Eosinophils in Health and Disease

Eosinophils in Health and Disease

James J. Lee
Mayo Clinic Arizona
Scottsdale, AZ

Helene F. Rosenberg
NIH
Laboratory of Allergic Diseases
Bethesda, MD

AMSTERDAM • BOSTON • HEIDELBERG • LONDON • NEW YORK • OXFORD
PARIS • SAN DIEGO • SAN FRANCISCO • SINGAPORE • SYDNEY • TOKYO

Academic Press is an Imprint of Elsevier

Academic Press is an imprint of Elsevier
32 Jamestown Road, London NW1 7BY, UK
225 Wyman Street, Waltham, MA 02451, USA
525 B Street, Suite 1800, San Diego, CA 92101-4495, USA

First edition 2013

Cover Credit: The cover illustration shows a scanning electron micrograph of an activated human peripheral blood eosinophil co-cultured with fibroblasts. Filopodia extend from the surface of the eosinophil in the direction of adjacent eosinophils and fibroblasts. Original magnification, 2,500X. Image provided courtesy of Paige Lacy, PhD, Department of Medicine, University of Alberta, Edmonton, Alberta, Canada, and Redwan Moqbel, PhD PRCPath, Department of Immunology, University of Manitoba, Winnipeg, Manitoba Canada.

Notice
No responsibility is assumed by the publisher for any injury and/or damage to persons or property as a matter of products liability, negligence or otherwise, or from any use or operation of any methods, products, instructions or ideas contained in the material herein

Because of rapid advances in the medical sciences, in particular, independent verification of diagnoses and drug dosages should be made

British Library Cataloguing-in-Publication Data
A catalogue record for this book is available from the British Library

Library of Congress Cataloging-in-Publication Data
A catalog record for this book is available from the Library of Congress

ISBN: 978-0-12-394385-9

For information on all Academic Press publications
visit our website at elsevierdirect.com

Typeset by TNQ Books and Journals Pvt Ltd.
www.tnq.co.in

Printed and bound in United States of America

12 13 14 15 16   10 9 8 7 6 5 4 3 2 1

Working together to grow
libraries in developing countries

www.elsevier.com | www.bookaid.org | www.sabre.org

ELSEVIER   BOOK AID
International   Sabre Foundation

This volume emerges from the intelligence and dedication put forth by innumerable investigators who came before us, as well as the ongoing efforts of our contemporary colleagues. All the ideas—the good *vs.* the bad, the correct *vs.* the incorrect, and the insightful *vs.* the just plain stupid—led the way and were the driving catalysts for much of the research presented herein. Consequently, our dedication is most simple: this book is a legacy and a gift to all. To this end, we dedicate *Eosinophils in Health and Disease* to the memories of those investigators who came before us, with a promise to those who will follow after we are gone. We acknowledge all of those who are working tirelessly to understand eosinophil biology, those who hope to determine what it is that these cells do, and how one can manage their helpful and harmful effects. These efforts have and will continue to contribute to improved health and well-being for everyone. More importantly, we hope and believe that this work will lead to a more fundamental understanding of the human condition, as none of us ever knows where our next ideas will come from or how far they will take us.

Contents

Contents

Had this book existed in the mid-1960s, my life with the eosinophil could have been very different. As it was, that life began when I was caring for patients with asthma at the Mayo Clinic in Rochester, Minnesota, USA. These patients had increased blood eosinophils and sputum packed with eosinophils. Autopsy studies had revealed that asthma is characterized by marked bronchial eosinophilia. Treatment with glucocorticoids, usually prednisone, dramatically reduced blood eosinophils, and patients improved. I sought more information, but there were no texts and the literature of the day was limited. Therefore, it gives me great pleasure to introduce this work. It is a splendid presentation of the eosinophil and takes our understanding to new levels.

We might regard this text as a kind of eosinophil anniversary, a birthday for the eosinophil! Of course, regarding the eosinophil as having a birthday presumes a date of birth, and this event is obscured in the fog of evolutionary history. Fortunately, even this issue is addressed in the Lee–Rosenberg compendium. A birthday also implies a celebration, and this book is a wonderful gift for current and future eosinophil lovers, perhaps best referred to as *eosinophilophiles*. It traces the origin and growth of the cell, its maturation, and its brief rite of passage through the bloodstream to its mature state in tissues. An adult eosinophil is called to sites of action, and these sites, as well as the mechanisms of transit, activation, and execution of effector functions, are presented. That is, once mobilized and ready to function, the eosinophil undergoes a series of activation states before fulfilling its mission. While the eosinophil's functions presently are not known fully, many are appreciated and these are reviewed. In particular, the book wonderfully summarizes our current understanding of eosinophil biology focusing the reader's attention to interactions with other cells, to antigen presentation, to production of cytokines, to fibrosis, and, not least, to the activities of its granule proteins. Knowledge of eosinophil associations with diseases, derived from both human medicine and from animal (mostly mouse) models, is presented along with current and novel approaches to therapy. The section on *Emerging Concepts* alone takes us from microRNA to zebrafish, to indoleamine 2,3-dioxy-genase, and to the actions of novel cytokines.

By the end of the book, I was struck by the remarkable advances over the past 50 years. Nonetheless, enigmas remain highlighted once again by the evolving role of the eosinophil in asthma, including current clinical conundrums such as nasal polyps, sensitivity to aspirin, and the occurrence of mixed pulmonary infiltrates. Moreover, what are the stimuli for the development of asthma and eosino-philia in the absence of immunoglobulin E antibodies to allergens? Despite our advances, we still do not know the underlying bases for the development of many eosinophil-associated diseases. Therefore, opportunities for research remain, which promise not only insights into fundamental disease pathophysiology, but also novel therapies. Thus, to readers of this wonderful compendium of our eosinophil knowledge, I say, "Enjoy the feast but remember that some of the courses are only hors d'oeuvres!"

Dr. Gerald J. Gleich,
University of Utah

As the 2009—2011 International Eosinophil Society (IES) President (J. J. L.) and as the Chair of the Scientific Program Committee for the 2011 Biennial IES meeting (H. F. R.), we set out to put in place several major academic and programmatic objectives for both the Society and for the eosinophil research community as a whole. Among these, we hoped to develop a scientific vision and to put eosinophils *on the map*. By *map*, we meant that we intended to define and to articulate a set of commonly accepted facts and ideas about eosinophils, to describe current research efforts and clinical investigations focused on eosinophils and eosinophil-related disorders, and to speculate on what the future might hold. In 2009, it was clear to us that one obstacle to this vision was the lack of an up-to-date encyclopedia—a one-stop reference book with all of the answers in one place at this moment in time. The idea for this reference book originated with the original go-to book entitled *Eosinophils: A Comprehensive Review and Guide to the Scientific and Medical Literature* by Christopher J. F. Spry (1988). It was a book that EVERYONE had and that EVERYONE would refer to when in doubt about anything having to do with eosinophils or eosinophil effector functions. Indeed, this was the first book nearly all of us purchased with those precious start-up funds we received with our first faculty position.

It has indeed been a remarkable journey for us. Our careers have been full of both painful reminders of the limits of our intellectual abilities and physical energies as well as the joys of discovery that have kept us in the laboratory and at the bench when common sense (and our friends and family) told us that it was time to go home. That journey has also led us to the interesting transition from junior research faculty to now elder (not too elder!!) statespersons of the eosinophil research community. However, we have found that, in this *information age*, our trainees rely primarily on research reports, reviews, and perspectives that were written within the last 5 years, and in many cases, those written in the last year or two. In fact, so do we. Our beloved Chris Spry book has become woefully, painfully out of date. Most of the nonhistorical information is now incomplete or, even worse, no longer exactly true. The scope of eosinophil research itself has expanded dramatically and includes methods, diseases, processes, and perspectives that were virtually unknown 23 years ago.

So, why a new book? We realized that discussions of eosinophil biology from smaller, individual perspectives had the tendency to fall into *pro forma* dogmatic patterns of ideas that were failing to explain recent discoveries. We envisioned this book as representing a snapshot of all things eosinophil at this single point in time, from basic and translational research activities to clinical studies and patient care. In addition, our desire was to create a book to serve as a portal into the minds of current researchers and clinicians, and to provide a solid foundation, and an opportunity for the next generation of researchers to build on the advances made by their predecessors. The irony of all of this has not escaped us. We have most definitely come full circle, as we hope that EVERYONE will have this book and that EVERYONE will refer to it when in doubt about anything related to eosinophils or eosinophil effector functions. We can only hope to be so fortunate!

Acknowledgements

Clearly, with a project of this scope and magnitude, the list of individuals who we would like to acknowledge, thank, and otherwise recognize is virtually endless. In many ways, it's like going to the movies. Once the movie is over, the cast and the major administrative folks (e.g., director and producer) are highlighted followed by an almost over-whelming list of movie credits noting the various contri-butions (what does a best boy or second grip do away?). This acknowledgement section is clearly going to follow a similar path, so we ask you to kindly stay in your seat until the movie is over and the lights come up!

It is probably wise and safe to start off by saying we may make mistakes and forget to acknowledge someone. Thus, let's begin by noting that if you were left out, it was unintended so don't feel hurt or ignored. It wouldn't be the first or last time we have misstepped. Having said this, we are going to divide our *thank you* remarks and notes of appreciation into two sections: first, personal acknowl-edgements by each of the editors, followed by general acknowledgements to those whose contributions were important for the success of this project.

James (Jamie) J. Lee: The organization and assembly of this book represent an exhausting project that has unfolded over a 2-year period. However, despite this effort there was a totally surprising and unexpected upside: I simply had too much damn fun!! The meetings, the discussions, and the debates with colleagues who contrib-uted sections and chapters to the book were spirited and at times direct, but they were always fun!! The book also gave me an opportunity to mentor young investigators and trainees on a scale that was far larger than anything even I ever tried before. The time and effort were, at times, overwhelming but it was fun!! I can't say enough about my coeditor Helene Rosenberg. She is a wonderful lady, a fantastic scientist, and an incredible academician. We made a very interesting team: Helene is very detail-oriented and timeline-driven. Me, I like the big picture and I will get to it when I get to it. We drove each other a little nuts, but you know … it was fun! The folks in Lee Laboratories past and present all need to be thanked over and over again for their collective efforts and patience. I have been blessed over the years with incredible folks who are enthusiastic, work hard, care about what they do, and want to do a good job—it doesn't get any better than this! I have also been

extraordinarily lucky to have the administrative assistant from heaven, Linda Mardel. Ms. Mardel single-handedly manages the lab, my schedule, and all of my professional activities. I would be lost without her! I would be remiss if I didn't also acknowledge the thing that kept me sane when issues at the Mayo Clinic and the efforts required to get this book completed got overwhelming—the children who play on my youth baseball club teams. Managing baseball teams and teaching children are activities that often extend beyond my love of science and research. The enthusiasm and innocence of children (even at age 14!) are incredibly motivating, especially when my real job (i.e., dealing with adults) gets difficult. Finally, no thank you would be complete without acknowledging the incredible efforts of my wife and colleague Nancy Lee. Nancy's support and encouragement over the years have singularly been the driving force in my life to be better than I currently am and to pursue ever-greater achievements. I am truly a lucky man.

Helene F. Rosenberg: This book has been an adven-ture, indeed an odyssey, and a heartfelt tribute to every-thing that makes scientific exploration amazing and wonderful. In that spirit, I would like to acknowledge first the members of my laboratory group, the Inflammation Immunobiology Section of the Laboratory of Allergic Diseases, National Institute of Allergy and Infectious Diseases, past and present, for their ongoing interest, energy, and excitement as we pursue this path together. Of particular note, I am deeply thankful to Kimberly Dyer and Caroline Percopo, whose efforts define sincerity, dedication, and scholarship. I also acknowledge and am profoundly grateful to my long-term collaborators, Joe Domachowske, Paul Foster, and the ever-exuberant Jamie Lee, for ongoing support, good humor, advice, mentor-ship, and the gift of true and everlasting friendship. And always, to my family: Joshua Rosenberg, Michael Rosenberg, and Lloyd Goodman, who bestow love and goodness upon me all the days of their lives.

General Acknowledgements: In no particular order (i.e., without correlating appearance on this list with the importance of the respective contribution to the book) the editors would like to thank the following individuals whose efforts were of significant assistance to us in preparing *Eosinophils in Health and Disease*:

Linda Mardel: The über-assistant whose tireless efforts made our collective jobs that much easier!

The 2009 Executive Council of the International Eosinophil Society: Amy D. Klion (President Elect), Steven J. Ackerman (Immediate Past President), Marc Rothenberg (Secretary Treasurer), Bruce Bochner, Redwan Moqbel, Francesca Levi-Schaffer, and Peter Weller. This group presided over what had been a simple collection of like-minded scientists, clinicians, and pathologist … and now … through their leadership, this *mom-and-pop* operation has become a true society and the focal point of all things eosinophil!

We wish to make a special note to Dr. Michael Lotze who encouraged us to pursue this project. His guidance, support, and encouragement to develop this project and bring it to fruition were singularly critical to the success of our efforts.

The Elsevier group should all take a bow. We don't know how other publishers operate, but Elsevier and their associates have been nothing but the most professional group of individuals who have made the publication of this book a reality. A special shout-out to Denise E.M. Penrose, Senior Acquisitions Editor, Mary Preap, Ashley Craig, Sonia Cutler and Caroline Johnson; we are in your debt.

David Abraham Thomas Jefferson University, Jefferson Medical College, Department of Microbiology and Immunology, 233 South Street, Philadelphia, PA 19107, USA

Seema S. Aceves Division of Allergy and Immunology, University of California, San Diego, Rady Children's Hospital, San Diego, 9500 Gilman Drive MC-0635, La Jolla, CA 92093, USA

Steven J. Ackerman University of Illinois at Chicago, Department of Biochemistry and Molecular Genetics (M/C 669), Molecular Biology Research Building, Rm. 2074, 900 S. Ashland Avenue, Chicago, IL 60607, USA

Darryl Adamko HMRC, University of Alberta, Edmonton, Alberta T6G 2S2, Canada

Koichi Akashi Medicine and Biosystemic Science, Kyushu University Graduate School of Medical Sciences, 3-1-1 Maidashi, Higashi-ku, Fukuoka 812-8582, Japan

Praveen Akuthota Beth Israel Deaconess Medical Center, 330 Brookline Ave., E/KSB-23 Pulmonary, Boston, MA 02215, USA

Rafeul Alam Division of Allergy & Immunology, Department of Medicine, National Jewish Health & University of Colorado, 1400 Jackson Street, Denver, CO 80206, USA

Judith A. Appleton Baker Institute for Animal Health, College of Veterinary Medicine, Cornell University, Ithaca, NY 14853, USA

Narcy Arizmendi Department of Immunology, University of Manitoba, Room 455 Apotex Centre, 750 McDermot Avenue, Winnipeg, MB R3E 0T5, Canada

Kewal Asosingh Department of Pathobiology/NC22, Cleveland Clinic, 9500 Euclid Avenue, Cleveland, OH 44195, USA

Keir M. Balla Department of Cellular & Molecular Medicine, Section of Cell & Developmental Biology, University of California, San Diego, 9500 Gilman Drive, Natural Sciences Building, Room 6107, La Jolla, CA 92093−0380, USA

Christianne Bandeira-Melo Laboratory of Inflammation, Carlos Chagas Filho Biophysics Institute, Federal University of Rio de Janeiro, Rio de Janeiro, Brazil

Marc Bartoli Aix-Marseille Université, UMR_S 910, Faculté de Médecine de la Timone, 13385, Marseille, France; INSERM UMR_S 910, 13385, Marseille, France; AP-HM, Hôpital d'Enfants de la Timone, Département de Génétique Médicale et de Biologie Cellulaire, 13385, Marseille, France

Fleur Samantha Benghiat Institute for Medical Immunology, Université Libre de Bruxelles, 8 rue A. Bolland, 1640 Charleroi, Belgium

Claudia Berek Deutsches Rheuma Forschungszentrum, Institut der Leibniz Geinschaft, Berlin, Germany

Utibe Bickham Department of Pathology and Laboratory Medicine, Madison, WI 53706, USA

Elizabeth R. Bivins-Smith Division of Pulmonary and Critical Care Medicine, Oregon Health & Science University, Portland, OR, USA

Carine Blanchard Nutrition and Health/Allergy Group, Nestlé Research Center, PO Box 44, CH-1000 Lausanne 26, Switzerland

Bruce S. Bochner Department of Medicine, Division of Allergy and Clinical Immunology, The Johns Hopkins University School of Medicine, 5501 Hopkins Bayview Circle, Baltimore, MD 21224, USA

Apostolos Bossios Krefting Research Centre, Department of Medicine and Clinical Nutrition, Sahlgrenska Academy, University of Gothenburg, SE-405 30 Göteborg, Sweden

Patricia T. Bozza Laboratory of Immunopharmacology, Oswaldo Cruz Institute, Fiocruz, Rio de Janeiro, Brazil

David Broide University of California San Diego, Biomedical Sciences Building, Room 5090, 9500 Gilman Drive, La Jolla, CA 92093−0635, USA

Miranda Buitenhuis Department of Hematology, Erasmus MC, Dr. Molewaterplein 50, Faculty Building Office H-Ee1330F, 3015 GE Rotterdam, The Netherlands

William W. Busse University of Wisconsin School of Medicine and Public Health, 600 Highland Avenue, Madison, WI 53792, USA

Joseph H. Butterfield Consultant, Division of Allergic Diseases, Department of Internal Medicine, Mayo Clinic, Rochester, MN, USA

Jose A. Cancelas Division of Experimental Hematology and Cancer Biology, Cincinnati Children's Hospital Medical Center, Hoxworth Blood Center, Research Division, University of Cincinnati College of Medicine, 3333 Burnet Avenue, Cincinnati, OH 45229-3039, USA

Monique Capron Inserm U995 − School of Medicine CHRU of Lille, Université Lille 2, Lille, France

Lars Olaf Cardell Division of ENT Diseases, Department of Clinical Sciences, Intervention and Technology, Karolinska Institutet, 141 86 Stockholm, Sweden

Van T. Chu Deutsches Rheuma Forschungszentrum, Institut der Leibniz Geinschaft, Berlin, Germany

Michael R. Comeau Department of Inflammation Research, Amgen, Seattle, Washington, Department of Inflammation Research, Amgen, 1201 Amgen Court West, Seattle, WA, 98119, USA

Joan M. Cook-Mills Allergy-Immunology Division, Northwestern University Feinberg School of Medicine, McGaw-M304, 240 E. Huron, Chicago, IL 60611, USA

Jan Cools Center for the Biology of Disease, VIB, Leuven, Belgium; Center for Human Genetics, KU Leuven, Leuven, Belgium

Olga V. Cravetchi Pulmonary Research Group, University of Alberta, Edmonton, Alberta, Canada

Benjamin P. Davis University of Cincinnati, Division of Allergy and Immunology, 3255 Eden Avenue #350 ML0563, Cincinnati, OH 45267−0563, USA

Judah A. Denburg Division of Clinical Immunology and Allergy, Department of Medicine, McMaster University, HSC 3V46, 1200 Main St West, Hamilton, ON L8N 3Z5, Canada

Joseph B. Domachowske Department of Pediatrics, Division of Infectious Diseases, State University of New York Upstate Medical University, 5400 University Hospital, 750 East Adams Street, Syracuse, NY 13210, USA

Virginie Driss Inserm U837, Institut de Recherche sur le Cancer de Lille, Université Lille 2, Lille, France

Jian Du University of Illinois at Chicago, Department of Biochemistry and Molecular Genetics, 900 S. Ashland (M/C 669), Chicago, IL 60607, USA

Ann M. Dvorak Department of Pathology, Beth Israel Deaconess Medical Center, Harvard Medical School, 330 Brookline Avenue, Boston, MA, 02215, USA

Kimberly D. Dyer Laboratory of Allergic Diseases, National Institute of Allergy and Infectious Diseases, National Institutes of Health, Bethesda, MA 20892 USA

Moran Elishmereni Department of Pharmacology and Experimental Therapeutics, Institute for Drug Research, School of Pharmacy, Faculty of Medicine, The Hebrew University of Jerusalem, POB 12065, Jerusalem, Israel

Michael J. Eppihimer Pre-Clinical Cell Biology, Boston Scientific Corporation, One Boston Scientific Place, Natick, MA, USA

Serpil C. Erzurum Lerner Research Institute and Respiratory Institute, Cleveland Clinic, 9500 Euclid Avenue, Cleveland, OH 44195, USA

Gary W. Falk Department of Medicine, Division of Gastroenterology, University of Pennsylvania School of Medicine, Philadelphia, PA, USA

Sophie Fillon Digestive Health Institute, Section of Pediatric Gastroenterology, Hepatology and Nutrition, Gastrointestinal Eosinophilic Diseases Program, The Children's Hospital, Denver, National Jewish Health, Department of Pediatrics, Mucosal Inflammation Program, University of Colorado Denver School of Medicine, Aurora, CO 80045, USA

Claudio Fiocchi Department of Pathobiology, Lerner Research Institute (NC22), and Department of Gastroenterology and Hepatology, Digestive Disease Institute, The Cleveland Clinic Foundation, 9500 Euclid Avenue, Cleveland, OH 44195, USA

Paul S. Foster Centre for Asthma and Respiratory Diseases, School of Biomedical Sciences and Pharmacy, Faculty of Health, University of Newcastle, Newcastle, NSW 2300, Australia

Allison D. Fryer Department of Pulmonary and Critical Care Medicine, Oregon Health and Science University, 3181 SW Sam Jackson Park Rd., L334, Portland, OR 97239, USA

Glenn T. Furuta Digestive Health Institute, Section of Pediatric Gastroenterology, Hepatology and Nutrition, Gastrointestinal Eosinophilic Diseases Program, The Children's Hospital, Denver, National Jewish Health, Department of Pediatrics, Mucosal Inflammation Program, University of Colorado Denver School of Medicine, Aurora, CO 80045, USA

Gail M. Gauvreau Division of Respirology, Department of Medicine, McMaster University, HSC 3U26, 1200 Main St West, Hamilton, ON L8N 3Z5, Canada

Nebiat G. Gebreselassie Baker Institute for Animal Health, College of Veterinary Medicine, Cornell University, Ithaca, NY 14853, USA

Erwin W. Gelfand Division of Cell Biology, Department of Pediatrics, National Jewish Health, Denver, CO 80206 and University of Colorado School of Medicine, Aurora, CO 80045, USA

Katrin Gentil Institute of Medical Microbiology, Immunology and Parasitology (IMMIP), University Clinic Bonn, Sigmund Freud Str. 25, 53105 Bonn, Germany

Gerald J. Gleich Department of Dermatology, University of Utah, 4A330 School of Medicine 30 North 1900 East, Salt Lake City, UT, USA

Pranab Haldar Department of Infection Immunity and Inflammation, Institute for Lung Health, University of Leicester, Glenfield Hospital, Groby Road, Leicester LE3 9QP, United Kingdom

Noriyasu Hirasawa Graduate School of Pharmaceutical Sciences, Tohoku University, 6-3 Aoba, Aramaki, Aoba-ku, Sendai, Miyagi 980-8578, Japan

Achim Hoerauf Institute of Medical Microbiology, Immunology and Parasitology (IMMIP), University Clinic Bonn, Sigmund Freud Str. 25, 53105 Bonn, Germany

Simon P. Hogan Division of Allergy and Immunology, Department of Pediatrics, University of Cincinnati, Cincinnati Children's Hospital Medical Center, Cincinnati, OH 45229, USA

Ramses Ilarraza Department of Immunology, University of Manitoba, Room 455 Apotex Centre, 750 McDermot Avenue, Winnipeg, MB R3E 0T5, Canada

Charles G. Irvin Vermont Lung Center, University of Vermont and Department of Medicine and Physiology, Room 226, HSRF, 149 Beaumont Avenue, Burlington, VT 05405-0075, USA

Kenji Ishihara Laboratory of Medical Science, Course for School Nurse Teacher, Faculty of Education, Ibaraki University, 2-1-1 Bunkyo, Mito, Ibaraki 310-8512, Japan

Hiromi Iwasaki Center for Cellular and Molecular Medicine, Kyushu University Hospital, 3-1-1 Maidashi, Higashi-ku, Fukuoka 812-8582, Japan

Elizabeth A. Jacobsen Research Associate, Division of Pulmonary and Critical Care Medicine, Scottsdale, AZ, USA

David B. Jacoby Division of Pulmonary and Critical Care Medicine, Oregon Health & Science University, Portland, OR, USA

Harsha H. Kariyawasam Department of Allergy and Medical Rhinology, Royal National Throat Nose Ear Hospital, University College London, London; Leukocyte Biology Section, National Heart and Lung Institute, Imperial College London, London, UK

Atsushi Kato Departments of Medicine and Otorhinolaryngology, Northwestern Feinberg School of Medicine, Chicago, IL 60611, USA

Howard R. Katz Department of Medicine, Harvard Medical School, Division of Rheumatology, Immunology and Allergy, Brigham and Women's Hospital, 1 Jimmy Fund Way, Room 638A, Boston, MA 02115, USA

A. Barry Kay Leukocyte Biology Section, National Heart and Lung Institute, Imperial College London, London, UK

Elizabeth A. Kelly Allergy, Pulmonary, and Critical Care Medicine Section of the Department of Medicine, University of Wisconsin School of Medicine and Public Health, 600 Highland Avenue, K4/928, Madison, WI 53792—9988, USA

Robert Kern Departments of Medicine and Otorhinolaryngology, Northwestern Feinberg School of Medicine, Chicago, IL 60611, USA

Paneez Khoury Eosinophil Pathology Unit, Laboratory of Parasitic Diseases, Bldg 4/Rm B1-28, Bethesda, MD 20892, USA

Hirohito Kita Consultant, Departments of Otorhinolaryngology, Immunology, and Division of Pediatric Allergy, Immunology, and Pulmonology, Department of Pediatric and Adolescent Medicine, Mayo Clinic, Rochester, MN, USA

Amy D. Klion Eosinophil Pathology Unit, Laboratory of Parasitic Diseases, Bldg 4/Rm B1-28, Bethesda, MD 20892, USA

Leo Koenderman Department of Pulmonary Diseases, E.03.511, University Medical Center Utrecht, Heidelberglaan 100, 3584CX Utrecht, The Netherlands

Roland Kolbeck Respiratory, Inflammation & Autoimmunity, MedImmune, LLC, One MedImmune Way, Gaithersburg, MD 20878, USA

Martin Krahn Aix-Marseille Université, UMR_S 910, Faculté de Médecine de la Timone, 13385, Marseille, France; INSERM UMR_S 910, 13385, Marseille, France; AP-HM, Hôpital d'Enfants de la Timone, Département de Génétique Médicale et de Biologie Cellulaire, 13385, Marseille, France

Paige Lacy 559 HMRC, Pulmonary Research Group, Department of Medicine, University of Alberta, Edmonton, Alberta T6G 2S2, Canada

Mark C. Lavigne Pre-Clinical Cell Biology, Boston Scientific Corporation, One Boston Scientific Place, Natick, MA, USA

Laura E. Layland Institute of Medical Microbiology, Immunology and Parasitology (IMMIP), University Clinic Bonn, Sigmund Freud Str. 25, 53105 Bonn, Germany and Institute of Medical Microbiology, Immunology and Hygiene an der Technischen Universität München, Trogerstrasse 30, 81675, München, Germany

Alain Le Moine Institute for Medical Immunology, Université Libre de Bruxelles, 8 rue A. Bolland, 1640 Charleroi, Belgium; Erasme Hospital, Université Libre de Bruxelles, Department of Nephrology 808 route de Lennik, 1070 Brussels, Belgium

James J. Lee Consultant, Division of Pulmonary and Critical Care Medicine, Department of Internal Medicine, Mayo Clinic, Scottsdale, AZ, USA

Nancy A. Lee Consultant, Division of Hematology and Oncology, Department of Internal Medicine, Mayo Clinic, Scottsdale, AZ, USA

Fanny Legrand Laboratoire d'Immunologie du CHRU de Lille, Université Lille 2, Centre de Biologie Pathologie, Lille, France

Philippe Lemaitre Institute for Medical Immunology, Université Libre de Bruxelles, 8 rue A. Bolland, 1640 Charleroi, Belgium

Séverine Letuve Inserm U700, Université Paris Diderot—Paris 7, Faculté de Médecine site Bichat 16, rue Henri Huchard, 75018 Paris, France

Francesca Levi-Schaffer Department of Pharmacology and Experimental Therapeutics, Institute for Drug Research, School of Pharmacy, Faculty of Medicine, The Hebrew University of Jerusalem, POB 12065, Jerusalem, Israel

Nicolas Levy Aix-Marseille Université, UMR_S 910, Faculté de Médecine de la Timone, 13385, Marseille, France; INSERM UMR_S 910, 13385, Marseille, France; AP-HM, Hôpital d'Enfants de la Timone, Département de Génétique Médicale et de Biologie Cellulaire, 13385, Marseille, France

Ramin Lotfi Institute for Transfusion Medicine, University of Ulm, and Institute of Clinical Transfusion Medicine and Immunogenetics Ulm, German Red Cross Blood Services Baden-Württemberg-Hessen, Helmholtzstr. 10, 89081 Ulm, Germany

Jan Lötvall Krefting Research Centre, Department of Medicine and Clinical Nutrition, Sahlgrenska Academy, University of Gothenburg, SE-405 30 Göteborg, Sweden

Michael Thomas Lotze G.27A Hillman Cancer Center of the University of Pittsburgh Cancer Institute, 5117 Centre Avenue, Pittsburgh, PA 15213, USA

Michael P. McGarry Visiting Scientist, Department of Biochemistry and Molecular Biology, Mayo Clinic, Scottsdale, AZ, USA

Kelly M. McNagny The Biomedical Research Centre, 2222 Health Sciences Mall, The University of British Columbia, Vancouver, BC, Canada, V6T 1Z3

Salahaddin Mahmudi-Azer Department of Medicine, University of Calgary and Foothills Medical Centre, 1403—29th Street NW, Calgary, Alberta, Canada

Steven Maltby The Biomedical Research Centre, 2222 Health Sciences Mall, The University of British Columbia, Vancouver, BC, Canada, V6T 1Z3

James S. Malter Waisman Center for Developmental Disabilities, Department of Pathology and Laboratory Medicine, University of Wisconsin School of Medicine and Public Health, Madison, WI 53705, USA

Anne M. Månsson Kvarnhammar Division of ENT Diseases, Department of Clinical Sciences, Intervention and Technology, Karolinska Institutet, 141 86 Stockholm, Sweden

Annick Massart Department of Nephrology, Dialysis and Transplantation, 808 route de Lennik, 1070 Brussels, Belgium

Joanne C. Masterson Digestive Health Institute, Section of Pediatric Gastroenterology, Hepatology and Nutrition, Gastrointestinal Eosinophilic Diseases Program, The Children's Hospital, Denver, National Jewish Health, Department of Pediatrics, Mucosal Inflammation Program, University of Colorado Denver School of Medicine, Aurora, CO 80045, USA

Kenji Matsumoto Department of Allergy and Immunology, National Research Institute for Child Health and Development, 2-10-1 Okura, Setagaya-ku, Tokyo 157-8535, Japan

Joerg Mattes Centre for Asthma and Respiratory Diseases, School of Biomedical Sciences and Pharmacy, Faculty of Health, University of Newcastle, Newcastle, NSW 2300, Australia

Rossana C.N. Melo Laboratory of Cellular Biology, Department of Biology, Federal University of Juiz de Fora, UFJF, Juiz de Fora, MG, 36036-900, Brazil

Nestor A. Molfino Clinical Development, MedImmune, LLC, One MedImmune Way, Gaithersburg, MD 20878, USA

Redwan Moqbel Department of Immunology, University of Manitoba, 471 Apotex Centre, 750 McDermot Avenue, Winnipeg, MB R3E 0T5, Canada

Yasuo Mori Medicine and Biosystemic Science, Kyushu University Graduate School of Medical Sciences, 3-1-1 Maidashi, Higashi-ku, Fukuoka 812-8582, Japan

Ariel Munitz Department of Microbiology and Clinical Immunology, The Sackler Faculty of Medicine, Tel-Aviv University, Tel-Aviv, Israel

Parameswaran Nair Firestone Institute for Respiratory Health, St Joseph's Healthcare, 50 Charlton Avenue East, Hamilton, ON L8N 4A6, Canada

Josiane Sabbadini Neves Institute of Biomedical Sciences, Federal University of Rio de Janeiro, 373 Carlos Chagas Filho Avenue, Centro de Ciências da Saúde (CCS), Room F 14, 1st floor, Ilha do Fundão, Zip Code 21941-590—Rio de Janeiro, RJ—Brazil

Thomas B. Nutman Helminth Immunology Section, Laboratory of Parasitic Diseases National Institute of Allergy and Infectious Diseases, National Institutes of Health, Bethesda, MD 20892, USA

Sergei I. Ochkur Research Associate, Division of Pulmonary Medicine, Department of Internal Medicine, Mayo Clinic, Scottsdale, AZ USA

S.O. (Wole) Odemuyiwa Department of Pathobiology, College of Veterinary Medicine, Nursing and Allied Health, Tuskegee University, AL 36088, USA

Kazuo Ohuchi Tohoku University, 6-3 Aoba, Aramaki, Aoba-ku, Sendai, Miyagi 980-8578, Japan

Kanami Orihara Department of Immunology, University of Manitoba, Room 455 Apotex Centre, 750 McDermot Avenue, Winnipeg, MB R3E 0T5, Canada

Ian D. Pavord Department of Infection Immunity and Inflammation, Institute for Lung Health, University of Leicester, Glenfield Hospital, Groby Road, Leicester LE3 9QP, United Kingdom

Caroline M. Percopo Inflammation Immunobiology Section, LAD, NIAID, NIH, Bethesda, MD, USA

Maximilian Plank Centre for Asthma and Respiratory Diseases, School of Biomedical Sciences and Pharmacy, Faculty of Health, University of Newcastle, Newcastle, NSW 2300, Australia

Marina Pretolani Inserm U700, Université Paris Diderot—Paris 7, Faculté de Médecine site Bichat 16, rue Henri Huchard, 75018 Paris, France

Calman Prussin Laboratory of Allergic Diseases, National Institute of Allergy and Infectious Diseases, NIH, Building 10, Room 11C207, National Institutes of Health, Bethesda, MD 20892-1881, USA

Catherine Ptaschinski Centre for Asthma and Respiratory Diseases, School of Biomedical Sciences and Pharmacy, Faculty of Health, University of Newcastle, Newcastle, NSW 2300, Australia

Madeleine Rådinger Krefting Research Centre, Department of Medicine and Clinical Nutrition, Sahlgrenska Academy, University of Gothenburg, SE-405 30 Göteborg, Sweden

Savita P. Rao Laboratory of Allergic Diseases and Inflammation, Department of Veterinary & Biomedical Sciences, and Medicine, University of Minnesota, 1971 Commonwealth Avenue, St. Paul, MN 55126, USA

Florian Rieder Department of Pathobiology, Lerner Research Institute (NC22), and Department of Gastroenterology and Hepatology, Digestive Disease Institute, The Cleveland Clinic Foundation, 9500 Euclid Avenue, Cleveland, OH 44195, USA

Douglas S. Robinson Leukocyte Biology Section, National Heart and Lung Institute, Imperial College London, London, UK; Laboratorios Leti SL, Madrid, Spain

Helene F. Rosenberg Laboratory of Allergic Diseases, National Institute of Allergy and Infectious Diseases, National Institutes of Health, Bethesda, MD 20892 USA

Marc E. Rothenberg Division of Allergy and Immunology, Cincinnati Center for Eosinophilic Disorders, Cincinnati Children's Hospital Medical Center, University of Cincinnati College of Medicine, 3333 Burnet Avenue, ML7028, Cincinnati, OH 45229-3039, USA

Florence Roufosse Institute for Medical Immunology, Université Libre de Bruxelles, 8 rue A. Bolland, 1640 Charleroi; Erasme Hospital, Université Libre de Bruxelles, Department of Internal Medicine, 808 route de Lennik, 1070 Brussels, Belgium

Robert P. Schleimer Departments of Medicine and Otorhinolaryngology, Northwestern Feinberg School of Medicine, Chicago, IL 60611, USA

Shauna Schroeder Digestive Health Institute, Section of Pediatric Gastroenterology, Hepatology and Nutrition, Gastrointestinal Eosinophilic Diseases Program, The Children's Hospital, Denver, National Jewish Health, Department of Pediatrics, Mucosal Inflammation Program, University of Colorado Denver School of Medicine, Aurora, CO 80045, USA

Gregory D. Scott Department of Pulmonary and Critical Care Medicine, Oregon Health and Science University, 3181 SW Sam Jackson Park Rd., L334, Portland, OR 97239, USA

Darren W. Sexton BioMedical Research Centre, Faculty of Health, University of East Anglia, Norwich, Norfolk NR4 7TJ, United Kingdom

Mi-Kyung Shin Department of Medicine, Division of Allergy and Clinical Immunology, The Johns Hopkins University School of Medicine, Baltimore, MD 21224, USA

Dagmar Simon Department of Dermatology, Inselspital, Bern University Hospital, CH-3010 Bern, Switzerland

Hans-Uwe Simon Institute of Pharmacology, University of Bern, Friedbuehlstrasse 49, CH-3010 Bern, Switzerland

Dirk E. Smith Department of Inflammation Research, Amgen, Seattle, Washington, Department of Inflammation Research, Amgen, 1201 Amgen Court West, Seattle, WA 98119, USA

Neal Spada G.27A Hillman Cancer Center of the University of Pittsburgh Cancer Institute, 5117 Centre Avenue, Pittsburgh, PA 15213, USA

Lisa A. Spencer Division of Allergy and Inflammation, Department of Medicine, Beth Israel Deaconess Medical Center, Harvard Medical School, 330 Brookline Ave., Boston, MA 02215, USA

P. Sriramarao Laboratory of Allergic Diseases and Inflammation, Department of Veterinary & Biomedical Sciences, and Medicine, University of Minnesota, 1971 Commonwealth Avenue, St. Paul, MN 55126, USA

Alex Straumann Department of Gastroenterology, University Hospital Basel, Chairman Swiss EoE Research Group, Roemerstrasse 7, CH-4600 Olten, Switzerland

Kiyoshi Takatsu Department of Immunobiology and Pharmacological Genetics, Graduate School of Medicine and Pharmaceutical Science, University of Toyama, Toyama 930-0194, Japan

Anastasya Teplinsky Department of Pharmacology and Experimental Therapeutics, Institute for Drug Research, School of Pharmacy, Faculty of Medicine, The Hebrew University of Jerusalem, POB 12065, Jerusalem, Israel

Alex Thomas University of Wisconsin School of Medicine and Public Health, Department of Medicine, Division of Allergy, Pulmonary and Critical Care Medicine, Madison, WI, USA

David Traver Department of Cellular & Molecular Medicine, Section of Cell & Developmental Biology, University of California, San Diego, 9500 Gilman Drive, Natural Sciences Building, Room 6107, La Jolla, CA 92093−0380, USA

Meri K. Tulic School of Paediatrics and Child Health, University of Western Australia, Princess Margaret Hospital for Children, Roberts Road, Subiaco, WA 6008, Australia

Per Venge Department of Medical Sciences, University of Uppsala, Uppsala, Sweden

Amanda Waddell Division of Allergy and Immunology, Department of Pediatrics, University of Cincinnati, Cincinnati Children's Hospital Medical Center, Cincinnati, OH 45229, USA

Lori A. Wagner University of Utah, Department of Dermatology, 30 North 1900 East, 4A330 Salt Lake City, UT, USA

Garry M. Walsh Division of Applied Medicine, School of Medicine and Dentistry, Institute of Medical Sciences, University of Aberdeen, Foresterhill, Aberdeen AB25 2ZD, United Kingdom

Haibin Wang Beth Israel Deaconess Medical Center, 330 Brookline Ave., CLS 930, Boston, MA 02215, USA

Andrew J. Wardlaw Department of Infection Immunity and Inflammation, Institute for Lung Health, University of Leicester, Glenfield Hospital, Groby Road, Leicester LE3 9QP, United Kingdom

Peter F. Weller Department of Medicine, Beth Israel Deaconess Medical Center, Harvard Medical School, 330 Brookline Ave., CLS 943, Boston, MA 02215, USA

Christine Wennerås Department of Medical Microbiology and Immunology, Göteborg University, Guldhedsgatan 10, 413 46 Göteborg, Sweden

Jason J. Xenakis Beth Israel Deaconess Medical Center, 330 Brookline Ave., CLS 930, Boston, MA 02215, USA

Yoshiyuki Yamada Division of Allergy and Immunology, Gunma Children's Medical Center, 779 Shimohakoda Hokkitsu, Shibukawa Gunma 377-8577, Japan

Shida Yousefi Institute of Pharmacology, University of Bern, Friedbuehlstrasse 49, CH-3010 Bern, Switzerland

Xiang Zhu Division of Allergy and Immunology, ML 7028, Children's Hospital Medical Center, 3333 Burnet Avenue, Cincinnati, OH 45229, USA

Nives Zimmermann Division of Allergy and Immunology, ML 7028, Department of Pediatrics, Children's Hospital Medical Center, 3333 Burnet Avenue, Cincinnati, OH 45229, USA

Historical Overview and Perspective on the Role of the Eosinophil in Health and Disease

Gerald J. Gleich

INTRODUCTION

Since the time of its discovery by Paul Ehrlich, the eosinophil has been a challenge. To some, it is the most attractive cell in the body because of the beauty and brilliance of its cytoplasmic granules. To others, it is of little relevance because it is not an easy cell to put into a functional framework. Comparison with the neutrophil is instructive. The absence of neutrophils results in a marked increase in susceptibility to bacterial infections. In contrast, eosinophils are commonly reduced, often to very low numbers, in the blood by the administration of glucocorticoids. Yet, no definite consequences follow from this pharmacological eosinophil ablation.

I first saw an eosinophil in 1954 as a medical student, and, in the early 1960s, John H. Vaughan, my mentor at the University of Rochester, New York, USA, sparked my interest by commenting that asthma associated with eosinophilia is especially challenging. When I joined the staff of the Mayo Clinic (Rochester, Minnesota) in 1965, one of my duties was caring for hospitalized patients with asthma. At that time, we had relatively little to offer for outpatient treatment, and many patients were hospitalized. Treating them consisted of assuring that asthma was the correct diagnosis and initiating care with hydration and oral glucocorticoids, usually prednisone. Most patients quickly responded with improved respiration. As many as 20–25 asthma patients might be hospitalized at a time and they stimulated my interest in the pathophysiology of the disease.

Some were allergic to environmental allergens, but others were not and had blood eosinophilia associated with nasal polyps and sensitivity to aspirin. I collected sputum and found many eosinophils. I read reports on the pathology of asthma and I realized that knowledge of the eosinophil, including how it might affect asthma, was sparse. Around 1968, I decided to direct research efforts at Mayo to investigating the eosinophil. In this overview, I will interweave elements of the efforts of our laboratory and the work of many others to tell the story of the eosinophil from its recognition to the present. In describing the evolution of knowledge of the eosinophil in health and disease, I arbitrarily divide these developments into five periods based on the scientific understanding of the times, with overlap in the third, fourth, and fifth period.

The first period, from 1879 until approximately 1914, highlights the contributions of Paul Ehrlich. The advances during this period were based on the knowledge of chemistry, especially dyes. Ehrlich brilliantly applied chemicals used in the German dye industry to the microscopic examination of human cells and tissues, and, from this work, discovered the eosinophil. The application of his methods for the analysis of blood provided clues to eosinophil function by showing associations with asthma and parasitic diseases. His contributions also established hematology as a specialized area of clinical knowledge.

The second period, from approximately 1915 to around 1970, was based on the association of eosinophilia with anaphylaxis and the recognition that eosinophils accumulate in tissues after this calamitous inflammatory event. These observations pointed to a role for eosinophils in tissue inflammation, possibly to counteract the damaging physiological effects from anaphylaxis. Investigators during most of this period were hampered by the lack of technology needed for protein analyses and were forced to rely on associations that were difficult to explain mechanistically.

The third period, from approximately 1970 until the present, exploited advances in protein chemistry such that eosinophil granule proteins were isolated, characterized, and their functions defined. The affinity of the eosinophil for eosin is due likely to these intensely cationic protein molecules. Investigation of the granule proteins immediately suggested mechanisms for eosinophil function in disease. By tracking the proteins, associations with asthma

FIGURE 1.1   Citations for *eosinophil* listed in PubMed per five-year period. This information suggests that very few publications occurred before about 1940. However, PubMed does not track the German literature of the early 1900s and before, and, therefore, many publications are not accounted for. As noted in the chapter, Schwarz published a review of the eosinophil in 1914 containing some 2758 references.[1]

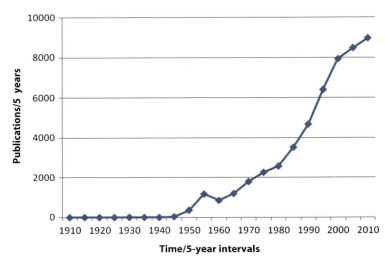

and other allergic disorders were strengthened, and better understanding of eosinophil activity emerged.

The fourth period, from approximately 1980 until the present, employed new knowledge of genetics to identify novel genes and to express proteins. A critical advance using this technology was the discovery of factors controlling eosinophil production, especially interleukin-5 (IL-5). The fifth period, beginning approximately in 1990, took advantage of the ability to create disease models by generating mice with defined attributes, transgenic mice, and other animals in which gene functions were eliminated or otherwise modified through a targeted mutation, gene knockout and knock-in mice, respectively. These technologies led to the development of mice with increased numbers of eosinophils and mice with no eosinophils. Investigation of these mice has strikingly increased our knowledge of eosinophil function. Importantly, progress in our understanding of the eosinophil has not been confined to studies of the cell itself, but also to diseases associated with eosinophils, and I comment on significant disease discoveries.

Fig. 1.1 shows the number of eosinophil citations from PubMed from 1910 to the present in 5-year periods and illustrates a striking increase from approximately 1960 until 2000 with the numbers of publications continuing to rise. These numbers are not simply a reflection of increased overall publications, because the percentage of eosinophil publications (total number of eosinophil citations in a 5-year period divided by the total number of publications in the same 5-year period) increased threefold from 1965–1969 to 1995–1999. This increased ratio reflects the increased attention paid to eosinophils by investigators over this time. The early literature on eosinophils is summarized by Schwarz in a monumental tome citing some 2758 references,[1] while Samter provides a scholarly and insightful analysis of advances from about 1915 until about 1970.[2]

PAUL EHRLICH AND THE DISCOVERY OF THE EOSINOPHIL

A superb account of Paul Ehrlich's life and his accomplishments was written by Hirsch and Hirsch.[3] They describe his life and career in detail, and a short synopsis is recounted here. Ehrlich was born in 1854 in Germany and was a good, though not outstanding, student as a child. He moved from school to school during his university education and published his first paper, *Beiträge zur Kenntnis der Anilinfärbungen und ihrer Verwendung in der mikroskopischen Technik* (translated from German as *Contributions to the Knowledge of Aniline Dyes and Their Use in Microscopic Techniques*), while still a medical student. After his graduation in 1878, he accepted a position at the Charité-Universitätsmedizin hospital in Berlin where he spent the next nine years in productive research making observations on clinical cases and on hematology and histochemistry. From 1888 to 1890, he developed a persistent productive cough and found tubercle bacilli in his own sputum. He spent the winter in southern Italy and Egypt to rest and recover. After returning to Berlin, he was jobless and, with support from his wife's parents, established a small laboratory in an apartment.

In 1890, Robert Koch, the discoverer of the tubercle bacillus, *Mycobacterium tuberculosis*, secured a post for him as head of a clinical observation station in Berlin, and, in 1896, Ehrlich was given government support and facilities. During the period from 1890 until 1905, he pursued work in immunology, developing methods for producing high levels of antibodies and for quantitating the potency of antisera, and on the toxophore and haptophore groups and the side–chain theory. Ehrlich shared the Nobel Prize in Physiology or Medicine with Ilya Mechnikov in 1908 for their contributions to immunology. In about 1905, Ehrlich turned his attention to chemotherapy, culminating in successful clinical trials of arsphenamine for the treatment of syphilis in 1910. He died in 1915 at the age of 61 of diabetes with cardiovascular and renal disease. With his death, the early period of eosinophil research ended.

Ehrlich had the good fortune to be the beneficiary of a remarkable blossoming of chemistry in Northern European countries such as England, France, and Germany. In 1824, Justus von Liebig had established the world's first major school of chemistry and taught a generation of chemists, including names that have become bywords for chemistry, such as August Kekulé, Emil Erlenmeyer, and August Wilhelm von Hofmann. Von Hofmann determined the nature of aniline and laid the scientific basis for the dyestuff industry. In 1865, Friedrich Engelhorn established the Badische Anilin- und Soda-Fabrik (Baden Aniline and Soda Factory; BASF), and that company came to exemplify a special symbiosis between business and scientific research.[4] Heinrich Caro was trained in the laboratory of Robert Bunsen and joined BASF in 1868. He is regarded as being responsible for the company's successes in the dye industry and developed methylene blue. He and his colleagues built the company from a small enterprise to an international giant on the basis of the synthesis of chemical dyes. In 1874, he synthesized eosin by the reaction of fluorescein with bromine in glacial acetic acid. Evidently, Caro knew Greek mythology because he named the dye, eosin, by reference to Eos, the Titan goddess of the dawn. Eos is described by Nonnus of Panopolis as opening the gates of heaven with *rosy fingers* and golden arms so that Helios, her brother, could ride his chariot across the sky every day.[5]

Ehrlich transferred the use of dyes from staining biological fabrics, such as wool, silk, and cotton, to staining cells and tissues. His special genius was the recognition that dyes could make chemical distinctions, and, from the earliest days of his career, he employed them to investigate cells in blood and tissues. He discovered that cells stained differentially and that he could distinguish among them by their staining properties. His doctoral thesis, submitted to the University of Leipzig in 1878, dealt with the chemical composition and classification of aniline dyes and their use in general histology. The thesis included a section on granulated connective tissue cells for which he proposed the name *mastzellen* or *mast cells* (from the German verb *mästen*, meaning to fatten) because of his belief that mast cells occurred at sites with enhanced blood flow and nutrition.

When Ehrlich accepted the position at the Charité-Universitätsmedizin in 1878, he turned his attention to blood. He started with a modification of the simple procedure that Koch used for the examination of bacteria in which blood or other fluid is spread as thinly as possible, rapidly dried at room temperature by exposure to air, and then stained. Hirsch and Hirsch comment that the use of

this simple air-drying procedure was fortunate because the integrity of neutrophils and eosinophils is destroyed by the chemical fixation procedures used at that time.[3] Ehrlich distinguished among the granules of leukocytes, and he referred to them as alpha and beta granules. Alpha granules bound acid coal tar dyes, especially eosin. In 1879, in a publication, Ehrlich referred to the cells containing alpha granules as *eosinophils*.[6] Ehrlich's work was seminal in that, in addition to discovering eosinophils, neutrophils, basophils, lymphocytes, and mast cells, he also established the methods for counting blood cells so that quantitative associations with disease states could be made.[7] While Thomas Wharton Jones likely first identified eosinophils by virtue of their refractile granules,[8] his observations did not provide the tools for the ready identification and quantification provided by Ehrlich.

Histology of the Blood, written by Ehrlich and Adolf Lazarus, is presently available in eBook format from the Project Gutenberg website.[7] In it, Ehrlich and Lazarus describe the methods for staining blood cells, the need for meticulously produced glass coverslips, the dyes employed and the care for their use and, in the case of the eosinophil, the diseases associated with increased numbers of blood eosinophils. For example, counting eosinophils per volume of blood provided us with the normal numbers of blood eosinophils in healthy subjects vs. increased numbers of eosinophils in asthma,[9] pemphigus (one wonders if the patient more likely suffered from bullous pemphigoid),[10] urticaria,[11] helminthiasis,[12] and in numerous other clinical situations, including postfebrile periods (for example, after an acute attack of malaria), in malignant tumors, after splenectomy, and after medications. Moreover, Ehrlich was aware of the association between Charcot–Leyden crystals in feces and eosinophilia, especially in patients with helminthic infection. He found that eosinophils develop in the bone marrow before migrating to the blood, and that they are a distinct cell line, such that a transition between neutrophils and eosinophils is not observed.

Ehrlich suggested that eosinophils and neutrophils possessed different *chemotactic irritability* and that eosinophils only migrate to sites where a *specific stimulating substance* is present; here, he anticipated the discovery of chemokines, including eotaxins. He commented on observations by Gollasch that the sputum of asthma patients contained Charcot–Leyden crystals and only eosinophils,[9] and that a 'material which attracts the eosinophils' exists and on observations showing a 'close connection' between the severity of asthma and eosinophilia and on similar observations in skin diseases. Ehrlich contended with theories that eosinophils are produced locally and argued strenuously against this opinion while noting that mast cells are produced locally. To explain selective eosinophilia, he cited observations by Leichtenstern that blood eosinophilia diminished after a bacterial infection only to rise again once the infection subsided.[7] He stressed the importance of determining the absolute number of eosinophils as opposed to the percentage and cited examples of how using just the percentages of eosinophils in the blood can lead the observer astray. In a lecture given in Paris in 1900, Ehrlich noted that 'the leukocyte granulations are in fact secretory products, which the cell dissolves and spreads to the environment as needed.' In *Histology of the Blood*, Ehrlich and Lazarus state that 'The final link of the chain of proof of the secretory nature of the granules would be the direct observation of secretion by a granular cell.'[7] Remarkably, this insight took almost a century to be realized and was shown by the release of eosinophil granule proteins with a coating of tissues in various diseases, especially bronchial asthma[13] and atopic dermatitis.[14]

However, functions of the eosinophilic leukocyte had to wait characterization of the molecules comprising the cell, so that observations showing the release of granule proteins in disease could be made, and the development of genetic molecular technology, so that animals with excessive numbers of eosinophils and animals devoid of eosinophils could be produced.

EARLY DAYS: THE EOSINOPHIL AND ANAPHYLAXIS

At the time of Ehrlich's death in 1915, the procedures for determining the numbers of blood eosinophils, performed much as he had described, and their associations with disease were established. Staining of tissues by hematoxylin and eosin became routine by the early 1900s,[15] and the associations between eosinophil tissue infiltration and diseases were further investigated, for example, the relationship between the occurrence of eosinophilia and asthma in 1889 by Gollasch[9] and between eosinophilia and trichinosis in 1898 by Brown.[16] However, the analyses of complex mixtures of proteins using procedures that we today take as routine were primitive. This period of eosinophil research is well described by Samter.[2]

Among these advances, several stand out. The first was the demonstration of marked eosinophil tissue infiltration in severe asthma.[17,18] Huber and Koessler[18] comment that the 'coincidence of sputum and blood eosinophilia in the same individual seems to be a pathognomonic symptom of the asthmatic state' and further state, 'This tissue eosinophilia is a phenomenon of far-reaching bearing, which it seems to us, if completely understood, would undoubtedly greatly elucidate the pathogenesis of asthma.' The findings of these authors[17,18] corroborated the earlier work of German workers showing that the presence of eosinophils and Charcot–Leyden crystals are characteristic of asthma. Another advance was the finding in 1912 by Schlecht and Schwenker of an association between eosinophilia and

anaphylaxis.[19] They examined the lungs of guinea pigs that survived anaphylaxis and described massive peribronchial eosinophilia. Because anaphylaxis in the guinea pig is often a lethal disease, it was reasonable to conjecture that the presence of eosinophils is related to the prior anaphylactic event. In 1931, Berger and Lang established that eosinophils also infiltrate the site of an immediate-type skin reaction in passively sensitized humans.[20] The concept that the eosinophil was associated with immediate-type, anaphylactic sensitivity held sway for much of the 20th century. This concept was tested by Archer and colleagues who found that injection of an eosinophil extract and histamine diminished the intensity of the edema produced by intradermal histamine alone.[21] In the absence of the eosinophil extract, a 10 μg histamine injection produced a wheal of 25 mm, whereas addition of eosinophils (between 0.8×10^6 and 25×10^6) reduced the wheal in a dose-related manner to 10 mm. Observations such as this seemed to point to the role of the eosinophil as a critical cell for modulating inflammation, especially that mediated by histamine. However, in these experiments, controls testing the abilities of other cells to neutralize histamine were not included. The concept of the eosinophil as an anti-inflammatory cell was further investigated using biochemical analyses and demonstrating that eosinophils contain a histaminase and an arylsulfatase that might function to degrade histamine and the slow-reacting substance of anaphylaxis [later characterized as leukotriene C_4 (LTC_4)].[22] However, a direct test of the hypothesis that eosinophils function to modulate immediate hypersensitivity reactions conducted in guinea pigs, and abolishing eosinophils with a specific antiserum failed to support this hypothesis.[23] The hypothesis that the eosinophil alters tissues undergoing anaphylaxis, while not supported by subsequent work on granule proteins, may yet be validated by current work on eosinophil cytokines, in that numerous molecules produced by the eosinophil could affect local tissue homeostasis.

When I joined the Mayo Clinic staff in 1965, differential leukocyte counts were still being performed by staining blood smears and counting 100–200 cells manually, and Mayo employed about 30 workers for this purpose. In essence, the technology was comparable to that introduced by Ehrlich 70 years earlier. In 1968, the seven-parameter Coulter Counter Model S was introduced, and in the next decade a series of innovations for automated blood counting proceeded. By the 1990s, automated counting was the norm and the accuracy and precision of leukocyte counting were markedly improved. Presently, counting of eosinophils is performed automatically and is ordinarily ordered by physicians as part of the complete blood count. Still, blood cell counts may be verified by visual inspection of a blood smear. I am often struck by the failure of physicians to order the enumeration of leukocytes in a differential cell count with the loss of information that might be helpful to their patients.

EOSINOPHIL GRANULE PROTEINS AND THEIR PROPERTIES

If the eosinophil does not have a dominant role as a cell responsible for modulating hypersensitivity reactions, then what are its functions? During the late 1960s and extending for three decades, our laboratory at the Mayo Clinic investigated eosinophil granules and their proteins. These studies and others, particularly by Per Venge and colleagues,[24] showed that the granule is composed of markedly cationic proteins, including eosinophil peroxidase (EPO), eosinophil granule major basic protein 1 (MBP-1) and its homologue (MBP-2), and the eosinophil ribonucleases, RNase2 [also referred to as eosinophil-derived neurotoxin (EDN)], and RNase 3 [the eosinophil cationic protein (ECP)].[25] On a molar basis, MBP-1 is the predominant granule protein while EPO predominates on a weight basis.[26] Miller working with Palade found that the eosinophil granule has a distinctive morphology, which consists of a core (or crystalloid) with a regular two-dimensional structure, as detected by electron microscopy, and a matrix.[27] MBP-1 is localized to the core of the granule[28] and appeared to be its sole constituent,[29] while the other granule proteins localize to the matrix. The cationic nature of the proteins at once explained the avidity of eosin, an acidic dye, for the eosinophil and also provided an explanation for certain eosinophil functions because the cationic granule proteins are toxins. For example, the granule proteins, well exemplified by MBP-1,[30,31] are able to kill helminths, such as the schistosomula of *Schistosoma mansoni*,[32] *Trichinella* larvae, and the microfilariae of *Brugia pahangi* and *Brugia malayi*,[33] and also damage tissues including the bronchial epithelium, keratinocytes, pneumocytes, and the gut epithelium.[34] The toxic granule proteins are deposited on tissues during the course of disease and coat the affected tissues.[35] These observations hark back to Ehrlich's speculation that 'the leukocyte granulations are in fact secretory products, which the cell dissolves and spreads to the environment as needed.'[3]

The links between diseases and eosinophil granule protein deposition suggested a protective role in helminthic diseases and an inimical role for the eosinophil in other diseases, such as asthma and hypereosinophilic syndrome (HES). However, this evidence is only associative, and it was not possible to specifically enhance or ablate eosinophils in disease until more recently (see the sections on IL-5 and transgenic and knockout mice later in this chapter). Continuing investigations of the granule proteins, again using MBP-1 as an example, confirmed their toxicities, but also showed that they activated cells including eosinophils

themselves,[36] basophils,[37] neutrophils,[38] and mast cells.[39] They interact with other molecules such as the M2 muscarinic receptors,[40] as well as clotting[41] and complement[42] components, altering their functions. They stimulate platelets with a potency comparable to thrombin,[43] implicating the granule proteins as important molecules in clotting through platelet activation. This finding is likely relevant to the formation of thrombi and emboli in HES. The demonstration that the granule proteins possess RNase activity, both RNase2 (EDN) and RNase3 (ECP),[44] and that they are able to neutralize viruses as a consequence of RNase activity[45] suggest that eosinophils protect against virus infection. Further evidence for this capability came from studies of animal models.[46]

Eosinophils and bronchial asthma have had an intimate connection since their discovery. Once the granule proteins were identified and became available, we used them to probe their relationship with asthma. The sputum from patients with asthma had elevated levels of MBP-1,[47] and it was diffusely deposited on damaged bronchi in patients dying of asthma.[13] A key experiment revealed that both EPO and MBP-1 provoked bronchospasm when instilled into the lungs of monkeys, and MBP-1 induced bronchial hyperresponsiveness.[48] Antibody to MBP-1 prevented the development of bronchial hyperresponsiveness in a guinea pig model of asthma.[49,50] Finally, EPO and MBP-1 stimulated the bronchial epithelium to synthesize factors that alter structural cells and modify extracellular matrix composition and turnover.[51] Overall, these findings indicate that eosinophil granule proteins have the capacity to produce many of the characteristic pathological changes seen in asthma.

While these discoveries related eosinophil granule proteins to asthma, other work found that LTC4, which powerfully contracts respiratory airways, is produced by eosinophils.[52,53] These discoveries ultimately lead to the development of new therapies for asthma to prevent LTC4 formation or block its activity.[54] Similarly, platelet-activating factor and prostaglandins are produced by eosinophils and likely are important in the pathophysiology of asthma.[55]

INTERLEUKIN-5, EOTAXIN, AND EOSINOPHIL-DERIVED CYTOKINES

With the development and application of recombinant DNA technology during the 1980s, it became possible to synthesize and purify proteins, such as cytokines, that are present in such low concentrations that their isolation and characterization are otherwise extremely difficult or virtually impossible. Moreover, cytokines have varied biological activities, and it may be similarly difficult to ascribe a given activity to a particular purified molecule because it may contain unrecognized impurities that mediate its biological activities. Therefore, the standard for the discovery and naming of cytokines, in general, is the discovery of the gene for the molecule and the expression of the protein free of other cytokine contaminants. IL-5 was independently discovered by three groups. The discoveries were based on three different biological activities illustrating the pleiotropic activities of the molecule.

While each of these discoveries is of interest, the most cogent instance is likely the identification of the factor responsible for eosinophilia following parasite infection, this association between eosinophilia and helminthic infection dating from Ehrlich's time. Later, T lymphocytes were implicated in the regulation of eosinophil production,[56] and soluble eosinophilopoietic factors were found to be produced by murine lymphocytes.[57] Sanderson and colleagues investigated a murine helminthic infection model[58] and established a murine thymoma cell line that produced a factor causing eosinophil differentiation, termed eosinophil differentiation factor (EDF).[59] However, the factor possessing EDF activity also stimulated murine B lymphocytes,[60] and, because of the difficulty obtaining pure factors so that their biological activities could be characterized, this finding raised the question whether the factor contained two proteins with two activities or whether one protein possessed both activities. Purification of the T-cell factor demonstrated that one factor had both activities, but the structure of the factor remained determined. Subsequently, Japanese investigators, led by Takatsu, expressed the protein from cells possessing murine B-cell stimulation activity.[61] The gene sequence was unique and was termed *interleukin 5 (colony-stimulating factor, eosinophil) IL5*. A sequence from murine *IL5* was then used to probe a human genomic library,[62] resulting in the isolation of the human counterpart of the murine B-cell growth factor. Sequencing of the human *IL5* gene showed that it was homologous to the murine B-cell growth factor and that the protein coded by the gene had exclusively eosinophilopoietic activity. Subsequently, *IL5* was independently discovered by analyses of a murine T-helper cell clone that expressed several activities, including the abilities to cause immunoglobulin A (IgA) enhancement[63] and eosinophil colony formation. A cDNA sequence was isolated from these cells by expression cloning, and a human clone was isolated by cross-hybridization. Sequencing of the human and murine genes showed that they were identical to the previously isolated *IL5* gene referred to earlier. Thus IL-5 was discovered by its biological activities as a murine B-cell growth factor, as EDF, and as a human IgA-enhancing factor and eosinophil colony-stimulating factor. Subsequent studies showed that the B-cell growth factor activity is confined to mice[64] and that human IL-5 possesses predominately the ability to stimulate bone marrow cells to produce eosinophils. Identification of the

IL-5 receptor showed that it shared molecular structures with the receptors for granulocyte-macrophage colony-stimulating factor (GM-CSF) and IL-3.[65,66] Presently, the clinical effects of monoclonal antibodies directed against human IL-5 and against the alpha chain of the IL-5 receptor[67] are under study and already have shown beneficial effects in the treatment of HES[68] and bronchial asthma.[69,70] However, regrettably, none of these drugs is presently registered for clinical use.

Another key discovery during this time period was eotaxin.[71] The infiltration of eosinophils into guinea pig lung after anaphylaxis established in 1912 initially identified the relationship between anaphylaxis and eosinophil tissue infiltration,[19] and this model was used to discover eotaxin as a chemokine that particularly attracts eosinophils. Subsequently, three eotaxins have been described, each derived from a different chromosomal gene but with sufficient three-dimensional similarity that they function through the same chemokine receptor, C-C chemokine receptor type 3. For a time IL-5 was also thought to have eosinophilotactic properties, but a direct test failed to demonstrate appreciable activity.[72] IL-5 did function in concert with eotaxin to increase infiltrating eosinophils by increasing the numbers of available circulating cells.

Lastly, the eosinophil is able to produce numerous cytokines, including: C-C motif chemokine 5 (CCL5/RANTES); eotaxin [C-C motif chemokine 11 (CCL11)], eotaxin-2 [C-C motif chemokine 24 (CCL24)], and eotaxin-3 [C-C motif chemokine 26 (CCL26)]; GM-CSF; interferon gamma (IFN-γ); IL-1 alpha, IL-2–6, IL-8–10, IL-12, and IL-16; macrophage inflammatory protein 1-alpha [MIP-1-alpha; C-C motif chemokine 3 (CCL3)]; nerve growth factor (NGF); platelet-derived growth factor subunit B (PDGF subunit B); stem cell factor; transforming growth factor alpha (TGF-α) and beta (TGF-β); and tumor necrosis factor (TNF-α) (reviewed in[55]). The capacity to produce these factors opens new horizons for the eosinophil in the instigation and control of physiological and pathological processes, and these roles are yet to be thoroughly defined.

TRANSGENIC AND KNOCKOUT MICE FOR THE INVESTIGATION OF EOSINOPHILS

Modification of the genetic makeup of experimental animals became a reality during the latter part of the 20th century with the production of transgenic and knockout mice. These animals constitute an invaluable resource for the study of the eosinophil because of the unique capacities to increase eosinophil production by the overexpression of IL-5[73,74] and to abolish eosinophil production by either deleting a high-affinity transcription factor binding site,

GATA,[75] or by *poisoning* eosinophil production by coupling the promotor for EPO to the gene for diphtheria toxin.[76] The latter animal strain, termed PHIL, is essentially devoid of eosinophils. When sensitized with antigen, PHIL fails to develop a characteristic feature of experimental asthma, namely bronchial hyperresponsiveness.[76] PHIL also fails to produce the expected T-helper type 2 (T_h2) cytokines, including IL-4 and IL-5, ordinarily seen in this asthma model, and to accumulate T lymphocytes in the lung.[77] Reconstitution of PHIL with eosinophils (and T_h2-polarized ovalbumin-specific T cells) restores its ability to accumulate pulmonary T cells and T_h2 cytokines. Therefore, the eosinophil appears able to control the accumulation of T cells. This capability could be expressed in numerous diseases and opens a new window on the possibilities for eosinophil participation in disease.

The studies on murine models of asthma, performed in the laboratories of James J. Lee and Nancy Lee at the Mayo Clinic in Arizona, have also provided strong evidence that the eosinophil has the capacity to inflict pulmonary damage that closely mimicks that seen in severe clinical asthma.[78] As noted above, PHIL failed to develop the characteristic airway hyperresponsiveness seen in wild-type animals. The absence of the eosinophil in this experimental model suggests that it is critical for the development of bronchial hyperresponsiveness. James J. Lee and Nancy Lee developed another murine asthma model by creating a double transgenic mouse expressing IL-5 systemically and eotaxin-2 (CCL24) in the respiratory epithelium.[78] Remarkably, these double transgenic mice developed pulmonary pathologies characteristic of severe asthma with marked local eosinophil lung infiltration, epithelial desquamation, and mucous hypersecretion, and structural remodeling with airway obstruction, subepithelial fibrosis, airway smooth muscle hyperplasia, and striking methacholine-induced airway responsiveness. Indeed, the mice showed such a dramatic sensitivity to methacholine that many died during challenge. These mice showed extensive eosinophil degranulation as judged by release of EPO and MBP-1. A key experiment was then conducted mating the double transgenic animals to PHIL; the resultant animals showed neither eosinophil pulmonary infiltration nor bronchial hyperresponsiveness or any of the other features seen in the double transgenic. This powerful genetic experiment is difficult to refute and strongly implicates the eosinophil as an inflammatory cell able to mediate most, if not all, of the features of severe asthma.

The role of the eosinophil in helminthic disease, namely a murine model of trichinosis, has suggested that the eosinophil, rather than being critical for parasite destruction, may actually be important for helminth survival.[79] This surprising result suggests that the helminth is able to modify its behavior as a consequence of its environment.

EOSINOPHIL-ASSOCIATED DISEASES

From the time of Ehrlich's death until the 1970s, our knowledge of eosinophil-associated diseases appeared to change very little. The definition of HES and the subsequent investigations by Wolff and colleagues at the National Institutes of Health in the 1970s and 1980s allowed the classification of patients with marked eosinophilia into a reasonably discrete group such that clinical investigations could be done.[80–82] A subset of HES patients showed mucosal ulcers and appeared to have a poor prognosis.[83] During the 1990s, clonal lymphocytes producing IL-5 defined a HES subset.[84,85] In the first decade of the 21st century, treatment of HES patients with imatinib mesylate, introduced for the treatment of chronic myeloid leukemia, produced dramatic and durable responses,[86] and a deletion on chromosome 4 defined a kinase responsible for eosinophil proliferation.[87,88] Elevated serum tryptase levels identified a subset of HES patients with tissue fibrosis, poor prognosis, and imatinib mesylate responsiveness.[89]

During the 1980s and 1990s, Spanish toxic oil syndrome[90] and eosinophilia myalgia syndrome[91] were discovered, the former as an epidemic linked to rapeseed oil intended for industrial use and adulterated with aniline added to olive oil and sold in Madrid and environs, and the latter to tainted tryptophan; diffuse fasciitis (Shulman syndrome), described in 1975, is often seen in these patients.[92] Patients with episodic angioedema showed striking eosinophilia and had cyclical elevations of IL-5.[93,94] Patients with subcutaneous nodules, eosinophilia, rheumatism, dermatitis, and swelling (NERDS) appeared to constitute a novel syndrome.[95] Eosinophilic esophagitis emerged as an entity related to food allergy and gastro-esophageal reflux disorder.[96,97] Overall, these advances broadened the scope of eosinophil-associated diseases to encompass patients with marked eosinophilia, and with novel cutaneous syndromes and gastrointestinal diseases.

CONCLUSION AND FUTURE PERSPECTIVES

The 1970s and 1980s were the decades of eosinophil granule proteins. They were isolated, characterized, and used to investigate eosinophil-associated diseases. The proteins are toxins, are deposited on damaged tissues in disease, and likely mediate damage. They activate numerous cells including other leukocytes, mast cells, and lung epithelial cells, to secrete their mediators and to produce factors altering tissues. The 1990s was the decade of eosinophil cytokine production, and a remarkable number of cytokines are attributable to the eosinophil. These cytokines are able to mediate a vast number of changes in eosinophils themselves, in other leukocytes, and in the cells of organs afflicted by eosinophil-associated diseases. The past decade might be termed the era of mouse-manipulated models for eosinophil mechanisms in that mice essentially devoid of eosinophils showed the role of the cell in disease models, especially asthma, and revealed capabilities not heretofore suspected, especially the ability to regulate the infiltration of T lymphocytes. It is noteworthy that the ability of eosinophils to regulate T lymphocytes was unexpected because prior studies showed that T cells regulate eosinophils.[77,98]

Moreover, eosinophils are able to present antigen to immunocompetent cells[99] and, thus, function comparably to dendritic cells and B cells as antigen-presenting cells, albeit perhaps not as efficiently as *professional antigen-presenting cells*. Overall, the past three decades have strikingly changed our views of the eosinophil from a cell that is primarily associated with helminths and asthma to a cell with the capacity to instigate and mediate inflammation and tissue damage and one associated with numerous novel syndromes. A beacon for the future may be the surprising finding of IL-4-positive eosinophils in murine adipose tissue implicating the eosinophil in the control of adipose tissue deposition.[100] This finding may point the way to further studies of eosinophils in health and disease. Examples of observations suggesting broad roles for eosinophils in health and disease are studies of human tissues from healthy persons dying suddenly,[101] which showed eosinophils in lymphatic tissues, including the thymus, and in gastrointestinal tissues. Eosinophil function in the gut may be critical for the defense against pathogens, especially with the formation of eosinophil traps composed of mitochondrial DNA, released in a catapult-like manner, and granule proteins.[102] Another possible site for eosinophil involvement is the bone marrow itself. Eosinophil promyelocytes appear to degranulate, releasing eosinophil-derived molecules, including Charcot—Leyden crystal protein and MBP-1 into the bone marrow itself.[103–105] However, the functions of eosinophil-derived molecules in hematopoiesis, and in other aspects of human biology, are obscure. Much has been discovered since Ehrlich's initial descriptions, yet much remains to be learned. The eosinophil remains a beautiful and challenging cell.

ACKNOWLEDGEMENTS

My work was supported by the Mayo Foundation for 36 years and allowed our group to commence and continue investigations of the eosinophil and eosinophil-associated diseases. I was also supported by grants from the National Institute of Allergy and Infectious Diseases for over 35 years, and this support was critical for the expansion of granule protein studies, and the studies on eosinophilia myalgia syndrome and HES. Many colleagues contributed to the work from our laboratory and merit mention, including David Loegering, who was a substantial contributor to the asthma studies, Gail Kephart, who participated in the critical studies of release of eosinophil granule proteins into tissues, Dan Lewis, Donald Wassom, Steven Ackerman, Evan Frigas, Larry Pease, David McKean, Frank Prendergast, Rosa Ten, Catherine Weiler, Joseph Butterfield, Kimm Hamann, Rob Barker, Hirohito Kita and, my wife and colleague, Kristin Leiferman.

REFERENCES

1. Schwarz E. Die Lehre von der allgemeinen und ortlichen 'Eosinophilie'. *Ergebn Allg Pathol Pathol Anat* 1914;**17**:137.
2. Samter M. Eosinophils: The First 90 Years. In: Mahmoud AAF, Austen KF, editors. *The Eosinophil in Health and Disease*. Grune and Stratton; 1980. p. 25—40.
3. Hirsch JG, Hirsch BI. Paul Ehrlich and the Discovery of the Eosinophil. In: Mahmoud AAF, Austen KF, editors. *The Eosinophil in Health and Disease*. Grune and Stratton; 1980. p. 3—23.
4. Abelshauser W. *BASF: innovation and adaptation in a German corporation since 1865*. Cambridge University Press; 2003.
5. Wikipedia. Eos. 2011; Available from: http://en.wikipedia.org/wiki/Eos.
6. Ehrlich P. Beitrage zur Kenntnis der granulierten Bindegewebszellen und der eosinophilen Leukocyten. *Arch Anat Physiol (Physiol Abt)*; 1879:166.
7. Ehrlich P, Lazarus A, Myers W. *Histology of the Blood*. Available from, http://www.gutenberg.org/ebooks/29842; 1900 (accessed 25 May 2012).
8. Wharton Jones T. The blood corpuscle considered in its different phases of development in the animal series. *Philos Trans R Soc Lond* 1846;**1**:82.
9. Gollasch. Zur Kenntniss des asthmatischen Sputums. *Fortschritte d Med*; 1889. VII.
10. Neusser. Klinisch-haematologische Mittheilungen (Pemphigus). *Wien klin Woch*; 1892. Nos. 3 and 4.
11. Canon. Uber eosinophilen Zellen und Mastzellen im Blut Gersunder und Kranker. *Deutsche Med Woch*; 1892. no. 10.
12. Zappert J. Neuerliche Beobachtungen uber das Vorkommen des Ankylostomum duodenale bei den Bergleuten. *Wiener Klin Woch*; 1892. no. 24.
13. Filley WV, et al. Identification by immunofluorescence of eosinophil granule major basic protein in lung tissues of patients with bronchial asthma. *Lancet* 1982;**2**(8288):11—6.
14. Leiferman KM, et al. Dermal deposition of eosinophil-granule major basic protein in atopic dermatitis. Comparison with onchocerciasis. *The New England Journal of Medicine* 1985;**313**(5):282—5.
15. Smith C. Our debt to the logwood tree: the history of hematoxylin. *The Free Library*; 2006.
16. Brown TR. Studies on Trichinosis, with Especial Reference to the Increase of the Eosinophilic Cells in the Blood and Muscle, the Origin of These Cells and Their Diagnostic Importance. *The Journal of Experimental Medicine* 1898;**3**(3):315—47.
17. Ellis A. The pathological anatomy of bronchial asthma. *Am J Med Sci* 1908;**136**:407—29.
18. Huber H, Koessler KK. The pathology of bronchial asthma. *Arch Intern Med* 1922;**30**:689—760.
19. Schlecht H, Schwenker G. Uber lokale Eosinophilie in der Bronchien und in der Lunge anaphylaktischer Meerschweinchen. *Arch Exp Pathol Pharamacol* 1912;**68**:163.

20. Berger W, Lang FJ. Zur Histopathologie der idiosynkrasischen Entzundung in der menshlichen Haut VI. *Beitr Pathol Anat Allgem Pathol* 1931;**87**:71.

21. Archer RK, editor. *The Eosinophil Leukocytes*. F.A. Davis; 1963. p. 205.

22. Goetzl EJ, Wasserman SI, Austen F. Eosinophil polymorphonuclear leukocyte function in immediate hypersensitivity. *Archives of Pathology* 1975;**99**(1):1—4.

23. Gleich GJ, Olson GM, Loegering DA. The effect of ablation of eosinophils on immediate-type hypersensitivity reactions. *Immunology* 1979;**38**(2):343—53.

24. Venge P. The eosinophil granulocyte in allergic inflammation. *Pediatric allergy and immunology: official publication of the European Society of Pediatric Allergy and Immunology* 1993;**4**(4 Suppl):19—24.

25. Hamann KJ, et al. The molecular biology of eosinophil granule proteins. *International Archives of Allergy and Applied Immunology* 1991;**94**(1—4):202—9.

26. Abu-Ghazaleh RI, et al. Eosinophil granule proteins in peripheral blood granulocytes. *Journal of Leukocyte Biology* 1992;**52**(6): 611—8.

27. Miller F, de Harven E, Palade GE. The structure of eosinophil leukocyte granules in rodents and in man. *The Journal of Cell Biology* 1966;**31**(2):349—62.

28. Peters MS, Rodriguez M, Gleich GJ. Localization of human eosinophil granule major basic protein, eosinophil cationic protein, and eosinophil-derived neurotoxin by immunoelectron microscopy. Laboratory investigation. *A Journal of Technical Methods and Pathology* 1986;**54**(6):656—62.

29. Lewis DM, et al. Localization of the guinea pig eosinophil major basic protein to the core of the granule. *The Journal of Cell Biology* 1978;**77**(3):702—13.

30. Gleich GJ, et al. Physiochemical and biological properties of the major basic protein from guinea pig eosinophil granules. *The Journal of Experimental Medicine* 1974;**140**(2):313—32.

31. Gleich GJ, et al. Comparative properties of the Charcot-Leyden crystal protein and the major basic protein from human eosinophils. *The Journal of Clinical Investigation* 1976;**57**(3):633—40.

32. Butterworth AE, et al. Damage to schistosomula of Schistosoma mansoni induced directly by eosinophil major basic protein. *Journal of Immunology* 1979;**122**(1):221—9.

33. Hamann KJ, et al. In vitro killing of microfilariae of Brugia pahangi and Brugia malayi by eosinophil granule proteins. *Journal of Immunology* 1990;**144**(8):3166—73.

34. Gleich GJ, et al. Cytotoxic properties of the eosinophil major basic protein. *Journal of Immunology* 1979;**123**(6):2925—7.

35. Gleich GJ, Adolphson CR. The eosinophilic leukocyte: structure and function. *Advances in Immunology* 1986;**39**:177—253.

36. Kita H, et al. Eosinophil major basic protein induces degranulation and IL-8 production by human eosinophils. *J Immunol* 1995; **154**(9):4749—58.

37. O'Donnell MC, et al. Activation of basophil and mast cell histamine release by eosinophil granule major basic protein. *The Journal of Experimental Medicine* 1983;**157**(6):1981—91.

38. Shenoy NG, Gleich GJ, Thomas LL. Eosinophil major basic protein stimulates neutrophil superoxide production by a class IA phosphoinositide 3-kinase and protein kinase C-zeta-dependent pathway. *J Immunol* 2003;**171**(7):3734—41.

39. Piliponsky AM, et al. Effects of eosinophils on mast cells: a new pathway for the perpetuation of allergic inflammation. *Molecular Immunology* 2002;**38**(16-18):1369.

40. Costello RW, et al. Eosinophils and airway nerves in asthma. *Histology and Histopathology* 2000;**15**(3):861—8.

41. Slungaard A, et al. Eosinophil cationic granule proteins impair thrombomodulin function. A potential mechanism for thromboembolism in hypereosinophilic heart disease. *The Journal of Clinical Investigation* 1993;**91**(4):1721—30.

42. Weiler JM, Gleich GJ. Eosinophil granule major basic protein regulates generation of classical and alternative-amplification pathway C3 convertases in vitro. *Journal of Immunology* 1988;**140**(5):1605—10.

43. Rohrbach MS, et al. Activation of platelets by eosinophil granule proteins. *The Journal of Experimental Medicine* 1990;**172**(4): 1271—4.

44. Gleich GJ, et al. Biochemical and functional similarities between human eosinophil-derived neurotoxin and eosinophil cationic protein: homology with ribonuclease. *Proceedings of the National Academy of Sciences of the United States of America* 1986;**83**(10): 3146—50.

45. Domachowske JB, et al. Eosinophil cationic protein/RNase 3 is another RNase A-family ribonuclease with direct antiviral activity. *Nucleic Acids Research* 1998;**26**(14):3358—63.

46. Adamko DJ, et al. Ovalbumin sensitization changes the inflammatory response to subsequent parainfluenza infection. Eosinophils mediate airway hyperresponsiveness, m(2) muscarinic receptor dysfunction, and antiviral effects. *The Journal of Experimental Medicine* 1999;**190**(10):1465—78.

47. Frigas E, et al. Elevated levels of the eosinophil granule major basic protein in the sputum of patients with bronchial asthma. *Mayo Clinic Proceedings Mayo Clinic* 1981;**56**(6):345—53.

48. Gundel RH, Letts LG, Gleich GJ. Human eosinophil major basic protein induces airway constriction and airway hyperresponsiveness in primates. *The Journal of Clinical Investigation* 1991;**87**(4):1470—3.

49. Evans CM, et al. Pretreatment with antibody to eosinophil major basic protein prevents hyperresponsiveness by protecting neuronal M2 muscarinic receptors in antigen-challenged guinea pigs. *J Clin Invest* 1997;**100**(9):2254—62.

50. Lefort J, et al. In vivo neutralization of eosinophil-derived major basic protein inhibits antigen-induced bronchial hyperreactivity in sensitized guinea pigs. *The Journal of Clinical Investigation* 1996;**97**(4):1117—21.

51. Pegorier S, et al. Eosinophil-derived cationic proteins activate the synthesis of remodeling factors by airway epithelial cells. *Journal of Immunology* 2006;**177**(7):4861—9.

52. Jorg A, et al. Leukotriene generation by eosinophils. *The Journal of Experimental Medicine* 1982;**155**(2):390—402.

53. Weller PF, et al. Generation and metabolism of 5-lipoxygenase pathway leukotrienes by human eosinophils: predominant production of leukotriene C4. *Proceedings of the National Academy of Sciences of the United States of America* 1983;**80**(24): 7626—30.

54. Leff AR. Discovery of leukotrienes and development of antileukotriene agents. *Annals of Allergy, Asthma & Immunology: official publication of the American College of Allergy, Asthma, & Immunology* 2001;**86**(Suppl. 1—6):4—8.

55. Moqbel R, Lacy P, Adamko DJ, Odemuyiwa SO. Biology of Eosinophils. Adkinson: Middleton's Allergy: Principles and Practice. In: Adkinson N, Bochner BS, Busse WW, Holgate ST, Lemanske Jr RF, Simons ER, editors. 7th ed. Mosby; 2008. p. 295–308.

56. Basten A, Beeson PB. Mechanism of eosinophilia. II. Role of the lymphocyte. *The Journal of Experimental Medicine* 1970;**131**(6): 1288–305.

57. Miller AM, McGarry MP. A diffusible stimulator of eosinophilopoiesis produced by lymphoid cells as demonstrated with diffusion chambers. *Blood* 1976;**48**(2):293–300.

58. Strath M, Sanderson CJ. Detection of eosinophil differentiation factor and its relationship to eosinophilia in Mesocestoides corti-infected mice. *Experimental Hematology* 1986;**14**(1):16–20.

59. Warren DJ, Sanderson CJ. Production of a T-cell hybrid producing a lymphokine stimulating eosinophil differentiation. *Immunology* 1985;**54**(4):615–23.

60. O'Garra A, et al. Interleukin 4 (B-cell growth factor II/eosinophil differentiation factor) is a mitogen and differentiation factor for preactivated murine B lymphocytes. *Proceedings of the National Academy of Sciences of the United States of America* 1986;**83**(14): 5228–32.

61. Kinashi T, et al. Cloning of complementary DNA encoding T-cell replacing factor and identity with B-cell growth factor II. *Nature* 1986;**324**(6092):70–3.

62. Campbell HD, et al. Molecular cloning, nucleotide sequence, and expression of the gene encoding human eosinophil differentiation factor (interleukin 5). *Proceedings of the National Academy of Sciences of the United States of America* 1987;**84**(19):6629–33.

63. Yokota T, et al. Isolation and characterization of lymphokine cDNA clones encoding mouse and human IgA-enhancing factor and eosinophil colony-stimulating factor activities: relationship to interleukin 5. *Proceedings of the National Academy of Sciences of the United States of America* 1987;**84**(21):7388–92.

64. Clutterbuck E, et al. Recombinant human interleukin 5 is an eosinophil differentiation factor but has no activity in standard human B cell growth factor assays. *European Journal of Immunology* 1987;**17**(12):1743–50.

65. Mita S, et al. Characterization of high-affinity receptors for interleukin 5 on interleukin 5-dependent cell lines. *Proceedings of the National Academy of Sciences of the United States of America* 1989;**86**(7):2311–5.

66. Murata Y, et al. Molecular cloning and expression of the human interleukin 5 receptor. *The Journal of Experimental Medicine* 1992;**175**(2):341–51.

67. Busse WW, et al. Safety profile, pharmacokinetics, and biologic activity of MEDI-563, an anti-IL-5 receptor alpha antibody, in a phase I study of subjects with mild asthma. *The Journal of Allergy and Clinical Immunology* 2010;**125**(6):1237–44. e2.

68. Rothenberg ME, et al. Treatment of patients with the hypereosinophilic syndrome with mepolizumab. *The New England Journal of Medicine* 2008;**358**(12):1215–28.

69. Haldar P, et al. Mepolizumab and exacerbations of refractory eosinophilic asthma. *The New England Journal of Medicine* 2009;**360**(10):973–84.

70. Nair P, et al. Mepolizumab for prednisone-dependent asthma with sputum eosinophilia. *The New England Journal of Medicine* 2009;**360**(10):985–93.

71. Jose PJ, et al. Eotaxin: a potent eosinophil chemoattractant cytokine detected in a guinea pig model of allergic airways inflammation. *J Exp Med* 1994;**179**(3):881–7.

72. Collins PD, et al. Cooperation between interleukin-5 and the chemokine eotaxin to induce eosinophil accumulation in vivo. *J Exp Med* 1995;**182**(4):1169–74.

73. Dent LA, et al. Eosinophilia in transgenic mice expressing interleukin 5. *The Journal of Experimental Medicine* 1990;**172**(5): 1425–31.

74. Lee JJ, et al. Interleukin-5 expression in the lung epithelium of transgenic mice leads to pulmonary changes pathognomonic of asthma. *The Journal of Experimental Medicine* 1997;**185**(12):2143–56.

75. Yu C, et al. Targeted deletion of a high-affinity GATA-binding site in the GATA-1 promoter leads to selective loss of the eosinophil lineage in vivo. *The Journal of Experimental Medicine* 2002; **195**(11):1387–95.

76. Lee JJ, et al. Defining a link with asthma in mice congenitally deficient in eosinophils. *Science* 2004;**305**(5691):1773–6.

77. Jacobsen EA, et al. Allergic pulmonary inflammation in mice is dependent on eosinophil-induced recruitment of effector T cells. *The Journal of Experimental Medicine* 2008;**205**(3):699–710.

78. Ochkur SI, et al. Coexpression of IL-5 and eotaxin-2 in mice creates an eosinophil-dependent model of respiratory inflammation with characteristics of severe asthma. *Journal of Immunology* 2007;**178**(12):7879–89.

79. Fabre V, et al. Eosinophil deficiency compromises parasite survival in chronic nematode infection. *Journal of Immunology* 2009; **182**(3):1577–83.

80. Chusid MJ, et al. The hypereosinophilic syndrome: analysis of fourteen cases and review of the literature. *Medicine (Baltimore)* 1975;**54**:1–27.

81. Fauci AS, et al. NIH conference. The idiopathic hypereosinophilic syndrome. Clinical, pathophysiologic, and therapeutic considerations. *Ann Intern Med* 1982;**97**(1):78–92.

82. Parrillo JE, Fauci AS, Wolff SM. Therapy of the hypereosinophilic syndrome. *Ann Intern Med* 1978;**89**(2):167–72.

83. Leiferman KM, et al. Recurrent incapacitating mucosal ulcerations. A prodrome of the hypereosinophilic syndrome. *Jama* 1982;**247**(7):1018–20.

84. Simon HU, et al. Abnormal clones of T cells producing interleukin-5 in idiopathic eosinophilia. *N Engl J Med* 1999;**341**(15): 1112–20.

85. Roufosse F, et al. Clonal Th2 lymphocytes in patients with the idiopathic hypereosinophilic syndrome. *British Journal of Haematology* 2000;**109**(3):540–8.

86. Gleich GJ, et al. Treatment of hypereosinophilic syndrome with imatinib mesilate. *Lancet* 2002;**359**(9317):1577–8.

87. Cools J, et al. A tyrosine kinase created by fusion of the PDGFRA and FIP1L1 genes as a therapeutic target of imatinib in idiopathic hypereosinophilic syndrome. *N Engl J Med* 2003;**348**(13): 1201–14.

88. Griffin JH, et al. Discovery of a fusion kinase in EOL-1 cells and idiopathic hypereosinophilic syndrome. *Proc Natl Acad Sci USA* 2003;**100**(13):7830–5.

89. Klion AD, et al. Elevated serum tryptase levels identify a subset of patients with a myeloproliferative variant of idiopathic hypereosinophilic syndrome associated with tissue fibrosis, poor prognosis, and imatinib responsiveness. *Blood* 2003;**101**(12):4660–6.

90. Gelpi E, et al. The Spanish toxic oil syndrome 20 years after its onset: a multidisciplinary review of scientific knowledge. *Environ Health Perspect* 2002;**110**(5):457—64.

91. Hertzman PA, et al. Association of the eosinophilia-myalgia syndrome with the ingestion of tryptophan. *N Engl J Med* 1990;**322**(13):869—73.

92. Shulman LE. Diffuse fasciitis with eosinophilia: a new syndrome? *Transactions of the Association of American Physicians* 1975;**88**:70—86.

93. Gleich GJ, et al. Episodic angioedema associated with eosinophilia. *The New England Journal of Medicine* 1984;**310**(25): 1621—6.

94. Butterfield JH, et al. Elevated serum levels of interleukin-5 in patients with the syndrome of episodic angioedema and eosinophilia. *Blood* 1992;**79**(3):688—92.

95. Butterfield JH, Leiferman KM, Gleich GJ. Nodules, eosinophilia, rheumatism, dermatitis and swelling (NERDS): a novel eosinophilic disorder. *Clinical and Experimental Allergy: Journal of the British Society for Allergy and Clinical Immunology* 1993;**23**(7): 571—80.

96. Liacouras CA. Eosinophilic esophagitis in children and adults. *Journal of Pediatric Gastroenterology and Nutrition* 2003;**37**(Suppl. 1):S23—8.

97. Rothenberg ME. Biology and treatment of eosinophilic esophagitis. *Gastroenterology* 2009;**137**(4):1238—49.

98. Jacobsen EA, et al. Eosinophils: singularly destructive effector cells or purveyors of immunoregulation? *The Journal of Allergy and Clinical Immunology* 2007;**119**(6):1313—20.

99. Shi HZ, et al. Lymph node trafficking and antigen presentation by endobronchial eosinophils. *J Clin Invest* 2000;**105**(7):945—53.

100. Wu D, et al. Eosinophils sustain adipose alternatively activated macrophages associated with glucose homeostasis. *Science* 2011; **332**(6026):243—7.

101. Kato M, et al. Eosinophil infiltration and degranulation in normal human tissue. *The Anatomical Record* 1998;**252**(3):418—25.

102. Yousefi S, et al. Catapult-like release of mitochondrial DNA by eosinophils contributes to antibacterial defense. *Nature Medicine* 2008;**14**(9):949—53.

103. Scott RE, Horn RG. Fine structural features of eosinophile granulocyte development in human bone marrow. Evidence for granule secretion. *Journal of Ultrastructure Research* 1970;**33**(1):16—28.

104. Hyman PM, et al. Secretion of primary granules from developing human eosinophilic promyelocytes. Proceedings of the Society for Experimental Biology and Medicine. *Society for Experimental Biology and Medicine* 1978;**159**(3):380—5.

105. Butterfield JH, et al. Evidence for secretion of human eosinophil granule major basic protein and Charcot-Leyden crystal protein during eosinophil maturation. *Experimental Hematology* 1984; **12**(3):163—70.

Chapter 2

The Evolutionary Origins and Presence of Eosinophils in Extant Species

Michael P. McGarry

INTRODUCTION

Reason suggests that in metazoans of increasing cellular/histological size and complexity, there need to be some basic and common functions performed by cellular elements that will be prerequisite to and necessary for survival and reproduction. These include, among others:

- systemic integration of function;
- nutrient availability;
- respiratory gas exchange;
- separation from the environment;
- elimination of programmed cell death and metabolic waste products; and
- protection from injury and invasion.

Evolutionary theories and the associated *trees* (i.e., cladograms) have been built, in part, based on the presumed conservation of diverse features that confer competitive advantages within a population or environmental niche. More recently this has been complemented, and in most cases confirmed, by the more robust methods of gene sequencing and protein function similarities.[1,2] One feature that occurs in such seemingly unrelated species as insects and amphibians is the transformation of body form known as metamorphosis. While generally identified with morphological features, this hormone-driven phenomenon (ecdysone and thyroxine for insects and amphibians, respectively) involves many more complex changes at the biochemical, metabolic, physiological, cellular, and histological levels. How these changes conferred selective advantages and contributed to the evolution of species is not defined, nor is it understood how vestiges of these processes may contribute to current issues of health and disease. Furthermore, the function of eosinophil granulocytes, or their precursors, in these transformational events may offer insights into both the evolution of contemporary species as well as offer clues as to the role of these unique hemocytes in health and disease.

INVERTEBRATE HEMOCYTES

The organ specialization evident in more complex metazoans is attended by a need for systemic mechanisms to integrate and coordinate intracorporeal structures and functions. This presented opportunities for the selective benefit of motile cells that could adapt for these tasks. Some of these generic qualities were well defined by Brehelin and colleagues[3] in their description of blood cells of select species of insects. The putative selective functions for their set of circulating cells included lineage elements for *blood cells*. Some had dense and spherical cytoplasmic granules and other organelles, others were actively phagocytic, while yet others contributed to capsule formation. Although stains for light microscopy were said not to be definitive, some of the cells had *eosinophilic* granulation.[3] These features reflect the qualities of blood cell populations for the successful extant species of higher vertebrates. That is, there are specialized cells for coagulation (hemostasis), protection from environmental irritants and pathogens (phagocytosis and elimination of foreign bodies), respiratory gas exchange, and for functions related to structural development and maintenance (e.g., encapsulation).

Complicating any efforts to categorize the various lineages of circulating *blood cells*, or *hemocytes*, of invertebrates is the fact that differentiation often occurs in circulation. While some localization of hemocytopoiesis occurs, morphological variation and lineage plasticity[4] result in a variety of developmental morphologies for the complement of cells that defies uniform definitions. In addition, many other metazoan invertebrate species also contain cells in the hemolymph that, to a greater or lesser extent, contain granulation of varying dimensions. The majority of morphological descriptions is for cells examined by electron microscopy and thus precludes description of eosin affinity. However, the ultrastructural qualities and the variety of granule structures may be reflective of more than one lineage of granulocyte, which is likely to include an *acidophilic* class. Thus, descriptions exist in the literature of such cells in moths[5] and a variety of other insects.[3]

Eosinophils in Health and Disease. http://dx.doi.org/10.1016/B978-0-12-394385-9.00002-X

They are also described in oysters[6] and clams.[7] Moreover, Stein and Cooper[8] cite eosinophil granulocytes in annelid roundworms. Ravindranath reviews arthropod eosinophil granulocytes[9] and Andrew gives descriptive reference to eosinophil granulocytes in a range of invertebrate species.[10]

The question of which invertebrate species is the best to exhibit *wandering* cells with acidophilic cytoplasmic granules in the hemolymph is not as important as the notion that they are, in fact, evolutionarily older than those representative vertebrate species that have definitive granulocytes of the distinct lineage which has come to be known as the eosinophil granulocyte. It is difficult, if not impossible, then, to infer if any of these cell types represent primitive cells on the evolutionary path to the vertebrate cell we collectively identify as the eosinophil (leukocyte). Biochemical and/or molecular confirmation of these eosinophilic granule-containing hemocytes as the evolutionary precursors of vertebrate eosinophils are sketchy at best. Eosinophil peroxidase is one of two myeloperoxidases and is specific to eosinophils. Daiyasu and Toh[11] constructed a family association to explore the relationship between these two peroxidases specifically associated with myeloid granulocytes. Based on protein sequence alignment, an unrooted phylogenetic family tree was constructed. Assessment of subfamily groups led them to conclude that myeloperoxidase (neutrophil granulocytes) and eosinophil peroxidase diverged before the mammalian divergence some 60−70 million years ago. However, these analyses were not robust enough to suggest where in earlier invertebrate/vertebrate evolution specialization of unique individual granulocyte lineages occurred.

VERTEBRATE EOSINOPHIL GRANULOCYTES

Vertebrate eosinophil granulocytes have been recognized as a unique lineage for more than a century.[12] They have been studied in representative species from all classes of vertebrates by light and electron microscopy. Much is known of their development. All are part of populations with self-renewal and are maintained within relatively stable ranges, even after substantial blood loss, by a process of dynamic compensatory proliferation. Nomenclature notwithstanding, the eosinophil granulocyte of mammals is the cell defined as having a multilobed nucleus with cytoplasmic granulation that is readily stained by eosin, a normal constituent of Romanowsky and Giemsa dye preparations. These granules frequently exhibit an internal crystalline structure by electron microscopy. Most, if not all, of these features are evident in cells of the eosinophil lineage from all vertebrates. Some are even apparent in transitional forms of other subphyla of Chordata. For example, Wright and Cooper,[13] in their studies of inflammatory reactions in

ascidians (subphylum: Urochordata), describe cells with cytoplasmic electron-dense granules, some of which are acidophilic. The granulocytes participate in encapsulation of experimentally induced reactions to foreign bodies. Presumptive eosinophil granulocytes are described in the Giemsa-stained peripheral blood of larval lampreys (ammocetes) (superclass: Cyclostomata), presumed to be contemporary examples of early vertebrates.[14] These authors conclude, however, that '... although lamprey and mammalian eosinophil granulocytes share a limited number of morphological and functional features no conclusive evidence exists in the findings to date for any evolutionary significance in their (sic) results.' They also quote Wittekind,[15] who suggested the unreliability of 'Romanowsky-type dyes ... to be a reliable indicator for any homology ... simply reflect(ing) a basic chemical similarity with which these stains react.' However, as one progresses from cyclostomes to cartilaginous fish, studies in a representative species of cartilaginous fish, the dogfish shark, reveal definitive peripheral blood eosinophil granulocytes—with an excentric, lobular nucleus and electron-dense cytoplasmic granules.[16]

If some extant teleosts (bony fishes) may be considered representative of those that may have been the initial candidates of this Class when they first appeared (circa 450 million years ago), then the eosinophil granulocyte has already achieved clear morphological lineage distinction. One might presume, therefore, that whatever selective pressures may have contributed, or may have been exerted, to result in its favorable retention/refinement in the successful evolution of more complex and adapted species, had already occurred. Light (Giemsa-stained) and electron microscopy were used by Weinreb[17] to describe eosinophil granulocytes which have an ' ... eccentric nucleus, occupying one-third to one-half of the cell ....' The nucleus is further described as perhaps being ' ... indented, sausage-shaped or bilobed ....' The caveat is made that ' ... unlike mammalian eosinophil granules ... those in the teleost are less distinct, smaller and less refractile.' However, the granules ' ... are relatively large ... enclosing one or more internal bodies ... sharply delineated ... are denser than the matrix of the granule ... large enough to distort the shape of the granule.' Similar descriptions have been made for the loach.[18] The distinctions are not so clear in other teleosts. Reite and Evensen[19] describe the mast cell/eosinophil granule cell in an excellent review of inflammatory cells of teleost fish, proposing that the 'rodlet cell [of salmonids (sic)] may represent a type of eosinophil granulocyte,' resurfacing the question of lineage identity of the purported eosinophil granulocyte of these vertebrates. More recently, with the zebrafish as the study subject, Lieschke and colleagues[20] have applied light and electron microscopy as well as genomics and proteomics to an analysis of granulocytes and macrophages in embryos and adults.

Eosinophil granulocytes were negative for myeloperoxidase, which typically is found in the primary granules of neutrophils. Eosinophil granules were positive for peroxidase. By catalytic domain analysis they determined that:

'zebrafish peroxidase lay basal to the 3 closely related mammalian peroxidases (myeloperoxidase, eosinophil peroxidase, and lactoperoxidase), thereby complicating the naming of the zebrafish peroxidase … isolated. This phylogeny suggests that the gene duplication and diversification that occurred in mammals to create these various peroxidases occurred after the evolutionary divergence of fish and tetrapods.[20]*'*

In perhaps a not unfamiliar commentary in a comprehensive review, Carradice and Lieschke[21] summarize that 'the function of this … [adult zebrafish-(sic)] … eosinophil granulocyte has not been demonstrated, there are no molecular markers for it, its ontogeny is not described, and one must not infer function from nomenclature.'

Although debate may surround the exact character of eosinophils from fish, no such concerns exist with the other classes of vertebrata. One of the most insightful descriptions of amphibian blood and bone marrow, with specific reference to eosinophils, is given by Jordan.[22] Some other examples of light and electron microscopic descriptions of eosinophil granulocytes of amphibians include work by Turpen and colleagues[23] and Frank[24,25] for frogs, as well as Tooze and Davies[26] and Hightower and Haar[27] for newts. Granulocytes with eosin-positive cytoplasmic granules were typically identified; eosinophils were identified by electron microscopy, but crystalloid bodies within the granules were not consistent findings. However, there seems to be no ambiguity of their lineage identity in both premetamorphic and postmetamorphic frogs. This is also true for the description of eosinophil granulocytes in the Class: Aves. That is, both light and electron microscopic morphology of eosinophil granulocytes of many avian species have been classically defined by Maxwell[28,29,30] and display an undeniable similarity with mammalian eosinophils.

Descriptions of mammalian eosinophils are readily available in any number of classical texts (see, for example, 31). In addition, the light microscopic similarities of eosinophils in Giemsa, Wright, May–Grünwald–Giemsa and Domenici-stained peripheral blood and marrow preparations of many mammalian species are striking (M. P. M.; unpublished observations) This has included the yellow and olive baboon, bat, cat, cow, deer, dog, goat, guinea pig, hamster, horse, monkeys—rhesus and stump-tailed macaques—pig, rabbit, and rat, as well as several inbred, hybrid, random-bred, wild, mutant, and genetically altered strains of the laboratory mouse (see Fig. 2.1). It continues to be an enigmatic cell as to its place both functionally and as an embedded clue to the evolution of higher vertebrates.

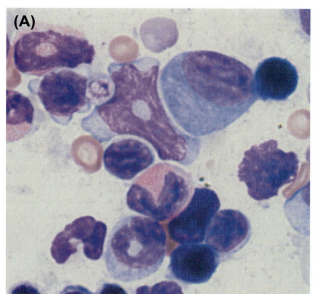

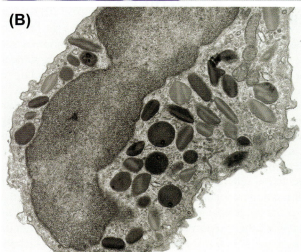

FIGURE 2.1 *(A)* Light photomicrograph of Romanowski-type stain of femoral marrow from a mature C57BL/6 mouse. Acidophilic granules, typical of mammalian eosinophils are evident in two cells; cells of other hemocytopoietic lineages are apparent. Morphologies and staining qualities are typical of most mammalian species, variations owing predominantly to granule size. *(B)* Electron micrograph of an eosinophil myelocyte from femoral marrow of a C57BL/6 mouse that exhibits the characteristic internal crystal structure of the specific cytoplasmic eosinophil granule. Its presence, considered definitive of the eosinophil lineage, is evident in mammals and several other non-mammalian species.

COMPARATIVE EOSINOPHILOPOIESIS

There are many attractive comparisons of hemocytopoiesis across the metazoan continuum. Even primitive invertebrates have hemocytopoiesis restricted to limited tissue sites.[32] Ghiretti-Magaldi and colleagues[33] concluded from their observations in the common littoral crab *Carcinus maenas* that three granulated and one agranular circulating hemocytes derive from a single, common stem cell. In vertebrates, the sites of hemocytopoiesis in general and eosinophilopoiesis in particular are especially conserved and are generally restricted to specialized organs. Even in the most primitive protochordates in which it has been studied, hemocytopoiesis is organ-specific.

As one ascends the vertebrate evolutionary *tree*, eosinophilopoiesis is found progressively in the liver, intestine, kidney, adrenal gland, thymus, spleen, and bone marrow (see 10). In mammals, this progression is followed developmentally, beginning in fetal liver, moving to the spleen, and eventually becoming established in the marrow of most of the bones of the mature body. Extramedullary hemocytopoiesis in mammals under stress (e.g., of excessive blood loss or other conditions related to overproduction or pathology) usually occurs in these developmental/more primitive hemocytopoietic sites.

SO, WHY EOSINOPHIL GRANULOCYTES?

The role and/or function of the eosinophil granulocyte in the vertebrate body have resisted definition. The arguments for a reconsideration of the classical understanding of this were recently put forward by Lee and colleagues.[34] It was proposed that '… accumulating tissue eosinophils are actually regulators of **L**ocal **I**mmunity **A**nd/or **R**emodeling/**R**epair in both health and disease—the **LIAR** hypothesis.' In it, the case is made that the role of eosinophils is not primarily in host defense. While there is ample evidence that these cells are well equipped for contributing to the elimination or containment of metazoan parasites and other select pathogens, consideration of other responses and the timing of their appearance in the evolutionary continuum would make it unreasonable to associate the eosinophil exclusively with conferring a selective advantage in the defense against parasitic infestation. Even the process of disgorging granule contents (i.e., degranulation) is not a consistently conserved function of all eosinophils in host defense mechanisms.[35] It is thus doubtful that this feature was a dominant determinant of a selectively advantageous function of evolving eosinophil granulocytes. In the LIAR hypothesis, eosinophil effector functions are proposed to be more closely related to remodeling events. The authors reference the observations of Jordan and Speidel[36] that the dramatic shortening of the frog tadpole gut at metamorphosis [M. P. M.; unpublished observation (see also[37])] is accompanied by a substantial infiltration of eosinophils. The metamorphosis of anuran amphibians is an elaborate process involving substantial histological and physiological changes: absorption of the extensive tadpole tail, development of complex multicellular skin glands, the aforementioned gut alterations as the animal changes from a herbivore to a carnivore, and gill resorption as gas

exchange occurs in newly developed lungs as the amphibian adapts from an aquatic to a terrestrial environment. These changes are dramatic and require metabolic, physiological, and anatomical alterations (i.e., remodeling). The role of eosinophils in these processes is not well defined, but warrants study. The evolutionary importance of the eosinophil may well be rooted in the selective advantages conferred on evolving species through its beneficial role in metamorphosis. Interestingly, Rheuben[5] observed large numbers of phagocytic granulocytic hemocytes accumulated on the surface of degenerating muscle fibers during the prepupal-to-first-day postecdysis in the moth, *Manduca sexta*. That is, granule-containing hemocytes are associated with tissue remodeling at metamorphosis in invertebrates!

Analysis of the eukaryotic orthologous groups (KOGs) of proteins has been applied to the dilemma of whether the foundational roots for the evolution of metazoans is based on Coelomata (with a true body cavity, e.g., arthropods and chordates) or Ecdysozoa (molting animals).[38] Although considerable controversy still prevails in the interpretation of data, suffice it to say that a significant base in the successful evolution of more complex metazoans occurred as earlier ancestors of contemporary vertebrates encountered selective pressures for survival. Either successive instars, which yielded intermediate periods for increased body growth among other refinements (Ecdysozoa), or possession of a body cavity, perhaps conferring advantages in organogenesis and anatomical composition (Coelomata), would be the determining structural advantage for the continued progressive evolution to early vertebrates. From that period through the emergence of reptiles, a midlife metamorphosis characterizes successfully evolved, more complex species. This process coincidently also reflects a substantial refinement of blood cell differentiation, especially as regards granulocytes and specifically the eosinophil. Primitive chordates also undergo a metamorphosis.[39] It seems reasonable to propose that evolution of higher vertebrates resulted from the successful adaptation of early vertebrate species that included a considerable body *makeover*, requiring remodeling events. Thus, could it be that residual functions of eosinophils have their roots in the breakdown and reconstruction of histological structures, especially those related to respiration and digestion?

It is enigmatic that the eosinophil granulocyte, so ubiquitous in vertebrate organisms, is still without a firm, exclusive function. Moreover, sophisticated genetic technologies have made it possible to extinguish the eosinophil lineage from at least one mammalian species, the mouse.[40] One might have anticipated that the absence of the lineage would result in some developmental or functional deficiency so as to make survival impossible. However, at least under the protective custody of contemporary husbandry for laboratory animals, no life-threatening abnormalities have become evident. This is even more enigmatic when one considers that all the most evolved vertebrate species have well-developed eosinophil granulocyte lineages! So the question remains: why eosinophils?

Though it may not be universally accepted, the most advanced vertebrates of our natural world are not the terminus of the evolutionary process, but simply a waypoint in which we reside. Recent data on structural and comparative biochemistry and the three-dimensional structure of eosinophil ribonucleases support the concept that evolutionary processes are still proceeding.[41,42] Perhaps, then, the more relevant question for contemporary eosinophilophiles is *'Eosinophils—Quo Vadis?'*

REFERENCES

1. Wang DY, Kumar S, Hedges SB. Divergence time estimates for the early history of animal phyla and the origin of plants, animals and fungi. *Proc R Soc Lond B* 1999;**266**:163—71.

2. Adoutte A, Balavoine G, Lartillot N, Lespinet O, Prud'homme B, de Rosa R. The new animal phylogeny: reliability and implications. *Proc Nat'l Acad Sci (USA)* 2000;**97**:4453—6.

3. Brehélin M, Zachary D, Hoffmann JA. A comparative ultrastructural study of blood cells from nine insect orders. *Cell Tissue Res* 1978;**195**:45—57.

4. Honti V, Csordás G, Márkus R, Kurucz E, Jankovics F, Andó I. Cell lineage tracing reveals the plasticity of the hemocyte lineages and of the hematopoietic compartments in Drosophila melanogaster. *Mol Immunol* 2010;**47**:1997—2004.

5. Rheuben MB. Degenerative changes in the muscle fibers of *Manduca sexta* during metamorphosis. *J Exp Biol* 1992;**167**:91—117.

6. Chang SJ, Tseng SM, Chou HY. Morphological characterization via light and electron microscopy of the hemacytes of two cultured bivalves: A comparison study between the hard clam (*Meretrix lustoria*) and Pacific oyster (*Crassostrea gigas*). *Zoological Studies* 2005;**44**:144—53.

7. Salimi L, Jamili S, Motalebi A, Eghtesadi-Araghi P, Rabbani M, Rostami-Beshman M. Morphological characterization and size of hemocytes in *Anodonta cygnea. J Invertebr Pathol* 2009;**101**:81—5.

8. Stein EA, Cooper EL. Inflammatory responses in annelids. *Amer Zool* 1983;**23**:145—56.

9. Ravindranath MH. The individuality of plasmatocytes and granular hemocytes of arthropods—a review. *Dev Comp Immunol* 1978;**2**:581—94.

10. Andrew W. *Comparative Hematology.* New York: Grune & Stratton; 1965.

11. Daiyasu H, Toh H. Molecular evolution of the myeloperoxidase family. *J Mol Evol* 2000;**51**:433—45.

12. Metchnikoff E. Varieties of leukocytes. Lecture VIII. In: *Lectures on the Comparative Pathology of Inflammation.* New York: Dover; 1891.

13. Wright RK, Cooper EL. Inflammatory reactions of the protochordata. *Amer Zool* 1983;**232**:205—11.

14. Rowley AF, Page M. Ultrastructural, cytochemical and functional studies on the eosinophilic granulocytes of larval lampreys. *Cell Tissue Res* 1985;**240**:705—9.

15. Wittekind DH. On the nature of Romanowsky—Giemsa staining and its significance for cytochemistry and histochemistry: an overall view. *Histochem J* 1983;**15**:1029—47.

16. Mainwaring G, Rowley AF. Separation of leucocytes in the dogfish (*Scyliorhinus canicula*) using density gradient centrifugation and differential adhesion to glass coverslips. *Cell Tissue Res* 1985;**241**:283—90.

17. Weinreb EL. Studies on the fine structure of teleost blood cells. I. Peripheral blood. *Anat Rec* 1963;**147**:219—38.

18. Ishizeki K, Nawa T, Tachibana T, Sakakura Y, Iida S. Hemopoietic sites and development of eosinophil granulocytes in the loach, *Misgurnus anguillicaudatus. Cell Tissue Res* 1984;**235**:419—26.

19. Reite OB, Evensen O. Inflammatory cells of teleostean fish: a review focusing on mast cells/eosinophilic granule cells and rodlet cells. *Fish Shellfish Immunol* 2006;**20**:192—208.

20. Lieschke GJ, Oates AC, Crowhurst MO, Ward AC, Layton JE. Morphologic and functional characterization of granulocytes and macrophages in embryonic and adult zebrafish. *Blood* 2001;**98**:3087—96.

21. Carradice D, Lieschke GJ. Zebrafish in hematology: sushi or science? *Blood* 2008;**111**:3331—42.

22. Jordan HE. The histology of the blood and the red bone marrow of the leopard frog, *Rana pipiens. Amer J Anat* 1919;**25**:437—80.

23. Turpen JB, Turpen CJ, Flajnik M. Experimental analysis of hematopoietic cell development in the liver of larval *Rana pipiens. Dev Biol* 1979;**69**:466—79.

24. Frank G. Granulopoiesis in tadpoles of *Rana esculenta.* Survey of the organs involved. *J Anat* 1988;**160**:59—66.

25. Frank G. Granulopoiesis in tadpoles of *Rana esculenta.* Ultrastructural observations on the developing granulocytes and on the development of eosinophil granules. *J Anat* 1989;**163**:97—105.

26. Tooze J, Davies HG. Light and electron microscopic observations on the spleen and the splenic leukocytes of the newt Triturus cristatus. *Am J Anat* 1968;**123**:521—56.

27. Hightower JA, Haar JL. A light and electron microscopic study of myelopoietic cells in the perihepatic subcapsular region of the liver in the adult aquatic newt, *Notophthalmus viridescens. Cell Tissue Res* 1975;**159**:63—71.

28. Maxwell MH. The development of eosinophils in the bone marrow of the fowl and the duck. *J Anat* 1978;**125**:387—400.

29. Maxwell MH. The fine structure of granules in eosinophil leucocytes from aquatic and terrestrial birds. *Tissue Cell* 1978;**10**:303—17.

30. Maxwell MH. Attempted induction of an avian eosinophilia using various agents. *Res Vet Sci* 1980;**29**:293—7.

31. Schalm OW, Jain NC, Carroll EJ. *Veterinary Hematology.* 3rd ed. Philadelphia: Lea & Febiger; 1975.

32. Jones JC. Hemocytopoiesis in Insects. In: Gordon AS, editor. *Regulation of Hematopoiesis*, vol. I. New York: Appleton Century Crofts; 1970. p. 7—75.

33. Ghiretti-Magaldi A, Milanesi C, Tognon G. Hemopoiesis in *crustacea decapoda*: Origin and evolution of hemacytes and cyanocytes of *Carcinus maenas. Cell Differentiation* 1977;**6**:167—86.

34. Lee JJ, Jacobsen EA, McGarry MP, Schleimer RP, Lee NA. Eosinophils in health and disease: the *LIAR* hypothesis. *Clin Exp Allergy* 2010;**40**:563—75.

35. Lee JJ, Lee NA. Eosinophil degranulation: an evolutionary vestige or a universally destructive effector function? *Clin Exp Allergy* 2005;**35**:986—94.

36. Jordan HE, Speidel CC. Blood cell formation and distribution in relation to the mechanism of thyroid-accelerated metamorphosis in the larval frog. *J Exp Med* 1923;**38**:529–41.

37. Marshall JA, Dixon KE. Cell specialization in the epithelium of the small intestine of feeding *Xenopus laevis* tadpoles. *J Anat* 1978;**126**:133–44.

38. Wolf YI, Rogozin IB, Koonin EV. Coelomata and not Ecdysozoa: evidence from genome-wide phylogenetic analysis. *Genome Res* 2004;**14**:29–36.

39. Nikitina N, Bronner-Fraser M, Sauka-Spengler T. *The sea lamprey* Petromyzon marinus: *a model for evolutionary and developmental*

biology. In: *Emerging Model Organisms: A Laboratory Manual*, vol. 1. Cold Spring Harbor: CSHL Press; 2009. Chapt. 16, p. 405–424.

40. Lee JJ, Dimina D, Macias MP, Ochkur SI, McGarry MP, O'Neill KR, et al. Defining a link with asthma in mice congenitally deficient in eosinophils. *Science* 2004;**305**:1773–6.

41. Zhang J, Dyer KD, Rosenberg HF. Evolution of the rodent eosinophil-associated RNase gene family by rapid gene sorting and positive selection. *Proc Nat'l Acad Sci (USA)* 2000;**97**:4701–6.

42. Rosenberg HF. RNase A ribonucleases and host defense: an evolving story. *J Leukoc Biol* 2008;**83**:1079–87.

Eosinophil Structure and Cell Surface Receptors

Introduction

Peter F. Weller

Eosinophils, like neutrophils and basophils, are bone marrow-derived granulocytes, so the granule compartments within each of these leukocytes are likely critical to their differential functional capabilities. For eosinophils, the rich content of cationic proteins within granules accounts for the cardinal tinctorial properties of these cells, properties responsible for the staining of eosinophils with acid aniline dyes such as eosin. The distinctive staining of eosinophils with eosin enabled the initial *discovery* of eosinophils by Paul Ehrlich in 1879[1] and anointed these distinct leukocytes with the name they now bear (*lovers of eosin*).

For the better part of a century following Ehrlich's cardinal recognition of eosinophils as distinct leukocytes, classic light microscopy was the sole modality to study and evaluate eosinophil involvement in health and disease. In 1974, Sheldon Cohen, a savvy clinician/investigator long interested in eosinophils, noted in an editorial in the *New England Journal of Medicine* that:

'The eosinophil is a hardy cell, completing its life cycle intact and morphologically unchanged. Thus, conventional light microscopy does not provide suitable criteria to indicate whether we are dealing with a cell of full functional potential or one that is spent.'[2]

The early applications of electron microscopy (EM) to visualize eosinophils documented that eosinophil-*specific* granules contained a cell lineage-specific *crystalline* core within their granules, an ultrastructural eosinophil signature perhaps more specific than Ehrlich's eosin staining. Initial EM and immunogold labeling EM studies focused on granules within eosinophils and their compartmentalized content of varied eosinophil proteins, and localized eosinophil granule major basic protein 1 (MBP-1) to the granule core and other cationic proteins [eosinophil cationic protein (ECP) and eosinophil peroxidase (EPO)] to

the matrix of the granules.[3] Another population of *small* granules was reported based on their ultrastructure and their cytochemically active arylsulfatase B activity.[4] These small granules are now no longer believed to be distinct granules but are vesicles derived from specific granules.[5,6] In fact, it is now recognized that eosinophils contain a morphologically distinct population of cytoplasmic large vesicles, centrally involved in secretion.[7] A population of *primary* granules that lack cores, as found in developing eosinophil progenitors, is likely a precursor of subsequently matured *specific* granules.[6] Thus, mature human eosinophils contain a single population of specific granules in contrast to neutrophils with their several types of granules.

With the recognitions in 1990 and thereafter that human eosinophils were sources of cytokines[8,9] and, in 1993, that eosinophil granules notably contained stores of preformed cytokine proteins,[10,11] attention has focused on the mechanisms whereby eosinophil granule-stored preformed proteins, including cationic proteins and cytokines, may be selectively and differentially secreted.[9] Notably, recent advances in EM methodologies, including better preservation of cytosolic vesicular compartments (lacking from most of the early EM studies), more sensitive EM immunonanogold labeling and detection strategies, and the applications of EM tomography, have revealed details about vesicle-mediated secretion [piecemeal degranulation (PMD)] by human eosinophils. These more sensitive mechanisms are not only recognizing granule-derived cationic proteins, including MBP-1, in likely granule-derived vesicles, but are also providing insights into the mechanisms of differential cytokine secretion by eosinophils, as reviewed in Chapter 3.2.

Likewise, an intracellular structure, often prominent in activated eosinophils and at times misjudged as a granule, the lipid body (also known as a lipid droplet), that has major roles in eicosanoid formation by eosinophils,[12] is having its onion-skin veneer unraveled to reveal its internal membranous ultrastructure in eosinophils and other leukocytes.[13]

In addition to the internal structure and function of human eosinophils, there has been a progressive increase in our understanding of eosinophil plasma membrane-expressed receptors. Following the report in 1974 of

Eosinophils in Health and Disease. http://dx.doi.org/10.1016/B978-0-12-394385-9.00003-1

a monospecific rabbit antihuman eosinophil serum,[14] there was the hope that eosinophils would express an eosinophil-unique cell surface marker.[2] To date, no unique eosinophil-specific cell surface marker has been identified, as noted in Chapter 3.3.

Nevertheless, the diversity of cell surface-expressed receptors provides evidence for the capacities of eosinophils to engage in multiple immune-mediated mechanisms. Of note, for many of the eosinophil surface-expressed receptors, eosinophils are known to elaborate ligands that engage these receptors, indicative either that they mediate autologous activation, or that they might contribute to secretion of the ligands from within eosinophils.[15]

ACKNOWLEDGMENTS

Studies were supported by National Institutes of Health grants AI020241, AI022571, and AI051645 to Peter F. Weller.

REFERENCES

1. Spry CJF. Eosinophils. *A comprehensive review and guide to the scientific and medical literature.* Oxford: Oxford Medical Publications; 1988.
2. Cohen SG. The eosinophil and eosinophilia (editorial). *N Engl J Med* 1974;**290**:457−9.
3. Egesten A, Aluments J, Meclenburg CV, Palmegren M, Olsson I. Localization of eosinophil cationic protein, major basic protein, and eosinophil peroxidase in human eosinophils by immunoelectron microscopic technique. *J Histochem Cytochem* 1986;**34**:1399−403.
4. Parmley RT, Spicer SS. Cytochemical and ultrastructural identification of a small type granule in human late eosinophils. *Lab Invest* 1974;**30**:557−67.
5. Egesten A, Weller PF, Olsson I. Arylsulfatase B is present in crystalloid-containing granules of human eosinophil granulocytes. *Int Arch Allergy Immunol* 1994;**104**:207−10.
6. Egesten A, Calafat J, Janssen H, Knol EF, Malm J, Persson T. Granules of human eosinophilic leucocytes and their mobilization. *Clin Exp Allergy* 2001;**31**:1173−88.
7. Melo RCN, Spencer LA, Dvorak AM, Weller PF. Mechanisms of eosinophil secretion: large vesiculotubular carriers mediate transport and release of granule-derived cytokines and other proteins. *J Leukoc Biol* 2008;**83**:229−36.
8. Wong DTW, Weller PF, Galli SJ, Rand TH, Elovic A, Chiang T, et al. Human eosinophils express transforming growth factor a. *J Exp Med* 1990;**172**:673−81.
9. Moqbel R, Coughlin JJ. Differential secretion of cytokines. *Sci STKE*; 2006. 2006:p.26.
10. Beil WJ, Weller PF, Tzizik DM, Galli SJ, Dvorak AM. Ultrastructural immunogold localization of tumor necrosis factor to the matrix compartment of human eosinophil secondary granules. *J Histochem Cytochem* 1993;**41**:1611−5.
11. Spencer LA, Szela CT, Perez SA, Kirchhoffer CL, Neves JS, Radke AL, et al. Human eosinophils constitutively express multiple Th1, Th2, and immunoregulatory cytokines that are secreted rapidly and differentially. *J Leukoc Biol* 2009;**85**:117−23.
12. Bozza PT, Yu W, Penrose JF, Morgan ES, Dvorak AM, Weller PF. Eosinophil lipid bodies: specific, inducible intracellular sites for enhanced eicosanoid formation. *J Exp Med* 1997;**186**:909−20.
13. Wan HC, Melo RCN, Jin Z, Dvorak AM, Weller PF. Roles and origins of leukocyte lipid bodies: proteomic and ultrastructural studies. *Faseb J* 2007;**21**:167−78.
14. Mahmoud AA, Kellermeyer RW, Warren KS. Production of monospecific rabbit antihuman eosinophil serums and demonstration of a blocking phenomenon. *N Engl J Med* 1974;**290**:417−20.
15. Spencer LA, Melo RCN, Perez SAC, Bafford SP, Dvorak AM, Weller PF. Cytokine receptor-mediated trafficking of preformed IL-4 in eosinophils identifies an innate immune mechanism of cytokine secretion. *Proc Natl Acad Sci U S A* 2006;**103**:3333−8.

Chapter 3.2

Eosinophil Ultrastructure

Rossana C.N. Melo, Ann M. Dvorak and Peter F. Weller

INTRODUCTION

Mature human eosinophils are readily distinguished by a major population of secretory granules termed secondary, crystalline, or specific granules. Eosinophils are also characterized by a prominent vesicular system, a polylobed nucleus with condensed, marginated nuclear chromatin, and osmiophilic lipid bodies (reviewed in[1−3]). Moreover, electron-dense glycogen particles and small numbers of primary granules, positive for Charcot−Leyden crystal protein, are seen in the eosinophil cytoplasm (reviewed in[3]).

SECRETORY GRANULES

Eosinophil-specific granules exhibit a distinctive morphology with a central crystalline core compartment and an outer matrix surrounded by a delimiting trilaminar membrane (Fig. 3.2.1). These granules store and secrete a large number of preformed proteins such as cytotoxic cationic proteins and an array of cytokines and chemokines (Fig. 3.2.1B). Because of their unique morphology, this granule population defines the eosinophil lineage in multiple species.[2,3]

Eosinophil-specific granules are not simply storage depots of preformed cationic and other proteins, but contain heretofore largely unrecognized internal membranotubular components within these compartmentalized organelles that undergo dynamic changes in their structure and contents in response to stimuli.[21] The contemporary application of techniques, including conventional electron microscopy (EM), immunogold labeling EM, and automated electron tomography, has revealed that eosinophil

(A)

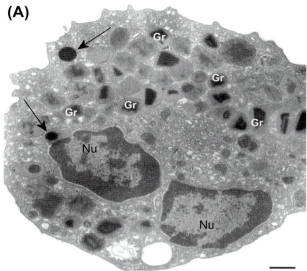

(B)

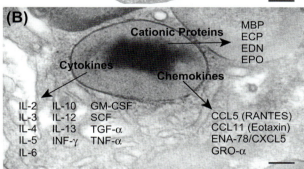

FIGURE 3.2.1   Ultrastructural image of a human eosinophil. *(A)* The eosinophil cytoplasm shows a major population of specific granules (Gr) which contain an internal, often electron-dense, crystalline core surrounded by an electron-lucent matrix, lipid bodies (arrows), transport vesicles, and a polylobed nucleus (Nu). *(B)* Multifunctional granule-stored products within human eosinophils. Eosinophil proteins documented within specific granules are listed. CCL5 (RANTES), C-C motif chemokine 5;[4] CCL11 (eotaxin), C-C motif chemokine 11;[5] ECP, eosinophil cationic protein;[6] EDN, eosinophil-derived neurotoxin;[7] CXCL5, C-X-C motif chemokine 5; ENA-78, epithelial-derived neutrophil-activating protein 78;[8] EPO, eosinophil peroxidase;[6] GM-CSF, granulocyte-macrophage colony-stimulating factor;[9,10] GRO-alpha, growth-regulated alpha protein;[11] IL-2, interleukin-2;[12] IL-3, interleukin-3;[10] IL-4, interleukin-4; IL-5, interleukin-5;[10,13] IL-6, interleukin-6;[14] IL-10, interleukin-10; IL-12, interleukin-12; IL-13, interleukin-13[15]; INF-γ, interferon gamma (INF-γ); MBP-1, eosinophil granule major basic protein 1;[16,17] NGF, nerve growth factor;[18] SCF, stem cell factor;[19] TGF-α, transforming growth factor-alpha; TNF-α tumor necrosis factor.[20] Bars: 1.2 μm *(A)*; 200 nm *(B)*.

granules have internal, membranous vesiculotubular domains[21] (Fig. 3.2.2). The presence of eosinophil intragranular membranes had been noted only occasionally in prior years in eosinophils from allergic subjects,[22] from individuals with Crohn disease,[23] and in platelet-activating factor (PAF)-stimulated human eosinophils.[24]

The presence of membrane domains within granules was confirmed more recently by immunonanogold labeling

for CD63 (Fig. 3.2.2A), a tetraspanin membrane protein.[21] CD63 is also consistently localized to the granule membranes[25] (Fig. 3.2.2A). Internal CD63-positive membranes have been recognized in other lysosome-related organelles such as platelet alpha granules[26] and major histocompatibility complex class II (MHC-II) compartments in dendritic cells.[27] However, the origin of the CD63-bearing membranes within eosinophil granules has not been ascertained. It is likely that these extensive intragranular membranous compartments emerge from endocytic recycling, from granule membranes, and/or from biosynthetic pathways; however, this remains to be delineated.

While conventional transmission EM (TEM) studies are usually performed on approximately 80-nm-thick sections, tomographic slices are only approximately 4 nm thick and offer a significant advantage over typical serial *thin sections* for tracking cell structures in three dimensions. Intragranular membranous subcompartments (Fig. 3.2.2B) were imaged in three-dimensional (3D) models, generated from electron tomography (Fig. 3.2.2C, D), as an aggregate of flattened tubular networks and tubules with interconnections in some planes and structural connections between the intragranular membranous network and the granule-limiting membrane. Intragranular structures appearing as round profiles in single, routine 80-nm sections of eosinophils were revealed by the 3D models to result from cross-sections of intragranular tubules rather than round vesicles.[21]

Tomographic reconstructions of human eosinophils also revealed that granule contents are rearranged within intragranular vesiculotubular compartments.[21,28] These findings have important functional implications, indicating that proteins may be specifically sorted and segregated within granule subcompartments before reaching the outer granule membrane to be delivered to the cell surface through vesicular compartments. It is now well established that differential release of cytokines occurs in response to specific stimuli[25,29] and the sorting of proteins into intragranular compartments makes them easier to be delivered to the cell surface through vesicular transport.

Activated eosinophils modulate immune responses and elicit effector functions through secretion. Hence, morphological changes in eosinophil secretory granules reflect the activation state of these granules. Eosinophil granules can show different stages of emptying which characterizes the secretory process of piecemeal degranulation (PMD), i.e., a *piece by piece* release of secretory granule contents in the absence of granule—granule or granule—plasma membrane fusion. This release is effected by the vesicular transport of small packets of materials from the cytoplasmic secretory granules to the cell surface (reviewed in[2,30]). Ultrastructural studies have demonstrated the presence of vesicles frequently attached to,

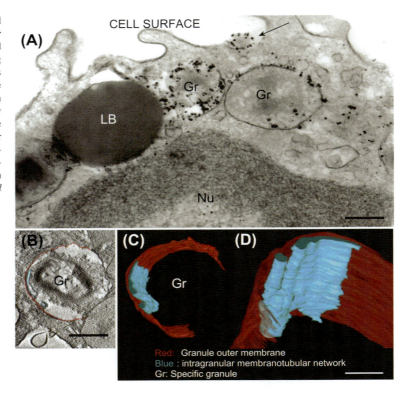

FIGURE 3.2.2 Secretory granules are highly labeled for CD63 and show membranes organized as a tubular network. *(A)* Immunogold electron microscopy revealed pools of CD63 within specific granules (Gr) undergoing depletion of their contents. Labeling for this tetraspanin is also observed at cell surface (arrow). *(B)* A representative tomographic slice (approximately 4 nm thick) obtained from a stimulated eosinophil secretory granule (GR) analyzed by automated electron tomography.[21] The outer granule membrane is partially traced in red and the intragranular vesiculotubular membranes in blue. *(C)* and *(D)* show three-dimensional models generated from serial tomographic slices. LB, lipid body; Nu, nucleus. Bars: 600 nm *(A)*; 500 nm *(B)*; 400 nm *(C)*; 180 nm *(D)*. Panels *(B–D) were reprinted from*[21] *with permission.*

surrounding, or budding from specific granules within activated human eosinophils.[2,30]

During PMD, secretory granules within eosinophils and other cells undergo a progressive emptying of their contents, as documented with TEM by the presence of lucent areas within their internal structure, reduced electron density (Fig. 3.2.3A), disassembled contents (Fig. 3.2.3A, asterisks) or membrane empty chambers (Fig. 3.2.3A, arrowheads).[21,31] Structural alterations within eosinophil secretory granules associated with PMD are described in diverse human inflammatory and allergic disorders including: asthma;[32,33] nasal polyposis;[33,34] allergic rhinitis;[33,35] ulcerative colitis;[33] Crohn disease;[33] atopic dermatitis;[36] gastric carcinoma;[37] shigellosis;[38] and cholera.[39] In this form of secretion, human eosinophils secrete granule matrix and/or core contents, but retain their intracellular granule containers. The results in EM images are eosinophils filled with partially empty, and/or fully empty, specific granules (reviewed in[2]) (Fig. 3.2.3A).

The numbers of emptying granules within human eosinophils increase when cells are activated, both *in vivo* and *in vitro*, in different conditions.[21,32,33,35] In addition, eosinophil-specific granules in the process of secreting their contents can be larger in size than resting granules in the same cell (Fig. 3.2.3A), a phenomenon likely due to the changes that occur within granules in response to degranulating stimuli.

Inflammatory stimuli, such as the classical eosinophil agonists, C-C motif chemokine 5 (CCL5/RANTES), C-C motif chemokine 11 (CCL11/eotaxin), or PAF (Fig. 3.2.3A), trigger PMD, and pretreatment with brefeldin A (BFA), an inhibitor of vesicular transport,[40] inhibits agonist-induced granule emptying.[21] Early aspects of stimulus-induced eosinophil PMD, when most granules did not yet show signs indicative of content losses, can be observed after 30 min of stimulation. At this time, granules developed into irregular structures with progressive protrusions from their surfaces, preferentially present on intact granules that had an ill-defined core and matrix.[21]

After 1 h of stimulation, specific granules show dramatic changes in their ultrastructure compared to those in unstimulated cells. In unstimulated eosinophils, granules are seen as round or elliptical structures with their classical morphology and full of contents. On stimulation, granule contents exhibit clear losses classically associated with PMD[21] (Fig. 3.2.3A).

Of interest, not all eosinophil-specific granules are uniformly, coordinately, and simultaneously responsive to stimuli. Studies using classical eosinophil agonists showed that whereas only 8% of granules in resting eosinophils had granules undergoing PMD, 25% (CCL5/RANTES), 43% (CCL11/eotaxin), and 34% (PAF)-activated eosinophils showed emptying granules.[21] Moreover, the responses of eosinophils were not uniformly distributed. For example, in

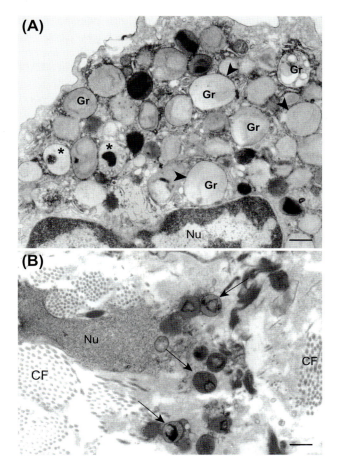

FIGURE 3.2.3 **Ultrastructural views of piecemeal degranulation and cytolysis (necrosis) in human eosinophils.** *(A)* Stimulation of blood eosinophils with platelet-activating factor (PAF) induced granule losses characteristic of piecemeal degranulation (PMD). Granules (Gr) showing reduced electron density, residual cores (asterisk), or membrane empty chambers (arrowheads) are clearly observed. *(B)* A tissue eosinophil undergoing cytolysis. Note the disintegrating nucleus and the entire membrane-bound secretory granules (arrows), which have spilled into the extracellular milieu. Blood eosinophils were isolated by negative selection from healthy donors, stimulated, immediately fixed, and prepared for transmission electron microscopy.[21] Tissue eosinophils were present in a skin biopsy performed on a patient with breast cancer who underwent treatment using recombinant human stem cell factor (SCF). CF, collagen fibrils; Nu, nucleus. Bars: 600 nm *(A)*; 800 nm *(B)*.

scoring the numbers of granules that exhibited loss of granule contents indicative of PMD, in resting cells, 70% of eosinophils had fewer than 10% of granules with losses whereas CCL11 (eotaxin) elicited a marked heterogeneity of granule emptying responses within eosinophils such that greater than 15% of eosinophils had more than 90% of their granules exhibiting content losses.[21] *In vivo*, the number of emptying eosinophil granules in nasal biopsies from patients with seasonal allergic rhinitis obtained before and after the pollen season has been evaluated.[35] Among the tissue eosinophils evaluated prior to the allergy season, an average of $37 \pm 3\%$ (mean $\pm$ SD) of the granules were

altered as a result of PMD, while the extent of PMD was increased to $87 \pm 2\%$ in association with seasonal pollen exposure.[35] Moreover, eosinophils showing signs of severe to complete loss of granule content were exclusively observed during the pollen season and the degree of eosinophil degranulation was correlated with levels of eosinophil cationic protein (ECP) in lavage fluids obtained at histamine challenge.[35]

Of note, eosinophils undergoing PMD sustain a pool of intact secretory granules. The presence of intact secretory granules intermingled with granules undergoing depletion of their contents has been described in different types of secretory cells, including other cells from the immune system such as mast cells and basophils and as such, this seems to be a general feature of PMD (reviewed in[41]). In eosinophils, this may contribute to the special capability of these cells, which undergo rapid release of their products under different or repetitive stimuli.[21,42]

Another mechanism by which granule-derived proteins may be specifically released is based on the responses of extracellular eosinophil granules. Eosinophil cytolysis is defined by EM by the presence of intact, membrane-bound granules released from eosinophils undergoing necrosis (Fig. 3.2.3B). This process has been recognized in diverse disorders (e.g., asthma, dermatitis). We demonstrated that cell-free eosinophil granules can function as independent secretory organelles capable of responding to cytokines (IFN-γ), chemokines (CCL11/eotaxin), and cysteinyl leukotrienes via cognate membrane-expressed receptors, topologically oriented with ligand-binding domains displayed externally on granule membranes.[43]

Eosinophils can also secrete through classical granule exocytosis, by which granules extrude their entire contents following granule fusion with the plasma membrane, including compound exocytosis, also involving intracellular granule–granule fusion before extracellular release (Fig. 3.2.4) (reviewed in[44]). Granule exocytosis is rarely documented during inflammatory responses and may occur when eosinophils interact with large targets, such as helminth parasites, while cytolysis in conjunction with PMD has been reported more frequently during human diseases. For example, in a recent study conducted in patients with nasal polyposis, 27.5% of eosinophils showed cytolysis, 30.7% were inactive, and 41.7% exhibited PMD.[34]

VESICULAR SYSTEM

Human eosinophils exhibit a prominent, morphologically distinct vesicular system. The eosinophil cytoplasm is filled with large tubular vesicles, termed eosinophil sombrero vesicles (EoSVs) (Fig. 3.2.5A), which in conjunction with small, classical round vesicles, represent pathways for the transport of granule products to the plasma membrane for extracellular release.[25,29,45]

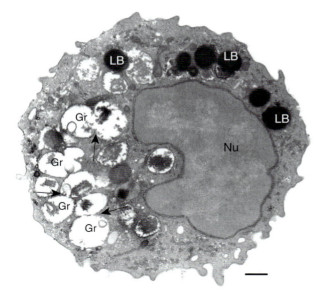

FIGURE 3.2.4 Compound granule exocytosis in a human, highly-activated eosinophil. This mode of secretion is characterized by granule–granule fusions (arrows) forming cytoplasmic channels. Note the presence of a great number of lipid bodies (LBs) with different sizes seen as very electron-dense organelles, while the nucleus and other organelles are less electron-dense. Cells were isolated from the peripheral blood, stimulated with the proinflammatory cytokine tumor necrosis factor-alpha (TNF-α) and processed for transmission electron microscopy using reduced osmium. Gr, secretory granules; Nu, nucleus. Bar: 600 nm.

The participation of large membrane-bound tubular compartments in the eosinophil secretory route was recognized only recently.[21,29,45,46] Although these vesiculotubular structures have long been identified by TEM in the cytoplasm of eosinophils,[3,22,47] little attention was given to them and their functional roles remained poorly understood for 30 years. In fact, EoSVs were previously reported as small granules, microgranules (reviewed in[3]), or even cup-shaped structures[47] in the early eosinophil literature.

While eosinophil secretory granules have been initially recognized for their content of cationic proteins and thereafter indicated as storage sites of preformed cytokines,[48] the ultrastructural immunolocalization of granule-stored proteins at transport vesicles was consistently documented only during the last 5 years.[21,25,45,46] The use of a pre-embedding immunolabeling approach—which is performed before standard EM processing and optimizes antigen preservation—and secondary antibodies conjugated to very small (1.4 nm) gold particles—to reach antigens at membrane microdomains such as vesicles—has enabled the identification of a consistent secretory vesicle traffic of granule-stored proteins from eosinophil-specific granules to the cell membrane.[25,45]

Immunogold EM studies have shown that both small round vesicles and EoSV compartments are positively

FIGURE 3.2.5 Ultrastructure of eosinophil sombrero vesicles. (A) Eosinophil sombrero vesicles (EoSVs; arrowheads) within an eotaxin-stimulated eosinophil are in contact with enlarged emptying granules (Gr) exhibiting reduced electron density. Intact, nonemptying granules (asterisk), with typical morphology are seen close to emptying granules. (B) EoSVs are labeled for eosinophil granule major basic protein 1 at their lumina. (C) and (D) Three-dimensional models of EoSVs generated from 4-nm-thick serial slices obtained by automated electron tomography. EoSVs are imaged as curved tubular and open structures surrounding a cytoplasmic center. Gr, granule; Nu, nucleus. Bar: 400 nm (A); 180 nm (B); 150 nm (C); and 100 nm (D). Panel (A) was published from,[46] panel (B) from,[57] and panels (C) and (D) from[45] with permission.

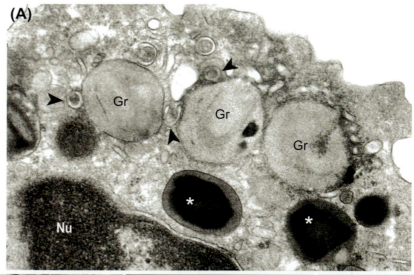

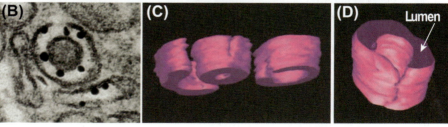

immunolabeled for typical eosinophil granule products such as interleukin-4 (IL-4) and eosinophil granule major basic protein 1 (MBP-1) (Fig. 3.2.5B),[21,25,45] recognized for a long time only within the cores of eosinophil granules.[49,50]

The recent EM identification of large vesicular structures as extragranular pools of MBP-1, even within resting cells, provides new insights into the complex task of protein secretion during immune responses.[25] These extragranular sites appear to be relevant for the rapid release of small concentrations of MBP-1 on cell activation without immediate disarrangement of the intricate crystalline cores within eosinophil-specific granules. This is important because it may underlie eosinophil functions as an immunoregulatory cell.[25] Interestingly, while MBP-1 is preferentially labeled within the vesicle lumen (Fig. 3.2.5B), IL-4 is clearly detected on the vesicle membrane,[45] a subcellular localization also described for transforming growth factor-alpha (TGF-α).[51] A functional implication of a membrane-bound vesicular transport of cytokines is that it adds support to the occurrence of selective release of products from eosinophils. In fact, the presence of selective, receptor-mediated mobilization and transport of cytokines is now recognized in the eosinophil secretory pathway.[29]

EoSVs present a typical morphology that resembles a *Mexican hat (sombrero)* when seen in conventional cross-sectional thin sections (Fig. 3.2.5A). These vesicles have a central area of cytoplasm and a brim of circular, membrane-delimited vesicle. EoSVs also exhibit a *C*-shaped morphology (Fig. 3.2.5A). Because of this particular morphology and large size (approximately 150–300 nm in diameter), EoSVs are easily identified by TEM in the cytoplasm of human eosinophils[28,46] (Fig. 3.2.5A).

EoSVs undergo a remarkable formation and redistribution within activated eosinophils. When eosinophils are stimulated with classical eosinophil agonists, such as CCL11 (eotaxin), there is an increase of the total number of cytoplasmic EoSVs.[45] In addition, EoSVs are more frequently observed surrounding and/or in contact with secretory granules[45] (Fig. 3.2.5A). By quantitative TEM, activation induced significant increases in the number of granule-attached EoSVs. Interestingly, the majority of these EoSVs (90%) were associated with granules showing ultrastructural changes related to the progressive release of their contents (Fig. 3.2.5A).

It was also demonstrated that the total number of EoSVs is significantly increased within eosinophils from a patient with a hypereosinophilic syndrome (HES).[25] Eosinophils from individuals with HES are typically activated,[52] compared to cells from normal donors. The identification of increased numbers of EoSVs in HES eosinophils is important because they would explain the reason for the loss of electron-dense cores (enriched in crystallized MBP-1) observed in tissue eosinophils from a range of disorders, such as Crohn disease, eosinophilic gastroenteritis, and HES.[53–55] In fact, deposition of MBP-1 can be demonstrated in affected tissues of patients with HES,[56] and vesicular trafficking is likely involved in this secretory mechanism.

Structural studies of EoSVs by EM revealed that these vesicles are curved tubular structures with remarkable plasticity (Fig. 3.2.5C, D). Along the length of EoSVs, both continuous, fully connected, cylindrical and circumferential domains, and incompletely connected and only partially circumferential curved domains, are observed.[45] These two domains explain both the *C*-shaped morphology of these vesicles and the presence of elongated tubular profiles very close to typical EoSVs, as frequently seen in two-dimensional, cross-sectional images of eosinophils (Fig. 3.2.5A). EoSVs present substantial membrane surfaces and are larger and more pleomorphic than the small, spherical vesicles (approximately 50 nm in diameter) classically involved in intracellular transport.[45,57] In fact, the findings using electron tomography highlight EoSVs as a dynamic system, with a remarkable ability to change their shape and to interact with secretory granules.[45,57] Another interesting finding revealed by TEM was the fact that EoSV formation can be inhibited by BFA,[21] a potential inhibitor of vesicular transport.[40] Studies using treatment with BFA showed that not only the total number of cytoplasmic EoSVs was reduced in the eosinophil cytoplasm, but also that membranes were clearly collapsed within secretory granules.[21] This observation pointed to a possible origin for EoSVs from secretory granules. To ascertain the budding of EoSVs from secretory granules, eosinophils were analyzed by electron tomography.[21] Tracking of vesicle formation by this technique revealed that EoSVs can indeed emerge from mobilized granules through a tubulation process.[21] This finding can explain the presence of these vesicles attached to and/or surrounding secretory granules, mainly granules undergoing the release of their contents (Fig. 3.2.5A).[46]

As noted, the morphology of EoSVs is quite distinct from that of conventional small transport vesicles and might offer several advantages to eosinophil secretion. First, it would provide a higher surface/volume ratio system for specific transport of membrane-bound proteins. Second, tubular carriers are more effective in dealing with the long distances that must be traversed in the cytoplasm until the cell surface is reached.[58] This fact might be particularly relevant for the rapid delivery of eosinophil preformed cytokines or other proteins. Third, EoSVs might serve as storage pools of specific proteins for rapid mobilization under stimulation and/or provide a more effective means to recycle granule membrane after the mobilization of granule products.[45]

The identification of granule-derived products at large transport carriers as revealed by EM adds support to a broader role for these large carriers in the intracellular trafficking and release of cytokines and other mediators from the immune system. In fact, while the functions of a number of cytokines and chemokines are well known, the intracellular pathways that lead to their secretion by different cells are only now beginning to emerge.[59] Immuno-EM studies have identified, for example, that the proinflammatory cytokines interleukin-6 (IL-6) and tumor necrosis factor-alpha (TNF-α) are loaded into large vesiculotubular structures budding from the trans Golgi network in activated macrophages.[60]

LIPID BODIES

Lipid bodies (LBs), also known as lipid droplets, are common organelles in the cytoplasm of human eosinophils (Fig. 3.2.1A and Fig. 3.2.2A). In these cells and others from the immune system, LBs are remarkably formed in response to inflammatory and infectious diseases and

different stimuli (Fig. 3.2.4), and are the focus for the catalyzed biosynthesis of inflammatory lipid mediators such as prostaglandins and leukotrienes.[61–64]

Eosinophil LBs appear as roughly round and very electron-dense organelles in TEM images (Fig. 3.2.6A). The high electron density of LBs in conjunction with the fact that LBs are not delimited by a true bilayer membrane enables their unambiguous ultrastructural identification within eosinophils (Fig. 3.2.1A, Fig. 3.2.2A, Fig. 3.2.4, and Fig. 3.2.6A). As a general feature of LBs in diverse cell types,[65,66] eosinophil LBs are composed of a neutral lipid core surrounded by a monolayer of phospholipids with associated proteins such as adipose differentiation-related protein (ADRP/adipophilin/perilipin-2) (Fig. 3.2.6B). Ultrastructural autoradiography and immunogold labeling have also demonstrated the incorporation of [³H]arachidonate by eosinophil LBs and the presence of TNF-α and enzymes involved in eicosanoid synthesis within them.[48,67,68]

Under favorable conditions, which may include postfixation with reduced osmium to increase contrast, LBs as small as 30 nm can be identified. In fact,

FIGURE 3.2.6 Lipid bodies in the cytoplasm of human eosinophils. *(A)* Fused (arrow), very electron-dense lipid bodies (LBs) are seen close to a secretory granule (Gr). Note that LBs are not surrounded by a typical bilayer membrane, thus differing from all other intracellular membranous organelles. *(B)* The LB periphery is heavily labeled by immuno-nanogold for adipose differentiation-related protein (ADRP/adipophilin/perilipin-2). *(C)* An organized membranous network is clearly seen within a LB. Note that typical endoplasmic reticulum cisternae (arrowheads) are seen in close association of another LB. An eosinophil sombrero vesicle (EoSV) is also closely associated with a LB. Bars: 400 nm *(A)* and *(B)*; 600 nm *(C)*; 300 nm. *Panel (C) was reprinted from[69] with permission.*

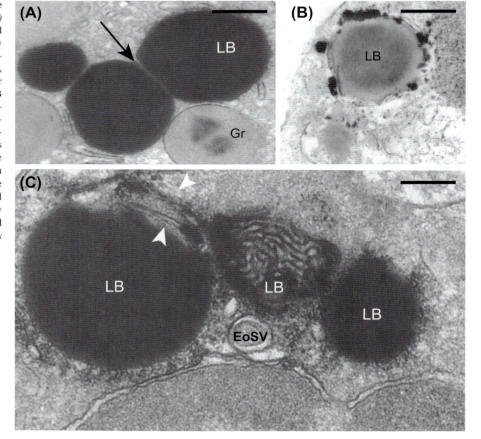

a morphological feature of LBs in most cells, including eosinophils, is their considerable variation in size. Large LBs can be formed in response to allergies, neoplasias, or *in vitro* stimulation with agonists (reviewed in[61]). Moreover, fused LBs are commonly seen in activated eosinophils (Fig. 3.2.6A).

A characteristic of LBs in eosinophils and other cells is their association with the endoplasmic reticulum (ER), which frequently appears around or even apparently intermingled in the periphery of LBs in conventional thin sections (Fig. 3.2.6C). Interestingly, ultrastructural analyses identified the presence of ribosomes attached to the circumferential surfaces of eosinophil LBs and even spread within their electron-dense core content.[69] The ribosomal localization at and within LBs in eosinophils may be linked to compartmentalized protein synthesis at LBs. This is fully consonant with prior ultrastructural localization studies of LBs in human mast cells:

1. [^{3}H]Uridine was shown to accumulate in LBs.
2. RNA was localized within LBs by hybridization with a ribonuclease-gold probe and by anti-RNA antibody immunogold labeling.
3. Poly(A) mRNA was detected within LBs by *in situ* hybridization with a poly(U) probe.
4. Several human autoimmune sera to ribosomal component proteins immunolabeled LBs.[70,71]

Eosinophil LBs were also observed by TEM in close association with the nucleus (Fig. 3.2.4) and EoSVs (Fig. 3.2.6C), but the meaning of these interactions remains to be delineated.

By conventional TEM, LBs from human eosinophils, as well as from other cells, generally appear as homogeneous organelles. However, recent studies indicate that the LB core is not a simple mass of lipid esters. Comprehensive examinations of thin sections prepared by conventional TEM[69] have pointed out that LBs in different cell types, including eosinophils, have a much more complex organization. Indeed, conventional TEM (Fig. 3.2.6C)[69] revealed the presence of membranes within LBs. The membranous system observed inside LBs may explain how membrane proteins stay associated with the LB core in different cell types. In eosinophils and other immune cells, for example, membrane-inserting proteins such as 5-lipoxygenase, leukotriene C$_4$ synthase, and prostaglandin-endoperoxide synthases (cyclooxygenases),[62,64,67] all linked to the synthesis of eicosanoid mediators, were localized fully within LBs. The identification of organized internal membranes within LBs may also be important to understand how LBs interact with membranous organelles and vesicles and manage intracellular trafficking.

The biogenesis of LBs is not fully understood. Although it is accepted that LBs originate from the ER,

there are two models to explain how they are formed. The classical model of LB biogenesis has neutral lipids accumulating between the cytoplasmic and luminal leaflets of the ER membranes followed by the budding off of LBs surrounded by a phospholipid monolayer derived from the cytoplasmic leaflets of the ER membranes.[72] In our newer, second model, proposed from studies using human eosinophils and other leukocytes, LBs would be formed by incorporating multiple loops of the ER membranes (both cytoplasmic and luminal leaflets of membranes within the developing LB). Accumulations of neutral lipids would develop among the membranes within LBs.[69]

The dynamic morphological aspects of LBs in eosinophils and other leukocytes as inducible, newly formed organelles, elicited in response to stimuli that lead to cellular activation contribute to the evolving understanding of LBs as organelles that are critical regulators of different inflammatory diseases, key markers of leukocyte activation, and attractive targets for novel anti-inflammatory therapies.

ACKNOWLEDGEMENTS

The work of the authors is supported by the National Institutes of Health, USA (grants AI020241, AI051645, and AI022571), the Conselho Nacional de Desenvolvimento Científico e Tecnológico (CNPq, Brazil), and the Fundação de Amparo a Pesquisa do Estado de Minas Gerais (FAPEMIG, Brazil). We thank Rita Monahan-Earley, Tracey Sciuto, Ellen Morgan (Electron Microscopy Unit, Department of Pathology, Beth Israel Deaconess Medical Center, Harvard Medical School), and Wim Voorhout of the FEI Company (Eindhoven, The Netherlands) for helpful discussions and previous electron microscopy assistance.

REFERENCES

1. Melo RCN, Dvorak AM, Weller PF. Contributions of electron microscopy to understand secretion of immune mediators by human eosinophils. *Microsc Microanal* 2010;**16**:653—60.
2. Melo RCN, Weller PF. Piecemeal degranulation in human eosinophils: a distinct secretion mechanism underlying inflammatory responses. *Histol Histopathol* 2010;**25**:1341—54.
3. Dvorak AM, Weller PF. Ultrastructural Analysis of Human Eosinophils. In: Marone G, editor. *Human Eosinophils: Biological and Chemical Aspects*. Basel: Karger; 2000. p. 1—28.
4. Lacy P, Mahmudi-Azer S, Bablitz B, Hagen SC, Velazquez JR, Man SF, et al. Rapid mobilization of intracellularly stored RANTES in response to interferon-gamma in human eosinophils. *Blood* 1999;**94**:23—32.
5. Nakajima T, Yamada H, Iikura M, Miyamasu M, Izumi S, Shida H, et al. Intracellular localization and release of eotaxin from normal eosinophils. *FEBS Lett* 1998;**434**:226—30.
6. Egesten A, Alumets J, von Mecklenburg C, Palmegren M, Olsson I. Localization of eosinophil cationic protein, major basic protein, and eosinophil peroxidase in human eosinophils

by immunoelectron microscopic technique. *J Histochem Cytochem* 1986;**34**:1399–403.

7. Peters MS, Rodriguez M, Gleich GJ. Localization of human eosinophil granule major basic protein, eosinophil cationic protein, and eosinophil-derived neurotoxin by immunoelectron microscopy. *Lab Invest* 1986;**54**:656–62.

8. Persson T, Monsef N, Andersson P, Bjartell A, Malm J, Calafat J, et al. Expression of the neutrophil-activating CXC chemokine ENA-78/CXCL5 by human eosinophils. *Clin Exp Allergy* 2003;**33**:531–7.

9. Levi-Schaffer F, Lacy P, Severs NJ, Newman TM, North J, Gomperts B, et al. Association of granulocyte-macrophage colony-stimulating factor with the crystalloid granules of human eosinophils. *Blood* 1995;**85**:2579–86.

10. Desreumaux P, Delaporte E, Colombel JF, Capron M, Cortot A, Janin A. Similar IL-5, IL-3, and GM-CSF syntheses by eosinophils in the jejunal mucosa of patients with celiac disease and dermatitis herpetiformis. *Clin Immunol Immunopathol* 1998;**88**:14–21.

11. Persson-Dajotoy T, Andersson P, Bjartell A, Calafat J, Egesten A. Expression and production of the CXC chemokine growth-related oncogene-alpha by human eosinophils. *J Immunol* 2003;**170**:5309–16.

12. Levi-Schaffer F, Barkans J, Newman TM, Ying S, Wakelin M, Hohenstein R, et al. Identification of interleukin-2 in human peripheral blood eosinophils. *Immunology* 1996;**87**:155–61.

13. Moller GM, de Jong TA, Overbeek SE, van der Kwast TH, Postma DS, Hoogsteden HC. Ultrastructural immunogold localization of interleukin 5 to the crystalloid core compartment of eosinophil secondary granules in patients with atopic asthma. *J Histochem Cytochem* 1996;**44**:67–9.

14. Lacy P, Levi-Schaffer F, Mahmudi-Azer S, Bablitz B, Hagen SC, Velazquez J, et al. Intracellular localization of interleukin-6 in eosinophils from atopic asthmatics and effects of interferon gamma. *Blood* 1998;**91**:2508–16.

15. Woerly G, Lacy P, Younes AB, Roger N, Loiseau S, Moqbel R, et al. Human eosinophils express and release IL-13 following CD28-dependent activation. *J Leukoc Biol* 2002;**72**:769–79.

16. Gleich GJ, Loegering DA, Maldonado JE. Identification of a major basic protein in guinea pig eosinophil granules. *J Exp Med* 1973;**137**:1459–71.

17. Lewis DM, Lewis JC, Loegering DA, Gleich GJ. Localization of the guinea pig eosinophil major basic protein to the core of the granule. *J Cell Biol* 1978;**77**:702–13.

18. Toyoda M, Nakamura M, Makino T, Morohashi M. Localization and content of nerve growth factor in peripheral blood eosinophils of atopic dermatitis patients. *Clin Exp Allergy* 2003;**33**:950–5.

19. Hartman M, Piliponsky AM, Temkin V, Levi-Schaffer F. Human peripheral blood eosinophils express stem cell factor. *Blood* 2001;**97**:1086–91.

20. Beil WJ, Weller PF, Tzizik DM, Galli SJ, Dvorak AM. Ultrastructural immunogold localization of tumor necrosis factor-alpha to the matrix compartment of eosinophil secondary granules in patients with idiopathic hypereosinophilic syndrome. *J Histochem Cytochem* 1993;**41**:1611–5.

21. Melo RCN, Perez SAC, Spencer LA, Dvorak AM, Weller PF. Intragranular vesiculotubular compartments are involved in piecemeal degranulation by activated human eosinophils. *Traffic* 2005;**6**:866–79.

22. Okuda M, Takenaka T, Kawabori S, Ogami Y. Ultrastructural study of the specific granule of the human eosinophil. *J Submicrosc Cytol* 1981;**13**:465–71.

23. Dvorak AM, Monahan RA, Osage JE, Dickersin GR. Crohn's disease: transmission electron microscopic studies. II. Immunologic inflammatory response. Alterations of mast cells, basophils, eosinophils, and the microvasculature. *Hum Pathol* 1980;**11**:606–19.

24. Kroegel C, Dewar A, Yukawa T, Venge P, Barnes PJ, Chung KF. Ultrastructural characterization of platelet-activating factor-stimulated human eosinophils from patients with asthma. *Clin Sci (Lond)* 1993;**84**:391–9.

25. Melo RCN, Spencer LA, Perez SA, Neves JS, Bafford SP, Morgan ES, et al. Vesicle-mediated secretion of human eosinophil granule-derived major basic protein. *Lab Invest* 2009;**89**:769–81.

26. Heijnen HF, Debili N, Vainchencker W, Breton-Gorius J, Geuze HJ, Sixma JJ. Multivesicular bodies are an intermediate stage in the formation of platelet-alpha-granules. *Blood* 1998;**91**:2313–25.

27. Barois N, de Saint-Vis B, Lebecque S, Geuze HJ, Kleijmeer MJ. MHC Class II compartments in human dendritic cells undergo profound structural changes upon activation. *Traffic* 2002;**3**:894–905.

28. Melo RCN, Dvorak AM, Weller PF. Electron tomography and immunonanogold electron microscopy for investigating intracellular trafficking and secretion in human eosinophils. *J Cell Mol Med* 2008;**12**:1416–9.

29. Spencer LA, Melo RCN, Perez SA, Bafford SP, Dvorak AM, Weller PF. Cytokine receptor-mediated trafficking of preformed IL-4 in eosinophils identifies an innate immune mechanism of cytokine secretion. *Proc Natl Acad Sci U S A* 2006;**103**:3333–8.

30. Melo RCN, Dvorak AM, Weller PF. New aspects of piecemeal degranulation in human eosinophils. In: Durand M, Morel CV, editors. *New Research in Innate Immunity*. Hauppauge: Nova Science Publishers; 2008. p. 329–48.

31. Dvorak AM. *Ultrastructure of Mast Cells and Basophils*. Basel: S. Karger; 2005.

32. Karawajczyk M, Seveus L, Garcia R, Bjornsson E, Peterson CG, Roomans GM, et al. Piecemeal degranulation of peripheral blood eosinophils: a study of allergic subjects during and out of the pollen season. *Am J Respir Cell Mol Biol* 2000;**23**:521–9.

33. Erjefalt JS, Greiff L, Andersson M, Adelroth E, Jeffery PK, Persson CG. Degranulation patterns of eosinophil granulocytes as determinants of eosinophil driven disease. *Thorax* 2001;**56**:341–4.

34. Armengot M, Garin L, Carda C. Eosinophil degranulation patterns in nasal polyposis: an ultrastructural study. *Am J Rhinol Allergy* 2009;**23**:466–70.

35. Ahlstrom-Emanuelsson CA, Greiff L, Andersson M, Persson CG, Erjefalt JS. Eosinophil degranulation status in allergic rhinitis: observations before and during seasonal allergen exposure. *Eur Respir J* 2004;**24**:750–7.

36. Cheng JF, Ott NL, Peterson EA, George TJ, Hukee MJ, Gleich GJ, et al. Dermal eosinophils in atopic dermatitis undergo cytolytic degeneration. *J Allergy Clin Immunol* 1997;**99**:683–92.

37. Caruso RA, Ieni A, Fedele F, Zuccala V, Riccardo M, Parisi E, et al. Degranulation patterns of eosinophils in advanced gastric carcinoma: an electron microscopic study. *Ultrastruct Pathol* 2005;**29**:29−36.

38. Raqib R, Moly PK, Sarker P, Qadri F, Alam NH, Mathan M, et al. Persistence of mucosal mast cells and eosinophils in Shigella-infected children. *Infect Immun* 2003;**71**:2684−92.

39. Qadri F, Bhuiyan TR, Dutta KK, Raqib R, Alam MS, Alam NH, et al. Acute dehydrating disease caused by Vibrio cholerae serogroups O1 and O139 induce increases in innate cells and inflammatory mediators at the mucosal surface of the gut. *Gut* 2004;**53**:62−9.

40. Nebenfuhr A, Ritzenthaler C, Robinson DG. Brefeldin A: deciphering an enigmatic inhibitor of secretion. *Plant Physiol* 2002;**130**:1102−8.

41. Crivellato E, Nico B, Mallardi F, Beltrami CA, Ribatti D. Piecemeal degranulation as a general secretory mechanism? *Anat Rec* 2003;**274A**:778−84.

42. Spencer LA, Szela CT, Perez SA, Kirchhoffer CL, Neves JS, Radke AL, et al. Human eosinophils constitutively express multiple Th1, Th2, and immunoregulatory cytokines that are secreted rapidly and differentially. *J Leukoc Biol* 2009;**85**:117−23.

43. Neves JS, Perez SA, Spencer LA, Melo RCN, Reynolds L, Ghiran I, et al. Eosinophil granules function extracellularly as receptor-mediated secretory organelles. *Proc Natl Acad Sci U S A* 2008;**105**:18478−83.

44. Moqbel R, Coughlin JJ. Differential secretion of cytokines. *Sci STKE*; 2006. 2006, pe26.

45. Melo RCN, Spencer LA, Perez SAC, Ghiran I, Dvorak AM, Weller PF. Human eosinophils secrete preformed, granule-stored interleukin-4 (IL-4) through distinct vesicular compartments. *Traffic* 2005;**6**:1047−57.

46. Melo RCN, Spencer LA, Dvorak AM, Weller PF. Mechanisms of eosinophil secretion: large vesiculotubular carriers mediate transport and release of granule-derived cytokines and other proteins. *J Leukoc Biol* 2008;**83**:229−36.

47. Komiyama A, Spicer SS. Microendocytosis in eosinophilic leukocytes. *J Cell Biol* 1975;**64**:622−35.

48. Beil WJ, Weller PF, Peppercorn MA, Galli SJ, Dvorak AM. Ultrastructural immunogold localization of subcellular sites of TNF-alpha in colonic Crohn's disease. *J Leukoc Biol* 1995;**58**:284−98.

49. Moqbel R, Ying S, Barkans J, Newman TM, Kimmitt P, Wakelin M, et al. Identification of messenger RNA for IL-4 in human eosinophils with granule localization and release of the translated product. *J Immunol* 1995;**155**:4939−47.

50. Moller GM, de Jong TA, van der Kwast TH, Overbeek SE, Wierenga-Wolf AF, Thepen T, et al. Immunolocalization of interleukin-4 in eosinophils in the bronchial mucosa of atopic asthmatics. *Am J Respir Cell Mol Biol* 1996;**14**:439−43.

51. Egesten A, Calafat J, Knol EF, Janssen H, Walz TM. Subcellular localization of transforming growth factor-alpha in human eosinophil granulocytes. *Blood* 1996;**87**:3910−8.

52. Ackerman SJ, Bochner BS. Mechanisms of eosinophilia in the pathogenesis of hypereosinophilic disorders. *Immunol Allergy Clin North Am* 2007;**27**:357−75.

53. Dvorak AM, Weller PF, Monahan-Earley RA, Letourneau L, Ackerman SJ. Ultrastructural localization of Charcot−Leyden crystal protein (lysophospholipase) and peroxidase in macrophages, eosinophils, and extracellular matrix of the skin in the hypereosinophilic syndrome. *Lab Invest* 1990;**62**:590−607.

54. Dvorak AM. Ultrastructural evidence for release of major basic protein-containing crystalline cores of eosinophil granules in vivo: cytotoxic potential in Crohn's disease. *J Immunol* 1980;**125**:460−2.

55. Torpier G, Colombel JF, Mathieu-Chandelier C, Capron M, Dessaint JP, Cortot A, et al. Eosinophilic gastroenteritis: ultrastructural evidence for a selective release of eosinophil major basic protein. *Clin Exp Immunol* 1988;**74**:404−8.

56. Tai PC, Ackerman SJ, Spry CJ, Dunnette S, Olsen EG, Gleich GJ. Deposits of eosinophil granule proteins in cardiac tissues of patients with eosinophilic endomyocardial disease. *Lancet* 1987;**1**:643−7.

57. Melo RCN, Dvorak AM, Weller PF. Electron tomography reveals the 3D structure of secretory organelles in eosinophils. *Microscopy & Analysis* 2007;**21**:15−7.

58. Bonifacino JS, Lippincott-Schwartz J. Coat proteins: shaping membrane transport. *Nat Rev Mol Cell Biol* 2003;**4**:409−14.

59. Stow JL, Low PC, Offenhauser C, Sangermani D. Cytokine secretion in macrophages and other cells: pathways and mediators. *Immunobiology* 2009;**214**:601−12.

60. Manderson AP, Kay JG, Hammond LA, Brown DL, Stow JL. Subcompartments of the macrophage recycling endosome direct the differential secretion of IL-6 and TNF{alpha}. *J Cell Biol* 2007;**178**:57−69.

61. Bozza PT, Melo RCN, Bandeira-Melo C. Leukocyte lipid bodies regulation and function: Contribution to allergy and host defense. *Pharmacol Ther* 2007;**113**:30−49.

62. Bozza PT, Yu W, Penrose JF, Morgan ES, Dvorak AM, Weller PF. Eosinophil lipid bodies: specific, inducible intracellular sites for enhanced eicosanoid formation. *J Exp Med* 1997;**186**:909−20.

63. Melo RCN, Fabrino DL, Dias FF, Parreira GG. Lipid bodies: Structural markers of inflammatory macrophages in innate immunity. *Inflamm Res* 2006;**55**:342−8.

64. D'Avila H, Melo RCN, Parreira GG, Werneck-Barroso E, Castro-Faria-Neto HC, Bozza PT. Mycobacterium bovis bacillus Calmette-Guerin induces TLR2-mediated formation of lipid bodies: intracellular domains for eicosanoid synthesis in vivo. *J Immunol* 2006;**176**:3087−97.

65. Brasaemle DL. Thematic review series: adipocyte biology. The perilipin family of structural lipid droplet proteins: stabilization of lipid droplets and control of lipolysis. *J Lipid Res* 2007;**48**:2547−59.

66. Tauchi-Sato K, Ozeki S, Houjou T, Taguchi R, Fujimoto T. The surface of lipid droplets is a phospholipid monolayer with a unique fatty acid composition. *J Biol Chem* 2002;**277**:44507−12.

67. Dvorak AM, Morgan ES, Tzizik DM, Weller PF. Prostaglandin endoperoxide synthase (cyclooxygenase): ultrastructural localization to nonmembrane-bound cytoplasmic lipid bodies in human eosinophils and 3T3 fibroblasts. *Int Arch Allergy Immunol* 1994;**105**:245−50.

68. Weller PF, Dvorak AM. Arachidonic acid incorporation by cytoplasmic lipid bodies of human eosinophils. *Blood* 1985;**65**:1269−74.

69. Wan HC, Melo RCN, Jin Z, Dvorak AM, Weller PF. Roles and origins of leukocyte lipid bodies: proteomic and ultrastructural studies. *FASEB J* 2007;**21**:167−78.

70. Dvorak AM, Morgan ES, Weller PF. RNA is closely associated with human mast cell lipid bodies. *Histol Histopathol* 2003;**18**:943−68.

71. Dvorak AM. Mast cell secretory granules and lipid bodies contain the necessary machinery important for the in situ synthesis of proteins. *Chem Immunol Allergy* 2005;**85**:252—315.

72. Robenek MJ, Severs NJ, Schlattmann K, Plenz G, Zimmer KP, Troyer D, et al. Lipids partition caveolin-1 from ER membranes into lipid droplets: updating the model of lipid droplet biogenesis. *FASEB J* 2004;**18**:866—8.

Chapter 3.3

Eosinophil Receptor Profile

Virginie Driss, Fanny Legrand and Monique Capron

INTRODUCTION

The first attempts at a characterization of eosinophil surface molecules began in the 1970s. Then, pioneering studies suggested that eosinophils expressed numerous cell surface markers, including adhesion molecules, receptors for complement, chemotactic factors, and immunoglobulins. Since then, new immune receptors have been discovered, such as toll-like receptors (TLRs), and inhibitory receptors, such as sialic acid-binding Ig-like lectins (Siglecs). Thanks to the development of new reagents and new animal models, the list has been extended, revealing a vast array of receptors and structures involved in the immune response expressed on the eosinophil surface (Fig. 3.3.1). The purpose of this chapter is to summarize what is known about the surface phenotype of eosinophils, with a focus on receptors involved in the immune response.

Cytokine Receptors

Cytokines are essential for hematopoietic cell development, differentiation, maturation, trafficking, and activation. As expected, eosinophils express receptors for the three major cytokines needed for their differentiation and maturation, specifically interleukin-3 receptor subunit alpha (IL-3RA/CD123), interleukin-5 receptor subunit alpha (IL-5RA/CD125), and granulocyte-macrophage colony-stimulating factor receptor subunit alpha (GM-CSF-R-alpha/CD116).[1] The major role of IL-5RA and eosinophils in allergen-induced airway remodeling has been reported in experimental models of allergic asthma, using mice lacking IL-5RA.[2] A humanized afucosylated immunoglobulin G1 (IgG1) anti-human IL-5RA monoclonal antibody (MEDI-563, formerly BIW-8405) that neutralizes IL-5RA activity and depletes tissue eosinophils in preclinical models is presently under phase I studies in patients with mild asthma.[3]

Cytokine receptors are present at relatively low levels. Studies have demonstrated receptors for interferon gamma (IFN-γ), both CD120a and CD120b for tumor necrosis factor (TNF-α), for CD124, CD132, and for CD129.[4] The existence of most cytokine receptors was first suggested by functional investigations, using cytokines that activated eosinophils *in vitro*; surface expression was subsequently confirmed by flow cytometry. Interestingly, receptors for IFN-γ and C-C chemokine receptor type 3 (CCR3/eosinophil eotaxin receptor) are also present on eosinophil granule membranes that function as receptor-mediated secretory organelles.[5] Eosinophils have been recently described as expressing receptors for interleukin-17A [interleukin-17 receptor A (IL-17RA)] and -17F (IL-17F), two members of the IL-17 family, and a receptor for interleukin-23 (IL-23R), three cytokines that play crucial roles in allergic inflammation. This receptor expression provides a new insight into T-helper type 17 (T$_h$17) lymphocyte-mediated activation of eosinophils via differential intracellular signaling cascades in allergic inflammation.[6]

Adhesion Molecules

Eosinophil accumulation at sites of inflammation is the result of several processes involving selective adhesion and migration. Transmigration of the eosinophil through the vascular endothelium is a multistep process involving rolling, tethering, firm adhesion, and transendothelial migration.

Eosinophils constitutively express L-selectin (CD62L), which interacts with CD34 and the mucosal addressin cell adhesion molecule 1 (MAdCAM-1). Eosinophils express also Sialyl-Lewis^X (CD15s) and selectin P ligand or P-selectin glycoprotein ligand 1 (PSGL-1/CD162), which regulate eosinophil tethering to the endothelium.[4]

In addition, eosinophils have been shown to express various integrins: $\alpha4\beta61$, $\alpha6\beta1$, $\alpha L\beta2$ [lymphocyte function-associated antigen 1(LFA-1); CD11a/CD18] $\alpha M\beta2$ [macrophage antigen-1 (MAC-1); CD11b/CD18)], $\alpha X\beta2$, $\alpha D\beta2$, and $\alpha4\beta7$. These integrin molecules are highly expressed by eosinophils and allow them to roll and adhere to endothelia expressing P-selectin and either vascular cell adhesion protein 1 (VCAM-1), intercellular adhesion molecule 1 (ICAM-1), -2 (ICAM-2), and -3 (ICAM-3), or MAdCAM-1. The specific interactions of integrins facilitate the migration of eosinophils to different tissue compartments during inflammation. Indeed, selective eosinophil recruitment during lung allergic inflammation is regulated by the interaction of very late activation antigen 4 (VLA-4) to VCAM-1.[7] Interestingly, anti-α_4 integrin and anti-VCAM-1 monoclonal antibodies inhibit eotaxin-induced eosinophil adhesion and extravasation within rat mesenteric venules.[8] Finally, eosinophils express platelet endothelial cell adhesion molecule (PECAM-1/CD31), ICAM-1, and ICAM-3.

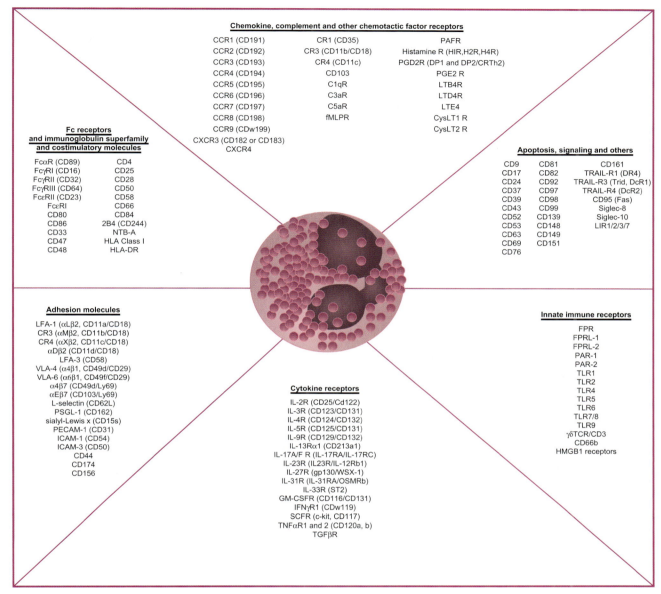

Chemokine, complement and other chemotactic factor receptors

CCR1 (CD191)	CR1 (CD35)	PAFR
CCR2 (CD192)	CR3 (CD11b/CD18)	Histamine R (HIR,H2R,H4R)
CCR3 (CD193)	CR4 (CD11c)	PGD2R (DP1 and DP2/CRTh2)
CCR4 (CD194)	CD103	PGE2 R
CCR5 (CD195)	C1qR	LTB4R
CCR6 (CD196)	C3aR	LTD4R
CCR7 (CD197)	C5aR	LTE4
CCR8 (CD198)	fMLPR	CysLT1 R
CCR9 (CDw199)		CysLT2 R
CXCR3 (CD182 or CD183)		
CXCR4		

Fc receptors and immunoglobulin superfamily and costimulatory molecules

FcαR (CD89)	CD4
FcγRI (CD16)	CD25
FcγRII (CD32)	CD28
FcγRIII (CD64)	CD50
FcεRII (CD23)	CD58
FcεRI	CD66
CD80	CD84
CD86	2B4 (CD244)
CD33	NTB-A
CD47	HLA Class I
CD48	HLA-DR

Apoptosis, signaling and others

CD9	CD81	CD161
CD17	CD82	TRAIL-R1 (DR4)
CD24	CD92	TRAIL-R3 (Trid, DcR1)
CD37	CD97	TRAIL-R4 (DcR2)
CD39	CD98	CD95 (Fas)
CD43	CD99	Siglec-8
CD52	CD139	Siglec-10
CD53	CD148	LIR1/2/3/7
CD63	CD149	
CD69	CD151	
CD76		

Adhesion molecules

LFA-1 (αLβ2, CD11a/CD18)
CR3 (αMβ2, CD11b/CD18)
CR4 (αXβ2, CD11c/CD18)
αDβ2 (CD11d/CD18)
LFA-3 (CD58)
VLA-4 (α4β1, CD49d/CD29)
VLA-6 (α6β1, CD49f/CD29)
α4β7 (CD49d/Ly69)
αEβ7 (CD103/Ly69)
L-selectin (CD62L)
PSGL-1 (CD162)
sialyl-Lewis x (CD15s)
PECAM-1 (CD31)
ICAM-1 (CD54)
ICAM-3 (CD50)
CD44
CD174
CD156

Innate immune receptors

FPR
FPRL-1
FPRL-2
PAR-1
PAR-2
TLR1
TLR2
TLR4
TLR5
TLR6
TLR7/8
TLR9
γδTCR/CD3
CD66b
HMGB1 receptors

Cytokine receptors

IL-2R (CD25/Cd122)
IL-3R (CD123/CD131)
IL-4R (CD124/CD132)
IL-5R (CD125/CD131)
IL-9R (CD129/CD132)
IL-13Rα1 (CD213a1)
IL-17A/F R (IL-17RA/IL-17RC)
IL-23R (IL23R/IL-12Rb1)
IL-27R (gp130/WSX-1)
IL-31R (IL-31RA/OSMRb)
IL-33R (ST2)
GM-CSFR (CD116/CD131)
IFNγR1 (CDw119)
SCFR (c-kit, CD117)
TNFαR1 and 2 (CD120a, b)
TGFβR

FIGURE 3.3.1 Classification of various types of cell surface molecules found on eosinophils. While many have been identified by flow cytometry, the presence of some receptors is based on cellular responsiveness to specific stimuli. 2B4, natural killer cell receptor 2B4; C1qR, complement component C1q receptor; C3aR, complement component 3a receptor; C5aR, complement component 5a receptor; CCR, C-C chemokine receptor; CR1, 3, 4, complement receptor type 1, 3, 4; CRTH2, chemoattractant receptor-homologous molecule expressed on T$_h$2 cells; CXC-R, C-X-C chemokine receptor; CysLT1, CysLT2, cysteinyl leukotriene receptor 1, 2; γδTCR, T cell receptor T3 delta chain; Fas, apoptosis-mediating surface antigen FAS; Fcα R immunoglobulin alpha Fc receptor; FcεRI, Fc-epsilon RI-alpha; fMLP, fMet-Leu-Phe; FPR, N-formyl peptide receptor; FPRL, formyl peptide receptor-like; GM-CSF-R-alpha, granulocyte-macrophage colony-stimulating factor receptor subunit alpha; H1R, histamine H1 receptor; HLA, human leukocyte antigen; HMGB1, high mobility group protein B1; ICAM-1, -3, intercellular adhesion molecule 1, 3; IFN, interferon gamma; IFN-gamma-R1, interferon gamma receptor 1; IL, interleukin; IL-2RA, interleukin-2 receptor subunit alpha, IL-3RA, interleukin-3 receptor subunit alpha, etc.; IL-13RA1, interleukin-13 receptor subunit alpha-1; IL-17A, interleukin-17A; LFA, lymphocyte function-associated antigen; LIR, leukocyte immunoglobulin-like receptor subfamily B; LTB$_4$, leukotriene B$_4$; LTD$_4$, leukotriene D$_4$; LTE4, leukotriene E$_4$; NTB-A, NK-T-B-antigen; OSMR, oncostatin-M-specific receptor subunit beta; PAF-R, platelet-activating factor receptor; PAR-1/2, proteinase-activated receptor 1, 2; PECAM-1, platelet endothelial cell adhesion molecule; PD2R2, prostaglandin D2 receptor 2; PGE2-R, prostaglandin E2 receptor EP3 subtype; PSGL-1, P-selectin glycoprotein ligand 1; SCFR, mast/stem cell growth factor receptor Kit; Siglec, sialic acid-binding Ig-like lectin; ST2, suppression of tumorigenicity 2; TGF, transforming growth factor; TLR, toll-like receptor; TNF, tumor necrosis factor; TNF-R, tumor necrosis factor receptor; TRAIL R, TNF-related apoptosis-inducing ligand receptor; VLA, very late antigen.

Chemokine Receptors

In physiological and inflammatory conditions, tropism of eosinophils for different tissues involves not only integrins but also chemokine production by different organs. The specific activity of chemokines is mediated by the selective expression of the seven transmembrane-spanning, G protein-coupled receptors. Human blood eosinophils express high levels of CCR3, a receptor responsible for both migration and degranulation.[9] Most chemokine receptors can bind multiple ligands, for example, CCR3 can bind C-C motif chemokine 5 (CCL5/RANTES), 7 (CCL7), 8 (CCL8), 11 (CCL11), 13 (CCL13), 24 (CCL24), and 26 (CCL26). Eosinophils respond to different ligands of C-C chemokine receptor type 1 (CCR-1) and CCR3, such as C-C motif chemokine 3 (CCL3), CCL5/RANTES, CCL7 [monocyte chemotactic protein 3 (MCP-3)], C-C motif chemokine 8 [CCL8/monocyte chemotactic protein 2 (MCP-2)], and CCL13 [monocyte chemotactic protein 4 (MCP-4)], eotaxin (CCL11), eotaxin-2 (CCL24), eotaxin-3 (CCL26), or CCL11, CCL24, and CCL26, and C-C motif chemokine 28 [CCL28/mucosae-associated epithelial chemokine (MEC)].[4] Interleukin-5 (IL-5) and CCL11, CCL24, and CCL26 are the main factors accounting for the maturation, recruitment, and chemotaxis of eosinophils.

Under physiological conditions, the key factor described for eosinophil recruitment to the gastrointestinal tract, but also to other organs like the thymus, mammary glands, and uterus is CCL11, a chemokine produced by epithelial cells, mast cells, alveolar macrophages, eosinophils themselves, bronchial smooth muscular fibers, and the vascular endothelium.

C-X-C motif chemokine 9 [CXCL9/monokine induced by interferon-gamma (MIG)], C-X-C motif chemokine 10 [CXCL10/10 kDa interferon gamma-induced protein (IP-10)], and C-X-C motif chemokine 12 [CXCL12/stromal cell-derived factor 1 (SDF-1)] induce strong eosinophil chemotaxis through the expression of their receptors, C-X-C chemokine receptor type 3 (CXC-R3, for MIG and IP-10)[11] and C-X-C chemokine receptor type 4 [CXC-R4 for CXCL12[10]]. C-C motif chemokine 4 [CCL4/macrophage inflammatory protein 1-beta (MIP-1-beta)] and C-C motif chemokine 1 (CCL1), C-C chemokine receptor type 5 (CCR5) and C-C chemokine receptor type 8 (CCR8) respectively, also induce eosinophil chemotaxis. There are conflicting reports regarding the expression of CCR1, C-C chemokine receptor type 2 (CCR2), C-C chemokine receptor type 6 (CCR6), C-X-C chemokine receptor type 1 (CXC-R1), 2 (CXC-R2), 3 (CXC-R3), and CXC-R4 by human blood eosinophils. Interestingly, CCR3 has been shown to deliver a negative signal in eosinophils, depending on the ligand engaged. In particular, the chemokine CXCL9 (MIG) inhibits eosinophil responses by CCR3.[12]

How chemokines and their receptors regulate eosinophil activation and recruitment into sites of allergic inflammation is not completely understood. To elucidate the role of chemokine receptors expressed by eosinophils, different strategies are used:

1. Using CCR gene-deleted mice, a number of receptors, including CCR3, CCR6, and CXC-R4, have been shown to be crucial for the development of eosinophilic airway inflammation and airway hyperresponsiveness. These studies, however, cannot differentiate between the requirement for the existence of a particular receptor on eosinophils versus its presence on other cell types that provide factors necessary for eosinophil migration to the airway. Results of two studies using CCR3 gene-deleted mice indicated a reduced eosinophil recruitment, but were contradictory with the development of bronchial hyperreactivity and seemed to depend on the allergen sensitization protocol[13,14] or the origin of the mice.[15]

2. Inhibition of chemokine receptor activity can be achieved either by using modified chemokines or small-molecule chemokine receptor antagonists, which function as antagonists, or by using neutralizing antibodies directed against the receptor. Two types of small-molecule antagonists used to block CCR3 significantly and selectively reduced eosinophil infiltration to the lung of mice.[16] To date, very few chemokine receptor antagonists have been developed for humans, but this area is of great interest for the pharmaceutical industry, and further antagonists are likely to emerge in the near future. Taken together, these studies suggest that some chemokines and their receptors at inflammatory sites play a pivotal role in the allergic inflammation process and other eosinophil-associated diseases through the initiation of eosinophil recruitment, activation, and regulation of disease severity.

Eosinophils carry intracellular stores of receptors that relocalize to the surface after stimulation with different molecules. After stimulation with IL-5, eosinophils express other chemokine receptors such as CCR5, CCR6, CCR8, CXC-R3, and CXC-R4.[4]

Other Receptors for Eosinophil Migration

Besides cytokines and chemokines, metabolites of arachidonic acid such as leukotrienes and prostaglandins are also involved in eosinophil chemotaxis and transmigration, through the expression of specific receptors for leukotriene B_4 receptor (LTB$_4$R, cysteinyl leukotriene) and for prostaglandin D2 receptor 2 [PD2R2/chemoattractant receptor-homologous molecule expressed on T$_h$2 cells (CRTH2)]. Clinical studies have demonstrated that, unlike cysteinyl leukotriene receptor 1 (CysLTR1) expressed on both eosinophils, cysteinyl leukotriene receptor 2 (CysLTR2) has been identified only on mature eosinophils. The function of these receptors on eosinophils has still to be defined. Interestingly, approaches based on the use of a CysLT1R antagonist or on the targeted deletion of LTB4R in mice reduce blood and lung eosinophilia.[17] In addition, a CysLTR1 antagonist blocks eosinophil differentiation and/or maturation.[18]

Eosinophils also express complement receptors, in particular complement receptor type 1 (CR1/CD35), complement receptor C3 subunit beta [CR3; integrin beta-2 (ITB2/CD18)], and integrin alpha-X (ITAX/CD11c). CR1, which binds to the complement fragments C3b, C4b, iC3b, and C1q, is regulated by different molecules such as LTB4.[19]

Immunoglobulin Receptors

Receptors for the Fc domain of immunoglobulin G (IgG) and E (IgE) were first identified on human eosinophils in the context of their effector function against parasites.[20] Next, increased Fc-epsilon RI-alpha (FcεRI) chain expression was shown on eosinophils from human allergen-induced atopic asthma, suggesting that this receptor could participate in eosinophil function during airway inflammation.[21] In contrast to mast cells, FcεRI expressed by human eosinophils is composed of the alpha and gamma chains but usually lacks the beta chain. However, the expression of the beta chain of FcεRI has been detected by flow cytometry on eosinophils from *Schistosoma mansoni*-infected patients,[22] in correlation with protection against reinfection. These studies suggest that this receptor is inducible, both *in vitro* and *in vivo* and that it participates in selective degranulation by inducing the release of eosinophil peroxidase (EPO) but not eosinophil cationic protein (ECP). Eosinophils also express low affinity immunoglobulin epsilon Fc receptor (FCER2/CD23).[23] Interestingly, both FCER1 and FCER2 receptors are only detected on human and rat eosinophils but are lacking on mouse eosinophils.

IgG antibodies bind to Fc-gamma RII-b (FcRII-b/CD32) and activate eosinophils, resulting in eosinophil degranulation with release of biologically active cationic proteins.[24] Eosinophils do not constitutively express Fc-gamma RI (CD64) or Fc-gamma RIII (Fc-RIII/CD16), but their expression can be upregulated by cytokines such as IFN-γ.[25]

Human eosinophils express functional polymeric immunoglobulin A (IgA), asialoglycoprotein receptor 1 and 2 (ASGPR 1, 2), and transferrin receptor (TfR), in addition to immunoglobulin alpha Fc receptor (IgA Fc receptor/CD89) and the receptor for the secretory component.[26] IgA Fc receptor (CD89) is probably the major receptor for IgA-mediated activation of eosinophils. Polymeric immunoglobulin receptor (PIGR) would be the second most effective receptor for IgA-induced eosinophil activation. Indeed, this receptor binds the J chain present in both IgA and immunoglobulin M (IgM) polymers.

Interestingly, these receptors for IgE or IgA are differentially expressed by eosinophils from different animal species. In addition to the question posed by the differences between mouse on the one hand and rat and human on the other, the absence of IgE receptors and IgA Fc receptor/CD89 on mouse eosinophils, in contrast to human and rat eosinophils, certainly prevent more in-depth studies on their functions.

Human eosinophil IgG (Fc) receptors are enhanced by chemotactic agents, such as LTB4 or fMet-Leu-Phe receptor (fMLP/FPR1), in a similar way to that previously described for complement receptors.

Sensors of Innate Immunity

Human eosinophils express two members of the fMLP and N-formyl peptide receptor 2 (FPR2) family, but not N-formyl peptide receptor 3 (FPR3).[27] In one study, FPR1 (FPR) and FPR2 mediated the activation of eosinophils by house dust mite. Interestingly, signals evoked via FPR caused the downregulation of CCR3-mediated chemotaxis but not respiratory burst in human eosinophils.[28] FPR is also known as the high-affinity receptor for fMLP, a bacterial by-product of protein synthesis, and could suggest a role of eosinophils in infectious diseases.

Several eosinophil-expressed receptors recently identified serve as pattern recognition receptors (PRRs), allowing eosinophils to be stimulated directly by 1 pathogen-associated molecular patterns (PAMPs) or 2 damage-associated molecular patterns (DAMPs), or by both 3:

1. Proteinase-activated receptors (PARs), in particular PAR-2, likely play a major role in eosinophil activation in response to proteases from allergens such as dust mites, fungi, or pollens.[29] Eosinophils express PAR-1 and PAR-2.[30] Eosinophil PAR-2 appears critical for eosinophil-mediated innate immune response to certain fungi.[31] Although the percentage of PAR-2-positive eosinophils is increased in allergic rhinitis,[32] further clarification of PAR function on eosinophils and their contribution to airway disease pathology is needed.

2. Protein expression of TLRs 1, 2, 4, 5, 6, 7/8, and 9 has been reported on human eosinophils.[33] This study shows that TLR2, 5, and 7 ligands differentially activate eosinophils leading to increased expression of some membrane molecules and/or differential release of cytokines/chemokines, superoxide ion, and granule proteins, depending on the given ligand. TLR3, 4, 8, and

9 ligands seemed to be less active or inactive. In a recent study, it has been demonstrated that eosinophils expressed inducible membrane TLR2 and TLR4 and that TLR2-dependent activation of human eosinophils led to α-defensin and ECP release, mediators involved in the decrease of mycobacterial growth.[34] Eosinophil activation via TLR7 and TLR9 might engender a link between viral infection and allergic exacerbations.[35] A recent study has shown intracellular expression of TLR3, 4, 5, and 7 in eosinophils isolated from the spleen of naive hypereosinophilic mice.[36]

3. DAMPs are typically released following necrotic and late, nonphagocytosed, apoptotic cell death (so-called aponecrosis). Interestingly, eosinophils are attracted by, and respond to, products released by necrotic cells or damaged tissues, including high mobility group protein B1 (HMGB1)[37] and crystalline uric acid.[38] Lotfi and colleagues showed that eosinophils express advanced glycosylation end product-specific receptor [AGER/receptor for advanced glycosylation end products (RAGE)], one of the receptors for HMGB1.[37] Hematopoietic cell signal transducer (HCST/DAP10) or TYRO protein tyrosine kinase-binding protein (TYOBP/DAP12)-associated receptors, as well as suppression of tumorigenicity 2 (ST2), the receptor for interleukin-33 (IL-33), are candidate molecules regulating the recruitment of inflammatory cells such as eosinophils in tumor immune responses.[39] In addition, the lectin family, including galectin-3, is considered as potential DAMP candidates that orchestrate innate immune responses alongside the PAMP system. CD66b [carcinoembryonic antigen-related cell adhesion molecule 8 (CEAM8)], which recognizes galectin-3, is highly expressed on the surface of human peripheral blood eosinophils and is involved in regulating the adhesion and activation of eosinophils.[40] These results suggest a role for eosinophils in sensing cell stress and damage, amplifying and regulating immune responses.

4. A restricted T-cell receptor (TCR) may be used as a PRR. For example, according to this paradigm, large numbers of gamma delta (γδ)T cells respond to common molecules produced by microbes and to stressed epithelial cells. Peripheral blood and cord blood-derived eosinophils express some receptors shared with T cells. Following the demonstration that eosinophils expressed CD4 and interleukin-2 receptor subunit alpha (IL-2RA/CD25), the alpha chain of the IL-2 receptor, as well as the co-stimulatory molecule CD28,[41] it has been recently demonstrated that human but not mouse eosinophils express a functional γδTCR/CD3 complex, with similar, but not identical, characteristics with γδTCR from γδT cells.[42] Agonist-dependent activation of this receptor induces the release of IFN-γ and lymphotoxin, and contributes to eosinophil innate responses against mycobacteria and tumors. Not only the expression of such a TCR represents an additional link between lymphoid and myeloid lineage, but it also confers on eosinophils new functions in immunosurveillance.[42] Indeed, due to their high cytotoxic potential and their γδTCR expression, eosinophils might thus represent very efficient innate effector cells in the defense against targets bearing γδTCR ligands, such as tumor cells, in particular when eosinophils surround or infiltrate solid tumors.

Receptors for Antigen Presentation

Eosinophils express major histocompatibility complex class II (MHC-II) and co-stimulatory molecules such as the B7 family members CD80, CD86, and CD40 necessary for T-cell activation and proliferation.[4] Although peripheral eosinophils of most normal donors are generally devoid of human leukocyte antigen, DR subregion (HLA-DR) protein expression, bronchoalveolar lavage (BAL) eosinophils from asthmatics have been shown to express HLA-DR.[43] In addition, eosinophils primed *ex vivo* or isolated from mouse tissues express detectable MHC-II.[44] Moreover, mature eosinophils express MHC-II when stimulated with granulocyte-macrophage colony-stimulating factor (GM-CSF).[45] Indeed, a recent study has demonstrated that recruited airway eosinophils serve as professional antigen-presenting cells (APCs) for ovalbumin antigen in mediating T-cell responses.[46] With their capacity to contribute to the initiation and amplification of immune responses to inhaled allergens, eosinophils acting as APCs may be important in mediating allergen-induced T_h2 responses. Finally, accumulated evidence revealed the important immunoregulatory role of eosinophils as APCs.

Recent studies have shown a restricted expression pattern of CD244 [natural killer cell receptor 2B4 (2B4)] subfamily of receptors on the surface of eosinophils.[47] Indeed, eosinophils express signaling lymphocytic activation molecule (SLAM) family member 6 [SLAF6/NK-T-B-antigen (NTB-A)], CD244 (2B4), CD84, CD58, and CD48. Cross-linking of CD244 on eosinophils induced EPO, IFN-γ, and IL-4 release. In addition, CD244 appears to be involved in tumor lysis by eosinophils against an Epstein–Barr virus (EBV)-positive B-cell lymphoma, expressing high CD48 levels.[47] CD48 is a high-affinity ligand for CD244 and is involved in the activation, co-stimulation, and adhesion of eosinophils. The expression of CD48 is increased on eosinophils from asthmatic donors and is upregulated by interleukin-3 (IL-3).[48] Therefore, eosinophils may interact either with themselves or with other cell types, such as T cells, via these CD2 subfamily receptors. A better understanding of the conditions in which eosinophils can interact with CD48 through CD244 (2B4) would define the factors that regulate eosinophil functions in health and disease. Importantly, Lee and colleagues demonstrated that CD244 (2B4) acts as an inhibitory receptor, rather than an activating one, on CD48 ligation.[49]

Inhibitory Receptors

As with natural killer (NK) cells, a new functional class of receptors has been recently discovered on eosinophils: inhibitory receptors (Table 3.3.1). Two classes of inhibitory receptors are described. First, receptors leading to the negative regulation of myeloid cell differentiation, proliferation, and survival after their activation have been described. Then, receptors implicated in the inhibition of cell migration were observed.

Although human eosinophils have been shown to express several inhibitory receptors, only a few of these have been examined thoroughly. One of these receptors is sialic acid-binding Ig-like lectin 8 (Siglec-8).[50] Siglec-8 activation leads to eosinophil apoptosis mediated by caspases and reactive oxygen species generation. Expression of Siglec-5 (or mSiglec-F) in purified mouse eosinophils has been detected.[51] Interestingly, activation of human eosinophils by Siglec-8 inhibits their survival by inducing apoptosis even in the presence of GM-CSF or IL-5, both key potent *eosinophil prosurvival cytokines*.[52]

CD300a [inhibitory receptor protein 60 (IRp60)] was also shown to suppress eosinophil survival.[53] However, in contrast to Siglec-8, CD300a does not actively induce apoptosis but rather prevents IL-5- and GM-CSF-mediated survival signals. The different outcome of Siglec-8 activation on eosinophils (induction of caspase-dependent apoptosis), as opposed to CD300a/IRp60 activation (inhibition of survival signals), may be due to the fact that Siglec-8 contains both an immunoreceptor tyrosine-based inhibition motif (ITIM) and an immunoreceptor tyrosine-based switch motif (ITSM), which may recruit adaptor molecules such as the signaling SLAM-associated protein containing an SH2 domain-containing protein 1A (SH21A) and/or SH2 domain-containing protein 1B [SH21B/ EWS/ FLI1-activated transcript 2 (EAT2)].[4]

Recruitment of eosinophils involves a signaling cascade where secreted chemokines interact with heterotrimeric G protein-coupled receptors (GPCRs) and especially with CCR3. Recent findings demonstrate a cross talk between GPCRs and inhibitory receptors signaling. Surprisingly, it has been shown that paired immunoglobulin-like type 2 receptor beta (PILRB) may actually exert a dual role in the regulation of eosinophil and neutrophil migration. Interestingly, a role for MIG (CXCL9) in the inhibition of murine eosinophil recruitment was demonstrated.[54] In their study, Fulkerson and colleagues reported that the binding of MIG to CCR3, a hallmark eosinophil chemokine receptor, activates an inhibitory cascade. In addition, both peripheral blood and nasal polyp eosinophils express the inhibitory receptors CD300a/IRp60, Siglec-7 [p75/adhesion inhibitory receptor molecule 1 (AIRM-1)], leukocyte immunoglobulin-like receptor subfamily B member 3 [LIR-3/ immunoglobulin-like transcript 5 (ILT5)], and Fc-gamma RII-b (FcγRIIB).[53] It is interesting to note that only 25% of human peripheral blood eosinophils express the inhibitory receptor p140.

TABLE 3.3.1 Inhibitory Receptors on Eosinophils

	Inhibitory Receptor	Alternative Name	Ligand
	FcRII-b	FCG2B	IgG
	CD300a, CD300LF	IRp60, CLM-1	Unknown
Inhibitory human Siglecs	Siglec-7	CD328, p75, AIRM-1	Sialic acid
	Siglec-8	No	Sialic acid
	Siglec-10	No	Sialic acid
Inhibitory mouse Siglecs	CD33	No	Sialic acid
	Siglec-5 (Siglec-F)	No	Sialic acid
Human LIR family	LIR-3	CD85a/ILT-5	Unknown
Mouse LIR family	LIR-4	Gp49B1	Unknown
	LIR-3	PIR-B	Unknown

AIRM, adhesion inhibitory receptor molecule 1; CLM-1, CMRF35-like molecule 1; IgG, immunoglobulin G; FcRII-b, Fc-gamma RII-b; FCG2B, low affinity immunoglobulin gamma Fc region receptor II-b; IgG, immunoglobulin G; ILT-5, immunoglobulin-like transcript 5; IRp60, inhibitory receptor protein 60; LIR-3, leukocyte immunoglobulin-like receptor subfamily B member 3; LIR-4, leukocyte immunoglobulin-like receptor 4; PIR-B, paired immunoglobulin-like receptor B; LIRA1, leukocyte immunoglobulin-like receptor subfamily A member 1; Siglec, sialic acid-binding Ig-like lectin.

CONCLUSION

This chapter provides an overview of some of the surface molecules expressed by eosinophils and involved in the immune response. Surprisingly, none of the molecules on the list is exclusively expressed by eosinophils. However, the growing collection of information does suggest that eosinophils express a unique pattern of cell surface markers. To date, multiple activation receptors have been described. These include receptors that mediate cellular functions such as adhesion, chemotaxis, cytokine signaling, mediator release, survival, and phagocytosis. In contrast to these activation pathways, an opposing and suppressive receptor system has evolved. These receptors can override the signals elicited by the activation pathways. It appears that a complex network of activating and inhibitory pathways can regulate the activities of eosinophils. Therefore, such receptors are potential future therapeutic targets, in particular in the settings of allergic inflammation. Surprisingly, no receptor is present on the surface of all eosinophils, suggesting some inherent flexibility and/or inducibility. Consequently, eosinophils might exert distinct functions, according to their membrane receptor profile expression. This provocative hypothesis has not been fully evaluated.

Finally, eosinophils express a vast number of surface molecules, including structures previously believed to be exclusively expressed by other cell types, such as FcεRI, thought to be mast cell-specific, or the TCR, γδTCR. Most studies dealing with eosinophil receptors have used eosinophils isolated from peripheral blood. We could guess that eosinophil migration from the peripheral blood to inflamed tissue sites, as well as signals from the tissue microenvironment, might both alter expression of receptor patterns. However, despite technological progress, little is known about the expression profile of membrane receptors on eosinophils at inflamed tissue sites.

REFERENCES

1. Rothenberg ME, Hogan SP. The eosinophil. *Annu Rev Immunol* 2006;**24**:147–74.
2. Tanaka H, Komai M, Nagao K, Ishizaki M, Kajiwara D, Takatsu K, et al. Role of interleukin-5 and eosinophils in allergen-induced airway remodeling in mice. *Am J Respir Cell Mol Biol* 2004;**31**:62–8.
3. Busse WW, Katial R, Gossage D, Sari S, Wang B, Kolbeck R, et al. Safety profile, pharmacokinetics, and biologic activity of MEDI-563, an anti-IL-5 receptor alpha antibody, in a phase I study of subjects with mild asthma. *J Allergy Clin Immunol* 2010;**125**:1237–44. e2.
4. Hogan SP, Rosenberg HF, Moqbel R, Phipps S, Foster PS, Lacy P, et al. Eosinophils: biological properties and role in health and disease. *Clin Exp Allergy* 2008;**38**:709–50.
5. Neves JS, Perez SA, Spencer LA, Melo RC, Reynolds L, Ghiran I, et al. Eosinophil granules function extracellularly as receptor-mediated secretory organelles. *Proc Natl Acad Sci U S A* 2008;**105**:18478–83.
6. Cheung PF, Wong CK, Lam CW. Molecular mechanisms of cytokine and chemokine release from eosinophils activated by IL-17A, IL-17F, and IL-23: implication for Th17 lymphocytes-mediated allergic inflammation. *J Immunol* 2008;**180**:5625–35.
7. Gonzalo JA, Lloyd CM, Kremer L, Finger E, Martinez AC, Siegelman MH, et al. Eosinophil recruitment to the lung in a murine model of allergic inflammation. The role of T cells, chemokines, and adhesion receptors. *J Clin Invest* 1996;**98**:2332–45.
8. Nagai K, Larkin S, Hartnell A, Larbi K, Razi Aghakhani M, Windley C. Human eotaxin induces eosinophil extravasation through rat mesenteric venules: role of alpha4 integrins and vascular cell adhesion molecule-1. *Immunology* 1999;**96**:176–83.
9. Ponath PD, Qin S, Ringler DJ, Clark-Lewis I, Wang J, Kassam N, et al. Cloning of the human eosinophil chemoattractant, eotaxin. Expression, receptor binding, and functional properties suggest a mechanism for the selective recruitment of eosinophils. *J Clin Invest* 1996;**97**:604–12.
10. Nagase H, Miyamasu M, Yamaguchi M, Fujisawa T, Ohta K, Yamamoto K, et al. Expression of CXCR4 in eosinophils: functional analyses and cytokine-mediated regulation. *J Immunol* 2000; **164**:5935–43.
11. Jinquan T, Jing C, Jacobi HH, Reimert CM, Millner A, Quan S, et al. CXCR3 expression and activation of eosinophils: role of IFN-gamma-inducible protein-10 and monokine induced by IFN-gamma. *J Immunol* 2000;**165**:1548–56.
12. Fulkerson PC, Zhu H, Williams DA, Zimmermann N, Rothenberg ME. CXCL9 inhibits eosinophil responses by a CCR3- and Rac2-dependent mechanism. *Blood* 2005;**106**:436–43.
13. Humbles AA, Lu B, Friend DS, Okinaga S, Lora J, Al-Garawi A, et al. The murine CCR3 receptor regulates both the role of eosinophils and mast cells in allergen-induced airway inflammation and hyperresponsiveness. *Proc Natl Acad Sci U S A* 2002;**99**:1479–84.
14. Ma W, Bryce PJ, Humbles AA, Laouini D, Yalcindag A, Alenius H, et al. CCR3 is essential for skin eosinophilia and airway hyperresponsiveness in a murine model of allergic skin inflammation. *J Clin Invest* 2002;**109**:621–8.
15. Pope SM, Zimmermann N, Stringer KF, Karow ML, Rothenberg ME. The eotaxin chemokines and CCR3 are fundamental regulators of allergen-induced pulmonary eosinophilia. *J Immunol* 2005;**175**:5341–50.
16. Das AM, Vaddi KG, Solomon KA, Krauthauser C, Jiang X, McIntyre KW, et al. Selective inhibition of eosinophil influx into the lung by small molecule CC chemokine receptor 3 antagonists in mouse models of allergic inflammation. *J Pharmacol Exp Ther* 2006;**318**:411–7.
17. Tager AM, Dufour JH, Goodarzi K, Bercury SD, von Andrian UH, Luster AD. BLTR mediates leukotriene B(4)-induced chemotaxis and adhesion and plays a dominant role in eosinophil accumulation in a murine model of peritonitis. *J Exp Med* 2000;**192**:439–46.
18. Thivierge M, Doty M, Johnson J, Stankova J, Rola-Pleszczynski M. IL-5 up-regulates cysteinyl leukotriene 1 receptor expression in HL-60 cells differentiated into eosinophils. *J Immunol* 2000; **165**:5221–6.
19. Fischer E, Capron M, Prin L, Kusnierz JP, Kazatchkine MD. Human eosinophils express CR1 and CR3 complement receptors for cleavage fragments of C3. *Cell Immunol* 1986;**97**:297–306.

20. Gounni AS, Lamkhioued B, Ochiai K, Tanaka Y, Delaporte E, Capron A, et al. High-affinity IgE receptor on eosinophils is involved in defence against parasites. *Nature* 1994;**367**:183—6.

21. Rajakulasingam K, Till S, Ying S, Humbert M, Barkans J, Sullivan M, et al. Increased expression of high affinity IgE (FcepsilonRI) receptor-alpha chain mRNA and protein-bearing eosinophils in human allergen-induced atopic asthma. *Am J Respir Crit Care Med* 1998;**158**:233—40.

22. Ganley-Leal LM, Mwinzi PN, Cetre-Sossah CB, Andove J, Hightower AW, Karanja DM, et al. Correlation between eosinophils and protection against reinfection with Schistosoma mansoni and the effect of human immunodeficiency virus type 1 coinfection in humans. *Infect Immun* 2006;**74**:2169—76.

23. Grangette C, Gruart V, Ouaissi MA, Rizvi F, Delespesse G, Capron A, et al. IgE receptor on human eosinophils (FcERII). Comparison with B cell CD23 and association with an adhesion molecule. *J Immunol* 1989;**143**:3580—8.

24. Kaneko M, Swanson MC, Gleich GJ, Kita H. Allergen-specific IgG1 and IgG3 through Fc gamma RII induce eosinophil degranulation. *J Clin Invest* 1995;**95**:2813—21.

25. Hartnell A, Kay AB, Wardlaw AJ. IFN-gamma induces expression of Fc gamma RIII (CD16) on human eosinophils. *J Immunol* 1992;**148**:1471—8.

26. Decot V, Woerly G, Loyens M, Loiseau S, Quatannens B, Capron M, et al. Heterogeneity of expression of IgA receptors by human, mouse, and rat eosinophils. *J Immunol* 2005;**174**:628—35.

27. Svensson L, Redvall E, Bjorn C, Karlsson J, Bergin AM, Rabiet MJ, et al. House dust mite allergen activates human eosinophils via formyl peptide receptor and formyl peptide receptor-like 1. *Eur J Immunol* 2007;**37**:1966—77.

28. Svensson L, Redvall E, Johnsson M, Stenfeldt AL, Dahlgren C, Wenneras C. Interplay between signaling via the formyl peptide receptor (FPR) and chemokine receptor 3 (CCR3) in human eosinophils. *J Leukoc Biol* 2009;**86**:327—36.

29. Miike S, McWilliam AS, Kita H. Trypsin induces activation and inflammatory mediator release from human eosinophils through protease-activated receptor-2. *J Immunol* 2001;**167**:6615—22.

30. Bolton SJ, McNulty CA, Thomas RJ, Hewitt CR, Wardlaw AJ. Expression of and functional responses to protease-activated receptors on human eosinophils. *J Leukoc Biol* 2003;**74**:60—8.

31. Matsuwaki Y, Wada K, White TA, Benson LM, Charlesworth MC, Checkel JL, et al. Recognition of fungal protease activities induces cellular activation and eosinophil-derived neurotoxin release in human eosinophils. *J Immunol* 2009;**183**:6708—16.

32. Dinh QT, Cryer A, Trevisani M, Dinh S, Wu S, Cifuentes LB, et al. Gene and protein expression of protease-activated receptor 2 in structural and inflammatory cells in the nasal mucosa in seasonal allergic rhinitis. *Clin Exp Allergy* 2006;**36**: 1039—48.

33. Wong CK, Cheung PF, Ip WK, Lam CW. Intracellular signaling mechanisms regulating toll-like receptor-mediated activation of eosinophils. *Am J Respir Cell Mol Biol* 2007;**37**:85—96.

34. Driss V, Legrand F, Hermann E, Loiseau S, Guerardel Y, Kremer L, et al. TLR2-dependent eosinophil interactions with mycobacteria: role of alpha-defensins. *Blood* 2009;**113**:3235—44.

35. Mansson A, Cardell LO. Role of atopic status in Toll-like receptor (TLR)7- and TLR9-mediated activation of human eosinophils. *J Leukoc Biol* 2009;**85**:719—27.

36. Phipps S, Lam CE, Mahalingam S, Newhouse M, Ramirez R, Rosenberg HF, et al. Eosinophils contribute to innate antiviral immunity and promote clearance of respiratory syncytial virus. *Blood* 2007;**110**:1578—86.

37. Lotfi R, Herzog GI, DeMarco RA, Beer-Stolz D, Lee JJ, Rubartelli A, et al. Eosinophils oxidize damage-associated molecular pattern molecules derived from stressed cells. *J Immunol* 2009;**183**:5023—31.

38. Kobayashi T, Kouzaki H, Kita H. Human eosinophils recognize endogenous danger signal crystalline uric acid and produce proinflammatory cytokines mediated by autocrine ATP. *J Immunol* 2010;**184**:6350—8.

39. Lanier LL. DAP10- and DAP12-associated receptors in innate immunity. *Immunol Rev* 2009;**227**:150—60.

40. Yoon J, Terada A, Kita H. CD66b regulates adhesion and activation of human eosinophils. *J Immunol* 2007;**179**:8454—62.

41. Woerly G, Roger N, Loiseau S, Dombrowicz D, Capron A, Capron M. Expression of CD28 and CD86 by human eosinophils and role in the secretion of type 1 cytokines (interleukin 2 and interferon gamma): inhibition by immunoglobulin a complexes. *J Exp Med* 1999;**190**:487—95.

42. Legrand F, Driss V, Woerly G, Loiseau S, Hermann E, Fournie JJ, et al. A functional gammadeltaTCR/CD3 complex distinct from gammadeltaT cells is expressed by human eosinophils. *PLoS One* 2009;**4**:e5926.

43. Hansel TT, Braunstein JB, Walker C, Blaser K, Bruijnzeel PL, Virchow Jr JC, et al. Sputum eosinophils from asthmatics express ICAM-1 and HLA-DR. *Clin Exp Immunol* 1991;**86**:271—7.

44. Mawhorter SD, Pearlman E, Kazura JW, Boom WH. Class II major histocompatibility complex molecule expression on murine eosinophils activated in vivo by Brugia malayi. *Infect Immun* 1993;**61**:5410—2.

45. Lucey DR, Nicholson-Weller A, Weller PF. Mature human eosinophils have the capacity to express HLA-DR. *Proc Natl Acad Sci U S A* 1989;**86**:1348—51.

46. Wang HB, Ghiran I, Matthaei K, Weller PF. Airway eosinophils: allergic inflammation recruited professional antigen-presenting cells. *J Immunol* 2007;**179**:7585—92.

47. Munitz A, Bachelet I, Fraenkel S, Katz G, Mandelboim O, Simon HU, et al. 2B4 (CD244) is expressed and functional on human eosinophils. *J Immunol* 2005;**174**:110—8.

48. Munitz A, Bachelet I, Eliashar R, Khodoun M, Finkelman FD, Rothenberg ME, et al. CD48 is an allergen and IL-3-induced activation molecule on eosinophils. *J Immunol* 2006;**177**:77—83.

49. Lee KM, McNerney ME, Stepp SE, Mathew PA, Schatzle JD, Bennett M, et al. 2B4 acts as a non-major histocompatibility complex binding inhibitory receptor on mouse natural killer cells. *J Exp Med* 2004;**199**:1245—54.

50. Floyd H, Ni J, Cornish AL, Zeng Z, Liu D, Carter KC, et al. Siglec-8. A novel eosinophil-specific member of the immunoglobulin superfamily. *J Biol Chem* 2000;**275**:861—6.

51. Aizawa H, Zimmermann N, Carrigan PE, Lee JJ, Rothenberg ME, Bochner BS. Molecular analysis of human Siglec-8 orthologs relevant to mouse eosinophils: identification of mouse orthologs of Siglec-5 (mSiglec-F) and Siglec-10 (mSiglec-G). *Genomics* 2003;**82**:521—30.

52. Nutku E, Aizawa H, Hudson SA, Bochner BS. Ligation of Siglec-8: a selective mechanism for induction of human eosinophil apoptosis. *Blood* 2003;**101**:5014—20.

53. Munitz A, Bachelet I, Eliashar R, Moretta A, Moretta L, Levi-Schaffer F. The inhibitory receptor IRp60 (CD300a) suppresses the effects of IL-5, GM-CSF, and eotaxin on human peripheral blood eosinophils. *Blood* 2006;**107**: 1996—2003.

54. Fulkerson PC, Zimmermann N, Brandt EB, Muntel EE, Doepker MP, Kavanaugh JL, et al. Negative regulation of eosinophil recruitment to the lung by the chemokine monokine induced by IFN-gamma (Mig, CXCL9). *Proc Natl Acad Sci U S A* 2004;**101**:1987—92.

Ex Vivo Models for the Study of Eosinophils

Introduction

Steven J. Ackerman

INTRODUCTION

Current understanding of eosinophil biochemistry, cellular, molecular, and immunobiology owes much of its initial origins and progress over the past 25 years or so to the availability of a variety of models developed for the *ex vivo* study of eosinophil biology. First and foremost, these have included the use of terminally differentiated, mature resting, or activated eosinophils purified by density gradient and/or magnetic-activated cell sorting approaches from the peripheral blood of normal subjects, and from patients with parasitic, asthmatic, allergic, and other eosinophil-associated diseases such as hypereosinophilic syndrome and chronic eosinophilic leukemia (CEL).

In particular, these models have been extremely important for early studies defining eosinophil surface receptors and signal transduction pathways pertinent to eosinophil migration and eosinophil-selective chemoattractant factors, and mechanisms of eosinophil activation and mediator release (cytokines, chemokines, and granule cationic proteins). In terms of animal models, this has included eosinophils purified from the peritoneal cavity of saline-lavaged guinea pigs,[1] which were initially used to generate antieosinophil antiserum for *in vivo* studies of eosinophil function[2] and to define a number of the major eosinophil secondary granule cationic proteins, such as eosinophil granule major basic protein 1 (MBP-1).[3]

More recently, peripheral blood eosinophils purified from hypereosinophilic interleukin-5 (IL-5) transgenic mouse strains[4,5] were used to clone, sequence, and purify murine granule cationic proteins[6,7] to develop antisera to these proteins, e.g., eosinophil peroxidase (EPO) and MBP-1, to detect eosinophils in tissues, both mouse and human,[8] and to purify eosinophils for reconstitution experiments in eosinophil-deficient (PHIL and ΔdblGATA) and other

transgenic and knockout mouse models to define the roles of eosinophils in the development of allergic inflammation, asthma, and host immune responses to parasitic infection.[9]

In addition to purified blood eosinophils from human and murine sources, transformed human leukemic cell lines that can be induced toward, are committed to, or partially differentiated to the eosinophil lineage have been extremely important for the initial cloning and characterization of eosinophil-specific genes, e.g., *interleukin 5 receptor, alpha* (*IL5RA*; 10), granule cationic proteins [eosinophil cationic protein (ECP), eosinophil-derived neurotoxin (EDN), EPO, and MBP-1], and Charcot–Leyden crystal protein (CLC)/galectin-10,[11–13] for studies of the mechanisms of eosinophil lineage commitment, transcription of eosinophil genes,[14–16] and terminal differentiation, and most recently as models for studying CEL as induced by constitutively active tyrosine kinase fusion proteins, e.g., FIP1-like 1 protein (FIP1L1) and platelet-derived growth factor receptor alpha (PDGF-R-alpha).[17] Unfortunately, similarly immortalized, eosinophil-committed, inducible, or differentiated cell lines from the mouse have not been developed, but would be extremely useful for developing novel methods and approaches for studies of mouse eosinophil biology. For example, studies that define regulatory regions of the promoters of the mouse granule cationic protein genes, as done for the development of the eosinophil-deficient PHIL mouse,[18] required testing of promoter/reporter gene constructs by this author in human eosinophilic cell lines, since equivalent mouse eosinophil lines were not available.

The availability of flow cytometric and MACS methods to purify the rare population of CD34+ hematopoietic stem cells (HSCs) and myeloid progenitors from human bone marrow aspirates, umbilical cord blood, and peripheral blood have provided an important source of HSCs and eosinophil lineage-committed progenitors (EoPs) that can be induced to proliferate and differentiate *ex vivo* by eosinophil lineage-active cytokines into nearly mature, terminally differentiated, and functionally active eosinophils. In addition to being extremely useful for studying the patterns of gene expression and the mechanisms that regulate gene transcription during eosinophil development,

Eosinophils in Health and Disease. http://dx.doi.org/10.1016/B978-0-12-394385-9.00004-3

39

these cells can be efficiently transduced with bicistronic retroviral or lentiviral vectors expressing green fluorescent protein (GFP) to ectopically overexpress transcription factors, signaling molecules, kinases, micro and other noncoding RNAs to study their roles in the processes of eosinophil lineage commitment and terminal differentiation. These cells also have considerable utility in studies using small interfering RNA or lentiviral, vector-mediated short hairpin RNA to selectively knock down eosinophil-specific gene products and analyze their roles in eosinophil development, lineage-specific gene transcription, and eosinophil cell biology in general.

The paucity of EoPs in the mouse bone marrow, the difficulties and costs involved in purifying them in large numbers and inducing them to proliferate and differentiate into useful numbers of functionally active eosinophils, has led to the recent development of *ex vivo* methods for doing this from whole (unselected) mouse bone marrow-derived mononuclear cells, providing a novel tool that has been adopted by many laboratories to generate mouse eosinophils for studies of eosinophil biology and functions.[19]

Chapter 4.2 in this section, by Ishihara and colleagues, reviews immortalized, eosinophil-differentiable cell line models, focusing on inducible eosinophilic subclones of the HL-60 promyelocytic leukemia line (HL-60 clone 15),[20] eosinophil-committed AML14 myeloblast and eosinophil-differentiated AML14.3D10 myelocyte lines,[21] and the inducible human eosinophilic leukemia (EoL-1) cell line[22–24] as models for studying the mechanisms of eosinophil differentiation, the transcriptional regulation of eosinophil-specific genes, and the roles of eosinophil-associated genes and proteins in eosinophil biology. The HL-60 clone 15 and EoL-1 lines can be induced to differentiate into immature eosinophils in response to butyric acid and/or alkaline pH culture conditions,[20] whereas the AML14 cell lines, including AML14.eos and AML14.3D10, possess signaling-competent cytokine receptors responsive to various eosinophil-active cytokines [e.g., granulocyte-macrophage colony-stimulating factor (GM-CSF), interleukin-3 (IL-3), and IL-5 for AML14 and AML14.eos] that either further induce or maintain their advanced state of differentiation (GM-CSF and IL-5 for AML14.3D10).[21] The EoL-1 line in particular possesses a gene deletion on chromosome 4 that generates a constitutively active tyrosine kinase fusion protein, termed FIP1L1−PDGF-R-alpha, that in part is responsible for the development of CEL,[17] and has been used as a model for studying this leukemic transformation and how it contributes to the hyperproliferation and dysregulated differentiation of eosinophil progenitors.[23]

Chapter 4.3 by C. M. Percopo reviews the history, methods, and uses of eosinophils purified from human peripheral blood, initially from patients with various hypereosinophilic syndromes, and subsequently from both normal subjects and patients with mild eosinophilia, e.g., due to allergy and/or asthma. The current negative selection method using a combination of Percoll or Ficoll-Hypaque density gradient centrifugation followed by MACS with anti-CD16b conjugated beads is now used routinely by most eosinophil research laboratories to obtain resting or *in vivo* activated blood eosinophils of high yield, purity, viability, and functional capacity for *ex vivo* studies of eosinophil cellular, molecular, and immunobiology. These include studies of eosinophil chemotaxis/migration, cytokine signaling, mechanisms of secretion and degranulation, and survival and apoptosis, to name just a few. Since eosinophil phenotypes and functional activities are particularly susceptible to modifications induced by different isolation methods, standardization of this purification procedure has significantly improved experimental reproducibility both within and between laboratories in the eosinophil and allied allergic and parasitic disease fields.

Chapter 4.4 in this section, by S. Ochkur, provides a review of the purification methods and *ex vivo* models currently in use for obtaining, using, and studying eosinophils from mouse peripheral blood. This typically has involved the use of various mouse strains transgenically overexpressing IL-5, either systemically or in a cell- or tissue-specific manner. Using methods similar to those for human eosinophils, mouse eosinophils were purified by the laboratory of Drs James and Nancy Lee from IL-5 transgenic mice in sufficiently large numbers to allow the routine purification of eosinophil secondary granules and consequently their cationic granule proteins, including EPO, MBP-1 and MBP-2, and the eosinophil-associated ribonucleases (EARs).[7] This approach enabled early studies of their protein biochemistry, enzymatic activities, and functions *in vitro*, and the gene cloning, characterization, and ultimately knockout of these proteins for studies of their *in vivo* activities in terms of tissue inflammation and damage, induction of tissue remodeling, bronchospasm and airways hyperreactivity, as well as the development of antibodies for neutralization studies in the mouse, and immunohistochemistry of mouse tissues to more efficiently identify tissue eosinophils, and their activation and secretion during eosinophilic inflammation.[8]

Chapter 4.5 in this section, by K. D. Dyer, provides a comprehensive review of the methods that have been developed for the generation of large numbers of essentially mature mouse eosinophils from minimally enriched bone marrow-derived mononuclear cells.[19] Historically, the purification and *ex vivo* culture of mouse bone marrow-derived myeloid progenitors, whether preselected or not, failed to routinely provide sufficiently large numbers of pure, phenotypically mature eosinophils for *in vitro* or mouse model studies. Dyer and colleagues devised an *ex vivo* culture system which routinely generates large numbers of high purity eosinophils from unselected mouse bone marrow progenitors that are phenotypically and functionally similar to eosinophils isolated from mouse peripheral blood. This novel approach has now been successfully adopted by many laboratories using both *in vitro* and mouse models to address important questions regarding the immunobiology and functions of the eosinophil in innate immunity, the development of T-helper type 2 (T_h2) polarized host allergic inflammatory responses and tissue remodeling, as well as host immune responses to parasitic infection.

Finally, in Chapter 4.6 by L. A. Wagner, we have a review of the past and current uses of purified human $CD34^+$ hematopoietic stem cell populations containing committed EoPs from bone marrow aspirates, umbilical cord, and peripheral blood for performing studies of eosinophil developmental biology, patterns of gene expression, and the transcriptional/post-transcriptional mechanisms that regulate these processes. In this *ex vivo* model, HSCs are identified and positively selected by MACS using antibodies to CD34, the sialomucin ligand present on approximately 0.5—2% of bone marrow or umbilical cord mononuclear cells, but <0.1% of peripheral blood cells. Once purified, this HSC population, which contains the recently redefined $CD34^+$ $CD45^+$/interleukin-5 receptor subunit alpha $(IL-5RA)^+$ EoP population,[25] can be induced to proliferate and terminally differentiate *ex vivo* in either bulk liquid culture or in colony-forming cell assays in semisolid media, providing useful numbers of relatively mature eosinophils by culture in IL-3 (to induce proliferation) and IL-5 (to drive terminal differentiation). This model system has now been used by multiple laboratories for studying the roles of transcription factors, kinases, and other proteins, by ectopic overexpression or gene knockdown, as well as microRNAs and other small, noncoding RNAs in eosinophil development.[26–29] Importantly, these $CD34^+$/$IL-5RA^+$ eosinophil progenitors can now be genetically engineered *ex vivo* to stably overexpress or to knock down eosinophil-specific genes of interest, to study the effects of these genetic manipulations on human eosinophil function *in vivo* in human/mouse chimeras generated by transplantation and reconstitution of the mouse hematopoietic system with the engineered human stem cells in β_2-microglobulin$^{-/-}$ nonobese diabetic/severe combined immunodeficient (NOD/SCID) mice.[30]

In summary, these *ex vivo* models have been and continue to be indispensible for studies of many aspects of eosinophil biology, functions, and roles in health and disease. They continue to be refined and improved upon, and currently represent the mainstay of most eosinophil-focused laboratories working on generating and testing novel hypotheses in this field that can ultimately be studied *in vivo* in either mouse or other animal models, or for translational research focused on human subjects.

ACKNOWLEDGEMENTS

Supported in part by National Institutes of Health grant R21AI079925, a Translational Research Award from the American Gastroenterology Association, the Campaign Urging Research on Eosinophilic Diseases, and the Thrasher Research Foundation and the American Partnership for Eosinophilic Diseases.

REFERENCES

1. Gleich GJ, Loegering D. Selective stimulation and purification of eosinophils and neutrophils from guinea pig peritoneal fluids. *J Lab Clin Med* 1973;**82**(3):522—8.

2. Gleich GJ, Loegering DA, Olson GM. Reactivity of rabbit antiserum to guinea pig eosinophils. *J Immunol* 1975;**115**(4):950—4.

3. Gleich GJ, Loegering DA, Maldonado JE. Identification of a major basic protein in guinea pig eosinophil granules. *J Exp Med* 1973;**137**(6):1459—71.

4. Dent LA, Strath M, Mellor AL, Sanderson CJ. Eosinophilia in transgenic mice expressing interleukin 5. *Journal of Experimental Medicine* 1990;**172**(5):1425—31.

5. Lee NA, McGarry MP, Larson KA, Horton MA, Kristensen AB, Lee JJ. Expression of IL-5 in thymocytes/T cells leads to the development of a massive eosinophilia, extramedullary eosinophilopoiesis, and unique histopathologies. *J Immunol* 1997;**158**(3):1332—44.

6. Larson KA, Horton MA, Madden BJ, Gleich GJ, Lee NA, Lee JJ. The identification and cloning of a murine major basic protein gene expressed in eosinophils. *J Immunol* 1995;**155**(6):3002—12.

7. Larson KA, Olson EV, Madden BJ, Gleich GJ, Lee NA, Lee JJ. Two highly homologous ribonuclease genes expressed in mouse eosinophils identify a larger subgroup of the mammalian ribonuclease superfamily. *Proc Natl Acad Sci U S A* 1996;**93**(22):12370—5.

8. Protheroe C, Woodruff SA, de Petris G, et al. A novel histologic scoring system to evaluate mucosal biopsies from patients with eosinophilic esophagitis. *Clin Gastroenterol Hepatol* 2009 Jul;**7**(7):749—55. e11.

9. Jacobsen EA, Ochkur SI, Pero RS, et al. Allergic pulmonary inflammation in mice is dependent on eosinophil-induced recruitment of effector T cells. *J Exp Med* 2008 Mar 17;**205**(3):699—710.

10. Tavernier J, Devos R, Cornelis S, et al. A human high affinity interleukin-5 receptor (IL5R) is composed of an IL5-specific alpha chain and a beta chain shared with the receptor for GM-CSF. *Cell* 1991;**66**(6):1175—84.

11. Ackerman SJ, Corrette SE, Rosenberg HF, et al. Molecular cloning and characterization of human eosinophil Charcot—Leyden crystal protein (lysophospholipase). Similarities to IgE binding proteins and the S-type animal lectin superfamily. *J Immunol* 1993;**150**(2):456—68.

12. Rosenberg HF, Ackerman SJ, Tenen DG. Human eosinophil cationic protein. Molecular cloning of a cytotoxin and helminthotoxin with ribonuclease activity. *J Exp Med* 1989;**170**(1):163—76.

13. Rosenberg HF, Tenen DG, Ackerman SJ. Molecular cloning of the human eosinophil-derived neurotoxin: a member of the ribonuclease gene family. *Proc Natl Acad Sci U S A* 1989;**86**(12):4460—4.

14. Yamaguchi Y, Zhang D-E, Sun Z-J, et al. Functional characterization of the promoter for the gene encoding human eosinophil peroxidase. *J Biol Chem* 1994;**269**(30):19410—9.

15. Sun Z, Yergeau DA, Wong IC, et al. Interleukin-5 receptor alpha subunit gene regulation in human eosinophil development: identification of a unique cis-element that acts like an enhancer in regulating activity of the IL-5R alpha promoter. *Current Topics in Microbiology and Immunology* 1996;**211**:173—87.

16. Du J, Stankiewicz MJ, Liu Y, et al. Novel combinatorial interactions of GATA-1, PU.1, and C/EBPepsilon isoforms regulate transcription of the gene encoding eosinophil granule major basic protein. *J Biol Chem* 2002 Nov 8;**277**(45):43481—94.

17. Gotlib J, Cools J. Five years since the discovery of FIP1L1-PDG-FRA: what we have learned about the fusion and other molecularly defined eosinophilias. *Leukemia* 2008 Nov;**22**(11):1999—2010.

18. Lee JJ, Dimina D, Macias MP, et al. Defining a link with asthma in mice congenitally deficient in eosinophils. *Science* 2004;**305**(5691):1773—6.

19. Dyer KD, Moser JM, Czapiga M, Siegel SJ, Percopo CM, Rosenberg HF. Functionally competent eosinophils differentiated ex vivo in high purity from normal mouse bone marrow. *J Immunol* 2008 Sep 15;**181**(6):4004—9.

20. Fischkoff SA, Rossi RM. Lineage directed HL-60 cell sublines as a model system for the study of early events in lineage determination of myeloid cells. *Leuk Res* 1990;**14**(11—12):979—88.

21. Baumann MA, Paul CC. The AML14 and AML14.3D10 cell lines: a long-overdue model for the study of eosinophils and more. *Stem Cells* 1998;**16**(1):16—24.

22. Saito H, Bourinbaiar A, Ginsburg M, et al. Establishment and characterization of a new human eosinophilic leukemia cell line. *Blood* 1985;**66**(6):1233—40.

23. Ishihara K, Kitamura H, Hiraizumi K, et al. Mechanisms for the proliferation of eosinophilic leukemia cells by FIP1L1-PDGFRalpha. *Biochem Biophys Res Commun* 2008 Feb 22;**366**(4):1007—11.

24. Ishihara K, Takahashi A, Kaneko M, et al. Differentiation of eosinophilic leukemia EoL-1 cells into eosinophils induced by histone deacetylase inhibitors. *Life Sci* 2007 Mar 6;**80**(13):1213—20.

25. Mori Y, Iwasaki H, Kohno K, et al. Identification of the human eosinophil lineage-committed progenitor: revision of phenotypic definition of the human common myeloid progenitor. *J Exp Med* 2009 Jan 16;**206**(1):183—93.

26. Buitenhuis M, Baltus B, Lammers JW, Coffer PJ, Koenderman L. Signal transducer and activator of transcription 5a (STAT5a) is required for eosinophil differentiation of human cord blood-derived CD34+ cells. *Blood* 2003 Jan 1;**101**(1):134—42.

27. Bedi R, Du J, Sharma AK, Gomes I, Ackerman SJ. Human C/EBP-epsilon activator and repressor isoforms differentially reprogram myeloid lineage commitment and differentiation. *Blood* 2009 Jan 8;**113**(2):317—27.

28. Wagner LA, Christensen CJ, Dunn DM, et al. EGO, a novel, non-coding RNA gene, regulates eosinophil granule protein transcript expression. *Blood* 2007 Jun 15;**109**(12):5191—8.

29. Hirasawa R, Shimizu R, Takahashi S, et al. Essential and instructive roles of GATA factors in eosinophil development. *J Exp Med* 2002 Jun 3;**195**(11):1379—86.

30. Buitenhuis M, van der Linden E, Ulfman LH, Hofhuis FM, Bierings MB, Coffer PJ. Protein kinase B (PKB/c-akt) regulates homing of hematopoietic progenitors through modulation of their adhesive and migratory properties. *Blood* 2010 Sep 30;**116**(13):2373—84.

Eosinophil Cell Lines

Kenji Ishihara, Noriyasu Hirasawa and Kazuo Ohuchi

HL-60 CLONE 15 CELLS

HL-60 clone 15 cells were established in 1988 by S. A. Fischkoff, through the long-term culture of HL-60 cells under alkaline conditions (pH 7.6).[1] The parental HL-60 cells were cloned from a 36-year-old female with acute promyelocytic leukemia and were established as a human promyelocytic leukemia cell line in 1977 by Collins and colleagues.[2] It has been shown that HL-60 cells are capable of differentiating into neutrophils,[3] macrophages,[4] and monocytes[5] in response to various chemical inducers and conditions. Fischkoff and colleagues indicated that the cells morphologically, histochemically, and cytogenetically differentiated into eosinophils or eosinophilic progenitors when cultured under mildly alkaline conditions (pH 7.6—7.8) and matured further in the presence of butyric acid at 0.5 mM.[6,7] The developed cells stained with Luxol fast blue and contained eosinophilic granule proteins, including eosinophil granule major basic protein 1 (MBP-1), eosinophil peroxidase (EPO), and Charcot—Leyden crystal protein (CLC).[6,7] However, when cultured in medium at pH 7.2, HL-60 cells were induced to differentiate into neutrophils by butyric acid.[7] Therefore, Fischkoff concluded that alkaline conditions act to alter the cells' ultimate lineage, directing them to differentiate into eosinophils, while butyric acid induces the simple commitment to mature.[1] Finally, a group of single-cell derived clones isolated from a population of HL-60 cells tending to differentiate into eosinophils was isolated and named *clone 15*.[1] A high percentage of cells differentiated into eosinophils when incubated in Roswell Park Memorial Institute medium (RPMI) 1640 medium under alkaline conditions (pH 7.6—7.8) with butyric acid at 0.5 mM for 7 days.[1]

The mechanisms by which HL-60 clone 15 cells differentiate into eosinophils on exposure to butyric acid remain to be elucidated. In 2004, butyric acid was revealed to induce differentiation by continuously inhibiting histone deacetylation.[8] Butyric acid is a histone deacetylase (HDAC) inhibitor along with apicidin and trichostatin A (TSA), all promoting the transactivation of various genes by causing the hyperacetylation of histones.[8—10] In general, histones sustain modifications such as phosphorylation, methylation, and acetylation during the regulation of gene expression.[9,10] The acetylation and deacetylation of specific lysine (Lys) residues is particularly important and reversibly catalyzed by histone acetyltransferase (HAT) and HDAC, respectively.[9,10] The acetylation by HAT neutralizes the positive charge of Lys residues, resulting in a decrease in their binding to negatively charged DNA, and promoting transcriptional activity.[9,10] On the other hand, the removal of acetate from acetylated Lys residues by HDAC restores their positive charge, resulting in a down-regulation of gene expression.[9,10] The balance of acetylation and deacetylation in histones regulates a cell's fate, differentiation, proliferation, or apoptosis, via gene expression. In HL-60 clone 15 cells, both butyric acid and apicidin induced differentiation into eosinophils via the continuous acetylation of histones.[8] In contrast, TSA did not induce eosinophilic differentiation due to a transient acetylation, though the continuous acetylation of histones achieved by repeated treatment with TSA did induce differentiation.[8] The induction was impaired by shortening the period of incubation with butyric acid or apicidin.[8] Therefore, the continuous acetylation of histones by butyric acid is necessary for HL-60 clone 15 cells to differentiate into eosinophils.

HL-60 clone 15 cells have contributed to the understanding of the molecular basis of the commitment of progenitors to the eosinophilic lineage and the mechanisms by which eosinophil-specific genes are regulated during differentiation. The levels of mRNA and protein, and the promoter region involved in the gene expression of C-C chemokine receptors,[11] CLC,[12] eosinophil cationic protein (ECP),[13] eosinophil-derived neurotoxin (EDN),[14,15] EPO,[16] the GATA transcription factors,[13,15] the interleukin-5 receptor subunit alpha (IL-5RA) chain,[17] and MBP-1,[18,19] have been analyzed. The expression of eosinophil granule proteins is regulated by the GATA and CCAAT/enhancer-binding protein (C/EBP) transcription factors.[13,15,18,19] Some of these findings were obtained by the transfection of genes encoding these factors into HL-60 clone 15 cells.[12,14–19] Another eosinophilic sub-line of HL-60 cells, HL-60 (3 + C-5) cells, was established in 1986 by Tomonaga and colleagues.[20] Eosinophilic differentiation of HL-60 (3 + C-5) cells was induced by IL-5 but not by butyric acid.[20] This cell line has also contributed to the understanding of eosinophils as well as HL-60 clone 15 cells. ECP,[21] EDN,[22] CLC,[23] and GATA-binding factor 1 (GATA-1)[13] were expressed concomitantly with eosinophilic differentiation using this cell line.

ACUTE MYELOID LEUKEMIA 14 CELLS

Acute myeloid leukemia 14 (AML14) cells were established in 1992 and first reported by Paul and colleagues in 1993,[24] then again by Baumann and Paul[25] in 1998. These cells were obtained by the long-term culture of the mononuclear fraction of peripheral blood from a 68-year-old man with acute myeloid leukemia. The doubling time of these cells is 24–36 h in culture.[24] It has been reported that the cells are large, with round nuclei, multiple prominent nucleoli, and a small amount of basophilic cytoplasm with occasional azurophilic granules; they are CD2$^-$ CD13$^+$ CD20$^-$, CD33$^+$ CD34$^-$, and weakly CD14$^+$.[24] AML14 cells are maintained in RPMI 1640 medium supplemented with 10% (v/v) fetal calf serum (FCS) in the absence of hematopoietic growth factors such as granulocyte-macrophage colony-stimulating factor (GM-CSF), interleukin-3 (IL-3), and IL-5.[24,26] They have been reported to express receptors for GM-CSF, IL-3, and IL-5, and to proliferate in response to cytokines.[24]

A combination of three cytokines also led to effective eosinophilic differentiation (up to 70% were stained with Luxol fast blue) within 21 days.[24] The mRNA and protein for CLC, ECP, EDN, EPO, and MBP-1 were expressed in AML14 cells induced to differentiate into eosinophils by these cytokines.[25,26] However, AML14 cells contain a subpopulation (10–20%) positive for integrin alpha-M (CD11b),[24] and had never actually been cloned from a single cell. Therefore, Paul and colleagues subcloned the AML14 cell line by limiting dilution, obtaining a subclone that both proliferated vigorously and maintained an advanced eosinophilic phenotype, which they named AML14.3D10.[25,27]

The AML14.3D10 cells were maintained in RPMI 1640 medium supplemented with 8% fetal bovine serum (FBS) in the absence of cytokines and proliferated with a doubling time of 48 h.[25] It was necessary to prevent the cells from exceeding a density of 1×10^6 cells/mL, which appeared to lead to degranulation, resulting in a decrease in cell viability.[25] On resurrection of the cells following thawing from frozen aliquots, initial viability was reported to be about 20–30%, likely due to the toxicity from the granule proteins.[25] AML14.3D10 cells are CD33$^+$, CD13$^+$, CD16$^-$, and weakly CD11b$^+$, and express CLC, ECP, EDN, EPO, and MBP-1 proteins.[27,28] It has been reported that the cells respond to GM-CSF and IL-5 via distinct receptors consisting of an alpha subunit and a beta common (βc) chain, but not to IL-3 due to the lack of an interleukin-3 receptor subunit alpha (IL-3RA) chain.[25,27,29] On the other hand, Du and colleagues found that the formation of Crk-like protein (CRKL)–signal transducer and activator of transcription 5 (STAT5) in AML14.3D10 cells was induced by IL-3 as well as GM-CSF and IL-5.[30]

AML14 cells and AML14.3D10 cells have also contributed to the understanding of eosinophils and the function of related molecules. In 1996, C-C chemokine receptor type 3 (CCR3) was isolated from human eosinophil RNA and reported to be a receptor for human eotaxin.[31] In this first report describing human CCR3,

AML14.3D10 cells were used for functional characterization by transfecting cDNA into these cells.[31] Neither C-C chemokine receptor type 1 (CCR1), nor CCR3, or C-C chemokine receptor type 5 (CCR5) were expressed on AML14.3D10 cells at baseline, but their functional expression was synergistically induced by butyric acid and IL-5 concomitant with cell differentiation.[28] Biological, functional, and pharmacological analyses of the CLC protein,[32] IL-5RA chain,[17] and the prostaglandin D2 receptor 2 (PD2R2)[33] have been performed using this cell line. The mechanisms by which transcription factors such as GATA1, transcription factor PU.1 (PU.1), and C/EBPε regulate the expression of MBP-1 have also been examined using this cell line.[34] The transfection of cDNA into AML14.3D10 cells has been performed in several laboratories.[17,28,31,34]

EOSINOPHILIC LEUKEMIA CELLS

Eosinophilic leukemia (EoL-1) cells were first described in 1985 by Saito and colleagues.[35] This human eosinophilic leukemia cell line was established from the peripheral blood of a 33-year-old man diagnosed with Philadelphia chromosome-negative eosinophilic leukemia.[35] EoL-1 cells are maintained in RPMI 1640 medium supplemented with 10% (v/v) FCS, with a saturation density of 1×10^6 cells/mL, and grown in single-cell suspensions with a doubling time of 48 h.[35] Surface markers of EoL-1 cells include the Ia antigen, myeloid antigen (IF10, MY9), and the interleukin-2 (IL-2) receptor (anti-tac).[35] The cells also express CD11b, CD49d, and CD95, the Fcγ receptor, low affinity immunoglobulin epsilon Fc receptor (FCER2/CD23), granulocyte-macrophage colony-stimulating factor receptor subunit alpha (GM-CSF-R-alpha), and interleukin 3 receptor subunit alpha (IL-3RA), and react weakly with the anti-T-cell antibody thyroxine 4 (T$_4$).[35−37] In the process of establishing the EoL-1 cell line, other eosinophilic cell lines (EoL-2 and EoL-3) were also cloned.[35] EoL-2 cells and EoL-3 cells grow with a doubling time of 72 h and have the same surface markers as EoL-1 cells.[35] Cytochemical staining showed that 10% of EoL-1 cells are positive for peroxidase, 6% for EPO, 100% for α-naphthyl acetate esterase, and 2% for Luxol fast blue.[35] Cultivation of EoL-1 cells in alkaline conditions (pH 8.0) increased the percentage of eosinophils to about 40%.[35] Nevertheless, EoL-1 cells cultured with dimethyl sulfoxide (DMSO) (0.8−1.2%) for 9 days did not differentiate into neutrophils, and retinoic acid (0.2−1.0 μM) had no effect on the cells.[35] Therefore, Saito and colleagues concluded that the EoL-1 cell line is already committed to eosinophilic differentiation.[35] In contrast to chemical induction, granulocyte colony-stimulating factor (G-CSF) produced few eosinophilic granule-positive cells, from 2% to 8%, and the induction was enhanced by tumor necrosis factor (TNF-α) (to 18%), but not IL-2, IL-3, interleukin-4 (IL-4), IL-5, GM-CSF, macrophage colony-stimulating factor 1 (M-CSF), or TNF-α.[36]

Platelet-activating factor (PAF) is known as a potent proinflammatory mediator. In 1991, the human PAF receptor was cloned from a human leukocyte cDNA library.[38] The expression of mRNA for the receptor was upregulated in EoL-1 cells concomitant with differentiation, when the cells were cultured with 0.5 mM of butyric acid,[39] or GM-CSF and butyric acid, for 2 weeks and then IL-5 and butyric acid.[38] On cultivation for eosinophilic differentiation, the cells also showed strong peroxidase activity and contained large spherical granules similar to those of normal eosinophils.[38] The eosinophilic differentiation of EoL-1 cells induced by butyric acid has been widely used to characterize eosinophils and related molecules as well as HL-60 clone 15 cells.[39,40] Butyric acid induces the eosinophilic differentiation of EoL-1 cells through the continuous acetylation of molecules such as histones by continuously inhibiting HDAC as well as HL-60 clone 15 cells.[8,40]

In 2003, it was reported that peripheral cells from some patients with hypereosinophilic syndrome and EoL-1 cells have a fusion gene generated by the interstitial deletion of the *FIP1 like 1 (S. cerevisiae) (FIP1L1)* gene and the *platelet-derived growth factor receptor, alpha polypeptide (PDGFRA)* gene, termed *FIP1L1−PDGFRA*.[41−44] The gene's product, FIP1L1−PDGFRA, is a constitutively activated tyrosine kinase and induces cell proliferation,[42,43] but it is not involved in the inhibition of differentiation toward mature eosinophils.[45] Furthermore, HDAC inhibitors such as butyric acid decrease the translation of FIP1L1−PDGFRA but not the transcription of its mRNA during the induction of differentiation into eosinophils.[46,47] Since the discovery of *FIP1L1−PDGFRA*, EoL-1 cells have also contributed to research as a model for chronic eosinophilic leukemia expressing FIP1L1−PDGFRA and the development of novel medications.[44] Conditions of electroporation for EoL-1 cells have been optimized.[48]

REFERENCES

1. Fischkoff SA. Graded increase in probability of eosinophilic differentiation of HL-60 promyelocytic leukemia cells induced by culture under alkaline conditions. *Leuk Res* 1988;**12**:679−86.
2. Collins SJ, Gallo RC, Gallagher RE. Continuous growth and differentiation of human myeloid leukaemic cells in suspension culture. *Nature* 1977;**270**:347−9.
3. Collins SJ, Ruscetti FW, Gallagher RE, Gallo RC. Terminal differentiation of human promyelocytic leukemia cells induced by dimethyl sulfoxide and other polar compounds. *Proc Natl Acad Sci USA* 1978;**75**:2458−62.

4. Rovera G, O'Brien TG, Diamond L. Induction of differentiation in human promyelocytic leukemia cells by tumor promoters. *Science* 1979;**204**:868−70.

5. McCarthy DM, San Miguel JF, Freake HC, Green PM, Zola H, Catovsky D, Goldman JM. 1,25-dihydroxyvitamin D3 inhibits proliferation of human promyelocytic leukaemia (HL60) cells and induces monocyte-macrophage differentiation in HL60 and normal human bone marrow cells. *Leuk Res* 1983;**7**:51−5.

6. Fischkoff SA, Pollak A, Gleich GJ, Testa JR, Misawa S, Reber TJ. Eosinophilic differentiation of the human promyelocytic leukemia cell line, HL-60. *J Exp Med* 1984;**160**:179−96.

7. Fischkoff SA, Condon ME. Switch in differentiative response to maturation inducers of human promyelocytic leukemia cells by prior exposure to alkaline conditions. *Cancer Res* 1985;**45**: 2065−9.

8. Ishihara K, Hong J, Zee O, Ohuchi K. Possible mechanism of action of the histone deacetylase inhibitors for the induction of differentiation of HL-60 clone 15 cells into eosinophils. *Br J Pharmacol* 2004;**142**:1020−30.

9. Ishihara K, Hong J, Zee O, Ohuchi K. Mechanism of the eosinophilic differentiation of HL-60 clone 15 cells induced by *n*-butyrate. *Int Arch Allergy Immunol* 2005;**137**(Suppl. 1):77−82.

10. Turner BM. Cellular memory and the histone code. *Cell* 2002;**111**:285−91.

11. Tiffany HL, Alkhatib G, Combadiere C, Berger EA, Murphy PM. CC chemokine receptors 1 and 3 are differentially regulated by IL-5 during maturation of eosinophilic HL-60 cells. *J Immunol* 1998;**160**:385−92.

12. Gomolin HI, Yamaguchi Y, Paulpillai AV, Dvorak LA, Ackerman SJ, Tenen DG. Human eosinophil Charcot−Leyden crystal protein: cloning and characterization of a lysophospholipase gene promoter. *Blood* 1993;**82**:1868−74.

13. Zon LI, Yamaguchi Y, Yee K, Albee EA, Kimura A, Bennett JC, et al. Expression of mRNA for the GATA-binding proteins in human eosinophils and basophils: potential role in gene transcription. *Blood* 1993;**81**:3234−41.

14. Tiffany HL, Handen JS, Rosenberg HF. Enhanced expression of the eosinophil-derived neurotoxin ribonuclease (RNS2) gene requires interaction between the promoter and intron. *J Biol Chem* 1996;**271**:12387−93.

15. Qiu Z, Dyer KD, Xie Z, Rådinger M, Rosenberg HF. GATA transcription factors regulate the expression of the human eosinophil-derived neurotoxin (RNase 2) gene. *J Biol Chem* 2009;**284**: 13099−109.

16. Yamaguchi Y, Zhang DE, Sun Z, Albee EA, Nagata S, Tenen DG, et al. Functional characterization of the promoter for the gene encoding human eosinophil peroxidase. *J Biol Chem* 1994;**269**:19410−9.

17. Sun Z, Yergeau DA, Tuypens T, Tavernier J, Paul CC, Baumann MA, et al. Identification and characterization of a functional promoter region in the human eosinophil IL-5 receptor α subunit gene. *J Biol Chem* 1995;**270**:1462−71.

18. Yamaguchi Y, Ackerman SJ, Minegishi N, Takiguchi M, Yamamoto M, Suda T. Mechanisms of transcription in eosinophils: GATA-1, but not GATA-2, transactivates the promoter of the eosinophil granule major basic protein gene. *Blood* 1998;**91**: 3447−58.

19. Yamaguchi Y, Nishio H, Kishi K, Ackerman SJ, Suda T. C/EBPβ and GATA-1 synergistically regulate activity of the eosinophil granule major basic protein promoter: implication for C/EBPβ activity in eosinophil gene expression. *Blood* 1999;**94**:1429−39.

20. Tomonaga M, Gasson JC, Quan SG, Golde DW. Establishment of eosinophilic sublines from human promyelocytic leukemia (HL-60) cells: demonstration of multipotentiality and single-lineage commitment of HL-60 stem cells. *Blood* 1986;**67**:1433−41.

21. Rosenberg HF, Ackerman SJ, Tenen DG. Human eosinophil cationic protein. Molecular cloning of a cytotoxin and helminthotoxin with ribonuclease activity. *J Exp Med* 1989;**170**:163−76.

22. Rosenberg HF, Tenen DG, Ackerman SJ. Molecular cloning of the human eosinophil-derived neurotoxin: a member of the ribonuclease gene family. *Proc Natl Acad Sci USA* 1989;**86**:4460−4.

23. Ackerman SJ, Corrette SE, Rosenberg HF, Bennett JC, Mastrianni DM, Nicholson-Weller A, et al. Molecular cloning and characterization of human eosinophil Charcot−Leyden crystal protein (lysophospholipase). Similarities to IgE binding proteins and the S-type animal lectin superfamily. *J Immunol* 1993;**150**:456−68.

24. Paul CC, Tolbert M, Mahrer S, Singh A, Grace MJ, Baumann MA. Cooperative effects of interleukin-3 (IL-3), IL-5, and granulocyte-macrophage colony-stimulating factor: a new myeloid cell line inducible to eosinophils. *Blood* 1993;**81**:1193−9.

25. Baumann MA, Paul CC. The AML14 and AML14.3D10 cell lines: a long-overdue model for the study of eosinophils and more. *Stem Cells* 1998;**16**:16−24.

26. Paul CC, Ackerman SJ, Mahrer S, Tolbert M, Dvorak AM, Baumann MA. Cytokine induction of granule protein synthesis in an eosinophil-inducible human myeloid cell line, AML14. *J Leukoc Biol* 1994;**56**:74−9.

27. Paul CC, Mahrer S, Tolbert M, Elbert BL, Wong I, Ackerman SJ, et al. Changing the differentiation program of hematopoietic cells: retinoic acid-induced shift of eosinophil-committed cells to neutrophils. *Blood* 1995;**86**:3737−44.

28. Zimmermann N, Daugherty BL, Stark JM, Rothenberg. ME. Molecular analysis of CCR-3 events in eosinophilic cells. *J Immunol* 2000;**164**:1055−64.

29. Paul CC, Mahrer S, McMannama K, Baumann MA. Autocrine activation of the IL-3/GM-CSF/IL-5 signaling pathway in leukemic cells. *Am J Hematol* 1997;**56**:79−85.

30. Du J, Alsayed YM, Xin F, Ackerman SJ, Platanias LC. Engagement of the CrkL adapter in interleukin-5 signaling in eosinophils. *J Biol Chem* 2000;**275**:33167−75.

31. Daugherty BL, Siciliano SJ, DeMartino JA, Malkowitz L, Sirotina A, Springer MS. Cloning, expression, and characterization of the human eosinophil eotaxin receptor. *J Exp Med* 1996;**183**: 2349−54.

32. Ackerman SJ, Liu L, Kwatia MA, Savage MP, Leonidas DD, Swaminathan GJ, et al. Charcot−Leyden crystal protein (galectin-10) is not a dual function galectin with lysophospholipase activity but binds a lysophospholipase inhibitor in a novel structural fashion. *J Biol Chem* 2002;**277**:14859−68.

33. Sawyer N, Cauchon E, Chateauneuf A, Cruz RP, Nicholson DW, Metters KM, et al. Molecular pharmacology of the human prostaglandin D2 receptor, CRTH2. *Br J Pharmacol* 2002;**137**:1163−72.

34. Du J, Stankiewicz MJ, Liu Y, Xi Q, Schmitz JE, Lekstrom-Himes JA, et al. Novel combinatorial interactions of GATA-1, PU.1, and C/EBPε isoforms regulate transcription of the gene encoding eosinophil granule major basic protein. *J Biol Chem* 2002;**277**:43481−94.

35. Saito H, Bourinbaiar A, Ginsburg M, Minato K, Ceresi E, Yamada K, et al. Establishment and characterization of a new human eosinophilic leukemia cell line. *Blood* 1985;**66**: 1233—40.

36. Morita M, Saito H, Honjo T, Saito Y, Tsuruta S, Kim KM, et al. Differentiation of a human eosinophilic leukemia cell line (EoL-1). by a human T-cell leukemia cell line (HIL-3)-derived factor. *Blood* 1991;**77**:1766—75.

37. Wong CK, Ho CY, Lam CW, Zhang JP, Hjelm NM. Differentiation of a human eosinophilic leukemic cell line, EoL-1: characterization by the expression of cytokine receptors, adhesion molecules, CD95 and eosinophilic cationic protein (ECP). *Immunol Lett* 1999;**68**: 317—23.

38. Nakamura M, Honda Z, Izumi T, Sakanaka C, Mutoh H, Minami M, et al. Molecular cloning and expression of platelet-activating factor receptor from human leukocytes. *J Biol Chem* 1991;**266**:20400—5.

39. Izumi T, Kishimoto S, Takano T, Nakamura M, Miyabe Y, Nakata M, et al. Expression of human platelet-activating factor receptor gene in EoL-1 cells following butyrate-induced differentiation. *Biochem J* 1995;**305**:829—35.

40. Ishihara K, Takahashi A, Kaneko M, Sugeno H, Hirasawa N, Hong J, et al. Differentiation of eosinophilic leukemia EoL-1 cells into eosinophils induced by histone deacetylase inhibitors. *Life Sci* 2007;**80**:1213—20.

41. Cools J, DeAngelo DJ, Gotlib J, Stover EH, Legare RD, Cortes J, et al. A tyrosine kinase created by fusion of the PDGFRA and FIP1L1 genes as a therapeutic target of imatinib in idiopathic hypereosinophilic syndrome. *N Engl J Med* 2003;**348**: 1201—14.

42. Griffin JH, Leung J, Bruner RJ, Caligiuri MA, Briesewitz R. Discovery of a fusion kinase in EOL-1 cells and idiopathic hypereosinophilic syndrome. *Proc Natl Acad Sci USA* 2003;**100**: 7830—5.

43. Cools J, Quentmeier H, Huntly BJ, Marynen P, Griffin JD, Drexler HG, et al. The EOL-1 cell line as an in vitro model for the study of FIP1L1-PDGFRA-positive chronic eosinophilic leukemia. *Blood* 2004;**103**:2802—5.

44. Gotlib J, Cools J. Five years since the discovery of FIP1L1-PDGFRA: what we have learned about the fusion and other molecularly defined eosinophilias. *Leukemia* 2008;**22**: 1999—2010.

45. Ishihara K, Kitamura H, Hiraizumi K, Kaneko M, Takahashi A, Zee O, et al. Mechanisms for the proliferation of eosinophilic leukemia cells by FIP1L1-PDGFRα. *Biochem Biophys Res Commun* 2008;**366**:1007—11.

46. Kaneko M, Ishihara K, Takahashi A, Hong J, Hirasawa N, Zee O, et al. Mechanism for the differentiation of EoL-1 cells into eosinophils by histone deacetylase inhibitors. *Int Arch Allergy Immunol* 2007;**143**(Suppl. 1):28—32.

47. Ishihara K, Kaneko M, Kitamura H, Takahashi A, Hong JJ, Seyama T, et al. Mechanism for the decrease in the FIP1L1-PDGFRα protein level in EoL-1 cells by histone deacetylase inhibitors. *Int Arch Allergy Immunol* 2008;**146**(Suppl. 1): 7—10.

48. Ohyama H, McBride J, Wong DT. Optimized conditions for gene transfection into the human eosinophilic cell line EoL-1 by electroporation. *J Immunol Methods* 1998;**215**:105—11.

Chapter 4.3

Isolation and Manipulation of Eosinophils from Human Peripheral Blood

Caroline M. Percopo

INTRODUCTION

Eosinophilic granulocytes are proinflammatory cells that develop in the bone marrow and are recruited from the bloodstream to tissue sites during asthma, allergic diseases, and parasitic infection.[1] The exact role of the eosinophil in each of these disease states is not clear, thus study *in vitro*, where their environment can be manipulated with various mediators and cell—cell interactions, would elucidate specific functions and capabilities. Nonetheless, the isolation of eosinophils has long been a challenge due to the fact that they represent only a small fraction (1—3%) of the total leukocytes in normal homeostatic human whole blood. The techniques for eosinophil isolation have employed the properties of unique cell density and peripheral blood cell marker specificity. With each advance in isolation technique, a host of issues addressing the potential for effect on the functional state of the eosinophil have arisen. This chapter provides a historical perspective leading up to the most current isolation and purification techniques, followed by a short review of the issues addressing eosinophil function and responses following each isolation procedure.

FEATURES OF PERIPHERAL BLOOD EOSINOPHILS

Eosinophils are easily identified and can be differentiated from other leukocytes in a Giemsa-stained preparation that reveals their characteristically (blue)-stained bilobed nuclei and numerous (red)-stained cytoplasmic granules (Fig. 4.3.1). Although these cells are easy to spot morphologically, the percentage of eosinophils in normal human peripheral blood is so low that it is hard to isolate these cells in great numbers. Due to the abundance of granules, eosinophils are characteristically denser than mononuclear cells, a feature that makes it easy to separate these cells from one another on a density gradient. However, neutrophils have similar densities to eosinophils, making it rather difficult to separate these two populations

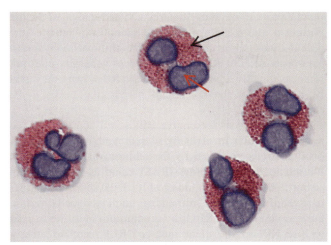

FIGURE 4.3.1 Freshly isolated eosinophils from human peripheral blood. Modified Giemsa-stained (Diff-Quik): bilobed nucleus (red arrow), granules (black arrow). Magnification 40×.

from each other. This latter point has driven the evolution of improved eosinophil isolation techniques.

EARLY ATTEMPTS AT EOSINOPHIL ISOLATION

Many researchers have worked tirelessly to find the optimal method to isolate eosinophils based on every possible way to exploit their unique characteristics. Initially, an obvious source for abundant eosinophils was from patients with some form of hypereosinophilia or parasitosis.[2-3] Although these sources were not functionally equivalent to normal donors and thus were not ideal, it was relatively easy to obtain high yields and thus this was a place to begin to delve into the study of eosinophils in isolation. Sher and colleagues[2] found that, following density gradient separation of lymphocytes and gelatin sedimentation to remove red blood cells, one could separate neutrophils from eosinophils by their differential phagocytosis of carbonyl iron. Parrillo and colleagues[3] found that, following removal of red blood cells and mononuclear cells, neutrophils could be separated from eosinophils by differential adhesion on a nylon wool column. However, these techniques were not optimal for consistent results for eosinophil isolation from donors with normal to low eosinophil counts in whole blood. An interesting observation following these early studies on eosinophil isolation was that on density gradient separation of eosinophils, there appeared to be a difference in the density of eosinophils from normal donors vs. hypereosinophilic donors.[4-8] These two subsets of eosinophils were isolated at slightly different densities and were termed normodense and hypodense eosinophils. The hypodense cells were more prominent in donors with hypereosinophilia, parasitic

diseases, or allergic diseases.[9-11] So, although hypereosinophilic donors were a good source of eosinophils in the early attempts to study eosinophils in isolation, the ability to establish a baseline of eosinophil function or activation in the normal state was thus clouded.

CONSIDERATIONS FOR THE ISOLATION OF EOSINOPHILS

Density Gradient Medium

The densities of human blood cells are as follows: neutrophils, eosinophils, and erythrocytes have densities >1.080 g/mL, whereas monocytes and lymphocytes (together known as mononuclear cells) and basophils have densities <1.080 g/mL (Fig. 4.3.2). Density differentials have historically been a useful characteristic and have allowed researchers to separate eosinophils and neutrophils from mononuclear cells in whole blood. Density gradients were prepared from sucrose, Ficoll-Hypaque, metrizamide [2-[[3-(Acetylamino)-5-(acetylmethylamino)-2,4,6-triiodobenzoyl]amino]-2-deoxy-D-glucose], Nycodenz [5-(N-2,3-dihydroxypropylacetamido)-2,4,6-tri-iodo-N, N′-bis (2,3-dihydroxypropyl)-isophthalamide], and Percoll (polyvinylpyrrolidone-coated silica gel).[12] Although each of these gradients was effective in separating eosinophils from mononuclear cells (and to some extent, eosinophils from neutrophils) by their density properties, each compound had some limitation or effect on eosinophil structure and/or function. The optimal compound would be non-toxic,

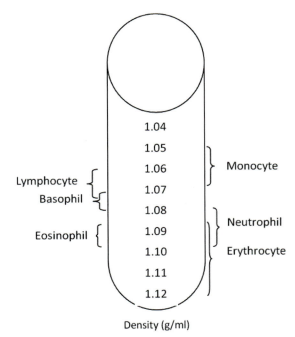

FIGURE 4.3.2 The densities of human blood cells.

metabolically inert, and form an isosmotic gradient. A side-by-side study of Nycodenz vs. metrizamide vs. Percoll one-step density gradient procedures was performed by Brattig and colleagues.[13] The distribution profiles of the gradients with the three media were found to be rather similar, but differences occurred in the degree of purity and recovery rates. For example, Nycodenz was originally introduced commercially for the separation of cells and subcellular particles; however, the application of Nycodenz to eosinophil isolation resulted in diminished purity and lower yield than metrizamide.[13] On the other hand, metrizamide caused minor morphological alterations of eosinophils, including cell shrinkage and in some instances cell aggregation. The compound that fulfilled the criteria for optimal eosinophil recovery with minimal alteration of its morphology was Percoll, and as such this compound is in routine use today, typically for eosinophil isolation within linear gradients, between 1.078 and 1.093 g/mL during centrifugation. The highest purity of eosinophils is found at $1.090-1.095$ g/mL. Pooled fractions within the range of $1.086-1.100$ g/mL can yield 86–99% eosinophils with a recovery rate of $38-56\%$.[14] Another interesting observation was the ability of the Percoll gradient to distinguish between normodense and hypodense eosinophils. Little and Casale[7] reported the separation of these populations of eosinophils, via a five-layer discontinuous density gradient of Percoll, with normodense eosinophils found at a density of ≥ 1.095 and hypodense eosinophils within the 1.090 and 1.095 range.

Density gradients are still difficult and time-consuming and the yields and purities can be inconsistent. To separate neutrophils from eosinophils most effectively, exposure of human neutrophils to the chemoattractant peptide fMet-Leu-Phe (fMLP) decreases their density, as demonstrated by centrifugation through a continuous gradient of Percoll. It was hypothesized that if anticoagulated whole blood, which was treated with fMLP, was then separated by a continuous Percoll gradient, the eosinophils would remain in the densest fraction while the neutrophils, which were now less dense, would migrate to a different layer. This in fact was the case and seemed a much-improved method of purifying eosinophils from neutrophils. Of note, it was reported that eosinophils were not activated by this treatment,[15,16] although subsequent studies observed that there was some response to fMLP by eosinophils[17]

Red Blood Cell Removal

Since red blood cells (RBCs) generally have a density similar to granulocytes, they readily sediment together when gradient methods are employed. There are several ways to remove RBCs, including differential sedimentation and differential hypotonic lysis.

An initial step in RBC removal is a presedimentation step with the polymer dextran sulfate. The RBCs sediment more rapidly through the material so that removal of the top portion of the solution results in retention of most of the granulocytes, while at the same time eliminating a significant proportion of RBCs. This process generally does not remove all RBCs, so some form of RBC lysis may still be necessary if the eosinophil preparation needs to be erythrocyte-free.

Hypotonic lysis to remove RBCs is a commonly used technique which involves brief treatment with nine parts distilled water followed by the addition of one part $10\times$ phosphate buffered saline (PBS) solution. If done with a short, 30-s incubation, this results in the differential lysis of RBCs, leaving the remaining nucleated cells viable. Differential lysis of RBCs by ammonium chloride is used to lyse RBCs. The membranes of RBCs are effectively permeable to ammonium chloride and cell lysis occurs due to the unbalanced osmotic pressure of their colloid content.[18] Although ammonium chloride is commonly used to lyse RBCs, it has been reported that during eosinophil isolation, it leads to changes in eosinophils, namely, viability, inhibition of cytokine-mediated survival, and changes in granule appearance on electron microscopic evaluation.[19]

Magnetic Bead Isolation

The advent of magnetic bead isolation opened up a new avenue for eosinophil purification based on cell-surface antigen recognition. This technology permits eosinophils to be separated from other cells by virtue of their different cell-surface antigens. When using magnetic bead cell technology there are two forms of isolation; cells can be separated by either negative or positive selection. Positive selection is the direct labeling of the target cell with a specified bead; those labeled cells are collected. Negative selection is the process whereby the target cell remains unlabeled and all unwanted cells are labeled and removed. Negative selection results in a population of cells *untouched* by magnetic beads and suggests they may have less of a chance of being inadvertently activated. A major cell-surface antigen, CD16b, is present on neutrophils but is absent on eosinophils. The absence of the CD16 antigen on most eosinophils allowed for their negative selection by magnetic bead isolation. Neutrophils are labeled with anti-CD16 antibody and are positively selected, whereas eosinophils are negatively selected for by virtue of the absence of the CD16 antigen.

Originally, two types of magnetic technologies were used for eosinophil isolation. These consisted of the large magnetic particles ($>0.5\,\mu m$), dynabeads (M-450 epoxy, $4.5\,\mu m$),[20,21] and the more commonly used today, magnetic microparticles (diameter $<0.5\,\mu m$). Not only did the magnetic beads evolve from large to small, but also magnetic sources evolved for better efficiency of isolation.

With dynabeads, the magnet used was a magnetic particle concentrator that could be reused after each application, following cleaning. This magnet worked well for larger particles but with the smaller particles, the binding time to the magnet was slow and thus not optimal for keeping cell viability high. This was a disadvantage for the smaller particles until the advent of small superparamagnetic microparticles and high gradient magnetic fields in a column framework, where the magnetically labeled cells are retained in the column and the unlabeled cells pass through a process called magnetic-activated cell sorting.[22] Another method for cell isolation using magnetic beads is where the tube containing the cells and magnetic beads are placed in a chamber, surrounded by a magnetic field (EasySep, STEMCELL Technologies Inc., Vancouver, BC, Canada). After a short incubation, the cells that are not magnetically labeled are poured off and the labeled ones remain in the tube.

Commercially Available Magnetic Bead-Based Eosinophil Isolation Kits

Currently, there are three kits available for the isolation of eosinophils from human peripheral blood.[23] All the kits require prior density centrifugation to isolate granulocytes from peripheral blood, followed by removal of RBCs. The kits are available from Miltenyi Biotec, Inc. (Auburn, CA, USA) and STEMCELL Technologies. The two kits available from Miltenyi Biotec, Inc. both use negative selection for eosinophils, meaning that the undesired cells are labeled and retained by the magnet, and the target cells, the eosinophils, flow freely through the column *untouched*. The first kit to be designed was one that uses the anti-CD16-coated magnetic beads, which bind to neutrophils that are then retained within the magnetic field, whereas the unlabeled eosinophils pass through. The second kit is a multibead kit that contains a cocktail of biotin-conjugated monoclonal antibodies against the leukocyte antigens CD2, CD14, CD16, CD19, interleukin-3 receptor subunit alpha (IL-3RA/CD123), and glycophorin-A (CD235a; MNS blood group), which after a short incubation is followed by the addition of microbeads conjugated to monoclonal antibiotin antibodies [isotype: mouse immunoglobulin G1 (IgG1)]. This multibead assay is also an untouched isolation, meaning cells that are not eosinophils are depleted, due to their retention on the column. The multibead assay is useful for removing stray mononuclear leukocytes, which may remain despite Ficoll-Hypaque separation prior to the magnetic bead methods, and is helpful when eosinophil purity is most important.

STEMCELL Technologies also produce an eosinophil isolation kit whereby the unwanted cells are specifically labeled with dextran-coated magnetic nanoparticles using bispecific tetrameric antibody complexes (TACs). The cocktail of antibodies are used against the following antigens: CD2, CD3ε, CD14, CD16, CD19, CD20 [membrane-spanning 4-domains subfamily A member 1 (MS4A1)], CD36, CD56 [neural cell adhesion molecule 1 (NCAM-1)], dextran, IL-3RA (CD123), and CD235a. The mouse monoclonal antibodies are subclass IgG1. The complexes recognize both the dextran and the cell-surface antigen expressed on the non-target cells. The small size of the magnetic dextran iron particles allows for efficient binding to the TAC-labeled cells. Following the STEMCELL Technologies protocol, magnetically labeled cells are then separated from unlabeled target cells by placing the tube containing labeled and unlabeled cells in the magnet, the unlabeled eosinophils are then poured off and the remaining unwanted labeled cells are retained in the tube.

EXPERIMENTAL APPLICATIONS

Eosinophil isolation techniques from human whole blood have evolved over the years. With each advance, isolation artifacts were found that affected eosinophil function. This section of the chapter will attempt to review some of the studies that illustrate these discrepancies in eosinophil function (Table 4.3.1). Some of the characteristics/functions of eosinophils that are studied *in vitro* include eosinophil adhesion, chemotaxis, migration, apoptosis, differentiation, superoxide production, cell–cell interactions, survival, degranulation, cytokine and protein secretion, and receptor studies. Sedgwick and colleagues[24] studied eosinophil functional differences between Percoll gradients alone versus immunomagnetic beads. No significant difference was observed for *in vitro* survival or adhesion. Casale and colleagues[25] reported that eosinophils isolated by magnetic cell sorting responded poorly to lipid chemoattractants. They compared leukotriene B$_4$ (LTB$_4$) and platelet-activating factor (PAF)-induced chemotactic responses of eosinophils isolated by Percoll versus the magnetic cell separation system techniques and found that although both methods of isolation yielded eosinophils capable of responding in a dose-dependent fashion to these mediators, for eosinophils isolated by the Percoll technique migration was observed to be greater at all doses of LTB$_4$ and PAF. Rozell and colleagues[26] reported that eosinophils isolated by MACS, even though they were in greater number and showed higher purity, did not migrate to interleukin-8 (IL-8) in contrast to eosinophils isolated by Percoll which did migrate to IL-8. A recent report by Schefzyk and colleagues[27] noted functional differences between eosinophils isolated by the new multibead isolation kit versus the original anti-CD16-conjugated magnetic beads kit. They observed changes in eosinophil survival in culture, apoptosis, release of superoxide anions, and

TABLE 4.3.1 Eosinophil Isolation Processes: Effect on Eosinophil Function

Isolation Process	Eosinophil Function (Characteristic Affected)	Reference
Source of eosinophils—patients with hypereosinophilia	Hypodense eosinophils from patients with hypereosinophilia exhibited a greater degree of cytoplasmic vacuolation and partial degranulation	7
Erythrocyte lysis	Moderate to extensive loss in the density of secondary granule core and/or matrix-altered granules	19, 29
Percoll vs. magnetic-activated cell sorting	Chemotactic responsiveness of eosinophils to leukotriene B_4 (LTB_4) and platelet-activating factor (PAF) Eosinophil migration in response to interleukin-8 (IL-8) CD16 (FCGR3)-increased baseline and stimulated leukotriene C_4; spontaneous O_2 generation; expression of specific cell-surface markers	24–26, 30
fMLP (FPR1) differential responsiveness	Enhances expression of Fc receptors and stimulates oxygen metabolite release as well as leukotriene C_4 generation	17
CD16 beads vs. multibead	Decreased survival; morphological features of apoptosis; increased release of superoxide anions; higher expression of CD69 and CD95 (apoptosis-mediating surface antigen FAS) Greater purity from granulocyte pack and interleukin-5 sustained viability; no difference in normal donors	27–28

FCGR3, Fc-gamma RIII; fMLP, fMet-Leu-Phe; FPR1, fMLP receptor.

expression of CD69 and CD95 (apoptosis-mediating surface antigen FAS). It was concluded that the new multibead eosinophil isolation kit should not be used for eosinophil investigation. However, our group also compared eosinophil isolation using these two kits. We found no significant differences with *in vitro* studies of eosinophil degranulation, interleukin-5 (IL-5)-supported eosinophil survival, or apoptosis. One noted difference from our group was an improved eosinophil purity using the multibead isolation kit of eosinophils from normal human donor granulocyte packs.[28]

In conclusion, a combination of density gradient separation as well as negative selection of human eosinophils with anti-CD16-coated magnetic beads can provide high yields at upwards of 99% eosinophil purity from both normal donors and hypereosinophilic patients. Studies on chemotaxis, migration, degranulation, survival, and apoptosis have revealed that careful consideration of mode of isolation is important when generating eosinophils for study *in vitro*.

REFERENCES

1. Dyer KD, Percopo CM, Fischer ER, Gabryszewski SJ, Rosenberg HF. Pneumoviruses infect eosinophils and elicit MyD88-dependent release of chemoattractant cytokines and interleukin-6. *Blood* 2009;**114**:2649–56.
2. Sher R, Glover A. Isolation of human eosinophils and their lymphocyte-like rosetting properties. *Immunology* 1976;**31**:337–41.
3. Parrillo JE, Fauci AS. Human eosinophils. Purification and cytotoxic capability of eosinophils from patients with the hypereosinophilic syndrome. *Blood* 1978;**51**:457–73.
4. De Simone C, Donneli G, Meli D, Rosati F, Sorice F. Human eosinophils and parasitic diseases. II. Characterization of two cell fractions isolated at different densities. *Clin Exp Immunol* 1982;**48**:249–55.
5. Frick WE, Sedgwick JB, Busse WW. Hypodense eosinophils in allergic rhinitis. *J Allergy Clin Immunol* 1988;**82**:119–25.
6. Fukuda T, Dunnette SL, Reed CE, Ackerman SJ, Peters MS, Gleich GJ. Increased numbers of hypodense eosinophils in the blood of patients with bronchial asthma. *Am Rev Respir Dis* 1985;**132**:981–5.
7. Little MM, Casale TB. Comparison of platelet-activating factor-induced chemotaxis of normodense and hypodense eosinophils. *J Allergy Clin Immunol* 1991;**88**:187–92.
8. Winqvist I, Olofsson T, Olsson I, Persson AM, Hallberg T. Altered density, metabolism and surface receptors of eosinophils in eosinophilia. *Immunology* 1982;**47**:531–9.
9. Gleich GJ, Adolphson CR. The eosinophilic leukocyte: structure and function. *Adv Immunol* 1986;**39**:177–253.
10. Hodges MK, Weller PF, Gerard NP, Ackerman SJ, Drazen JM. Heterogeneity of leukotriene C4 production by eosinophils from asthmatic and from normal subjects. *Am Rev Respir Dis* 1988;**138**:799–804.
11. Weller PF. Cytokine regulation of eosinophil function. *Clin Immunol Immunopathol* 1992;**62**:S55–9.
12. Vadas MA, David JR, Butterworth A, Pisani NT, Siongok TA. A new method for the purification of human eosinophils and neutrophils, and a comparison of the ability of these cells to damage schistosomula of Schistosoma mansoni. *J Immunol* 1979;**122**:1228–36.

13. Brattig NW, Medina-De la Garza CE, Tischendorf FW. Comparative study of eosinophil purification on Nycodenz, Metrizamide and Percoll density gradients. *Eur J Haematol* 1987;**39**:148—53.

14. Gartner I. Separation of human eosinophils in density gradients of polyvinylpyrrolidone-coated silica gel (Percoll). *Immunology* 1980;**40**:133—6.

15. Roberts RL, Gallin JI. Rapid method for isolation of normal human peripheral blood eosinophils on discontinuous Percoll gradients and comparison with neutrophils. *Blood* 1985;**65**:433—40.

16. Laviolette M, Bosse M, Rocheleau H, Lavigne S, Ferland C. Comparison of two modified techniques for purifying blood eosinophils. *J Immunol Methods* 1993;**165**:253—61.

17. Yazdanbakhsh M, Eckmann CM, Koenderman L, Verhoeven AJ, Roos D. Eosinophils do respond to fMLP. *Blood* 1987;**70**: 379—83.

18. Phillips WA, Hosking CS, Shelton MJ. Effect of ammonium chloride treatment on human polymorphonuclear leucocyte iodination. *J Clin Pathol* 1983;**36**:808—10.

19. Ide M, Weiler D, Kita H, Gleich GJ. Ammonium chloride exposure inhibits cytokine-mediated eosinophil survival. *J Immunol Methods* 1994;**168**:187—96.

20. Hansel TT, Pound JD, Pilling D, Kitas GD, Salmon M, Gentle TA, et al. Purification of human blood eosinophils by negative selection using immunomagnetic beads. *J Immunol Methods* 1989;**122**: 97—103.

21. Hansel TT, De Vries IJ, Iff T, Rihs S, Wandzilak M, Betz S, et al. An improved immunomagnetic procedure for the isolation of highly purified human blood eosinophils. *J Immunol Methods* 1991;**145**: 105—10.

22. Miltenyi S, Muller W, Weichel W, Radbruch A. High gradient magnetic cell separation with MACS. *Cytometry* 1990;**11**: 231—8.

23. Munoz NM, Leff AR. Highly purified selective isolation of eosinophils from human peripheral blood by negative immunomagnetic selection. *Nat Protoc* 2006;**1**:2613—20.

24. Sedgwick JB, Shikama Y, Nagata M, Brener K, Busse WW. Effect of isolation protocol on eosinophil function: Percoll gradients versus immunomagnetic beads. *J Immunol Methods* 1996;**198**: 15—24.

25. Casale TB, Erger RA, Rozell MD. Eosinophils isolated by magnetic cell sorting respond poorly to lipid chemoattractants. *Ann Allergy Asthma Immunol* 1999;**83**:127—31.

26. Rozell MD, Erger RA, Casale TB. Isolation technique alters eosinophil migration response to IL-8. *J Immunol Methods* 1996;**197**:97—107.

27. Schefzyk M, Bruder M, Schmiedl A, Stephan M, Kapp A, Wedi B, et al. Eosinophil granulocytes: functional differences of a new isolation kit compared to the isolation with anti-CD16-conjugated MicroBeads. *Exp Dermatol* 2009;**18**:653—5.

28. Percopo CM, Dyer KD, Killoran KE, Rosenberg HF. Isolation of human eosinophils: microbead method has no impact on IL-5 sustained viability. *Exp Dermatol* 2010;**19**:467—9.

29. Malm-Erjefält M, Stevens TR, Persson CG, Erjefalt JS. Discontinuous Percoll gradient centrifugation combined with immunomagnetic separation obviates the need for erythrocyte lysis and yields isolated eosinophils with minimal granule abnormalities. *J Immunol Methods* 2004;**288**:99—109.

30. Adamko DJ, Wu Y, Gleich GJ, Lacy P, Moqbel R. The induction of eosinophil peroxidase release: improved methods of measurement and stimulation. *J Immunol Methods* 2004;**291**:101—8.

Chapter 4.4

Induction of Eosinophilia in Mice and the Isolation of Eosinophils for *Ex Vivo* Manipulations

Sergei I. Ochkur

INTRODUCTION

Several excellent and comprehensive reviews on eosinophil biology have been published recently[1-7] and serve as a foundation for the methods used to purify and manipulate these granulocytes *ex vivo*. This chapter focuses on the practical aspects of eosinophil-related studies and, specifically, studies of eosinophil biology using mouse eosinophils *ex vivo*. It describes the methods used to induce eosinophilia in mice to isolate eosinophils from mouse tissues, as well as *ex vivo* assays directed at the elucidation of eosinophil function.

INDUCTION OF EOSINOPHILIA IN THE MOUSE

Eosinophils comprise only 1—3% of circulating leukocytes, but an increase in this population in the peripheral blood and tissue is often a diagnostic feature associated with a variety of inflammatory and disease states, including parasitic and fungal infections, allergic diseases (e.g., asthma, rhinitis, and sensitivities to specific foods), cancer, and transplant rejection. Mouse models provide sources of eosinophils, often from tissue-specific and/or pathology-specific environments. However, for *ex vivo* studies, the ease and the level of purification are of the utmost importance. The common tissue sources of mouse eosinophils are peripheral blood (PB), peritoneal lavage (PL), and bronchoalveolar lavage (BAL) fluids. Allergic inflammation, cytokine administration, and transgenic approaches are employed to induce eosinophilia so as to facilitate the isolation of these cells from these locations.

ALLERGIES AND INFECTIONS

Several mechanisms of eosinophilia in allergic inflammatory conditions have been proposed.[8] On allergen exposure, the epithelium, dendritic cells, mast cells, and T cells release factors that directly or indirectly induce eosinophilia in the bone marrow (BM). Thus, one of the most common ways to elicit eosinophilia in the mouse is via exposure to allergens such as house dust mites (HDMs), the pollen of ragweed (*Ambrosia*), fungus (*Aspergillus*, *Alternaria*) extracts, ovalbumin (OVA), or a combination of several agents. The protocols usually consist of three or more allergen exposures: two applications 1–2 weeks apart, commonly called sensitizations, followed by local allergen provocation, called challenges, which are administered 1–2 weeks later for an acute protocol and for a longer time for a chronic protocol. In general, sensitizations are usually systemic; local allergen provocation is performed by aerosolized allergen endotracheal (e.t.) or intranasal (i.n.) inoculation, if eosinophilia is to be elicited in the lung, or intraperitoneally (i.p.) if eosinophilia is to be elicited in the peritoneal cavity.

Ovalbumin

Several detailed descriptions of OVA sensitization and challenge protocols can be found in the literature.[9–12] In its most common form, sensitization consists of i.p. injection of 100–200 μL of phosphate buffered saline (PBS) solution containing 20–100 μg OVA with 2 mg alum on day 1 and 14, with allergen provocation on days 28, 29, and 30 with an aerosol derived from 1% OVA in saline, followed by evaluation on day 32. Variations of this protocol have different time intervals, different doses of OVA, and different routes of airway challenge (aerosol, e.t., or i.n. instillations). A version of this protocol produced a 10% eosinophilia and over a million eosinophils in BAL.[13] Using a similar protocol, over 9×10^6 cells were elicited in BAL, over 44% of which were eosinophils.[14]

Shorter protocols were also successful. For example, two i.p. sensitizations 5 days apart with 0.5 mL solution containing 8 μg OVA bound to 4 mg of aluminum hydroxide [Al(OH)$_3$], then 8 days later for five consecutive days i.n. OVA challenge with 100 μg OVA in 25 μL PBS yielded approximately a 40% BAL eosinophilia and approximately 10^7 mature eosinophils in the lungs.[15]

Peritoneal eosinophilia can be antigen-induced. The following OVA-based protocol has been described.[16,17] BALB/c mice were sensitized on day 0 and day 7 by subcutaneous injection of 100 μg OVA adsorbed in 1.6 mg Al(OH)$_3$ in 0.2 mL of saline, and challenged a week later by i.p. injection of 0.2 mL of 5 mg/mL OVA. Eosinophil accumulation was noticeable even 6 h after challenge and reached a plateau at approximately 10^6 cell/mL between 24 and 48 h. Use of a similar protocol resulted in approximately 10^7 of eosinophils recoverable from the peritoneal cavity.[18] Another approach is to sensitize mice by implanting subcutaneous, heat-coagulated egg white

pellets followed by challenge with i.p. injection of 10 μg OVA 15 days after OVA implantation surgery, and collecting cells 48 h after challenge. Over 10^6 eosinophils were recovered from the peritoneal cavity using this method.[19]

Ragweed

Substantial eosinophil recruitment to the lung is induced by ragweed sensitization, by i.p. injection of 50 μg ragweed extract (Greer Laboratories, Inc., Lenoir, NC, USA) with alum on day 0 and 5 and challenge with 50 μg ragweed in 30 μL PBS e.t. on day 14 and i.n. on day 15. This protocol resulted in over 10^6 cells in the BAL and over half of these were eosinophils.[20] Eosinophils can also be attracted into the peritoneal cavity after ragweed immunization in an interleukin (IL)-5 dependent fashion. Immunization protocols consisting of five subcutaneous injections of variously diluted ragweed pollen extract on days 0, 1, 6, 8, 14, and 20 yielded over 1.5×10^6 or 22% eosinophils in the peritoneal cavity.[21]

Fungi

A mouse model of allergic bronchopulmonary aspergillosis, which was created by i.n. immunization with 200 μg of *Aspergillus fumigatus* antigen in 50 μL of PBS twice a week for 4 weeks, featured approximately 40×10^4 eosinophils/mL.[22] The same *A. fumigatus* antigen used i.n. at 75 μg of protein in 50 μL of PBS for 7 consecutive days resulted in eosinophil counts of 120, 17, and 35×10^4 cells/mL in the PB, BM, and BAL respectively.[23] Havaux and colleagues presented a model of asthma via i.p. sensitization with 2×10^6 *Alternaria alternata* spores in PBS/alum twice and followed by i.n. challenge with 2×10^5 spores once or three times.[24] After three challenges, the number of eosinophils in BAL was over 4×10^5. *Alternaria* may also act as an adjuvant and cause significant eosinophilia even when administered i.n. with doses of OVA or ragweed that normally elicit very minimal response. For example, 8×10^4 eosinophil/BAL sample were detected when *A. alternata* was used in addition to OVA, and 10^5 eosinophils/BAL when used in combination with ragweed.[25]

House Dust Mite

Acute and chronic HDM inoculation protocols were used to create allergic inflammation in mice.[26] In an acute protocol, 25 μg of HDM extract without adjuvant was injected i.p. on day 0 and challenged with 25 μg of the extract on day 10. About 7×10^5 cells were counted in BAL of which over 20% were eosinophils. A chronic protocol using 40 μg HDM via i.n. administration 5 days a week for 3 weeks led to the accumulation of approximately 6×10^6 cells/BAL sample, half of which were eosinophils. Another chronic protocol was established by using HDM extract at 25 μg/ 10 μL i.n. 5 days a week for up to 5 weeks. After 3 weeks of this protocol, about 7% of blood and 40% of BAL leukocytes were eosinophils:[27] after 5 weeks of this protocol about of 10^6 cells in BAL were eosinophils.[28]

Cat

Mouse models of cat allergy have been generated via i.p. injections on days 0 and 14 of purified Fel d 1, the major antigen responsible for cat allergies in humans (Indoor Biotechnology Inc., Charlottesville, VA, USA). The protocol includes sensitization with 5 μg protein in 2 mg of $Al(OH)_3$ followed by e.t. challenge on days 28, 29, 30, 33, 40, 47, and 54 with 1 mg of native Fel d 1.[29] This protocol was designed to study immunotherapeutic approaches and led to 20% eosinophils in BAL.

Cockroach

Acute allergic lung inflammation was established by sensitizing BALB/c mice with i.p. injection of 10 μg of cockroach mix antigens in 2.25 mg of alum on days 0 and 14, followed by challenge with 1% aerosolized antigens for 30 min on days 28, 29, and 30, and studied on day 32.[30] BAL contained about 10^6 cells, about 65% of which were eosinophils.

Parasites

Allergic disorders are often thought of as a *misfired immune response*, which was intended to protect against metazoan parasitic infections.[31] A classical example is eosinophilia caused by infection with *Schistosoma mansoni 1*.[32] PB and peritoneal eosinophilia could reach 1000 cells/mm³ and as much as 70% of BM progenitors would be eosinophilic.[33] Using the same parasite and its soluble egg antigen, 50% of circulating granulocytes were eosinophils.[34] A combination of *S. mansoni* and thioglycolate produced approximately 50% eosinophilia in the peritoneum.[35] Parasitic helminths also cause tropical pulmonary eosinophilia, a severe asthmatic response due to entrapment of this blood parasite in the microvasculature of the lung. A mouse model replicates many of the characteristics of the human disease.[36] Mice immunized with three weekly subcutaneous injections of 10^5 killed *Brugia malayi* microfilariae and challenged 10 days later intravenously with 2×10^5 live microfilariae exhibit severe pulmonary eosinophilia with BAL eosinophil counts of 5.3×10^5 (85% of cells). After using the same helminth antigen with an adjuvant for i.p. sensitization and i.p. inoculation of 10^4 live microfilariae as a challenge, peritoneal eosinophilia reached as high as 55%.[37] *Mesocestoides corti* infection[38] leads to blood, peritoneal, BM, and spleen eosinophilia.

Multiple Allergens

House dust extract (HDE), which contained high levels of the cockroach antigens Bla g1 and Bla g2 and very low levels of HDM Der p 1 or Der p 2 and feline Fel d 1 or Fel d 2 antigens, was used to induce allergic pulmonary inflammation. Mice were sensitized by an i.p. injection of 50 μL of HDE emulsified in 50 μL TiterMax Gold (CytRx, Atlanta, GA, USA) on day 0. On days 14 and 21, mice were given an e.t. challenge of 50 μL of HDE, and by 36−48 h after the last challenge the number of eosinophils in BAL reached 10^5. Dual or multiple allergen administration[39,40] is reported to enhance allergic response, especially if there is an interest in eosinophils conditioned by multiple allergen chronic pulmonary inflammation.[40] A combination of HDM and OVA caused no higher eosinophilia in BAL than HDM alone, but greater airway hyperresponsiveness (AHR).[39] Combination of HDM, ragweed, and *Aspergillus* helped to break the tolerance after a chronic allergen protocol; however, an extent of eosinophilia was not reported.[40] Even though such protocols do not result in higher eosinophil numbers, those eosinophils may have unique properties that would be of interest for a particular study.

CYTOKINES AND CHEMOKINES

A number of cytokines and chemokines are known to be important for eosinophil proliferation and recruitment to the sites of inflammation. They include C-C motif chemokine 11 (CCL11; eotaxin), C-C motif chemokine 24 (CCL24; eotaxin-2), granulocyte-macrophage colony-stimulating factor (GM-CSF), interleukin-3 (IL-3), IL-5, and interleukin-33 (IL-33), which play distinct roles in the maturation, chemotaxis, and activation of mouse eosinophils.

Interleukin-5

IL-5, an *eosinophil differentiation factor*[41] and the most universally acknowledged eosinophilopoetin, is necessary for eosinophil proliferation in the BM and activation *in vitro* and *in vivo*. On exposure to allergens or helminths, various cells [e.g., CD4$^+$ T-helper type 2 (T$_h$2) polarized lymphocytes, mast cells, gamma delta T cells, natural killer cells, natural killer T cells, and nonhematopoietic cells] produce IL-5. A practical way to ensure a sustainable level of IL-5 in the mouse is to overexpress it using a transgenic approach. Several reports describing various IL-5 transgenic mice are summarized in Table 4.4.1. Constitutive overexpression of eosinophil-active cytokines can be

TABLE 4.4.1 Interleukin-5 Transgenic Mice are Characterized by Profound Eosinophilia

Reference	Regulatory Element	IL-5 Serum Concentration	Peripheral Blood Eosinophilia, 10^3 Eosinophils/μL	Tissue Eosinophilia
38	CD2	39 U[a]	53 at 9–12 weeks	Spleen, peritoneum
45	Metallothionein-1	16.6 ng/mL at 25 weeks[b]	14 at 25 weeks	Spleen, peritoneum
43	CD3ε	0.8 ng/mL at 4 weeks, 0.4 ng/mL at older age	26.5 at 4 weeks, 160 at 28 weeks	Spleen, liver, peritoneum
46	CC10	1.7 ng/mL		Lung
47	FABPL	Not reported	Circa 90	BM, intestine

BM, bone marrow; CC10, Clara cells 10 kDa secretory protein; FABPL, fatty acid-binding protein, liver.
[a]*Interleukin-5 (IL-5) was assayed by the eosinophil differentiation assay. The level of IL-5 in the serum of IL-5 transgenic mice was approximately the same as in Mesocestoides corti-infected mice.*
[b]*Found in homozygous mice; hemizygous mice had approximately half of that amount.*

restricted to specific tissues [see, for example,[42]]. Ochkur and colleagues[42] described a mouse model of severe asthma in mice that overexpress both IL-5 and CCL24 (eotaxin-2); expression of CCL24 was directed by the uteroglobin [Clara cells 10 kDa secretory protein (CC10)] promoter, and IL-5 from the T-cell CD3 delta molecule (CD3δ) promoter. These mice had the same level of IL-5 and blood eosinophilia as the original IL-5 transgenic mice,[43] but massive eosinophilia in the lung with eosinophil numbers $>10^7$/total BAL. The eosinophils from these mice were activated and showed distinct signs of degranulation.

Allergen exposure in transgenic mice has been studied.[44] In this study, OVA sensitization and challenge did not alter the amount of IL-5 in the airways, but caused substantial lung eosinophilia (5.8×10^6 eosinophils/BAL).

Another example of successful induction of eosinophilia and creation of a model of murine hypereosinophilic syndrome was reported by using hematopoietic stem cells/progenitors from IL-5 transgenic mice that were retrovirally transduced with the *FIP1 like 1 (S. cerevisiae) (FIP1L1)−platelet-derived growth factor receptor, alpha polypeptide (PDGFRA)* fusion gene.[48] Lethally irradiated mice intravenously received $2.5−6.5 \times 10^6$ transduced cells. Four weeks later, they developed eosinophilia in the PB ($35.0 \pm 9.9 \times 10^3$ eosinophils) and multiple organs, as well as developing splenomegaly.

Eotaxin

Matthews and colleagues discovered that eotaxin (CCL11) is needed for maintaining a baseline level of eosinophils.[49] Rothenberg and colleagues administered eotaxin intravenously to transgenic mice that overexpressed IL-5 under

the CD2 promoter and saw a remarkable eosinophil influx, reaching 88% and 2×10^6 eosinophils in BAL after 5 nmol of intranasal instillation, and significant lung tissue eosinophilia.[50] Subcutaneous injection of eotaxin brought eosinophil infiltration into the skin even 1 h after injection. Similar results were obtained in other studies.[51] After e.t. instillation of 5 μg eotaxin in the same IL-5 transgenic mice, the authors recovered 16.8×10^5 eosinophils/mL BAL.[42] Ochkur and colleagues instilled mouse eotaxin (Ccl11) and Ccl24 (eotaxin-2) or human CCL24, and C-C motif chemokine 26 (eotaxin-3) into the trachea of IL-5 transgenic mice [T-cell surface glycoprotein CD3 delta chain (CD3δ)] and showed that mouse eotaxin (Ccl11) and eotaxin-2 (Ccl24), and human CCL24 are the most effective chemokines in bringing mouse eosinophils into the airway lumen from the circulation and that eosinophil influx peaks at 6 h after instillation. Ccl24 (eotaxin-2) also elicits the recruitment of eosinophil progenitor cells [CD34$^+$and C-C chemokine receptor type 3 (CCR3)$^+$] in the lungs when instilled e.t. into IL-5 transgenic mice.[15]

Overexpression of eotaxin driven by intestinal fatty acid-binding protein, liver (FABPL) promoter did not lead to an increase of eosinophils in the BM or blood.[47] Eotaxin could stimulate the development of functional eosinophils from mouse embryonic stem cells when cultured in the presence of either IL-3 or IL-5 on a neonatal, calvaria-derived stromal cell feeder layer; up to 50% of the cells became eosinophils.[52] Eotaxin stimulated differentiation of embryonic stem cells *in vitro* into eosinophils, but only in the presence of IL-3 or IL-5.[52] *In vivo*, eotaxin also did not stimulate eosinophilia if it was overexpressed alone.

Double transgenic mice[47] that overexpress eotaxin and IL-5 under the same FABPL promoter had the same eosinophil count as single eotaxin transgenic mice.

However, substantially increased eosinophilia (10^7/BAL) was attained in double transgenic mice where eotaxin-2 overexpression was driven by the uteroglobin (CC10) promoter and IL-5 by the CD3δ promoter compared to the respective single transgenic mice.[42]

Interleukin-33

IL-33 [interleukin-1 family member 11 (IL-1F11)] is an alarmin that is released during cell necrosis and is associated with T_h2 bias. The IL-33 receptor, interleukin-1 receptor-like 1 (ILRL1), is found on multiple cell types, including eosinophils, and can induce airway inflammation even in the absence of allergy.[53] When mice were treated daily with 0.4 or 4 μg of IL-33 for 7 days, they developed splenomegaly and significant blood eosinophilia of up to 2000 cells/μL.[54] Intranasal administration of 4 μg of IL-33 to allergen-naive mice for 6 consecutive days resulted in an asthma-like pathology; the number of eosinophils in BAL reached 10^6 cells/mL. IL-33 remarkably enhanced airway eosinophilia when it was given in addition to OVA.[55] IL-33 also stimulated $CD117^+$ (c-Kit) hematopoietic progenitor cells differentiation into eosinophils.[56]

The EL4 mouse lymphoma cell line that secretes the active form of IL-33 (termed EL-4-IL-33),[57] when injected into the peritoneal cavity at 10^7 cells/mouse caused massive eosinophil accumulation. In this study, eosinophils were defined as having a glutathione reductase (Gr)-1^{int} sialic acid-binding Ig-like lectin (Siglec)-F^+ surface phenotype by means of flow cytometry. They were also identified microscopically, expressed eosinophil-associated genes and proteins, and migrated *in vitro* in response to eotaxin.

Interleukin-9

Transgenic mice overexpressing interleukin-9 (IL-9)[58] 3 days after bleomycin sulfate administration have 10^5 eosinophils in BAL. When peritoneal inflammation was induced by sodium thioglycolate, 72 h after injection, the PL contained 81%, 16×10^6 eosinophils.[59] Doxycycline-inducible IL-9, driven by the lung epithelium-specific CC10 promoter,[60] recruited 3×10^5 eosinophils in BAL by day 5 of doxycycline administration.[59]

Interleukin-17

Overexpression of human interleukin-17E (IL-17E) under the apolipoprotein E hepatic promoter in BDF mice[61] resulted in high levels of circulating eosinophils (about 2000 cells/μL), eosinophilia in the spleen (68×10^6 cells), lymph nodes (28×10^6 cells), and BM (approximately 0.8×10^6 cells).

ISOLATION AND PURIFICATION OF EOSINOPHILS

Peripheral Blood

In our laboratory, we routinely use the following method[62]: 300 μL of blood from a tail vein of IL-5 transgenic mice is collected into a tube filled with 5 mL PBS supplemented with 100 units of heparin sodium. The diluted blood is carefully layered over a Percoll solution, containing 60% Percoll, ρ = 1.084 g/mL, and 1 × Hanks balanced salt solution (HBSS) (1×), 15 mM 4-(2-hydroxyethyl)-1-piperazineethanesulfonic acid (HEPES) acid (pH 7.4), and then centrifuged for 45 min, at 2000 g, 4°C. The buffy coat is removed and washed twice with PBS that contains 2% fetal calf serum (FCS). The cell pellet is resuspended and residual red blood cells (RBCs) are lysed by 20-s incubation in 20 mL of cold water followed by dilution with 2 × PBS. Then the cells are washed twice in PBS/FCS. At this point the cell suspension consists of eosinophils and lymphocytes. B and T cells are removed using magnetic-activated cell sorting (MACS) microbeads (Miltenyi Biotec, Inc.) conjugated to CD45 and CD90, according to the manufacturer's protocol. The purified eosinophils (usually >98% pure) are washed twice, resuspended in Roswell Park Memorial Institute (RPMI) 1640 medium with 2% FBS and can be kept refrigerated for up to 24 h prior to analysis.

Spleen

Isolation and purification of eosinophils from spleen differs from the PB protocol chiefly in the method used to obtain the cell suspension. The process starts with sacrificing a mouse, excising the spleen, cleaning excess fatty tissue from the outside of the spleen. The spleen is then sliced with a scalpel into approximately 3-mm pieces, which are then placed onto a 45-μm mesh placed over a 45 mL conical tube. Using the rubber end of the 3 cc syringe plunger, the spleen tissue is gently worked against the screen with an occasional flush with PBS/FCS to wash single cells through the screen. The cells are washed twice with PBS/FCS, the RBCs are lysed, and this is followed by Percoll separation and magnetic bead purification as described earlier. Eosinophil purity is usually close to 98%.

Dyer and colleagues made a single-cell suspension by cutting spleen into small pieces in HBSS containing 1% FBS and 10 mM HEPES acid. The tissue is then passed through a 100-μm strainer and a 21-gauge needle. RBCs are then lysed and the eosinophils are purified using magnetic microbeads conjugated to antilymphocyte CD90.2 and CD45R (B220 isoform) antibodies according to the manufacturer's instructions. The resulting purity was about 80%.[63]

Peritoneal Cavity

Purification of eosinophils from peritoneal cavity exudates is similar to the one from the PB and spleen. The difference is mainly in the first steps. Zuany-Amorim and colleagues used the following procedure.[17] Mice were sacrificed, the peritoneal cavity was opened and washed with 3 mL of heparinized saline (10 U/mL), and 90% of the original volume is recovered. PL cells are counted and analyzed by the differential analysis of Diff-Quick-stained slides obtained after the cytocentrifugation of PL fluid and/or by flow cytometry. About 50–60% of PL cells are eosinophils.[18] Eosinophils can be further purified by MACS (Miltenyi Biotec GmbH, Bergisch Gladbach, Germany) using CD45 (B220 isoform), CD90, and major histocompatibility complex class II (MHC-II) magnetic beads. With this method, over 10^7 eosinophils per mouse can be obtained with 90–98% purity. A good description of the isolation and purification of mouse eosinophils from the peritoneal cavity can be found in Legrand et al.[64]

Bronchoalveolar Lavage

BAL eosinophils can be purified in the same way as blood and spleen eosinophils. An original way to purify eosinophils from BAL has been described by Shinagawa and Anderson:[14] 13 plant lectins were screened for their utility in the purification of BAL eosinophils from mice sensitized and challenged with OVA. Initially, BAL contained 45% eosinophils, 44% macrophages, 12% lymphocytes, and 0.1% neutrophils. The best yield (about 40%) and purity (up to 98.5%) was achieved with Jacalin and *Griffonia (Bandeiraea) simplicifolia* isolectin I at 10 µg/mL with 1 h incubation. Minor cell populations were CD4, CD8, and B220 lymphocytes.

Stem Cells

Eosinophils may be obtained from stem cells derived from mouse embryos[52] cultured in the presence of IL-5 and either GM-CSF or IL-13. Eotaxin stimulated eosinophil development either with GM-CSF or IL-5. Another source of mouse eosinophils is the BM.[65] Dyer and colleagues developed a method of generating large numbers of mouse eosinophils from the BM. BM cells were cultured for 4 days in augmented RPMI 1640 medium containing stem cell factor (SCF) and receptor-type tyrosine-protein kinase FLT3 (FLT3) ligand, after which the last two reagents were replaced with IL-5. By day 17 of culture, the number of eosinophils was 10 times greater than the original number of BM cells. Eosinophils were >90% pure and functionally competent. Later, the same authors[66] were also able to obtain

eosinophils from the BM of several knockout mice on two genetic backgrounds; eosinophils were 100% pure essentially and thus required no further purification.

Eosinophilopoiesis in BM cultures may be influenced by IL-13 and eotaxin, with cysteinyl leukotriene receptor 1 (CysLTR1) playing an important role.[67]

EX VIVO ASSAYS

Ex vivo assays are used to study eosinophil biology in isolation of the natural microenvironment and in strictly controlled culture conditions. Under these conditions eosinophils are stimulated to release various mediators (Table 4.4.2), display cell-surface molecules, change shape, adhere to a surface, move (chemotaxis assays; Table 4.4.3), and affect other cell types.

Mobilization of Intracellular Ca^{2+}

This can be used as a convenient indication of eosinophil stimulation for *ex vivo* assays. Many chemokines act on their receptors, which results in the mobilization of intracellular Ca^{2+}, leading to an increase in cytoplasmic free Ca^{2+}. Many chemokine and hormone receptors that are present on eosinophils belong to a very broad and ancient family of G protein-coupled receptors, seven transmembrane domain receptors that work in concert with trimeric guanosine triphosphate (GTP)-binding proteins (G proteins). These receptors relate their signals in one of two major ways: via an increase in the concentration of cyclic adenosine monophosphate (cAMP) or free Ca^{2+}. Those receptors that are coupled to Gq proteins use Ca^{2+} as a second messenger. Normally, cytosolic Ca^{2+} concentrations are kept at extremely low levels ($\leq 10^{-7}$ M) by Ca^{2+}-ATPases located in the plasma membrane and the membranes of the endoplasmic reticulum and mitochondria. On stimulation of the G protein-coupled receptor, the Gq protein activates phospholipase Cβ, which cleaves phosphatidylinositol 4,5-biphosphate (PIP_2) to produce diacylglycerol and inositol 1,4,5-triphosphate (IP_3). The latter binds to tetrameric, IP_3-gated Ca^{2+} release channels in the endoplasmic reticulum. The opening of these channels leads to the increase in cytosolic Ca^{2+} concentration to micromolar levels. This increase can be readily measured with the help of Ca^{2+}-binding dyes. Borchers and colleagues,[62] and Rothenberg and colleagues,[50] showed that eosinophils mobilize intracellular Ca^{2+} in a dose-dependent fashion at concentrations of eotaxin up to 100 g/mL, reaching a plateau thereafter. A slight response to C-C motif chemokine 3 [CCL3/macrophage inflammatory protein 1-alpha (MIP-1-alpha)] was also detected.[34]

TABLE 4.4.2 Mouse Eosinophils' Response to Stimulation *Ex Vivo*

Molecule Tested	Stimulus	Detected	Method of Detection	Eosinophil Source (Mouse Line)	Reference
Ca^{2+}	MIP-1-alpha, eotaxin	Yes	Fura-2-acetoxy methyl ester	Peritoneum (IL-5 transgenic, thioglycolate; *Schistosoma mansoni*)	34
Ca^{2+}	MCP-1, MDC, TARC, MIP-1-beta, TCA-3	No	Fura-2-acetoxy methyl ester	Peritoneum (IL-5 transgenic, thioglycolate; *Schistosoma mansoni*)	34
Ca^{2+}	Eotaxin	Yes	Fura-2 dye	Spleen (CD2-IL-5 transgenic)	50
Ca^{2+}	mEotaxin, mEotaxin-2, mTARC	Yes	Fura-2 dye	Blood (CD3δ-IL-5 transgenic)	62
Ca^{2+}	Eotaxin, Eotaxin-2, PAF	Yes	Fura-2-acetoxy methyl ester	Spleen (CD2-IL-5 transgenic)	71
Ca^{2+}	mMIP-1-beta, KC	No	Fura-2 dye	Blood (CD3D-IL-5 transgenic)	62
Mitochondrial DNA	LPS, C5a, Eotaxin primed with IL-5 or IFN-γ	Yes	SYTO 13, SYTOX Orange	Blood (CD3D-IL-5 transgenic)	68
CCL11, CCL24, IL-4, CXCL10, CCL3, IL-17, CXCL1, CXCL9, FGF, IL-1 alpha, IL-2, IL-5, IL-10, IL-12, IFN-γ, IL-1 beta, TNF-α	IL-33	No	ELISA, Luminex	BM cultures (WT)	56
RANTES, MDC, MIP-1-beta, C10	SCF	Yes	ELISA	Peritoneum (IL5 transgenic, thioglycolate; *Schistosoma mansoni*)	35
TARC, CXC-R3, CXC-R5, RANTES	SCF	Yes	RT-PCR	Peritoneum (IL-5 transgenic, thioglycolate; *Schistosoma mansoni*)	35
CD11b	Eotaxin, LTB_4	Yes	Fluorescence-assisted cell sorting	Blood (WT, mettallo thionein-IL-5 transgenic)	72
CD29, CD80	*Strongyloides stercoralis* antigens	No	Fluorescence-assisted cell sorting	Spleen (CD3D-IL-5 transgenic)	43
CD69, CD86, MHC-II, CD62L	*Strongyloides stercoralis* antigens	Yes	Fluorescence-assisted cell sorting	Spleen (CD3D-IL-5 transgenic)	43
IFN-γ, IL-4, IL-6, MIP-1-alpha, IL-9, IL-10	Non-BM culture	Yes	Bio-Plex suspension array system (Bio-Rad)	BM culture (WT)	65
EPO	Calcium ionophore (A23187)	Yes	OPD	Spleen (CD2-IL-5 transgenic)	70
EPO	Eotaxins, IL-5, IgA beads, PMA-induced protein 1, PMA-induced protein 1 + myristate	No	OPD	Spleen (CD2-IL-5 transgenic)	70
EPO	PAF, lyso-PAF	Yes	OPD	BM culture, spleen (IL-5 transgenic)	63
EPO	IL-5, IL-3, GM-CSF, Eotaxin, IL-5/IL-3, IL-5/GM-SCF, IL-3/GM-CSF, FMLF, PMA-induced protein 1	No	OPD	BM culture (IL-5 transgenic)	63

(Continued)

TABLE 4.4.2 Mouse Eosinophils' Response to Stimulation *Ex Vivo*—cont'd

Molecule Tested	Stimulus	Detected	Method of Detection	Eosinophil Source (Mouse Line)	Reference
EPO	PAF	Yes	OPD	BM culture (WT or TLR3$^{-/-}$)	66
EPO	SCF	Yes	OPD	Peritoneum (IL-5 transgenic, thioglycolate; *Schistosoma mansoni*)	35
EPO	IFN-γ	Yes	Luminol, Luciferin	Spleen (CD2-IL-5 transgenic)	73
MBP-1	PMA-induced protein 1	Yes	Immunoblot	Peritoneum (CD2-IL-5 transgenic)	12
F-actin	Eotaxin, LTB$_4$, MIP-1-alpha	Yes	FACS	Blood (WT, metallo thionein-IL-5 transgenic)	72
F-actin	Eotaxin-1, PAF	Yes	FACS	Spleen (CD2-IL-5 transgenic)	71
F-actin	RANTES, PGD2, LTD$_4$	No	FACS	Blood (WT, metallo thionein-IL-5 transgenic)	72
IL-13, IL-6, CCL17, TGF-β	IL-33	Yes	ELISA	BM cultures (WT)	56
IL-1 beta, IL-12, IFN-γ, IL-9, TNF-α, Mcp-1 (CCL2)	IL-6	Yes	ELISA, Milliplex (Millipore), Bioplex (Bio-Rad)	BM culture (WT)	63
IL-1 beta, IL-12, IFN-γ, IL9, TNF-α, Mcp-1 (CCL2)	Eotaxin	No	ELISA, Milliplex (Millipore), Bioplex (Bio-Rad)	BM culture (WT)	63
LTC$_4$	Eotaxin, MIP-1-alpha, MIP-1-beta, TCA-3	Yes	ELISA	Peritoneum (IL-5 transgenic, thioglycolate; *Schistosoma mansoni*)	34
LTC$_4$	IL-5, PMA-induced protein 1 + ionomycin	Yes	ELISA	Peritoneum (WT, OVA)	18
LTC$_4$	Calcium ionophore A23187	Yes	ELISA	Peritoneum (IL-5 transgenic CH3/HeN)	44
LTC$_4$	Extract of *M. corti*	Yes	HPLC	Peritoneum (WT, *M. corti*)	74
LTC$_4$	SCF	Yes	ELISA	Peritoneum (IL-5 transgenic, thioglycolate; *Schistosoma mansoni*)	35
MHC-II, CD80, CD86	GM-CSF	Yes	FACS	Blood (WT, metallo thionein-IL-5 transgenic)	75
MHC-II, CD80, CD86, CD40	GM-CSF	Yes	FACS	Blood (WT, metallo thionein-IL-5 transgenic)	76
MHC-II	GM-CSF; increased with IL-4, decreased with IFN-γ, IL-5 had no effect	Yes	FACS	Peritoneum (WT *Brugia malayi* infected)	37
O$_2^-$	fMLP	No	Cytochrome C	Spleen (CD2-IL-5 transgenic)	70
O$_2^-$	PMA-induced protein 1, 500 ng/mL	Yes	Cytochrome C	Spleen (CD2-IL-5 transgenic)	70
O$_2^-$	Eotaxin-2, PAF, IL-5 had no effect	Yes	FACS	Spleen (CD2-IL-5 transgenic)	70

TABLE 4.4.2 Mouse Eosinophils' Response to Stimulation *Ex Vivo*—cont'd

Molecule Tested	Stimulus	Detected	Method of Detection	Eosinophil Source (Mouse Line)	Reference
O_2^-	PMA-induced protein 1	Yes	DH-23	BAL (WT)	14
O_2^-	PMA-induced protein 1	Yes	Cytochrome C	Peritoneum (IL-5 transgenic CH3/HeN)	44
O_2^-	PAF, IL-5, eotaxin	No	Cytochrome C	Peritoneum (IL-5 transgenic CH3/HeN)	44
O_2^-	Extract of *M. corti*	Yes	Cytochrome C	Peritoneum (WT, *M. corti*)	74
TGF-beta-1	IL-5, IL-3	Yes	ELISA	Peritoneum (IL-5 transgenic CH3/HeN)	44
TGF-beta-1	Eotaxin	No	ELISA	Peritoneum (IL-5 transgenic CH3/HeN)	44

BAL, bronchoalveolar lavage; C5a, complement component C5a; C10, protein C10; CCL2, C-C motif chemokine 2; CCL3, C-C motif chemokine 3; CCL11, C-C motif chemokine 11 (eotaxin); CCL17, C-C motif chemokine 17; CCL24, C-C motif chemokine 24 (eotaxin-2); CXCL1, C-X-C motif chemokine 1; CXCL9, C-X-C motif chemokine 9; CXCL10, C-X-C motif chemokine 10; CXC-R3, C-X-C chemokine receptor type 3; CXC-R5, C-X-C chemokine receptor type 5; EPO, eosinophil peroxidase; ELISA, enzyme-linked immunosorbent assay; FACS, fluorescence-activated cell sorting; FGF, fibroblast growth factor; fMLP, fMet-Leu-Phe; GM-CSF, granulocyte-macrophage colony-stimulating factor; HPLC, high-performance liquid chromatography; IFN-γ, interferon gamma; IL-, interleukin; KC, keratinocyte chemoattractant; LPS, lipopolysaccharide; LTB$_4$, leukotriene B$_4$; LTC$_4$, leukotriene C$_4$; LTD$_4$, leukotriene D$_4$; MBP-1, eosinophil granule major basic protein 1; *M. corti*, *Mesocestoides corti*; MCP-1, monocyte chemoattractant protein-1; MDC, macrophage-derived chemokine; MHC-II, major histocompatibility complex class II; MIP-1-alpha, macrophage inflammatory protein 1-alpha; MIP-1-beta, macrophage inflammatory protein 1-beta; OPD, o-phenylenediamine; PAF, platelet-activating factor; PGD2, prostaglandin-D2; RANTES, regulated upon activation, normally T-expressed, and presumably secreted; RT-PCR, reverse transcription polymerase chain reaction; SCF, stem cell factor; SYTO13, green fluorescent nucleic acid stain; TARC, thymus and activation-regulated chemokine; TCA-3, T-cell activation protein 3; TGF-β, transforming growth factor beta; TGF-beta-1, transforming growth factor beta-1; TLR, toll-like receptor; TNF-α, tumor necrosis factor; WT, wild-type.

Degranulation

Degranulation is the release of mediators stored in eosinophil granules. Traditionally, it is thought of as secretion of eosinophil-associated ribonucleases (EARs), eosinophil peroxidase (EPO), and eosinophil granule major basic protein 1 (MBP-1), although other components are also released in this process. When studying degranulation in response to stimuli *ex vivo*, it is piecemeal degranulation that is examined. Normally, eosinophils from various tissues are incubated in a culture medium with and without a stimulant and the presence of EPO in the culture medium is measured. EPO is chosen because it is relatively stable in the solution and because it is eosinophil-specific. MBP-1 is too basic (pI > 11) and easily adheres to surfaces (although methods to detect this granule protein by immunoblot have been described[12]), EARs are not strictly specific for eosinophils. The following methods are used to measure EPO in solution: (1) the most common method is based on measuring the enzymatic activity of EPO using o-Phenylenediamine (OPD) as a substrate; (2) immunoblot; and (3) enzyme-linked immunosorbent assay (ELISA).

Reactive Oxygen Species

These are secreted by mouse eosinophils. Kobayashi and colleagues stimulated mouse eosinophils isolated from the peritoneal cavity of IL-5 transgenic mice or from the BAL of these mice after OVA sensitization.[44] Eosinophils were stimulated *ex vivo* with platelet-activating factor (PAF), IL-5, eotaxin, phorbol-12-myristate-13-acetate-induced protein 1 (PMA-induced protein 1), and immobilized human and mouse immunoglobulin G (IgG). Only PMA-induced protein 1 induced a robust superoxide production. Shinagawa and colleagues detected robust superoxide production from BAL eosinophils isolated from OVA-sensitized and -challenged mice.[14]

Leukotriene C$_4$

Leukotriene C$_4$ (LTC$_4$) is produced and secreted by eosinophils on activation. Kobayashi and colleagues stimulated the eosonophils from the murine peritoneal cavity isolated from IL-5 transgenic mice with IL-5, immobilized immunoglobulin G2a (IgG2a), and calcium ionophore A23187:[44] only calcium ionophore A23187 elicited a vigorous response. Oliveira and colleagues demonstrated LTC$_4$ release from eosinophils in response to CCL4 (MIP-1-beta) and C-C motif chemokine 1 [CCL1/T-cell activation protein 3 (TCA3)].[34] Eosinophils isolated from the peritonea of OVA-sensitized and -challenged BALB/c mice release LTC$_4$ when stimulated with IL-5 in a dose-dependent manner.[18]

TABLE 4.4.3 Chemotaxis of the Mouse Eosinophil

Chemokine	Eosinophil Source (Mouse Line and Treatment)	Chemotaxis Observed	Reference
Eotaxin (CCL11), eotaxin-2 (CCL4)	Blood or BM	Yes	15
Eotaxin (CCL11)	Spleen (CD2-IL-5 transgenic)	Yes	50
MIP-1-alpha, eotaxin (CCL11)	Peritoneum (IL-5 transgenic, thioglycolate)	Yes	34
MIP-1-beta, TCA-3	Peritoneum (IL-5 transgenic, thioglycolate)	No	34
MIP-1-alpha, eotaxin (CCL11)	Blood (IL-5 transgenic, thioglycolate)	Yes	34
MIP-1-beta, TCA-3	Blood (IL-5 transgenic, thioglycolate)	No	34
Eotaxin (CCL11), MIP-1-alpha, MIP-1-beta, TCA-3	Peritoneum (IL-5 transgenic, thioglycolate; *Schistosoma mansoni*)	Yes	34
Eotaxin (CCL11), PAF	Peritoneum (WT, OVA)	Yes	18
TARC	Peritoneum (WT, OVA)	No	18
Eotaxin (CCL11)	BM culture (WT)	Yes	65
Eotaxin (CCL11)	Peritoneum (WT, EL-4-IL-33 cell inoculated)	Yes	57
mKC, mMIP-1-beta, mTARC, eotaxin (CCL11), MCP-4, eotaxin-2 (CCL24), mEotaxin	Blood (CD3D-IL-5 transgenic)	Yes	62
mRantes, mMIP-1-alpha, RANTES, mMIP-3-beta, mMIP-1-gamma, mTECK, mBLC, mC10, mMDC, eotaxin-3 (CCL26)	Blood (CD3D-IL-5 transgenic)	No	62
Eotaxin (CCL11) + IFN-γ	Peritoneum (CD2-IL-5 transgenic)	Yes	73
PAF, LTB$_4$	Spleen (CD2-IL-5 transgenic)	Yes	71

BLC, B lymphocyte chemoattractant; BM, bone marrow; C10, protein C10; CCL2, C-C motif chemokine 2; CCL4, C-C motif chemokine 4; CCL24, C-C motif chemokine 24; CCL26, C-C motif chemokine 26; IFN-γ, interferon gamma; KC, keratinocyte chemoattractant; LTB$_4$, leukotriene B$_4$; MCP-4, monocyte chemotactic protein 4; MDC, macrophage-derived chemokine; MIP-1-alpha, macrophage inflammatory protein 1-alpha; MIP-1-beta, macrophage inflammatory protein 1-beta; MIP-3-beta, macrophage inflammatory protein 3 beta; MIP-1-gamma, macrophage inflammatory protein 1-gamma; PAF, platelet-activating factor; OVA, ovalbumin; RANTES, regulated upon activation, normally T-expressed, and presumably secreted; TARC, thymus and activation-regulated chemokine; TCA-3, T-cell activation protein 3; TECK, thymus-expressed chemokine; WT, wild-type.

DNA Traps

DNA traps may be used by eosinophils as a means to protect the host from bacterial infections. Eosinophils release mitochondrial DNA on priming with IL-5 or interferon gamma (IFN-γ) and when subsequently stimulated with lipopolysaccharide (LPS), complement component C5a (C5a), or eotaxin[68]

Antigen Processing

Antigen processing *ex vivo* has been described by Mackenzie and colleagues,[13] although the functional signature of this activity has been called into question.[69] When cultured with DQ OVA, which exhibits bright green fluorescence on proteolytic degradation, eosinophils sequester and process this molecule within 30 min and fully process it within 4 h.

Shape Change

Shape change occurs in response to CCL24 (eotaxin-2) and PAF.[70] It can be detected by confocal microscopy using a filamentous actin (F-actin) phalloidin probe conjugated to rhodamine and 4′,6-diamidino-2-phenylindole (DAPI).

Actin Polymerization

Actin polymerization was detected by flow cytometry after stimulation of eosinophils *ex vivo* with CCL24 (eotaxin-2) and PAF.[70]

Chemotaxis

Chemotaxis, related receptors, and their ligands have been described in a comprehensive review.[77]

Eosinophils express C-C chemokine receptor type 1 (CCR1), 2 (CCR2), CCR3, 5 (CCR5), and 8 (CCR8), and C-X-C chemokine receptor type 2 (CXC-R2) and 4 (CXC-R4) mRNA, and migrate predominantly using the CCR3 receptor.[62] Cytokines, media, and drugs, and 24-well tissue-culture treated polystyrene plates with Transwell polycarbonate membranes with 3–5 μm pore size are commonly used in our laboratory to assess the ability of mouse eosinophils to migrate *ex vivo* in response to various reagents, including chemokines.[62] The upper chamber (insert) is first preincubated with 200 μL of RPMI 1640 medium containing 5% FCS for 1 h at room temperature. The medium in the upper chamber is replaced with 200 μL of RPMI/FSC containing 30 ng/mL of IL-5 and 10^6 purified eosinophils. The lower well is filled with 500 μL of medium with or without chemoattractant and the insert is placed on top of the lower well. The plates are placed in a 5% CO_2 incubator at 37°C for 90 min. Cells from the upper chamber are removed and the cells that passed though the mesh but remained attached to it are recovered as follows. The top chamber is wiped with a cotton swab, the insert is placed in the well filled with 500 μL ice-cold piperazine-N,N′-bis(2-ethanesulfonic acid) (PIPES) buffer containing 5 mM ethylenediaminetetraacetic acid (EDTA) and tapping the plates lightly. The total number of cells and cell differentials is determined. Using a similar chemotaxis assay, Rothenberg and colleagues[50] studied the properties of recombinant mouse eotaxin at concentrations of 0–1000 ng/mL and showed that eotaxin is an eosinophil-specific chemokine that stimulates dose-dependent migration, reaching a plateau at a concentration of 50 ng/mL. This assay is also used to show the functional integrity of eosinophils derived from BM.[66]

Adhesion Assay

An adhesion assay based on EPO activity was developed by Clark and colleagues.[12] Eosinophils were stimulated with PMA-induced protein 1 to adhere to flat-bottomed wells coated with 2.5% bovine serum albumin (BSA). Nonadhering cells were washed away at various time points. Adhering cells were lysed and the amount of EPO was assessed and expressed as a percentage of the total value of initially introduced eosinophils.

Co-cultures

Co-cultures of eosinophils with other cell types are used to study cell–cell interactions. Eosinophils and macrophage co-cultures were used to study the effect of IL-33-activated eosinophils on macrophage phenotype switching to M2.[56] A co-culture of *Strongyloides stercoralis* antigens-pulsed eosinophils with primed CD4[+] T cells resulted in MHC-II-dependent IL-5 production from the T cells.[78] Co-culture of eosinophils purified from IL-5 transgenic mice with CD4[+] T cells polarized *in vitro* to T_h2 phenotype caused the production of IL-4, IL-5, and IL-13 in an antigen-specific manner, though not IFN-γ from those T cells, which confirmed the ability of eosinophils for antigen presentation. Likewise, eosinophils isolated from the BAL of allergic mice co-cultured with *in vivo* polarized CD4[+] T cells isolated from the spleens of the same mice caused T cells to produce IL-5 in an antigen-dependent manner.[13] Eosinophils caused antigen-dependent/specific T-cell proliferation *ex vivo*. Airway-derived eosinophils from OVA-, BSA-, or human gamma globulin-treated mice were cultured with respective or different antigen-sensitized T cells. Only when antigen exposure was substantial and the same for the eosinophils and the T cells, T-cell proliferation was observed. This proliferation was partially inhibited when anti-CD80 or anti-CD86 antibodies were added, thus confirming the antigen-presenting capability of eosinophils.[79] Eosinophils from allergen-naive mice can stimulate T-cell proliferation in the presence of the cognate antigen, however, at lower levels compared to dendritic cells.[80]

CONCLUSION

Multiple approaches exist to elicit eosinophilia in the mouse, ranging from allergen treatment, specific cytokine administration, and directing progenitor cell development, to using genetic manipulation and transformed eosinophilic cell lines. Various methods have been developed to isolate and purify a substantial number of eosinophils from the mouse. The stimulation of eosinophils outside the body under strict experimental conditions has been used to study eosinophil activation, chemotaxis, secretion of specific granule proteins, reactive oxygen species, cytokines, and other inflammatory mediators. An exploration of the biology of mouse eosinophils *ex vivo* provides a wealth of information about the biology and significance of these cells in homeostatic as well as pathological conditions.

REFERENCES

1. Rothenberg ME, Hogan SP. The eosinophil. *Annu Rev Immunol* 2006;**24**:147–74.
2. Ackerman SJ, Bochner BS. Mechanisms of Eosinophilia in the Pathogenesis of Hypereosinophilic Disorders. *Immunology and Allergy Clinics of North America Hypereosinophilic Syndromes* 2007;**27**(3):357–75.

3. Hogan SP, et al. Eosinophils: Biological Properties and Role in Health and Disease doi:10.1111/j.1365−2222.2008.02958.x. *Clinical & Experimental Allergy* 2008;**38**(5):709−50.

4. Blanchard C, Rothenberg ME. Biology of the eosinophil. *Adv Immunol* 2009;**101**:81−121.

5. Park YM, Bochner BS. Eosinophil survival and apoptosis in health and disease. *Allergy Asthma Immunol Res* 2010;**2**(2):87−101.

6. Shamri R, Xenakis J, Spencer LA. Eosinophils in innate immunity: an evolving story. *Cell and Tissue Research* 2010:1−27. Springer Berlin/Heidelberg.

7. Kita H. Eosinophils: multifaceted biological properties and roles in health and disease. *Immunol Rev* 2011;**242**(1):161−77.

8. Barnes PJ. Pathophysiology of allergic inflammation. *Immunol Rev* 2011;**242**(1):31−50.

9. Pichavant M, et al. Animal models of airway sensitization. *Curr Protoc Immunol*; 2007. Chapter 15: p. Unit 15 18.

10. Conrad ML, et al. Comparison of adjuvant and adjuvant-free murine experimental asthma models. *Clin Exp Allergy* 2009;**39**(8): 1246−54.

11. Takeda K, et al. Strain dependence of airway hyperresponsiveness reflects differences in eosinophil localization in the lung. *Am J Physiol Lung Cell Mol Physiol* 2001;**281**(2):L394−402.

12. Clark K, et al. Eosinophil degranulation in the allergic lung of mice primarily occurs in the airway lumen. *J Leukoc Biol* 2004;**75**(6): 1001−9.

13. MacKenzie JR, et al. Eosinophils promote allergic disease of the lung by regulating CD4(+) Th2 lymphocyte function. *J Immunol* 2001;**167**(6):3146−55.

14. Shinagawa K, Anderson GP. Rapid isolation of homogeneous murine bronchoalveolar lavage fluid eosinophils by differential lectin affinity interaction and negative selection. *J Immunol Methods* 2000;**237**(1−2):65−72.

15. Radinger M, et al. Local proliferation and mobilization of CCR3(+) CD34(+) eosinophil-lineage-committed cells in the lung. *Immunology* 2011;**132**(1):144−54.

16. Zuany Amorim C, et al. Characterization and pharmacological modulation of antigen-induced peritonitis in actively sensitized mice. *British Journal of Pharmacology* 1993;**110**(2):917−24.

17. Zuany-Amorim C, et al. Modulation by IL-10 of antigen-induced IL-5 generation, and CD4+ T lymphocyte and eosinophil infiltration into the mouse peritoneal cavity. *J Immunol* 1996;**157**(1): 377−84.

18. Ebihara S, Kurachi H, Watanabe Y. A simple preparation method for mouse eosinophils and their responses to anti-allergic drugs. *Inflamm Res* 2007;**56**(3):112−7.

19. Cheraim AB, et al. Leukotriene B4 is essential for selective eosinophil recruitment following allergen challenge of CD4+ cells in a model of chronic eosinophilic inflammation. *Life Sci* 2008; **83**(5−6):214−22.

20. Nemeth K, et al. Bone marrow stromal cells use TGF-beta to suppress allergic responses in a mouse model of ragweed-induced asthma. *Proc Natl Acad Sci U S A* 2010;**107**(12):5652−7.

21. Kaneko M, et al. Role of interleukin-5 in local accumulation of eosinophils in mouse allergic peritonitis. *Int Arch Allergy Appl Immunol* 1991;**96**(1):41−5.

22. Kurup VP, et al. A murine model of allergic bronchopulmonary aspergillosis with elevated eosinophils and IgE. *J Immunol* 1992;**148**(12):3783−8.

23. Murali PS, et al. Development of bone marrow eosinophilia in mice induced by Aspergillus fumigatus antigens. *Clin Immunol Immunopathol* 1997;**84**(2):216−20.

24. Havaux X, et al. A new mouse model of lung allergy induced by the spores of Alternaria alternata and Cladosporium herbarum molds. *Clin Exp Immunol* 2005;**139**(2):179−88.

25. Kobayashi T, et al. Asthma-related environmental fungus, Alternaria, activates dendritic cells and produces potent Th2 adjuvant activity. *J Immunol* 2009;**182**(4):2502−10.

26. Simeone-Penney MC, et al. Airway epithelial STAT3 is required for allergic inflammation in a murine model of asthma. *J Immunol* 2007;**178**(10):6191−9.

27. Fattouh R, et al. Eosinophils are dispensable for allergic remodeling and immunity in a model of house dust mite-induced airway disease. *Am J Respir Crit Care Med* 2011;**183**(2):179−88.

28. Botelho FM, et al. Cigarette Smoke Differentially Impacts Eosinophilia and Remodeling in a House Dust Mite Asthma Model. *Am J Respir Cell Mol Biol*; 2011. p. 2010−0404OC.

29. Terada T, et al. A chimeric human-cat Fcgamma-Fel d1 fusion protein inhibits systemic, pulmonary, and cutaneous allergic reactivity to intratracheal challenge in mice sensitized to Fel d1, the major cat allergen. *Clin Immunol* 2006;**120**(1): 45−56.

30. McGee HS, Edwan JH, Agrawal DK. Flt3-L increases CD4+CD25+Foxp3+ICOS+ cells in the lungs of cockroach-sensitized and -challenged mice. *Am J Respir Cell Mol Biol* **42**(3): p. 331−340.

31. Rajan TV. The Gell-Coombs classification of hypersensitivity reactions: a re-interpretation. *Trends Immunol* 2003;**24**(7): 376−9.

32. Lenzi HL, Sobral AC, Lenzi JA. In vivo kinetics of eosinophils and mast cells in experimental murine schistosomiasis. *Mem Inst Oswaldo Cruz* 1987;**82**(Suppl. 4):67−76.

33. Lenzi HL, Lenzi JA. Comparative distribution of eosinophils in bone marrow, blood and peritoneal cavity in murine schistosomiasis. *Braz J Med Biol Res* 1990;**23**(10):989−94.

34. Oliveira SH, et al. Increased responsiveness of murine eosinophils to MIP-1beta (CCL4) and TCA-3 (CCL1) is mediated by their specific receptors, CCR5 and CCR8. *J Leukoc Biol* 2002;**71**(6): 1019−25.

35. Oliveira SHP, et al. Stem cell factor induces eosinophil activation and degranulation: mediator release and gene array analysis. *Blood* 2002;**100**(13):4291−7.

36. Hall LR, et al. An essential role for interleukin-5 and eosinophils in helminth-induced airway hyperresponsiveness. *Infection & Immunity* 1998;**66**(9):4425−30.

37. Mawhorter SD, et al. Class II major histocompatibility complex molecule expression on murine eosinophils activated in vivo by Brugia malayi. *Infect Immun* 1993;**61**(12):5410−2.

38. Dent LA, et al. Eosinophilia in transgenic mice expressing interleukin 5. *J Exp Med* 1990;**172**(5):1425−31.

39. DiGiovanni FA, et al. Concurrent dual allergen exposure and its effects on airway hyperresponsiveness, inflammation and remodeling in mice. *Dis Model Mech* 2009;**2**(5−6):275−82.

40. Goplen N, et al. Combined sensitization of mice to extracts of dust mite, ragweed, and Aspergillus species breaks through tolerance and establishes chronic features of asthma. *J Allergy Clin Immunol* 2009;**123**(4):925−32. e11.

41. Lopez AF, et al. Murine eosinophil differentiation factor. An eosinophil-specific colony-stimulating factor with activity for human cells. *J Exp Med* 1986;**163**(5):1085−99.

42. Ochkur SI, et al. Coexpression of IL-5 and Eotaxin-2 in Mice Creates an Eosinophil-Dependent Model of Respiratory Inflammation with Characteristics of Severe Asthma. *J Immunol* 2007; **178**(12):7879−89.

43. Lee NA, et al. Expression of IL-5 in thymocytes/T cells leads to the development of a massive eosinophilia, extramedullary eosinophilopoiesis, and unique histopathologies. *J Immunol* 1997;**158**(3):1332−44.

44. Kobayashi T, Iijima K, Kita H. Marked airway eosinophilia prevents development of airway hyper-responsiveness during an allergic response in IL-5 transgenic mice. *J Immunol* 2003;**170**(11): 5756−63.

45. Tominaga A, et al. Transgenic mice expressing a B cell growth and differentiation factor gene (interleukin 5) develop eosinophilia and autoantibody production. *J Exp Med* 1991;**173**(2):429−37.

46. Lee JJ, et al. Interleukin-5 expression in the lung epithelium of transgenic mice leads to pulmonary changes pathognomonic of asthma. *J Exp Med* 1997;**185**(12):2143−56.

47. Mishra A, et al. Enterocyte expression of the eotaxin and interleukin-5 transgenes induces compartmentalized dysregulation of eosinophil trafficking. *J Biol Chem* 2002;**277**(6):4406−12.

48. Yamada Y, et al. The FIP1L1-PDGFRA fusion gene cooperates with IL-5 to induce murine hypereosinophilic syndrome (HES)/chronic eosinophilic leukemia (CEL)-like disease 10.1182/blood-2005-08-3153. *Blood* 2006;**107**(10):4071−9.

49. Matthews AN, et al. Eotaxin is required for the baseline level of tissue eosinophils. *Proc Natl Acad Sci U S A* 1998;**95**(11):6273−8.

50. Rothenberg ME, et al. Eotaxin triggers eosinophil-selective chemotaxis and calcium flux via a distinct receptor and induces pulmonary eosinophilia in the presence of interleukin 5 in mice. *Mol Med* 1996;**2**(3):334−48.

51. Hisada T, et al. Cysteinyl-leukotrienes partly mediate eotaxin-induced bronchial hyperresponsiveness and eosinophilia in IL-5 transgenic mice. *Am J Respir Crit Care Med* 1999;**160**(2): 571−5.

52. Hamaguchi-Tsuru E, et al. Development and functional analysis of eosinophils from murine embryonic stem cells. *Br J Haematol* 2004;**124**(6):819−27.

53. Kurowska-Stolarska M, et al. IL-33 amplifies the polarization of alternatively activated macrophages that contribute to airway inflammation. *J Immunol* 2009;**183**(10):6469−77.

54. Schmitz J, et al. IL-33, an interleukin-1-like cytokine that signals via the IL-1 receptor-related protein ST2 and induces T helper type 2-associated cytokines. *Immunity* 2005;**23**(5):479−90.

55. Kurowska-Stolarska M, et al. IL-33 Induces Antigen-Specific IL-5+ T Cells and Promotes Allergic-Induced Airway Inflammation Independent of IL-4. *J Immunol* 2008;**181**(7):4780−90.

56. Stolarski B, et al. IL-33 exacerbates eosinophil-mediated airway inflammation. *J Immunol* 2010;**185**(6):3472−80.

57. Kim W, et al. A novel method for procuring a large quantity of mature murine eosinophils in vivo. *Journal of Immunological Methods* 2010;**363**(1):90−4.

58. Steenwinckel V, et al. IL-13 mediates in vivo IL-9 activities on lung epithelial cells but not on hematopoietic cells. *J Immunol* 2007;**178**(5):3244−51.

59. Louahed J, et al. Interleukin 9 promotes influx and local maturation of eosinophils. *Blood* 2001;**97**(4):1035−42.

60. Temann UA, Ray P, Flavell RA. Pulmonary overexpression of IL-9 induces Th2 cytokine expression, leading to immune pathology. *J Clin Invest* 2002;**109**(1):29−39.

61. Kim MR, et al. Transgenic overexpression of human IL-17E results in eosinophilia, B-lymphocyte hyperplasia, and altered antibody production. *Blood* 2002;**100**(7):2330−40.

62. Borchers MT, et al. In vitro assessment of chemokine receptor-ligand interactions mediating mouse eosinophil migration. *J Leukoc Biol* 2002;**71**(6):1033−41.

63. Dyer KD, et al., Mouse and Human Eosinophils Degranulate in Response to Platelet-Activating Factor (PAF) and LysoPAF via a PAF-Receptor-Independent Mechanism: Evidence for a Novel Receptor. 2010;**184**(11):6327−6334.

64. Legrand F, et al. Innate immune function of eosinophils: from antiparasite to antitumor cells. *Methods Mol Biol* 2008;**415**:215−40.

65. Dyer KD, et al. Functionally competent eosinophils differentiated ex vivo in high purity from normal mouse bone marrow. *J Immunol* 2008;**181**(6):4004−9.

66. Dyer KD, Percopo CM, Rosenberg HF. Generation of eosinophils from unselected bone marrow progenitors: wild-type, TLR- and eosinophil-deficient mice. *Open Immunol J* 2009;**2**:163−7.

67. Queto T, et al. Cysteinyl-leukotriene type 1 receptors transduce a critical signal for the up-regulation of eosinophilopoiesis by interleukin-13 and eotaxin in murine bone marrow. *J Leukoc Biol* 2010;**87**(5):885−93.

68. Yousefi S, et al. Catapult-like release of mitochondrial DNA by eosinophils contributes to antibacterial defense. *Nat Med* 2008;**14**(9):949−53.

69. Jacobsen EA, et al. Eosinophils regulate dendritic cells and Th2 pulmonary immune responses following allergen provocation. *Journal of Immunology*; 2011. In press.

70. Lacy P, et al. Agonist Activation of F-Actin-Mediated Eosinophil Shape Change and Mediator Release is Dependent on Rac2. *Int Arch Allergy Immunol* 2011;**156**(2):137−47.

71. Fulkerson PC, et al. CXCL9 inhibits eosinophil responses by a CCR3- and Rac2-dependent mechanism. *Blood* 2005;**106**(2):436−43.

72. Kudlacz E, et al. Functional effects of eotaxin are selectively upregulated on IL-5 transgenic mouse eosinophils. *Inflammation* 2002;**26**(3):111−9.

73. Kanda A, et al. Eosinophil-derived IFN-gamma induces airway hyperresponsiveness and lung inflammation in the absence of lymphocytes. *J Allergy Clin Immunol*; 2009.

74. de Andres B, et al. Release of O2- and LTC4 by murine eosinophils: role of intra- and extracellular calcium. *Immunology* 1990;**69**(2):271−6.

75. Tamura N, et al. Requirement of CD80 and CD86 molecules for antigen presentation by eosinophils. *Scand J Immunol* 1996;**44**(3):229−38.

76. Wang H-B, et al. Airway Eosinophils: Allergic Inflammation Recruited Professional Antigen-Presenting Cells. *J Immunol* 2007;**179**(11):7585−92.

77. Zimmermann N, et al. Chemokines in asthma: cooperative inter-action between chemokines and IL-13. *J Allergy Clin Immunol* 2003;**111**(2):227−42. quiz 243.

78. Padigel UM, et al. Eosinophils can function as antigen-presenting cells to induce primary and secondary immune responses to Strongyloides stercoralis. *Infect Immun* 2006;**74**(6):3232−8.

79. Shi HZ, et al. Lymph node trafficking and antigen presentation by endobronchial eosinophils. *J Clin Invest* 2000;**105**(7):945—53.

80. Rose Jr CE, et al. Murine lung eosinophil activation and chemokine production in allergic airway inflammation. *Cell Mol Immunol* 2010;**7**(5):361—74.

Chapter 4.5

Culture and Characterization of Mouse Bone Marrow-Derived Eosinophils

Kimberly D. Dyer

Among the difficulties encountered by eosinophil biologists is the fact that eosinophil-specific events represent only a fraction of the ongoing hematopoietic activity in bone marrow at any given time, even under profound T-helper type 2 (T$_h$2) stimulation such as *Schistosoma mansoni* infection.[1,2] Historically, cytokine-stimulated *ex vivo* culture conditions have not been very effective at generating large quantities of pure, phenotypically mature eosinophils from unselected mouse bone marrow progenitors. Solid (i.e., agar, methylcellulose) cultures have been used to identify mediators that stimulate progenitor colony production and differentiation of eosinophils. Using solid culture systems, interleukin-5 (IL-5) was identified as the major regulator of eosinophils.[3] Interestingly, IL-5 shares the common beta chain of its membrane receptor with the receptors for granulocyte-macrophage colony-stimulating factor (GM-CSF) and interleukin-3 (IL-3),[4,5] and both of these cytokines can act synergistically with IL-5 to promote eosinophil production.[6] Agar culture systems were also used to elucidate the role of stem cell factor (SCF) in the proliferation of stem cells and early hematopoietic progenitors[7] via proto-oncogene c-Kit (c-Kit/CD117). Solid culture systems continue to be good models for studying the roles of exogenous growth factors in eosinophil hematopoiesis, but do not produce eosinophils in sufficient numbers for further functional experiments. The use of cytokines whose role in hematopoiesis was elucidated in solid culture systems has allowed researchers to develop liquid *ex vivo* culture systems that allow for the production of eosinophils from mouse bone marrow.

Ishihara and colleagues[8] cultured rat bone marrow progenitors in the presence of recombinant rat IL-5 and obtained 90% eosinophils but with a reduced total cell number, demonstrating that in the rat, IL-5 was important for the proliferation and terminal differentiation of the eosinophil. Culturing unselected BALB/c bone marrow progenitors in a cytokine regimen that includes GM-CSF, IL-3, and IL-5 yielded increased cell numbers, but only 5—6% eosinophils.[9] Dyer and colleagues[10] used SCF acting through c-Kit (CD117)[7] and fms-related tyrosine kinase 3 ligand (Flt3L) acting on the receptor-type tyrosine-protein kinase FLT3 (FLT3)[11] in combination to support the proliferation of the progenitor cells, which resulted in an enhanced response to IL-5. This *ex vivo* culture system generates large numbers of eosinophils at high purity (>90%) from unselected mouse bone marrow progenitors (Fig. 4.5.1). This method has been used to generate eosinophils from both BALB/c and C57BL/6 strains of mice and a variety of gene-deleted mice.[12] In response to four days of culture with recombinant mouse (rm)Flt3L and rmSCF followed by IL-5 alone thereafter, the resulting bone marrow-derived eosinophils look like eosinophils (Fig. 4.5.2) and express immunoreactive eosinophil granule major basic protein 1 (MBP-1), sialic acid binding Ig-like lectin 5 (Siglec-5/Siglec-F), IL-5 receptor alpha, and transcripts encoding mouse eosinophil peroxidase, C-C motif chemokine 3 (Ccl3), the IL-3/IL-5/GM-CSF receptor common beta chain (βc), and the transcription factor GATA-binding factor 1 (Gata-1).[10] The bone marrow eosinophils are functionally competent in that they undergo chemotaxis toward mouse chemokine eotaxin [C-C motif chemokine 11 (CCL11)] and produce characteristic cytokines (Table 4.5.1), including Ccl3 [macrophage inflammatory protein 1-alpha (Mip-1-alpha)], interferon gamma (IFN-γ), interleukin-4 (IL-4), and interleukin-6 (IL-6).[10]

Bone marrow eosinophils generated *ex vivo* from this culture system have been used to demonstrate that eosinophils can be infected with the mouse respiratory pathogen, pneumonia virus of mice (PVM), and that infective PVM is released from these cells and bone

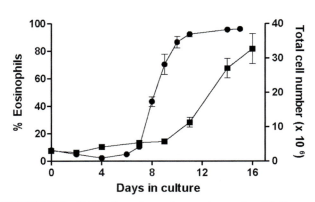

FIGURE 4.5.1 **Total cell number and percentage eosinophils from *ex vivo* culture of unselected mouse bone marrow.** Percentage of eosinophils (circles, left axis) and total cell number (squares, right axis) after 4 days in stem cell factor and fms-related tyrosine kinase 3 ligand (Flt3L) followed by interleukin-5 alone from day 4 onward. *Reprinted with permission from reference[10].*

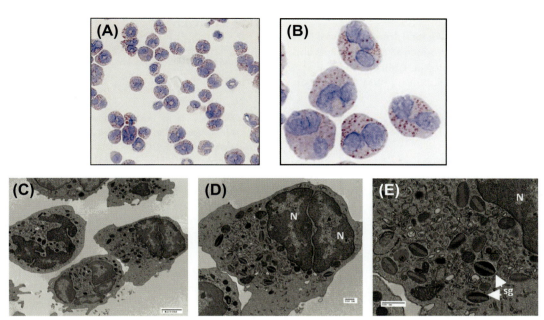

FIGURE 4.5.2 *Ex vivo* **differentiated bone marrow eosinophils are phenotypically normal.** *(A)* and *(B)* Light microscopic image of bone marrow eosinophils at day 9 of cytokine stimulation stained with Diff-Quik (a modified Giemsa stain). *(A)* Original image magnification 20×; *(B)* original magnification 63×. *(C−E)* Electron micrographs of bone marrow eosinophils demonstrate morphological features characteristic of eosinophils including the bilobed nucleus (N) and the specific granules [sg, arrowheads in *(E)*]. Magnification *(C)* 2500×, bar = 2 μm; *(D)* 12,000× and *(E)* 5000×, bar = 500 nm. *Reprinted with permission from references[10] and[12].*

marrow eosinophils derived from mice devoid of the toll-like receptor (TLR) signaling adaptor protein myeloid differentiation primary response gene (88) (MyD88[−/−] mice) were used to elucidate the role of MyD88 in bone marrow eosinophil response to PVM infection.[12] The question of whether or not mouse eosinophils degranulate has also been addressed using this model system, leading to the identification of platelet-activating factor (PAF) and lyso-PAF as secretagogues for eosinophils.[13] Human peripheral eosinophils, eosinophils isolated from the spleen of IL-5 transgenic mice,[14] and bone marrow eosinophils release eosinophil peroxidase (EPO) in response to PAF and lyso-PAF stimulation (Fig. 4.5.3) independent of the characterized G protein-coupled PAF receptor,[15−17] suggesting the presence of a non-classical receptor or, perhaps, receptor-independent degranulation in eosinophils in response to stimulation with these phospholipids. Additionally the authors demonstrated the release of cytokines from bone marrow eosinophils in response to IL-6 stimulation (Table 4.5.1). Interestingly, PAF and lyso-PAF do not stimulate the release of cytokines from bone marrow eosinophils suggesting that piecemeal degranulation (PMD) occurs in these cells similar to PMD characterized in human eosinophils.[18] Rankin and colleagues[19] differentiated bone marrow eosinophils from wild-type and devoid of the IL-33 receptor subunit, interleukin-1 receptor 4 (IL-1R4/protein ST2; ST2[−/−] mice) and showed that IL-33 stimulated the

release of IL-13 from bone marrow eosinophils in an ST2-dependent manner, suggesting a role for eosinophil-derived IL-13 in the development of IL-33-induced cutaneous fibrosis. These findings highlight the utility of this protocol as a method by which eosinophils can be obtained in sufficient numbers to perform biologically meaningful experiments.

The ability to generate eosinophils from the bone marrow of gene-deleted mice will allow researchers to investigate the role of diverse molecules in eosinophil development and activation. For instance, does the TLR family of pattern recognition molecules play a role in eosinophil biology during and following viral infections and if so, which ones are required? Now that it is known that mouse eosinophils exhibit PMD, the questions of which molecules allow for the selective release of granule mediators can be asked and perhaps answered using bone marrow from mice deficient in molecules associated with vesicular trafficking and degranulation. This model of eosinophil hematopoiesis may provide insight into the emergence of transcription factors during eosinophil development. Finally, bone marrow eosinophils can be generated in sufficient numbers to reconstitute the eosinophil compartment of eosinophil-deficient mice, such as transgenic PHIL[20] or ΔdblGATA[21] mice, allowing researchers to investigate the molecules responsible for the trafficking of eosinophils, and examine the role of various molecules *in vivo*.

TABLE 4.5.1 Eosinophils are a Rich Source of Cytokines Bone Marrow Eosinophils Produce Many Cytokines and Release these Mediators into the Culture Media at Baseline (unstimulated) or Following PVM Infection or IL-6 Stimulation

Cytokine	Stimulus for Secretion	Reference
IFN-α	PVM infection	12
IFN-γ	Unstimulated	10
	IL6 stimulation	13
IL-1 beta	IL-6 stimulation	13
IL-4	Unstimulated	10
IL-6	Unstimulated	10
	PVM infection	12
IL-9	Unstimulated	10
	IL6 stimulation	13
IL-12 (p70)	IL-6 stimulation	13
IP-10/CXCL10	Unstimulated	10
	PVM infection	12
MCP-1/CCL2	Unstimulated	12
	IL6 stimulation	13
MIP-1-alpha/CCL3	Unstimulated	10
	PVM infection	12
TNF-α	IL-6 stimulation	13

CCL2, C-C motif chemokine 2; CCL3, C-C motif chemokine 3; CXCL10, C-X-C motif chemokine 10; IFN-α, interferon alpha; IFN-γ, interferon gamma; IL-1β, interleukin-1 beta; IL-4, interleukin-4; IL-6, interleukin-6; IL-9, interleukin-9; IL-12, interleukin-12; IP-10, 10 kDa interferon gamma-induced protein; MCP-1, monocyte chemotactic protein 1; MIP-1-alpha, macrophage inflammatory protein 1-alpha; PVM, pneumonia virus of mice; TNF-α, tumor necrosis alpha.

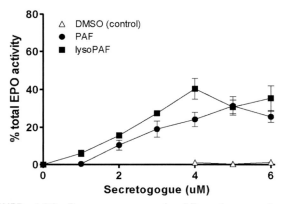

FIGURE 4.5.3 Bone marrow eosinophils release eosinophil peroxidase in response to stimulation with platelet-activating factor and lyso-platelet-activating factor. DMSO, dimethyl sulfoxide; PAF, platelet-activating factor. *Reprinted with permission from reference[13].*

REFERENCES

1. Bystrom J, Dyer KD, Ting-De Ravin SS, Naumann N, Stephany DA, Foster PS, et al. Interleukin-5 does not influence differential transcription of transmembrane and soluble isoforms of IL-5R alpha in vivo. *Eur J Haematol* 2006;**77**:181—90.

2. Swartz JM, Dyer KD, Cheever AW, Ramalingam T, Pesnicak L, Domachowske JB, et al. Schistosoma mansoni infection in eosinophil lineage-ablated mice. *Blood* 2006;**108**:2420—7.

3. Lopez AF, Begley CG, Williamson DJ, Warren DJ, Vadas MA, Sanderson CJ. Murine eosinophil differentiation factor. An eosinophil-specific colony-stimulating factor with activity for human cells. *J Exp Med* 1986;**163**:1085—99.

4. Kitamura T, Sato N, Arai K, Miyajima A. Expression cloning of the human IL-3 receptor cDNA reveals a shared beta subunit for the human IL-3 and GM-CSF receptors. *Cell* 1991;**66**:1165—74.

5. Tavernier J, Devos R, Cornelis S, Tuypens T, Van der Heyden J, Fiers W, et al. A human high affinity interleukin-5 receptor (IL5R) is composed of an IL5-specific alpha chain and a beta chain shared with the receptor for GM-CSF. *Cell* 1991;**66**:1175—84.

6. Takamoto M, Sugane K. Synergism of IL-3, IL-5, and GM-CSF on eosinophil differentiation and its application for an assay of murine IL-5 as an eosinophil differentiation factor. *Immunol Lett* 1995;**45**:43—6.

7. Metcalf D, Mifsud S, Di Rago L. Stem cell factor can stimulate the formation of eosinophils by two types of murine eosinophil progenitor cells. *Stem Cells* 2002;**20**:460—9.

8. Ishihara K, Satoh I, Mue S, Ohuchi K. Generation of rat eosinophils by recombinant rat interleukin-5 in vitro and in vivo. *Biochim Biophys Acta* 2000;**1501**:25—32.

9. Dyer KD, Czapiga M, Foster B, Foster PS, Kang EM, Lappas CM, et al. Eosinophils from lineage-ablated Delta dblGATA bone marrow progenitors: the dblGATA enhancer in the promoter of GATA-1 is not essential for differentiation ex vivo. *J Immunol* 2007;**179**:1693—9.

10. Dyer KD, Moser JM, Czapiga M, Siegel SJ, Percopo CM, Rosenberg HF. Functionally competent eosinophils differentiated ex vivo in high purity from normal mouse bone marrow. *J Immunol* 2008;**181**:4004—9.

11. Matthews W, Jordan CT, Wiegand GW, Pardoll D, Lemischka IR. A receptor tyrosine kinase specific to hematopoietic stem and progenitor cell-enriched populations. *Cell* 1991;**65**:1143—52.

12. Dyer KD, Percopo CM, Fischer ER, Gabryszewski SJ, Rosenberg HF. Pneumoviruses infect eosinophils and elicit MyD88-dependent release of chemoattractant cytokines and interleukin-6. *Blood* 2009;**114**:2649—56.

13. Dyer KD, Percopo CM, Xie Z, Yang Z, Kim JD, Davoine F, et al. Mouse and human eosinophils degranulate in response to platelet-activating factor (PAF) and lyso-PAF via a PAF-receptor-independent mechanism: evidence for a novel receptor. *J Immunol* 2010;**184**:6327—34.

14. Dent LA, Strath M, Mellor AL, Sanderson CJ. Eosinophilia in transgenic mice expressing interleukin 5. *J Exp Med* 1990;**172**:1425—31.

15. Honda Z, Nakamura M, Miki I, Minami M, Watanabe T, Seyama Y, et al. Cloning by functional expression of platelet-activating factor receptor from guinea-pig lung. *Nature* 1991;**349**:342—6.

16. Ishii S, Kuwaki T, Nagase T, Maki K, Tashiro F, Sunaga S, et al. Impaired anaphylactic responses with intact sensitivity to endotoxin in mice lacking a platelet-activating factor receptor. *J Exp Med* 1998;**187**:1779−88.

17. Ishii S, Shimizu T. Platelet-activating factor (PAF) receptor and genetically engineered PAF receptor mutant mice. *Prog Lipid Res* 2000;**39**:41−82.

18. Melo RC, Spencer LA, Perez SA, Ghiran I, Dvorak AM, Weller PF. Human eosinophils secrete preformed, granule-stored interleukin-4 through distinct vesicular compartments. *Traffic* 2005;**6**:1047−57.

19. Rankin AL, Mumm JB, Murphy E, Turner S, Yu N, McClanahan TK, et al., IL-33 induces IL-13-dependent cutaneous fibrosis. *J Immunol* **184**:1526−1535.

20. Lee JJ, Dimina D, Macias MP, Ochkur SI, McGarry MP, O'Neill KR, et al. Defining a link with asthma in mice congenitally deficient in eosinophils. *Science* 2004;**305**:1773−6.

21. Yu C, Cantor AB, Yang H, Browne C, Wells RA, Fujiwara Y, et al. Targeted deletion of a high-affinity GATA-binding site in the GATA-1 promoter leads to selective loss of the eosinophil lineage in vivo. *J Exp Med* 2002;**195**:1387−139521.

Chapter 4.6

Modeling Eosinophil Development *Ex Vivo* from Human CD34+ Cells

Lori A. Wagner

INTRODUCTION

Eosinophils are myeloid cells derived from hematopoietic stem cells in the bone marrow. Hematopoietic stem cells are identified by the expression of CD34, a sialomucin ligand. CD34 is a marker for pluripotent stem cells that can stably reconstitute both myelopoiesis and lymphopoiesis in irradiated bone marrow. Several *ex vivo* models have shed light on the cytokine requirements and pathways leading to eosinophilopoiesis. CD34+ cells are present at 0.5−2% in bone marrow mononuclear cells and in umbilical cord mononuclear blood, and have been used as model systems for eosinophil development.[1] Peripheral blood, in contrast, is readily available but contains only small numbers (about 0.1%) of CD34+ hematopoietic progenitors.[2] The addition of specific cytokines is required to grow large numbers of nearly pure eosinophils or eosinophil colonies in liquid or semisolid culture, respectively. All three of these *ex vivo* model systems for eosinophil development, bone marrow, umbilical cord blood, and peripheral blood, have been used to elucidate the transcription factors and patterns of gene expression in eosinophil development.

EX VIVO MODELS

Ex Vivo Eosinophilopoiesis: Bone Marrow or Umbilical Cord Blood

In 1988, human interleukin-5 (IL-5), formerly called eosinophil differentiation factor, was cloned and shown to induce specific differentiation of eosinophils from umbilical cord blood mononuclear cells (UCMCs) or bone marrow mononuclear cells (BMMCs).[3] UCMCs in liquid culture for 21 days in the presence of IL-5 developed into 92% eosinophils. Bone marrow cells also differentiated into 83% eosinophils after 28 days of liquid culture in IL-5. The addition of interleukin-4 (IL-4) to BMMC cultures resulted in minimal levels of eosinophils and mainly yielded lymphocyte (85%), macrophage, and neutrophil development. Interleukin-3 (IL-3) was also less eosinophil-specific; addition of this cytokine resulted in 64% eosinophils after 28 days. Eosinophils were morphologically normal, although electron-dense secondary granule cores were not formed. Secondary granule cores, which contain crystalline eosinophil granule major basic protein 1 (MBP-1), apparently do not form in any *ex vivo* human model system investigated.[3−5]

Culture of BMMCs in semisolid media produced similar results as compared to liquid culture; the addition of IL-5 resulted in mostly eosinophil colonies (up to 98%).[5] In contrast, IL-3 resulted in a greater overall number of eosinophil colonies; however, it was less specific for eosinophil formation, resulting only in up to 61% eosinophil colonies. Granulocyte-macrophage colony-stimulating factor (GM-CSF) also induced eosinophil colony formation but it was even less specific, with about 35% eosinophil colonies formed. Early experiments showed that nonadherent BMMCs stimulated with IL-5 in semisolid media resulted in pure eosinophil colonies; however, the addition of GM-CSF or IL-3 produced five times as many eosinophil colonies in addition to other cell types.[6]

The majority of cells seen in cultures grown only in GM-CSF or IL-3 are myelocytes and promyelocytes.[6] Bone marrow, nonadherent, low-density, T-cell-depleted buffy coats produced similar results in semisolid media. Greater than 94% pure eosinophil colonies were produced with murine IL-5 stimulation alone in both serum and serum-free conditions. Without the addition of cytokines, no colonies were formed. IL-3 addition alone produced slightly fewer eosinophil colonies, with many more colonies and clusters of neutrophils and macrophages. IL-3 and IL-5 stimulation resulted in more overall eosinophil colonies but less specificity (79% eosinophils), while interleukin-1 (IL-1) and interleukin-6 (IL-6) had no effect on eosinophil colony formation.[7] The effects of stem cell factor (SCF) on CD34+ lineage nonadherent BMMCs were also examined.[8] SCF increases colony formation of all myeloid colonies. SCF in combination with GM-CSF or SCF with GM-CSF and IL-3 produced the greatest number of eosinophil colonies; however, macrophage and neutrophil colonies were also increased. The overall conclusion from this work is that IL-3 supports the growth of multipotent progenitors, and IL-5 expands committed eosinophil progenitors.[6]

Peripheral Blood Models

Eosinophils can also be derived from hematopoietic stem cells (CD34[+]) in peripheral blood by the addition of GM-CSF, IL-3, and IL-5. In this case, the addition of all three cytokines resulted in more specific eosinophil development (>80%).[2] Peripheral blood CD34[+] cells will not support colony growth in semisolid media, such as methylcellulose, if plated with IL-5 alone. The addition of IL-3 resulted in 65% pure eosinophil colonies; the results in response to GM-CSF alone were similar, with 63% pure eosinophil colonies. A combination of GM-CSF and IL-3 increased eosinophil colonies to 78% and the addition of IL-5 only slightly augmented eosinophil colony purity. It was shown that pre-culturing CD34[+] cells with GM-CSF and IL-3 for 7 days generated an IL-5-responsive population in liquid culture, with 83% eosinophils after 28 days (2 weeks following a shift to IL-5 only).[9] Pre-culture of eosinophils with GM-CSF and IL-3, followed by a shift to IL-5 in methylcellulose colonies, also induces a significant number of eosinophil colonies.

In conclusion, multiple experiments with three models of eosinophil development all show nonspecific proliferation of myeloid cell types in the presence of GM-CSF or IL-3. IL-5 induces the lineage-specific development of eosinophils and/or eosinophil colonies. This supports a model in which GM-CSF and IL-3 expand multipotent common myeloid progenitors (CMPs), whereas IL-5 induces the proliferation and terminal differentiation of IL-5R[+] committed eosinophil progenitors.

Eosinophil Progenitors as Defined by Ex Vivo Models

The identification of the committed human eosinophil progenitor supports the earlier model described above. The human common myeloid progenitor (hCMP) has been previously described; lineage CD34[+] CD38[+] human progenitors were fractionated into hCMPs, human granulocyte/macrophage progenitors (hGMPs), and human megakaryocyte/erythrocyte progenitors (hMEPs).[10] The hCMP fraction was sorted as IL-3R[+] CD45RA[−].[10] Human CMPs grown in semisolid media in a cytokine cocktail generate eosinophils and basophils as well as a variety of other myelomonocytic and megakaryocyte/erythroid cells. The IL-5R[+] CMP fraction gives rise to pure eosinophil colonies, but never differentiates into basophils or neutrophils and is thus considered to be the human

eosinophil lineage-committed progenitor (hEoP).[10] The IL-5R[−] fraction of the hCMP gives rise to a variety of colony types. IL-5R[+] cells are thought to develop from IL-5R[−] CMPs as pretreatment of cells with multicolony stimulating factors such as IL-3 cause an increase in eosinophil numbers. The authors showed that the hEoP expresses GATA-binding protein 2 (GATA-2) at higher levels than GATA-binding protein 1 (GATA-1), and, as expected, does not express friend of GATA protein 1 (FOG-1). In addition, the hEoP has high levels of the transcription factor CCAAT/enhancer-binding protein alpha (C/EBPα) and expresses CCAAT/enhancer-binding protein epsilon (C/EBPε). IL-3R and IL-5R are expressed, as well as low levels of the GM-CSF receptor. Granule proteins expressed in the progenitor include eosinophil peroxidase (EPO), Charcot—Leyden crystal protein/galectin-10 (CLC/Gal-10), MBP-1, and low levels of myeloperoxidase. Commitment is thought to proceed through an uncommitted hematopoietic stem cell, to a multipotent hCMP which is responsive to IL-3. The hCMP then splits into an IL-5-responsive hEoP as well as a GMP (neutrophil, monocyte, and basophil) and MEP (megakaryocyte/erythrocyte) (Fig. 4.6.1).[10]

Reprogramming of Eosinophil Development

Hematopoietic progenitors can be reprogrammed to become eosinophils by bypassing the natural pathways of development via cytokines and directly activating transcription factors or signal transduction pathways. For example, intermediate to high levels of the transcription factor, GATA-1 are required for eosinophil development. Enforced expression of GATA-1 in CD34[+] UCMCs

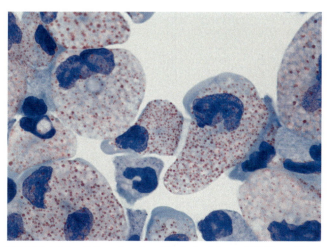

FIGURE 4.6.1   Cultured eosinophils derived from CD34[+] umbilical cord blood mononuclear cells as described in[13]. *Photograph kindly donated by Zhijun Qiu and Helene Rosenberg.*

cultured without IL-5 in myeloid conditions developed exclusively into eosinophils. GATA-2 transfected cells yielded similar results.[11] CD34+ cells transduced with GATA-1 were IL-5 responsive, whereas cells transduced with a dominant/negative GATA-1 did not respond to IL-5.

Another prominent transcription factor that is required for terminal eosinophil development is C/EBPε. Gene-deleted mice lacking C/EBPε have no mature granulocytes.[12] C/EBPε has four isoforms produced by alternative translational start sites, alternative mRNA splicing, and different promoters: the 32 kDa, 30 kDa, 27 kDa, and 14 kDa isoforms. The 32/30 kDa forms of C/EBPε, transduced into CD34+ UCMCs prestimulated for 18 h in SCF, fms-related tyrosine kinase 3 ligand (Flt3L) and thrombopoietin have been reported to induce the eosinophil lineage in the absence of IL-5 (Fig. 4.6.1).[13] Eosinophil-derived neurotoxin (EDN/RNase 2), EPO, IL-5R (both soluble and transmembrane forms), MBP-1, as well as the transcription factor GATA-1, are expressed in 32/30 kDa transduced cells, as is expected in developing eosinophils. In contrast the 27 kDa isoform of C/EBPε strongly inhibits eosinophil differentiation and gene expression, including GATA-1 expression, and instead promotes neutrophil and macrophage development. The 14 kDa isoform, which lacks a transactivation domain, also inhibits eosinophil differentiation, but promotes erythroid development.

IL-5 activates signal transducer and activator of transcription 5A (STAT5A), a signal transducer required for eosinophil development.[14] STAT5A is expressed highly at 7−10 days of eosinophil differentiation from CD34+ UCMCs stimulated for 3 days with SCF, Flt3L, GM-CSF, IL-3, and IL-5, then subsequently with IL-3 and IL-5. CD34+ UCMCs infected with retrovirus expressing wild-type STAT5A resulted in 27-fold more eosinophils after 14 days. Dominant/negative STAT5A, a mutant form of STAT5A that blocks function, decreased the number of eosinophils, as expected.[15]

DNA-binding protein inhibitor ID-1 (ID1) and 2 (ID2) are basic helix-loop-helix (bHLH) inhibitory transcriptional factors that play a role in lymphopoiesis. ID proteins block the binding of bHLH transcription factors to enhancer box (E-box) DNA-binding sites during development. Investigation of ID expression during myelopoiesis showed that ID1 is initially upregulated during granulopoiesis and ID2 is expressed later in maturation. ID1 and ID2 were retrovirally transduced into the previously mentioned model system of eosinophil development. Constitutive expression of ID1 inhibited eosinophil differentiation but enhanced neutrophil development. Inhibition of ID2 expression blocked both lineages from developing.[16] Therefore, reprogramming can also occur fairly late in the developmental pathway.

Gene Expression During Eosinophil Development as Defined by Ex Vivo Models

Gene expression during eosinophil development has been characterized by the use of ex vivo models. CD34+ hematopoietic stem cells derived from umbilical cord or peripheral blood contain a variety of eosinophil transcripts and proteins that increase during eosinophil development.

CD34+ cells derived from UCMCs, cultured with IL-3 and IL-5, yielded 84% developing eosinophils by day 14. CD34 gene expression was lost by day 3 with the acquisition of C-C chemokine receptor type 3 (CCR3), granulocyte-macrophage colony-stimulating factor receptor subunit alpha (GM-CSF-R-alpha), IL-3R, and IL-5R by 21 days of culture.[17] CD34+ UCMCs were cultured in serum-free conditions with GM-CSF, IL-3, IL-5, and SCF for 5 days. Cells were cultured in IL-3 and IL-5 until day 10 and then in IL-5 until day 22. Ninety percent eosinophils were observed after 3 weeks by light microscopy. Reverse transcription polymerase chain reaction (RT-PCR) showed EDN and MBP-1 transcripts peaking at 15 days, IL-5R mRNA at 18 days, IL-3R transcripts peaking at 18−21 days, and GM-CSF-R-alpha peaking at 18 days. CCR3, the eotaxin receptor, increased throughout the experiment.[18] UCMC stimulated with IL-5 or conditioned murine EL-4 media was shown to result in about 80% eosinophils after 60 days.[19] Eosinophils were shown to produce the granule proteins eosinophil cationic protein (ECP/RNase 3), EDN, EPO, and MBP-1, and the expression of these proteins was blocked by anti-IL-5. Therefore, IL-5 was shown to induce expression of eosinophil-specific granule proteins as well as cytokine and chemokine receptors ex vivo.

Granule protein expression was shown to be in the granule matrix of UCMC CD34+ grown in methylcellulose in the presence of IL-3 and IL-5. Immunoreactive ECP, IL-6, MBP-1 and RANTES were not present in freshly purified cells, but between days 16 and 28 of culture, all of these proteins were detected by confocal laser microscopy. Staining appeared in the periphery of granules initially, and by day 28 they were present throughout eosinophil granule structure. Therefore, developing eosinophils produce and store granule proteins and cytokines simultaneously.[20]

Similar results were obtained with peripheral CD34+ cells stimulated with GM-CSF, Flt3L, IL-3, IL-5, and SCF for 3 days, then with GM-CSF, IL-3, and IL-5 for 7 days. After day 7, GM-CSF was omitted from the media and after day 21, IL-3 was omitted. By day 7, granularity was observed and by day 14, a bilobed nucleus, characteristic of eosinophils, was present. By four weeks, 85% pure eosinophils were present (>200-fold amplification). Positive immunostaining for EDN and MBP-1 was detected in

15—21% of cells as early as day 3 of culture. By day 28, 80% of eosinophils expressed the previously cited granule proteins.[21]

Gene expression of EPO mRNA in peripheral blood CD34$^+$ cells stimulated with GM-CSF, IL-3, and IL-5 for 28 days followed a similar pattern. At initial isolation, no eosinophil granule protein transcripts were present, but by day 3, small amounts of transcript encoding CLC, ECP, EDN, EPO, and MBP-1 could be observed. These transcripts increased until day 7 and then remained stable.[22]

In concurrence with the above experiments, GM-CSF-, IL-3-, IL-5-, IL-6-, and SCF-stimulated peripheral blood CD34$^+$ cells expressed transcripts encoding CLC, ECP, EDN, and EPO at 2—3 days of culture. MBP-1 increased after 3—4 days. Protein expression of CLC, ECP, EPO, and MBP-1 was highest at day 21 of culture.[23]

Finally, an early study evaluated the expression of mRNAs encoding the eosinophil granule proteins in blood eosinophils purified from patients with eosinophilia and in eosinophils differentiated from mononuclear cells by culture with GM-CSF, IL-3, and IL-5. By Northern blot, MBP-1 mRNA was absent in mature blood eosinophils, while EPO mRNA was detectable in only approximately 25% of patients with eosinophilia. In contrast, ECP and EDN mRNAs were present in the blood eosinophils of all patients studied. In developing cord blood-derived eosinophils, EPO and MBP-1 mRNAs were highly abundant in the immature cells, but the amounts decreased significantly as the eosinophils matured; ECP and EDN mRNAs followed the same patterns but were less abundant overall. The loss of mRNA expression was not due to changes in mRNA stability; rather the transcription of the eosinophil granule genes was apparently silenced as the eosinophil progenitors terminally differentiated, consistent with their lack of expression in the mature blood eosinophil.[24]

CONCLUSION

Three models, CD34$^+$ cells derived from peripheral blood, or CD34$^+$ or mononuclear cells derived from bone marrow or umbilical cord blood have been shown to be useful for studying human eosinophil development *ex vivo*. Bone marrow as a tissue source mimics the natural process most closely and has become more accessible recently due to commercially available tissue (AllCells, LLC, Emeryville, CA, USA; ReachBio LLC, Seattle, WA, USA; STEMCELL Technologies Inc., Vancouver, BC, Canada). UCMCs and peripheral blood, however, are more easily acquired by many researchers and have given similar results to bone marrow cultures in most studies. Semisolid culture media is useful for identifying how single progenitor cells develop into eosinophils. In contrast, liquid culture produces much larger numbers of developing eosinophils and may mimic the interaction of many cell types and cytokines in the bone marrow microenvironment. Researchers mainly choose a model system in which one combines IL-3 to expand the multipotent progenitor cell population with IL-5 to stimulate eosinophil progenitor cell proliferation and terminal differentiation following IL-3 expansion.

REFERENCES

1. Nimgaonkar MT, Roscoe RA, Persichetti J, Rybka WB, Winkelstein A, Ball ED. A unique population of CD34+ cells in cord blood. *Stem Cells* 1995;**13**:158—66.
2. Shalit M, Sekhsaria S, Malech HL. Modulation of growth and differentiation of eosinophils from human peripheral blood CD34+ cells by IL5 and other growth factors. *Cell Immunol* 1995;**160**:50—7.
3. Saito H, Hatake K, Dvorak AM, Leiferman KM, Donnenberg AD, Arai N, et al. Selective differentiation and proliferation of hematopoietic cells induced by recombinant human interleukins. *Proc Natl Acad Sci U S A* 1988;**85**:2288—92.
4. Clutterbuck EJ, Sanderson CJ. Human eosinophil hematopoiesis studied in vitro by means of murine eosinophil differentiation factor (IL5): production of functionally active eosinophils from normal human bone marrow. *Blood* 1988;**71**:646—51.
5. Clutterbuck EJ, Hirst EM, Sanderson CJ. Human interleukin-5 (IL-5) regulates the production of eosinophils in human bone marrow cultures: comparison and interaction with IL-1, IL-3, IL-6, and GMCSF. *Blood* 1989;**73**:1504—12.
6. Sonoda Y, Arai N, Ogawa M. Humoral regulation of eosinophilopoiesis in vitro: analysis of the targets of interleukin-3, granulocyte/macrophage colony-stimulating factor (GM-CSF), and interleukin-5. *Leukemia* 1989;**3**:14—8.
7. Lu L, Lin ZH, Shen RN, Warren DJ, Leemhuis T, Broxmeyer HE. Influence of interleukins 3,5, and 6 on the growth of eosinophil progenitors in highly enriched human bone marrow in the absence of serum. *Exp Hematol* 1990;**18**:1180—6.
8. Ichihara M, Hotta T, Asano H, Inoue C, Murate T, Kobayashi M, et al. Effects of stem cell factor (SCF) on human marrow neutrophil, neutrophil/macrophage mixed, macrophage and eosinophil progenitor cell growth. *Int J Hematol* 1994;**59**:81—9.
9. Shalit M. Growth and differentiation of eosinophils from human peripheral blood CD 34+ cells. *Allerg Immunol (Paris)* 1997;**29**:7—10.
10. Mori Y, Iwasaki H, Kohno K, Yoshimoto G, Kikushige Y, Okeda A, et al. Identification of the human eosinophil lineage-committed progenitor: revision of phenotypic definition of the human common myeloid progenitor. *J Exp Med* 2009;**206**:183—93.
11. Hirasawa R, Shimizu R, Takahashi S, Osawa M, Takayanagi S, Kato Y, et al. Essential and instructive roles of GATA factors in eosinophil development. *J Exp Med* 2002;**195**:1379—86.
12. Lekstrom-Himes JA. The role of C/EBP(epsilon) in the terminal stages of granulocyte differentiation. *Stem Cells* 2001;**19**:125—33.
13. Bedi R, Du J, Sharma AK, Gomes I, Ackerman SJ. Human C/EBP-epsilon activator and repressor isoforms differentially reprogram myeloid lineage commitment and differentiation. *Blood* 2009;**113**:317—27.
14. Caldenhoven E, van Dijk TB, Tijmensen A, Raaijmakers JA, Lammers JW, Koenderman L, et al. Differential activation of functionally distinct STAT5 proteins by IL-5 and GM-CSF during eosinophil and neutrophil differentiation from human CD34+ hematopoietic stem cells. *Stem Cells* 1998;**16**:397—403.

15. Buitenhuis M, Baltus B, Lammers JW, Coffer PJ, Koenderman L. Signal transducer and activator of transcription 5a (STAT5a) is required for eosinophil differentiation of human cord blood-derived CD34+ cells. *Blood* 2003;**101**:134−42.

16. Buitenhuis M, van Deutekom HW, Verhagen LP, Castor A, Jacobsen SE, Lammers JW, et al. Differential regulation of granulopoiesis by the basic helix-loop-helix transcriptional inhibitors Id1 and Id2. *Blood* 2005;**105**:4272−81.

17. Robinson DS, North J, Zeibecoglou K, Ying S, Meng Q, Rankin S, et al. Eosinophil development and bone marrow and tissue eosinophils in atopic asthma. *Int Arch Allergy Immunol* 1999;**118**:98−100.

18. Hashida R, Ogawa K, Miyagawa M, Sugita Y, Matsumoto K, Akasawa A, et al. Gene expression accompanied by differentiation of cord blood-derived CD34+ cells to eosinophils. *Int Arch Allergy Immunol* 2001;**125**(Suppl. 1):2−6.

19. Ten RM, Butterfield JH, Kita H, Weiler DA, Fischkoff S, Ishizaka T, et al. Eosinophil differentiation of human umbilical cord mononuclear cells and prolonged survival of mature eosinophils by murine EL-4 thymoma cell conditioned medium. *Cytokine* 1991;**3**:350−9.

20. Mahmudi-Azer S, Velazquez JR, Lacy P, Denburg JA, Moqbel R. Immunofluorescence analysis of cytokine and granule protein expression during eosinophil maturation from cord blood-derived CD34 progenitors. *J Allergy Clin Immunol* 2000;**105**:1178−84.

21. Al-Rabia MW, Blaylock MG, Sexton DW, Thomson L, Walsh GM. Granule protein changes and membrane receptor phenotype in maturing human eosinophils cultured from CD34+ progenitors. *Clin Exp Allergy* 2003;**33**:640−8.

22. Shalit M, Sekhsaria S, Mauhorter S, Mahanti S, Malech HL. Early commitment to the eosinophil lineage by cultured human peripheral blood CD34+ cells: messenger RNA analysis. *J Allergy Clin Immunol* 1996;**98**:344−54.

23. Rosenberg HF, Dyer KD, Li F. Characterization of eosinophils generated in vitro from CD34+ peripheral blood progenitor cells. *Exp Hematol* 1996;**24**:888−93.

24. Gruart V, Truong MJ, Plumas J, Zandecki M, Kusnierz JP, Prin L, et al. Decreased expression of eosinophil peroxidase and major basic protein messenger RNAs during eosinophil maturation. *Blood* 1992;**79**:2592−7.

Eosinophilopoiesis

Introduction

Steven Maltby and Kelly M. McNagny

Since their discovery by Paul Ehrlich over a century ago, eosinophils have been recognized as a highly distinctive, but low-frequency, subset of hematopoietic cells in the blood of essentially all vertebrate species. Despite their early discovery, the true function of eosinophils in tissue homeostasis and pathology has remained a point of contention. This is likely due to their paucity in normal tissues and, until recently, the lack of well-defined biochemical and genetic tools to evaluate their functional significance. Similarly, a detailed understanding of the factors governing their genesis is only now coming to light. Eosinophilopoiesis encompasses all the steps involved in the initial commitment, differentiation, and maturation along the eosinophil lineage in the bone marrow, as well as proliferation, survival, and migration of cells into peripheral tissues. In this section of the book we review the latest developments and current *state-of-the-art* knowledge of factors governing eosinophil formation and development under steady state and pathological conditions.

Ultimately, as with all blood cells, eosinophils are derived from common hematopoietic stem cells and multilineage progenitors in the bone marrow. Research over the last 20 years has greatly improved our understanding of the role of specific transcription factors in eosinophil lineage commitment and function, as described in detail in Chapter 5.2 by Dr. Steven Ackerman and colleagues. Early work from the 1980s and 1990s quickly established a hierarchy of transcription factors governing myeloid and erythroid lineage specification.[1] Macrophage and neutrophil differentiation was shown to be regulated primarily by the CCAAT/enhancer-binding protein (C/EBP) family of transcription factors and the E-Twenty-Six (ETS) family transcription factor PU.1, while erythroid and megakaryocyte differentiation was regulated by GATA factors and the GATA-interacting protein, friend of GATA protein 1 (FOG-1).[1,2] The fact that these factors mutually antagonize each other's functions provided a mechanism for reinforcing fate decisions. Eosinophils, on the other hand, appeared to

break the rules, since these cells were shown to coexpress both PU.1 and GATA, as well as the C/EBP factors. Through elegant ectopic gene expression studies, it was subsequently shown that the eosinophil's fate represents a delicate combinatorial balance between the threshold level expression of these factors, and that a subtle disruption of this balance can tip cells toward either an erythroid or myeloid fate.[2,3] Of note, these same transcription factors (C/EBP, ETS, and GATA family members) also regulate the expression of eosinophil-specific genes.[2] In particular, *double GATA* sites are features of several eosinophil gene promoters, including *the Gata1 gene* itself (in the mouse) and *genes encoding the interleukin 5 receptor, alpha chain (IL5RA), proteoglycan 2, bone marrow/eosinophil granule major basic protein (PRG2/MBP)* and *chemokine (C-C motif) receptor 3 (CCR3* in humans; reviewed in Rothenberg and Hogan).[4] The levels of these factors change with eosinophil maturation, and recent studies are revealing, in much more intricate detail, how these complicated transcription factor interactions regulate eosinophil lineage commitment, differentiation, and maturation.

While *in vitro* studies have provided an insight into the transcription factors that are sufficient to drive eosinophil commitment and differentiation, *in vivo* studies have shaped our understanding of how progenitor populations give rise to eosinophils. In Chapter 5.3, Dr. Koichi Akashi and colleagues expand on the current understanding and characterization of eosinophil lineage-committed progenitors (EoPs). Eosinophil differentiation occurs in the bone marrow where hematopoietic stem cells give rise to multipotent progenitor cells and, ultimately, committed EoPs.

Expression of the two surface markers, CD34 and interleukin-5 receptor subunit alpha (IL-5RA), in addition to characteristic stem cell markers, are used to define EoPs and to distinguish lineage-committed EoPs from other progenitors. An increased frequency of CD34$^+$ IL-5RA$^+$ progenitors correlates with an increase in the production of mature eosinophil numbers and the expansion of eosinophil precursors is a consistent observation in asthmatic patients. Thus, expanded eosinophilopoiesis may be a prerequisite for full asthmatic disease and may offer a focal point for therapeutic intervention. Interestingly, data from our own laboratory has revealed that in mice, CD34, beyond its role as a marker for eosinophil precursors, is also expressed on mature eosinophils under inflammatory conditions and

Eosinophils in Health and Disease. http://dx.doi.org/10.1016/B978-0-12-394385-9.00005-5

serves an important functional role in promoting eosinophil migration into inflamed peripheral tissues in both asthma and ulcerative colitis.[5,6]

With regard to their bone marrow origins, eosinophils are usually placed on their own branch of the mammalian hematopoietic developmental tree, diverging early from the other myeloid lineages; there is continued controversy over whether EoPs arise from an eosinophil/basophil bipotent progenitor (reviewed in Gauvreau),[7] or arise directly from the common myeloid progenitor (CMP) population,[8,9] and there also appear to be differences between their origins in humans and the mouse.[9] In human hematopoiesis, it has been shown conclusively that human EoPs can develop from CMPs, although it remains to be determined whether they can also develop directly from earlier multipotent progenitors or hematopoietic stem cells.[9] In light of the fact that relatively minor manipulations in transcription factor expression can promote the *transdifferentiation* of eosinophils from a number of progenitor subsets, including both granulocyte/macrophage progenitors and megakaryocyte/ erythroid precursors,[2] it is worth considering whether certain *in vivo* inflammatory stimuli may lead to altered eosinophilopoiesis. Several recent high-profile reports have revealed the expansion of previously undescribed multilineage precursors in response to T-helper type 2 (T_h2) cell inflammatory disease, particularly for mast cells and basophils.[10,11] Thus, there is a precedent to suggest that inflammation/disease leads to shifts in the multipotent progenitor pathways used to respond to disease. Regardless, improved characterization of *lineage-committed* EoPs allows for the more accurate characterization and study of EoP development *in vivo* and will facilitate the development of therapeutics targeting early eosinophil lineage commitment and differentiation.

Once committed, further maturation and proliferation of eosinophils is guided largely by extrinsic cues from within the eosinophil's microenvironment. Cytokine receptors including those for granulocyte-macrophage colony-stimulating factor (GM-CSF), interleukin-3 (IL-3), and interleukin-5 (IL-5) (all of which share the 160 kDa common beta chain), are widely known to play an important role in eosinophil maturation and effector functions.[12] Of these cytokines, IL-5 is relatively specific to eosinophils due to the selective expression of the IL-5R α-chain and, indeed, much of our knowledge on the transcription factors that govern eosinophil development has been gleaned through the analysis of the gene encoding this receptor.

In Chapter 5.4, Dr. Kiyoshi Takatsu focuses on the role of IL-5 as the key cytokine driving eosinophil expansion and survival in tissues, and the expression of the IL-5R α-chain (which forms the IL-5R in conjunction with the βc chain), within the eosinophil lineage. Mice overexpressing IL-5 exhibit profound eosinophilia in the bone marrow, blood, and tissues. Conversely, mice lacking either the

interleukin 5 (colony-stimulating factor, eosinophil) (IL5) or the *IL5RA* gene, or mice treated with anti-IL-5 antibodies, exhibit decreased eosinophil numbers and impaired eosinophil function.[12] The interaction of IL-5 with the IL-5R complex activates a number of pathways, including the mitogen-activated protein kinase 3 (MAPK3) and tyrosine-protein kinase JAK (JAK)—signal transducer and activator of transcription (STAT) signaling pathways, and modulates eosinophil survival/apoptosis, proliferation, and degranulation.[12] Interestingly, recent evidence suggests that IL-5 has no effect on initial EoP lineage commitment, but rather drives expansion and survival of committed eosinophils.[8] Thus, while current intervention strategies targeting IL-5 signaling have been proposed to decrease eosinophil numbers, these interventions will likely not affect lineage commitment or EoP numbers in the bone marrow.

As mentioned, eosinophil development has traditionally been thought to occur in the bone marrow, but as Dr. Madeleine Rådinger and colleagues note in Chapter 5.5, emerging evidence suggests that eosinophilopoiesis also occurs at extramedullary sites, particularly during inflammatory disease.[13] Peripheral blood eosinophil numbers are normally tightly regulated and at steady state they comprise roughly 1—3% of the total white blood cells. However, under certain pathological conditions (particularly during allergic inflammation), peripheral blood eosinophilia develops, which corresponds closely with increased numbers of lineage-committed EoPs in the bone marrow and periphery. Thus, inflammatory signals at peripheral sites have a potent effect on the rate of eosinophilopoiesis in the bone marrow.

$CD34^+$ $IL-5RA^+$ (EoP) cells are increased in asthmatic patients, and in mouse models of asthma, within the bone marrow and the peripheral blood.[14–16] These cells give rise to eosinophil/basophil colonies in colony-forming assays *in vitro*, when cultured with GM-CSF, IL-3, and IL-5.[17] $CD34^+$ cell populations with colony-forming unit (CFU)-eosinophil/basophil potential are also present within lung tissue in models of asthmatic disease, suggesting that either EoPs or early progenitor populations migrate from the bone marrow to local sites of inflammation and produce mature eosinophils *in situ*. These emerging findings suggest that interventions aimed solely at limiting eosinophil production in the bone marrow once a disease state is established will not prevent eosinophil accumulation at peripheral disease sites. However, novel approaches aimed at inhibiting EoP seeding of tissues early may limit disease pathology in chronic conditions.

Finally, in Chapter 5.6, Dr. Nancy Lee and colleagues highlight the current mouse models of eosinophil proliferation and deficiency, which have greatly advanced eosinophil research. As IL-5 drives eosinophil proliferation and survival, IL-5-overexpressing transgenic mice exhibit profound eosinophilia, serving as valuable models of human eosinophilia and also providing a valuable source of primary eosinophils for functional *in vitro* studies and eosinophil adoptive transfer models. More recently, two eosinophil-deficient mouse strains were also generated by the eosinophil-specific expression of the diphtheria toxin A, *heparin-binding EGF-like growth factor (HBEGF/DTR)* gene (PHIL; EPO-DTA mice)[18] and deletion of the presumed autoregulatory *double GATA1* site in the GATA1 promoter (ΔdblGATA mice).[19] These mice have provided an unprecedented opportunity to evaluate the effects of eosinophil ablation on disease progression in a wide variety of mouse models of inflammatory disease and have already begun to clarify the role (or lack thereof) of eosinophils in antiparasite immunity.[20,21] These mouse models have, and will continue to provide, insight into factors regulating eosinophilopoiesis and the function of eosinophils in general health and disease.[22,23]

With our increasing understanding of eosinophilopoiesis, and an ever-increasing list of eosinophilic disorders and conditions, the aim of future research must be to regulate and control eosinophil differentiation to dampen disease pathology in patients. While initial anti-IL-5 antibody trials have yielded controversial results, many unexplored avenues remain open to suppress or modulate eosinophil expansion. An improved understanding of factors, both intrinsic and extrinsic, that regulate all stages of eosinophil lineage commitment, differentiation, proliferation, and survival may lead to clinical success in this area. Novel therapeutics may include small-molecule or small interfering RNA (siRNA)-based therapies targeting specific signaling pathways downstream of IL-5/IL-5R activation, modulating eosinophil transcription factor levels or treatments that block either the production of EoPs or the migration of cells with eosinophilopoietic potential to extramedullary sites in disease. Currently, however, no therapies directly targeting either initial eosinophil commitment or EoP expansion have been developed for general use in the clinic. We have reached the stage where we understand some of the regulatory complexity of eosinophilopoiesis; the future of the field will be in using this knowledge to alter eosinophil function during disease conditions to ultimately improve patient health.

ACKNOWLEDGEMENTS

S. M. holds a Canadian Institutes of Health Research (CIHR) and Heart and Stroke Foundation of Canada Transfusion Science Fellowship from the Centre for Blood Research (CBR). K. M. M. is a Michael Smith Foundation for Health Research Scholar (Senior) and member of the CBR, the Stem Cell Network, and AllerGen Networks of Centres of Excellence (NCEs).

REFERENCES

1. Graf T. Differentiation plasticity of hematopoietic cells. *Blood* 2002;**99**:3089–101.
2. McNagny K, Graf T. Making eosinophils through subtle shifts in transcription factor expression. *J Exp Med* 2002;**195**:F43–7.
3. Akashi K, Traver D, Miyamoto T, Weissman IL. A clonogenic common myeloid progenitor that gives rise to all myeloid lineages. *Nature* 2000;**404**:193–7.
4. Rothenberg ME, Hogan SP. The eosinophil. *Annu Rev Immunol* 2006;**24**:147–74.
5. Blanchet MR, Maltby S, Haddon DJ, Merkens H, Zbytnuik L, McNagny KM. CD34 facilitates the development of allergic asthma. *Blood* 2007;**110**:2005–12.
6. Maltby S, Wohlfarth C, Gold M, Zbytnuik L, Hughes MR, McNagny KM. CD34 is required for infiltration of eosinophils into the colon and pathology associated with DSS-induced ulcerative colitis. *Am J Pathol* 2010;**177**:1244–54.
7. Gauvreau GM, Ellis AK, Denburg JA. Haemopoietic processes in allergic disease: eosinophil/basophil development. *Clin Exp Allergy* 2009;**39**:1297–306.
8. Iwasaki H, Mizuno S, Mayfield R, Shigematsu H, Arinobu Y, Seed B, et al. Identification of eosinophil lineage-committed progenitors in the murine bone marrow. *J Exp Med* 2005;**201**:1891–7.
9. Mori Y, Iwasaki H, Kohno K, Yoshimoto G, Kikushige Y, Okeda A, et al. Identification of the human eosinophil lineage-committed progenitor: revision of phenotypic definition of the human common myeloid progenitor. *J Exp Med* 2009;**206**:183–93.
10. Saenz SA, Siracusa MC, Perrigoue JG, Spencer SP, Urban Jr JF, Tocker JE, et al. IL25 elicits a multipotent progenitor cell population that promotes T(H)2 cytokine responses. *Nature* 2010;**464**:1362–6.
11. Neill DR, Wong SH, Bellosi A, Flynn RJ, Daly M, Langford TK, et al. Nuocytes represent a new innate effector leukocyte that mediates type-2 immunity. *Nature* 2010;**464**:1367–70.
12. Kouro T, Takatsu K. IL-5- and eosinophil-mediated inflammation: from discovery to therapy. *Int Immunol* 2009;**21**:1303–9.
13. Radinger M, Lotvall J. Eosinophil progenitors in allergy and asthma—do they matter? *Pharmacol Ther* 2009;**121**:174–84.
14. Sehmi R, Wood LJ, Watson R, Foley R, Hamid Q, O'Byrne PM, et al. Allergen-induced increases in IL-5 receptor alpha-subunit expression on bone marrow-derived CD34+ cells from asthmatic subjects. A novel marker of progenitor cell commitment towards eosinophilic differentiation. *J Clin Invest* 1997;**100**:2466–75.
15. Inman MD, Ellis R, Wattie J, Denburg JA, O'Byrne PM. Allergen-induced increase in airway responsiveness, airway eosinophilia, and bone-marrow eosinophil progenitors in mice. *Am J Respir Cell Mol Biol* 1999;**21**:473–9.
16. Sergejeva S, Johansson AK, Malmhall C, Lotvall J. Allergen exposure-induced differences in CD34+ cell phenotype: relationship to eosinophilopoietic responses in different compartments. *Blood* 2004;**103**:1270–7.
17. Wood LJ, Inman MD, Watson RM, Foley R, Denburg JA, O'Byrne PM. Changes in bone marrow inflammatory cell progenitors after inhaled allergen in asthmatic subjects. *Am J Respir Crit Care Med* 1998;**157**:99–105.
18. Lee JJ, Dimina D, Macias MP, Ochkur SI, McGarry MP, O'Neill KR, et al. Defining a link with asthma in mice congenitally deficient in eosinophils. *Science* 2004;**305**:1773–6.

19. Humbles AA, Lloyd CM, McMillan SJ, Friend DS, Xanthou G, McKenna EE, et al. A critical role for eosinophils in allergic airways remodeling. *Science* 2004;**305**:1776–9.

20. Swartz JM, Dyer KD, Cheever AW, Ramalingam T, Pesnicak L, Domachowske JB, et al. Schistosoma mansoni infection in eosinophil lineage-ablated mice. *Blood* 2006;**108**:2420–7.

21. Fabre V, Beiting DP, Bliss SK, Gebreselassie NG, Gagliardo LF, Lee NA, et al. Eosinophil deficiency compromises parasite survival in chronic nematode infection. *J Immunol* 2009;**182**:1577–83.

22. Lee JJ, Jacobsen EA, McGarry MP, Schleimer RP, Lee NA. Eosinophils in health and disease: the LIAR hypothesis. *Clin Exp Allergy* 2010;**40**:563–75.

23. Jacobsen EA, Taranova AG, Lee NA, Lee JJ. Eosinophils: singularly destructive effector cells or purveyors of immunoregulation? *J Allergy Clin Immunol* 2007;**119**:1313–20.

Chapter 5.2

Transcriptional Regulation of Eosinophil Lineage Commitment and Differentiation

Steven J. Ackerman and Jian Du

INTRODUCTION

This chapter reviews the basic aspects of the transcriptional regulation and control of the processes of eosinophil lineage commitment and terminal differentiation from pluripotent hematopoietic stem cells (HSCs) and eosinophil lineage-committed progenitors (EoPs). These processes are highly pertinent to understanding the overall mechanisms that regulate basal eosinophilopoiesis relative to the development of eosinophilia in the bone marrow, blood, and tissues, in response to innate and adaptive host immune responses in allergic, parasitic, and other eosinophil-associated diseases. The transcription factor *codes* that specify eosinophil lineage commitment have been deciphered in part through *in vitro* studies of eosinophil-specific gene transcription in eosinophil-committed human leukemic cell lines, *ex vivo* studies of *authentic* bone marrow and umbilical cord blood HSCs and early myeloid progenitors, and *in vivo* studies using transgenic (both gain of function and gene mutation) and knockout (loss of function) approaches in the mouse.

Studies of the regulation of blood cell development over the past 15–20 years, including the eosinophil, have provided valuable models for determining how hematopoietic programs initially become established and then executed in terms of cell fate decisions and terminal differentiation. Identification of common myeloid (CMP) and lymphoid (CLP) progenitors in the mouse and man allowed the direct assessment of the regulatory mechanisms of lineage commitment, importantly showing that multiple markers of the various hematopoietic lineages are initially coexpressed in HSCs and early progenitors, a phenomenon now referred to as lineage priming.[1,2] In the current paradigm, *promiscuous* expression of multiple lineage-associated genes, including key hematopoietic transcription factors of interest to eosinophil development,[2,3] precedes lineage commitment without altering the developmental potential of pluripotent HSCs and multipotent myeloid or lymphoid progenitors; this allows flexibility in cell fate commitment (and reprogramming) at these early stages of hematopoietic development through a mechanism by which transcription factor antagonism is thought to resolve lineage promiscuity in the transition from multipotency to single-lineage commitment.[4–7] The roles played by lineage priming, combinatorial transcription factor networks, and the hierarchy and level of transcription factor expression in resolving hematopoietic stem cell pluripotency into myeloid progenitor cell multipotency, commitment to the eosinophil lineage, and finally differentiation under homeostatic conditions in health will be contrasted to eosinophil lineage-specific cytokine [interleukin-5 (IL-5)]-mediated mechanisms that induce and amplify eosinophilopoiesis to generate blood and tissue eosinophilia as part of disease pathogenesis. The ability of ectopically expressed transcription factors, signaling factors, and kinases key to eosinophil development to reprogram HSCs to the eosinophil lineage will be reviewed.

EOSINOPHILOPOIESIS IN THE BONE MARROW

Human eosinophils develop from a recently redefined population of lineage-committed, stem cell-derived CD34$^+$, interleukin-5 receptor subunit alpha (IL-5RA)$^+$ eosinophil lineage-committed progenitors (hEoPs) in the bone marrow that are an *offshoot* of the CMP and not derived from the granulocyte/macrophage progenitor (GMP) as originally thought (Fig. 5.2.1).[8] In contrast, in murine hematopoiesis, eosinophils differentiate from a mEoP downstream of the mGMP.[9] Thus, the hEoPs present in adult bone marrow, previously thought to develop from the hGMP, develop from an IL-5RA$^+$-committed progenitor, an independent component of the human (h)CMP population that originally included both the IL-5R$^-$ and IL-5R$^+$ populations.[10] These IL-5RA$^+$ CD34$^+$ CD38$^+$ interleukin-3 receptor subunit alpha (IL-3RA)$^+$ CD45RA$^-$ hEoPs give rise exclusively to eosinophils and

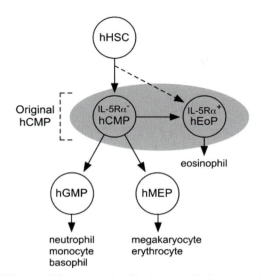

FIGURE 5.2.1 **Lineage relationship between the human eosinophil progenitor and other myeloid progenitor populations.** The original hCMP (shaded) contained the hEoP. The revised hCMP is now defined as the IL-5RA$^+$ fraction (IL-5RA$^+$ CD34$^+$ CD38$^+$ IL-3RA$^+$ CD45RA$^-$) of the original hCMP. The hEoP develops from the hCMP or its upstream multipotent progenitor, independent of the hGMP and the hMEP. hCMP, human common myeloid progenitor; hEoP, human eosinophil lineage-committed progenitor; hGMP, human granulocyte/macrophage progenitor; hHSC, human hematopoietic stem cell; hMEP, human megakaryocyte/erythrocyte progenitor; IL-5RA, interleukin-5 receptor subunit alpha. *Modified with permission from Mori Y et al. J Exp Med 2009;206:183—93.*[8]

not to basophils or neutrophils, whereas hGMPs or human megakaryocyte/erythroid progenitors (hMEPs) never differentiate into eosinophils (Fig. 5.2.1). Thus, the hCMP population has been phenotypically redefined to exclude the IL-5RA$^+$ hEoP.[8] Of note, the number of hEoPs increases to comprise upward of 10—20% of the hCMP population in the bone marrow of patients with blood eosinophilia of various known etiologies or idiopathic origins, suggesting that hEoPs participate in the expansion of eosinophilopoiesis in allergic and other diseases and hypereosinophilic syndromes. Importantly, the committed hEoPs already express mRNA encoding a number of the key transcription factors required for eosinophil development, as well as eosinophil-associated cytokine receptors required for EoP proliferation, and specific granule cationic and other proteins that define the lineage. Relative to other myeloid progenitors, the hEoP expresses moderate levels of the GATA family of transcription factors [GATA-binding factor 1 (GATA-1) < GATA-binding factor 2 (GATA-2)], little to no friend of GATA protein 1 (FOG-1), along with moderate levels of the E-twenty-six (ETS) family transcription factor PU.1 (PU.1), and high levels of the CCAAT/enhancer-binding protein (C/EBP) family, including both C/EBP alpha (C/EBPα) and epsilon (C/EBPε).

Of note, although morphologically still undifferentiated, the hEoP expresses mRNA encoding eosinophil secondary granule cationic proteins [eosinophil peroxidase (EPO) and eosinophil granule major basic protein 1 (MBP-1)] and the Charcot—Leyden crystal protein (CLC)/galectin-10.

In addition to being IL-5RA$^+$, the hEoP expresses the IL-3RA and granulocyte-macrophage colony-stimulating factor receptor subunit alpha (GM-CSF-R-alpha) along with their beta common (βc) chain that forms the high-affinity receptors for granulocyte-macrophage colony-stimulating factor (GM-CSF), IL-3, and IL-5, respectively, cytokines that further regulate eosinophil differentiation and function at a number of levels including: (1) EoP proliferation and terminal differentiation in the bone marrow; (2) priming, activation, and/or survival of eosinophils in blood and tissues; and (3) recruitment of eosinophils into allergic reactions.

IL-5, principally derived from activated T-helper type 2 (T$_h$2) lymphocytes[11] and mast cells,[12,13] regulates the development of blood and tissue eosinophilia *in vivo*.[14-16] While GM-CSF and IL-3 are multipotent cytokines with activities on many other myeloid lineages, IL-5 is eosinophil-selective and plays a crucial role in driving lineage-committed EoP cell proliferation, terminal differentiation, and postmitotic activation in the development of eosinophilia.[17] IL-5 is maximally active on the IL-5RA$^+$ EoP pool that is initially expanded by the multipotent GM-CSF and IL-3 cytokines.[17] Although IL-5 is both necessary and sufficient for the development of eosinophilia,[17,18] basal eosinophil differentiation itself, at least in the mouse, does not require any IL-5.[19] IL-5-mediated, upregulated expression of IL-5RA on the EoP is a prerequisite for IL-5-induced eosinophil terminal differentiation and the development of eosinophilia, and IL-5 overexpression underlies many eosinophil-associated diseases and syndromes, such as hypereosinophilic syndrome (HES).[20-22] IL-5-overexpressing transgenic mice develop profound eosinophilias,[23,24] further evidence that IL-5 is key in promoting the expanded production of eosinophils. Conversely, IL-5-deficient (gene knockout) mice[19,25] do not develop significant blood or tissue eosinophilia,[25] nor in response to infection with helminth parasites.[19] The finding that IL-5-deficient mice still generate basal numbers of bone marrow eosinophils fits the current paradigm that basal blood cell development: (1) occurs independently of lineage-specific cytokines [i.e., erythropoietin, IL-5, and macrophage colony-stimulating factor 1 (M-CSF)]; and (2) is regulated at the level of gene transcription through combinatorial networks of transcription factors, coactivators, and corepressors that act to resolve lineage-promiscuous gene expression patterns in early, uncommitted hematopoietic progenitors.[4,5,26]

FIGURE 5.2.2 **Combinatorial transcription factor** *codes* **that specify eosinophil lineage commitment and terminal differentiation.** GATA-1 is both necessary and sufficient to drive eosinophil development, and C/EBPε is required for eosinophil terminal differentiation. Eosinophils have recently been shown to develop from a distinct CD34+/IL-5RA+ eosinophil lineage-committed progenitor (EoP) derived from the common myeloid progenitor (CMP) pool and not from the granulocyte/macrophage progenitor (GMP) as previously thought.[8] C/EBPα, CCAAT/enhancer-binding protein alpha; C/EBPε, CCAAT/enhancer-binding protein epsilon; EoP, eosinophil lineage-committed progenitor; Ery, erythrocyte; FOG-1, friend of GATA protein 1; GATA-1, GATA-binding factor 1; GMP, granulocyte/macrophage progenitor; HSC, hematopoietic stem cell; IL-5RA, interleukin-5 receptor subunit alpha; Mac, macrophage; Meg, megakaryocyte; MEP, megakaryocyte/erythroid progenitor; PMN, polymorphonuclear leukocyte; PU.1, transcription factor PU.1. *Modified and updated with permission from: McNagny K, Graf T. J. Exp. Med. 2002;195:[11]F43−47.*[3]

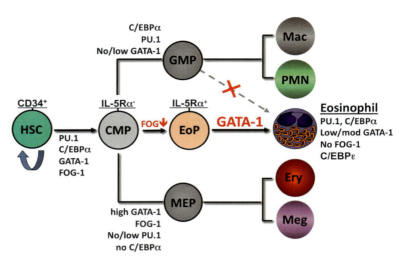

TRANSCRIPTIONAL REGULATION OF EOSINOPHIL LINEAGE COMMITMENT

Avian, mouse, and human studies have shown that only a handful of transcription factors and their functional interactions are really key to specifying eosinophil lineage commitment and terminal differentiation (Fig. 5.2.2).[3] The combinatorial activities of C/EBPα, GATA-1, and PU.1 are required, with GATA-1 being the pivotal factor that determines if CMPs (GMPs in the mouse) will commit and differentiate to the eosinophil lineage (GATA-1 is required), or neutrophil and macrophage lineages (GATA-1 is absent) (Fig. 5.2.2). Additionally, the GATA-1 coactivator FOG-1, which is required for erythroid and megakaryocyte differentiation, functions instead as a corepressor of GATA-1 within the eosinophil lineage[27] and must be

downregulated for eosinophil differentiation to proceed, most likely through the repression by C/EBPα [as for C/EBP beta (C/EBPβ) in the chicken] (Fig. 5.2.3).[28] Of note, enforced expression of FOG-1 in eosinophil progenitor cell lines induced them to *dedifferentiate* into MEPs.[28]

Notably, eosinophil development in fetal liver is deficient in *GATA binding protein 1 (globin transcription factor 1) (GATA1)* gene-deleted mice,[29] and the transgenic deletion of a high-affinity double-GATA site in the HS2 (DNase I hypersensitivity) region of the murine Gata1 promoter itself was shown to serendipitously block eosinophil development,[30] providing an eosinophil-deficient mouse model now used in multiple studies of eosinophil function in allergic and parasitic diseases.[31−34] Of note, we previously identified similar, high-affinity double-GATA sites in the promoters of a number of

FIGURE 5.2.3 **Resolution of lineage promiscuity in eosinophil lineage-committed progenitor leading to the induction of eosinophil gene transcription.** A model for the induction of the eosinophil-specific gene expression program during eosinophil lineage commitment is shown in terms of the cross-antagonistic roles of transcription factors including C/EBPs, FOG-1, GATA-1, and PU.1. C/EBPα (βin chicken) represses *zinc finger protein, multitype 1 (ZFPM1/FOG1)* gene expression at the transcriptional level. *(A)* Eosinophil C/EBP target genes contain cis-acting regulatory sequences that mediate the binding of C/EBPs, the E-twenty-six (ETS) factor PU.1, and the GATA factors. *(B)* In contrast to the erythroid lineage, FOG-1 acts as a transcriptional corepressor in the eosinophil lineage through protein−protein interaction with GATA-1, and abrogates C/EBPβ-mediated activation of eosinophil gene expression, so it must be downregulated for eosinophil development to proceed. C/EBPα, CCAAT/enhancer-binding protein alpha; EoP, eosinophil lineage-committed progenitor; Eos, eosinophil; FOG-1, friend of GATA protein 1; GATA-1, GATA-binding factor 1; MEP, avian Myb-ETS transformed multipotent hematopoietic progenitor; PMN, polymorphonuclear leukocyte; MØ, macrophage; PU.1, transcription factor PU.1. *Modified and updated with permission from Querfurth et al. Genes and Development 2000;14:2515.*[28]

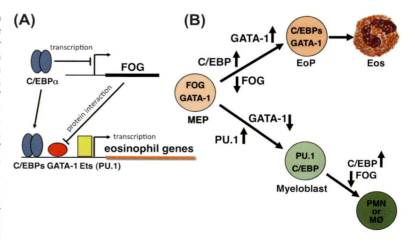

Eosinophil Genes	Double GATA consensus binding sites	
	WGATAW T/AGATAA/G	WGATAW T/AGATAA/G
IL-5Rα-P1 promoter	**AGATAGCCATTATCT**GATAA	
IL-5Rα-P1 (chr3-P2925)	AGATAA..AGG...AGATAA	
MBP1-P2 promoter double GATA-1	AGATAGCAAGG..CTGATAA	
MBP1-P2 (PRG2-chr11-P556)	AGATAGG.A.G...AGATAT	
CLC/Gal-10 promoter (chr19-1-P324)	AGATAT.AA....TTGATAT	
EPX (chr17-P1440)	AGATAA...GG...AGATAT	
C/EBPε (chr14-2-P282)	AGATAT.AA....TTGATAT	

FIGURE 5.2.4 Sequence alignments of high-affinity double-GATA-binding sites in the regulatory regions of eosinophil gene promoters. Alignments are shown for cis-oriented double-GATA sites identified in the upstream regulatory regions for eosinophil lineage-specific [IL-5 receptor subunit alpha (IL-5R)], eosinophil granule major basic protein 1 (MBP-1), Charcot—Leyden crystal protein/galectin-10 (CLC/Gal-10), eosinophil peroxidase (EPO)] and eosinophil lineage-associated [CCAAT/enhancer-binding protein epsilon (C/EBPε)] promoters. Five of six eosinophil-specific genes contained one or more cis-oriented, high-affinity double-GATA binding motifs in their promoter regions (shaded sequences). The IL-5RA P1 promoter also contains a trans-oriented double-GATA palindromic site (underlined and bold), which is essentially identical to the murine Gata-1 HS2 domain double-GATA palindrome that, when transgenically deleted, resulted in the eosinophil-deficient ΔdblGATA mouse strain.[30]

hallmark human eosinophil-*specific* genes including *proteoglycan 2, bone marrow (natural killer cell activator, eosinophil granule major basic protein) (PRG2/MBP)* (its eosinophil-specific P2 promoter), eosinophil peroxidase (*EPO*), *Charcot—Leyden crystal protein (CLC/galectin 10), interleukin 5 receptor, alpha (IL5RA),*[35] the eotaxin receptor *chemokine (C-C motif) receptor 3 (CCR3),*[36] and the transcription factor *CCAAT/enhancer binding protein (C/EBP), epsilon (CEBPE)*, the latter required for eosinophil terminal differentiation[37] (Fig. 5.2.4), which may serve as functional regulators. Finally, in contrast to the deletion of the murine Gata-1 HS2 double-GATA site enhancer, resulting in a block in eosinophil development, deletion of the mouse HS1 region active in erythropoiesis did not affect eosinophil Gata-1 mRNA expression or eosinophil differentiation, nor chromatin histone acetylation of the eosinophil HS2 Gata-1 locus.[38] In eosinophils, GATA-1 and C/EBPε did not bind the HS1 enhancer, but instead were found to bind selectively to a novel cis-regulatory element in the first GATA-1 intron.[38] Thus, the GATA-1 HS1 enhancer is not required for eosinophil GATA expression, further highlighting eosinophil lineage-specific regulation of GATA-1 transcription, and the role of a GATA-1 intronic element for this purpose, which requires further investigation.

The current consensus for the combinatorial transcription factor *code* that specifies the eosinophil relative to the other hematopoietic lineages is outlined in Fig. 5.2.2 and Fig. 5.2.3. Importantly, regulation of both the timing and levels of expression of these transcription factors is important and required to generate eosinophils, such that commitment and terminal differentiation of eosinophils from EoPs requires the coordinated expression of C/EBPα and PU.1, a moderate level of GATA-1, and the absence of FOG-1,[3] with later stages, including the promyelocyte to myelocyte transition, terminal differentiation, and functional maturation of the eosinophil being regulated by the combinatorial activities of C/EBPε's various activator and repressor isoforms,[39,40] these also being expressed in a temporally regulated fashion[41] (Fig. 5.2.5).

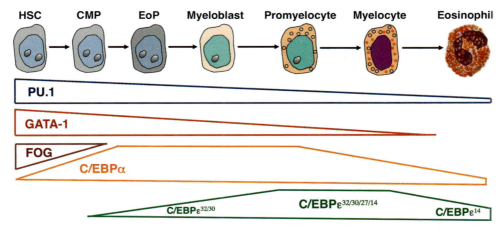

FIGURE 5.2.5 Temporal expression patterns of transcription factors specifying the eosinophil lineage during eosinophil differentiation. Comparative timing and levels of expression of the eosinophil lineage's differentiation program-associated transcription factors, including PU.1, GATA-1, FOG-1, C/EBPα, and C/EBPε [with its activator ($\varepsilon^{32/30}$) and repressor ($\varepsilon^{27/14}$) isoforms] is shown schematically during eosinophilopoiesis from lineage-committed EoPs. Downregulation of FOG-1 during lineage commitment of the common myeloid progenitor (CMP) to the EoP is a prerequisite for the expression of GATA-1-regulated eosinophil target genes. GATA-1 expression ceases in the mature, terminally differentiated blood eosinophil. C/EBPα, CCAAT/enhancer-binding protein alpha; C/EBPε, CCAAT/enhancer-binding protein epsilon; EoP, eosinophil lineage-committed progenitor; FOG-1, Friend of GATA protein 1; GATA-1, GATA-binding factor 1; HSC, hematopoietic stem cell; PU.1, transcription factor PU.1.

TRANSCRIPTION FACTOR PU.1 IS REQUIRED FOR THE TERMINAL DIFFERENTIATION OF EOSINOPHILS, BUT NOT FOR THE SPECIFICATION OF EOSINOPHIL PROGENITORS

PU.1 is a member of the ETS transcription factor family expressed in various hematopoietic lineages, principally myeloid (monocyte, granulocyte) and lymphocyte (B cell). Gene disruption of *spleen focus forming virus (SFFV) proviral integration oncogene spi1 (SPI1/PU.1$^{-/-}$)* in two knockout mouse strains developed in the mid-1990s severely impaired the development of both myeloid and lymphoid lineages, leading either to embryonic lethality by day 16–18 of gestation[42] or to viable mice with no detectable monocyte/macrophages, B cells, and delayed/reduced neutrophil development.[43] *SPI1 (PU.1)* gene disruption caused a severe reduction, but not elimination, of myeloid progenitors, which were still capable of responding to multilineage cytokines [IL-3, interleukin-6 (IL-6), and stem cell factor (SCF)], but not to myeloid-specific cytokines [granulocyte colony-stimulating factor (G-CSF), GM-CSF, and MCSF] *ex vivo*.[44] The *SPI1 (PU.1)$^{-/-}$* null progenitors could also undergo limited differentiation into neutrophils, but not into monocytes. Eosinophil progenitors (both EoP and developing eosinophils) also express *SPI1 (PU.1)*, and it continues to be expressed at low levels by mature eosinophils. As the effect of *SPI1 (PU.1)* knockout on eosinophilopoiesis had not been analyzed, we used reverse transcription polymerase chain reaction (RT-PCR) to analyze the expression of eosinophil lineage-specific genes including *EPO, IL5RA,* and *PRG2 (MBP)* in the two *SPI1 (PU.1)$^{-/-}$* mouse strains in fetal liver, spleen, and bone marrow cells.[45] Our results showed that the EPO, IL-5RA, and MBP-1 mRNA was not expressed in the fetal liver of PU.1$^{-/-}$ embryos, but could be detected equivalently in PU.1$^{+/-}$ and PU.1$^{+/+}$ control RNA (Fig. 5.2.6). In contrast, *IL5RA* gene expression was reduced by approximately three to fourfold in day 9 spleens of viable PU.1$^{-/-}$ mice, whereas MBP-1 and EPO expression levels were identical in PU.1$^{+/+}$ and PU.1$^{+/-}$ mice (Fig. 5.2.6E). Histochemical evaluations using Fast Green/Neutral Red staining for eosinophils in PU.1$^{-/-}$ fetal liver, bone marrow, and spleen cells showed eosinophils in the bone marrow and spleens of PU.1$^{+/+}$ and PU.1$^{+/-}$ mice, but none in the viable PU.1$^{-/-}$ mice. Eosinophils could not be detected in (d14.5) PU.1$^{+/+}$, PU.1$^{+/-}$, and PU.1$^{-/-}$ fetal livers, since few granulocytes could be identified at this stage. In contrast, small numbers of eosinophils were identified in the d16.5 fetal livers of PU.1$^{+/+}$ and PU.1$^{+/-}$ mice, but not in the PU.1$^{-/-}$ embryonic lethal or viable null mice. These data indicated that eosinophilopoiesis did not occur in PU.1 embryonic lethal knockouts, whereas eosinophil gene expression was still detectable in the absence of eosinophil terminal differentiation in the viable PU.1 null mice.[45] The defect in *IL5RA* gene expression suggests a role for PU.1 in regulating its expression during eosinophil differentiation, perhaps contributing to the

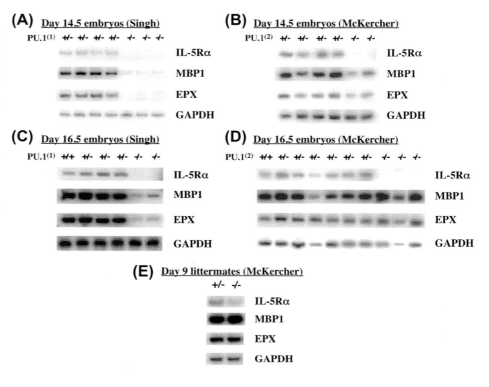

FIGURE 5.2.6 Effects of *SPI1/PU.1* gene knockout on eosinophil development. Assessment of mRNA expression levels of eosinophil lineage-specific genes in two different PU.1 knockout mouse strains[42,43] using semi-quantitative reverse transcription polymerase chain reaction (RT-PCR). *(A)* mRNA expression of eosinophil lineage-specific genes in day 14.5 fetal livers of PU.1$^{+/+}$, $^{+/-}$, and null embryos (Singh);[42] *(B)* Day 14.5 fetal livers of PU.1$^{+/+}$, $^{+/-}$, and null$^{-/-}$ embryos (McKercher);[43] *(C)* Day 16.5 fetal livers of PU.1$^{+/+}$, $^{+/-}$, and null$^{-/-}$ embryos (Singh);[42] *(D)* Day 16.5 fetal livers of PU.1$^{+/+}$, $^{+/-}$, and null$^{-/-}$ embryos (McKercher);[43] and *(E)* Day 9 littermates of PU.1$^{+/-}$ and null$^{-/-2}$ mice (McKercher).[43] EPO, eosinophil peroxidase; GAPDH, glyceraldehyde-3-phosphate dehydrogenase; IL-5Rα, interleukin-5 receptor subunit α; MBP-1, eosinophil granule major basic protein 1; PU.1, transcription factor PU.1.

absence of terminally differentiated eosinophils in PU.1$^{-/-}$ null mice, which may require IL-5 signaling for proliferation and survival, despite the expression of a number of eosinophil secondary granule protein genes. These findings support the concept that PU.1 is not essential for the specification of granulocyte (including EoP) progenitors (in the mouse), but instead controls their proliferation and terminal differentiation in a cell-autonomous manner by regulating expression of growth factor receptor (i.e., IL-5RA in the eosinophil) and other lineage-specific genes. [44,46–48]

TRANSCRIPTIONAL REGULATION OF EOSINOPHIL TERMINAL DIFFERENTIATION—THE ROLE OF CCAAT/ENHANCER-BINDING PROTEIN EPSILON

C/EBPε is a member of the basic leucine zipper (bZIP) family of C/EBP transcription factors that also includes C/EBPα, β, γ, δ, and ζ, [49] is preferentially expressed in granulocytes and is required for the promyelocyte to myelocyte transition of terminally differentiating eosinophils and neutrophils. [37,50,51] C/EBPε, expressed mainly during the promyelocyte to myelocyte transition, is required for eosinophil terminal differentiation. [37,49] Studies of C/EBPε knockout mice showed that eosinophil (and neutrophil) differentiation both require C/EBPε, since these mice lack terminally differentiated and functionally mature eosinophils and neutrophils and die by 3–5 months of age due to opportunistic infections, despite treatment with antibiotics. [37,49] The absence of both eosinophils and neutrophils in C/EBPε-null mice may reflect their shared development from the same progenitor cell population (GMP) in the mouse. Similarly, patients with specific granule deficiency (SGD) disease have a mutation in their *C/EBPE* gene that notably leads to a failure of both neutrophil and eosinophil terminal differentiation, with failed expression of important secondary granule protein genes in both granulocyte populations. [52] Of note, these SGD patients have a small population of GM-CSF-responsive, EPO$^+$ cells that lack secondary granules, express CLC/galectin-10, but are deficient in their expression of eosinophil-derived neurotoxin (EDN/RNase 2), eosinophil cationic protein (ECP) and MBP-1, despite the presence of some mRNA transcripts for these proteins. [52]

In the mouse, C/EBPε is expressed as two activator isoforms of approximately 36 kDa and 34 kDa, [53] compared to humans in which four distinct isoforms are expressed as proteins of 32, 30, 27, and 14 kDa through alternative splicing, differential promoter usage, and translational start sites. [50,51,54,55] The human C/EBPε isoforms are all identical in sequence at their C-terminus, which encodes the

basic DNA-binding and bZIP domains. [50,51] Human neutrophils and eosinophils express all four C/EBPε isoforms, [35,50] with the highest levels of expression reported during the promyelocyte to myelocyte transition in both neutrophilic cell lines [50,51,56,57] and granulocyte progenitors purified from human bone marrow. [58] Additionally, developing EoPs and eosinophil myeloblast and myelocyte cell lines (AML14 and AML14.3D10, respectively) express all four C/EBPε isoforms. On the other hand, mature blood eosinophils express high levels of mainly the 14 kDa repressor isoform. [35,41] The C/EBPε^{32} and C/EBPε^{30} isoforms are weak transcriptional activators that interact with and require coactivators for full functional activity. [37,59] The activities of the shorter 27 kDa and 14 kDa isoforms was unclear [51] until transactivation studies in cell lines suggested they function as transcriptional repressors of GATA-1 (C/EBPε^{27}) or other C/EBPs (C/EBPε^{14}). [35,60]

The full length human C/EBPε^{32} and shorter ε^{30} isoforms contain well-defined transactivation, repression and DNA-binding and dimerization domains, [51,54,61] but their function as transcriptional activators of myeloid promoters requires transcriptional cofactors, particularly transcriptional activator Myb (c-Myb). [59] Cotransfection of c-Myb with C/EBPε^{30} or C/EBPε^{32} cooperatively transactivates both the c-Myb-induced myeloid protein 1 (MIM-1) and neutrophil elastase promoters. [59] C/EBPε^{32} also cooperatively interacts with the nuclear factor NF-kappa-B p65 subunit [(NF-κB)/p65 (RelA)] in binding to C/EBP sites, such that transcriptional activity of the C/EBPε-dependent MIM-1 promoter was significantly reduced in the absence of p65 (RelA). [62] The p65 (RelA)-C/EBPε interaction was also shown to be important for the activation of neutrophil secondary granule genes and for C/EBPε-induced differentiation of murine 32Dc13 myeloid cells. [62] Activity of murine C/EBPε has been shown to be post-transcriptionally dependent on small ubiquitin-like modifier (SUMO)ylation of a lysine residue in a *VKEEP* (valine-lysine-glutamic acid-glutamic acid-proline) motif present in the RDI repressor domain, a highly conserved sequence present in the human C/EBPε 30, 32, and 27 kDa isoforms and three other C/EBP family members, including C/EBPα, C/EBPβ, and C/EBPδ. [63,64]

The C/EBPε^{27} isoform contains a unique N-terminus (RD27) and RDI repressor domain, [54] both of which we showed contributed to its inhibition of GATA-1 activity. [35,60] Our prior studies showed that C/EBPε^{27} (but not the other C/EBPε isoforms) specifically antagonizes GATA-1 transactivation of the hallmark eosinophil-specific promoter, the MBP-1 P2 promoter, [35] and transduction of an eosinophil myelocyte cell line (AML14.3D10) with a human immunodeficiency virus (HIV) Tat-specific factor 1(Tat-SF1)−C/EBPε^{27} fusion protein potently inhibits endogenous *PRG2 (MBP)* gene expression. [60,65] The AML14.3D10 line expresses all of the C/EBPε isoforms and GATA-1. Co-immunoprecipitation with antibodies to

GATA-1 or C/EBPε showed that C/EBPε[27] physically interacts with GATA-1.[35] Importantly, we showed that C/EBPε[27] is a potent repressor of GATA-1 activity in developing EoPs that potently inhibits both GATA-1 expression itself and GATA-1-dependent eosinophil gene transcription.[41]

The C/EBPε[14] isoform contains only DNA-binding and bZIP dimerization domains and was hypothesized to function as a dominant negative repressor either as a heterodimer with other C/EBP family members (e.g., C/EBPε,[32/30] C/EBPα, or C/EBPβ) or by direct competition as a homodimer for C/EBP sites in granulocyte promoters;[49] our reports, which showed that C/EBPε[14] inhibited transactivation of the eosinophil MBP-1 P2 promoter by C/EBPα and C/EBPβ in a dose-dependent manner,[35] and inhibited endogenous *PRG2 (MBP)* gene transcription in AML14.3D10 eosinophils transduced with a Tat-SF1−C/EBPε[14] fusion protein,[60,65] supported a likely repressor role for this isoform in downregulating eosinophil gene transcription during eosinophil terminal differentiation.

As reviewed in the *Transcriptional Reprogramming of Hematopoietic Progenitors* section of this chapter, there is now novel evidence showing that these human C/EBPε isoforms have distinct activities and roles in eosinophilopoiesis, and are capable of reprogramming eosinophil (and other myeloid) lineage commitment and terminal differentiation when ectopically expressed in umbilical cord blood-derived human CD34[+] hematopoietic progenitors.[41] These studies were the first to characterize individual functional activities for the C/EBPε activator and repressor isoforms in eosinophil and myeloid development. Surprisingly, overexpression of the C/EBPε activator isoforms in HSCs drove them exclusively to the eosinophil lineage at the expense of other myeloid programs in a cytokine (IL-5)-independent manner, and regardless of the presence of other lineage-specific cytokines (Fig. 5.2.7). In contrast, the C/EBPε[27/14] repressor isoforms were capable of inhibiting differentiation to the eosinophil lineage, even in the presence of IL-5 to drive the process, findings reflected in the induction (by C/EBPε[32/30]) or repression (by C/EBPε[27] and C/EBPε[14]) of lineage-specific genes that uniquely define the eosinophil (Fig. 5.2.7). Of note, enforced overexpression of the C/EBPε[32/30] activator isoforms, in the context of endogenous C/EBPε isoform expression, both maintained CMP proliferation and potently induced their differentiation, regardless of whether IL-5 or other lineage-specific cytokines were provided, suggesting that early expression of C/EPBε[32/30] in CMPs defaults their hematopoietic program to EoP commitment and eosinophil differentiation. Importantly, it implies that subtle changes in the balance between activator vs. repressor isoforms of C/EBPε (increased C/EBPε[27] and C/EBPε[14] relative to C/EBPε[32/30]) observed during eosinophil development impacts the outcome of their combinatorial activities on

eosinophil gene transcription and terminal differentiation. Of note, that eosinophils differentiate from C/EBPε[32/30] transduced CD34[+] progenitors in the absence of IL-5 is consistent with the IL-5 and IL-5RA knockout (null) mouse phenotype, in which basal eosinophilopoiesis proceeds normally in the bone marrow, but the mice fail to develop significant blood or tissue eosinophilia in response to allergic stimuli or parasitic infections.[66,67] These findings support the current concept that basal eosinophilopoiesis occurs in a transcription factor-dependent, but IL-5-independent, manner and that IL-5 signaling may only be required for blood and tissue eosinophilia in response to innate or T-helper type 2 (T$_h$2)-mediated allergic or antiparasite immune responses.[68,69]

Eosinophils are clearly unique in their commitment and differentiation from the CMP-derived EoP population, as they require combinatorial expression of transcription factors that include C/EBPα, C/EBPε, GATA-1, and PU.1.[3] Additionally, their developmental program is clearly different than that of GMP-derived neutrophils, since GATA-1 is an absolute requirement for eosinophil development[29] and a pivotal difference between these two granulocytes.[3,35] However, the role of C/EBPε had largely been considered ancillary to C/EBPα, which regulates C/EBPε expression in granulocyte development.[70] The four human C/EBPε isoforms are expressed early and concurrently during granulopoiesis in the bone marrow,[58] although expression levels vary temporally during eosinophil differentiation,[41] suggesting their expression ratios and combinatorial interactions serve to finely regulate one another's transcriptional activities and the activities of other C/EBPs, e.g., C/EBPα or C/EBPβ (C/EBPε[14]), or GATA-1 (C/EBPε[27]),[35] or serve to downregulate secondary granule protein gene expression during terminal eosinophil differentiation.[71] Consistent with the latter, it was shown that mature blood eosinophils continue to express high levels of mainly the C/EBPε[14] repressor isoform.[35]

Continued studies defining the combinatorial interactions and activities of the C/EBPε activator and repressor isoforms should further inform our understanding of their roles in regulating transcriptional control of the eosinophil developmental program and gene expression in this process.

TRANSCRIPTIONAL REPROGRAMMING OF HEMATOPOIETIC PROGENITORS TO THE EOSINOPHIL LINEAGE

Reprogramming by GATA-Binding Factor 1 and 2

As discussed earlier, GATA-1 is pivotal in determining whether human CMPs (GMPs in the mouse) differentiate into the eosinophil (GATA-1 required) or

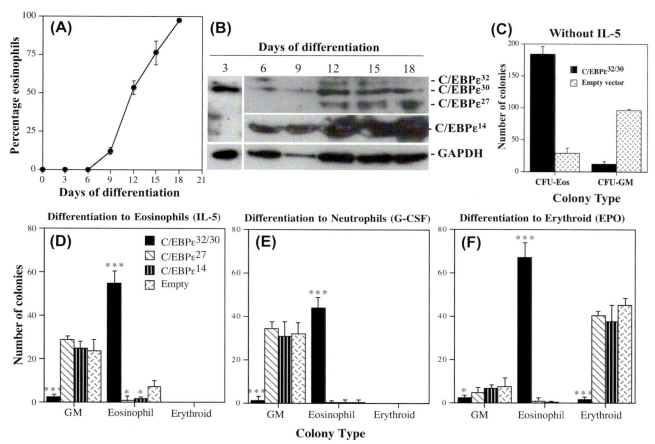

FIGURE 5.2.7 **Roles of the C/EBP epsilon activator and repressor isoforms in eosinophil terminal differentiation.** Panels *(A)* and *(B)* show temporal changes in the expression of the C/EBPε isoforms *(B)* during eosinophilopoiesis *(A)*. CD34$^+$ progenitors were differentiated to eosinophils by suspension culture in Fms-related tyrosine kinase 3 ligand (Flt3L), granulocyte-macrophage colony-stimulating factor (GM-CSF), interleukin-3 (IL-3) and -5 (IL-5), and stem cell factor (SCF) for 3 days, followed by only IL-3 and IL-5 thereafter. The percentage of eosinophils *(A)* in the cultures, based on differential counts using Fast Green/Neutral Red staining to distinguish secondary granule formation, is shown. C/EBPε isoform protein expression was analyzed by Western blot of whole cell lysates using combined anti-C/EBPε C- and N-terminus antibodies, compared to the glyceraldehyde-3-phosphate dehydrogenase (GAPDH) loading control *(B)*. Panels *(D–F)* show the effects of the C/EBPε isoforms on eosinophil, erythroid, and neutrophil differentiation in the presence of lineage-specific cytokines. CD34$^+$ cord blood progenitors were transduced with retroviral vectors encoding each of the C/EBPε isoforms. After 72 h, CD34$^+$/green fluorescent protein (GFP)$^+$ cells were sorted by fluorescence-activated cell sorting (FACS), plated in Collagen Cult medium colony assays, and the cells were induced to differentiate toward the eosinophil *(D:* IL-3, IL-5, and SCF), neutrophil *(E:* G-CSF, IL-3, and SCF), and erythroid *(F:* EPO, IL-3, and SCF) lineages. Significant differences are indicated: *$p \leq 0.05$; **$p \leq 0.01$, and ***$p \leq 0.001$. Panel *(C)* shows that C/EBPε$^{32/30}$ enhancement of eosinophil differentiation does not require interleukin-5 (IL-5). CD34$^+$ cord blood progenitors were transduced as above with the retroviral vector encoding the C/EBPε$^{32/30}$ activator isoforms or empty vector control. CD34$^+$/GFP$^+$ transduced cells were sorted by FACS, plated in Collagen Cult medium containing SCF and IL-3, and cells differentiated for 14 days without any IL-5. Eosinophil and granulocyte/macrophage (GM) colonies were enumerated using Alkali Fast Green 10 G/Neutral Red staining. The mean number of eosinophil colony-forming units (CFU-Eos) and GM colony-forming units (CFU-GMs) is shown for triplicate determinations. C/EBPε, CCAAT/enhancer-binding protein epsilon; G-CSF, granulocyte colony-stimulating factor. *Figures modified and reproduced with permission from Bedi et al. Blood 2009;113:317–27.*[41]

granulocyte-macrophage (in the absence of GATA-1) lineages (Fig. 5.2.2).[3,45] In addition to eosinophil development, the GATA transcription factors are major regulators of hematopoietic and immune system development, with GATA-1, GATA-2, and GATA-3 playing crucial roles in the development of the erythroid, megakaryocytic, and mast cell lineages (GATA-1), hematopoietic stem and early progenitor cells (GATA-2), and T$_h$2 T cells (GATA-3).[72] Eosinophils express high levels of both GATA-1 and GATA-2,[73] and our initial studies suggested that GATA-2 functioned as

a transcriptional repressor rather than activator of human eosinophil genes, such as the MBP-1 P2 promoter,[74] consistent with a requirement for the downregulation of GATA-2 expression as HSCs commit to the erythroid and other myeloid lineages.[75]

Importantly, Hirasawa and colleagues[29] showed that retroviral vector-enforced ectopic expression of GATA-1 *ex vivo* in primary human cord blood CD34$^+$ hematopoietic progenitors completely reprogrammed their myeloid cell fate into eosinophils, such that expression of GATA-1 exclusively promoted the development and terminal maturation of eosinophils at the expense of the other myeloid lineages, regardless of the presence of other lineage-specific cytokines during culture. Functional domain analyses showed that the C-terminus zinc finger of GATA-1 was essential whereas the N-terminus zinc finger, which binds FOG-1, is not required for the switch and was dispensable for inducing eosinophil differentiation.[29] Surprisingly, GATA-2 was similarly instructive for eosinophil differentiation in cord blood CD34+ HSCs, comparable to GATA-1, and efficiently rescued eosinophil development *in vivo* in the eosinophil-deficient, GATA-1-deficient (knockout) mouse, indicating that at least in the mouse, both GATA factors are capable of reprograming HSCs to the eosinophil lineage[2] (see the next section of this chapter). Thus, these studies elegantly demonstrated the essential and instructive roles of GATA-1 and GATA-2 in eosinophil development, and their ability to reprogram HSCs to eosinophil differentiation.

Reprogramming by Hierarchical Transcription Factor Expression: GATA-Binding Factor 2 vs. CCAAT/Enhancer-Binding Protein Alpha in the Mouse

The timing and level of transcription factor expression can also determine the lineage specification of multipotent HSCs to the eosinophil vs. other myeloid lineages. Iwasaki and colleagues[2] isolated murine progenitors with eosinophil,[9] basophil, or mast cell developmental potential, which in the mouse all originate from GMPs[9] rather than CMPs as for human eosinophils. They showed that the timing of expression of GATA-2 relative to C/EBPα differentially controlled lineage commitment to the eosinophil lineage, such that ectopic overexpression of GATA-2 instructed C/EBPα-expressing GMPs to commit exclusively to the eosinophil lineage, while it induced basophil and/or mast cell lineage commitment only if C/EBPα expression was suppressed in the GMPs. Switching the order of C/EBPα and GATA-2 transduction, even in progenitors committed to the lymphoid lineage, also reprogrammed these cells into eosinophil, basophil, and mast cell lineages, demonstrating the plasticity of hematopoietic EoPs, e.g., GMPs and CLPs. Both the order and level of expression of transcription factors in HSCs and more committed hematopoietic progenitors is therefore crucial for their hierarchical interactions to selectively induce the developmental programs of eosinophils and other myeloid lineages. Thus, at least in the mouse, both GATA-1 and GATA-2 participate in GMP commitment upstream of C/EBPα in eosinophil lineage specification.

Reprogramming by CCAAT/Enhancer-Binding Protein Epsilon Activator vs. Repressor Isoforms

To further define the functional activities for the activator and repressor isoforms of C/EBPε in myelopoiesis in general and eosinophilopoiesis in particular, we transduced human CD34+ progenitors with internal ribosome entry site (IRES)-enhanced green fluorescent protein (eGFP) retroviral vectors encoding the 32/30, 27, and 14 kDa isoforms, purified the retrovirus-transduced cells by fluorescence-activated cell sorting (FACS), and analyzed them in colony-forming assays and suspension culture.[41] Cord blood CD34+ progenitors transduced with C/EBPε[32/30] were shown to default exclusively to eosinophil differentiation and gene expression in an IL-5-independent manner, and regardless of the inclusion of cytokines, to induce neutrophil or other myeloid lineage differentiation (Fig. 5.2.7). As expected of developing eosinophils, *GATA binding protein 1 (globin transcription factor 1) (GATA1)* and the genes encoding the IL-5RA forms (both soluble and transmembrane alternative splice forms) and the secondary granule cationic protein genes encoding EDN (RNase 2), EPO, and MBP-1 were strongly induced. In contrast, the repressor C/EBPε[27] isoform was shown to strongly inhibit eosinophil differentiation and the expression of EDN (RNase 2), EPO, GATA-1 autoregulated expression itself, IL-5RA (both soluble and transmembrane isoforms), and MBP-1, thus promoting granulocyte (neutrophil)-macrophage differentiation (Fig. 5.2.7). Of note, the C/EBPε[14] repressor isoform was shown to strongly inhibit eosinophil development and gene expression, promoting erythropoiesis. Thus, the various C/EBPε isoforms have the unique capacity to reprogram myelopoiesis toward or away from eosinophil differentiation and gene expression in a manner consistent with their predicted activities based on their activator and repressor domains and *in vitro* functional activities.[41] Further *ex vivo* studies in CD34+ HSCs designed to define the combinatorial interactions of the four C/EBPε activator and repressor isoforms in eosinophil differentiation, and *in vivo* studies using retrovirally transduced human cord blood progenitors ectopically expressing these activator/repressor isoforms transplanted into the immunodeficient nonobese diabetic (NOD)/severe combined immunodeficient or immunodeficiency (SCID) mouse model, should further define the transcriptional roles of C/EBPε in regulating the eosinophil's developmental program, including lineage commitment and terminal maturation.

Reprogramming by Signal Transducer and Activator of Transcription 5

Signal transducer and activator of transcription 5 (STAT5), both STAT5A and STAT5B isoforms, are expressed by human eosinophils and their signaling pathways are induced by GM-CSF and IL-5.[76–78] IL-5 signals through its high-affinity receptor, IL-5Rα, on the mature eosinophil to induce the phosphorylation and consequent nuclear translocation of STAT5 and the activation of the *signal transducer and activator of transcription 5A (STAT5A)* target genes.[77–79] STAT5A is highly expressed by CD34+ cord blood progenitors early in eosinophilopoiesis, after 7–10 days of eosinophil development induced *ex vivo* by culture with IL-3 and IL-5,[80] and protein expression is downregulated during the terminal stages of eosinophil differentiation. Ectopic overexpression of STAT5A in CD34+ progenitors by retroviral transduction resulted in both enhanced eosinophil progenitor cell proliferation and accelerated eosinophil terminal differentiation by approximately 27-fold.[80] Additionally, forced expression of STAT5A also enhanced *STAT5A* target gene expression, including apoptosis regulator Bcl-2 (BCL2) and cyclin-dependent kinase inhibitor 1 (p21), suggesting roles in STAT5-mediated eosinophil differentiation. In contrast, retroviral expression of dominant negative STAT5A or B mutants blocked both the progenitor proliferative and eosinophil differentiation responses. These results demonstrate that STAT5 plays a key role in eosinophil differentiation of primary human hematopoietic EoPs downstream of IL-5 signaling, is capable of reprogramming their proliferation and differentiation potential, and thus likely participates in the IL-5-mediated amplification of EoP differentiation leading to eosinophilia in response to allergic or parasitic host immune responses.

Reprogramming by DNA-binding Protein Inhibitors

DNA-binding protein inhibitors (IDs) have been shown to act as repressors of the family of basic helix-loop-helix (HLH) transcription factors, and to play a key role in regulating lymphocyte differentiation. Roles for the ID proteins in regulating granulopoiesis have also been addressed by Buitenhuis and colleagues,[81] who studied the roles of DNA-binding protein inhibitor ID-1 (ID-1) and ID-2 in eosinophil and neutrophil development, using both *ex vivo* transduction of CD34+ hematopoietic progenitors with retroviral expression vectors, and the establishment of human/mouse bone marrow chimeras in immunodeficient mice. Expression of ID-1 was found to be initially upregulated during the earliest stages of granulopoiesis, followed by decreasing expression during terminal maturation. In contrast, expression of ID-2 was upregulated in more terminally differentiated granulocytes. To determine whether ID-1 or ID-2 plays a role in regulating granulocyte development, they were ectopically overexpressed in umbilical cord blood-derived CD34+ HSCs by

transduction with bicistronic, ID/green fluorescent protein (GFP) retroviral expression vectors. Using FACS-sorted CD34+/GFP+ transduced cells, this approach showed that constitutive expression of ID-1 inhibited eosinophil differentiation in response to IL-5, whereas neutrophil development was modestly enhanced. In contrast, constitutive ID-2 expression accelerated the terminal maturation of both eosinophils and neutrophils, whereas knockdown of ID-2 expression blocked the differentiation of both lineages. Transplantation of NOD/SCID mice with retrovirally transduced human CD34+/GFP+ cells ectopically expressing ID-1 enhanced neutrophil development, compared to ectopically overexpressed ID-2, which induced both eosinophil and neutrophil development. Taken together, these *ex vivo* and *in vivo* models indicate that: (1) both ID-1 and ID-2 participate in regulating granulopoiesis, albeit through differential roles in eosinophil vs. neutrophil development; and (2) reprogramming of the eosinophil's differentiation program can occur fairly late in its developmental pathway.[81]

Reprogramming by Phosphatidylinositol 3-Kinase Signaling and CCAAT/Enhancer-Binding Protein Alpha Phosphorylation

Phosphatidylinositol 3-kinase (PI3K) signaling participates in the regulation of normal hematopoiesis, with dysregulated PI3K signaling identified in a number of leukemias. Using similar approaches as for STAT5 and the ID proteins, Buitenhuis and colleagues[82] also explored roles for PI3K signaling and its downstream target, protein kinase B/proto-oncogene c-Akt (PKB/c-Akt), in eosinophil (and neutrophil) development. In cord blood CD34+ HSCs, PI3K activity was found essential for hematopoietic progenitor cell survival. In cord blood CD34+ HSCs transduced with a PKB/c-Akt retroviral expression vector, the ectopically elevated PKB activity promoted neutrophil and monocyte development, whereas a reduction in PKB activity optimally induced eosinophil differentiation. Furthermore, bone marrow transplantation of NOD/SCID mice with human cord blood CD34+ cells ectopically expressing constitutively active PKB/c-Akt enhanced both neutrophil and monocyte, in contrast to an inactive dominant negative PKB/c-Akt, which induced eosinophil differentiation. The transcription factor C/EBPα is required for both eosinophil and neutrophil development, as the development of both lineages is impaired in the C/EBPα knockout mouse.[83] Of note, ectopic expression of PKB/c-Akt in CD34+ hematopoietic progenitors also abrogated threonine 222/226 inhibitory phosphorylation of C/EBPα, while expression of a non-phosphorylatable C/EBPα mutant inhibited eosinophil differentiation and induced neutrophil development, identifying a role for PKB/

c-Akt-mediated C/EBPα phosphorylation in regulating eosinophil vs. neutrophil development.[82] These studies have demonstrated a novel role for the PKB/c-Akt kinase in regulating cell fate decisions, including eosinophil development, in hematopoietic lineage commitment and differentiation.[84]

Reprogramming Eosinophil Lineage Commitment by the FIP1-Like 1 Protein-Platelet-Derived Growth Factor Receptor Alpha Fusion Protein in Chronic Eosinophilic Leukemia

HES was historically confused with chronic eosinophilic leukemia (CEL) because distinguishing between the malignant vs. nonmalignant causes of hypereosinophilia is often quite difficult. Molecular explanations for the development of HES and CEL were elusive until it was reported that some patients with HES or CEL responded to a tyrosine kinase inhibitor—imatinib mesylate—with complete hematological and cytogenetic remissions at fourfold lower doses than typically used to treat chronic myeloid leukemia.[85,86] This suggested that these patients might possess abnormal gene fusions or activating mutations that generate novel tyrosine kinase targets of imatinib mesylate, sparking a *reverse* bedside-to-bench translational research effort that identified an activated tyrosine kinase gene fusion on chromosome 4q12, producing a novel interstitial chromosomal deletion of the cysteine-rich hydrophobic domain 2 protein (CHIC2) domain that fused an uncharacterized human gene *FIP1 like 1 (S. cerevisiae) (FIP1L1)* to the *platelet-derived growth factor receptor, alpha polypeptide (PDGFRA)* gene.[87] The mechanism by which the *FIP1L1−PDGFRA* gene fusion leads to the proliferation/differentiation of eosinophils over other myeloid lineages has been investigated by a number of groups, with initial investigations suggesting that STAT5 may be one of the downstream targets of FIP1L1−PDGF-R-alpha tyrosine kinase activity.[88] As discussed previously, retrovirus-mediated ectopic overexpression of STAT5A in cord blood-derived CD34[+] HSCs selectively amplified their commitment, proliferation, and terminal differentiation to the eosinophil lineage *ex vivo* in the presence of IL-5,[80] and STAT5 activation by tyrosine kinases such as active breakpoint cluster region-related protein (ABR)−tyrosine-protein kinase ABL1 (ABL1) could contribute to the transformation of leukemic cells.[89] In a murine transgenic model, the *FIP1L1−PDGFRA* fusion gene was not sufficient to induce an HES/CEL-like disease independently, but required a second event, i.e., the overexpression of IL-5.[90]

Fukushima and colleagues[91] went on to define the molecular mechanism underlying the development of HES/CEL by FIP1L1−PDGFRA. When introduced into lineage-negative (Lin[−]) stem cell antigen 1 (Sca-1[+]) proto-oncogene c-Kit (c-Kit[high]) murine progenitors, FIP1L1−PDGFRA conferred cytokine-independent growth and enhanced self-renewal, failed to immortalize the transduced CMPs, but significantly enhanced the development of IL-5R[+] EoPs from Lin[−] Sca-1[+]c-Kit[high] progenitors and promoted eosinophil development. Of note, when the *FIP1L1−PDGFRA* fusion was ectopically overexpressed in MEP progenitors or CLPs, it inhibited their differentiation to these lineages and reprogrammed their development to eosinophils. The mechanism for this FIP1L1−PDGFRA-induced eosinophil differentiation involved the activation of the kinase's dual specificity mitogen-activated protein kinase kinase 1 (MAPKK 1) and 2 (MAPKK 2) and p38 mitogen-activated protein kinase (p38 MAPK), augmented expression of C/EBPα, GATA-1, and GATA-2, and rat sarcoma (Ras)-mediated inhibition of PU.1 transcriptional activity. Additionally, small hairpin RNAs against C/EBPα and GATA-2, and a dominant negative GATA-3 KRR (lysine-arginine-arginine) mutant that inhibits all GATA family members, inhibited FIP1L1−PDGFRA-induced eosinophil development. Thus, the *FIP1L1−PDGFRA* fusion enhances eosinophil development by modifying the expression and activity of eosinophil lineage-specific transcription factors that include the C/EBPs (C/EBPα and C/EBPε), GATA-1 and -2, and PU.1 through the modulation of the Ras/MAPKK1/2 and p38 MAPK cascades in hematopoietic progenitors;[91] this provides yet another important example for how the eosinophil's basal developmental program is regulated in a cytokine (IL-5)-independent manner through the induction of an eosinophil lineage-specific transcription factor network.

CONCLUSION

The network of transcription factors, their coactivators, and corepressors that interact in HSCs and early myeloid progenitors to resolve lineage promiscuity toward eosinophil development through transcription factor antagonism, autoregulation, and synergy[5−7,26] have now been defined at a basic level.[3] However, the question of how the hierarchical expression of transcription factors is initially regulated in HSCs and early progenitors has not been resolved. Whether this process is simply stochastic or is selectively regulated at the level of the hematopoietic stem cell niche, through interactions with bone marrow microenvironmental or other factors, remains to be determined. Greater understanding of the complex combinatorial and functional interactions of the transcription factors that regulate eosinophil lineage commitment from the CMP to the EoP, eosinophil terminal differentiation, functional maturation, postmitotic activation in disease, and lineage-specific gene expression during these processes may lead to novel targets for ablating eosinophil development in general, or selectively knocking down eosinophil expression of key inflammatory mediators, e.g., granule cationic proteins or the eotaxin receptor C-C chemokine receptor type 3 (CCR3), as novel therapeutic approaches to treating eosinophil-mediated allergic diseases including asthma, eosinophilic esophagitis, and other hypereosinophilic syndromes.

ACKNOWLEDGEMENTS

This work has been supported in part by National Institutes of Health (NIH) grant R21AI079925, a Translational Research Award from the American Gastroenterology Association (AGA), the Campaign Urging Research for Eosinophilic Diseases (CURED), and the Thrasher Research Foundation.

REFERENCES

1. Miyamoto T, Iwasaki H, Reizis B, et al. Myeloid or lymphoid promiscuity as a critical step in hematopoietic lineage commitment. *Dev Cell* 2002 Jul;**3**(1):137–47.

2. Iwasaki H, Mizuno S, Arinobu Y, et al. The order of expression of transcription factors directs hierarchical specification of hematopoietic lineages. *Genes Dev* 2006 Nov 1;**20**(21):3010–21.

3. McNagny K, Graf T. Making eosinophils through subtle shifts in transcription factor expression. *J Exp Med* 2002 Jun 3;**195**(11):F43–7.

4. Cantor AB, Orkin SH. Hematopoietic development: a balancing act. *Curr Opin Genet Dev* 2001 Oct;**11**(5):513–9.

5. Miyamoto T, Akashi K. Lineage promiscuous expression of transcription factors in normal hematopoiesis. *Int J Hematol* 2005 Jun;**81**(5):361–7.

6. Singh H, Medina KL, Pongubala JM. Contingent gene regulatory networks and B cell fate specification. *Proc Natl Acad Sci U S A* 2005 Apr 5;**102**(14):4949–53.

7. Singh H. Shaping a helper T cell identity. *Nat Immunol* 2007 Feb;**8**(2):119–20.

8. Mori Y, Iwasaki H, Kohno K, et al. Identification of the human eosinophil lineage-committed progenitor: revision of phenotypic definition of the human common myeloid progenitor. *J Exp Med* 2009 Jan 16;**206**(1):183–93.

9. Iwasaki H, Mizuno S, Mayfield R, et al. Identification of eosinophil lineage-committed progenitors in the murine bone marrow. *Journal of Experimental Medicine* 2005;**201**(12):1891–7.

10. Manz MG, Miyamoto T, Akashi K, Weissman IL. Prospective isolation of human clonogenic common myeloid progenitors. *Proc Natl Acad Sci U S A* 2002 Sep 3;**99**(18):11872–7.

11. Takatsu K, Tominaga A, Harada N, et al. T cell-replacing factor (TRF)/interleukin 5 (IL-5): molecular and functional properties. *Immunol Rev* 1988;**102**:107–35.

12. Plaut M, Pierce JH, Watson CJ, Hanley-Hyde J, Nordan RP, Paul WE. Mast cell lines produce lymphokines in response to cross-linkage of Fc epsilon RI or to calcium ionophores. *Nature* 1989;**339**(6219):64–7.

13. Galli SJ, Gordon JR, Wershil BK, Elovic A, Wong DT, Weller PF. Mast cell and eosinophil cytokines in allergy and inflammation. In: Kay AB, Gleich GJ, editors. *Eosinophils in Allergy and Inflammation*. New York: Marcel Dekker; 1994. p. 255–80.

14. Palacios R, Karasuyama H, Rolink A. Ly1+ PRO-B lymphocyte clones. Phenotype, growth requirements and differentiation in vitro and in vivo. *Embo J* 1987;**6**(12):3687–93.

15. Takatsu K, Yamaguchi N, Hitoshi Y, Sonoda E, Mita S, Tominaga A. Signal transduction through interleukin-5 receptors. *Cold Spring Harb Symp Quant Biol* 1989;**2**:745–51.

16. Yamaguchi Y, Suda T, Suda J, et al. Purified interleukin 5 supports the terminal differentiation and proliferation of murine eosinophilic precursors. *J Exp Med* 1988;**167**(1):43–56.

17. Sanderson CJ. Eosinophil differentiation factor (interleukin-5). *Immunol Ser* 1990;**49**:231–56.

18. Sanderson CJ. Control of eosinophilia. *Int Arch Allergy Appl Immunol* 1991;**94**(1-4):122–6.

19. Kopf M, Brombacher F, Hodgkin PD, et al. IL-5-deficient mice have a developmental defect in CD5+ B-1 cells and lack eosinophilia, but have normal antibody and cytotoxic T cell responses. *Immunity* 1996;**4**:15–24.

20. Owen WF, Rothenberg ME, Petersen J, et al. Interleukin 5 and phenotypically altered eosinophils in the blood of patients with the idiopathic hypereosinophilic syndrome. *J Exp Med* 1989;**170**(1):343–8.

21. Owen Jr W, Petersen J, Sheff DM, et al. Hypodense eosinophils and interleukin 5 activity in the blood of patients with the eosinophilia-myalgia syndrome. *Proc Natl Acad Sci U S A* 1990;**87**(21):8647–51.

22. Sanderson CJ. Interleukin-5, eosinophils, and disease. *Blood* 1992;**79**(12):3101–9.

23. Dent LA, Strath M, Mellor AL, Sanderson CJ. Eosinophilia in transgenic mice expressing interleukin-5. *J Exp Med* 1990;**172**(5):1425–31.

24. Tominaga A, Takaki S, Koyama N, et al. Transgenic mice expressing a B cell growth and differentiation factor gene (interleukin 5) develop eosinophilia and autoantibody production. *J Exp Med* 1991;**173**(2):429–37.

25. Foster PS, Hogan SP, Ramsay AJ, Matthaei KI, Young IG. Interleukin 5 deficiency abolishes airways eosinophilia, airways hyper-reactivity, and lung damage in mouse asthma model. *J Exp Med* 1996;**183**:195–201.

26. Akashi K. Lineage promiscuity and plasticity in hematopoietic development. *Ann N Y Acad Sci* 2005 Jun;**1044**:125–31.

27. Yamaguchi Y, Nishio H, Kishi K, Ackerman SJ, Suda T. C/EBPbeta and GATA-1 synergistically regulate activity of the eosinophil granule major basic protein promoter: implication for C/EBPbeta activity in eosinophil gene expression. *Blood* 1999;**94**(4):1429–39.

28. Querfurth E, Schuster M, Kulessa H, et al. Antagonism between C/EBPbeta and FOG in eosinophil lineage commitment of multipotent hematopoietic progenitors. *Genes and Development* 2000;**14**(19):2515–25.

29. Hirasawa R, Shimizu R, Takahashi S, et al. Essential and instructive roles of GATA factors in eosinophil development. *J Exp Med* 2002 Jun 3;**195**(11):1379–86.

30. Yu C, Cantor AB, Yang H, et al. Targeted deletion of a high-affinity GATA-binding site in the GATA-1 promoter leads to selective loss of the eosinophil lineage in vivo. *J Exp Med* 2002;**195**(11):1387–95.

31. Simson L, Ellyard JI, Dent LA, et al. Regulation of carcinogenesis by IL-5 and CCL11: a potential role for eosinophils in tumor immune surveillance. *J Immunol* 2007 Apr 1;**178**(7):4222–9.

32. Mishra A, Wang M, Pemmaraju VR, et al. Esophageal remodeling develops as a consequence of tissue specific IL-5-induced eosinophilia. *Gastroenterology* 2008 Jan;**134**(1):204—14.

33. Humbles AA, Lloyd CM, McMillan SJ, et al. A critical role for eosinophils in allergic airways remodeling. *Science* 2004;**305**(5691):1776—9.

34. Percopo CM, Qiu Z, Phipps S, Foster PS, Domachowske JB, Rosenberg HF. Pulmonary eosinophils and their role in immunopathologic responses to formalin-inactivated pneumonia virus of mice. *J Immunol* 2009 Jul 1;**183**(1):604—12.

35. Du J, Stankiewicz MJ, Liu Y, et al. Novel combinatorial interactions of GATA-1, PU.1, and C/EBPepsilon isoforms regulate transcription of the gene encoding eosinophil granule major basic protein. *J Biol Chem* 2002 Nov 8;**277**(45):43481—94.

36. Zimmermann N, Colyer JL, Koch LE, Rothenberg ME. Analysis of the CCR3 promoter reveals a regulatory region in exon 1 that binds GATA-1. *BMC Immunol* 2005;**6**(1):7.

37. Yamanaka R, Barlow C, Lekstrom-Himes J, et al. Impaired granulopoiesis, myelodysplasia, and early lethality in CCAAT/enhancer binding protein epsilon-deficient mice. *Proc Natl Acad Sci U S A* 1997;**94**(24):13187—92.

38. Guyot B, Valverde-Garduno V, Porcher C, Vyas P. Deletion of the major GATA1 enhancer HS 1 does not affect eosinophil GATA1 expression and eosinophil differentiation. *Blood* 2004 Jul 1;**104**(1):89—91.

39. Lekstrom-Himes JA. The role of C/EBP(epsilon) in the terminal stages of granulocyte differentiation. *Stem Cells* 2001;**19**(2):125—33.

40. Gombart AF, Kwok SH, Anderson KL, Yamaguchi Y, Torbett BE, Koeffler HP. Regulation of neutrophil and eosinophil secondary granule gene expression by transcription factors C/EBP epsilon and PU.1. *Blood* 2003 Apr 15;**101**(8):3265—73.

41. Bedi R, Du J, Sharma AK, Gomes I, Ackerman SJ. Human C/EBP-epsilon activator and repressor isoforms differentially reprogram myeloid lineage commitment and differentiation. *Blood* 2009 Jan 8;**113**(2):317—27.

42. Scott EW, Simon MC, Anastasi J, Singh H. Requirement of transcription factor PU.1 in the development of multiple hematopoietic lineages. *Science* 1994;**265**(5178):1573—7.

43. McKercher SR, Torbett BE, Anderson KL, et al. Targeted disruption of the PU.1 gene results in multiple hematopoietic abnormalities. *Embo J* 1996;**15**(20):5647—58.

44. Anderson KL, Smith KA, Conners K, McKercher SR, Maki RA, Torbett BE. Myeloid development is selectively disrupted in PU.1 null mice. *Blood* 1998;**91**(10):3702—10.

45. Ackerman SJ, Du J, Xin F, et al. Eosinophilopoeisis: To be or not to be (an eosinophil)? That is the question: transcriptional mechanisms regulating eosinophil genes and development. *Respiratory Medicine* 2000;**94**(11):1135—40.

46. Scott EW, Fisher RC, Olson MC, Kehrli EW, Simon MC, Singh H. PU.1 functions in a cell-autonomous manner to control the differentiation of multipotential lymphoid-myeloid progenitors. *Immunity* 1997;**6**(4):437—47.

47. Fisher RC, Scott EW. Role of PU.1 in hematopoiesis. *Stem Cells* 1998;**16**(1):25—37.

48. Simon MC, Olson M, Scott E, Hack A, Su G, Singh H. Terminal myeloid gene expression and differentiation requires the transcription factor PU.1. *Curr Top Microbiol Immunol* 1996;**211**:113—9.

49. Yamanaka R, Lekstrom-Himes J, Barlow C, Wynshaw-Boris A, Xanthopoulos KG. CCAAT/enhancer binding proteins are critical components of the transcriptional regulation of hematopoiesis (Review). *Int J Mol Med* 1998;**1**(1):213—21.

50. Yamanaka R, Kim GD, Radomska HS, et al. CCAAT/enhancer binding protein epsilon is preferentially up-regulated during granulocytic differentiation and its functional versatility is determined by alternative use of promoters and differential splicing. *Proc Natl Acad Sci U S A* 1997;**94**(12):6462—7.

51. Lekstrom-Himes JA. The role of C/EBP(epsilon) in the terminal stages of granulocyte differentiation. *Stem Cells* 2001;**19**(2):125—33.

52. Rosenberg HF, Gallin JI. Neutrophil-specific granule deficiency includes eosinophils. *Blood* 1993;**82**(1):268—73.

53. Williams SC, Du Y, Schwartz RC, et al. C/EBPepsilon is a myeloid-specific activator of cytokine, chemokine, and macrophage-colony-stimulating factor receptor genes. *J Biol Chem* 1998 May 29;**273**(22):13493—501.

54. Williamson EA, Xu HN, Gombart AF, et al. Identification of transcriptional activation and repression domains in human CCAAT/enhancer-binding protein epsilon. *J Biol Chem* 1998;**273**(24):14796—804.

55. Chumakov AM, Grillier I, Chumakova E, Chih D, Slater J, Koeffler HP. Cloning of the novel human myeloid-cell-specific C/EBP-epsilon transcription factor. *Mol Cell Biol* 1997;**17**(3):1375—86.

56. Chih DY, Chumakov AM, Park DJ, Silla AG, Koeffler HP. Modulation of mRNA expression of a novel human myeloid-selective CCAAT/enhancer binding protein gene (C/EBP epsilon). *Blood* 1997;**90**(8):2987—94.

57. Morosetti R, Park DJ, Chumakov AM, et al. A novel, myeloid transcription factor, C/EBP epsilon, is upregulated during granulocytic, but not monocytic, differentiation. *Blood* 1997;**90**(7):2591—600.

58. Bjerregaard MD, Jurlander J, Klausen P, Borregaard N, Cowland JB. The in vivo profile of transcription factors during neutrophil differentiation in human bone marrow. *Blood* 2003 Jun 1;**101**(11):4322—32.

59. Verbeek W, Gombart AF, Chumakov AM, Muller C, Friedman AD, Koeffler HP. C/EBPepsilon directly interacts with the DNA binding domain of c-myb and cooperatively activates transcription of myeloid promoters. *Blood* 1999 May 15;**93**(10):3327—37.

60. Stankiewicz MJ, Du J, Ackerman SJ. CCAAT/Enhancer-binding protein epsilon[27] antagonism of GATA-1 transcriptional activity is mediated by a unique N-terminal repression domain, is independent of sumoylation and does not require DNA-binding. *Exp Hematol*; 2011. Manuscript in preparation.

61. Angerer ND, Du Y, Nalbant D, Williams SC. A short conserved motif is required for repressor domain function in the myeloid-specific transcription factor CCAAT/enhancer-binding protein epsilon. *J Biol Chem* 1999;**274**(7):4147—54.

62. Chumakov AM, Silla A, Williamson EA, Koeffler HP. Modulation of DNA binding properties of CCAAT/enhancer binding protein epsilon by heterodimer formation and interactions with NFkappaB pathway. *Blood* 2007 May 15;**109**(10):4209—19.

63. Kim J, Sharma S, Li Y, Cobos E, Palvimo JJ, Williams SC. Repression and coactivation of CCAAT/enhancer-binding protein epsilon by sumoylation and protein inhibitor of activated STATx proteins. *J Biol Chem* 2005 Apr 1;**280**(13):12246—54.

64. Kim J, Cantwell CA, Johnson PF, Pfarr CM, Williams SC. Transcriptional activity of CCAAT/enhancer-binding proteins is controlled by a conserved inhibitory domain that is a target for sumoylation. *J Biol Chem* 2002 Oct 11;**277**(41):38037—44.

65. Stankiewicz MJ, Du J, Ackerman SJ. Use of HIV TAT-transcription factor fusion proteins to study mechanisms of myeloid gene

regulation: Transcriptional repression by C/EBPε[27] and C/EBPε[14] in the eosinophil lineage. *Blood* 2003;**102**(11):573.

66. Kopf M, Brombacher F, Hodgkin PD, et al. IL-5-deficient mice have a developmental defect in CD5+ B-1 cells and lack eosinophilia but have normal antibody and cytotoxic T *cell responses. Immunity* 1996 Jan;**4**(1):15−24.

67. Yoshida T, Ikuta K, Sugaya H, et al. Defective B-1 cell development and impaired immunity against Angiostrongylus cantonensis in IL-5R alpha-deficient mice. *Immunity* 1996 May;**4**(5):483−94.

68. Foster PS, Mould AW, Yang M, et al. Elemental signals regulating eosinophil accumulation in the lung. *Immunol Rev* 2001 Feb;**179**:173−81.

69. Shen HH, Ochkur SI, McGarry MP, et al. A causative relationship exists between eosinophils and the development of allergic pulmonary pathologies in the mouse. *J Immunol* 2003 Mar 15;**6**(170): 3296−305.

70. Wang QF, Friedman AD. CCAAT/enhancer-binding proteins are required for granulopoiesis independent of their induction of the granulocyte colony-stimulating factor receptor. *Blood* 2002 Apr 15;**99**(8):2776−85.

71. Gruart V, Truong MJ, Plumas J, et al. Decreased expression of eosinophil peroxidase and major basic protein messenger RNAs during eosinophil maturation. *Blood* 1992;**79**(10):2592−7.

72. Weiss MJ, Orkin SH. GATA transcription factors: key regulators of hematopoiesis. *Exp Hematol* 1995 Feb;**23**(2):99−107.

73. Zon LI, Yamaguchi Y, Yee K, et al. Expression of mRNA for the GATA-binding proteins in human eosinophils and basophils: potential role in gene transcription. *Blood* 1993;**81**(12):3234−41.

74. Yamaguchi Y, Ackerman SJ, Minegishi N, Takiguchi M, Yamamoto M, Suda T. Mechanisms of transcription in eosinophils: GATA-1, but not GATA-2, transactivates the promoter of the eosinophil granule major basic protein gene. *Blood* 1998;**91**(9):3447−58.

75. Tsai FY, Orkin SH. Transcription factor GATA-2 is required for proliferation/survival of early hematopoietic cells and mast cell formation, but not for erythroid and myeloid terminal differentiation. *Blood* 1997;**89**(10):3636−43.

76. Hall DJ, Cui J, Bates ME, et al. Transduction of a dominant-negative H-Ras into human eosinophils attenuates extracellular signal-regulated kinase activation and interleukin-5-mediated cell viability. *Blood* 2001;**98**(7):2014−21.

77. Bhattacharya S, Stout BA, Bates ME, Bertics PJ, Malter JS. Granulocyte macrophage colony-stimulating factor and interleukin-5 activate STAT5 and induce CIS1 mRNA in human peripheral blood eosinophils. *Am J Respir Cell Mol Biol* 2001;**24**(3):312−6.

78. Stout BA, Bates ME, Liu LY, Farrington NN, Bertics PJ. IL-5 and granulocyte-macrophage colony-stimulating factor activate STAT3 and STAT5 and promote Pim-1 and cyclin D3 protein expression in human eosinophils. *J Immunol* 2004 Nov 15;**173**(10):6409−17.

79. Caldenhoven E, van Dijk TB, Tijmensen A, et al. Differential activation of functionally distinct STAT5 proteins by IL-5 and GM-CSF during eosinophil and neutrophil differentiation from human CD34+ hematopoietic stem cells. *Stem Cells* 1998;**16**(6):397−403.

80. Buitenhuis M, Baltus B, Lammers JW, Coffer PJ, Koenderman L. Signal transducer and activator of transcription 5a (STAT5a) is required for eosinophil differentiation of human cord blood-derived CD34+ cells. *Blood* 2003 Jan 1;**101**(1):134−42.

81. Buitenhuis M, van Deutekom HW, Verhagen LP, et al. Differential regulation of granulopoiesis by the basic helix-loop-helix transcriptional inhibitors Id1 and Id2. *Blood* 2005 Jun 1;**105**(11):4272−81.

82. Buitenhuis M, Verhagen LP, van Deutekom HW, et al. Protein kinase B (c-akt) regulates hematopoietic lineage choice decisions during myelopoiesis. *Blood* 2008 Jan 1;**111**(1):112−21.

83. Zhang DE, Zhang P, Wang ND, Hetherington CJ, Darlington GJ, Tenen DG. Absence of granulocyte colony-stimulating factor signaling and neutrophil development in CCAAT enhancer binding protein alpha-deficient mice. *Proceedings of the National Academy of Sciences of the United States of America* 1997;**94**(2):569−74.

84. Buitenhuis M, Coffer PJ. The role of the PI3K-PKB signaling module in regulation of hematopoiesis. *Cell Cycle* 2009 Feb 15;**8**(4):560−6.

85. Gleich GJ, Leiferman KM, Pardanani A, Tefferi A, Butterfield JH. Treatment of hypereosinophilic syndrome with imatinib mesilate. *Lancet* 2002 May 4;**359**(9317):1577−8.

86. Gotlib J, Cools J, Malone 3rd JM, Schrier SL, Gilliland DG, Coutre SE. The FIP1L1-PDGFRalpha fusion tyrosine kinase in hypereosinophilic syndrome and chronic eosinophilic leukemia: implications for diagnosis, classification, and management. *Blood* 2004 Apr 15;**103**(8):2879−91.

87. Cools J, DeAngelo DJ, Gotlib J, et al. A tyrosine kinase created by fusion of the PDGFRA and FIP1L1 genes as a therapeutic target of imatinib in idiopathic hypereosinophilic syndrome. *N Engl J Med* 2003 Mar 27;**348**(13):1201−14.

88. Gotlib J, Cools J. Five years since the discovery of FIP1L1-PDGFRA: what we have learned about the fusion and other molecularly defined eosinophilias. *Leukemia* 2008 Nov;**22**(11):1999−2010.

89. de Groot RP, Raaijmakers JA, Lammers JW, Jove R, Koenderman L. STAT5 activation by BCR-Abl contributes to transformation of K562 leukemia cells. *Blood* 1999;**94**(3):1108−12.

90. Yamada Y, Rothenberg ME, Lee AW, et al. The FIP1L1-PDGFRA fusion gene cooperates with IL-5 to induce murine hypereosinophilic syndrome (HES)/chronic eosinophilic leukemia (CEL)-like disease. *Blood* 2006 May 15;**107**(10):4071−9.

91. Fukushima K, Matsumura I, Ezoe S, et al. FIP1L1-PDGFRalpha imposes eosinophil lineage commitment on hematopoietic stem/progenitor cells. *J Biol Chem* 2009 Mar 20;**284**(12):7719−32.

Chapter 5.3

Eosinophil Lineage-Committed Progenitors

Yasuo Mori, Hiromi Iwasaki and Koichi Akashi

INTRODUCTION

Eosinophils normally constitute only 1−3% of nucleated blood cells. They mainly reside in the gastrointestinal mucosa and may contribute not only to the host defense against microorganisms, but also toward the regulation of T-helper type 1 and 2 (T_h1, T_h2) balance, as well as the negative selection of developing thymocytes in the post-natal period.[1] Eosinophils are also involved in particular pathological conditions, such as allergic disorders (e.g., asthma, atopic dermatitis, and inflammatory bowel disease) and myeloproliferative neoplasms [e.g., *clonal chronic eosinophilic leukemia (CEL)* and *idiopathic hypereosinophilic syndrome (HES)*]. In these disease settings, eosinophils produce and release chemical mediators and cytokines, resulting in inflammatory tissue damage. Although corticosteroids and/or neutralizing anti-interleukin-5 (IL-5) antibodies have been used to control eosinophil numbers, these drugs are not fully effective at inhibiting eosinophil-mediated tissue inflammation. Thus, more powerful therapeutic strategies that control amplified eosinophil development are required. To this end, it is critical to identify the eosinophil lineage-committed progenitors (EoPs).

Eosinophils originate from hematopoietic stem cells (HSCs) in the bone marrow, whereas their downstream developmental pathway remains controversial. Some investigators reported the existence of hybrid eosinophil/basophil granulocytes, which suggests a close relationship between the eosinophil and basophil lineages.[2] On the other hand, according to *in vitro* colony assays, eosinophils develop simultaneously with erythrocytes[3] or myelomonocytes[4] in single colonies. However, retrospective assessments of progenitor functions based on lineage read-outs of colony assays do not necessarily reflect their full lineage potential, because multipotent progenitors (MPPs) could decide lineage fates in a stochastic manner at least *in vitro*.[5] Again, it is critical to isolate the EoP population prospectively in the hematopoietic hierarchy.

The differentiation, survival, and proliferation of hematopoietic cells are supported, at least in part, by extrinsic signals through a variety of cytokine receptors or adhesion molecules. In eosinophil lineage development, cytokines including granulocyte-macrophage colony-stimulating factor (GM-CSF), interleukin-3 (IL-3), and IL-5, whose receptors share a common β subunit (β chain), play pivotal roles. With regard to eosinophil lineage specificity, IL-5 signaling is particularly critical. IL-5 selectively supports eosinophil generation from MPPs *in vitro*,[6] and also promotes terminal differentiation, survival, chemotaxis, and superoxide production by mature eosinophils. Therefore, it is conceivable that putative EoPs already express receptors for IL-5 on their surface to proliferate and differentiate in response to IL-5 signaling.

The lineage commitment of HSCs/MPPs is determined largely in a cell-autonomous manner by altering the expression of lineage-specific transcription factors. Transcription factors of the CCAAT/enhancer-binding protein (C/EBP) family are predominantly expressed in granulocytes, including eosinophils. C/EBPα and C/EBPβ activate the eosinophil-specific *antigen p97 (melanoma associated) identified by monoclonal antibodies 133.2 and 96.5 (MFI2)* gene in chickens.[7,8] C/EBPα-deficient mice lack neutrophils as well as eosinophils[9] and C/EBPε-deficient mice also lack eosinophils.[10] These observations suggest that C/EBP transcription factors play an essential role in eosinophil development and function. GATA-binding factor 1 (GATA-1), which is identified as an indispensable transcription factor for megakaryocyte/erythrocyte development, is also expressed in eosinophils.[11] GATA-1-deficient mice lack eosinophils[12] and the forced expression of GATA-1 or GATA-binding factor 2 (GATA-2) in myeloid progenitors results in the formation of eosinophil colonies at the expense of neutrophil/macrophage colonies.[13] These data also suggest that the expression of GATA factors is a cue for eosinophil lineage commitment from myeloid progenitors.

In this chapter, we summarize the recent advances in delineating the developmental pathway of murine and human eosinophils, highlighting the EoP population. The mechanism of eosinophil lineage commitment is also discussed in view of the transcription factor network. Finally, the possibility of novel therapeutic strategies for eosinophil-mediated disorders that target the EoP population is discussed.

OVERVIEW OF THE DEVELOPMENTAL PATHWAY FOR EOSINOPHILS BASED ON COLONY ASSAYS

In murine hematopoiesis, the most primitive HSCs with a long-term self-renewal potential can be purified as cells with the antigen profile CD90[lo] or CD34[−] cells within the lineage-negative (Lin)[−] stem cell antigen 1 (Sca-1)[+] proto-oncogene c-Kit (c-Kit)[+] (LSK) fraction.[14,15] CD90[−] or CD34[+] LSK cells are short-term HSCs or MPPs that are capable of reconstituting multilineage hematopoiesis only for approximately 3 months. Common lymphoid progenitors (CLPs), which can produce all lymphoid cell types but not myeloid cells, can be isolated as the interleukin-7 receptor subunit alpha (IL-7RA)[+] Lin[−] Sca-1[lo] c-Kit[lo] population.[16] In the myeloid pathway, common myeloid progenitors (CMPs) first emerge in the IL-7RA[−] Lin[−] Sca-1[−/lo] c-Kit[+] fraction with the CD34[+] Fc-gamma RIIa (Fc-gamma-RIIa)[lo] phenotype[17,18] and give rise to downstream granulocyte/macrophage progenitors [GMPs: IL-7R[−] Lin[−] Sca-1[−] c-Kit[+] CD34[+] Fc-gamma RIIb (FcRII-b)/Fc-gamma RIIIa (Fc-gamma RIIIa)[hi]] and megakaryocyte/erythrocyte progenitors [MEPs: IL-7R[−] Lin[−] Sca-1[−] c-Kit[+] CD34[−] Fc-gamma RII-a/IIIa[lo]).[17] CMPs are capable of generating all types of myeloid colonies but not lymphoid cells, whereas GMPs and MEPs are committed to the granulocyte/macrophage and the megakaryocyte/erythrocyte lineages, respectively.[17]

Thereafter, distinct progenitors which possess granulocyte/macrophage and lymphoid potentials but lose megakaryocyte/erythrocyte read-outs are newly identified in the Lin⁻ Sca-1^{lo/+} c-Kit⁺ MPP fraction as the lymphoid-primed multipotent progenitors [LMPPs; Lin⁻ Sca-1⁺ c-Kit⁺ CD34⁺ fms-like tyrosine kinase 3 (Flt3)⁺][19] or the granulocyte/monocyte/lymphoid-restricted progenitors [Lin⁻ Sca-1^{lo} c-Kit⁺ CD34⁺ transcription factor PU.1 (PU.1)⁺][18] (Fig. 5.3.1A).

In limiting dilution assays, one in four HSCs, one in 22 CMPs, and one in 72 GMPs gave rise to eosinophils *in vitro*, whereas eosinophils were not developed in cultures of MEPs or CLPs.[20] Therefore, the eosinophil potential exists along with the granulocyte-macrophage (GM) differentiation pathway from HSCs (Fig. 5.3.1A). In methylcellulose single-cell colony assays, eosinophils were found in approximately 1.5% of single GMP-derived colonies. Approximately half of these colonies were pure eosinophil colonies, whereas the remaining had bipotentiality for eosinophils and neutrophils or tripotentiality for eosinophils, neutrophils, and macrophages.[20] Basophils and mast cells also developed downstream of GMPs independent of the eosinophil lineage via a common basophil/mast cell progenitor (BMCP) population and monopotent progenitors for basophils or mast cells [basophil lineage-committed progenitor (BaP) and mast cell progenitor (MCP)], respectively[21] (Fig. 5.3.1A).

In human hematopoiesis, the Lin⁻ CD34⁺ bone marrow mononuclear cell fraction was subdivided into the CD38⁻ human HSCs (hHSCs) and the CD38⁺ progenitor population.[22] We have reported that the CD34⁺ CD38⁺ myeloid progenitor population can be further fractionated into the human CMPs (hCMPs), the hGMPs, and the hMEPs according to the expression patterns of IL-3RA and CD45RA; hCMPs, hGMPs, and hMEPs were defined as the IL-3RA⁺ CD45RA⁻, the IL-3RA⁺ CD45RA⁺, and the IL-3RA⁻ CD45RA⁻ populations, respectively[23] (Fig. 5.3.1B). In liquid cultures supplemented with a cytokine cocktail, hHSCs and hCMPs generated eosinophils as well as basophils, other GM and megakaryocyte/erythrocyte cells, whereas hGMPs and hMEPs gave rise mainly to neutrophils/monocytes/macrophages and megakaryocyte/erythrocyte cells, respectively, but did not differentiate into eosinophils.[24] In methylcellulose assays, a few proportion (approximately 5%) of hHSC- or hCMP-derived colonies contained eosinophils, whereas hGMP- or hMEP-derived colonies did not. Moreover, a fraction of hCMPs generated pure eosinophil colonies,[24] suggesting that, in human hematopoiesis, progenitors restricted to the eosinophil lineage reside within the hCMP fraction. On the other hand,

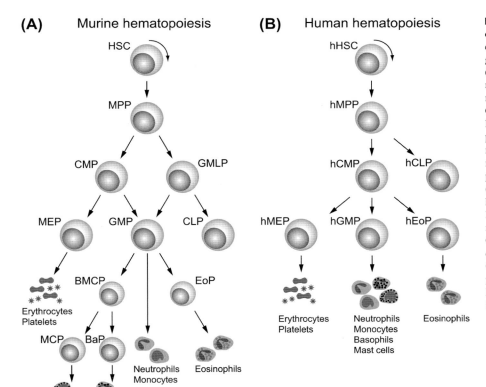

(A) Murine hematopoiesis

(B) Human hematopoiesis

FIGURE 5.3.1 Developmental pathway of eosinophils. *(A)* In murine hematopoiesis, eosinophils develop downstream of granulocyte/macrophage progenitors (GMPs) together with neutrophils and monocytes/macrophages. Basophils and mast cells also develop downstream of GMPs independently of the eosinophil lineage via a common basophil/mast cell progenitor (BMCP) population and monopotent progenitors for basophils or mast cells [basophil lineage-committed progenitor (BaP); mast cell progenitor (MCP)], respectively. *(B)* In contrast, human GMPs (hGMPs) lose the eosinophil lineage potential, whereas common myeloid progenitors (hCMPs) possess it. CLP, common lymphoid progenitor; EoP, eosinophil lineage-committed progenitor; GMLP, granulocyte-monocyte-lymphoid progenitor; HSC, hematopoietic stem cell; MEP, megakaryocyte/erythrocyte progenitor; MPP, multipotent progenitor.

human basophils and mast cells developed downstream of hGMPs (Fig. 5.3.1B).

THE ROLE OF INTERLEUKIN-5 SIGNALING IN EOSINOPHIL DEVELOPMENT

IL-5, initially described as a B-cell differentiation factor that stimulates the production of immunoglobulin A (IgA) and M (IgM) in the mouse, also plays a critical role in eosinophil development in both murine and human hematopoiesis.[1] Mice carrying an *interleukin 5 (colony-stimulating factor, eosinophil) (IL5)* transgene showed an expansion of total white blood cells in the peripheral blood with persistent eosinophilia.[25,26] However, a relatively normal basal eosinophil level was maintained in mice deficient for IL-5[27]. Likewise, in an *in vitro* colony assay, bone marrow progenitors could generate mature eosinophils in the absence of IL-5.[24] In contrast, bone marrow cells expressing ectopic IL-5RA gave rise to various types of myeloid colonies in response to IL-5 alone, without other cytokine signals.[28] These observations suggest that IL-5/IL-5RA signaling does not play an instructive role in eosinophil lineage commitment but likely provides permissive signals for eosinophil differentiation. In this context, IL-5 can specifically stimulate eosinophil development simply because IL-5RA is expressed on condition that the myeloid cells are already committed to the eosinophil lineage. These data strongly suggest that the expression of IL-5RA might be a key positive marker for the purification of progenitors committed to the eosinophil lineage.

IDENTIFICATION OF MOUSE EOSINOPHIL-LINEAGE COMMITTED PROGENITORS

In murine hematopoiesis, purified murine GMPs (mGMPs) can generate a minor population of eosinophils, basophils, and mast cells in addition to neutrophils, monocytes, and macrophages.[20,21] GATA-1 is expressed in mature eosinophils[11] and can accumulate gene expression for eosinophil lineage commitment at the myeloid progenitor stage.[29] Moreover, the nucleotide sequence of the 5′ flanking region of the murine *interleukin 5 receptor, alpha (IL5RA)* gene contains potential binding sites for GATA-1.[30] Thus, murine eosinophil-lineage committed progenitors (mEoPs) were searched downstream of mGMPs using the activation of GATA-1 transcription as a positive marker. Green fluorescent protein (GFP)-tagged GATA-1 transgenic mice were used to purify the early myeloid progenitors activating GATA-1.[20] A small fraction of GATA-1/GFP+ cells was isolated within the progeny of mGMPs after 3-day cultures.

These cells expressed CD34, IL-5RA, and low levels of c-Kit, but not Sca-1. Purified GATA-1/GFP+ IL-5RA+ c-Kit^lo Sca-1− CD34+ cells gave rise exclusively to eosinophils and were regarded as the mEoP population[20] (Fig. 5.3.2A).

Based on the phenotypic characteristics of GATA-1/GFP+ mEoPs, mEoPs could be isolated in normal murine bone marrow by using IL-5RA expression as a positive marker. The IL-5RA+ Lin− c-Kit^lo Sca-1− CD34+ mEoPs constitute approximately 0.05% of total murine bone marrow and can generate exclusively pure eosinophil colonies.[20] The mEoP population expands in response to helminth infection, suggesting that the EoP is the developmental stage for physiological eosinophil production.[20]

IDENTIFICATION OF HUMAN EOSINOPHIL-LINEAGE COMMITTED PROGENITORS

IL-5RA upregulation also occurs in concert with eosinophil development in human hematopoiesis. Committed eosinophil precursors in the human bone marrow were identified by the expression of IL-5RA and C-C chemokine receptor type 3 (CCR3) in addition to CD34.[31] In allergic patients, the CD34+ IL-5RA+ population was considerably increased in their bone marrow.[32] Recently, we evaluated the expression of IL-5RA with human myeloid progenitors.[23] Interestingly, IL-5RA-expressing progenitors resided only within the hCMP population and the IL-5RA+ hCMPs expressed CCR3, though at a low level, in agreement with a previous report.[31] Notably, purified IL-5RA+ hCMPs gave rise only to pure eosinophil colonies in methylcellulose assays. In contrast, the IL-5RA− hCMP population generated a variety of myeloid colonies including rare eosinophil-containing colonies. Thus, the hEoP was defined as the IL-5RA+ CD34+ CD38+ IL-3RA+ CD45RA− population[24] (Fig. 5.3.3A). Mature human basophils also expressed IL-5RA on their surface, and hCMPs and hGMPs gave rise to basophils *in vitro*.[24] Importantly, the IL-5RA+ CD34+ CD38+ population did not give rise to basophils at least *in vitro*, and we still do not know at what stage basophils upregulate IL-5RA. Because all IL-5RA+ CD34+ CD38+ cells have committed to the eosinophil lineage, whereas hGMPs were capable of producing basophils, putative human basophil progenitors (hBaPs) may exist within or downstream of the hGMP population.

This study also showed that the conventional hCMP population defined by our earlier work[23] was heterogeneous and contained at least the hEoP. The true hCMP (i.e., the revised hCMP) resides in the IL-5RA− fraction of conventional hCMPs. The IL-5RA− hCMP gave rise to IL-5RA+ hEoPs, suggesting that eosinophil lineage

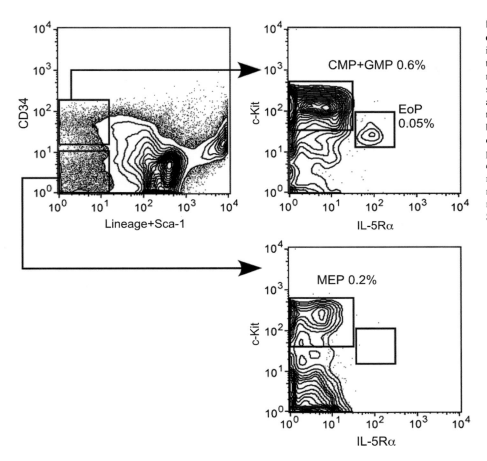

FIGURE 5.3.2 Isolation of the murine eosinophil-committed progenitor. Murine EoPs (mEoPs) can be isolated using the Lin$^-$ Sca-1$^-$ CD34$^+$ (LSK) bone marrow fraction by their specific expression of interleukin-5 receptor subunit alpha (IL-5RA). MEoPs normally constitute approximately 0.05% of the whole bone marrow cell population. c-Kit, proto-oncogene c-Kit; CMP, common myeloid progenitor; EoP, eosinophil lineage-committed progenitor; GMP, granulocyte/macrophage progenitor; Lin$^-$, lineage-negative; Lin$^+$, lineage-positive; MEP, megakaryocyte/erythrocyte progenitor; Sca-1, stem cell antigen 1.

commitment occurs within the IL-5RA$^-$-revised hCMP stage[24] (Fig. 5.3.3B). The hGMP might not completely correspond to the mGMP either, since they possess different lineage potentials; the hGMP lacks the eosinophil potential (Fig. 5.3.1), whereas the mGMP is capable of producing all granulocyte subclasses, including eosinophils and basophils via their progenitor populations, the mEoP and the mBaP, respectively.[21] Therefore, it is also possible that the true hGMP with all granulocyte subclass potentials may reside within the IL-5RA$^-$-revised hCMP fraction (Fig. 5.3.3B). Further studies are required to completely delineate the human myeloid developmental pathways.

MECHANISMS OF EOSINOPHIL LINEAGE COMMITMENT

As described above, IL-5 signaling can stimulate the proliferation and differentiation of cells expressing its receptors, but cannot play an instructive role in eosinophil lineage specification. The enforced IL-5 signaling in mGMP transduced with murine IL-5RA could induce differentiation into the neutrophil/monocyte/macrophage lineages, but not specifically into the eosinophil lineage.[20] Consistent with

these data, both mEoPs and hEoPs generated only eosinophils in the presence of cytokines without mIL-5 or hIL-5, respectively.[24,33] Thus, IL-5R signaling provides permissive but not instructive signals for eosinophil development, as in the case of granulocyte colony-stimulating factor (G-CSF) or GM-CSF signaling in multipotent progenitors to develop myeloid lineage cells.[34,35]

Lineage commitment and subsequent differentiation of multipotent progenitors is likely to involve the selective activation and silencing of a set of genes.[36] Such programs could be triggered by extrinsic signals, intrinsic signals, or by both at different developmental stages, which are ultimately controlled by transcription factors. Transcription factors can play a key role in activating lineage-specific programs dependent on their expression and timing,[37] and multiple transcription factors exert collaborative or competitive actions for cell fate decisions.[36] Chapter 5.2 provides a more detailed description of transcriptional regulation of eosinophil lineage commitment. Here, we describe the transcription factor regulation within the scheme of inter-relationships between myeloid progenitors.

Eosinophil lineage specification is dictated by the interplay of at least three classes of transcription factors

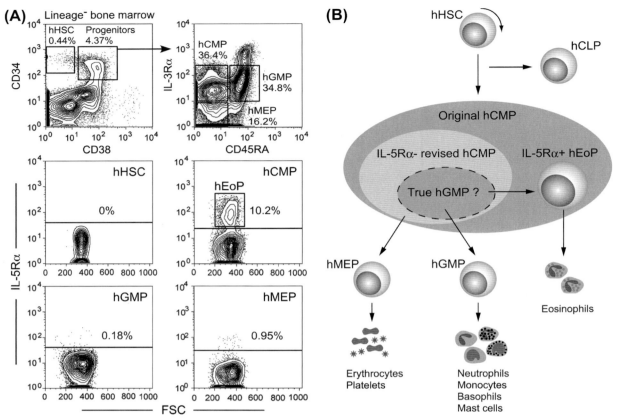

FIGURE 5.3.3   Isolation of the human eosinophil lineage-committed progenitor and redefinition of the human common myeloid progenitor. *(A)* The hEoPs are distinguished from the conventional hCMP by their upregulation of interleukin-5 receptor subunit alpha (IL-5RA). *(B)* The revised hCMP is redefined as an IL-5RA⁻ fraction of the conventional hCMP and gives rise to the hEoP. It is speculated that the true hGMP, which corresponds to the mGMP with regard to its lineage potential, might exist in the revised hCMP population. hCMP, human common myeloid progenitor; EoP, eosinophil lineage-committed progenitor; hGMP, human granulocyte/macrophage progenitor; hHSC, human hematopoietic stem cell; hMEP, human megakaryocyte/erythrocyte progenitor.

including GATAs (zinc finger protein family member), PU.1 [E-twenty-six (ETS)-domain family member], and C/EBPs (CCAAT/enhancer-binding protein family).[8,38] Gene-deleted mice of one of these transcription factors lacked eosinophil differentiation.[9,33,39] In contrast, the enforced expression of GATA-1 or GATA-2 using retroviral vectors could induce eosinophil lineage commitment from myeloid progenitors in both human and mouse.[13,40] Prospectively purified mEoPs and hEoPs expressed both GATA-1 and GATA-2 at high levels.[20,24] Friend of GATA protein 1 (FOG-1) is a critical cofactor of the GATA family of transcription factors and plays an indispensable role in megakaryocyte/erythrocyte lineage development,[41] but in turn, exerts inhibitory effects on eosinophil development.[42] Neither mEoPs nor hEoPs expressed FOG-1.[20,24]

PU.1 is a master transcription factor for both myeloid and lymphoid development. GATA-1 and PU.1 antagonize each other at the bifurcation of the megakaryocyte/erythrocyte GM lineages,[36] but they exert synergistic activities when regulating eosinophil lineage development.[43] The

high-level coexpression of GATA-1 and PU.1 was observed in both mEoPs and hEoPs.[20,24]

C/EBPα is a well-known critical transcription factor for granulocyte development and enhances the transcription of multiple GM-affiliated genes including the *colony stimulating factor 2 receptor, beta, low-affinity (granulocyte-macrophage) (CSF2RB)* gene.[44] The expression level of C/EBPα is elevated along with the GM commitment from CMPs to GMPs, but is shut off in MEPs both in murine and human hematopoiesis.[17,23] C/EBPα-deficient mice lacked eosinophils as well as neutrophils[9] and a cell line study revealed that GATA factors and C/EBPα can cooperatively instruct eosinophil lineage commitment.[38] In murine hematopoiesis, even common lymphoid progenitors can be reprogrammed to EoPs by the sequential upregulation of C/EBPα, which is then followed by the upregulation of GATA-2.[37] Consistent with these observations, hEoPs also express both GATA factors and C/EBPα at high levels.[24]

The cooperative dynamics of lineage-instructive transcription factors during murine myeloerythroid

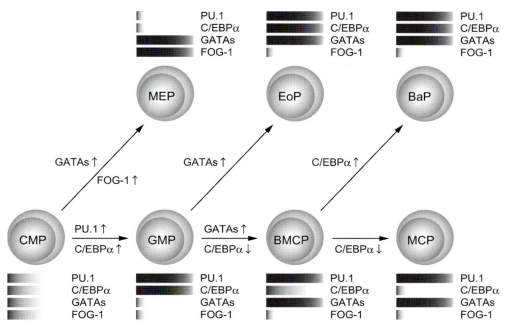

FIGURE 5.3.4 Changes in transcription factor expression during myeloid lineage commitment. Common myeloid progenitors (CMPs) express myeloerythroid lineage-affiliated transcription factors at low levels. Simultaneous upregulation of GATAs (GATA-1 and GATA-2) and FOG-1 accompanied by downregulation of PU.1 and C/EBPα results in the commitment to MEPs. Reciprocal upregulation of PU.1 and C/EBPα leads to the development of GMPs silencing GATAs and FOG-1. After the GMP stage, upregulation of GATAs, but not FOG-1, with maintained expression of PU.1 and C/EBPα instructs to become EoPs. On the other hand, the same upregulation of GATAs accompanied by the downregulation of C/EBPα induces basophil/mast cell lineage development. Furthermore, after the BMCP stage, reactivation or maintained silencing of C/EBPα leads to the basophil or the mast cell lineage specification, respectively. BaP, basophil lineage-committed progenitor; BMCP, basophil/mast cell progenitor; C/EBPα, CCAAT/enhancer-binding protein alpha; EoP, eosinophil lineage-committed progenitor; FOG-1, friend of GATA protein 1; GATA 1, GATA-binding factor 1, GATA-2, GATA-binding factor 2; GMP, granulocyte/macrophage progenitor; MCP, mast cell progenitor; MEP, megakaryocyte/erythrocyte progenitor; PU.1, transcription factor PU.1.

differentiation are summarized in Fig. 5.3.4. The expression patterns exhibited by the major transcription factor families are well conserved between mice and humans, at least as far as the EoP is concerned. In brief, CMPs widely express myeloerythroid-affiliated transcription factors at low levels, reflecting their priming status with a multilineage differentiation potential. Simultaneous upregulation of GATAs (GATA-1 and GATA-2) and FOG-1, accompanied by downregulation of PU.1 and C/EBPα, results in megakaryocyte/erythrocyte lineage commitment. Reciprocal upregulation of PU.1 and C/EBPα leads to the granulocyte/monocyte/macrophage lineage differentiation, silencing the GATAs and FOG-1. After the GMP stage, upregulation of GATAs but not FOG-1 with maintained expression of PU.1 and C/EBPα instructs the eosinophil-lineage differentiation. In contrast, upregulation of GATAs accompanied by downregulation of C/EBPα induces basophil/mast cell lineage development. Furthermore, after the BMCP stage, reactivation or maintained silencing of C/EBPα leads to basophil or mast cell lineage specification, respectively. Further analysis of each purified progenitor population might help understand more critical issues as to how such changes in the expression level of these

transcription factors are regulated to decide commitment into each lineage.

EOSINOPHIL PROGENITORS AS A THERAPEUTIC TARGET FOR EOSINOPHIL-MEDIATED DISORDERS

The newly identified hEoPs appear to contribute toward abnormal eosinophil production in humans. The hEoP population significantly expanded (three- to fourfold) in eosinophilia patients suffering from various diseases, including HES, T-cell malignancies, chronic myeloid leukemia (CML) and allergic granulomatosis (Churg–Strauss syndrome).[24] Moreover, the hEoPs isolated from HES patients generated very large eosinophil colonies, with higher plating efficiency when compared to those from healthy individuals (unpublished data). More recently, we analyzed a CEL patient with a novel fusion gene involving the *platelet-derived growth factor receptor, alpha polypeptide (PDGFRA)* locus. The hEoPs in this patient's bone marrow increased approximately fivefold in numbers when compared to those in normal bone marrow (unpublished data). These observations strongly suggest that the newly identified hEoP plays a critical role in the expansion of eosinophils in both *reactive* and *neoplastic* disorders. Therefore, the hEoP population may be a critical cellular target useful for the control of both reactive and malignant eosinophilia.

The majority of CEL patients are characterized as carrying a deletion or an insertion in chromosome 4q12, which results in the formation of constitutive active *PDGFRA* fusions such as *FIP1 like 1 (S. cerevisiae) FIP1L1—PDGFRA*.[45] In general, CEL patients with *PDGFRA* fusions respond well to the administration of a tyrosine kinase inhibitor, imatinib mesylate (IM), and have a favorable prognosis.[45] However, some patients relapse after the cessation of IM treatment. Since abnormal eosinophil production in these patients is likely to occur at the hEoP stage, it could be critical to target the hEoP population. To this end, it is reasonable to try to identify molecules especially expressed in transformed hEoPs but not in normal myeloid progenitors, as it has been done in leukemic stem cells in acute myeloid leukemia.[46–48] In this context, IL-5RA is a promising molecule because it is expressed strongly in hEoPs but not in other myeloid progenitors. In fact, antihuman IL-5RA antibodies with a strong neutralizing activity[49] have been developed to target hEoPs in both reactive and neoplastic eosinophilia, and phase I studies for asthmatic patients are ongoing. The further profiling and characterization of human EoPs might of course be useful to identify additional critical targets to control the production of eosinophils.

REFERENCES

1. Hogan SP, et al. Eosinophils: biological properties and role in health and disease. *Clin Exp Allergy* 2008;**38**(5):709—50.
2. Boyce JA, et al. Differentiation in vitro of hybrid eosinophil/basophil granulocytes: autocrine function of an eosinophil developmental intermediate. *J Exp Med* 1995;**182**(1):49—57.
3. Nakahata T, Spicer SS, Ogawa M. Clonal origin of human erythro-eosinophilic colonies in culture. *Blood* 1982;**59**(4):857—64.
4. Clutterbuck EJ, Hirst EM, Sanderson CJ. Human interleukin-5 (IL-5) regulates the production of eosinophils in human bone marrow cultures: comparison and interaction with IL-1, IL-3, IL-6, and GMCSF. *Blood* 1989;**73**(6):1504—12.
5. Suda T, Suda J, Ogawa M. Disparate differentiation in mouse hemopoietic colonies derived from paired progenitors. *Proc Natl Acad Sci U S A* 1984;**81**(8):2520—4.
6. Yamaguchi Y, et al. Highly purified murine interleukin 5 (IL-5) stimulates eosinophil function and prolongs in vitro survival. IL-5 as an eosinophil chemotactic factor. *J Exp Med* 1988;**167**(5):1737—42.
7. McNagny KM, et al. Regulation of eosinophil-specific gene expression by a C/EBP-Ets complex and GATA-1. *Embo J* 1998;**17**(13):3669—80.
8. Nerlov C, et al. Distinct C/EBP functions are required for eosinophil lineage commitment and maturation. *Genes Dev* 1998;**12**(15):2413—23.
9. Zhang DE, et al. Absence of granulocyte colony-stimulating factor signaling and neutrophil development in CCAAT enhancer binding protein alpha-deficient mice. *Proc Natl Acad Sci U S A* 1997;**94**(2):569—74.
10. Yamanaka R, et al. Impaired granulopoiesis, myelodysplasia, and early lethality in CCAAT/enhancer binding protein epsilon-deficient mice. *Proc Natl Acad Sci U S A* 1997;**94**(24):13187—92.
11. Zon LI, et al. Expression of mRNA for the GATA-binding proteins in human eosinophils and basophils: potential role in gene transcription. *Blood* 1993;**81**(12):3234—41.
12. Fujiwara Y, et al. Arrested development of embryonic red cell precursors in mouse embryos lacking transcription factor GATA-1. *Proc Natl Acad Sci U S A* 1996;**93**(22):12355—8.
13. Hirasawa R, et al. Essential and instructive roles of GATA factors in eosinophil development. *J Exp Med* 2002;**195**(11):1379—86.
14. Morrison SJ, Weissman IL. The long-term repopulating subset of hematopoietic stem cells is deterministic and isolatable by phenotype. *Immunity* 1994;**1**(8):661—73.
15. Osawa M, et al. Long-term lymphohematopoietic reconstitution by a single CD34-low/negative hematopoietic stem cell. *Science* 1996;**273**(5272):242—5.
16. Kondo M, Weissman IL, Akashi K. Identification of clonogenic common lymphoid progenitors in mouse bone marrow. *Cell* 1997;**91**(5):661—72.
17. Akashi K, et al. A clonogenic common myeloid progenitor that gives rise to all myeloid lineages. *Nature* 2000;**404**(6774):193—7.
18. Arinobu Y, et al. Reciprocal activation of GATA-1 and PU.1 marks initial specification of hematopoietic stem cells into myeloerythroid and myelolymphoid lineages. *Cell Stem Cell* 2007;**1**(4):416—27.
19. Adolfsson J, et al. Identification of Flt3+ lympho-myeloid stem cells lacking erythro-megakaryocytic potential a revised road map for adult blood lineage commitment. *Cell* 2005;**121**(2):295—306.
20. Iwasaki H, et al. Identification of eosinophil lineage-committed progenitors in the murine bone marrow. *J Exp Med* 2005;**201**(12):1891—7.
21. Arinobu Y, et al. Developmental checkpoints of the basophil/mast cell lineages in adult murine hematopoiesis. *Proc Natl Acad Sci U S A* 2005;**102**(50):18105—10.
22. Bhatia M, et al. Purification of primitive human hematopoietic cells capable of repopulating immune-deficient mice. *Proc Natl Acad Sci U S A* 1997;**94**(10):5320—5.
23. Manz MG, et al. Prospective isolation of human clonogenic common myeloid progenitors. *Proc Natl Acad Sci U S A* 2002;**99**(18):11872—7.
24. Mori Y, et al. Identification of the human eosinophil lineage-committed progenitor: revision of phenotypic definition of the human common myeloid progenitor. *J Exp Med* 2009;**206**(1):183—93.
25. Dent LA, et al. Eosinophilia in transgenic mice expressing interleukin 5. *J Exp Med* 1990;**172**(5):1425—31.
26. Tominaga A, et al. Transgenic mice expressing a B cell growth and differentiation factor gene (interleukin 5) develop eosinophilia and autoantibody production. *J Exp Med* 1991;**173**(2):429—37.
27. Kopf M, et al. IL-5-deficient mice have a developmental defect in CD5+ B-1 cells and lack eosinophilia but have normal antibody and cytotoxic T cell responses. *Immunity* 1996;**4**(1):15—24.

28. Takagi M, et al. Multi-colony stimulating activity of interleukin 5 (IL-5) on hematopoietic progenitors from transgenic mice that express IL-5 receptor alpha subunit constitutively. *J Exp Med* 1995;**181**(3):889—99.

29. Heyworth C, et al. Transcription factor-mediated lineage switching reveals plasticity in primary committed progenitor cells. *Embo J* 2002;**21**(14):3770—81.

30. Imamura F, et al. The murine interleukin-5 receptor alpha-subunit gene: characterization of the gene structure and chromosome mapping. *DNA Cell Biol* 1994;**13**(3):283—92.

31. Sehmi R, et al. Allergen-induced fluctuation in CC chemokine receptor 3 expression on bone marrow CD34+ cells from asthmatic subjects: significance for mobilization of haemopoietic progenitor cells in allergic inflammation. *Immunology* 2003;**109**(4):536—46.

32. Sehmi R, et al. Allergen-induced increases in IL-5 receptor alpha-subunit expression on bone marrow-derived CD34+ cells from asthmatic subjects. A novel marker of progenitor cell commitment towards eosinophilic differentiation. *J Clin Invest* 1997;**100**(10):2466—75.

33. Iwasaki H, et al. Distinctive and indispensable roles of PU.1 in maintenance of hematopoietic stem cells and their differentiation. *Blood* 2005;**106**(5):1590—600.

34. Nishijima I, et al. A human GM-CSF receptor expressed in transgenic mice stimulates proliferation and differentiation of hemopoietic progenitors to all lineages in response to human GM-CSF. *Mol Biol Cell* 1995;**6**(5):497—508.

35. Yang FC, et al. Human granulocyte colony-stimulating factor (G-CSF) stimulates the in vitro and in vivo development but not commitment of primitive multipotential progenitors from transgenic mice expressing the human G-CSF receptor. *Blood* 1998;**92**(12):4632—40.

36. Iwasaki H, Akashi K. Myeloid lineage commitment from the hematopoietic stem cell. *Immunity* 2007;**26**(6):726—40.

37. Iwasaki H, et al. The order of expression of transcription factors directs hierarchical specification of hematopoietic lineages. *Genes Dev* 2006;**20**(21):3010—21.

38. McNagny K, Graf T. Making eosinophils through subtle shifts in transcription factor expression. *J Exp Med* 2002;**195**(11):F43—7.

39. Yu C, et al. Targeted deletion of a high-affinity GATA-binding site in the GATA-1 promoter leads to selective loss of the eosinophil lineage in vivo. *J Exp Med* 2002;**195**(11):1387—95.

40. Iwasaki H, et al. GATA-1 converts lymphoid and myelomonocytic progenitors into the megakaryocyte/erythrocyte lineages. *Immunity* 2003;**19**(3):451—62.

41. Tsang AP, et al. FOG, a multitype zinc finger protein, acts as a cofactor for transcription factor GATA-1 in erythroid and mega-karyocytic differentiation. *Cell* 1997;**90**(1):109—19.

42. Querfurth E, et al. Antagonism between C/EBPbeta and FOG in eosinophil lineage commitment of multipotent hematopoietic progenitors. *Genes Dev* 2000;**14**(19):2515—25.

43. Du J, et al. Novel combinatorial interactions of GATA-1, PU.1, and C/EBPepsilon isoforms regulate transcription of the gene encoding eosinophil granule major basic protein. *J Biol Chem* 2002;**277**(45):43481—94.

44. van Dijk TB, et al. A composite C/EBP binding site is essential for the activity of the promoter of the IL-3/IL-5/granulocyte-macrophage colony-stimulating factor receptor beta c gene. *J Immunol* 1999;**163**(5):2674—80.

45. Cools J, et al. A tyrosine kinase created by fusion of the PDGFRA and FIP1L1 genes as a therapeutic target of imatinib in idiopathic hypereosinophilic syndrome. *N Engl J Med* 2003;**348**(13):1201—14.

46. Hosen N, et al. CD96 is a leukemic stem cell-specific marker in human acute myeloid leukemia. *Proc Natl Acad Sci U S A* 2007;**104**(26):11008—13.

47. Jin L, et al. Targeting of CD44 eradicates human acute myeloid leukemic stem cells. *Nat Med* 2006;**12**(10):1167—74.

48. Kikushige Y, et al. TIM-3 is a promising target to selectively kill acute myeloid leukemia stem cells. *Cell Stem Cell* 2010;**7**(6):708—17.

49. Koike M, et al. Establishment of humanized anti-interleukin-5 receptor alpha chain monoclonal antibodies having a potent neutralizing activity. *Hum Antibodies* 2009;**18**(1—2):17—27.

Chapter 5.4

Interleukin-5 and its Receptor Molecules

Kiyoshi Takatsu

INTRODUCTION

Allergic diseases, including asthma and atopic diseases, are characterized by inflammation with pronounced infiltration of T-helper type 2 (T_H2) cells and granulocytes such as mast cells, basophils, eosinophils, and neutrophils.[1] With regard to the inflammatory cells implicated in asthma, immunoglobulin E-mediated degranulation of mast cells contributes to inflammatory infiltrates and acute broncho-constriction in the early phase of allergic inflammation, whereas recruitment of CD4+ T cells and eosinophils is a central feature of the late-phase response.[1,2] Pulmonary allergen exposure results in both increased output of eosinophils from hematopoietic tissues and increased migration to the lung in both mice and humans. Eosinophils are produced and proliferate in the bone marrow under the influence of cytokines and eosinophilia is associated with a wide variety of conditions, including allergic diseases, helminth infection, drug hypersensitivity, and neoplastic disorders. Classical T cell-derived cytokines and eosinophils are thought to play critical roles in the induction of airway hyperreactivity and the development of lesions that underpin chronic airway wall remodeling.[3]

In the 1980s, numerous attempts were made to uncover the molecular nature of T cell-derived cytokines that are involved in eosinophil production. Surprisingly, it was found that T cell-derived B-cell differentiation factor acts on eosinophils to induce proliferation and differentiation.[4]

A murine T-cell hybrid was reported to promote the differentiation of eosinophils [eosinophil differentiation factor (EDF)] and cause the proliferation of murine B-cell leukemia (BCL1). Both activities seemed to be associated with a protein of 44 kDa.[5]

Next, a factor that induces terminal differentiation of activated B cells to Ig-secreting cells was analyzed. Known as T-cell replacing factor (TRF), its activity was originally screened by its ability to support the anti-hapten, immunoglobulin G plaque-forming cell response of T cell-depleted mouse B cells,[6] and TRF activity was found in supernatants of lymph node cell cultures of *Mycobacterium tuberculosis*-primed mice after stimulation with purified protein derivatives (PPD)-presenting cells. The establishment of a TRF-producing T-cell hybrid B151K12 (B151) demonstrated that TRF was a novel cytokine distinct from other cytokines.[7] Subsequently, murine chronic BCL1 leukemia cells growing *in vivo* were shown to differentiate into immunoglobulin M (IgM)-secreting cells when stimulated with TRF-containing B151 supernatant.[8] A homogeneous TRF preparation purified from B151 supernatants was an acidic glycoprotein with a molecular mass of 50–60 kDa, which had a smaller mass (25–30 kDa) under reducing conditions,[8] indicating that bioactive T cell-derived TRF consisted of a homodimer. The purified TRF preparation contained undetectable interleukin-1–4 (IL-1–4) and interferon gamma (IFN-γ) activity, and the growth-promoting activity [=B-cell growth factor II (BCGF-II) activity] of BCL1 cells.[9] BCGF-II activity always resided in the same fraction where TRF activity was detected, suggesting that a single molecule was able to induce B-cell growth and differentiation. It was then possible to generate monoclonal antibodies against mouse TRF.[10]

The molecular cloning of cDNA-encoding mouse Trf and expression and functional studies have convincingly demonstrated that a single molecule is responsible for both TRF and BCGF-II activity.[11] Furthermore, recombinant TRF has been shown to exert pleiotropic activities on various target cells besides B cells, including T cells and eosinophils.[8] Anti-TRF monoclonal antibodies (mAbs) could neutralize TRF activity and immunoprecipitate TRF-active molecules.[10] It was therefore proposed that TRF should be called interleukin-5 (IL-5).[11] While initially identified by its ability to support the growth and differentiation of activated B cells and eosinophils in the mouse, IL-5 is synonymous with BCGF-II, EDF, immunoglobulin A (IgA)-enhancing factor, and TRF, and exerts pleiotropic effects on various cell types involved in the immune system and inflammation.[8]

IL-5 activates mouse B cells and the proliferation and differentiation of eosinophils. In particular, IL-5 is indispensable for maintaining homeostatic proliferation and survival of mouse B-1 cells.[12,13] In humans, the biological effects of IL-5 are best characterized for eosinophils.[14,15] In addition to inducing terminal maturation of eosinophils, IL-5 prolongs eosinophil survival, modulates eosinophil chemotaxis, increases eosinophil adhesion to endothelial cells, and enhances eosinophil effector functions.[14–18] IL-5 appears to be involved in the pathogenesis of hypereosinophilic syndromes and eosinophil-dependent inflammatory diseases.[18] However, it remains unclear whether IL-5 is indispensable in asthma, as recent attempts to inhibit the accumulation of eosinophils in the airways of asthma patients by targeting IL-5 have had only limited success.[19,20]

Now that we have explored the role of IL-5 in eosinophil activation and disease control,[21–23] we will focus on the genetic and protein structure of IL-5, on its receptor, interleukin-5 receptor subunit alpha (IL-5RA), and on IL-5 signal transduction cascades.

INTERLEUKIN-5

Complementary DNA and Gene Expression

Mouse IL-5 cDNA (mTRF, pSP6K clone) was isolated from cDNA libraries constructed from poly(A)+ mRNA of the alloreactive T-cell clone, 2.19.[11] Mouse IL-5 (mIL-5) cDNA encodes a mature polypeptide of 113 amino acids with a calculated molecular mass of 13 kDa that contains three potential N-linked glycosylation sites and three cysteine residues. N-terminus amino acid sequencing of T cell-derived mIL-5 revealed a sequence identical to that deduced from the nucleotide sequence of mIL-5 cDNA and confirms the inferred N-terminus of the mature protein.[8] Using the mIL-5 cDNA as a probe, cDNA encoding human IL-5 (hIL-5) was also isolated,[24] and this codes for a mature polypeptide consisting of 115 amino acids. The cDNA encoding human EDF was isolated and found to be identical to hIL-5.[15]

The *interleukin 5 (colony-stimulating factor, eosinophil) (IL5)* gene includes three introns and has numerous repetitive sequences. A common possible regulatory element, including a conserved decamer, lies adjacent to the TATA boxes (5′-TATAAA-3′) upstream of the transcription initiation sites. The exon-intron organization and the location of the cysteine codons of the *IL5* gene resemble those of the *colony stimulating factor 2 (granulocyte-macrophage) (CSF2/GM-CSF)*, *interleukin-2 (IL2)*, and *interleukin-4 (IL4)* genes, but are quite distinct from those of the *colony stimulating factor 3 (granulocyte) (CSF3/G-CSF)* gene. The human *IL5* gene is mapped on chromosome 5q23-q31 on which the *IL4*, *CSF2 (GM-CSF)*, *interferon regulatory factor 1 (IRF1)*, *interleukin-19 (IL19)*, and *interleukin 3 (colony-stimulating factor, multiple) (IL3)* genes are mapped, and their similar exon-intron organization suggests an evolutionary relationship.

Mouse *IL5* genes cluster within a 230-kb region of mouse chromosome 11.

Both mouse and human IL-5 proteins are in the range of 50−60 kDa when expressed in mammalian cells, but migrate to 25−30 kDa when treated with reducing and alkylating agents,[4,10] indicating that recombinant IL-5 is a disulfide-linked dimer. A single polypeptide with a molecular mass of 14 kDa is secreted when mIL-5 mRNA is transiently translated in rabbit reticulocyte lysates. The large variation in the molecular mass of IL-5 is predominantly a result of the heterogeneous addition of carbohydrate in the post-translational biosynthetic state.[25] Dimer formation is essential for expressing biological activity, as the monomeric IL-5 produced by reduction and alkylation of IL-5 loses its biological activity.

Structurally, IL-5 is a unique member of the short-chain helical-bundle subfamily of cytokines whose canonical motif includes four helices (A−D) arranged in an up−up−down−down topology. In contrast to other subfamily members, IL-5 forms a pair of helical bundles of two identical monomers that contribute a D helix to the other A−C helices.[26]

It is predicted that the lack of bioactivity by an IL-5 monomer is due to a short loop between helices C and D, which physically prevents unimolecular folding of helix D into a functionally obligate structural motif. By using an insertional mutant of IL-5 leading to monomeric IL-5, Dickason and colleagues demonstrated that all of the structural features necessary for IL-5 to function are contained within a single helical bundle.[27]

Cells Types Producing Interleukin-5

The major cellular sources of IL-5 are T_H2 lymphocytes: type 2 cytotoxic T (T_c2) cells, mast cells, eosinophils, and gamma delta ($\gamma\delta$)T cells. In particular, T_H2 cells produce IL-5 on stimulation with antigens such as *M. tuberculosis* and *Toxocara canis* or with allergens, and mast cells secrete IL-5 in response to stimulation with allergen/IgE complex or calcium ionophore.[4,6,8,12] Natural killer (NK) and natural killer T (NKT) cells or nonhematopoietic cells, including epithelial cells, can also produce IL-5. Significant IL-5 mRNA expression is detected in the lungs, spleen, and small intestine of wild-type, V(D)J recombination-activating protein 2 (Rag-2)$^{-/-}$ and T-cell receptor (TCR)β' $^{-}\delta^{-/-}$ transgenic mice. Furthermore, high levels of IL-5 mRNA expression are detected in proto-oncogene c-Kit (c-Kit)$^-$ IL-5RA$^{-/-}$ cells in the lungs and small intestine of Rag-2$^{-/-}$ mice,[13] implying that IL-5 can be produced by non-T/non-mast/non-eosinophil cells. Interestingly, CD4$^-$ c-kit$^-$ CD3e$^-$ interleukin-2 receptor subunit alpha (IL-2RA)$^+$ cells in the mouse Peyer patch (designated PP CD3$^-$ IL-2R$^+$ cells) are reported to produce high levels of IL-5 when stimulated with IL-2 or phorbol myristate acetate (PMA) plus the mobile ion carrier A23187.[28] PP CD3$^-$ IL-2R$^+$ cells do not express TCRβ and TCR$\gamma\delta$, CD23, CD49b, or NK1.1 antigen. The cells express several Toll-like receptors (TLRs), respond to poly I:C (TLR3 ligand), and secrete IL-5, suggesting that they may belong to a novel subset of IL-5-producing cells.

Recently, natural helper cells, a new type of innate lymphocyte subset of lineage-negative (Lin)$^-$ c-Kit$^+$ stem cell antigen 1 (Sca-1)$^+$ interleukin-7 receptor subunit alpha (IL-7RA)$^+$ interleukin-33 receptor subunit alpha (IL-33RA)$^+$ cells, have been identified.[29] These cells present in a novel lymphoid structure associated with adipose tissue in the peritoneal cavity and proliferate in response to IL-2, resulting in the production of large amounts of IL-5 together with other T_H2 cytokines such as interleukin-6 (IL-6) and interleukin-13 (IL-13). Taken together, cells involved in both innate and acquired immunity, regardless of hematopoietic and nonhematopoietic lineage, are able to secrete IL-5.

THE INTERLEUKIN-5 RECEPTOR

Gene and Protein Structure

In the mouse, IL-5-responsive B cells and eosinophils express relatively small numbers (approximately 50) of high-affinity IL-5R (kDa, 10−150 pM) and larger numbers (around 1000) of low-affinity IL-5R (kDa, 2−10 nM).[4,12,30,31] Chemical cross-linking of mIL-5R with IL-5 reveals that the high-affinity mIL-5R consists of two distinct polypeptide chains [α subunit, 60 kDa; β chain (βc) subunit, 130 kDa]. IL-5 binds to the 60 kDa subunit on B cells and eosinophils. The anti-mIL-5R mAbs H7 and R52.120 recognize the α subunit with 60 kDa and the βc subunit with 130 kDa, respectively.[32,33] The H7 mAb specifically inhibits IL-5-induced proliferation, while the R52.120 mAb partially inhibits both IL-5- and IL-3-induced proliferation.[34]

Interleukin-5 Receptor Subunit Alpha

Expression cloning of mIL-5RA cDNA was carried out with the use of H7 and T21 mAbs.[35] Results revealed that the mIL-5RA subunit is a type 1-membrane protein of 415 amino acids, containing a glycosylated extracellular domain, a single transmembrane segment, and a cytoplasmic tail. The extracellular domain of the mIL-5RA contains two motifs that are conserved in a set of type 1 cytokine receptor family molecules; a particular spacing of four cysteine residues in the N-terminus half of the region and the WSXWS (tryptophan-serine-X-tryptophan-serine) motifs located close to the transmembrane domain. In addition, the extracellular region comprises three tandemly repeated sets of a fibronectin type III domain. The cytoplasmic domain does not contain the consensus sequences for either a tyrosine kinase domain or a catalytic domain of protein kinases. Intriguingly, the cytoplasmic domain has a motif rich in proline, the PPXP (proline-proline-X-proline) motif following the transmembrane domain that is well conserved in IL-5RA and receptors for granulocyte-macrophage colony-stimulating factor (GM-CSF), growth hormone, IL-3, and prolactin.

The entire nucleotide sequence of hIL-5RA cDNA shows considerable similarity to the coding sequence of mIL-5RA cDNA. The cytoplasmic regions, which are rich in proline residues and which follow the transmembrane domain, are also well conserved. Complementary DNA coding for soluble forms of both mIL-5RA and hIL-5RA subunits have been isolated.[36,37] CD99-cos7 transfectants expressing recombinant mIL-5RA bind IL-5 with low affinity. The recombinant hIL-5RA subunit expressed on CD99-cos7 cells binds hIL-5 with high affinity (kDa, 300−450 pM).

The mIL-5RA genomic gene is divided into eleven exons and ten introns and spans more than 35 kb.[4] Gene organization revealed a pattern of considerable structural homology with genomic genes encoding for other cytokine genes such as *interleukin 2 receptor, beta (IL2RB), interleukin 3 receptor, alpha (low affinity) (IL3RA), interleukin 4 receptor (IL4R), interleukin 7 receptor (IL7R)*, and *erythropoietin receptor (EPOR)*. Chromosomal localization of the *mIL5RA* and *hIL5RA* genes was mapped on the distal half of mouse chromosome 6 and human chromosome 3 (3p24−3p26), respectively.

In humans, Tavernier and colleagues[38] have shown that two forms of soluble hIL-5RA, soluble (s)1hIL-5RA and s2hIL-5RA, are generated by a different mechanism from the mouse. One is generated by a normal splicing event using the first splice acceptor site coupled with the extracellular domain to a specific small exon, and the other is generated without splicing. In contrast to smIL-5RA, recombinant shIL-5RA binds hIL-5 with high affinity and inhibits the biological activities of hIL-5. Tavernier and colleagues[38] also showed that shIL-5RA can bind dimerized hIL-5 in a 1:1 ratio. They also speculated that shIL-5RA may act as a potent regulator for hIL-5, if the protein can be secreted enough for it to compete with membrane-bound IL-5RA. The differences between smIL-5RA and shIL-5RA in the mechanism for production and ligand-binding affinity might suggest different functional roles for shIL-5RA and smIL-5RA.

Interleukin-5 Receptor Subunit Beta

With regard to the βc, anti-mIL-5R mAb (R52.120) and anti-mIL-3R (anti-Aic-2) mAb could immunoprecipitate identical proteins with a molecular mass of 130 kDa and partially inhibit IL-5-dependent proliferation,[34] although they did not react with the mIL-5RA subunit. IL-3-dependent FDC-P1 transfectants expressing recombinant mIL-5RA bind IL-5 with high affinity and proliferate in response to IL-5.[39] The homologue protein [interleukin-3 receptor-like protein (AIC2B)] of low-affinity mIL-3R (AIC2A) was identified as the IL-5Rβc subunit by transfection of AIC2B cDNA into recombinant mIL-5RA-expressing L cells.[39] The βc subunit contains motifs conserved among cytokine receptor families and contributes to the signaling molecule for mouse granulocyte-macrophage colony-stimulating factor receptor subunit alpha (GM-CSF-R-alpha) and mIL-3R. The human βc is also identified as the βc of GM-CSF-R-alpha and IL-3R.[15,37,40,41]

The X-ray structure of the βc subunit is an intertwined homodimer in which each chain contains four domains with approximate fibronectin type III topology. The two βc subunits that compose the homodimer are interlocked by virtue of the swapping of β strands between domain 1 of one subunit and domain 3 of the other subunit.[42] Site-directed mutagenesis analyses have revealed that the interface between domains 1 and 4 is predicted to form the functional epitope. Using fluorescence resonance energy transfer imaging, the βc subunit was shown to exist as preformed homo-oligomers and IL-5 stimulation induces βc assembly in the presence of IL-5RA.[43]

Martinez-Moczygemba and colleagues demonstrated that following cytokine ligation, βc signaling is terminated partially by ubiquitination and proteasome degradation of its cytoplasmic region, resulting in the generation of truncated βc products, a process termed βc intracytoplasmic proteolysis (βIP).[44] The truncated IL-5R complex (IL-5RA and βIP) is degraded in the lysosome.

As described, IL-5 acts on target cells through IL-5R, which consists of the IL-5RA and βc subunits. IL-5RA is expressed on a subset of mouse B cells, B-1 cells, and naturally activated B-2 cells. Mouse and human eosinophils and basophils also express significant numbers of IL-5RA, and the βc is constitutively expressed on hematopoietic progenitor cells.[45−47] The IL-5/IL-5RA complex induces the recruitment of βc to the complex and the βc/IL-5/IL-5RA complex can transduce IL-5 signals through the βc, while βc alone does not bind IL-5.[47] Both IL-5RA and βc subunits have the conserved motifs of the hematopoietic receptor superfamily. The cytoplasmic domains of IL-5RA and βc are indispensable for signal transduction. The membrane proximal, proline-rich sequence (the PPXP motif) of the cytoplasmic domain of both IL-5RA and βc are essential for IL-5 signal transduction.[35,48−50] GM-CSF, IL-3, and IL-5 are proinflammatory cytokines that control the production and function of myeloid and lymphoid cells. Their receptors are composed of a ligand-specific α subunit and a shared βc as a signal-transducing subunit.[41] Both GM-CSF and IL-3RA conserve the membrane proximal PPXP motif of the cytoplasmic domain that is essential for signal transduction.

Regulation of Interleukin-5 Receptor Expression on Progenitors for B cells and Eosinophils

IL-5RA expression in wild-type mice is readily detectable by anti-IL-5RA mAb on B-1 cells that reside largely in the peritoneum and pleural cavities. B-1 cells respond to IL-5 for survival, proliferation, and differentiation to antibody-secreting cells (ASCs).[45,50] IL-5 transgenic mice show a marked increase in the proportion and number of B-1 cells in the spleen with concomitant hyper-gammaglobulinemia and autoantibody production.[51] Likewise, IL-5-responsive B-1 cells are increased in the spontaneously autoimmune NZB and (NZB × NZW) F1 mice.

In contrast, most resting B-2 cells constitutively express βc, but not IL-5RA. Once B cells are activated by T_H cells and antigen through the B-cell receptor, they express IL-5RA and become responsive to IL-5 resulting in the integration into the plasma cell differentiation program. CD38 ligation of B-2 cells induces an increase in the expression of IL-5RA. A complex of transcription factors including CCAAT/enhancer-binding protein beta, E12, E47, octamer-binding protein 2 (Oct-2), and transcription factor Sp1 coordinately bind to the promoter of the mouse *IL5RA* gene and regulate IL-5RA expression.[4] IL-5RA expression is also inducible on B-2 cells on stimulation with CD40 or interferon regulatory factor 6 (IRF-6). IL-4 upregulates IL-5RA expression on CD40- and IRF-6-stimulated B cells,[52] while IL-4 by itself does not induce IL-5RA. Corcoran and colleagues reported that Oct-2 contributes to, but does not entirely control, IL-5RA levels. Oct-2 binds directly to the promoter of the *IL5RA* gene to activate its transcription specifically in B cells.

Eosinophilic progenitors and mature eosinophils in the mouse and human constitutively express IL-5RA.[46,53] Ackerman and colleagues identified a unique cis-element that acts as an enhancer in regulating the activity of the IL-5RA promoter.[54] Iwama and colleagues identified an enhancer-like cis-element in the IL-5RA promoter that is important for both full promoter function and lineage-specific activity.[55] They also showed that the major histocompatibility complex class II regulatory factor X 1 (RFX1), X 2 (RFX2), and X 3 (RFX3) homodimers and heterodimers specifically bind to the cis-element of the IL-5RA promoter and contribute to the activity and lineage specificity of the IL-5RA promoter through the activation and repression domains. At this moment, we are yet to uncover the entire regulatory mechanisms of *IL5RA* gene expression in eosinophils.

By examining IL-5RA expression in B-1 cell progenitors using fluorescence-activated cell sorting (FACS), we found IL-5RA$^+$ cells in the Lin$^-$ fraction in fetal bone marrow, but not among CD19$^-$ B cell isoform of 220 kDa (B220)$^+$ cells. Cell transfer experiments into severe combined immuno-deficient or immunodeficiency (SCID) mice revealed that IL-5RA$^+$ cells in fetal liver are unable to differentiate into B-1 cells in SCID mice, while IL-5RA$^-$ CD19$^+$ B220$^-$ fetal liver cells become B-1 cells. Rather, IL-5RA$^+$ cells in fetal liver are able to differentiate into eosinophils in *in vitro* culture under the influence of cytokine cocktails.[56] These results suggest that CD19$^+$ B220$^-$ B-1 progenitors in fetal liver do not express a detectable level of IL-5RA, while eosinophil progenitors do express IL-5RA.

Among GM-CSF, IL-3, and IL-5, IL-5 plays a role in the development and proliferation of eosinophils. Iwasaki and colleagues have demonstrated that in normal bone marrow, the Lin$^-$ Sca-1$^-$ CD34$^+$ fraction contains a small number of cells expressing IL-5RA and a low level of c-Kit. Purified Lin$^-$ Sca-1$^-$ IL-5RA$^+$ CD34$^+$ c-Kit^{lo} cells, which are blastic cells with scattered eosinophilic granules, respond to IL-5 alone or GM-CSF, eosinophil peroxidase, IL-3, IL-5, interleukin-9, stem cell factor, and thyroid peroxidase, leading to differentiate exclusively into eosinophils.[53] Thus, Lin$^-$ Sca-1$^-$ IL-5RA$^+$ CD34$^+$ c-Kit^{lo} cells appear to be eosinophil lineage-committed progenitors (EoPs). In the bone marrow of mice infected with *Trichinella spiralis*, Lin$^-$ Sca-1$^-$ IL-5RA$^+$ CD34$^+$ c-Kit^{lo} EoPs significantly expanded by approximately threefold in numbers, while numbers of granulocyte/macrophage progenitors and common myeloid progenitors were not affected. EoPs are therefore involved in the physiological development of eosinophils.

INTERLEUKIN-5 RECEPTOR-MEDIATED SIGNALING

IL-5 stimulation induces rapid tyrosine phosphorylation of cellular proteins, including the βc, hematopoietic SH2/SH3 domain-containing proteins, such as guanine nucleotide exchange factor VAV and SHC-transforming protein, Bruton tyrosine kinase (Btk) and the Btk-associated molecules, tyrosine-protein kinase JAK1 and JAK2, signal transducer and activator of transcription 1-alpha/beta (STAT1) and signal transducer and activator of transcription A (STAT5), phosphatidylinositol 3-kinase (PI3K), and mitogen-activated protein kinases (MAPKs), and activates downstream signaling molecules.[47–49] Treatment of IL-5-dependent cells with herbimycin A completely inhibits IL-5-dependent cell growth, indicating the indispensable role of tyrosine kinases in IL-5 signaling. Activated STAT5 can induce the gene expression of a cytokine-inducible SH2-containing protein (CIS) and a JAK2-binding protein [JAB, also called suppressor of cytokine signaling 1 (SOCS-1)] in eosinophils that are one of the feedback loops of negative regulation of IL-5 signaling.[57] IL-5 also activates the NF-kappa-B repressing factor (NF-κB-repressing factor) that is dependent on TNF receptor-associated factor 6 (TRAF6).[58]

IL-5 enhances gene expression of *v-myc myelocytomatosis viral oncogene homolog (avian) (MYK/c-Myc)*, *FBJ murine osteosarcoma viral oncogene homolog (FOS/c-Fos)*, jun proto-oncogene *(JUN/c-Jun)*, *cytokine inducible SH2-containing protein (CISH/CIS)*, *suppressor of cytokine signaling 1 (SOCS1/JAB)*, and *pim-1 oncogene (pim-1)* in B cells.[22,48] IL-5-induced cell proliferation and anti-apoptosis depends on the Janus kinase (JAK)—c-Myc pathway, and IL-5-induced upregulation of c-Myc is dependent on JAK1 and JAK2 activation.[59]

The JAK and STAT Pathway

Binding of IL-5 to IL-5R on mouse B cells and eosinophils activates JAK1 and JAK2 and STAT5. The activation of JAK2 and STAT5 is essential for IL-5-dependent signal transduction in human eosinophils.[17,60,61] We have already elucidated the hIL-5RA cytoplasmic domain regulating JAK activation for IL-5 in human eosinophils. Results revealed that JAK2 is constitutively associated with hIL-5RA, regardless of IL-5 stimulation.[47] In contrast, JAK1 is constitutively associated with βc and is associated with hIL-5RA only after cells are stimulated with IL-5.[61] As in the IL-5R system, JAK2 and JAK1 are constitutively associated with hIL-3RA and βc, respectively.[59] Both JAK1 and JAK2 are activated by a relevant cytokine, either IL-5 or IL-3. The region of hIL-5RA necessary for JAK2 binding is located in amino acid residues 346—387, including the proline-rich sequences of the cytoplasmic domain. JAK1 at the N-terminus binds to the conserved regions of the proline-rich motif (Box 1 and Box 2 regions) of βc, but signal activation requires JAK1 at the C-terminus.[59]

By using CD99-cos7 transfectants expressing intact βc and a kinase-negative form of JAK1 and JAK2, (DN) JAK1 and (DN) JAK2, respectively, we found that overexpression of (DN) JAK2 inhibits IL-5-induced activation of JAK2 and JAK1 and cell proliferation.[4] In contrast, (DN) JAK1 overexpression inhibits JAK1 activation, but does not suppress JAK2 activation. IL-5-induced tyrosine phosphorylation of βc is diminished only when (DN) JAK2 is overexpressed. Thus, JAK2 activation is essential and sufficient to transduce IL-5 signals.

Inhibition of βc proteasome degradation results in the prolonged activation of βc, JAK2, STAT5, and tyrosine-protein phosphatase non-receptor type 11 (PTPN11/SHP-2), in which JAK activity is required for the direct ubiquitination of the βc cytoplasmic domain and proteasome degradation.[62]

Bruton Tyrosine Kinase Activation

Bruton agammaglobulinemia tyrosine kinase (BTK) is the gene responsible for human X-linked agammaglobulinemia (XLA), which is characterized by a near absence of peripheral B cells, low concentrations of serum immunoglobulins, and varying degrees of bacterial infections. Btk is a cytoplasmic tyrosine kinase expressed in myeloid, erythroid, and B-lineage cells except plasma cells. A spontaneous Btk mutation (R28C) in mice produces X-linked immunodeficiency (XID). The B cells from XID mice as well as Btk$^{-/-}$ mice show impaired B-cell development and function. XID B cells are hyporesponsive to IL-5, IL-10, and IRF-6 and fail to proliferate in response to stimulation via the B cell antigen receptor or CD38.[22]

There are several pieces of evidence that support the involvement of *BTK* in IL-5 signaling in B cells. IL-5 stimulation enhances Btk activity in a murine early B-cell line.[63] Activated B cells from XID mice respond poorly to IL-5. Murine activated splenic B cells expressing recombinant wild-type IL-5RA are unable to respond to IL-5.[22] To elucidate the function of Btk at the biochemical level, we isolated Btk-PH-domain binding protein (BAM11) that binds to the PH-domain of Btk.[64] BAM11 is a murine homologue of human protein ENL, a fusion partner of the myeloid/lymphoid or *mixed-lineage leukemia (trithorax homolog, Drosophila) (MLL/ALL-1/HRX)* gene in infantile leukemia cells. Btk localizes both in the nucleus and cytoplasm, and shows nucleocytoplasmic shuttling.[65] Forced expression of BAM11 in B-cell progenitors inhibits IL-5-induced proliferation and the activity of Btk.

BAM11 has transcriptional activity that is enhanced by Btk through the intact pleckstrin homology (PH) and kinase domain of Btk.[65] This *positive—negative mutual regulation system* between BAM11 and Btk may elucidate a novel mechanism of B-cell signaling through IL-5R. Furthermore, general transcription factor II-I (GTFII-I/TFII-I), another Btk-binding protein with transcriptional activity, together with BAM11 and Btk augments BAM11- and Btk-dependent transcriptional coactivation. As BAM11 is co-immunoprecipitated with the SWI/SNF-related matrix-associated actin-dependent regulator of chromatin subfamily B member 1 (SNF5) protein, a member of the SWItch/Sucrose NonFermentable (SWI/SNF) complex, IL-5 may regulate gene transcription in B cells by activating Btk, BAM11, and the SWI/SNF transcriptional complex via GTFII-I/TFII-I activation.

Ras/Extracellular Signal-Regulated Kinase Activation

In addition to the JAK2/STAT5 pathway, the Ras/extracellular signal-regulated kinase (ERK) pathway has also been implicated in the signaling of IL-5 and other cytokines for maintaining eosinophil survival, proliferation, and differentiation, such as degranulation.[66] The Sprouty (Spry) proteins are identified as negative regulators for several types of growth factor-induced ERK activation including fibroblast growth factor and epidermal growth factor. Yoshimura and colleagues cloned the Sprouty-related, EVH1 domain-containing protein 1 (Spred-1) and identified it as a negative regulator of growth factor-mediated, Ras-dependent ERK activation. Inoue and colleagues[67] demonstrated that Spred-1 negatively regulates allergen-induced airway eosinophilia and hyperresponsiveness, without affecting T_H cell differentiation. Biochemical assays indicate that Spred-1 suppresses IL-5-dependent cell proliferation and ERK activation. Moreover, Spred-1 deficiency shows overexpression of IL-13 in eosinophils. These data indicate that Spred-1 is a negative regulator of ERK activation and modulates eosinophil activation normally mediated by IL-5.

CONCLUSION

The cytokine IL-5 is a critical regulator of inflammation and immune responses in mammals. It controls the production and function of myeloid and lymphoid cells. IL-5 exerts its effects on target cells via receptors that comprise an IL-5-specific subunit and βc, which is shared with the cytokine-specific subunit of the GM-CSF and IL-3 receptors. IL-5RA is expressed on mature eosinophils in both mouse and human. IL-5 activates several kinases including Btk, JAK2, tyrosine-protein kinase Lyn, and RAF proto-oncogene serine/threonine-protein kinase, and the phosphatase PTPN11/SHP-2. In addition, the ERK pathway is also implicated in IL-5 signaling, which is important for IL-5-dependent cell survival, proliferation, and function of B cells and eosinophils. IL-5 induces the activation of STAT5 and NF-κB-repressing factor, which play critical roles in IL-5-dependent proliferation, differentiation, cell survival, and function.

Structural, functional, and clinical studies described herein provide an insight into the role of IL-5 and its receptor system in the immune response, in inflammation, and in disease control. We have also emphasized a strong impetus for investigating the biological functions of IL-5 regarding the link between natural and adaptive immunity specific to the epitopes of natural ligands and exogenous allergens. As the role of hIL-5 in health and disease is best characterized in human eosinophils, we may emphasize the efficacy of anti-IL-5 and anti-IL-5RA antibody therapy in the pathogenesis of asthma and hypereosinophilic syndromes.

REFERENCES

1. Kay B. Allergy and allergic diseases. First of two parts. *N Eng J Med* 2001;**344**:30–7.
2. Galli SJ, Tsai M, Piliponsky AM. The development of allergic inflammation. *Nature* 2008;**453**:445–54.
3. Nakajima H, Takatsu K. Role of cytokines in allergic inflammation. *Int Archives Allergy Immunol* 2007;**142**:265–73.
4. Takatsu K, Takaki S, Hitoshi Y. Interleukin 5 and its receptor system: Implications in the immune response and inflammation. *Adv Immunol* 1994;**54**:134–70.
5. O'Garra Warren DJ, Holman M, Popham AM, Sanderson CJ, Klaus GG. Interleukin 4 (B-cell growth factor II/eosinophil differentiation factor) is a mitogen and differentiation factor for pre-activated murine B lymphocytes. *Proc Natl Acad Sci USA* 1986;**83**:5228–32.
6. Takatsu K, Tominaga A, Hamaoka T. Antigen-induced T cell-replacing factor (TRF). I. Functional characterization of a TRF-producing helper T cell subset and genetic studies on TRF production. *J Immunol* 1980;**124**:2414–22.
7. Takatsu K, Tanaka K, Tominaga A, Kumahara Y, Hamaoka T. Antigen-induced T-cell replacing factor (TRF). III. Establishment of T cell hybridoma producing TRF and analysis of released TRF. *J Immunol* 1980;**125**:2646–53.
8. Takatsu K, Tominaga A, Harada N, Mita S, Matsumoto M, Takahashi T, et al. T cell-replacing factor (TRF)/interleukin 5 (IL-5): molecular and functional properties. *Immunol Rev* 1988;**102**:107–35.
9. Harada N, Kikuchi Y, Tominaga A, Takaki S, Takatsu K. BCGFII Activity on activated B cells of a purified murine T cell-replacing factor (TRF) from a T cell hybridoma (B151K12). *J Immunol* 1985;**134**:3944–51.
10. Harada N, Takahashi T, Matsumoto M, Kinashi T, Kikuchi Y, Koyama N, et al. Production of a monoclonal antibody useful in the molecular characterization of murine T-cell-replacing factor/B-cell growth factor II. *Proc Natl Acad Sci USA* 1987;**84**:4581–5.
11. Kinashi T, Harada N, Severinson E, Tanabe T, Sideras P, Konishi M, et al. Cloning of complementary DNA encoding T-cell replacing factor and identity with B-cell growth factor II. *Nature* 1986;**324**:70–3.
12. Takatsu K, Dickason R, Huston D. Interleukin-5. In: LeRoith D, Bondy C, editors. *Growth Factors and Cytokines in Health and Disease*, vol. 2 PART A. London England: JAI Press Inc.; 1997. p. 143–200.
13. Moon BG, Takaki S, Miyake K, Takatsu K. The role of IL-5 for mature B-1 cells in homeostatic proliferation, cell survival, and Ig production. *J Immunol* 2004;**172**:6020–9.
14. Sanderson CJ. Interleukin-5, eosinophils, and disease. *Blood* 1992;**79**:3101–9.
15. Campbell HD, Tucker WQJ, Hort Y, Martinson ME, Mayo G, Clutterbuck EJ, et al. Molecular cloning, nucleotide sequence, and expression of the gene encoding human eosinophil differentiation factor (interleukin-5). *Proc Natl Acad Sci USA* 1987;**84**:6629–33.
16. Yamaguchi YY, Hayashi Y, Sugama Y, Miura Y, Kasahara T, Torisu T, et al. *J Exp Med* **167**, 1737–1742.
17. Yamaguchi Y, Suda T, Suda J, Eguchi M, Miura Y, Harada N, et al. Purified interleukin-5 supports the terminal differentiation and proliferation of murine eosinophilic precursors. *J Exp Med* 1988;**167**:43–56.

18. Rothenberg ME, Hogan SP. The eosinophil. *Annu Rev Immunol* 2006;**24**:147—74.

19. Rothenberg ME, Klion AD, Roufosse FE, Emmanuel Kahn J, Weller PF, Simon H-U, et al. Mepolizumab HES Study Group. Treatment of patients with the hypereosinophilic syndrome with mepolizumab. *N Engl J Med* 2008;**358**:1215—28.

20. Flood-Page P, Swenson C, Faiferman I, Matthews J, Williams M, Brannick L, et al. International Mepolizumab Study Group: A study to evaluate safety and efficacy of mepolizumab in patients with moderate persistent asthma. *Am J Respir Crit Care Med* 2007;**176**:1062—71.

21. Takatsu K, Nakajima H. IL-5 and eosinophilia. *Curr Opin Immunol* 2008;**20**:288—94.

22. Takatsu K. Interleukin 5 and B cell differentiation. *Cytokines Growth Factor Rev* 1998;**9**:25—35.

23. Takatsu K, Kouro T, Nagai Y. Interleukin 5 in the link between innate and acquired immune response. *Adv Immunol* 2009;**101**:191—236.

24. Azuma C, Tanabe T, Konishi M, Kinashi T, Noma T, Matsuda R, et al. Cloning of cDNA encoding human T-cell replacing factor (Interleukin-5) and comparison with the murine homologue. *Nucleic Acids Res* 1986;**14**:9149—58.

25. Tominaga A, Takahashi T, Kikuchi Y, Mita S, Naomi S, Harada N, et al. Role of carbohydrate moiety of IL-5. Effect of tunicamycin on the glycosylation of IL-5 and the biologic activity of deglycosylated IL-5. *J Immunol* 1990;**144**:1345—52.

26. Milburn MV, Hassell AM, Lambert MH, Jordan SR, Proudfoot AEI, Graber P, et al. A novel dimer configuration revealed by the crystal structure at 2.4 A resolution of human interleukin-5. *Nature* 1993;**363**:172—6.

27. Dickason RR, Huston DP. Creation of a biologically active interleukin-5 monomer. *Nature* 1996;**379**:652—5.

28. Kuraoka M, Hashiguchi M, Hachimura S, Kaminogawa S. CD4— c-kit— CD3ε— IL-2Rα+ Peyer's patch cells are a novel cell subset which secrete IL-5 in response to IL-2: implications for their role in IgA production. *Eur J Immunol* 2004;**34**:1920—9.

29. Moro K, Yamada T, Tanabe M, Takeuchi T, Ikawa T, Kawamoto H, et al. Innate production of TH2 cytokines by adipose tissue-associated c-Kit1Sca-11 lymphoid cells. *Nature* 2010;**463**:540—5.

30. Mita S, Harada N, Naomi N, Hitoshi Y, Sakamoto K, Akagi M, et al. Receptors for T cell-replacing factor/interleukin 5. Specificity, quantitation, and its implication. *J Exp Med* 1988;**168**:863—78.

31. Mita S, Tominaga A, Hitoshi Y, Sakamoto K, Honjo T, Akagi M, et al. Characterization of high-affinity receptors for interleukin 5 on interleukin 5-dependent cell lines. *Proc Natl Acad Sci USA* 1989a;**86**:2311—5.

32. Yamaguchi N, Hitoshi Y, Mita S, Hosoya Y, Kikuchi Y, Tominaga A, et al. Characterization of the murine interleukin 5 receptor by using a monoclonal antibody. *Int Immunol* 1990;**2**:181—7.

33. Rolink AG, Melchers F, Palacios R. Monoclonal antibodies reactive with the mouse interleukin 5 receptor. *J Exp Med* 1989;**169**:1693—701.

34. Mita S, Takaki S, Hitoshi Y, Rolink AG, Tominaga A, Yamaguchi N, et al. Molecular characterization of the β chain of the murine interleukin-5 receptor. *Int Immunol* 1991;**3**:665—72.

35. Takaki S, Tominaga A, Hitoshi Y, Mita S, Sonoda E, Yamaguchi N, et al. Molecular cloning and expression of the murine interleukin-5 receptor. *EMBO J* 1990;**9**:4367—74.

36. Murata Y, Takaki S, Migita M, Kikuchi Y, Tominaga A, Takatsu K. Molecular cloning and expression of the human interleukin-5 receptor. *J Exp Med* 1992;**175**:341—51.

37. Tavernier J, Devos R, Cornelis S, Tuypens T, Van der Heyden J, Fiers W, et al. A human high affinity interleukin-5 receptor (IL5R) is composed of an IL5-specific alpha chain and a beta chain shared with the receptor for GM-CSF. *Cell* 1991;**66**:1175—84.

38. Tavernier J, Tuypens T, Verhee A, Plaetinck G, Devos R, Van der Heyden J, et al. Identication of receptor-binding domains on human interleukin 5 and design of an interleukin 5-derived receptor antagonist. *Proc Natl Acad Sci USA* 1995;**92**:5194—8.

39. Takaki S, Mita S, Kitamura T, Yonehara S, Yamaguchi N, Tominaga A, et al. Identification of the second subunit of the murine interleukin-5 receptor: interleukin-3 receptor-like protein, AIC2B is a component of the high affinity interleukin-5 receptor. *EMBO J* 1991;**10**:2833—8.

40. Takaki S, Murata Y, Kitamura T, Miyajima A, Tominaga A, Takatsu K. Reconstitution of the functional receptors for murine and human interleukin 5x. *J Exp Med* 1993;**177**:1523—9.

41. Miyajima A, Kitamura T, Harada N, Yokota T, Arai K. Cytokine receptors and signal transduction. *Annu Rev Immunol* 1992;**10**:295—331.

42. Mirza S, Chen J, Ewens CL, Dai J, Murphy JM, Young IG. Two modes of beta-receptor recognition are mediated by distinct epitopes on mouse and human interleukin-3. *J Biol Chem* 2010;**285**:22370—81.

43. Zaks-Zilberman M, Hamington AE, Ishino T, Chaikaen IM. Interleukin-5 receptor subunit oligomerization and rearrangement revealed by fluorescence resonance energy transfer imaging. *J Biol Chem* 2008;**283**:13398—406.

44. Martinez-Moczygemba M, Huston DP. Proteasome regulation of βc signaling reveals a novel mechanism for heterotypic desensitization. *J Clin Invest* 2001;**108**:1797—806.

45. Hitoshi Y, Yamaguchi N, Mita S, Takaki S, Tominaga A, Takatsu K. Distribution of IL-5 receptor-positive B cells. Expression of IL-5 receptor on Ly-1(CD5)+ B cells. *J Immunol* 1990;**144**:4218—25.

46. Hitoshi Y, Sonoda E, Kikuchi Y, Yonehara S, Nakauchi H, Tominaga A, et al. Interleukin-5 receptor positive B cells, but not eosinophils are functionally and numerically influenced in the mice carried with X-linked immunodeficiency. *Int Immunol* 1993;**5**:1183—90.

47. Ogata N, Kouro T, Yamada A, Koike M, Hanai N, Ishikawa T, et al. JAK2 and JAK1 constitutively associate with an interleukin 5 (IL-5) receptor and βc subunit, respectively and are activated upon IL-5 stimulation. *Blood* 1998;**91**:2264—71.

48. Takaki S, Kanazawa H, Shiiba M, Takatsu K. A critical cytoplasmic domain of the interleukin-5 (IL-5) receptor alpha chain and its function in IL-5-mediated growth signal transduction. *Mol Cell Biol* 1994;**14**:7404—13.

49. Kouro T, Kikuchi Y, Kanazawa H, Hirokawa K, Harada N, Shiiba M, et al. Critical proline residues of the cytoplasmic domain of the IL-5 receptor alpha chain and its function in IL-5-mediated activation of JAK kinase and STAT5. *Int Immunol* 1996;**8**:237—45.

50. Moon BG, Yoshida T, Shiiba M, Nakao K, Katsuki M, Takaki S, et al. Functional dissection of the cytoplasmic subregions of the interleukin-5 receptor alpha chain in growth and immunoglobulin G1 switch recombination of B cells. *Immunology* 2001;**102**:289—300.

51. Tominaga A, Takaki S, Koyama N, Katoh S, Matsumoto R, Migita M, et al. Transgenic mice expressing a B cell growth and differentiation factor gene (interleukin-5) develop eosinophilia and autoantibody production. *J Exp Med* 1991;**173**:429—41.

52. Emslie D, D'Costa K, Hasbold J, Metcalf D, Takatsu K, Hodgkin PO, et al. Oct2 enhances antibody-secreting cell differentiation through regulation of IL-5 receptor alpha chain expression on activated B cells. *J Exp Med* 2008;**205**:409—21.

53. Iwasaki H, Mizuno S, Mayfield R, Shigematsu H, Arinobu Y, Seed B, et al. Identification of eosinophil lineage-committed progenitors in the murine bone marrow. *J Exp Med* 2005;**201**:1891—7.

54. Sun Z, Yergeau DA, Tuypens T, Tavernier J, Paul CC, Baumann MA, et al. Identification and characterization of a functional promoter region in the human eosinophil IL-5 receptor alpha subunit gene. *J Biol Chem* 1995;**270**:1462—71.

55. Iwama A, Pan J, Zhang P, Reith W, Mach B, Tenen DG, et al. Dimeric RFX proteins contribute to the activity and lineage specificity of the interleukin-5 receptor alpha promoter through activation and repression domains. *Mol Cell Biol* 1999;**19**:3940—50.

56. Kouro T, Ikutani M, Kariyone A, Takatsu K. Expression of IL-5Rα on B-1 cell progenitors in mouse fetal liver and involvement of Bruton's tyrosine kinase in their development. *Immunol Lett* 2009;**123**:169—78.

57. Zahn S, Godillot P, Yoshimura A, Chaiken I. IL-5-induced JAB—JAK2 interaction. *Cytokine* 2000;**12**:299—306.

58. Meads MB, Li ZW, Dalton WS. A novel TNF receptor-associated factor 6 binding domain mediates NF-kB signaling by the common cytokine receptor beta subunit. *J Immunol* 2010;**185**:1606—15.

59. Huang HM, Lee YL, Chang TW. JAK1 N-terminus binds to conserved Box 1 and Box 2 motifs of cytokine receptor common beta subunit but signal activation requires JAK1 C-terminus. *J Cell Biochem* 2006;**99**:1078—84.

60. Kagami S, Nakajima H, Kumano K, Suzuki K, Suto A, Imada K, et al. Both Stat5a and Stat5b are required for antigen-induced eosinophil and T-cell recruitment into the tissue. *Blood* 2000;**95**:1370—7.

61. Horikawa K, Kaku H, Nakajima H, Davey HW, Hennighausen L, Iwamoto I, et al. Essential role of Stat5 for IL-5-dependent IgH switch recombination in mouse B cells. *J Immunol* 2001;**167**:5018—26.

62. Martinez-Moczygemba M, Huston DP, Lei JT. JAK kinases control IL-5 receptor ubiquitination, degradation and internalization. *J Leuk, Biol* 2007;**81**:1137—48.

63. Satoh S, Katagiri T, Takaki S, Kikuchi Y, Hitoshi Y, Yonehara S, et al. IL-5 receptor-mediated tyrosine phosphorylation of SH2/SH3-containing proteins and activation of Bruton's tyrosine and Janus-2 kinases. *J Exp Med* 1995;**180**:2101—11.

64. Kikuchi Y, Hirano M, Seto M, Takatsu K. Identification and characterization of a molecule, BAM11, that associates with the pleckstrin homology domain of mouse Btk. *Int Immunol* 2000;**12**:1397—408.

65. Hirano M, Kikuchi Y, Nisitani S, Yamaguchi A, Sato A, Ito T, et al. Bruton's tyrosine kinase (Btk) enhances transcriptional co-activation activity of BAM11, a Btk-associated molecule of a subunit of SWI/SNF complexes. *Int Immunol* 2004;**16**:747—57.

66. Pazdrak K, Schreiber D, Forsythe P, Justement L, Alam R. The intracellular signal transduction mechanism of interleukin 5 in eosinophils: the involvement of lyn tyrosine kinase and the Ras-Raf-1-MEK-microtubule-associated protein kinase pathway. *J Exp Med* 1995;**181**:1827—34.

67. Inoue H, Kato R, Fukuyama S, Nonami A, Taniguch K, Matsumoto K, et al. Spred-1 negatively regulates allergen-induced airway eosinophilia and hyperresponsiveness. *J Exp Med* 2005;**201**:73—82.

Chapter 5.5

Extramedullary Recruitment and Proliferation of Eosinophil Progenitors

Madeleine Rådinger, Apostolos Bossios and Jan Lötvall

EOSINOPHIL PROGENITORS

Circulating hematopoietic stem cells (HSCs) and progenitors play important roles in the physiology and homeostasis of the hematopoietic system. Hematopoietic progenitors are defined as undifferentiated pluripotent stem cells that are capable of self-renewal and can differentiate into all blood cell types, as well as lineage-committed progenitors that have a limited capacity for self-renewal and are thus committed to a specific cell lineage. Human hematopoietic progenitors express the hematopoietic progenitor cell antigen CD34 on their surface and form colony-forming units in culture. The majority of the hematopoietic activity takes place in the bone marrow, under the influence of resident stromal cells, monocytes, T lymphocytes, and their secretory products.[1] Eosinophil progenitors are currently defined by the cell surface antigen profile interleukin-5 receptor subunit alpha (IL-5RA)$^+$ lineage-negative (Lin)$^-$ stem cell antigen 1 (Sca-1)$^-$CD34$^+$ (c-KIT)^{lo}.

Eosinophil development primarily occurs under the influence of granulocyte-macrophage colony-stimulating factor (GM-CSF), interleukin-3, and interleukin-5 (IL-5), and eosinophilopoiesis is mainly restricted to the bone marrow compartment, although eosinophilopoiesis may also occur in the spleen, thymus, and lymph nodes to some degree.[3,4] Of note, IL-5 is not essential for homeostatic eosinophil development, as IL-5- and IL-5R-deficient mice have eosinophil levels close to wild-type mice under homeostatic conditions.[5,6] Nevertheless, IL-5 is the most specific for the eosinophil lineage and regulates the differentiation, maturation, and survival of eosinophils, and has been shown to provide an essential signal for both expansion and mobilization of eosinophils from the bone marrow in response to allergen exposure.[7] IL-5 has also been shown to prime eosinophil progenitors to respond to eotaxin [C-C motif chemokine 11 (CCL11)], most likely by the upregulation of the eotaxin receptor, C-C chemokine receptor type 3 (CCR3).[8]

EXTRAMEDULLARY RECRUITMENT OF EOSINOPHIL PROGENITORS

Small numbers of HSCs are detected in the bloodstream of mammals under homeostatic conditions. The mechanisms controlling their trafficking under steady-state conditions are to a large extent unknown. However, a recent study pointed out that mobilization of HSCs into the circulation involves complex mechanisms, which in part are regulated by proteolytic cleavage of stromal cell-derived factor-1 (SDF-1), leading to a decrease in SDF-1 in the bone marrow microenvironment and egress of HSCs into the circulation.[5] The downregulation of SDF-1 expression in the bone marrow is orchestrated by the central nervous system and controlled by circadian fluctuations in norepinephrine (noradrenaline) secretion, which in turn result in an attraction of HSCs to specific bone marrow niches.[9] Little is known about the trafficking of eosinophil progenitors under homeostatic conditions. However, many mature eosinophils traffic to the gastrointestinal tract, but also to the thymus, mammary gland, and uterus, a process that is further regulated by a constitutive expression of the chemokine eotaxin (CCL11).[10,11] Eotaxin (CCL11) also promotes eosinophil hematopoiesis and is directly responsible for mobilizing eosinophil progenitors (i.e., CD34$^+$ IL-5RA$^+$cells) in the bone marrow for exit into the peripheral circulation.[12–14] Many studies have suggested a role for progenitor cells in inflammatory disorders such as, for example, allergy. This is largely based on the fact that CD34$^+$ cells as well as CD34$^+$ IL-5RA$^+$cells are found in the circulation and recruited to sites of inflammation in allergic rhinitis and asthma.[15–18]

Adhesion molecules are critical for maintaining progenitor-related adhesive interactions in the bone marrow. Human CD34$^+$ cells express members of the beta-1 (β1) integrin family, VLA-4 subunit beta (VLA-4) and the beta-2 (β2) integrin family, lymphocyte function-associated antigen 1 (LFA-1). Both of these integrins are expressed at lower levels on mobilized blood-derived progenitors as compared to bone marrow-derived progenitors.[19–23] From these observations it was assumed that mobilization of progenitors from the bone marrow in part results from an altered expression and/or function of adhesion molecules and their corresponding ligands. Consistent with these findings, VLA-4 expression is downregulated on bone marrow-derived CD34$^+$ CD45$^+$ progenitor cells in atopic asthmatics after allergen challenge. The decrease in VLA-4 was further shown to correspond to a reduction of the *in vitro* adhesive properties of these cells, suggesting that adhesive changes in response to allergen challenge facilitates progenitor cell mobilization and egress to the peripheral circulation during the allergic inflammatory response.[24] In addition, studies performed on a human eosinophilic cell line have pointed to a distinct adhesion phenotype during eosinophil differentiation, characterized by the upregulation of beta-7 integrin and downregulation of β1 and beta-5 integrin.[25,26] Thus, VLA-4 has been suggested to be a key adhesive receptor for HSCs since administration of VLA-4 blocking antibodies increases the number of hematopoietic progenitor cells in the circulation.[27] Likewise, the homing selectin ligand, P-selectin glycoprotein ligand 1 (PSGL-1/Selectin P ligand), is also known to play a key role in homing and adhesion of HSCs in the bone marrow through the interaction with E- and P-selectin on endothelial cells; however, its specific role in extramedullary recruitment of eosinophil progenitor remains unknown.[28] In addition, functionally, CD34 has been proposed to orchestrate the recruitment of hematopoietic cells from the bone marrow through the regulation of integrin-mediated adhesion to stromal cells,[29–31] and Blanchet and colleagues recently showed that CD34 itself appears to act like *Teflon* to enhance invasiveness and cell motility.[30] Thus, in a mouse model of allergic asthma, a reduced mobility of bone marrow-derived cells was evident in cells lacking CD34 expression. These experiments also suggest that CD34 regulates the motility of progenitors by inhibiting cell–cell adhesion.[30] Furthermore, a dramatic reduction in allergen-induced eosinophil and mast cell infiltration into the airways was shown, suggesting that CD34 expression is a prerequisite signal for the development of allergic airway inflammation. In addition to adhesion molecules, mouse studies have shown that sustained exposure to increased concentrations of IL-5 (in IL-5 transgenic mice) results in the increased sensitivity of progenitors to chemotactic factors that are elaborated at sites of tissue inflammation, leading to directed migration of progenitors to sites of inflammation.[4,32]

COMMUNICATION BETWEEN AIRWAYS AND BONE MARROW

Progenitor cells also express receptors that bind chemokines, including CCR3, C-C motif chemokine 3 (CCL3/MIP-1-alpha), and as previously mentioned, SDF-1, all of which have been implicated in different aspects of the mobilization and motility of progenitor cells.[33,34] Thus, extramedullary recruitment of eosinophil progenitors is most likely regulated by a coordinated network and by interactions between adhesion molecules and their receptors on the endothelium in a way that is regulated

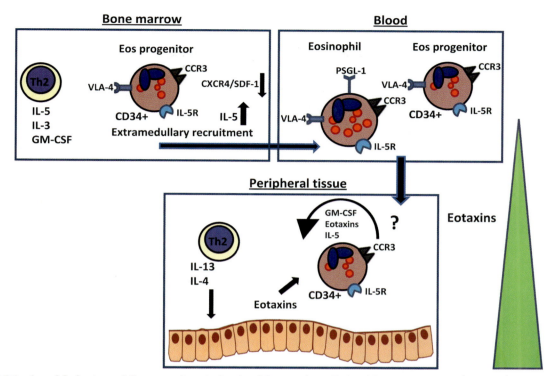

FIGURE 5.5.1 A model of extramedullary recruitment of eosinophil progenitors. Eosinophil progenitors (CD34[+] IL-5RA[+] cells) differentiate in the bone marrow under the influence of T-helper type 2 (T$_h$2) lymphocyte-derived granulocyte-macrophage colony-stimulating factor (GM-CSF), interleukin-3 (IL-3), and interleukin-5 (IL-5). In response to airway allergen exposure, both mature eosinophils and their progenitors exit the bone marrow. This process is mainly regulated by IL-5, which most likely primes the cells for extramedullary recruitment by increasing C-C chemokine receptor type 3 (CCR3) on their surface. In addition, a decrease in C-X-C chemokine receptor type 4 (CXC-R4/SDF-1) interactions in the bone marrow, as well as a decrease in VLA-4 subunit beta (VLA-4), might also result in the egress of progenitors from the bone marrow in response to allergen exposure. In the bloodstream, eosinophil progenitors respond to chemokine signaling (eotaxins/CCR3) and adhesion molecules and their receptors. Thus, eosinophil progenitors most likely exit the bloodstream by interaction with the endothelium, which is mediated through selectins, such as P-selectin glycoprotein ligand 1 (PSGL-1/selectin P ligand) and integrins, such as VLA-4 and vascular cell adhesion protein 1 (VCAM-1) in a direction determined by an eotaxin gradient. Furthermore, interleukin-4 (IL-4) and 13 (IL-13) produced by T$_h$2, epithelial, and endothelial cells promote tissue eosinophilia by regulating adhesion pathways as well as local production of eotaxin. Eosinophil progenitors in the peripheral tissue proliferate and undergo *in situ* differentiation, which is presumably driven by GM-CSF, the eotaxins, and IL-5. IL-5RA, interleukin-5 receptor subunit alpha; SDF-1, stromal cell-derived factor 1.

via cytokine and chemokine gradients (Fig. 5.5.1 and Table 5.5.1).

Allergen-induced airway inflammation results in enhanced bone marrow eosinophilopoiesis in asthmatic subjects and in animal models. Although the nature of bone marrow activation after allergen exposure is unknown, several pathways have been suggested, including stimulation of resident bone marrow cells and release of hematopoietic growth factors. Animal studies suggest that bone marrow eosinophilia is initiated in a complex manner, involving T cell-independent and T cell-dependent mechanisms. Thus, it has been shown that inhaled allergen induces trafficking of IL-5-producing T cells to the bone marrow and thus regulates the events occurring in the bone marrow.[35] Furthermore, adoptive transfer of CD3[+] T cells overexpressing IL-5 causes increased bone marrow eosinophilia in sensitized and allergen-challenged mice.[36] Several lines of evidence exist to demonstrate an increase of IL-5RA expression on CD34[+] cells in subjects with

allergic asthma with a coincident expression of CD3[+] IL-5 mRNA[+] cells in bone marrow.[37] Taken together, in response to allergic stimuli in the lung, IL-5 regulates the expansion of eosinophil progenitors in the bone marrow and potentiates mobilization into the blood. In the blood, eosinophils and their progenitors respond to chemotactic eotaxin (CCL11) signals from the respiratory epithelial cells. Thus, a second signal operates in the allergic lung that regulates the transmigration of these cells into the tissue. This is mainly coordinated by the T-helper type 2 (T$_h$2) cytokines interleukin-4 and interleukin-13 (IL-13), most likely by regulating adhesion molecules and by regulating the local production of the eotaxin chemokines (Fig. 5.5.1).

A recent study showed that both human primitive bone marrow CD34[+] cells and eosinophil-lineage committed progenitor cells (CD34[+] IL-5RA[+] cells) express the eotaxin (CCL11) receptor, CCR3.[13] Importantly, this study also revealed that CD34[+] CCR3[+] cells are significantly increased in the bone marrow of atopic subjects who

TABLE 5.5.1 Inflammatory Receptors and their Corresponding Ligands Expressed on Eosinophil Progenitors

Receptor	Ligand
Integrin alpha-4 (CD49d)/ integrin beta-1 (CD29)	Fibronectin; VCAM-1
Integrin alpha-4/integrin beta-7	Fibronectin, VCAM-1, MAdCAM-1
CD11b/CD18, CR3	ICAM-1
CCR3/CD193	Eotaxins
CD44	Hyaluronic acid
c-KIT/CD117	SCF
CXC-R4/CD184	SDF-1
IL-5RA/CDw125	IL-5
PSGL-1/CD162	E-Selectin, P-Selectin

c-Kit; CR3, complement receptor C3 subunit beta; CCR3, C-C chemokine receptor type 3; CXC-R4, C-X-C chemokine receptor type 4; ICAM-1, intercellular adhesion molecule 1; IL-5, interleukin-5; IL-5RA, interleukin-5 receptor subunit alpha; MAdCAM-1, mucosal addressin cell adhesion molecule 1; PSGL-1, P-selectin glycoprotein ligand 1; SCF, stem cell factor; SDF-1, stromal cell-derived factor 1; VCAM-1, vascular cell adhesion protein 1.

develop an asthmatic response, such as increased airway responsiveness to methacholine as well as blood and sputum eosinophilia, following airway allergen challenge. In contrast, those atopic subjects who did not develop asthma following allergen challenge showed attenuation of CD34[+] CCR3[+] bone marrow cells compared with baseline conditions. Furthermore, the upregulation in CCR3 expression was associated with an increased migration response by bone marrow CD34[+] cells toward eotaxin (CCL11) in vitro. These data suggest that upregulation of CCR3 on CD34[+] cells may facilitate the mobilization of progenitor cells from the bone marrow in response to allergen challenge, thus facilitating the development of pulmonary eosinophilia, a hallmark of allergen-induced asthma. Moreover, data have suggested that IL-5 may prime CD34[+] progenitor cells to respond to eotaxin (CCL11), presumably by the upregulation of CCR3, and that eotaxin (CCL11) may induce the expression of IL-5RA in CD34[+] cells, thus priming them to respond to IL-5.[8] By evaluating CCR3 receptor density using fluorescence-activated cell sorting (FACS), we have been able to determine that the CD34[+] cells in the bone marrow of allergen-challenged mice have higher numbers of receptors per cell when compared to mature bone marrow eosinophils, quantified as the mean fluorescent index (Rådinger et al., unpublished data), which may explain the greater migration

of these cells in response to eotaxin (CCL11) and eotaxin-2 [C-C motif chemokine 24 (CCL24)].[33]

A study using normal subjects and atopic asthmatic subjects with either early or early and late responses to allergen exposure suggests that fluctuations in C-X-C chemokine receptor type 4 (CXC-R4) and CCR3 expression on bone marrow CD34[+] cells, together with changes in SDF-1 and eotaxin (CCL11) levels, may regulate the release of progenitors from the bone marrow into the blood.[17] Allergen challenge decreased CXC-R4 expression and SDF-1-alpha levels in bone marrow in those with late asthmatic responses to allergen. By contrast, the intensity of CCR3 expression on CD34[+] bone marrow cells was significantly increased in these subjects, as well as enhanced migrational responses toward the eotaxin (CCL11) of bone marrow CD34[+] cells. Consistent with these findings, CCR3 expression has been found on mouse CD34[+] bone marrow cells and eotaxin (CCL11) has been shown to mobilize bone marrow-derived CD34[+] cells in vivo from blood into the skin in mice.[8] Furthermore, intravenous infusion of eotaxin (CCL11) has been shown to stimulate a rapid mobilization of eosinophil progenitors from the bone marrow into the peripheral circulation in guinea pigs.[12] Interestingly, the percentage of both CD34[+] cells and lineage-committed progenitors (CD34[+] IL-5RA[+] cells), as well as mature eosinophils expressing CCR3, are increased in the bone marrow of asthma subjects compared to controls, suggesting that there is a larger pool of bone marrow CCR3[+] cells available for rapid mobilization.[13,16,38]

Taken together, these studies suggest a significant role for the eotaxin/CCR3 pathway in promoting eosinophil progenitor release from the bone marrow in allergen-induced asthma. To date, however, it is not clear at which stage of eosinophil differentiation and recruitment eotaxin (CCL11), eotaxin-2 (CCL24), or IL-5 exert their effect on CD34+ progenitor cells. Even though the trafficking of eosinophils from the bone marrow to sites of inflammation is dependent on a variety and coordinated action of chemokines, cytokines, and adhesion molecules, as well as leukotrienes, only IL-5 and the eotaxins selectively regulate eosinophil trafficking. Thus, during either steady-state traffic, or in response to potent systemic inflammatory signals, eosinophil progenitors are equipped with the appropriate chemokine receptors as well as adhesion and homing molecules for trafficking to sites of an ongoing inflammation.

EVIDENCE FOR EXTRAMEDULLARY DIFFERENTIATION AND PROLIFERATION OF EOSINOPHIL PROGENITORS

Differentiation and maturation of hematopoietic progenitors have traditionally been thought to be restricted to the bone marrow environment. However, accumulating evidence suggests that eosinophil progenitors present in the allergic tissue may contribute to the process of allergic inflammation by both giving rise to mature eosinophils within the bone marrow, and by trafficking as undifferentiated cells via the peripheral circulation to the allergen-exposed tissue, where they differentiate under the control of specific chemokines and cytokines.[39,40] This latter process has been termed *in situ* hematopoiesis. The fact that hematopoietic progenitors can expand in the local tissue of allergic subjects is further supported by an increased number of IL-5RA mRNA$^+$ CD34$^+$cells found in biopsies from bronchial mucosa in subjects with atopic asthma.[41] Likewise, we have recently shown that subjects with allergic rhinitis have an increased number of CD34$^+$ progenitor cells and eosinophils expressing CD34 in the nasal mucosa during the pollen season.[42]

One of the first studies that suggested that eosinophilic and basophilic progenitors might traffic from the bone marrow to the airways, where they undergo *in situ* hematopoiesis, was by Otsuka and colleagues.[43] This study demonstrated that subjects developing symptoms of rhinitis during the allergic season have decreased numbers of circulating progenitors during the height of the season, compared to before and after the season, suggesting trafficking of these cells into tissues in response to allergen challenge. Several studies have subsequently provided further evidence for *in situ* hematopoiesis, including a study by Cameron and colleagues,[39] which clearly demonstrated that allergen-induced eosinophil differentiation could be blocked by soluble IL-5RA.[39] A recent study showed a decline in IL-5RA$^+$ CD34$^+$ mRNA cells within the bronchial mucosa of asthmatics 24 h after the inhalation of IL-5,[44] which is in agreement with the work by Cameron and colleagues, showing that in mucosal tissue incubated with IL-5 for 24 h, the number of IL-5RA mRNA$^+$ CD34$^+$ cells significantly decreased with a subsequent increase in mature eosinophils [i.e., eosinophil granule major basic protein (MBP-1)$^+$ cells].

Furthermore, we have previously shown that CD34$^+$ cells release markedly more IL-5 compared with CD34$^-$ eosinophils, suggesting that airway CD34$^+$ cells may play an autocrine role in their final maturation to eosinophils.[42] In addition to the role of IL-5 in eosinophil differentiation and maturation, studies in humans have revealed a role for eotaxin (CCL11) in the differentiation of CD34$^+$ cells toward eosinophils, because cord blood-derived CD34$^+$ cells cultured in the presence of eotaxin (CCL11) differentiate into eosinophils.[8] Human endothelial cell-derived eotaxin (CCL11) was recently shown to be acting as an eosinophil survival factor by altering the capacity of peripheral blood-derived eosinophils to undergo apoptosis.[45] Finally, we have recently shown that CD34$^+$ IL-5RA$^+$ eosinophil progenitors are increased in the

airways of allergen-challenged mice, confirming previous published data in mice and humans.[18,46−49] However, in our study, we also demonstrated a significant increase in proliferating IL-5RA$^+$ cells *in vivo* in the lung in response to allergen challenge. Interestingly, most of the IL-5RA$^+$ cells in the airways of allergen-exposed mice coexpressed CCR3, implicating that these receptors may have complementary functions in cell proliferation and maturation of the local lung eosinophil progenitor.[33] Indeed, studies have demonstrated that *in vitro* colony formation of lung CD34$^+$ cells is increased by IL-5.[42,49] However, our study also suggests a role for eotaxin-2 (CCL24) in this process, since in addition to IL-5, eotaxin-2 (CCL24) increased in the *in vitro* colonies of lung CD34$^+$ cells. Taken together, these data suggest that changes in eosinophil numbers after allergen exposure may be caused at least in part from local differentiation of progenitor cells via both eotaxin-2 (CCL24) and IL-5-dependent mechanisms.

EOSINOPHIL PROGENITORS AS TRUE EFFECTOR CELLS IN ALLERGIC INFLAMMATION

A growing body of evidence points out a role for stem cells in chronic inflammation and emerging evidence points out that stem cells themselves could augment the degree of tissue inflammation via the production of inflammatory and chemotactic mediators, thereby promoting recruitment of both other precursors and mature inflammatory cells. Thus, HSCs and progenitors have been shown to respond to inflammatory stimuli in both the bone marrow and in peripheral tissue. We, and others, have shown that CD34$^+$ cells themselves respond to allergen, since it was shown that allergen stimulates bone marrow CD34$^+$ cells to release IL-5 *in vitro*.[42,50] Furthermore, circulating CD34$^+$ cells from asthmatic subjects expressed intracellular IL-5 and likewise demonstrated high expression of IL-5 mRNA.[51] A recent study showed that human CD34$^+$ progenitor cells can act as effector cells in allergy, as sputum CD34$^+$ cells from allergic subjects contained both IL-5 and IL-13.[52] The same study revealed that human CD34$^+$ cells express two main receptors involved in driving the T$_h$2-dependent allergic response, namely the thymic stromal lymphopoietin (TSLP) and interleukin-33 (IL-33) receptors. Furthermore, CD34$^+$ cells were capable to release GM-CSF, IL-5, and IL-13 in response to TSLP and IL-33 stimulation. Thus, the above studies clearly demonstrate that CD34$^+$ progenitors may participate in and promote chronic inflammation.

In summary, eosinophil progenitors are equipped with adhesion and homing molecules, as well as the appropriate chemokine and cytokine receptors, for either steady-state trafficking or in response to systemic inflammatory signals.

Moreover, eosinophil progenitors at the site of inflammation most likely proliferate and potentially augment the degree of inflammation by amplifying the source of effector cells, such as eosinophils, at the site of inflammation, but also in maintaining a constant supply of inflammatory cytokines that drives the ongoing inflammation.

REFERENCES

1. Mayani H, Guilbert LJ, Janowska-Wieczorek A. Biology of the hemopoietic microenvironment. *Eur J Haematol* 1992;**49**:225−33.
2. Iwasaki H, Mizuno S, Mayfield R, Shigematsu H, Arinobu Y, Seed B, et al. Identification of eosinophil lineage-committed progenitors in the murine bone marrow. *J Exp Med* 2005;**201**:1891−7.
3. McEwen BJ. Eosinophils: a review. *Vet Res Commun* 1992;**16**:11−44.
4. Lee NA, McGarry MP, Larson KA, Horton MA, Kristensen AB, Lee JJ. Expression of IL-5 in Thymocytes/T Cells Leads to the Development of a Massive Eosinophilia, Extramedullary Eosinophilopoiesis, and Unique Histopathologies. *Journal of Immunology* 1997;**158**:1332−44.
5. Kopf M, Brombacher F, Hodgkin PD, Ramsay AJ, Milbourne EA, Dai WJ, et al. IL-5-deficient mice have a developmental defect in CD5+ B-1 cells and lack eosinophilia but have normal antibody and cytotoxic T cell responses. *Immunity* 1996;**4**:15−24.
6. Yoshida T, Ikuta K, Sugaya H, Maki K, Takagi M, Kanazawa H, et al. Defective B-1 cell development and impaired immunity against Angiostrongylus cantonensis in IL-5R alpha-deficient mice. *Immunity* 1996;**4**:483−94.
7. Collins PD, Marleau S, Griffiths-Johnson DA, Jose PJ, Williams TJ. Cooperation between interleukin-5 and the chemokine eotaxin to induce eosinophil accumulation in vivo. *J Exp Med* 1995;**182**:1169−74.
8. Lamkhioued B, Abdelilah SG, Hamid Q, Mansour N, Delespesse G, Renzi PM. The CCR3 receptor is involved in eosinophil differentiation and is up-regulated by Th2 cytokines in CD34+ progenitor cells. *J Immunol* 2003;**170**:537−47.
9. Mendez-Ferrer S, Lucas D, Battista M, Frenette PS. Haematopoietic stem cell release is regulated by circadian oscillations. *Nature* 2008;**452**:442−7.
10. Gouon-Evans V, Rothenberg ME, Pollard JW. Postnatal mammary gland development requires macrophages and eosinophils. *Development* 2000;**127**:2269−82.
11. Rothenberg ME, Mishra A, Brandt EB, Hogan SP. Gastrointestinal eosinophils in health and disease. *Adv Immunol* 2001;**78**:291−328.
12. Palframan RT, Collins PD, Williams TJ, Rankin SM. Eotaxin induces a rapid release of eosinophils and their progenitors from the bone marrow. *Blood* 1998;**91**:2240−8.
13. Sehmi R, Dorman S, Baatjes A, Watson R, Foley R, Ying S, et al. Allergen-induced fluctuation in CC chemokine receptor 3 expression on bone marrow CD34+ cells from asthmatic subjects: significance for mobilization of haemopoietic progenitor cells in allergic inflammation. *Immunology* 2003;**109**:536−46.
14. Quackenbush EJ, Wershil BK, Aguirre V, Gutierrez-Ramos JC. Eotaxin modulates myelopoiesis and mast cell development from embryonic hematopoietic progenitors. *Blood* 1998;**92**:1887−97.
15. Li J, Saito H, Crawford L, Inman MD, Cyr MM, Denburg JA. Haemopoietic mechanisms in murine allergic upper and lower airway inflammation. *Immunology* 2005;**114**:386−96.
16. Zeibecoglou K, Ying S, Yamada T, North J, Burman J, Bungre J, et al. Increased mature and immature CCR3 messenger RNA+ eosinophils in bone marrow from patients with atopic asthma compared with atopic and nonatopic control subjects. *J Allergy Clin Immunol* 1999;**103**:99−106.
17. Dorman SC, Babirad I, Post J, Watson RM, Foley R, Jones GL, et al. Progenitor egress from the bone marrow after allergen challenge: role of stromal cell-derived factor 1alpha and eotaxin. *J Allergy Clin Immunol* 2005;**115**:501−7.
18. Dorman SC, Efthimiadis A, Babirad I, Watson RM, Denburg JA, Hargreave FE, et al. Sputum CD34+IL-5Ralpha+ cells increase after allergen: evidence for in situ eosinophilopoiesis. *Am J Respir Crit Care Med* 2004;**169**:573−7.
19. Leavesley DI, Oliver JM, Swart BW, Berndt MC, Haylock DN, Simmons PJ. Signals from platelet/endothelial cell adhesion molecule enhance the adhesive activity of the very late antigen-4 integrin of human CD34+ hemopoietic progenitor cells. *J Immunol* 1994;**153**:4673−83.
20. Mohle R, Murea S, Kirsch M, Haas R. Differential expression of L-selectin, VLA-4, and LFA-1 on CD34+ progenitor cells from bone marrow and peripheral blood during G-CSF-enhanced recovery. *Exp Hematol* 1995;**23**:1535−42.
21. Imamura R, Miyamoto T, Yoshimoto G, Kamezaki K, Ishikawa F, Henzan H, et al. Mobilization of human lymphoid progenitors after treatment with granulocyte colony-stimulating factor. *J Immunol* 2005;**175**:2647−54.
22. Gigant C, Latger-Cannard V, Bensoussan D, Feugier P, Bordigoni P, Stoltz JF. Quantitative expression of adhesion molecules on granulocyte colony-stimulating factor-mobilized peripheral blood, bone marrow, and cord blood CD34+ cells. *J Hematother Stem Cell Res* 2001;**10**:807−14.
23. Sovalat H, Racadot E, Ojeda M, Lewandowski H, Chaboute V, Henon P. CD34+ cells and CD34+CD38− subset from mobilized blood show different patterns of adhesion molecules compared to those from steady-state blood, bone marrow, and cord blood. *J Hematother Stem Cell Res* 2003;**12**:473−89.
24. Catalli AE, Thomson JV, Babirad IM, Duong M, Doyle TM, Howie KJ, et al. Modulation of beta1-integrins on hemopoietic progenitor cells after allergen challenge in asthmatic subjects. *J Allergy Clin Immunol* 2008;**122**:803−10.
25. Lundahl J, Sehmi R, Hayes L, Howie K, Denburg JA. Selective upregulation of a functional beta7 integrin on differentiating eosinophils. *Allergy* 2000;**55**:865−72.
26. Lundahl J, Sehmi R, Moshfegh A, Hayes L, Howie K, Upham J, et al. Distinct phenotypic adhesion molecule expression on human cord blood progenitors during early eosinophilic commitment: upregulation of beta(7) integrins. *Scand J Immunol* 2002;**56**:161−7.
27. Craddock CF, Nakamoto B, Andrews RG, Priestley GV, Papayannopoulou T. Antibodies to VLA4 integrin mobilize long-term repopulating cells and augment cytokine-induced mobilization in primates and mice. *Blood* 1997;**90**:4779−88.
28. Katayama Y, Hidalgo A, Furie BC, Vestweber D, Furie B, Frenette PS. PSGL-1 participates in E-selectin-mediated progenitor homing to bone marrow: evidence for cooperation between E-selectin ligands and alpha4 integrin. *Blood* 2003;**102**:2060−7.

29. Drew E, Merkens H, Chelliah S, Doyonnas R, McNagny KM. CD34 is a specific marker of mature murine mast cells. *Exp Hematol* 2002;**30**:1211.

30. Blanchet MR, Maltby S, Haddon DJ, Merkens H, Zbytnuik L, McNagny KM. CD34 facilitates the development of allergic asthma. *Blood* 2007;**110**:2005—12.

31. Nielsen JS, McNagny KM. Influence of host irradiation on long-term engraftment by CD34-deficient hematopoietic stem cells. *Blood* 2007;**110**:1076—7.

32. Khaldoyanidi S, Sikora L, Broide DH, Rothenberg ME, Sriramarao P. Constitutive overexpression of IL-5 induces extramedullary hematopoiesis in the spleen. *Blood* 2003; **101**:863—8.

33. Radinger M, Bossios A, Sjostrand M, Lu Y, Malmhall C, Dahlborn AK, et al. Local proliferation and mobilization of CCR3(+) CD34(+) eosinophil-lineage-committed cells in the lung. *Immunology*, http://dx.doi.org/10.1111/j.1365-2567.2010.03349.x; 2010.

34. Broxmeyer HE, Kim CH, Cooper SH, Hangoc G, Hromas R, Pelus LM. Effects of CC, CXC, C, and CX3C chemokines on proliferation of myeloid progenitor cells, and insights into SDF-1-induced chemotaxis of progenitors. *Ann N Y Acad Sci* 1999;**872**:142—62. discussion 163.

35. Minshall EM, Schleimer R, Cameron L, Minnicozzi M, Egan RW, Gutierrez-Ramos JC, et al. Interleukin-5 expression in the bone marrow of sensitized Balb/c mice after allergen challenge. *Am J Respir Crit Care Med* 1998;**158**:951—7.

36. Johansson AK, Sergejeva S, Sjostrand M, Lee JJ, Lotvall J. Allergen-induced traffic of bone marrow eosinophils, neutrophils and lymphocytes to airways. *Eur J Immunol* 2004;**34**:3135—45.

37. Wood LJ, Sehmi R, Dorman S, Hamid Q, Tulic MK, Watson RM, et al. Allergen-induced increases in bone marrow T lymphocytes and interleukin-5 expression in subjects with asthma. *Am J Respir Crit Care Med* 2002;**166**:883—9.

38. Robinson DS, North J, Zeibecoglou K, Ying S, Meng Q, Rankin S, et al. Eosinophil development and bone marrow and tissue eosinophils in atopic asthma. *Int Arch Allergy Immunol* 1999;**118**:98—100.

39. Cameron L, Christodoulopoulos P, Lavigne F, Nakamura Y, Eidelman D, McEuen A, et al. Evidence for local eosinophil differentiation within allergic nasal mucosa: inhibition with soluble IL-5 receptor. *J Immunol* 2000;**164**:1538—45.

40. Kim YK, Uno M, Hamilos DL, Beck L, Bochner B, Schleimer R, et al. Immunolocalization of CD34 in nasal polyposis. Effect of topical corticosteroids. *Am J Respir Cell Mol Biol* 1999; **20**:388—97.

41. Robinson DS, Damia R, Zeibecoglou K, Molet S, North J, Yamada T, et al. CD34(+)/interleukin-5Ralpha messenger RNA+ cells in the bronchial mucosa in asthma: potential airway eosinophil progenitors. *Am J Respir Cell Mol Biol* 1999;**20**:9—13.

42. Sergejeva S, Johansson AK, Malmhall C, Lotvall J. Allergen exposure-induced differences in CD34+ cell phenotype: relationship to eosinophilopoietic responses in different compartments. *Blood* 2004;**103**:1270—7.

43. Otsuka H, Dolovich J, Befus AD, Telizyn S, Bienenstock J, Denburg JA. Basophilic cell progenitors, nasal metachromatic cells, and peripheral blood basophils in ragweed-allergic patients. *J Allergy Clin Immunol* 1986;**78**:365—71.

44. Menzies-Gow AN, Flood-Page PT, Robinson DS, Kay AB. Effect of inhaled interleukin-5 on eosinophil progenitors in the bronchi and bone marrow of asthmatic and non-asthmatic volunteers. *Clin Exp Allergy* 2007;**37**:1023—32.

45. Farahi N, Cowburn AS, Upton PD, Deighton J, Sobolewski A, Gherardi E, et al. Eotaxin-1/CC chemokine ligand 11: a novel eosinophil survival factor secreted by human pulmonary artery endothelial cells. *J Immunol* 2007;**179**:1264—73.

46. Sehmi R, Wood LJ, Watson R, Foley R, Hamid Q, O'Byrne PM, et al. Allergen-induced increases in IL-5 receptor alpha-subunit expression on bone marrow-derived CD34+ cells from asthmatic subjects. A novel marker of progenitor cell commitment towards eosinophilic differentiation. *J Clin Invest* 1997;**100**:2466—75.

47. Dorman SC, Sehmi R, Gauvreau GM, Watson RM, Foley R, Jones GL, et al. Kinetics of bone marrow eosinophilopoiesis and associated cytokines after allergen inhalation. *Am J Respir Crit Care Med* 2004;**169**:565—72.

48. Radinger M, Lotvall J. Eosinophil progenitors in allergy and asthma—do they matter? *Pharmacol Ther* 2009;**121**:174—84.

49. Southam DS, Widmer N, Ellis R, Hirota JA, Inman MD, Sehmi R. Increased eosinophil-lineage committed progenitors in the lung of allergen-challenged mice. *J Allergy Clin Immunol* 2005; **115**:95—102.

50. Johansson AK, Sjostrand M, Tomaki M, Samulesson AM, Lotvall J. Allergen stimulates bone marrow CD34(+) cells to release IL-5 in vitro; a mechanism involved in eosinophilic inflammation? *Allergy* 2004;**59**:1080—6.

51. Kuo HP, Wang CH, Lin HC, Hwang KS, Liu SL, Chung KF. Interleukin-5 in growth and differentiation of blood eosinophil progenitors in asthma: effect of glucocorticoids. *Br J Pharmacol* 2001;**134**:1539—47.

52. Allakhverdi Z, Comeau MR, Smith DE, Toy D, Endam LM, Desrosiers M, et al. CD34+ hemopoietic progenitor cells are potent effectors of allergic inflammation. *J Allergy Clin Immunol* 2009;**123**:472—8.

Chapter 5.6

Mouse Models Manipulating Eosinophilopoiesis

Nancy A. Lee

INTRODUCTION

Eosinophils have long been thought to be the *bad guys* in inflammation. Who, what, where, when? The answers make these granulocytes the prime suspect. Who? Eosinophils are often seen at sites of inflammation, correlating with observed pathologies. What? Eosinophils express many potentially toxic proteins and generate copious amounts of reactive oxygen species. Where? Locations are often linked with environmental interfaces or sites of defined immune-mediated inflammatory events. When? More so when there is active inflammation or disease. In the human analogy of *film noir*, eosinophils are often thought of as ominously lurking in a shadowy blood vessel, smoking a Tareyton, and looking malevolent—the eosinophil version of William Holden carrying an eosinophil effector gun that shoots at unsuspecting innocent tissues.

This classic view of eosinophils has dominated thought for many years. However, more recent studies have shown that eosinophils may be a functional link in the immune system. These mysterious granulocytes appear to mediate intricate behind-the-scenes networks controlling the immune system, sometimes leading to inflammation, leaving only the faintest hint of their presence. In particular, these new studies suggest that eosinophils can secrete specific cytokines, interact with T cells, and possibly dendritic cells, as part of cascades leading to inflammation and other pathologies.

The advent of transgenic and gene knockout/knock-in mouse technology has allowed researchers to modify and manipulate circulating eosinophil numbers as well as to add, modify, or delete specific genes involved in eosinophil biology. This chapter on mouse models of eosinophil proliferation and deficiency is divided into two sections. The first section discusses models of mice that show *increased* levels of eosinophils. The second section discusses mouse models that show *decreased* numbers of eosinophils.

Specifically, there are multiple transgenic mouse models using tissue-specific promoters to express interleukin-5 (IL-5) in transgenic mice. The one consensus taken from these multiple studies is that increasing eosinophils in specific areas affects the local immune microenvironment and leads to pathologies associated with T-helper type 2 (T_h2)-mediated diseases. Alternatively, eosinophil deficiency and models of eosinophil ablation in mice helped to prove that eosinophils are required for a specific pathology. Thus, together with add-back strategies using eosinophils with various gene deletions (created from knockout mice crossed with IL-5 transgenic mice), a very powerful strategy emerged for elucidating the role of eosinophils and specific gene products in disease.

MOUSE MODELS WITH INCREASED NUMBERS OF EOSINOPHILS

CD2/Interleukin-5 (Transgenic 5C1 and 5C2)[1]
Metallothionein 1/Interleukin-5 Mice (Inducible Increased Interleukin-5 Expression)[2]
NJ.1638 (CD3δ/Interleukin-5 Mice)[3]
NJ.1726 (Clara Cells 10 kDa Secretory Protein/Interleukin-5)[4]
NJ.692 Keratin-14/Interleukin-5 Transgenic Mice[5]
I5E2 (CD3δ/Interleukin-5 and Clara Cells 10 kDa Secretory Protein/Eotaxin-2 Double Transgenic Mice)[6]

MOUSE MODELS WITH DECREASED NUMBERS OF EOSINOPHILS

Interleukin-5 Knockout Mice[7]
ΔdblGATA Mice[8]
PHIL[9]
Inducible PHIL (our unpublished data)
CD70 Transgenic Mice[10]

MOUSE MODELS WITH INCREASED NUMBERS OF EOSINOPHILS

CD2/Interleukin-5 (Transgenic 5C1 and 5C2)

Dent and colleagues published their work generating transgenic mice that expressed mouse IL-5 using the cis-regulatory elements from the human *CD2 molecule (CD2)* gene.[1] In their CD2/IL-5 transgenic construct, the authors used the human CD2 enhancer and promoter element inserted before a mouse *interleukin 5 (colony-stimulating factor, eosinophil) IL5* genomic gene, leaving the 5′ regulatory regions intact from the IL-5 locus. The CD2 dominant control region and promoter has been shown to restrict gene expression to B cells, thymocytes, and T cell subpopulations in transgenic mice in a copy number-dependent manner.[11] Four independent lines of mice were created, each expressing a different level of IL-5, and two of these four lines are described in detail here.

Levels of serum IL-5 in these two transgenic lines (Tg5C1 and Tg5C2) were found to be 65-fold and 265-fold elevated over wild-type, respectively, with *Mesocestoides corti*-infected wild-type mice showing a further 35-fold elevation of serum IL-5. The investigators demonstrated that the CD2/IL-5 transgenic construct results in constitutive expression of the *IL5* gene and concomitant high eosinophil levels. These authors reported that eosinophil levels (relative to wild-type) increased 130% and 530% in the peripheral blood of the two independent transgenic lines relative to wild-type mice. This level of eosinophils in the blood was comparable to that found in sera from mice infected with *M. corti*, but not as high as other transgenic IL-5 mice generated subsequently (discussed later in this chapter). In particular, white blood cell (WBC) counts in CD2/IL-5 mice were increased only about threefold over those of wild-type animals, suggesting that the presence of the 5′ regulatory regions from the IL-5 promoter may have affected the levels of expression obtained and/or the specificity of transgene expression of IL-5. In summary, these CD2/IL-5 transgenic mice were the first *hypereosinophilic* mice showing that overexpression of IL-5 could increase the number of eosinophils. However, the presence of the endogenous IL-5 promoter elements may have resulted in only moderately increased levels of eosinophils, as opposed to the dramatic increases seen in later studies in which the endogenous IL-5 promoter elements were removed (discussed later in this chapter).

Metallothionein 1/Interleukin-5 Mice (Inducible Increased Interleukin-5 Expression)

Tominaga and colleagues reported that they had generated transgenic mice expressing IL-5 from a metallothionein 1 (MT-1) promoter.[2] The authors used a mouse (m)MT-1 promoter fragment to express a mouse IL-5 cDNA, creating an expression construct free of endogenous IL-5 regulatory elements. The MT-1 promoter is induced by the administration of cadmium or zinc. Two independent founder animals were generated and studied. Serum IL-5 levels following heavy metal administration in the water were measured and found to be elevated, between 2.3 and 16.6 ng/mL for homozygotes whereas wild-type mice had levels below 0.01 ng/mL. Moreover, cadmium sulfate injection into transgenic mice increased the serum IL-5 levels within 24 h. Transgenic IL-5 production was demonstrated in both liver and spleen tissues by Northern blot analysis. The increased production of IL-5 led to eosinophil levels that were increased over wild-type littermates, from 20-fold in hemizygotes to 70-fold in homozygotes. Eosinophil levels and serum immunoglobulin levels were both elevated in transgenic mice, demonstrating effects on both B cells and eosinophils.

NJ.1638 (CD3δ/Interleukin-5 Mice)

Lee and colleagues used the CD3δ promoter and enhancer regions to express IL-5 from a construct in which all the IL-5 regulatory regions had been eliminated.[3] Specifically, this was accomplished by fusing a mouse cDNA to the 5′ end of a mouse genomic clone, creating a genomic/cDNA *minigene*. This placed the minigene under the control of a T cell-specific regulatory element, the CD3δ enhancer and promoter, restricting expression to all T cells and thymocytes.[12]

Unlike previous IL-5 transgenic models, the resulting NJ.1638 mice displayed a remarkable increase in circulating eosinophils. Out of three independently derived transgenic lines, each showing increased peripheral eosinophil numbers and elevated WBC counts, one, NJ.1638, was chosen for extensive analysis. Northern blot analysis of total RNA showed the tissue-specific expression of IL-5 in the thymus and other tissues with significant numbers of mature T cells, such as spleen, bone marrow, and blood. Circulating serum levels of IL-5 were elevated consistently to levels of 400−800 pg/mL. These transgenic mice not only showed dramatically increased eosinophil levels, but also a significant increase in other WBC numbers, especially B cells. The increase in both eosinophils and WBC numbers was truly remarkable. For example, WBC counts in these transgenic mice averaged greater than $250,000$ cells/mm^3 of blood. Mice with WBC counts greater than 1 million cells/mm^3 of blood were not uncommon. Concomitantly, the circulating eosinophil levels in these mice increased to greater than 40% of all WBC. Thus, unlike wild-type mice which have $60-180$ eosinophils/mm^3 in blood, NJ.1638 mice display levels nearly 1000 times higher or approximately $150,000$ eosinophils/mm^3 of blood. Even the spleen and liver in transgenic mice were greatly enlarged due to the increased WBC numbers and increased levels of eosinophils (Fig. 5.6.1).

Surprisingly, it was subsequently shown that bone development was also affected by the transgenic IL-5 expression.[13] Thus, these data indicated that IL-5 overexpression in transgenic mice not only elevates eosinophil numbers but also mobilizes, in an eosinophil-dependent fashion, the formation of marrow-derived osteogenic progenitors or the inhibition of osteoclasts.

NJ.1726 (Clara Cells 10 kDa Secretory Protein/Interleukin-5)

Lee and colleagues expressed IL-5 specifically in the lung using the Clara cells 10 kDa secretory protein (CC10) promoter in transgenic mice.[4] The CC10 promoter directs expression to the non-ciliated Clara cells of the central airways of the lung including within the trachea, bronchi, and bronchioles.[14] The NJ.1726 mice developed eosinophil-mediated pathologies specifically in the lung tissue, including an increase in bronchoalveolar lymphoid tissue (BALT), increased goblet cell metaplasia, and increases in airway wall thickness and airway smooth muscle. Blood and bronchoalveolar lavage eosinophil numbers were increased relative to wild-type mice but did not increase further after antigen challenge. Pulmonary function after methacholine challenge showed significant increases in airway hyperresponsiveness, a response not observed in NJ.1638 mice, despite the much higher eosinophil numbers found in the T cell-specific line of mice. Significantly, no eosinophil degranulation was observed in the lungs of NJ.1726 mice, representing a significant difference from that observed in human asthmatics. In summary, this IL-5 transgenic line demonstrates the importance of site-specific expression of this cytokine—i.e., paracrine-mediated IL-5 activities.

Interestingly, pathologies arising from site-specific expression appeared to occur even within the lung, as not all lung-specific promoters produced the same phenotype

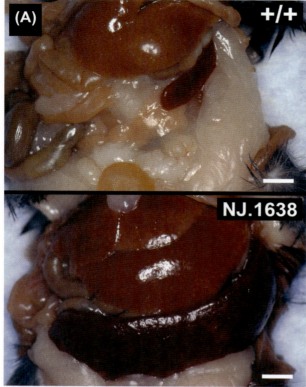

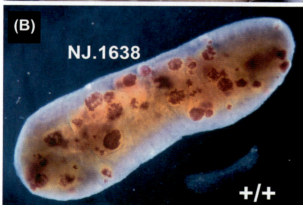

FIGURE 5.6.1 **Perturbations in spleen size and the *de novo* appearance of bone resulting from the overexpression of interleukin-5 in thymocytes/T cells.** *(A)* Ventral views of the abdominal cavities of a 7-month-old[+/+] male, and age- and sex-matched NJ.1638 mouse. The scale bar represents 0.5 cm. *(B)* The multifocal nodules of NJ.1638 spleens exhibit extensive staining for bone but not cartilage. Dark-field photomicrograph of a wild-type ([+/+]) and NJ.1638 spleen after whole-mount staining with alizarin red S/alcian blue to identify mineralized bone/cartilage, respectively. In contrast to wild-type ([+/+]) spleen, which displayed no staining with either reagent, numerous and randomly distributed nodules were evident in the NJ.1638 spleen as heavily alizarin red S-staining structures representing centers of mineralized bone.

as the CC10/IL-5 transgenic mice. The pulmonary surfactant-associated protein C (SP-C) promoter, a different lung-specific promoter[15] was also used to generate transgenic mice overexpressing IL-5 in the lung (N. Lee, unpublished

data). SP-C directs gene expression to the alveolar type II epithelial cells of the distal lung instead of the more proximal Clara cells.[16] The mice generated with the SP-C/IL-5 construct showed eosinophil infiltration into the lung tissues in all three independently derived founder lines with >30% of the transgenic mice developing fatal expansions of pulmonary lymphoid tissue.

NJ.692 Keratin-14/Interleukin-5 Transgenic Mice

Eosinophil proliferation and granule protein deposition are also key features of dermatological diseases, including atopic dermatitis or atopic eczema. Transgenic mice were created using the human keratin-14 promoter to express the mouse *IL5* minigene described earlier.[6] One of these transgenic lines (NJ.692) was chosen for further analysis.[6] The skin-specific expression of IL-5 was demonstrated by *in situ* hybridization of IL-5 to a section of skin from an NJ.692 transgenic mouse (Fig. 5.6.2). The number of eosinophils in the peripheral blood in NJ.692 is elevated tenfold relative to wild-type mice (37% vs. 3.5%, respectively). These mice develop skin-specific pathologies, including dermatitis and hair loss consistent with our site-specific hypothesis of IL-5-mediated activities. The NJ.692 mouse shows signs of skin pathology that appear to be caused by the influx of eosinophils into the skin. However, the increased eosinophils in the skin did not replicate many of the pathologies observed in human allergic skin diseases. NJ.692 more closely reproduced eosinophilic fasciitis,[17] with the majority of the eosinophils located within the fascia layer adjacent to the dermis.

Most of the eosinophils in NJ.692 appear intact and not degranulated. However, degranulation in NJ.692 was observed when skin tissue sections were analyzed from areas that the animal was scratching, suggesting that moderate skin injury may physically induce degranulation. It is also of note that eosinophil infiltration into the skin of transgenic mice accelerated nerve growth, suggesting that the pathophysiological itching in these mice may, in part, be the result of increased nerve growth.[6] In contrast to NJ.1726/IL-5 expression in the lung, and similar to NJ.1638 T cell-specific IL-5 expression, this transgenic line does not show any increased airway hyperresponsiveness (AHR), suggesting once again that the increased infiltration of eosinophils can cause tissue-specific pathologies, whether the tissue is blood, lung, or skin.

I5E2 (CD3δ/Interleukin-5 and Clara Cells 10 kDa Secretory Protein/Eotaxin-2 Double Transgenic Mice)

Many of the mouse models that were studied prior to 2007 did not show any evidence of eosinophil degranulation.

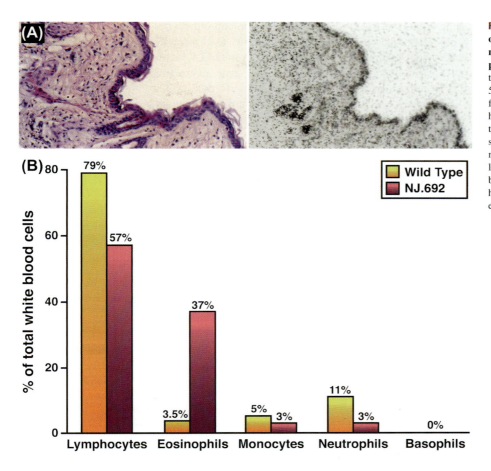

FIGURE 5.6.2 **Transgenic expression of interleukin-5 in the skin of NJ.692 mice leads to a selective increase of peripheral blood eosinophils.** *(A)* NJ.692 transgenic mice express interleukin 5 (IL-5) from a skin-specific promoter derived from the *keratin 14 gene (KRT14)*. *In situ* hybridization with an IL-5 probe showed transgene expression selectively in the skin keratinocytes of these transgenic mice. *(B)* NJ.692 mice have increased levels of eosinophils in the peripheral blood. This cell differential analysis highlights the selective eosinophilopoietic effect of ectopic IL-5 expression.

Malm-Erjefält and colleagues discussed the lack of evidence of eosinophil degranulation in mice as a fundamental distinction between the mouse models of allergic airway inflammation and human asthma.[18] Kobayashi and colleagues used the MT-1/IL-5 transgenic mice[2] and induced airway eosinophils with ovalbumin (OVA) sensitization and challenge;[19] however, they observed no increase in AHR in these antigen-stimulated IL-5 transgenic mice. They speculated that the lack of eosinophil degranulation in their study could be responsible for the lack of increased AHR. In addition, Clark and colleagues described that they only saw evidence of low level piecemeal degranulation in the airway lumen of IL-5 transgenic mice after allergen provocation.[20] Together, these studies suggested a serious flaw with the existing mouse models of eosinophil activities—they failed to replicate the eosinophil degranulation observed in human subjects.

The consistent and problematic inability to reproduce eosinophil degranulation in animal models was rectified by Ochkur and colleagues, who generated transgenic mice expressing IL-5 from the CD3δ promoter, which express IL-5 in all T cells, and the eosinophil agonist chemokine, eotaxin-2 [C-C motif chemokine 24 (CCL24)], which is expressed from the CC10 promoter only in lung epithelial Clara cells.[5] The resulting mice, I5/E2, not only showed all the signs of pathology present in NJ.1726,[4] but also significant levels of eosinophil degranulation, epithelial desquamation, as well as increases in AHR similar to what is seen in human asthma patients. The importance of this model is difficult to overestimate as these authors showed that the observed pathologies were eosinophil-specific. That is, crosses of these mice with eosinophil-less PHIL transgenic animals[9] (see later in this chapter) produced animals expressing IL-5 and eotaxin in the absence of eosinophils with *no* induced pathologies.

MOUSE MODELS WITH DECREASED NUMBERS OF EOSINOPHILS

Interleukin-5 Knockout Mice

A genetic deletion of the *IL5* gene was created in knockout mice by Kopf and colleagues.[7] These IL-5 knockout mice have abnormal B cell numbers, especially the B-1 cell population in younger animals. In addition, they showed no increases in eosinophils when mice are infested with parasites (*M. corti*). However, they displayed normal (i.e., similar to wild-type) levels of eosinophils at homeostatic base line, demonstrating that eosinophil lineage commitment and basal production occur by an IL-5-independent mechanism. In a mouse model of acute allergen challenge, the IL-5 knockout animals developed an airway eosinophilia that was only 10% of the allergen-challenged wild-type. Nonetheless, this decrease was enough to ablate lung pathologies characteristic of T_h2 inflammation,[21] although some strain variability is linked with induced pathologies.[22] IL-5 expression from a recombinant vaccinia virus infection of the IL-5 knockout mice restored AHR and lung pathologies that were abolished by the IL-5 knockout, providing the first indication that T_h2 pathologies were in part dependent on the presence of eosinophils. In summary, eosinophils are present in the IL-5 knockout mouse; however, their numbers are insufficient to cause histologically evident lung pathologies in OVA-induced airway inflammation.

ΔdblGATA Mice

Yu and colleagues reported that deletion of a GATA-binding factor 1 (GATA-1) binding site in knockout mice surprisingly resulted in mice with no eosinophils.[23] GATA-1 is a transcription factor involved in the hematopoietic regulation of erythroid cells, megakaryocytes, and eosinophils. The transcriptional control of the GATA-1 locus is complex, with several regulatory regions identified by DNase I hypersensitivity that identify binding sites for transcriptional regulatory proteins. A DNase I hypersensitive site upstream of the GATA-1 locus was targeted through a knockout approach by these authors. The resulting ΔdblGATA mice appear relatively normal in most aspects; however, they lack staining for eosinophils by conventional hematoxylin-eosin (H&E) staining. Fluorescence-activated cell sorting (FACS) analysis of ΔdblGATA mice showed that eosinophils, identified by their forward scatter and peroxidase staining, were absent in the knockout mice, although the authors noted an increased presence of *large unclassified cells*. To highlight the loss of eosinophils by FACS analysis, the ΔdblGATA mice were crossed with the CD2/IL-5 transgenic mice created by Dent and colleagues, which revealed that even under the eosinophilopoietic pressure of IL-5 expression the loss of eosinophils was nearly absolute.[23]

Gerard and colleagues[8] showed that the allergen challenge of ΔdblGATA mice reduced remodeling events in a chronic allergen challenge protocol with little to no effects observed in acute allergen models. However, it is noteworthy that subsequent studies have shown that this lack of pathologies following acute allergen challenge was limited only to mice of the BALB/c background, suggesting strain variability associated with these eosinophil-less

mice.[24] Fulkerson and colleagues also provided additional insights regarding this line of eosinophil-less animals, examining the ΔdblGATA mice as part of their study of eosinophils and gene expression during allergic inflammation.[25] They performed a microarray analysis of genes expressed in allergic airway inflammation using several different knockout mice, including the ΔdblGATA mice, as well as mice deficient in both eotaxin (CCL11) and eotaxin-2 (CCL24; eotaxin-1/2 double knockout). Their results show that *Aspergillus* challenge of the ΔdblGATA mice resulted in lower levels of *interleukin-4 (IL4)* and *interleukin-13 (IL13)* gene expression than wild-type. Thus, Fulkerson and colleagues reported that the ΔdblGATA mice were associated with a significant reduction of pulmonary T_h2 gene expression [69% of interleukin-4 (IL-4) and 30% of IL-13], an issue exploited by others in future studies.[26,27]

PHIL

PHIL is a transgenic line of mice that are deficient in eosinophils by the expression of diphtheria toxin (DT) specifically in eosinophils via the eosinophil peroxidase (EPO) promoter.[9] DT is a single-chain polypeptide with two functional domains, A and B, and is known to be lethal to cells by binding a protein necessary for protein synthesis [i.e., elongation factor 2 (EF-2)].[28] DT domain B binds a receptor [proheparin-binding EGF-like growth factor (HB-EGF)[29]] on primate cells including humans, while DT toxin domain A is released in the cytosol and is lethal to cells. DT domain A is known to be lethal even at extremely low levels, and thus, this becomes a very sensitive method of targeted cell ablation in transgenic mice.[30] Moreover, this avoids the obvious obstacle that wild-type mice are resistant to DT due to the lack of a receptor capable of binding this bacterial toxin. In PHIL, because DT domain A was expressed specifically inside eosinophil lineage-committed cells, this obviated the need for a DT receptor. The induced DT domain A expression in eosinophil lineage-committed marrow progenitors in PHIL mice led to their specific death and the subsequent absence of mature eosinophils in circulation.

The first indication that PHIL mice lacked eosinophils were blood smears confirming the lack of eosinophils. More extensive confirmation of the absence of eosinophils in PHIL was obtained by immunohistochemistry with antibodies specific for eosinophil granule major basic protein 1 (MBP-1) as well as the absence of eosinophils in PHIL mice constitutively expressing IL-5.[9] Significantly, it was also found that PHIL mice did not develop AHR or any of the other pathologies associated with acute allergen challenge by a strain-independent mechanism(s) (data not shown).

Further studies by Jacobsen and colleagues[26] highlighted the importance of eosinophils in the development of AHR and allergic airway inflammation in mice through the modulation of pulmonary T_h2 immune responses. Jacobsen and colleagues showed that adoptive transfer of both T cells *and* eosinophils are necessary for the T_h2 inflammatory lung responses in OVA-treated PHIL mice. These analyses definitively underscored the critical importance of eosinophils in creating T_h2-mediated pathologies in the lung and highlighted two critical immune regulatory mechanisms: (1) eosinophil-mediated recruitment of effector T cells to the lung and (2) eosinophil-T cell cross talk via a dendritic cell-dependent mechanism.[27]

Inducible PHIL

The iPHIL mouse represents an inducible form of the eosinophil-less mouse, PHIL (*i*nducible PHIL). This novel gene knock-in line of mice provides an essentially wild-type mouse that can inducibly become eosinophil-less, thus avoiding the potential ontological effects of eosinophils on the immune system. iPHIL is created using genetic knock-in technology by inserting the human DT receptor into the endogenous EPO locus, thus expressing the human DT receptor specifically in eosinophils (our unpublished data). DT is then administered to the mice to ablate the eosinophils specifically, which are the only cell type expressing the human DT receptor.

The success and utility of this strategy is demonstrated in the DT administration study presented in Fig. 5.6.3. Two intraperitoneal injections of DT (15 ng/g of body weight) were given on consecutive days to a cohort of mice ($n = 14$) and the presence of eosinophils was determined in the blood and bone marrow as a function of time. These data dramatically showed that, within 4 days of the first DT injection, circulating eosinophils levels were reduced to absolute zero and remained at this level for 4–5 days before returning to preadministration levels (Fig. 5.6.3A).

This time course has been shown to be invariant and reproducible. (Control animals were either iPHIL mice that received vehicle injections or wild-type mice that were administered DT.) Equally important, cell differential analyses of the blood and bone marrow of DT-treated mice following ablation of eosinophils showed that only the eosinophil lineage was affected, with no changes occurring in either total blood cell counts or the composition of the remaining WBCs. Significantly, recent preliminary data ablating eosinophils during a 1-month-long acute OVA model demonstrates our present ability to ablate eosinophils from the lungs of allergen-sensitized/challenged mice by an inducibly reversible mechanism (Fig. 5.6.3B).

CD70 Transgenic Mice

De Bruin and colleagues describe that CD70 transgenic mice have greatly increased numbers of interferon gamma (IFN-γ)-producing T cells and therefore significantly decreased numbers of eosinophils.[31] These mice were generated by expressing CD70 via the CD19B cell-specific promoter.[10] The T cell expression of IFN-γ inhibits eosinophil differentiation in the bone marrow. Even the induction of allergic airway inflammation by OVA sensitization and challenge fails to produce an elevation in eosinophils. This study again highlights the importance of IFN-γ and T cell/eosinophil cross talk in immune homeostasis. Moreover, Liu and colleagues have described studies in which they show that IFN-γ and tumor necrosis factor (TNF-α) stimulate the production of T-helper type 1 (T_h1) chemokines from eosinophils.[32] However, IL-4 and TNF-α stimulate the production of T_h2 chemokines from eosinophils. Thus, as we suggested in our local immunity and/or remodeling/repair (*LIAR*) hypothesis of eosinophil-mediated activities,[33] depending on the immune microenvironment (i.e., whether IFN-γ or IL-4 is prevalent) eosinophils may elaborate the T_h1 or T_h2 character of the immune response by T_h1 or T_h2 chemokine secretion, respectively.

(A)

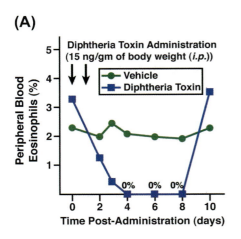

(B)

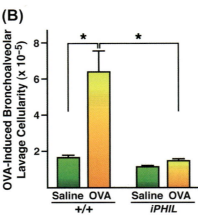

FIGURE 5.6.3 Intraperitoneal administration of diphtheria toxin (15 ng/g of body weight) results in the ablation of eosinophils from peripheral blood that is both absolute and reversible. *(A)* Time course of eosinophil ablation from circulation of allergen-naive iPHIL mice administered diphtheria toxin on two consecutive days. Eosinophil ablation occurs for 4–5 days before returning to wild-type levels. *(B)* Diphtheria toxin administration of iPHIL mice during a 1-month ovalbumin (OVA) acute sensitization/challenge protocol also leads to the ablation of eosinophils and a correspondingly significant decrease in OVA-induced bronchoalveolar lavage cellularity. i.p., intraperitoneal. *$p < 0.05$.

CONCLUSION

Eosinophils are often overlooked but have emerged as a critically important cell in immune homeostasis. Specifically, eosinophils are often linked to T_h2-mediated inflammation, including asthma (allergic airway inflammation in mice), atopic disease, and rhinosinusitis. In recent years, however, the importance of eosinophil-mediated pathways has grown as eosinophilic esophageal disease and eosinophil gastroenteritis have become widespread, especially among atopic and allergic children. Indeed, the prevalence of food allergies has dramatically increased; however, the reasons for the increase in allergic and eosinophil-mediated diseases are unclear. The mouse models that have been described in this chapter have one commonality: eosinophils contribute to, and in some circumstances are the sole cause of, T_h2-mediated pathology. That is, the expression of particular genes or cytokines by eosinophils may be an important mechanism for the immune system to regulate T_h2 inflammation and pathophysiology. Moreover, eosinophil—T cell cross talk is likely an important immune regulatory pathway in allergic diseases. The availability of these and future transgenic and knockout/knock-in mice will allow investigators to answer the *who*, *what*, *when*, and *where* questions of eosinophil biology. More importantly, mouse models that increase or decrease in eosinophils will remain powerful tools that can be combined with specific genetic deletions of particular genes and pathways.

We suggest that the future of eosinophil research using these mouse models is bright and we look forward to the ever-increasing data and future studies of eosinophils in health and disease.

REFERENCES

1. Dent LA, Strath M, Mellor AL, Sanderson CJ. Eosinophilia in transgenic mice expressing interleukin 5. *J Exp Med* 1990;**172**(5):1425—31.
2. Tominaga A, Takaki S, Koyama N, Katoh S, Matsumoto R, Migita M, et al. Transgenic mice expressing a B cell growth and differentiation factor gene (interleukin 5) develop eosinophilia and autoantibody production. *J Exp Med* 1991;**173**(2):429—37.
3. Lee NA, McGarry MP, Larson KA, Horton MA, Kristensen AB, Lee JJ. Expression of IL-5 in thymocytes/T cells leads to the development of a massive eosinophilia, extramedullary eosinophilopoiesis, and unique histopathologies. *J Immunol* 1997;**158**(3):1332—44.
4. Lee JJ, McGarry MP, Farmer SC, Denzler KL, Larson KA, Carrigan PE, et al. Interleukin-5 expression in the lung epithelium of transgenic mice leads to pulmonary changes pathognomonic of asthma. *Journal of Experimental Medicine* 1997;**185**(12):2143—56.
5. Foster EL, Simpson EL, Fredrikson LJ, Lee JJ, Lee NA, Fryer AD, et al. Eosinophils increase neuron branching in human and murine skin and in vitro. *PLoS One* 2011;**6**(7). e22029.
6. Ochkur SI, Jacobsen EA, Protheroe CA, Biechele TL, Pero RS, McGarry MP, et al. Co-Expression of IL-5 and Eotaxin-2 in Mice Creates an Eosinophil-Dependent Model of Respiratory Inflammation with Characteristics of Severe Asthma. *Journal of Immunology* 2007;**178**(12):7879—89.
7. Kopf M, Brombacher F, Hodgkin PD, Ramsay AJ, Milbourne EA, Dai WJ, et al. IL-5-deficient mice have a developmental defect in CD5+ B-1 cells and lack eosinophilia but have normal antibody and cytotoxic T cell responses. *Immunity* 1996;**4**(1):15—24.
8. Humbles AA, Lloyd CM, McMillan SJ, Friend DS, Xanthou G, McKenna EE, et al. A critical role for eosinophils in allergic airways remodeling. *Science* 2004;**305**(5691):1776—9.
9. Lee JJ, Dimina D, Macias MP, Ochkur SI, McGarry MP, O'Neill KR, et al. Defining a link with asthma in mice congenitally deficient in eosinophils. *Science* 2004;**305**(5691):1773—6.
10. Arens R, Tesselaar K, Baars PA, van Schijndel GM, Hendriks J, Pals ST, et al. Constitutive CD27/CD70 interaction induces expansion of effector-type T cells and results in IFNgamma-mediated B cell depletion. *Immunity* 2001;**15**(5):801—12.
11. Greaves DR, Wilson FD, Lang G, Kioussis D. Human CD2 3'-flanking sequences confer high-level, T cell-specific, position-independent gene expression in transgenic mice. *Cell* 1989;**56**(6):979—86.
12. Lee NA, Loh DY, Lacy E. CD8 surface levels alter the fate of ab TCR expressing thymocytes in transgenic mice. *J Exp Med* 1992;**175**:1013—25.
13. Macias MP, Fitzpatrick LA, Brenneise I, McGarry MP, Lee JJ, Lee NA. Expression of IL-5 alters bone metabolism and induces ossification of the spleen in transgenic mice. *J Clin Invest* 2001;**107**(8):949—59.
14. Hackett BP, Gitlin JD. 5' flanking region of the Clara cell secretory protein gene specifies a unique temporal and spatial pattern of gene expression in the developing pulmonary epithelium. *Am J Respir Cell Mol Biol* 1994;**11**(2):123—9.
15. Glasser SW, Burhans MS, Eszterhas SK, Bruno MD, Korfhagen TR. Human SP-C gene sequences that confer lung epithelium-specific expression in transgenic mice. *Am J Physiol Lung Cell Mol Physiol* 2000;**278**(5):L933—45.
16. Glasser SW, Eszterhas SK, Detmer EA, Maxfield MD, Korfhagen TR. The murine SP-C promoter directs type II cell-specific expression in transgenic mice. *Am J Physiol Lung Cell Mol Physiol* 2005;**288**(4):L625—32.
17. Jarratt M, Bybee JD, Ramsdell W. Eosinophilic fasciitis: an early variant of scleroderma. *J Am Acad Dermatol* 1979;**1**(3):221—6.
18. Malm-Erjefält M, Persson CGA, Erjefalt JS. Degranulation status of airway tissue eosinophils in mouse models of allergic airway inflammation. *American Journal of Respiratory Cell and Molecular Biology* 2001;**24**(3):352—9.
19. Kobayashi T, Iijima K, Kita H. Marked airway eosinophilia prevents development of airway hyper-responsiveness during an allergic response in IL-5 transgenic mice. *J Immunol* 2003;**170**(11):5756—63.
20. Clark K, Simson L, Newcombe N, Koskinen AM, Mattes J, Lee NA, et al. Eosinophil degranulation in the allergic lung of mice primarily occurs in the airway lumen. *J Leukoc Biol* 2004;**75**(6):1001—9.
21. Foster PS, Hogan SP, Ramsay AJ, Matthaei KI, Young IG. Interleukin 5 deficiency abolishes eosinophilia, airways hyperreactivity, and lung damage in a mouse asthma model [see comments]. *J Exp Med* 1996;**183**(1):195—201.

22. Hogan SP, Koskinen A, Foster PS. Interleukin-5 and eosinophils induce airway damage and bronchial hyperreactivity during allergic airway inflammation in BALB/c mice. *Immunol Cell Biol* 1997;**75**(3):284—8.

23. Yu C, Cantor AB, Yang H, Browne C, Wells RA, Fujiwara Y, et al. Targeted deletion of a high-affinity GATA-binding site in the GATA-1 promoter leads to selective loss of the eosinophil lineage in vivo. *J Exp Med* 2002;**195**(11):1387—95.

24. Walsh ER, Sahu N, Kearley J, Benjamin E, Kang BH, Humbles A, et al. Strain-specific requirement for eosinophils in the recruitment of T cells to the lung during the development of allergic asthma. *J Exp Med* 2008;**205**(6):1285—92.

25. Fulkerson PC, Fischetti CA, McBride ML, Hassman LM, Hogan SP, Rothenberg ME. A central regulatory role for eosinophils and the eotaxin/CCR3 axis in chronic experimental allergic airway inflammation. *Proc Natl Acad Sci U S A* 2006;**103**(44): 16418—23.

26. Jacobsen EA, Ochkur SI, Pero RS, Taranova AG, Protheroe CA, Colbert DC, et al. Allergic Pulmonary Inflammation in Mice is Dependent on Eosinophil-induced Recruitment of Effector T Cells. *Journal of Experimental Medicine* 2008;**205**(3):699—710.

27. Jacobsen EA, Zellner KR, Colbert D, Lee NA, Lee JJ. Eosinophils regulate dendritic cells and Th2 pulmonary immune responses following allergen provocation. *Journal of Immunology* 2011;**187**(11): 6059—68.

28. Collier RJ. Understanding the mode of action of diphtheria toxin: a perspective on progress during the 20th century. *Toxicon* 2001;**39**(11):1793—803.

29. Saito M, Iwawaki T, Taya C, Yonekawa H, Noda M, Inui Y, et al. Diphtheria toxin receptor-mediated conditional and targeted cell ablation in transgenic mice. *Nat Biotechnol* 2001;**19**(8):746—50.

30. Palmiter RD, Behringer RR, Quaife CJ, Maxwell F, Maxwell IH, Brinster RL. Cell lineage ablation in transgenic mice by cell-specific expression of a toxin gene [published erratum appears in Cell 1990 Aug 10;62(3):following 608]. *Cell* 1987;**50**(3):435—43.

31. de Bruin AM, Buitenhuis M, van der Sluijs KF, van Gisbergen KP, Boon L, Nolte MA, et al. Eosinophil differentiation in the bone marrow is inhibited by T cell-derived IFN-gamma. *Blood* 2010;**116**(14):2559—69.

32. Liu LY, Bates ME, Jarjour NN, Busse WW, Bertics PJ, Kelly EA. Generation of Th1 and Th2 chemokines by human eosinophils: evidence for a critical role of TNF-alpha. *J Immunol* 2007;**179**(7): 4840—8.

33. Lee JJ, Jacobsen EA, McGarry MP, Schleimer RP, Lee NA. Eosinophils in Health and Disease: The *LIAR* Hypothesis. *Clinical and Experimental Allergy* 2010;**40**(4):563—75.

Eosinophil Trafficking

Introduction

Christine Wennerås

In healthy humans, eosinophils are mainly found in the gastrointestinal tract, and to a lesser extent in lymphoid organs, e.g., the thymus (of neonates), spleen, and lymph nodes. Eosinophils can be recruited to other organs, such as the lungs, in response to parasitic and/or allergic inflammatory disease states. Eosinophils are present in the intestine of the embryo and attain adult levels soon after birth. Moreover, intestinal eosinophil levels are the same in most major inbred mouse strains, and studies in germ-free mice reveal that the existence of enteric microflora is not a requirement for the establishment of intestinal eosinophils.

What regulates the trafficking of eosinophils into the tissues? How is recruitment achieved?

In their chapter, Drs Zhu and Zimmermann discuss the mechanisms of eosinophil chemotaxis and the signals eliciting eosinophil trafficking into peripheral tissues. They identify the steps involved in the transit of eosinophils from the bloodstream and discuss how these processes promote the role of eosinophils in promoting homeostasis and disease processes.

The vascular bed likewise plays an important role in inflamed tissue by slowing down blood flow and allowing leukocytes to marginate from the bloodstream to the blood vessel wall, along which they roll and tether until signals from the endothelium make them stop. Endothelial cells indicate the point of exit by exhibiting receptors that are normally hidden and whose counter-receptors are found on the eosinophilic cell surface. The adhesion molecule P-selectin glycoprotein ligand 1 is stored in the Weibel–Palade bodies of endothelial cells and is rapidly exposed on the cell surface. Drs Matsumoto and Bochner discuss the role of specific eosinophil adhesion molecules, regulated by both *inside out* and *outside in* signaling in modulating these specific processes.

Likewise, the chapter by Dr. Cook-Mills addresses the interactions between these cells more directly and considers how eosinophil–endothelial cell interactions further modify the functions of both eosinophils and endothelial cells thereafter.

Another interesting topic addressed in this chapter is that of circulating endothelial progenitor cells. Drs Asosingh and Erzurum describe these eotaxin-positive cells, which may recruit eosinophils as a component of the pathogenesis of asthma.

Finally, in answer to the question *where do eosinophils go?*, Drs Rao and Sriramarao address this question directly with an update on the technologies directed toward understanding eosinophil trafficking using intravital microscopic methods.

Overall, these contributions, which promote our understanding of the cells and signals that direct the trafficking of eosinophils to particular tissues, may contribute to our understanding of what roles these fascinating cells play in health and disease.

Eosinophil Chemotaxis

Xiang Zhu and Nives Zimmermann

INTRODUCTION

Eosinophils are multifunctional granulocytes that accumulate in various tissues and are implicated in the pathogenesis of numerous inflammatory processes, including parasitic helminth infections and allergic diseases. Thus, elucidating the processes that regulate eosinophil tissue accumulation has the potential to advance knowledge and the clinical care of patients with eosinophil-mediated disease. The recruitment of eosinophils to inflamed tissue is a multistage process in which eosinophils undergo the following: (1) priming; (2) rolling along endothelial cells; (3) firm adhesion to the endothelium; (4) transendothelial diapedesis; and (5) chemotaxis to the inflammatory site. A substantial effort has been made to understand the mechanisms for selective accumulation of eosinophils and the pathways regulating these responses. This chapter focuses primarily on eosinophil chemotaxis, particularly on the eotaxin subfamily of chemokines and their receptors since these chemokines specifically target eosinophils.

Eosinophils in Health and Disease. http://dx.doi.org/10.1016/B978-0-12-394385-9.00006-7

Furthermore, we discuss the inhibitory signals and signal integration *in vivo*, as well as the involvement of eosinophil chemotaxis in disease states.

CHEMOKINE FAMILY

Chemokines constitute a large family of chemotactic cytokines that are divided into four groups, designated CC, CX3C, CXC, and XC, depending on the spacing of conserved cysteine motifs (where X represents any amino acid). The CC chemokines target a variety of cell types, including macrophages, eosinophils, and basophils, while the CXC chemokines mainly target neutrophils and lymphocytes. Due to the complexity of the previous naming system, a new nomenclature was adopted in 2000.[1] This new chemokine nomenclature is based on the chemokine receptor nomenclature, which uses CC, CX3C, CXC, or XC to designate the chemokine group followed by R (for receptor) and then a number. The chemokine nomenclature replaces R with L (for ligand) and uses CC for the *SCYA* gene family, CXC for *SCYB*, CX3C for *SCYD*, and XC for *SCYC*. The numbering system is the same as that from the now-retired small inducible cytokine (SCY) nomenclature, which already has numbers assigned to the genes known to encode chemokines. Thus, a given gene will have the same number as its protein ligand. Since the SCY nomenclature was retired, the gene and chemokine symbols match; for example, the gene encoding eotaxin is *chemokine (C-C motif) ligand 11 (CCL11)*, and the chemokine symbol is CCL11.

For the most part, chemokines are 8−12 kDa and contain 1−3 (usually 2) disulfide bonds that stabilize the overall topology. Although their sequence identity ranges from <20 to 90%, investigators have revealed by X-ray crystallography and/or nuclear magnetic resonance that all chemokines share very similar tertiary structures.[2] After translation, most chemokines are secreted from the cell, with the exceptions of C-X3-C motif chemokine 1 (CX3CL1/fractalkine) and C-X-C motif chemokine 16 (CXCL16) that are tethered to the extracellular surface. In secreted chemokines, the disordered N-terminus functions as a key signaling domain. The N-terminus loop region that follows the first two cysteine residues and connects the N-terminus to three antiparallel beta-pleated sheets is the major receptor-binding site, and the sequence therein confers receptor specificity. In most CC and CXC chemokines, the C-terminus is speculated to be the glycosaminoglycan binding site, which is important for gradient generation *in vivo* but is often dispensable for *in vitro* chemotaxis. Finally, while many chemokines form dimers or other higher-order structures, it appears that most chemokines interact with receptors as monomers. For more details of the structure-function relations of chemokines and chemokine receptors, as well as implications for

development of agents targeting these interactions, we refer the reader to a recent review by Allen and colleagues.[2]

CHEMOKINES ACTIVE ON EOSINOPHILS

Eotaxins

The eotaxins are a subfamily of chemokines that specifically target eosinophils. Eotaxin (CCL11) was first described in a guinea pig model of allergic lung inflammation.[3,4] The bronchoalveolar lavage fluid (BALF) of allergen-challenged guinea pigs was reported to promote the local accumulation of eosinophils when injected into the skin of naive guinea pigs. The chemoattractant in the BALF was selective for eosinophils; it was subsequently purified and identified as a new guinea pig CC chemokine, accordingly named eotaxin. Later, the human[5−7] and mouse[8] eotaxins were cloned. Eotaxin is expressed constitutively in multiple tissues, with the highest constitutive expression in the gastrointestinal tract (except for the esophagus).[7−10] Indeed, gastrointestinal eosinophils represent the single largest population of these granulocytes outside of the bone marrow. The marked decrease of this eosinophil population in eotaxin-deficient mice confirms the significance of eotaxin-dependent gastrointestinal eosinophil recruitment by eotaxin.[10] Constitutive eotaxin expression has been demonstrated in epithelial cells,[11] endothelial cells,[9] T cells, macrophages, and eosinophils themselves.[7] Eotaxin mRNA expression can be induced by several proinflammatory and T-helper type 2 (T_h2) cytokines. For example, the proinflammatory cytokines interleukin-1 (IL-1) alpha and tumor necrosis factor (TNF-α) upregulate eotaxin mRNA expression in human dermal fibroblasts.[12] Similarly, TNF-α and IL-1 beta induce the accumulation of eotaxin mRNA in the pulmonary epithelial cell lines A549 and BEAS 2B in a dose-dependent manner.[13] Of the T_h2 cytokines, interleukin-4 (IL-4) and -13 (IL-13) are both potent inducers of eotaxin mRNA and protein expression in the BEAS 2B cell line.[14,15] Moreover, intranasal challenge of recombinant mouse IL-13 can strongly induce pulmonary eosinophilia that is associated with increased eotaxin in the tissue.[16]

Interestingly, interleukin-5 (IL-5) and eotaxin can cooperatively promote tissue eosinophilia.[17−19] Administration of eotaxin intranasally or subcutaneously in mice induces rapid and selective tissue accumulation of eosinophils in the lung or skin, respectively, in the presence of high levels of IL-5 (using IL-5 transgenic mice).[17] IL-5 synergizes with eotaxin in promoting tissue eosinophilia by increasing the pool of circulating eosinophils (by stimulating eosinophilopoiesis and bone marrow release) and by priming eosinophils to have enhanced responsiveness to eotaxin.

Single nucleotide polymorphism (SNP) analysis in the *eotaxin/[chemokine (C-C motif) ligand 11 (CCL11)]* gene revealed that a naturally occurring mutation ($+123G \rightarrow A$) causes an amino acid substitution in the last amino acid of the signal peptide (alanine $\rightarrow$ threonine) that results in less effective cellular secretion of eotaxin *in vivo* and *in vitro*. Notably, this SNP ($+123G \rightarrow A$) is associated with significantly lower plasma eotaxin levels, lower circulating eosinophil levels, and improved lung function in patients with asthma.[20] More importantly, this SNP has been associated with disease in more than one patient population,[21,22] suggesting that it may be an important genetic determinant of a subphenotype of patients with asthma.

The second member of the eotaxin subfamily, eotaxin-2 [chemokine (C-C motif) ligand 24 (CCL24)], is functionally very similar to eotaxin despite having only 39% amino acid identity and differing almost completely in the amino terminal region.[23] *In vitro* chemotaxis assays demonstrated that eotaxin-2 is a chemoattractant for eosinophils and basophils (primed with interleukin-3). Experimental induction of cutaneous and pulmonary late-phase responses in atopic subjects revealed that eotaxin is induced at an early time point (6 h) and correlates with early eosinophil recruitment, whereas eotaxin-2 correlates with eosinophil accumulation 24 h post-allergen challenge.[24,25] Eotaxin-2 can also synergize with IL-5 to promote lung eosinophilia and production of IL-13.[26] By using eotaxin and eotaxin-2 single- and double gene-deficient mice, investigators established that both chemokines have non-overlapping roles in regulating the temporal and regional distribution of eosinophils at allergic inflammatory sites.[27–29] These studies demonstrated that:

1. eotaxin enhances the magnitude of the early, but not late, eosinophil recruitment after antigen challenge;[27]
2. in contrast to observations in eotaxin-deficient mice, eotaxin-2-deficient mice have normal baseline eosinophil levels in the hematopoietic tissues and gastrointestinal tract;[28]
3. cooperative interactions between eotaxin-2 and IL-13 promote eosinophil recruitment in response to allergen challenge;[28] and
4. specific synergy between eotaxin and eotaxin-2, with a dominant role for eotaxin-2, is responsible for the accumulation of eosinophils in the airway lumen in response to allergen challenge.[29]

An SNP analysis revealed that the SNP ($+1272A \rightarrow G$) and two haplotypes, haplotype 2 (ht2) and 6 (ht6) of eotaxin-2 are strongly associated with higher plasma eotaxin-2 levels in patients with asthma ($+1272A \rightarrow G$, $p = 0.006$; ht2, $p = 0.006$; and ht6, $p = 0.002$).[30]

The third member of the eotaxin subfamily, C-C motif chemokine 26 (CCL26/eotaxin-3), has 37% amino acid identity with eotaxin and was first discovered in IL-4-stimulated human vascular endothelial cells.[31] Subsequently, its expression was reported in dermal fibroblasts,[32] airway epithelial cells,[33] airway smooth muscle cells,[34] and intestinal epithelial cells.[35] Interestingly, a putative mouse *eotaxin-3 [chemokine (C-C motif) ligand 26 (Ccl26)]* gene sequence with homology to human eotaxin-3 was found approximately 7 kb downstream from eotaxin-2 on mouse chromosome 5. However, the expression of this gene has not been demonstrated either at baseline or with allergen challenge,[28] suggesting that this sequence may be a pseudogene in the mouse genome. Similar to eotaxin and eotaxin-2, human eotaxin-3 can promote chemotaxis of eosinophils; however, it seems to be approximately 90% less potent than eotaxin.[36] Interestingly, eotaxin-3 is expressed at later stages (24–48 h) following allergen challenge and most likely supports prolonged eosinophil recruitment.[37,38] Studies have demonstrated that eotaxin-2 and eotaxin-3, like eotaxin, are both upregulated by the T_h2 cytokines IL-4 and IL-13[33,35] via a signal transducer and activator of transcription 6-dependent pathway.[35] Most recently, eotaxin-3 has been identified as a biomarker for the allergic disorder eosinophilic esophagitis.[39] One SNP ($+2496T \rightarrow G$) in the *eotaxin-3* gene, which may disrupt an adenylate-uridylate (AU)-rich RNA-stability regulatory element, has been identified in association with susceptibility to eosinophilic esophagitis.[39] In summary, the three eotaxins regulate eosinophil accumulation in a temporally and spatially restricted manner.

Other Eosinophil-Active Chemokines

Monocyte chemotactic proteins (MCPs) and eotaxin constitute a subfamily of CC chemokines that share approximately 65% amino acid identity.[40,41] The genomic organization of MCPs and eotaxin is also similar; MCPs and eotaxin are localized on human chromosome 17q11.2-q21.2, suggesting that these genes arose from a common ancestral gene. As their name implies, all MCPs have strong chemoattractant activity for monocytes. Moreover, they display overlapping chemoattractant activity on basophils and eosinophils in humans.[42,43] Monocyte chemotactic protein 1 (MCP-1) is not active on eosinophils because MCP-1 only binds to the C-C chemokine receptor type 2 (CCR2) receptor, which eosinophils do not express. In contrast, monocyte chemotactic protein 2–4 (MCP-2–4) can bind to both CCR2 and C-C chemokine receptor type 3 (CCR3) receptors, with the latter providing activity on eosinophils. Interestingly, even though full-length MCP-1 is not active on eosinophils, deletion of the N-terminus residue converts MCP-1 from an activator of basophil mediator release to an eosinophil chemoattractant.[44] Members of the MCP family also have other effects on eosinophils aside from chemoattraction. For example, MCP-3 has been shown to differentially regulate beta-1 (β1) and beta-2 (β2) integrin avidity. MCP-3-induced upregulation of VLA-4 (α4β1) avidity is dependent on the actin cytoskeleton whereas the effect of MCP-3 on cell surface glycoprotein MAC-1 subunit alpha [MAC-1 (αMβ2)] avidity is dependent on conformational changes.[45] The main stimuli for the production of MCPs appear to be similar to those for eotaxin (i.e., early proinflammatory cytokines such as IL-1 and TNF-α and bacterial or viral products).[40,41] Additionally, the T-helper type 1 (Th1) cytokine interferon gamma (IFN-γ) and the Th2 cytokine IL-4 can induce the production of MCPs individually as well as synergize with IL-1 and TNF-α to induce chemokine secretion.[41]

C-C motif chemokine 5 (CCL5), also known as regulated upon activation, normally T-cell-expressed, and presumably secreted (RANTES), is expressed in T cells,[46] thrombin-activated platelets,[47] endothelial cells,[48] bronchial epithelial cells,[49] and eosinophils themselves.[50] In humans, CCL5 can attract monocytes and eosinophils *in vitro*, which is consistent with its receptor binding to C-C chemokine receptor type 1 (CCR1) and C-C chemokine receptor type 5 (CCR5) on monocytes and CCR3 on eosinophils. Other than attracting eosinophil migration, CCL5 can activate eosinophils by inducing exocytosis of eosinophil cationic proteins and by eliciting the respiratory burst.[51] In addition, injection of CCL5 into human skin induces eosinophil and monocyte recruitment, and this induction is augmented in allergic subjects.[52]

C-C motif chemokine 3 (CCL3), otherwise known as macrophage inflammatory protein 1-alpha (MIP-1-alpha), induces eosinophil migration in humans to a lesser extent than CCL5 (RANTES) *in vitro* and also promotes transient changes in intracellular free calcium concentration and eosinophil cationic protein release.[51] In *Schistosoma mansoni* egg antigen-induced airway inflammation, a significant increase of CCL3 expression was found in BALF cells.[53] Further studies demonstrated CCL3 protein expression in airway epithelial cells, macrophages, and recruited eosinophils.[53] In the same model, pretreatment of mice with a neutralizing anti-CCL3 antibody reduced eosinophilia by >50%.[53] However, in ovalbumin (OVA)-induced lung inflammation, blocking CCL3 with neutralizing antibody only modestly inhibits the accumulation of eosinophils.[54,55]

Interleukin-8 (IL-8), otherwise known as C-X-C motif chemokine 8 (CXCL8), is primarily active on neutrophils. However, its chemotactic activity for eosinophils was discovered by injection of human IL-8 into guinea pigs, which resulted in eosinophil infiltration.[56] Eosinophils from atopic patients respond to IL-8 whereas eosinophils from nonallergic individuals are generally unresponsive. *In vitro*, IL-5 primes eosinophils to respond to IL-8, probably by upregulating the IL-8 receptor, C-X-C chemokine receptor type 2 (CXC-R2), on their surface.[57]

Other Eosinophil Chemoattractants

There are several other *classic eosinophil chemoattractants* based on their ability to attract eosinophils into tissue, including platelet-activating factor (PAF), complement component C5a, leukotriene B4 (LTB4), and formyl-Methionyl-Leucyl-Phenylalanine (fMLP). PAF and complement component C5a are more potent than LTB4 and fMLP for eosinophil attraction. All of these mediators also attract neutrophils; however, PAF exhibits greater potency for eosinophils.[58] *In vivo*, PAF induces local eosinophilia after intracutaneous administration into atopic subjects[59] and promotes infiltration of eosinophils into the airways in a guinea pig model of asthma.[60]

EOSINOPHIL CHEMOKINE RECEPTORS

Chemokines induce leukocyte migration and activation by binding to specific, seven-transmembrane-spanning, G protein-coupled receptors (GPCRs). So far, seven CXC-chemokine receptors [C-X-C chemokine receptor type 1−7 (CXC-R1−7)] and 11 C-C chemokine receptors (CCR1−11) have been identified. The chemokine and leukocyte selectivity of chemokine receptors overlap extensively; a given leukocyte often expresses multiple chemokine receptors, and more than one chemokine typically binds to the same receptors. For example, eosinophils express both CCR1 and CCR3, and chemokines eotaxin, eotaxin-2 and eotaxin-3, MCP-2−4, CCL5 (RANTES), and C-C motif chemokine 28 (CCL28) can bind CCR3. While eosinophils mainly express CCR1 and CCR3, they have been shown to express other chemokine receptors under certain conditions (Fig. 6.2.1).

Eosinophils from most healthy donors express CCR3 at the highest level (4 × 10^5 receptors per cell) and CCR1 at a significantly lower level (0.5−2 × 10^4 receptors per cell).[61] However, eosinophils from a subset of individuals (approximately 19% of the donor pool), designated as CCL3 highly responsive donors, express high levels of CCR1.[62] Overall, and consistent with the expression of CCR1 and CCR3, human eosinophils respond to CCR1- and CCR3-relevant chemokines, including CCL3, CCL5 (RANTES), CCL28, eotaxin, eotaxin-2, eotaxin-3, MCP-2, and MCP-3. CCR3 appears to function as the predominant eosinophil chemokine receptor since CCR3 ligands are generally more potent eosinophil chemoattractants. As mentioned earlier, eosinophils have the capacity to express CXC-R2 (the

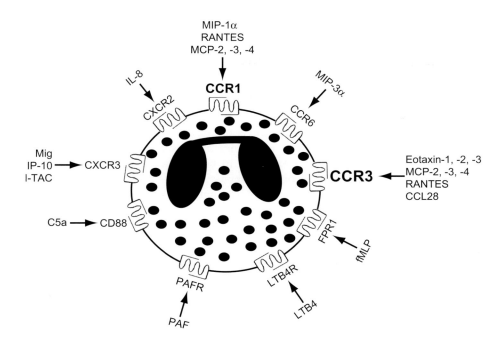

FIGURE 6.2.1 Schematic diagram of eosinophil chemokine receptors and their ligands. Several classic eosinophil chemoattractants and their receptors are also shown. C5a, complement component 5a; CCL5, C-C motif chemokine 5; CCL28, C-C motif chemokine 28; CCR1, C-C chemokine receptor type 1; CCR3, C-C chemokine receptor type 3; CCR6, C-C chemokine receptor type 6; CXC-R2, C-X-C chemokine receptor type 2; CXC-R3, C-X-C chemokine receptor type 3; fMLP, fMet-Leu-Phe; FPR1, N-formyl peptide receptor; IL-8, interleukin-8; IP-10, IFN-γ inducible protein of 10 kDa; I-TAC, interferon-inducible T-cell alpha chemoattractant; LTB$_4$, leukotriene B$_4$; LTB$_4$-R1, leukotriene B$_4$ receptor 1; MCP, monocyte chemotactic protein; MIG, monokine induced by interferon gamma; MIP-1-alpha, macrophage inflammatory protein 1-alpha; PAF, platelet-activating factor; PAFR, platelet-activating factor receptor; RANTES, regulated upon activation, normally T-expressed, and presumably secreted.

low-affinity IL-8 receptor) when cultured in IL-5, and this may explain the ability of IL-5 to prime eosinophils to respond to IL-8.[57] Lastly, eosinophils have been shown to express or respond to ligands of C-C chemokine receptor type 6 (CCR6), C-X-C chemokine receptor type 3 (CXC-R3) and 4 (CXC-R4). For example, eosinophils isolated from allergic donors responded to MIP-3-alpha (a ligand of CCR6) in chemotaxis and calcium mobilization assays, and fluorescence-activated cell sorting (FACS) analysis revealed that approximately 20% of eosinophils express low levels of CCR6.[63] In contrast, another group of researchers could not detect the expression of CCR6 on either blood or BALF eosinophils from allergic subjects 48 h after segmental bronchoprovocation with antigen.[64] In terms of CXC-R3, FACS analysis found that 50% of eosinophils from nonallergic donors express CXC-R3 and that these cells respond in a functional assay. Moreover, CXC-R3 mRNA and protein expression in eosinophils is upregulated and downregulated by interleukin-2 and -10, respectively.[65] In addition, eosinophils isolated from nonallergic donors express CXC-R4 on their surface following incubation *in vitro*. Interestingly, CXC-R4 expression can be upregulated by IFN-γ, TNF-α, and transforming growth factor beta (TGF-β) and downregulated by IL-4 and IL-5.[66] Taken together, these findings indicate the significance of these chemokine receptors in eosinophil accumulation in health and disease and substantiate the need for further investigation.

Gene Structure and Polymorphisms

The majority of *CCR* genes, with the exception of *chemokine (C-C motif) receptor 6* (*CCR6*) (chromosome 6q27) and 7 (*CCR7*) (chromosome 17q12-q21.2), are located on human chromosome 3p21—24. Similar to the other chemoattractant receptor genes, *CCR* genes are composed of a single coding exon, and the 5′ untranslated region is separated by at least one large intron. For example, the *chemokine (C-C motif) receptor 3* (*CCR3*) gene consists of four exons with multiple 5′ untranslated exons differentially used for alternative splicing and/or as promoters.[67] The open reading frame of the human *CCR3* gene (exon 4) is polymorphic and contains multiple nucleotide variants. One variant (+824G$\rightarrow$A, allele frequency = 0.01) encodes for a change of arginine to glutamine at position 275 of the protein; this is a nonconservative amino acid change in the third extracellular domain of the receptor, a region implicated in ligand binding to CCR3. Stratification of DNA samples by the history of asthma yielded no change in this allele's frequency; however, the frequency may change in other inflammatory conditions or populations.[68] Another polymorphism (+1052T$\rightarrow$C, allele frequency = 0.005) encodes for a change of leucine to proline at position 351 in the serine/threonine-rich cytoplasmic tail.[68] This occurs in a putative G protein-coupled receptor kinase 2 (GPRK2)-phosphorylation site and therefore may have consequences on receptor signaling. The polymorphism

$+51T \rightarrow C$ in *CCR3* was identified in Japanese (Asian) and British (white) subjects.[69] *CCR3* $+51T \rightarrow C$ demonstrated a significant association with the diagnosis of asthma in the British population (odds ratio 2.35, $p < 0.01$), but not in the Japanese population. In addition, a genetic association between *CCR3* polymorphisms and the number of circulating eosinophils was revealed as a novel finding; the *CCR3* ht2 had a negative gene dose effect on the eosinophil count ($p = 0.003 - 0.009$).[70] Moreover, CCR3 protein expression was higher on the eosinophils of asthmatic patients without ht2 than of those with ht2. These associations between *CCR3* polymorphisms and the number of circulating eosinophils were more pronounced when the *CCR3* polymorphisms were paired with polymorphisms in *interleukin 5 receptor, alpha (IL5RA)*. More recently, it was shown that the $+971T \rightarrow C$ SNP, which results in a leucine to proline substitution at amino acid 324 in the *CCR3* C-terminus region, gives rise to a protein that is expressed but is subsequently degraded since it cannot traffic to the cell membrane. This leads to a complete loss of chemotactic function and points the way to alternative approaches for antagonizing CCR3 function in the treatment of eosinophil-associated disease.[71] In conclusion, CCR3 contains several genetic variations that may have consequences on disease processes that involve the use of this receptor.

Signal Transduction

Chemokine receptors are members of the large family of GPCRs. GPCR activation initiates signaling cascades involving a number of pathways, thus leading to a variety of biological responses. Here we summarize what is known about eosinophil signal transduction following chemokine binding (primarily to CCR3) in the context of the general knowledge of GPCR signal transduction. Most chemokine receptors couple to the pertussis toxin-sensitive $G\alpha_i$ family of heterotrimeric G proteins. Activated $G\alpha_i$ classically inhibits adenylate cyclase; however, this does not occur following CCR3 activation in eosinophils, despite the presence of pertussis toxin-sensitive calcium transients.[72] Activated $G\alpha_i$ also activates the Src family of tyrosine kinases. Indeed, in eosinophils, tyrosine-protein kinase HCK (HCK) and tyrosine-protein kinase Fgr (FGR) were shown to associate with CCR3 following activation and are postulated to be involved in the rapid cell shape changes required for eosinophil migration.[73] In contrast, the liberated $G\beta\gamma$ subunit activates phospholipase C beta (PLCB) leading to inositol 1,4,5-triphosphate (IP3) formation and a transient rise in the concentration of intracellular free calcium (Ca^{2+}). The rise in intracellular calcium is measured by a simple assay and is often used to demonstrate chemokine activation of eosinophils. Furthermore, the calcium flux leads

to actin polymerization and eosinophil chemotaxis. In addition, mitogen-activated protein kinases have been shown to be phosphorylated and activated after exposure of eosinophils to CCR3 ligands, which is required for eotaxin-induced eosinophil chemotaxis and degranulation.[74] Notably, on triggering intracellular signaling pathways, CCR3 appears to undergo internalization.[75] While internalization has classically been viewed as a negative regulatory process, CCR3 internalization in eosinophils seems to be required for positive signaling and biological responses such as shape change and actin polymerization, but not desensitization and calcium mobilization.[76] In summary, a complex array of signaling molecules is activated upon chemokine binding that collectively leads to biological responses, including eosinophil chemotaxis. However, most of these studies are conducted in reductionist scenarios, in which eosinophils are activated with one stimulus at a time. The *in vivo* system provides additional layers of complexity wherein eosinophils are stimulated by multiple positive and negative signals in a temporal and spatial manner.

INHIBITORY CHEMOKINES AND RECEPTORS REGULATING EOSINOPHIL CHEMOTAXIS

A paradigm has emerged implicating the T_h2 cytokines, IL-4 and IL-13, in the induction of eosinophil-active chemokines that mainly signal through the CCR3 receptor expressed on eosinophils. In contrast, T_h1 cytokines (e.g., IFN-γ) induce a different set of chemokines [e.g., C-X-C motif chemokine 9 (CXCL9), also known as monokine induced by interferon gamma (MIG), C-X-C motif chemokine 10 (CXCL10), also known as 10 kDa interferon gamma-induced protein (IP-10), and C-X-C motif chemokine 11 (CXCL11), also known as interferon-inducible T-cell alpha chemoattractant (I-TAC)]. These chemokines uniquely signal through the CXC-R3 receptor expressed on activated T cells, preferentially of the T_h1 phenotype. A few studies suggest that these T_h1- and T_h2-associated chemokines may inhibit CCR3 and CXC-R3, respectively.[77–79] For example, human CXC-R3 ligands have been demonstrated to be CCR3 antagonists, inhibiting the action of CCR3 ligands on human eosinophils and CCR3$^+$ cells *in vitro*.[77,78] In addition, eotaxin has been reported to be an antagonist for CXC-R3.[79] These results suggest a feedback loop by which T_h1- and T_h2-associated chemokines coordinately regulate eosinophil responses. In support of this, we found that both T_h1- and T_h2-associated chemokines were upregulated in an experimental asthma model.[80] Further studies revealed that the T_h1-associated chemokine CXCL9 plays an inhibitory role in regulating eosinophil chemoattraction.[80] CXCL9 not only potently inhibits eosinophil migration *in vitro* but also markedly attenuates eosinophil recruitment to the lung in response to diverse stimuli *in vivo*, including T_h2-associated chemokines, IL-13, and allergen. Thus, these results demonstrate the existence of naturally occurring eosinophil inhibitory chemokines, such as CXCL9, identifying a pathway with potential therapeutic significance.

Furthermore, a recent study indicated that paired immunoglobulin-like receptor B (PIR-B), which contains immunoreceptor tyrosine-based inhibitory motifs (ITIMs) in its cytoplasmic portion, negatively regulates eotaxin-dependent eosinophil chemotaxis *in vivo* and *in vitro* (i.e., PIR-B$^{-/-}$ eosinophils have increased eotaxin responsiveness).[81] In contrast, PIR-B$^{-/-}$ eosinophils had decreased LTB$_4$-dependent chemotactic responses *in vitro*. Mechanistic analysis revealed that PIR-B recruits activating kinases after LTB$_4$, but not eotaxin, stimulation.[81] These data challenge the conventional wisdom that inhibitory receptors are restricted to inhibitory signals and are further discussed in section 7 (Eosinophil Signal Transduction), Chapter 7.3.

In summary, these studies illustrate the complexity of signal integration resulting in eosinophil chemotaxis *in vivo*. For example, in allergen-challenged lungs, eosinophils are exposed to multiple stimuli. Some prime for chemotaxis (IL-5), some activate chemotaxis (chemokines and classical chemoattractants), and others inhibit or decrease signaling (CXCL9, PIR-B). The mechanisms of signal integration are underexplored and a challenge for future research (Fig. 6.2.2).

EOSINOPHIL CHEMOKINES IN ALLERGY MODELS AND HUMAN DISEASES

The importance of eosinophils and chemokines in the pathogenesis of allergic disease has been an active research area in animal models of eosinophilic pulmonary inflammation. Gene targeting and neutralizing monoclonal antibodies have been used to address the role of individual chemokines in allergic responses. For example, *Cd11* (*eotaxin*) gene-targeted mice demonstrated approximately 70% reduction in BALF eosinophils 18 h after OVA challenge, whereas the numbers of neutrophils, lymphocytes, and macrophages were comparable to those from wild-type mice.[27] However, 48 h after OVA challenge, there was no longer a reduction in eosinophil numbers in the BALF,[27] indicating that eotaxin is important in the early recruitment of eosinophils to the lung in this model. In a later report, a significant reduction of BALF eosinophils was observed in OVA-challenged *chemokine (C-C motif) ligand 24 (Ccl24/eotaxin-2)* gene-deficient mice, suggesting a predominant role of CCL24 (eotaxin-2) for late-phase eosinophil accumulation in the lung.[29] Moreover, it was

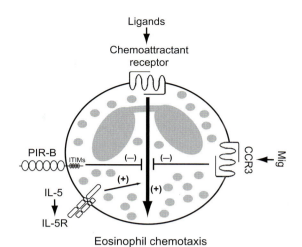

FIGURE 6.2.2 Schematic diagram of examples of positive and negative signals leading to eosinophil chemotaxis. CCR3, C-C chemokine receptor type 3; interleukin-5 (IL-5), IL-5RA, interleukin-5 receptor subunit alpha; ITIM, immunoreceptor tyrosine-based inhibitory motif; Mig, monokine induced by interferon gamma; PIR-B, paired immunoglobulin-like receptor B.

demonstrated that eotaxin and eotaxin-2 can synergistically regulate allergen-induced airway eosinophilia.[29] Consistently, *Ccr3*-deficient mice had a significant reduction of BALF eosinophil numbers.[29] Additionally, administration of neutralizing antibody against eotaxin transiently reduced eosinophil infiltration and airway hyperresponsiveness after each allergen challenge, whereas neutralization of monocyte chemotactic protein 5 (MCP-5; C-C motif chemokine 12) abolished airway hyperresponsiveness apparently by altering the trafficking of eosinophils through the lung interstitium.[54] In contrast, neutralization of MCP-1 blocked the development of airway hyperresponsiveness, even though eosinophil recruitment was unchanged in these mice.[54] Taken together, these studies provide the evidence for promising therapeutic strategies that target the chemokines or chemokine receptor pathways involved in allergic inflammation.

The key involvement of eosinophil chemokines in allergic animal models strongly suggests that these chemokines play an important role in human allergic diseases. Substantial preclinical evidence supports a role for the eotaxin chemokines in human allergic disease. For example, experimental induction of cutaneous and pulmonary late-phase responses in humans has revealed that eotaxin is induced early (within 6 h) and correlates with early eosinophil recruitment; in contrast, eotaxin-2 and -3 correlate with eosinophil accumulation at 24 h.[76] Notably, constitutive expression of eotaxin was found in nasal biopsies from normal individuals; however, the number of eotaxin-expressing cells was significantly increased in tissue from patients with chronic sinusitis or allergic rhinitis.[82] Likewise, protein expression of eotaxin, eotaxin-2, and -3 is significantly higher in eosinophilic nasal polyps than that in controls and is significantly correlated to the amount of eosinophilia.[83] Additionally, the levels of CCL5 (RANTES), MCP-1, and MIP-1-alpha are also significantly higher in the BALF of patients with asthma compared with those in healthy individuals,[84] and MCP-3 and MCP-4 expression in tissue from patients with chronic sinusitis is significantly increased compared with those from normal controls.[85] Notably, transcriptome analysis of biopsies from patients with eosinophilic esophagitis revealed that *eotaxin-3 [chemokine (C-C motif) ligand 26 (CCL26)]* is among the most highly induced genes and may be a biomarker for this emerging allergic disorder.[39] Collectively, these studies suggest the importance of chemokines in the pathogenesis of airway inflammation in humans.

CONCLUSION

Chemokines, particularly the eotaxin subfamily, are involved in eosinophil recruitment in homeostasis and disease. A combination of mouse and human studies defined a critical role for chemokines in eosinophil-associated diseases and their potential as therapeutic targets. The challenge for the future will be to understand the role of individual chemokines and chemokine receptors and their polymorphic variants in the pathophysiology of eosinophilic diseases. Furthermore, a better understanding of the mechanisms by which various signals are integrated and the identification of signal transduction and integration mediators will expand the knowledge of eosinophil recruitment decision-making at baseline and in disease. Once this is accomplished, neutralization of chemokine function may have therapeutic value in the treatment of eosinophil-associated diseases.

REFERENCES

1. Zlotnik A, Yoshie O. Chemokines: a new classification system and their role in immunity. *Immunity* 2000;**12**:121–7.
2. Allen SJ, Crown SE, Handel TM. Chemokine: receptor structure, interactions, and antagonism. *Annu Rev Immunol* 2007;**25**:787–820.
3. Griffiths-Johnson DA, Collins PD, Rossi AG, Jose PJ, Williams TJ. The chemokine, eotaxin, activates guinea-pig eosinophils in vitro and causes their accumulation into the lung in vivo. *Biochem Biophys Res Commun* 1993;**197**:1167–72.
4. Jose PJ, Griffiths-Johnson DA, Collins PD, Walsh DT, Moqbel R, Totty NF, et al. Eotaxin: a potent eosinophil chemoattractant cytokine detected in a guinea pig model of allergic airways inflammation. *J Exp Med* 1994;**179**:881–7.
5. Ponath PD, Qin S, Ringler DJ, Clark-Lewis I, Wang J, Kassam N, et al. Cloning of the human eosinophil chemoattractant, eotaxin.

6. Kitaura M, Nakajima T, Imai T, Harada S, Combadiere C, Tiffany HL, et al. Molecular cloning of human eotaxin, an eosinophil-selective CC chemokine, and identification of a specific eosinophil eotaxin receptor, CC chemokine receptor 3. *J Biol Chem* 1996;**271**:7725–30.
7. Garcia-Zepeda EA, Rothenberg ME, Ownbey RT, Celestin J, Leder P, Luster AD. Human eotaxin is a specific chemoattractant for eosinophil cells and provides a new mechanism to explain tissue eosinophilia. *Nat Med* 1996;**2**:449–56.
8. Rothenberg ME, Luster AD, Leder P. Murine eotaxin: an eosinophil chemoattractant inducible in endothelial cells and in interleukin 4-induced tumor suppression. *Proc Natl Acad Sci U S A* 1995;**92**:8960–4.
9. Rothenberg ME, Luster AD, Lilly CM, Drazen JM, Leder P. Constitutive and allergen-induced expression of eotaxin mRNA in the guinea pig lung. *J Exp Med* 1995;**181**:1211–6.
10. Mishra A, Hogan SP, Lee JJ, Foster PS, Rothenberg ME. Fundamental signals that regulate eosinophil homing to the gastrointestinal tract. *J Clin Invest* 1999;**103**:1719–27.
11. Cook EB, Stahl JL, Lilly CM, Haley KJ, Sanchez H, Luster AD, et al. Epithelial cells are a major cellular source of the chemokine eotaxin in the guinea pig lung. *Allergy Asthma Proc* 1998;**19**:15–22.
12. Bartels J, Schluter C, Richter E, Noso N, Kulke R, Christophers E, et al. Human dermal fibroblasts express eotaxin: molecular cloning, mRNA expression, and identification of eotaxin sequence variants. *Biochem Biophys Res Commun* 1996;**225**:1045–51.
13. Lilly CM, Nakamura H, Kesselman H, Nagler-Anderson C, Asano K, Garcia-Zepeda EA, et al. Expression of eotaxin by human lung epithelial cells: induction by cytokines and inhibition by glucocorticoids. *J Clin Invest* 1997;**99**:1767–73.
14. Fujisawa T, Kato Y, Atsuta J, Terada A, Iguchi K, Kamiya H, et al. Chemokine production by the BEAS-2B human bronchial epithelial cells: differential regulation of eotaxin, IL-8, and RANTES by TH2- and TH1-derived cytokines. *J Allergy Clin Immunol* 2000;**105**:126–33.
15. Matsukura S, Stellato C, Georas SN, Casolaro V, Plitt JR, Miura K, et al. Interleukin-13 upregulates eotaxin expression in airway epithelial cells by a STAT6-dependent mechanism. *Am J Respir Cell Mol Biol* 2001;**24**:755–61.
16. Li L, Xia Y, Nguyen A, Lai YH, Feng L, Mosmann TR, et al. Effects of Th2 cytokines on chemokine expression in the lung: IL-13 potently induces eotaxin expression by airway epithelial cells. *J Immunol* 1999;**162**:2477–87.
17. Rothenberg ME, Ownbey R, Mehlhop PD, Loiselle PM, van de Rijn M, Bonventre JV, et al. Eotaxin triggers eosinophil-selective chemotaxis and calcium flux via a distinct receptor and induces pulmonary eosinophilia in the presence of interleukin 5 in mice. *Mol Med* 1996;**2**:334–48.
18. Collins PD, Marleau S, Griffiths-Johnson DA, Jose PJ, Williams TJ. Cooperation between interleukin-5 and the chemokine eotaxin to induce eosinophil accumulation in vivo. *J Exp Med* 1995;**182**:1169–74.
19. Mould AW, Matthaei KI, Young IG, Foster PS. Relationship between interleukin-5 and eotaxin in regulating blood and tissue eosinophilia in mice. *J Clin Invest* 1997;**99**:1064–71.

Expression, receptor binding, and functional properties suggest a mechanism for the selective recruitment of eosinophils. *J Clin Invest* 1996;**97**:604–12.

20. Nakamura H, Luster AD, Nakamura T. In KH, Sonna LA, Deykin A, et al. Variant eotaxin: its effects on the asthma phenotype. *J Allergy Clin Immunol* 2001;**108**:946−53.

21. Chae SC, Lee YC, Park YR, Shin JS, Song JH, Oh GJ, et al. Analysis of the polymorphisms in eotaxin gene family and their association with asthma, IgE, and eosinophil. *Biochem Biophys Res Commun* 2004;**320**:131−7.

22. Shin HD, Kim LH, Park BL, Jung JH, Kim JY, Chung IY, et al. Association of Eotaxin gene family with asthma and serum total IgE. *Hum Mol Genet* 2003;**12**:1279−85.

23. Forssmann U, Uguccioni M, Loetscher P, Dahinden CA, Langen H, Thelen M, et al. Eotaxin-2, a novel CC chemokine that is selective for the chemokine receptor CCR3, and acts like eotaxin on human eosinophil and basophil leukocytes. *J Exp Med* 1997;**185**:2171−6.

24. Ying S, Meng Q, Zeibecoglou K, Robinson DS, Macfarlane A, Humbert M, et al. Eosinophil chemotactic chemokines (eotaxin, eotaxin-2, RANTES, monocyte chemoattractant protein-3 (MCP-3), and MCP-4), and C-C chemokine receptor 3 expression in bronchial biopsies from atopic and nonatopic (intrinsic) asthmatics. *J Immunol* 1999;**163**:6321−9.

25. Ying S, Robinson DS, Meng Q, Barata LT, McEuen AR, Buckley MG, et al. C-C chemokines in allergen-induced late-phase cutaneous responses in atopic subjects: association of eotaxin with early 6-hour eosinophils, and of eotaxin-2 and monocyte chemoattractant protein-4 with the later 24-hour tissue eosinophilia, and relationship to basophils and other C-C chemokines (monocyte chemoattractant protein-3 and RANTES). *J Immunol* 1999;**163**:3976−84.

26. Yang M, Hogan SP, Mahalingam S, Pope SM, Zimmermann N, Fulkerson P, et al. Eotaxin-2 and IL-5 cooperate in the lung to regulate IL-13 production and airway eosinophilia and hyperreactivity. *J Allergy Clin Immunol* 2003;**112**:935−43.

27. Rothenberg ME, MacLean JA, Pearlman E, Luster AD, Leder P. Targeted disruption of the chemokine eotaxin partially reduces antigen-induced tissue eosinophilia. *J Exp Med* 1997;**185**:785−90.

28. Pope SM, Fulkerson PC, Blanchard C, Akei HS, Nikolaidis NM, Zimmermann N, et al. Identification of a cooperative mechanism involving interleukin-13 and eotaxin-2 in experimental allergic lung inflammation. *J Biol Chem* 2005;**280**:13952−61.

29. Pope SM, Zimmermann N, Stringer KF, Karow ML, Rothenberg ME. The eotaxin chemokines and CCR3 are fundamental regulators of allergen-induced pulmonary eosinophilia. *J Immunol* 2005;**175**:5341−50.

30. Min JW, Lee JH, Park CS, Chang HS, Rhim TY, Park SW, et al. Association of eotaxin-2 gene polymorphisms with plasma eotaxin-2 concentration. *J Hum Genet* 2005;**50**:118−23.

31. Shinkai A, Yoshisue H, Koike M, Shoji E, Nakagawa S, Saito A, et al. A novel human CC chemokine, eotaxin-3, which is expressed in IL-4-stimulated vascular endothelial cells, exhibits potent activity toward eosinophils. *J Immunol* 1999;**163**:1602−10.

32. Dulkys Y, Schramm G, Kimmig D, Knoss S, Weyergraf A, Kapp A, et al. Detection of mRNA for eotaxin-2 and eotaxin-3 in human dermal fibroblasts and their distinct activation profile on human eosinophils. *J Invest Dermatol* 2001;**116**:498−505.

33. Komiya A, Nagase H, Yamada H, Sekiya T, Yamaguchi M, Sano Y, et al. Concerted expression of eotaxin-1, eotaxin-2, and eotaxin-3 in human bronchial epithelial cells. *Cell Immunol* 2003;**225**:91−100.

34. Zuyderduyn S, Hiemstra PS, Rabe KF. TGF-beta differentially regulates TH2 cytokine-induced eotaxin and eotaxin-3 release by human airway smooth muscle cells. *J Allergy Clin Immunol* 2004;**114**:791−8.

35. Blanchard C, Durual S, Estienne M, Emami S, Vasseur S, Cuber JC. Eotaxin-3/CCL26 gene expression in intestinal epithelial cells is up-regulated by interleukin-4 and interleukin-13 via the signal transducer and activator of transcription 6. *Int J Biochem Cell Biol* 2005;**37**:2559−73.

36. Kitaura M, Suzuki N, Imai T, Takagi S, Suzuki R, Nakajima T, et al. Molecular cloning of a novel human CC chemokine (Eotaxin-3) that is a functional ligand of CC chemokine receptor 3. *J Biol Chem* 1999;**274**:27975−80.

37. Berkman N, Ohnona S, Chung FK, Breuer R. Eotaxin-3 but not eotaxin gene expression is upregulated in asthmatics 24 hours after allergen challenge. *Am J Respir Cell Mol Biol* 2001;**24**:682−7.

38. Ravensberg AJ, Ricciardolo FL, van Schadewijk A, Rabe KF, Sterk PJ, Hiemstra PS, et al. Eotaxin-2 and eotaxin-3 expression is associated with persistent eosinophilic bronchial inflammation in patients with asthma after allergen challenge. *J Allergy Clin Immunol* 2005;**115**:779−85.

39. Blanchard C, Wang N, Stringer KF, Mishra A, Fulkerson PC, Abonia JP, et al. Eotaxin-3 and a uniquely conserved gene-expression profile in eosinophilic esophagitis. *J Clin Invest* 2006;**116**:536−47.

40. Proost P, Wuyts A, Van Damme J. Human monocyte chemotactic proteins-2 and -3: structural and functional comparison with MCP-1. *J Leukoc Biol* 1996;**59**:67−74.

41. Garcia-Zepeda EA, Combadiere C, Rothenberg ME, Sarafi MN, Lavigne F, Hamid Q, et al. Human monocyte chemoattractant protein (MCP)-4 is a novel CC chemokine with activities on monocytes, eosinophils, and basophils induced in allergic and nonallergic inflammation that signals through the CC chemokine receptors (CCR)-2 and -3. *J Immunol* 1996;**157**:5613−26.

42. Dahinden CA, Geiser T, Brunner T, von Tscharner V, Caput D, Ferrara P, et al. Monocyte chemotactic protein 3 is a most effective basophil- and eosinophil-activating chemokine. *J Exp Med* 1994;**179**:751−6.

43. Weber M, Uguccioni M, Ochensberger B, Baggiolini M, Clark-Lewis I, Dahinden CA. Monocyte chemotactic protein MCP-2 activates human basophil and eosinophil leukocytes similar to MCP-3. *J Immunol* 1995;**154**:4166−72.

44. Weber M, Uguccioni M, Baggiolini M, Clark-Lewis I, Dahinden CA. Deletion of the NH2-terminal residue converts monocyte chemotactic protein 1 from an activator of basophil mediator release to an eosinophil chemoattractant. *J Exp Med* 1996;**183**:681−5.

45. Weber C, Kitayama J, Springer TA. Differential regulation of beta 1 and beta 2 integrin avidity by chemoattractants in eosinophils. *Proc Natl Acad Sci U S A* 1996;**93**:10939−44.

46. Schall TJ, Simpson NJ, Mak JY. Molecular cloning and expression of the murine RANTES cytokine: structural and functional conservation between mouse and man. *Eur J Immunol* 1992;**22**:1477−81.

47. Kameyoshi Y, Dorschner A, Mallet AI, Christophers E, Schroder JM. Cytokine RANTES released by thrombin-stimulated platelets is a potent attractant for human eosinophils. *J Exp Med* 1992;**176**:587−92.

48. Terada N, Maesako K, Hamano N, Ikeda T, Sai M, Yamashita T, et al. RANTES production in nasal epithelial cells and endothelial cells. *J Allergy Clin Immunol* 1996;**98**:S230—7.

49. Wang JH, Devalia JL, Xia C, Sapsford RJ, Davies RJ. Expression of RANTES by human bronchial epithelial cells in vitro and in vivo and the effect of corticosteroids. *Am J Respir Cell Mol Biol* 1996;**14**:27—35.

50. Ying S, Meng Q, Taborda-Barata L, Corrigan CJ, Barkans J, Assoufi B, et al. Human eosinophils express messenger RNA encoding RANTES and store and release biologically active RANTES protein. *Eur J Immunol* 1996;**26**:70—6.

51. Rot A, Krieger M, Brunner T, Bischoff SC, Schall TJ, Dahinden CA. RANTES and macrophage inflammatory protein 1 alpha induce the migration and activation of normal human eosinophil granulocytes. *J Exp Med* 1992;**176**:1489—95.

52. Beck LA, Dalke S, Leiferman KM, Bickel CA, Hamilton R, Rosen H, et al. Cutaneous injection of RANTES causes eosinophil recruitment: comparison of nonallergic and allergic human subjects. *J Immunol* 1997;**159**:2962—72.

53. Lukacs NW, Strieter RM, Shaklee CL, Chensue SW, Kunkel SL. Macrophage inflammatory protein-1 alpha influences eosinophil recruitment in antigen-specific airway inflammation. *Eur J Immunol* 1995;**25**:245—51.

54. Gonzalo JA, Lloyd CM, Wen D, Albar JP, Wells TN, Proudfoot A, et al. The coordinated action of CC chemokines in the lung orchestrates allergic inflammation and airway hyperresponsiveness. *J Exp Med* 1998;**188**:157—67.

55. Das AM, Ajuebor MN, Flower RJ, Perretti M, McColl SR. Contrasting roles for RANTES and macrophage inflammatory protein-1 alpha (MIP-1 alpha) in a murine model of allergic peritonitis. *Clin Exp Immunol* 1999;**117**:223—9.

56. Burrows LJ, Piper PJ, Lindley ID, Westwick J. Intraperitoneal injection of human recombinant neutrophil-activating factor/interleukin 8 (hrNAF/IL-8) produces a T cell and eosinophil infiltrate in the guinea pig lung. Effect of PAF antagonist WEB2086. *Ann N Y Acad Sci* 1991;**629**:422—4.

57. Heath H, Qin S, Rao P, Wu L, LaRosa G, Kassam N, et al. Chemokine receptor usage by human eosinophils. The importance of CCR3 demonstrated using an antagonistic monoclonal antibody. *J Clin Invest* 1997;**99**:178—84.

58. Morita E, Schroder JM, Christophers E. Differential sensitivities of purified human eosinophils and neutrophils to defined chemotaxins. *Scand J Immunol* 1989;**29**:709—16.

59. Henocq E, Vargaftig BB. Skin eosinophilia in atopic patients. *J Allergy Clin Immunol* 1988;**81**:691—5.

60. Lellouch-Tubiana A, Lefort J, Simon MT, Pfister A, Vargaftig BB. Eosinophil recruitment into guinea pig lungs after PAF-acether and allergen administration. Modulation by prostacyclin, platelet depletion, and selective antagonists. *Am Rev Respir Dis* 1988;**137**:948—54.

61. Daugherty BL, Siciliano SJ, DeMartino JA, Malkowitz L, Sirotina A, Springer MS. Cloning, expression, and characterization of the human eosinophil eotaxin receptor. *J Exp Med* 1996;**183**:2349—54.

62. Phillips RM, Stubbs VE, Henson MR, Williams TJ, Pease JE, Sabroe I. Variations in eosinophil chemokine responses: an investigation of CCR1 and CCR3 function, expression in atopy, and identification of a functional CCR1 promoter. *J Immunol* 2003;**170**:6190—201.

63. Sullivan SK, McGrath DA, Liao F, Boehme SA, Farber JM, Bacon KB. MIP-3alpha induces human eosinophil migration and activation of the mitogen-activated protein kinases (p42/p44 MAPK). *J Leukoc Biol* 1999;**66**:674—82.

64. Liu LY, Jarjour NN, Busse WW, Kelly EA. Chemokine receptor expression on human eosinophils from peripheral blood and bronchoalveolar lavage fluid after segmental antigen challenge. *J Allergy Clin Immunol* 2003;**112**:556—62.

65. Jinquan T, Jing C, Jacobi HH, Reimert CM, Millner A, Quan S, et al. CXCR3 expression and activation of eosinophils: role of IFN-gamma-inducible protein-10 and monokine induced by IFN-gamma. *J Immunol* 2000;**165**:1548—56.

66. Nagase H, Miyamasu M, Yamaguchi M, Fujisawa T, Ohta K, Yamamoto K, et al. Expression of CXCR4 in eosinophils: functional analyses and cytokine-mediated regulation. *J Immunol* 2000;**164**:5935—43.

67. Zimmermann N, Daugherty BL, Kavanaugh JL, El-Awar FY, Moulton EA, Rothenberg ME. Analysis of the CC chemokine receptor 3 gene reveals a complex 5′ exon organization, a functional role for untranslated exon 1, and a broadly active promoter with eosinophil-selective elements. *Blood* 2000;**96**:2346—54.

68. Zimmermann N, Bernstein JA, Rothenberg ME. Polymorphisms in the human CC chemokine receptor-3 gene. *Biochim Biophys Acta* 1998;**1442**:170—6.

69. Fukunaga K, Asano K, Mao XQ, Gao PS, Roberts MH, Oguma T, et al. Genetic polymorphisms of CC chemokine receptor 3 in Japanese and British asthmatics. *Eur Respir J* 2001;**17**:59—63.

70. Lee JH, Chang HS, Kim JH, Park SM, Lee YM, Uh ST, et al. Genetic effect of CCR3 and IL5RA gene polymorphisms on eosinophilia in asthmatic patients. *J Allergy Clin Immunol* 2007;**120**:1110—7.

71. Wise EL, Bonner KT, Williams TJ, Pease JE. A single nucleotide polymorphism in the CCR3 gene ablates receptor export to the plasma membrane. *J Allergy Clin Immunol* 2010;**126**:150—7.

72. Zimmermann N, Daugherty BL, Stark JE, Rothenberg ME. Molecular analysis of CCR-3 events in eosinophilic cells. *J Immunol* 2000;**164**:1055—64.

73. El-Shazly A, Yamaguchi N, Masuyama K, Suda T, Ishikawa T. Novel association of the src family kinases, hck and c-fgr, with CCR3 receptor stimulation: A possible mechanism for eotaxin-induced human eosinophil chemotaxis. *Biochem Biophys Res Commun* 1999;**264**:163—70.

74. Kampen GT, Stafford S, Adachi T, Jinquan T, Quan S, Grant JA, et al. Eotaxin induces degranulation and chemotaxis of eosinophils through the activation of ERK2 and p38 mitogen-activated protein kinases. *Blood* 2000;**95**:1911—7.

75. Zimmermann N, Conkright JJ, Rothenberg ME. CC chemokine receptor-3 undergoes prolonged ligand-induced internalization. *J Biol Chem* 1999;**274**:12611—8.

76. Zimmermann N, Hershey GK, Foster PS, Rothenberg ME. Chemokines in asthma: cooperative interaction between chemokines and IL-13. *J Allergy Clin Immunol* 2003;**111**:227—42.

77. Loetscher P, Pellegrino A, Gong JH, Mattioli I, Loetscher M, Bardi G, et al. The ligands of CXC chemokine receptor 3, I-TAC, Mig, and IP10, are natural antagonists for CCR3. *J Biol Chem* 2001;**276**:2986—91.

78. Xanthou G, Duchesnes CE, Williams TJ, Pease JE. CCR3 functional responses are regulated by both CXCR3 and its ligands CXCL9, CXCL10 and CXCL11. *Eur J Immunol* 2003;**33**:2241−50.

79. Weng Y, Siciliano SJ, Waldburger KE, Sirotina-Meisher A, Staruch MJ, Daugherty BL, et al. Binding and functional properties of recombinant and endogenous CXCR3 chemokine receptors. *J Biol Chem* 1998;**273**:18288−91.

80. Fulkerson PC, Zimmermann N, Brandt EB, Muntel EE, Doepker MP, Kavanaugh JL, et al. Negative regulation of eosinophil recruitment to the lung by the chemokine monokine induced by IFN-gamma (Mig, CXCL9). *Proc Natl Acad Sci U S A* 2004;**101**:1987−92.

81. Munitz A, McBride ML, Bernstein JS, Rothenberg ME. A dual activation and inhibition role for the paired immunoglobulin-like receptor B in eosinophils. *Blood* 2008;**111**:5694−703.

82. Minshall EM, Cameron L, Lavigne F, Leung DY, Hamilos D, Garcia-Zepada EA, et al. Eotaxin mRNA and protein expression in chronic sinusitis and allergen-induced nasal responses in seasonal allergic rhinitis. *Am J Respir Cell Mol Biol* 1997;**17**:683−90.

83. Olze H, Forster U, Zuberbier T, Morawietz L, Luger EO. Eosinophilic nasal polyps are a rich source of eotaxin, eotaxin-2 and eotaxin-3. *Rhinology* 2006;**44**:145−50.

84. Alam R, York J, Boyars M, Stafford S, Grant JA, Lee J, et al. Increased MCP-1, RANTES, and MIP-1alpha in bronchoalveolar lavage fluid of allergic asthmatic patients. *Am J Respir Crit Care Med* 1996;**153**:1398−404.

85. Wright ED, Frenkiel S, Ghaffar O, al-Ghamdi K, Luster A, Miotto D, et al. Monocyte chemotactic protein expression in allergy and non-allergy-associated chronic sinusitis. *J Otolaryngol* 1998;**27**:281−7.

Chapter 6.3

Adhesion Molecules

Kenji Matsumoto and Bruce S. Bochner

INTRODUCTION

Eosinophil-selective accumulation into specific tissue sites is one of the key events for the onset, maintenance, exacerbation, or perhaps even repair of pathological and physiological responses in which eosinophils play a role, such as allergic inflammation, helminth infections, and cancer regulation. This local eosinophil accumulation is the result of a series of binding and detaching events to different cells and extracellular matrix (ECM) proteins mediated by several different adhesion molecules.[1,2] Accumulating reports have elucidated that eosinophil-selective recruitment is elegantly orchestrated by interactions of chemokines and adhesion molecules.[3] Of note, such translocations of eosinophils to the tissue sites at the same time *prime* eosinophils to become even more susceptible to subsequent stimuli (so-called *outside-in signaling*). This mechanism not only facilitates eosinophil functions in the tissue but also probably prevents systemic overactivation of circulating eosinophils.

ADHESION MOLECULES ON EOSINOPHILS

Selectins

Selectins are glycoproteins composed of multiple domains: one homologous to lectins, one homologous to epidermal growth factor, and two or more homologous to the consensus repeat units found in complement component 3/4 (C3/C4) binding proteins. Freshly isolated eosinophils express L-selectin (CD62L), which plays a critical role in the rolling of eosinophils on endothelial cells under the sheath flow[4] through binding to carbohydrates [Sialyl-Lewis^X (sLe^x), 6-sulfated sLe^x, and fucoidan), proteoglycans, E- and P-selectins, and others, including nucleolin. Impairment of L-selectin reduces eosinophil rolling on endothelial cells and reduces the chances of establishing firm adhesion leading to extravasation. Some proteoglycans are known to bind to chemokines and create gradients, both in the tissue interstitium and on the endothelium. Binding to proteoglycans through L-selectin allows eosinophils to be exposed to proteoglycan-binding chemokines more efficiently.[5] L-selectin has a short intracellular tail insufficient to transduce signals directly; however, an L-selectin association molecule, urokinase plasminogen activator surface receptor (U-PAR/CD87), mediates the activation of β_2 integrins through Ca^{2+} mobilization after L-selectin binding to its ligands.[6]

Eosinophils also express P-selectin and P-selectin-mediated adhesion is comparable between eosinophils and neutrophils in that both express similar levels of CD162 [P-selectin glycoprotein ligand 1 (PSGL-1)], both bind P-selectin equally well, both lose their P-selectin binding activity following treatment with sialidase or protease, and both use P-selectin for binding *in vivo*.[7–11] An exception to this similarity was reported under conditions where endothelial cells were treated with interleukin-13, resulting in P-selectin-dependent adhesion of eosinophils but not neutrophils under physiological conditions of shear stress.[12]

Integrins

Integrins are composed of two non-covalently associated transmembrane glycoprotein subunits called α and β, both of which contribute to the binding function of the adhesion molecules to matrix proteins and other ligands (Fig. 6.3.1).[13] The α subunit contains three or four divalent, cation-binding domains in the large extracellular part that regulates binding affinity of integrins on eosinophils.[14] The β subunit has talin

TABLE 6.3.1 Adhesion Molecules Found on Peripheral Blood Eosinophils

Adhesion Molecule		Alternative Name/s		Cluster of Differentiation Molecule/s	Ligands
Selectins	L-selectin	SELL	CD62L	P- and E-selectin, 6-sulfo-sLe^x	
	P-selectin	SELP	CD62P	L- and E-selectin, PSGL-1	
Integrins					
	α4β1	VLA-4	CD49d/CD29	VCAM-1, FN	
	α6β1	VLA-6	CD49f/CD29	LN	
	αLβ2	LFA-1	CD11a/CD18	ICAM-1, 2, 3, 5	
	αMβ2	MAC-1, CR3	CD11b/CD18	iC3b, FG, ICAM-1, others	
	αxβ2	p150, 95, CR4	CD11c/CD18	iC3b, FG, ICAM-1, others	
	αDβ2	LPAM-1	CD11d/CD18	VCAM-1, ICAM-3,	
	α4β7		CD49d/CD29	MAdCAM-1, VCAM-1, FN	
Others					
	Siglec-8	SAF-2		6'-sulfo-sLe^x	
	CD44		CD44	Hyaluronic acid	
	Galectin-3			β-galactosides, VCAM-1	
	CD66b	CEACAM8		Galectin-3	
	CD31	PECAM-1	CD31	CD31	
	PSGL-1		CD162	P-selectin	
	sLe^x			E-selectin	

CEACAM8, carcinoembryonic antigen-related cell adhesion molecule-8; FG, fibrinogen; FN, fibronectin; ICAM, intercellular adhesion molecule; LFA-1, lymphocyte function-associated antigen-1; LN, laminin; LPAM-1, lymphocyte Peyer patch adhesion molecule-1; MAC-1, cell surface glycoprotein MAC-1 subunit alpha; MAdCAM-1, mucosal addressin cell adhesion molecule-1; PECAM-1, platelet endothelial cell adhesion molecule 1; PSGL-1, P-selectin glycoprotein ligand-1; SAF-2, sialoadhesin family member 2; Siglec-8, sialic acid-binding Ig-like lectin 8; sLe^x: Sialyl-Lewis X, VLA, very late antigen; VCAM-1, vascular cell adhesion molecule-1.

and α-actinin binding sites in the intracellular tail that mediate the assembly of actin filaments in the cell and critically regulate morphological changes of eosinophils on binding to other cells and ECM proteins.

To date, at least seven integrin molecules have been reported to be expressed by fresh eosinophils; α4β1, α6β1, αLβ2, αMβ2, αXβ2, αDβ2, and α4β7 integrin. Eosinophil binding to vascular cell adhesion protein 1 (VCAM-1) and an ECM protein, fibronectin (FN), is mediated by α4β1 integrin. Differences in the binding affinities of integrin α4β1 for VCAM-1 and FN suggest that the binding sites of α4β1 for VCAM-1 and FN may be distinct.[15] An ECM protein, laminin (LN),[16] which participates in eosinophilopoiesis and in the recognition of eosinophil apoptosis by macrophages,[18] binds to α6β1 integrin. Three β2 integrins, αLβ2, αMβ2, and αXβ2, conserve high binding affinity to intercellular adhesion molecule 1 (ICAM-1) and critically regulate the transendothelial migration of eosinophils,[19] whereas αDβ2 integrin can only bind to VCAM-1 and intercellular adhesion molecule 3 (ICAM-3).[20] However, several eosinophil functions require eosinophil binding to ECM proteins such as fibrinogen, complement fragments, and other degenerative proteins through αMβ2 integrin.[21–26] Finally, α4β7 integrin binds to mucosal addressin cell adhesion molecule 1 (MAdCAM-1), VCAM-1, and FN, and probably regulates eosinophil migration to mucosal surfaces.[27]

Other Carbohydrate Molecules and their Ligands

Eosinophils express a variety of glycosylated surface structures involved in cell adhesion. For example, eosinophils express low levels (compared to neutrophils) of the E-selectin ligands sLe^x and a sialylated dimeric form of Le^x.[28] The levels of their expression were the same on eosinophils from different types of donors and remained the same even after cell activation. For both eosinophils and neutrophils, removal of terminal sialic acids by treatment with sialidase resulted in the complete loss of detectable sLe^x and sialyl-

Integrin

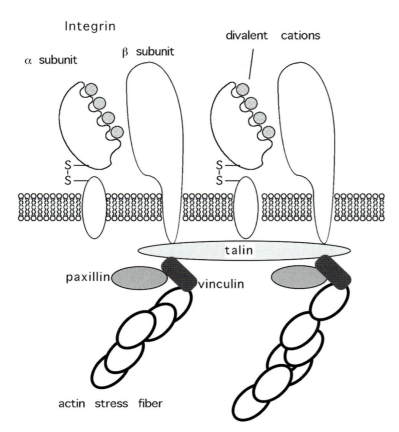

FIGURE 6.3.1 The structure of typical α and β integrin subunits is shown along with their intracellular association molecules.

dimeric Le^x surface expression, and eliminated the cells' adhesion to E-selectin. Treatment of eosinophils and neutrophils with endo-β-galactosidase, which removes the $β_{1,4}$ galactose from glycans such as that found in sialyl-dimeric Le^x, also inhibited cell adhesion and eliminated the expression of sialyl-dimeric Le^x as detected by flow cytometry. Endo-β-galactosidase treatment also reduced sLe^x expression on eosinophils, but not neutrophils, suggesting that the major proportion of the sLe^x-containing E-selectin ligand on the surface of eosinophils, but not neutrophils, is in the form of sialyl-dimeric Le^x. For both cell types, treatment with proteases failed to alter either the expression of these E-selectin ligands or adhesion to E-selectin. Indeed, subsequent studies with human neutrophils confirmed that the primary ligand for E-selectin is a sialylated glycolipid, not a glycoprotein.[29] These marked differences between eosinophil and neutrophil sLe^x-containing surface molecules may explain why E-selectin contributes little if anything to eosinophil recruitment during inflammatory responses *in vivo.*[10,11,30,31]

Further insight into glycans for selectin-mediated adhesion was provided by studies in which the role of fucosyltransferases (FUTs) in mouse eosinophil

selectin ligand activity was explored. Both alpha-(1,3)-fucosyltransferase 4 (FUT4) and 7 (FUT7) mRNAs were increased by transforming growth factor beta-1 (TGF-β1), but the FUT4 transcript was consistently predominant in eosinophils. While P-selectin bound identically to eosinophils from FUT4-null and wild-type mice, E-selectin binding was partially reduced in these mice.[9]

Very recently, high-sensitivity mass spectrometry-based glycomics methodologies were applied to the analysis of N-linked glycans derived from isolated populations of human eosinophils (North and colleagues, manuscript in preparation). Studies revealed considerable quantities of terminal N-acetylglucosamine-containing structures on the eosinophil surface similar to those known to bind the C-type lectin (CLEC) LSECtin, with a wide range of high-mannose N-glycans and a number of complex glycans consistent with bi-, tri-, and tetra-antennary forms. Sialylation was relatively low as a terminal modification compared to neutrophils. Some low-level antennal fucosylation was detected, together with small amounts of simple hybrid structures. Overall, the most distinctive feature of the eosinophil carbohydrate phenotype was a high level of truncated or unsubstituted and partly processed glycans. How this distinctive glycan pattern contributes to eosinophil biology will require further study.

Other Adhesion Molecules on Eosinophils

Sialic acid-binding Ig-like lectins (Siglecs) are cell surface proteins of the immunoglobulin superfamily, composed of an N-terminus immunoglobulin V (IgV) domain which can recognize sugars as I-type (Ig-type) lectins, a variable number of C2-type Ig (IgC2) domains extracellularly, and one or more immunoreceptor tyrosine-based inhibitory motifs (ITIMs) intracellularly. Among various Siglecs, sialic acid-binding Ig-like lectin 8 (Siglec-8) is most selectively and prominently expressed in eosinophils, but also basophils and mast cells.[10] For more information on eosinophil Siglecs, see the Anti-Eosinophil Therapeutics section of this book, Chapter 15.4 on Therapeutic Approaches Targeting Siglecs.

CD44 is a receptor for hyaluronic acid and can also interact with other ligands, including osteopontin, collagens, and matrix metalloproteinases (MMPs). Expression of CD44 on eosinophils is upregulated on exposure to eosinophil-activating cytokines such as granulocyte-macrophage colony-stimulating factor (GM-CSF), interleukin-3, and interleukin-5 (IL-5), even though eosinophils showed only marginal binding to hyaluronic acid in vitro.[11] However, binding to hyaluronic acid may play a role in eosinophilopoiesis[32] and TGF-β production by eosinophils.[33]

Galectin-3 is a member of the lectin family and contains a carbohydrate-recognition-binding domain that enables the specific binding of β-galactosides. Eosinophils from allergic donors expressed elevated levels of galectin-3 that colocalized with α4 integrins and reportedly contribute to eosinophil rolling on VCAM-1 under flow conditions.[34]

CD66b [carcinoembryonic antigen-related cell adhesion molecule 8 (CEAM8)] is a single chain, glycosylphosphatidylinisotol (GPI)-anchored, highly glycosylated protein belonging to the carcinoembryonic antigen supergene family. On binding of CD66b to its natural ligand, galectin-3, eosinophil activation was observed. However, this signal transduction is likely to be mediated by αMβ2 integrin because CD66b was constitutively and physically associated with αMβ2 integrin.[35]

CD31 [platelet endothelial cell adhesion molecule (PECAM-1)] is expressed on eosinophils. Homotypic binding between CD31 on the endothelial cells (expressed near the endothelial cell—cell junctions) and CD31 on eosinophils is supposed to be a critical step in transendothelial migration. However, administration of anti-CD31 monoclonal antibody failed to block eosinophil accumulation in the lung in a mouse model of asthma.[36,37]

Cadherins are a class of type-1 transmembrane proteins and include cadherins, protocadherins, desmogleins, desmocollins, and others. Cadherins play important roles in the formation of tissue structures by contributing to tight adhesion of cells in a Ca^{2+} ion-dependent manner. According to an eosinophil transcriptome database (the Gene Expression Omnibus database; available at: http://www.ncbi.nlm.nih.gov/geo/query/acc.cgi?acc=GSE12837), eosinophils express at least mRNA of the cadherin family molecules, but no functional information regarding these structures in eosinophils is available.

CHANGES IN AVIDITY OF ADHESION MOLECULES ON EOSINOPHILS
Number of Adhesion Molecules

Elevated expression levels of several adhesion molecules, including αLβ2 integrin,[38] αMβ2 integrin,[39] α1β1 integrin and α2β1 integrin,[40] CD44,[31] and galectin-3[34] have been demonstrated for eosinophils from allergic donors and after activation in vitro. However, with regard to the roles of integrins in eosinophil extravasation to inflammatory sites, such changes in the expression levels of integrins may not be clinically important because regulation of binding affinity by conformational changes is far more critical for the function of integrins. On the other hand, the increased expression levels of those adhesion molecules are probably important in emphasizing activation signals after eosinophils have reached their target tissues.

Conversely, cell surface expression of L-selectin is decreased on exposure to eosinophil-activating cytokines and following transendothelial migration of eosinophils in vitro and in vivo.[14,39] This shedding of L-selectin is mediated by proteolysis by a disintegrin and metalloproteinase domain-containing protein 8 (ADAM 8).[41] L-selectin shedding is not only a marker of activated eosinophils, but also probably renders eosinophils no longer capable of binding to their ligands to interrupt the movement of eosinophils in the tissue interstitium.

Distribution of Adhesion Molecules

Adhesion molecules are not always evenly distributed on the surface of eosinophils. Most of the adhesion molecules accumulate on the cell surface to form focal patches similar to the lipid rafts in T-cell and B-cell interactions.[42–44] Such homotypic accumulations of adhesion molecules not only increase binding efficiency, but also provide several association molecules with intracellular enzymes and their substrates with more chances to interact with one another. This facilitates intracellular signaling, especially

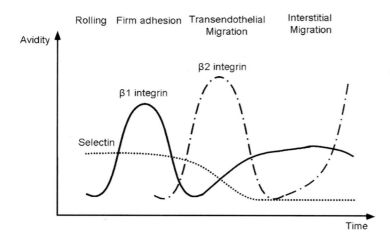

FIGURE 6.3.2 Schema of avidity changes in adhesion molecules on eosinophils that are felt to occur during recruitment into tissues. The x-axis indicates the time course, while the y-axis indicates the avidity. *Reproduced from reference[2] with permission.*

phosphorylation, which, in turn, activates eosinophils (so-called outside-in signaling).[45] Interestingly, the accumulated adhesion molecules move on the cell surface to regulate cell tethering during transendothelial migration.

Regulation of the Affinity of Adhesion Molecules

During eosinophil transendothelial migration and movement into the tissue interstitium, eosinophils must regulate the avidity of adhesion molecules serially and dynamically. Unlike monomeric adhesion molecules, integrins are composed of α and β subunits and are responsible for the rapid and dynamic changes in binding affinity, presumably within seconds.[46] The affinity of integrins on eosinophils is known to be critically regulated by their activation with cytokines, chemokines, and even the process of adhesion itself through other molecules (this affinity regulation is another example of *inside-out signaling*);[47] the binding through integrins accompanies cytosolic gel movement and cytoskeleton changes and leads to cell movement and shape changes.

The activation signals that increase the affinity of β1[48] and β2 integrins[49] are distinct, suggesting that the intracellular molecules regulating integrin affinity differ between different β integrins. In eosinophils, cytokines such as IL-5 increase the affinity of β2 integrins through the activation of phosphatidylinositol 3-kinase (PI3K), cytosolic phospholipase A2 (cPLA2)[50] and NF-kappa-B-repressing factor (NF-κB-repressing factor).[51–54] In contrast, chemokines such as eotaxin [C-C motif chemokine 11 (CCL11)] increase the affinity of β1 integrins through the activation of a tyrosine kinase[10] and MAPK/ERK kinase 1 (MEK 1).[55] Other chemical mediators such as cysteinyl leukotriene D4 (LTD4)[56] and histamine[57] activate mainly αMβ2 integrin.

Importantly, however, such increases in the affinities of integrins should be transient but sufficient enough to lead eosinophils to the target tissues (Fig. 6.3.2). For instance, once eosinophils firmly adhere to endothelial cells on exposure to chemokines, they start to migrate into tissues through the endothelial cell layer. This process requires not only firm adhesion of eosinophils through α4β1 integrin-VCAM-1 interactions but also upregulation of the affinities of β2 integrins to adhere to ICAM-1 as well as detachment of α4β1 integrin-VCAM-1 binding.[58–61] If such detachment is inhibited, eosinophils will remain stuck to the endothelium and will not undergo transendothelial migration.[58]

To date, several monoclonal antibodies have been developed that react specifically with active forms of β1 [15/7[62]] or β2 [HUTS-21, N29, 24, CBRM1/5[8]] integrins. Not surprisingly, these antibodies are useful tools for adhesion studies.

ACTIVATION OF EOSINOPHILS BY ADHESION

As mentioned earlier, several functions are enhanced after eosinophils adhere to other cells or ECM proteins via outside-in signaling. One of the most sensitive assays for measuring such activation is the prolongation of eosinophil survival. On binding via α4β1, α4β7, and αMβ2 integrins, eosinophils survive longer, mainly as a result of autocrine/paracrine secretion of GM-CSF.[63–65] In addition, binding of eosinophils via integrins, in particular αMβ2 integrin, enhances various eosinophil functions such as degranulation, superoxide generation, and chemokine and cytokine production.[16–21] Production of eosinophil-recruiting chemokines such as eotaxins and C-C motif chemokine 5 (CCL5/RANTES) by eosinophils facilitates further eosinophil migration from the bloodstream and at the same time allows activated eosinophils to stay in the tissue. Most of these enhanced eosinophil functions are mediated by autocrine/paracrine secretion of GM-CSF. However, some direct effects of integrin signaling through the activation of focal adhesion kinase 1 (FADK 1), protein kinase C (PKC), and actin assembly have been demonstrated.[66]

ADHESION MOLECULES AS POTENTIAL THERAPEUTIC TARGETS

Most of what we know about the contributions of cell adhesion molecules to eosinophil recruitment responses comes from mouse studies. For instance, mice deficient in P-selectin, L-selectin, or ICAM-1, or hypomorphic for VCAM-1 show varying degrees of reduced eosinophil accumulation.[26,67−71] E-selectin and CD31 make little if any contribution to eosinophil recruitment responses, while CD44 was shown to make a modest contribution in some models.[37,67,72] Most consistent is the contribution of the $\alpha4\beta1$ and $\alpha4\beta7$ integrins, mediated via VCAM-1 and MAdCAM-1, to eosinophil accumulation into various sites, including the lung, skin, and gastrointestinal tract ([73,74] and reviewed in[75]).

Studies using early glycomimetic antagonists suggested that human eosinophils, as well as neutrophils, would be expectedly susceptible to blockade of selectin function.[76,77] However, to date, no glycomimetics have been tried in the treatment of eosinophilic disorders. Studies in humans using the $\alpha L\beta2$ antibody efalizumab, which is no longer commercially available in the United States, found a disappointingly modest effect on allergen-induced eosinophil recruitment into the airways.[78] Natalizumab, an $\alpha4$ integrin antibody approved for use in a highly restricted manner (due to the risk of developing progressive multifocal leukoencephalopathy) for refractory relapsing remitting multiple sclerosis (RRMS), increased circulating eosinophil counts, presumably by blocking eosinophil extravasation,[79] but it has never been formally tested in eosinophilic disorders. Unfortunately, attempts to date to use inhaled, small-molecule $\alpha4\beta1$ antagonists to treat eosinophilic airway inflammation, which appeared promising based on *in vitro* studies,[80,81] have met with failure *in vivo*.[82−84] While natalizumab would be predicted to be quite effective in treating eosinophilic disorders, concerns regarding immunosuppression will likely prevent its use for these disorders. Perhaps some of the newer orally available $\alpha4\beta1$ antagonists that were reportedly well tolerated in phase I clinical trials[85] will one day prove useful for asthma.

Another strategy for treating diseases in which eosinophils play a role is induction of apoptosis. Ligation of two adhesion molecules, ICAM-3[86] and Siglec-8,[87] was demonstrated to induce eosinophil apoptosis *in vitro* (see the Anti-Eosinophil Therapeutics section of this book, Chapter 15.4 on Therapeutic Approaches Targeting Siglecs).

CONCLUSION

Eosinophils express various types of adhesion molecules that play critical roles in rolling and firm adhesion to endothelial cells under shear stress of blood flow, transendothelial migration, and movement in the tissue interstitium. In addition, eosinophil binding itself critically regulates various cellular functions. Many of the mechanisms involved in the regulation of adhesion molecule expression or alteration in cellular functions via adhesion have been well elucidated for eosinophils and other leukocytes. The differences between the expression and function of adhesion molecules on eosinophils and on other cells like neutrophils are thought to be a potential therapeutic opportunity for diseases in which eosinophils play a role, but so far no trials in humans have been done successfully.

REFERENCES

1. Kitayama J, Fuhlbrigge RC, Puri KD, Springer TA. P-selectin, L-selectin, and alpha 4 integrin have distinct roles in eosinophil tethering and arrest on vascular endothelial cells under physiological flow conditions. *J Immunol* 1997;**159**:3929−39.

2. Yoshikawa M, Matsumoto K, Iida M, Akasawa A, Moriyama H, Saito H. Effect of extracellular matrix proteins on platelet-activating factor-induced eosinophil chemotaxis. *Int Arch Allergy Immunol* 2002;**128**(Suppl. 1):3−11.

3. Ebnet K, Kaldjian EP, Anderson AO, Shaw S. Orchestrated information transfer underlying leukocyte endothelial interactions. *Annu Rev Immunol* 1996;**14**:155−77.

4. Knol EF, Tackey F, Tedder TF, Klunk DA, Bickel CA, Sterbinsky SA, et al. Comparison of human eosinophil and neutrophil adhesion to endothelial cells under nonstatic conditions. Role of L-selectin. *J Immunol* 1994;**153**:2161−7.

5. Shikata K, Suzuki Y, Wada J, Hirata K, Matsuda M, Kawashima H, et al. L-selectin and its ligands mediate infiltration of mononuclear cells into kidney interstitium after ureteric obstruction. *J Pathol* 1999;**188**:93−9.

6. Sitrin RG, Pan PM, Blackwood RA, Huang J, Petty HR. Cutting edge: evidence for a signaling partnership between urokinase receptors (CD87) and L-selectin (CD62L) in human polymorphonuclear neutrophils. *J Immunol* 2001;**166**:4822−5.

7. Bochner BS. Road signs guiding leukocytes along the inflammation superhighway. *J Allergy Clin Immunol* 2000;**106**:817−28.

8. Bochner BS, Schleimer RP. Mast cells, basophils, and eosinophils: distinct but overlapping pathways for recruitment. *Immunol Rev* 2001;**179**:5−15.

9. Satoh T, Kanai Y, Wu MH, Yokozeki H, Kannagi R, Lowe JB, et al. Synthesis of {alpha}(1,3) fucosyltransferases IV- and VII-dependent eosinophil selectin ligand and recruitment to the skin. *Am J Pathol* 2005;**167**:787−96.

10. Kikly KK, Bochner BS, Freeman SD, Tan KB, Gallagher KT, D'Alessio K, et al. Identification of SAF-2, a novel siglec expressed on eosinophils, mast cells, and basophils. *J Allergy Clin Immunol* 2000;**105**:1093−100.

11. Matsumoto K, Appiah-Pippim J, Schleimer RP, Bickel CA, Beck LA, Bochner BS. CD44 and CD69 represent different types of cell-surface activation markers for human eosinophils. *Am J Respir Cell Mol Biol* 1998;**18**:860−6.

12. Woltmann G, McNulty CA, Dewson G, Symon FA, Wardlaw AJ. Interleukin-13 induces PSGL-1/P-selectin-dependent adhesion of eosinophils, but not neutrophils, to human umbilical vein endothelial cells under flow. *Blood* 2000;**95**:3146−52.

13. Barthel SR, Johansson MW, McNamee DM, Mosher DF. Roles of integrin activation in eosinophil function and the eosinophilic inflammation of asthma. *J Leukoc Biol* 2008;**83**:1−12.

14. Werfel SJ, Yednock TA, Matsumoto K, Sterbinsky SA, Schleimer RP, Bochner BS. Functional regulation of beta 1 integrins on human eosinophils by divalent cations and cytokines. *Am J Respir Cell Mol Biol* 1996;**14**:44−52.

15. Matsumoto K, Sterbinsky SA, Bickel CA, Zhou DF, Kovach NL, Bochner BS. Regulation of alpha 4 integrin-mediated adhesion of human eosinophils to fibronectin and vascular cell adhesion molecule-1. *J Allergy Clin Immunol* 1997;**99**:648−56.

16. Georas SN, McIntyre BW, Ebisawa M, Bednarczyk JL, Sterbinsky SA, Schleimer RP, et al. Expression of a functional laminin receptor (alpha 6 beta 1, very late activation antigen-6) on human eosinophils. *Blood* 1993;**82**:2872−9.

17. Tourkin A, Anderson T, LeRoy EC, Hoffman S. Eosinophil adhesion and maturation is modulated by laminin. *Cell Adhes Commun* 1993;**1**:161−76.

18. Seton K, Hakansson L, Venge P. Enhanced adhesion to laminin by apoptotic eosinophils. *Scand J Immunol* 2003;**58**:412−8.

19. Ebisawa M, Bochner BS, Georas SN, Schleimer RP. Eosinophil transendothelial migration induced by cytokines. I. Role of endothelial and eosinophil adhesion molecules in IL-1 beta-induced transendothelial migration. *J Immunol* 1992;**149**:4021−8.

20. Grayson MH, Van der Vieren M, Sterbinsky SA, Michael Gallatin W, Hoffman PA, Staunton DE, et al. alphadbeta2 integrin is expressed on human eosinophils and functions as an alternative ligand for vascular cell adhesion molecule 1 (VCAM-1). *J Exp Med* 1998;**188**:2187−91.

21. Horie S, Kita H. CD11b/CD18 (Mac-1) is required for degranulation of human eosinophils induced by human recombinant granulocyte-macrophage colony-stimulating factor and platelet-activating factor. *J Immunol* 1994;**152**:5457−67.

22. Kaneko M, Horie S, Kato M, Gleich GJ, Kita H. A crucial role for beta 2 integrin in the activation of eosinophils stimulated by IgG. *J Immunol* 1995;**155**:2631−41.

23. Nagata M, Sedgwick JB, Bates ME, Kita H, Busse WW. Eosinophil adhesion to vascular cell adhesion molecule-1 activates superoxide anion generation. *J Immunol* 1995;**155**:2194−202.

24. Kita H, Horie S, Gleich GJ. Extracellular matrix proteins attenuate activation and degranulation of stimulated eosinophils. *J Immunol* 1996;**156**:1174−81.

25. Kato M, Abraham RT, Okada S, Kita H. Ligation of the beta2 integrin triggers activation and degranulation of human eosinophils. *Am J Respir Cell Mol Biol* 1998;**18**:675−86.

26. Munoz NM, Hamann KJ, Rabe KF, Sano H, Zhu X, Leff AR. Augmentation of eosinophil degranulation and LTC(4) secretion by integrin-mediated endothelial cell adhesion. *Am J Physiol* 1999;**277**:L802−10.

27. Walsh GM, Symon FA, Lazarovils AL, Wardlaw AJ. Integrin alpha 4 beta 7 mediates human eosinophil interaction with MAdCAM-1, VCAM-1 and fibronectin. *Immunology* 1996;**89**:112−9.

28. Bochner BS, Sterbinsky SA, Bickel CA, Werfel S, Wein M, Newman W. Differences between human eosinophils and neutrophils in the function and expression of sialic acid-containing counterligands for E-selectin. *J Immunol* 1994;**152**:774−82.

29. Nimrichter L, Burdick MM, Aoki K, Laroy W, Fierro MA, Hudson SA, et al. E-selectin receptors on human leukocytes. *Blood* 2008;**112**:3744−52.

30. Bochner BS, Schleimer RP. The role of adhesion molecules in human eosinophil and basophil recruitment. *J Allergy Clin Immunol* 1994;**94**:427−38. quiz 439.

31. Sriramarao P, Norton CR, Borgstrom P, DiScipio RG, Wolitzky BA, Broide DH. E-selectin preferentially supports neutrophil but not eosinophil rolling under conditions of flow in vitro and in vivo. *J Immunol* 1996;**157**:4672−80.

32. Ohkawara Y, Tamura G, Iwasaki T, Tanaka A, Kikuchi T, Shirato K. Activation and transforming growth factor-beta production in eosinophils by hyaluronan. *Am J Respir Cell Mol Biol* 2000;**23**:444−51.

33. Ohashi H, Takei M, Ide Y, Ishii H, Kita H, Gleich GJ, et al. Effect of interleukin-3, interleukin 5 and hyaluronic acid on cultured eosinophils derived from human umbilical cord blood mononuclear cells. *Int Arch Allergy Immunol* 1999;**118**:44−50.

34. Rao SP, Wang Z, Zuberi RI, Sikora L, Bahaie NS, Zuraw BL, et al. Galectin-3 functions as an adhesion molecule to support eosinophil rolling and adhesion under conditions of flow. *J Immunol* 2007;**179**:7800−7.

35. Yoon J, Terada A, Kita H. CD66b regulates adhesion and activation of human eosinophils. *J Immunol* 2007;**179**:8454−62.

36. Chiba R, Nakagawa N, Kurasawa K, Tanaka Y, Saito Y, Iwamoto I. Ligation of CD31 (PECAM-1) on endothelial cells increases adhesive function of alphavbeta3 integrin and enhances beta1 integrin-mediated adhesion of eosinophils to endothelial cells. *Blood* 1999;**94**:1319−29.

37. Miller M, Sung KL, Muller WA, Cho JY, Roman M, Castaneda D, et al. Eosinophil tissue recruitment to sites of allergic inflammation in the lung is platelet endothelial cell adhesion molecule independent. *J Immunol* 2001;**167**:2292−7.

38. Lantero S, Alessandri G, Spallarossa D, Scarso L, Rossi GA. LFA-1 expression by blood eosinophils is increased in atopic asthmatic children and is involved in eosinophil locomotion. *Eur Respir J* 1998;**12**:1094−8.

39. Georas SN, Liu MC, Newman W, Beall LD, Stealey BA, Bochner BS. Altered adhesion molecule expression and endothelial cell activation accompany the recruitment of human granulocytes to the lung after segmental antigen challenge. *Am J Respir Cell Mol Biol* 1992;**7**:261−9.

40. Bazan-Socha S, Bukiej A, Pulka G, Marcinkiewicz C, Musial J. Increased expression of collagen receptors: alpha1beta1 and alpha2beta1 integrins on blood eosinophils in bronchial asthma. *Clin Exp Allergy* 2006;**36**:1184−91.

41. Matsuno O, Miyazaki E, Nureki S, Ueno T, Kumamoto T, Higuchi Y. Role of ADAM8 in experimental asthma. *Immunol Lett* 2006;**102**:67−73.

42. Tourkin A, Bonner M, Mantrova E, LeRoy EC, Hoffman S. Dot-like focal contacts in adherent eosinophils, their redistribution into peripheral belts, and correlated effects on cell migration and protected zone formation. *J Cell Sci* 1996;**109**(Pt 8):2169—77.

43. Johansson MW, Lye MH, Barthel SR, Duffy AK, Annis DS, Mosher DF. Eosinophils adhere to vascular cell adhesion molecule-1 via podosomes. *Am J Respir Cell Mol Biol* 2004;**31**:413—22.

44. Meliton AY, Munoz NM, Leff AR. Blockade of avidity and focal clustering of beta 2-integrin by cysteinyl leukotriene antagonism attenuates eosinophil adhesion. *J Allergy Clin Immunol* 2007;**120**: 1316—23.

45. Myou S, Zhu X, Boetticher E, Myo S, Meliton A, Lambertino A, et al. Blockade of focal clustering and active conformation in beta 2-integrin-mediated adhesion of eosinophils to intercellular adhesion molecule-1 caused by transduction of HIV TAT-dominant negative Ras. *J Immunol* 2002;**169**:2670—6.

46. Sung KL, Li Y, Elices M, Gang J, Sriramarao P, Broide DH. Granulocyte-macrophage colony-stimulating factor regulates the functional adhesive state of very late antigen-4 expressed by eosinophils. *J Immunol* 1997;**158**:919—27.

47. Barthel SR, Jarjour NN, Mosher DF, Johansson MW. Dissection of the hyperadhesive phenotype of airway eosinophils in asthma. *Am J Respir Cell Mol Biol* 2006;**35**:378—86.

48. Ulfman LH, Joosten DP, van der Linden JA, Lammers JW, Zwaginga JJ, Koenderman L. IL-8 induces a transient arrest of rolling eosinophils on human endothelial cells. *J Immunol* 2001;**166**:588—95.

49. Weber C, Kitayama J, Springer TA. Differential regulation of beta 1 and beta 2 integrin avidity by chemoattractants in eosinophils. *Proc Natl Acad Sci U S A* 1996;**93**:10939—44.

50. Zhu X, Munoz NM, Kim KP, Sano H, Cho W, Leff AR. Cytosolic phospholipase A2 activation is essential for beta 1 and beta 2 integrin-dependent adhesion of human eosinophils. *J Immunol* 1999;**163**:3423—9.

51. Wong CK, Ip WK, Lam CW. Interleukin-3, -5, and granulocyte macrophage colony-stimulating factor-induced adhesion molecule expression on eosinophils by p38 mitogen-activated protein kinase and nuclear factor-[kappa] B. *Am J Respir Cell Mol Biol* 2003;**29**:133—47.

52. Liu J, Munoz NM, Meliton AY, Zhu X, Lambertino AT, Xu C, et al. Beta2-integrin adhesion caused by eotaxin but not IL-5 is blocked by PDE-4 inhibition and beta2-adrenoceptor activation in human eosinophils. *Pulm Pharmacol Ther* 2004;**17**:73—9.

53. Ip WK, Wong CK, Wang CB, Tian YP, Lam CW. Interleukin-3, -5, and granulocyte macrophage colony-stimulating factor induce adhesion and chemotaxis of human eosinophils via p38 mitogen-activated protein kinase and nuclear factor kappaB. *Immuno-pharmacol Immunotoxicol* 2005;**27**:371—93.

54. Sano M, Leff AR, Myou S, Boetticher E, Meliton AY, Learoyd J, et al. Regulation of interleukin-5-induced beta2-integrin adhesion of human eosinophils by phosphoinositide 3-kinase. *Am J Respir Cell Mol Biol* 2005;**33**:65—70.

55. Tachimoto H, Kikuchi M, Hudson SA, Bickel CA, Hamilton RG, Bochner BS. Eotaxin-2 alters eosinophil integrin function via mitogen-activated protein kinases. *Am J Respir Cell Mol Biol* 2002;**26**:645—9.

56. Nagata M, Saito K, Tsuchiya K, Sakamoto Y. Leukotriene D4 upregulates eosinophil adhesion via the cysteinyl leukotriene 1 receptor. *J Allergy Clin Immunol* 2002;**109**:676—80.

57. Wu P, Mitchell S, Walsh GM. A new antihistamine levocetirizine inhibits eosinophil adhesion to vascular cell adhesion molecule-1 under flow conditions. *Clin Exp Allergy* 2005;**35**:1073—9.

58. Kuijpers TW, Mul EP, Blom M, Kovach NL, Gaeta FC, Tollefson V, et al. Freezing adhesion molecules in a state of high-avidity binding blocks eosinophil migration. *J Exp Med* 1993;**178**:279—84.

59. Sung KP, Yang L, Kim J, Ko D, Stachnick G, Castaneda D, et al. Eotaxin induces a sustained reduction in the functional adhesive state of very late antigen 4 for the connecting segment 1 region of fibronectin. *J Allergy Clin Immunol* 2000;**106**:933—40.

60. Tachimoto H, Burdick MM, Hudson SA, Kikuchi M, Konstantopoulos K, Bochner BS. CCR3-active chemokines promote rapid detachment of eosinophils from VCAM-1 in vitro. *J Immunol* 2000;**165**:2748—54.

61. Alblas J, Ulfman L, Hordijk P, Koenderman L. Activation of Rhoa and ROCK are essential for detachment of migrating leukocytes. *Mol Biol Cell* 2001;**12**:2137—45.

62. Picker LJ, Treer JR, Nguyen M, Terstappen LW, Hogg N, Yednock T. Coordinate expression of beta 1 and beta 2 integrin 'activation' epitopes during T cell responses in secondary lymphoid tissue. *Eur J Immunol* 1993;**23**:2751—7.

63. Anwar AR, Moqbel R, Walsh GM, Kay AB, Wardlaw AJ. Adhesion to fibronectin prolongs eosinophil survival. *J Exp Med* 1993;**177**: 839—43.

64. Meerschaert J, Vrtis RF, Shikama Y, Sedgwick JB, Busse WW, Mosher DF. Engagement of alpha4beta7 integrins by monoclonal antibodies or ligands enhances survival of human eosinophils in vitro. *J Immunol* 1999;**163**:6217—27.

65. Chihara J, Kakazu T, Higashimoto I, Saito N, Honda K, Sannohe S, et al. Signaling through the beta2 integrin prolongs eosinophil survival. *J Allergy Clin Immunol* 2000;**106**:S99—103.

66. Suzuki M, Kato M, Hanaka H, Izumi T, Morikawa A. Actin assembly is a crucial factor for superoxide anion generation from adherent human eosinophils. *J Allergy Clin Immunol* 2003;**112**: 126—33.

67. Gonzalo JA, Lloyd CM, Kremer L, Finger E, Martinez AC, Siegelman MH, et al. Eosinophil recruitment to the lung in a murine model of allergic inflammation. The role of T cells, chemokines, and adhesion receptors. *J Clin Invest* 1996;**98**:2332—45.

68. Broide DH, Humber D, Sullivan S, Sriramarao P. Inhibition of eosinophil rolling and recruitment in P-selectin- and intracellular adhesion molecule-1-deficient mice. *Blood* 1998;**91**:2847—56.

69. Broide DH, Sullivan S, Gifford T, Sriramarao P. Inhibition of pulmonary eosinophilia in P-selectin- and ICAM-1-deficient mice. *Am J Respir Cell Mol Biol* 1998;**18**:218—25.

70. Sriramarao P, von Andrian UH, Butcher EC, Bourdon MA, Broide DH. L-selectin and very late antigen-4 integrin promote eosinophil rolling at physiological shear rates in vivo. *J Immunol* 1994;**153**:4238—46.

71. Sriramarao P, DiScipio RG, Cobb RR, Cybulsky M, Stachnick G, Castaneda D, et al. VCAM-1 is more effective than MAdCAM-1 in supporting eosinophil rolling under conditions of shear flow. *Blood* 2000;**95**:592—601.

72. Katoh S, Matsumoto N, Kawakita K, Tominaga A, Kincade PW, Matsukura S. A role for CD44 in an antigen-induced murine model of pulmonary eosinophilia. *J Clin Invest* 2003;**111**: 1563—70.

73. Fryer AD, Costello RW, Yost BL, Lobb RR, Tedder TF, Steeber DA, et al. Antibody to VLA-4, but not to L-selectin, protects neuronal M2 muscarinic receptors in antigen-challenged guinea pig airways. *J Clin Invest* 1997;**99**:2036–44.

74. Soler D, Chapman T, Yang LL, Wyant T, Egan R, Fedyk ER. The binding specificity and selective antagonism of vedolizumab, an anti-alpha4beta7 integrin therapeutic antibody in development for inflammatory bowel diseases. *J Pharmacol Exp Ther* 2009;**330**:864–75.

75. Bochner BS. Adhesion molecules as therapeutic targets. *Immunol Allergy Clin North Am* 2004;**24**:615–30. vi.

76. Kim MK, Brandley BK, Anderson MB, Bochner BS. Antagonism of selectin-dependent adhesion of human eosinophils and neutrophils by glycomimetics and oligosaccharide compounds. *Am J Respir Cell Mol Biol* 1998;**19**:836–41.

77. Davenpeck KL, Berens KL, Dixon RA, Dupre B, Bochner BS. Inhibition of adhesion of human neutrophils and eosinophils to P-selectin by the sialyl Lewis antagonist TBC1269: preferential activity against neutrophil adhesion in vitro. *J Allergy Clin Immunol* 2000;**105**:769–75.

78. Gauvreau GM, Becker AB, Boulet LP, Chakir J, Fick RB, Greene WL, et al. The effects of an anti-CD11a mAb, efalizumab, on allergen-induced airway responses and airway inflammation in subjects with atopic asthma. *J Allergy Clin Immunol* 2003;**112**:331–8.

79. Miller DH, Khan OA, Sheremata WA, Blumhardt LD, Rice GP, Libonati MA, et al. A controlled trial of natalizumab for relapsing multiple sclerosis. *N Engl J Med* 2003;**348**:15–23.

80. Bochner BS, Hudson SA, Wasserman M, Stilz H-U, Wehner V, Wolos J. Effect of eight VLA-4 antagonists on human eosinophil adhesion to immobilized human VCAM-1 in vitro. *J Allergy Clin Immunol* 2000;**105**:S258–9.

81. Sedgwick JB, Jansen KJ, Kennedy JD, Kita H, Busse WW. Effects of the very late adhesion molecule 4 antagonist WAY103 on human peripheral blood eosinophil vascular cell adhesion molecule 1-dependent functions. *J Allergy Clin Immunol* 2005;**116**:812–9.

82. Norris V, Choong L, Tran D, Corden Z, Boyce M, Arshad H, et al. Effect of IVL745, a VLA-4 antagonist, on allergen-induced bronchoconstriction in patients with asthma. *J Allergy Clin Immunol* 2005;**116**:761–7.

83. Diamant Z, Kuperus J, Baan R, Nietzmann K, Millet S, Mendes P, et al. Effect of a very late antigen-4 receptor antagonist on allergen-induced airway responses and inflammation in asthma. *Clin Exp Allergy* 2005;**35**:1080–7.

84. Ravensberg AJ, Luijk B, Westers P, Hiemstra PS, Sterk PJ, Lammers JW, et al. The effect of a single inhaled dose of a VLA-4 antagonist on allergen-induced airway responses and airway inflammation in patients with asthma. *Allergy* 2006;**61**:1097–103.

85. Muro F, Iimura S, Sugimoto Y, Yoneda Y, Chiba J, Watanabe T, et al. Discovery of trans-4-[1-[[2,5-Dichloro-4-(1-methyl-3-indolylcarboxamido)phenyl]acetyl]-(4S)-me thoxy-(2S)-pyrrolidinylmethoxy]cyclohexanecarboxylic acid: an orally active, selective very late antigen-4 antagonist. *J Med Chem* 2009;**52**:7974–92.

86. Kessel JM, Sedgwick JB, Busse WW. Ligation of intercellular adhesion molecule 3 induces apoptosis of human blood eosinophils and neutrophils. *J Allergy Clin Immunol* 2006;**118**:831–6.

87. Bochner BS. Siglec-8 on human eosinophils and mast cells, and Siglec-F on murine eosinophils, are functionally related inhibitory receptors. *Clin Exp Allergy* 2009;**39**:317–24.

Chapter 6.4

Eosinophil-Endothelial Cell Interactions during Inflammation

Joan M. Cook-Mills

INTRODUCTION TO EOSINOPHIL-ENDOTHELIAL CELL INTERACTIONS (ROLLING, ADHESION, AND MIGRATION)

During inflammation, eosinophils from the blood are recruited across the postcapillary venule endothelium and into tissues. The specificity for recruitment of eosinophils is mediated by the combination of chemokines in the microenvironment, adhesion molecules on the endothelium that mediate eosinophil binding, and eosinophil receptors for chemokines and endothelial adhesion molecules.[1] The expression of adhesion molecules on endothelial cells is induced by several mediators including cytokines, high levels of reactive oxygen species (ROS), turbulent blood flow, or activation of endothelial cell Toll-like receptors.[2–7] Eosinophil binding to endothelial cell adhesion molecules activates signals within the endothelial cells that allow the opening of small intercellular gaps, forming narrow vascular passageways, through which eosinophils migrate (Fig. 6.4.1). Eosinophil movement through these passageways is stimulated by chemokines that are produced by the endothelium and the tissue (Fig. 6.4.1). The majority of leukocytes migrate through intercellular gaps but, under conditions of high levels of inflammation, a small percentage of some leukocyte cell types have been reported to also migrate through individual endothelial cells by transcellular migration.[8] When inhibitors are used to block the signals that open small passageways in the endothelium, eosinophils bind to the endothelium but do not complete their transendothelial migration.[9] Eosinophils that bind to the endothelium but do not complete transendothelial migration are often released from the endothelium and continue in the blood flow. Thus, the adhesion molecules and their intracellular signals are a source for intervention in leukocyte recruitment.

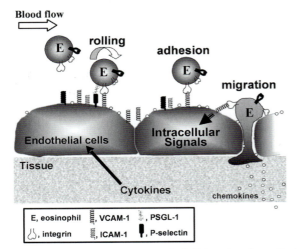

FIGURE 6.4.1 Leukocyte transendothelial migration. During inflammation, cytokines produced in the tissue induce endothelial cell adhesion molecule expression. In addition, chemoattractants released by both tissue cells and endothelial cells increase leukocyte adhesion molecule affinity as well as provide direction for leukocyte migration. This vascular recruitment of leukocytes is a three-step process involving low-affinity rolling of leukocytes on the endothelium, arrest of leukocytes on the endothelium through high-affinity adhesion, and then transmigration of leukocytes through the endothelium. Low-affinity adhesion is mediated by eosinophil PSGL-1 or $\alpha 4\beta_1$ integrin binding to endothelial cell P-selectin and VCAM-1, respectively. There is also some contribution of $\alpha L\beta_2$ integrin binding to ICAM-1. Chemokines upregulate the affinity of integrins for firm adhesion of eosinophils to endothelial cells. The endothelial cell adhesion molecules, including VCAM-1, activate signals in endothelial cells for localized endothelial cell retraction and the formation of small passageways through which the eosinophils migrate. ICAM, intercellular adhesion molecule 1; PSGL1, P-selectin glycoprotein ligand 1; VCAM-1, vascular cell adhesion protein 1.

EOSINOPHIL BINDING TO ENDOTHELIAL CELLS

The recruitment of eosinophils is a multistep process involving the rolling of eosinophils on the endothelium through low-affinity receptors, arrest of the eosinophils on the endothelium through high-affinity receptors, and then transmigration through the endothelium (Fig. 6.4.1). Eosinophil rolling followed by firm adhesion to the endothelium is primarily mediated by the binding of the eosinophil receptors P-selectin glycoprotein ligand 1 (PSGL-1) and $\alpha 4\beta 1$ integrin to endothelial cell P-selectin and vascular cell adhesion protein 1 (VCAM-1), respectively. However, additional receptors also contribute.

Selectins and P-Selectin Glycoprotein Ligand 1 Mediate Low-Affinity Eosinophil Binding

The rolling of eosinophils on the luminal side of the activated endothelium is mediated primarily by the low-affinity

receptor P-selectin on the endothelium and its heavily glycosylated ligand PSGL-1 on the eosinophils.[10,11] In contrast, the adhesion molecules L-selectin and platelet endothelial cell adhesion molecule (PECAM-1) are not critical for eosinophil recruitment.[12,13] The combination of anti-P-selectin and anti-E-selectin antibodies completely inhibits eosinophil rolling on endothelial cells *in vitro*.[14] *In vivo*, mice deficient in P-selectin and/or E-selectin have reduced lung eosinophils.[14,15] In skin, allergen-induced recruitment of eosinophils is blocked with anti-P-selectin antibodies and to a lesser extent with anti-E-selectin antibodies, whereas anti-$\alpha 4$ integrin antibodies completely blocked eosinophil recruitment to the skin.[16] Human eosinophils from allergic-asthmatics have increased PSGL-1 expression and increased recruitment on interleukin-4 (IL-4)-stimulated human umbilical vein endothelial cells (HUVECs) expressing P-selectin and VCAM-1.[17] Therefore, P-selectin and VCAM-1 have important roles in eosinophil recruitment in animal models and in humans.

High-Affinity Eosinophil-Endothelial Cell Binding through Integrins, Galectin-3, SPARC (Osteonectin), Intercellular Adhesion Molecule 1 and Vascular Cell Adhesion Protein 1

It is reported that, in lieu of P-selectin interactions with eosinophil PSGL-1, eosinophil rolling can also be mediated by eosinophil $\alpha 4\beta 1$ integrin in its low-affinity state interacting with VCAM-1 (CD106) on the endothelium.[3] The low-affinity integrin receptors are induced to a high-affinity state by signals from the selectins, PECAM-1, chemokine receptors, and cations. The CC chemokines, C-C motif chemokine 5 (CCL5/RANTES), eotaxin, eotaxin-2, and monocyte chemotactic protein 3 (MCP-3) and 4 (MCP-4) increase eosinophil migration across recombinant human VCAM-1[18] and tumor necrosis factor (TNF-α) or interleukin-1β (IL-1β)-stimulated HUVECs.[19] The high-affinity $\alpha 4\beta 1$ and $\alpha L\beta 2$ integrins bind to the endothelial cell adhesion molecules VCAM-1 (CD106) and intercellular adhesion molecule 1 (ICAM-1) (CD54), respectively (Figure 6.4.2). In summary, activation for high-affinity binding to VCAM-1 is determined by cell type-specific chemokines in the microenvironment.

Eosinophils express several receptors that bind to VCAM-1. In humans, VCAM-1 is expressed as a seven-immunoglobulin-like-domain form as well as a six-domain form that is missing domain 4. Mice express the seven-domain form of VCAM-1 and a truncated glyco-sylphosphatidylinisotol-linked three-domain form of VCAM-1. The $\alpha 4\beta 1$ integrins on eosinophils bind to domains 1 and 4 of VCAM-1 and the $\alpha M\beta 2$ integrins on eosinophils bind to domain 4 of recombinant VCAM-1

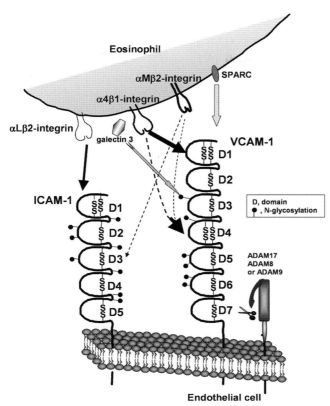

FIGURE 6.4.2 Eosinophil receptor binding to endothelial cell adhesion molecules. Eosinophil integrin binding to VCAM-1 is regulated by the integrin activation state; α4β1 integrin binds readily to domain 1 of VCAM-1 but requires higher-affinity activation for binding to domain 4 of VCAM-1. Galectin 3 binds to N-linked glycosylation sites on VCAM-1. VCAM-1 has six N-linked glycosylation sites that may participate in Gal-3 binding. αLβ2 integrin binds to domain 1 of ICAM-1. αMβ2 integrin can bind to domain 4 of VCAM-1 or domain 3 of ICAM-1. SPARC (osteonectin) has also been reported to bind to VCAM-1. VCAM-1 is clipped from the cell surface by disintegrin and metalloproteinase domain-containing protein 8 (ADAM 8), 9 (ADAM 9), and 17 (ADAM 17). The arrows indicate ligand binding sites on VCAM-1 and ICAM-1. ICAM-1, intercellular adhesion molecule 1; SPARC, secreted protein acidic and rich in cysteine; VCAM-1, vascular cell adhesion protein 1.

(Figure 6.4.2).[20] In addition to integrin adhesion to VCAM-1, eosinophil galectin-3 (Gal-3) mediates rolling and adhesion on recombinant VCAM-1 (Figure 6.4.2).[21] Gal-3 is secreted and then binds to cell surface extracellular matrix laminin, elastin, fibronectin, VCAM-1, α4 integrin, and Gal-3 itself.[21] Gal-3-deficient mice have reduced eosinophil recruitment to the lung and lung lavage after ovalbumin (OVA) challenge.[22] In addition, secreted protein acidic and rich in cysteine (SPARC), also called osteonectin, on leukocytes binds to recombinant VCAM-1 (Fig. 6.4.2)[23] and mediates eosinophil recruitment *in vivo* as determined using SPARC-deficient mice.[23]

Mice deficient in α4 integrin or β2 integrin have reduced eosinophil recruitment after OVA challenge[24] and anti-α4 integrin antibodies block >85% of OVA-induced

eosinophil recruitment in allergic conjunctivitis[25] and OVA-induced lung eosinophilia.[26] In contrast to this predominant role for VCAM-1 in the recruitment of eosinophils, recruitment of lymphocytes is much less dependent on VCAM-1. The concentrations of anti-α4 integrin antibodies that block eosinophil recruitment do not alter lymphocyte recruitment; yet, a high concentration of anti-α4 integrin antibodies results in a 20% inhibition of lymphocyte recruitment to the lung after OVA challenge.[26]

Induction of Adhesion Molecule Expression on Endothelial Cells for the Recruitment of Eosinophils

Cytokines and chemokines expressed during allergic inflammation [IL-4, interleukin-13 (IL-13), TNFα, and eotaxin] induce expression of the vascular adhesion molecules P-selectin and VCAM-1 during eosinophil recruitment. Endothelial cells store P-selectin, as well as eotaxin-3, interleukin-8 (IL-8), and angiopoietin-2 (ANG-2), in specialized endothelial cell organelles, the Weibel—Palade-bodies. IL-4 increases the endothelial cell expression of P-selectin, VCAM-1, and eotaxin-3 but reduces the expression of IL-8 and ANG-2.[27,28] VCAM-1 expression on pulmonary microvascular endothelial cells and eosinophil adhesion to these endothelial cells is induced by cytokines with the potency of TNF-α + IL-4 > TNF-α > IL-1β.[29] In addition to cytokine stimulation, VCAM-1 expression is induced in inflammatory sites by high levels of ROS, turbulent shear stress, and microbial stimulation of endothelial cell Toll-like receptors (TLRs).[2–7] TNF-α induces endothelial cell expression of E-selectin, ICAM-1, and VCAM-1 but not P-selectin, whereas IL-13 induces endothelial cell expression of P-selectin and VCAM-1, but not E-selectin or ICAM-1.[30,31] *In vivo* administration of IL-4 induces VCAM-1 expression and eosinophil rolling on cremaster postcapillary venules and this eosinophil rolling is blocked by antibodies to P-selectin or α4 integrin but not E-selectin or L-selectin.[32] Interferon beta (IFN-β) augments TNF-α-induced expression of VCAM-1 and ICAM-1 on HUVECs and this elevates α4 integrin-dependent and β2 integrin-dependent adhesion of eosinophils.[33] Eosinophil chemotactic factors such as eotaxin and complement component C3a and C5a can also induce VCAM-1 and ICAM-1 expression on endothelial cells.[34–36] The nucleotide adenosine triphosphate (ATP)/uridine triphosphate (UTP) P2Y purinoceptor 2 (P2Y2) is expressed on endothelial cells and eosinophils; ATP/UTP induces endothelial cell VCAM-1 expression and regulates allergic inflammation.[37–39]

VCAM-1 can be released from the endothelial cell surface through cleavage by thrombin[40] or the metalloproteinases disintegrin and metalloproteinase domain-containing protein 17 (ADAM 17),[41] ADAM 8,[42,43] or

ADAM 9 (Fig. 6.4.2).[44] Eosinophils bind to VCAM-1 through α4 integrin on eosinophil podosomes that contain ADAM 8. The ADAM 8 in eosinophil podosomes can cleave underlying recombinant VCAM-1.[45] The soluble form of VCAM-1 (sVCAM-1) in the plasma is used as a predictive biomarker of disease, including asthma and severe allergic rhinitis.[46–48] The sVCAM-1 in plasma is thought to either limit leukocyte integrin binding to endothelial VCAM-1 by binding to leukocytes or stimulate leukocyte chemotaxis in vivo.[49,50]

ENDOTHELIAL CELL PRODUCTION OF CYTOKINES AND CHEMOKINES REGULATES EOSINOPHIL RECRUITMENT

Activated endothelial cells produce granulocyte-macrophage colony-stimulating factor (GM-CSF) that functions in eosinophil hematopoiesis and increases α4β1 integrin affinity on eosinophils, indicating that endothelial cell-derived cytokines regulate eosinophilic inflammation.[51,52] Endothelial cells also produce chemokines that regulate eosinophil chemotaxis including CCL5 (RANTES), eotaxin, eotaxin-2, eotaxin-3, monocyte chemotactic protein 1 (MCP-1), and MCP-4; these chemokines bind to the heparin and heparin sulfate that coats the endothelial cell surface and regulates recruitment of leukocytes, including eosinophils.[53–56] Mice deficient in heparin sulfate have reduced lung eosinophilia after OVA challenge.[57] Chemokines also modulate endothelial cell function during eosinophil recruitment. C-C chemokine receptor type 3 (CCR3) is expressed by endothelial cells, and eotaxin [C-C motif chemokine 11 (CCL11)] binding to the endothelial cells suppresses TNF-α-induced endothelial cell production of IL-8,[58] suggesting a role for endothelial cells in switching from neutrophilic to eosinophilic inflammation.

EOSINOPHIL BINDING TO ENDOTHELIAL CELLS MODIFIES EOSINOPHIL FUNCTION

Eosinophils that have migrated through endothelial cell monolayers on VCAM-1 or ICAM-1 exhibit increased production of leukotriene C_4 (LTC_4), eosinophil peroxidase, and respiratory burst ROS.[59–61] Eosinophil binding to VCAM-1 in the presence leukotriene D_4 (LTD_4) increases eosinophil superoxide generation and eosinophil-derived neurotoxin (EDN) release, whereas exposure to either VCAM-1 or LTD_4 alone does not induce eosinophil superoxide generation.[62] The ICAM-1 and VCAM-1 activation of eosinophil superoxide and EDN release is also augmented by GM-CSF, a cytokine produced by endothelial cells.[63,64] Eosinophil exposure to GM-CSF, as well as eosinophil transendothelial migration across IL-1β-activated HUVECs increases eosinophil CD13 (aminopeptidase N).[65] Migration through the endothelial basement membrane extracellular matrix increases eosinophil CD44 expression and CD44 mediates binding to hyaluronic acid in tissue matrix.[59] Therefore, eosinophil function is altered by exposure to adhesion receptor-ligand engagement.

EOSINOPHIL MEDIATORS REGULATE ENDOTHELIAL CELL FUNCTION

IL-1β and TNF-α in supernatants from cultured human eosinophils activate the expression of VCAM-1, ICAM-1, and E-selectin on HUVECs.[66,67] Eosinophils also regulate vascular angiogenesis (vessel growth). Vascular angiogenesis is induced by eosinophil granule major basic protein 1 (MBP-1), at subcytotoxic concentrations.[68] Human eosinophils also express preformed vascular endothelial growth factor (VEGF), which increases endothelial angiogenesis.[69] Eosinophils can release their VEGF after stimulation with GM-CSF.[70] Eosinophil sonicates also induce VEGF mRNA expression in endothelial cells.[69] Interestingly, VEGF expression is elevated in eosinophilic bronchitis[71] and is found in the induced sputum of asthmatic patients.[72] In addition, the VEGF receptor fms-like tyrosine kinase 1 (FLT-1) is expressed by human eosinophils and VEGF induces eosinophil chemotaxis.[73] In summary, VEGF, which is derived from eosinophils and the activated endothelium, may regulate eosinophil transendothelial migration by activating signals in endothelium and in eosinophils.

CLINICAL INHIBITORS THAT BLOCK EITHER ADHESION MOLECULE BINDING OR INDUCTION OF ADHESION MOLECULE EXPRESSION

Inhibitors of selectins and integrins have been used to block eosinophilia and inflammation in research studies and clinical trials. The small-molecule pan-selectin antagonist bimosiamose (TBC-1269) blocks leukocyte recruitment mediated by E-selectin and attenuates late asthmatic reactions after allergen challenge to mild asthmatics.[74–79] A small-molecule antagonist of α4β1 integrin, valategrast (R411), is a potential strategy for asthma.[75] The anti-α4 integrin antibody natalizumab has been used in clinical trials.[80,81] However, treatment with natalizumab is complicated by the rare occurrence of progressive multifocal leukoencephalopathy[80,81] and natalizumab can induce allergic reactions to the treatment.[82,83] In vitro pretreatment of endothelial cells with theophylline, a bronchodilator that attenuates eosinophilia in atopic asthmatics, inhibits IL-4 and TNF-α-stimulated VCAM-1 and ICAM-1 expression.[84] Inhibitors of leukotriene D_4 block LTD_4-induced expression of eosinophil $\beta2$ integrin, eosinophil adhesion to recombinant human ICAM-1, and eosinophil transmigration across endothelial cells.[85] The cysteinyl leukotriene receptor 1 (CysLTR1) antagonist montelukast inhibits eosinophil transmigration across HUVECs in response to platelet-activating factor (PAF) or LTD_4.[85,86] The CysLTR1 antagonist pranlukast blocks eosinophil transendothelial migration in response to LTD_4 but not eotaxin, CCL5 (RANTES), or PAF.[87] The histamine H1 receptor antagonist levocabastine binds to $\alpha4\beta_1$ integrin and prevents adhesion to VCAM-1.[88] The antihistamine levocetirizine inhibits eosinophil adhesion to VCAM-1.[89,90] The small-compound antagonist of $\alpha4\beta_1$ integrin, compound A, blocks eosinophil accumulation in the OVA-challenged lungs of mice.[91] Another small-compound antagonist of $\alpha4\beta_1$ integrin, WAY103, blocks human eosinophil binding to recombinant human VCAM-1 and eosinophil transendothelial migration.[60] An orally active small-molecule antagonist of the $\alpha4\beta_1/\alpha4\beta_7$ integrins, TR14035, reduces eosinophil infiltration into the rat lung after antigen challenge by reducing leukocyte rolling and adhesion.[92] Therefore, inhibitors of induction of adhesion molecule expression or inhibitors of integrins block eosinophil recruitment *in vitro* and in animal models.

VASCULAR CELL ADHESION PROTEIN 1 SIGNALS OPEN VASCULAR JUNCTIONS FOR TRANSENDOTHELIAL MIGRATION

VCAM-1 is located on the luminal surface and lateral surface of endothelial cells but not on the basal surface of endothelial cells. As eosinophils bind to VCAM-1, VCAM-1 activates intracellular signals (Fig. 6.4.3A). The localized VCAM-1 signals induce changes in endothelial cell shape for the opening of passageways through which eosinophils migrate. The VCAM-1 signals are transient and occur within minutes; this is consistent with the transient, rapid nature of leukocyte transendothelial migration.

An overview of VCAM-1 signaling is introduced here before discussing the contribution of VCAM-1 signal transduction to eosinophil recruitment *in vivo*. VCAM-1 stimulates calcium channels, intracellular calcium release, the G protein $G\alpha i2$, and the low molecular weight G protein Ras-related C3 botulinum toxin substrate 1 (Rac1) (Fig. 6.4.3A).[93–95] The calcium flux and Rac1 activate the nicotinamide adenine dinucleotide phosphate (NADPH) dual oxidase 2 (NOX2) (Fig. 6.4.3A),[93,94] but not other enzymes that generate ROS.[94] The activated NOX2

generates superoxide that then dismutates to generate $1\,\mu M$ H_2O_2.[93,96] This $1\,\mu M$ H_2O_2 is relatively low compared to the $50-200\,\mu M$ H_2O_2 produced by macrophages or neutrophils.[97,98] It is also much lower than the exogenous $100-1000\,\mu M$ H_2O_2 in studies on the oxidative damage of the endothelium or for induction of VCAM-1 expression.[99,100] These differences in H_2O_2 levels are important in understanding the functions of oxidation, as several reports indicate that $1\,\mu M$ H_2O_2 and $>50\,\mu M$ H_2O_2 have opposing effects on signal transduction.[101–104]

During VCAM-1 signaling, the $1\,\mu M$ H_2O_2 oxidizes the prodomain of endothelial cell-associated matrix metalloproteinases (MMPs), causing autocatalytic cleavage of the prodomain and activation of the MMPs (Fig. 6.4.3A).[101] The $1\,\mu M$ H_2O_2 also diffuses through cell membranes at $100\,\mu m/s$[105] to directly oxidize and transiently activate intracellular protein kinase C alpha (PKCα) in endothelial cells (Fig. 6.4.3A).[103] This activated PKCα induces phosphorylation and activation of protein-tyrosine phosphatase 1B (PTP-1B) (Fig. 6.4.3A).[106] Interestingly, the PTP-1B that has an oxidizable cysteine in its catalytic domain is not oxidized during VCAM-1 signaling in endothelial cells,[106] indicating specificity of targets for oxidation by the low concentrations of ROS generated during VCAM-1 signaling. VCAM-1 signals through NOX2 also modify endothelial cell actin polymerization and intercellular gap formation.[94,107,108] Most importantly, the signals shown in Fig. 6.4.3 have been demonstrated to function in the regulation of VCAM-1-dependent leukocyte transendothelial migration *in vitro* and *in vivo*.[9,93–95,101,103,106,109,110] Thus, VCAM-1 is not simply a scaffold for leukocyte adhesion, since it also activates endothelial signals.

VASCULAR CELL ADHESION PROTEIN 1 SIGNALS DURING EOSINOPHIL RECRUITMENT *IN VIVO*

G Protein $G\alpha_{i2}$ Regulation of Vascular Cell Adhesion Protein 1-dependent Eosinophilia *In Vivo*

The G protein $G\alpha_{i2}$ functions in VCAM-1-dependent leukocyte recruitment *in vitro* and *in vivo* (Fig. 6.4.3). In OVA-challenged $G\alpha i2^{-/-}$ mice, VCAM-1-dependent eosinophil recruitment is inhibited, blood leukocyte numbers are elevated, and there is an accumulation of leukocytes on the luminal surface of blood vessels.[95] *In vitro*, inhibition of endothelial cell $G\alpha_i$ blocks VCAM-1-dependent leukocyte transendothelial migration.[95] Interestingly, there is a reduced recruitment of wild-type eosinophils transferred into $G\alpha_{i2}^{-/-}$ mice during allergic responses and it is reported that the $G\alpha_{i2}^{-/-}$ eosinophils respond to chemotactic factors,[95] indicating that signaling

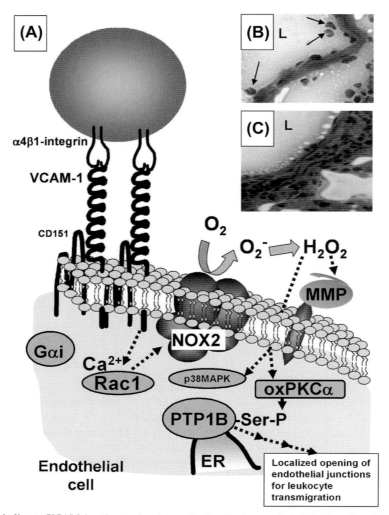

FIGURE 6.4.3 Eosinophil binding to VCAM-1 activates signal transduction for transendothelial migration. *(A)* Model for VCAM-1 signaling. VCAM-1 cell surface expression requires associated tetraspanins CD151 or CD9. Ligand cross-linking of VCAM-1 activates calcium fluxes and Ras-related C3 botulinum toxin substrate 1 (Rac1), which then activates endothelial cell NADPH dual oxidase 2 (NOX2). NOX2 catalyzes the production of superoxide that then dismutates to H_2O_2. VCAM-1 induces the production of only 1 µM of H_2O_2. H_2O_2 activates endothelial cell-associated matrix metalloproteinases (MMPs) that degrade the extracellular matrix and endothelial cell surface receptors in cell junctions. H_2O_2 also diffuses through membranes at 100 µm/s to activate mitogen-activated protein kinase (MAPK). H_2O_2 also oxidizes and transiently activates endothelial cell protein kinase C alpha (PKCα). PKCα phosphorylates and activates protein-tyrosine phosphatase 1B (PTP-1B) on the endoplasmic reticulum (ER). PTP-1B is not oxidized. These signals through reactive oxygen species (ROS), MMPs, PKCα, and PTP-1B are required for VCAM-1-dependent leukocyte trans-endothelial migration. The G protein Gαi is also involved in VCAM-1 signaling. *(B)* Adapted from 9. Cybb mice that lack NOX2 activity were irradiated and received a bone marrow transplant with wild-type bone marrow. Thus, the leukocytes expressed wild-type NOX2 but the nonhematopoietic cells, including endothelial cells, were NOX2 deficient. The mice were sensitized with ovalbumin (OVA)/alum intraperitoneally and challenged intranasally with OVA in saline. Lung tissue sections were collected and stained with hematoxylin and eosin. The arrows indicate an accumulation of eosinophils bound to the luminal surface of the endothelium in representative lung tissue sections from OVA-challenged chimeric NOX2-deficient mice. *(C)* Representative lung tissue section from OVA-challenged chimeric wild-type control mice; the leukocytes had crossed the endothelium and accumulated in the lung tissue. L, vessel lumen; NADPH, nicotinamide adenine dinucleotide phosphate; VCAM-1, vascular cell adhesion protein 1.

in a resident cell of the lung is required for the accumulation of eosinophils.[95] This occurs without altering lung levels of the T-helper type 2 (T_h2) cytokines IL-4 and interleukin-5 (IL-5), and without inducing the T-helper type 1 (T_h1) cytokine interferon gamma (IFN-γ).[95] These data are consistent with $Gα_{i2}$-mediated signaling in endothelial cells for the extravasation of eosinophils.

Nonhematopoietic NOX2 Knockout Regulates Vascular Cell Adhesion Protein 1-Dependent Eosinophil Recruitment

VCAM-1 signals through NOX2 regulate eosinophil recruitment *in vivo*.[9] The gene that encodes the catalytic subunit of NOX2 is *cytochrome b-245, beta polypeptide*

(CYBB/gp91-phox). Cybb (gp91-phox)-deficient mice that were irradiated and transplanted with wild-type bone marrow have reduced (68% inhibition) eosinophils in the bronchoalveolar lavage (BAL) and lung tissue after OVA challenge; the infiltration of other leukocytes, which migrate on other adhesion molecules, is not altered in these mice.[9] This is consistent with reports that 70% of OVA-induced eosinophil infiltration is VCAM-1-dependent.[26] There are no alterations in OVA-induced cytokines, chemokines, VCAM-1 expression, OVA-specific immuno-globulin E, or blood eosinophil numbers.[9] Most interestingly, there is an accumulation of eosinophils on the luminal surface of the endothelial cells in the lung tissue of OVA-challenged chimeric Cybb (gp91-phox)-deficient mice (Fig. 6.4.3B, C) suggesting that eosinophils adhere to the endothelium but that they cannot undergo VCAM-1-dependent transmigration.[9]

The chimeric Cybb (gp91-phox)-deficient mice also exhibit a 70% reduction in airway hyperresponsiveness (AHR).[9] Intratracheal administration of purified eosino-phils into the chimeric Cybb (gp91-phox)-deficient mice recovers the AHR,[9] suggesting that (1) bypassing the endothelium overcomes the reduced AHR in the chimeric Cybb (gp91-phox)-deficient mice and (2) that Cybb (gp91-phox) expression by other lung cells, such as fibroblasts, is not critical for the reduced AHR in the nonhematopoietic, Cybb (gp91-phox)-deficient mice. Thus, eosinophils were critical to the AHR.[9] In summary, these studies provide support for the in vivo relevance of VCAM-1 signals during eosinophil recruitment in allergic inflammation and experimental asthma.

ANTIOXIDANT REGULATION OF VASCULAR CELL ADHESION PROTEIN 1 SIGNALS AND LUNG EOSINOPHILIA

The Antioxidant Bilirubin Inhibits VCAM-1 Signaling and Eosinophilia

Antioxidant Bilirubin Inhibits Vascular Cell Adhesion Protein 1-Induced Signals In Vitro

Bilirubin and biliverdin are membrane permeable and are taken up by endothelial cells;[109,111,112] endothelial cells also express hemoxygenase-1 for bilirubin synthesis and express biliverdin reductase for bilirubin redox cycling.[109,113] Pretreatment of endothelial cells with bilirubin blocks VCAM-1-dependent leukocyte migration without altering leukocyte adhesion in vitro and blocks VCAM-1 activation of MMPs.[109] Consistent with an antioxidant function for bilirubin, VCAM-1-dependent leukocyte migration is not blocked by the stable bilirubin conjugate ditaurobilirubin, which cannot scavenge ROS.[109]

Antioxidant Bilirubin Inhibits Vascular Cell Adhesion Protein 1-Dependent Eosinophilia in Allergic Lung Inflammation

Intraperitoneal administration of upper physiological levels of bilirubin to OVA-sensitized mice reduces eosinophils in the BAL and lung tissue by >90% and reduces lymphocytes in the BAL by 60%.[109] The reduction in eosinophil and lymphocyte infiltration is consistent with the VCAM-1 dependence of eosinophil migration and the partial VCAM-1 dependence of lymphocyte migration in response to OVA.[26] There is no effect of bilirubin administration on the OVA-induced infiltration of monocytes or neutrophils,[109] which is independent of VCAM-1. Bilirubin treatment does not alter endothelial VCAM-1 expression, chemokines, and T_h2 cytokines in OVA-challenged mice,[109] and bilirubin does not induce T_h1 cytokines.[109] Although ROS can induce VCAM-1 expression, the lack of an antioxidant effect on VCAM-1 expression is consistent with compensatory mechanisms by proinflammatory mediator induction of VCAM-1 expression. Therefore, bilirubin blocks VCAM-1 signals through ROS in vitro and inhibits VCAM-1-dependent eosinophilia in allergic responses in mice.

Vitamin E Isoforms Differentially Regulate Vascular Cell Adhesion Protein 1 Signaling and Eosinophilia

Vitamin E is commonly used as an antioxidant. However, there are contradictory outcomes for clinical and experimental studies that focus on one form of vitamin E, α-tocopherol, even though multiple forms of vitamin E are present in the studies. We recently reported that isoforms of vitamin E have opposing functions in the recruitment of eosinophils during allergic inflammation by, at least, a direct effect on VCAM-1 signaling during leukocyte transendothelial migration.[110]

Vitamin E Isoforms

The natural forms of vitamin E include α-tocopherol, β-tocopherol, γ-tocopherol, and δ-tocopherol as well as the tocotrienol forms of each of these. The α-tocopherol and γ-tocopherol isoforms (Fig. 6.4.4A) are the most abundant in diets, supplements, and tissues. However, the α-tocopherol isoform in tissues is about tenfold higher than γ-tocopherol since there is preferential transfer of the α-tocopherol isoform of vitamin E by α-tocopherol transfer protein.[114] At equal molar concentrations, the α-tocopherol and γ-tocopherol isoforms have relatively similar capacity to scavenge ROS during lipid oxidation.[115] Thus, in vivo, there is likely more ROS scavenging by α-tocopherol than γ-tocopherol because there is a tenfold higher

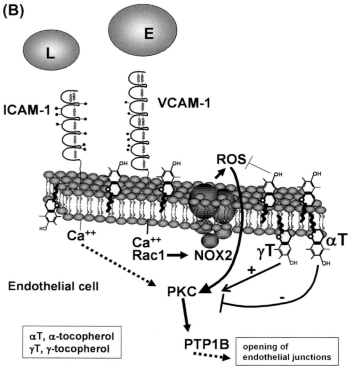

FIGURE 6.4.4 The α-tocopherol and γ-tocopherol forms of vitamin E regulate VCAM-1 signal transduction during eosinophil recruitment. *(A)* α-tocopherol differs from γ-tocopherol by one methyl group (arrows). *(B)* Tocopherols are lipids in the plasma membrane. The tocopherol head group is hydrophilic and is thus poised in membranes for scavenging of reactive oxygen species (ROS). There is approximately tenfold more α-tocopherol (αT) than γ-tocopherol (γT) in membranes *in vivo*.[110] Nevertheless, αT inhibits leukocyte recruitment and this inhibition is ablated by γT.[110] Tocopherols regulate migration of leukocytes (L) and eosinophils (E) on ICAM-1 and VCAM-1, both of which signal through PKCα.[110] Leukocyte recruitment is regulated by tocopherol scavenging of ROS and tocopherol regulation of VCAM-1 activation of PKCα. Moreover, we have recently determined that the tocopherols can directly regulate recombinant PKCα.[119] Thus, *in vivo* in endothelial cells, tocopherols that demonstrate opposing regulatory functions during eosinophil recruitment likely do so through both antioxidant mechanisms and the regulation of PKCα signaling. ICAM-1, intercellular adhesion molecule 1; NOX2, NADPH dual oxidase 2; PKCα, protein kinase C alpha; PTP-1B, protein-tyrosine phosphatase 1B; Rac1, Ras-related C3 botulinum toxin substrate 1; VCAM-1, vascular cell adhesion protein 1.

concentration of α-tocopherol in the tissues. In contrast to α-tocopherol, γ-tocopherol can also react with reactive nitrogen species.[116] When tocopherols are oxidized, they are recycled by reduction by vitamin C.[117] In addition to the antioxidant functions of tocopherols, they also have nonantioxidant functions during inflammation.[110,118]

Regulation of Vascular Cell Adhesion Protein 1-Dependent Leukocyte Transmigration by Vitamin E Isoforms In Vitro

In vitro, α-tocopherol blocks whereas γ-tocopherol elevates VCAM-1-dependent leukocyte transmigration through a direct effect of the tocopherols on endothelial cells.[110] In contrast, pretreatment of leukocytes with tocopherols has no effect on VCAM-1-dependent leukocyte transmigration.[110] Moreover, endothelial cell treatment with γ-tocopherol ablates the inhibition by α-tocopherol, such that leukocyte transendothelial migration is the same as the vehicle-treated control.[110] Thus, γ-tocopherol ablates the effects of α-tocopherol, even though it is at a physiological concentration in tissues that is 1:10 that of α-tocopherol.[110] Tocopherol modulation of VCAM-1-dependent transmigration occurs by altering VCAM-1-induced activation of endothelial cell PKCα.[110] Tocopherols not only regulate the oxidative activation of PKCα (Fig. 6.4.4B), but they can also directly bind and regulate PKCα.[119] Therefore, since PKCα is a signaling intermediate for several adhesion molecules including VCAM-1 and ICAM-1,[103,120] tocopherols regulate endothelial cell function during recruitment of eosinophils, monocytes, neutrophils, and lymphocytes that migrate on VCAM-1 and ICAM-1 *in vivo* (Fig. 6.4.4B).[110] This is in contrast to bilirubin, which exhibits predominantly antioxidant functions and preferentially blocks VCAM-1-dependent eosinophil and lymphocyte recruitment to OVA-challenged lungs.[109]

Vitamin E Isoform-Specific Regulation of Eosinophil Recruitment In Vivo

Okamoto and colleagues[121] reported that feeding mice α-tocopherol, starting 2 weeks before sensitization with OVA, reduces the number of eosinophils in the BAL. In our study, tocopherols were administered after OVA antigen sensitization to determine whether tocopherols modulate eosinophil recruitment during the OVA antigen challenge phase.[110] In this study, d-γ-tocopherol elevates the accumulation of eosinophils and other leukocytes in the BAL and lung tissue, whereas d-α-tocopherol inhibits accumulation of eosinophils and other leukocytes in the BAL and lung tissue.[110] Moreover, d-γ-tocopherol (at as little as 10% the tissue concentration of d-α-tocopherol) ablates the anti-inflammatory benefit of the d-α-tocopherol isoform *in vivo* in response to OVA.[110] D-α-tocopherol or d-γ-tocopherol did not alter blood eosinophils or OVA-induced expression of several cytokines, chemokines, prostaglandin E2, or adhesion molecules that regulate leukocyte recruitment.[110] Tocopherol modulation of leukocyte infiltration in allergic inflammation, without alteration of adhesion molecules, cytokines, or chemokines, is similar to several previous reports of *in vivo* inhibition of intracellular signals in endothelial cells without alteration of expression of these immune modulators of leukocyte recruitment.[9,95,109] In summary, the tocopherol regulatory function in allergic responses is, at least in part, by regulation of endothelial cell VCAM-1 activation of PKCα and leukocyte transendothelial migration (Fig. 6.4.4B). The opposing

functions of tocopherol isoforms have important implications for the interpretation of clinical reports and animal studies of vitamin E regulation of eosinophil recruitment and allergic inflammation.

Reinterpretation of Animal Studies and Clinical Reports on Vitamin E Regulation of Eosinophil Accumulation and Allergic Inflammation

Many studies on vitamin E *in vitro* or in animals indicate that vitamin E was administered but the form, source, and purity of tocopherols are often not reported. In addition, tissue levels of tocopherol isoforms after administration are often not determined. There also needs to be consideration for tocopherol isoforms that are present in dietary oils or in the oil vehicles used for the delivery of tocopherols. We and other investigators have determined the levels of α-tocopherol and γ-tocopherol in dietary oils (Fig. 6.4.5A).[110,122] In some reports, α-tocopherol is administered in oil vehicles that contain γ-tocopherol; investigators found no major effect of α-tocopherol on eosinophil recruitment, immune parameters, or lung airway responsiveness in rats challenged with OVA.[123] An interpretation of this is that γ-tocopherol in the vehicle antagonizes the function of the α-tocopherol that was administered. In another report, γ-tocopherol reduced the number of eosinophils and lymphocytes in the BAL of γ-tocopherol-treated mice after OVA challenge,[124] but the leukocyte infiltration was, unexpectedly, predominantly neutrophils and the purity of the γ-tocopherol once it was placed in the corn oil vehicle was not reported. Reports also conflict as to whether the antioxidant tocopherols modulate mediators of leukocyte recruitment during inflammation including prostaglandins, cytokines, chemokines, and adhesion molecules. We suggest that the variations in reports on outcomes of tocopherol treatments *in vitro* and *in vivo* result, at least in part, from differences in isoforms and purity of tocopherols, in physiological versus high pharmacological tocopherol concentrations, in cell type-specific tocopherol effects, and in experimental systems.

There are also conflicting reports for clinical studies on vitamin E regulation of allergic inflammation. For interpretation of clinical studies, it is especially important to take into consideration the dietary contribution of tocopherols because γ-tocopherol is more abundant in Western diets. The average plasma concentration of α-tocopherol is the same among many countries (Fig. 6.4.5B),[122] whereas in the United States and in The Netherlands, the average plasma γ-tocopherol level is 2−6 times higher than that reported for six European countries, China and Japan (Fig. 6.4.5B).[122] This is a result of the American diet which is rich in γ-tocopherol-containing soybean oil. In contrast, γ-tocopherol is low in other oils (sunflower and olive oil) that are commonly used in some European countries

FIGURE 6.4.5 **Alpha-tocopherol and gamma-tocopherol in dietary oils and human plasma.** *(A)* Adapted from[110]. Tocopherols were extracted from dietary oils and measured by high-pressure liquid chromatography with an electrochemical detector. *(B)* Human plasma α-tocopherol and γ-tocopherol concentrations are given in several reports from Asian countries, European countries, and the USA. Plasma α-tocopherol levels are the same among these countries. However, plasma γ-tocopherol is 2–6 times higher in the USA and The Netherlands than in other Asian and European countries. αT, alpha tocopherol; γT, gamma tocopherol.

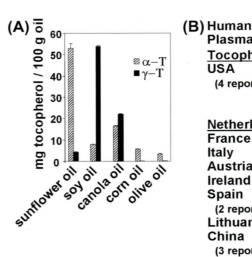

(B) Human Plasma

Tocopherol:	γT(μM)	αT(μM)	reference
USA	2.5	22	(125)
(4 reports)	5.4	22	(136)
	5.2	27	(137)
	7	20	(138)
Netherlands	2.3	25	(139)
France	1.2	26	(140)
Italy	1.2	24	(141)
Austria	1.4	21	(142)
Ireland	1.8	26	(140)
Spain	1.7	27	(140)
(2 reports)	1.7	27	(143)
Lithuania	1.6	22	(144)
China	1.4	19	(145)
(3 reports)	2.4	19	(146)
		22	(147)
Japan	1.7	23	(148)
(2 reports)	2.0	23	(149)

(Fig. 6.4.5A).[110,122,125] In clinical studies of asthma, it is reported that α-tocopherol supplementation of asthmatic patients is beneficial in Finland and Italy but, disappointingly, α-tocopherol is not beneficial for asthmatic patients in studies in the United States or The Netherlands.[126–130] These clinical outcomes are consistent with an interpretation that there is little benefit of α-tocopherol for inflammation in the presence of elevated plasma γ-tocopherol because γ-tocopherol is elevated two- to sixfold in people in the United States and The Netherlands (Fig. 6.4.4C). Although there are other environmental and genetic differences affecting the people in these countries, clinical data are consistent with animal studies demonstrating the opposing functions of the tocopherol isoforms on eosinophil accumulation during allergic inflammation.[110] Clinical literature does caution that tocopherols should be at supplemental, not high, levels since high doses of vitamin E may increase hemorrhagic stroke, all-cause mortality, or post-trial cerebral infarction.[131–133] Therefore, since α-tocopherol levels are low in asthmatics[134,135] and since α-tocopherol can reduce inflammation, supplemental levels of α-tocopherol in the presence of low γ-tocopherol may be beneficial in limiting leukocyte recruitment during allergic inflammation.

CONCLUSION

Eosinophils, which are recruited from the blood, bind to and migrate across the endothelium and into tissues during inflammation. The specificity of this recruitment is mediated by the combination of chemokines in the microenvironment, adhesion molecules induced on the endothelium

that mediate eosinophil binding, and eosinophil receptors for chemokines and adhesion molecules.[1] During eosinophil recruitment across the endothelium, PSGL-1 and integrins on eosinophils and their ligands P-selectin and VCAM-1 on endothelial cells have a major role in the recruitment of leukocytes. The activated endothelium also produces chemokines that regulate eosinophil recruitment, and eosinophils produce mediators that regulate endothelial cell function and angiogenesis. The binding of eosinophils to the adhesion molecules on endothelial cells activates signals within the endothelial cells that allow the opening of narrow vascular passageways through which eosinophils migrate. Eosinophil movement through these passageways is stimulated by chemokines that are produced by the endothelium and underlying cells in the tissue. Inhibition of adhesion molecule binding or adhesion molecule signals limits eosinophil recruitment and thus blocks eosinophilic inflammation. The eosinophils that bind to the endothelium but do not complete transendothelial migration are often released from the endothelium and continue in the blood flow. Adhesion molecules and their intracellular signals are a source for intervention in eosinophil recruitment. Clinical inhibitors modulate these pathways and limit inflammation, although there is a need for inhibitors that have reduced off-target effects, increased effectiveness, and increased selectivity for eosinophil recruitment. Since eosinophils migrate on VCAM-1, VCAM-1 and its signals are additional potential targets for intervention. VCAM-1 signals through ROS are regulated by antioxidants, including vitamin E isoforms. Importantly, isoforms of vitamin E have opposing functional effects on eosinophil recruitment and, therefore, future clinical studies of vitamin E regulation of inflammatory diseases should include a systematic design to examine opposing functions of the isoforms and levels of vitamin E on inflammation, leukocyte recruitment, and disease parameters. In summary, inhibiting eosinophil interactions with endothelial cells or endothelial cell intracellular signals blocks eosinophilic tissue inflammation.

REFERENCES

1. Lalor PF, Shields P, Grant A, Adams DH. Recruitment of lymphocytes to the human liver. *Immunol Cell Biol* 2002;**80**:52−64.
2. Marui N, Offermann MK, Swerlick R, Kunsch C, Rosen CA, et al. Vascular cell adhesion molecule-1 (VCAM-1) gene transcription and expression are regulated through an antioxidant-sensitive mechanism in human vascular endothelial cells. *J Clin Invest* 1993;**92**:1866−74.
3. Reinhardt PH, Kubes P. Differential leukocyte recruitment from whole blood via endothelial adhesion molecules under shear conditions. *Blood* 1998;**92**:4691−9.
4. Lee YW, Kuhn H, Hennig B, Neish AS, Toborek M. IL-4-induced oxidative stress upregulates VCAM-1 gene expression in human endothelial cells. *J Mol Cell Cardiol* 2001;**33**:83−94.
5. Mackay F, Loetscher H, Stueber D, Gehr G, Lesslauer W. Tumor necrosis factor alpha (TNF-alpha)-induced cell adhesion to human endothelial cells is under dominant control of one TNF receptor type, TNF-R55. *J Exp Med* 1993;**177**:1277−86.
6. Hortelano S, Lopez-Fontal R, Traves PG, Villa N, Grashoff C, et al. ILK mediates LPS-induced vascular adhesion receptor expression and subsequent leucocyte trans-endothelial migration. *Cardiovasc Res* 2010;**86**:283−92.
7. O'Keeffe LM, Muir G, Piterina AV, McGloughlin T. Vascular cell adhesion molecule-1 expression in endothelial cells exposed to physiological coronary wall shear stresses. *J Biomech Eng* 2009;**131**:081003.
8. Cernuda-Morollon E, Gharbi S, Millan J. Discriminating between the paracellular and transcellular routes of diapedesis. *Methods Mol Biol* 2010;**616**:69−82.
9. Abdala-Valencia H, Earwood J, Bansal S, Jansen M, Babcock G, et al. Nonhematopoietic NADPH oxidase regulation of lung eosinophilia and airway hyperresponsiveness in experimentally induced asthma. *Am J Physiol Lung Cell Mol Physiol* 2007;**292**:L1111−25.
10. Edwards BS, Curry MS, Tsuji H, Brown D, Larson RS, et al. Expression of P-selectin at low site density promotes selective attachment of eosinophils over neutrophils. *J Immunol* 2000;**165**:404−10.
11. Satoh T, Kanai Y, Wu MH, Yokozeki H, Kannagi R, et al. Synthesis of α(1,3) fucosyltransferases IV- and VII-dependent eosinophil selectin ligand and recruitment to the skin. *Am J Pathol* 2005;**167**:787−96.
12. Fiscus LC, Van Herpen J, Steeber DA, Tedder TF, Tang ML. L-Selectin is required for the development of airway hyper-responsiveness but not airway inflammation in a murine model of asthma. *J Allergy Clin Immunol* 2001;**107**:1019−24.
13. Miller M, Sung KL, Muller WA, Cho JY, Roman M, et al. Eosinophil tissue recruitment to sites of allergic inflammation in the lung is platelet endothelial cell adhesion molecule independent. *J Immunol* 2001;**167**:2292−7.
14. Lukacs NW, John A, Berlin A, Bullard DC, Knibbs R, et al. E- and P-selectins are essential for the development of cockroach allergen-induced airway responses. *J Immunol* 2002;**169**:2120−5.
15. Broide DH, Sullivan S, Gifford T, Sriramarao P. Inhibition of pulmonary eosinophilia in P-selectin- and ICAM-1-deficient mice. *Am J Respir Cell Mol Biol* 1998;**18**:218−25.
16. Teixeira MM, Hellewell PG. Contribution of endothelial selectins and alpha 4 integrins to eosinophil trafficking in allergic and nonallergic inflammatory reactions in skin. *J Immunol* 1998;**161**:2516−23.
17. Dang B, Wiehler S, Patel KD. Increased PSGL-1 expression on granulocytes from allergic-asthmatic subjects results in enhanced leukocyte recruitment under flow conditions. *J Leukoc Biol* 2002;**72**:702−10.
18. Yamamoto H, Nagata M, Sakamoto Y. CC chemokines and transmigration of eosinophils in the presence of vascular cell adhesion molecule 1. *Ann Allergy Asthma Immunol* 2005;**94**:292−300.
19. Shahabuddin S, Ponath P, Schleimer RP. Migration of eosinophils across endothelial cell monolayers: interactions among IL-5, endothelial-activating cytokines, and C-C chemokines. *J Immunol* 2000;**164**:3847−54.
20. Barthel SR, Annis DS, Mosher DF, Johansson MW. Differential engagement of modules 1 and 4 of vascular cell adhesion molecule-1 (CD106) by integrins alpha4beta1 (CD49d/29) and alphaMbeta2 (CD11b/18) of eosinophils. *J Biol Chem* 2006;**281**:32175−87.
21. Rao SP, Wang Z, Zuberi RI, Sikora L, Bahaie NS, et al. Galectin-3 functions as an adhesion molecule to support eosinophil rolling and adhesion under conditions of flow. *J Immunol* 2007;**179**:7800−7.
22. Ge XN, Bahaie NS, Kang BN, Hosseinkhani MR, Ha SG, et al. Allergen-induced airway remodeling is impaired in galectin-3-deficient mice. *J Immunol* 2010;**185**:1205−14.
23. Kelly KA, Allport JR, Yu AM, Sinh S, Sage EH, et al. SPARC is a VCAM-1 counter-ligand that mediates leukocyte transmigration. *J Leukoc Biol* 2007;**81**:748−56.
24. Banerjee ER, Jiang Y, Henderson Jr WR, Latchman Y, Papayannopoulou T. Absence of alpha 4 but not beta 2 integrins restrains development of chronic allergic asthma using mouse genetic models. *Exp Hematol* 2009;**37**:715−27. e713.
25. Ebihara N, Yokoyama T, Kimura T, Nakayasu K, Okumura K, et al. Anti VLA-4 monoclonal antibody inhibits eosinophil infiltration in allergic conjunctivitis model of guinea pig. *Curr Eye Res* 1999;**19**:20−5.
26. Chin JE, Hatfield CA, Winterrowd GE, Brashler JR, Vonderfecht SL, et al. Airway recruitment of leukocytes in mice is dependent on alpha4-integrins and vascular cell adhesion molecule-1. *Am J Physiol* 1997;**272**:L219−29.
27. Inomata M, Into T, Nakashima M, Noguchi T, Matsushita K. IL-4 alters expression patterns of storage components of vascular endothelial cell-specific granules through STAT6- and SOCS-1-dependent mechanisms. *Mol Immunol* 2009;**46**:2080−9.
28. Patel KD. Eosinophil tethering to interleukin-4-activated endothelial cells requires both P-selectin and vascular cell adhesion molecule-1. *Blood* 1998;**92**:3904−11.

29. Yamamoto H, Sedgwick JB, Busse WW. Differential Regulation of Eosinophil Adhesion and Transmigration by Pulmonary Microvascular Endothelial Cells. *J Immunol* 1998;**161**:971–7.

30. Tachimoto H, Ebisawa M. Effect of interleukin-13 or tumor necrosis factor-alpha on eosinophil adhesion to endothelial cells under physiological flow conditions. *Int Arch Allergy Immunol.* 2007;**143**(Supp. 1):33–7.

31. Woltmann G, McNulty CA, Dewson G, Symon FA, Wardlaw AJ. Interleukin-13 induces PSGL-1/P-selectin-dependent adhesion of eosinophils, but not neutrophils, to human umbilical vein endothelial cells under flow. *Blood* 2000;**95**:3146–52.

32. Hickey MJ, Granger DN, Kubes P. Molecular mechanisms underlying IL-4-induced leukocyte recruitment in vivo: a critical role for the alpha 4 integrin. *J Immunol* 1999;**163**:3441–8.

33. Broide DH, Campbell K, Gifford T, Sriramarao P. Inhibition of eosinophilic inflammation in allergen-challenged, IL-1 receptor type 1-deficient mice is associated with reduced eosinophil rolling and adhesion on vascular endothelium. *Blood* 2000;**95**:263–9.

34. Hohki G, Terada N, Hamano N, Kitaura M, Nakajima T, et al. The effects of eotaxin on the surface adhesion molecules of endothelial cells and on eosinophil adhesion to microvascular endothelial cells. *Biochem Biophys Res Commun* 1997;**241**:136–41.

35. Nagai K, Larkin S, Hartnell A, Larbi K, Razi Aghakhani M, et al. Human eotaxin induces eosinophil extravasation through rat mesenteric venules: role of alpha4 integrins and vascular cell adhesion molecule-1. *Immunol* 1999;**96**:176–83.

36. DiScipio RG, Schraufstatter IU. The role of the complement anaphylatoxins in the recruitment of eosinophils. *Int Immunopharmacol* 2007;**7**:1909–23.

37. Seye CI, Yu N, Jain R, Kong Q, Minor T, et al. The P2Y2 nucleotide receptor mediates UTP-induced vascular cell adhesion molecule-1 expression in coronary artery endothelial cells. *J Biol Chem* 2003;**278**:24960–5.

38. Vanderstocken G, Bondue B, Horckmans M, Di Pietrantonio L, Robaye B, et al. P2Y2 receptor regulates VCAM-1 membrane and soluble forms and eosinophil accumulation during lung inflammation. *J Immunol* 2010;**185**:3702–7.

39. Muller T, Robaye B, Vieira RP, Ferrari D, Grimm M, et al. The purinergic receptor P2Y(2) receptor mediates chemotaxis of dendritic cells and eosinophils in allergic lung inflammation. *Allergy* 2010;**65**:1545–53.

40. Barthel SR, Johansson MW, Annis DS, Mosher DF. Cleavage of human 7-domain VCAM-1 (CD106) by thrombin. *Thromb Haemost* 2006;**95**:873–80.

41. Garton KJ, Gough PJ, Philalay J, Wille PT, Blobel CP, et al. Stimulated shedding of vascular cell adhesion molecule 1 (VCAM-1) is mediated by tumor necrosis factor-alpha-converting enzyme (ADAM 17). *J Biol Chem* 2003;**278**:37459–64.

42. Matsuno O, Miyazaki E, Nureki S, Ueno T, Ando M, et al. Elevated soluble ADAM8 in bronchoalveolar lavage fluid in patients with eosinophilic pneumonia. *Int Arch Allergy Immunol* 2007;**142**:285–90.

43. Matsuno O, Miyazaki E, Nureki S, Ueno T, Kumamoto T, et al. Role of ADAM8 in experimental asthma. *Immunol Lett* 2006;**102**:67–73.

44. Peduto L. ADAM9 as a potential target molecule in cancer. *Curr Pharm Des* 2009;**15**:2282–7.

45. Johansson MW, Lye MH, Barthel SR, Duffy AK, Annis DS, et al. Eosinophils adhere to vascular cell adhesion molecule-1 via podosomes. *Am J Respir Cell Mol Biol* 2004;**31**:413–22.

46. Ohashi Y, Nakai Y, Tanaka A, Kakinoki Y, Ohno Y, et al. Soluble vascular cell adhesion molecule-1 in perennial allergic rhinitis. *Acta Otolaryngol* 1998;**118**:105–9.

47. Robroeks CM, Rijkers GT, Jobsis Q, Hendriks HJ, Damoiseaux JG, et al. Increased cytokines, chemokines and soluble adhesion molecules in exhaled breath condensate of asthmatic children. *Clin Exp Allergy* 2010;**40**:77–84.

48. Hamzaoui A, Ammar J, El Mekki F, Borgi O, Ghrairi H, et al. Elevation of serum soluble E-selectin and VCAM-1 in severe asthma. *Mediators Inflamm* 2001;**10**:339–42.

49. Tokuhira M, Hosaka S, Volin MV, Haines 3rd GK, Katschke Jr KJ, et al. Soluble vascular cell adhesion molecule 1 mediation of monocyte chemotaxis in rheumatoid arthritis. *Arthrit Rheumat* 2000;**43**:1122–33.

50. Ueki S, Kihara J, Kato H, Ito W, Takeda M, et al. Soluble vascular cell adhesion molecule-1 induces human eosinophil migration. *Allergy* 2009;**64**:718–24.

51. Yamamoto H, Nagata M. Regulatory mechanisms of eosinophil adhesion to and transmigration across endothelial cells by alpha4 and beta2 integrins. *Int Arch Allergy Immunol* 1999;**120**(Suppl. 1):24–6.

52. Nagata M, Yamamoto H, Tabe K, Sakamoto Y, Matsuo H. Eosinophil-adhesion-inducing activity produced by antigen-stimulated mononuclear cells involves GM-CSF. *Int Arch Allergy Immunol* 2000;**122**(Suppl. 1):15–9.

53. Cuvelier SL, Patel KD. Shear-dependent eosinophil transmigration on interleukin 4-stimulated endothelial cells: a role for endothelium-associated eotaxin-3. *J Exp Med* 2001;**194**:1699–709.

54. Farahi N, Cowburn AS, Upton PD, Deighton J, Sobolewski A, et al. Eotaxin-1/CC chemokine ligand 11: a novel eosinophil survival factor secreted by human pulmonary artery endothelial cells. *J Immunol* 2007;**179**:1264–73.

55. Kanda A, Adachi T, Kayaba H, Yamada Y, Ueki S, et al. Red blood cells regulate eosinophil chemotaxis by scavenging RANTES secreted from endothelial cells. *Clin Exp Allergy* 2004;**34**:1621–6.

56. Ellyard JI, Simson L, Bezos A, Johnston K, Freeman C, et al. Eotaxin selectively binds heparin. An interaction that protects eotaxin from proteolysis and potentiates chemotactic activity in vivo. *J Biol Chem* 2007;**282**:15238–47.

57. Zuberi RI, Ge XN, Jiang S, Bahaie NS, Kang BN, et al. Deficiency of endothelial heparan sulfates attenuates allergic airway inflammation. *J Immunol* 2009;**183**:3971–9.

58. Cheng SS, Lukacs NW, Kunkel SL. Eotaxin/CCL11 suppresses IL-8/CXCL8 secretion from human dermal microvascular endothelial cells. *J Immunol* 2002;**168**:2887–94.

59. Dallaire MJ, Ferland C, Lavigne S, Chakir J, Laviolette M. Migration through basement membrane modulates eosinophil expression of CD44. *Clin Exp Allergy* 2002;**32**:898–905.

60. Sedgwick JB, Jansen KJ, Kennedy JD, Kita H, Busse WW. Effects of the very late adhesion molecule 4 antagonist WAY103 on human peripheral blood eosinophil vascular cell adhesion molecule 1-dependent functions. *J Allergy Clin Immunol* 2005;**116**:812–9.

61. Munoz NM, Hamann KJ, Rabe KF, Sano H, Zhu X, et al. Augmentation of eosinophil degranulation and LTC(4) secretion by integrin-mediated endothelial cell adhesion. *Am J Physiol* 1999;**277**:L802–10.

62. Mori M, Takaku Y, Kobayashi T, Hagiwara K, Kanazawa M, et al. Eosinophil superoxide anion generation induced by adhesion molecules and leukotriene D4. *Int Arch Allergy Immunol* 2009;**149**(Suppl. 1):31—8.

63. Nagata M, Sedgwick JB, Kita H, Busse WW. Granulocyte macrophage colony-stimulating factor augments ICAM-1 and VCAM-1 activation of eosinophil function. *Am J Respir Cell Mol Biol* 1998;**19**:158—66.

64. Nagata M, Sedgwick JB, Busse WW. Synergistic activation of eosinophil superoxide anion generation by VCAM-1 and GM-CSF. Involvement of tyrosine kinase and protein kinase C. *Int Arch Allergy Immunol* 1997;**114**:S78—80.

65. Braun RK, Foerster M, Workalemahu G, Haefner D, Kroegel C, et al. Differential regulation of aminopeptidase N (CD13) by transendothelial migration and cytokines on human eosinophils. *Exp Lung Res* 2003;**29**:59—77.

66. Molet S, Gosset P, Vanhee D, Tillie-Leblond I, Wallaert B, et al. Modulation of cell adhesion molecules on human endothelial cells by eosinophil-derived mediators. *J Leukoc Biol* 1998; **63**:351—8.

67. Chihara J, Yamamoto T, Kayaba H, Kakazu T, Kurachi D, et al. Degranulation of eosinophils mediated by intercellular adhesion molecule-1 and its ligands is involved in adhesion molecule expression on endothelial cells-selective induction of VCAM-1. *J Allergy Clin Immunol* 1999;**103**:S452—6.

68. Puxeddu I, Berkman N, Nissim Ben Efraim AH, Davies DE, Ribatti D, et al. The role of eosinophil major basic protein in angiogenesis. *Allergy* 2009;**64**:368—74.

69. Puxeddu I, Alian A, Piliponsky AM, Ribatti D, Panet A, et al. Human peripheral blood eosinophils induce angiogenesis. *Int J Biochem Cell Biol* 2005;**37**:628—36.

70. Horiuchi T, Weller PF. Expression of vascular endothelial growth factor by human eosinophils: upregulation by granulocyte macrophage colony-stimulating factor and interleukin-5. *Am J Respir Cell Mol Biol* 1997;**17**:70—7.

71. Kanazawa H, Nomura S, Yoshikawa J. Role of microvascular permeability on physiologic differences in asthma and eosinophilic bronchitis. *Am J Respir Crit Care Med* 2004;**169**:1125—30.

72. Asai K, Kanazawa H, Kamoi H, Shiraishi S, Hirata K, et al. Increased levels of vascular endothelial growth factor in induced sputum in asthmatic patients. *Clin Exp Allergy* 2003;**33**:595—9.

73. Feistritzer C, Kaneider NC, Sturn DH, Mosheimer BA, Kahler CM, et al. Expression and function of the vascular endothelial growth factor receptor FLT-1 in human eosinophils. *Am J Respir Cell Mol Biol* 2004;**30**:729—35.

74. Beeh KM, Beier J, Meyer M, Buhl R, Zahlten R, et al. Bimosiamose, an inhaled small-molecule pan-selectin antagonist, attenuates late asthmatic reactions following allergen challenge in mild asthmatics: a randomized, double-blind, placebo-controlled clinical cross-over-trial. *Pulm Pharmacol Ther* 2006;**19**:233—41.

75. Woodside DG, Vanderslice P. Cell adhesion antagonists: therapeutic potential in asthma and chronic obstructive pulmonary disease. *BioDrugs* 2008;**22**:85—100.

76. Kogan TP, Dupre B, Bui H, McAbee KL, Kassir JM, et al. Novel synthetic inhibitors of selectin-mediated cell adhesion: synthesis of 1,6-bis[3-(3-carboxymethylphenyl)-4-(2-alpha-D-mannopyranosyloxy)phenyl]hexane (TBC1269). *J Med Chem* 1998;**41**: 1099—111.

77. Aydt E, Wolff G. Development of synthetic pan-selectin antagonists: a new treatment strategy for chronic inflammation in asthma. *Pathobiol* 2002;**70**:297—301.

78. Hicks AE, Abbitt KB, Dodd P, Ridger VC, Hellewell PG, et al. The anti-inflammatory effects of a selectin ligand mimetic, TBC-1269, are not a result of competitive inhibition of leukocyte rolling in vivo. *J Leukoc Biol* 2005;**77**:59—66.

79. Romano SJ. Selectin antagonists: therapeutic potential in asthma and COPD. *Treat Respir Med* 2005;**4**:85—94.

80. Clifford DB, DeLuca A, Simpson DM, Arendt G, Giovannoni G, et al. Natalizumab-associated progressive multifocal leukoencephalopathy in patients with multiple sclerosis: lessons from 28 cases. *Lancet Neurol* 2010;**9**:438—46.

81. O'Connor PW. Use of natalizumab in multiple sclerosis patients. *Can J Neurol Sci* 2010;**37**:98—104.

82. Cohen M, Rocher F, Vivinus S, Thomas P, Lebrun C. Giant urticaria and persistent neutralizing antibodies after the first natalizumab infusion. *Neurol* 2010;**74**:1394—5.

83. Leussink VI, Lehmann HC, Hartung HP, Gold R, Kieseier BC. Type III systemic allergic reaction to natalizumab. *Arch Neurol* 2008;**65**:851—2. author reply 852.

84. Choo JH, Nagata M, Sutani A, Kikuchi I, Sakamoto Y. Theophylline attenuates the adhesion of eosinophils to endothelial cells. *Int Arch Allergy Immunol* 2003;**131**(Suppl. 1):40—5.

85. Saito K, Nagata M, Kikuchi I, Sakamoto Y. Leukotriene D4 and eosinophil transendothelial migration, superoxide generation, and degranulation via beta2 integrin. *Ann Allergy Asthma Immunol* 2004;**93**:594—600.

86. Virchow Jr JC, Faehndrich S, Nassenstein C, Bock S, Matthys H, et al. Effect of a specific cysteinyl leukotriene-receptor 1-antagonist (montelukast) on the transmigration of eosinophils across human umbilical vein endothelial cells. *Clin Exp Allergy* 2001;**31**:836—44.

87. Nagata M, Saito K, Kikuchi I, Hagiwara K, Kanazawa M. Effect of the cysteinyl leukotriene antagonist pranlukast on transendothelial migration of eosinophils. *Int Arch Allergy Immunol* 2005;**137**(Suppl. 1):2—6.

88. Qasem AR, Bucolo C, Baiula M, Sparta A, Govoni P, et al. Contribution of alpha4beta1 integrin to the antiallergic effect of levocabastine. *Biochem Pharmacol* 2008;**76**:751—62.

89. Wu P, Mitchell S, Walsh GM. A new antihistamine levocetirizine inhibits eosinophil adhesion to vascular cell adhesion molecule-1 under flow conditions. *Clin Exp Allergy* 2005;**35**:1073—9.

90. Thomson L, Blaylock MG, Sexton DW, Campbell A, Walsh GM. Cetirizine and levocetirizine inhibit eotaxin-induced eosinophil transendothelial migration through human dermal or lung microvascular endothelial cells. *Clin Exp Allergy* 2002;**32**:1187—92.

91. Koo GC, Shah K, Ding GJ, Xiao J, Wnek R, et al. A small molecule very late antigen-4 antagonist can inhibit ovalbumin-induced lung inflammation. *Am J Respir Crit Care Med* 2003;**167**:1400—9.

92. Cortijo J, Sanz MJ, Iranzo A, Montesinos JL, Nabah YN, et al. A small molecule, orally active, alpha4beta1/alpha4beta7 dual antagonist reduces leukocyte infiltration and airway hyperresponsiveness in an experimental model of allergic asthma in Brown Norway rats. *Br J Pharmacol* 2006;**147**:661—70.

93. Cook-Mills JM, Johnson JD, Deem TL, Ochi A, Wang L, et al. Calcium mobilization and Rac1 activation are required for VCAM-1

(vascular cell adhesion molecule-1) stimulation of NADPH oxidase activity. *Biochem J* 2004;**378**:539−47.

94. Matheny HE, Deem TL, Cook-Mills JM. Lymphocyte Migration through Monolayers of Endothelial Cell Lines involves VCAM-1 Signaling via Endothelial Cell NADPH Oxidase. *J Immunol* 2000;**164**:6550−9.

95. Pero RS, Borchers MT, Spicher K, Ochkur SI, Sikora L, et al. Galphai2-mediated signaling events in the endothelium are involved in controlling leukocyte extravasation. *Proc Natl Acad Sci U S A* 2007;**104**:4371−6.

96. Tudor KSRS, Hess KL, Cook-Mills JM. Cytokines Modulate Endothelial Cell Intracellular Signal Transduction Required for VCAM-1-dependent Lymphocyte Transendothelial Migration. *Cytokine* 2001;**15**:196−211.

97. DeLeo FR, Quinn MT. Assembly of the phagocyte NADPH oxidase: molecular interaction of oxidase proteins. *J Leuk Biol* 1996;**60**:677−91.

98. Cook-Mills JM, Wirth JJ, Fraker PJ. Possible Roles for Zinc in Destruction of {UTrypanosoma} {Ucruzi} by Toxic Oxygen Metabolites Produced by Mononuclear Phagocytes. In: Bendich A, Phillips M, Tengerdy R, editors. *Antioxidant Nutrients and Immune Function*. New York, NY: Plenum Press; 1990. p. 111−21.

99. Hogg N, Browning J, Howard T, Winterford C, Fitzpatrick D, et al. Apoptosis in vascular endothelial cells caused by serum deprivation, oxidative stress and transforming growth factor-beta. *Endothelium* 1999;**7**:35−49.

100. Usatyuk PV, Natarajan V. Role of mitogen-activated protein kinases in 4-hydroxy-2-nonenal-induced actin remodeling and barrier function in endothelial cells. *J Biol Chem* 2004;**279**:11789−97.

101. Deem TL, Cook-Mills JM. Vascular cell adhesion molecule-1 (VCAM-1) activation of endothelial cell matrix metalloproteinases: role of reactive oxygen species. *Blood* 2004;**104**:2385−93.

102. Rajagopalan S, Meng XP, Ramasamy S, Harrison DG, Galis ZS. Reactive oxygen species produced by macrophage-derived foam cells regulate the activity of vascular matrix metalloproteinases in vitro. Implications for atherosclerotic plaque stability. *J Clin Invest* 1996;**98**:2572−9.

103. Abdala-Valencia H, Cook-Mills JM. VCAM-1 signals activate endothelial cell protein kinase Cα via oxidation. *J Immunol* 2006;**177**:6379−87.

104. Fialkow L, Chan CK, Downey GP. Inhibition of CD45 during neutrophil activation. *J Immunol* 1997;**158**:5409−17.

105. Mathai JC, Sitaramam V. Stretch sensitivity of transmembrane mobility of hydrogen peroxide through voids in the bilayer. Role of cardiolipin. *J Biol Chem* 1994;**269**:17784−93.

106. Deem TL, Abdala-Valencia H, Cook-Mills JM. VCAM-1 activation of PTP1B in endothelial cells. *J Immunol* 2007;**178**:3865−73.

107. Carman CV, Springer TA. A transmigratory cup in leukocyte diapedesis both through individual vascular endothelial cells and between them. *J Cell Biol* 2004;**167**:377−88.

108. van Wetering S, van den Berk N, van Buul JD, Mul FPJ, Lommerse I, et al. VCAM-1-mediated Rac signaling controls endothelial cell-cell contacts and leukocyte transmigration. *Am J Physiol Cell Physiol* 2003;**285**:C343−52.

109. Keshavan P, Deem TL, Schwemberger SJ, Babcock GF, Cook-Mills JM, et al. Unconjugated bilirubin inhibits VCAM-1-mediated transendothelial leukocyte migration. *J Immunol* 2005;**174**: 3709−18.

110. Berdnikovs S, Abdala-Valencia H, McCary C, Somand M, Cole R, et al. Isoforms of vitamin E have opposing immunoregulatory funcitons during inflammation by regulating leukocyte recruitment. *J Immunol* 2009;**182**:4395−405.

111. Ostrow JD, Mukerjee P, Tiribelli C. Structure and binding of unconjugated bilirubin: relevance for physiological and pathophysiological function. *J Lipid Res* 1994;**35**:1715−37.

112. Stocker R, McDonagh AF, Glazer AN, Ames BN. Antioxidant activities of bile pigments: biliverdin and bilirubin. *Methods Enzymol* 1990;**186**:301−9.

113. Sedlak TW, Snyder SH. Bilirubin benefits: cellular protection by a biliverdin reductase antioxidant cycle. *Pediatrics* 2004;**113**: 1776−82.

114. Wolf G. How an increased intake of alpha-tocopherol can suppress the bioavailability of gamma-tocopherol. *Nutr Rev* 2006;**64**: 295−9.

115. Yoshida Y, Saito Y, Jones LS, Shigeri Y. Chemical reactivities and physical effects in comparison between tocopherols and tocotrienols: physiological significance and prospects as antioxidants. *J Biosci Bioeng* 2007;**104**:439−45.

116. Christen S, Woodall AA, Shigenaga MK, Southwell-Keely PT, Duncan MW, et al. Gamma-tocopherol traps mutagenic electrophiles such as NO(X) and complements alpha-tocopherol: physiological implications. *Proc Natl Acad Sci U S A* 1997;**94**:3217−22.

117. Buettner GR. The pecking order of free radicals and antioxidants: lipid peroxidation, alpha-tocopherol, and ascorbate. *Arch Biochem Biophys* 1993;**300**:535−43.

118. Azzi A, Stocker A. Vitamin E: non-antioxidant roles. *Prog Lipid Res* 2000;**39**:231−55.

119. McCary CA, Yoon Y, Panagabko C, Cho W, Atkinson J, et al. Vitamin E isoforms directly bind PKCalpha and differentially regulate activation of PKCalpha. *Biochem J* 2012;**441**:189−98.

120. Etienne-Manneville S, Manneville JB, Adamson P, Wilbourn B, Greenwood J, et al. ICAM-1-coupled cytoskeletal rearrangements and transendothelial lymphocyte migration involve intracellular calcium signaling in brain endothelial cell lines. *J Immunol* 2000;**165**:3375−83.

121. Okamoto N, Murata T, Tamai H, Tanaka H, Nagai H. Effects of alpha tocopherol and probucol supplements on allergen-induced airway inflammation and hyperresponsiveness in a mouse model of allergic asthma. *Int Arch Allergy Immunol* 2006;**141**:172−80.

122. Wagner KH, Kamal-Eldin A, Elmadfa I. Gamma-tocopherol—an underestimated vitamin? *Ann Nutr Metab* 2004;**48**:169−88.

123. Suchankova J, Voprsalova M, Kottova M, Semecky V, Visnovsky P. Effects of oral alpha-tocopherol on lung response in rat model of allergic asthma. *Respirol* 2006;**11**:414−21.

124. Wagner JG, Jiang Q, Harkema JR, Ames BN, Illek B, et al. Gamma-tocopherol prevents airway eosinophilia and mucous cell hyperplasia in experimentally induced allergic rhinitis and asthma. *Clin Exp Allergy* 2008;**38**:501−11.

125. Jiang Q, Christen S, Shigenaga MK, Ames BN. Gamma-tocopherol, the major form of vitamin E in the US diet, deserves more attention. *Am J Clin Nutr* 2001;**74**:714−22.

126. Weiss ST. Diet as a risk factor for asthma. *Ciba Foundation Symposium* 1997;**206**:244−57.

127. Troisi RJ, Willett WC, Weiss ST, Trichopoulos D, Rosner B, et al. A prospective study of diet and adult-onset asthma. *Am J Respir Critical Care Med* 1995;**151**:1401—8.
128. Dow L, Tracey M, Villar A, Coggon D, Margetts BM, et al. Does dietary intake of vitamins C and E influence lung function in older people? *Am J Respir Critical Care Med* 1996;**154**:1401—4.
129. Smit HA, Grievink L, Tabak C. Dietary influences on chronic obstructive lung disease and asthma: a review of the epidemiological evidence. *Proc Nutr Soc* 1999;**58**:309—19.
130. Tabak C, Smit HA, Rasanen L, Fidanza F, Menotti A, et al. Dietary factors and pulmonary function: a cross sectional study in middle aged men from three European countries. *Thorax* 1999;**54**:1021—6.
131. Saremi A, Arora R. Vitamin E and cardiovascular disease. *Am J Ther* 2010;**17**:e56—65.
132. Leppala JM, Virtamo J, Fogelholm R, Huttunen JK, Albanes D, et al. Controlled trial of alpha-tocopherol and beta-carotene supplements on stroke incidence and mortality in male smokers. *Arterioscler Thromb Vasc Biol* 2000;**20**:230—5.
133. Miller 3rd ER, Pastor-Barriuso R, Dalal D, Riemersma RA, Appel LJ, et al. Meta-analysis: high-dosage vitamin E supplementation may increase all-cause mortality. *Ann Intern Med* 2005;**142**:37—46.
134. Kalayci O, Besler T, Kilinc K, Sekerel BE, Saraclar Y. Serum levels of antioxidant vitamins (alpha tocopherol, beta carotene, and ascorbic acid) in children with bronchial asthma. *Turk J Pediatrics* 2000;**42**:17—21.
135. Kelly FJ, Mudway I, Blomberg A, Frew A, Sandstrom T. Altered lung antioxidant status in patients with mild asthma. *Lancet* 1999;**354**:482—3.
136. Block G, Norkus E, Hudes M, Mandel S, Helzlsouer K. Which plasma antioxidants are most related to fruit and vegetable consumption? *Am J Epidemiol* 2001;**154**:1113—8.
137. Sato R, Helzlsouer KJ, Alberg AJ, Hoffman SC, Norkus EP, et al. Prospective study of carotenoids, tocopherols, and retinoid concentrations and the risk of breast cancer. *Cancer Epidemiol Biomarkers Prev* 2002;**11**:451—7.
138. Cooney RV, Custer LJ, Okinaka L, Franke AA. Effects of dietary sesame seeds on plasma tocopherol levels. *Nutr Cancer* 2001;**39**:66—71.
139. Sen CK, Khanna S, Roy S. Tocotrienols in health and disease: the other half of the natural vitamin E family. *Mol Aspects Med* 2007;**28**:692—728.
140. Olmedilla B, Granado F, Southon S, Wright AJ, Blanco I, et al. Serum concentrations of carotenoids and vitamins A, E, and C in control subjects from five European countries. *Br J Nutr* 2001;**85**:227—38.
141. Palli D, Masala G, Vineis P, Garte S, Saieva C, et al. Biomarkers of dietary intake of micronutrients modulate DNA adduct levels in healthy adults. *Carcinogenesis* 2003;**24**:739—46.
142. Tomasch R, Wagner KH, Elmadfa I. Antioxidative power of plant oils in humans: the influence of alpha- and gamma-tocopherol. *Ann Nutr Metab* 2001;**45**:110—5.
143. Ruiz Rejon F, Martin-Pena G, Granado F, Ruiz-Galiana J, Blanco I, et al. Plasma status of retinol, alpha- and gamma-tocopherols, and main carotenoids to first myocardial infarction: case control and follow-up study. *Nutrition* 2002;**18**:26—31.
144. Kristenson M, Kucinskiene Z, Schafer-Elinder L, Leanderson P, Tagesson C. Lower serum levels of beta-carotene in Lithuanian men are accompanied by higher urinary excretion of the oxidative DNA adduct, 8-hydroxydeoxyguanosine. The LiVicordia study. *Nutrition* 2003;**19**:11—5.
145. Taylor PR, Qiao YL, Abnet CC, Dawsey SM, Yang CS, et al. Prospective study of serum vitamin E levels and esophageal and gastric cancers. *J Natl Cancer Inst* 2003;**95**:1414—6.
146. Ratnasinghe D, Tangrea JA, Forman MR, Hartman T, Gunter EW, et al. Serum tocopherols, selenium and lung cancer risk among tin miners in China. *Cancer Causes Control* 2000;**11**:129—35.
147. Kumagai Y, Pi JB, Lee S, Sun GF, Yamanushi T, et al. Serum antioxidant vitamins and risk of lung and stomach cancers in Shenyang, China. *Cancer Lett* 1998;**129**:145—9.
148. Persson C, Sasazuki S, Inoue M, Kurahashi N, Iwasaki M, et al. Plasma levels of carotenoids, retinol and tocopherol and the risk of gastric cancer in Japan: a nested case-control study. *Carcinogenesis* 2008;**29**:1042—8.
149. Fukui K, Ostapenko VV, Abe K, Nishide T, Miyano M, et al. Changes in plasma alpha and gamma tocopherol levels before and after long-term local hyperthermia in cancer patients. *Free Radic Res* 2006;**40**:893—9.

Chapter 6.5

Eosinophils and Circulating Endothelial Progenitor Cells

Kewal Asosingh and Serpil C. Erzurum

ENDOTHELIAL PROGENITOR CELLS

Endothelial progenitor cells (EPCs) in postnatal neovascularization is a groundbreaking concept introduced by Asahara and colleagues.[1] This discovery has added tremendously to our understanding of physiological and pathological new blood vessel formation. In the meantime, new studies are revealing multiple functions and types of endothelial progenitors in novel blood vessel development.[2,3] These cells are derived from the vascular wall or from the bone marrow (Fig. 6.5.1). Bone marrow-derived EPCs are derived from the hemangioblast (i.e., the common hematopoietic and endothelial stem cell), and are also referred to as *early outgrowth* EPCs or proangiogenic progenitors. Whether this heterogeneous group of cells differentiates into endothelial cells is uncertain. Different biological assays and cell surface markers are used to define subsets of these progenitors. Proangiogenic progenitors influence neovascularization through paracrine effects. Indeed, there is conclusive evidence that these cells release high levels of potent angiogenic cytokines such as vascular endothelial growth factor (VEGF) and angiogenic proteases such as matrix metalloproteinases (MMPs).[4] Under steady-state physiological conditions, proangiogenic cells are rare cells in the peripheral blood circulation but are

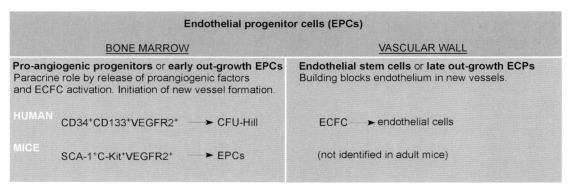

FIGURE 6.5.1 Endothelial progenitor cells. Endothelial progenitor cells (EPCs) define a heterogeneous group of rare blood-borne cells with crucial roles in angiogenesis. Bone marrow-derived proangiogenic progenitors originate from hematopoietic stem cells and give rise to potent paracrine actors of new vessel formation. Different cell surface markers are used to detect these cells in both human and murine circulation. In humans, *in vitro* cultures of these cells are called colony-forming units of endothelial-like cells (CFU-ECs; also known as CFU-Hill cells), which are distinguished from the vascular wall-derived endothelial colony-forming cells (ECFCs). ECFCs differentiate into true endothelial cells, which serve as structural cells of the endothelium during angiogenesis. C-Kit, proto-oncogene c-Kit; Sca-1, stem cell antigen 1; VEGFR-2, vascular endothelial growth factor receptor 2.

rapidly mobilized from the bone marrow in the presence of angiogenic stimuli.[5] They can be enumerated by flow cytometric analysis using a combination of hematopoietic stem cell markers and endothelial cell antigens, such as CD34, CD133 [stem cell antigen 1 (Sca-1) and proto-oncogene c-Kit (c-Kit) in mice] and vascular endothelial growth factor receptor 2 (VEGFR-2). In humans, progeny of these cells *in vitro* gives rise to colony-forming units of endothelial-like cells (CFU-ECs) or CFU-Hill colonies.[2,6,7] CFU-ECs are heterogeneous colonies comprised of proliferating progenitors in the presence of T cells.[8] Immunophenotypic analysis indicates low to intermediate expression of the hematopoietic stem cell markers CD34 and CD133. These cells are also positive for endothelial cell-like features such as uptake of acetylated low-density lipoprotein, lectin binding, and the expression of CD31, von Willebrand factor, and vascular endothelial cadherin. In addition, CFU-ECs also express the pan-hematopoietic marker CD45, myeloid marker CD33, CD11b, and α-smooth muscle actin; thus, they are not considered true endothelial cells.[9,10] CFU-ECs have extensive proangiogenic effects in *in vitro* tube formation assays and ischemic hind limb models mainly attributable to the release of paracrine factors.[11,12] Further investigations have shown that these cells are mainly of myeloid origin.[2]

In addition to CFU-ECs, there is a newly described endothelial progenitor, termed endothelial colony-forming cells (ECFCs) or late outgrowth EPCs because they appear after 3−4 weeks of culture.[13,14] ECFCs are considered true endothelial stem cells and are most probably derived from the vascular wall. Little is known regarding this very recently described ECFCs, which are extremely infrequent in the circulation but can expand numerous times when needed for new vessel formation. These ECFCs serve as structural building blocks in the formation of angiogenic

tubes *in vitro* and blood vessels in animal models. Interestingly, CFU-ECs enhance—through the release of paracrine factors—angiogenic tube formation by ECFCs.[15] Support for proangiogenic progenitors as the initiators of the angiogenic switch (the onset of new angiogenesis) comes from several animal models of disease such as cancer, asthma, and hind limb ischemia[10,16−18]. In all these different systems, the arrival of proangiogenic progenitors marks imminent angiogenesis. In fact, increased levels of circulating proangiogenic progenitors is a valid cellular biomarker for an increased angiogenic state.[5]

ENDOTHELIAL PROGENITORS AND EOSINOPHIL RECRUITMENT

Allergic Lung Disease

Increased airway angiogenesis in asthma has been well documented historically,[19] but the *understanding* of this process in the origin of asthma remains unexplored. Asthma patients have elevated levels of circulating CD34[+] CD133[+] progenitors compared to healthy controls and patients with allergic rhinitis (i.e., controls for upper airway inflammation).[9] *In vitro*, CFU-Hill assays demonstrated that these asthmatic proangiogenic progenitor cells are highly proliferative. In addition, angiogenesis assays showed increased incorporation of the asthmatic proangiogenic progenitors into endothelial tubular networks.

To further elucidate the roles of these EPCs, complementary experiments were performed using the mouse ovalbumin (OVA) model of allergic airway inflammation. Data obtained from this model indicated that increased mobilization of proangiogenic progenitors occurs within hours after allergen exposure. Interestingly, these progenitors homed specifically to allergen-challenged lungs and not into other organs, suggesting that a very early response to airway allergen exposure is an organ-specific mobilization and migration of proangiogenic progenitors from the bone marrow pool into the lungs. Quantification of lung microvessel density during kinetic experiments showed that the proangiogenic progenitors induced lung angiogenesis (angiogenic switch) already during the primordial stages of allergic lung disease, before the initiation of eosinophilic inflammation. Strikingly, the migration of proangiogenic progenitors to allergen-challenged lungs, as well as the resulting angiogenic switch, preceded the bulk recruitment of eosinophils. There was an exponential correlation between eosinophilic infiltration and proangiogenic progenitors arriving in the lungs. This strong temporal association between migration of proangiogenic progenitors and eosinophils lead us to the hypothesis that proangiogenic progenitors are pro-eosinophilic.

In the search for a causal relationship between the migration patterns of these two cell types, the expression of eotaxins—the main eosinophilic chemoattractants—by proangiogenic progenitors were analyzed. In mice, *eotaxin [chemokine (C-C motif) ligand 11 (CCL11)]* and *eotaxin-2 [chemokine (C-C motif) ligand 24 (CCL24)]* are expressed,[20,21] while *eotaxin-3 [chemokine (C-C motif) ligand 26 (CCL26)]* is a pseudogene. Recent findings suggest that *eotaxin (CCL11)* is the functional paralogue of human *eotaxin-3*.[22] Proangiogenic progenitors in the OVA mouse model, as well as asthma patients, showed increased levels of *eotaxin* expression (Fig. 6.5.2). Since proangiogenic progenitors are a heterogeneous group of

cells, careful characterization of the eotaxin-expressing proangiogenic progenitors was necessary, and they were immunophenotyped as angiogenic progenitors in the monocytic myeloid lineage, as indicated by the expression of hematopoietic progenitor cell antigens (CD34 and CD133 in humans and c-Kit and Sca-1 in mice) and by the expression of the myeloid markers CD45, CD33, and CD11b (Fig. 6.5.2). These cells also expressed low levels of *eotaxin-2* (unpublished observation).

During allergen challenge, eotaxin[+] proangiogenic progenitors progressively homed to the lungs.[23] The finding that proangiogenic progenitors specifically migrated to the lungs indicated selective interactions with lung vascular cells, which also appeared to be involved in triggering the release of the proangiogenic progenitor eotaxin content. Lung endothelial and smooth muscle cells from allergen-sensitized and airway allergen-exposed animals, but not from control lungs or other organs such as the liver and heart, were able to induce eotaxin release from proangiogenic progenitors *in vitro*. *In vivo*, antibody-blocking experiments further confirmed that proangiogenic progenitor-derived eotaxin is functionally active in the recruitment of eosinophils into allergen-exposed lungs. Together, these findings point toward a causal relationship between lung recruitment of proangiogenic progenitors and subsequent allergic airway inflammation (Fig. 6.5.3).

Other Angiogenic Conditions

Since proangiogenic progenitors are involved in most, if not all, postnatal angiogenesis, eosinophil recruitment at angiogenic sites would have been expected to be a common

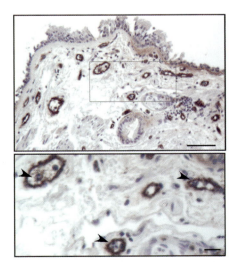
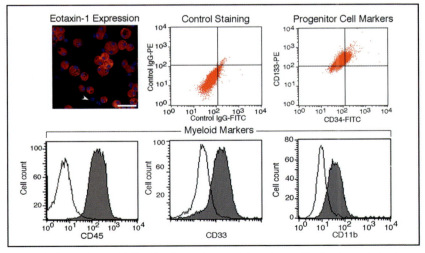

FIGURE 6.5.2   Characterization of Eotaxin[+] Proangiogenic Progenitors. *(A)* Von Willebrand factor staining of bronchial blood vessels in a human asthmatic airway. The arrows indicate inflammatory cells adherent to the endothelium. *(B)* Confocal imaging of the eotaxin content of proangiogenic progenitors. The arrowhead indicates a negative granulocyte for comparison (scale bar = 50 μm). Flow cytometry profiles depict the expression of stem cell and myeloid antigens. FITC, fluorescein isothiocyanate; IgG, immunoglobulin G. *Reprinted with permission from Asosingh et al.,* Journal of Immunology, ***vol 168,*** *pp. 6482−6494, 2007 and from Asosingh et al.,* Am J Pathology, ***vol 172,*** *pp. 615−627, 2008.*

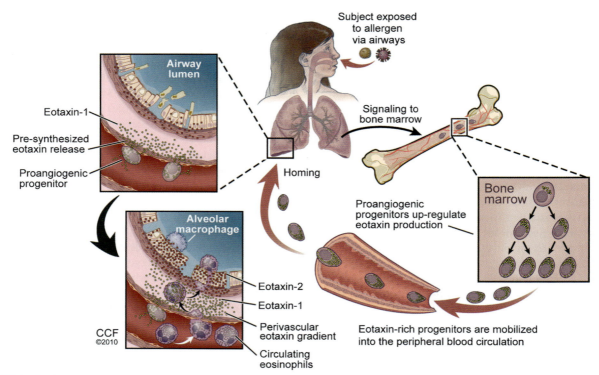

FIGURE 6.5.3 Proangiogenic progenitors and migration of eosinophils in allergic airway inflammation. Bone marrow-derived proangiogenic progenitors express constitutive low levels of eotaxin, which is upregulated by allergen sensitization. On subsequent allergen exposure these cells are rapidly mobilized into the peripheral blood circulation and recruited specifically into the lungs. Interactions with lung vascular cells induce the release of the presynthesized eotaxin by proangiogenic progenitors, which increases the perivascular gradient of eotaxin in the lung and attracts circulating eosinophils into the lung tissue. Eotaxin released by airway epithelial cells, other structural lung cells, and alveolar macrophages further attract the eosinophils into the airways. *Illustration by David Schumick, BS, CMI. Reprinted with the permission of the Cleveland Clinic Center for Medical Art & Photography © 2010. All Rights Reserved.*

observation. Direct support of this comes from tumor angiogenesis in carcinomas, gastrointestinal tumors, and Hodgkin lymphoma where increased numbers of eosinophils, among other hematopoietic cells, in the inflammatory infiltrate in growing tumor foci seems to be common.[18] Furthermore, angiogenesis during inflammation and fibrosis has also been linked to eosinophilia.[24] Eosinophils themselves are also considered to promote angiogenesis via the expression and secretion of several angiogenic factors including: VEGF, basic fibroblast growth factor (bFGF), interleukin-6 and 8, granulocyte-macrophage colony-stimulating factor (GM-CSF), platelet-derived growth factor (PDGF), transforming growth factor beta (TGF-β), and matrix metalloproteinase-9 (MMP-9).[25] These factors are stored in the secretory granules in mature eosinophils and are rapidly activated by interleukin-5 or tumor necrosis factor. Evidence for the angiogenesis, however, comes mainly from *in vitro* studies in the rat aortic ring and chorioallantoic membrane assays.[26] Nevertheless, proangiogenic, progenitor-mediated eosinophil recruitment could be an additional mechanism to fuel vascularization during specific angiogenic responses. The expression of

eotaxin by proangiogenic progenitors in tumor angiogenesis and other eosinophilic inflammatory disorders remains to be determined.

Implications

Several specialized tissue-resident cells have been reported to be involved in the recruitment of eosinophils. The bone marrow-derived proangiogenic progenitor is a newcomer in this process and is the only mobile cell type. What is the evolutionary selective advantage of such a mechanism in addition to resident cells? Perhaps because proangiogenic progenitors preloaded with eotaxin provide a mechanism of rapid eosinophilia as opposed to resident cells, which often have to synthesize eotaxins on microenvironmental stimulation. Proangiogenic progenitors circulate through the whole body, but only specific interactions between proangiogenic progenitors and activated vascular cells allows the release of presynthesized eotaxin, creating a localized increased eotaxin gradient and thus selective, site-specific recruitment of eosinophils. Vascular cells may be activated by local stimuli to upregulate proangiogenic progenitor-specific adhesion molecules and/or chemokines, though this is yet to be determined.

This novel insight that proangiogenic progenitors mediate eosinophil recruitment is in complete accordance with the new paradigm of eosinophils in health and disease: the local immunity and/or remodeling/repair (LIAR) hypothesis.[27] In this hypothesis eosinophils are postulated as regulators of LIAR under physiological and pathological conditions at sites with increased cell turnover or increased local stem/progenitor cell activities. Indeed, asthmatic airway disease is characterized by increased cell death and epithelial cell turnover, in addition to the early migration of bone marrow proangiogenic progenitors. This hypothesis also helps to explain why, in naive mice, proangiogenic progenitor-derived eotaxin fails to accumulate eosinophils in the lungs. That is, under steady-state conditions, the lack of allergen-induced expression of eosinophil chemo-attractants and survival factors are probably necessary to promote the accumulation of eosinophils in the lung. The LIAR hypothesis also raises the possibility that proangiogenic progenitors may exert immunomodulatory and remodeling/repair activities in asthma by controlling the recruitment of eosinophils.

While these findings add exciting new insights into eosinophil recruitment, several unanswered questions are raised. Eotaxin expression by resident cells is regulated by T-helper type 2 (T_h2) responses.[28,29] Whether this also holds true for proangiogenic progenitors and the specific cytokines involved remains to be examined however, it is conceivable that during allergen exposure T_h2 cells signal to the bone marrow inducing the upregulation of eotaxin by proangiogenic progenitors. Future studies also need to address whether eotaxin expression by proangiogenic progenitors is necessary and sufficient for airway eosinophilia. Specific roles have been ascribed to specific eotaxins in the migration of eosinophils to the lungs.[21,30] Eotaxin, expressed by proangiogenic progenitors, endothelial cells, and smooth muscle cells attracts eosinophils from the circulation into the lung tissue. Once in the lung tissue, the eotaxin-2 gradient generated by airway epithelial cells and alveolar macrophages further attracts the eosinophils toward the airway lumen. Recently, eotaxin-3 has also been reported to be expressed by airway epithelial cells in humans, but its exact role(s) in the eosinophil migration pathway is unknown. Thus, the intriguing question of whether proangiogenic progenitor-derived eotaxin is a universal mechanism of eosinophil recruitment or a mechanism specific for allergic lung disease remains to be determined. Understanding these processes will hopefully identify new biology-based targets and novel, more effective, therapeutic approaches.

SUMMARY

Bone marrow-derived EPCs:

- are a heterogeneous group of proangiogenic myeloid hematopoietic progenitors;
- are rare cells in the peripheral blood circulation, but have crucial paracrine effects on vascular homeostasis and new blood vessel formation;
- upregulate eotaxin expression on allergen sensitization; and
- home to allergen-challenged lungs where they secrete their presynthesized eotaxin, which recruits circulating eosinophils into the lungs.

ONGOING DISCUSSION

- How crucial is proangiogenic progenitor-derived eotaxin in the recruitment of eosinophils?
- What are the molecular mechanisms of eotaxin upregulation in proangiogenic progenitors?
- Is proangiogenic, progenitor-derived eotaxin a universal mechanism of eosinophil recruitment?

CONCLUSION

The involvement of proangiogenic progenitors in the migration of eosinophils is a novel concept highlighting close interactions between stem cell activity and eosinophilia. This view opens new perspectives in the understanding of how bone marrow progenitor cells regulate tissue homeostasis, repair, and remodeling via eosinophil recruitment. Most importantly, further exploration of this paradigm will be helpful in designing more effective clinical intervention strategies.

REFERENCES

1. Asahara T, Murohara T, Sullivan A, Silver M, van der Zee R, Li T, et al. Isolation of putative progenitor endothelial cells for angiogenesis. *Science* 1997;**275**:964—7.
2. Prater DN, Case J, Ingram DA, Yoder MC. Working hypothesis to redefine endothelial progenitor cells. *Leukemia* 2007;**21**:1141—9.
3. Yoder MC. Defining human endothelial progenitor cells. *J Thromb Haemost* 2009;**1**(Suppl. 7):49—52.
4. Urbich C, Aicher A, Heeschen C, Dernbach E, Hofmann WK, Zeiher AM, et al. Soluble factors released by endothelial progenitor cells promote migration of endothelial cells and cardiac resident progenitor cells. *J Mol Cell Cardiol* 2005;**39**:733—42.
5. Shaked Y, Bertolini F, Man S, Rogers MS, Cervi D, Foutz T, et al. Genetic heterogeneity of the vasculogenic phenotype parallels angiogenesis; Implications for cellular surrogate marker analysis of antiangiogenesis. *Cancer Cell* 2005;**7**:101—11.

6. Dimmeler S. Regulation of bone marrow-derived vascular progenitor cell mobilization and maintenance. *Arterioscler Thromb Vasc Biol* **30**:1088–93.

7. Hill JM, Zalos G, Halcox JP, Schenke WH, Waclawiw MA, Quyyumi AA, et al. Circulating endothelial progenitor cells, vascular function, and cardiovascular risk. *N Engl J Med* 2003;**348**:593–600.

8. van Beem RT, Noort WA, Voermans C, Kleijer M, ten Brinke A, et al. The presence of activated CD4(+) T cells is essential for the formation of colony-forming unit-endothelial cells by CD14(+) cells. *J Immunol* 2008;**180**:5141–8.

9. Asosingh K, Swaidani S, Aronica M, Erzurum SC. Th1- and Th2-dependent endothelial progenitor cell recruitment and angiogenic switch in asthma. *J Immunol* 2007;**178**:6482–94.

10. Asosingh K, Erzurum SC. Angioplasticity in asthma. *Biochem Soc Trans* 2009;**37**:805–10.

11. Hur J, Yoon CH, Kim HS, Choi JH, Kang HJ, Hwang KK, et al. Characterization of two types of endothelial progenitor cells and their different contributions to neovasculogenesis. *Arterioscler Thromb Vasc Biol* 2004;**24**:288–93.

12. Urbich C, Heeschen C, Aicher A, Dernbach E, Zeiher AM, Dimmeler S. Relevance of monocytic features for neovascularization capacity of circulating endothelial progenitor cells. *Circulation* 2003;**108**:2511–6.

13. Ingram DA, Mead LE, Tanaka H, Meade V, Fenoglio A, Mortell K, et al. Identification of a novel hierarchy of endothelial progenitor cells using human peripheral and umbilical cord blood. *Blood* 2004;**104**:2752–60.

14. Yoder MC, Mead LE, Prater D, Krier TR, Mroueh KN, Li F, et al. Redefining endothelial progenitor cells via clonal analysis and hematopoietic stem/progenitor cell principals. *Blood* 2007;**109**:1801–9.

15. Sieveking DP, Buckle A, Celermajer DS, Ng MK. Strikingly different angiogenic properties of endothelial progenitor cell subpopulations: insights from a novel human angiogenesis assay. *J Am Coll Cardiol* 2008;**51**:660–8.

16. Tongers J, Roncalli JG, Losordo DW. Role of endothelial progenitor cells during ischemia-induced vasculogenesis and collateral formation. *Microvasc Res* **79**:200–6.

17. Gao D, Nolan D, McDonnell K, Vahdat L, Benezra R, Altorki N, et al. Bone marrow-derived endothelial progenitor cells contribute to the angiogenic switch in tumor growth and metastatic progression. *Biochim Biophys Acta* 2009;**1796**:33–40.

18. Murdoch C, Muthana M, Coffelt SB, Lewis CE. The role of myeloid cells in the promotion of tumour angiogenesis. *Nat Rev Cancer* 2008;**8**:618–31.

19. Dunnill MS. The pathology of asthma, with special reference to changes in the bronchial mucosa. *J Clin Pathol* 1960;**13**:27–33.

20. Rothenberg ME. Eotaxin. An essential mediator of eosinophil trafficking into mucosal tissues. *Am J Respir Cell Mol Biol* 1999;**21**:291–5.

21. Pope SM, Fulkerson PC, Blanchard C, Akei HS, Nikolaidis NM, Zimmermann N, et al. Identification of a cooperative mechanism involving interleukin-13 and eotaxin-2 in experimental allergic lung inflammation. *J Biol Chem* 2005;**280**:13952–61.

22. Zuo L, Fulkerson PC, Finkelman FD, Mingler M, Fischetti CA, Blanchard C, et al. IL-13 induces esophageal remodeling and gene expression by an eosinophil-independent, IL-13R alpha 2-inhibited pathway. *J Immunol* 185:660–9.

23. Asosingh K, Hanson JD, Cheng G, Aronica MA, Erzurum SC. Allergen-induced, eotaxin-rich, proangiogenic bone marrow progenitors: a blood-borne cellular envoy for lung eosinophilia. *J Allergy Clin Immunol* **125**:918–25.

24. Ge XN, Bahaie NS, Kang BN, Hosseinkhani MR, Ha SG, Frenzel EM, et al. Allergen-induced airway remodeling is impaired in galectin-3-deficient mice. *J Immunol* **185**:1205–14.

25. Puxeddu I, Berkman N, Nissim Ben Efraim AH, Davies DE, Ribatti D, Gleich GJ, et al. The role of eosinophil major basic protein in angiogenesis. *Allergy* 2009;**64**:368–74.

26. Puxeddu I, Alian A, Piliponsky AM, Ribatti D, Panet A, Levi-Schaffer F. Human peripheral blood eosinophils induce angiogenesis. *Int J Biochem Cell Biol* 2005;**37**:628–36.

27. Lee JJ, Jacobsen EA, McGarry MP, Schleimer RP, Lee NA. Eosinophils in health and disease: the LIAR hypothesis. *Clin Exp Allergy* **40**:563–75.

28. Li L, Xia Y, Nguyen A, Lai YH, Feng L, Mosmann TR, et al. Effects of Th2 cytokines on chemokine expression in the lung: IL-13 potently induces eotaxin expression by airway epithelial cells. *J Immunol* 1999;**162**:2477–87.

29. Li L, Xia Y, Nguyen A, Feng L, Lo D. Th2-induced eotaxin expression and eosinophilia coexist with Th1 responses at the effector stage of lung inflammation. *J Immunol* 1998;**161**:3128–35.

30. Smit JJ, Lukacs NW. A closer look at chemokines and their role in asthmatic responses. *Eur J Pharmacol* 2006;**533**:277–88.

Chapter 6.6

In Vivo Assessment of Eosinophil Trafficking by Intravital Microscopy

Savita P. Rao and P. Sriramarao

INTRODUCTION

The interaction of eosinophils with the vascular endothelium in inflamed blood vessels is a critical event in the recruitment of eosinophils to sites of inflammation. Until 15 years ago, it was not possible to assess trafficking of eosinophils in real time *in vivo* due to the lack of appropriate animal model systems and imaging tools. Interpretation of cell recruitment has largely been based on the retrospective assessment of postmortem, histopathological tissue specimens or bronchial biopsies. Further, assessment of dynamic cellular interactions relevant to eosinophil trafficking has largely depended on observations from *in vitro* studies carried out with isolated eosinophils from peripheral blood or bronchoalveolar lavage fluid (BALF). As such, there has been a compelling need for the adaptation of emerging imaging tools and *in vivo* animal models to study eosinophil trafficking in a more physiological setting, especially in the context of recruitment to the gut or lung parenchyma. Exploitation of noninvasive imaging techniques such as magnetic resonance imaging and positron emission tomography, has certainly expanded our knowledge about various molecular and cellular responses in the lung, but these studies have failed to delineate the mechanism of eosinophil trafficking in the pulmonary blood stream. Intravital microscopy (IVM) is a powerful imaging tool that has facilitated the assessment of cell behavior within inflamed blood vessels under conditions of blood flow in real time. Here we outline studies conducted by us and other researchers using IVM in different tissue settings that have resulted in the current awareness of how eosinophils traffic intravascularly prior to emigration to sites of inflammation.

IN VIVO ASSESSMENT OF EOSINOPHIL TRAFFICKING

Intravital Microscopy

Advancements in the field of IVM have provided scientists with the ability to visualize the fate of leukocytes in a *dynamic and live animal* setting. The earliest IVM studies in the blood vessels of animals using bright-field illumination was carried out more than 170 years ago[1,2] and form the basis of current concepts of leukocyte trafficking and the 'multistep paradigm of leukocyte adhesion',[3] (described later in this chapter). Rapid advances and improvements in imaging and optical tools, as well as availability of more easily adaptable digital technology, have led to a surge in our knowledge of the detailed concepts of leukocyte trafficking especially in the last 15 years. Combining these tools and technologies with IVM has resulted in the expansion of our understanding of the cellular mechanisms that orchestrate the delicate balance of leukocyte interactions within inflamed blood vessels, their emigration across the endothelium, and finally their extravascular migration within inflamed tissues.

Compared to bright-field microscopy, which is limited to the visualization of total circulating leukocytes within the bloodstream, assessment of the dynamic interaction of individual leukocytes and their subsets has been facilitated by combining IVM with epifluorescence or multiphoton microscopy. Between these two methods, epifluorescence has been more widely used and established. However, the coupling of two-photon and multiphoton technology with IVM has resulted in tremendous excitement in the field of *in vivo* assessment of leukocyte migration in deep tissue.[4] Multiphoton IVM uses infrared pulsed laser excitation to generate fluorescent signals several hundreds of micrometers below the surface of solid tissues to assess cell migration via time-lapse recordings of three-dimensional tissue reconstructions.[4,5] As such, multiphoton IVM has, by

and large, been deployed to study the interaction of leukocytes (predominantly T cells and their subsets, B cells, natural killer cells, and dendritic cells) in deeper tissue settings to observe immune responses at subcellular resolution *in situ*.[4–7] In contrast to visualizing cell–cell interactions in deep tissue by multiphoton IVM, epifluorescence IVM has been used extensively to study intravascular interactions and trafficking of fluorescently labeled distinct leukocyte subpopulations in blood vessels in a two-dimensional setting. This technique typically uses a video-triggered xenon arc stroboscope with long working distance water immersion objectives for close access to vasculature in surgically exposed tissues. In this section we focus predominantly on how epifluorescence IVM (from now on referred to as IVM) has been used in multiple tissues to understand the multistep adhesion cascade and trafficking of eosinophils in the context of allergic inflammation.

The Multistep Paradigm of Eosinophil Trafficking

Eosinophils traffic from the bone marrow through the systemic blood circulation to sites such as the skin, lungs, or gastrointestinal (GI) tract during inflammation where they participate in the initiation and propagation of inflammatory responses.[8] During episodes of allergic inflammation, including asthma, recruitment of eosinophils to extravascular tissue sites, is mediated by a series of adhesive interactions between circulating eosinophils and endothelial cells within inflamed blood vessels under conditions of physiological blood flow.[9] These interactions (summarized in Fig. 6.6.1) involve discrete sequential steps, comprising leukocyte or eosinophil tethering/rolling followed by firm adhesion and transmigration, that are mediated by specific extracellular adhesion receptors, intracellular signaling molecules, as well as soluble inflammatory mediators.[10] The first step involves P-selectin glycoprotein ligand 1 (PSGL-1), L- and P-selectin-mediated tethering to the endothelium followed by α4 integrin-dependent rolling along the vessel wall. Recent studies have also identified a role for eosinophil-expressed galectin-3 in promoting rolling on vascular cell adhesion protein 1 (VCAM-1) *in vitro*[11,12] and *in vivo* (unpublished observations; Ge and colleagues). Following activation by selective chemokines [eotaxin, C-C motif chemokine 5 (CCL5/RANTES), complement component C5a], rolling eosinophils engage both α4 (α4β1, α4β7) integrins and β2 (αMβ2, αLβ2) integrins to firmly adhere to vascular endothelial cells. Finally, in the presence of chemotactic gradients, adherent eosinophils transmigrate between endothelial cells (paracellular), a process involving junctional rearrangement of the endothelium and proteolysis, to facilitate the subsequent extravasation/migration of

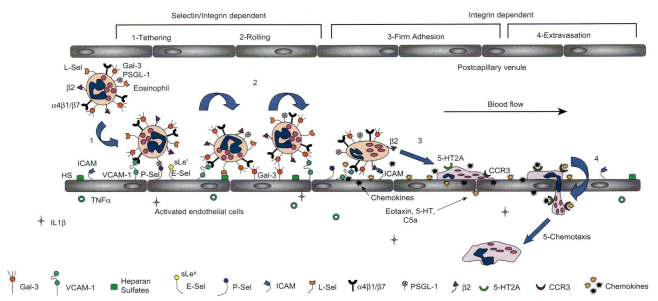

FIGURE 6.6.1 **Multistep paradigm of eosinophil trafficking.** Eosinophils initially undergo tethering to the endothelium followed by rolling in a selectin/integrin-dependent manner. This is followed by activation-dependent, integrin-mediated firm adhesion and transmigration to extravascular spaces in response to chemotactic gradients. 5-HT, 5-hydroxytryptamine; 5-HT-2A, 5-hydroxytryptamine receptor 2A; β2, β2 integrins; CCR3, C-C chemokine receptor type 3; E-Sel, E-selectin; Gal-3, galectin-3; ICAM, intercellular adhesion molecule; IL-1-beta, interleukin-1 beta; L-Sel, L-selectin; P-Sel, P-selectin; PSGL-1, P-selectin glycoprotein ligand 1; sLe^x, Sialyl-Lewis^x; TNF-α, tumor necrosis factor; VCAM-1, vascular cell adhesion protein 1.

eosinophils into tissues. Chapter 6.2 in this section of the book discusses the mechanisms of eosinophil chemotaxis, while Chapter 6.3 addresses in detail the gamut of adhesion molecules expressed by eosinophils, and Chapter 6.4 considers how eosinophil-endothelial cell interactions further modify the functions of both eosinophils and endothelial cells. In this chapter, we outline specific examples of how IVM has been used to assess individual steps of eosinophil trafficking in various vascular beds of live, anesthetized animals.

Models for the Assessment of Eosinophil Trafficking

The Rabbit Mesentery as a Model System to Evaluate Human Eosinophil Trafficking

Because of interspecies homology and conservation of function between several human and rabbit adhesion molecules,[13] the rabbit mesenteric microcirculation has been extensively used by us to evaluate human eosinophil rolling on the inflamed vascular endothelium.[14–17] This *in vivo* technique, originally described for human neutrophils,[18] enables the differentiation of fluorescently labeled interacting vs. noninteracting human eosinophils and the assessment of cellular interactions in arteries, venules, or capillaries (Fig. 6.6.2). Using function-blocking monoclonal antibodies (mAbs), studies from our laboratory have

demonstrated that L-selectin, as well as α4 integrins, can support the rolling of human eosinophils on interleukin-1 (IL-1)-stimulated postcapillary venules *in vivo*.[14] These studies not only led to the identification of α4 integrins as rolling receptors for eosinophils in addition to selectins, an observation subsequently confirmed with other cells,[19,20] but also supported the earlier observations that L-selectin may play a role during homing of eosinophils in the asthmatic lung after allergen challenge.[21] IVM studies in postcapillary venules of the rabbit mesentery also demonstrated that the functional status of α4β1 integrin determines the firm adhesion of eosinophils to the vascular endothelium and that activation of β1 integrins can rapidly induce eosinophil adhesion *in vivo* as well as resistance to detachment from VCAM-1.[17] In addition, a role for α4β7 integrin in supporting eosinophil adhesion to both VCAM-1 and mucosal addressin cell adhesion molecule 1 was also identified using this model.[17] Activation-dependent stable adhesion of eosinophils to the vascular endothelium mediated by β2 integrins was also established by IVM.[16]

The ability to evaluate the effect of specific antibodies and inhibitors or even inflammatory mediators such as cytokines and chemokines on eosinophil rolling on the venular endothelium has made this model invaluable to identify vascular adhesion receptors as well as potential targets for the blockade of human eosinophil recruitment in an *in vivo* setting. Along these lines, the inducible rolling of human eosinophils in the rabbit mesenteric circulation was

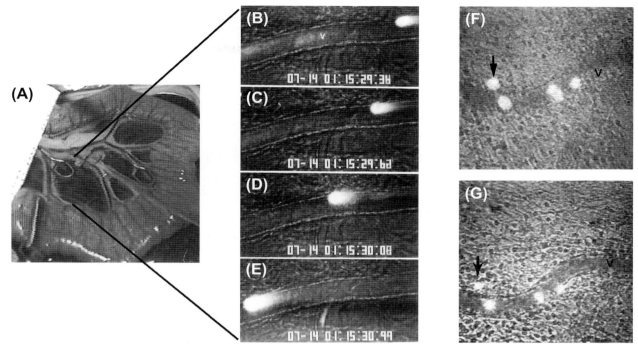

FIGURE 6.6.2   Trafficking of human eosinophils in postcapillary venules of the rabbit mesentery. *(A)* Photomicrograph of the exteriorized rabbit mesentery. Carboxyfluorescein diacetate succinimidyl ester (CFDA-SE)-labeled eosinophils were injected into the terminal mesentery artery bloodstream 6 h after interleukin-1 (IL-1) administration. Black lines indicate the area of mesenteric microcirculation observed by intravital microscopy (IVM). *(B–E)* Rolling of a single fluorescently labeled eosinophil in an IL-1-stimulated venule (blood flow from right to left).[14] *(F–G)* Emigration of adherent, CFDA-SE-labeled eosinophils in response to complement component C5a (10^{-7} M) at 5 min *(F)* and 30 min *(G)*. Black arrowheads point to individual eosinophils that migrated into the extravascular space in response to complement component C5a. V indicates a postcapillary venule.[16] *Reproduced with permission, Copyright 1994 and 1999, The American Association of Immunologists, Inc.*

significantly inhibited when blood vessels were pretreated with an anti-VCAM-1 mAb, thus establishing a role for VCAM-1 in supporting human eosinophil rolling *in vivo*.[17]

Yet another study using this model demonstrated that E-selectin is likely to function as a major vascular adhesion receptor in predominantly mediating neutrophil, but not eosinophil, rolling in inflamed postcapillary venules.[15] Using this model, we demonstrated that 5-hydroxytryptamine (5-HT/serotonin), which is elevated during asthma, not only functions as a chemotactic agent for eosinophils but also induces human eosinophil rolling on the inflamed vascular endothelium that could be prevented by blocking type 2 5-HT receptors (5-HT-2) on eosinophils.[22] In addition, the rabbit mesentery has been used to compare differences in trafficking between human eosinophils and neutrophils with respect to adhesion molecules. For instance, while L-selectin is a prerequisite for neutrophil rolling,[18] eosinophils were found to use both L-selectin and α4 integrins to roll on the venular endothelium.[14] Additionally, granulocyte-macrophage colony-stimulating factor, a cytokine generated in the lungs in response to allergen exposure, was found to exert a differential effect on human neutrophil versus eosinophil adhesive interactions, consistently inhibiting rolling of neutrophils but not

eosinophils in inflamed postcapillary venules.[23] Subsequent steps of eosinophil trafficking, i.e., stable adhesion and chemokine-induced transmigration, can also be assessed in the rabbit mesenteric microcirculation by IVM. For eosinophils, this has been accomplished by superfusion of the vascular bed with chemoattractants such as complement component C3a and C5a (Fig. 6.6.2).[16] These studies have revealed that exposure to complement component C3a induces stable adhesion of rolling eosinophils, while complement component C5a can induce both stable adhesion as well as transmigration. Overall, this xenogeneic model is currently the only method available that allows for the visualization and assessment of human eosinophil trafficking under conditions of physiological blood flow in an *in vivo* setting by IVM.

Rodent (Mouse and Rat) Models for Evaluating Eosinophil Trafficking

Most studies describing mechanisms of eosinophil trafficking have used murine models. Although there are significant differences between rodent and human eosinophils, they share several common features,[10] including the ability of rodent eosinophils to engage with the vascular endothelium as part of the multistep adhesion cascade. Accordingly, IVM studies have been carried out to evaluate eosinophil trafficking in various vascular beds of mice such as the mesentery, cremaster muscle, bone marrow, and lung in the context of inflammation, including that induced by exposure to allergens. An advantage of working with mouse eosinophils and mouse models of inflammation is the availability of a wide range of reagents such as specific antibodies, inhibitors, and recombinant proteins as well as gene-targeted mice.

Mesenteric Microcirculation

Using a combination of gene-targeted mice and specific antibodies, the mouse mesenteric microcirculation was used to evaluate eosinophil trafficking in real time by IVM in a model of peritoneal eosinophilic inflammation induced by ragweed challenge.[24] These studies demonstrated that eosinophil rolling and firm adhesion to the inflamed vascular endothelium is significantly reduced in P-selectin-deficient mice and that P-selectin, intercellular adhesion molecule 1 (ICAM-1), and VCAM-1 are important for the peritoneal recruitment of eosinophils after ragweed challenge. Importantly, IVM studies such as this enable mechanistic validation at a cellular level for the reduced pulmonary eosinophilic inflammation observed in mice deficient for P-selectin or ICAM-1[25] and support the finding that selectin interactions drive allergic inflammation in the lung and skin.[26,27]

The critical role of proinflammatory cytokines that are expressed during allergic inflammation, such as IL-1 and tumor necrosis factor (TNF-α),[28] in inducing endothelial adhesiveness and thus promoting eosinophil rolling and firm adhesion contributing to overall recruitment of circulating eosinophils has been established using IVM and interleukin-1 receptor type 1 (IL-1R-1), tumor necrosis factor (TNF) receptor 1 p55/p75 and TNF receptor p55-deficient mice.[29,30] These latter studies underscore an important aspect of the potential effects of elevated levels of these cytokines in the lungs during allergic airway inflammation and asthma. The mouse mesentery has also been used to examine the contribution of signaling molecules to leukocyte trafficking in the context of allergic inflammation.[31] Using mice deficient for Gα_i, a subunit of the heterotrimeric G-protein complex, IVM studies have demonstrated a significant role for Gα_i in leukocyte diapedesis. Leukocytes deficient in Gα_i exhibited significantly increased adhesion to Gα_i-deficient vascular endothelium compared to wild-type endothelium leukocytes (Fig. 6.6.3), resulting in the inhibition of leukocyte extravasation and contributing to decreased recruitment of inflammatory leukocytes to the lung during allergic inflammation.

In addition to the mouse mesentery, IVM studies in rat mesenteric microcirculation have been used to establish

FIGURE 6.6.3 Intravital microscopy of leukocyte trafficking in the postcapillary venules of LPS-exposed mouse mesentery showing that loss of Gα_{i2} results in a significant increase in the number of stationary leukocytes adherent (stable adhesion) to the vascular endothelium. The numbered white arrows indicate individual rolling leukocytes and the black arrows identify stationary leukocytes.[31] LPS, lipopolysaccharide. *Reproduced with permission, Copyright 2007, National Academy of Sciences, USA.*

that eotaxin-induced migration of eosinophils *in vivo* is dependent on the α4/VCAM-1 adhesion pathway[32] and to demonstrate the feasibility of an α4β1/α4β7 dual antagonist, TR14035, as an inhibitor of airway leukocyte recruitment and hyperresponsiveness in a rat model of allergic asthma based on its ability to significantly inhibit leukocyte rolling flux, adhesion, and emigration *in vivo*.[33] Further, using this model, the distinct roles of P- and L-selectin in endotoxin-induced leukocyte infiltration were identified; while P-selectin plays a role in lipopolysaccharide (LPS)-induced rolling and is essential for LPS-induced leukocyte adhesion, L-selectin participates in LPS-induced rolling, but not in adhesion.[34]

Cremaster Muscle Microcirculation

The organized vascular network of the cremaster muscle is another alternative for the assessment of cell trafficking *in vivo* under flow conditions by IVM. This model has been widely used to evaluate trafficking of leukocytes, particularly neutrophils,[35–37] and to some extent lymphocytes.[38,39] These studies have led to the understanding of important aspects of leukocyte trafficking including slow rolling, intraluminal crawling, and pathways of transmigration.[40] IVM studies in the cremaster muscle of mice deficient for P-selectin identified a regulatory role for P-selectin in interleukin-13-induced leukocyte transmigration that was tissue specific.[41] Another study using this model demonstrated a critical role for α4 integrins during interleukin-4 (IL-4)-induced leukocyte recruitment (eosinophils and mononuclear cells) due to its ability to initiate leukocyte-endothelial cell interactions even in the absence of selectins under shear conditions *in vivo*.[42] While

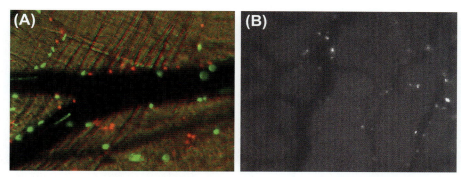

FIGURE 6.6.4 **Intravital microscopy of cell trafficking in different mouse tissues.** *(A)* Interaction of mouse eosinophils within microvessels of the cremaster muscle. Two different mouse eosinophil populations (wild-type and gene-deficient, for example) were fluorescently labeled with carboxy-fluorescein succinimidyl ester (CFSE, green) or 5-(and-6)-(((4-chloromethyl)benzoyl)amino)tetramethylrhodamine (orange) and infused simultaneously by intravenous injection. Eosinophil trafficking was evaluated by IVM. *(B)* Interaction of CFSE-labeled bone marrow cells from chronic allergen-challenged mice within bone marrow microcirculation of the skull by IVM.

there are no published reports of trafficking of eosinophils *per se* in cremaster muscle microvessels, with our capability to culture functionally competent eosinophils from mouse bone marrow using recently described methods,[43] we have demonstrated the feasibility of evaluating rolling and adhesion of eosinophils within the cremaster muscle vascular network by IVM (Fig. 6.6.4A). Furthermore, using dual-color IVM, the interaction of differing eosinophil populations [control vs. treated (e.g., inflammatory mediators), gene-deleted vs. wild-type, etc.] tagged with different fluorescent dyes with the vascular endothelium can be evaluated simultaneously to identify the effect of a specific treatment or condition on trafficking *in vivo*.

Bone Marrow Microcirculation

Another vascular bed important in the development of inflammation at the level of initiation as well as ongoing maintenance of cellular inflammation is that of the bone marrow. Elevated T-helper type 2 (T_h2) cytokines, particularly interleukin-5, associated with exposure to allergens or parasitic infections promote progenitor differentiation leading to increased production of mature eosinophils that are subsequently recruited to sites of inflammation.[44] A novel technique was developed by Mazo and colleagues to visualize the interactions of hematopoietic progenitor cells (HPCs) with bone marrow microvessels in the frontoparietal skulls of mice, which indicated that rolling of HPCs was mediated by α4-VCAM-1 interactions with partial involvement of E- and P-selectin, but did not involve L-selectin.[45]

Subsequently, this model has been used in our laboratory to evaluate the interaction of bone marrow leukocytes (containing both mature and progenitor cells) from allergen vs. non-allergen-exposed mice with the bone marrow microvasculature of naive mice to understand the effect of allergen challenge on the trafficking of these cells (Fig. 6.6.4B).[46] This model can be used to evaluate

trafficking of isolated eosinophils (including progenitor cells) as well as other relevant cell types within the bone marrow microvasculature of naive mice or mice that have been manipulated (allergen-exposed, genetically modified, treated with specific agents, etc.) to further understand cellular trafficking within the hematopoietic compartment at a molecular level.

Lung Microcirculation

Eosinophils constitute the major inflammatory cells recruited to the lungs during allergic airway inflammation and asthma. Thus, identifying how these cells traffic within the dense pulmonary vascular network during airway recruitment is vital to the understanding of the pathophysiology of allergic asthma. However, assessing and imaging the trafficking of leukocytes within the lungs of live mice by IVM presents a significant challenge due to the anatomical location of this organ in the thoracic rib cage, which is constantly moving. To overcome these challenges, we developed a mouse model to evaluate leukocyte trafficking under conditions of continuous blood flow within a network of microvessels comprising arterioles, capillaries, and postcapillary venules of implanted lung allografts placed in dorsal skinfold chambers in mice (Fig. 6.6.5, left panel).[47] A unique feature of this model is the ability to evaluate the effect of inflammatory mediators such as cytokines and chemokines on leukocyte-endothelial interactions in lung microvessels.

Studies using this model demonstrated that leukocyte rolling and adhesion is induced in response to cytokine (TNF-α) stimulation and occurs in both arterioles and postcapillary venules of the lung microvascular network;[47] that is, as opposed to the systemic circulation where rolling is observed predominantly in postcapillary venules and less frequently in arterioles.[24] Likewise, superfusion of implanted lung allografts with an inflammatory agent such as

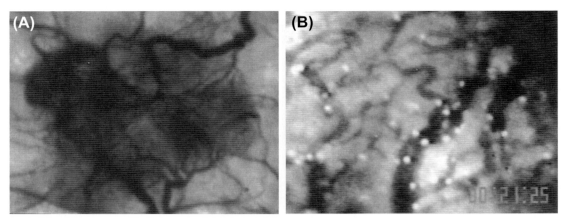

FIGURE 6.6.5   Intravital microscopy of cell trafficking in lung microcirculation. *(A)* Photomicrograph of a completely revascularized lung allograft in the dorsal skinfold chamber of a nude mouse that has established connections with the recipient cutaneous vessels all around the circumference of the allograft. (Pending permission.) *(B)* Interaction of carboxyfluorescein succinimidyl ester (CFSE)-labeled murine eosinophils within tumor necrosis factor (TNF-α)-stimulated microvessels of a revascularized lung allograft.

nicotine significantly enhanced rolling and adhesion of leukocytes within lung microvessels in a selectin-dependent manner.[48] Another advantage of this model is the ability to assess the contribution of the lung vascular endothelium to leukocyte trafficking by implanting lung tissue from mice that have been manipulated, i.e., allergen-challenged, gene-targeted, or even in the context of chronic obstructive pulmonary disease. For example, lung allografts of mice exposed to chronic cigarette smoke demonstrate increased intravascular neutrophil rolling and adhesion compared to the lungs of mice exposed to fresh air in this model.[49]

In the context of allergic inflammation, using lung allografts from allergen-challenged mice deficient in endothelial and leukocyte *N*-deacetylase/*N*-sulfotransferase-1, a role was ascribed to endothelial-expressed, but not leukocyte-expressed, heparan sulfates in regulating the airway recruitment of inflammatory cells,[50] most likely through L-selectin binding as previously described for neutrophils.[51] In addition to circulating leukocytes or infused, fluorescently labeled neutrophils, trafficking of mouse eosinophils within TNF-α-stimulated microvessels of the lung vasculature can be assessed by IVM using peripheral blood- or bone marrow-derived mouse eosinophils (Fig. 6.6.5, right panel). This represents a significant advancement in the capability to evaluate the outcome of eosinophil-endothelial interactions within the lung microcirculation that hitherto has not been feasible.

Why Assess Trafficking of Eosinophils *In Vivo*?

The need to study trafficking of isolated eosinophils rather than total leukocytes under conditions of physiological flow in an *in vivo* setting is in part based on the cell type-specific requirement of adhesion receptors for interaction with the

vascular endothelium. For example, in mice deficient for both E- and P-selectin, while a 98% inhibition of leukocyte rolling in response to IL-4 stimulation was observed, a clearly discernible population of leukocytes continued to roll in postcapillary venules of the cremaster muscle microcirculation and are most likely to be eosinophils since eosinophil rolling is dependent on the engagement of α4 integrins. This could in part explain the similar extent of tissue eosinophil recruitment in both wild-type and E- and P-selectin double knockout mice exposed to IL-4.[42] Moreover, while *in vitro* studies provide valuable information, validation of *in vitro* inferences in an *in vivo* setting is crucial. For example, while *in vitro* static adhesion assays indicated that E-selectin can support eosinophil adhesion,[52] *in vivo* IVM studies demonstrated that human eosinophils do not preferentially use E-selectin as an adhesion molecule.[15] Similarly, complement component C3a, which was found to act as a potent and selective chemoattractant for eosinophils in *in vitro* chemotaxis studies, induced stable adhesion but not transmigration of eosinophils across inflamed blood vessels *in vivo* by IVM.[16] Overall, working with isolated human and mouse eosinophils under conditions of physiological blood flow in different tissues *in vivo* by IVM will further increase our understanding of eosinophil trafficking in various disease states.

CONCLUSION

Access to new molecular and genetic tools coupled with improved imaging techniques are empowering researchers to address unresolved and important questions relevant to inflammation and immunity in living animals. Using IVM, researchers are able to dissect leukocyte movement and interaction within tissues at a cellular and molecular level. IVM studies have demonstrated that eosinophils both share and distinguish themselves from other leukocytes in their ability to interact within inflamed blood vessels. In addition to assessing human eosinophil trafficking in the rabbit mesentery, IVM has been used to evaluate the trafficking of mouse eosinophils in multiple vascular beds such as the mesentery, cremaster, and bone marrow, as well as more complicated and difficult to access intrathoracic organs such as the lung, albeit as explants grafted onto skin chambers implanted in mice. Overall, IVM studies are delineating new paradigms of eosinophil trafficking hitherto restricted to the realm of *in vitro* experimentation or postmortem samples and tissue analysis. Over the next few years, it is anticipated that with access to more sophisticated imaging tools including the ability to specifically tag eosinophils *in vivo*, IVM will be able to identify the close interplay between cell surface receptors, the cytoskeleton, and intracellular signaling mechanisms that regulate eosinophil-endothelial interactions and the fate of eosinophils once they migrate into extravascular tissue spaces. Further, with the advent of multiphoton IVM, which has been used extensively to examine lymphocyte trafficking, it is envisioned that adaption of such technology to an allergic setting will facilitate the rapid understanding of how eosinophils potentially interact with other cells extravascularly within tissues such as the nasal mucosa, trachea, lungs, and GI tract, including the esophagus and the gut.

REFERENCES

1. Wagner R. *Erlauterungstafeln zur Physiologie und Entwicklungsgeschichte*. Leipzig: Leopold Voss; 1839.
2. Cohnheim J. Lectures on general pahology, *A Handbook for Practioners and Students*. London: The New Sydenham Society; 1889.
3. Butcher EC. Leukocyte-endothelial cell recognition: three (or more) steps to specificity and diversity. *Cell* 1991;**67**:1033–6.
4. Mempel TR, Scimone ML, Mora JR, von Andrian UH. *In vivo* imaging of leukocyte trafficking in blood vessels and tissues. *Curr Opin Immunol* 2004;**16**:406–17.
5. Cahalan MD, Parker I, Wei SH, Miller MJ. Two-photon tissue imaging: seeing the immune system in a fresh light. *Nat Rev Immunol* 2002;**2**:872–80.
6. Sumen C, Mempel TR, Mazo IB, von Andrian UH. Intravital microscopy: visualizing immunity in context. *Immunity* 2004;**21**:315–29.
7. Huppa JB, Davis MM. T-cell-antigen recognition and the immunological synapse. *Nat Rev Immunol* 2003;**3**:973–83.
8. Simon D, Wardlaw A, Rothenberg ME. Organ-specific eosinophilic disorders of the skin, lung, and gastrointestinal tract. *J Allergy Clin Immunol* 2010;**126**:3–13.
9. Broide DH, Sriramarao P. Eosinophil trafficking to sites of allergic inflammation. *Immunol Reviews* 2001;**179**:163–72.
10. Rosenberg HF, Phipps S, Foster PS. Eosinophil trafficking in allergy and asthma. *J Allergy Clin Immunol* 2007;**119**:1303–10.
11. Rao SP, Wang Z, Zuberi RI, Sikora L, Bahaie NS, Zuraw BL, et al. Galectin-3 functions as an adhesion molecule to support eosinophil rolling and adhesion under conditions of flow. *J Immunol* 2007;**179**: 7800–7.
12. Ge XN, Bahaie NS, Kang BN, Hosseinkhani RM, Ha SG, Frenzel EM, et al. Allergen-induced airway remodeling is impaired in galectin-3 deficient mice. *J. Immunol* 2010;**185**:1205–14.
13. Bevilacqua MP, Nelson RD. Selectins. *J Clin Invest* 1993;**91**: 379–87.
14. Sriramarao P, von Andrian UH, Butcher EC, Bourdon MA, Broide DH. L-selectin and very late antigen-4 integrin promote eosinophil rolling at physiological shear rates *in vivo*. *J Immunol* 1994;**53**:4238–46.
15. Sriramarao P, Norton CR, Borgström P, DiScipio RG, Wolitzky BA, Broide DH. E-selectin preferentially supports neutrophil but not eosinophil rolling under conditions of flow *in vitro* and *in vivo*. *J Immunol* 1996;**157**:4672–80.
16. DiScipio RG, Daffern PJ, Jagels MA, Broide DH, Sriramarao P. A comparison of C3a and C5a mediated stable adhesion of rolling eosinophil in postcapillary venules and transendothelial migration in vitro and in vivo. *J Immunol* 1999;**162**:1127–36.
17. Sriramarao P, DiScipio RG, Cobb RR, Cybulsky M, Stachnick G, Castenada D, et al. VCAM-1 is more effective than MAdCAM-1 in supporting eosinophil rolling under conditions of flow. *Blood* 2000;**95**:592–601.
18. von Andrian UH, Hansell P, Chambers JD, Berger EM, Torres Filho I, et al. L-selectin function is required for β2 integrin-mediated neutrophil adhesion at physiological shear rates *in vivo*. *Am J Physiol* 1992;**263**:H1034–44.
19. Alon R, Kassner PD, Carr MW, Finger EB, Hemler ME, Springer TA. The integrin VLA-4 supports tethering and rolling in flow on VCAM-1. *J Cell Biol* 1995;**128**:1243–53.
20. Berlin C, Bargatze RF, Campbell JJ, von Andrian UH, Szabo MC, et al. α4 integrins mediate lymphocyte attachment and rolling under physiologic flow. *Cell* 1995;**80**:413–22.
21. Mengelers HJ, Maikoe T, Hooibrink B, Kuypers TW, Kreukniet J, Lammers J, et al. Down modulation of L-selectin on eosinophils recovered from bronchoalveolar lavage fluid after allergen provocation. *Clin Exp Allergy* 1993;**23**:196–204.
22. Boehme SA, Lio FM, Sikora L, Pandit TS, Lavrador K, Rao SP, et al. Cutting edge: serotonin is a chemotactic factor for eosinophils and functions additively with eotaxin. *J Immunol* 2004;**173**:3599–603.
23. Sheikh Bahaie N, Rao SP, Massoud A, Sriramarao P. GM-CSF differentially regulates eosinophil and neutrophil adhesive interactions with vascular endothelium in vivo. *Iran J Allergy Asthma Immunol* 2010;**9**:207–17.
24. Broide DH, Humber D, Sriramarao P. Inhibition of eosinophil rolling and recruitment in P-selectin- and intracellular adhesion molecule-1-deficient mice. *Blood* 1998;**91**:2847–56.
25. Broide DH, Sullivan S, Gifford T, Sriramarao P. Inhibition of pulmonary eosinophilia in P-selectin and ICAM-1 deficient mice. *Am J Respir Cell Mol Biol* 1998;**18**:218–25.
26. Jiao A, Fish SC, Mason LE, Schelling SH, Goldman SJ, Williams CM. A role for endothelial selectins in allergic and nonallergic inflammatory disease. *Ann Allergy Asthma Immunol* 2007;**98**:83–8.
27. Teixeira MM, Hellewell PG. Contribution of endothelial selectins and alpha 4 integrins to eosinophil trafficking in allergic and

nonallergic inflammatory reactions in skin. *J Immunol* 1998;**161**: 2516–23.

28. Thomas PS. Tumor necrosis factor-alpha: the role of this multifunctional cytokine in asthma. *Immunol Cell Biol* 2001;**79**:132–40.

29. Broide D, Campbell K, Sriramarao P. Inhibition of eosinophilic inflammation in allergen challenged IL-1 receptor type I deficient mice is associated with reduced eosinophil rolling and adhesion on vascular endothelium. *Blood* 2000;**95**:263–9.

30. Broide DH, Stachnick G, Castaneda D, Nayar J, Sriramarao P. Inhibition of eosinophilic inflammation in allergen-challenged TNF receptor p55/p75- and TNF receptor p55-deficient mice. *Am J Respir Cell Mol Biol* 2001;**24**:304–11.

31. Pero RS, Borchers MT, Spicher K, Ochkur SI, Sikora L, Rao SP, et al. Galpha(i2)-mediated signaling events in the endothelium are involved in controlling leukocyte extravasation. *Proc Natl Acad Sci U S A* 2007;**104**:4371–6.

32. Nagai K, Larkin S, Hartnell A, Larbi K, Razi Aghakhani M, Windley C, et al. Human eotaxin induces eosinophil extravasation through rat mesenteric venules: role of alpha4 integrins and vascular cell adhesion molecule-1. *Immunology* 1999;**96**:176–83.

33. Cortijo J, Sanz MJ, Iranzo A, Montesinos JL, Nabah YN, Alfón J, et al. A small molecule, orally active, alpha4beta1/alpha4beta7 dual antagonist reduces leukocyte infiltration and airway hyperresponsiveness in an experimental model of allergic asthma in Brown Norway rats. *Br J Pharmacol* 2006;**147**:661–70.

34. Davenpeck KL, Steeber DA, Tedder TF, Bochner BS. P- and L-selectin mediate distinct but overlapping functions in endotoxininduced leukocyte-endothelial interactions in the rat mesenteric microcirculation. *J Immunol* 1997;**159**:1977–86.

35. Ferri LE, Swartz D, Christou NV. Soluble L-selectin at levels present in septic patients diminishes leukocyte-endothelial cell interactions in mice in vivo: a mechanism for decreased leukocyte delivery to remote sites in sepsis. *Crit Care Med* 2001;**29**.

36. Smith ML, Olson TS, Ley K. CXCR2- and E-selectin-induced neutrophil arrest during inflammation in vivo. *J Exp Med* 2004;**200**: 935–9.

37. Zarbock A, Ley K. Neutrophil adhesion and activation under flow. *Microcirculation* 2009;**16**:31–42.

38. Singbartl K, Thatte J, Smith ML, Wethmar K, Day K, Ley K. A CD2-green fluorescence protein-transgenic mouse reveals very late antigen-4-dependent CD8+ lymphocyte rolling in inflamed venules. *J Immunol* 2001;**166**:7520–6.

39. Yakubenia S, Frommhold D, Schölch D, Hellbusch CC, Körner C, Petri B, et al. Leukocyte trafficking in a mouse model for leukocyte adhesion deficiency II/congenital disorder of glycosylation IIc. *Blood* 2008;**112**:1472–81.

40. Zarbock A, Ley K. New insights into leukocyte recruitment by intravital microscopy. *Curr Top Microbiol Immunol* 2009;**334**:129–52.

41. Larbi KY, Dangerfield JP, Culley FJ, Marshall D, Haskard DO, Jose PJ, et al. P-selectin mediates IL-13-induced eosinophil transmigration but not eotaxin generation in vivo: a comparative study with IL-4-elicited responses. *J Leukoc Biol* 2003;**73**: 67–73.

42. Hickey MJ, Granger DN, Kubes P. Molecular mechanisms underlying IL-4-induced leukocyte recruitment in vivo: a critical role for the alpha 4 integrin. *J Immunol* 1999;**163**:3441–8.

43. Dyer KD, Moser JM, Czapiga M, Siegel SJ, Percopo CM, Rosenberg HF. Functionally competent eosinophils differentiated ex vivo in high purity from normal mouse bone marrow. *J Immunol* 2008;**181**:4004–9.

44. Gauvreau GM, Ellis AK, Denburg JA. Haemopoietic processes in allergic disease: eosinophil/basophil development. *Clin Exp Allergy* 2009;**39**:1297–306.

45. Mazo IB, Gutierrez-Ramos JC, Frenette PS, Hynes RO, Wagner DD, von Andrian UH. Hematopoietic progenitor cell rolling in bone marrow microvessels: parallel contributions by endothelial selectins and vascular cell adhesion molecule 1. *J Exp Med* 1998;**188**:465–74 [published erratum appears in *J Exp Med* 1998 Sep 1997; **1188**(1995):1001].

46. Pandit TS, Hosseinkhan RM, Kang BN, Bahaie NS, Ge XN, Rao SP, et al. Chronic allergen challenge induces pulmonary extramedullary hematopoiesis. *Exp Lung Res* 2011;**37**:279–90.

47. Sikora L, Johansson ACM, Rao SP, Hughes GK, Broide DH, Sriramarao P. A murine model to study leukocyte rolling and intravascular trafficking in lung microvessels. *Am J Pathol* 2003;**162**:2019–28.

48. Sikora L, Rao SP, Sriramarao P. Selectin-dependent rolling and adhesion of leukocytes in nicotine-exposed microvessels of lung allografts. *Am J Physiol Lung Cell Mol Biol* 2003;**285**: L654–63.

49. Rao SP, Sikora L, Hosseinkhani MR, Pinkerton KE, Sriramarao P. Exposure to environmental tobacco smoke induces angiogenesis and leukocyte trafficking in lung microvessels. *Exp Lung Res* 2009;**35**:119–35.

50. Zuberi RI, Ge X, Jiang S, Bahaie NS, Kang BN, Hosseinkhani RM, et al. Deficiency of endothelial heparan sulfates attenuates allergic airway inflammation. *J Immunol* 2009;**183**:3971–9.

51. Wang L, Fuster MM, Sriramarao P, Esko JD. Endothelial heparan sulfate deficiency impairs L-selectin- and chemokine-mediated neutrophil trafficking during inflammatory responses. *Nat Immunol* 2005;**6**:902–10.

52. Bochner BS, Luscinskas FW, Gimbrone Jr MA, Newman W, Sterbinsky SA, Derse-Anth C, et al. Adhesion of human basophils, eosinophils and neutrophils to IL-1 activated human vascular endothelial cells: contributions of endothelial cell adhesion molecules. *J Exp Med* 1991;**173**:1553–7.

Eosinophil Signal Transduction

Introduction

Rafeul Alam and Francesca Levi-Schaffer

EARLY CHALLENGES WITH EOSINOPHIL SIGNALING STUDIES

Signal transduction studies, unlike many other cellular studies, require pure cell populations. Since eosinophils represent less than 5% of the peripheral blood cell population, signaling studies with this cell type were exceptionally challenging. In the late 1970s and early 1980s, multiple gradients based on density centrifugation with Ficoll-Hypaque or Percoll were the primary means of eosinophil purification. The introduction of antibody-bound, magnetic bead-based purification methods was an important step that allowed the high-quality purification of eosinophils and accelerated progress in this field.

Signaling studies with eosinophils remain a challenge because of other limiting factors. One major limiting factor is the short half-life and relative metabolic quiescence of eosinophils. This severely limits genetic manipulations (gene overexpression, mutation, knockdown with small hairpin RNA, and so on) of signaling pathways. There are cell lines that mimic many cellular and functional aspects of eosinophils. They are certainly useful in some initial screening studies. Given that they are not terminally differentiated cells, conclusions drawn from these cell lines require confirmation with freshly isolated eosinophils.

There are problems in interpreting the results from signaling studies. Cells isolated from organs can be at different states of activation. Another drawback is that most of the signaling studies on eosinophils are performed on peripheral blood cells that might differ from those isolated from one or more peripheral tissues. Most investigators initially focus on optimizing the isolation and culture conditions to render quiescence. Although this is an important step that ensures that the signaling results are easily interpretable, it raises concerns about the fidelity of the results. This problem is overcome by demonstrating the biological relevance of the study conclusion through *in vivo* experiments. Finally, signaling studies are typically conducted in a reductionist model. This allows a clear interpretation of the study results. However, the reductionist models may not represent physiological conditions. For this reason, it is imperative that the results from reductionist studies are confirmed in more integrated, holistic models.

EARLY SIGNALING STUDIES

Signal transduction studies are aimed at elucidating biochemical processes that generate and transduce signals from the receptor to various internal compartments to elicit a cellular response. Some early signaling studies examined the role of calcium[1] and cyclic adenosine monophosphate (cAMP)[2] signaling in eosinophils. These studies confirmed an important role of calcium signaling in eosinophil degranulation and superoxide generation. Subsequent studies linked calcium mobilization to inositol triphosphate and the relevance of lipid mediators in eosinophil signaling.[3] In contrast to calcium signaling, cAMP inhibited immunoglobulin A-induced eosinophil degranulation. The cross talk and cross-regulation between calcium signaling and the cAMP pathway remains an important research subject.

Later research in eosinophil signaling was dominated by the recognition of the importance of kinases and phosphatases in signal generation and propagation.[4] It was recognized that some of the tyrosine kinases, especially those belonging to the Janus and Src families, were rapidly activated following the stimulation of cells by cytokines and other inflammatory mediators.[5,6] It is important to note that most receptors do not have intrinsic enzymatic activity, so they rely on associated kinases or other enzymes to trigger a biochemical reaction. Kinases that belong to the Src and Janus families are associated with many receptors and are among the first to be activated following receptor engagement. Most kinases are basally inactive and require an activation step. For Src kinases this activation involves a stepwise dephosphosphorylation of the C-terminus phosphotyrosine and dissociation of the linker region from the Src homology 3 (SH3) domain. The transmembrane phosphatase CD45 and certain SH3 ligands, such as protein unc-119 homolog A,[7] play a critical role in the activation of Src family kinases.

Eosinophils in Health and Disease. http://dx.doi.org/10.1016/B978-0-12-394385-9.00007-9

CURRENT STATE OF THE ART

The thrust of signaling research addresses the mechanism of eosinophil activation by common activating agents. These include chemokines, granulocyte-macrophage colony-stimulating factor (GM-CSF), interleukin-3 (IL-3) and 5 (IL-5), and other chemotactic agents (e.g., platelet-activating factor, fMet-Leu-Phe, and complement component C5a), immobilized immunoglobulins, integrin ligands, and inhibitory receptors. Some of the major players of eosinophil activation include phosphatidylinositol 3-kinase (PI3K), extracellular signal-regulated kinase 1 and 2 (ERK1, ERK2), and p38 mitogen-activated protein kinase (p38 MAPK), nuclear factor kappa-light-chain-enhancer of activated B cells, and the tyrosine-protein kinase JAK (JAK)—signal transducer and activator of transcription signaling pathway.[8] Signals generated by growth factor receptors, cytokine receptors, and G protein-coupled receptors seem to converge on one of the foregoing signaling pathways. There are limited studies on the role of the notch and beta-catenin signaling pathways in eosinophils and one would like to see more.

MECHANISM OF EOSINOPHIL PRIMING

Cellular priming has been observed in many disease conditions. Primed cells function like memory T cells. These cells have a lower threshold for activation. They may manifest features of partial activation. Hematopoietic growth factors, such as GM-CSF, IL-3, and IL-5, prime eosinophils for a multitude of functions. Following exposure to these agents eosinophils change their physical properties and become hypodense. Eosinophil recovered from the airways of patients with asthma display the characteristics of primed and hypodense cells. How the priming signal differs from the activation signal is unclear. Dr. Leo Koenderman elucidates the signaling mechanism of eosinophil priming and its relevance in Chapter 7.2.

EOSINOPHIL INHIBITORY RECEPTORS

Cells of the innate immune system typically express multiple inhibitory receptors. This system may have evolved to prevent overreaction to innocuous environmental encounters and to maintain a balance between activating and inhibitory signals. As a member of the innate immune system, the eosinophil expresses many inhibitory receptors including Fc-gamma RII-b, leukocyte immunoglobulin-like receptor subfamily B member 3/immunoglobulin-like transcript 5, CD33, sialic acid-binding Ig-like lectin 7 (Siglec-7/p75/adhesion inhibitory receptor molecule 1), Siglec-8, Siglec-10, killer cell immunoglobulin-like receptor 3DL2/p140, and CD300a (inhibitory receptor protein 60).[9] We have a better understanding of the role of inhibitory receptors on natural killer cells and macrophages. We know less about the function of these receptors on eosinophils. Dr. Ariel Munitz examines the expression and the signaling mechanism of these inhibitory receptors and their possible use as target of antiallergic therapy in Chapter 7.3.

MECHANISM OF EOSINOPHIL APOPTOSIS

Shortly after terminal differentiation, eosinophils die by apoptosis. Apoptotic cell death is particularly important for eosinophils since they contain toxic granular proteins. How the cells are programmed to die shortly after differentiation is a fascinating topic. In Chapter 7.4 Dr. Utibe Bickham and Dr. James Malter elucidate the mechanism of spontaneous eosinophil apoptosis and its prevention by pro-survival cytokines.[10] In the absence of cytokines or under conditions of low expression of peptidyl-prolyl cis-trans isomerase NIMA-interacting 1 (Pin1), the proapoptotic molecule apoptosis regulator BAX (Bax) is cleaved at the N-terminus by calpain-like proteases. This allows the translocation of Bax to the mitochondria, which causes cytochrome c release, activation of caspase-3 and -9, and apoptotic cell death. In response to cytokine signaling, Pin1 is activated while Bax is phosphorylated at threonine 167 (Thr167) by ERK1/2. This phosphorylation enhances Pin1 binding to, and the likely isomerization of, the phospho-Thr167-proline 168 peptide bond. This isomerization constrains Bax in an inactive conformation that is resistant to calpain-mediated cleavage and translocation to the mitochondria.

MECHANISM OF EOSINOPHIL DIFFERENTIATION

Eosinophil differentiation from myeloid progenitor cells remains a fascinating subject. Early studies focused on the role of transcription factors in determining the fate of myeloid cells. Much less is known about the signal transduction pathways leading to eosinophil differentiation. In Chapter 7.5, Dr. Miranda Buitenhuis examines the signaling pathways that regulate the differentiation-associated transcription factors in eosinophils. Signal transduction molecules involved in the development of eosinophils include PI3K and its downstream effector protein kinase B (PKB/Akt), p38MAPK, and the interleukin-5 receptor subunit alpha-associated proteins JAK2, tyrosine-protein kinase Lyn, and syntenin. Interestingly, two ubiquitously regulated signal transduction pathways, the PI3K/PKB pathway and the p38MAPK pathway, seem to counteract each other in terms of eosinophil differentiation, even though both are activated by all cytokines, suggesting that changes in the balance of activating and inhibitory signals induced by cytokines determines whether or not a progenitor cell becomes a mature eosinophil.

MECHANISM OF EOSINOPHIL DEGRANULATION

One of the important aspects of eosinophil function is its degranulation. Eosinophils are highly granulated cells. The process of degranulation is a complex event, requiring tightly controlled intracellular signaling pathways allowing the release of granules and vesicles from cells. Despite its important role in eosinophil biology, little is known regarding the precise molecular and intracellular mechanisms that regulate degranulation and, ultimately, the secretion of eosinophil-derived granule products. Eosinophil degranulation can occur through classic exocytosis, piecemeal degranulation, and cytolysis. The signaling mechanism for these processes is likely to be different. In Chapter 7.6, Dr. Paige Lacy and Dr. Redwan Moqbel examine the mechanism of granulogenesis, the types of degranulation, and the involvement of GTPases, Ca^{2+}-dependent kinases, SNAP soluble NSF attachment protein (SNAP) REceptor (SNARE) isoforms, and Sec/Munc proteins in eosinophil degranulation.

EOSINOPHILS AND INNATE IMMUNE RECEPTORS

Eosinophils are cells of the innate immune system that can respond to pathogen-associated molecular patterns and damage-associated molecular patterns. They are activated by many pattern and danger recognition receptors. Examples of some pattern recognition receptors include toll-like receptors, nucleotide oligomerization domain (NOD)-like receptors, RIG-like receptors, C-type lectin receptors, and the advanced glycosylation end product-specific receptor. In Chapter 7.7, Dr. Anne Månsson Kvarnhammar and Dr. Lars Olaf Cardell present the current knowledge on eosinophil expression of these molecules and the effect of these receptors' engagement on eosinophil survival, degranulation, and superoxide and cytokine/chemokine and adhesion molecule expression production. These findings underscore the importance of eosinophils in innate immunity.

CONCLUSION AND FUTURE CHALLENGES

Recent advances in signaling studies include dynamic live cell imaging[11,12] and high-resolution subcellular imaging.[13] Direct observation of single fluorophores now enables scientists to gather extensive molecular information. By following temporal and spatial trajectories, one can calculate diffusion constants and binding kinetics of signaling molecules. The analyses of fluorescence lifetime, intensity, polarization, and spectra provide chemical and conformational information about molecules in live cells. Another area of progress is high-resolution subcellular imaging. A variety of microscopic approaches including transmission electron microscopy and atomic force microscopy with fluorescent imaging have been used to obtain spatiotemporally resolved images of receptors and signaling molecules.[12] Adaptation of these approaches will help develop a comprehensive understanding of signaling processes in eosinophils.

REFERENCES

1. Kroegel C, Yukawa T, Westwick J, Barnes PJ. Evidence for two platelet activating factor receptors on eosinophils: dissociation between PAF-induced intracellular calcium mobilization degranulation and superoxides anion generation in eosinophils. *Biochem Biophys Res Commun* 1989;**162**:511−21.
2. Kita H, Abu-Ghazaleh RI, Gleich GJ, Abraham RT. Regulation of Ig-induced eosinophil degranulation by adenosine 3′,5′-cyclic monophosphate. *J Immunol* 1991;**146**:2712−8.
3. Kroegel C, Chilvers ER, Giembycz MA, Challiss RA, Barnes PJ. Platelet-activating factor stimulates a rapid accumulation of inositol (1,4,5)trisphosphate in guinea pig eosinophils: relationship to calcium mobilization and degranulation. *J Allergy Clin Immunol* 1991;**88**:114−24.
4. Kita H, Kato M, Gleich GJ, Abraham RT. Tyrosine phosphorylation and inositol phosphate production: are early events in human eosinophil activation stimulated by immobilized secretory IgA and IgG? *J Allergy Clin Immunol* 1994;**94**:1272−81.
5. Yousefi S, Green DR, Blaser K, Simon HU. Protein-tyrosine phosphorylation regulates apoptosis in human eosinophils and neutrophils. *Proc Natl Acad Sci USA* 1994;**91**:10868−72.
6. van der Bruggen T, Caldenhoven E, Kanters D, Coffer P, Raaijmakers JA, Lammers JW, et al. Interleukin-5 signaling in human eosinophils involves JAK2 tyrosine kinase and Stat1 alpha. *Blood* 1995;**85**:1442−8.
7. Cen O, Gorska MM, Stafford SJ, Sur S, Alam R. Identification of UNC119 as a novel activator of SRC-type tyrosine kinases. *J Biol Chem* 2003;**278**:8837−45.
8. Goplen N, Gorska MM, Stafford SJ, Rozario S, Guo L, Liang Q, et al. A phosphosite screen identifies autocrine TGF-beta-driven activation of protein kinase R as a survival-limiting factor for eosinophils. *J Immunol* 2008;**180**:4256−64.
9. Munitz A, Levi-Schaffer F. Inhibitory receptors on eosinophils: a direct hit to a possible Achilles heel? *J Allergy Clin Immunol* 2007;**119**:1382−7.
10. Shen ZJ, Esnault S, Schinzel A, Borner C, Malter JS. The peptidyl-prolyl isomerase Pin1 facilitates cytokine-induced survival of eosinophils by suppressing Bax activation. *Nat Immunol* 2009;**10**:257−65.
11. Douglass AD, Vale RD. Single-molecule microscopy reveals plasma membrane microdomains created by protein-protein networks that exclude or trap signaling molecules in T cells. *Cell* 2005;**121**:937−50.

12. Rink J, Ghigo E, Kalaidzidis Y, Zerial M. Rab conversion as a mechanism of progression from early to late endosomes. *Cell* 2005;**122**:735–49.

13. Hsieh MY, Yang S, Raymond-Stinz MA, Edwards JS, Wilson BS. Spatio-temporal modeling of signaling protein recruitment to EGFR. *BMC Syst Biol* 2010;**4**:57.

Chapter 7.2

Priming: A Critical Step in the Control of Eosinophil Activation

Leo Koenderman

INTRODUCTION

The immune system faces an important dilemma. Fast and robust responses are necessary to combat invading microorganisms, but sufficient compensatory mechanisms should prevent the induction of hyperactivation. Such hyperresponsiveness of the immune system can cause damage to the host's tissues, which is typically found in chronic inflammatory lesions. Despite the consensus that the treatment of chronic inflammatory diseases should be aimed at the mechanisms underlying chronic activation of the immune system, surprisingly little success has been reached with currently available anti-inflammatory drugs. This lack of success is now a major concern in the long-term treatment or cure of these chronic diseases and new therapeutics are badly needed. Many lines of evidence point at a complex interplay between environment and pathogenic mechanisms, which are beyond the scope of this short chapter. However, a consistent finding is the importance of hyperactivation of the innate immune system during the chronic phase of these disorders. Limiting the activation of the innate immune system has proven difficult using currently available anti-inflammatory drugs such as glucocorticosteroids (Fig. 7.2.1).

THE CONCEPT OF PRIMING

Priming is a complex mechanism that is defined at the level of individual cellular responses rather than at the level of the whole cell.[1,2] A cellular response is primed when an agent does not evoke the response itself but amplifies the response by a heterologous stimulus.[3,4]

Priming *In Vitro*

It is important to emphasize that mediators can prime certain responses and activate others. A clear example is platelet-activating factor (PAF), which is a very potent activator of eosinophil chemotaxis,[5] but which is only a priming agent for the activation of the respiratory burst evoked by opsonized particles.[6]

The activation of human eosinophils with physiologically relevant stimuli is typically controlled by priming. Several soluble priming agents have been described for eosinophils: chemotactic lipids (e.g., PAF) and chemokines (e.g., eotaxin), cytokines [e.g., granulocyte-macrophage colony-stimulating factor; interleukin-3 (IL-3) and 5 (IL-5)], complement fragments (e.g., complement component C5a), and pathogen-associated molecular patterns (PAMPs).[3,4] Many of these priming agents can activate responses that facilitate the differentiation and homing of eosinophils to the tissues, including: differentiation, adhesion, transendothelial migration, and chemotaxis.[3,4] On the other hand, these agents are potent primers of cytotoxicity-associated events such as activation of phagocytosis, respiratory burst, and degranulation.[3,4]

Apart from the aforementioned soluble mediators, immobilized ligands can also potently prime cellular functions such as the induction of survival and cytotoxicity-associated responses. Kaneko and colleagues[7] have shown

FIGURE 7.2.1 **The importance of priming for eosinophil functions.** Normal eosinophils isolated from the blood of normal control donors are characterized by a resting phenotype that has limited responsiveness toward many immunological stimuli. On interaction with priming agents produced by other cells or microorganisms, that eosinophil phenotype quickly changes toward a primed cell with upregulated effector functions. GM-CSF, granulocyte-macrophage colony-stimulating factor; PAF, platelet-activating factor; TLR, Toll-like receptor; TNF-α, tumor necrosis factor.

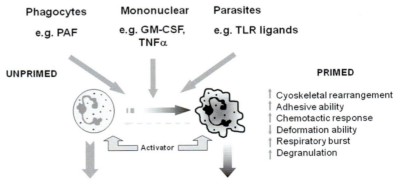

Priming is Essential for Activation of Eosinophils

that ligation of β2 integrins clearly modulates the response of eosinophils activated by coated immunoglobulins, a finding that points to the complex interplay between adhesion and immunoglobulin (Fc) receptors. Ulfman and colleagues[8] have shown that only eosinophils that adhere to a physiologically relevant surface respond toward interleukin-8 (IL-8) with an increase in cytosolic free Ca^{2+} concentration $[(Ca^{2+})_i]$. Also, survival of eosinophils *in vitro* is clearly modulated by adhesion.[9,10]

Not only adhesion, but also other poorly defined environmental factors (e.g., heat shock, endotoxin, and so on) can specifically prime granulocytes during isolation and *in vitro* handling.[11] This has led to a lack of consensus in the literature on how potent certain stimuli are and whether these stimuli need priming to activate eosinophils. A clear illustration is finding that untouched eosinophils are unresponsive to the PAMP fMet-Leu-Phe (fMLP) in the context of activation of the respiratory burst, whereas fMLP is an activator of this response in primed cells.[12] A similar situation is found with serum-treated zymosan (STZ), which only activates eosinophils when primed with added cytokines/chemokines[6] or via paracrine means, by the production of PAF.[13]

These data clearly indicate that adhesion *in vitro* or other environmental factors should always be considered as important to interpret priming responses *in vitro*.

Priming *In Vivo*

Several lines of evidence show that priming of eosinophils occurs *in vivo* in patients with allergies and other chronic inflammatory diseases.[14−19] These priming responses are found in the context of adhesion,[142] transendothelial migration,[15] chemotaxis,[16,17] activation of the respiratory burst,[18] and degranulation.[19] Therefore, eosinophil priming could be an interesting diagnostic tool to determine and study systemic inflammation in patients *in vivo*. However, all these responses require the isolation of cells, which makes the measurement of these responses difficult to apply in clinical practice. Apart from these practical considerations, the occurrence of isolation artifacts also makes it difficult to interpret and compare different studies. Nevertheless, these studies clearly show that eosinophils are modulated systemically by signals originating from the tissues affected by allergic or parasitic inflammation.

MECHANISMS UNDERLYING THE PRIMING OF EOSINOPHILS

Expression of Receptors

Many studies focusing on the (pre)activation of granulocytes have shown that activation of these cells is associated with modulated expression of receptors such as cell surface integrin alpha-M (CD11b), L-selectin (CD62L), carcinoembryonic antigen-related cell adhesion molecule 8 (CEAM8; CD66b), and lysosomal-associated membrane protein 3 (LAMP-3; CD63).[19−22] Although many studies show low but significant induction of these markers on peripheral blood granulocytes, no appreciable correlation with priming has been demonstrated. A likely hypothesis that can explain this lack of success is the localization of these proteins/receptors in the resting cell. They are typically found associated with specific granules and secretory vesicles.[23] This means that enhanced expression is associated with a limited degranulation response, i.e., associated with activation rather than with priming. There is very little evidence that degranulation of specific granules is initiated in the peripheral blood of patients with chronic inflammatory diseases. Also, the finding that eosinophil cationic protein is found in enhanced concentrations in serum is largely based on the release of the protein *ex vivo*, during the blood clotting process in the tube.[24] Degranulation does occur in the tissue, as sampling at inflammatory loci has shown multiple granule-associated proteins found in tissue fluids such as sputum, bronchoalveolar lavage (BAL), and synovial fluid.[24]

Inside-Out Control and Affinity of Receptors

Recent studies show that the functionality of several very important receptors associated with cytotoxic responses in eosinophils is controlled by intracellular signals induced by inflammatory mediators. This mechanism,[25−27] which is generally referred to as *inside-out control*, was first described clearly for integrin receptors. These adhesion molecules consist of heterodimeric receptors made up of an alpha chain non-covalently bound to a beta chain. These integrins are very important receptors in the adhesion of cells to other cells and to the extracellular matrix. Five important integrins are expressed on the surface of eosinophils: integrin α4/β1 [CD49d; VLA-4 subunit beta (VLA-4/CD29)], α4/β7 (CD49d/integrin beta-7), αM/β2 [cell surface glycoprotein MAC-1 subunit alpha (MAC-1): CD11b/CD18), αL/β2 [leukocyte adhesion glycoprotein LFA-1 (LFA-1); CD11a, CD18) α/β2 (p150, 95 subunit beta/CD11c; CD18).

Two of these integrins, VLA-4 and MAC-1, have been studied in detail in the context of inside-out control on eosinophils. These integrins are typically expressed on the surface of quiescent, untouched eosinophils. However, their functionality is very low and does not facilitate binding to their ligands, vascular cell adhesion protein 1 (VCAM-1) and fibronectin for VLA-4,[28] and the complement protein fragment C3bi for MAC-1.[29] The interaction of eosinophils with inflammatory mediators rapidly leads to rapid upregulation of the functionality of these receptors and the cells start binding to the different ligands.

The functionality of receptors for immunoglobulin G [IgG; Fcγ RII-a (CD32)] and immunoglobulin A [IgA; FcαR (CD89)] on eosinophils is controlled by similar mechanisms as integrins.[30] As eosinophils only express one Fcγ RII-a (CD32) and one FcαR (CD89), these cells are ideal to study these receptors in their physiological primary cell context. Eosinophils bind poorly to IgA- and IgG-coated ligands, but rapidly upregulate this function after interaction with inflammatory mediators.[30] The expression of the receptors changes minimally under these conditions, which is indicative of an inside-out type of regulation. Upregulation of the functionality of these receptors occurs both *in vitro* and *in vivo* in patients with allergic asthma.[31]

The fact that both integrins and Fc receptors on eosinophils are under an inside-out type of control suggests that the interaction with opsonized particles might also be under similar control. This is indeed found in studies applying opsonized particles such as serum-treated zymosan.[6] The main opsonins on these yeast particles are immunoglobulins and complement component fragments such as C3bi. Many studies have shown that the interaction with and activation by these particles is very sensitive for priming with inflammatory mediators such as chemotactic agents and cytokines. These data indicate that inside-out control of receptors is an important mechanism in the priming of eosinophils.

Mechanisms Underlying Priming and Inside-Out Control Mechanisms

The mechanisms underlying priming and inside-out control in eosinophils are still poorly defined, but recent studies have shed some light on the putative mechanisms:

1. *Affinity versus valency of eosinophil receptors.* Part of the priming response, which is mediated by inside-out control, is the switching of the functionality of receptors between low-affinity, high-affinity, and high-valency/avidity (the interaction of multiple receptors with their ligands, Fig. 7.2.2).[32] The consensus is that for optimal functionality of both integrins and Fc receptors, these receptors need to be both in a high-affinity and high-valency state. The processes controlling affinity and valency are likely to be different but poorly defined. Several lines of evidence show that, as a minimum, phosphorylation of integrins and Fc receptors is important in controlling at least the affinity of the receptors. In the intracellular tails of both the α4 integrin chain (CD49d)[33] and the immunoglobulin alpha Fc receptor (CD89),[34] critical serine residues have been identified for the control of the affinity of these receptors. Point mutations of serine 263 (Ser263S → A) in the tail of the immunoglobulin alpha Fc receptor, as well as the point mutation of serine 488 in the intracellular tail of CD49d, are characterized by a marked dysfunctional phenotype. Up to now, no kinases have been identified that control these receptors on eosinophils.

2. *Formation of supramolecular complexes.* As mentioned previously, several opsonin receptors are controlled by similar inside-out control mechanisms.

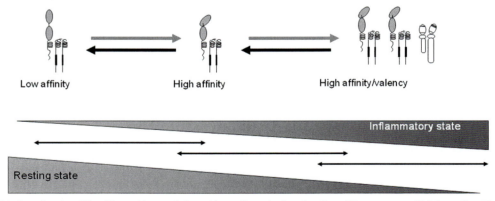

FIGURE 7.2.2 Priming of eosinophils with cytokines and chemokines affects the functionality of Fc receptors, which is mediated by two different but connected mechanisms: enhanced affinity and valence. Enhanced affinity is mediated by an intramolecular change that facilitates ligand binding. Increased valence is mediated by the grouping of receptors.[30–36]

Therefore, several groups have tested the hypothesis that these receptors, under conditions of priming and/or activation, can cluster and/or functionally interact. Important work has shown that Fc receptors can interact with MAC-1 on the surface of phagocytes and that this interaction is facilitated by several adaptor molecules such as urokinase receptor[35] and L-plastin.[32] The consequence of these interactions is the formation of large protein complexes in the membranes of these cells.

3. The functional consequences of these multimeric interactions are only now emerging. Pioneering work by van Spriel and colleagues[36] has shown that the formation of a so-called immunological synapse is necessary for a stable interaction of phagocytes with their opsonized targets. These authors showed that the absence of MAC-1 in CD11b$^{-/-}$ mice resulted in an unstable interaction with IgG-coated targets. These findings implicate an important intellectual challenge, as the absence of one protein will affect the functionality of another. Analysis of eosinophils with blocking antibodies directed against certain membrane receptors can induce the impairment of binding of other receptors to their ligands and, as a consequence, downstream signaling of these receptors. Therefore, the finding that MAC-1-blocking antibodies inhibit the activation of granulocytes is not necessarily an indication that MAC-1 is the signaling receptor. Blocking MAC-1 might just affect the signaling of a heterologous receptor. An illustration of the latter hypothesis comes from work by Kaneko and colleagues, which shows that blocking antibodies directed against MAC-1 completely block immunoglobulin-induced activation.[7]

4. An even higher order of complexity is the finding that the membranes of mammalian cells are organized by specialized regions such as lipid rafts.[37] The distribution of many receptors is, therefore, not random but associated with these specialized domains. Little is known regarding priming and changes in function and/or distribution of these regions.

5. *Signal transduction.* Several studies have been published regarding signal transduction pathways in eosinophils initiated by priming and activating agonists. Many signaling pathways have been described (see[38] for a review). Little attention has been given to the hypothesis that priming not only potentiates the signals initiated by heterologous receptors, but also changes this process. A few studies have suggested that this process at least can occur. Bates and colleagues provided evidence that mitogen-activated protein kinases (MAPKs) are particularly activated in primed cells compared to non-primed cells.[39] Apart from MAPKs, the opsonized particle-induced

production of 1,2-diacylglycerol and changes in cytosolic free Ca^{2+} concentrations $[Ca^{2+}]_i$ are different in primed eosinophils.[6] The importance of these changes remains to be established.

6. *Negative signaling.* A very important mechanism in controlling priming in eosinophils is mediated by signals that actively suppress activation in non-primed cells. This paradigm is referred to as *negative signaling.* Surprisingly, little is known regarding this mechanism but recent studies have identified several of these signaling modules. Early work applying several protein kinase C (PKC) inhibitors has shown that one of the PKC isoforms is involved in suppressing the activation of eosinophils by opsonized particles. As described previously, non-primed eosinophils are relatively unresponsive to opsonized targets such as STZ. Short-term treatment with staurosporine and several more PKC-specific staurosporine derivatives provides a very potent priming signal.[40] This indicates that certain kinases are operational in non-primed eosinophils, actively keeping the opsonin receptor complement receptor C3 subunit beta (CR3; CD11b/CD18) and/or Fc-gamma RII-a (CD32) in an inactive form. Priming can then be accomplished by inhibiting the inhibitory signal.

7. Investigation of the underlying mechanisms of negative signaling and the signaling enzymes involved has only recently been occurring. Studies focusing on the immunoglobulin alpha Fc receptor (CD89) have identified a mechanism by which negative signals can control the functionality of the receptor. Bracke and colleagues[34] identified a serine (Ser265) at the C-terminus region in the intracellular domain of the immunoglobulin alpha Fc receptor, which is instrumental in the functionality of the receptor. Stable transfection of immunoglobulin alpha Fc receptor in the IL-3-dependent murine pre-B cell line, Ba/F3, resulted in the expression of the receptor in the correct context. As in eosinophils, the receptor bound to the ligand only in the presence of the priming stimulus, murine IL-3. Interestingly, the requirement for priming was lost in the immunoglobulin alpha Fc receptor containing a point mutation at position 263 (S→A). This receptor could no longer be phosphorylated at this position, which coincided with maximal functionality in terms of ligand binding. The reverse approach, by introducing glutamic acid as a *phosphomimetic* at this position, rendered the receptor non-functional and not sensitive to inside-out activation. These findings are consistent with the hypothesis that the inhibition of a kinase involved in suppressing the functionality of immunoglobulin alpha Fc receptor leads to priming of this Fc receptor.

8. Since these early findings, more studies indicated that inhibition of constitutive signals leads to modulation of granulocyte responses.[41] This poses, however, a very important conceptual problem in targeting the priming process as an innovative new target for the treatment of eosinophil-based diseases. Several kinases can be controlled by different mechanisms and the function of these kinases can be dependent on the upstream signals. Whereas constitutively active PKC is involved in anti-inflammatory pathways by the negative control of opsonin receptors, this kinase is also involved in proinflammatory pathways through the proinflammatory, mediator-induced activation of PKC by the phosphatidylinositol 4,5-biphosphate (PIP_2)-diacylglycerol $(2DG)-Ca^{2+}$ pathway. The dilemma is clear; inhibition of such kinases will lead to both anti- as well as proinflammatory processes dependent on the time and context of inhibition of the kinase. A similar finding regarding the inhibition and/or activation of the transcription factor, nuclear factor kappa-light-chain-enhancer of activated B cells (NFκB), in human granulocytes was recently published.[42]

9. *Inhibitory receptors.* Human granulocytes express inhibitory receptors. These receptors seem to particularly modulate proinflammatory signals initiated by tyrosine kinases. Two types of receptors have been characterized to some detail: receptors expressing immunoreceptor tyrosine-based inhibition motifs (ITIMs) or an intracellular tyrosine phosphatase activity. Leukocyte-associated immunoglobulin-like receptor 1 (LAIR-1) and signal inhibitory receptor on leukocytes-1 (SIRL-1) are homologous receptors expressing an ITIM domain in the intracellular part of the receptor.[43,44] These ITIM domains are involved in the downregulation of signaling paradigms that involve several tyrosine kinases. Several studies show that cross-linking of these receptors leads to the activation of these domains by phosphorylation. Recent studies have identified human collagens as ligands for at least the LAIR-1 protein. These findings are consistent with the hypothesis that matrix proteins such as collagen can affect eosinophils once they are in the tissue. Another relevant receptor in this respect is CD45, which is a transmembrane tyrosine phosphatase.[45] Data in cells other than eosinophils indicate that CD45 is involved in the downregulation of the signaling pathways mediated by the activation of tyrosine kinases.

10. *A change in specificity and/or responsiveness for chemotaxins.* Eosinophils are relatively refractory to multiple, soluble, proinflammatory stimuli, such as chemokines, and several end-organ chemotaxins such as PAF and IL-8. On priming with cytokines such as IL-5, eosinophils switch their phenotype and become very sensitive to PAF and even responsive to IL-8.[5] The underlying mechanism for this switch remains elusive. Particularly, the responsiveness to IL-8 is difficult to explain as neither C-X-C chemokine receptor type 1 (CXC-R1/CD181) nor C-X-C chemokine receptor type 2 (CXC-R2/CD182) are expressed by eosinophils even after priming.[8] Nonetheless, primed eosinophils are responsive to IL-8 as illustrated by the fact that (i) IL-8 can induce a transient arrest of rolling cells on tumor necrosis factor (TNF-α)-activated endothelial cells and (ii) IL-8 induces changes in $[Ca^{2+}]_i$ in single adherent eosinophils.[8] The receptor(s) mediating these effects remain(s) to be identified.

DETERMINATION OF PRIMING OF EOSINOPHILS *IN VIVO*

Priming of Functional Eosinophil Responses *In Vivo*

Many studies have tested the hypothesis that eosinophils are primed in the peripheral blood of allergic asthmatics at level of (1) adhesion/transmigration/chemotaxis, (2) respiratory burst, (3) degranulation, and (4) cytokine release. All these studies have come to the same conclusion that eosinophils are primed in the peripheral blood, tissue, and BAL fluid in patients with allergic diseases. Therefore, these studies are clear proof of principle, but cannot be applied in clinical practice, because the assays require isolation of relatively rare cells and laborious *in vitro* experiments.

An important new development is the application of antibodies that recognize only primed cells.[46,47] These antibodies allow the rapid analysis of eosinophil priming in whole blood by flow cytometric assays. Initial studies have shown that these antibodies can very sensitively detect primed cells in blood and tissue and have identified subpopulations of eosinophils with distinct functionalities.

Occurrence of Eosinophil Subpopulations *In Vivo*

The interpretation of studies on the priming of granulocytes is still hampered by the possibility that multiple functional phenotypes of these cells exist. A functional phenotype in this case is a type of cell which has distinct functional characteristics not shared with other phenotypes. The consensus in the field is that the majority of eosinophils are formed in the bone marrow as a homogeneous population. Hereafter, the cells are distributed through the body and possibly altered by environmental factors such as those found in the tissues. Although very little is known regarding eosinophils, some key studies have been published showing the first indications of subpopulations. Liu and colleagues[48] analyzed eosinophils in the BAL fluid of asthma patients after segmental allergen challenge. These eosinophils have several unique characteristics not found in eosinophils obtained from the peripheral blood. One very relevant change is the downregulation of the interleukin-5 receptor subunit alpha (IL-5Rα) chain (CD125), whereas the granulocyte-macrophage colony-stimulating factor receptor subunit alpha (GM-CSF-R-α) chain (CD116) is not affected. This renders the cells not sensitive to IL-5. This type of modulation of receptor expression can lead to a redirection of the redistribution and activation of these cells. When these cells are redistributing to other tissues, they are not targeted by innovative new drugs, such as the humanized anti-IL-5 antibody, mepolizumab. Apart from expression of the IL-5RA chain, several additional characteristics have been shown for BAL eosinophils compared to blood cells, which include enhanced signaling on stimulation with inflammatory mediators, increased adhesion/degranulation, and modulated expression of several proteins expressed on the cell surface.[39] It is uncertain at the moment whether these cells should be targeted with different therapeutics as compared to *normal* eosinophils that are newly mobilized from the bone marrow.

ACTIVATION EPITOPES

Many studies have shown that several markers on the surface of granulocytes can be used as readout for activation of these cells. These markers include carcinoembryonic antigen-related cell adhesion molecule 8 (CD66b), intercellular adhesion molecule 1 (CD54), L-selectin (CD62L), LAMP-3 (CD63), and MAC-1 (CD11b/CD18). These markers have been particularly studied in *in vitro* activation model systems.[20] None of these markers have been proven applicable in the analysis of priming of eosinophils in the peripheral blood *in vivo*, although small changes in expression of these markers in patients have been published. Interestingly, these markers are upregulated in cells analyzed in tissues.[49] These findings are consistent with the hypothesis that activation characterized by upregulation of these markers, and downregulation of L-selectin, typically does not occur in the peripheral blood, but much more in the tissues. Therefore, these markers are not very suitable for the study of eosinophil priming in the peripheral blood.

Analysis of priming *in vivo* by flow cytometry requires antibodies recognizing very subtle changes of receptors on

the cell membrane, which are associated with cellular processes occurring in eosinophils in the peripheral blood. Ideal targets for these antibodies are those receptors that are under the control of inside-out signaling (see earlier in the chapter). Two antibodies have been studied in some detail in the context of eosinophils and allergic asthma:

1. *Expression of activated integrins.* Johansson and colleagues[46] showed that the application of an antibody directed against the active form of beta-1 (β1) (CD29, clone N29) integrins can be used to study the activation of eosinophils in different types of asthma. In particular, the difficult-to-treat type of asthma studied in the Severe Asthma Research Program (SARP) initiative is associated with eosinophils in the peripheral blood and relatively high amounts of active β1 integrins.[46] The authors showed that it was not so much an increase in the total number of receptors, but mainly the active confirmation of these receptors.

2. *Expression of activated IgG Fc receptor II-a (CD32).* We have developed two antibodies that recognize Fc-gamma RII-a (CD32) only in the active configuration. As Fc-gamma RII-a is expressed on all phagocytes, this antibody allows the detection of two important processes: (i) the extent of priming and (ii) the identification of cell types that are primed. Applications of these antibodies have shown that priming of eosinophils in the peripheral blood is a common feature of normal, human asthma responsive to treatment. This priming characteristic can be potentiated by allergen challenge, and fully primed cells are found in the lung of atopic patients after segmental allergen challenge.[50]

PRIMING AS A DIAGNOSTIC TOOL

As described in this chapter, eosinophil priming can be evoked by a multitude of inflammatory mediators ranging from the interaction of cytokines, to adhesion to certain surfaces. The markers available at present do not discriminate between the different signals that prime eosinophils *in vivo*. However, priming can be seen as a final common pathway for a large array of priming mediators. It is, therefore, tempting to speculate that probes recognizing primed cells can be applied in the diagnosis of eosinophil-driven diseases.

Identification of Type of Inflammatory Response

The different antibodies recognizing primed cells allow the discrimination between primed and non-primed cells to be made. When primed cells are detected, the pathogenesis of the disease is associated with peripheral

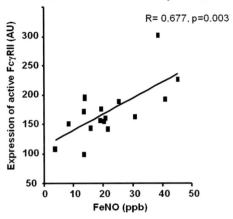

Priming of blood eosinophils correlates with exhaled NO in asthma patients

R= 0.677, p=0.003

FIGURE 7.2.3   Correlation of blood eosinophil priming and exhaled nitric oxide in asthma patients. The severity of allergic asthma, measured by the fraction of exhaled nitric oxide (FeNO) in patients with stable disease, positively correlates with the activation of peripheral eosinophils characterized by enhanced expression of Fc-gamma RII-a [measured by fluorescein isothiocyanate (FITC)-labeled antibody A27]. The expression of active Fc-gamma RII-a is visualized by fluorescence-activated cell sorting (FACS) analysis and FeNO was determined as described by Ravensberg and colleagues.[52] This figure is adapted from[53].

priming of these cells, which is indicative of the nature of the inflammatory response. Stable asthma that is responsive to therapy is typically associated with priming of eosinophils in the peripheral blood. In these patients, no priming of neutrophils is found. These data suggest that determination of eosinophil priming in the systemic circulation can help to discriminate between eosinophil and neutrophil asthma.

The Extent of Eosinophil Priming

In peripheral blood, the extent of eosinophil priming correlates with several markers of disease severity, such as exhaled nitric oxide [fraction of exhaled NO (FeNO), Fig. 7.2.3]. Therefore, determination of the extent of priming of peripheral blood cells can help in diagnosing the extent of the systemic inflammatory response at the time of blood sampling.

This type of analysis can have consequences in clinical decision-making. Normal, stable asthma of the eosinophilic type is typically responsive toward treatment with corticosteroids. Therefore, the assessment of specific eosinophil priming in the peripheral blood can be used to support the decision for treatment with glucocorticosteroids (GCS). More important is when neutrophil priming is found in the blood of asthma patients. Neutrophils are rather unresponsive to GCS in terms of downregulation of activation mechanisms. In fact, GCS have been shown to increase

neutrophil survival and several effector mechanisms are even upregulated.[50] Until now, this type of asthma is difficult to diagnose, therefore determination of neutrophil priming in the peripheral blood may help.

Lack of Priming in the Peripheral Blood is Associated with Severe Systemic Inflammation

The concept of priming of granulocytes in the peripheral blood *in vivo* is complicated by the fact that priming of these cells in the peripheral blood can only be detected at a certain level and/or time in the inflammatory response. Recent work[51] has shown that severe systemic inflammation, such as that found after major trauma and sepsis, is associated with the downregulation of priming-associated markers. This finding can be explained by the fact that primed cells under these conditions are mobilized to the tissue. This hypothesis awaits experimental confirmation.

The result, however, is that a low priming signal in the blood can be associated with both the absence of inflammation or a very marked inflammatory response (Fig. 7.2.4). This situation seems to preclude the application of these markers to clinical practice as the cells from both situations seem to be similar. This is, however, not the case. It turns out that cells in the peripheral blood during severe systemic inflammation, which is characterized by a low expression of priming markers, have lost the capacity to upregulate these markers in response to *ex vivo* activation. These so-called *refractory cells* cannot upregulate the expression of the active form of Fc-gamma RII-a (CD32) on these cells in response to the innate immune stimulus fMLP (see Fig. 7.2.4). Therefore, diagnostic tests focused on the priming phenotype of granulocytes should always take place in the absence and presence of an activator. Indeed, when primed cells have left the circulation, refractory cells will stay behind. These latter cells show a low responsive to, e.g., fMLP.

CONCLUSION

Priming of eosinophils *in vivo* is a critical process in the control of these cells both in the context of distribution and activation of cytotoxic processes. A multitude of inflammatory mediators is involved in the priming of certain eosinophil responses and different mechanisms underlie this priming mechanism. Priming is, therefore, a critical step in the control of the most cytotoxic cell in the body, namely the eosinophil.

Complex priming phenotype of blood eosinophils during systemic inflammation

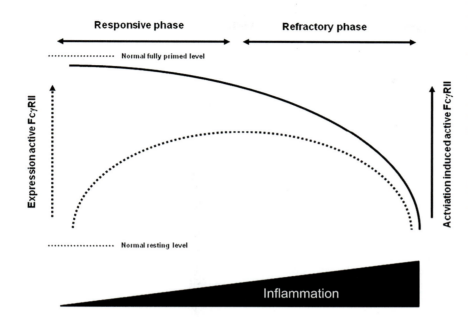

FIGURE 7.2.4 Hypothetical model for the complex priming phenotype of eosinophils found in the peripheral blood during systemic inflammation. Under normal healthy conditions, eosinophils, like other granulocytes, have a resting, non-primed phenotype that is very sensitive to priming by multiple priming mediators (see, e.g., Fig. 7.2.1). During mild-to-moderate systemic inflammation, the expression of active Fc-gamma RII-a is enhanced, but the cells still respond adequately to additional priming mediators (*responsive phase*). The situation changes during moderate-to-severe systemic inflammation, because the cells seem unprimed, though they become refractory to priming mediators (*refractory phase*). During this phase cells should be studied in the context of responsiveness to activation rather than expression of priming epitopes.

REFERENCES

1. Koenderman L, Bruijnzeel PL. Increased sensitivity of the chemo-attractant-induced chemiluminescence in eosinophils isolated from atopic individuals. *Immunology* 1989;**67**(4):534–6.

2. Hakansson L, Carlson M, Stalenheim G, Venge P. Migratory responses of eosinophil and neutrophil granulocytes from patients with asthma. *J Allergy Clin Immunol* 1990;**85**(4):743–50.

3. Koenderman L, van der BT, Schweizer RC, et al. Eosinophil priming by cytokines: from cellular signal to in vivo modulation. *Eur Respir J Suppl* 1996;**22**:119s–25s.

4. Coffer PJ, Koenderman L. Granulocyte signal transduction and priming: cause without effect? *Immunol Lett* 1997;**57**(1-3):27–31.

5. Warringa RA, Koenderman L, Kok PT, Kreukniet J, Bruijnzeel PL. Modulation and induction of eosinophil chemotaxis by granulocyte-macrophage colony-stimulating factor and interleukin-3. *Blood* 1991;**77**(12):2694–700.

6. Koenderman L, Tool AT, Roos D, Verhoeven AJ. Priming of the respiratory burst in human eosinophils is accompanied by changes in signal transduction. *J Immunol* 1990;**145**(11):3883–8.

7. Kaneko M, Horie S, Kato M, Gleich GJ, Kita H. A crucial role for beta 2 integrin in the activation of eosinophils stimulated by IgG. *J Immunol* 1995;**155**(5):2631–41.

8. Ulfman LH, Joosten DP, van der Linden JA, Lammers JW, Zwaginga JJ, Koenderman L. IL-8 induces a transient arrest of rolling eosinophils on human endothelial cells. *J Immunol* 2001;**166**(1):588–95.

9. Chihara J, Kakazu T, Higashimoto I, et al. Signaling through the beta2 integrin prolongs eosinophil survival. *J Allergy Clin Immunol* 2000;**106**(1 Pt 2):S99–103.

10. Meerschaert J, Vrtis RF, Shikama Y, Sedgwick JB, Busse WW, Mosher DF. Engagement of alpha4beta7 integrins by monoclonal antibodies or ligands enhances survival of human eosinophils in vitro. *J Immunol* 1999;**163**(11):6217–27.

11. Kuijpers TW, Tool AT, van der Schoot CE, et al. Membrane surface antigen expression on neutrophils: a reappraisal of the use of surface markers for neutrophil activation. *Blood* 1991;**78**(4):1105–11.

12. Koenderman L, Kok PT, Hamelink ML, Verhoeven AJ, Bruijnzeel PL. An improved method for the isolation of eosinophilic granulocytes from peripheral blood of normal individuals. *J Leukoc Biol* 1988;**44**(2):79–86.

13. Tool AT, Koenderman L, Kok PT, Blom M, Roos D, Verhoeven AJ. Release of platelet-activating factor is important for the respiratory burst induced in human eosinophils by opsonized particles. *Blood* 1992;**79**(10):2729–32.

14. Hakansson L, Heinrich C, Rak S, Venge P. Priming of eosinophil adhesion in patients with birch pollen allergy during pollen season: effect of immunotherapy. *J Allergy Clin Immunol* 1997;**99**(4):551–62.

15. Moser R, Fehr J, Bruijnzeel PL. IL-4 controls the selective endo-thelium-driven transmigration of eosinophils from allergic individuals. *J Immunol* 1992;**149**(4):1432–8.

16. Sehmi R, Wardlaw AJ, Cromwell O, Kurihara K, Waltmann P, Kay AB. Interleukin-5 selectively enhances the chemotactic response of eosinophils obtained from normal but not eosinophilic subjects. *Blood* 1992;**79**(11):2952–9.

17. Warringa RA, Mengelers HJ, Kuijper PH, Raaijmakers JA, Bruijnzeel PL, Koenderman L. In vivo priming of platelet-activating factor-induced eosinophil chemotaxis in allergic asthmatic individuals. *Blood* 1992;**79**(7):1836–41.

18. Sannohe S, Adachi T, Hamada K, et al. Upregulated response to chemokines in oxidative metabolism of eosinophils in asthma and allergic rhinitis. *Eur Respir J* 2003;**21**(6):925–31.

19. Torsteinsdottir I, Arvidson NG, Hallgren R, Hakansson L. Enhanced expression of integrins and CD66b on peripheral blood neutrophils and eosinophils in patients with rheumatoid arthritis, and the effect of glucocorticoids. *Scand J Immunol* 1999;**50**(4):433–9.

20. Liu L, Hakansson L, Ridefelt P, Garcia RC, Venge P. Priming of eosinophil migration across lung epithelial cell monolayers and upregulation of CD11b/CD18 are elicited by extracellular Ca2+. *Am J Respir Cell Mol Biol* 2003;**28**(6):713–21.

21. Mengelers HJ, Maikoe T, Hooibrink B, et al. Down modulation of L-selectin expression on eosinophils recovered from bronchoalveolar lavage fluid after allergen provocation. *Clin Exp Allergy* 1993;**23**(3):196–204.

22. Mahmudi-Azer S, Downey GP, Moqbel R. Translocation of the tetraspanin CD63 in association with human eosinophil mediator release. *Blood* 2002;**99**(11):4039–47.

23. Calafat J, Kuijpers TW, Janssen H, Borregaard N, Verhoeven AJ, Roos D. Evidence for small intracellular vesicles in human blood phagocytes containing cytochrome b558 and the adhesion molecule CD11b/CD18. *Blood* 1993;**81**(11):3122–9.

24. Koh GC, Shek LP, Goh DY, Van BH, Koh DS. Eosinophil cationic protein: is it useful in asthma? A systematic review. *Respir Med* 2007;**101**(4):696–705.

25. Luo BH, Carman CV, Springer TA. Structural basis of integrin regulation and signaling. *Annu Rev Immunol* 2007;**25**:619–47.

26. Laudanna C, Kim JY, Constantin G, Butcher E. Rapid leukocyte integrin activation by chemokines. *Immunol Rev* 2002;**186**:37–46.

27. Ginsberg MH, Partridge A, Shattil SJ. Integrin regulation. *Curr Opin Cell Biol* 2005;**17**(5):509–16.

28. Ulfman LH, Kamp VM, van Aalst CW, et al. Homeostatic intracellular-free Ca2+ is permissive for Rap1-mediated constitutive activation of alpha4 integrins on eosinophils. *J Immunol* 2008;**180**(8):5512–9.

29. Carlson M, Peterson C, Venge P. The influence of IL-3, IL-5, and GM-CSF on normal human eosinophil and neutrophil C3b-induced degranulation. *Allergy* 1993;**48**(6):437–42.

30. Bracke M, Dubois GR, Bolt K, et al. Differential effects of the T helper cell type 2-derived cytokines IL-4 and IL-5 on ligand binding to IgG and IgA receptors expressed by human eosinophils. *J Immunol* 1997;**159**(3):1459–65.

31. Bracke M, van de GE, Lammers JW, Coffer PJ, Koenderman L. In vivo priming of FcalphaR functioning on eosinophils of allergic asthmatics. *J Leukoc Biol* 2000;**68**(5):655–61.

32. Jones SL, Wang J, Turck CW, Brown EJ. A role for the actin-bundling protein L-plastin in the regulation of leukocyte integrin function. *Proc Natl Acad Sci U S A* 1998;**95**(16):9331–6.

33. Han J, Liu S, Rose DM, Schlaepfer DD, McDonald H, Ginsberg MH. Phosphorylation of the integrin alpha 4 cytoplasmic domain regulates paxillin binding. *J Biol Chem* 2001;**276**(44):40903–9.

34. Bracke M, Lammers JW, Coffer PJ, Koenderman L. Cytokine-induced inside-out activation of FcalphaR (CD89) is mediated by a single serine residue (S263) in the intracellular domain of the receptor. *Blood* 2001;**97**(11):3478–83.

35. Wei Y, Lukashev M, Simon DI, et al. Regulation of integrin function by the urokinase receptor. *Science* 1996;**273**(5281):1551–5.

36. van Spriel AB, Leusen JH, van EM, et al. Mac-1 (CD11b/CD18) is essential for Fc receptor-mediated neutrophil cytotoxicity and immunologic synapse formation. *Blood* 2001;**97**(8):2478–86.

37. Kannan KB, Barlos D, Hauser CJ. Free cholesterol alters lipid raft structure and function regulating neutrophil Ca2+ entry and respiratory burst: correlations with calcium channel raft trafficking. *J Immunol* 2007;**178**(8):5253–61.

38. Gorska MM, Alam R. Signaling molecules as therapeutic targets in allergic diseases. *J Allergy Clin Immunol* 2003;**112**(2):241–50.

39. Bates ME, Sedgwick JB, Zhu Y, et al. Human airway eosinophils respond to chemoattractants with greater eosinophil-derived neurotoxin release, adherence to fibronectin, and activation of the Ras-ERK pathway when compared with blood eosinophils. *J Immunol* 2010;**184**(12):7125–33.

40. van der Bruggen T, Kok PT, Blom M, et al. Transient exposure of human eosinophils to the protein kinase C inhibitors CGP39-360, CGP41-251, and CGP44-800 leads to priming of the respiratory burst induced by opsonized particles. *J Leukoc Biol* 1993;**54**(6):552–7.

41. Dent G, Munoz NM, Ruhlmann E, et al. Protein kinase C inhibition enhances platelet-activating factor-induced eicosanoid production in human eosinophils. *Am J Respir Cell Mol Biol* 1998;**18**(1):136–44.

42. Langereis JD, Raaijmakers HA, Ulfman LH, Koenderman L. Abrogation of NF-kappaB signaling in human neutrophils induces neutrophil survival through sustained p38-MAPK activation. *J Leukoc Biol* 2010;**88**(4):655–64.

43. Meyaard L. LAIR and collagens in immune regulation. *Immunol Lett* 2010;**128**(1):26–8.

44. Steevels TA, Lebbink RJ, Westerlaken GH, Coffer PJ, Meyaard L. Signal inhibitory receptor on leukocytes-1 is a novel functional inhibitory immune receptor expressed on human phagocytes. *J Immunol* 2010;**184**(9):4741–8.

45. Hermiston ML, Xu Z, Weiss A. CD45: a critical regulator of signaling thresholds in immune cells. *Annu Rev Immunol* 2003;**21**:107–37.

46. Johansson MW, Barthel SR, Swenson CA, et al. Eosinophil beta 1 integrin activation state correlates with asthma activity in a blind study of inhaled corticosteroid withdrawal. *J Allergy Clin Immunol* 2006;**117**(6):1502–4.

47. Kanters D, ten Hove W, Luijk B, et al. Expression of activated Fc gamma RII discriminates between multiple granulocyte-priming phenotypes in peripheral blood of allergic asthmatic subjects. *J Allergy Clin Immunol* 2007;**120**(5):1073–81.

48. Liu LY, Sedgwick JB, Bates ME, et al. Decreased expression of membrane IL-5 receptor alpha on human eosinophils: I. Loss of membrane IL-5 receptor alpha on airway eosinophils and increased soluble IL-5 receptor alpha in the airway after allergen challenge. *J Immunol* 2002;**169**(11):6452–8.

49. Mengelers HJ, Maikoe T, Brinkman L, Hooibrink B, Lammers JW, Koenderman L. Immunophenotyping of eosinophils recovered from blood and BAL of allergic asthmatics. *Am J Respir Crit Care Med* 1994;**149**(2 Pt 1):345–51.

50. Luijk B, Lindemans CA, Kanters D, et al. Gradual increase in priming of human eosinophils during extravasation from peripheral blood to the airways in response to allergen challenge. *J Allergy Clin Immunol* 2005;**115**(5):997–1003.

51. Pillay J, Hietbrink F, Koenderman L, Leenen LP. The systemic inflammatory response induced by trauma is reflected by multiple phenotypes of blood neutrophils. *Injury* 2007;**38**(12):1365–72.

52. Ravensberg AJ, Luijk B, Westers P, Hiemstra PS, Sterk PJ, Lammers JW, et al. *Allergy* 2006;**61**(9):1097–103.

53. Luijk B. *Systemic and airway inflammation after allergen and adenosine inhalation in asthma.* Thesis University of Utrecht; 2005. page 69.

Chapter 7.3

Eosinophil Receptor-Mediated Inhibition

Ariel Munitz

INTRODUCTION

The main purpose of the immune system is to track down and eliminate pathogens that endanger our health. As such, complex mechanisms of innate and adaptive immunity collaborate to accomplish this task, which is primarily achieved by efficient generation, recruitment, and activation of immune cells. Traditional studies focusing on immune cells have largely concentrated on pathways that activate their function in response to antigens and pathogens. Extensive research has revealed that immune cell activation is regulated by a complex interplay between positive signals (leading to cellular activation) and negative signals (leading to the suppression of activation). These opposing signals are the basis for immune cell homeostasis

in health and disease and are often achieved through a combination of signals mediated by cell surface receptors, intracellular signaling intermediates, and transcribed gene products.[1]

In this chapter, we focus on one emerging family of regulatory receptors termed *inhibitory receptors*. Lanier first coined the term *inhibitory receptor superfamily* in 1998, when he described receptors that suppress activation of natural killer (NK) cells.[2] The traditional view is that these receptors possess three distinct features to inhibit cellular activation (Fig. 7.3.1):

- They suppress cellular activation of adjacent activation receptors within the same cell type which contain immunoreceptor tyrosine-based activation motifs (ITAMs).
- They all contain either one or more immunoreceptor tyrosine-based inhibitory motifs (ITIMs) that are necessary for their function.
- They bind intracellular phosphatases for their inhibitory function.[2]

It is important to note that inhibitory receptors are capable of counter-regulating multiple receptor-mediated activation pathways and usually display basal inhibitory activity that can be monitored by baseline tyrosine phosphorylation and physical interaction with phosphatases. These biochemical responses and interactions regulate basal cellular activity.

Much of the data on inhibitory receptor function has emerged from studies relating to cells from the lymphoid lineages, including NK cells, B cells and T cells.[2,3] However, over the past few years it has become apparent that the expression of inhibitory receptors is not

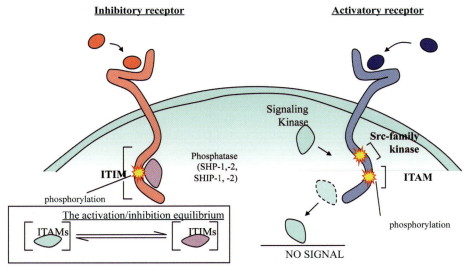

Inhibitory receptor

Activatory receptor

Signaling Kinase

Src-family kinase

Phosphatase (SHP-1,-2, SHIP-1, -2)

ITIM

phosphorylation

ITAM

phosphorylation

The activation/inhibition equilibrium

ITAMs → ITIMs

NO SIGNAL

Consensus ITIM:

I/V/L/SxYxxL/V

Consensus ITAM:

DxxYxxL/Ix$_8$YxxL/I

FIGURE 7.3.1 The *classical* view of inhibitory receptor signaling. Traditionally, inhibitory receptors have been described as functioning via immunoreceptor tyrosine-based activation motifs (ITAMs), via the recruitment of intracellular phosphatases (SHP-1, SHP-2, SHIP-1) to their immunoreceptor tyrosine-based inhibitory motif (ITIM), which often displays constitutive low-level phosphorylation. ITIM receptors are perceived to inhibit in *trans* to an activation receptor. SHP, Src homology phosphatase; SHIP-1, SH2 domain-containing inositol 5′-phosphatase 1.

limited to cells of the lymphoid lineage, as many cells from the myeloid lineage express functional inhibitory receptors as well. Moreover, the broad expression of these inhibitory receptors on cells from the lymphoid and myeloid lineages indicates that such receptors likely possess important roles in the regulation of myeloid cell function, particularly in settings where these cells are recruited and activated.[4]

This chapter discusses recent findings on the expression and function of inhibitory receptors in eosinophils. The basic structure and mechanism of action will be illustrated and current knowledge obtained from experimental settings will be summarized. Finally, rationalizing inhibitory receptors as pharmacological targets for treatment of eosinophilic-related diseases will be outlined.

STRUCTURE OF INHIBITORY RECEPTORS

The inhibitory receptor superfamily can be divided into two groups: the immunoglobulin superfamily (IgSF) and the C-type lectin superfamily.[2,3]

The IgSF receptor is characterized by a single variable V-type Ig-like domain in the extracellular portion and may contain several constant C-type domains. Eosinophils express various IgSF receptors, including leukocyte immunoglobulin-like receptor subfamily B member (LIR)/immunoglobulin-like transcript (ILT),[5] leukocyte-associated immunoglobulin-like receptor (LAIR),[6] leukocyte immunoglobulin-like receptor subfamily B member 3 (LILRB3) and 4 (LILRB4)[7], CD300 family members [such as inhibitory receptor protein 60 (IRp60/CD300a)],[8] and the sialic acid-binding Ig-like lectins (Siglecs)[9] (Table 7.3.1).

The term *C-type lectin* stems from the presence of a calcium-dependent, carbohydrate-binding protein motif that is found in C-type lectin receptors. Despite the presence of such a motif, many C-type lectin inhibitory receptors do not possess any calcium binding or carbohydrate specificity.[10] The C-type lectin inhibitory receptor group also includes proteins with a structural domain that was originally identified as a protein fold in the carbohydrate recognition domain of mannose-binding lectin.[10] The prototypical C-type lectin inhibitory receptor is the NKG2-A/B type II integral membrane protein/CD94 complex or Ly49.[11] The expression and function of C-type lectin inhibitory receptors on various myeloid cells and especially on dendritic cells and macrophages has been studied in depth. However, only limited knowledge exists on their expression and/or function in eosinophils. Consequently, the next sections of this chapter focus on IgSF receptors, which have received considerably more attention.

INHIBITORY SIGNALING: MECHANISM

The *hallmark* immune inhibitory receptor (either an IgSF or C-type lectin) can be identified by a consensus amino acid sequence, the ITIM, which is present in the cytoplasmic domains of these receptors. The ITIM sequence is composed of six amino acids (Ile/Val/Leu/Ser)-X-Tyr-X-X-(Leu/Val), where X denotes any amino acid (Fig. 7.3.1). Inhibitory receptors can express either one or more ITIM domains and their maximal inhibitory activity is not necessarily dependent on all motifs.

Activation signals can generally be divided into four signaling steps (Fig. 7.3.2). The initial, proximal signaling step is carried through by juxtamembranous signaling steps (often involving Src family kinases and other tyrosine kinases). The second, interfacing step involves recruitment of signaling intermediates such as adaptor molecules and lipid mediators. The third step involves a cytosolic compartment of signaling intermediates such as the mitogen-activated protein kinase (MAPK) family and involves amplification of the activation signal. Finally, efficient activation leads to the fourth and final stage involving binding of transcription factors and initiation of gene transcription. The general paradigm for inhibitory receptor signaling is that inhibitory receptors elicit suppression of cellular functions by the inhibition of proximal juxtamembranous activation signaling events (Fig. 7.3.2). This proximal inhibitory signal enables potent suppression of cellular activation prior to the establishment of signal amplification steps (mediated by cytosolic compartment signaling intermediates).

Mechanistically, the inhibitory signaling cascade can be also divided into three stages. First, and on engagement of an inhibitory receptor with its ligand(s), the intracellular ITIMs undergo tyrosine phosphorylation (often by an Src family kinase). Second, the phosphorylated ITIM provides a docking site for the recruitment of cytoplasmic phosphatases having an Src homology 2 (SH2) domain such as tyrosine-protein phosphatase non-receptor type 6 [PTPN6; Src homology phosphatase (SHP)] and SHP-2 (PTPN11) but not the lipid phosphatase phosphatidylinositol 3,4,5-triphosphate 5-phosphatase (SHIP)-1 or -2[12] (Fig. 7.3.1). Finally, these aforementioned phosphatases are perceived to dephosphorylate the tyrosine residues that provide docking sites for signaling kinases, which are recruited by activation receptors (and therefore elicit inhibition), or hydrolyze phosphate groups of phosphatidylinositol, thereby affecting multiple signaling pathways.

Most inhibitory receptors recruit the tyrosine phosphatases SHP-1 (PTN6) or phosphatidylinositol 3,4,5-trisphosphate 5-phosphatase 2 (SHIP-2) but not the lipid phosphatase SHIP. Fc-gamma RII-b is an exceptional inhibitory receptor since it primarily recruits SHIPs but not SHPs.[13] However, there are some inhibitory receptors that are capable of recruiting both SHP-1, -2 and SHIP such as CD300a.[14–16]

Although inhibitory receptors are capable of binding SH2-containing phosphatases (i.e., SHP-1, SHP-2,

TABLE 7.3.1 Itim and Itam-bearing Receptors that are Expressed by Eosinophils

Receptor Superfamily	Receptor Name	Expressed on Eosinophils?
Human CD300	CD300a	Yes
	CD300b	n.d.
	CD300c	n.d.
	CD300d	n.d.
	CD300f	Yes
	CD300g	n.d.
Mouse Clm	Clm-1	Yes (Munitz et al., unpublished data)
	Clm-2	Yes
	Clm-3	n.d.
	Clm-4	Yes
	Clm-5	Yes
	Clm-6	n.d.
	Clm-7	Yes
	Clm-8	Yes
	Clm-9	n.d.
Human CD85	CD85a/ILT5/LIR3/LILRB3	Yes
	CD85b/ILT8/LILRA6	n.d.
	CD85c/LIR8/LILRB5	n.d.
	CD85d/ILT4/LIR2/LILRB2	Yes*
	CD85e/ILT6/LIR4/LILRA3	No
	CD85f/ILT11/LIR9/LILRA5	n.d.
	CD85g/ILT7/LILRA4	n.d.
	CD85h/ILT1/LIR7/LILRA2	Yes
	CD85i/LIR6/LILRA1	No
	CD85j/ILT2/LIR1/LILRB1	Yes*
	CD85k/ILT3/LIR5/LILRB4	Yes
	CD85l/ILT9/LILRP1	n.d.
	CD85m/ILT10/LILRP2	n.d.
Mouse	Pir-b/Lilrb3	Yes
	gp49B1/Lilrb4	Yes
Human Siglecs		
	Siglec-1/sialoadhesin/CD169	No
	Siglec-2/CD22	No
	Siglec-3/CD33	Yes

(Continued)

TABLE 7.3.1 Itim and Itam-bearing Receptors that are Expressed by Eosinophils—cont'd

Receptor Superfamily	Receptor Name	Expressed on Eosinophils?
	Siglec-4/MAG	n.d.
	Siglec-5/CD170	Yes
	Siglec-6/CD327	n.d.
	Siglec-7/CD328/p75/AIRM	Yes
	Siglec-8	Yes
	Siglec-9/CD329	No
	Siglec-10	Yes
Mouse Siglecs	Siglec-1/sialoadhesin/CD169	n.d.
	Siglec-2/CD22	Yes
	Siglec-3/CD33	Yes
	Siglec-4/MAG	n.d.
	Siglec-5	Yes (mRNA)
	Siglec-10	Yes (mRNA)
	Siglec-E	Yes (mRNA)
	Siglec-5	Yes
	Siglec-G	No
	Siglec-H	n.d.
Human LAIR	LAIR-1	Yes
	LAIR-2	n.d.
Mouse LAIR	LAIR-1	Munitz et al., unpublished data
	LAIR-2	n.d.

AIRM, adhesion inhibitory receptor molecule; CLM, CMRF35-like molecule; ILT, immunoglobulin-like transcript; LAIR, leukocyte-associated immunoglobulin-like receptor; LILR, leukocyte immunoglobulin-like receptor subfamily; LIR, leukocyte immunoglobulin-like receptor; MAG, myelin-associated glycoprotein; Pir-b, paired immunoglobulin-like receptor B; Siglec, sialic acid-binding Ig-like lectin.
*Found in 33% of eosinophil donors. n.d., not determined.

SHIP-1), the precise inhibitory mechanism is not fully understood and the actual requirement for SHP-1 or SHP-2 molecules may occasionally be redundant. One interesting example of an inhibitory effect in the absence of SHP molecules comes from the inhibitory receptor LAIR-1, which is also expressed by eosinophils. On its phosphorylation, LAIR-1 can bind tyrosine-protein kinase CSK (CSK), which is a negative regulator of the Src family kinases and inhibits B-cell receptor-induced activation even in the absence of SHP-1 and -2.[17] Additional inhibitory receptors, such as ILT-2 and signal-regulatory protein alpha (SIRP-alpha) can also bind CSK (Fig. 7.3.2). Thus, the binding of inhibitory receptors to CSK and consequent SHP-independent inhibition is likely a shared phenomenon at least between several inhibitory receptor families.[18] In addition to CSK, inhibitory receptors can also recruit adaptor molecules with negative regulation capabilities such as the docking protein 1 (DOK1) and 2 (DOK2) family members.[19] In this scenario, inhibitory receptors may recruit DOK proteins, which elicit inhibition that is either mediated by forming a complex with SHIP/SHPs or independently through direct binding of DOK to Ras GTPase-activating protein (RasGAP).[20,21]

It is important to note that the simplistic and traditional view of inhibitory receptor function initially described these receptors as counter-regulators of receptors which contain ITAMs[22] (Fig. 7.3.1). Such motifs are present in various adaptor molecules that facilitate ITAM receptor-mediated activation, including DNAX-activation protein 10 (DAP-10) and DNAX-activation protein 12 (DAP-12), as well as on the Fc-gamma RII-b chain.[23] Contributing to this traditional viewpoint are the findings that many inhibitory receptors are expressed in pairwise fashion with activation receptors. For example, the inhibitory receptor paired immunoglobulin-like receptor B (PIR-B/LILRB3) is expressed in cells that also express paired immunoglobulin-like receptor A (PIR-A/LILRA) (except for B cells, which express PIR-B but do not express PIR-A) and functions as a counter-regulator of PIR-A-induced activation signals.[24] In addition, many inhibitory receptor families comprise both inhibitory and activation receptors such as the CD300 and the ILT receptor families.[8,25] In contrast to the traditional view, emerging data indicate that inhibitory receptors can regulate cellular functions in a broader fashion that is not limited to: (1) their paired activation receptor or (2) an ITAM-containing receptor.[26–28] This claim is also likely true for eosinophils, since several inhibitory receptors can regulate G protein-coupled receptor (GPCR) and interleukin-5 (IL-5) signaling, which do not use ITAMs for their activity. For example, CD300a was found to inhibit both GPCR signaling and cytokine receptor signaling.[15] PIR-B can inhibit chemokine receptor activation, integrin signaling, and innate immune receptor-mediated responses.[29–33] In addition, and supporting this notion, inhibitory receptors are relatively outnumbered by activation pathways (including adhesion molecules, complement component receptors, co-stimulatory molecules, cytokine/chemokine receptors, Fc receptors, pattern recognition receptors, and so on) and a single inhibitory receptor can suppress multiple activation pathways. Thus, these findings suggest that inhibitory receptor signaling dominates activation signals and that inhibitory receptors can counter-regulate several activation pathways.

LIGANDS FOR INHIBITORY RECEPTORS

Early studies in the NK and cytotoxic T-cell field have illustrated the *missing self* hypothesis, in which inhibitory receptors recognize a repertoire of major histocompatibility complex class I (MHC-I) molecules that prevent cellular activation on recognition of a *self* cell.[2,3,34] However, on viral infection or on acquiring a malignant phenotype, MHC-I molecules are significantly downregulated and therefore the cytotoxic cell *loses* its inhibition, rendering the cell active by the presence of activation receptors that are expressed on its surface.

While this paradigm is well understood for cells that are specialized in cell-mediated cytotoxicity, the biological rationale for MHC-I recognition by inhibitory receptors on eosinophils is not quite clear,[27] especially since eosinophils express several MHC-I-binding inhibitory receptors including leukocyte immunoglobulin-like receptor subfamily B member 2 (LIR-2), protein p140, and paired immunoglobulin-like receptor B (PIR-B).[15,35] Other than MHC-I-binding inhibitory receptors, the majority of the inhibitory receptors that are expressed by eosinophils do not bind MHC-I molecules. Therefore, it is evident that another class of MHC-I-independent inhibitory recognition is present in eosinophils (and other myeloid cells).[36] For example, the ligand for gp49B1 has been reported to be αVβ3 integrin and collagens have been shown to serve as functional, high-affinity ligands for the inhibitory immune receptor LAIR-1.[37,38]

Interestingly, even PIR-B, which was originally identified as a myeloid-specific inhibitory receptor that binds MHC-I molecules, has been recently shown to recognize and bind various bacteria and inhibit bacterial-induced cellular activation.[30,39,40] Of note, *Pirb*$^{-/-}$ mice display increased gastrointestinal eosinophilia at baseline that may be due to increased reactivity of *Pirb*$^{-/-}$ eosinophils to eotaxin or to some (yet-to-be-defined) innate stimulus. Substantiating a role for non-MHC-restricted inhibitory signaling are also recent reports demonstrating that the inhibitory receptor PILRA recognizes CD99, which is widely expressed on all leukocytes.[41]

A ligand for CD300a and CD300f has been identified (see note added in proof). In fact, CD300a was considered at one point to be a potential NK cell inhibitory molecule, but it failed to bind to MHC-I molecules,[8] making it unlikely that MHC-I molecules are its natural ligand. Future studies defining the ligands of inhibitory receptors that are expressed by eosinophils are likely to contribute to our understanding of the physiological role for such receptors in immune regulation of eosinophil functions and are likely to shed light on the potential interactions of eosinophils with other cell types or extracellular matrix components.

INHIBITION OF EOSINOPHIL RESPONSES

Eosinophil Survival

As mentioned previously, functional studies on inhibitory receptor activity in eosinophils are more limited. While eosinophils have been shown to express several inhibitory receptors, only a few of them have been examined thoroughly.

Siglec-8 is expressed in eosinophils, mast cells, and basophils.[42,43] Co-ligation of Siglec-8 by anti-Siglec-8 antibodies, or a polymer expressing its ligand with

FIGURE 7.3.2 Inhibitory receptors as an opposing arm of immune cell signaling. Activation of cellular responses via the activation of receptors such as tyrosine kinases and/or G protein-coupled receptors, signal transduction can occur in four main steps namely: (1) juxtamembranous signaling steps (i.e., tyrosine phosphorylation); (2) binding of kinases and/or adaptor proteins; (3) amplification of the signaling cascade (activation of MAPKs, STATs, etc.); and finally (4) targeted gene transcription by multiple transcription factors. Inhibitory receptors can also undergo several steps that initiate their signaling: (1) phosphorylation of ITIMs; (2) binding to intracellular phosphatases; and (3) dephosphorylation of activation receptor-mediated pathways. Of note, all signaling steps of inhibitory receptors occur at a proximal stage of signaling, therefore resulting in the suppression of activation prior to signal transduction amplification. ITIM, immunoreceptor tyrosine-based inhibitory motif; MAPK, mitogen-activated protein kinase; STAT, signal transducer and activator of transcription.

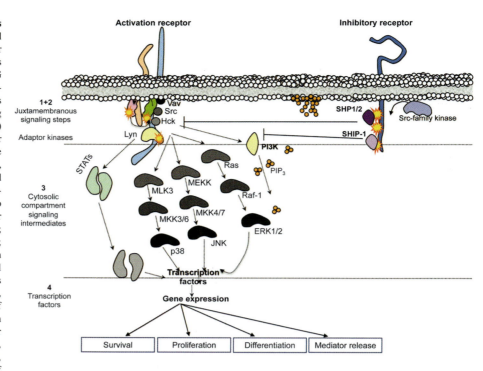

a secondary antibody, has been shown to mediate eosinophil apoptosis by initiating mitochondrial injury, generation of reactive oxygen species (ROS), and rapid cleavage of caspase-3 (CASP-3), -8 (CASP-8) and -9 (CASP-9)[44–46] (Fig. 7.3.3). Of even greater interest is the finding that in the presence of the *hallmark eosinophil survival cytokines* IL-5 and granulocyte-macrophage colony-stimulating factor (GM-CSF), Siglec-8 activity was further enhanced and requirement for co-ligation of the receptor by a secondary antibody was not needed.[45,46]

Furthermore, administration of anti-Siglec-F antibody (the murine functional paralogue of Siglec-8) in a model of experimental asthma significantly reduced peribronchial eosinophilic inflammation and subepithelial fibrosis.[47] These results correlated with increased eosinophil apoptosis in the lung and bone marrow.[47] In other studies, a single dose of anti-Siglec-F antibody to IL-5 transgenic mice resulted in rapid reduction of peripheral blood eosinophilia, which was accompanied by a reduction in tissue eosinophilia.[48] Interestingly, and in contrast to its role in eosinophils, Siglec-8 was unable to induce mast cell apoptosis but rather significantly dampened Fc-epsilon RI-alpha-dependent histamine and prostaglandin D2 receptor 2 (PD2R2) secretion, as well as Fc-epsilon RI-alpha-dependent contraction of human airway bronchial rings. Importantly, point mutation analysis revealed that this effect is dependent on the proximal ITIM of Siglec-8 (Fig. 7.3.3).[44] In this regard, it is noteworthy that the functional mechanism of Siglecs can differ between various

cell types. For example, Siglec-9 has been shown to actively induce neutrophil death via a caspase-dependent (apoptotic) and -independent pathway.[49] Results from experiments using scavengers of ROS or neutrophils that were unable to generate ROS indicated that both Siglec-9-mediated, caspase-dependent and caspase-independent forms of neutrophil death depend on ROS. Interestingly, the caspase-independent apoptotic pathway in neutrophils was characterized by cytoplasmic vacuolization and several other nonapoptotic morphological features.[49]

Similar to studies with Siglec-8, CD300a/IRp60 was also shown to suppress eosinophil survival (Fig. 7.3.3).[15] However, in contrast to Siglec-8, CD300a does not actively induce apoptosis but rather prevents IL-5- and GM-CSF-mediated survival signals. Indeed, CD300a was shown to suppress tyrosine-protein kinase JAK2 (JAK2) phosphorylation and therefore may be suppressing the function of the common beta chain (βc) of the GM-CSF, interleukin-3 (IL-3), and IL-5 complex.[15] Remarkably (and analogous to Siglec-8), increasing concentrations of GM-CSF and IL-5 enhanced the inhibitory activity of CD300a.[15]

The different outcome of Siglec-8 activation in eosinophils (induction of caspase-dependent apoptosis) as opposed to IRp60/CD300 activation (inhibition of survival signals) may be due to the fact that Siglec-8 contains both ITIMs and immunoreceptor tyrosine-based switch motifs (ITSMs), which may recruit adaptor molecules such as the signaling lymphocytic activation molecule-associated protein (SLAM-associated protein) containing SH2

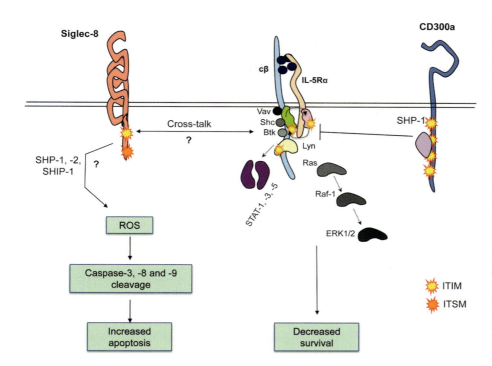

FIGURE 7.3.3 Regulation of eosinophil survival by CD300a and Siglec-8. Activation of the inhibitory receptors CD300 and Siglec-8 results in eosinophil death. Despite the common outcome, CD300a and Siglec-8 use distinct molecular mechanisms to achieve cell death. Siglec-8 activation induces the generation of reactive oxygen species (ROS) and cleavage of caspases ultimately leading to the induction of active apoptosis in eosinophils. In contrast to Siglec-8, ligation of CD300a suppresses survival signals that are delivered by the IL-5 receptor (IL-5R) complex via SHP-1 and suppression of JAK2, p38, and MAPK 1 and 2 phosphorylation. Interestingly, IL-5 and other eosinophil survival cytokines (i.e., IL-3 and GM-CSF) amplify the proapoptotic and anti-survival effects elicited by Siglec-8 and CD300a, respectively. GM-CSF, granulocyte-macrophage colony-stimulating factor; IL-3, interleukin-3; MAPK, mitogen-activated protein kinase; SHP, Src homology phosphatase, Siglec, sialic acid-binding Ig-like lectin.

domain-containing protein 1A (SH2D1A/SAP) and/or EWS/FLI1-activated transcript 2 (EAT-2/SH2D1B).[50] Nonetheless, this hypothesis needs to be examined as the signaling mechanisms of both Siglec-8 and CD300a in eosinophils have not been fully described and the actual function and necessity of the ITSM domain of Siglec-8 is currently unknown.

EOSINOPHIL MIGRATION

Besides the role of inhibitory receptors in eosinophil survival, their role in the regulation of eosinophil migration has received some attention. Recruitment of eosinophils involves a signaling cascade in which secreted chemokines interact with heterotrimeric GPCRs and especially with C-C chemokine receptor type 3 (CCR3), the receptor for eotaxins. In response to eotaxin stimulation, CCR3 induces a signaling cascade that is accompanied by Ca^{2+} mobilization and the activation of Ras—MAPK, especially extracellular signal-regulated kinase 1 and 2 (ERK-1/2)-dependent pathways, and can associate with Src family kinases such as tyrosine-protein kinase Fgr (FGR) and tyrosine-protein kinase HCK (HCK).[51] Recent findings demonstrate cross talk between GPCR signaling and inhibitory receptors signaling. We have recently shown that CD300a can negatively regulate human eosinophil migration in response to eotaxin *in vitro* and that PIR-B is a negative regulator of murine eosinophil migration both *in vivo* and *in vitro* in response to eotaxin. Unexpectedly,

we observed that in response to leukotriene B4 (LTB4) stimulation, PIR-B could deliver activation signals rather than inhibitory ones. The exact mechanism for this dual response is currently unknown. Yet, it is likely to involve differential Src family kinase recruitment to the inhibitory receptor. Specifically, the kinases HCK and FGR may be involved. In fact, HCK and FGR can phosphorylate the ITIMs of PIR-B in neutrophils, which in turn recruits SHP-1 and SHP-2, resulting in the suppression of cell activation.[26] Furthermore, HCK and FGR, as well as SHP-1, play key roles in the regulation of myeloid leukocyte migration. Neutrophils and dendritic cells that lack PIR-B or SHP-1 display enhanced chemokine signaling and functional responses, as do FGR- and HCK-deficient cells.[26,29] Although HCK interacts with PIR-B in eosinophils, its involvement in phosphorylation of PIR-B is unclear since HCK—PIR-B interactions were demonstrated following SHP recruitment and phosphorylation and not preceding them. Thus, in eosinophils, alternative molecular switches may regulate the regulatory function of PIR-B.

Mechanistically, both CD300a and PIR-B have been shown to regulate the phosphorylation of ERK-1 and -2 and to interact with phosphatases. However, the role of these phosphatases and the downstream molecular events leading to negative regulation of MAPK 1 and 2 activation are yet to be defined.

Interestingly, a novel role for C-X-C motif chemokine 9/monokine induced by interferon-gamma (CXCL9/MIG) in the inhibition of murine eosinophil recruitment was recently demonstrated.[52] In their study, Fulkerson and colleagues reported that the binding of CXCL9/MIG to CCR3, activates an inhibitory cascade by a yet-to-be-defined mechanism.[53] Although this study was not conducted on a classical ITIM-bearing receptor, it suggests that different chemokines and perhaps other agonists can use CCR3 to inhibit eosinophil functions. Mechanistically, these findings could imply that a substantial cross talk occurs between inhibitory receptors and *eosinophil-specific* cytokine receptors such as CCR3.

EOSINOPHIL EFFECTOR FUNCTIONS

Eosinophils have been attributed various effector functions, some opposing to others. Given that inhibitory receptors can counter-regulate eosinophil activation, an important question arises on whether they are capable of regulating eosinophil effector functions. Several lines of evidence indicate that this may indeed be the case. *In vitro*, we have shown that CD300a cross-linking inhibits profibrogenic mediator release from eosinophils resulting in decreased fibroblast proliferation. *In vivo*, a bispecific antibody targeting CCR3 and CD300a was capable of activating the inhibitory machinery of CD300 and suppressing the chronic remodeling effects that were observed in experimental asthma settings including mucus production, fibrosis, and transforming growth factor beta (TGF-β) expression. In addition, allergen-challenged mice that lack the inhibitory receptor gp49B1 display increased lung inflammation and eosinophil recruitment that is correlated with a more severe phenotype of allergic inflammation.[54,55] While the latter study did not directly examine the role of gp49B1 in eosinophils, gp49B1 expression in eosinophils was likely responsible at least in part for the increased disease phenotype that was observed in the gene-targeted mice.

TARGETING INHIBITORY RECEPTORS AS A THERAPEUTIC APPROACH

Despite limited data on the function of inhibitory receptors on eosinophils (compared to other cells such as NK cells and B cells) various approaches have been already taken to target such receptors as a therapeutic approach. Evidently, the fact that several independent groups have used similar strategies to target these receptors supports the notion that inhibitory receptors are potent targets for future drug design.

As previously described, IRp60/CD300a is a potent negative regulator of eosinophil responses.[14,15] Bispecific antibody fragments, capable of recognizing CCR3 and CD300a/IRp60 or immunoglobulin E (IgE) and CD300a were designed and administered *in vivo* in murine models of allergic peritonitis, and a chronic model of established allergic eosinophilic airway inflammation. These studies demonstrate the potential of targeting inhibitory receptors for the specific suppression of eosinophils in allergic responses and perhaps even reversal of the inflammatory process and associated remodeling[16,56] (Fig. 7.3.4). Furthermore, they highlight the potential of inhibitory receptors as pharmacological targets in other eosinophil-associated diseases, including hypereosinophilic syndromes and eosinophilic gastrointestinal disorders.

FIGURE 7.3.4 Targeting inhibitory receptors on eosinophils as a therapeutic strategy. Inhibitory receptors that are expressed by eosinophils could serve as potential therapeutic targets in eosinophil-associated diseases such as allergy, hypereosinophilic syndromes, and eosinophil gastrointestinal disorders. Such receptors could potentially suppress multiple pathways in eosinophils including differentiation, proliferation, chemotaxis, survival, degranulation/mediator release, and consequent tissue effector functions. CCR3, C-C chemokine receptor type 3.

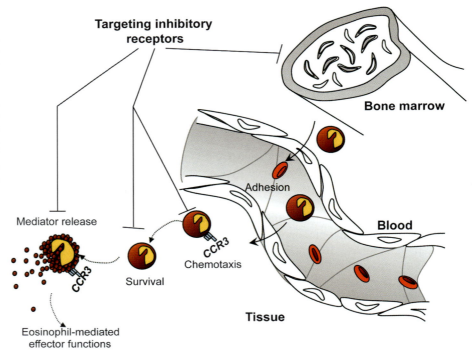

Nevertheless and most importantly, when targeting these receptors one should take into account various factors including cell specificity, general immune suppression, and the inflammatory context, which may induce ITIM-dependent coactivation rather than inhibition. In fact, recent data suggest that ITIM-bearing receptors may actually mediate activation and that ITAM-bearing receptors could induce inhibition.[57,58] Although the subject of activation through ITIMs is not fully understood, there are now various reports on the reciprocal activity for such receptors. We have recently shown that in response to LTB$_4$ stimulation, PIR-B can serve as a docking site for various adaptor kinases that can promote cellular activation. In fact, our data support a model in which, following LTB$_4$ activation, PIR-B can interact with several activating kinases, such as JAK1, JAK2, SHC-transforming protein (SHC) and adaptor molecule crk (CRK).[31] Similar to PIR-B, the inhibitory receptor SIRP-alpha, was recently reported to stimulate nitric oxide production in macrophages via the JAK—signal transducer and activator of transcription (STAT) and phosphatidylinositol 3-kinase (PI3K)—/Ras-related protein Rac1 (Rac1) pathways, thereby suggesting an activating role for SIRP-alpha in a similar molecular mechanism to that of PILRB.[59] Furthermore, the inhibitory receptors Trem-like transcript 1 protein (TLT-1) and cholecystokinin (CCK) receptor have been also shown to enhance cellular activation.[57] For example, TLT-1 amplifies Fc-epsilon RI-alpha-mediated calcium signaling and CCK (via the recruitment of SHP-2 to its ITIM) activates the Akt [protein kinase B (PKB)] pathway.[60] Finally, studies assessing the role of CMRF35-like molecule 1 (CLM-1/CD300f) in mast cells reveal that CLM-1 cross-linking enhanced cytokine production of bone marrow-derived mast cells stimulated by lipopolysaccharide, while suppressing their production stimulated by other Toll-like receptor agonists or by stem cell factor.

Thus, although it appears that the primary role of inhibitory receptors is to indeed suppress cellular activation, under specific circumstances they can also induce cellular activation.

CONCLUSION AND FUTURE PERSPECTIVES

Despite ongoing progress in the field of inhibitory receptor function and signaling in eosinophils, our understanding of these receptors' activities is still lacking and further investigations studying these receptors are needed. One specific interest is to define whether IL-5 signaling is regulated by inhibitory receptors. Given the dependency of eosinophils on IL-5 signaling *in vitro* and *in vivo*, this task may be technically challenging. However, it will likely provide substantial information on the role of inhibitory receptors in eosinophil maturation and differentiation.

In summary, future studies on the roles of inhibitory receptors will likely define:

- the precise role of inhibitory receptors in various inflammatory settings involving eosinophils;
- whether inhibitory receptors differentially regulate eosinophil function and whether eosinophils from different anatomical sources (especially blood vs. lung vs. gastrointestinal tract) are differentially regulated by a given inhibitory receptor;
- the molecular pathways that mediate ITIM-dependent and -independent inhibition, especially defining the kinase complexes that can phosphorylate ITIM domains in eosinophils;
- and characterize potential ligands for inhibitory receptors;
- the pathways that regulate inhibitory receptor expression on eosinophils in homeostasis and in inflammatory settings.

Addressing these questions will enhance our understanding of the molecular regulation of eosinophil function in health and disease. Furthermore, they may provide novel tools for targeting inhibitory receptors on eosinophils as a therapeutic approach in eosinophil-associated diseases.

NOTE ADDED IN PROOF

Recently, two independent laboratories[61-63] have identified that CD300a/CLM-8 binds phosphatidylserine (PS) and phosphatidylethanolamine on apoptotic cells. Studies using $Clm8^{-/-}$ mice revealed that CD300a:PS interactions critically regulate mast cell inflammatory responses to microbial infections. The relevance of these findings to CD300a expression in eosinophils is yet to be determined.

ACKNOWLEDGEMENTS

I would like to thank Drs Bruce Bochner and Nives Zimmermann for the helpful discussions and exchange of information. In addition, I would like to thank Dr. Marc E. Rothenberg for his mentorship, support, and willingness to share some unpublished results. Finally, I would like to thank Dr. Francesca Levi-Schaffer, who introduced me to the fascinating world of eosinophil research.

A.M. is supported by grants from the FP7 Marie-Curie Reintegration program (PIRG-GA-2009-256311) and by the Bi-National US-Israel Science Foundation (#2009222).

REFERENCES

1. Daeron M, Jaeger S, Du Pasquier L, Vivier E. Immunoreceptor tyrosine-based inhibition motifs: a quest in the past and future. *Immunological Reviews* 2008;**224**:11—43.
2. Lanier LL. NK cell receptors. *Annual Review of Immunology* 1998;**16**:359—93.
3. Ravetch JV, Lanier LL. Immune inhibitory receptors. *Science* 2000;**290**:84—9.

4. Munitz A, Levi-Schaffer F. Inhibitory receptors on eosinophils: a direct hit to a possible Achilles heel? *The Journal of Allergy and Clinical Immunology* 2007;**119**:1382−7.

5. Cella M, Dohring C, Samaridis J, Dessing M, Brockhaus M, Lanzavecchia A, et al. A novel inhibitory receptor (ILT3) expressed on monocytes, macrophages, and dendritic cells involved in antigen processing. *The Journal of Experimental Medicine* 1997;**185**:1743−51.

6. Meyaard L, Adema GJ, Chang C, Woollatt E, Sutherland GR, Lanier LL, et al. LAIR-1, a novel inhibitory receptor expressed on human mononuclear leukocytes. *Immunity* 1997;**7**:283−90.

7. Katz HR. Inhibition of pathologic inflammation by leukocyte Ig-like receptor B4 and related inhibitory receptors. *Immunological Reviews* 2007;**217**:222−30.

8. Clark GJ, Ju X, Tate C, Hart DN. The CD300 family of molecules are evolutionarily significant regulators of leukocyte functions. *Trends in Immunology* 2009;**30**:209−17.

9. Crocker PR, Paulson JC, Varki A. Siglecs and their roles in the immune system. *Nature Reviews* 2007;**7**:255−66.

10. Robinson MJ, Sancho D, Slack EC, LeibundGut-Landmann S, Reis e Sousa C. Myeloid C-type lectins in innate immunity. *Nature Immunology* 2006;**7**:1258−65.

11. Brooks AG, Posch PE, Scorzelli CJ, Borrego F, Coligan JE. NKG2A complexed with CD94 defines a novel inhibitory natural killer cell receptor. *The Journal of Experimental Medicine* 1997;**185**:795−800.

12. Long EO. Regulation of immune responses through inhibitory receptors. *Annual Review of Immunology* 1999;**17**:875−904.

13. Fong DC, Malbec O, Arock M, Cambier JC, Fridman WH, Daeron M. Selective in vivo recruitment of the phosphatidylinositol phosphatase SHIP by phosphorylated Fc gammaRIIB during negative regulation of IgE-dependent mouse mast cell activation. *Immunology Letters* 1996;**54**:83−91.

14. Bachelet I, Munitz A, Moretta A, Moretta L, Levi-Schaffer F. The inhibitory receptor IRp60 (CD300a) is expressed and functional on human mast cells. *J Immunol* 2005;**175**:7989−95.

15. Munitz A, Bachelet I, Eliashar R, Moretta A, Moretta L, Levi-Schaffer F. The inhibitory receptor IRp60 (CD300a) suppresses the effects of IL-5, GM-CSF, and eotaxin on human peripheral blood eosinophils. *Blood* 2006;**107**:1996−2003.

16. Bachelet I, Munitz A, Levi-Schaffer F. Abrogation of allergic reactions by a bispecific antibody fragment linking IgE to CD300a. *The Journal of Allergy and Clinical Immunology* 2006;**117**:1314−20.

17. Verbrugge A, Rijkers ES, de Ruiter T, Meyaard L. Leukocyte-associated Ig-like receptor-1 has SH2 domain-containing phosphatase-independent function and recruits C-terminal Src kinase. *European Journal of Immunology* 2006;**36**:190−8.

18. Sayos J, Martinez-Barriocanal A, Kitzig F, Bellon T, Lopez-Botet M. Recruitment of C-terminal Src kinase by the leukocyte inhibitory receptor CD85j. *Biochemical and Biophysical Research Communications* 2004;**324**:640−7.

19. Zhang S, Cherwinski H, Sedgwick JD, Phillips JH. Molecular mechanisms of CD200 inhibition of mast cell activation. *J Immunol* 2004;**173**:6786−93.

20. Berg KL, Siminovitch KA, Stanley ER. SHP-1 regulation of p62(DOK) tyrosine phosphorylation in macrophages. *The Journal of Biological Chemistry* 1999;**274**:35855−65.

21. Robson JD, Davidson D, Veillette A. Inhibition of the Jun N-terminal protein kinase pathway by SHIP-1, a lipid phosphatase that interacts with the adaptor molecule Dok-3. *Molecular and Cellular Biology* 2004;**24**:2332−43.

22. Long EO. Negative signaling by inhibitory receptors: the NK cell paradigm. *Immunological Reviews* 2008;**224**:70−84.

23. Wu J, Cherwinski H, Spies T, Phillips JH, Lanier LL. DAP10 and DAP12 form distinct, but functionally cooperative, receptor complexes in natural killer cells. *The Journal of Experimental Medicine* 2000;**192**:1059−68.

24. Kubagawa H, Burrows PD, Cooper MD. A novel pair of immunoglobulin-like receptors expressed by B cells and myeloid cells. *Proceedings of the National Academy of Sciences of the United States of America* 1997;**94**:5261−6.

25. Cella M, Nakajima H, Facchetti F, Hoffmann T, Colonna M. ILT receptors at the interface between lymphoid and myeloid cells. *Current Topics in Microbiology and Immunology* 2000;**251**:161−6.

26. Brown EJ. Leukocyte migration: dismantling inhibition. *Trends in Cell Biology* 2005;**15**:393−5.

27. Sinclair NR. Why so many coinhibitory receptors? *Scandinavian Journal of Immunology* 1999;**50**:10−3.

28. Dietrich J, Nakajima H, Colonna M. Human inhibitory and activating Ig-like receptors which modulate the function of myeloid cells. *Microbes and Infection/Institut Pasteur* 2000;**2**:323−9.

29. Zhang H, Meng F, Chu CL, Takai T, Lowell CA. The Src family kinases Hck and Fgr negatively regulate neutrophil and dendritic cell chemokine signaling via PIR-B. *Immunity* 2005;**22**:235−46.

30. Nakayama M, Underhill DM, Petersen TW, Li B, Kitamura T, Takai T, et al. Paired Ig-like receptors bind to bacteria and shape TLR-mediated cytokine production. *J Immunol* 2007;**178**:4250−9.

31. Munitz A, McBride ML, Bernstein JS, Rothenberg ME. A dual activation and inhibition role for the paired immunoglobulin-like receptor B in eosinophils. *Blood* 2008;**111**:5694−703.

32. Torii I, Oka S, Hotomi M, Benjamin Jr WH, Takai T, Kearney JF, et al. PIR-B-deficient mice are susceptible to Salmonella infection. *J Immunol* 2008;**181**:4229−39.

33. Munitz A, Cole ET, Beichler A, Groschwitz K, Ahrens R, Steinbrecher K, et al. Paired immunoglobulin-like receptor B (PIR-B) negatively regulates macrophage activation in experimental colitis. *Gastroenterology* 2010;**139**:530−41.

34. Moretta A, Bottino C, Vitale M, Pende D, Biassoni R, Mingari MC, et al. Receptors for HLA class-I molecules in human natural killer cells. *Annual Review of Immunology* 1996;**14**:619−48.

35. Tedla N, Bandeira-Melo C, Tassinari P, Sloane DE, Samplaski M, Cosman D, et al. Activation of human eosinophils through leukocyte immunoglobulin-like receptor 7. *Proceedings of the National Academy of Sciences of the United States of America* 2003;**100**:1174−9.

36. Lebbink RJ, Meyaard L. Non-MHC ligands for inhibitory immune receptors: novel insights and implications for immune regulation. *Molecular Immunology* 2007;**44**:2153−64.

37. Castells MC, Klickstein LB, Hassani K, Cumplido JA, Lacouture ME, Austen KF, et al. gp49B1-alpha(v)beta3 interaction inhibits antigen-induced mast cell activation. *Nature Immunology* 2001;**2**:436−42.

38. Lebbink RJ, de Ruiter T, Adelmeijer J, Brenkman AB, van Helvoort JM, Koch M, et al. Collagens are functional, high affinity ligands for the inhibitory immune receptor LAIR-1. *The Journal of Experimental Medicine* 2006;**203**:1419−25.

39. Takai T. Paired immunoglobulin-like receptors and their MHC class I recognition. *Immunology* 2005;**115**:433−40.

40. Masuda A, Nakamura A, Maeda T, Sakamoto Y, Takai T. Cis binding between inhibitory receptors and MHC class I can regulate mast cell activation. *The Journal of Experimental Medicine* 2007;**204**:907−20.

41. Shiratori I, Ogasawara K, Saito T, Lanier LL, Arase H. Activation of natural killer cells and dendritic cells upon recognition of a novel CD99-like ligand by paired immunoglobulin-like type 2 receptor. *The Journal of Experimental Medicine* 2004;**199**:525−33.

42. Floyd H, Ni J, Cornish AL, Zeng Z, Liu D, Carter KC, et al. Siglec-8. A novel eosinophil-specific member of the immunoglobulin superfamily. *The Journal of Biological Chemistry* 2000;**275**:861−6.

43. Kikly KK, Bochner BS, Freeman SD, Tan KB, Gallagher KT, D'Alessio KJ, et al. Identification of SAF-2, a novel siglec expressed on eosinophils, mast cells, and basophils. *The Journal of Allergy and Clinical Immunology* 2000;**105**:1093−100.

44. Yokoi H, Choi OH, Hubbard W, Lee HS, Canning BJ, Lee HH, et al. Inhibition of FcepsilonRI-dependent mediator release and calcium flux from human mast cells by sialic acid-binding immunoglobulin-like lectin 8 engagement. *The Journal of Allergy and Clinical Immunology* 2008;**121**:499−505. e491.

45. Nutku E, Aizawa H, Hudson SA, Bochner BS. Ligation of Siglec-8: a selective mechanism for induction of human eosinophil apoptosis. *Blood* 2003;**101**:5014−20.

46. Nutku E, Hudson SA, Bochner BS. Mechanism of Siglec-8-induced human eosinophil apoptosis: role of caspases and mitochondrial injury. *Biochemical and Biophysical Research Communications* 2005;**336**:918−24.

47. Song DJ, Cho JY, Lee SY, Miller M, Rosenthal P, Soroosh P, et al. Anti-Siglec-F antibody reduces allergen-induced eosinophilic inflammation and airway remodeling. *J Immunol* 2009;**183**:5333−41.

48. Zimmermann N, McBride ML, Yamada Y, Hudson SA, Jones C, Cromie KD, et al. Siglec-F antibody administration to mice selectively reduces blood and tissue eosinophils. *Allergy* 2008;**63**:1156−63.

49. von Gunten S, Yousefi S, Seitz M, Jakob SM, Schaffner T, Seger R, et al. Siglec-9 transduces apoptotic and nonapoptotic death signals into neutrophils depending on the proinflammatory cytokine environment. *Blood* 2005;**106**:1423−31.

50. Engel P, Eck MJ, Terhorst C. The SAP and SLAM families in immune responses and X-linked lymphoproliferative disease. *Nature Reviews* 2003;**3**:813−21.

51. El-Shazly A, Yamaguchi N, Masuyama K, Suda T, Ishikawa T. Novel association of the src family kinases, hck and c-fgr, with CCR3 receptor stimulation: A possible mechanism for eotaxin-induced human eosinophil chemotaxis. *Biochemical and Biophysical Research Communications* 1999;**264**:163−70.

52. Fulkerson PC, Zimmermann N, Brandt EB, Muntel EE, Doepker MP, Kavanaugh JL, et al. Negative regulation of eosinophil recruitment to the lung by the chemokine monokine induced by IFN-gamma (Mig, CXCL9). *Proceedings of the National Academy of Sciences of the United States of America* 2004;**101**:1987−92.

53. Fulkerson PC, Zhu H, Williams DA, Zimmermann N, Rothenberg ME. CXCL9 inhibits eosinophil responses by a CCR3- and Rac2-dependent mechanism. *Blood* 2005;**106**:436−43.

54. Breslow RG, Rao JJ, Xing W, Hong DI, Barrett NA, Katz HR. Inhibition of Th2 adaptive immune responses and pulmonary inflammation by leukocyte Ig-like receptor B4 on dendritic cells. *J Immunol* 2010;**184**:1003−13.

55. Norris HH, Peterson ME, Stebbins CC, McConchie BW, Bundoc VG, Trivedi S, et al. Inhibitory receptor gp49B regulates eosinophil infiltration during allergic inflammation. *Journal of Leukocyte Biology* 2007;**82**:1531−41.

56. Munitz A, Bachelet I, Levi-Schaffer F. Reversal of airway inflammation and remodeling in asthma by a bispecific antibody fragment linking CCR3 to CD300a. *The Journal of Allergy and Clinical Immunology* 2006;**118**:1082−9.

57. Barrow AD, Trowsdale J. You say ITAM and I say ITIM, let's call the whole thing off: the ambiguity of immunoreceptor signalling. *European Journal of Immunology* 2006;**36**:1646−53.

58. Pinheiro da Silva F, Aloulou M, Benhamou M, Monteiro RC. Inhibitory ITAMs: a matter of life and death. *Trends in Immunology* 2008;**29**:366−73.

59. Alblas J, Honing H, de Lavalette CR, Brown MH, Dijkstra CD, van den Berg TK. Signal regulatory protein alpha ligation induces macrophage nitric oxide production through JAK/STAT- and phosphatidylinositol 3-kinase/Rac1/NAPDH oxidase/H2O2-dependent pathways. *Molecular and Cellular Biology* 2005;**25**:7181−92.

60. Vatinel S, Ferrand A, Lopez F, Kowalski-Chauvel A, Esteve JP, Fourmy D, et al. An ITIM-like motif within the CCK2 receptor sequence required for interaction with SHP-2 and the activation of the AKT pathway. *Biochimica et Biophysica Acta* 2006;**1763**:1098−107.

61. Nakahashi-Oda C, Tahara-Hanaoka S, Shoji M, Okoshi Y, Nakano-Yokomizo T, et al. Apoptotic cells suppress mast cell inflammatory responses via the cd300a immunoreceptor. *J Exp Med* 2012;**209**:1493−503.

62. Simhadri VR, Andersen JF, Calvo E, Choi SC, Coligan JE, Borrego F. Human cd300a binds to phosphatidylethanolamine and phosphatidylserine, and modulates the phagocytosis of dead cells. *Blood* 2012;**119**:2799−809.

63. Nakahashi-Oda C, Tahara-Hanaoka S, Honda S, Shibuya K, Shibuya A. Identification of phosphatidylserine as a ligand for the cd300a immunoreceptor. *Biochem Biophys Res Commun* 2012;**417**:646−50.

Chapter 7.4

Apoptotic and Survival Signaling in Eosinophils

Utibe R. Bickham and James S. Malter

INTRODUCTION

The mammalian immune system is composed of a variety of effector cells predominantly generated in the bone marrow. Despite a common developmental location and overlapping lineage designations, the disparate functions of immune cells necessitate considerable variation in their physiology. One important discriminating feature, which has recently gained considerable attention, are differences in how and under what circumstances populations of immune cells die. Alterations in this fundamental process can be an appropriate or dysregulated response to physiological or pathological stimuli, respectively. In this chapter, we discuss the unique features of eosinophil survival and death decisions with emphasis on the intracellular signals that mediate these processes.

NORMAL EOSINOPHIL PHYSIOLOGY

In the bone marrow, eosinophils develop from pluripotent stem cells, which differentiate into eosinophil lineage committed progenitors (EoPs). EoPs are more extensively differentiated progenitors than granulocyte/macrophage progenitors (GMPs)[1] and are characterized by the surface antigen profile IL-5RA[+] CD34[+] CD38[+] IL-3RA[+] CD45RA[−]. Such progenitors give rise, nearly exclusively, to eosinophils.[2] After stimulation by cytokines, particularly interleukin-3 (IL-3), interleukin-5 (IL-5) and/or granulocyte-macrophage colony stimulating factor (GM-CSF), the progenitor cells differentiate into eosinophils, exit the bone marrow and enter the peripheral blood circulation. Peripheral counts are positively influenced by chemokines such as eotaxin and eotaxin-2, as well as IL-5. Eosinophils typically circulate for approximately 8–12 h[2] before entering targeted resident tissues, including the gastrointestinal tract, thymus, uterus, and mammary glands.[1] Eosinophil egress from the circulation requires interactions with endothelial surfaces, the extracellular matrix, as well as the stroma of resident tissues; these interactions likely contribute to eosinophil death decisions and are discussed in more detail later in this chapter.

DEATH SIGNALS

In the absence of inflammatory signaling, circulating or tissue-based eosinophils survive briefly with an average life span of 3–4 days.[2] Peripheral blood eosinophils isolated by negative selection from normal or mildly atopic individuals and cultured *in vitro* show a similar life span.[3] Thus, peripheral blood eosinophils are physiologically programmed to undergo cell death in the absence of exogenous stimuli, making them an excellent experimental system to study these events in normal, primary human cells.

While there have been reports of autophagic bodies within dying granulocytes, apoptosis is clearly the predominant pathway by which eosinophils die. Like other cells, eosinophil apoptosis generally, but not always, requires the activation of pro-death and the suppression of pro-survival B-cell lymphoma 2 protein (Bcl-2) family members, caspase activation, loss of plasma and nuclear membrane integrity, and DNA and protein fragmentation.[4,5]

Caspases are cysteine proteases typically with specificity for aspartate and are subclassified as apoptotic initiators or effectors. Initiators include caspase-2 (CASP-2), -8 (CASP-8), -9 (CASP-9) and -10 (CASP-10).

These caspases contain a long prodomain containing either a caspase recruitment domain-containing protein (CARD) or a death effector domain (DED). Initiator caspases are characterized by their ability to autoactivate within specialized complexes.[6] CASP-3, CASP-6, and CASP-7 are effectors. Effector caspases contain a short prodomain whose activating cleavage is performed by other proteases (such as caspases or calpain). Caspase-cleaved proteins show aberrant function but are not catabolized.[7] CASP-3, CASP-8, and CASP-9 are constitutively expressed and all have been implicated in eosinophil apoptosis induced by dexamethasone as well as growth factor depletion.[8,9]

Eosinophil apoptosis can result from the activation of either intrinsic or extrinsic pathways. These pathways differ in the identity of initiator caspases as well as the involvement of mitochondria. Death signaling through the intrinsic pathway requires mitochondrial disruption and the release of proapoptotic proteins including apoptosis-inducing factor (AIF), second mitochondria-derived activator of caspase (Smac/Diablo), high temperature requirement protein A2 (HtrA2/Omi) or cytochrome c into the cytoplasm. Cytochrome c binds to the WD40 repeat domain of apoptotic protease-activating factor 1 (APAF-1) while the CARD domains of APAF-1 interact with those of procaspase-9 to form the apoptosome.[9] In the presence of adenosine triphosphate (ATP), these interactions trigger both a conformational change and oligomerization. Oligomerized APAF-1 recruits additional molecules of procaspase-9, which are then activated by cleavage.

The activity of the apoptosome can be modulated by proteins other than those involved in the formation of the complex. For instance, the mitogen-activated protein kinase (MAPK) c-Jun N-terminal kinase (JNK) interacts with the pre-apoptosome and delays CASP-9 recruitment and activation.[10] However, JNK is activated by Fas antigen ligand (FasL) or reactive oxygen species in eosinophils and is required for apoptosis.[11] Thus, a single signaling intermediate can have divergent actions, possibly depending on the activation conditions as well as the cell type. Other examples include the inhibitor of apoptosis (IAP) family [baculoviral IAP repeat-containing protein 3 (also known as c-IAP2), X-linked inhibitor of apoptosis protein, and baculoviral IAP repeat-containing protein 5 (also known as apoptosis inhibitor survivin)], which bind to and modulate the activity of caspases via their ubiquitin ligase activity. Recent data indicate that eosinophils from patients with hypereosinophilic syndrome (HES) overexpress c-IAP2 and survivin.[12] Similar changes in gene expression were observed in normal eosinophils treated with IL-5 or GM-CSF as well as anti-CD40.[2] Thus, the prolonged survival of eosinophils from HES patients may reflect a constitutive upregulation in anti-apoptotic regulators. On the other hand, proteins such as histone H1.2 promote apoptosome activation[13] and the cleavage of CASP-3 and CASP-9.

The extrinsic pathway is initiated by the cell surface activation of membrane-bound receptors. On eosinophils, these include the tumor necrosis factor (TNF-α) receptor superfamily members, TNF-related apoptosis-inducing ligand receptor 1, 3, and 4 (TRAIL-R1, R3, and R4) as well as apoptosis-mediating surface antigen (Fas). Apoptotic signaling TNF receptors (TNFRs) are trans-membrane proteins that may have an intracellular death domain (DD), CARD domain, death effector domain (DED) domain or a TNFR-associated factor (TRAF) binding domain.[14] Some TNFRs can bind the ligand but lack the DD; these receptors are unable to transduce death signals and are referred to as decoy receptors (DcRs). DD-containing receptors, when bound to their ligands, typically induce apoptosis directly by recruiting *adaptor* proteins such as TRAF, as well as procaspase-8. After segmental allergen challenge, pulmonary eosinophils showed elevated levels of TRAIL-DcR2 and decreased levels of TRAIL receptor 1 (death receptor 4) and TRAIL receptor 2 (death receptor 5). Rather than inducing apoptosis, the interaction of TRAIL-Rs with extracellular TNF-α triggered a pro-survival response.[15]

Fas, also known as Apo-1 antigen or CD95, is a member of the TNFR superfamily. FasL (CD95-L) is a homotri-meric transmembrane protein that is highly expressed on immune cells including T-helper type 2 (T_h2) cells, monocytes, and airway epithelial cells. Fas is highly expressed on the surface of eosinophils,[5] which is consis-tent with a functional role. *In vitro* studies with FasL or anti-CD95-L on purified populations of unstimulated eosinophils have not shown dramatic effects on life span.[16] In addition, the expression of FasL is induced on inflam-matory cells[17] during allergic inflammation. In aggregate, these data suggest Fas/FasL interactions do not play a major role in the programmed cell death of circulating, unstimulated eosinophils but may be quite important in resolving eosinophilic inflammation by inducing the apoptosis of tissue-based, activated cells.

Mechanistically, the Fas receptor—Fas ligand interac-tion triggers receptor oligomerization and the formation of the death-inducing signaling complex (DISC[6,18]). The complex consists of Fas, FasL, the adaptor molecule, FAS-associated death domain protein (FADD), and procaspase-8.[7] DISC assembly requires homotypic interactions between DDs and DEDs. The DD of FADD binds to the DD of Fas while the DED of FADD binds to the DED of pro-caspase-8.[19] The association with FADD clusters and activates caspase-8, which is then subsequently released. In addition to FADD, the DISC also contains other DED-containing proteins such as cellular FLICE-like inhibitory protein (c-FLIP),[20] which at low concentrations promotes procaspase-8 incorporation and cleavage but has the

opposite effect when overexpressed. It is currently unclear if c-FLIP is expressed by eosinophils.

Once activated via the DISC complex, CASP-8 can lead to mitochondria-independent (extrinsic) or mitochondria-dependent (intrinsic) eosinophil cell death.[21] In the extrinsic pathway, activated CASP-8 directly cleaves procaspase-3 and -7, leading to cell death. However, CASP-8 also provides an important link to the intrinsic pathway by cleaving the BH3-interacting domain death agonist (Bid).[22]

The Bcl-2 Family and Apoptosis

The Bcl-2 family regulates apoptosis in nearly all organ-isms. There are three distinct groups that are characterized by their Bcl-2 homology (BH) domains. The anti-apoptotic members include apoptosis regulator Bcl-2 (Bcl-2), Bcl2 antagonist of cell death (Bad), Bcl-2 related protein A1 (B2LA1), and induced myeloid leukemia cell differentia-tion protein Mcl-1 homolog (Mcl-1), and contain four BH domains. The pro-apoptotic members such as Bcl-2 homologous antagonist/killer (BAK), apoptosis regulator BAX (Bax), and Bcl-2-related ovarian killer protein (Bok) possess BH1, BH2, and BH3 domains, while Bid, Bcl-2 interacting mediator of cell death (Bim), Bcl-2-interacting killer (Bik), and p53 up-regulated modulator of apoptosis (PUMA) contain a single BH3 domain.[9] BH domains mediate the formation of homo- and heterodimers between members of this family. BH1−3 domains form a hydro-phobic pocket where the BH3 domain of another Bcl-2 family member can dock.

Eosinophils express a variety of Bcl-2 family members including Bad, Bax, and Bid,[5,22,23] with Mcl-1 and Bcl-2 barely detectable,[4] although they may be elevated in eosinophils from patients treated with IL-5 or GM-CSF.[23] However, the overall expression of BH proteins by eosin-ophils appears to be relatively unaffected by cytokine stimulation, suggesting that the steady-state ratio of pro-versus anti-apoptotic protein and their post-translational regulation likely determines whether eosinophils live or die.

The mechanism of action of BH1−3 or BH3 proteins has been studied intensively but remains extremely controversial. Nearly all of these studies have been done in tumor cell lines with only a handful in primary cells such as eosinophils. Bax is highly expressed in peripheral blood eosinophils as well as those from bronchoalveolar lavage (BAL) after segmental allergen challenge.[24] Peripheral blood eosinophils cultured *in vitro* for 48 h without pro-survival cytokines show significantly increased punctate Bax staining,[25] consistent with trans-location to the mitochondria. Within 6 h of culture, the Bax activation epitope, as detected with monoclonal antibody 6A7, was also increased and associated with the mitochondria. As these events precede the appearance of active CASP-9, Bax activation and translocation is likely a critical, early step in the apoptotic cascade in unstimulated eosinophils. This hypothesis is strengthened by data showing that Bax remains inactivated and largely cytosolic in IL-5- or GM-CSF-treated cells,[25,26] and inhibitory peptides that antagonize Bax function reduce eosinophil apoptosis.[26]

Once activated and translocated, Bax mediates mitochondrial outer membrane pore formation. The predominant view is that on commitment to apoptosis, Bax undergoes conformational changes and oligomerization;[9] its activation may also be preceded by cleavage, probably by calpain.[26] Bax assembly and pore formation is facilitated by truncated Bid (tBid).[27] TBid, created by procaspase-8 cleavage after Fas/FasL signaling, translocates to the mitochondria where it facilitates active Bax oligomerization and mitochondrial pore assembly. TBID may also promote death indirectly by binding to and titrating away anti-apoptotic Bad, Mcl-1, or Bcl-2 from pro-apoptotic Bax. Consistent with the importance of these proteins, BAL eosinophils from allergen-challenged *BH3 interacting domain death agonist* (*BID*) or *BCL2-associated X protein* (*BAX*) gene-deleted mice show resistance to FasL or steroid-induced apoptosis,[22] demonstrating their physiological relevance.

Pro-apoptotic Bcl-2 family members can be activated through a variety of signaling cascades. Calcium influx sensitive to 1,2-bis(o-aminophenoxy)ethane-N,N,N',N'-tetraacetic acid (BAPTA) rapidly (approximately 3−6 h) triggers mitochondrial disruption and CASP-3 activation in eosinophils,[28] conceivably through the activation of Bax. Calcium flux can activate calpain-1, a non-lysosomal cysteine protease that is highly expressed in eosinophils and participates in the control of cytoskeletal remodeling, cell cycle progression, gene expression, and apoptotic cell death.[29] Calpains have been reported to regulate eosinophil apoptosis,[26] probably through the cleavage and activation of pro-apoptotic Bax.

Glucocorticosteroids

Glucocorticosteroids (GCs) induce apoptosis of unstimulated peripheral blood eosinophils as well as those present in the lung after allergen challenge.[30] Mechanistically, GCs reduce the transcription and enhance the decay of cytokine mRNAs encoding pro-survival cytokines, including GM-CSF, IL-3, and IL-5,[31] effectively blocking their production and release. In addition, GCs directly activate eosinophil apoptosis by inducing CASP-8 activation.[5] High concentrations of exogenous GM-CSF or IL-5 antagonized GC-induced CASP-8 activation demonstrating that pro-death and pro-survival signaling must overlap at or before this early step in the cascade. The effect of GCs on other components of pro-death signaling, such as Bax activation, remains unknown.

Transforming Growth Factor Beta

Transforming growth factor beta (TGF-β) is produced by activated eosinophils in the asthmatic lung where it contributes to airway remodeling. In addition to a *trans* effect, recent evidence suggests that TGF-β also modulates eosinophil function and survival. Eosinophils express type I and II TGF-β receptors, as well as downstream SMAD proteins, implying a fully functional signaling cascade. TGF-β inhibited IL-5-induced tyrosine phosphorylation of tyrosine-protein kinase Lyn (Lyn), tyrosine-protein kinase JAK2 (JAK-2), signal transducer and activator of transcription 1, and extracellular signal-regulated kinase (ERK),[32] and reversed IL-5-mediated dephosphorylation of protein kinase RNA-activated (PKR), a RNA-dependent protein kinase.[33] PKR inhibition promoted eosinophil survival and was downregulated in long-lived, pulmonary eosinophils induced by allergen challenge. Apoptosis induced by PKR usually activates the FADD−CASP-8 pathway, although APAF/CASP-9 activation has also been observed during PKR-induced apoptosis.

Other Apoptotic Stimuli

A variety of other stimuli are capable of accelerating eosinophil apoptosis. These include cross-linking of the sialic acid-binding Ig-like lectin (Siglec) family (Siglec-5/Siglec-F, Siglec-8, -12, and -G). These plasma membrane proteins are highly expressed by eosinophils and bind sialic acid.[34] Cross-linking of Siglec-8 triggers CASP-3 activation as well as ROS production and mitochondrial damage.[34] Curiously, pro-survival cytokines such as IL-5 synergize with rather than antagonize Siglec-8 cross-linking and accelerate eosinophil death,[34] although a non-apoptotic pathway involving ROS generation was responsible. *In vivo* administration of anti-Siglec-5 resolved eosinophilic gastritis and allergic pulmonary inflammation[35] in models of asthma by accelerating eosinophil apoptosis.

Cross-linking of CD30, a TNF-R superfamily member generates rapid eosinophil death, which was prevented by inhibitors of phosphatidylinositol 3-kinase (PI3K) and the MAPKs ERK and p38.[36] As this process was insensitive to a variety of broad caspase inhibitors, it is unclear if apoptosis or another process mediated cell death. Conversely, CD30 knockout mice show reduced levels of pulmonary eosinophilia, which may have been due to reduced levels of interleukin-13 (IL-13) after allergen challenge.[37] Histone deacetylase-like amidohydrolase (HDAC-like amidohydrolase) antagonized the pro-survival effects of GM-CSF and IL-5 and accelerated spontaneous eosinophil apoptosis, probably through the intrinsic pathway.[13] Nitric oxide is increased in the asthmatic lung and also induces eosinophil death through a CASP-6-dependent mechanism.[38]

SURVIVAL SIGNALS

As discussed earlier, in the absence of stimulation, eosinophils are programmed to apoptose in 2—3 days, maintaining peripheral counts at between 1—3% of nucleated cells. However, under inflammatory conditions such as allergic asthma, tissue based-eosinophils can survive for weeks. These data suggest eosinophils are exposed to anti-apoptotic signals during their circulation, migration from the vasculature, and/or while in residence in the lung or other inflamed tissue. Indeed, multiple cytokines and extracellular matrix components have been shown, both *in vivo* and *in vitro*, to antagonize eosinophil programmed cell death and promote cell longevity.

Granulocyte-Macrophage Colony-Stimulating Factor, Interleukin-3, and Interleukin-5

Bone marrow and circulating eosinophils express the unique alpha (α) and shared beta (β) chains that comprise the heterodimeric granulocyte-macrophage colony-stimulating factor subunit alpha (GM-CSF-R-alpha) (CD116), interleukin-3 receptor subunit alpha (IL-3RA) (CD123), and interleukin-5 receptor subunit alpha (IL-5RA) (CD125).[2] The cytokine-specific α chain mediates ligand binding while the common β chain initiates downstream signaling. Despite their considerable overlap, each cytokine also shows distinctive biological properties suggesting that signaling through the common β chain is uniquely modified by the α chains,[2] as well as the formation of higher-order, hexameric and dodecameric structures.[39]

These cytokines are widely expressed by immune and parenchymal cells as well as activated eosinophils. After receptor ligation, β chain-associated JAK-2 trans-phosphorylates the cytoplasmic tail of nearby β chains, permitting the recruitment of Src homology 2 (SH2) and phosphotyrosine-binding domain proteins and the initiation of tyrosine phosphorylation-dependent signaling pathways. These include the JAK—signal transducer and activator of transcription (STAT), Ras—MAPK, and PI3K—RAC-alpha serine/threonine-protein kinase (Akt) pathways. STAT translocation induces the transcriptional upregulation of multiple anti-apoptotic proteins including Bcl-2, Bad, GM-CSF, IL-5, serine/threonine-protein kinase pim-1 (Pim1), and others,[5,23] likely enabling long-term resistance to cell death. Differential activation of serine vs. tyrosine phosphorylation signaling pathways depends on the ligand concentration. GM-CSF at picomolar concentrations induces β-chain phosphorylation at

serine 585 (Ser585) and promotes cell survival signaling in the absence of tyrosine 577 (Tyr577) phosphorylation, while higher concentrations of GM-CSF selectively induces β-chain Tyr577 phosphorylation without Ser585 phosphorylation.[40] Thus, GM-CSF/IL-3/IL-5 can selectively suppress apoptosis at nearly undetectable cytokine concentrations (femtomolar).[3]

Signal bifurcation likely involves the adaptor protein 14-3-3 which, after tyrosine 179 (Tyr179) phosphorylation by Src or Lyn, recognizes phosphorylated β chain Ser585,[41] binds to Src homology collagen-like protein and generates a molecular bridge for PI3K/Akt activation and survival signaling.[41] Interestingly, blockade of either tyrosine or serine signaling accelerates apoptosis, suggesting both are essential. The effects of serine blockade occur within hours while the inhibition of tyrosine signaling is somewhat slower. These observations suggest that serine signaling maintains eosinophil survival in the short term (hours) while tyrosine signaling, by inducing new, pro-survival proteins, is required for longer-term (days—weeks) viability.

PI3K/Akt, as well as the downstream MAPKs, are essential intermediates in pro-survival cytokine signaling in eosinophils as well as other cells. Within 5 min of GM-CSF stimulation, PI3K was transiently activated, with return to baseline after 10—15 min.[41] Akt is the major downstream target of PI3K. It is recruited to the plasma membrane through binding to inositol 1,4,5-triphosphate (IP3), and becomes activated by 3-phosphoinositide-dependent protein kinase 1 (PDPK1)-mediated phosphorylation at threonine 308 (Thr308) and Ser473. Active Akt phosphorylates apoptotic regulators such as CASP-3 and CASP-9, pro-apoptotic Bad and Bax, glycogen synthase kinase-3 beta (GSK-3-beta), which regulates proteins involved in cell survival and proliferation, and tuberous sclerosis 2 protein [TSC2, which is an inhibitor of the mammalian target of rapamycin (mTOR) that is involved in several cell processes including proliferation and survival].[42] In neutrophils, Akt likely phosphorylates Bax at Ser184, which prevents its activation and translocation to the mitochondria.

JNK has been reported to play a role in eosinophil apoptosis. Conversely, treatment of eosinophils with PI3K, Akt, or ERK inhibitors blunts the effects of GM-CSF or IL-5 and induces apoptosis.[26] P38 or JNK show some pro-apoptotic effects but are less potent. Thus, a variety of cytokine-responsive kinase cascades including PI3K—Akt and MAPK mediate survival signaling, likely by specifically phosphorylating Bcl-2 family members, particularly Bax.

Many but not all of the aforementioned phosphorylated serine or threonine moieties in Bcl-2 or caspase family members are immediately N-terminus to prolines. This is not surprising as many kinases involved in survival signaling are thought to be proline-directed and preferentially phosphorylate serine or threonine adjacent to a proline. This dipeptide

(Ser/Thr—Pro) is recognized by a group of enzymes with peptidyl-prolyl cis-trans isomerase-like (PPIase) activity, which includes the cyclophilin A and peptidyl-prolyl cis-trans isomerase FK506 binding protein families as well as Pin1. The former are targeted by cyclosporin (CsA) and FK506, respectively, after solid organ transplant and are required for nuclear factor of activated T-cells (NFAT)-mediated transcription that culminates in interleukin-2 (IL-2) production. Both agents have been used as anti-T cell drugs in asthma with limited success.

PPIases interact with and isomerize the Ser/Thr—Pro bond from *cis* to *trans*, altering the conformational and hence biological properties of the isomerized protein. Pin1 is a highly conserved, ubiquitously expressed, bipartite isomerase with an N-terminus tryptophan-tryptophan (WW) target binding domain and C-terminus PPIase domain. The protein is insensitive to CsA or FK506 but can be inhibited by low micromolar concentrations of the naphthoquinone, 5-hydroxynaphthoquinone (juglone). Pin1-mediated isomerization of the Ser/Thr—Pro peptide bond is increased by approximately 1000-fold if either the serine or the threonine is phosphorylated. This suggests that Pin1 might be a molecular adaptor, akin to 14-3-3, which senses site-specific Ser/Thr phosphorylation, but then modifies target function via isomerization.

The first indication that Pin1 played a role in eosinophil biology were *in vitro* experiments showing that juglone accelerated the apoptosis of purified, unstimulated, or GM-CSF-treated peripheral blood eosinophils as well as BAL eosinophils obtained after allergen challenge.[43] Higher concentrations of juglone were necessary to kill GM-CSF- or IL-5-treated eosinophils suggesting that Pin1 was involved in both the autocrine production of pro-survival cytokines as well as the signaling they induced. Consistent with these results, Pin1 PPIase activity was rapidly increased in eosinophils treated *in vitro* with GM-CSF or IL-5, or activated *in vivo*,[43] suggesting a possible interaction with and isomerization of target proteins modified by pro-survival cytokine signaling.

Molecular analysis revealed Pin1 modulated GM-CSF and IL-5 production by activated eosinophils by influencing the decay of their coding mRNAs. Many cytokine mRNAs including GM-CSF, IL-2, IL-3, IL-5, interferon gamma, and TNF-α contain 3′ untranslated region (UTR) adenylate-uridylate (AU)-rich elements (AREs) that interact with mRNA binding proteins including heterogeneous nuclear ribonucleoprotein C (hnRNP C), ribonucleoprotein D (hnRNP D), Hu-antigen R (HuR), and tristetraprolin (TTP).[44] Under resting conditions, the 3′ UTR is occupied by hnRNP D or TTP, causing cytokine mRNAs to be rapidly degraded.[45] Eosinophil activation through the α4/β7 integrins, cytokine receptors, or CD44 reduces the affinity of destabilizing proteins for the AREs in favor of

stabilizers HuR or hnRNP C,[45] rapidly increasing the steady-state levels of cytokine mRNAs. In eosinophils, fibroblasts, and T cells, Pin1 mediated this process by binding to and isomerizing hnRNP D on cell activation.[43,46] Isomerization reduced hnRNP D binding affinity for GM-CSF mRNA and initiated hnRNP D degradation via the proteasome. The ARE was then occupied by hnRNP C, which facilitated stabilization of GM-CSF mRNA. While not directly shown, it is likely that other ARE-containing cytokine mRNAs produced by eosinophils (IL-3, IL-5, and TNF-α) show similar regulation.

As discussed previously, eosinophils exposed to GM-CSF or IL-5 show extended life spans. Under these conditions, Bax and downstream caspases remain inactive. How Bax is constrained by pro-survival signaling has been under intense investigation in many cell systems. Bax, CASP-8, and c-IAP all contain Ser/Thr—Pro motifs, which could interact with Pin1. This suggested that Pin1 could bind to and potentially isomerize critical apoptotic effectors after phosphorylation induced by pro-survival signaling. Immunoprecipitation revealed that CASP-8 and c-IAP indeed interacted with Pin1 in resting or cytokine-stimulated eosinophils while Bax interacted with Pin1 only after cell stimulation with GM-CSF or IL-5. Interestingly, Pin1 PPIase activity was rapidly enhanced by cytokines and blockade of that activity with juglone reduced the Bax—Pin1 interaction and accelerated apoptosis. These results suggested that active Pin1 could prevent Bax translocation after cytokine signaling.

Bax contains two potential Pin1 interaction sites, Ser87—Pro88 and Thr167—Pro168. The Ser87—Pro88 potential site is largely inaccessible within the hydrophobic interior while the Thr167—Pro168 is located on the exterior face in the hinge region adjacent to the C-terminus activation domain.[47] GM-CSF or IL-5 treatment stimulated the phosphorylation of Thr167, which was sensitive to ERK but not JNK or p38 inhibitors. Mutation of Thr167 to alanine (T167A) significantly reduced the Bax—Pin1 interaction, while substitution of a phosphomimetic (threonine to glutamic acid) increased it. Consistent with a regulatory role, T167A mutant Bax no longer bound Pin1 and was not suppressed by pro-survival signaling after its transduction into eosinophils. Therefore, pro-survival signaling induces ERK-mediated phosphorylation of Bax at Thr167, facilitating the binding of active Pin1 to this site. Under such conditions, Bax remains inactive and mitochondrial targeting and disruption are prevented.

These data suggest that eosinophil agonists which activate the PI3K/Akt/ERK cascade are likely to be pro-survival while those that suppress it are likely to be pro-death. The former include leptin, which activates PI3K and Akt,[2] while hypoxia, interleukin-27,[48] thymic stromal lymphopoietin,[49] eotaxin, and fibroblast growth factor all activate ERK.[50] Anti-CD30 induces eosinophil death by antagonizing PI3K,[36] while TGF-β suppresses Akt.

CONCLUSION

Many agonists including cytokines, the extracellular matrix, and cell surface receptor ligands modulate eosinophil apoptotic decisions. In the absence of outside-in signaling, these cells die quickly in the circulation, likely through the default activation of Bax and downstream caspases. Exposure to pro-survival cytokines, including GM-CSF, IL-3, and IL-5, which trigger PI3K/Akt/ERK signaling, prevents Bax activation through a Pin1-dependent process. NFκB-mediated transcriptional upregulation of anti-apoptotic Bcl-2 family members, as well as inhibitors of apoptosis (Mcl-1 and c-IAP), contributes to long-term eosinophil survival. A critical unresolved issue remains how the apoptotic pathway can be selectively engaged to reduce the numbers and functionality of pathogenic, tissue-based eosinophils without reliance on corticosteroids. Despite the involvement of Pin1 in diverse cell metabolism, aerosol delivery of specific inhibitors should be considered as a possible new direction for drug development.

ACKNOWLEDGEMENTS

The chapter was made possible by the work of numerous authors many of whom could not be cited due to space limitations. In addition, we would like to acknowledge other members of the Malter laboratory for their helpful insights and the UW-Asthma Research Group. U. R. B. received support from UW-SciMed Graduate Research Scholars and 5T32GM08349 and J. S. M. from National Institutes of Health (P01 HL088594 and R01HL087950) and the Waisman Center (P30HD003352).

REFERENCES

1. Rothenberg M, Hogan S. The eosinophil. *Annu Rev Immunol* 2006;**24**:47–174.
2. Bochner B, Park Y. Eosinophil survival and apoptosis in health and disease. *Allergy Asthma Immunol Res* 2010;**2**(2):87–101.
3. Begley C, Lopez A, Nicola N, Warren D, Vadas M, Sanderson C, et al. Purified colony-stimulating factors enhance the survival of human neutrophils and eosinophils in vitro: a rapid and sensitive microassay for colony-stimulating factors. *Blood* 1986;**68**,(1):162–6.
4. Druilhe A, Arock M, Le Goff L, Pretolani M. Human eosinophils express Bcl-2 family proteins: Modulations of Mcl-1 expression by IFN-γ. *Am J Respir Cell Mol Biol* 1998;**18**(3):315–22.
5. Zangrilli J, Robertson N, Shetty A, Wu J, Hastie A, Fish J, et al. Effect of IL-5, glucocorticoid, and Fas ligation on Bcl-2 homologue expression and caspase activation in circulating human eosinophils. *Clin Exp Immunol* 2000;**120**(1):12–21.
6. Bao Q, Shi Y. Apoptosome: a platform for the activation of initiator caspases. *Cell Death Differ* 2007;**14**(1):56–65.
7. Salvesen G, Dixit V. Caspase activation: The induced-proximity model. *Proc Natl Acad Sci USA;* 1999;**96**:10964–7.
8. Simon H, Alam R. Regulation of eosinophil apoptosis: transduction of survival and death signals. *Int. Arch. Allergy Immunol* 1999;**118**:7–14.
9. Tait S, Green D. Mitochondria and cell death: outer membrane permeablization and beyond. *Nature Reviews: Molecular Cell Biology* 2010;**11**:621–31.
10. Tran T, Andreka P, Rodrigues C, Webster K, Bishopric N. Jun kinase delays caspase-9 activation by interaction with the apoptosome. *J Biol Chem* 2007;**282**(28):20340–50.
11. Shrivastava P, Pantano C, Watkin R, McElhinney B, Guala A, Poynter M, et al. Reactive nitrogen species-induced cell death requires Fas-dependent activation of c-Jun N-terminal kinase. *Mol Cell Biol* 2004;**24**(15):6372–763.
12. Vassina E, Yousefi S, Simon D, Zwicky C, Conus S, Simon H. cIAP-2 and survivin contribute to cytokine-mediated delayed eosinophil apoptosis. *Eur J Immunol* 2006;**36**(7):1975–84.
13. Kankaanranta H, Janka-Junttila M, Ilmarinen-Salo P, Jalonen U, Ito M, Adcock I, et al. Histone deacetylase inhibitors induce apoptosis in human eosinophils and neutrophils. *J Inflamm (Lond)* 2010;**7**:9.
14. Thomas L, Johnson R, Reed J, Thorburn A. The C terminal tails of tumor necrosis factor-related apoptosis inducing ligand (TRAIL) and Fas receptors have opposing functions in Fas-associated death domain (FADD) recruitment and can regulate agonist-specific mechanisms of receptor activation. *J Biol Chem* 2004;**279**:52479–86.
15. Robertson N, Zangrilli J, Steplewski A, Hastie A, Lindemeyer R, Planeta M, et al. Differential expression of TRAIL AND TRAIL receptors in allergic asthmatics following segmental antigen challenge: evidence for a role of TRAIL in eosinophil survival. *J Immunol* 2002;**169**:5986–96.
16. Liles W, Kiener P, Ledbetter J, Aruffo A, Klebanoff S. Differential expression of Fas (CD95) and Fas ligand on normal human phagocytes: implications for the regulation of apoptosis in neutrophils. *J Exp Med* 1996;**184**:429–40.
17. Peter M, Kischkel F, Scheuerpflug C, Medema J, Debatin K, Krammer P. Resistance of cultured peripheral T cells towards activation-induced cell death involves a lack of recruitment of FLICE (MACH/caspase 8) to the CD95 death-inducing signaling complex. *Eur J Immunol* 1997;**27**:1207–12.
18. Matsumoto K, Schleimer R, Saito H, Likura Y, Bochner B. Induction of apoptosis in human eosinophils by anti-Fas antibody treatment. *Blood* 1995;**86**(4):1437–43.
19. Reed J, Doctor K, Godzik A. The domains of apoptosis: a genomics perspective. *Sci STKE* 2004;**239**. re9.
20. Krueger A, Baumann S, Krammer P, Kirchhoff S. FLICE-inhibitory proteins: regulators of death receptor-mediated apoptosis. *Mol Cell Biol* 2001;**21**:8247–54.
21. Simon H, Yousefi S, Dibbert B, Hebestreit H, Weber M, Branch D, et al. Role of tyrosine-phosphorylation and Lyn tyrosine kinase in Fas receptor-mediated apoptosis in eosinophils. *Blood* 1998;**92**(2):547–57.
22. Maret M, Ruffié C, Séverine L, Phelep A, Thidaudeau O, Marchal J, et al. A role for Bid in eosinophil apoptosis and in allergic airway reaction. *J Immunol;* 2009;**182**:5740–7.
23. Dibbert B, Daigle I, Braun D, Schranz C, Weber M, Blaser K, et al. Role for Bcl-x$_L$ in delayed eosinophil apoptosis mediated by

granulocyte-macrophage colony-stimulating factor and interleukin-5. *Blood* 1998;**92**(3):778−83.

24. Bates M, Liu L, Esnault S, Stout B, Fonkem E, Kung V, et al. Expression of interleukin-5-and granulocyte macrophage−colony-stimulating factor-responsive genes in blood and airway eosinophils. *Am J Respir Cell Mol Bio* 2004;**30**:736−43.

25. Dewson G, Cohen G, Wardlaw A. Interleukin-5 inhibits translocation of Bax to the mitochondria, cytochrome c release, and activation of caspases in human eosinophils. *Blood* 2001;**98**(7):2239−47.

26. Shen Z, Esnault S, Schinzel A, Borner C, Malter J. The peptidyl-prolyl isomerase Pin1 facilitates cytokine-induced survival of eosinophils by suppressing Bax activation. *Nat Immunology* 2009;**10**(3):57−265.

27. Lovell J, Billen L, Bindner S, Shamas-Din A, Fradin C, Leber B, et al. Membrane binding by tBid initiates an ordered series of events culminating in membrane permeablization by Bax. *Cell* 2008;**135**(6):1074−84.

28. Yan H, Xue G, Mei Q, Ding F, Wang Y, Sun S. Calcium-dependent proapoptotic effect of *Taenia solium* metacestodes annexin B1 on human eosinophils: A novel strategy to prevent host immune response. *Int J Biochem Cell Biol* 2008;**40**(10):2151−63.

29. Franco S, Huttenlocher A. Regulating cell migration: calpains make the cut. *J Cell Sci* 2005;**118**:3829−38.

30. Meagher L, Cousin J, Seckl J, Haslett C. Opposing effects of glucocorticoids on the rate of apoptosis in neutrophilic and eosinophilic granulocytes. *J Immunol* 1996;**156**:4422−8.

31. Masuyama K, Till S, Jacobson M, Kamil A, Cameron L, Juliusson S, et al. Nasal eosinophilia and IL-5 mRNA expression in seasonal allergic rhinitis induced by natural allergen exposure: effect of topical corticosteroids. *J Allergy Clin Immunol* 1998;**102**(4 Pt 1):610−7.

32. Pazdrak K, Justement L, Alam R. Mechanism of inhibition of eosinophil activation by transforming growth factor-beta. Inhibition of Lyn, MAP, Jak2 kinases and STAT1 nuclear factor. *J Immunol* 1995;**155**(9):4454−8.

33. Goplen N, Gorska M, Stafford S, Rozario S, Guo L, Alam R. A phosphosite screen identifies autocrine TGf-β-driven activation of protein kinase R as a survival-limiting factor for eosinophils. *J Immunol* 2008;**180**:4256−64.

34. Nutku E, Hudson SA, Bochner BS. Mechanism of Siglec-8-induced human eosinophil apoptosis: role of caspases and mitochondrial injury. *Biochem Biophys Res Commun* 2005;**336**:918−24.

35. Song D, Cho J, Miller M, Rosenthal P, Soroosh P, Croft M, et al. Anti-Siglec-F antibody reduces allergen-induced eosinophilic inflammation and airway remodeling. *J Immunol* 2009;**183**:5333−41.

36. Matsumoto K, Terakawa M, Fukuda S, Saito H. Analysis of signal transduction pathways involved in anti-CD30 mAb-induced human eosinophil apoptosis. *Int Arch Allergy Immunol Suppl* 2010;**152**(1):2−8.

37. Nam S, Kim Y, Do J, Choi Y, Seo H, Yi H, et al. CD30 supports lung inflammation. *Int Immunol* 2008;**20**(2):177−84.

38. Ilmarinen-Salo P, Moilanen E, Kankaanranta H. Nitric oxide induces apoptosis in GM-CSF treated eosinophils via caspase-6-dependent lamin and DNA fragmentation. *Pulmonary Pharmacology & Therapeutics* 2010;**23**(4):365−71.

39. Hansen G, Hercus T, McClure B, Stomski F, Dottore M, Powell J, et al. The structure of the GM-CSF receptor complex reveals a distinct mode of cytokine receptor activation. *Cell* 2008;**134**:496−507.

40. Guthridge M, Stomski F, Barry E, Winnall W, Woodcock J, McClure B, et al. Site-specific serine phosphorylation of the IL-3 receptor is required for hemopoietic cell survival. *Mol Cell* 2000;**6**(1):99−108.

41. Barry E, Felquer F, Powell J, Biggs L, Stomski F, Urbani A, et al. 14-3-3:Shc scaffolds integrate phosphoserine and phosphotyrosine signaling to regulate phosphatidylinositol 3-kinase activation and cell survival. *J Biol Chem* 2009;**284**(18):12080−90.

42. Stiles B. PI-3-K and AKT: Onto the mitochondria. *Advanced Drug Delivery Reviews* 2009;**61**(14):1276−82.

43. Shen Z, Esnault S, Malter J. The peptidyl-prolyl isomerase Pin1 regulates the stability of granulocyte-macrophage colony stimulating factor mRNA in activated eosinophils. *Nat Immunology* 2005;**6**:1280−7.

44. Capowski E, Esnault S, Bhattacharya S, Malter J. Y Box-binding factor promotes eosinophil survival by stabilizing granulocyte-macrophage-stimulating factor mRNA. *J Immunol* 2001;**167**:5970−6.

45. Ross J. mRNA stability in mammalian cells. *Microbiol Rev* 1995;**59**:423−50.

46. Esnault S, Shen Z, Whitesel E, Malter J. The pepitdyl-prolyl isomerase Pin1 regulates granulocyte-macrophage colony-stimulating factor mRNA stability in T lymphocytes. *J Immunol* 2006;**177**:6999−7006.

47. Schinzel A, Kaufmann T, Schuler M, Martinalbo J, Grubb D, Borner C. Conformational control of Bax localization and apoptotic activity by pro168. *J Cell Biol* 2004;**164**(7):1021−32.

48. Hu S, Wong C, Lam C. Activation of eosinophils by IL-12 family cytokine IL-27: Implications of the pleiotropic roles of IL-27 in allergic responses. *Immunobiology* 2011;**216**(1−2):54−65. Epub 2010 Mar 16.

49. Wong C, Hu S, Cheung P, Lam C. Thymic stromal lymphopoietin induces chemotacti and prosurvival effects in eosinophils: implications in allergic inflammation. *Am J Respir Cell Mol Biol* 2010;**43**(3):305−15.

50. Huang J, John J, Fuentebella J, Patel A, Nguyen T, Seki S, et al. Eotaxin and FGF enhance signaling through an extracellular signal-related kinase (ERK)-dependent pathway in the pathogenesis of eosinophilic esophagitis. *Allergy, Asthma & Clin Immunol* 2010;**6**:25.

FURTHER READING

Normal Eosinophil Physiology

Akashi K, Traver D, Miyamoto T, Weissman IL. A clonogenic common myeloid progenitor that gives rise to all myeloid lineages. *Nature* 2000;**404**:193−7.

Mori Y, Iwasaki H, Kohno K, Yoshimoto G, Kikushige Y, Okeda A, et al. Identification of the human eosinophil lineage-committed progenitor: revision of phenotypic definition of the human common myeloid progenitor. *JEM* 2008;**206**(1):183−93.

Death Signals

Esnault S, Malter J. Minute quantities of granulocyte-macrophage colony-stimulating factor prolong eosinophil survival. *J Interferon Cytokine Res* 2001;**21**:117−24.

Letuve S, Druilhe A, Grandsaigne M, Aubier M, Pretolani. Critical role of mitochondria, but not caspases, during glucocorticosteroid-induced human eosinophil apoptosis. *Am J Respir Cell Mol Biol* 2002;**26**:565−71.

Ruiz-Vela A, Korsmeyer S. Proapoptotic histone H1.2 induces CASP-3 and -7 activation by forming a protein complex with CYT c, APAF-1 and CASP-9. *FEBS Lett* 2007;**18**:3422−8. 581.

De Rose V, Cappello P, Sorbello V, Ceccarini B, Gani F, Bosticardo M, et al. IFN-gamma inhibits the proliferation of allergen-activated T lymphocytes from atopic, asthmatic patients by inducing Fas/FasL-mediated apoptosis. *J Leukoc Biol* 2004;**76**(2):423−32.

Segal M, Niazi S, Simons M, Galati S, Zangrilli J. Bid activation during induction of extrinsic and intrinsic apoptosis in eosinophils. *Immunology and Cell Biology* 2007;**85**:518−24.

Ogawa K, Hashida R, Miyagawa M, Kagaya S, Sugita Y, Matsumoto K, et al. Analysis of gene expression in peripheral blood eosinophils from patients with atopic dermatitis and in vitro cytokine-stimulated blood eosinophils. *Clin Exp Immunol* 2003;**131**(3):436−45.

Interleukin-5, Interleukin-3, and Granulocyte-Macrophage Colony-Stimulating Factor

Guthridge M, Stomski F, Barry E, Winnall W, Woodcock J, McClure B, et al. Site-specific serine phosphorylation of the IL-3 receptor is required for hemopoietic cell survival. *Mol Cell* 2000;**6**(1):99−108.

Olayioye M, Guthridge M, Stomski F, Lopez A, Visvader J, Linderman G. Threonine 391 phosphorylation of the human prolactin receptor mediates a novel interaction with 14-3-3 proteins. *J Biol Chem* 2003;**278**(35):32929−35.

Kim B, Ryu S, Song B. JNK- and p38 kinase-mediated phosphorylation of Bax leads to its activation and mitochondrial translocation and to apoptosis of human hepatoma HepG2 cells. *J Biol Chem* 2006;**281**(30):21256−65.

Yang W, Inouye C, Seto E. Cyclophilin A and FKBP12 interact with YY1 and alter its transcriptional activity. *J Biol Chem* 1995;**270**(23):15187−93.

Eckstein J, Fung J. A new class of cyclosporine analogues for the treatment of asthma. *Expert Opin Investig Drugs* 2003;**12**(4):647−53.

Efraim A, Eliashar R, Levi-Schaffer F. Hypoxia modulates human eosinophil function. *Clin Mol Allergy* 2010;**8**:10.

Chapter 7.5

Molecular Mechanisms Underlying Eosinophil Hematopoiesis

Miranda Buitenhuis

INTRODUCTION

The production of blood cells, or *hematopoiesis*, is a complex series of events that results in the formation of all blood lineages. Adult hematopoiesis predominantly occurs in the bone marrow of the trabecular bones and is dependent on the correct function of the bone marrow microenvironment. Lineage development is specifically directed by a variety of cytokines including erythropoietin (EPO), granulocyte colony-stimulating factor (G-CSF), granulocyte-macrophage colony-stimulating factor (GM-CSF), interleukin-3 (IL-3), interleukin-5 (IL-5), macrophage colony-stimulating factor (M-CSF), and thrombopoietin (TPO). Whereas GM-CSF and IL-3 are cytokines regulating proliferation and survival during myeloid differentiation of various lineages, EPO, G-CSF, IL-5, M-CSF, and TPO are required for the final maturation of erythrocytes,[1] megakaryocytes,[2] neutrophils,[3] monocytes,[4] and eosinophils,[5−6] respectively. On binding of cytokines to their cognate receptors, the activity of intracellular signal transduction pathways is regulated leading to the modulation of gene expression. Although our appreciation of the transcriptional regulators of hematopoiesis has developed considerably, until recently the roles of specific intracellular signal transduction modules were completely unknown. This chapter focuses on recent studies that have extended our understanding of how signal transduction pathways can mediate signals from cytokines to transcription factors, resulting in the regulation of hematopoiesis.

THE PHOSPHATIDYLINOSITOL 3-KINASE/PROTEIN KINASE B SIGNALING MODULE

An important mediator of cytokine signals is the phosphatidylinositol 3-kinase (PI3K)/protein kinase B (PKB/Akt) signaling module. The PI3K family consists of three distinct subclasses that can be characterized by their structure and substrate specificity. Thus far, the role of these lipid kinases in modulating hematopoiesis has only been examined in terms of regulation of the activity and expression of class I isoforms. To date, four distinct catalytic class I isoforms have been identified; p110α, β, δ, and γ.[7] These isoforms are predominantly activated by protein tyrosine kinases and form heterodimers with a group a regulatory adaptor molecules, including p85α and β, p50α, p55α and γ, and p101γ.[7] The most important substrate for these class I PI3Ks is phosphatidylinositol 4,5-bisphosphate [PI(4,5)P2] which can be phosphorylated at the D3 position of the inositol ring on extracellular stimulation, resulting in the formation of phosphatidylinositol 3,4,5 triphosphate [PI(3,4,5)P3].[8] PI(3,4,5)P3 subsequently serves as an anchor for pleckstrin homology (PH) domain-containing proteins, such as PKB/Akt.[9] PKB/Akt itself subsequently regulates the activity of multiple downstream substrates, including the serine/threonine kinase glycogen synthase kinase-3

(GSK-3),[10] members of the forkhead box protein (FoxO) subfamily of forkhead transcription factors, forkhead box protein O1 (FoxO1), O3 (FoxO3), and O4 (FoxO4),[11] and the serine/threonine kinase mammalian target of rapamycin (mTOR) as part of the mTOR complex 1 (mTORC1). In contrast to GSK-3 and the FoxO transcription factors that are inhibitory phosphorylated by PKB/Akt, activation of mTOR is positively regulated by PKB.[12-13] The activity of PI3K can be inhibited by both phosphatidylinositol 3,4,5-trisphosphate 3-phosphatase and dual-specificity protein phosphatase PTEN (PTEN),[14] a ubiquitously expressed tumor suppressor protein that dephosphorylates phosphoinositide-binding protein PIP3 (PIP3), resulting in the formation of PI(4,5)P$_2$,[14] and SH2 domain-containing inositol-5′-phosphatase (SHIP-1),[15] a protein predominantly expressed in hematopoietic cells,[16] which hydrolyzes PIP3 to generate PI(3,4)P2.[15]

Mutations in both the gene encoding for *phosphatase and tensin homolog (PTEN)* and *inositol polyphosphate-5-phosphatase, 145 kDa (INPP5D/SHIP)* have been observed in a wide variety of malignancies, including leukemia,[17-19] suggesting that correct regulation of PI3K activity plays an important role in the regulation of hematopoiesis (reviewed in[20]). Modulation of the activity of the PI3K/PKB signaling module in both transgenic mice and human umbilical cord blood-derived hematopoietic progenitors revealed that this pathway indeed plays an important role in the expansion of hematopoietic progenitors and cell fate decisions during hematopoiesis. The role of PI3K class I isoforms was initially examined using knockout mice deficient for one or multiple regulatory or catalytic subunits. Combined deletion of p85α, p55α, and p50α resulted in a complete block in B cell development.[21] Similarly, introduction of a mutated, catalytically inactive p110δ (p110δ D910A) in the normal p110δ locus resulted in a block in early B cell development, while T cell development was unaffected.[22-23] These results indicate that PI3K activity is essential for normal B lymphocyte development. In addition, conditional deletion of *PTEN* or *INPP5D (SHIP)* in mature hematopoietic stem cells (HSCs), resulting in the activation of the PI3K pathway, not only reduced the level of B-lymphocytes but also enhanced the level of myeloid cells.[24-26] Furthermore, these mice developed a myeloproliferative disorder that progressed to leukemia.[24-26] In addition, *Pten* heterozygote ($^{+/-}$) *Ship* null ($^{-/-}$) mice have also been generated and this was found to result in a more severe myeloproliferative phenotype compared to the individual mutant mice,[27] displayed by reduced erythrocyte and platelet numbers and enhanced white blood cell counts including elevated levels of neutrophils and monocytes in the peripheral blood. Interestingly, inhibition of either Pten or Ship in the HSCs of adult mice resulted in an initial expansion of HSCs

followed by a depletion of long-term repopulating HSCs,[15,28] suggesting that strict control of PI3K activity is not only essential for correct lineage development but is also required for stem cell maintenance. A role for PI3K in the regulation of eosinophil differentiation has not yet been described using these mouse models. However, pharmacological inhibition of PI3K activity in human umbilical cord blood-derived hematopoietic stem and progenitor cells, which are defined by the expression of the extracellular marker CD34, revealed that inhibition of the activity of PI3K in these cells is sufficient to completely abrogate both proliferation and differentiation during *ex vivo* eosinophil and neutrophil development, eventually leading to cell death.[29] Taken together, these studies suggest that correct temporal regulation of PI3K activity by both PTEN and SHIP is critical for HSC maintenance by the induction of quiescence and plays an important role in the regulation of lineage development.

PKB/Akt, an important effector of PI3K signaling, has been demonstrated to play an important role in the regulation of cell survival and proliferation in a variety of systems.[10] To date, three highly homologous PKB isoforms have been described to be expressed in mammalian cells; PKBα, β, and γ. Analysis of HSCs derived from PKBα/β double-knockout mice revealed that PKB plays an important role in the maintenance of long-term repopulating HSCs. These PKBα/β double-deficient HSCs were found to persist in the G$_0$ phase of the cell cycle, suggesting that the long-term functional defects observed in these mice were caused by increased quiescence.[30] In contrast, ectopic expression of constitutively active PKB in mouse HSCs resulted in transient expansion and increased cycling of HSCs, which was also associated with impaired engraftment,[31] again demonstrating the importance of PKB in HSC maintenance. Using an *ex vivo* human granulocyte differentiation system, it has recently been demonstrated that PKB not only plays a role in the expansion of hematopoietic progenitors, but also has an important function in the regulation of cell fate decisions during hematopoietic lineage commitment.[29] Inhibition of PKB activity in CD34$^+$ cells with a pharmacological inhibitor was sufficient to reduce proliferation during both *ex vivo* eosinophil and neutrophil differentiation, but, in contrast to inhibition of PI3K, did not affect the survival of progenitor cells. High PKB activity was found to promote neutrophil development *ex vivo*, while conversely, reduction of PKB activity is required to induce optimal eosinophil differentiation. In addition, transplantation of β2-microglobulin ($^{-/-}$) nonobese diabetic (NOD)/severe combined immunodeficient or immunodeficiency (SCID) mice with hematopoietic progenitors expressing constitutively active PKB not only resulted in a higher percentage of human neutrophils in the mouse bone marrow *in vivo*, but also induced monocyte development and reduced the percentage of B-lymphocytes. Furthermore, ectopic expression of dominant-negative Pkb was sufficient to enhance the percentage of differentiated eosinophils *in vivo*.[29] Similarly, the development of a myeloproliferative disease, T-cell lymphoma, or acute myeloid leukemia has recently been observed in mice transplanted with mouse HSCs ectopically expressing constitutively active PKB.[31] Although the inhibition of PKB activity appears to be essential to induce terminal maturation of eosinophils, inhibition of PKB activity is not sufficient to induce eosinophil differentiation in the absence of IL-5.[29] This suggests that IL-5 induces specific, currently unknown, signal transduction pathways that are essential for eosinophil differentiation.

To understand the molecular mechanisms underlying the PKB-mediated regulation of hematopoiesis, the roles of its downstream effectors in hematopoiesis have been investigated. An *ex vivo* granulocyte differentiation system was used to investigate the role of mTOR, one of the known downstream effectors of PKB, in the regulation of eosinophil and neutrophil development.[32] In contrast to the inhibition of PKB activity, inhibition of mTOR with the pharmacological inhibitor rapamycin dramatically reduced the expansion of hematopoietic progenitors, during both eosinophil and neutrophil differentiation, without altering levels of apoptosis or maturation.[32] In addition, as described previously, activation of the PI3K signaling pathway by conditional deletion of Pten in adult murine HSCs resulted in an initial expansion followed by exhaustion of long-term HSCs. Inhibition of mTOR in these cells with rapamycin was sufficient to revert this phenotype, suggesting that mTORC1 signaling plays a role in the proliferation of both HSCs and myeloid progenitors.[33]

In addition to mTOR, members of the family of FoxO transcription factors are also known to play an important role in the regulation of hematopoiesis. Conditional deletion of FoxO1, 3, and 4 in the adult hematopoietic system, using mice harboring the interferon-inducible transgene *Mx1-cre* in a *forkhead box O1/2/3 (FoxO1/3/4^{LoxP/LoxP})* background, was, for example, sufficient to induce leukocytosis characterized by relative neutrophilia and lymphocytopenia.[34] In addition, these mice displayed a similar phenotype in terms of HSC expansion and subsequent exhaustion compared to *Pten*-deficient mice in which PKB activity was constitutively upregulated, suggesting that FoxO is a major PI3K—PKB effector in the maintenance of hematopoietic stem cells. Ectopic expression of a constitutively active, non-phosphorylatable, FoxO3 mutant in mouse hematopoietic progenitors conversely results in a decrease in the formation of both myeloid and erythroid colonies.[35] An effect of the modulation of FoxO expression and/or activity on eosinophil differentiation has not been described and remains to be investigated. However, since

ectopic expression of the transcription factor DNA-binding protein inhibitor ID-1 (Id1), which is transcriptionally downregulated by FoxO3,[36] induces neutrophil development and inhibits eosinophil differentiation,[37] thereby mimicking the PKB effect, it could be hypothesized that FoxO3 may indeed be involved in the regulation of eosinophil differentiation.

It has recently been demonstrated that GSK-3, in a way similar to FoxO transcription factors, plays an important role in the maintenance of HSC homeostasis in mice; disruption of GSK-3 activity resulted in a transient expansion of the HSC population.[38] A role for GSK-3, which is also inhibitory phosphorylated by PKB, in the regulation of lineage development has been demonstrated using human CD34[+] hematopoietic progenitors. The ectopic expression of active GSK-3 in CD34[+] cells resulted in the inhibition of neutrophil development *ex vivo*, whereas eosinophil differentiation was enhanced, thereby mimicking the effect of PKB inhibition on hematopoiesis.[29] In addition, it has recently been shown that inhibitory phosphorylation of GSK-3 by PKB results in dephosphorylation and subsequent activation of CCAAT/enhancer-binding protein alpha (C/EBPα), itself a key regulator of hematopoiesis. In contrast, activation of GSK-3 on inhibition of PKB activity results in the phosphorylation of C/EBPα on Thr222/226. Dephosphorylation of C/EBPα on these amino acid residues in human hematopoietic progenitors has been demonstrated to be sufficient to induce neutrophil differentiation and inhibit eosinophil maturation,[29] suggesting that the inhibition of GSK-3 activity, on activation of PI3K/PKB, affects lineage development, at least in part, through the regulation of C/EBPα transcriptional activity.

Taken together, it is evident that the PI3K signaling module plays a critical role in the regulation of both maintenance of HSCs and maturation during lineage development[20] (Fig. 7.5.1). Although expansion of eosinophil progenitors is positively regulated by PKB and mTOR, differentiation of hematopoietic progenitors to mature eosinophils requires the inhibition of PKB activity and subsequent activation of its downstream effector GSK-3.

THE MITOGEN-ACTIVATED PROTEIN KINASE FAMILY

In addition to the PI3K/PKB signaling module, a plethora of external stimuli including hematopoietic growth factors and cytokines mediate their intracellular responses through members of the mitogen-activated protein kinase (MAPK) family. This group of kinases consists of three major subtypes: the extracellular signal-regulated kinases (ERKs), c-Jun-amino-terminal kinase-interacting protein

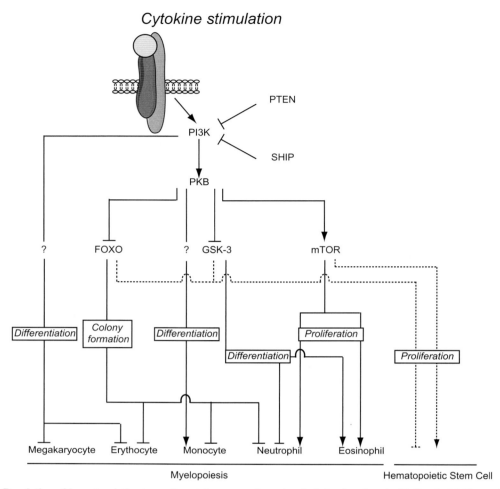

FIGURE 7.5.1 **Regulation of hematopoietic stem cell maintenance and myelopoiesis by the phosphatidylinositol 3-kinase signaling module.** Modulation of the activity of PI3K and its downstream effector PKB plays a critical role in the regulation of HSC maintenance. The activation of PKB by PI3K and subsequent inhibition of the downstream PKB effectors FoxO and GSK-3, as well as activation of mTOR, results in a transient expansion of HSCs followed by exhaustion. In addition, this pathway plays an important role in the regulation of myelopoiesis. Activation of PI3K/PKB results in an induction of neutrophil and monocyte development, while eosinophil differentiation and the development of megakaryocytes and erythrocytes is inhibited. FoxO, forkhead box protein; GSK-3, glycogen synthase kinase-3; HSC, hematopoietic stem cell; mTOR, rapamycin; PI3K, phosphatidylinositol 3-kinase; PKB, protein kinase B; PTEN, phosphate and tensin homologue; SHIP, SH2 domain-containing inositol-5′-phosphatase.

(JNK), and p38 MAPKs. All MAPKs are activated through a similar kinase cascade: MAP kinase kinase kinase (MAP3K or MEKK) activates a MAP kinase kinase (MKK or MEK), which in turn activates MAPK via dual phosphorylation of a Thr-X-Tyr (threonine-X-tyrosine) motif. Whereas ERK activity is induced on phosphorylation of MEK 1/2 (MAPK/ERK kinase) in response to various cytokines and growth factors, members of the JNK and p38 family of MAPKs are predominantly activated by MKK4/MKK7 and MKK3/MKK6, respectively. MKK6 directly phosphorylates all four described p38 isoforms (p38α, β,γ, and δ), while MKK3 is more restricted and activates only p38α, p38γ, and p38δ. MKK4 is even further restricted and only activates p38α (reviewed in[39]).

The constitutive activation of both the ERK and p38 MAPK signaling pathways has been observed in a wide variety of myeloid malignancies,[39] suggesting that these molecules play an important role in the regulation of hematopoiesis. Although the homologous deletion of Jnk1 or Jnk2 in mice has no apparent effect on hematopoiesis, which might be due to redundancy of the different JNK isoforms, several studies have demonstrated that the JNK signaling module does play an important role in the regulation of proliferation and survival during erythropoiesis.[40–41] Similarly, activation of ERK1/2, also known as p42 MAPK and p44 MAPK, has also been shown to be essential for the survival of erythrocyte progenitors.[42] In contrast, during neutrophil differentiation, optimal progenitor expansion, but not survival, depends on the correct regulation of the MEK/ERK1/2 pathway.[42] Whereas the molecular mechanism underlying MEK/ERK1/2-mediated expansion of neutrophil progenitors was found to involve the regulation of the expression of various cell cycle-modulating proteins, including transcriptional regulator Myc (c-Myc), proto-oncogene c-Fos (c-Fos), p21,[WAF/CIP1] and cyclin D1 and D3, during erythrocyte development survival may be regulated by MEK/ERK1/2-mediated modulation of Bcl2 antagonist of cell death (Bad) expression.[42] Although these studies demonstrate that both JNK and ERK1/2 are involved in the regulation of hematopoiesis, their function in eosinophil development is currently unknown. Recent studies indicate that activation of p38 MAPK, the third major MAPK subtype, induces the expansion of HSCs thereby negatively regulating HSC homeostasis.[43] In addition, p38 MAPKs have been demonstrated to differentially regulate both proliferation and maturation during the development of different lineages. It has, for example, been demonstrated that inhibition of p38 MAPK using the pharmacological inhibitor SB 203580 in human hematopoietic progenitors enhances both the expansion and maturation of neutrophil progenitors ex vivo, but conversely reduces proliferation and maturation during eosinophil differentiation. In contrast, transplantation of β-2 microglobulin ($^{-/-}$) in NOD/SCID mice with CD34$^+$ cells ectopically expressing constitutively active MKK3 resulted in reduced neutrophil differentiation in vivo, whereas eosinophil development was enhanced.[44] Similar to eosinophil differentiation, proliferation of CD34$^+$-derived erythroid progenitors was found to be positively regulated by p38 MAPK.[45] Interestingly, distinct stage-specific mRNA expression patterns of the four p38 MAPK isoforms have been described during erythroid differentiation.[46] Although it remains to be investigated whether a similar expression profile is present during differentiation of other hematopoietic lineages, this suggests that the p38 MAPK isoforms are not completely redundant and may also have distinct functions during hematopoiesis.

It has been demonstrated that p38 MAPK regulates the phosphorylation of various proteins, including the kinases MAP kinase-activated protein kinase 2 and 3 (MAPKAPK-2 and -3), the small 27 kDa heat shock protein (Hsp27), and several transcription factors, including C/EBPβ, activating transcription factor 1 and 2 (ATF-1/2), C/EBP-homologous protein (CHOP), signal transducer and activator of transcription 1-alpha/beta (STAT1) and signal transducer and activator of transcription 3 (STAT3) (reviewed in[39]). However, the importance of phosphorylation of these molecules in the regulation of hematopoiesis remains to be determined. In addition to these proteins, it has recently been demonstrated that activation of p38 MAPK can inhibitory phosphorylate C/EBPα on serine 21.[44] Similar to ectopic expression of a C/EBPα

mutant, in which phosphorylation on Thr222/226 was prevented resulting in an induction of neutrophil differentiation and inhibition of eosinophil development,[29] expression of a C/EBPα mutant that could no longer be phosphorylated on serine 21 was sufficient to abrogate MKK3-induced inhibition of neutrophil development.[44] Taken together, this indicates that MKK3−p38 MAPK plays an important role in the regulation of lineage choices during myelopoiesis, at least in part through the modulation of C/EBPα activity. However, C/EBPα can also be phosphorylated on serine 21 by ERK1/2.[47] Although both p38 MAPK and ERK1/2 regulate the phosphorylation of C/EBPα on the same residue, these proteins appear to have different roles in hematopoiesis. This suggests that although C/EBPα is a key regulator in hematopoiesis additional proteins must be involved in determination of cell fate decisions.

Taken together, although the three major subtypes of the MAPK family, ERK, JNK, and p38 MAPK, all appear to be involved in regulation of hematopoiesis (Fig. 7.5.2), p38 MAPK is thus far the only member of the MAPK family implicated in the regulation of eosinophil differentiation; p38 MAPK positively regulates both proliferation and maturation of eosinophil precursors at least in part via the regulation of the activity of C/EBPα.

INTERLEUKIN-5 RECEPTOR SUBUNIT ALPHA-ASSOCIATING PROTEINS

Many cytokine receptors consist of multiple subunits, one of which is unique and specific for a certain cytokine while the other subunit is shared by multiple cytokines. Although cytokines can exhibit a wide variety of functions in different cell types and tissues, there is also some redundancy between cytokines. The fact that different cytokines exhibit, in part, similar functions can be explained by the presence of shared receptor subunits. The GM-CSF, IL-3, and IL-5 cytokine receptors family, for example, consist of two subunits; a cytokine-specific α chain, which is responsible for ligand binding, and a common β chain. The β chain is not capable of ligand binding by itself but forms a high-affinity receptor on association with the ligand-bound α chain. The β chain is thought to mediate intracellular signaling through the activation of a variety of signal transduction pathways. Receptor activation induces overlapping signal transduction pathways that depend on both α and β chains, which are pre-associated with tyrosine-protein kinase JAK (JAK) that activate STAT. In addition, both the PI3K/PKB signaling module and members of the MAPK family are activated by all three receptors. Although these cytokines have overlapping functions, GM-CSF, IL-3, or IL-5 can also regulate distinct hematopoietic stem cell fate

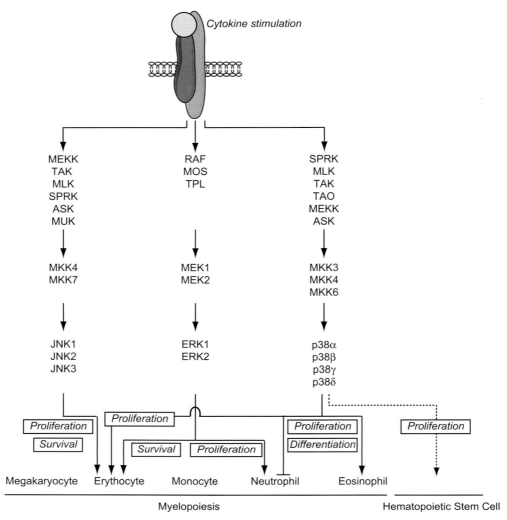

FIGURE 7.5.2   Regulation of hematopoietic stem cell maintenance and myelopoiesis by members of the mitogen-activated protein kinase family.
Modulation of the activity of p38 MAPK plays an important role in the regulation of HSC maintenance. Activation of p38 MAPK results in a transient expansion of HSCs followed by exhaustion. In addition, members of the MAPK family also play an important role in the regulation of myelopoiesis. Proliferation of erythrocyte progenitors is positively regulated by all three major subtypes of the MAPK family, ERK1/2, JNK, and p38 MAPK. Similarly, ERK1/2 also induces proliferation of neutrophil progenitors. In contrast, p38 MAPK inhibits both the proliferation and differentiation of neutrophil progenitors. The development of eosinophils is conversely induced by p38 MAPK. ASK, apoptosis signal-regulating kinase; ERK, extracellular signal-regulated kinase; JNK, c-Jun-amino-terminal kinase; MAPK, mitogen-activated protein kinase; MEK, MAPK/ERK kinase; MEKK, MAPK/ERK kinase kinase; MKK, dual specificity mitogen-activated protein kinase kinase; MLK, mixed-lineage kinase; MOS, proto-oncogene serine/threonine-protein kinase mos; MUK, MAPK upstream kinase; RAF, RAF proto-oncogene serine/threonine-protein kinase; SPRK, Src-homology 3 domain-containing proline-rich kinase; TAK, TGF-beta-activated kinase 1 and MAP3K7-binding protein 1; TAO, thousand and one amino acid protein; TPL, tumor progression locus.

decisions during myelopoiesis. In contrast to GM-CSF and IL-3, IL-5 plays an essential role in eosinophil differentiation. Since the intracellular domains of the cytokine receptor-specific α chains are unique, these are excellent candidates to mediate specific signal transduction pathways. To understand the molecular mechanisms underlying IL-5 mediated regulation of eosinophil differentiation it is essential to identify proteins that are specifically associated with the interleukin-5 receptor subunit alpha (IL-5RA) chain.

Although both JAK1 and JAK2 are activated on stimulation of cells with IL-5, the mechanism of activation is different. Analysis of the cytoplasmic tail of IL-5RA revealed that JAK2, but not JAK1, is constitutively associated with human IL-5RA (hIL-5RA) regardless of IL-5 stimulation. In contrast, JAK1, which is constitutively associated with the β chain, only associates with IL-5RA on stimulation with IL-5.[48] Detailed study of the C-terminus-truncated cytoplasmic domain of hIL-5RA revealed that the cytoplasmic stretch at position 346–387, containing the proline-rich region, is necessary for JAK2 binding. Furthermore, tyrosine phosphorylation of JAK1 was dependent on the activation of JAK2. These observations suggest that the activation of hIL-5RA-associated JAK2 is indispensable for IL-5 signaling. The role of JAK2 in the regulation of eosinophil development has been shown using a pharmacological inhibitor of JAK2. Those investigations showed that inhibition of JAK2 activity is sufficient to reduce the development of mouse bone marrow cells to mature eosinophils.[49] However, since signal transducer and activator of transcription 5 (STAT5), a downstream effector of JAK2, has been demonstrated to play an important role in proliferation and maturation during both eosinophil and neutrophil differentiation,[50] it remains to be investigated whether JAK2 inhibition specifically affects eosinophil differentiation.

The proto-oncogene tyrosine-protein kinase Src (Src), tyrosine-protein kinase SYK (SYK), and tyrosine-protein kinase Tec (Tec) family kinases are three of the best characterized, non-receptor tyrosine kinase families found in the human genome. Members of these kinase families function downstream of a wide variety of receptors in hematopoietic cells and induce calcium mobilization, altered gene expression, cytokine production, and cell proliferation. It has recently been demonstrated that from the Src family members present in eosinophils, tyrosine-protein kinase Lyn (Lyn), tyrosine-protein kinase HCK (HCK), tyrosine-protein kinase Fgr (FGR), and tyrosine-protein kinase Lck (LCK), only Lyn can associate with IL-5RA. On stimulation with IL-5, Lyn is capable of phosphorylating both IL-5RA and IL-5RB. However, since both GM-CSF and IL-3 can also upregulate Lyn kinase activity, suggesting the β chain is involved,[51] it remains to be investigated whether IL-5-mediated activation of Lyn results in the regulation of IL-5-specific downstream effectors. The importance of Lyn kinase for eosinophil differentiation has been studied using both antisense oligodeoxynucleotides[49] and an antagonistic peptide that prevents association of Lyn with the β chain.[52] Both studies showed that the inhibition of Lyn is sufficient to block eosinophil differentiation from stem cells in a dose-dependent manner. In addition, the importance of Lyn for eosinophil differentiation was further studied using Lyn knockout mice. Although Lyn-deficient mice exhibited a mildly enhanced basal production of neutrophils and reduced levels of B cells,[53] basal eosinophilopoiesis was unaffected.[53] However, IL-5-induced eosinophil differentiation of bone marrow cells was significantly inhibited in Lyn-deficient mice as compared to controls.[49]

In addition to JAK2 and Lyn, a direct interaction has been demonstrated between IL-5RA and the two PSD-95/Discs-Large/ZO-1 (PDZ) domains containing the adaptor protein syntenin.[54] Deletion of either of the PDZ domains

is sufficient to completely abrogate the interaction with IL-5RA. In addition, a single-point mutant in the C-terminus of IL-5RA completely abrogates the interaction between syntenin and IL-5RA, demonstrating that syntenin associates with the C-terminus of IL-5RA. No interaction between syntenin and granulocyte-macrophage colony-stimulating factor receptor subunit alpha and interleukin-3 receptor subunit alpha could be detected.[54]

Since adaptor proteins are non-catalytic proteins that consist of multiple protein interaction domains that serve to regulate the assembly of protein complexes, it is likely that syntenin serves to recruit additional signaling molecules to the IL-5R. Indeed, it has, for example, been demonstrated that transcription factor SOX-4 (SOX-4) directly interacts with syntenin, but does not bind to IL-5RA itself.[54] In addition, it has recently been demonstrated that syntenin does enhance IL-5-mediated phosphorylation of ERK-1 and STAT5, suggesting that syntenin is also involved in the modulation of IL-5 signaling in general.[55] Moreover, it has also been demonstrated that syntenin enhances the IL-5-mediated phosphorylation of JAK2, but not Lyn kinase.[55] The formation of complexes including syntenin and multiple IL-5RA chains suggest that the syntenin-enhanced IL-5R output may result from stabilization of an IL-5-induced oligomeric receptor complex.[55] Although it has been demonstrated that, in addition to SOX-4 and IL-5RA, syntenin can also associate with other proteins,[56] a role for these molecules in hematopoietic progenitors and eosinophils remains to be determined. Syntenin expression is upregulated during eosinophil differentiation of human umbilical cord blood-derived $CD34^+$ cells, but downregulated during neutrophil development, suggesting that syntenin may specifically regulate eosinophil differentiation.[55] Indeed, manipulation of syntenin levels by ectopic expression or knockdown affects IL-5-induced eosinophil differentiation. Ectopic expression of syntenin in primary human $CD34^+$ hematopoietic progenitor cells was sufficient to enhance IL-5-induced eosinophil differentiation, whereas reduction of syntenin with, for example, small hairpin RNA resulted in reduced eosinophil development.[55] It remains to be investigated whether syntenin-mediated regulation of eosinophil differentiation is due to a general enhancement of IL-5 signaling or whether regulation of the activity of specific syntenin binding partners, including SOX4, is essential.

Taken together, to date, three IL-5RA interacting proteins, JAK2, Lyn, and syntenin, have been described to play an important role in the regulation of eosinophil differentiation (Fig. 7.5.3). However, since also other cytokines, including GM-CSF and IL-3, activate both JAK2 and Lyn, it is likely that these molecules are essential for receptor activation in general. Although syntenin does not interact with the GM-CSF and IL-3 receptors, suggesting

FIGURE 7.5.3 Schematic representation of the interleukin-5 receptor and its binding partners. Under basal condition, Jak2, Lyn, and syntenin are all associated with IL-5RA. Jak1, in contrast, is constitutively associated with the beta chain (βc). Upon ligand binding, the βc is recruited to IL-5RA resulting in the activation of multiple signal transduction pathways. ERK, extracellular signal-regulated kinase; IL-5, interleukin-5; IL-5RA, interleukin-5 receptor subunit alpha; JAK, tyrosine-protein kinase JAK; JNK, c-Jun-amino-terminal kinase-interacting protein; Lyn, tyrosine-protein kinase Lyn; MAPK, mitogen-activated protein kinase; PI3K, phosphatidylinositol 3-kinase; SOX4, transcription factor SOX-4; STAT5, signal transducer and activator of transcription A.

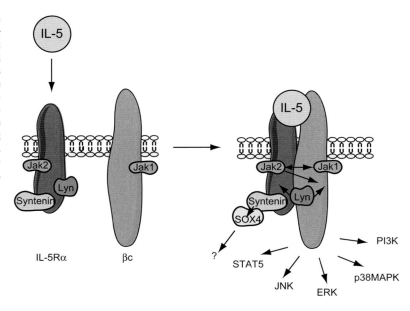

a higher level of specificity compared to JAK2 and Lyn, it remains to be investigated whether syntenin specifically induces eosinophil differentiation.

CONCLUSION

Cytokines play an important role in directing cell fate decisions during lineage development, suggesting that the activity of cytokine-specific effectors must be regulated. However, most of the currently known effectors that are involved in hematopoiesis, including PKB and p38 MAPK, are regulated by cytokines in a similar manner. In addition, although the activity of both PKB and p38 MAPK is induced on cytokine stimulation, these molecules appear to counteract each other in the regulation of some hematopoietic processes. Whereas, for example, neutrophil differentiation is positively regulated by PKB, it is inhibited by p38 MAPK (Fig. 7.5.4). In contrast, both activation of p38 MAPK and/or inhibition of PKB activity induce

eosinophil differentiation. This differential effect on granulocyte development suggests that the balance between inhibitory and activating signals induced by cytokines determines whether or not a progenitor cell becomes a mature eosinophil. Differences in signal strength and length could potentially change this balance in inhibitory and activating signals. In addition, it could also be hypothesized that the activity of p38 MAPK and/or PKB could be differentially regulated in hematopoietic stem and progenitors cells on interaction of those cells with stromal cells in the bone marrow microenvironment.

In addition to these generally regulated molecules, it has also been shown that the IL-5RA-associated proteins JAK2, Lyn, and syntenin are involved in eosinophil development (Fig. 7.5.4). However, since the activity of these molecules is regulated by a variety of cytokines, it is unlikely that JAK2, Lyn, and syntenin specifically regulate eosinophil differentiation. In contrast, syntenin does not interact with the GM-CSF-R-alpha and IL-3 receptors, suggesting

FIGURE 7.5.4 Signal transduction molecules involved in the regulation of eosinophil differentiation. Proliferation of eosinophil progenitors is positively regulated by both PKB and its downstream effectors mTOR and p38 MAPK. The differentiation of eosinophils requires the activation of Jak2, Lyn, p38 MAPK, and syntenin. In contrast, the inhibition of PKB activity is essential for normal eosinophil maturation. GSK-3, glycogen synthase kinase-3; JAK2, tyrosine-protein kinase JAK2; Lyn, tyrosine-protein kinase Lyn; MAPK, mitogen-activated protein kinase, mTOR, rapamycin; PKB, protein kinase B.

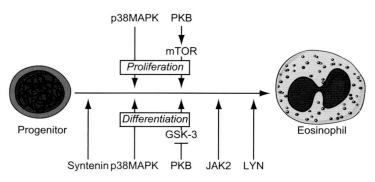

a higher level of specificity compared to JAK2 and Lyn. It would be of great interest to investigate whether syntenin indeed specifically induces eosinophil differentiation. In addition, to date, the molecular mechanisms underlying syntenin-mediated regulation of eosinophil differentiation are completely unclear and should be investigated.

Taken together, although our knowledge of cytokine-mediated regulation of hematopoiesis has increased over the last few years, many questions remain to be answered.

Research should, for example, focus on the identification of transcription factors that are regulated by the signaling molecules involved in the regulation of hematopoiesis. In addition, since several signal transduction pathways, which are regulated in a similar manner on cytokine stimulation, can counteract one another in terms of regulation of lineage development, the balance between inhibitory and activating signals should also be investigated.

REFERENCES

1. Klingmuller U, Wu H, Hsiao JG, Toker A, Duckworth BC, Cantley LC, et al. Identification of a novel pathway important for proliferation and differentiation of primary erythroid progenitors. *Proc Natl Acad Sci U S A* 1997;**94**:3016—21.

2. Kaushansky K, Broudy VC, Grossmann A, Humes J, Lin N, Ren HP, et al. Thrombopoietin expands erythroid progenitors, increases red cell production, and enhances erythroid recovery after myelosuppressive therapy. *J Clin Invest* 1995;**96**:1683—7.

3. Panopoulos AD, Watowich SS. Granulocyte colony-stimulating factor: molecular mechanisms of action during steady state and 'emergency' hematopoiesis. *Cytokine* 2008;**42**:277—88.

4. Rohrschneider LR, Bourette RP, Lioubin MN, Algate PA, Myles GM, Carlberg K. Growth and differentiation signals regulated by the M-CSF receptor. *Mol Reprod Dev* 1997;**46**:96—103.

5. Yamaguchi Y, Hayashi Y, Sugama Y, Miura Y, Kasahara T, Kitamura S, et al. Highly purified murine interleukin 5 (IL-5) stimulates eosinophil function and prolongs in vitro survival. IL-5 as an eosinophil chemotactic factor. *J Exp Med* 1988;**167**:1737—42.

6. Sanderson CJ, Warren DJ, Strath M. Identification of a lymphokine that stimulates eosinophil differentiation in vitro. Its relationship to interleukin 3, and functional properties of eosinophils produced in cultures. *J Exp Med* 1985;**162**:60—74.

7. Vanhaesebroeck B, Leevers SJ, Ahmadi K, Timms J, Katso R, Driscoll PC, et al. Synthesis and function of 3-phosphorylated inositol lipids. *Annu Rev Biochem* 2001;**70**:535—602.

8. Hawkins PT, Anderson KE, Davidson K, Stephens LR. Signalling through Class I PI3Ks in mammalian cells. *Biochem Soc Trans* 2006;**34**:647—62.

9. Burgering BM, Coffer PJ. Protein kinase B (c-Akt) in phosphatidylinositol-3-OH kinase signal transduction. *Nature* 1995;**376**:599—602.

10. Manning BD, Cantley LC. AKT/PKB signaling: navigating downstream. *Cell* 2007;**129**:1261—74.

11. Kops GJ, de Ruiter ND, De Vries-Smits AM, Powell DR, Bos JL, Burgering BM. Direct control of the Forkhead transcription factor AFX by protein kinase B. *Nature* 1999;**398**:630—4.

12. Nave BT, Ouwens M, Withers DJ, Alessi DR, Shepherd PR. Mammalian target of rapamycin is a direct target for protein kinase B: identification of a convergence point for opposing effects of insulin and amino-acid deficiency on protein translation. *Biochem J* 1999;**344**(Pt 2):427—31.

13. Inoki K, Li Y, Zhu T, Wu J, Guan KL. TSC2 is phosphorylated and inhibited by Akt and suppresses mTOR signalling. *Nat Cell Biol* 2002;**4**:648—57.

14. Maehama T, Dixon JE. The tumor suppressor, PTEN/MMAC1, dephosphorylates the lipid second messenger, phosphatidylinositol 3,4,5-trisphosphate. *J Biol Chem* 1998;**273**:13375—8.

15. Damen JE, Liu L, Rosten P, Humphries RK, Jefferson AB, Majerus PW, et al. The 145-kDa protein induced to associate with Shc by multiple cytokines is an inositol tetraphosphate and phosphatidylinositol 3,4,5-triphosphate 5-phosphatase. *Proc Natl Acad Sci U S A* 1996;**93**:1689—93.

16. Liu Q, Nozari G, Sommer SS. Single-tube polymerase chain reaction for rapid diagnosis of the inversion hotspot of mutation in hemophilia A. *Blood* 1998;**92**:1458—9.

17. Leupin N, Cenni B, Novak U, Hugli B, Graber HU, Tobler A, et al. Disparate expression of the PTEN gene: a novel finding in B-cell chronic lymphocytic leukaemia (B-CLL). *Br J Haematol* 2003;**121**:97—100.

18. Liu TC, Lin PM, Chang JG, Lee JP, Chen TP, Lin SF. Mutation analysis of PTEN/MMAC1 in acute myeloid leukemia. *Am J Hematol* 2000;**63**:170—5.

19. Luo JM, Yoshida H, Komura S, Ohishi N, Pan L, Shigeno K, et al. Possible dominant-negative mutation of the SHIP gene in acute myeloid leukemia. *Leukemia* 2003;**17**:1—8.

20. Buitenhuis M, Coffer PJ. The role of the PI3K-PKB signaling module in regulation of hematopoiesis. *Cell Cycle* 2009;**8**:560—6.

21. Fruman DA, Mauvais-Jarvis F, Pollard DA, Yballe CM, Brazil D, Bronson RT, et al. Hypoglycaemia, liver necrosis and perinatal death in mice lacking all isoforms of phosphoinositide 3-kinase p85 alpha. *Nat Genet* 2000;**26**:379—82.

22. Okkenhaug K, Bilancio A, Farjot G, Priddle H, Sancho S, Peskett E, et al. Impaired B and T cell antigen receptor signaling in p110delta PI 3-kinase mutant mice. *Science* 2002;**297**:1031—4.

23. Jou ST, Carpino N, Takahashi Y, Piekorz R, Chao JR, Wang D, et al. Essential, nonredundant role for the phosphoinositide 3-kinase p110delta in signaling by the B-cell receptor complex. *Mol Cell Biol* 2002;**22**:8580—91.

24. Zhang J, Grindley JC, Yin T, Jayasinghe S, He XC, Ross JT, et al. PTEN maintains haematopoietic stem cells and acts in lineage choice and leukaemia prevention. *Nature* 2006;**441**:518—22.

25. Helgason CD, Damen JE, Rosten P, Grewal R, Sorensen P, Chappel SM, et al. Targeted disruption of SHIP leads to hemopoietic perturbations, lung pathology, and a shortened life span. *Genes Dev* 1998;**12**:1610—20.

26. Liu Q, Sasaki T, Kozieradzki I, Wakeham A, Itie A, Dumont DJ, et al. SHIP is a negative regulator of growth factor receptor-mediated PKB/Akt activation and myeloid cell survival. *Genes Dev* 1999;**13**:786—91.

27. Moody JL, Xu L, Helgason CD, Jirik FR. Anemia, thrombocytopenia, leukocytosis, extramedullary hematopoiesis, and impaired

progenitor function in Pten+/−SHIP−/− mice: a novel model of myelodysplasia. *Blood* 2004;**103**:4503−10.

28. Helgason CD, Antonchuk J, Bodner C, Humphries RK. Homeostasis and regeneration of the hematopoietic stem cell pool are altered in SHIP-deficient mice. *Blood* 2003;**102**:3541−7.

29. Buitenhuis M, Verhagen LP, van Deutekom HW, Castor A, Verploegen S, Koenderman L, et al. Protein kinase B (c-akt) regulates hematopoietic lineage choice decisions during myelopoiesis. *Blood* 2008;**111**:112−21.

30. Juntilla MM, Patil VD, Calamito M, Joshi RP, Birnbaum MJ, Koretzky GA. AKT1 and AKT2 maintain hematopoietic stem cell function by regulating reactive oxygen species. *Blood* 2010;**115**:4030−8.

31. Kharas MG, Okabe R, Ganis JJ, Gozo M, Khandan T, Paktinat M, et al. Constitutively active AKT depletes hematopoietic stem cells and induces leukemia in mice. *Blood* 2010;**115**:1406−15.

32. Geest CR, Zwartkruis FJ, Vellenga E, Coffer PJ, Buitenhuis M. Mammalian target of rapamycin activity is required for expansion of CD34+ hematopoietic progenitor cells. *Haematologica* 2009;**94**:901−10.

33. Yilmaz OH, Valdez R, Theisen BK, Guo W, Ferguson DO, Wu H, et al. Pten dependence distinguishes haematopoietic stem cells from leukaemia-initiating cells. *Nature* 2006;**441**:475−82.

34. Tothova Z, Kollipara R, Huntly BJ, Lee BH, Castrillon DH, Cullen DE, et al. FoxOs are critical mediators of hematopoietic stem cell resistance to physiologic oxidative stress. *Cell* 2007;**128**:325−39.

35. Engstrom M, Karlsson R, Jonsson JI. Inactivation of the forkhead transcription factor FoxO3 is essential for PKB-mediated survival of hematopoietic progenitor cells by kit ligand. *Exp Hematol* 2003;**31**:316−23.

36. Birkenkamp KU, Essafi A, van der Vos KE, da Costa M, Hui RC, Holstege F, et al. FOXO3a induces differentiation of Bcr-Abl-transformed cells through transcriptional down-regulation of Id1. *J Biol Chem* 2007;**282**:2211−20.

37. Buitenhuis M, van Deutekom HW, Verhagen LP, Castor A, Jacobsen SE, Lammers JW, et al. Differential regulation of granulopoiesis by the basic helix-loop-helix transcriptional inhibitors Id1 and Id2. *Blood* 2005;**105**:4272−81.

38. Huang J, Zhang Y, Bersenev A, O'Brien WT, Tong W, Emerson SG, et al. Pivotal role for glycogen synthase kinase-3 in hematopoietic stem cell homeostasis in mice. *J Clin Invest* 2009;**119**:3519−29.

39. Geest CR, Coffer PJ. MAPK signaling pathways in the regulation of hematopoiesis. *J Leukoc Biol* 2009;**86**:237−50.

40. Nagata Y, Takahashi N, Davis RJ, Todokoro K. Activation of p38 MAP kinase and JNK but not ERK is required for erythropoietin-induced erythroid differentiation. *Blood* 1998;**92**:1859−69.

41. Jacobs-Helber SM, Sawyer ST. Jun N-terminal kinase promotes proliferation of immature erythroid cells and erythropoietin-dependent cell lines. *Blood* 2004;**104**:696−703.

42. Geest CR, Buitenhuis M, Groot Koerkamp MJ, Holstege FC, Vellenga E, Coffer PJ. Tight control of MEK-ERK activation is essential in regulating proliferation, survival, and cytokine production of CD34+-derived neutrophil progenitors. *Blood* 2009;**114**:3402−12.

43. Ito K, Hirao A, Arai F, Takubo K, Matsuoka S, Miyamoto K, et al. Reactive oxygen species act through p38 MAPK to limit the lifespan of hematopoietic stem cells. *Nat Med* 2006;**12**:446−51.

44. Geest CR, Buitenhuis M, Laarhoven AG, Bierings MB, Bruin MC, Vellenga E, et al. p38 MAP kinase inhibits neutrophil development through phosphorylation of C/EBPalpha on serine 21. *Stem Cells* 2009;**27**:2271−82.

45. Somervaille TC, Linch DC, Khwaja A. Different levels of p38 MAP kinase activity mediate distinct biological effects in primary human erythroid progenitors. *Br J Haematol* 2003;**120**:876−86.

46. Uddin S, Ah-Kang J, Ulaszek J, Mahmud D, Wickrema A. Differentiation stage-specific activation of p38 mitogen-activated protein kinase isoforms in primary human erythroid cells. *Proc Natl Acad Sci U S A* 2004;**101**:147−52.

47. Ross SE, Radomska HS, Wu B, Zhang P, Winnay JN, Bajnok L, et al. Phosphorylation of C/EBPalpha inhibits granulopoiesis. *Mol Cell Biol* 2004;**24**:675−86.

48. Ogata N, Kouro T, Yamada A, Koike M, Hanai N, Ishikawa T, et al. JAK2 and JAK1 constitutively associate with an interleukin-5 (IL-5) receptor alpha and betac subunit, respectively, and are activated upon IL-5 stimulation. *Blood* 1998;**91**:2264−71.

49. Stafford S, Lowell C, Sur S, Alam R. Lyn tyrosine kinase is important for IL-5-stimulated eosinophil differentiation. *J Immunol* 2002;**168**:1978−83.

50. Buitenhuis M, Baltus B, Lammers JW, Coffer PJ, Koenderman L. Signal transducer and activator of transcription 5a (STAT5a) is required for eosinophil differentiation of human cord blood-derived CD34+ cells. *Blood* 2003;**101**:134−42.

51. Torigoe T, O'Connor R, Santoli D, Reed JC. Interleukin-3 regulates the activity of the LYN protein-tyrosine kinase in myeloid-committed leukemic cell lines. *Blood* 1992;**80**:617−24.

52. Adachi T, Stafford S, Sur S, Alam R. A novel Lyn-binding peptide inhibitor blocks eosinophil differentiation, survival, and airway eosinophilic inflammation. *J Immunol* 1999;**163**:939−46.

53. Chan VW, Meng F, Soriano P, DeFranco AL, Lowell CA. Characterization of the B lymphocyte populations in Lyn-deficient mice and the role of Lyn in signal initiation and down-regulation. *Immunity* 1997;**7**:69−81.

54. Geijsen N, Uings IJ, Pals C, Armstrong J, McKinnon M, Raaijmakers JA, et al. Cytokine-specific transcriptional regulation through an IL-5Ralpha interacting protein. *Science* 2001;**293**:1136−8.

55. Beekman JM, Verhagen LP, Geijsen N, Coffer PJ. Regulation of myelopoiesis through syntenin-mediated modulation of IL-5 receptor output. *Blood* 2009;**114**:3917−27.

56. Beekman JM, Coffer PJ. The ins and outs of syntenin, a multifunctional intracellular adaptor protein. *J Cell Sci* 2008;**121**:1349−55.

Chapter 7.6

Signaling and Degranulation

Paige Lacy and Redwan Moqbel

DEGRANULATION DEFINED

Eosinophils possess a remarkable series of membrane-bound secretory granules and vesicles that are readily detectable by light and electron microscopy.[1] Their most distinctive organelle is the unique crystalloid granule, also known as the secondary granule. The crystalloid granule, a modified lysosome, contains an electron-dense core that is highly cationic and is intensely stained by acidic dyes such as eosin. Along with these are the multitudes of electron-translucent granules and tubulovesicular carriers, which can be visualized by electron microscopy. These intracellular membrane-bound organelles are generated through granulogenesis, which occurs during the development of eosinophils in the bone marrow and other sites of hematopoiesis.[2,3]

The crystalloid granules form the major intracellular site of storage of all of the eosinophil's highly cationic granule proteins, including eosinophil granule major basic protein 1 (MBP-1), eosinophil peroxidase (EPO), and the various ribonucleases [eosinophil cationic protein (ECP) and eosinophil-derived neurotoxin (EDN) in humans; eosinophil-associated ribonucleases in mice].[1,4] In addition, a range of cytokines, chemokines, and growth factors may be found in crystalloid granules.[5,6] These proteins are synthesized and stored within granules for the purpose of controlled release following receptor—stimulus coupling.

There are a significant number of crystalloid granules present in each eosinophil (approximately 300—400 per cell)[7] and their contents are frequently found in tissues from patients with allergy, asthma, viral diseases, and parasitic worm diseases. By comparison, healthy individuals show a complete absence of eosinophil granule proteins in similar tissue samples. Thus, eosinophil degranulation must occur during the pathogenesis of these various diseases. How do eosinophils release their granule proteins? Is it enough for eosinophils to simply arrive at inflammatory foci and somehow *shed* their proteins? This chapter discusses how granules are derived, the pathways for eosinophil degranulation, and how eosinophils release their tissue-damaging cationic proteins along with cytokines and chemokines.

Granulogenesis

The formation of the crystalloid granule occurs through granulogenesis at the early stages of eosinophil progenitor development in hematopoietic tissues. Each granule has a lipid bilayer to contain the granule proteins and ensure their delivery to the cell surface. The crystalloid granule contains crystallized MBP-1 at its core and is surrounded by a matrix that is filled with chemokines, cytokines, ECP, EDN, and other granule contents (Table 7.6.1).[1,6,8] The development of the crystalloid granule involves a maturation process in which immature granules form via membrane trafficking from the Golgi apparatus. Patch

clamp analysis of human eosinophils has shown that immature crystalloid granules derive from unit granules (1—4 unit granule per crystalloid) that bud off from the Golgi apparatus and fuse together.[7] At the core of the immature crystalloid granule, the precursor form of MBP-1 (pro-MBP-1) gradually condenses until a crystalline core containing the mature MBP-1 is developed.[2] The appearance of the crystalline core in these granules is the hallmark of the mature crystalloid granule of the eosinophil.

WHAT DEGRANULATION MEANS

Eosinophils typically release their granule products through degranulation in response to receptor stimulation. Many different types of receptors converge through separate signaling pathways to stimulate the release of eosinophil granule products. These include G protein-coupled receptors, such as complement component (C5a) receptors, chemoattractants [f-Met-Leu-Phe (fMLP), platelet-activating factor (PAF)], and chemokine receptors, which signal through phospholipase C and induce the release of diacylglycerol, inositol phosphates, and Ca^{2+}, that act in concert with protein serine/threonine kinases to elicit granule mobilization and release.[1,9] Similarly, receptors belonging to immunoglobulin-binding families [such as Fcgamma receptor II and immunoglobulin alpha Fc receptor (IgA Fc receptor) in human eosinophils], on cross-linking by appropriate ligands, trigger a cascade of tyrosine kinase phosphorylation that leads to granule secretion.[3,9]

The process of degranulation describes the release of granules, either whole or ruptured, or their granule proteins from cells possessing specialized secretory granules. Degranulation is an umbrella term that defines the release of granules without reference to the specific mechanism that regulates their release. Hence, degranulation can describe the release of granule proteins from viable cells that tightly regulate their secretion, or the release of intact/ruptured granules from dying, necrotic cells.

There is a substantial body of evidence from *in vitro* and *in vivo* studies showing that eosinophils degranulate when they are recruited and activated at sites of inflammation. In allergic disease, eosinophil degranulation is thought to be an essential component of the late phase mucosal tissue response to allergen challenge. Tissue eosinophil degranulation is found in severe asthma, allergic rhinitis, and cutaneous allergic reactions.[1,6,8] Moreover, eosinophil granule products have been detected in viral and parasitic worm infections.[10—12] In many of these disorders or infections, eosinophil degranulation is associated with a deteriorating clinical picture. Therefore, it is critically important to identify and characterize key regulatory proteins associated with degranulation in eosinophils, with

TABLE 7.6.1   Proteins and Enzymes in Human Eosinophil Granules and Secretory Vesicles

Crystalloid Granules	Primary Granules	Small Granules	Lipid Bodies	Secretory Vesicles
Core				
Catalase				
Cathepsin D				
Enoyl-CoA-hydrolase				
β-glucuronidase				
Eosinophil granule major basic protein				
Matrix				
Acid phosphatase	Charcot–Leyden crystal protein (galectin-10)	Acid phosphatase	Arachidonic acid	Plasma proteins (albumin)
Acyl-CoA oxidase		Arylsulfatase B (active)	Cyclooxygenase	
Arylsulfatase B (inactive)		Catalase	Eosinophil peroxidase	
Acid β-glycerophosphatase		Elastase	Esterase	
Bactericidal/permeability-increasing protein		Eosinophil cationic protein	5-Lipoxygenase	
Catalase			15-Lipoxygenase	
Cathepsin D			LTC$_4$ synthase	
Collagenase				
Elastase				
Enoyl-CoA-hydrolase				
Eosinophil granule cationic protein				
Eosinophil-derived neurotoxin				
Eosinophil peroxidase				
FAD				
β-Glucuronidase				
β-Hexosaminidase				
3-Ketoacyl-CoA thiolase				
Lysozyme				
Eosinophil granule major basic protein				
Phospholipase A$_2$ (type II)				
Nonspecific esterases				
Membrane				
CD63		VAMP-7		Cytochrome b$_{558}$ [p22phox]
V-type H$^+$-ATPase		VAMP-8		VAMP-2
VAMP-7				VAMP-7
VAMP-8				VAMP-8

FAD, Flavin adenine dinucleotide; VAMP, vesicle-associated membrane protein.

the ultimate aim of developing novel therapeutic strategies directed at controlling or inhibiting this process.

TYPES OF DEGRANULATION

One of the principal ways in which degranulation occurs is through exocytosis, an exquisitely regulated process of granule membrane fusion with the plasma membrane. The first step of exocytosis is the contact of the outer leaflet of the lipid bilayer membrane surrounding the granule with the inner leaflet of the plasma membrane. This process is known as docking and occurs following the tethering of granules. After docking, the granule and plasma membrane form a continuous lipid bilayer that causes the development of a fusion pore, a reversible structure that can spontaneously open and close. Expansion of the fusion pore leads to a complete integration of the granule membrane with the plasma membrane, which directly increases the surface area of the cell. At the same time, the interior surface of the granule (the lumenal surface) becomes exposed to the exterior of the cell, and the contents of the granule are expelled to the extracellular milieu.

There are four modes of eosinophil degranulation, as determined by *in vitro* and *in vivo* imaging of cells prepared from peripheral blood or tissue samples. These are:[1] classical exocytosis, in which granules are released individually following membrane fusion;[2] compound exocytosis, in which secretory granules undergo homotypic fusion to allow clusters of granules to release their contents through a single fusion pore;[13,14,3] piecemeal degranulation, which involves the shuttling of granular components containing selective crystalloid granule contents to the plasma membrane;[15,16] and[4] cytolysis, which is the necrotic disintegration of eosinophils leading to the loss of membrane integrity and subsequently causing the dispersal of intact, membrane-bound crystalloid granules (cell-free granules) into the tissues (Fig. 7.6.1A).

Regulated and Constitutive Exocytosis

Classical exocytosis can be further differentiated as either constitutive or regulated exocytosis. Constitutive exocytosis describes the pathway of continuous vesicular trafficking and release of newly synthesized protein cargo. For example, all cells including eosinophils have the ability to constitutively synthesize and secrete extracellular matrix proteins. Constitutively released proteins are synthesized in the endoplasmic reticulum and transported through the Golgi complex to be packaged at the trans-Golgi network into vesicles, which are immediately released at the cell surface (Fig. 7.6.1B). Constitutive exocytosis can be stimulated indirectly by receptor stimulation through enhanced nuclear transcription of vesicular cargo.

Conversely, regulated exocytosis involves receptor-mediated stimulation of granule mobilization and fusion with the plasma membrane. In this case, cargo proteins are synthesized in advance by the cell and packaged into granules for later release on appropriate stimulation (Fig. 7.6.1B). Eosinophils have the capacity to undergo both constitutive and regulated exocytosis by virtue of their heterogeneous population of tubulovesicular structures and secretory granules.[16,17]

Eosinophils release their granule contents through regulated exocytosis. In one form of the latter, eosinophils release their granules one at a time by fusing these individually with the plasma membrane. Patch-clamp analysis of degranulating human eosinophils has demonstrated that a stepwise increment in cell capacitance can be observed when eosinophils are patched in the whole cell configuration with a pipette containing the nonhydrolyzable analog for guanosine triphosphate, guanosine $5'$-O-γ-thio-triphosphate (GTPγS), and Ca^{2+}.[18] Each step in capacitance corresponds to the exocytotic fusion event associated with a single secretory granule. However, clinical samples do not show evidence of classical exocytosis in tissue eosinophils, and this type of exocytosis has only been observed *in vitro* with purified eosinophil preparations.

Another form of regulated exocytosis in eosinophils is compound exocytosis. Here, cells release their granule products from the intracellular fusion of multiple granules in a highly focused manner onto target surfaces after cell adherence. An example of this is eosinophil granule release following intimate attachment to appropriately opsonized larvae of the parasitic helminth *Schistosoma mansoni* (Fig. 7.6.2A).[12,19,20] Compound exocytosis from eosinophils has also been characterized *in vitro* by patch-clamp analysis using whole-cell configuration, showing homotypic granule fusion preceding the release of granule contents through individual fusion pores.[7] Fusion pores associated with granules undergoing compound exocytosis have been visualized in tannic acid-arrested degranulating eosinophils.[21]

Although classical and compound exocytosis have been shown to occur in eosinophils from patients with inflammatory bowel disease and tissue-invasive infections, it is unusual to see morphological evidence of these types of exocytosis in eosinophilic inflammation associated with asthma or allergy.

Piecemeal Degranulation

The most commonly observed physiological mode of degranulation in eosinophils in allergy is piecemeal degranulation, which is a form of regulated exocytosis. This type of degranulation was first identified by electron microscopy in eosinophils differentiated in culture from human peripheral blood mononuclear cells.[22,23] In

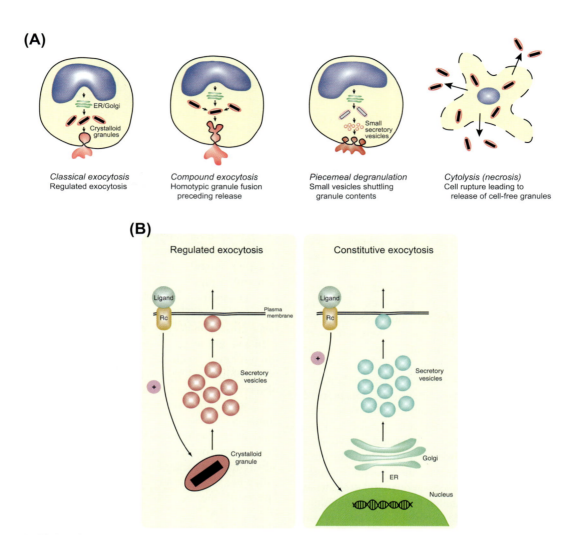

FIGURE 7.6.1   Modes of eosinophil degranulation. *(A)* Eosinophils have been observed to undergo at least four different modes of degranulation following *in vitro* stimulation by secretagogues or in tissue specimens. Tissue eosinophils have been observed to exhibit each of these types of degranulation, except for classical exocytosis, in biopsies from a range of inflammatory diseases. *(B)* Cells may undergo either regulated or constitutive exocytosis. In regulated exocytosis, receptor stimulation leads to the release of granule-stored mediators, shown in this example as piecemeal degranulation from an eosinophil crystalloid granule. Constitutive exocytosis is associated with either continuous vesicular trafficking of protein cargo of newly synthesized protein deriving from mRNA translation, or receptor stimulation of *de novo* protein synthesis, which is subsequently packaged and trafficked. In tissue eosinophils from allergic subjects, eosinophils frequently show regulated exocytosis through piecemeal degranulation. ER, endoplasmic reticulum; Rc, receptor.

piecemeal degranulation, numerous small electron-lucent secretory vesicles are found in abundance in the cytoplasm, adjacent to translucent, *hollowed-out* crystalloid granules containing partially eroded core components that are normally solid and electron-dense in nonstimulated cells.[20] Eosinophils exhibiting the characteristics of piecemeal degranulation are found in association with allergy,[24,25] particularly in nasal polyps seen in allergic rhinitis.[26,27] In nasal polyp sections, piecemeal degranulation was observed in 67—83% of tissue eosinophils by morphological examination using transmission electron microscopy,[26,27] but was not evident in circulating eosinophils.[28]

Piecemeal degranulation is hypothesized to be a mechanism by which selective, piecemeal release of granule-derived mediators occurs in response to differential receptor stimulation. This mode of degranulation is considered a secretory method used by immune cells and other cell types in the body, including enteroendocrine cells of the gastrointestinal tract, chromaffin cells of the adrenal medulla, and chief cells of the parathyroid gland.[29] In eosinophils, piecemeal degranulation is associated with the differential release of the granule proteins ECP and EDN, during stimulation by complexes of different immunoglobulin subclasses.[30,31]

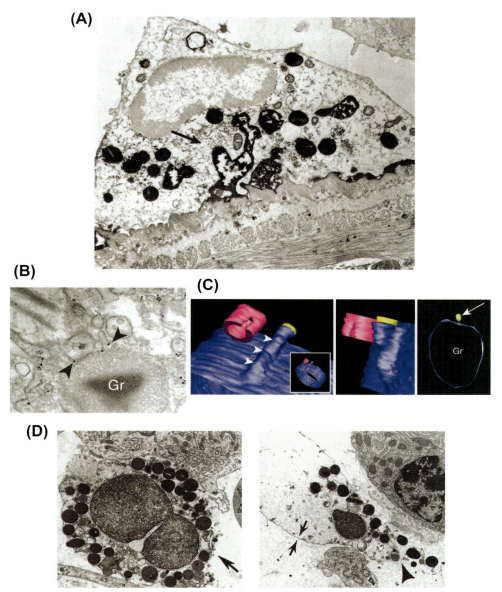

FIGURE 7.6.2   Electron microscopic evidence of compound exocytosis and piecemeal degranulation in eosinophils. *(A)* Compound exocytosis of eosinophil peroxidase-positive granules (indicated by the arrow) directed toward the surface of an opsonized parasite. Transmission electron micrograph (magnification ×14,800), reproduced with permission pending, McLaren et al.[19] *(B)* Piecemeal degranulation of human eosinophils. Vesicle budding from a crystalloid granule containing IL-4RA on its membrane (arrowheads). These vesicles are also enriched in IL-4, suggesting a selective trafficking mechanism for cytokine release from crystalloid granules. Gr, crystalloid granule. Transmission electron micrograph reproduced with permission pending, Spencer et al.[34] Copyright (2006) National Academy of Sciences, USA. *(C)* Meshed models of an eosinophil granule with sombrero vesicles budding from it, using contours generated by automated electron tomography. A protrusion can be seen progressively emerging from the granule, in blue (arrowheads). In the center panel, a vesicle is shown in yellow to indicate the way in which sombrero vesicles appear in sections through the cell. In the right panel, the arrow indicates the same vesicle from a different perspective near the granule (Gr). This model was generated by meshes of contours arising from serial sections of the same granule in a single cell. Cells were fixed and processed for conventional transmission electron microscopy following stimulation with eotaxin. Images reproduced with permission pending from Melo et al.[35] Reprinted with permission from *Traffic* 2005;6:1047—1057, published by John Wiley and Sons. *(D)* Cytolysis in tissue eosinophils from allergic subjects. Cytolysis is evident in a tissue eosinophil in which the plasma membrane is lost (arrow); the panel on the right shows late-stage cytolysis with eosinophil granules being released into the extracellular matrix (arrowhead), with parts of the cell membrane remaining (arrows) but lacking organelles except for crystalloid granules. *Transmission electron micrograph reproduced with permission pending from Erjefält et al.[26] Reprinted from Erjefält JS, Andersson M, Greiff L, Korsgren M, Gizycki M, Jeffery PK, Persson CG. Cytolysis and piecemeal degranulation as distinct modes of activation of airway mucosal eosinophils. J Allergy Clin Immunol 102:286—294. Copyright (1998), with permission from Elsevier.*

Cytokine release from eosinophils also occurs through piecemeal degranulation in a differential and selective manner. While interleukin-6 is stored in the crystalloid granules,[32] the chemokine C-C motif chemokine 5 (CCL5/ RANTES) is present in two different secretory compartments in human eosinophils.[33] The first is the crystalloid granule and the second is a pool of small, rapidly mobilizable secretory vesicles. These secretory vesicles form a part of the eosinophil's tubulovesicular network that shuttles crystalloid granule contents to the cell surface.[34] The tubulovesicular carriers that form this network appear to be involved in the selective release of cytokines and chemokines from their sites of storage within crystalloid granules, since differing kinetics of release of CCL5/ RANTES and β-hexosaminidase were observed in eosinophils stimulated by interferon gamma (IFN-γ) (Fig. 7.6.3).[33] In support of this, a mechanism for selective cytokine release from crystalloid granules was proposed in which the cytokine receptor, in this case interleukin-4 receptor subunit alpha (IL-4RA), binds to granular IL-4 and carries granule-derived IL-4 to the cell membrane for secretion.[34] In this work, elegant imaging using transmission electron microscopy of sequential sections demonstrated that IL-4-containing tubulovesicular carriers eosinophil sombrero vesicles[16] could be seen budding from the surface of secretory granules (Fig. 7.6.2B and C). These findings are supportive of the notion of piecemeal degranulation in eosinophils, leading to the selective release of cytokines and chemokines in response to different agonists. The process associated with the secretion of cytokines by piecemeal degranulation is discussed in greater detail within Chapter 8.2 (Release of Cytokines and Chemokines from Eosinophils).

Cytolysis (Necrosis)

The fourth mode of degranulation observed in eosinophils is cytolysis, which is the process by which cells rupture during necrotic cell death. Eosinophils exhibit cytolysis at sites of inflammation, with a loss of plasma membrane integrity associated with the spilling of cellular contents, including intact and disrupted secretory granules, into the surrounding tissues. The release of eosinophil secretory granules into the extracellular matrix has a major impact on tissue functions. Cytolysis is the second most commonly observed mode of degranulation in tissue eosinophils from allergic subjects, with 10−33% of tissue eosinophils exhibiting cytolytic degranulation in nasal polyps from subjects with allergic rhinitis (Fig. 7.6.2D).[26,27] This is likely to be the main source of membrane-bound, intact, cell-free granules from eosinophils accumulating at inflammatory foci. Cell-free granules from eosinophils have been observed in blood and tissues from a range of diseases including allergy,[36] primary biliary cirrhosis,[37]

inflammatory bowel disease,[38] and in the bone marrow of patients with neoplastic disorders.[39] They are readily detectable by fluorescence microscopy analysis of tissue sections due to their autofluorescence, caused by an enrichment of flavin adenine dinucleotide (FAD).[40] These granules have the potential to act as secretory organelles as they express a range of functional cell receptors and release their contents during receptor stimulation,[41] a recent area of eosinophil research that is discussed in detail elsewhere in this book (see Chapter 8.6, Cell-Free Granules are Functional Secretory Organelles).

INTRACELLULAR AND MOLECULAR EVENTS REGULATING EXOCYTOSIS

The process of granule mobilization and exocytosis occurs through a highly regulated series of intracellular signaling events in response to receptor stimulation. Precise coordination of molecular signals enables: (1) the mobilization of granules or vesicles to the cell periphery; (2) tethering of these membrane-bound organelles to the inner leaflet of the plasma membrane; (3) docking of granules/vesicles to the plasma membrane, finally resulting in (4) fusion pore formation as a result of membrane fusion between the granule and plasma membrane. The fusion pore is a reversible structure that is ultimately responsible for the discharge of granule contents on its expansion. Early studies demonstrated that Ca^{2+} and guanosine triphosphate (GTP) were the two essential effectors required to induce exocytosis using permeabilized or patch-clamped eosinophils.[13,42] The exocytotic response was enhanced by the addition of the energy molecule adenosine triphosphate (ATP).[43] The obligatory requirement of eosinophils for GTP suggested that at least one GTP-binding protein was required for the process of degranulation.[44] These findings demonstrated that, similar to many other secretory tissues, both Ca^{2+} and GTP are important second messengers for receptor-mediated degranulation in eosinophils, and implicate a role for both Ca^{2+}- and GTP-binding proteins in exocytotic fusion. Receptor stimulation by a range of ligands, particularly from G protein-coupled receptors, leads to increased Ca^{2+} levels, which then acts on a cohort of Ca^{2+}-binding proteins to initiate exocytosis. Parallel with this, numerous GTP-binding proteins are activated to bind to GTP after the dissociation of guanosine diphosphate (GDP), and consequently stimulate their respective effector proteins for exocytotic trafficking.

GTPases

There is a plethora of GTP-binding proteins expressed in cells, many of which are GTPases that hydrolyze protein-bound GTP to GDP to form an inactive GDP-bound structure. Regulatory GTPases can be broadly classified

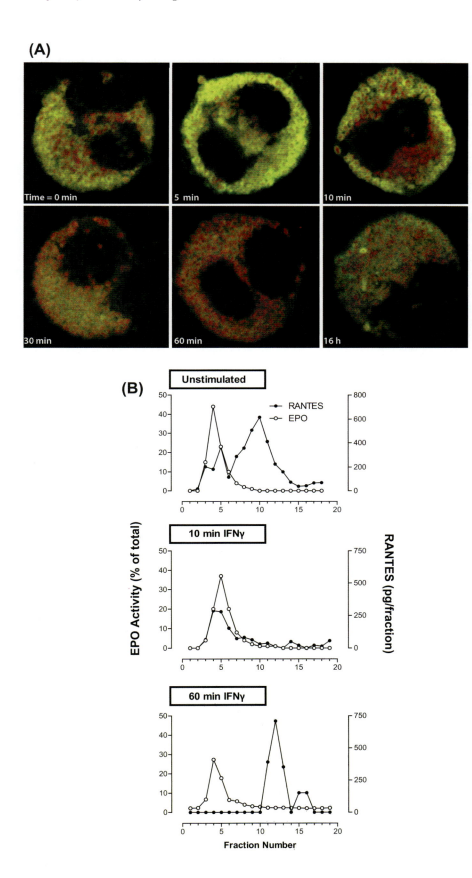

(A)

Time = 0 min | 5 min | 10 min
30 min | 60 min | 16 h

(B)

Unstimulated

10 min IFNγ

60 min IFNγ

EPO Activity (% of total)

RANTES (pg/fraction)

Fraction Number

● RANTES
○ EPO

FIGURE 7.6.3 Confocal imaging of piecemeal degranulation of C-C motif chemokine 5 (RANTES) in eosinophils. Shown is *(A)* a time course of IFN-γ-induced C-C motif chemokine 5 (CCL5/RANTES) release from human eosinophils determined by confocal microscopy of immunofluorescence. CCL5/RANTES (green) selectively traffics to the cell membrane during IFN-γ stimulation away from granules containing MBP-1 (red). Some colocalization of MBP-1 and CCL5/RANTES is evident at the cell periphery, seen as a yellow color resulting from the combination of red and green fluorescence. *(B)* Subcellular fractionation of resting and IFN-γ-stimulated eosinophils (5×10^7 per fractionation) across a linear density gradient. A time course analysis of 500 U/ml IFN-γ-induced CCL5/RANTES release was carried out using eosinophils from the same donor. Immunoreactivity to CCL5/RANTES shifted from higher to lower density fractions during stimulation by IFN-γ, suggesting selective mobilization of this chemokine. EPO activity indicates the location of crystalloid granules in the gradient. EPO, eosinophil peroxidase; IFN-γ, interferon gamma; MBP-1, eosinophils granule major basic protein 1. *Reproduced from Lacy et al.[33] This research was originally published in Blood. Lacy P, Mahmudi-Azer S, Bablitz B, Hagen SC, Velazquez JR, Man SF, Moqbel R. Rapid mobilization of intracellularly stored RANTES in response to interferon-gamma in human eosinophils. Blood 1999 Jul 1;94(1):23-32. © The American Society of Hematology.*

into two classes: one is the receptor-associated hetero-trimeric G proteins, which contain three subunits (α, β, and γ),[44] and the second is the vast family of rat sarcoma (Ras)-related GTPases, which consist of a single monomer of about 21 kDa.[45] Of the Ras-related GTPases, there are five subfamilies that have been defined based on homologous amino acid sequences (Ras, Rho, Rab, Arf, and Ran). Three of these subfamilies have been associated with the regulation of exocytosis in many cells (Rho, Rab, and Arf).[17,46] Rho GTPases are composed of 20 members that regulate actin remodeling and cell motility, with unique functions and localizations identified for each protein.[47] Rab GTPases constitute a diverse subfamily (>40 isoforms) that localize to membranes of distinct intracellular compartments and which are engaged in membrane trafficking at all levels.[48]

Eosinophils express heterotrimeric G protein subunits ($G\alpha_{q11}$, $G\alpha_s$, $G\alpha_{i2/i3}$, and $G\beta$) and the small monomeric GTPases Rab1, Rab2, Rab4, and Rab11.[49] In addition, they express Ras-related protein Rac1 (Rac1) and 2 (Rac2), which are members of the Rho-related GTPases.[50–52] Rac1 and Rac2 are important for the activation of the nicotinamide adenine dinucleotide phosphate (NADPH) oxidase enzyme complex, which generates superoxide during receptor or phorbol ester stimulation, and eosinophils employ Rac1 and Rac2 for their extracellular release of superoxide.[52] Rac2, which possesses 92% sequence homology to Rac1, is a regulator of actin remodeling in hematopoietic cells including eosinophils.[53] We have shown that Rac2 is also required for exocytosis of azurophilic granules induced by fMLP and leukotriene B$_4$ in neutrophils.[54] Similarly, Rac2 is essential for Ca^{2+}-mediated exocytosis of EPO-containing crystalloid granules in eosinophils.[55]

A potential role for membrane-bound Ras-related protein Rab-27A (Rab27) in eosinophil activation and degranulation has recently been explored, in which Rab27 was shown to be activated during stimulation by secretory immunoglobulin A-coated beads.[56] Eosinophils from atopic donors also showed higher levels and accelerated activation of Rab27 compared with normal healthy eosinophils.[56] These observations show that eosinophils may utilize several GTPases, specifically Rac2 and Rab27, for the mobilization and docking of granules as a prelude to exocytosis (Fig. 7.6.4).

Soluble NSF Attachment Protein Receptor Isoforms in Regulated Exocytosis

For the past two decades, investigators have focused intently on a highly homologous group of membrane-associated proteins known as SNAP (Soluble NSF Attachment Protein) REceptors (SNAREs), which are critical for granule or vesicle docking and fusion.[57] Originally these proteins were separated into vesicle (v-SNARE) and target (t-SNARE) labels to describe their cellular localization. However, due to the overlap in the functions and localization of several v- and t-SNAREs, SNAREs are now classified on the basis of their expression of a conserved arginine (R) or glutamine (Q) residue in the core sequence of the conserved SNARE docking complex.[58]

In exocytosis, the SNARE complex is composed of one R-SNARE and two Q-SNAREs bound together to form a four-helix coiled-coil bundle. The exocytotic R-SNARE, usually a vesicle-associated membrane protein (VAMP) isoform, provides one coiled-coil domain, and the two plasma membrane-associated Q-SNAREs contribute three coiled-coil strands. The three coiled-coil strands from the Q-SNAREs consist of one sequence from a syntaxin isoform, and two sequences from a synaptosomal-associated protein of 23, 25, or 29 kDa (SNAP-23, SNAP-25, or SNAP-29; Fig. 7.6.4). The cytosolic protein's soluble N-ethylmaleimide-sensitive factor (NSF)-attachment protein (α- and β-SNAP, which are unrelated to SNAP-23, SNAP-25, and SNAP-29) and NSF, for which the SNAREs are named, are recruited to the SNARE complex during fusion and are responsible for the rapid disassembly and recycling of the SNARE components to their original compartments.[57]

The formation of SNARE complexes is a critical event preceding membrane fusion and granule mediator release in almost all cell types. This has been demonstrated using tetanus neurotoxin, which specifically cleaves certain R-SNAREs (VAMP-1, VAMP-2, and VAMP-3/cellubrevin), and botulinum neurotoxins, which cleave several Q-SNARE isoforms.[58] Cleavage of SNAREs leads to inhibition of their assembly into granule docking complexes, and prevents the subsequent fusion of granules with the plasma membrane.[59] The highly specific actions of these toxins have led to the exploitation of botulinum neurotoxin type A (BOTOX) in cosmetic applications to reduce wrinkling and furrowing in the skin, as well as for clinical use to treat headaches. In these applications, BOTOX acts to decrease neurotransmitter release at neuromuscular junctions in the scalp and forehead, causing nonpermanent but long-lasting paralysis of muscle activity.[60,61]

One of the main hypotheses put forward originally by Rothman and colleagues in their original paper describing SNAREs[62] was that v-SNAREs (R-SNAREs) and t-SNAREs (Q-SNAREs) confer specificity of binding between intracellular sites, so that a given donor compartment will recognize and bind to its appropriate target compartment. In other words, SNAREs localized to specific granules/vesicles are predicted to form complexes with their binding partners in the plasma membrane (cognate SNAREs) and not with SNAREs localized to other cellular compartments (noncognate

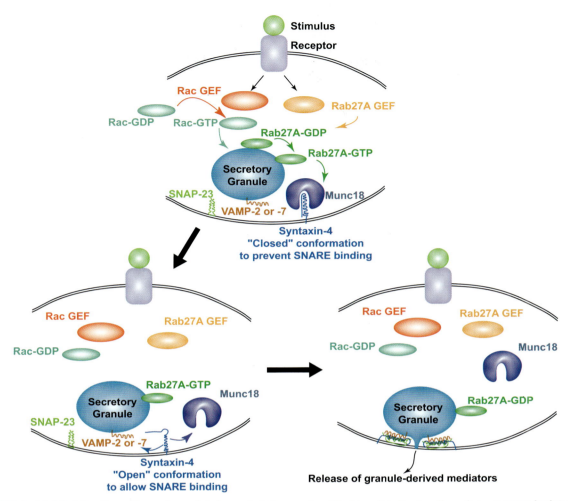

FIGURE 7.6.4 **Model of signaling pathways in regulated exocytosis from eosinophils.** Ligand binding to cell membrane receptors leads to activation of an intracellular signaling cascade that results in GTP binding of Rac and Rab27 through their respective guanine exchange factors (GEFs). Rac-GTP and Rab27-GTP are proposed to stimulate the mobilization of the secretory granule to the cell membrane and initiate SNARE binding. SNARE binding is initiated by Rab27-mediated dissociation of syntaxin-bound Munc18, leading to an *open* conformation for syntaxin-4. The *open* conformation of syntaxin-4 allows its association with the R-SNARE on the secretory granule, VAMP-2 or -7, and with the second Q-SNARE, SNAP-23. Binding of the SNARE complex causes the formation of a four-helix coiled-coil structure that pulls the granule membrane to the plasma membrane and permits the development of a fusion pore, leading to the release of granule-derived mediators. Munc, mammalian uncoordinated (protein); GDP, guanosine diphosphate; GTP, guanosine triphosphate; Rab, Ras-related protein Rab; Rac, Ras-related protein; SNAP, synaptosomal-associated protein; SNARE, SNAP (Soluble NSF Attachment Protein) Receptor, VAMP, vesicle-associated membrane protein.

SNAREs). This was a reasonable proposal that was based on the diversity of SNARE isoforms found in each cell type. However, this is now a controversial concept, as noncognate SNARE isoforms readily associate *in vitro*,[63] and individual R-SNAREs can interact with high affinity with multiple intracellular syntaxins at different sites, at least in yeast studies.[64] Moreover, multiple R-SNARE isoforms promoted equally efficient fusion events on binding with plasma membrane Q-SNAREs, suggesting that SNARE isoforms cannot account for specificity on their own.[65]

Human eosinophils express R-SNARE, VAMP-2, VAMP-7, and VAMP-8, as determined by subcellular fractionation and confocal microscopy analysis of fluorescent antibody labeling.[66,67] Tetanus toxin-sensitive VAMP-2 localized predominantly to small secretory vesicles, based on differential ultracentrifugation of light membranes, while no VAMP-2 could be detected in crystalloid granules.[66,68,69] However, immunoreactivity to VAMP-2 colocalized with a pool of small secretory vesicles containing CCL5/RANTES, and was similarly mobilized to the cell surface on stimulation with IFN-γ.[66] Thus, VAMP-2 may provide a mechanism for piecemeal degranulation of CCL5/RANTES release from crystalloid granules by shuttling small secretory vesicles from the granules to the cell surface.

The crystalloid granules are instead enriched in VAMP-7 and VAMP-8, which are proposed to serve as vesicular SNAREs to allow binding and fusion of granules with the plasma membrane.[67] Using permeabilized eosinophils, it was shown that neutralizing antibody to VAMP-7 reduced EDN and EPO secretion following GTPγS- and Ca^{2+}-triggered exocytosis, while anti-VAMP-8 was without effect at large doses. Administration of anti-VAMP-2 partially suppressed EPO release, suggesting that piecemeal degranulation of EPO is dependent on VAMP-2. Interestingly, these findings support the notion that SNARE isoforms do not confer specificity to granule release, since anti-VAMP-7 completely abrogated all forms of granule release not only from eosinophils, but also from neutrophils.[67]

The Q-SNAREs expressed by eosinophils so far identified are SNAP-23 and syntaxin-4, which are predominantly found in the plasma membrane.[70] Eosinophils do not appear to express the neuronal isoforms SNAP-25 or syntaxin-1.[48] Both SNAP-23 and syntaxin-4 appear to function as cognate t-SNAREs for VAMP-2 in piecemeal degranulation of CCL5/RANTES (Fig. 7.6.4).[70] In summary, these findings suggest that SNARE binding plays a crucial role in regulated exocytosis in eosinophils.

Sec/Munc Proteins

The Sec/mammalian uncoordinated (Munc) (SM) proteins constitute a family of conserved syntaxin-binding proteins, which has been directly implicated in the regulation of SNARE complex assembly. Three mammalian homologs of Munc18 have been identified: Munc18-1, -2, and -3.[71] Similar to SNAREs, SM proteins are essential for secretion. The current model for SM proteins in exocytosis is that they selectively bind to a *closed* conformation of syntaxin on the inner leaflet of the plasma membrane (Fig. 7.6.4). Thus, when bound to Munc18, syntaxin is unable to bind to other SNAREs to form docking complexes.[72] The dissociation of Munc18 is catalyzed by kinases such as cyclin-dependent kinase 5 (CDK5). Eosinophils express Munc18-3 on their crystalloid granules and require CDK5 for secretion.[73] However, the specific details of the role that these SM proteins play in secretion are not yet elucidated.

Ca^{2+}-Dependent Kinases

Among the numerous effector molecules stimulated by transient Ca^{2+} increases in response to receptor binding, protein kinase C (PKC) is regarded as a key regulatory molecule in the coupling of stimulus to secretion in granule-derived mediator release. This phosphorylating enzyme is activated by membrane-derived diacylglycerol and by inositol phosphate-mediated Ca^{2+} transients. Eosinophils express the conventional, novel, and atypical isoforms of PKC.[74] A role for PKC in mediator release from eosinophils is supported by observations of ECP and EPO secretion induced by PAF or allergen.[75-77]

Several investigations have shown important links between PKC and the regulation of SNARE partnering with other SNAREs and/or SNARE-associated accessory proteins. Syntaxin, SNAP-25, and SM proteins are phosphorylated by PKC *in vitro*.[78,79] Direct phosphorylation of SM proteins by PKC leads to their reduced affinity for syntaxins.[80,81] An attractive hypothesis is that stimulus-coupled Ca^{2+} activation of PKC leads to the phosphorylation of Munc18, leading to an *open* conformation of syntaxin and allowing SNARE binding. However, evidence for a role for PKC in the activation of exocytosis in eosinophils, either at the level of SNARE phosphorylation or at other sites in the cell, has not yet been reported.

CONCLUSION

Our understanding of degranulation processes has developed exponentially since the discovery of the eosinophil and its involvement in inflammatory diseases. The signaling pathways described here pertain only to what we have learned in receptor-mediated exocytosis of small secretory vesicles and crystalloid granules, and how this is important in piecemeal degranulation. However, there are still several areas of degranulation that are poorly defined. For example, signaling events associated with necrotic cell death, leading to the release of intact or ruptured granules and vesicles, are not defined in eosinophils. Since cytolytic degranulation from eosinophils is frequently witnessed in tissue samples from a range of inflammatory diseases, it is important to understand how cytolysis occurs and how it may be modulated to prevent inflammation and the damaging sequelae of eosinophilic inflammation. We anticipate that current studies using genetically modified animal models will generate many new insights into how eosinophil degranulation is linked to the pathogenesis of disease, and the specific molecular mechanisms required for this inflammatory process.

REFERENCES

1. Hogan SP, Rosenberg HF, Moqbel R, Phipps S, Foster PS, Lacy P, et al. Eosinophils: biological properties and role in health and disease. *Clin Exp Allergy* 2008;**38**:709—50.
2. Popken-Harris P, Checkel J, Loegering D, Madden B, Springett M, Kephart G, et al. Regulation and processing of a precursor form of

eosinophil granule major basic protein (ProMBP) in differentiating eosinophils. *Blood* 1998;**92**:623−31.

3. Mahmudi-Azer S, Velazquez JR, Lacy P, Denburg JA, Moqbel R. Immunofluorescence analysis of cytokine and granule protein expression during eosinophil maturation from cord blood-derived CD34 progenitors. *J Allergy Clin Immunol* 2000;**105**:1178−84.

4. Rosenberg HF. RNase A ribonucleases and host defense: an evolving story. *J Leukoc Biol* 2008;**83**:1079−87.

5. Lacy P, Moqbel R. Eosinophil cytokines. *Chem Immunol* 2000;**76**:134−55.

6. Moqbel R, Odemuyiwa SO, Lacy P, Adamko DJ. The Human Eosinophil. In: Greer JP, Foerster J, Rodgers GM, Paraskevas F, Glader B, Arber DA, Means RT, editors. *Wintrobe's Clinical Hematology*. Philadelphia: Wolters Kluwer/Lippincott Williams & Wilkins; 2009. p. 214−35.

7. Hartmann J, Scepek S, Lindau M. Regulation of granule size in human and horse eosinophils by number of fusion events among unit granules. *The Journal of Physiology* 1995;**483**:201−9.

8. Rothenberg ME, Hogan SP. The eosinophil. *Annu Rev Immunol* 2006;**24**:147−74.

9. Giembycz MA, Lindsay MA. Pharmacology of the eosinophil. *Pharmacol Rev* 1999;**51**:213−340.

10. Gleich GJ, Adolphson CR. The eosinophilic leukocyte: structure and function. *Adv Immunol* 1986;**39**:177−253.

11. Garofalo R, Kimpen JL, Welliver RC, Ogra PL. Eosinophil degranulation in the respiratory tract during naturally acquired respiratory syncytial virus infection. *J Pediatr* 1992;**120**:28−32.

12. Scepek S, Moqbel R, Lindau M. Compound exocytosis and cumulative degranulation by eosinophils and their role in parasite killing. *Parasitol Today* 1994;**10**:276−8.

13. Scepek S, Lindau M. Focal exocytosis by eosinophils—compound exocytosis and cumulative fusion. *EMBO J* 1993;**12**:1811−7.

14. Scepek S, Coorssen JR, Lindau M. Fusion pore expansion in horse eosinophils is modulated by Ca^{2+} and protein kinase C via distinct mechanisms. *EMBO J* 1998;**17**:4340−5.

15. Moqbel R, Coughlin JJ. Differential secretion of cytokines. *Sci STKE* 2006;**338**. pe26.

16. Melo RC, Spencer LA, Dvorak AM, Weller PF. Mechanisms of eosinophil secretion: large vesiculotubular carriers mediate transport and release of granule-derived cytokines and other proteins. *J Leukoc Biol* 2008;**83**:229−36.

17. Lacy P. The role of Rho GTPases and SNAREs in mediator release from granulocytes. *Pharmacol Ther* 2005;**107**:358−76.

18. Nusse O, Lindau M, Cromwell O, Kay AB, Gomperts BD. Intracellular application of guanosine-5′-O-(3-thiotriphosphate) induces exocytotic granule fusion in guinea pig eosinophils. *J Exp Med* 1990;**171**:775−86.

19. McLaren DJ, Mackenzie CD, Ramalho-Pinto FJ. Ultrastructural observations on the in vitro interaction between rat eosinophils and some parasitic helminths (*Schistosoma mansoni, Trichinella spiralis* and *Nippostrongylus brasiliensis*). *Clin Exp Immunol* 1977;**30**:105−18.

20. McLaren DJ, Ramalho-Pinto FJ, Smithers SR. Ultrastructural evidence for complement and antibody-dependent damage to schistosomula of *Schistosoma mansoni* by rat eosinophils *in vitro*. *Parasitology* 1978;**77**:313−24.

21. Newman TM, Tian M, Gomperts BD. Ultrastructural characterization of tannic acid-arrested degranulation of permeabilized guinea pig eosinophils stimulated with GTPγS. *Eur J Cell Biol* 1996;**70**:209−20.

22. Dvorak AM, Furitsu T, Letourneau L, Ishizaka T, Ackerman SJ. Mature eosinophils stimulated to develop in human cord blood mononuclear cell cultures supplemented with recombinant human interleukin-5. Part I. Piecemeal degranulation of specific granules and distribution of Charcot-Leyden crystal protein. *Am J Pathol* 1991;**138**:69−82.

23. Dvorak AM, Ackerman SJ, Furitsu T, Estrella P, Letourneau L, Ishizaka T. Mature eosinophils stimulated to develop in human-cord blood mononuclear cell cultures supplemented with recombinant human interleukin-5. II. Vesicular transport of specific granule matrix peroxidase, a mechanism for effecting piecemeal degranulation. *Am J Pathol* 1992;**140**:795−807.

24. Karawajczyk M, Seveus L, Garcia R, Bjornsson E, Peterson CG, Roomans GM, et al. Piecemeal degranulation of peripheral blood eosinophils: a study of allergic subjects during and out of the pollen season. *Am J Respir Cell Mol Biol* 2000;**23**:521−9.

25. Ahlstrom-Emanuelsson CA, Greiff L, Andersson M, Persson CG, Erjefalt JS. Eosinophil degranulation status in allergic rhinitis: observations before and during seasonal allergen exposure. *Eur Respir J* 2004;**24**:750−7.

26. Erjefalt JS, Andersson M, Greiff L, Korsgren M, Gizycki M, Jeffery PK, et al. Cytolysis and piecemeal degranulation as distinct modes of activation of airway mucosal eosinophils. *J Allergy Clin Immunol* 1998;**102**:286−94.

27. Erjefalt JS, Greiff L, Andersson M, Matsson E, Petersen H, Linden M, et al. Allergen-induced eosinophil cytolysis is a primary mechanism for granule protein release in human upper airways. *Am J Respir Crit Care Med* 1999;**160**:304−12.

28. Malm-Erjefalt M, Greiff L, Ankerst J, Andersson M, Wallengren J, Cardell LO, et al. Circulating eosinophils in asthma, allergic rhinitis, and atopic dermatitis lack morphological signs of degranulation. *Clin Exp Allergy* 2005;**35**:1334−40.

29. Crivellato E, Nico B, Mallardi F, Beltrami CA, Ribatti D. Piecemeal degranulation as a general secretory mechanism? *Anat Rec A Discov Mol Cell Evol Biol* 2003;**274**:778−84.

30. Capron M, Tomassini M, Torpier G, Kusnierz JP, MacDonald S, Capron A. Selectivity of mediators released by eosinophils. *Int Arch Allergy Appl Immunol* 1989;**88**:54−8.

31. Tomassini M, Tsicopoulos A, Tai PC, Gruart V, Tonnel AB, Prin L, et al. Release of granule proteins by eosinophils from allergic and nonallergic patients with eosinophilia on immunoglobulin-dependent activation. *J Allergy Clin Immunol* 1991;**88**:365−75.

32. Lacy P, Levi-Schaffer F, Mahmudi-Azer S, Bablitz B, Hagen SC, Velazquez J, et al. Intracellular localization of interleukin-6 in eosinophils from atopic asthmatics and effects of interferon gamma. *Blood* 1998;**91**:2508−16.

33. Lacy P, Mahmudi-Azer S, Bablitz B, Hagen SC, Velazquez JR, Man SF, et al. Rapid mobilization of intracellularly stored RANTES in response to interferon-γ in human eosinophils. *Blood* 1999;**94**:23−32.

34. Spencer LA, Melo RC, Perez SA, Bafford SP, Dvorak AM, Weller PF. Cytokine receptor-mediated trafficking of preformed IL-4 in eosinophils identifies an innate immune mechanism of cytokine secretion. *Proc Natl Acad Sci U S A* 2006;**103**:3333−8.

35. Melo RC, Spencer LA, Perez SA, Ghiran I, Dvorak AM, Weller PF. Human eosinophils secrete preformed, granule-stored

interleukin-4 through distinct vesicular compartments. *Traffic* 2005;**6**:1047−57.

36. Erjefalt JS, Persson CG. New aspects of degranulation and fates of airway mucosal eosinophils. *Am J Respir Crit Care Med* 2000;**161**:2074−85.

37. Yamazaki K, Suzuki K, Nakamura A, Sato S, Lindor KD, Batts KP, et al. Ursodeoxycholic acid inhibits eosinophil degranulation in patients with primary biliary cirrhosis. *Hepatology* 1999;**30**:71−8.

38. Rubio CA. A method for the detection of eosinophilic granulocytes in colonoscopic biopsies from IBD patients. *Pathol Res Pract* 2003;**199**:145−50.

39. Samoszuk MK, Espinoza FP. Deposition of autofluorescent eosinophil granules in pathologic bone marrow biopsies. *Blood* 1987;**70**:597−9.

40. Mayeno A, Hamann K, Gleich G. Granule-associated flavin adenine dinucleotide (FAD) is responsible for eosinophil autofluorescence. *J Leukoc Biol* 1992;**51**:172−5.

41. Neves JS, Perez SA, Spencer LA, Melo RC, Reynolds L, Ghiran I, et al. Eosinophil granules function extracellularly as receptor-mediated secretory organelles. *Proc Natl Acad Sci U S A* 2008;**105**:18478−83.

42. Cromwell O, Bennett JP, Hide I, Kay AB, Gomperts BD. Mechanisms of granule enzyme secretion from permeabilized guinea pig eosinophils. Dependence on Ca^{2+} and guanine nucleotides. *J Immunol* 1991;**147**:1905−11.

43. Gomperts BD. G_E: a GTP-binding protein mediating exocytosis. *Annu Rev Physiol* 1990;**52**:591−606.

44. Tesmer JJG. The quest to understand heterotrimeric G protein signaling. *Nat Struct Mol Biol* 2010;**17**:650−2.

45. Takai Y, Sasaki T, Matozaki T. Small GTP-binding proteins. *Physiol Rev* 2001;**81**:153−208.

46. Watson EL. GTP-binding proteins and regulated exocytosis. *Crit Rev Oral Biol Med* 1999;**10**:284−306.

47. Bokoch GM. Regulation of innate immunity by Rho GTPases. *Trends Cell Biol* 2005;**15**:163−71.

48. Zerial M, McBride H. Rab proteins as membrane organizers. *Nat Rev Mol Cell Biol* 2001;**2**:107−17.

49. Lacy P, Thompson N, Tian M, Solari R, Hide I, Newman TM, et al. A survey of GTP-binding proteins and other potential key regulators of exocytotic secretion in eosinophils. Apparent absence of rab3 and vesicle fusion protein homologues. *J Cell Sci* 1995;**108**(Pt 11):3547−56.

50. Someya A, Nishijima K, Nunoi H, Irie S, Nagaoka I. Study on the superoxide-producing enzyme of eosinophils and neutrophils—comparison of the NADPH oxidase components. *Arch Biochem Biophys* 1997;**345**:207−13.

51. Lacy P, Mahmudi-Azer S, Bablitz B, Gilchrist M, Fitzharris P, Cheng D, et al. Expression and translocation of Rac2 in eosinophils during superoxide generation. *Immunology* 1999;**98**:244−52.

52. Lacy P, Abdel-Latif D, Steward M, Musat-Marcu S, Man SF, Moqbel R. Divergence of mechanisms regulating respiratory burst in blood and sputum eosinophils and neutrophils from atopic subjects. *J Immunol* 2003;**170**:2670−9.

53. Fulkerson PC, Zhu H, Williams DA, Zimmermann N, Rothenberg ME. CXCL9 inhibits eosinophil responses by a CCR3- and Rac2-dependent mechanism. *Blood* 2005;**106**:436−43.

54. Abdel-Latif D, Steward M, Macdonald DL, Francis GA, Dinauer MC, Lacy P. Rac2 is critical for neutrophil primary granule exocytosis. *Blood* 2004;**104**:832−9.

55. Lacy P, Willetts L, Kim JD, Lo AN, Lam B, MacLean EI, et al. *Int Arch Allergy Immunol* 2011;**156**:137−47.

56. Coughlin JJ, Odemuyiwa SO, Davidson CE, Moqbel R. Differential expression and activation of Rab27A in human eosinophils: relationship to blood eosinophilia. *Biochem Biophys Res Commun* 2008;**373**:382−6.

57. Jahn R, Scheller RH. SNAREs—engines for membrane fusion. *Nat Rev Mol Cell Biol* 2006;**7**:631−43.

58. Fasshauer D, Sutton RB, Brunger AT, Jahn R. Conserved structural features of the synaptic fusion complex: SNARE proteins reclassified as Q- and R-SNAREs. *Proc Natl Acad Sci U S A* 1998;**95**:15781−6.

59. Schiavo G, Matteoli M, Montecucco C. Neurotoxins affecting neuroexocytosis. *Physiol Rev* 2000;**80**:717−66.

60. Glogau RG. Review of the use of botulinum toxin for hyperhidrosis and cosmetic purposes. *Clin J Pain* 2002;**18**:S191−7.

61. Loder E, Biondi D. Use of botulinum toxins for chronic headaches: A focused review. *Clin J Pain* 2002;**18**:S169−76.

62. Sollner T, Whiteheart SW, Brunner M, Erdjument-Bromage H, Geromanos S, Tempst P, et al. SNAP receptors implicated in vesicle targeting and fusion. *Nature* 1993;**362**:318−24.

63. Scales SJ, Bock JB, Scheller RH. The specifics of membrane fusion. *Nature* 2000;**407**:144−6.

64. von Mollard GF, Nothwehr SF, Stevens TH. The yeast v-SNARE Vti1p mediates two vesicle transport pathways through interactions with the t-SNAREs Sed5p and Pep12p. *J Cell Biol* 1997;**137**:1511−24.

65. McNew JA, Parlati F, Fukuda R, Johnston RJ, Paz K, Paumet F, et al. Compartmental specificity of cellular membrane fusion encoded in SNARE proteins. *Nature* 2000;**407**:153−9.

66. Lacy P, Logan MR, Bablitz B, Moqbel R. Fusion protein vesicle-associated membrane protein 2 is implicated in IFN-gamma-induced piecemeal degranulation in human eosinophils from atopic individuals. *J Allergy Clin Immunol* 2001;**107**:671−8.

67. Logan MR, Lacy P, Odemuyiwa SO, Steward M, Davoine F, Kita H, et al. A critical role for vesicle-associated membrane protein-7 in exocytosis from human eosinophils and neutrophils. *Allergy* 2006;**61**:777−84.

68. Feng D, Flaumenhaft R, Bandeira-Melo C, Weller P, Dvorak A. Ultrastructural localization of vesicle-associated membrane protein(s) to specialized membrane structures in human pericytes, vascular smooth muscle cells, endothelial cells, neutrophils, and eosinophils. *J Histochem Cytochem* 2001;**49**:293−304.

69. Hoffmann HJ, Bjerke T, Karawajczyk M, Dahl R, Knepper MA, Nielsen S. SNARE proteins are critical for regulated exocytosis of ECP from human eosinophils. *Biochem Biophys Res Commun* 2001;**282**:194−9.

70. Logan MR, Lacy P, Bablitz B, Moqbel R. Expression of eosinophil target SNAREs as potential cognate receptors for vesicle-associated membrane protein-2 in exocytosis. *J Allergy Clin Immunol* 2002;**109**:299−306.

71. Rizo J, Sudhof TC. Snares and Munc18 in synaptic vesicle fusion. *Nat Rev Neurosci* 2002;**3**:641−53.

72. Misura KMS, Scheller RH, Weis WI. Three-dimensional structure of the neuronal-Sec1-syntaxin 1a complex. *Nature* 2000;**404**:355−62.

73. Logan MR, Odemuyiwa SO, Lacy P, Moqbel R. Eosinophil SNAREs are critical components for eosinophil peroxidase (EPO) release and are associated with the Sec1/Munc18 (SM) regulator, Munc18c. *J Allergy Clin Immunol* 2003;**111**:S211.

74. Bates ME, Bertics PJ, Calhoun WJ, Busse WW. Increased protein kinase C activity in low density eosinophils. *J Immunol* 1993;**150**:4486–93.

75. Kroegel C, Yukawa T, Dent G, Venge P, Chung KF, Barnes PJ. Stimulation of degranulation from human eosinophils by platelet-activating factor. *J Immunol* 1989;**142**:3518–26.

76. Evans DJ, Lindsay MA, Webb BL, Kankaanranta H, Giembycz MA, O'Connor BJ, et al. Expression and activation of protein kinase C-zeta in eosinophils after allergen challenge. *Am J Physiol* 1999;**277**:L233–9.

77. Takizawa T, Kato M, Kimura H, Suzuki M, Tachibana A, Obinata H, et al. Inhibition of protein kinases A and C demonstrates dual modes of response in human eosinophils stimulated with platelet-activating factor. *J Allergy Clin Immunol* 2002;**110**:241–8.

78. Risinger C, Bennett MK. Differential phosphorylation of syntaxin and synaptosome-associated protein of 25 kDa (SNAP-25) isoforms. *J Neurochem* 1999;**72**:614–24.

79. de Vries KJ, Geijtenbeek A, Brian EC, de Graan PN, Ghijsen WE, Verhage M. Dynamics of munc18-1 phosphorylation/dephosphorylation in rat brain nerve terminals. *Eur J Neurosci* 2000;**12**:385–90.

80. Reed GL, Houng AK, Fitzgerald ML. Human platelets contain SNARE proteins and a Sec1p homologue that interacts with syntaxin 4 and is phosphorylated after thrombin activation: implications for platelet secretion. *Blood* 1999;**93**:2617–26.

81. Fujita Y, Sasaki T, Fukui K, Kotani H, Kimura T, Hata Y, et al. Phosphorylation of Munc-18/n-Sec1/rbSec1 by protein kinase C: its implication in regulating the interaction of Munc-18/n-Sec1/rbSec1 with syntaxin. *J Biol Chem* 1996;**271**:7265–8.

Chapter 7.7

Eosinophil Responses to Pathogen-Associated and Damage-Associated Molecular Patterns

Anne Månsson Kvarnhammar and Lars Olof Cardell

PATTERN- RECOGNITION RECEPTORS

In 1997, Medzhitov and colleagues identified a human homologue of the Toll protein in *Drosophila melanogaster*, later designated Toll-like receptor 4 (TLR4).[1] Since then, the TLR family has grown rapidly. This finding has led to the discovery of pattern recognition receptors (PRRs), a diverse set of proteins that recognize specific molecular signatures on pathogens called pathogen-associated molecular patterns (PAMPs) or endogenous molecules produced by injured or dying cells in response to microbial invasion termed damage-associated molecular patterns (DAMPs).[2,3] The recognition of PAMPs and DAMPs enables the innate immune system to distinguish between pathogenic and non-pathogenic microorganisms and subsequently signal invasion by pathogens that can cause tissue damage.[4] The PRR system contains four major receptor families: TLRs, nucleotide oligomerization domain (NOD)-like receptors (NLRs), RIG-like receptors (RLRs), and C-type lectin receptors (CLRs).[2] The TLRs are positioned on the cell surface (TLR1, 2, 4, 5, 6, and 10) where they primarily detect bacterial proteins, lipoproteins, and polysaccharides, as well as in endosomes (TLR3, 7, 8, and 9) where they detect viral nucleic acids.[5-7] In contrast, the NLRs and RLRs are positioned in the cytosol where they sense mainly bacterial peptidoglycan and viral double-stranded RNA (dsRNA), respectively.[8,9] The CLRs are cell membrane-associated proteins that recognize various carbohydrate moieties, particularly on fungi.[4,10]

The advanced glycosylation end product-specific receptor (AGER/RAGE) is another receptor included in the PRR family that deserves the attention in the context of eosinophil responses to DAMPs. AGER (RAGE) is a transmembrane receptor responsible for the recognition of endogenous ligands generated on cell death and tissue injury.[11] A schematic outline of the PRR members, their ligands, and signaling pathways is presented in Fig. 7.7.1.

The Toll-Like Receptor Family

The human TLR family is made up of 10 members (TLR1–TLR10), whereas it comprises 12 receptors in mice.[3] TLR2 acts as a heterodimer in concert with TLR1 or TLR6, and mediates the responses to lipoproteins, lipoteichoic acids, peptidoglycan, and zymosan. Depending on the heterodimer, TLR2 can discriminate between diacyl (TLR1/2) and triacyl (TLR2/6) lipopeptides. There are indications that TLR10 also has the ability to form heterodimers with TLR1 and TLR2, but the specific ligands have not yet been identified. TLR3 is involved in the recognition of dsRNA from viruses and the synthetic dsRNA analogue polyinosinic:polycytidylic acid [poly (I:C)]. TLR4 is the main lipopolysaccharide (LPS) receptor. However, it is dependent on the cooperation with the LPS binding protein, CD14 and lymphocyte antigen 96 (MD-2). TLR5 recognizes flagellin, an essential component of the bacterial flagella. TLR7 and TLR8 mediate responses to viral single-stranded RNA (ssRNA), but also to immunomodulatory drugs such as imiquimod (R-837) and resiquimod (R-848). TLR9 responds to bacterial and viral DNA containing unmethylated CpG

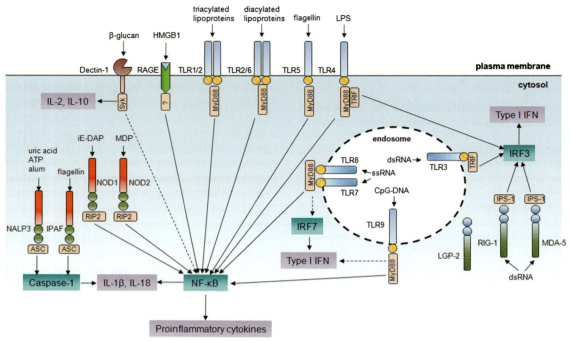

FIGURE 7.7.1 **Schematic outline of the pattern recognition receptors, their ligands, and signaling pathways.** The TLRs are located at the cell surface or in endosomes where they sense bacterial proteins and viral nucleic acids, respectively. The NLRs and RLRs are positioned in the cytoplasm where they detect bacterial structures, danger signals, and viral RNA, whereas the CLR family member dectin-1 is a transmembrane receptor that responds to β-glucan from yeast cell walls. AGER (RAGE) is a transmembrane receptor responsible for the recognition of HMGB1. All PRRs activate downstream signaling cascades involving NF-κB, IRFs, and caspase-1 with the resultant release of cytokines. AGER/RAGE, advanced glycosylation end product-specific receptor; ASC, apoptosis-associated speck-like protein containing a CARD; CpG, cytosine-phosphodiester-guanine; HMGB1, high mobility group protein B1; iE-DAP, γ-D-glutamyl-*meso*-diaminopimelic acid; IFN; interferon; IL-1 beta, interleukin-1 beta; IL-2, interleukin-2; IL-10, interleukin-10; IL-18, interleukin-18; Ipaf, ice protease-activating factor; IPS, interferon beta promoter stimulator protein 1; IRF, IFN regulatory factor; LGP2, probable ATP-dependent helicase LGP2; LPS, lipopolysaccharide; MDA-5, melanoma differentiation-associated protein 5; MDP, muramyl dipeptide; MyD88, myeloid differentiation primary response protein MyD88; NALP3, NACHT, LRR and PYD domains-containing protein 3; NF-κB, nuclear factor-kappaB; NOD, nucleotide oligomerization domain; RIG-1, retinoic acid-inducible gene 1 protein; RIP-2, receptor-interacting protein 2; Syk, spleen tyrosine kinase; TLR, toll-like receptor.

(cytosine-phosphodiester-guanine) motifs.[5,12−15] All TLRs except TLR3 signal through the myeloid differentiation primary response protein MyD88 (MyD88)-dependent pathway to activate nuclear factor-kappaB (NF-κB) repressing factor, which ultimately regulates the transcription of proinflammatory genes. To some extent, TLR7, 8, and 9 also trigger the production of type I interferons (IFNs) through MyD88-dependent activation of IFN regulatory factors (IRFs). In contrast, the MyD88-independent pathway, selectively used by TLR3 (and to some extent TLR4), uses Toll-receptor-associated activator of interferon (TRIF) as an adaptor to activate IRF-3 leading to the transcription of type I IFNs.[2,5]

The Nucleotide Oligomerization Domain-Like Receptor Family

The NLR family consists of more than 20 cytoplasmic pathogen and danger sensors.[16] Based on the N-terminus domain, they can be divided into three major

families: caspase recruitment domain-containing proteins (CARDs), NACHT, LRR and PYD domains-containing proteins (NALPs), and NLR family, apoptosis inhibitory protein (NAIPs).[17] The ligands for most NLRs are unknown and their physiological functions are therefore poorly understood. The best-characterized proteins so far are NOD1, NOD2, NALP3, and ice protease-activating factor (Ipaf). NOD1 and NOD2 recognize the peptidoglycan substructures γ-D-glutamyl-*meso*-diaminopimelic acid (iE-DAP) and muramyl dipeptide (MDP), respectively,[16,18] and activate NF-κB through the recruitment of receptor-interacting protein 2 (RIP-2).[9] In contrast, NALP3 and Ipaf respond to host-derived danger signals and flagellin, respectively, and activate the caspase-1-dependent inflammasome through apoptosis-associated speck-like protein containing a CARD (ASC).[2,9] Additionally, recent studies demonstrate a role for the NALP3 inflammasome in the recognition of aluminum adjuvants (alum), although it is unclear whether this interaction is NALP3-specific.[19,20]

The Rig-Like Receptor Family

Unlike TLRs that recognize viruses in the endosomal compartment (i.e., *outside* the cell), the RLRs are located in the cytoplasm, where they mediate responses against viruses that replicate inside the cell.[8,9] The RLR family comprises three members: probable ATP-dependent helicase LGP2 (LGP2), melanoma differentiation-associated protein 5 (MDA-5), and retinoic acid-inducible gene 1 protein (RIG-1). RIG-1 recognizes ssRNA that has a triphosphate moiety in its 5'-terminus, along with short blunt dsRNA, whereas long dsRNA and long poly(I:C) are detected preferentially by MDA-5. No ligand is yet defined for LGP2, but it has been described as a positive regulator of the RIG-1- and MDA-5-mediated immune responses.[2,3] All RLRs transmit their signal through a common adaptor protein, interferon beta promoter stimulator protein 1 (IPS-1), to induce type I IFN production.[8]

The C-Type Lectin Receptor Family

The CLRs are transmembrane receptors characterized by the presence of a carbohydrate-binding domain. The best known members are dendritic cell-associated C-type lectin 1 and 2 (dectin-1, dectin-2) and macrophage-inducible C-type lectin [Mincle; also known as C-type lectin domain family 4 member E (Clec4e/Clecsf9)], but other receptors that will not be further discussed here, such as dendritic cell-specific ICAM-3-grabbing non-integrin 1 (DC-SIGN), langerin, sialic acid-binding Ig-like lectins (Siglecs), and the mannose receptor, are also included in this family of receptors.[2-4,10,21] Dectin-1 and dectin-2 recognize β-glucan, the major fungal cell wall component,[22] whereas Mincle responds to Sin3-associated polypeptide p130 (SAP130), a component of a small nuclear ribonucleoprotein that is released from damaged cells.[23] In contrast to murine dectin-1, the human homologue, termed β-glucan receptor (βGR), can be divided into two major (and several minor) isoforms—βGRA and βGRB.[24] CLR signaling is largely mediated by a spleen tyrosine kinase (SYK)-dependent activation of mitogen-activated protein kinase (MAPK) and NF-κB with the resultant generation of proinflammatory cytokines.[3]

Advanced Glycosylation End Product-Specific Receptor

AGER (RAGE) is a transmembrane receptor responsible for the recognition of advanced glycosylation end-products (AGEs), non-enzymatically glycated or oxidized proteins, lipids, and nucleic acids, which are formed in the environment of oxidant stress and hyperglycemia. One prototypic DAMP recognized by AGER (RAGE) is the high mobility group protein B1 (HMGB1). On binding, AGER (RAGE) initiates cellular signals that activate NF-κB, with the subsequent induction of proinflammatory cytokines. However, the adaptor protein transmitting the signals is unknown.[11]

PATTERN RECOGNITION RECEPTORS IN EOSINOPHILS

The first studies to examine TLRs in eosinophils were published in the early 2000s, but since then the field has developed surprisingly slowly. Overall, the expression of PRRs in eosinophils is minimal compared to neutrophils that express high levels of most PRRs.[25-29] What follows next is a review of what is known to date about eosinophil responses to PAMPs and DAMPs exerted by the various PRRs. A summary of PRR expression in eosinophils is provided in Table 7.7.1.

Toll-Like Receptors and Their Response

Eosinophils have been found to express mRNA and/or protein for most TLRs, although at varying intensity. There has been some controversy regarding the presence or absence of TLR3, TLR4, and TLR6, while TLR7 is the only receptor that has been found highly active in all studies.[25,26,30-35]

TLR2 forms heterodimers with either TLR1 or TLR6 and discriminates between different PAMPs depending on which protein it associates with. Three groups have shown that expression of TLR2 in eosinophils is absent, and that Pam$_3$CSK$_4$ (the ligand for TLR1/TLR2) and peptidoglycan (TLR2) are unable to induce activation.[26,30,36] In contrast, Wong and colleagues have detected mRNA and proteins for TLR1, TLR2, and TLR6, and shown that the TLR2 agonist peptidoglycan can activate eosinophils by upregulating cell surface expression of intercellular adhesion molecule 1 (ICAM-1) and CD18, and induce the release of interleukin-1 beta (IL-1 beta), interleukin-6 and -8 (IL-6, IL-8), growth-regulated protein homolog alpha (GRO-alpha), eosinophil cationic protein (ECP), and superoxides. These effects are shown to be mediated by a combined action of extracellular signal-regulated kinase (ERK), phosphatidylinositol 3-kinase (PI3K), and NF-κB pathways.[25] In another study by the same group, peptidoglycan was found to enhance eosinophil survival, induce the expression of ICAM-1 and CD18, suppress the expression of ICAM-3 and L-selectin, and enhance C-C motif chemokine 5 (CCL5)-induced migration. It has also been demonstrated that the signals transmitted through ERK are mediated by the focal adhesion kinase 1 (FAK).[37] Moreover, Driss and colleagues have reported that *Mycobacterium bovis* bacillus Calmette–Guérin (BCG) activates human eosinophils and that TLR2 plays a major

TABLE 7.7.1 Pattern Recognition Receptor Expression in Human Eosinophils and Their Activators

PRR	Expression	Activator	Origin of Activator	Reference
TLRs				
TLR1	++	Triacylated lipopeptides	Bacteria	25,26
TLR2	+	Lipoprotein	Bacteria, viruses, fungi, self	25,26,30,36
TLR3	−/+	Poly (I:C)	Viruses	25,26,31,33
TLR4	−/+	LPS	Bacteria	25,26,30,31,34,36,39,40
TLR5	++	Flagellin	Bacteria	25,26
TLR6	−/+	Diacylated lipopeptides	Bacteria, viruses	25,26
TLR7	+++	ssRNA	Viruses	25,26,31,32,41
TLR8	−	ssRNA	Viruses	25,26,41
TLR9	++	CpG-DNA	Bacteria, viruses	25,26,32,35,42,44
TLR10	+	Unknown	Unknown	26
NLRs				
NOD1	+	iE-DAP	Bacteria	28
NOD2	++	MDP	Bacteria	28
NALP3	−	Uric acid, alum, ATP	Self	28
Ipaf	n.d.	Flagellin	Bacteria	−
RLRs				
RIG-1	++	Short dsRNA, 5′-triphosphate dsRNA	RNA viruses, DNA viruses	28
MDA-5	−	Long dsRNA	RNA viruses	28
LGP2	n.d.	Unknown	RNA viruses	−
CLRs				
Dectin-1	−/++	β-glucan, zymosan	Fungi	29,45,46
Dectin-2	n.d.	β-glucan, mannan	Fungi	−
Mincle	n.d.	SAP130	Self, fungi	−
Others				
AGER/RAGE	++	HMGB1	Self	48

+, ++, +++ and − indicate the intensity of the expression of each pattern recognition receptor. AGER/RAGE, advanced glycosylation end product-specific receptor; CLR, C-type lectin receptor; CpG, cytosine-phosphodiester-guanine; iE-DAP, γ-D-glutamyl-*meso*-diaminopimelic acid; Ipaf, ice protease-activating factor; HMGB1, high mobility group protein B1; LGP2, probable ATP-dependent helicase LGP2; LPS, lipopolysaccharide; MDA-5, melanoma differentiation-associated protein 5; MDP, muramyl dipeptide; Mincle, macrophage-inducible C-type lectin; NALP3, NACHT, LRR and PYD domains-containing protein 3; n.d., not determined; NLR, nucleotide oligomerization domain (NOD)-like receptor; NOD, nucleotide oligomerization domain; RIG-1, retinoic acid-inducible gene 1 protein; RLR, RIG-like receptor; SAP130, Sin3-associated polypeptide p130; TLR, Toll-like receptor.

role in the activation process. They show that both live BCG and lipomannan purified from *M. bovis* BCG attract eosinophils and promote the synthesis of alpha-defensin, ECP, eosinophil peroxidase (EPO), reactive oxygen species (ROS), and tumor necrosis factor (TNF-α) by activating MyD88-, p38 MAPK-, and NF-κB-dependent pathways. In addition, they observed a heterogeneous donor-dependent and inducible expression of TLR2 and TLR4.[38]

In contrast to the majority of studies showing no or low levels of TLR3,[25,26,37] one report has demonstrated the presence of TLR3 mRNA and protein in eosinophils from bone marrow and peripheral blood.[33] In this study, TLR3 expression was found to be higher in bone marrow-derived cells than in circulating cells and was downregulated in both compartments during symptomatic allergic rhinitis and in the presence of interleukin-4 (IL-4) and interleukin-5 (IL-5). The same study also showed that poly(I:C) has the ability to induce an increase in the expression of CD11b and the release of IL-8, effects found to be mediated via the p38 MAPK and NF-κB signaling pathways. Further, IL-5 was proven to augment the poly(I:C)-induced response.[33] In support of this, TLR3 mRNA has been detected in murine splenic eosinophils.[31] However, it should be emphasized that the responses reported to be induced by poly(I:C) are relatively weak compared to the effects seen by other TLRs known to be functionally active in eosinophils.

The immunostimulatory properties of LPS have been known long before the discovery of its major receptor, TLR4. In an early study, an activation of eosinophils in response to LPS was demonstrated, manifested by an induction in survival and secretion of granulocyte-macrophage colony-stimulating factor (GM-CSF), IL-8, and TNF-α.[39] A more recent study revealed the expression of TLR4 mRNA and that LPS-induced eosinophil activation in terms of ECP and TNF-α release is dependent on CD14.[34] The presence of Tlr4 mRNA has also been detected in eosinophils isolated from the spleen of naive hypereosinophilic mice.[31] In contrast, another group has demonstrated that eosinophils are unresponsive to LPS and do not express TLR4 or CD14 protein.[30] Moreover, by comparing CD14-depleted and non-CD14-depleted eosinophil preparations, LPS activation is proven to be monocyte-dependent,[40] suggesting an indirect effect of LPS on eosinophil function. In line with this, the most recent data on the topic show no eosinophil activating capability of LPS alone.[25,26,37]

While one study found no evidence for TLR5 mRNA in eosinophils,[26] another detected the TLR5 protein using flow cytometry and Western blot.[25] The latter also demonstrated that the TLR5 ligand flagellin is capable of inducing eosinophil activation in terms of chemokine and cytokine secretion (GRO-alpha, IL-1 beta, IL-6, IL-8), superoxide generation, prolongation of eosinophil survival, and activation of the adhesion system (CD18, ICAM-1, ICAM-3, L-selectin) with a subsequent increase in CCL5-induced migration. These effects are mediated by NF-κB and a FAK-dependent activation of ERK signaling pathways.[25,37]

TLR7 is the most prominent TLR in eosinophils, highly expressed at both mRNA and protein level, whereas expression of TLR8 is lacking.[25,26,32] This is in contrast to neutrophils that selectively express TLR8 but not TLR7.[26,41] The first study to investigate TLRs in eosinophils demonstrated that resiquimod (R-848) was the main PAMP to activate eosinophils via the TLR system. This report was based on the ability of resiquimod (R-848) to regulate the expression of CD11b and L-selectin, enhance viability, increase the generation of superoxides, and activate the p38 MAPK pathway. They further showed that interferon gamma (IFN-γ) upregulated the mRNA levels of several TLRs and synergistically affected CD11b and L-selectin expression.[26] Following this study, several investigators have confirmed and further developed these findings. One group has shown that imiquimod (R-837) regulates the expression of CD18, ICAM-1, ICAM-3, and L-selectin, and subsequently enhances CCL5-induced migration, induces the release of GRO-alpha, IL-1 beta, IL-6, IL-8, and superoxides, and prolongs survival by activating the NF-κB, PI3K, p38 MAPK, and ERK signaling pathways.[25,37] Another group verified imiquimod (R-837)-induced IL-8 secretion, prolongation in survival, and activation of the adhesion system with a subsequent increase in chemotactic migration. They also showed that priming with IL-5 sensitizes eosinophils for TLR7 activation and that the responses of eosinophils to imiquimod (R-837) are higher in atopic patients with allergic rhinitis than in healthy, non-allergic controls.[32] It should also be mentioned that even though imiquimod (R-837) affects several eosinophil functions, two independent studies have failed to detect release of the granule proteins ECP and eosinophil-derived neurotoxin (EDN) in response to stimulation.[25,32] Moreover, Phipps and colleagues showed that murine eosinophils express TLR7 that can be activated by ssRNA stimulation. By infecting the airways of hypereosinophilic (IL-5 transgenic) mice with respiratory syncytial virus (RSV) or by the adoptive transfer of MyD88-sufficient, but not MyD88-deficient, eosinophils to wild-type mice, they showed that eosinophils mediate accelerated RSV clearance via MyD88-dependent pathways.[31]

TLR9 recognizes bacterial and viral DNA that is characterized by unmethylated CpG sequences.[3] Even though eosinophil expression of TLR9 has been demonstrated by several investigators,[25,26,32,35,42] there has been some controversy as to the ability of CpG to activate eosinophils. This might be explained by the great variation in the structure of CpG, with three different classes (type A, B, and C, based on the backbone composition), each characterized by numerous sequence variations.[43] While three studies were unable to see any effect of CpG,[25,26,37] Ilmarinen and colleagues found that bacterial DNA delays eosinophil apoptosis in a TLR9-dependent fashion by activating PI3K and NF-κB signaling cascades.[35] In support of this, another group has shown that CpG prolongs survival, regulates the expression of CD11b, CD69, and L-selectin, leading to facilitated IL-8-induced migration, and enhances the release of IL-8 and EDN. Also, conditioned media from CpG-treated eosinophils is found to promote airway epithelial damage by inducing cell death and cytokine release (IL-6 and IL-8). It has also been demonstrated that priming with IL-5 increases the responsiveness to CpG stimulation and that the TLR9-induced responses are higher in patients with allergic rhinitis compared to healthy, non-allergic controls in terms of IL-8 and EDN release.[32] Moreover, Lotfi and colleagues took this one step further by showing that CpG promotes eosinophil-induced dendritic cell (DC) maturation [upregulation of CD80, CD83, CD86, and human leukocyte antigen, DR subregion (HLA-DR)] by, e.g., releasing eosinophil granule major basic protein 1 (MBP-1) that is taken up and internalized by the DCs. Direct cell—cell interactions appear to enhance maturation, but this is not obligatory.[44]

Nucleotide Oligomerization Domain-Like Receptors and Rig-Like Receptors Responses

NLRs and RLRs represent the cytosolic PRR members. In contrast to the TLR system, not much information is available regarding their eosinophil expression. Our personal, so far unpublished, observations suggest that eosinophils express mRNA and protein for NOD1, NOD2, and RIG-1, but not for NALP3 and MDA-5. Stimulation with iE-DAP and MDP, ligands for NOD1 and NOD2, respectively, induces the secretion of IL-8, regulates the expression of CD11b, CD69, and L-selectin and facilitates IL-8- and eotaxin-induced migration. MDP also promotes the release of EDN, while no effects are seen on stimulation with alum or poly(I:C)/LyoVec, agonists for NALP3 and RIG-1/MDA-5, respectively. Moreover, NOD1- and NOD2-induced activation is augmented by GM-CSF and IL-5, but not by IFN-γ.[28] Notably, the responses to iE-DAP and MDP are quite modest compared to TLR-induced activation, but they can be enhanced in the presence of eosinophil-activating cytokines.

C-Type Lectin Receptor Response

Dectin-1 is the most prominent and well-studied type, detecting primarily β-glucans derived from the yeast cell wall.[10] While Yoon and colleagues found no signs of dectin-1 in human eosinophils either on the cell surface or in the intracellular compartment,[45] two other groups detected several dectin-1 isoforms, including βGRA and βGRB.[29,46] One of the groups also showed that nontypeable *Haemophilus influenzae* activates eosinophils to produce IL-8, induces oxidative burst, and upregulates several genes regulating, e.g., signal transduction, cell division, and cytokine/chemokine release. Blocking experiments with the soluble glucan derivates laminarin and scleroglucan shows this activation to be dependent on β-glucan receptors.[46] Moreover, an early study conducted before the identification of dectin-1, showed that unopsonized zymosan and glucan particles activate eosinophils to produce leukotriene C_4 through a glucan recognition mechanism.[47]

Advanced Glycosylation End Product-Specific Receptor Mediates Responses to Damage-Associated Molecular Patterns

AGER (RAGE) is one of the most prominent DAMP sensors, which responds to a range of signals derived from dying or damaged cells. Lotfi and colleagues have demonstrated that eosinophils express AGER (RAGE) and that necrotic material from tumor cell lysates induces oxidative burst and the release of EPO and MBP-1. In addition, they showed that HMGB1 induces degranulation and oxidative burst and serves as a chemoattractant and survival factor. Blocking experiments revealed that HMGB1-induced activation is AGER (RAGE)-dependent.[48]

CONCLUSION

Eosinophils express a range of PRRs that, on recognition of pathogens or host-derived danger signals, trigger an immune reaction by inducing the release of proinflammatory cytokines, chemokines, cytotoxic granule proteins, and ROS, and cause delayed apoptosis and the activation of the adhesion system and cellular trafficking. An overview of the eosinophil response to PAMPs and DAMPs is illustrated in Fig. 7.7.2. There are discrepancies between different studies in terms of PRR expression and the ability of their agonists to induce activation. These might be donor- or compartment-related, or may be due to regulation in expression caused by the presence of microbes, cytokines, and environmental stresses.[33,38,42] GM-CSF, IFN-γ, IL-4, and IL-5 have, for example, been shown to regulate PRR expression and augment the responsiveness to ligand stimulation.[26,28,32,33]

The function of each PRR in eosinophils has mainly been studied in isolated single-cell systems that do not reflect the natural interactive biological environment. In real life, the detection and clearance of an infection is dependent on the simultaneous triggering of several signals and events. Recent data also suggest that microbial recognition involves a cross talk between several pathogen sensors. The overlapping detection of given ligands among TLRs, NLRs, RLRs, and CLRs in conjunction with their diverse compartmental distribution further emphasize how the different components of innate immunity cooperate in a synergistic way to eradicate an invading pathogen.

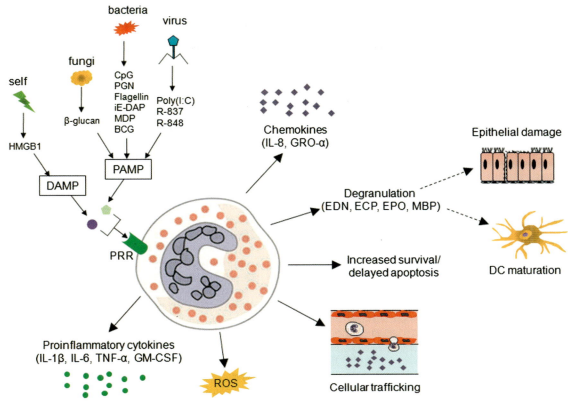

FIGURE 7.7.2 Eosinophil responses to pathogen-associated and damage-associated molecular patterns. Eosinophils are activated by various stimuli, leading to increased survival/delayed apoptosis, increased cellular trafficking and production of reactive oxygen species (ROS), proinflammatory cytokines, chemokines, and granule proteins that subsequently influence other cells. BCG, Bacillus of Calmette and Guérin; CpG, cytosine-phospho-diester-guanine; DAMP, damage-associated molecular pattern; DC, dendritic cell; ECP, eosinophil cationic protein; EDN, eosinophil-derived neurotoxin; EPO, eosinophil peroxidase; GM-CSF, granulocyte-macrophage colony-stimulating factor; GRO-alpha, growth-regulated protein homolog alpha; HMGB1, high mobility group protein B1; iE-DAP, γ-D-glutamyl-*meso*-diaminopimelic acid; IL-1 beta, interleukin-1 beta; IL-6, interleukin-6; IL-8, interleukin-8; MDP, muramyl dipeptide; PAMP, pathogen-associated molecular pattern; PRR, pattern recognition receptor; R-837, imiquimod; R-848, resiquimod; TNF-α, tumor necrosis factor.

Traditionally, eosinophils have been regarded as a key player in the immune response against parasitic helminths by the release of cytotoxic granule proteins, cytokines, and lipid mediators.[49] However, the recent discovery of a broad set of functional PRRs in eosinophils suggests that the innate immunity exerted by eosinophils reaches well beyond parasites to include bacteria, viruses, and filamentous fungi. The relationships between eosinophils and PAMPs and DAMPs summarized here also make it clear that eosinophils not only have a direct effector function in our microbial defense, but also play an instrumental role in modulating immunity by interacting with other cells, such as inducing epithelial damage and DC maturation.

Eosinophils are accumulated both in the asthmatic airway and within tumors where they are associated with inflammation and tissue damage. In the former, their numbers correlate with disease severity,[49] whereas in the latter they are associated with a good prognosis.[50] Hence, the destructive effector functions exerted by eosinophils

on PAMP and DAMP activation might be detrimental in allergic disease, by their ability to cause exacerbations, and beneficial in cancer by causing suppression of tumor growth. Careful manipulation of the eosinophil-based PRR system might therefore be of therapeutic use in future management of both atopic and cancer-related diseases.

REFERENCES

1. Medzhitov R, Preston-Hurlburt P, Janeway Jr CA. A human homologue of the Drosophila Toll protein signals activation of adaptive immunity. *Nature* 1997;**388**:394−7.
2. Kumagai Y, Akira S. Identification and functions of pattern-recognition receptors. *J Allergy Clin Immunol* 2010;**125**:985−92.
3. Takeuchi O, Akira S. Pattern recognition receptors and inflammation. *Cell* 2010;**140**:805−20.
4. Sato S, St-Pierre C, Bhaumik P, Nieminen J. Galectins in innate immunity: dual functions of host soluble beta-galactoside-binding lectins as damage-associated molecular patterns (DAMPs) and as

receptors for pathogen-associated molecular patterns (PAMPs). *Immunol Rev* 2009;**230**:172−87.

5. Parker LC, Prince LR, Sabroe I. Translational mini-review series on Toll-like receptors: networks regulated by Toll-like receptors mediate innate and adaptive immunity. *Clin Exp Immunol* 2007;**147**:199−207.

6. Iwasaki A, Medzhitov R. Toll-like receptor control of the adaptive immune responses. *Nat Immunol* 2004;**5**:987−95.

7. Pandey S, Agrawal DK. Immunobiology of Toll-like receptors: emerging trends. *Immunol Cell Biol* 2006;**84**:333−41.

8. Kawai T, Akira S. The roles of TLRs, RLRs and NLRs in pathogen recognition. *Int Immunol* 2009;**21**:317−37.

9. Creagh EM, O'Neill LA. TLRs, NLRs and RLRs: a trinity of pathogen sensors that co-operate in innate immunity. *Trends Immunol* 2006;**27**:352−7.

10. Lee MS, Kim YJ. Pattern-recognition receptor signaling initiated from extracellular, membrane, and cytoplasmic space. *Mol Cells* 2007;**23**:1−10.

11. Lin L. RAGE on the Toll Road? *Cell Mol Immunol* 2006;**3**:351−8.

12. Akira S, Takeda K, Kaisho T. Toll-like receptors: critical proteins linking innate and acquired immunity. *Nat Immunol* 2001;**2**:675−80.

13. Akira S, Hemmi H. Recognition of pathogen-associated molecular patterns by TLR family. *Immunol Lett* 2003;**85**:85−95.

14. Heine H, Lien E. Toll-like receptors and their function in innate and adaptive immunity. *Int Arch Allergy Immunol* 2003;**130**:180−92.

15. Takeda K, Akira S. Toll receptors and pathogen resistance. *Cell Microbiol* 2003;**5**:143−53.

16. Fritz JH, Ferrero RL, Philpott DJ, Girardin SE. Nod-like proteins in immunity, inflammation and disease. *Nat Immunol* 2006;**7**:1250−7.

17. Kaparakis M, Philpott DJ, Ferrero RL. Mammalian NLR proteins; discriminating foe from friend. *Immunol Cell Biol* 2007;**85**:495−502.

18. Le Bourhis L, Benko S, Girardin SE. Nod1 and Nod2 in innate immunity and human inflammatory disorders. *Biochem Soc Trans* 2007;**35**:1479−84.

19. Kool M, Petrilli V, De Smedt T, Rolaz A, Hammad H, van Nimwegen M, et al. Cutting edge: alum adjuvant stimulates inflammatory dendritic cells through activation of the NALP3 inflammasome. *J Immunol* 2008;**181**:3755−9.

20. Eisenbarth SC, Colegio OR, O'Connor W, Sutterwala FS, Flavell RA. Crucial role for the Nalp3 inflammasome in the immunostimulatory properties of aluminium adjuvants. *Nature* 2008;**453**:1122−6.

21. Willment JA, Brown GD. C-type lectin receptors in antifungal immunity. *Trends Microbiol* 2008;**16**:27−32.

22. Brown GD. Dectin-1: a signalling non-TLR pattern-recognition receptor. *Nat Rev Immunol* 2006;**6**:33−43.

23. Yamasaki S, Ishikawa E, Sakuma M, Hara H, Ogata K, Saito T. Mincle is an ITAM-coupled activating receptor that senses damaged cells. *Nat Immunol* 2008;**9**:1179−88.

24. Willment JA, Gordon S, Brown GD. Characterization of the human beta-glucan receptor and its alternatively spliced isoforms. *J Biol Chem* 2001;**276**:43818−23.

25. Wong CK, Cheung PF, Ip WK, Lam CW. Intracellular signaling mechanisms regulating toll-like receptor-mediated activation of eosinophils. *Am J Respir Cell Mol Biol* 2007;**37**:85−96.

26. Nagase H, Okugawa S, Ota Y, Yamaguchi M, Tomizawa H, Matsushima K, et al. Expression and function of Toll-like receptors

in eosinophils: activation by Toll-like receptor 7 ligand. *J Immunol* 2003;**171**:3977−82.

27. Ekman A-K, Cardell L-O. The expression and function of Nod-like receptors in neutrophils. *Immunology* 2010 May;**130**(1):55−63.

28. Månsson Kvarnhammar A, Petterson T, Cardell LO. NOD-like receptors and RIG-I-like receptors in human eosinophils: activation by NOD1 and NOD2 agonists. *Immunology* 2011 Nov;**134**(3):25−314.

29. Willment JA, Marshall AS, Reid DM, Williams DL, Wong SY, Gordon S, et al. The human beta-glucan receptor is widely expressed and functionally equivalent to murine Dectin-1 on primary cells. *Eur J Immunol* 2005;**35**:1539−47.

30. Sabroe I, Jones EC, Usher LR, Whyte MK, Dower SK. Toll-like receptor (TLR)2 and TLR4 in human peripheral blood granulocytes: a critical role for monocytes in leukocyte lipopolysaccharide responses. *J Immunol* 2002;**168**:4701−10.

31. Phipps S, Lam CE, Mahalingam S, Newhouse M, Ramirez R, Rosenberg HF, et al. Eosinophils contribute to innate antiviral immunity and promote clearance of respiratory syncytial virus. *Blood* 2007;**110**:1578−86.

32. Mansson A, Cardell LO. Role of atopic status in Toll-like receptor (TLR)7- and TLR9-mediated activation of human eosinophils. *J Leukoc Biol* 2009;**85**:719−27.

33. Mansson A, Fransson M, Adner M, Benson M, Uddman R, Bjornsson S, et al. TLR3 in human eosinophils: functional effects and decreased expression during allergic rhinitis. *Int Arch Allergy Immunol* 2010;**151**:118−28.

34. Plotz SG, Lentschat A, Behrendt H, Plotz W, Hamann L, Ring J, et al. The interaction of human peripheral blood eosinophils with bacterial lipopolysaccharide is CD14 dependent. *Blood* 2001;**97**:235−41.

35. Ilmarinen P, Hasala H, Sareila O, Moilanen E, Kankaanranta H. Bacterial DNA delays human eosinophil apoptosis. *Pulm Pharmacol Ther* 2009;**22**:167−76.

36. Mattsson E, Persson T, Andersson P, Rollof J, Egesten A. Peptidoglycan induces mobilization of the surface marker for activation marker CD66b in human neutrophils but not in eosinophils. *Clin Diagn Lab Immunol* 2003;**10**:485−8.

37. Cheung PF, Wong CK, Ip WK, Lam CW. FAK-mediated activation of ERK for eosinophil migration: a novel mechanism for infection-induced allergic inflammation. *Int Immunol* 2008;**20**:353−63.

38. Driss V, Legrand F, Hermann E, Loiseau S, Guerardel Y, Kremer L, et al. TLR2-dependent eosinophil interactions with mycobacteria: role of alpha-defensins. *Blood* 2009;**113**:3235−44.

39. Takanaski S, Nonaka R, Xing Z, O'Byrne P, Dolovich J, Jordana M. Interleukin 10 inhibits lipopolysaccharide-induced survival and cytokine production by human peripheral blood eosinophils. *J Exp Med* 1994;**180**:711−5.

40. Meerschaert J, Busse WW, Bertics PJ, Mosher DF. CD14(−) cells are necessary for increased survival of eosinophils in response to lipopolysaccharide. *Am J Respir Cell Mol Biol* 2000;**23**:780−7.

41. Janke M, Poth J, Wimmenauer V, Giese T, Coch C, Barchet W, et al. Selective and direct activation of human neutrophils but not eosinophils by Toll-like receptor 8. *J Allergy Clin Immunol* 2009;**123**:1026−33.

42. Fransson M, Benson M, Erjefalt JS, Jansson L, Uddman R, Bjornsson S, et al. Expression of Toll-like receptor 9 in nose,

peripheral blood and bone marrow during symptomatic allergic rhinitis. *Respir Res* 2007;**8**:17.

43. Krieg AM. Therapeutic potential of Toll-like receptor 9 activation. *Nat Rev Drug Discov* 2006;**5**:471−84.

44. Lotfi R, Lotze MT. Eosinophils induce DC maturation, regulating immunity. *J Leukoc Biol* 2008;**83**:456−60.

45. Yoon J, Ponikau JU, Lawrence CB, Kita H. Innate antifungal immunity of human eosinophils mediated by a beta 2 integrin, CD11b. *J Immunol* 2008;**181**:2907−15.

46. Ahren IL, Eriksson E, Egesten A, Riesbeck K. Nontypeable Haemophilus influenzae activates human eosinophils through beta-glucan receptors. *Am J Respir Cell Mol Biol* 2003;**29**:598−605.

47. Mahauthaman R, Howell CJ, Spur BW, Youlten LJ, Clark TJ, Lessof MH, et al. The generation and cellular distribution of leukotriene C4 in human eosinophils stimulated by unopsonized zymosan and glucan particles. *J Allergy Clin Immunol* 1988;**81**:696−705.

48. Lotfi R, Herzog GI, DeMarco RA, Beer-Stolz D, Lee JJ, Rubartelli A, et al. Eosinophils oxidize damage-associated molecular pattern molecules derived from stressed cells. *J Immunol* 2009;**183**:5023−31.

49. Hogan SP, Rosenberg HF, Moqbel R, Phipps S, Foster PS, Lacy P, et al. Eosinophils: biological properties and role in health and disease. *Clin Exp Allergy* 2008;**38**:709−50.

50. Lotfi R, Lee JJ, Lotze MT. Eosinophilic granulocytes and damage-associated molecular pattern molecules (DAMPs): role in the inflammatory response within tumors. *J Immunother* 2007;**30**:16−28.

Eosinophil Secretory Functions

Introduction

Darryl Adamko

The presence of eosinophils in diseased or healthy tissue is well described. The exact role that eosinophils play has become somewhat controversial. In some cases, gene-deleted mouse models and human anti-interleukin-5 data have created more questions than answers on the exact role played by eosinophils in disease. Such disconnections between eosinophil presence and effector function could be related to many factors, but one key element is the state of the eosinophil while in the tissue. Eosinophil presence in tissue is important, but ultimately eosinophil mediator release is needed for it to be a true effector cell.

In our work, we have described how virus-induced airway hyperresponsiveness in allergic models is a complex process requiring not only eosinophil recruitment to the lung, but more specifically recruitment to airway nerves.[1] Further, these eosinophils, once positioned, must still be stimulated by the virus through an interaction with T cells. We have described the release of granule-stored mediators and the production of leukotrienes,[2] but we acknowledge that the full spectrum of virus-induced release remains to be determined. It is fascinating how one cell could know when to be active and when to stay silent. In this section of the book, the authors review not only the many mediators created by eosinophils, but equally important they endeavor to explain how this cell can control all these factors.

To understand how release could be controlled, one must first understand that the position of the mediators within the eosinophil is variable. For example, the eosinophil granules store an array of cytokines and chemokines. Dr. Lisa Spencer endeavors to unravel the complex interplay of these granule products and their impact in a variety of tissue states. This includes highlighting the key functions of eosinophil cytokines and chemokines, and the accepted pathways for their release. She also highlights our understanding of the eosinophil's fascinating ability to release such mediators in a selective fashion.

In addition to stored mediators, the eosinophil has the ability to synthesize and to secrete distinctive lipid mediators. Dr. Christianne Bandeira-Melo and Dr. Patricia Bozza highlight the unique pattern of oxidative metabolism in eosinophils, especially in the context of a diverse number of inflammatory diseases, such as infection, asthma, and cancer.

The eosinophil was named in part because of the unique staining properties of its many granule-stored proteins. Dr. Joseph Butterfield gives an in-depth review of the biology of two well-described granule-stored proteins, eosinophil granule major basic protein 1 and eosinophil peroxidase. Equally important, he reviews the data regarding the association or lack of association of these proteins with disease. While the association with disease is strong in terms of eosinophil presence in humans, he discusses how gene-deleted models for these proteins in mice sometimes fail to show much correlation. He does highlight an immunomodulatory effect of eosinophils in mice with selective loss of eosinophil granule proteins. In addition, the cationic nature of eosinophil granule proteins is an important factor to consider. Dr. Charles Irvin reviews some of the pathophysiology of asthma, but also focuses on the diseases in the context of these potentially damaging cationic proteins.

Finally, the ability to see an eosinophil in tissue traditionally has largely relied on our ability to stain the intact cell. If the eosinophil dies by cytolysis, what happens to the granules? The cell is not seen, but could function remain? In our virus models, we suspect that cytolysis was a plausible mechanism for eosinophil degranulation, as the staining for eosinophil presence was absent despite clear association with eosinophil effector function. Dr. Josiane Neves and colleagues review eosinophil cytolysis, but more importantly share some exciting findings showing that even outside the cell, eosinophil granules are regulated: yet another level of complexity.

While our understanding of the mediators eosinophils produce has grown substantially, the mechanisms underlying the release of such mediators still could be improved. In this section, the reader will be shown the latest opinion on the breadth factors produced by eosinophils, and the accepted mechanisms underlying their release. Understanding the mechanisms of release will help us better understand the role that eosinophils play in health and disease.

Eosinophils in Health and Disease. http://dx.doi.org/10.1016/B978-0-12-394385-9.00008-0

REFERENCES

1. Adamko DJ, Fryer AD, Bochner BS, Jacoby DB. CD8+ T-lympho-cytes in viral hyperreactivity and M2 muscarinic receptor dysfunc-tion. *Am J Respir Crit Care Med* 2003;**167**:550−6.
2. Davoine F, Cao M, Wu Y, Ajamian F, Ilarraza R, Kokaji AI, et al. Virus-induced eosinophil mediator release requires antigen-present-ing and CD4+ T cells. *J Allergy Clin Immunol* 2008;**122**:69−77. e1-2.

Chapter 8.2

Release of Cytokines and Chemokines from Eosinophils

Lisa A. Spencer

EFFECTOR FUNCTIONS OF EOSINOPHIL-DERIVED CYTOKINES AND CHEMOKINES

Many of the roles of eosinophils in health and disease derive from their capacity to secrete numerous cytokines, chemokines, and growth factors with varied biological activities. Eosinophil-derived cytokines, chemokines, and growth factors identified to date are included in Table 8.2.1. The recognition that in addition to granule-derived cationic proteins, eosinophils secrete a vast array of biologically relevant cytokines was a driving force in the appreciation of eosinophils as more than tissue-destructive effector cells. The three dozen and more cytokines derived from eosinophils elicit strong proinflammatory [e.g., tumor necrosis factor (TNF-α)], anti-inflammatory [e.g., interleukin-10 (IL-10)], immunomodulatory [e.g., interleukin-4 and -12 (IL-4, IL-12)], and tissue remodeling [e.g., transforming growth factor (TGF-β)] effects, and many are critical mediators in helminth infections, allergic asthma, rhinitis, atopic dermatitis, eosinophilic esophagitis, and other eosinophilic inflammatory disorders. Discussed in this chapter and summarized in Fig. 8.2.1 are specific examples of the recognized effects of eosinophil-derived cytokines in health and disease. More detailed discussions of the roles of eosinophils in the processes described herein are found in other chapters within this book.

Modulating Adaptive Immunity

Eosinophil secretion of T-helper type 2 (T_h2) polarizing cytokines very early in helminth infections and in allergic inflammation contributes to the immunomodulatory potential of eosinophils. Murine eosinophils express IL-4 and interleukin-13 (IL-13) mRNA early in ontogeny,[1] and maintain a constitutively active *interleukin 4 (IL4)* gene

TABLE 8.2.1

Cytokines and Growth Factors	Chemokines
IL-1 beta	CCL2 (MCP-1)
IL-2	CCL3 (MIP-1-alpha)
IL-3	CCL5 (RANTES)
IL-4	CCL6 (C10)
IL-5	CCL7 (MCP-3)
IL-6	CCL9 (MIP-1-alpha)
IL-10	CCL11 (Eotaxin)
IL-12	CCL17 (TARC)
IL-13	CCL22 (MDC)
IL-16	CCL24 (Eotaxin-2)
IL-18	CCL13 (MCP-4)
IL-25 (IL-17E)	CCL18 (PARC)
IFN-γ	CXCL1 (GRO-alpha)
GM-CSF	CXCL8 (IL-8)
SCF	CXCL9 (MIG)
TGF-α	CXCL10 (IP-10)
TGF-β	CXCL12 (SDF-1)
TNF-α	MIF
TNF-β	
LIF	
NGF	
PDGF	
VEGF	

CCL, C-C motif chemokine; CXCL, C-X-C motif chemokine; GM-CSF, granulocyte-macrophage colony-stimulating factor; GRO-alpha, growth-regulated alpha protein; IFN, interferon; IL, interleukin; IP-10, 10 kDa interferon gamma-induced protein; LIF, leukemia inhibitory factor; MCP, monocyte chemotactic protein; MDC, macrophage-derived chemokine; MIF, macrophage migration inhibitory factor; MIG, monokine induced by interferon gamma; MIP, macrophage inflammatory protein; NGF, nerve growth factor; PARC, pulmonary and activation-regulated chemokine; PDGF, platelet-derived growth factor; CCL5 (RANTES), regulated upon activation, normal T-cell expressed, and secreted; SCF, stem cell factor; SDF, stromal cell-derived factor; TARC, thymus and activation-regulated chemokine; TGF, transforming growth factor; TNF, tumor necrosis factor; VEGF, vascular endothelial growth factor.

locus while in circulation.[2,3] Eosinophils are rapidly recruited to tissue sites following exposure to T_h2-eliciting pathogens or allergens, independent of, and often preceding, an adaptive immune response.[3,4] In mice, eosinophils are the predominant IL-4-producing cell type recruited to the lungs following exposure to *Nippostrongylus brasiliensis*[2] or intranasal administration of ovalbumin,[5] and within the peritoneal cavity in response to

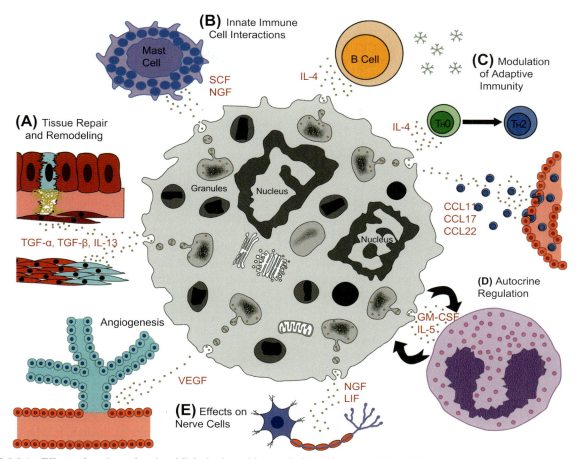

FIGURE 8.2.1 **Effector functions of eosinophil-derived cytokines and chemokines.** *(A)* Eosinophil-derived cytokines and chemokines promote tissue repair and remodeling, and vascular angiogenesis through effects on epithelial cells, fibroblasts, smooth muscle cells, and endothelial cells, and exert effects on innate immune cells, including mast cells and nerve cells *(B, E)*. *(C)* Eosinophil-derived IL-4 lowers the activation threshold required for B cell production of IgM. Eosinophil-derived cytokines may act indirectly through dendritic cells, or directly on naive T cells to influence T helper cell differentiation, and eosinophil-secreted chemokines promote the recruitment of specific effector T-cell subsets. *(D)* Eosinophils also secrete cytokines and chemokines with autocrine effects to propagate eosinophil recruitment and activation. Light blue coloring indicates proliferation. CCL, C-C motif chemokine; GM-CSF, granulocyte-macrophage colony-stimulating factor; IgM, immunoglobulin M; IL-4, interleukin-4; IL-5, interleukin-5; IL-13, interleukin-13; LIF, leukemia inhibitory factor; NGF, nerve growth factor; SCF, stem cell factor; TGF, transforming growth factor; T_h, T-helper (cell); VEGF, vascular endothelial cell growth factor.

inoculation with *Schistosoma mansoni* eggs.[6] Similarly, within the nasal mucosa of allergic rhinitis patients 6 h after allergen challenge, 76% and 77% of cells positive for IL-4 or interleukin-5 (IL-5) mRNA, respectively, were identified as eosinophils.[7] Once an established T_h2 immune response has been generated, cytokines such as eosinophil-derived interleukin-25 may augment T_h2 memory cell function.[8] Despite their general association with T_h2 immune responses, eosinophils also secrete T-helper type 1 (T_h1)-related cytokines, including interleukin-2, interferon gamma (IFN-γ),[9] and IL-12.[10]

Eosinophil-derived chemokines are implicated in the specific recruitment of innate and adaptive immune cells in mouse models of allergic airway inflammation. Coincident with the infiltration of C-C chemokine receptor type 1 (CCR1)-expressing dendritic cells into allergic airways,

eosinophils are the dominant source of the CCR1 ligands C-C motif chemokine 6 (CCL6; protein C10) and C-C motif chemokine 9 [CCL9; macrophage inflammatory protein 1-gamma (MIP-1-gamma)] within the lungs.[11] Moreover, in mice made deficient of eosinophils, generated either through genetic ablation of eosinophils or through C-C chemokine receptor type 3 (CCR3) deficiency, T_h2 effector cell recruitment and accumulation in response to airway allergen challenge is diminished, which is attributable to the absence of eosinophil-derived chemokines, specifically eotaxin (CCL11), C-C motif chemokine 22 [CCL22; macrophage-derived chemokine (MDC)] and thymus- and activation-regulated chemokine C-C motif chemokine 17 [CCL17; CC chemokine TARC (TARC)].[12–15] Similar to the capacity of eosinophils to secrete cytokines participating in T_h1 or T_h2 immunity,

T_h1-associated chemokines, such as C-X-C motif chemokine 9 [CXCL9; monokine induced by interferon gamma (MIG)] and C-X-C motif chemokine 10 [CXCL10; 10 kDa interferon gamma-induced protein (IP-10)], are also included in the eosinophil repertoire.[16] Thus, immune-biased cytokines and chemokines elaborated from eosinophils, in concert with the direct antigen presentation functions of eosinophils, contribute to the development of immune-polarized responses.

Although the dynamic synergism between eosinophils and T cells, specifically in the development of a T_h2 immune response, is well established, eosinophil-mediated effects on adaptive immunity are not limited to interactions with T cells. Studies of *Strongyloides stercoralis* infections in eosinophil-deficient mice reveal a loss of immunoglobulin M (IgM)-mediated immunity[17] that is restored on reconstitution with eosinophils,[18,19] suggesting eosinophils influence B-cell activities. In a different setting, early priming of splenic B cells following the administration of the adjuvant alum requires a population of IL-4-secreting, Gr-1[+] leukocytes,[20] now identified to be eosinophils. In mice, intraperitoneal (i.p.) injection of alum elicits the splenic recruitment of eosinophils, triggering eosinophil-dependent priming of B cells for major histocompatibility complex class II (MHC-II) signaling-mediated Ca^{2+} flux and subsequent production of IgM.[21]

Tissue Homeostasis, Repair, and Remodeling

Eosinophil-derived cytokines exert profound effects on the structure, composition, and integrity of the tissue microenvironment. Cytokines secreted by eosinophils with demonstrated tissue effects include interleukin-13 (IL-13), interleukin-1 beta (IL-1 beta), platelet-derived growth factor (PDGF), transforming growth factor alpha (TGF-α), and transforming growth factor beta (TGF-β). In health, baseline eosinophils may contribute to the maintenance of epithelial barrier functions within the gastrointestinal tract through the secretion of cytokines including TGF-β,[22,23] and eosinophils recruited to sites of tissue injury promote wound healing through the secretion of cytokines such as TGF-α.[24] Under conditions of excessive eosinophilic infiltration, eosinophil-derived cytokines, particularly TGF-β, contribute to aspects of tissue remodeling, including the induction of smooth muscle cell hyperplasia, fibroblast proliferation, and the deposition of extracellular matrix materials. Eosinophils have been implicated as effector cells in fibrotic processes associated with several allergic diseases with marked eosinophilia, including allergic asthma,[25,26] idiopathic pulmonary fibrosis,[27] eosinophilic esophagitis,[28] eosinophil myalgia syndrome,[29] scleroderma,[29] and eosinophilic endomyocardial fibrosis,[30]

and are the predominant cellular source of TGF-β within the lungs of asthmatics[26] and within the esophagus of pediatric eosinophilic esophagitis patients.[28]

In addition to tissue remodeling, eosinophil-derived cytokines, including angiogenin [ribonuclease 5 (RNase 5)], basic fibroblast growth factor (bFGF), granulocyte-macrophage colony-stimulating factor (GM-CSF), interleukin-6 (IL-6), interleukin-8, PDGF, TGF-β, and vascular endothelial growth factor (VEGF) promote remodeling of the vascular tree, a common manifestation in several diseases associated with high eosinophil counts (reviewed in[31] and[32]). Using the chick embryo chorioallantoic membrane model, eosinophils promoted endothelial cell proliferation and the subsequent formation of new vessels, in part through the secretion of VEGF.[33] Likewise, in human subjects, increases in peribronchial and leaky vessels within asthmatic airways parallel the enhanced pulmonary expression of angiogenin (RNase5), bFGF, and VEGF-A, colocalizing predominantly with CD34[+] cells (27−28%), eosinophils (21−23%), and macrophages (21−23%).[34,35]

INTERCELLULAR INTERACTIONS

Significant evidence to date highlights the importance of an eosinophil–mast cell axis in several allergic inflammatory diseases. Eosinophil secretion of granule-derived cationic proteins and several cytokines, including stem cell factor (SCF) and nerve growth factor (NGF), promotes mast cell growth, survival and activation (reviewed in[36]). Conversely, mast cell-derived proteases and cytokines are involved in eosinophil recruitment and promote eosinophil activation and survival.[37−39] A recent *in vitro* ultrastructural study proposed a model of mediator passage between eosinophils and mast cells through a mechanism mediated by direct cell–cell contact.[40]

Eosinophil-mediated effects on nerve cells also involve secretion of eosinophil-derived cytokines. Eosinophils adhere directly to nerve cells through the intercellular adhesion molecule 1 (ICAM-1) and vascular cell adhesion protein 1 (VCAM-1) molecules and are stimulated to degranulate.[41] Human eosinophils constitutively express mRNA and maintain preformed stores of leukemia inhibitory factor (LIF) and neurotrophins, including NGF, available for secretion.[42,43]

Autocrine Signaling

In addition to effects on neighboring tissue and immune cells, eosinophils engage in significant autocrine activation. Included among the many cytokines synthesized and stored by eosinophils are the eosinophilopoietins GM-CSF, IL-3, and IL-5, critical for the growth, differentiation, activation, and survival of eosinophils. In addition, eosinophils secrete chemokines, such as eotaxin and C-C motif chemokine 5 (CCL5/RANTES), promoting their own recruitment into tissue sites in a positive feedback loop. Of note, receptors for most, if not all, of the cytokines secreted by eosinophils are also expressed by eosinophils. Expression of cognate receptors for eosinophil-derived cytokines allows for significant autocrine regulation. However, as will be discussed later, eosinophil-expressed cytokine receptors also participate in the mobilization of their cognate ligands from preformed intragranular stores.

PATHWAYS OF EOSINOPHIL CYTOKINE SECRETION

Eosinophil secretion of cytokines and chemokines may occur through pathways using new gene transcription, intracellular stores of mRNA, or secretion from arsenals of cytokines stored preformed within intracellular granules and vesicles. Secretion of cytokines from the preformed, intragranular pool can be further divided into three ultra-structurally distinct mechanisms of degranulation (Fig. 8.2.2): (1) classical exocytosis, whereby intracellular granules may or may not fuse with one another before fusing with the plasma membrane, eliciting wholesale release of granule contents (hereafter referred to as exocytosis); (2) piecemeal degranulation (PMD), whereby small packets of proteins are released from granules into cytoplasmic vesicles and carried to the plasma membrane for extracellular secretion (Fig. 8.2.2C, E); and (3) eosinophil cytolysis (ECL), whereby the integrity of the plasma membrane is breached and intact granules are released extracellularly prior to or in association with cell death (Fig. 8.2.2B, D). Electron microscopy (EM) has proven to be a powerful tool in understanding eosinophil secretory processes (reviewed in[44]).

De Novo Cytokine Synthesis

As with other nucleated cells, *de novo* synthesis of cytokines within the endoplasmic reticulum (ER) and secretion through the Golgi apparatus occurs within eosinophils. Likewise, the capacity of eosinophils to secrete cytokines lacking signaling peptides required for ER—Golgi secretion, including IL-1 beta and macrophage migration inhibitory factor (MIF), suggest eosinophils may also engage in nonconventional mechanisms of *de novo* cytokine secretion (reviewed in[45]), although few studies to date have addressed this question. Constitutive and inducible mRNA expression within eosinophils has been noted for several cytokines. The extent to which eosinophil effector functions rely on classic *de novo* cytokine synthesis or maintenance of an intracellular mRNA pool for direct cytokine translation and secretion is unclear. Eosinophils may use pathways of *de novo* synthesis to restock the

preformed cytokine arsenal within intracellular granules following degranulation. Further studies are needed to determine the extent to which PMD in particular depletes intracellular cytokine concentrations, and to monitor the capacity of eosinophils to replenish granule stores of cytokines.

Mechanisms of stabilizing mRNA transcripts have also been implicated in the stimulus-induced production of at least some cytokines, including GM-CSF and TGF-β. GM-CSF promotes the growth and differentiation of myeloid cells, including eosinophils. In eosinophil-associated diseases, GM-CSF plays a critical role in priming, activating, and prolonging the survival of eosinophils within tissues. As noted previously, GM-CSF is also a product of eosinophils, providing eosinophils with an autocrine mechanism of perpetuating an eosinophilic inflammatory response. Studies using eosinophils isolated from mild asthmatic patients revealed intracellular stores of preformed GM-CSF protein colocalized with granule-enriched subcellular fractions, suggesting eosinophil secretion of GM-CSF may occur through PMD and/or cytolysis.[46] However, in addition to preformed protein stores, additional studies identified constitutive expression of GM-CSF mRNA that is rapidly degraded within resting eosinophils. In contrast, eosinophils recovered from the bronchoalveolar lavage (BAL) fluid of asthmatics exhibited stabilized GM-CSF mRNA, a condition replicated *in vitro* in human blood eosinophils through the activation with ionomycin[47,48] and hyaluronic acid, or ligation of β_7 integrins in the presence of TNF-α.[49] Stabilization of GM-CSF mRNA was accompanied by GM-CSF protein secretion and prolonged survival of activated eosinophils, and was dependent on the extracellular signal-regulated kinase (ERK)—mitogen-activated protein kinase (MAPK) intracellular signaling pathway.[50]

GM-CSF mRNA stability is regulated through 3'-untranslated region adenylate-uridylate (AU)-rich elements (AREs), whose binding partners dictate the rate of decay or stabilization of associated mRNA molecules. In the case of eosinophil-expressed GM-CSF mRNA, peptidyl-prolyl cis-trans isomerase NIMA-interacting 1 (Pin1) has been identified as a critical regulator of ARE binding partners, and consequently GM-CSF mRNA stability. In nonstimulated eosinophils, GM-CSF AREs are bound by the ARE binding protein AU-rich element RNA-binding protein 1 (AUF1) in association with Pin1 in its phosphorylated, catalytically inactive state. AUF1 recruits a multisubunit ribonuclease complex (the exosome), promoting rapid 3'—5' decay of the GM-CSF mRNA. In contrast, eosinophil activation elicits the dephosphorylation of Pin1, converting the prolyl isomerase (PPIase) to its catalytically active state, in combination with hyperphosphorylation of AUF1, leading to an exchange of ARE binding partners, with heterogeneous nuclear ribonucleoprotein C (hnRNP C) replacing

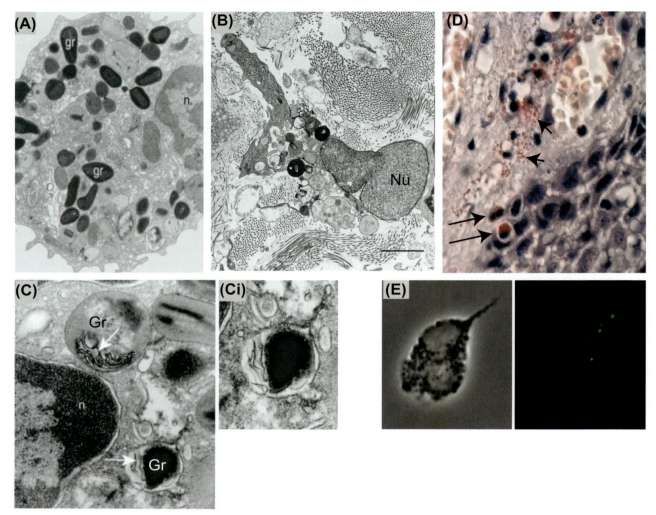

FIGURE 8.2.2 **Secretion of preformed, granule-stored cytokines.** Electron micrographs of eosinophils at rest *(A)*, or undergoing ECL *(B)* or PMD *(C and Ci)* forms of degranulation. In *(B)*, evidence of ECL includes nuclear chromatolysis and the absence of an intact plasma membrane. In *(C)*, evidence of PMD includes the appearance of a dynamic vesiculotubular compartment within granules (arrow, also see *Ci*) and an increase in small spherical and larger tubular secretory vesicles within the cytoplasm. In *(D)*, intact eosinophils (long arrows) and cell-free extracellular granules (short arrows) are apparent within human dermal tissue. In *(E)*, secretion of IL-4 by PMD from eotaxin-stimulated eosinophils is visualized as discrete, punctate spots using the EliCell assay technique. ECL, eosinophil cytolysis; IL-4, interleukin-4; PMD, piecemeal degranulation. (A), (C), and (Ci) *are reproduced from*[77]; (B) *is reproduced from*[78], *with permission;* (D) *is reproduced from*[57]; *and* (E) *is reproduced from*[79].

AUF1, and a consequent stabilization of GM-CSF mRNA.[51] In full support of these findings, eosinophils isolated from the BAL fluid of asthmatics exhibited GM-CSF mRNA transcripts preferentially associated with hnRNP C over AUF1.[51]

Similarly, studies in rodents recently identified Pin1 regulation of TGF-β mRNA decay within eosinophils. Of important clinical significance, *in vivo* studies targeting Pin1 revealed the attenuation of eosinophilic inflammation, transforming growth factor beta-1 (TGF-β1) protein and mRNA expression, and collagen deposition within the BAL fluid and lung tissue of allergen-challenged rats following specific inhibition of Pin1.[52] These studies suggest Pin1 may provide an attractive therapeutic target for the treatment of eosinophilic inflammatory diseases, and provide compelling evidence for the contribution of *de novo* protein translation to eosinophil functions *in vivo*.

Exocytosis

Classic exocytosis, as defined by whole granules fusing with the plasma membrane, is the form of degranulation best described in basic biology textbooks, and is therefore erroneously considered a predominant mode of secretion from granule-containing innate immune cells. In classic, whole granule exocytosis, limiting membranes of intracellular granules fuse with the cell surface membrane, creating a fusion pore. Granule membranes are incorporated into the plasma membrane, and the entire granule contents are released extracellularly. Under some circumstances, two or more intracellular granules fuse prior to fusion with the surface membrane, resulting in compound exocytosis of the contents of multiple granules. Fusion pore formation and whole granule exocytosis can be measured using the patch-clamp technique, where increases in the surface area of the cell membrane resulting from the incorporation of granule membranes is detected as an increase in the membrane capacitance. Whole-cell patch clamping of eosinophils reveals that the exocytosis of granules results in a definable, stepwise increase in membrane capacitance, while vesicle-mediated degranulation is measurable by capacitance charting as a gradual incline. Using the intracellular addition of guanosine 5'-O-γ-thiotriphosphate (GTPγS) as an artificial stimulus for the complete exocytotic release of granules, exocytotic events are observed in differentiating eosinophils as early as the third week in cultures of umbilical cord blood-derived eosinophils.[53] Other non-physiological stimuli, such as treatment with Ca^{2+} ionophores, also induce morphological characteristics consistent with granule–granule and granule–plasma membrane fusions.

However, evidence for classic exocytosis under physiological conditions and *in vivo* is less compelling. Ultrastructurally, granule–granule and granule–plasma membrane fusions are observed only very rarely within tissues. One exception may be in the direct killing of parasitic helminths by eosinophils. EM analyses of *in vitro* co-cultures of rat eosinophils and *Schistosoma mansoni*, *Trichinella spiralis*, or *Nippostrongylus brasiliensis* revealed ultrastructural evidence of granule–granule fusions and compound exocytosis, discharging large amounts of granule-derived materials onto parasite surfaces. Intracellular granules within rat eosinophils forming intimate interactions with the parasite cuticle were observed to exhibit signs of activation, with matrix material appearing more coarsely granular. Some of these activated granules fused with one another, forming intracellular vacuoles before fusing with the plasma membrane, releasing the entire granule-derived vacuole contents onto the surface of the helminth.[54] This pathway of compound exocytosis would enable the secretion of granule contents to be concentrated onto a defined area, magnifying the potential cytotoxic effects of granule cationic proteins on parasitic pathogens.

Cytolysis

The process of ECL differs from classic exocytosis in that granules are released extracellularly through breaches in the cell plasma membrane and deposited within tissues in their entirety, often with granule-limiting membranes still intact.[55,56] Cytokines presumably remain sequestered within the intact granules until the granules are acted on by exogenous agonists to promote stimulus-dependent secretion. ECL culminates in eosinophil death and as such may be viewed as an alternative to apoptosis that retains a capacity for degranulation. Factors propelling eosinophil fates toward either apoptosis or ECL have yet to be elucidated. The ultrastructural characteristics of ECL include rupturing of the plasma membrane and chromatolysis (Fig. 8.2.2B). Within tissues, these ultrastructural observations are observed together with the deposition of intact eosinophil granules (Fig. 8.2.2D). Recently, collaborative studies from the laboratories of Dr. Peter Weller and Dr. Redwan Moqbel demonstrated that cell-free, extracellular granules express functional receptors on the granule membrane surface and maintain the capacity to function extracellularly as secretory-competent organelles, including the stimulus-dependent secretion of cationic proteins and cytokines.[57] It is likely that degranulation of cell-free extracellular granules within tissues occurs through a process akin to the PMD observed in intact cells. Extracellularly released eosinophil granules are the subject of a later section in this chapter.

Piecemeal Degranulation

The process of PMD was proposed over 30 years ago based on ultrastructural observations by EM of basophilic leukocytes,[58] and has since been identified as an important secretory process in a variety of cells, including eosinophils, mast cells, and endocrine cells.[59] PMD is readily induced by physiological stimuli *in vitro* and is the mode of degranulation most often observed *in vivo* by eosinophils within tissue sections.

In PMD, specific proteins residing within intracellular granules emerge from the granule within spherical vesicles and elongated tubules, and are shuttled to the plasma membrane for extracellular release. Therefore, unlike exocytosis, where whole granule contents are simultaneously released, PMD delivers discrete packets of granule-derived proteins to the cell surface (Fig. 8.2.2E). Ultrastructural evidence for PMD includes a disordering of the granule crystalline core, observable as an overall tendency toward electron lucency and less well-defined core, the appearance of a network of membranous compartments within granules, and the association of granule-limiting membranes with spherical and tubular vesicles (Fig. 8.2.2C). Intracytoplasmic evidence includes a significant increase in the number of secretory vesicles, including a subset of distinct tubular vesicles termed eosinophil sombrero vesicles (EoSVs) in reference to the concentric *doughnut*-shaped configuration they exhibit ultrastructurally in cross section.[60] Importantly, eosinophils remain fully viable throughout PMD and maintain their secretory capacities, allowing for sequential stimulations.

Based on ultrastructural observations of eosinophils from blood and tissues in a variety of disease settings, PMD and ECL are, respectively, the first and second most commonly observed secretory pathways *in vivo*. In light of the strong *in vivo* evidence for PMD as the major physiological pathway of eosinophil cytokine secretion, the remainder of this chapter focuses primarily on the mechanisms regulating this mode of degranulation. Cytolysis, and in particular its consequent extracellular release of intact eosinophil granules, is the subject of a subsequent section in this chapter.

EOSINOPHILS STORE PREFORMED CYTOKINES, AVAILABLE FOR RAPID AND DIFFERENTIAL SECRETION

Eosinophils are distinguished from many other innate immune secretory cells and most lymphocytes by their capacity to secrete cytokines within only minutes of stimulation. The very rapid cytokine secretion from eosinophils, enabling the effector functions described previously, is made possible by their differential degranulation of cytokines from premade intracellular stores through PMD, bypassing the time required for new gene transcription or protein translation of nascent mRNA. Preformed intracellular stores of cytokines are generated within eosinophils during bone marrow development.[61] Therefore, eosinophils emerge into the circulation already equipped for rapid cytokine secretion. Although eosinophils tend to be associated primarily with T_h2 immunity, the cache of preformed cytokines within eosinophils available for immediate mobilization is not restricted to the repertoire of T_h2-associated proteins, but also includes cytokines normally associated with T_h1 (IFN-γ, IL-12), T-regulatory (T_{reg}; IL-10, TGF-β), and T_h17 cells. We recently analyzed the relative preformed concentrations of seven diverse cytokines (IFN-γ, IL-4, IL-6, IL-10, IL-12, IL-13, TNF-α, and p70) within human blood eosinophils from healthy subjects, and subjects with mild atopy. These specific cytokines were chosen based on their relationships with T_h1, T_h2, or T_{reg} immunity, or for their pro- or anti-inflammatory properties. All seven cytokines were detected within eosinophils from all donors analyzed, in relative concentrations remarkably well conserved among donors with the exception of TNF-α, which displayed a notable degree of donor-dependent variability[10] (Fig. 8.2.3A).

The highest intracellular content of preformed cytokines within human eosinophils colocalizes with granule-enriched fractions, as evidenced by subcellular fractionation and immunological methods using light and EM. In addition to the granular stores, some cytokines and cationic proteins, including IFN-γ,[10] eosinophil granule major basic protein 1,[62] CCL5/RANTES,[63] TGF-α,[64] and IL-4,[65] have been identified within secretory vesicles of human blood-derived eosinophils (Fig. 8.2.3B). Therefore, vesicular compartments may serve as a second, potentially even more rapidly accessible, reservoir for preformed proteins within human eosinophils. Whether vesicular stores of cytokines within freshly isolated eosinophils represent a mobilization of cytokines from granules into vesicles, or originate through *de novo* synthesis into Golgi-derived vesicles is yet to be determined. Also unknown is whether vesicle loading is constitutively active in the absence of exogenous stimulations, requires an *in vivo* priming event, or represents an artifact of cell isolation.

A wide range of soluble and cell-associated stimuli derived from immune and tissue cells elicit PMD of vesicle- and granule-stored cytokines from eosinophils recovered from the blood. These agonists include cross-linking surface CD9 or leukocyte immunoglobulin-like receptor 7 (LIR-7),[66] exposure to cytokines, including IFN-γ or TNF-α,[10] or stimulation by chemokines, including CCL5 (RANTES) and eotaxin.[67] Different stimuli elicit very rapid and distinctive patterns of cytokine secretion by eosinophils.[10] Evidence of intracellular cytokine mobilization is observed within minutes of stimulation,[63] and secreted cytokines are detected as early as 5 min after stimulation.[67]

Eosinophils recruited into allergic airways and tissues exhibit enhanced secretory capacities and often display a lowered threshold requirement for activation.[68] Accounting for the enhanced activation of tissue compared with blood eosinophils, interaction with vascular endothelial cells and subsequent transmigration across the endothelium activates eosinophils and promotes degranulation.[69,70] Moreover, in many allergic diseases, blood and tissue expression of eosinophilopoietins such as GM-CSF and IL-5, and inflammatory mediators such as TNF-α, is augmented. Eosinophils primed with GM-CSF or IL-5 resemble airway eosinophils,[71] and TNF-α synergizes with cytokine stimuli to elicit robust secretion of T_h1- and T_h2-associated chemokines.[16]

MECHANISMS REGULATING EOSINOPHIL SECRETION OF PREFORMED CYTOKINES THROUGH PIECEMEAL DEGRANULATION

The clinical significances of eosinophil-derived cytokine secretion make a full delineation of molecular and biochemical mechanisms regulating PMD an attractive

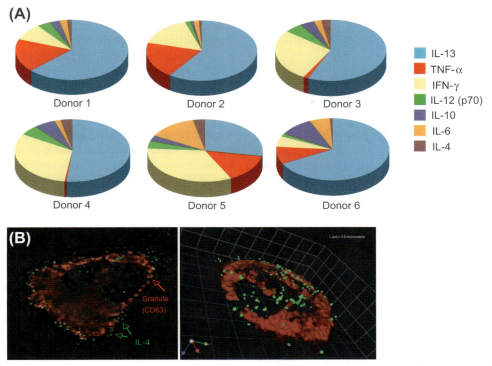

FIGURE 8.2.3 Preformed intracellular caches of cytokines stored within human blood eosinophils. *(A)* Relative concentrations of preformed cytokines stored within blood eosinophils isolated from six different donors, as determined by multiplex analyses of eosinophil lysates. *(B)* Vesicle-stored IL-4 is detected within nonstimulated eosinophils by immunofluorescence microscopy. Treatment of cells with saponin resulted in the permeabilization of plasma and vesicle membranes, but not granule membranes. IL-4 reactivity (green) is shown outside of CD63-labeled granules (red) in a two-dimensional image of a single focal plane (left panel) and in a three-dimensional re-rendering of an image stack following deconvolution (right panel). IFN-γ, interferon gamma; IL-4, interleukin-4; IL-6, interleukin-6; IL-10, interleukin-10; IL-12 (p70), interleukin-12; IL-13, interleukin-13; TNF-α, tumor necrosis factor.

goal in the development of novel therapeutic approaches. Some of these mechanisms are beginning to be unraveled.

Receptor-Chaperoned Trafficking of Preformed Cytokines

Secretion of cytokines from eosinophil intragranular stores is highly agonist-specific, raising the question of how stimulus-induced parsimony of granule-derived cytokines is achieved through PMD. Indeed, intracellular mechanisms governing differential trafficking of cytokines in all cells remain poorly delineated. We recently identified a novel mechanism by which such parsimonious and differential sorting and trafficking may occur through cytokine chaperoning by cognate receptor ligands.[72]

The first evidence of receptor chaperoning of cognate cytokine ligands was gleaned through ultrastructural observations of human eosinophils undergoing PMD. Eotaxin stimulation elicits rapid secretion of IL-4, mobilized from preformed granule stores, as demonstrated by the detection of secreted IL-4 within 30 min of eotaxin stimulation, irrespective of pretreatment with actinomycin

D or cycloheximide to block new gene transcription or protein translation, respectively.[67] Eotaxin is a selective agonist, as several other granule-stored cytokines (e.g., IL-12) are not mobilized or secreted.[67] Characteristic of PMD, by immunoelectron microscopy IL-4 was visualized within small spherical vesicles and tubular vesicles, emerging from intracellular granules and trafficking within the cytoplasm of eotaxin-stimulated cells. Intriguingly, the pattern of IL-4 immunoreactivity revealed IL-4 to be associated primarily with the luminal surface of secretory vesicle membranes.[65] Lacking a transmembrane domain of its own, this observation suggested that IL-4 within secretory vesicles was tethered to the vesicle membrane by a carrier molecule. Of note, in a little-appreciated correlation, eosinophils express receptors specific for many, if not all, of the cytokines and chemokines secreted by eosinophils, including interleukin-4 receptor subunit alpha (IL-4RA) (Fig. 8.2.4A) and common gamma (γc) chains, components of the type 1 IL-4 receptor complex. Western blot of eosinophil subcellular fractions revealed that, in addition to nominal surface expression, IL-4RA chains colocalized predominantly with granule-enriched fractions (Fig. 8.2.4B). Immunoelectron microscopy of

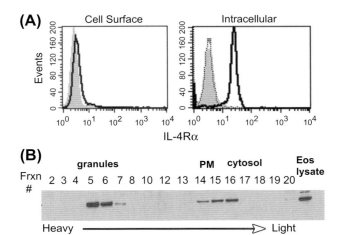

FIGURE 8.2.4 Interleukin-4 receptor subunit alpha chains are associated predominantly with intracellular granules. *(A)* Surface and intracellular staining of human blood-derived eosinophils for expression of interleukin-4 receptor subunit alpha (IL-4RA) chains. Shaded histogram, immunoglobulin G (IgG) control; dotted line, staining in the presence of blocking peptide. *(B)* Intracellular IL-4RA chains colocalize predominantly with granule-enriched subcellular fractions. Frxn, fraction. *Adapted from*[72].

eotaxin-stimulated eosinophils revealed mobilization of IL-4RA chains out of granules and into secretory vesicles in a pattern resembling that observed for IL-4. Flow cytometry of eotaxin-stimulated cells in which plasma membrane and vesicular membranes (but not granule membranes) were permeabilized confirmed the mobilization of the IL-4RA chains (but not the γc chains) out of granules and into vesicles. In contrast, an antibody that is excluded from IL-4RA chains when IL-4 is bound failed to detect the mobilized IL-4RA, suggesting that IL-4RA chains within vesicles emerging from the granules of eotaxin-stimulated cells are bound by IL-4.[72] That γc chains do not appear to be mobilized in parallel with IL-4RA implies that IL-4RA might chaperone bound IL-4 without the stabilizing effect of heterodimer formation with the γc chains, favoring the release of bound IL-4 on fusion of the vesicle and plasma membranes, and avoiding activation of intracellular signaling cascades by secretory vesicles destined for extracellular cytokine release.

Interpreting these data within the context of granule ultrastructural changes characteristic of PMD, a model for differential cytokine secretion has been proposed whereby exogenous stimulation elicits mobilization of cytokine receptor chains associated with intragranular membranes (Fig. 8.2.5A, B). Thus, mobilized receptors could conceivably draw their specific cytokine ligands out from the mixture of granule-stored cytokines and escort bound cytokines into newly formed vesicles arising from granules. Substantial intracellular expression of other cytokine- and chemokine-binding receptor chains, including interleukin-13 receptor subunit alpha, interleukin-6 receptor subunit

alpha, and CCR3, supports the hypothesis that receptor-mediated trafficking may be a mechanism broadly applicable to the mobilization and secretion of preformed, eosinophil-derived cytokines and chemokines.

Moreover, the recognition of receptor-mediated trafficking of cognate ligands in eosinophils provides a new appreciation of the EoSVs prominently observed within eosinophils undergoing PMD. The high surface area to volume ratio inherent in the tubular structures of these vesicles is ideally suited for a transport mechanism relying on membrane tethering of cargo proteins.[60] Of note, since the publication of the aforementioned study in eosinophils, another study reported the transport of interleukin-15 (IL-15)—interleukin-15 receptor complexes from the Golgi to the plasma membrane of dendritic cells,[73] suggesting receptor-mediated cytokine trafficking may represent a common mode of cytokine secretion. Moreover, other innate immune cells secreting granule-derived cytokines express receptors in association with intracellular granules. For example, neutrophils can secrete IL-10[74] and contain IL-10R+-specific granules.[75] Likewise, the CCR3 ligands eotaxin and CCL5 (RANTES) are secreted by mast cells, and CCR3 expression has been localized to mast cell granules.[76]

Granule Release and Trafficking of Cytokine-Loaded Secretory Vesicles

What are the processes by which cytokine-laden vesicles emerge from and are released by mobilized granules, and ultimately dock and fuse with the plasma membrane? Significant progress has been made over the past several years toward answering these questions. Although early EM analyses of vesicles protruding from granules in eosinophils within tissues and in agonist-stimulated eosinophils *in vitro* suggested that cargo-laden secretory vesicles arise directly from granules, more recent studies using electron tomography confirm that both small spherical vesicles and elongated tubule structures are actively formed from granules, and reveal a dynamic intragranular membranous network that functions to reorganize and sequester granule-derived products for packaging and secretion.[79] Specifically, the use of electron tomography enabled the three-dimensional visualization of tubulovesicular structures physically interconnected with membranous networks within granule matrices and with the granule-limiting membrane while emerging from emptying granules.

Use of specific staining and fixation methods optimized for the preservation of membrane bilayers has enabled a more careful study of the intragranular microdomains.[79] Following 1 h of agonist stimulation of human blood eosinophils, a complex network of membranes is evident,

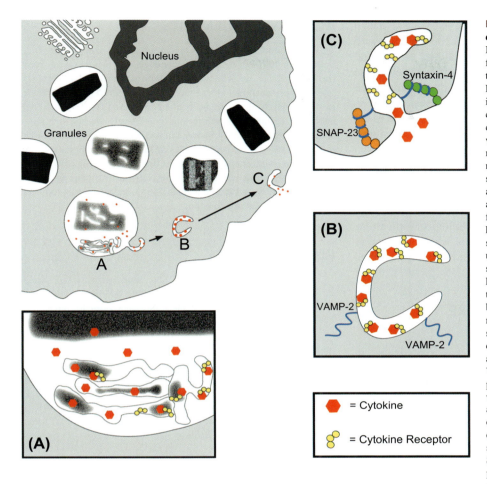

FIGURE 8.2.5 Model of piecemeal degranulation in human eosinophils. Based on recent published findings, the following model of piecemeal degranulation is proposed for human eosinophils. Following cell stimulation, eosinophil intracellular granules undergo marked changes, including a loss of electron density of the crystalline core in association with a relocalization of electron-dense materials, and the appearance of a complex network of intragranular membrane-bound structures. (A) Specific cytokine receptors associated with intragranular membranes are mobilized in a stimulus-dependent fashion, and sequester their cognate cytokines from the preformed protein pool stored within eosinophil intracellular granules. Receptor chains facilitate the specific sorting and packaging of respective cytokines into secretory vesicles emerging from the granules. (B) Cytokines are chaperoned by their respective membrane-attached receptors to the cell periphery within small spherical vesicles and elongated tubular carriers (EoSVs) that express the vesicle-associated SNARE molecule VAMP-2. (C) VAMP-2-expressing vesicles fuse with the plasma membrane through interactions with the membrane SNAREs syntaxin-4 and SNAP-23, and release their cytokine cargo into the extracellular space. EoSV, eosinophil sombrero vesicle; SNAP-23, synaptosomal-associated protein 23; SNARE, SNAP (Soluble NSF attachment protein) Receptor; VAMP-2, vesicle-associated membrane protein 2.

primarily within granule matrices. This membranous compartment takes on the form of tubules and vesicles, and is associated with an apparent reordering of the electron-dense material within granules, suggestive of a stimulus-driven sorting and/or packaging of granule-derived proteins occurring within the granule itself (Fig. 8.2.5A). Pretreatment with brefeldin A (BFA), an inhibitor of vesicular transport, is known to prevent the secretion of cytokines by PMD. Ultrastructural analyses of eosinophils pretreated with BFA prior to stimulation with eotaxin revealed a collapse of the membranous network within mobilized granules, resulting in the appearance of electron-dense lipid deposits within granule matrices.[77]

Vesicles released from granules must traffic through the cytoplasmic space, and dock and fuse with the cell surface membrane. Molecular mechanisms controlling these steps of PMD involve the coordinated efforts of complexes comprised of synaptosomal-associated protein (SNARE) isoforms. Human eosinophils express the SNARE isoforms vesicle-associated membrane protein 2 (VAMP-2), 7 (VAMP-7), and 8 (VAMP-8).[80–82] Expression of VAMP-7 and VAMP-8 is enriched on eosinophil-specific granules, while VAMP-2 is associated predominantly with secretory vesicles.[82–84] The SNARE molecules SNAP-23 and syntaxin-4 colocalize with the plasma membrane and appear to function as binding targets for the VAMP-2-expressing secretory vesicles[85] (Fig. 8.2.5C). Molecular mechanisms regulating PMD, including the involvement of SNARE isoforms, is discussed in greater detail elsewhere in this volume.

DO MOUSE EOSINOPHILS ENGAGE IN PMD?

Although numerous studies describe the functions of eosinophils in the development of allergic inflammatory diseases in mice, and many of these functions are ascribed to the secretion of eosinophil-derived cytokines and chemokines, differences in the secretory potential of eosinophils are observed between species. Of note, our discussion in this chapter regarding the pathways of degranulation and mechanisms regulating PMD is based on *in vitro* and *in vivo* studies of human eosinophils. In contrast to findings with human cells, ultrastructural examinations of murine eosinophils within tissues do not reveal extensive evidence of degranulation.[86] Studies are currently under way which are attempting to rationalize these differences. However, a combined effort from several laboratories interested in the immunobiology of eosinophils has offered some insights, concluding that murine eosinophils are induced to undergo PMD, albeit with a higher threshold for agonist-induced activation. Moreover, using a model of allergen sensitization and airway challenge, these authors demonstrated that although eosinophil degranulation is not observed within the submucosa of allergic lungs, morphological indications of PMD were exhibited by eosinophils within the airway lumen, corresponding to the detection of granule-derived proteins.[87]

CONCLUSIONS AND FUTURE DIRECTIONS

An appreciation of eosinophils as active participants in diverse processes including immunomodulation and tissue repair and remodeling, using aspects of their immunobiology distinct from cationic protein-mediated destructive effector functions, has highlighted the importance of the cytokine secretory processes of eosinophils. As such, there is potential therapeutic value in fully elucidating the mechanisms governing the generation, storage, and selective secretion of eosinophil-derived cytokines and chemokines. Toward this end, the field has seen significant recent advances in the demonstration of SNARE molecules directing fusion events between granules, vesicles, and surface membranes, and in the identification of receptor-mediated chaperoning of cognate cytokine ligands. Significant questions still remain, including:

- What are the signaling events upstream of intracellular granule mobilization that link external stimuli to movement of intragranular cytokine receptor chains?
- Which cytokines and chemokines are secreted through mechanisms using receptor chaperoning?
- How are receptors recycled following the delivery of cytokine cargo to the cell periphery?
- To what extent might eosinophils participate in polarized cytokine secretion, and what are the molecular and biochemical mechanisms regulating polarized secretion?

Answers to these questions hold promise not only for the full elucidation of cytokine secretion from eosinophils, but also will likely be more broadly applicable to our understanding of cytokine secretion from other innate immune cells.

ACKNOWLEDGEMENTS

This work and the perspectives herein draw upon years of conversations and input from numerous individuals, across several laboratories. In particular, we acknowledge members of the Weller laboratory (past and present), including Peter Weller, Sandra Perez-Rodriguez, Rossana Melo, and Josiane Sabbadini Neves; Ann Dvorak, for a long history of collaborative efforts and consistently sharing her ultrastructural expertise; and Linying Liu and Paige Lacy for the critical reading of this work. Special thanks to Jason J. Xenakis for creating the computer-generated illustrations included as Fig. 8.2.1 and Fig. 8.2.5. Funding was provided by National Institutes of Health grants HL095699, AI020241, and AI051645, and a Grant-in-Aid from the American Heart Association.

REFERENCES

1. Gessner A, Mohrs K, Mohrs M. Mast cells, basophils, and eosinophils acquire constitutive IL-4 and IL-13 transcripts during lineage differentiation that are sufficient for rapid cytokine production. *J Immunol* 2005;**174**(2):1063–72.
2. Voehringer D, Shinkai K, Locksley RM. Type 2 immunity reflects orchestrated recruitment of cells committed to IL-4 production. *Immunity* 2004;**20**(3):267–77.
3. Shinkai K, Mohrs M, Locksley RM. Helper T cells regulate type-2 innate immunity in vivo. *Nature* 2002;**420**(6917):825–9.
4. Sabin EA, Pearce EJ. Early IL-4 production by non-CD4+ cells at the site of antigen deposition predicts the development of a T helper 2 cell response to Schistosoma mansoni eggs. *J Immunol* 1995;**155**(10):4844–53.
5. Voehringer D, et al. Type 2 immunity is controlled by IL-4/IL-13 expression in hematopoietic non-eosinophil cells of the innate immune system. *J Exp Med* 2006;**203**(6):1435–46.
6. Sabin EA, Kopf MA, Pearce EJ. Schistosoma mansoni egg-induced early IL-4 production is dependent upon IL-5 and eosinophils. *J Exp Med* 1996;**184**(5):1871–8.
7. Nouri-Aria KT, et al. Cytokine expression during allergen-induced late nasal responses: IL-4 and IL-5 mRNA is expressed early (at 6 h) predominantly by eosinophils. *Clin Exp Allergy* 2000;**30**(12):1709–16.
8. Wang YH, et al. IL-25 augments type 2 immune responses by enhancing the expansion and functions of TSLP-DC-activated Th2 memory cells. *J Exp Med* 2007;**204**(8):1837–47.
9. Woerly G, et al. Expression of Th1 and Th2 immunoregulatory cytokines by human eosinophils. *Int Arch Allergy Immunol* 1999;**118**(2-4):95–7.
10. Spencer LA, et al. Human eosinophils constitutively express multiple Th1, Th2, and immunoregulatory cytokines that are secreted rapidly and differentially. *J Leukoc Biol* 2009;**85**(1):117–23.
11. Rose Jr CE, et al. Murine lung eosinophil activation and chemokine production in allergic airway inflammation. *Cell Mol Immunol* 2010;**7**(5):361–74.
12. Fulkerson PC, et al. A central regulatory role for eosinophils and the eotaxin/CCR3 axis in chronic experimental allergic airway inflammation. *Proc Natl Acad Sci USA* 2006;**103**(44):16418–23.

13. Jacobsen EA, et al. Allergic pulmonary inflammation in mice is dependent on eosinophil-induced recruitment of effector T cells. *J Exp Med* 2008;**205**(3):699−710.

14. Lee JJ, et al. Defining a link with asthma in mice congenitally deficient in eosinophils. *Science* 2004;**305**(5691):1773−6.

15. Walsh ER, et al. Strain-specific requirement for eosinophils in the recruitment of T cells to the lung during the development of allergic asthma. *J Exp Med* 2008;**205**(6):1285−92.

16. Liu LY, et al. Generation of Th1 and Th2 chemokines by human eosinophils: evidence for a critical role of TNF-alpha. *J Immunol* 2007;**179**(7):4840−8.

17. Abraham D, et al. Strongyloides stercoralis: protective immunity to third-stage larvae inBALB/cByJ mice. *Exp Parasitol* 1995;**80**(2):297−307.

18. Herbert DR, et al. Role of IL-5 in innate and adaptive immunity to larval Strongyloides stercoralis in mice. *J Immunol* 2000;**165**(8): 4544−51.

19. Galioto AM, et al. Role of eosinophils and neutrophils in innate and adaptive protective immunity to larval strongyloides stercoralis in mice. *Infect Immun* 2006;**74**(10):5730−8.

20. Jordan MB, et al. Promotion of B cell immune responses via an alum-induced myeloid cell population. *Science* 2004;**304**(5678): 1808−10.

21. Wang HB, Weller PF. Pivotal advance: eosinophils mediate early alum adjuvant-elicited B cell priming and IgM production. *J Leukoc Biol* 2008;**83**(4):817−21.

22. Stenfeldt AL, Wenneras C. Danger signals derived from stressed and necrotic epithelial cells activate human eosinophils. *Immunology* 2004;**112**(4):605−14.

23. Fillon S, et al. Epithelial function in eosinophilic gastrointestinal diseases. *Immunol Allergy Clin North Am* 2009;**29**(1):171−8. xii-xiii.

24. Todd R, et al. The eosinophil as a cellular source of transforming growth factor alpha in healing cutaneous wounds. *Am J Pathol* 1991;**138**(6):1307−13.

25. Kay AB. The role of eosinophils in the pathogenesis of asthma. *Trends Mol Med* 2005;**11**(4):148−52.

26. Minshall EM, et al. Eosinophil-associated TGF-beta1 mRNA expression and airways fibrosis in bronchial asthma. *Am J Respir Cell Mol Biol* 1997;**17**(3):326−33.

27. Gharaee-Kermani M, Phan SH. Molecular mechanisms of and possible treatment strategies for idiopathic pulmonary fibrosis. *Curr Pharm Des* 2005;**11**(30):3943−71.

28. Aceves SS, et al. Esophageal remodeling in pediatric eosinophilic esophagitis. *J Allergy Clin Immunol* 2007;**119**(1):206−12.

29. Varga J, Kahari VM. Eosinophilia-myalgia syndrome, eosinophilic fasciitis, and related fibrosing disorders. *Curr Opin Rheumatol* 1997;**9**(6):562−70.

30. Spry CJ. The pathogenesis of endomyocardial fibrosis: the role of the eosinophil. *Springer Semin Immunopathol* 1989;**11**(4): 471−7.

31. Aceves SS, Broide DH. Airway fibrosis and angiogenesis due to eosinophil trafficking in chronic asthma. *Curr Mol Med* 2008;**8**(5): 350−8.

32. Detoraki A, et al. Angiogenesis and lymphangiogenesis in bronchial asthma. *Allergy* 2010;**65**(8):946−58.

33. Puxeddu I, et al. Human peripheral blood eosinophils induce angiogenesis. *Int J Biochem Cell Biol* 2005;**37**(3):628−36.

34. Hoshino M, Takahashi M, Aoike N. Expression of vascular endothelial growth factor, basic fibroblast growth factor, and angiogenin immunoreactivity in asthmatic airways and its relationship to angiogenesis. *J Allergy Clin Immunol* 2001;**107**(2):295−301.

35. Hoshino M, Nakamura Y, Hamid QA. Gene expression of vascular endothelial growth factor and its receptors and angiogenesis in bronchial asthma. *J Allergy Clin Immunol* 2001; **107**(6):1034−8.

36. Piliponsky AM, et al. Effects of eosinophils on mast cells: a new pathway for the perpetuation of allergic inflammation. *Mol Immunol* 2002;**38**(16-18):1369−72.

37. Wong CK, et al. Signalling mechanisms regulating the activation of human eosinophils by mast-cell-derived chymase: implications for mast cell-eosinophil interaction in allergic inflammation. *Immunology* 2009;**126**(4):579−87.

38. Shakoory B, et al. The role of human mast cell-derived cytokines in eosinophil biology. *J Interferon Cytokine Res* 2004;**24**(5):271−81.

39. Levi-Schaffer F, et al. Mast cells enhance eosinophil survival in vitro: role of TNF-alpha and granulocyte-macrophage colony-stimulating factor. *J Immunol* 1998;**160**(11):5554−62.

40. Matsuba-Kitamura S, et al. Contribution of IL-33 to induction and augmentation of experimental allergic conjunctivitis. *Int Immunol* 2010;**22**(6):479−89.

41. Sawatzky DA, et al. Eosinophil adhesion to cholinergic nerves via ICAM-1 and VCAM-1 and associated eosinophil degranulation. *Am J Physiol Lung Cell Mol Physiol* 2002;**282**(6): L1279−88.

42. Kobayashi H, et al. Human eosinophils produce neurotrophins and secrete nerve growth factor on immunologic stimuli. *Blood* 2002;**99**(6):2214−20.

43. Zheng X, et al. Leukemia inhibitory factor is synthesized and released by human eosinophils and modulates activation state and chemotaxis. *J Allergy Clin Immunol* 1999;**104**(1):136−44.

44. Melo RC, Dvorak AM, Weller PF. Contributions of electron microscopy to understand secretion of immune mediators by human eosinophils. *Microsc Microanal* 2010;**16**(6):653−60.

45. Stanley AC, Lacy P. Pathways for cytokine secretion. *Physiology (Bethesda)* 2010;**25**(4):218−29.

46. Levi-Schaffer F, et al. Association of granulocyte-macrophage colony-stimulating factor with the crystalloid granules of human eosinophils. *Blood* 1995;**85**(9):2579−86.

47. Esnault S, Malter JS. Primary peripheral blood eosinophils rapidly degrade transfected granulocyte-macrophage colony-stimulating factor mRNA. *J Immunol* 1999;**163**(10):5228−34.

48. Esnault S, Jarzembowski JA, Malter JS. Stabilization of granulocyte-macrophage colony-stimulating factor RNA in a human eosinophil-like cell line requires the AUUUA motifs. *Proc Assoc Am Physicians* 1998;**110**(6):575−84.

49. Esnault S, Malter JS. Granulocyte macrophage-colony-stimulating factor mRNA is stabilized in airway eosinophils and peripheral blood eosinophils activated by TNF-alpha plus fibronectin. *J Immunol* 2001;**166**(7):4658−63.

50. Esnault S, Malter JS. Extracellular signal-regulated kinase mediates granulocyte-macrophage colony-stimulating factor messenger RNA stabilization in tumor necrosis factor-alpha plus fibronectin-activated peripheral blood eosinophils. *Blood* 2002;**99**(11): 4048−52.

51. Shen ZJ, Esnault S, Malter JS. The peptidyl-prolyl isomerase Pin1 regulates the stability of granulocyte-macrophage colony-stimulating

factor mRNA in activated eosinophils. *Nat Immunol* 2005;**6**(12): 1280−7.

52. Shen ZJ, et al. Pin1 regulates TGF-beta1 production by activated human and murine eosinophils and contributes to allergic lung fibrosis. *J Clin Invest* 2008;**118**(2):479−90.

53. Scepek S, Lindau M. Exocytotic competence and intergranular fusion in cord blood-derived eosinophils during differentiation. *Blood* 1997;**89**(2):510−7.

54. McLaren DJ, Mackenzie CD, Ramalho-Pinto FJ. Ultrastructural observations on the in vitro interaction between rat eosinophils and some parasitic helminths (Schistosoma mansoni, Trichinella spiralis and Nippostrongylus brasiliensis). *Clin Exp Immunol* 1977;**30**(1): 105−18.

55. Persson CG, Erjefalt JS. Eosinophil lysis and free granules: an in vivo paradigm for cell activation and drug development. *Trends Pharmacol Sci* 1997;**18**(4):117−23.

56. Erjefalt JS, et al. Cytolysis and piecemeal degranulation as distinct modes of activation of airway mucosal eosinophils. *J Allergy Clin Immunol* 1998;**102**(2):286−94.

57. Neves JS, et al. Eosinophil granules function extracellularly as receptor-mediated secretory organelles. *Proc Natl Acad Sci USA* 2008;**105**(47):18478−83.

58. Dvorak HF, Dvorak AM. Basophilic leucocytes: structure, function and role in disease. *Clin Haematol* 1975;**4**(3):651−83.

59. Crivellato E, et al. Piecemeal degranulation as a general secretory mechanism? *Anat Rec A Discov Mol Cell Evol Biol* 2003;**274**(1): 778−84.

60. Melo RC, et al. Mechanisms of eosinophil secretion: large vesiculotubular carriers mediate transport and release of granule-derived cytokines and other proteins. *J Leukoc Biol* 2008;**83**(2): 229−36.

61. Mahmudi-Azer S, et al. Immunofluorescence analysis of cytokine and granule protein expression during eosinophil maturation from cord blood-derived CD34 progenitors. *J Allergy Clin Immunol* 2000;**105**(6 Pt 1):1178−84.

62. Melo RC, et al. Vesicle-mediated secretion of human eosinophil granule-derived major basic protein. *Lab Invest* 2009;**89**(7): 769−81.

63. Lacy P, et al. Rapid mobilization of intracellularly stored RANTES in response to interferon-gamma in human eosinophils. *Blood* 1999;**94**(1):23−32.

64. Egesten A, et al. Subcellular localization of transforming growth factor-alpha in human eosinophil granulocytes. *Blood* 1996;**87**(9): 3910−8.

65. Melo RC, et al. Human eosinophils secrete preformed, granule-stored interleukin-4 through distinct vesicular compartments. *Traffic* 2005;**6**(11):1047−57.

66. Tedla N, et al. Activation of human eosinophils through leukocyte immunoglobulin-like receptor 7. *Proc Natl Acad Sci USA* 2003; **100**(3):1174−9.

67. Bandeira-Melo C, et al. Cutting edge: eotaxin elicits rapid vesicular transport-mediated release of preformed IL-4 from human eosinophils. *J Immunol* 2001;**166**(8):4813−7.

68. Bates ME, et al. Human airway eosinophils respond to chemoattractants with greater eosinophil-derived neurotoxin release, adherence to fibronectin, and activation of the Ras-ERK pathway when compared with blood eosinophils. *J Immunol* 2010;**184**(12): 7125−33.

69. Dallaire MJ, et al. Endothelial cells modulate eosinophil surface markers and mediator release. *Eur Respir J* 2003;**21**(6):918−24.

70. Yamamoto H, et al. The effect of transendothelial migration on eosinophil function. *Am J Respir Cell Mol Biol* 2000;**23**(3):379−88.

71. Sedgwick JB, et al. Effect of interleukin-5 and granulocyte-macrophage colony stimulating factor on in vitro eosinophil function: comparison with airway eosinophils. *J Allergy Clin Immunol* 1995;**96**(3):375−85.

72. Spencer LA, et al. Cytokine receptor-mediated trafficking of preformed IL-4 in eosinophils identifies an innate immune mechanism of cytokine secretion. *Proc Natl Acad Sci U S A* 2006;**103**(9):3333−8.

73. Mortier E, et al. IL-15Ralpha chaperones IL-15 to stable dendritic cell membrane complexes that activate NK cells via trans presentation. *J Exp Med* 2008;**205**(5):1213−25.

74. Piskin G, Bos JD, Teunissen MB. Neutrophils infiltrating ultraviolet B-irradiated normal human skin display high IL-10 expression. *Arch Dermatol Res* 2005;**296**(7):339−42.

75. Elbim C, et al. Intracellular pool of IL-10 receptors in specific granules of human neutrophils: differential mobilization by proinflammatory mediators. *J Immunol* 2001;**166**(8):5201−7.

76. Price KS, et al. CC chemokine receptor 3 mobilizes to the surface of human mast cells and potentiates immunoglobulin E-dependent generation of interleukin 13. *Am J Respir Cell Mol Biol* 2003;**28**(4):420−7.

77. Melo RC, et al. Intragranular vesiculotubular compartments are involved in piecemeal degranulation by activated human eosinophils. *Traffic* 2005;**6**(10):866−79.

78. Cheng JF, et al. Dermal eosinophils in atopic dermatitis undergo cytolytic degeneration. *J Allergy Clin Immunol* 1997;**99**(5): 683−92.

79. Spencer LA, et al. A gel-based dual antibody capture and detection method for assaying of extracellular cytokine secretion: EliCell. *Methods Mol Biol* 2005;**302**:297−314.

80. Logan MR, et al. A critical role for vesicle-associated membrane protein-7 in exocytosis from human eosinophils and neutrophils. *Allergy* 2006;**61**(6):777−84.

81. Stow JL, Manderson AP, Murray RZ. SNAREing immunity: the role of SNAREs in the immune system. *Nat Rev Immunol* 2006;**6**(12):919−29.

82. Lacy P, et al. Fusion protein vesicle-associated membrane protein 2 is implicated in IFN-gamma-induced piecemeal degranulation in human eosinophils from atopic individuals. *J Allergy Clin Immunol* 2001;**107**(4):671−8.

83. Feng D, et al. Ultrastructural localization of vesicle-associated membrane protein(s) to specialized membrane structures in human pericytes, vascular smooth muscle cells, endothelial cells, neutrophils, and eosinophils. *J Histochem Cytochem* 2001;**49**(3):293−304.

84. Hoffmann HJ, et al. SNARE proteins are critical for regulated exocytosis of ECP from human eosinophils. *Biochem Biophys Res Commun* 2001;**282**(1):194−9.

85. Logan MR, et al. Expression of eosinophil target SNAREs as potential cognate receptors for vesicle-associated membrane protein-2 in exocytosis. *J Allergy Clin Immunol* 2002;**109**(2): 299−306.

86. Malm-Erjefalt M, Persson CG, Erjefalt JS. Degranulation status of airway tissue eosinophils in mouse models of allergic airway inflammation. *Am J Respir Cell Mol Biol* 2001;**24**(3):352−9.

87. Clark K, et al. Eosinophil degranulation in the allergic lung of mice primarily occurs in the airway lumen. *J Leukoc Biol* 2004;**75**(6): 1001−9.

Release of Lipid Mediators: the Role of Lipid Bodies

Christianne Bandeira-Melo and Patricia T. Bozza

DISTINCTIVE EOSINOPHIL-DERIVED EICOSANOIDS

Eosinophils have the potential to generate and release diverse lipid mediators critical to the development and perpetuation of a number of inflammatory diseases. Among the biologically active lipids that are rapidly generated at sites of inflammation by eosinophils, eicosanoids are derived from the oxidative metabolism of arachidonic acid (AA) and are key molecular targets of commercially available anti-inflammatory drugs, and have been the focus of development of new therapeutic interventions. In eosinophils, as in other leukocytes, to produce eicosanoids, free AA molecules must be metabolized through one of two distinct pathways mediated by either the cyclooxygenase (COX; PGH synthase) or lipoxygenase (LO) enzymes. It has long been established that eosinophils display both enzymatic cascades of AA metabolism and therefore are capable of generating both the LO metabolites, which, in addition to leukotrienes in eosinophils, also include lipoxins and eoxins, as well as COX-driven prostanoids.[1–3]

As the substrate for eicosanoid biosynthesis, AA is found in very low levels in its free form; most is found esterified to the sn-2 position of phospholipids. On cell activation, this pool of AA can be rapidly mobilized through a requisite event performed by a superfamily of enzymes collectively known as secretory phospholipase A2 (PLA2). Of the more than 20 different PLA2s described so far, cytosolic (c)PLA2-alpha, also known as group IVA PLA2, is the only entity known to be selective for AA in the sn-2 position. cPLA2-alpha is the most extensively studied PLA2 and has a central role in AA mobilization in conditions ranging from parturition to allergic reactions. (For a review of PLA2 nomenclature and functions, see[4].) Human eosinophils, in addition to expressing cPLA2-alpha,[5] contain high amounts of the secretory group II PLA2,[6] but lack the endogenous secretory group V PLA2.[7] Although eosinophils contain at least two structurally distinct forms of PLA2, cPLA2-alpha appears to represent the central AA-mobilizing enzyme involved in eicosanoid synthesis within eosinophils. However, it has been also postulated that eosinophils may represent the only cell type in which AA release can also happen by a mechanism entirely independent of cPLA2-alpha activity, but via a paracrine action of group V PLA2.[7–8]

The oxidative metabolism of free AA mediated by one of two cyclooxygenases [COX, prostaglandin G/H synthase (PGH)] or a family of LO enzymes culminate with the generation of a variety of eicosanoids according to cell type and stimulus. Specifically, within human eosinophils, a newly synthesized eicosanoid array may include prostanoids, cysteinyl leukotrienes, eoxins [14,15-leukotrienes (LTs)], lipoxins, and platelet-activating factor.

Prostanoids

Both COX enzymes, nominally the constitutive COX-1 and the inducible COX-2, catalyze the same two reactions: a COX reaction that inserts two molecules of oxygen into substrate AA to form prostaglandin G2 (PGG2) and a subsequent endoperoxidase reaction that reduces PGG2 to its 15-hydroxy analogue, prostaglandin G/H synthase 2 (PGH2; Fig. 8.3.1). Of note, eosinophils contain active COX-1 and may express COX-2 under proper stimulatory conditions. The specific COX pathway-derived eicosanoids synthesized from PGH2 include a series of prostanoids that are defined by the terminal enzymes differentially expressed by different cell types. For instance, platelets express thromboxane A synthase and therefore preferentially synthesize thromboxane B2, endothelial cells produce mostly prostaglandin I2 (prostacyclin) synthase (PGI2) as they express high levels of PGI synthase. Mast cells generate large amounts of prostaglandin-D2 synthase (PGD2 synthase), but not prostaglandin E2 (PGE2), due to selective expression of hematopoietic PGD2 synthase (H-PGD2), while within macrophages, PGH2 is isomerized to PGE2 by a specific terminal PGES enzyme that preferentially combines with the inducible form of COX-2 to generate PGE2 under a range of inflammatory conditions.[9]

What about eosinophils? As illustrated in Fig. 8.3.1, among prostanoids, although direct demonstration of thromboxane-A synthase expression within eosinophils remains to be demonstrated, it has been clearly established that eosinophils produce thromboxane B2. Similarly, there is no report so far documenting the synthesis of any of the three known PGES enzymes—inducible microsomal PGES-1 (mPGES-1) or either of the constitutive forms, mPGES-2 and cPGES[10]—within eosinophils. However, a variety of stimulatory conditions confer on eosinophils the well-documented ability to synthesize PGE2, confirming that, while remaining to be identified, eosinophils likely express at least one terminal PGES enzyme. On the other hand, the presumed lack of PGD2-synthesizing capability by eosinophils relies on anecdotal evidence of lack of hematopoietic prostaglandin D synthase (H-PGDS) expression. It is now clear that PGD2 synthesis is not restricted to mast cells, as it seems to be

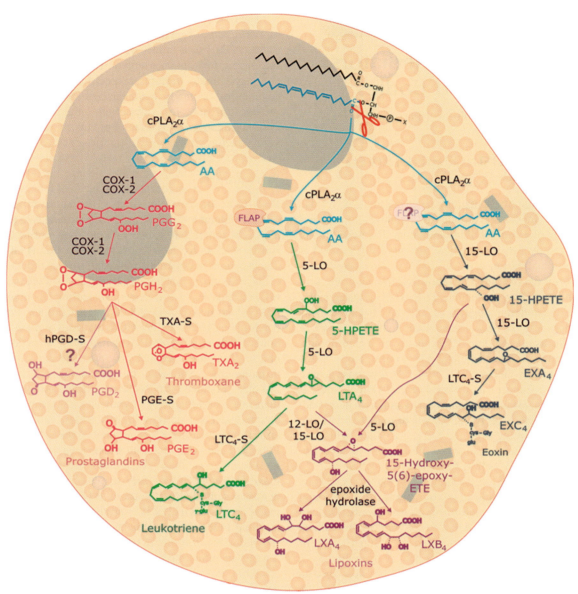

FIGURE 8.3.1 Pathways of eicosanoid synthesis within eosinophils. Eicosanoid synthesis within any cell type begins with the release of arachidonic acid (AA) from a membrane phospholipid by the action of cPLA2-alpha. Within eosinophils, the oxidative metabolism of free AA can be mediated by either COX or LOs, culminating with the generation of large amounts of cysteinyl LTs, in addition to EXC4, LXA4, PGE2, and TXB2. 5-HPETE, 5S-hydroperoxyeicosatetraenoic acid; COX, cyclooxygenase; EXA4, eoxin A4; EXC4, eoxin C4; FLAP (AL5AP), 5-lipoxygenase-activating protein; LO, lipoxygenase; LTC4; leukotriene C4; LXA4, lipoxin A4; LXB4, lipoxin B4; PGD, prostaglandin-D2 synthase; PGE2, prostaglandin E2; PLA2, phospholipase A2; TXA2, thromboxane A2 TXB2, thromboxane B2.

a common feature of other allergy-associated cell types. Of note, it has also been reported that T-helper type 2 (Th2) lymphocytes and skin dendritic cells are capable of synthesizing significant amounts of PGD2.[11] While definitive demonstration of eosinophil-mediated production of significant amounts of biologically relevant PGD2 is lacking, recently H-PGDS was detected by immunological methods within eosinophils found infiltrating human nasal polyps.[12]

Cysteinyl Leukotrienes

As an alternative pathway to COX-mediated oxidation, free AA can be metabolized within eosinophils by two mammalian lipoxygenases—5-LO and 15-LO. These LOs belong to a family of structurally related, lipid-peroxidizing enzymes that are implicated in the pathogenesis of asthma and other inflammatory disorders. Of note, eosinophils are major cellular sources of cysteinyl LTs,[13] which are well known as mediators of the allergic response, causing bronchoconstriction, mucous hypersecretion, increased microvascular permeability, and bronchial hyper-responsiveness, as well as autocrine actions on eosinophils themselves, which serve to regulate infiltration and activation.[14]

Among these LOs, 5-LO is the limiting enzyme of LT synthesis. Like the COX enzymes, 5-LO also catalyzes a two-step reaction (Fig. 8.3.1). First, 5-LO targets free AA in concert with the arachidonate 5-lipoxygenase-activating protein (FLAP) to insert one oxygen molecule into the 5 position of AA and form 5S-hydroperoxyeicosatetraenoic acid (5-HPETE), then transforms 5-HPETE into an unstable allylic epoxide, named leukotriene A_4 hydrolase (LTA$_4$ hydrolase). Similar to cell type-specific differential prostanoid production, the subsequent metabolism of LTA$_4$ hydrolase also differs between leukocytes. In neutrophils, for instance, LTA$_4$ hydrolase enzymatically hydrolyzes 5-LO-metabolite LTA$_4$ to leukotriene B_4 (LTB$_4$). In contrast, in eosinophils, which do not express LTA$_4$ hydrolase and are therefore are incapable of LTB$_4$ synthesis, a specific glutathione S-transferase, named leukotriene C_4 synthase (LTC$_4$ synthase) catalyzes the incorporation of reduced glutathione (a tripeptide composed by glutamic acid, glycine and cysteine) to LTA$_4$ hydrolase to form LTC$_4$ synthase. Regarding LTC$_4$ synthase, it is noteworthy that: (1) LTC$_4$ synthase is an 18 kDa integral membrane protein that belongs to a superfamily of membrane-associated proteins in eicosanoid and glutathione metabolism (MAPEG) that also includes AL5AP (FLAP) and microsomal prostaglandin E synthase 1 (mPGES-1)[15]; (2) it has a more restricted distribution than other eicosanoid-forming enzymes like 5-LO, appearing to be limited to eosinophils, mast cells, basophils, mononuclear phagocytes, and platelets; and (3) in vivo, eosinophils play a central role in LTC$_4$ synthase secretion during allergic inflammatory reactions,[16] in part because of allergy-related eosinophil LTC$_4$ synthase overexpression.[1]

After energy-dependent export, LTC$_4$ synthase is converted to cysteinyl leukotriene D_4 (LTD$_4$) and leukotriene E_4 (LTE$_4$) through sequential enzymatic removal of the glutamic acid by γ-glutamyl transpeptidases and then the glycine by dipeptidases. Therefore, because these LTs share a cysteine, LTC$_4$ synthase and its extracellular derivatives LTD$_4$ and LTE$_4$ are collectively called cysteinyl LTs.

Eoxins (14,15-Leukotrienes)

Contrasting with 5-LO, little is known about the biological function of human 15-LO. With an amino acid sequence similarity of only approximately 40%, two forms of 15-LO exist, named 15-lipoxygenase-1 (15-LO-1) and -2 (15-LO-2). Human eosinophils, similar to airway epithelial cells

and subsets of mast cells and dendritic cells, contain high numbers of 15-LO-1.[17] Via 15-LO-1, the major AA metabolite formed is 15-hydroxyeicosatetraenoic acid (15-HETE), which has been used as a marker for 15-LO activity in vivo, as well as in various cells and tissues in vitro. To form 15-HETE, 15-LO-1 targets free AA and inserts one oxygen molecule into position 15 of AA to form 15(S)-hydroperoxyeicosatetraenoic acid (15-HPETE), which may then undergo reduction to 15-HETE. Alternatively, like 5-LO, 15-LO-1 may also catalyze a second reaction, transformed by dehydration of 15-HPETE into 14,15-epoxy-eicosatetraenoic acid (14,15-LTA$_4$).[18]

15-LO-1 appears to be highly active particularly within eosinophils. In the 1990s Schwenk and colleagues[19] demonstrated that, besides 15-HETE, eosinophils can synthesize 5-oxo-15-hydroxy-6,8,11,13-eicosatetraenoic acid, a distinct 15-LO-1-derived metabolite with potent eosinophil chemotactic activity. More recently, Feltenmark and colleagues[20] reported that eosinophils display an additional 15-LO-initiated biosynthetic pathway that forms the most recently described class of eicosanoids, the cysteinyl 14,15-LTs. As illustrated in Fig. 8.3.1, following the two-step reactions catalyzed by 15-LO-1, 14,15-LTA$_4$ can also be metabolized by LTC$_4$ synthase, that pivotal enzyme for LTC$_4$ biosynthesis. LTC$_4$ synthase catalyzes the incorporation of a glutathione to 14,15-LTA$_4$ to form a cysteinyl 14,15-LTC$_4$. Just like LTC$_4$, 15-LO-1-driven 14,15-LTC$_4$ is actively exported from cells and then metabolized to 14,15-LTD$_4$ and 14,15-LTE$_4$. To avoid confusion with 5-LO-derived cysteinyl LTs, cysteinyl 14,15-LTs are collectively known as eoxins (EXs). Coined after eosinophils, which represent a major cell source of these 15-LO-1-derived cysteinyl products, such nomenclature includes EX(C$_4$), EX(D$_4$), and EX(E$_4$) instead of 14,15-LTC$_4$, 14,15-LTD$_4$, and 14,15-LTE$_4$, respectively.[20]

Little is known about the regulatory events controlling the synthesis of EXs. However, it is already clear that the regulation of 5-LO and 15-LO within eosinophils differs, and that the specific LO pathway to be triggered will depend on stimulatory conditions. It is known that: (1) receptor-mediated activation of human eosinophils is able to trigger the production and release of both EXs and cysteinyl LTs, indicating that, like cysteinyl LTs, EXs can be produced from the endogenous pool of AA in eosinophils; (2) either cysteinyl LTs or EXs are generated in response to exogenous AA; however, such stimulatory condition appears to favor the 15-LO over the 5-LO pathway; and (3) increasing intracellular calcium levels by calcium ionophores, while promptly triggering 5-LO-mediated cysteinyl LTs, fail to activate 15-LO-driven synthesis.

Alternatively, although human 15-LO-1 displays mainly 15-LO activity, like its animal orthologue that is commonly referred to as 12/15-LO (earlier called leukocyte-type 12-LO) and possesses predominantly a 12-LO activity, human 15-LO-1 also exhibit a relatively low 12-LO activity. The 12-LO pathway converts AA, by rearrangement of 12-HPETE, into the hydroxy epoxides hepoxilin A3 (HxA3) and B31 (HxB3). Recently, it has been shown that the 15-LO-1 expressed on human eosinophils also exhibits intrinsic 12-LO-driven HxA3 and HxB31 synthase activities, even though metabolites derived from its 15-LO activity still predominate after incubation with AA.[21]

The physiological and pathological functions played by the 15-LO pathway are poorly defined and emerge as highly complex. While 15-HETE displays well-established chemotactic activities and EXs are emerging as 15-LO-1-derived proinflammatory mediators capable of inducing edema, there is also evidence suggesting that other 15-LO-derived metabolites may display anti-inflammatory functions. The potential 15-LO-1 anti-inflammatory impact appears to depend on 15-LO-1-driven formation of lipoxins.[22]

Lipoxins

Lipoxin A_4 (LXA4) and its positional isomer lipoxin B_4 (LXB4) were identified by Serhan and colleagues[23] as the products of interactions between lipoxygenases.

Lipoxins (LXs) are generated from free AA through the sequential actions of distinct lipoxygenases and subsequent epoxide hydrolase reaction that provides specific trihydroxytetraene-containing eicosanoids. Two major LO-driven enzymatic routes of lipoxin synthesis have been established (Fig. 8.3.1):[1] 15-LO inserts molecular oxygen into AA at the carbon 15 position to produce 15-HETE, which is then metabolized by 5-LO to form first an unstable epoxy intermediate, i.e., 5S,6S-epoxy-15S-hydroxy-ETE and then, depending on the cell type, by activity of specific hydrolases to form either LXA4 or LXB4; or (2) 5-LO converts AA to the epoxide product LTA4, which is then sequentially transformed via a second lipoxygenation by 12- or 15-LO followed by the activity of lipoxin hydrolase enzymes to form LXA4 or LXB4.[24]

Based on the mechanism involving the sequential activities of distinct LOs, it is clear that cells that do not express simultaneously two complementary LO enzymes and are not able to produce lipoxins on their own. Lipoxin synthesis, in contrast to intracellular biosynthesis of prostanoids and LTs, largely depends on cell–cell associations by a process known as transcellular biosynthesis. Among distinct cellular cross talk mechanisms controlling lipoxin transcellular biosynthesis, some are well established: (1) the neutrophil–platelet interaction, by which the LTA4 synthesized within neutrophils is promptly taken up by platelets (which do not express 5-LO) in a transcellular manner and converted by platelet 12-LO to LXA4; (2)

interactions between leukocytes and mucosal surfaces, where concerted actions of epithelial 15-LO-1 (such as in the airway) acting on AA followed by leukocytic 5-LO, generate LXs that are released from mucosal infiltrating leukocytes; and (3) intracellular pathogen and phagocyte, where LTA4 derived from phagocytic 5-LO is catalyzed by pathogen-secreted 15-LO into LXA4, which is then secreted by the infected leukocyte.[22]

There is also an aspirin-triggered pathway leading to the synthesis of the epimeric lipoxins, also called aspirin-triggered lipoxins (ATLs). By this route, aspirin-mediated irreversible acetylation of the active site of COX-2 results in the conversion of AA to 15(R)-HETE, which, when released by endothelial or epithelial cells, may be transformed by leukocyte 5-LO to generate endogenously 15-epi-LXA4 or 15-epi-LXB4, which share the potent anti-inflammatory actions of LXs. Therefore, such an aspirin-driven pathway can also be achieved by cell–cell interactions, which are initiated when activated leukocytes (mainly neutrophils) adhere to the vascular endothelium or to epithelial cells as a consequence of inflammation, resulting in the 15-LO/5-LO transcellular biosynthesis of epi-LXs.

Distinct from neutrophils, platelets, or other cell types, human eosinophils can synthesize LXs and epi-LXs on their own without any need for transcellular biosynthesis mechanisms. As mentioned previously, eosinophils contain both 5-LO and 15-LO-1, and can express COX-2 when activated. Thus, as illustrated in Fig. 8.3.1, both enzymatic pathways of lipoxin synthesis, featuring either 5-LO/15-LO or 12-LO/15-LO/acetylated COX-2 to 5-LO activity routes, may take place within eosinophils. It is noteworthy that although actual demonstration that eosinophils express epoxide hydrolases is still missing, eosinophils have long been recognized as cell sources of LXs. Serhan and colleagues reported that eosinophil-enriched leukocytes from eosinophilic donors when challenged in vitro with the ionophore A23187 produced several LO-derived compounds, including LXA4.[25]

Platelet-Activating Factor

Though not an eicosanoid, platelet-activating factor (PAF) is another potent lipid mediator synthesized by a range of cell types, including monocytes/macrophages, mast cells, platelets, neutrophils, endothelial cells, and eosinophils. PAF is an ether analogue of phosphatidylcholine and contains an acetyl group at its sn-2 position. Under inflammatory conditions PAF is biosynthesized through a two-step remodeling pathway. First, alkyl-phosphatidylcholine is cleaved from membrane phospholipids at its sn-2 position by phospholipase A2 (PLA2), generating lyso-PAF. Within the PLA2 family and akin to the AA-mobilizing activity, it is cPLA2-alpha that plays a major role in inflammatory cell production by PAF. PAF is then synthesized from lyso-PAF by acetyltransferase.[26]

In addition to eosinophils, PAF is produced by many other cells associated with asthmatic inflammation. The biological effects of PAF are mediated by the activation of specific receptors expressed on effector cell surfaces, limited by PAF-specific acetylhydrolase degradation, and include some of the clinical hallmarks of asthma, such as bronchoconstriction, mucus production, and airway hyperresponsiveness (AHR). In addition, PAF has profound chemoattractant properties for eosinophils and neutrophils and promotes an increase in microvascular permeability and edema formation within the airways. As yet unavailable, a specific inhibitor of PAF synthesis could be a valuable and promising complementary therapeutic tool in the future.

REGULATION OF EICOSANOID SYNTHESIS

Distinct from protein molecules, eicosanoid structures are conserved across species, and are not stored but, instead, are synthesized *de novo* after cell activation. It is now well established that effective AA mobilization and metabolism, ensuring successful eicosanoid synthesis, are not merely determined by cellular AA availability and proper expression of the relevant enzymes; eicosanoid-forming enzymes need to be both activated and mobilized toward specific sites where they can interact with their substrates in a cell- and stimulus-dependent manner. Much effort has been made toward understanding the regulatory mechanisms of eicosanoid formation. It is now clear that eicosanoid synthesis is a highly regulated event governed by a variety of both activating and limiting processes, some of which may involve:

1. *Translocation of cytosolic enzymes*: PLA2-alpha and 5-LO are cytosolic enzymes that on cell activation translocate to have access to their selective substrates. Translocation depends on a rise in cytosolic calcium levels that allow these enzymes to be targeted to cellular membranes mediated by their C2 domains.[27]
2. *Enzyme phosphorylation*: Both activation and substrate targeting by at least two key enzymes of eicosanoid synthesis, PLA2-alpha and 5-LO, are controlled by phosphorylation via mitogen-activated protein kinase 2 (MAPK 2).[28] Eosinophil cPLA2 can be regulated both by phosphorylation and increases in intracellular Ca^{2+}.[29] Whether 5-LO becomes phosphorylated in eosinophils as in other cells is not known.
3. *Induced expression of non-constitutive enzymes*: The enhanced cellular capacity of synthesizing eicosanoids is strongly controlled by inducible mechanisms like the induction of expression of COX-2—the non-constitutive isoform of the PGH2 synthesizing enzyme. COX-2 is thought to display inflammatory roles and is expressed only following cellular activation by inflammatory stimuli, such as allergic reactions. While controversy exists, clinical studies showed that the number of eosinophils expressing COX-2 and PGH synthase increases in asthmatic subjects.[30]

4. *Increased expression of constitutive enzymes*: Another regulatory mechanism involved in increased eicosanoid synthesis characteristic of inflammatory conditions is the overexpression of key eicosanoid-forming enzymes. Concerning eosinophils, while several studies aimed to demonstrate altered expression of 5-LO and AL5AP (FLAP) in asthmatic patients, it was the enhanced expression of LTC4 synthase within eosinophils that was striking and which may explain the high levels of cysteinyl LTs found in those patients.[1,31]
5. *Macromolecular organization of multiprotein complexes*: Successful LT synthesis is largely reliant on prearranged multiprotein complexes. Assembly of such functional heteroligomers by the MAPEG AL5AP (FLAP) and LTC4 synthase allows the efficient transfer of LTA4 from the 5-LO-driven reaction to the downstream enzyme. Therefore, the LTC4 synthase-associated AL5AP (FLAP), besides presenting AA to 5-LO, which prevents the diffusion of AA through the cellular membranes, also favors a shift of LTA4 from 5-LO to LTC4 synthase, restricting LTA4 diffusion and/or degradation by other enzymes. AL5AP (FLAP) regulatory functions may also include the ability to *dock* 5-LO after translocation to specific sites of LTC4 synthase activity.[32] It is also noteworthy that similar oligomer-driven organization capable of controlling prostanoid synthesis has not yet been identified.
6. *Differential intracellular compartmentalization of eicosanoid biosynthesis*: In addition to molecular regulation, there is now an evolving understanding that eicosanoid synthesis varies based on specific localization at distinct intracellular compartments; this has emerged as a key regulatory aspect governing eicosanoid formation, which enables functional interactions among biosynthetic enzymes involved in AA release and metabolism, and ultimately controls efficient eicosanoid synthesis and, likely, subsequent eicosanoid functions.

DISTINCT INTRACELLULAR DOMAINS OF EICOSANOID SYNTHESIS

Compartmentalization of eicosanoid-forming enzymes in non-stimulated cells varies depending on the cell type and can be altered by distinct stimulatory conditions. On cell activation, four distinct intracellular domains may compartmentalize such macromolecular machinery:

1. *Nuclear envelope (and contiguous endoplasmic reticulum)*: Whereas plasma membranes were initially assumed to be the sites of the paracrine mediator of eicosanoid synthesis in leukocytes, the finding that the integral membrane protein AL5AP (FLAP) and 5-LO were localized by immunogold electron microscopy to perinuclear membranes in activated human neutrophils and monocytes directed attention to the nuclear envelope as sites of LTs synthesis.[33] Subsequently, LTC$_4$ synthase was also localized to the nuclear envelope.[34] Specifically within eosinophils, such perinuclear localization may also play a role in eicosanoid synthesis. It has been shown that following the activation of eosinophils with A23187, 5-LO protein was immunolocalized in a perinuclear pattern, likely in the nuclear envelope and endoplasmic reticulum.[35] In another study, Cowburn and colleagues reported that the eosinophilopoietic cytokine, interleukin-5 (IL-5), which is known to enhance eosinophil LTC$_4$ release, not only elicited enhanced synthesis of AL5AP (FLAP) in eosinophils, but also elicited increased LTC$_4$ release in response to A23187 and induced the nuclear localization of 5-LO.[36] Indeed, translocation of cytoplasmic 5-LO to the nuclear envelope has emerged as a major candidate mechanism potentially involved in regulating LT formation within leukocytes[37]; however, it is noteworthy that, particularly within eosinophils, this nuclear translocation has been associated with both increased and decreased LTC$_4$ formation.[35,36]

2. *Phagosomal membrane*: By studying the role of group V secretory (s)PLA2 on zymosan phagocytosis by mouse peritoneal macrophages, Balestrieri and colleagues[38] unveiled an unpredicted function of phagosomes. They showed that, though within resting peritoneal macrophages group V, sPLA2 localizes at the Golgi apparatus and recycling endosome. On zymosan intake, group V sPLA$_2$ moves to the phagosomal membrane where cPLA2-alpha, 5-LO, AL5AP (FLAP), and LTC$_4$ synthase are compartmentalized. While such phagosomal macromolecular machinery appears to regulate macrophage phagocytic function, as a non-phagocytic cell, eicosanoid synthesis within eosinophils does not seem to be regulated by this kind of intracellular compartment.

3. *Cytoplasmic lipid body*: Eosinophil lipid bodies contain arachidonyl phospholipids.[39] Enzymes needed to release AA from phospholipids, including cPLA2 and activating MAP kinases, can also be found within lipid bodies.[40] In addition, eosinophil lipid bodies may compartmentalize several of those enzymes needed for eicosanoid synthesis by eosinophils including 5-LO, 15-LO, COX and LTC$_4$ synthase.[16,41] As illustrated in Fig. 8.3.2, although most conventionally thin-sectioned lipid bodies show no sign of internal structure, a honeycombed, membranous matrix or individual scattered membranous structures can be imaged inside eosinophil lipid bodies intermingling with the lipid content[41] (Melo et al., unpublished data). In support of the presence of an internal structure, concentrically arranged lamellar structures or layers were recently revealed by freeze-fracture at varying depths within lipid body cores from cultured smooth muscle cells.[42] The functional meaning of internal membranes within lipid bodies remains to be defined, but may explain how key integral membrane proteins, like AL5AP (FLAP) and LTC$_4$ synthase, enzymes with obligatory membrane anchorage, like COX-1 and COX-2, or inducible membrane translocation, like cPLA2-alpha and 5-LO, can be localized within lipid bodies (Fig. 8.3.2).

In agreement with the role played by lipid body bodies in eicosanoid synthesis, studies using enucleated eosinophils (eosinophilic cytoplasts) ruled out the functional relevance of perinuclear eicosanoid synthesis within eosinophils.[41] Eosinophilic cytoplasts were capable of assembling new lipid bodies in response to PAF stimulation, and these lipid bodies contained immunolocalized 5-LO, COX, and LTC$_4$ synthase. Moreover, in PAF-primed, nucleus-free eosinophilic cytoplasts, the increased numbers of lipid bodies strongly correlated with the increased production of both LTC$_4$ and PGE2 following stimulation with calcium ionophore.

EOSINOPHIL LIPID BODIES AS ACTIVE DOMAINS OF EICOSANOID SYNTHESIS

As mentioned previously, intracellular compartmentalization of the eicosanoid synthesis machinery has emerged as a key component of the regulation of eicosanoid synthesis (reviewed in[43]). However, the direct evaluation of specific subcellular locales of eicosanoid synthesis has been elusive, as eicosanoids are newly formed, not stored, and often rapidly released on cell stimulation. Indeed, in the majority of studies, intracellular sites of eicosanoid synthesis have been inferred based on the identification of eicosanoid-forming enzymes localization. Therefore, just by detecting eicosanoid-forming enzymes within discrete subcellular structures, one cannot assume that those sites are indeed accountable for the efficient and enhanced eicosanoid synthesis observed during inflammatory responses. The immunolocalization of eicosanoid-forming proteins does not necessarily ascertain that the localized protein is functional and activated to synthesize a specific eicosanoid lipid at an intracellular site. Using a strategy that cross-links, captures, and localizes newly formed eicosanoids at their sites of synthesis, a more direct approach to

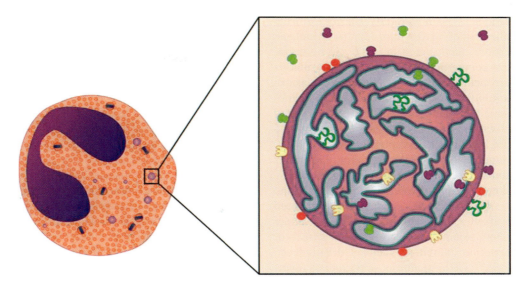

SYMBOL	PROTEIN	REFERENCE
(red dot)	ADRP	Vieira-de-Abreu *et al.*, 2005 Mesquita-Santos *et al.*, 2006 Vieira-de-Abreu *et al.*, 2010
(purple)	5-LO	Bozza *et al.*, 1997 Bozza *et al.*, 1998 Brock *et al.*, 1999 Vieira-de-Abreu *et al.*, 2005
(green)	15-LO	Bozza *et al.*, 1998
(green)	COX	Dvorak *et al.*, 1994 Bozza *et al.*, 1997
(yellow)	LTC₄-S	Bozza *et al.*, 1997

FIGURE 8.3.2 **Compartmentalization and organization of eicosanoid-forming proteins within eosinophil lipid bodies.** Lipid bodies are composed of an outer phospholipid monolayer and a lipid-rich core, currently thought to be composed of a system of endoplasmic reticulum (ER)-like invaginations comprising double-layer membranes that enclose hydrophilic lumina. Within the organelle, distinct proteins target different subcompartments. PAT (perilipin, ADRP and TIP) family member adipose differentiation-related protein (ADRP) is mostly found on the external lipid body membrane. When intracellular calcium levels rise, cytosolic proteins like cPLA2-alpha, 5-LO, and 15-LO are translocated to the lipid body membrane, but can also be found within the organelle's core. Other proteins involved in eicosanoid synthesis, like the membrane-anchored COX and the membrane-spanning structurally related LTC₄ synthase, reside not only on the outer membrane, but also within the core of the lipid body. COX, cyclooxygenase; LO, lipoxygenase; LTC₄, leukotriene C₄; PLA2, phospholipase.

detect the intracellular sites of AA-derived lipid mediator formation in leukocytes and other cell types has been developed. Building on such a novel strategy for *in situ* immunolocalization of newly formed eicosanoids to ascertain the intracellular compartmentalization of their synthesis—the EicosaCell microscopic assay—a modification of a prior technique was used.[44] The EicosaCell rationale relies on the specific features of the hetero-bifunctional cross-linker 1-ethyl-3-[3-dimethylamino-propyl]carbodiimide (C8H17N3-HCl; EDAC). EDAC immobilizes newly synthesized eicosanoids by cross-linking the eicosanoid carboxyl groups to the amines of adjacent proteins localized at the eicosanoid-synthesizing compartment. Besides the coupling of an immunodetectable eicosanoid at its site of formation, EDAC enables: (1) the ending of the cell stimulation step; (2) cell fixation; (3) cell permeabilization, allowing the penetration of both antieicosanoid and detecting fluorochrome-conjugated antibodies into cells; and, importantly, (4) the relative preservation of lipid domains, such as membranes and droplets (such as lipid bodies), which dissipate with air-drying or commonly used alcohol fixation.

By employing the EicosaCell technique, it has been possible to confirm the dynamic aspect involved in the localization of eicosanoid synthesis, providing direct evidence of compartmentalization within three distinct sites. The perinuclear envelope,[45] phagosomes,[38] or lipid bodies[45,46] were found to compartmentalize eicosanoid synthesis, in accord to the specific cell type and stimulatory condition studied. So far, the EicosaCell assay has been used to identify LTC production,[16,38,45,46] LTB₄,[46,47] PGE2[45,46] and PGD2 (unpublished observations) in different cell types and under different stimulatory conditions.

Specifically within eosinophils, the EicosaCell assay shown that human eosinophils stimulated with A23187 synthesize LTC₄ and that synthesis takes place at the perinuclear membrane,[45,48] which is fully in accord with the perinuclear localization of 5-LO within A23187-stimulated eosinophils. Notably, there are also more physiological stimuli that lead to compartmentalization of LTC$_4$ synthesis to the perinuclear membrane of human eosinophils, as can be seen after cross-linking surface leukocyte immunoglobulin-like receptor 7 (LIR-7) or CD9.[45] In contrast, eosinophils stimulated with eotaxin or C-C motif chemokine 5 (CCL5/RANTES), rather than supraphysiological A23187, were shown to synthesize LTC$_4$ mainly within lipid bodies and only occasionally around the perinuclear membrane.[45] After co-stimulation of eosinophils with A23187 and eotaxin or CCL5 (RANTES), LTC$_4$ synthesis was enhanced but remained localized mainly within lipid bodies.[45] Under *in vivo* inflammatory conditions, such as allergic, PGD2- or macrophage migration inhibitory factor (MIF)-induced pleurisy models,[16,46] LTC$_4$ synthesis appeared compartmentalized to lipid bodies and did not occur in any other subcellular compartment. Interestingly, and as mention previously, EicosaCell confirmed that zymosan-stimulated peritoneal macrophages synthesize LTC$_4$ at the phagosomal membrane rather than lipid bodies or nuclear envelope.[38] Eicosanoids other than LTC$_4$ have also been detected inside the lipid bodies of other cell types by EicosaCell assay, together with the respective eicosanoid-forming enzymes, contributing to the appreciation that eicosanoid synthesis within these organelles is subject to tight regulation of cell-specific signaling (for a review see[48]).

The high sensitivity of the EicosaCell assay allows the detection of low levels of intracellularly generated eicosanoids even when extracellularly released eicosanoids could not be detected by conventional eicosanoid enzyme immune assay. For that reason, by using the EicosaCell assay, it was possible to discover that lipid body-derived LTC$_4$ has intracellular functions in controlling cytokine release from eosinophils.[48] Therefore, by identifying compartmentalized levels of eicosanoids, besides providing new insights into the regulation of eicosanoid biosynthesis, the EicosaCell assay may contribute to the identification of likely intracrine functions of newly synthesized eicosanoids.

LIPID BODY BIOGENESIS AS REGULATORY EVENT OF EICOSANOID SYNTHESIS

In addition to localization of eicosanoid-forming enzymes within eosinophil lipid bodies and the definitive EicosaCell-driven demonstration that these organelles are active sites of eicosanoid synthesis, it has also been shown that increases in lipid body numbers correlated with increases in LTC$_4$ and PGE2 released by these cells. For instance, eosinophils first primed to form lipid bodies with increasing concentrations of PAF or *cis*-unsaturated fatty acids

released correspondingly increased amounts of LTC$_4$ and PGE2 upon submaximal calcium ionophore activation.[41,49] Accordingly, in a murine model of allergen-driven eosinophilic inflammation with increased cysteinyl LTs levels, infiltrating eosinophils displayed elevated numbers of cytoplasmic lipid bodies.[16] Indeed, lipid bodies were shown to increase both in characteristic size and number *in vivo* in cells associated with human inflammatory diseases; this included blood eosinophils from patients with hypereosinophilic syndrome (HES) and tissue eosinophils in biopsies from Crohn's disease.[41,50]

Therefore, the evolving understanding of lipid body biogenic features, including the distinction between different cell types, specific stimuli, key signaling pathways, and induction during *in vivo* inflammatory reactions, is critical to the increased understanding of the regulation of eicosanoid synthesis. In this regard, it is noteworthy that much of the current knowledge on lipid body role in eicosanoid synthesis and key aspects of leukocyte lipid body biogenesis were achieved using human eosinophils as the cell model. Of special relevance to allergic inflammation, key mediators of the allergic response, in addition to PAF, are capable of eliciting newly formed lipid bodies within human eosinophils. IL-5 alone or combined with granulocyte-macrophage colony-stimulating factor (GM-CSF) in the absence of exogenous lipids, as well as immobilized immunoglobulin G (IgG), lead to a significant increase in lipid body numbers,[41,51] suggesting that receptor-mediated stimuli result in intracellular lipid remodeling and lipid body formation. Moreover, chemokines acting via C-C chemokine receptor type 3 (CCR3) receptors, including CCL5 (RANTES), eotaxin, eotaxin-2, and eotaxin-3, can initiate intracellular signaling in eosinophils, but not neutrophils, that trigger the formation of lipid bodies.[45,52] More recently, it was also shown that PGD2 and MIF also trigger lipid body biogenesis within both *in vitro*-stimulated human eosinophils and eosinophils recruited *in vivo* to sites of allergic inflammation.[53]

Lipid bodies assembled in response to all these diverse stimuli display similar size, number, localization, and morphology within eosinophils. However, an important observation made by studying lipid body biogenesis within eosinophils uncovered that the intracellular signaling pathways leading to rapid lipid body assembly varies according to the different receptor engaged. As summarized in Fig. 8.3.3, PAF, acting via its G protein-coupled receptor, induces lipid body formation via downstream signaling involving protein kinase C (PKC) activation.[41,54] Of note, other agonists of the G protein-coupled receptors IL-8, component complement C5a, and LTB$_4$ did not induce eosinophil lipid body formation.[41,54] Unlike the PAF-engaged signaling pathway, eotaxin- and CCL5 (RANTES)-activating CCR3 receptor signals through phosphatidylinositol 3-kinase (PI3K) and the extracellular signal-regulated kinase 1 and 2 (ERK1/2), and p38 MAP kinases.[45] While the MIF-related signaling pathway that induces lipid body assembly is still unknown, it has been shown that a cross talk between the MIF receptor CD74 and CCR3 activation by an autocrine/paracrine activity of eotaxin takes place. It is noteworthy that although due to different types of stimulation and subsequent signaling pathways, increases in lipid body numbers triggered by PAF, eotaxins, CCL5 (RANTES), or MIF correlated with increased LTC_4 released by these cells.

In contrast to such correlative increases in lipid body formation and LTC_4 release, under PGD2 stimulation, a new and unexpected intracellular pathway was recently found to control lipid body biogenesis within eosinophils. PGD2 signaling through its prostaglandin D2 receptor (DP1) coupled to $G\alpha_s$ proteins leads to increases in cyclic adenosine monophosphate (cAMP) and protein kinase A (PKA) activity—an intracellular signal transducing cascade that is classically related to the inhibition of chemoattractant-induced eosinophil motility,[55] and consistent with the idea of cAMP elevating agents as powerful anti-inflammatory[56] or proresolution[57] tools for the treatment of diseases in which eosinophil accumulation is thought to play a key role.[57] In this context and in contrast to PAF- or eotaxin-induced LTC_4-synthesizing lipid bodies, it was congruent that DP1-driven, PKA-dependent newly formed lipid bodies were found to be non-competent LTC_4 synthesizing organelles. The enhanced PGD2-elicited LTC_4 synthesis, besides being dependent on the activation of DP1-elicited PKA-regulated lipid bodies, needs equally

important and concomitant DP2-elicited $G\alpha_i$/calcium-regulated signaling pathway, which prompts DP1-driven newly formed lipid bodies to synthesize enhanced amounts of LTC_4. Collectively, these findings demonstrate that eicosanoid synthesis and, perhaps, other lipid body functions are directly regulated by a specific type of stimulation and signaling, which controls their assembly.

Eosinophil lipid bodies are crucial contributors to allergic inflammatory disorders, such as asthma, since lipid bodies control key aspects of the prevalent pathology, i.e., the synthesis of cysteinyl leukotrienes. Therefore, lipid body biogenesis within eosinophils should emerge as a key target for anti-allergic treatment. In the future, lipid body-specific inhibitors may provide further therapeutic options in the management of allergic airway diseases.

CONCLUSIONS

In the last years, a growing body of data has been generated that has to some degree delineated the mechanisms governing eosinophil-driven differential synthesis and release of eicosanoids with functions ranging from pro-inflammatory to anti-inflammatory or proresolution. Throughout this effort, it was learned that the generation of eicosanoids within eosinophils is important not only for the roles of eicosanoids as extracellular paracrine mediators of inflammation, but also for their roles as intracrine signal transducing molecules that may regulate fundamental cellular responses within eosinophils. Currently, the main therapeutic agents used in eosinophil-related pathologies are inhibitors of 5-LO or cysteinyl LT1 receptors that may

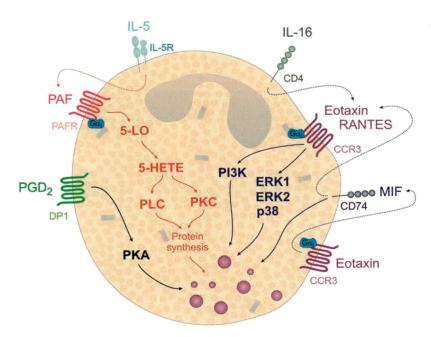

FIGURE 8.3.3 **Receptor-induced signaling pathways of eosinophil lipid body biogenesis.** Different stimuli, including protein and lipid agonists, by triggering a variety of downstream signaling leads to lipid body formation within eosinophils. 5-HETE, 15-hydroxyeicosatetraenoic acid; RANTES (CCL5), C-C motif chemokine 5; CCR3, C-C chemokine receptor type 3; DP1, prostaglandin D2 receptor; ERK1/2, extracellular signal-regulated kinase 1 and 2; IL-5, interleukin-5; IL-5RA, interleukin-5 receptor subunit alpha; IL-16, interleukin-16; LO, lipoxygenase; MIF, macrophage migration inhibitory factor; PAF; platelet-activating factor; PAF-R, platelet-activating factor receptor; PGD_2, prostaglandin-D2 synthase; phosphatidylinositol 3-kinase; PI3K, phosphatidylinositol 3-kinase; PKA, protein kinase A; PKC, protein kinase C; PLC, phospholipase C.

have activities that are broader than simply blocking the paracrine mediator activities of cysteinyl LTs. Moreover, it is clear that newer therapeutic targets directed against other bioactive eosinophil-derived eicosanoids may be unveiled, with the potential to have an impact on asthma pathology and the pathology of other allergic diseases, rather than just eosinophil-derived LTC$_4$. Improved therapeutics will be developed once we have a better understanding of the complete eicosanoid array derived from eosinophils, the intracellular sites of eosinophil eicosanoid synthesis, and the functional consequences of eicosanoid interactions with their functional receptors.

REFERENCES

1. Cowburn AS, Sladek K, Soja J, Adamek L, Nizankowska E, Szczeklik A, et al. *J Clin Invest* 1998;**101**:834.
2. Bandeira-Melo C, Singh Y, Cordeiro RS, e Silva PM, Martins MA. *Br J Pharmacol* 1996;**118**:2192.
3. Bandeira-Melo C, Woods LJ, Phoofolo M, Weller PF. *J Exp Med* 2002;**196**:841.
4. Schaloske RH, Dennis EA. *Biochim Biophys Acta* 2006;**1761**:1246.
5. Zhu X, Munoz NM, Rubio N, Herrnreiter A, Mayer D, Douglas I, et al. *J Immunol Methods* 1996;**199**:119.
6. Blom M, Tool AT, Wever PC, Wolbink GJ, Brouwer MC, Calafat J, et al. *Blood* 1998;**91**:3037.
7. Munoz NM, Kim YJ, Meliton AY, Kim KP, Han SK, Boetticher E, et al. *J Biol Chem* 2003;**278**:38813.
8. Wijewickrama GT, Kim JH, Kim YJ, Abraham A, Oh Y, Ananthanarayanan B, et al. *J Biol Chem* 2006;**281**:10935.
9. Fahmi H. *Curr Opin Rheumatol* 2004;**16**:623.
10. Hara S, Kamei D, Sasaki Y, Tanemoto A, Nakatani Y, Murakami M. *Biochimie* 2010;**92**:651.
11. Matsuoka T, Hirata M, Tanaka H, Takahashi Y, Murata T, Kabashima K, et al. *Science* 2000;**287**:2013.
12. Hyo S, Kawata R, Kadoyama K, Eguchi N, Kubota T, Takenaka H, et al. *Arch Otolaryngol Head Neck Surg* 2007;**133**:693.
13. Weller PF, Lee CW, Foster DW, Corey EJ, Austen KF, Lewis RA. *Proc Natl Acad Sci USA* 1983;**80**:7626.
14. Austen KF, Maekawa A, Kanaoka Y, Boyce JA. *J Allergy Clin Immunol* 2009;**124**:406.
15. Bresell A, Weinander R, Lundqvist G, Raza H, Shimoji M, Sun TH, et al. *FEBS J* 2005;**272**:1688.
16. Vieira-de-Abreu A, Assis EF, Gomes GS, Castro-Faria-Neto HC, Weller PF, Bandeira-Melo C, et al. *Am J Respir Cell Mol Biol* 2005;**33**:254.
17. Turk J, Maas RL, Brash AR, Roberts 2nd LJ, Oates JA. *J Biol Chem* 1982;**257**:7068.
18. Maas RL, Brash AR. *Proc Natl Acad Sci USA* 1983;**80**:2884.
19. Schwenk U, Morita E, Engel R, Schroder JM. *J Biol Chem* 1992;**267**:12482.
20. Feltenmark S, Gautam N, Brunnstrom A, Griffiths W, Backman L, Edenius C, et al. *Proc Natl Acad Sci USA* 2008;**105**:680.
21. Nigam S, Zafiriou MP, Deva R, Ciccoli R, Roux-Van der Merwe R. *FEBS J* 2007;**274**:3503.
22. Bafica A, Scanga CA, Serhan C, Machado F, White S, Sher A, et al. *J Clin Invest* 2005;**115**:1601.
23. Serhan CN, Hamberg M, Samuelsson B. *Proc Natl Acad Sci USA* 1984;**81**:5335.
24. Romano M, Serhan CN. *Biochemistry* 1992;**31**:8269.
25. Serhan CN, Hirsch U, Palmblad J, Samuelsson B. *FEBS Lett* 1987;**217**:242.
26. Prescott SM, Zimmerman GA, Stafforini DM, McIntyre TM. *Annu Rev Biochem* 2000;**69**:419.
27. Leslie CC, Gangelhoff TA, Gelb MH. *Biochimie* 2010;**92**:620.
28. Werz O, Burkert E, Fischer L, Szellas D, Dishart D, Samuelsson B, et al. *FASEB J* 2002;**16**:1441.
29. Zhu X, Sano H, Kim KP, Sano A, Boetticher E, Munoz NM, et al. *J Immunol* 2001;**167**:461.
30. Sousa A, Pfister R, Christie PE, Lane SJ, Nasser SM, Schmitz-Schumann M, et al. *Thorax* 1997;**52**:940.
31. Seymour ML, Rak S, Aberg D, Riise GC, Penrose JF, Kanaoka Y, et al. *Am J Respir Crit Care Med* 2001;**164**:2051.
32. Mandal AK, Skoch J, Bacskai BJ, Hyman BT, Christmas P, Miller D, et al. *Proc Natl Acad Sci USA* 2004;**101**:6587.
33. Woods JW, Evans JF, Ethier D, Scott S, Vickers PJ, Hearn L, et al. *J Exp Med* 1993;**178**:1935.
34. Penrose JF, Spector J, Lam BK, Friend DS, Xu K, Jack RM, et al. *Am J Respir Crit Care Med* 1995;**152**:283.
35. Brock TG, Anderson JA, Fries FP, Peters-Golden M, Sporn PH. *J Immunol* 1999;**162**:1669.
36. Cowburn AS, Holgate ST, Sampson AP. *J Immunol* 1999;**163**:456.
37. Peters-Golden M, Brock TG. *Am J Respir Crit Care Med* 2000;**161**:S36.
38. Balestrieri B, Hsu VW, Gilbert H, Leslie CC, Han WK, Bonventre JV, et al. *J Biol Chem* 2006;**281**:6691.
39. Weller PF, Monahan-Earley RA, Dvorak HF, Dvorak AM. *Am J Pathol* 1991;**138**:141.
40. Yu W, Bozza PT, Tzizik DM, Gray JP, Cassara J, Dvorak AM, et al. *Am J Pathol* 1998;**152**:759.
41. Bozza PT, Yu W, Penrose JF, Morgan ES, Dvorak AM, Weller PF. *J Exp Med* 1997;**186**:909.
42. Robenek MJ, Severs NJ, Schlattmann K, Plenz G, Zimmer KP, Troyer D, et al. *Faseb J* 2004;**18**:866.
43. Bozza PT, Magalhaes KG, Weller PF. *Biochim Biophys Acta* 2009;**1791**:540.
44. Liu LX, Buhlmann JE, Weller PF. *Am J Trop Med Hyg* 1992;**46**:520.
45. Tedla N, Bandeira-Melo C, Tassinari P, Sloane DE, Samplaski M, Cosman D, et al. *Proc Natl Acad Sci USA* 2003;**100**:1174.
46. Mesquita-Santos FP, Vieira-de-Abreu A, Calheiros AS, Figueiredo IH, Castro-Faria-Neto HC, Weller PF, et al. *J Immunol* 2006;**176**:1326.
47. Silva AR, Pacheco P, Vieira-de-Abreu A, Maya-Monteiro CM, D'Alegria B, Magalhaes KG, et al. *Biochim Biophys Acta* 2009;**1791**:1066.
48. Bandeira Melo C, Weller PF, Bozza PT. *Methods Molec Biol* 2010. in press.
49. Bozza PT, Payne JL, Morham SG, Langenbach R, Smithies O, Weller PF. *Proc Natl Acad Sci USA* 1996;**93**:11091.
50. Solley GO, Maldonado JE, Gleich GJ, Giuliani ER, Hoagland HC, Pierre RV, et al. *Mayo Clin Proc* 1976;**51**:697.
51. Bartemes KR, McKinney S, Gleich GJ, Kita H. *J Immunol* 1999;**162**:2982.

52. Bandeira-Melo C, Herbst A, Weller PF. *Am J Respir Cell Mol Biol* 2001;**24**:653.

53. Vieira-de-Abreu A, Calheiros AS, Mesquita-Santos FP, Magalhaes ES, Mourao-Sa D, Castro-Faria-Neto HC, et al. *Am J Respir Cell Mol Biol*; 2010.

54. Bozza PT, Payne JL, Goulet JL, Weller PF. *J Exp Med* 1996;**183**:1515.

55. Hirai H, Tanaka K, Yoshie O, Ogawa K, Kenmotsu K, Takamori Y, et al. *J Exp Med* 2001;**193**:255.

56. Teixeira MM, Williams TJ, Hellewell PG. *Eur J Pharmacol* 1995;**272**:185.

57. Sousa LP, Carmo AF, Rezende BM, Lopes F, Silva DM, Alessandri AL, et al. *Biochem Pharmacol* 2009;**78**:396.

58. Bozza PT, Yu W, Cassara J, Weller PF. *J Leukoc Biol* 1998;**64**:563.

59. Dvorak AM, Morgan ES, Tzizik DM, Weller PF. *Int Arch Allergy Immunol* 1994;**105**:245.

Chapter 8.4

Gene-Knockout Mice: What Can They Teach Us about Sequelae Resulting from the Absence of Specific Eosinophil Cationic Proteins?

Joseph H. Butterfield

EOSINOPHIL SECRETORY PRODUCTS AND GRANULE-DERIVED PROTEINS

Eosinophils produce a wealth of secretory products that include 13 interleukins, three reactive oxygen intermediates, 10 lipid mediators, four enzymes, seven chemokines, six growth factors, and others, including granulocyte-macrophage colony-stimulating factor (GM-CSF), interferon gamma (IFN-γ), and tissue necrosis factor alpha (TNF-α).[1,2] These products promote eosinophil participation in multiple inflammatory processes. Eosinophils possess receptors for many of the cytokines that they synthesize, such as GM-CSF, interleukin-3 (IL-3) and 5 (IL-5), and platelet-activating factor (PAF), thereby establishing autostimulatory loops enhancing eosinophil activation, survival, migration, and hematopoiesis above and beyond the other effects of these cytokines in the extracellular milieu.[3,4]

In addition to these products, eosinophils synthesize and secrete unique cationic proteins that are localized in the eosinophil secondary granules. These include eosinophil-derived neurotoxin (EDN), eosinophil peroxidase (EPO), eosinophil cationic protein (ECP), eosinophil granule major basic protein (MBP-1), and major basic protein homolog.[1,5] Among the 17 most abundant transcripts of IL-5-differentiated umbilical cord mononuclear cells, five were transcripts of granule proteins, and MBP-1 messenger RNA was the most abundant (8.12%).[5]

The cytoplasm of eosinophils contains the coarse secondary granules that stain bright orange-red with acidic dyes such as eosin, whence derives the cell's name—eosin + phil (*lover of*). On an ultrastructural level, eosinophil's secondary granules are between 0.5 and 0.8 μm in diameter and are comprised of an electron-dense core surrounded by an electron-lucent matrix, an appearance that is unique on electron microscopy.[6] MBP-1 composes the core of the eosinophil granule, and the other cationic proteins reside in the surrounding matrix.

MAJOR BASIC PROTEIN-1

MBP-1 is an arginine-rich, 13.8 kDa, 117 amino acid protein with an isoelectric point unable to be accurately measured because of the extreme basic nature of the protein.[2,7] MBP-1 is first synthesized in a prepro form in eosinophil promyelocytes with subsequent processing to pro-MBP-1, transport to immature granules, cleavage to MBP-1, and condensation to develop the mature crystalline core.[8,9] MBP-1 makes up an enormous fraction, up to 50%, of an eosinophil's total cellular protein.[7] Naturally, interest in this protein's functions has been intense for years.

Hardly a month goes by that a new injurious or inflammatory effect of MBP-1 is not discovered. Beginning with guinea pig MBP-1,[10] the cytotoxic effect of this protein has repeatedly been reported in *in vitro* assays against various targets. Additionally, by the use of immunofluorescent staining, extracellular MBP-1 deposition has been detected in pathological specimens from diverse clinical diseases, a finding thereby providing evidence for eosinophil degranulation at sites where intact eosinophils are absent by hematoxylin-eosin staining. This staining technique serves to confirm eosinophil participation that otherwise could have been overlooked by the use of traditional stains.[11] Much of the interest in the effects of MBP-1 has centered on its contribution to the pathogenesis of asthma, its involvement in the defense against parasites, and its role in intestinal disorders.

During the past 40 years, the role of eosinophils in the pathogenesis of asthma has undergone substantial revision. Before 1980, eosinophils were thought to be important for neutralizing mast cell mediators of anaphylaxis.[12] Subsequent studies beginning with the study of MBP-1 seemed to contradict this role by showing that, among other actions, MBP-1 was toxic to the respiratory epithelium,[10] was deposited in the lung tissue of patients with asthma,[11] was increased in the sputum of patients with asthma,[13] caused hyperreactivity of respiratory smooth muscle,[14] and allosterically antagonized inhibitory muscarinic M2 receptors in the guinea pig airway and thereby increased vagal responsiveness and bronchial hyperreactivity.[15,16] Neutralization of MBP-1 *in vivo* inhibited antigen-induced bronchial hyperreactivity in a sensitized guinea pig model.[17] Moreover, substantial hypereosinophilia is found in Churg–Strauss syndrome in which asthma, often severe, is a prominent component.[18] Eosinophil-associated inflammation has been comprehensively reviewed.[1]

Helminth infections of humans include infestations by nematodes (roundworms), trematodes (flukes), and cestodes (tapeworms) and are host specific. Marked eosinophilia, increased total and parasite-specific immunoglobulin E levels, and mastocytosis are hallmarks of infection with helminths, especially during acute infection and tissue migratory phases.[19] Eosinophils are capable of killing the larvae of many helminths.[20,21] *In vitro*, MBP-1 has been shown to be a potent helminthotoxin to newborn larvae of *Trichinella spiralis*,[22] microfilariae of *Brugia pahangi* and *Brugia malayi*,[23] and schistosomula of *Schistosoma mansoni*.[24] Nonetheless, the role of the eosinophil *in vivo* as an effector cell in helminth infections and, within that context, the role of eosinophil MBP-1 have been re-examined. Recent evidence suggests that the combination of eosinophils, antibody, and complement has a role in the killing of the infective stages of helminth parasites; however, eosinophils do not seem to have a major role in the rejection of adult parasites.[25]

The gastrointestinal tract is a location in which resident eosinophils and eosinophil degranulation are normally observed; however, tissue procurement methods can affect the degranulation that is seen. More degranulation is present in specimens obtained with endoscopic forceps than in those obtained with a scalpel.[26] Eotaxin is an eosinophil-specific chemoattractant constitutively expressed in the normal large and small intestines, and eotaxin messenger RNA is upregulated in lesions of ulcerative colitis and Crohn's disease; as such, it provides a mechanism for the increased eosinophil infiltration common to these disorders.[27] Deposition of eosinophil granule products has been found in celiac disease (MBP-1)[28,29] and eosinophilic gastroenteritis.[29,30] Eosinophilia may accompany inflammatory bowel disease or sclerosing cholangitis.[31] In an *in vitro* model using T84 human colonic carcinoma epithelial cells, eosinophils reduced colonic epithelial barrier function; in this same model, MBP-1 was shown to cause a time-dependent loss of occludin RNA, an apical tight junction protein.[32]

Consequences of the Absence of Eosinophil Granule Major Basic

Protein-1

The eosinophil's role as an end-stage cell primarily causing cytotoxicity of various targets by degranulation and release of its cationic granule proteins is currently being reassessed.[33,34] This change in perspective to a nuanced view of the eosinophil's function in disease pathogenesis has come about for several reasons. These, in part, include the failure of anti-IL-5 treatment to benefit patients with asthma,[35,36] the difficulty of reconciling the *in vitro* demonstrated ability of eosinophils and eosinophil granule proteins to kill early parasitic forms (e.g., the schistosomula of *S. mansoni*), and the inability of anti-IL-5 treatment to affect the immunity to schistosome or *T. spiralis* infections in mice.[25,37,38] A more overarching problem, however, may be the use of mice as *in vivo* models for examining the eosinophil's role in parasite immunity. Helminth parasites are host specific, and mice are not the natural host species of most nematode parasites. As nonpermissive hosts, mice can make enhanced innate immune responses, such as complement activation or generation of T-helper type 1 (T_h-1) responses, which can abrogate parasitic infections.[25]

There are no clinical reports of patients who have eosinophils that lack MBP-1. However, the murine knockout strain lacking MBP (mMBP$^{-/-}$) has afforded an insight into the contribution of eosinophils and, specifically, MBP-1 to the pathophysiology of several disorders. The characteristics of eosinophils from mMBP$^{-/-}$ mice are listed in Table 8.4.1.

Asthma

A murine model of allergic asthma using ovalbumin (OVA) sensitization by intraperitoneal injection with OVA plus an adjuvant followed by several OVA aerosol challenges has been used by many investigators to examine the participation of eosinophils in the pathogenesis of allergic asthma.[39,40] The use of this model with wild-type and mMBP-1$^{-/-}$ mice found that both strains had indistinguishable selective recruitment of eosinophils to the airways. Trafficking to the interstitial regions of the lung in mMBP-1$^{-/-}$ mice was also unaffected. Moreover, mMBP-1$^{-/-}$ mice had histopathological findings identical to those of wild-type mice in response to OVA challenge (bronchiolar-associated leukocyte aggregates, airway epithelial cell hypertrophy, and goblet cell hyperplasia) and identical airway epithelial mucous content. Methacholine dose response assays showed that wild-type mice needed a lower threshold dose of methacholine to induce equivalent airflow changes than did mMBP-1$^{-/-}$ mice. Additionally, the absolute magnitude of the airflow changes was significantly ($p < 0.05$) higher in wild-type animals than in mMBP-1$^{-/-}$ mice.[41] In contrast, there was no difference between the two mouse groups in the response to another nonspecific stimulus, serotonin. The findings in this model of allergic asthma suggest that establishing a causal link between

TABLE 8.4.1 Comparison of Eosinophils from Homozygous mMBP-1$^{-/-}$ Knockout Mice with those from Wild-Type Mice

Observation	Mouse Type	
	mMBP-1$^{-/-}$ Knockout	Wild-Type
Number of secondary granules	Unaffected	Normal
Size of secondary granules	Unaffected	Normal
Intensity of Wright–Giemsa staining color	Less intense than wild-type	Intense
Electron-dense cores in secondary granules	Absent	Present

MBP-1, eosinophil granule major basic protein 1.

eosinophil degranulation and the development of multiple measures of airway hyperresponsiveness (eosinophil infiltration, inflammatory histopathological changes, epithelial mucous content, and response to nonspecific stimuli), at least in the mouse, continues to prove to be an elusive goal.[41]

Parasites

MBP-1$^{-/-}$ mice have also been used to examine host protection in parasitic infections. In one report, the larval recoveries of animals injected intraperitoneally with 50 *Brugia pahangi* third-stage (L3) larvae organisms were, after 2 weeks, comparable to wild-type mice and mMBP-1$^{-/-}$ knockout mice. Although mice lacking EPO$^{-/-}$ had an approximately 50% reduction in their eosinophil numbers in this system, there was no defect in eosinophil numbers in the mMBP-1$^{-/-}$ mice.[42]

In contrast, mMBP-1$^{-/-}$ mice deficient in MBP-1 (or EPO—see below) had significantly ($p < 0.05$) higher worm burdens than wild-type 129/SvJ mice with a murine filarial infection (*Litomosoides sigmodontis*). Additionally, thoracic cavity macrophages from mMBP-1$^{-/-}$ mice produced more interleukin-10 (IL-10) than 129/SvJ wild-type cells and an equivalent amount of IL-5. This finding suggests that a lack of mMBP-1 can modulate cytokine responses that may, in turn, lead to different levels of permissiveness for parasite infection. MBP-1 deficiency in mice was not associated with either a decrease of eosinophil peroxidase in either the thoracic cavity cells of infected animals or the thoracic cavity fluid. Conversely, cytospin slides of thoracic cavity cells from EPO$^{-/-}$ mice were positive for MBP-1,[43] a finding showing that production of these two granule proteins is not mutually dependent.

Gastrointestinal Diseases

MBP-1 has not been detected in the lavage fluid from patients with inflammatory bowel disease,[44] although deposition of this cationic protein has been found in celiac disease[28,29] and eosinophilic gastroenteritis.[29,30] *In vitro*, MBP-1 but not EDN significantly ($p < 0.05$), though reversibly, decreased epithelial barrier function in a concentration-dependent fashion and downregulated occludin expression on T84 human colonic carcinoma epithelial lines. In an extension of these studies to a T-helper type 2 (T$_h$2)-mediated model of colitis, oxazolone colitis, mMBP-1$^{-/-}$ mice were significantly ($p < 0.05$) protected from colitis-induced weight loss and were protected from increased permeability compared with wild-type mice.[32] These studies suggest that MBP-1 can contribute to experimental colitis by decreasing mucosal barrier function.

Miscellaneous Inflammatory Diseases: Muscular Dystrophy

Tissue eosinophilia is a prominent cellular component of the mdx mouse model of Duchenne muscular dystrophy.[45] Eosinophils from both wild type and mMBP$^{-/-}$ mice both lyse myotubes, but this effect is enhanced by the addition of PAF only to wildtype eosinophils and not to eosinophils from mMBP$^{-/-}$ mice.

In mdx mice that were also mMBP$^{-/-}$, the eosinophil granule contribution to muscle damage *in vivo* was examined. In contrast to eosinophil-depleted mice, mMBP$^{-/-}$ mice showed no significant ($p = 0.44$) reduction in the percentage of necrotic muscle fibers, and the concentration of creatine kinase in the serum was unaffected by the mMBP$^{-/-}$ mutation. The degree of muscle membrane lysis in mdx mice soleus muscles and diaphragmatic muscle membrane at 4 weeks of age was not reduced in mdx mice with the mMBP$^{-/-}$ mutation, nor was the diaphragm damage reduced at 3 months of age. The mMBP$^{-/-}$ mutation had no effect on mdx muscle regeneration. However, the mMBP$^{-/-}$ mutation was associated with a significantly reduced proportion of limb muscles, diaphragm muscles, and hearts occupied by collagen types I, III, IV, and V; reduced hydroxyproline content as an index of fibrosis in the diaphragms and limb muscle at 14 months of age; and an 89% reduction in pathological fibrosis in 18-month-old mdx hearts. Lastly, mdx mMBP-1$^{-/-}$ mice had increased numbers of cytotoxic lymphocytes in hind-limb muscles, diaphragms, and hearts at 4 weeks of age that were comparable to the increase in cytotoxic lymphocytes in eosinophil-depleted mice of the same age.[46]

Taken together, these results suggest that mMBP-1 can cause lysis of muscle cells *in vitro*. *In vivo*, ablation of mMBP-1 does not reduce the lysis of muscle cells or influence regeneration of muscle fibers after damage, but it is associated with reduced fibrotic changes in muscle. As with the findings in murine filarial infections,[43] in which increased IL-10 levels were associated with mMBP$^{-/-}$, the association of increased numbers of cytotoxic lymphocytes in mMBP$^{-/-}$ mice suggests that eosinophils can serve immunomodulatory roles in diverse inflammatory disorders.

EOSINOPHIL PEROXIDASE

Human EPO, a cationic heme-containing protein located in the matrix of eosinophil secondary granules, is a 67-kDa protein comprised of two subunits with molecular weights of 52 kDa and 15 kDa by sodium dodecyl sulfate polyacrylamide gel electrophoresis (SDS-PAGE). Amino acid analysis has shown that arginine and aspartic acid constitute high proportions (13.6% and 13.4%, respectively) of the total molecular weight of the protein. EPO shows no immunological reactivity on double immunodiffusion with antibodies against chymotrypsin-like cationic protein, ECP, EDN, elastase, lactoferrin, lysozyme, or myeloperoxidase. The cell content of EPO is about 15.0 µg/10^6 eosinophils. EPO has an isoelectric point above 11, similar to that of other eosinophil granule proteins.[47] EPO preferentially oxidizes bromine (Br$^-$) to hypobromous acid (HOBr)[48] and, like other peroxidases, has potent antimicrobial activity in the presence of H_2O_2 and thiocyanate, iodide, chloride, or bromide ions.[49]

The combination of EPO, H_2O_2, and halides is capable of causing pneumocyte injury[50] and endocardial toxicity in the rat heart.[48] EPO is the most potent of the eosinophil granule proteins, on a molar basis, in its ability to kill microfilariae of *B. pahangi* and *B. malayi*.[23] EPO activates platelets[51] and tightly binds to mast cell granules.[52] The

EPO-mast cell granule complex retains the capacity of EPO to catalyze the iodination reaction when supplemented with iodide, H_2O_2, and a protein receptor, and to kill microorganisms when supplemented with H_2O_2 and a halide.[52] At low concentrations, the EPO-H_2O_2-halide complex causes noncytotoxic mast cell degranulation, whereas the EPO-H_2O_2-halide complex at higher concentrations of EPO causes nonselective loss of mast cell cytoplasmic constituents.[53] Post-translational tyrosine nitration of eosinophil granule proteins detected as immunostaining for 3-nitrotyrosine occurs during the maturation of eosinophils independent of inflammation. Tyrosine nitration is mediated by EPO in the presence of nicotinamide adenine dinucleotide phosphate oxidase and small amounts of mono-nitrogen oxides.[54] EPO can utilize nitrite (NO$_2^-$), an end product of NO metabolism, as a substrate to nitrate protein tyrosyl residues and is a more efficient catalyst of the formation of 3-nitrotyrosine than myeloperoxidase.[55]

Human EPO Deficiency

Unlike MBP-1, for which there are no clinical reports of selective deficiency in humans, there have been, since the report by Presentey in 1968,[56] case reports of patients and families with partial or complete EPO deficiency.[56–58] Human EPO deficiency remains extremely rare. It is inherited in an autosomal recessive manner.[59,60] Eosinophils of EPO-deficient patients may have hypersegmentation of the nucleus, a scarcity of specific granules, and negative peroxidase and phospholipid staining of eosinophilic granules.[57] There is an increased ratio of the secondary granule core volume to the total granule volume, due chiefly to decreased matrix volume. Two other matrix constituents, ECP and EDN, are present in normal amounts in EPO-deficient granules.[61] There has been no clear association between EPO deficiency and other diseases, and subjects with EPO deficiency are otherwise healthy.[62]

Analysis of complementary DNA from eosinophil precursors of an EPO-deficient patient showed a compound heterozygosity for the defect. This consisted of (1) a base transition leading to an amino acid substitution (arginine286→histidine) on one allele and (2) an insertion in an intron-exon junction causing a reading frame shift and the generation of a premature stop codon on the other allele. EPO-deficient eosinophils have no spectroscopic evidence of the EPO heme group, and, although synthesis of EPO could be verified in eosinophil precursor cells, EPO could no longer be detected in mature eosinophils, a finding suggesting that synthesized EPO was unstable or was degraded.[63] In another patient with EPO deficiency, a novel point mutation (g.2060G>A) causing an amino acid change from aspartic acid to asparagine (D648N) was reported. This mutation was found to cause loss of the electrostatic interaction of Asn648 with Arg146 that is crucial for disulfide bonds of the light chain in the N-terminus.[64]

EPO-Deficient Murine Strains

New Zealand White Mice

New Zealand white (NZW) mice with EPO deficiency, confirmed both by staining for peroxidase activity and by measuring EPO activity in extracts of eosinophil-rich cell suspensions, were reported in 1996. Similar to staining results in eosinophils from humans with EPO deficiency, eosinophils from NZW mice showed negative phospholipid staining (Sudan black) and smaller granules (whole granule and core) in morphometric studies, than control BDF1 eosinophils.[65] NZW mice do not have any hematological abnormalities, with the exception of the deletion of a T-cell receptor β chain.[66]

NZW mice have been used in the OVA sensitization challenge model[39,40] to study allergic asthma. NZW mice produced a significant ($p < 0.05$) eosinophilia in both bronchoalveolar lavage fluid (BALF) and lung tissue at day 22. The NZW mice had twofold higher eosinophil numbers in BALF than in control BALB/c or C57BL/6 mice; however, BALB/c and C57BL/6 mice had two and a half and fourfold, respectively, more eosinophils around the bronchi than NZW mice. The basal pulmonary perfusion pressure was similar in all groups of mice and, like BALB/c and C57BL/6 mice, sensitized NZW mice had greater responses to methacholine than control NZW mice. BALF from NZW mice had a strikingly higher concentration of vascular endothelial growth factor than the other strains. Both NZW and C57BL/6 mice had antigen challenge-induced, time-dependent increases in BALF levels of transforming growth factor beta (TGF-β) at day 22 that lasted until day 28. NZW mice also had less pulmonary mucus and, after antigen challenge, no change in peri-bronchial collagen compared with BALB/c and C57BL/6 mice.

These findings suggest that despite the production of vascular endothelial growth factor and TGF-β, tissue remodeling was not a feature of NZW mice in this model of allergic asthma. This finding could be due to the lower degree of pulmonary parenchymal eosinophilic infiltration in NZW mice than in the other strains.[67] NZW mice also exhibit markedly decreased 3-nitrotyrosine staining around the airways after OVA challenge compared with control (A/J and C57BL/6) strains.[68]

EPO Knockout Mice

Mice with a null allele at the EPO locus (EPO$^{-/-}$) have no peroxidase activity, whereas heterozygous EPO$^{+/-}$ mice

have eosinophils with peroxidase activity approximately 50% that of wild-type mice. Eosinophils from the peritoneal cavity of EPO knockout mice after sensitization challenge with a whole protein extract from the helminth *Mesocestoides corti* showed no loss of granule core or surrounding matrix nor was the average number of granules per cross-sectional area different from that in wild-type mice[66] (Table 8.4.2). Similar to that in NZW mice, the granule size in EPO knockout mice is smaller than that in wild-type mice (by about 57%), and, likewise, 3-nitrotyrosine is not detectable in eosinophils from EPO$^{-/-}$ mice[54] (Table 8.4.3).

Asthma

EPO$^{-/-}$ mice have been used in several model systems to examine specific effector functions mediated by this granule protein. In the OVA sensitization challenge model of allergic asthma, there was no difference in the localization and number of eosinophils following OVA challenge in the lungs of wild-type or EPO$^{-/-}$ mice.[69] 3-Nitrotyrosine levels in whole lung homogenates were undetectable in saline-treated control or OVA-treated EPO$^{-/-}$ mice, findings consistent with the role of EPO in tyrosine nitration. Similar to the findings in NZW mice, OVA-induced pulmonary inflammation resulted in equivalent numbers of wild-type and EPO$^{-/-}$ eosinophils recruited to BALF and no effect of the EPO knockout phenotype on the recruitment of any leukocyte types compared with OVA-treated wild-type controls. No measurable EPO was detected in BALF from EPO$^{-/-}$ mice. Airway epithelial mucous content was identical in EPO$^{-/-}$ mice and controls. OVA-sensitized EPO$^{-/-}$ and wild-type mice had identical increases in methacholine-induced airflow changes compared with saline controls.[69] These studies suggest that in the OVA model of allergic asthma in mice, EPO release is not critical to or necessary for the observed pathophysiological changes. In combination with the results of

TABLE 8.4.2 Characteristics of Eosinophils from EPO$^{-/-}$ Knockout Mice

Nuclear hypersegmentation
Scarcity of specific granules
Negative peroxidase staining
Negative phospholipid staining
Increased granule core volume/total granule volume
Normal quantity of eosinophil cationic protein, eosinophil-derived neurotoxin
EPO, eosinophil peroxidase.

TABLE 8.4.3 Comparison of New Zealand White Mice and EPO$^{-/-}$ Knockout Mice

Observation	Mouse Type	
	New Zealand White	EPO$^{-/-}$ Knockout
EPO activity by		
Stain	Absent	Absent
Enzyme activity	Absent	Absent
Granule cores	Intact	Intact
Phospholipid staining	Absent	Not tested
Number of granules/cross-sectional area	Not tested	Normal
Granule size	Smaller than wild-type	Smaller than wild-type
Post-translational tyrosine nitration	Absent	Absent
Response in OVA sensitization challenge model:		
BALF eosinophils after challenge	2 × increase versus BALB/c, C57BL/6	Equal to wild-type
Bronchial tissue after challenge	Fewer peribronchial eosinophils than in BALB/c, C57BL/6 mice	No change in number or localization versus wild-type
Postchallenge bronchial mucus	Decreased versus control	Increased, same as wild-type
Airway reactivity to methacholine	Greater response of NZW sensitized mice than control NZW mice	Increase in airflow changes equal to that in wild-type

BALF, bronchoalveolar lavage fluid; EPO, eosinophil peroxidase; NZW, New Zealand White; OVA, ovalbumin.

studies of MBP-1 knockout mice,[41] pulmonary pathological changes observed in this model of allergic asthma in mice appear not to depend on the two major eosinophil granule cationic proteins.

Parasites

EPO$^{-/-}$ knockout mice have been used to examine the contribution of eosinophils and EPO to protective immunity against parasites. In one model, naive and EPO$^{-/-}$ knockout mice immunized with irradiated *Onchocerca volvulus* L3 larvae received implanted diffusion chambers containing L3 larvae. Three weeks later, the L3 recovery rates were the same for naive 129/SvJ wild-type mice and for naive EPO$^{-/-}$ knockout mice, a finding showing that innate antilarval immunity was not affected by EPO deficiency. Subsequently, equal levels of adaptive antilarval immunity developed in the immunized wild-type and EPO$^{-/-}$ knockout mice.[70]

In a study mentioned previously in this chapter,[42] EPO$^{-/-}$ and MBP-1$^{-/-}$ mice were examined for their ability to eliminate infection after intraperitoneal injection of *B. pahangi* L3 larvae. There was no difference in larval recoveries in either knockout strain compared with that in wild-type mice. In the same study, eosinophil depletion using anti-C-C chemokine receptor type 3 (CCR3) antibody resulted in higher parasite numbers than isotype control-treated mice,[42] a suggestion that other actions, direct or indirect, of eosinophils, not related to mMBP-1 or EPO, were responsible for reducing larval numbers.

Study of infection with the rodent filaria *L. sigmodontis* has shown both EPO and MBP-1 to be important components of immunity to this helminth. Helminth numbers in EPO$^{-/-}$ knockout mice were increased compared with those in wild-type 129/SvJ mice and equal to those in the fully permissive BALB/c strain. In contrast to MBP-1$^{-/-}$ knockout mice, EPO$^{-/-}$ knockout mice had a twofold increase in intrathoracic eosinophils, a suggestion that EPO can regulate eosinophil influx. Thoracic cavity cells from EPO$^{-/-}$ mice and BALB/c mice produced significantly ($p <$ 0.05) more IL-5 and IL-10 than those from MBP-1$^{-/-}$ or wild-type mice. In the case of IL-10, macrophages of both deficient strains showed significantly ($p < 0.05$) higher IL-10 levels in response to worm antigen than those of the wild-type mice.

Production of IL-10 by CD4 T cells from EPO$^{-/-}$ mice did not differ from that in either wild-type or MBP-1$^{-/-}$ mice. However, EPO$^{-/-}$ CD4 T cells produced two to three times more IL-5 than wild-type and MBP-1$^{-/-}$ cells. Splenocytes from uninfected EPO$^{-/-}$ and MBP-1$^{-/-}$ mice produced less IL-4 and equivalent amounts of IL-10 after challenge with various stimuli (medium, *L. sigmodontis* antigen, anti-CD3, conconavalin A) than cells from wild-type animals.[43]

Considered together, these studies suggest that, depending on the parasite chosen and the permissiveness of the host animal, eosinophils can be shown to contribute to parasite immunity. Depletion of EPO does not seem to affect innate antilarval immunity to *O. volvulus*. In another system, eosinophil depletion but not MBP-1 or EPO depletion led to the impaired elimination of *B. pahangi* larvae, whereas in a third system, immunity to *L. sigmodontis* was dependent on the presence of MBP-1 and EPO. In this third system, the absence of MBP-1 and EPO seemed to increase the production of IL-10 by thoracic cavity macrophages, whereas CD4 T cells from EPO$^{-/-}$ knockout mice produced more IL-5 than those from wild-type or MBP-1$^{-/-}$ mice, a finding that demonstrates a modulatory effect of granule protein deficiency on the immune system.

Gastrointestinal Diseases

A murine model of ulcerative colitis that uses dextran sodium sulfate (DSS) to induce inflammatory changes has been used to examine the contribution of eosinophils to the immunopathogenesis of this disorder. The administration of DSS to mice induces an acute inflammation of the colon, weight loss, and bloody diarrhea followed subsequently by diarrhea, rectal bleeding, and shortening of the colon. Histological examination of the inflamed bowel wall using a polyclonal antibody to MBP-1 showed the presence of eosinophils throughout the mucosa and submucosa of animals treated with DSS. Luminal levels of EPO were about 1000-fold higher in DSS-treated wild-type mice than in control animals.

DSS challenge of MBP-1$^{-/-}$, EPO$^{-/-}$, and wild-type control mice showed results that were quite different between the two knockout strains. Whereas MBP-1$^{-/-}$ mice showed inflammatory changes comparable to those of wild-type mice, colitis in EPO$^{-/-}$ mice was attenuated

compared with that in strain-matched wild-type mice or MBP-1$^{-/-}$ mice despite equivalent eosinophil intestinal accumulation and pattern of distribution. This attenuation was associated with the loss of EPO activity.

Resorcinol, an EPO inhibitor, also attenuated experimental colitis, colonic shortening, and luminal EPO levels in DSS-treated mice compared with DSS-treated mice subsequently injected with the (resorcinol) control vehicle, although eosinophil levels were comparable in both groups.[71] These results parallel findings in human gut lavage fluid from patients with inflammatory bowel disease, which showed that the concentrations of ECP, EDN, and EPO, but not MBP-1, were significantly ($p < 0.05$) higher in gut lavage fluid from patients with inflammatory bowel disease than in healthy control subjects.[44] Considered together, the available data suggest different roles for MBP-1 and EPO in experimental colitis.

REFERENCES

1. Gleich GJ. Mechanisms of eosinophil-associated inflammation. *J Allergy Clin Immunol* 2000;**105**:651–63.
2. Moqbel R, Lacy P, Adamko DJ, Odemuyiwa SO. Biology of eosionphils. In: Adkinson, Jr. NF, Bochner BS, Busse WW, Holgate ST, Lemanske, Jr. RF, Simons FER, editors. *Middleton's Allergy: Principles & Practices*. 7th ed, Vol. I. Philadelphia: Mosby Elsevier; 2009. p. 295–310.
3. Yamada T, Sun Q, Zeibecoglou K, Bungre J, North J, Kay AB, et al. IL-3, IL-5, granulocyte-macrophage colony-stimulating factor receptor alpha-subunit, and common beta-subunit expression by peripheral leukocytes and blood dendritic cells. *J Allergy Clin Immunol* 1998;**101**:677–82.
4. Kroegel C, Yukawa T, Westwick J, Barnes PJ. Evidence for two platelet activating factor receptors on eosinophils—Dissociation between PAF-induced intracellular calcium mobilization degranulation and superoxides anion generation in eosinophils. *Biochem Biophys Res Commun* 1989;**162**:511–21.
5. Plager DA, Loegering DA, Weiler DA, Checkel JL, Wagner JM, Clarke NJ, et al. A novel and highly divergent homolog of human eosinophil granule major basic protein. *J Biol Chem* 1999;**274**:14464–73.
6. Gleich GJ, Loegering DA, Maldonado JE. Identification of a major basic protein in guinea pig eosinophil granules. *J Exp Med* 1973;**137**:1459–71.
7. Hamann KJ, Barker RL, Ten RM, Gleich GJ. The molecular biology of eosinophil granule proteins. *Int Arch Allergy Appl Immunol* 1991;**94**:202–9.
8. Popken-Harris P, McGrogan M, Loegering DA, Checkel JL, Kubo H, Thomas LL, et al. Expression, purification, and characterization of the recombinant proform of eosinophil granule major basic protein. *J Immunol* 1995;**155**:1472–80.
9. Popken-Harris P, Checkel J, Loegering D, Madden B, Springett M, Kephart G, et al. Regulation and processing of a precursor form of eosinophil granule major basic protein (ProMBP) in differentiating eosinophils. *Blood* 1998;**92**:623–31.
10. Frigas E, Loegering DA, Gleich GJ. Cytotoxic effects of the guinea pig eosinophil major basic protein on tracheal epithelium. *Lab Invest* 1980;**42**:35–43.

11. Filley WV, Holley KE, Kephart GM, Gleich GJ. Identification by immunofluorescence of eosinophil granule major basic protein in lung tissues of patients with bronchial asthma. *Lancet* 1982;**2**:11−6.

12. Goetzl EJ, Wasserman SI, Austen F. Eosinophil polymorphonuclear leukocyte function in immediate hypersensitivity. *Arch Pathol* 1975;**99**:1−4.

13. Frigas E, Loegering DA, Solley GO, Farrow GM, Gleich GJ. Elevated levels of the eosinophil granule major basic protein in the sputum of patients with bronchial asthma. *Mayo Clin Proc* 1981;**56**:345−53.

14. Flavahan NA, Slifman NR, Gleich GJ, Vanhoutte PM. Human eosinophil major basic protein causes hyperreactivity of respiratory smooth muscle—Role of the epithelium. *Am Rev Respir Dis* 1988;**138**:685−8.

15. Jacoby DB, Gleich GJ, Fryer AD. Human eosinophil major basic protein is an endogenous allosteric antagonist at the inhibitory muscarinic M2 receptor. *J Clin Invest* 1993;**91**:1314−8.

16. Fryer AD, Jacoby DB. Function of pulmonary M2 muscarinic receptors in antigen-challenged guinea pigs is restored by heparin and poly-L-glutamate. *J Clin Invest* 1992;**90**:2292−8.

17. Lefort J, Nahori MA, Ruffie C, Vargaftig BB, Pretolani M. In vivo neutralization of eosinophil-derived major basic protein inhibits antigen-induced bronchial hyperreactivity in sensitized guinea pigs. *J Clin Invest* 1996;**97**:1117−21.

18. Guillevin L, Cohen P, Gayraud M, Lhote F, Jarrousse B, Casassus P. Churg-Strauss syndrome—Clinical study and long-term follow-up of 96 patients. *Medicine (Baltimore)* 1999;**78**:26−37.

19. Kay AB, Moqbel R, Durham SR, MacDonald AJ, Walsh GM, Shaw RJ, et al. Leucocyte activation initiated by IgE-dependent mechanisms in relation to helminthic parasitic disease and clinical models of asthma. *Int Arch Allergy Appl Immunol* 1985;**77**:69−72.

20. Butterworth AE. Cell-mediated damage to helminths. *Adv Parasitol* 1984;**23**:143−235.

21. Aiyar S, Zaman V, Ha CS. Mechanism of destruction of *Brugia malayi* microfilariae in vitro—The role of antibody and leucocytes. *Acta Trop* 1982;**39**:225−36.

22. Wassom DL, Gleich GJ. Damage to *Trichinella spiralis* newborn larvae by eosinophil major basic protein. *Am J Trop Med Hyg* 1979;**28**:860−3.

23. Hamann KJ, Gleich GJ, Checkel JL, Loegering DA, McCall JW, Barker RL. In vitro killing of microfilariae of *Brugia pahangi* and *Brugia malayi* by eosinophil granule proteins. *J Immunol* 1990;**144**:3166−73.

24. Ackerman SJ, Gleich GJ, Loegering DA, Richardson BA, Butterworth AE. Comparative toxicity of purified human eosinophil granule cationic proteins for schistosomula of Schistosoma mansoni. *Am J Trop Med Hyg* 1985;**34**:735−45.

25. Meeusen EN, Balic A. Do eosinophils have a role in the killing of helminth parasites? *Parasitol Today* 2000;**16**:95−101.

26. Kato M, Kephart GM, Morikawa A, Gleich GJ. Eosinophil infiltration and degranulation in normal human tissues—Evidence for eosinophil degranulation in normal gastrointestinal tract. *Int Arch Allergy Immunol* 2001;**125**:55−8.

27. Garcia-Zepeda EA, Rothenberg ME, Ownbey RT, Celestin J, Leder P, Luster AD. Human eotaxin is a specific chemoattractant for eosinophil cells and provides a new mechanism to explain tissue eosinophilia. *Nat Med (NY, NY, US)* 1996;**2**:449−56.

28. Colombel JF, Torpier G, Janin A, Klein O, Cortot A, Capron M. Activated eosinophils in adult coeliac disease—Evidence for a local release of major basic protein. *Gut* 1992;**33**:1190−4.

29. Talley NJ, Kephart GM, McGovern TW, Carpenter HA, Gleich GJ. Deposition of eosinophil granule major basic protein in eosinophilic gastroenteritis and celiac disease. *Gastroenterology* 1992;**103**:137−45.

30. Torpier G, Colombel JF, Mathieu-Chandelier C, Capron M, Dessaint JP, Cortot A, et al. Eosinophilic gastroenteritis: ultrastructural evidence for a selective release of eosinophil major basic protein. *Clin Exp Immunol* 1988;**74**:404−8.

31. Mir-Madjlessi SH, Sivak Jr MV, Farmer RG. Hypereosinophilia, ulcerative colitis, sclerosing cholangitis, and bile duct carcinoma. *Am J Gastroenterol* 1986;**81**:483−5.

32. Furuta GT, Nieuwenhuis EE, Karhausen J, Gleich G, Blumberg RS, Lee JJ, et al. Eosinophils alter colonic epithelial barrier function—Role for major basic protein. *American Journal of Physiology Gastrointestinal and Liver Physiology* 2005;**289**:G890−7.

33. Lee JJ, Jacobsen EA, McGarry MP, Schleimer RP, Lee NA. Eosinophils in health and disease—The LIAR hypothesis. *Clin Exp Allergy* 2010;**40**:563−75.

34. Jacobsen EA, Taranova AG, Lee NA, Lee JJ. Eosinophils—Singularly destructive effector cells or purveyors of immunoregulation? *J Allergy Clin Immunol* 2007;**119**:1313−20.

35. Leckie MJ, ten Brinke A, Khan J, Diamant Z, O'Connor BJ, Walls CM, et al. Effects of an interleukin-5 blocking monoclonal antibody on eosinophils, airway hyper-responsiveness, and the late asthmatic response. *Lancet* 2000;**356**:2144−8.

36. Flood-Page PT, Menzies-Gow AN, Kay AB, Robinson DS. Eosinophil's role remains uncertain as anti-interleukin-5 only partially depletes numbers in asthmatic airway. *Am J Respir Crit Care Med* 2003;**167**:199−204.

37. Sher A, Coffman RL, Hieny S, Cheever AW. Ablation of eosinophil and IgE responses with anti-IL-5 or anti-IL-4 antibodies fails to affect immunity against *Schistosoma mansoni* in the mouse. *J Immunol* 1990;**145**:3911−6.

38. Herndon FJ, Kayes SG. Depletion of eosinophils by anti-IL-5 monoclonal antibody treatment of mice infected with *Trichinella spiralis* does not alter parasite burden or immunologic resistance to reinfection. *J Immunol* 1992;**149**:3642−7.

39. Tomaki M, Zhao LL, Lundahl J, Sjostrand M, Jordana M, Linden A, et al. Eosinophilopoiesis in a murine model of allergic airway eosinophilia—Involvement of bone marrow IL-5 and IL-5 receptor alpha. *J Immunol* 2000;**165**:4040−50.

40. Shen H, O'Byrne PM, Ellis R, Wattie J, Tang C, Inman MD. The effects of intranasal budesonide on allergen-induced production of interleukin-5 and eotaxin, airways, blood, and bone marrow eosinophilia, and eosinophil progenitor expansion in sensitized mice. *Am J Respir Crit Care Med* 2002;**166**:146−53.

41. Denzler KL, Farmer SC, Crosby JR, Borchers M, Cieslewicz G, Larson KA, et al. Eosinophil major basic protein-1 does not contribute to allergen-induced airway pathologies in mouse models of asthma. *J Immunol* 2000;**165**:5509−17.

42. Ramalingam T, Porte P, Lee J, Rajan TV. Eosinophils, but not eosinophil peroxidase or major basic protein, are important for host protection in experimental *Brugia pahangi* infection. *Infect Immun* 2005;**73**:8442−3.

43. Specht S, Saeftel M, Arndt M, Endl E, Dubben B, Lee NA, et al. Lack of eosinophil peroxidase or major basic protein impairs defense against murine filarial infection. *Infect Immun* 2006;**74**:5236−43.

44. Levy AM, Gleich GJ, Sandborn WJ, Tremaine WJ, Steiner BL, Phillips SF. Increased eosinophil granule proteins in gut lavage fluid from patients with inflammatory bowel disease. *Mayo Clin Proc* 1997;**72**:117−23.

45. Cai B, Spencer MJ, Nakamura G, Tseng-Ong L, Tidball JG. Eosinophilia of dystrophin-deficient muscle is promoted by perforin-mediated cytotoxicity by T cell effectors. *Am J Pathol* 2000;**156**:1789−96.

46. Wehling-Henricks M, Sokolow S, Lee JJ, Myung KH, Villalta SA, Tidball JG. Major basic protein-1 promotes fibrosis of dystrophic muscle and attenuates the cellular immune response in muscular dystrophy. *Hum Mol Genet* 2008;**17**:2280−92.

47. Carlson MG, Peterson CG, Venge P. Human eosinophil peroxidase—Purification and characterization. *J Immunol* 1985;**134**:1875−9.

48. Slungaard A, Mahoney Jr JR. Bromide-dependent toxicity of eosinophil peroxidase for endothelium and isolated working rat hearts—A model for eosinophilic endocarditis. *J Exp Med* 1991;**173**:117−26.

49. Jong EC, Henderson WR, Klebanoff SJ. Bactericidal activity of eosinophil peroxidase. *J Immunol* 1980;**124**:1378−82.

50. Agosti JM, Altman LC, Ayars GH, Loegering DA, Gleich GJ, Klebanoff SJ. The injurious effect of eosinophil peroxidase, hydrogen peroxide, and halides on pneumocytes in vitro. *J Allergy Clin Immunol* 1987;**79**:496−504.

51. Rohrbach MS, Wheatley CL, Slifman NR, Gleich GJ. Activation of platelets by eosinophil granule proteins. *J Exp Med* 1990;**172**:1271−4.

52. Henderson WR, Jong EC, Klebanoff SJ. Binding of eosinophil peroxidase to mast cell granules with retention of peroxidatic activity. *J Immunol* 1980;**124**:1383−8.

53. Henderson WR, Chi EY, Klebanoff SJ. Eosinophil peroxidase-induced mast cell secretion. *J Exp Med* 1980;**152**:265−79.

54. Ulrich M, Petre A, Youhnovski N, Promm F, Schirle M, Schumm M, et al. Post-translational tyrosine nitration of eosinophil granule toxins mediated by eosinophil peroxidase. *J Biol Chem* 2008;**283**:28629−40.

55. Wu W, Chen Y, Hazen SL. Eosinophil peroxidase nitrates protein tyrosyl residues—Implications for oxidative damage by nitrating intermediates in eosinophilic inflammatory disorders. *J Biol Chem* 1999;**274**:25933−44.

56. Presentey BZ. A new anomaly of eosinophilic granulocytes. *Tech Bull Registr Med Technol* 1968;**38**:131−4.

57. Presentey B. Cytochemical characterization of eosinophils with respect to a newly discovered anomaly. *Am J Clin Pathol* 1969;**51**:451−7.

58. Hoffmann JJ, Tielens AG. Partial deficiency of eosinophil peroxidase. *Blut* 1987;**54**:165−9.

59. Presentey B, Szapiro L. Hereditary deficiency of peroxidase and phospholipids in eosinophilic granulocytes. *Acta Haematol* 1969;**41**:359−62.

60. Presentey B. Partial and severe peroxidase and phospholipid deficiency in eosinophils—Cytochemical and genetic considerations. *Acta Haematol* 1970;**44**:345−54.

61. Zabucchi G, Soranzo MR, Menegazzi R, Vecchio M, Knowles A, Piccinini C, et al. Eosinophil peroxidase deficiency—Morphological and immunocytochemical studies of the eosinophil-specific granules. *Blood* 1992;**80**:2903−10.

62. Kutter D, Mueller-Hagedorn S, Forges T, Glaesener R. A case of eosinophil peroxidase deficiency. *Ann Hematol* 1995;**71**:315−7.

63. Romano M, Patriarca P, Melo C, Baralle FE, Dri P. Hereditary eosinophil peroxidase deficiency—Immunochemical and spectroscopic studies and evidence for a compound heterozygosity of the defect. *Proc Natl Acad Sci USA* 1994;**91**:12496−500.

64. Nakagawa T, Ikemoto T, Takeuchi T, Tanaka K, Tanigawa N, Yamamoto D, et al. Eosinophilic peroxidase deficiency—Identification of a point mutation (D648N) and prediction of structural changes. *Hum Mutat* 2001;**17**:235−6.

65. Ohmori J, Tokunaga H, Ezaki T, Maruyama H, Nawa Y. Eosinophil peroxidase deficiency in New Zealand white mice. *Int Arch Allergy Immunol* 1996;**111**:30−5.

66. Kotzin BL, Barr VL, Palmer E. A large deletion within the T-cell receptor beta-chain gene complex in New Zealand white mice. *Science (NY, NY)* 1985;**229**:167−71.

67. Vasquez-Pinto LM, Landgraf RG, Bozza PT, Jancar S. High vascular endothelial growth factor levels in NZW mice do not correlate with collagen deposition in allergic asthma. *Int Arch Allergy Immunol* 2007;**142**:19−27.

68. Duguet A, Iijima H, Eum SY, Hamid Q, Eidelman DH. Eosinophil peroxidase mediates protein nitration in allergic airway inflammation in mice. *Am J Respir Crit Care Med* 2001;**164**:1119−26.

69. Denzler KL, Borchers MT, Crosby JR, Cieslewicz G, Hines EM, Justice JP, et al. Extensive eosinophil degranulation and peroxidase-mediated oxidation of airway proteins do not occur in a mouse ovalbumin-challenge model of pulmonary inflammation. *J Immunol* 2001;**167**:1672−82.

70. Abraham D, Leon O, Schnyder-Candrian S, Wang CC, Galioto AM, Kerepesi LA, et al. Immunoglobulin E and eosinophil-dependent protective immunity to larval *Onchocerca volvulus* in mice immunized with irradiated larvae. *Infect Immun* 2004;**72**:810−7.

71. Forbes E, Murase T, Yang M, Matthaei KI, Lee JJ, Lee NA, et al. Immunopathogenesis of experimental ulcerative colitis is mediated by eosinophil peroxidase. *J Immunol* 2004;**172**:5664−75.

Chapter 8.5

The Eosinophil and Airways Hyperresponsiveness: the Role and Mechanisms of Cationic Proteins

Charles G. Irvin

INTRODUCTION

The eosinophil is a multifunctional granulocytic leukocyte that shares in common many general abilities and capacities with the neutrophil and other adaptive response inflammatory cells. As discussed in other chapters in this book, the eosinophil has been long linked to the pathogenesis of asthma and is commonly present within the inflammatory lesions found in this disease. However, it is important to note that eosinophils are also present in many other inflammatory lung diseases, but these eosinophilic diseases sometimes do and sometimes do not exhibit airway hyperresponsiveness (AHR). Nonetheless, there is no debate that, under the right conditions, eosinophils can cause considerable structural damage and can also promote subsequent repair or scar formation.[1] The armementarium of the eosinophil in terms of potential contribution to the physiological dysfunction associated with asthma are extensive, but in general these consist of granule-associated and highly charged proteins, lipid mediators, protein cytokines, and reactive oxygen species.[1,2] Among the best known eosinophilic mediators are the eosinophil granule proteins eosinophil granule major basic protein 1 (MBP-1) and eosinophil cationic protein [ECP; ribonuclease 3 (RNase 3)]. The highly cationic nature of these proteins is due to the many negatively charged arginine residues. In this chapter, the focus will be on these cationic proteins and the postulated complex mechanisms by which they cause lung dysfunction and AHR in particular.

ASTHMA AND AIRWAY HYPERRESPONSIVENESS

Asthma is a clinical syndrome characterized by three fundamental aspects of lung dysfunction that include: (1) reversible airflow limitation; (2) periodic airway dysfunction; and (3) AHR.[3,4] However, none of the clinical manifestations of this disease are specific to asthma, with the possible exception of periodic airflow limitation. Variation in peak expiratory flow is known to be related to the presence of eosinophilic inflammation. In this regard asthma is usually diagnosed clinically when a patient presents with periodic exacerbations in their disease or exhibits a phenomenon best described as *twitchy lung*, i.e., hyperresponsiveness to known triggers (e.g., protein allergens). Periodic worsening of the disease can be more spontaneous in nature, such as occurring in patients with intrinsic asthma, or can be a very regular daily presentation, such as the patient with nocturnal asthma. For some time it was thought that AHR was a hallmark of asthma but subsequent investigations have now shown that most inflammatory disorders of the lung will exhibit AHR.[5] As such, AHR is a sensitive but nonspecific indicator of the disease. However, the prototypical presentation of asthma is a degree of AHR that is often much more severe than other lung diseases. In atopic patients, AHR is associated with eosinophilic inflammation and it is therefore used in clinical studies as an outcome measure that most physicians accept as indirectly assessing inflammation.[6]

In animal models of asthma (or perhaps better described as animal systems of airway inflammation), changes in baseline lung function (i.e., bronchospasm) are often not observed in response to inflammation or a specific inflammatory mediator.[7] Cationic proteins in particular have little or no effect on baseline lung caliber in laboratory animals, whereas the leukotrienes, for example, have been shown to be potent bronchoconstrictors in animals and humans.[8] These observations suggest that multiple mediators with different effector functions work in concert to produce the functional phenotype of asthma. In small laboratory rodents, in part because they have such large airway lumens relative to the parenchyma,[9] bronchoconstriction is difficult to appreciate. There are a number of means by which AHR can be assessed.[10] First, it can be obtained noninvasively with a whole-body plethysmography technique, but this approach has now been largely discredited.[11] Second, the use of several different invasive measurement approaches that requires that one introduces a catheter into the trachea to attach the animal to a ventilator; this is currently the preferred approach. Invasive measurement approaches include just simple pressure measurement or very sophisticated analysis of forced oscillation responses providing a more detailed assessment of lung function, as will be described more fully later in this chapter.[10]

INFLAMMATION, ASTHMA, AND AIRWAY HYPERRESPONSIVENESS

Prior to the 1980s, the pathogenesis of asthma was thought to be largely due to a dysfunction of the automatic neural axis that controls airway caliber. This hypothesis held sway until two important observations were made. First, by using the then newly developed technique of bronchoalveolar lavage (BAL), it was shown by several investigators that the asthmatic lung was significantly inflamed with the presence of *both* eosinophils and neutrophils.[12,13] These investigations suggested that rather than a neural dysfunction, asthma might in fact be an inflammatory lung disease. This idea was rapidly accepted because it formed a rational basis for therapy and a plausible mechanism of action of anti-inflammatory therapies for asthma and, in particular, the use of steroids. Early clinical investigations into the association between inflammation and airway dysfunction linked infiltrating eosinophils to AHR by the use of BAL, and more recently this has been complemented by the induced sputum technique.[2,8,14,15] Moreover, later investigations demonstrated that products from the eosinophil, e.g., leukotriene C_4 (LTC_4) were also linked to AHR.[16] Other studies using BAL measurements showed a significant relationship between eosinophil numbers (dose response) and the degree of AHR,[17] and this relationship has been extended to biopsy studies[18] or induced sputum studies.[19]

Next, a series of studies were published that used a widely divergent series of animal models of induced lung inflammation and that found clear associations between inflammation and airway dysfunction.[20] Inflammation could be induced by endotoxin,[21] ozone,[22] complement fragments,[23] or antigen in a sensitized animal[24] and in each instance caused AHR. These pioneering animal studies were instrumental in linking inflammation as caused by either neutrophils or eosinophils to AHR.[20] Moreover, when both neutrophils and eosinophils were depleted,[22,23,25] the increases in hyperresponsiveness were ameliorated. However, the most convincing evidence came from studies that used reconstitution with purified granulocytes[26] to restore AHR, especially since the degree of AHR was also correlated to the number of transferred cells. It was years later that similar results were reported with T cells in mouse models of allergic airway inflammation.[7] Taken together, these results indicate that both types of granulocytes as well as other cell types, e.g., lymphocytes are capable of promoting both bronchospasm and AHR that are both considered to be clinical hallmarks of the asthma phenotype; that is, there is a case to be made for cellular redundancy. Moreover, it was not at all unreasonable to postulate that eosinophils released some mediator(s) that then caused AHR (Fig. 8.5.1); however, the very limited success of anti-interleukin-5 (IL-5) therapies[27] has brought this long held hypothesis into question in recent years. Nevertheless, therapies targeted at reducing eosinophil numbers are currently associated with improved asthma control in those patients with eosinophilia.[28,29]

ESTABLISHING THE SPECIFIC ROLE OF THE EOSINOPHIL

More recently the role of the eosinophil has been more firmly established as sufficient to cause both the pathology and pathophysiology associated with asthma. This evidence comes from the use of genetic manipulation of the mammalian genome. First, Lee and colleagues[30] showed that the entire allergic phenotype could be abrogated by the deletion of eosinophils using a strategy that targets eosinophil progenitors in the mouse bone marrow. By contrast, Humbles and colleagues[31] showed that eosinophil gene deletion in the ΔdblGATA mouse model likewise reduced many of the pathological features of asthma, although AHR was not diminished. Given that GATA-1 signaling is needed for secondary granule formation[32,33] the data of Humbles might be explained by a stored mediator founded in the cytoplasm (e.g., LTC$_4$) that is responsible for causing AHR.

CATIONIC PROTEINS

The animal studies discussed thus far primarily associated the eosinophil with airway structural changes (remodeling) and lung dysfunction as caused by allergic inflammation. The next question then is: What is released by the eosinophil that would best explain the physiological phenotype associated with eosinophil diapedesis into the airway lumen? If eosinophils were somehow specific for asthma (which of course is not strictly true) then perhaps

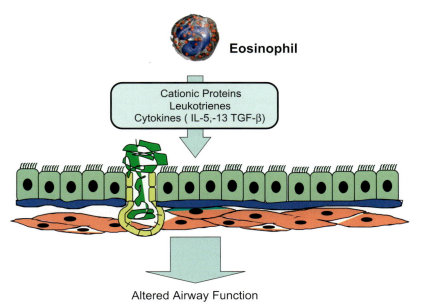

Eosinophil

Cationic Proteins
Leukotrienes
Cytokines (IL-5,-13 TGF-β)

Altered Airway Function

FIGURE 8.5.1 The general and widely held, but simplistic, view of the key role the eosinophil might have in causing airway dysfunction such as observed in asthma. IL-5, interleukin-5; IL-13, interleukin-13, TGF-β, transforming growth factor beta.

investigating the granular proteins found within the large secondary granules of the eosinophil might prove insightful.

MBP-1 when instilled into the rat lung did not cause bronchospasm but did increase, in a markedly dose-dependent fashion, the responsiveness to inhaled methacholine.[34,35] At the time most workers in this field thought that MBP-1 worked through a receptor,[34] so as a control other large polycationic synthetic proteins such as poly-L-arginine or poly-L-lysine (PLL) were instilled. Surprisingly, an identical degree of AHR in response to repetitive polycations as with MBP-1 was observed. It was determined that these artificial mimics of naturally occurring cationic proteins needed to be minimally >1 kDa in size. Moreover, if the cationic sites on poly-L-lysine were acetylated, this completely abrogated the increase in airway responsiveness hence demonstrating the importance of charge. These results also suggested cationic proteins are potent mediators of AHR through a mechanism that does not involve a receptor and one that is very much related to the degree of cationic charge.

Eosinophils are not the only cells that release cationic protein mediators of inflammation. As an example, neutrophils release small, highly cationic proteins called defensins but also produce and take up ECP.[36] In addition, we have shown that other cationic proteins such as platelet factor 4 and cathepsin G also caused similar degrees of AHR.[37] These results supported the idea that cationic proteins may be a more general class of inflammatory mediator having common and redundant mechanisms. These results have further significance as they explain why so many inflammatory cell types have been implemented in asthma (Fig. 8.5.2). Indeed, virtually all adaptive inflammatory cells, when activated, release mediators that have low molecular weight and are highly charged cationic proteins, and as such the total *cationic protein burden* should be considered. These studies suggest the feasibility of several diverse strategies to neutralize the highly charged nature of these proteins. Of relevant interest are two reports in patients with asthma where the use of heparin was effective in blocking the asthmatic response to exercise,[38] but not the asthmatic response to antigen,[39] and the intriguing report that asthmatic patients have a low level of endogenous circulating heparin.[40] Investigating ways of neutralizing the cationic protein burden from inflammatory cells, and especially eosinophils, with charge neutralizing therapies may be warranted.[41]

The airway epithelium is known to have on its apical surface a number of highly anionic proteoglycans such as heparan sulfate and glycosaminoglycans[42] and so it is tempting to speculate that this *anionic barrier* is present to protect the lung from cationic proteins released from the adaptive inflammatory responses (Fig. 8.5.2). Moreover, the mucous hypersecretion so commonly observed in

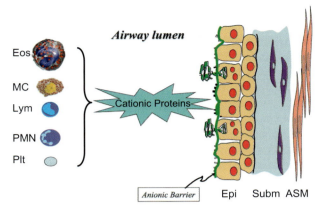

FIGURE 8.5.2 An array of inflammatory cells including: platelet factor 4; polymorphonuclear leukocytes (PMN, i.e., neutrophils), e.g., defensins; lymphocytes (Lyms), e.g., granulysin, granzyme A; mast cells (MCs), e.g., chymase, cathepsin D; and the eosinophil (Eos) with the release of cationic proteins such as MBP-1 on ECP which in aggregate would be expected to compromise the function of the epithelium. The anionic barrier on the apical side of the epithelium appears to account for why luminal but not mucosal presence of cationic protein leads to AHR. ASM, airway smooth muscle; ECP, eosinophil cationic protein; Epi, epithelium; Subm, submucosa; MBP-1, eosinophil granule major basic protein 1; PF-4, platelet factor 4;

asthma might in fact serve a similar protective function by binding up released cationic proteins from the inflammatory process within the airway lumen.

The dose-response characteristic of the AHR that is caused by cationic proteins is also unlike other inflammatory mediators. Rather than a gradual relationship between the dose of endotracheally instilled protein and responsiveness, the dose-response to cationic proteins is characterized by a sharp, abrupt increase,[35] i.e., a threshold effort. This observation may explain why, even though cationic proteins can be found in the tissue or sampled fluids (e.g., BAL) and cellular degranulation is detected,[43] there may not be a direct relationship to subsequent dysfunction if the total concentration of cationic charge is not sufficient to trigger a response. Moreover, given that mucus is anionic, it would be expected that in more chronic animals where mucous hyperplasia occurs, this increased mucus would neutralize cationic proteins. In this instance, the dose-response curve would be shifted to the right and no effect be observed. Lastly, this observation may also explain why MBP-1 knockout mice are not significantly different from wild-type ones when sensitized and challenged with antigen,[44] and would also explain the shift in the dose-response curve to cationic proteins observed in antigen-sensitized and -challenged animals.[45]

SECONDARY MEDIATOR CASCADE

Do cationic proteins cause AHR by a direct mechanism, such as the postulated destruction or death of epithelial cells[17,46] or by some other mechanism? One mechanism long thought to be the structural cause of asthma is the opening of intercellular spaces and the increase in penetration of the bronchoconstricting agent (e.g., methacholine) to the underlying airway smooth muscle (Fig. 8.5.3). To investigate this possible role for cationic proteins, it was postulated that kininogen-1 (KNG1) might be involved, since it was generated by kallikreins from a plasma protein precursor, kininogen, in the presence of a positively charged surface.[47] In addition, Venge and colleagues[48] reported that ECP can activate bradykinin and then generate the kallikreins so it seemed reasonable to postulate that any cationic protein, even a repetitive polycation such as PLL, could do the same by a similar mechanism.

To address this question, we first investigated the effects of NPC17713, a selective BRK-2 receptor antagonist by first demonstrating its ability to block the lung effects caused by aerosolizing BRK-2 and then showing that NPC17713 completely blocked the increase in airways responsiveness caused by both MBP-1 and poly-L-lysine.[49] We reasoned that BRK-2 was an early step in the response since NPC17713 did not mitigate the response if given immediately after PLL. We next showed that PLL or MBP-1 caused the release of kallikrein-like activity and that detectable levels of bradykinin appeared within the BAL. Coinstallation of PLL with heparin to neutralize the cationic charge did not cause kallikrein nor bradykinin release into the airway lumen. These results are consistent

with reports that ECP activates plasma kallikrein by a factor XII-dependent mechanism[48] and increases in bradykinin following nasal allergen challenge in humans are associated with the presence of ECP.[50]

As the tachykinins (e.g., substance P and neurokinin-A) have been implicated in several asthma-like functional abnormalities, such as smooth muscle constriction, increased vascular permeability, or mucous secretions, and the tachykinins are known to be released during allergic inflammatory processes, we investigated the role of the tachykinin in the PLL-induced AHR system.[51] We first showed that neonatal capsaicin treatment that permanently depletes neural sensory endings of their tachykinin mediators abrogated the increase in airway responsiveness caused by PLL. Phosphoramidon, an inhibitor of the neutral endopeptides that normally break down tachykinins, enhanced the effects of PLL. To test the system in a loss of function, the neurokinin 1 (NK1) antagonists, CP-96, 345, and RP-67580 in a dose-dependent fashion, blocked PLL-induced airways responsiveness. We then showed that like bradykinin, PLL caused the release of tachykinins as measured by the release of calcitonin gene-related peptide (CGRP), and lastly we linked PLL to increased plasma protein extravasation. More recently,[52] we have extended these observations and have determined that cationic proteins cause a decrease in epithelial function via a c-Jun N-terminal kinase (JNK)-dependent process underscoring the complexity of the mechanisms involved.

Taken together, these findings established that rather than causing the destruction of the epithelium as previously postulated, cationic proteins elicit a complex cascade of bradykinin and sensory ending release of tachykinins that in turn increase airway permeability and AHR (Fig. 8.5.3).

ROLE OF THE EPITHELIUM

Apical but not basolateral application of cationic proteins causes AHR. The airway lumen is lined with an anionic barrier comprised of heparan sulfate and glycosaminoglycans that display a basically charged surface to the airway lumen. Accordingly, it was hypothesized that cationic proteins increase airway responsiveness by a mechanism that involved the airway epithelium. To assess this hypothesis, several investigations were conducted into the role of the epithelia by using several different *in vitro* strategies. An *in vitro* tracheal tube system was developed[53] that allows for agents to be presented to either the lumenal or mucosal side of the airway tube system was used to directly implicate the epithelium. Denudation of the airway tube preparation caused a marked increase in both the lumenal agonist dose-response relationship and the epithelial-derived relaxation response. Cationic proteins caused almost an identical response to epithelial removal, i.e., an increase in lumenal dose-response to bronchoconstricting agents and a loss of

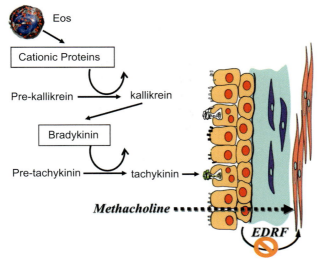

FIGURE 8.5.3 The introduction of cationic proteins into the airway lumen elicits a complex cascade of signaling events that include the conversion of pro-kallikreins into kinins such as bradykinin. Bradykinin in turn converts the pre-tachykinin into tachykinins such as substance P and protachykinin-1 (TKN1). The functional effect caused by this complex mediator cascade includes: (1) enhanced permeability and (2) loss of factors released by the epithelium that inhibit airway smooth muscle function such as epithelial-derived relaxation factor (EDRF).

the epithelial-derived relaxation response factor. On the other hand, extra lumenal application of cationic proteins was ineffective. As in the intact animal, when the charge was neutralized with either albumin or heparin, these cationic protein responses were blocked. Of interest, a microscopic examination of the tracheal epithelia after treatment with cationic proteins failed to reveal any overt structural changes or epithelial shedding.

The next study investigated the effect of cationic proteins on cultured epithelial cells, exploring the direct effects on epithelial function and, in particular, changes in epithelial permeability.[54] Using primary cell cultures from rabbit airways, we assessed the effects of cationic proteins on both transepithelial resistance and mannitol permeability. Structural integrity was assessed with keratin staining. PLL caused a dose-dependent and marked reduction in transepithelial resistance when it was applied to both the basolateral and apical sides of the confluent epithelial cultures, but the apical side was much more sensitive to the effects of cationic proteins. Cationic proteins caused a significant increase in mannitol permeability. Increasing Ca^{2+} markedly reduced these effects following the treatment of cultures with PLL. We speculated that the cationic proteins might in some unknown way displace calcium at the intracellular junctions leading to disruption of the tight junctions and thus the observed increase in macromolecule passage.

Collectively these results show that cationic proteins clearly cause a complex change in the functioning of the airway epithelium that compromises its function in a number of ways. There is an increase in the permeability of macromolecules, a decrease in epithelial integrity, and a loss of function specifically defined here as a decrease in the release of epithelium-derived relaxation factor. Like the intact airway, the apical as opposed to the basolateral surface seems the most sensitive to the effects of cationic proteins consistent with the location of glycosaminoglycans. These effects of cationic proteins are sustained (>1 h) but can be reversed if exposed to increased presence of anion proteins, suggesting a reversible mechanism. A recent study by Fan and colleagues[55] showed that intravenous injection of ECP could be detected specifically on the apical surface of the airway epithelium. These *in vivo*, *in situ*, and *in vitro* results are contrary to the widely held belief that cationic proteins cause asthma-like changes solely through a mechanism dependent on a specific receptor engagement, cell death, and tissue necrosis.

ROLE OF CATIONIC PROTEINS IN DISEASE SEVERITY

The hypothesis that the role of cationic proteins released from completely activated eosinophils may be responsible

or involved in more severe forms of asthma was posited. The degree of AHR that is observed in animal models is often not as severe as that observed in clinical cases of human asthma. In particular, the physiological response in humans with stable disease or animals is dominated by the closure of peripheral airways whereas the narrowing of central airways is modest.[56] On the other hand, when cationic proteins are directly instilled into the inflamed lung, there is a marked and exaggerated increase in central airway narrowing,[57,58] which we previously found to be difficult to cause in the mouse. The time course of the temporal response of hyperresponsiveness to inhaled methacholine was also markedly different (Fig. 8.5.4). In the animals that are only sensitized and challenged with antigen, the response of the central airway to methacholine is characterized by a modest but significant increase in resistance and the peak response was temporarily delayed to the right (longer time to reach peak response); however, after cationic protein exposure, the response of the central airway to inhaled methacholine is both highly exaggerated

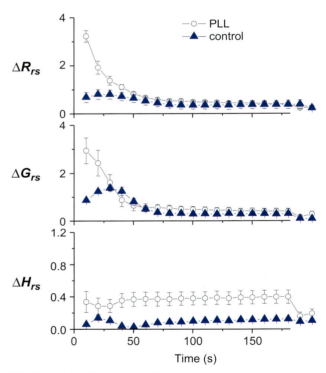

FIGURE 8.5.4 The lung mechanical response to methacholine. Methacholine is delivered to the lung and the mechanical response is assessed with forced oscillation using the constant phase model of analysis[9,10,52] that yields changes (Δ) in three variables: Rrs (resistance, of the respiratory system), Grs (tissue resistance—here a measure of small airway function), and Hrs (elastance of the lung). Control naive mice are the solid triangle and the poly-L-lysine (PLL) charged variables are indicated by the open circles. PLL alters only R and G—an airways effect. Note, too, that the peak response occurred during the methacholine exposure period (pre-time) consistent with an increase in epithelial permeability. *Adapted from reference[58].*

and so rapid that the peak response occurs during the exposure to methacholine. An *in silico* model was then used to create an improved understanding of the mechanism behind this remarkable response and required that we radically increased the time profile and magnitude for fractional airway narrowing of the conducting airways. These findings are consistent with the notion that cationic proteins greatly compromise the unregulated and protective functions of the epithelium, and by doing so the epithelium is essential to preventing maximal activation and shortening of airway smooth muscle (Fig. 8.5.3). The findings are also consistent with the notion that cationic proteins increase the transit of methacholine to the underlying smooth muscle. Compromising the function of the epithelium as a biophysical barrier seems to be at the epicenter of this pathophysiological mechanism.

Linking maximal activation of the eosinophil to a severe functional phenotype required the extensive genetic manipulation and was guided by the observation that mice deficient in both MBP-1 and EPO did not show a diminished asthma-like phenotype and therefore suggested that more vigorous or complete activation of the eosinophil might be required. Maximal chemotaxis of eosinophils into the BAL of IL-5 transgenic mice resulted when human eotaxin-2 (or mouse eotaxin) was instilled into a transgenic mouse that had high circulating numbers of eosinophils (IL-5 transgenic); hence, a mouse that overexpressed

eotaxin (hE2) via the Clara cells' 10 kDa secretory protein (CC10) promoter to originate the chemotactic signal to the epithelia was crossed with an IL-5 hypereosinophilic mouse.[59] This new mouse line (IL5/E2) had a profound phenotype characterized by: (1) high numbers of eosinophils in the airway lumen and the tissue; (2) marked eosinophil activation assessed by CD69[+], EPO release, and electron microscope evidence of maximal degranulation (Fig. 8.5.5). These mice exhibited altered airway structure (i.e., remodeling) with evidence of mucous hyperplasia, impressive collagen deposition, and epithelial cell shedding. However, the most impressive phenotype was the increase in AHR to such a degree that mice frequently expired at low doses of methacholine. Moreover, all died at higher methacholine doses that only caused little lung function change in even the antigen-challenged wild-type counterparts (Fig. 8.5.5).

This significant result suggests that if properly activated, the eosinophil is capable of the profound pathophysiological derangements that are associated with severe asthma. Novel insights will be forthcoming if an understanding as to why most asthmatics do not have this degree of eosinophil activation is obtained. It is also important to note that the strain of mouse (C57BL/6) is not one that commonly exhibits much of a functional phenotype to antigen in the first place.[43] These findings demonstrate that overactivation of the eosinophil is capable of overcoming

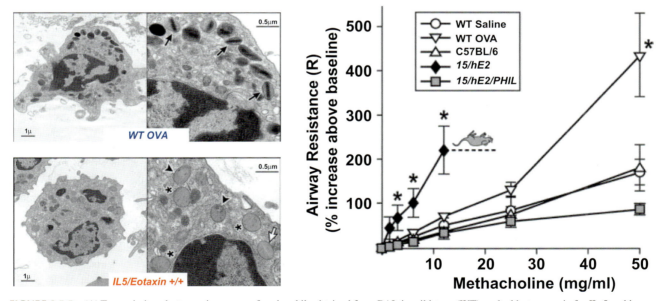

FIGURE 8.5.5 *(A)* Transmission electron microscopy of eosinophils obtained from BAL in wild-type (WT) or double transgenic for IL-5 and human eotaxin-2 (IL-5/hE2) mice. The wild-type mice eosinophil shows signs of cytoplasmic activation but little degranulation (arrowed in the upper section). On the other hand, the IL-5/hE2 eosinophils display piecemeal degranulation and near complete loss of the central core of the secondary granules (asterisks) or partial loss (arrowheads). *(B)* Double transgenic mice (IL-5/hE2) exhibit extreme methacholine-induced responses. Airway resistance was assessed with the single compartment model. As a comparison, wild-type mice (WT), antigen control (silver), wild-type immunized, and antigen-challenged (WT OVA), C57BL/6 mice not challenged (C57BL6), and IL-5/hE2 crossed to an eosinophil-less mouse (PHIL). The last mouse shows the dependency of the phenotype on eosinophils. $*p < 0.05$. *Adapted from reference*[59].

even significant genetic deficiencies that limit the response to antigen. These results also underscore the complex and multiple signals that go into activating inflammatory effector cells and debunk the common assumption that single mediators or mechanisms operate in isolation.

The evidence up to this point suggests that antigen sensitization and challenge leads to AHR by a very complex mechanism and involves a pantheon of mediators, cells, and downstream signaling mechanisms. On the other hand, the mechanisms and mediators of the AHR caused by cationic proteins are also complex, but different.[60] To investigate the role of multiple mechanisms, we decided to treat antigen-sensitized and -challenged animals with cationic proteins to determine if an upregulated (synergistic) response occurred.[61] We used a dose of PLL that native mice previously tolerated well, but now, when administered to sensitized mice challenged with antigen, all these animals died at the first dose of inhaled methacholine (Fig. 8.5.6). When the dose of PLL was reduced by half a log, the animals survived but the dose-response was moved significantly left and upwards, consistent with the changes observed in patients with more severe disease[3] and exhibited a pattern consistent with profound airway closure. Central airway responses, if anything, seem to be protected at least with regards to the effects of PLL but there was synergy in the responses observed in the small airways. Synergy also occurred in another strain of mouse (A/J) that exhibits a naturally occurring AHR that is determined by a genetic mechanism of increased contractibility of airway smooth muscle.[60] It is noteworthy that synergy was not present in C57BL/6 mice, and since we had previously shown in this mouse strain a defect in NF-kappaB repressing factor (NF-κB) signaling,[62] this might explain these results. Since asthmatic patients, especially those with severe disease, are thought to have an exuberant NF-κB-dependent response, these findings might have clinical relevance.

To summarize and to propose a working hypothesis, we would suggest that profound protective mechanisms or genetic differences located within the epithelium protect the allergic airway smooth muscle from the ravages of cationic proteins and in general prevent or diminish asthma attacks from being fatal. Moreover, in the situation of maximal eosinophil activation or in the case where there is participation by other cationic protein releasing effector cells (e.g., neutrophils), synergy and fatality are more likely.

CONCLUSIONS

Cationic proteins cause a profound increase in airway responsiveness. While this clearly implicates the eosinophil, it also suggests other inflammatory cells could be involved. The mechanism of lung dysfunction caused by cationic

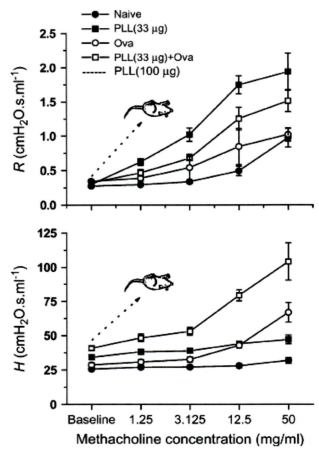

FIGURE 8.5.6 Methacholine dose-response relationships for airway resistance (R) and by elastance (H) for four groups of BALB/c mice. Naive, untreated; PLL, PLL given endotracheally at the dose indicated; OVA, animals are immunized and challenged with the antigen ovalbumin. Antigen-challenged animals given 100 μg PLL died immediately when challenged with methacholine. Reducing the dose of PLL to 33 μg showed a decreased response to methacholine compared to PLL alone in antigen-naive mice, indicating a reduction in synergy of the central airway and this is likely due to enhanced protection afforded by epithelial remodeling. On the other hand, the peripheral response (elastance or H) is greatly enhanced. The latter suggest the complex role of multiple mechanisms in causing AHR—especially in severe situations. *Taken from reference*[61].

proteins is not limited to a single mediator but involves a cascade of secondary mediators that include bradykinin and the release of neural tachykinins. The effects cationic proteins have on the functioning of the airway are also complex, involving both an increased permeability and loss of airway smooth muscle inhibitors. More recent studies suggest that eosinophil cationic proteins may be more important in more severe phenotypes of asthma. There is also the likelihood that cationic proteins synergize with other mediators (e.g., leukotrienes) and lead to the complex phenotypes that are collectively known as the syndrome of asthma and these various scenarios may also account for the diverse phenotypes associated with this disease.

REFERENCES

1. Hogan SP, Rosenberg HF, Moqbel R, Phipps S, Foster PS, Lacy P, et al. Eosinophils: biological properties and role in health and disease. *Clin Exp Allergy* 2008;**38**(5):709–50.

2. Kay AB. The role of eosinophils in the pathogenesis of asthma. *Trends Mol Med* 2005;**11**(4):148–52.

3. Bosse Y, Riesenfeld EP, Pare PD, Irvin CG. It's not all smooth muscle: non-smooth-muscle elements in control of resistance to airflow. *Annu Rev Physiol* 2010;**72**:437–62.

4. Irvin CG. Development, Structure and Physiology in Normal and Asthmatic Lung. In: Nguyen T, Scott J, editors. *Middleton's Allergy*. 7th ed. New York: Elsevier, Inc; 2008.

5. ATS Committee on Proficiency Standards. Guidelines for methacholine and exercise challenge testing–1999. *Am J Respir Crit Care Med* 2000;**161**(1):50–6. PMID: 10619836.

6. Reddel HK, Taylor DR, Bateman ED, et al. An official American Thoracic Society/European Respiratory Society statement: asthma control and exacerbations: standardizing endpoints for clinical asthma trials and clinical practice. *Am J Respir Crit Care Med* 2009;**180**:59–99.

7. Bates JH, Rincon M, Irvin CG. Animal models of asthma. *Am J Physiol Lung Cell Mol Physiol* 2009;**297**:L401–10.

8. Kay AB, Phipps S, Robinson DS. A role for eosinophils in airway remodelling in asthma. *Trends Immunol* 2004;**25**(9):477–82.

9. Irvin CG, Bates JH. Measuring the lung function in the mouse: the challenge of size. *Respir Res* 2003;**4**:4.

10. Bates JH, Irvin CG. Measuring lung function in mice: the phenotyping uncertainty principle. *J Appl Physiol* 2003;**94**:1297–306.

11. An SS, Bai TR, Bates JH, et al. Airway smooth muscle dynamics: a common pathway of airway obstruction in asthma. *Eur Respir J* 2007;**29**:834–60.

12. Diaz P, Galleguillos FR, Gonzalez MC, Pantin CF, Kay AB. BAL in asthma: the effect of disodium cromoglycate (cromolyn) on leukocyte counts, immunoglobulins, and complement. *J Allergy Clin Immunol* 1984;**74**(1):41–8.

13. Metzger WJ, Nugent K, Richerson HB, Moseley P, Lakin R, Zavala D, et al. Methods for BAL in asthmatic patients following bronchoprovocation and local antigen challenge. *Chest* 1985;**87**(1 Suppl):16S–9S.

14. Rosenberg HF, Phipps S, Foster PS. Eosinophil trafficking in allergy and asthma. *J Allergy Clin Immunol* 2007;**119**(6):1303–10.

15. Wardlaw AJ, Brightling CE, Green R, Woltmann G, Bradding P, Pavord ID. New insights into the relationship between airway inflammation and asthma. *Clin Sci (Lond)* 2002;**103**(2):201–11.

16. Schauer U, Daume U, Muller R, Riedel F, Gemsa D, Rieger CH. Relationship between LTC4 generation of hypodense eosinophils and bronchial hyperreactivity in asthmatic children. *Int Arch Allergy Appl Immunol* 1990;**92**:82–7.

17. Gleich GS, Adolphsom CR, Kia H. The Eosinophil and Asthma. In: Asthma & Rhinitis, Busse WW, Holgate ST, editors. *Blackwell Science, Oxford, UK* 2000;Vol. I. p. 429–79.

18. Bradley BL, Azzawi M, Jacobson M, et al. Eosinophils, T-lymphocytes, mast cells, neutrophils, and macrophages in bronchial biopsy specimens from atopic subjects with asthma: comparison with biopsy specimens from atopic subjects without asthma and normal control subjects and relationship to bronchial hyperresponsiveness. *J Allergy Clin Immunol* 1991;**88**:661–74.

19. Pin I, Freitag AP, O'Byrne PM, et al. Changes in the cellular profile of induced sputum after allergen-induced asthmatic responses. *Am Rev Respir Dis* 1992;**145**:1265–9.

20. Irvin CG, Bethel R, Hensen P. Granulocyte mediators and the induction of increased airways reactivity. In: Ray AB, editor. *Alergy and Inflammation*. London: Academic Press; 1987.

21. Hutchison AA, Hinson Jr JM, Brigham KL, Snapper JR. Effect of endotoxin on airway responsiveness to aerosol histamine in sheep. *J Appl Physiol* 1983;**54**(6):1463–8.

22. Holtzman MJ, Fabbri LM, O'Byrne PM, Gold BD, Aizawa H, Walters EH, et al. Importance of airway inflammation for hyperresponsiveness induced by ozone. *Am Rev Respir Dis* 1983;**127**(6):686–90.

23. Irvin CG, Berend N, Henson PM. Airways hyperreactivity and inflammation produced by aerosolization of human C5A des arg. *Am Rev Respir Dis* 1986;**134**(4):777–83.

24. Marsh WR, Irvin CG, Murphy KR, Behrens BL, Larsen GL. Increases in airway reactivity to histamine and inflammatory cells in BAL after the late asthmatic response in an animal model. *Am Rev Respir Dis* 1985;**131**(6):875–9.

25. O'Byrne PM, Walters EH, Gold BD, Aizawa HA, Fabbri LM, Alpert SE, et al. Neutrophil depletion inhibits airway hyperresponsiveness induced by ozone exposure. *Am Rev Respir Dis* 1984;**130**(2):214–9.

26. Murphy KR, Wilson MC, Irvin CG, Glezen LS, Marsh WR, Haslett C, et al. The requirement for polymorphonuclear leukocytes in the late asthmatic response and heightened airways reactivity in an animal model. *Am Rev Respir Dis* 1986;**134**(1):62–8.

27. Flood-Page PT, Menzies-Gow AN, Kay AB, Robinson DS. Eosinophil's role remains uncertain as anti-interleukin-5 only partially depletes numbers in asthmatic airway. *Am J Respir Crit Care Med* 2003;**167**:199–204.

28. Green RH, Brightling CE, McKenna S, Hargadon B, Parker D, Bradding P, et al. Asthma exacerbations and sputum eosinophil counts: a randomised controlled trial. *Lancet* 2002;**360**(9347):1715–21.

29. Nair P, Pizzichini MM, Kjarsgaard M, et al. Mepolizumab for prednisone-dependent asthma with sputum eosinophilia. *N Engl J Med* 2009;**360**:985–93.

30. Lee JJ, Dimina D, Macias MP, et al. Defining a link with asthma in mice congenitally deficient in eosinophils. *Science* 2004;**305**:1773–6.

31. Humbles AA, Lloyd CM, McMillan SJ, et al. A critical role for eosinophils in allergic airways remodeling. *Science* 2004;**305**:1776–9.

32. McGarry MP, Prtheroe CA, Lee JJ. *Mouse Hematology: A Laboratory Manual*. Cold Spring Harbor, New York: Cold Spring Harbor Laboratory Press; 2010.

33. Dyer KD, Czapiga M, Foster B, et al. Eosinophils from lineage-ablated Delta dblGATA bone marrow progenitors: the dblGATA enhancer in the promoter of GATA-1 is not essential for differentiation ex vivo. *J Immunol* 2007;**179**:1693–9.

34. Gundel RH, Letts LG, Gleich GJ. Human eosinophil major basic protein induces airway constriction and airway hyperresponsiveness in primates. *J Clin Invest* 1991;**87**(4):1470–3.

35. Uchida DA, Ackerman SJ, Coyle AJ, et al. The effect of human eosinophil granule major basic protein on airway responsiveness in

the rat in vivo. A comparison with polycations. *Am Rev Respir Dis* 1993;**147**:982–8.

36. Monteseirin J, Vega A, Chacon P, Camacho MJ, El Bekay R, Asturias JA, et al. Neutrophils as a novel source of eosinophil cationic protein in IgE-mediated processes. *J Immunol* 2007;**179**(4):2634–41.

37. Coyle AJ, Ackerman SJ, Irvin CG. Cationic proteins induce airway hyperresponsiveness dependent on charge interactions. *Am Rev Respir Dis* 1993;**147**:896–900.

38. Ahmed T, Garrigo J, Danta I. Preventing bronchoconstriction in exercise-induced asthma with inhaled heparin. *N Engl J Med* 1993;**329**:90–5.

39. Diamant Z, Timmers MC, van der Veen H, Page CP, van der Meer FJ, Sterk PJ. Effect of inhaled heparin on allergen-induced early and late asthmatic responses in patients with atopic asthma. *Am J Respir Crit Care Med* 1996;**153**:1790–5.

40. Davids H, Ahmed A, Oberholster A, van der Westhuizen C, Mer M, Havlik I. Endogenous heparin levels in the controlled asthmatic patient. *S Afr Med J* 2010;**100**:307–8.

41. Young E. The anti-inflammatory effects of heparin and related compounds. *Thromb Res* 2008;**122**:743–52.

42. Simionescu D, Simionescu M. Differentiated distribution of the cell surface charge on the alveolar-capillary unit. Characteristic paucity of anionic sites on the air-blood barrier. *Microvasc Res* 1983;**25**(1):85–100.

43. Takeda K, Haczku A, Lee JJ, Irvin CG, Gelfand EW. Strain dependence of airway hyperresponsiveness reflects differences in eosinophil localization in the lung. *Am J Physiol Lung Cell Mol Physiol* 2001;**281**:L394–402.

44. Denzler KL, Farmer SC, Crosby JR, et al. Eosinophil major basic protein-1 does not contribute to allergen-induced airway pathologies in mouse models of asthma. *J Immunol* 2000;**165**:5509–17.

45. Bates JH, Cojocaru A, Haverkamp HC, Rinaldi LM, Irvin CG. The synergistic interactions of allergic lung inflammation and intra-tracheal cationic protein. *Am J Respir Crit Care Med* 2008;**177**:261–8.

46. Motojima S, Frigas E, Loegering DA, Gleich GJ. Toxicity of eosinophil cationic proteins for guinea pig tracheal epithelium in vitro. *Am Rev Respir Dis* 1989;**139**(3):801–5.

47. Proud D, Kaplan AP. Kinin formation: mechanisms and role in inflammatory disorders. *Annu Rev Immunol* 1988;**6**:49–83.

48. Venge P, Dahl R, Hallgren R. Enhancement of factor XII dependent reactions by eosinophil cationic protein. *Thromb Res* 1979;**14**:641–9.

49. Coyle AJ, Ackerman SJ, Burch R, Proud D, Irvin CG. Human eosinophil-granule major basic protein and synthetic polycations induce airway hyperresponsiveness in vivo dependent on bradykinin generation. *J Clin Invest* 1995;**95**:1735–40.

50. Svensson C, Andersson M, Persson CG, Venge P, Alkner U, Pipkorn U. Albumin, bradykinins, and eosinophil cationic protein on the nasal mucosal surface in patients with hay fever during natural allergen exposure. *J Allergy Clin Immunol* 1990;**85**:828–33.

51. Coyle AJ, Perretti F, Manzini S, Irvin CG. Cationic protein-induced sensory nerve activation: role of substance P in airway hyper-responsiveness and plasma protein extravasation. *J Clin Invest* 1994;**94**:2301–6.

52. van der Velden J, Brown AL, McElhinney B, Irvin CG, Janssen-Heininger YM. c-Jun N-terminal kinase 1 is required for the development of pulmonary fibrosis. *Am J Respir Cell Mol Biol* 2009;**40**:422–32.

53. Coyle AJ, Mitzner W, Irvin CG. Cationic proteins alter smooth muscle function by an epithelium-dependent mechanism. *J Appl Physiol* 1993;**74**:1761–8.

54. Uchida DA, Irvin CG, Ballowe C, Larsen G, Cott GR. Cationic proteins increase the permeability of cultured rabbit tracheal epithelial cells: modification by heparin and extracellular calcium. *Exp Lung Res* 1996;**22**:85–99.

55. Fan TC, Fang SL, Hwang CS, et al. Characterization of molecular interactions between eosinophil cationic protein and heparin. *J Biol Chem* 2008;**283**:25468–74.

56. Wagers S, Lundblad LK, Ekman M, Irvin CG, Bates JH. The allergic mouse model of asthma: normal smooth muscle in an abnormal lung? *J Appl Physiol* 2004;**96**:2019–27.

57. Homma T, Bates JH, Irvin CG. Airway hyperresponsiveness induced by cationic proteins in vivo: site of action. *Am J Physiol Lung Cell Mol Physiol* 2005;**289**:L413–8.

58. Bates JH, Wagers SS, Norton RJ, Rinaldi LM, Irvin CG. Exaggerated airway narrowing in mice treated with intratracheal cationic protein. *J Appl Physiol* 2006;**100**:500–6.

59. Ochkur SI, Jacobsen EA, Protheroe CA, et al. Coexpression of IL-5 and eotaxin-2 in mice creates an eosinophil-dependent model of respiratory inflammation with characteristics of severe asthma. *J Immunol* 2007;**178**:7879–89.

60. Wagers SS, Haverkamp HC, Bates JH, et al. Intrinsic and antigen-induced airway hyperresponsiveness are the result of diverse physiological mechanisms. *J Appl Physiol* 2007;**102**:221–30.

61. Bates JH, Cojocaru A, Haverkamp HC, Rinaldi LM, Irvin CG. The synergistic interactions of allergic lung inflammation and intra-tracheal cationic protein. *Am J Respir Crit Care Med* 2008;**177**:261–8.

62. Alcorn JF, et al. Strain-dependent activation of NF-kappaB in the airway epithelium and its role in allergic airway inflammation. *(Translated from eng) Am J Physiol Lung Cell Mol Physiol* 2010;**298**(1):L57–66 (in eng).

Chapter 8.6

Cell-Free Granules are Functional Secretory Organelles

Josiane S. Neves, Salahaddin Mahmudi-Azer, Redwan Moqbel and Peter F. Weller

GRANULES ARE HIGHLY STRUCTURED ORGANELLES AND SITES OF EOSINOPHIL PROTEIN STORAGE

Mature eosinophils have a single population of secondary (or specific, or crystalloid) granules ultrastructurally characterized as membrane-bound organelles that contain a crystalloid core surrounded by a matrix. It is now recognized that human eosinophils synthesize and store cationic proteins such as eosinophil cationic protein (ECP), eosinophil-derived neurotoxin (EDN), eosinophil peroxidase (EPO), eosinophil granule major basic protein 1 (MBP-1), enzymes and over 36 cytokines, chemokines, and growth factors and secrete these in response to a range of stimuli and agonists.[1-7] Diverse eosinophil electron microscopy and subcellular fractionation studies have revealed that granules are the predominant intracellular sites of storage of preformed cationic proteins, growth factors, and cytokines/chemokines within eosinophils. These include: C-C motif chemokine 5 (CCL5/RANTES), C-X-C motif chemokine 5 (CXCL5/ENA-78), eotaxin (CCL11), granulocyte-macrophage colony-stimulating factor (GM-CSF), growth-regulated alpha protein (GRO-α), interferon gamma (IFN-γ), interleukins-2—6 (ILs-2—6), interleukin-10 (IL-10), 12 (IL-12), 13 (IL-13), nerve growth factor (NGF), stem cell factor (SCF), transforming growth factor alpha (TGF-α), and tumor necrosis factor alpha (TNF-α).[8] Although the dogma is that the proteins that are pre-stored in eosinophil crystalloid granules are imported from the cytoplasm after messenger RNA translation and cell nucleus transcription, pre-existing and new ultrastructural evidence indicates that protein synthesis may take place within the eosinophil granule.[9-11] Previously, using transmission electron microscopy (TEM) techniques, it was shown that radiolabeled [³H]uridine can be incorporated into RNA in specific eosinophil granules.[9,10] Recently, these studies were extended and the results indicated that eosinophil granules may be sites of RNA and DNA localization.[11]

In addition, TEM and electron tomography studies followed by three-dimensional reconstituted images of human eosinophils revealed that eosinophil secretory granules are not only intracellular sites of protein storage, but are highly structured organelles limited by a lipid bilayer membrane containing an elaborate interconnected internal membranous vesiculotubular network able to selectively segregate, sequester, and relocate granule products during eosinophil activation in response to different stimuli.[6,7,12] It was also shown that this membranous network within granules exhibits points of continuity with the membrane surrounding the granule.[12] These findings might suggest that these intragranular membranous subcompartments may have important functional implications regarding granule protein sorting and mobilization in the early stages of selective eosinophil secretion. Robust intragranular stores of preformed proteins combined with highly compartmentalized granule morphology uniquely enable eosinophils to rapidly and selectively secrete their variety of cytokines in response to different stimuli, distinguishing eosinophils from most other leukocytes that require new protein synthesis prior to

cytokine secretion. Interestingly, eosinophil crystalloid granules are also rich sites of localization of chemokine [e.g., C-C chemokine receptor type 3 (CCR3)], cytokine [e.g., interleukin-4 receptor subunit alpha (IL-4RA)], and leukotriene receptors [e.g., cysteinyl leukotriene receptor 1 (CysLT1R) and 2 (CysLT2R), and P2Y purinoceptor 12 (P2Y12)][5,13,14] (Fig. 8.6.1).

CYTOLYSIS AS A MECHANISM FOR THE RELEASE AND DEPOSITION OF MEMBRANE-BOUND GRANULES

The mechanisms governing the selective secretion of granule-derived proteins underlie the biological activities of eosinophils in health and disease. Unlike basophils or mast cells that undergo acute exocytotic degranulation on cross-linking of their Fc epsilon receptors (FcεRs), a physiological mechanism to elicit comparable degranulation of eosinophils has not been reported. In fact, three different mechanisms are known to facilitate the release of eosinophil granule-stored proteins. Compound exocytosis, whereby the entire granule contents are released extracellularly following fusion of granules with plasma membranes, occurs when eosinophils interact with large targets, such as helminthic parasites, but is otherwise not commonly observed *in vivo*. Instead, secretion of granule contents from within intact eosinophils occurs mainly by a mechanism termed piecemeal degranulation (PMD)

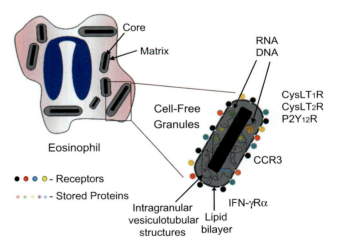

FIGURE 8.6.1 Schematic representation of eosinophil granule components. Eosinophil granules contain a crystalloid core surrounded by a matrix and an extensive vesiculotubular network, as well as a variety of preformed proteins, and likely RNA and DNA. Granules express receptors topologically oriented to engage eotaxin (CCL11)—CCR3, IFN-γ—IFN-γ receptor α chain, two cysteinyl leukotriene receptors—CysLT1R, CysLT2R—and the purinergic receptor P2Y12. CCR3, C-C chemokine receptor type 3; CysLT1R, cysteinyl leukotriene receptor 1; CysLT2R, cysteinyl leukotriene receptor 2; IFN-γ, interferon gamma; P2Y12, P2Y purinoceptor 12.

whereby granule contents are selectively mobilized into spherical and tubular vesicles that need to disengage from granules, cross the cytoplasm, and fuse with the plasma membrane to release their specific granule-derived protein cargo at the cell surface.[8,12,15]

A third pattern of eosinophil *degranulation* is associated with cytolysis. Following lysis of an eosinophil with loss of its plasma membrane integrity, there is the release and deposition of intact, cell-free, membrane-bound granules. Although PMD has been considered the predominant mechanism underlying eosinophil degranulation and secretion, analyses of the published literature reveal that cytolysis is a common mechanism for cell-free eosinophil granule release and deposition in tissue biopsies of patients with different allergic manifestations, including atopic dermatitis,[16] nasal allergy,[17,18] nasal polyps,[19] and other upper airway respiratory mucosal disorders.[20] The relevance of cell-free eosinophil granules and their capacity to mobilize their contents of preformed proteins independently of the cell have been neglected over the years, but recent findings have shed new light on the functional roles of cell-free eosinophil granules.

ROLE OF FREE EXTRACELLULAR EOSINOPHIL GRANULES IN DISEASE

The intriguing presence of membrane-bound cell-free granules extruded from eosinophils has been long recognized in tissues associated with eosinophilia, including allergic diseases and immune and inflammatory responses to helminths. The first recognition of free eosinophil granules in the sputum of asthmatics was reported in the early 1880s,[21] but the relevance of this observation was unknown and over the years this finding was at times attributed to *crush artifacts*. However, to date, signs of eosinophil cytolysis and the presence of membrane-bound free eosinophil granules in tissue sites associated with eosinophilic disorders have repeatedly been reported in the literature.[16,19,22−31]

With their unique ultrastructural morphology, clusters of intact extracellular eosinophil granules were described in asthmatic sputum samples[21] and in association with sinus tissue or nasal mucus of patients with chronic rhinosinusitis[23,29] and nasal allergy,[18] respectively. The skin of patients with chronic idiopathic urticaria[30] revealed the existence of cytolytic eosinophils and extracellular eosinophil granules, while TEM analyses of skin biopsies of patients with atopic dermatitis showed the presence of cell-free membrane-bound eosinophil granules in the dermis without recognizable adjacent eosinophils.[16] In patients with eosinophilic esophagitis, blind histological evaluation of esophageal biopsy specimens revealed significantly increased numbers of degranulated eosinophils (defined as free eosinophil granules) in the esophageal epithelium.[28] In a different study, after analyzing biopsies of patients with eosinophilic esophagitis, eosinophil granules were also extensively found extracellularly.[27] The presence of cell-free eosinophil granules was also recognized in host response to helminths. It was shown that after amocarzine anthelminthic chemotherapy, granules from necrotic eosinophils were regularly found on the surface of damaged microfilariae of *Onchocerca volvulus*.[25] Extracellular cell-free granules were also often found in tissue sites of other diseases other than allergic and anthelminthic responses. In patients presenting with advanced gastric carcinoma, TEM images revealed extracellular deposition of clusters of membrane-bound extracellular granules mixed with lipid bodies adjacent to late apoptotic eosinophils.[26] Furthermore, free eosinophil granules have also been found in histopathological analyses of subcutaneous fat necrosis lesions in newborns.[31]

The clinical significance of these observations in terms of disease severity or prognosis is not yet known. However, there is one report showing that the presence of free extracellular granules correlates with the severity of urticaria.[30] Moreover, another study demonstrated that anti-Fas (CD95) antibody-induced cytolysis of murine airway eosinophils and the release of cell-free granules aggravated rather than resolved experimental airway inflammation.[32] The release of free extracellular granules might potentially explain the noted enhancement of the inflammatory response, but an underlying mechanism by which extracellular eosinophil granules might aggravate inflammation had not been investigated.

EOSINOPHIL GRANULES AS SECRETORY-COMPETENT ORGANELLES

Despite the fact that intact, membrane-bound granules extruded from eosinophils have been extensively found in tissue sites associated to eosinophilic inflammation, the functional roles and capacities of eosinophil granules to secrete their preformed cationic and cytokine proteins has only recently become clear. Notably, it has been shown that isolated human eosinophil granules exert an extracellular function as secretion-competent organelles.[13,14,33] It was demonstrated that purified cell-free eosinophil granules, isolated by subcellular fractionation from disrupted human peripheral blood eosinophils,[34] express receptors, including CCR3, IFN-γ receptor α (CD119) chain, CysLT1R and CysLT2R, and the purinergic receptor P2Y12, topologically oriented to engage their ligands.[13,14]

G protein-coupled receptors (GPCRs) have traditionally been considered to transduce signals at the plasma membrane[35] and less often at the nuclear membrane.[36−38] However, it has been suggested that certain receptors might signal in other subcellular compartments[39] and that cytokine receptors may have critical roles in mediating selective and rapid secretion in different cell types.[5,40] Previously, CCR3 expression has been demonstrated on intracellular mast cell granules[41] and IL-10 receptors have been localized on neutrophil granules[42]; however, no functional roles for the granule localizations of these receptors have been suggested. Interestingly, on eosinophil cell-free granules, the topology of receptors [CCR3, cysteinyl leukotriene receptors (CysLTRs), IFN-γ receptor α chain], with ligand-binding domains displayed on the outer granule membranes, not only allows granules to function extracellularly, but also suggests that granule-expressed receptors might potentially function intracellularly to mediate eosinophil granule protein mobilization and secretion.

In response to stimulation with IFN-γ and eotaxin (CCL11), purified cell-free granules elicited intragranular signal transduction that lead to differential secretion of granule-stored cytokines and other proteins. These stimuli induced a dose-dependent release of ECP, EPO, and hydrolytic enzymes such as beta hexosaminidase subunit alpha. The treatment of isolated granules with IFN-γ also induced differential secretion of cytokines, thus eliciting IL-4 and IL-6, but not IL-13 release. Inhibitors of intracellular kinases, including tyrosine kinases, protein kinase C, and p38 mitogen-activated protein kinase (p38 MAPK), inhibited IFN-γ-dependent release of effector proteins from the granules. Eotaxin-dependent granule content release was sensitive to a p38 MAPK inhibitor, a phosphatidylinositol 3-kinase (PI3K) inhibitor, and pertussis toxin, an inhibitor of $G_i\alpha$-coupled GPCRs. The presence and the activation of these kinases within purified cell-free granules were confirmed by immunoblotting assays.[14]

In response to leukotriene C4 (LTC4) and the extracellularly generated leukotriene D4 (LTD4) and E4 (LTE4),

granules secreted ECP but not cytokines.[13] Notably, the dose-response to the three cysteinyl leukotrienes varied. LTC4 and LTE4 elicited ECP secretion only in lower (subnanomolar) concentrations, which was fully consistent with the high-dose inhibition characteristic of the GPCRs. Intriguingly, LTD4 elicited ECP secretion at low and high, but not at intermediate concentrations. This dose-response might suggest engagements of two receptors sensitive to LTD4; the first responds to low LTD4 levels and then exhibits higher dose inhibition, while a second receptor putatively mediates secretion by higher concentrations of LTD4. In fact, homo- and heterodimerization of CysLTRs and purinergic receptors have been widely suggested.[43,44] However, whether dimerization of receptors is involved in this response, remains to be elucidated.

A notable finding was the capacity of low concentrations of LTE4 to elicit granule ECP secretion. LTE4 is a major extracellularly generated CysLTR commonly found in the biological fluids of allergic inflammatory sites.[45] LTE4 is a weak stimulus for both CysLTRs, but has been recognized to elicit responses that are unlikely based on CysLT1R and CysLT2R[45–48]; this may imply the existence of another receptor sensitive to LTE4. Recently, the purinergic P2Y12 has been indicated as an additional LTE4- and LTD4-sensitive receptor, as identified by *in silico* and *in vitro* methods.[49] Further, *in vivo* studies have characterized the relevance of P2Y12 in the pulmonary allergic response mediated by LTE4[50] (Fig. 8.6.2).

In response to all three CysLTRs, a P2Y12 antagonist and montelukast, not likely acting only as a classic inhibitor of CysLT1R, effectively inhibited ECP secretion by cell-free eosinophil granules stimulated with LTC4, LTD4, and LTE4.[13] In fact, off-target actions of montelukast have been demonstrated lately; however, whether broader inhibitory activities of montelukast suggest functional

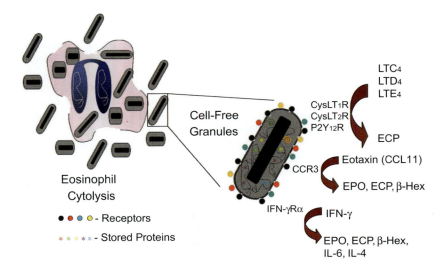

FIGURE 8.6.2 Eosinophil post-cytolytic mechanisms lead to granule extrusion and deposition. Extracellular eosinophil granules express receptors, topologically oriented to engage their ligands, for CCR3, IFN-γreceptor α chain, two cysteinyl leukotriene receptors (CysLT1R, CysLT2R), and the purinergic receptor P2Y12. On activation by eotaxin (CCL11), IFN-γ and the cysteinyl leukotrienes (LTC4, LTD4, and LTE4) granules respond by secreting their preformed proteins as indicated. β-hex, beta hexosaminidase subunit alpha; CCR3, C-C chemokine receptor type 3; CysLT1R, cysteinyl leukotriene receptor 1; CysLT2R, cysteinyl leukotriene receptor 2; ECP, eosinophil cationic protein; EPO, eosinophil peroxidase; IFN-γ, interferon gamma; IL-4, interleukin-4; IL-6, interleukin-6; LTC4, leukotriene C4; LTD4, leukotriene D4; LTE4, leukotriene E4; P2Y12, P2Y purinoceptor 12.

heterodimerization between CysLT1R, CysLT2R, and P2Y12 remains unclear. These findings highlight the capacity of CysLTRs to stimulate cell-free granule secretory responses and identify novel mechanisms whereby LTC_4 and the extracellularly generated LTD_4 and LTE_4 may serve as intracrine mediators of eosinophil granule-derived secretion.

Although the understanding of the mechanism by which cell-free granule proteins are released is still evolving, there is evidence that the impressive intragranular membranous network has a crucial role during cell-free granule protein mobilization and secretion. As previously described,[33] granule protein release was dramatically suppressed when the intragranular membranotubular network was collapsed following granule treatment with brefeldin A. In a similar fashion, brefeldin A was shown to act within granules in intact eosinophils.[6]

All these findings are remarkable since they provide additional understanding about the capacity of eosinophils to contribute to modulating host and inflammatory responses after eosinophil cytolysis. Cytolytic release of intact eosinophil granules yields extracellular organelles fully capable of ligand-elicited active secretory responses, and able to act as functional *cluster bombs* amplifying the differential secretory properties of eosinophils and likely contributing to the persistence and exacerbation of the inflammatory response.

CONCLUSION AND FUTURE PERSPECTIVES

Current knowledge is only beginning to explore the functional biology and responses of eosinophil granules. Eosinophils contain morphologically unique cytoplasmic secondary (crystalloid) granules. Evolving knowledge has established that eosinophil granules incorporate a secretory machinery that enables them to selectively mobilize and secrete specific protein contents intracellularly, as well as to function as secretory *organelles* when extruded, deposited, or released extracellularly during eosinophil cytolysis *in vivo*. However, many cell biology questions remain unanswered. Most notably, it is yet to be ascertained how specific granule-derived proteins are selectively mobilized for secretion either from isolated eosinophil granules or from intracellular granules by PMD within intact eosinophils. The molecular mechanisms that regulate eosinophil granule protein mobilization and secretion continue to be intriguing and requiring further delineation.

ACKNOWLEDGEMENTS

This work was supported by the National Institutes of Health Grants AI 020241, AI 022571, and AI 051645 (PFW), and the Canadian Institutes of Health Research (RM).

REFERENCES

1. Gleich GJ. Mechanisms of eosinophil-associated inflammation. *J Allergy Clin Immunol* 2000;**105**:651−63.
2. Blanchard C, Rothenberg ME. Biology of the eosinophil. *Adv Immuno* 2009;**101**:81−121.
3. Spencer LA, Szela CT, Perez SA, Kirchhoffer CL, Neves JS, Radke AL, et al. Human eosinophils constitutively express multiple Th1, Th2, and immunoregulatory cytokines that are secreted rapidly and differentially. *J Leukoc Biol* 2009;**85**:117−23.
4. Moqbel R, Coughlin JJ. Differential secretion of cytokines. *Sci STKE* 2006;**2006**. pe26.
5. Spencer LA, Melo RC, Perez SA, Bafford SP, Dvorak AM, Weller PF. Cytokine receptor-mediated trafficking of preformed IL-4 in eosinophils identifies an innate immune mechanism of cytokine secretion. *Proc Natl Acad Sci USA* 2006;**103**:3333−8.
6. Melo RC, Spencer LA, Perez SA, Ghiran I, Dvorak AM, Weller PF. Human eosinophils secrete preformed, granule-stored interleukin-4 through distinct vesicular compartments. *Traffic* 2005;**6**:1047−57.
7. Melo RC, Spencer LA, Perez SA, Neves JS, Bafford SP, Morgan ES, et al. Vesicle-mediated secretion of human eosinophil granule-derived major basic protein. *Lab Invest* 2009;**89**:769−81.
8. Melo RC, Spencer LA, Dvorak AM, Weller PF. Mechanisms of eosinophil secretion: large vesiculotubular carriers mediate transport and release of granule-derived cytokines and other proteins. *J Leukoc Biol* 2008;**83**:229−36.
9. Wickramasinghe SN, Hughes M. High resolution autoradiographic studies of RNA, protein and DNA synthesis during human eosinophil granulocytopoiesis: evidence for the presence of RNA on or within eosinophil granules. *Br J Haematol* 1978;**38**:179−83.
10. Dvorak AM, Morgan ES, Lichtenstein LM, Schleimer R. Ultrastructural autoradiographic analysis of RNA in isolated human lung mast cells during secretion and recovery from secretion. *Int Arch Allergy Immunol* 2000;**122**:124−36.
11. Behzad AR, Walker DC, Abraham T, McDonough J, Mahmudi-Azer S, Chu F, et al. Localization of DNA and RNA in eosinophil secretory granules. *Int Arch Allergy Immunol* 2010;**152**: 12−27.
12. Melo RC, Perez SA, Spencer LA, Dvorak AM, Weller PF. Intragranular vesiculotubular compartments are involved in piecemeal degranulation by activated human eosinophils. *Traffic* 2005;**6**:866−79.
13. Neves JS, Radke AL, Weller PF. Cysteinyl leukotrienes acting via granule membrane-expressed receptors elicit secretion from within cell-free human eosinophil granules. *J Allergy Clin Immunol* 2010;**125**:477−82.
14. Neves JS, Perez SA, Spencer LA, Melo RC, Reynolds L, Ghiran I, et al. Eosinophil granules function extracellularly as receptor-mediated secretory organelles. *Proc Natl Acad Sci USA* 2008;**105**:18478−83.
15. Melo RC, Weller PF. Piecemeal degranulation in human eosinophils: a distinct secretion mechanism underlying inflammatory responses. *Histol Histopathol* 2010;**25**: 1341−1354.
16. Cheng JF, Ott NL, Peterson EA, George TJ, Hukee MJ, Gleich GJ, et al. Dermal eosinophils in atopic dermatitis undergo cytolytic degeneration. *J Allergy Clin Immunol* 1997;**99**:683−92.
17. Erjefalt JS, Greiff L, Andersson M, Matsson E, Petersen H, Linden M, et al. Allergen-induced eosinophil cytolysis is a primary mechanism for granule protein release in human upper airways. *Am J Respir Crit Care Med* 1999;**160**:304−12.

18. Watanabe K, Misu T, Inoue S, Edamatsu H. Cytolysis of eosinophils in nasal secretions. *Ann Otol Rhinol Laryngol* 2003;**112**:169−73.

19. Armengot M, Garin L, Carda C. Eosinophil degranulation patterns in nasal polyposis: an ultrastructural study. *Am J Rhinol Allergy* 2009;**23**:466−70.

20. Erjefalt JS, Persson CG. New aspects of degranulation and fates of airway mucosal eosinophils. *Am J Respir Crit Care Med* 2000;**161**:2074−85.

21. Persson CG, Erjefalt JS. 'Ultimate activation' of eosinophils in vivo: lysis and release of clusters of free eosinophil granules (Cfegs). *Thorax* 1997;**52**:569−74.

22. Persson CG, Erjefalt JS. Eosinophil lysis and free granules: an in vivo paradigm for cell activation and drug development. *Trends Pharmacol Sci* 1997;**18**:117−23.

23. Greiff L, Erjefalt JS, Andersson M, Svensson C, Persson CG. Generation of clusters of free eosinophil granules (Cfegs) in seasonal allergic rhinitis. *Allergy* 1998;**53**:200−3.

24. Farinelli P, Gattoni M, Delrosso G, Boggio P, Raselli B, Merlo E, et al. Eosinophilic granules in subcutaneous fat necrosis of the newborn: what do they mean? *J Cutan Pathol* 2008;**35**:1073−4.

25. Gutierrez-Pena EJ, Knab J, Buttner DW. Immunoelectron microscopic evidence for release of eosinophil granule matrix protein onto microfilariae of Onchocerca volvulus in the skin after exposure to amocarzine. *Parasitol Res* 1998;**84**:607−15.

26. Caruso RA, Ieni A, Fedele F, Zuccala V, Riccardo M, Parisi E, et al. Degranulation patterns of eosinophils in advanced gastric carcinoma: an electron microscopic study. *Ultrastruct Pathol* 2005;**29**:29−36.

27. Chehade M, Sampson HA, Morotti RA, Magid MS. Esophageal subepithelial fibrosis in children with eosinophilic esophagitis. *J Pediatr Gastroenterol Nutr* 2007;**45**:319−28.

28. Aceves SS, Newbury RO, Dohil R, Bastian JF, Broide DH. Esophageal remodeling in pediatric eosinophilic esophagitis. *J Allergy Clin Immunol* 2007;**119**:206−12.

29. Ponikau JU, Sherris DA, Kephart GM, Kern EB, Congdon DJ, Adolphson CR, et al. Striking deposition of toxic eosinophil major basic protein in mucus: implications for chronic rhinosinusitis. *J Allergy Clin Immunol* 2005;**116**:362−9.

30. Toyoda M, Maruyama T, Morohashi M, Bhawan J. Free eosinophil granules in urticaria: a correlation with the duration of wheals. *Am J Dermatopathol* 1996;**18**:49−57.

31. Tajirian A, Ross R, Zeikus P, Robinson-Bostom L. Subcutaneous fat necrosis of the newborn with eosinophilic granules. *J Cutan Pathol* 2007;**34**:588−90.

32. Uller L, Rydell-Tormanen K, Persson CG, Erjefalt JS. Anti-Fas mAb-induced apoptosis and cytolysis of airway tissue eosinophils aggravates rather than resolves established inflammation. *Respir Res* 2005;**6**:90.

33. Neves JS, Weller PF. Functional extracellular eosinophil granules: novel implications in eosinophil immunobiology. *Curr Opin Immunol* 2009;**21**:694−9.

34. Neves JS, Perez SA, Spencer LA, Melo RC, Weller PF. Subcellular fractionation of human eosinophils: isolation of functional specific granules on isoosmotic density gradients. *J Immunol Methods* 2009;**344**:64−72.

35. Koelle MR. Heterotrimeric G protein signaling: Getting inside the cell. *Cell* 2006;**126**:25−7.

36. Gobeil F, Fortier A, Zhu T, Bossolasco M, Leduc M, Grandbois M, et al. G-protein-coupled receptors signalling at the cell nucleus: an emerging paradigm. *Can J Physiol Pharmacol* 2006;**84**:287−97.

37. Nielsen CK, Campbell JI, Ohd JF, Morgelin M, Riesbeck K, Landberg G, et al. A novel localization of the G-protein-coupled CysLT1 receptor in the nucleus of colorectal adenocarcinoma cells. *Cancer Res* 2005;**65**:732−42.

38. Jiang Y, Borrelli LA, Kanaoka Y, Bacskai BJ, Boyce JA. CysLT2 receptors interact with CysLT1 receptors and down-modulate cysteinyl leukotriene dependent mitogenic responses of mast cells. *Blood* 2007;**110**:3263−70.

39. Calebiro D, Nikolaev VO, and Lohse MJ. Imaging of persistent cAMP signaling by internalized G protein-coupled receptors. *J Mol Endocrinol* 45, 1−8.

40. Duitman EH, Orinska Z, Bulanova E, Paus R, Bulfone-Paus S. How a cytokine is chaperoned through the secretory pathway by complexing with its own receptor: lessons from interleukin-15 (IL-15)/IL-15 receptor alpha. *Mol Cell Biol* 2008;**28**:4851−61.

41. Price KS, Friend DS, Mellor EA, De Jesus N, Watts GF, Boyce JA. CC chemokine receptor 3 mobilizes to the surface of human mast cells and potentiates immunoglobulin E-dependent generation of interleukin 13. *Am J Respir Cell Mol Biol* 2003;**28**:420−7.

42. Elbim C, Reglier H, Fay M, Delarche C, Andrieu V, El Benna J, et al. Intracellular pool of IL-10 receptors in specific granules of human neutrophils: differential mobilization by proinflammatory mediators. *J Immunol* 2001;**166**:5201−7.

43. Milligan G. G protein-coupled receptor hetero-dimerization: contribution to pharmacology and function. *Br J Pharmacol* 2009;**158**:5−14.

44. Capra V, Thompson MD, Sala A, Cole DE, Folco G, Rovati GE. Cysteinyl-leukotrienes and their receptors in asthma and other inflammatory diseases: critical update and emerging trends. *Med Res Rev* 2007;**27**:469−527.

45. Peters-Golden M, Henderson Jr WR. Leukotrienes. *N Engl J Med* 2007;**357**:1841−54.

46. Lee TH, Woszczek G, Farooque SP. Leukotriene E4: perspective on the forgotten mediator. *J Allergy Clin Immunol* 2009;**124**:417−21.

47. Gauvreau GM, Parameswaran KN, Watson RM, O'Byrne PM. Inhaled leukotriene E(4), but not leukotriene D(4), increased airway inflammatory cells in subjects with atopic asthma. *Am J Respir Crit Care Med* 2001;**164**:1495−500.

48. Austen KF, Maekawa A, Kanaoka Y, Boyce JA. The leukotriene E4 puzzle: finding the missing pieces and revealing the pathobiologic implications. *J Allergy Clin Immunol* 2009;**124**:406−14. quiz 415−406.

49. Nonaka Y, Hiramoto T, Fujita N. Identification of endogenous surrogate ligands for human P2Y12 receptors by in silico and in vitro methods. *Biochem Biophys Res Commun* 2005;**337**:281−8.

50. Paruchuri S, Tashimo H, Feng C, Maekawa A, Xing W, Jiang Y, et al. Leukotriene E$_4$-induced pulmonary inflammation is mediated by the P2Y12 receptor. *J Exp Med* 2009;**206**:2543−55.

Eosinophils and Anti-Pathogen Host Defense

Introduction

Gerald J. Gleich and Kristin Leiferman

In the beginning, the eosinophil was a mystery leukocyte, brilliantly staining but without obvious function. The earliest clues came from the application of Paul Ehrlich's then new (in the latter part of the 19th century) technology for staining blood revealing associations with asthma and helminthiasis. By the end of Ehrlich's life in 1915, the eosinophil had been further linked to guinea pig anaphylaxis and, subsequently, human hypersensitivity reactions. However, for a time, the quest for more discoveries seemed stuck, and anaphylaxis, asthma, and helminth infections appeared to be the major diseases in which eosinophils had a role. This view was not to last. Once eosinophil granule proteins and reagents to localize them were available, it became clear that eosinophils participate in diseases that would not be predicted based on blood counts. Along with many inflammatory disorders, a role for eosinophils in infectious diseases beyond parasitosis has emerged. This section discusses eosinophil function relating to bacterial, fungal, and viral diseases, and strong cases are made for important eosinophil contributions to some of these diseases.

Observations dating to 1893 reported that blood eosinophils were reduced during bacterial infections. Much later, in the mid-1970s, studies in murine models showed that pyelonephritis caused by *Escherichia coli* and early subcutaneous pneumococcal abscesses produced eosinopenia, whereas trichinosis infections were accompanied by eosinophilia.[1] Remarkably, establishment of pyelonephritis or pneumococcal abscesses suppressed the eosinophilia induced by trichinosis infection.[2] A factor that caused eosinopenia was identified and partially characterized, but its molecular identity was not determined.[3] Eosinopenia was produced by the injection of chemotactic factors, such as complement component C5a,[4] so that they could account for at least part of the eosinopenia observed during bacterial infection. Numerous studies have compared the phagocytic and bactericidal activities of eosinophils and neutrophils,[5–9]

and, overall, the eosinophil emerges as less able to ingest and kill bacteria than the neutrophil. However, experimental conditions in these studies varied, and the activation status of the cell, i.e., whether derived from a healthy, normal subject or a patient with eosinophilia, was important because cells from patients with eosinophilia were more active.[10] Recognition that eosinophil granule proteins function as toxins prompted studies to determine their ability to kill bacteria, and results of the studies clearly showed that granule proteins kill both gram-negative and gram-positive bacteria.[11] Nonetheless, few investigations pointed to an important role for eosinophils in bacterial disease. More recent studies indicate a mechanism by which eosinophils are able to kill bacteria, summarized in this chapter by Simon and Yousefi. They focus, in particular, on the formation of extracellular DNA traps generated by eosinophils, and the demonstration that these traps are able to bind bacteria and kill them. The traps are formed by the extrusion of mitochondrial DNA, referred to as *catapult-like* because of its rapidity, and by the deposition of granule proteins on the extruded DNA. Earlier studies had shown that DNA avidly binds eosinophil granule major basic protein,[12] and, most probably, this complex is stable. Simon and colleagues found that eosinophil DNA traps are present in eosinophil-associated inflammatory diseases such as skin diseases and bronchial asthma. In other work, studies on eosinophil-deficient mice show reduced ability to clear *Pseudomonas* species from the peritoneal cavity and increased protection in the presence of added eosinophils.[13] Therefore, the old literature's teaching that eosinophils are not important in bacterial diseases must be questioned on the basis of these new findings.

Concerning fungal diseases and eosinophilia and in contrast to the comments above on the relationship between bacterial diseases and eosinophils, the literature is bereft of reports on eosinophil-fungus interactions. While numerous observations show that mucin derived from the sinus cavities of patients suffering from chronic rhinosinusitis (CRS) contains fungal elements, eosinophils, and Charcot—Leyden crystals, no studies had explored the mechanisms of eosinophil-fungus interactions. In this chapter, Kita presents a summary of the mechanisms by which the

Eosinophils in Health and Disease. http://dx.doi.org/10.1016/B978-0-12-394385-9.00009-2

immune system responds to fungi. Extracts from numerous fungal species activate eosinophils from normal individuals with release of eosinophil-derived neurotoxin (RNase2); interestingly, *Alternaria* extracts do not correspondingly induce neutrophil activation. Investigation of the mechanism by which *Alternaria* extracts activate eosinophils concludes that the G protein-coupled protease-activated receptor is critical in the process. Eosinophils interact with living *Alternaria alternata* and release granule proteins onto the surface of the organism with death of the fungus. This interaction is mediated by the adherence of eosinophils through the β2 integrin, integrin alpha-M (ITAM/CD11b), possibly through recognition of β-glucans. Thus, two key factors appear important for the interactions of eosinophils with fungi, namely PAR and ITAM (CD11b). Kita then reviews the mechanisms of fungal-mediated eosinophil inflammation *in vivo*, and he stresses the importance of chitin and fungal proteases possibly via respiratory epithelial-derived molecules, such as thymic stromal lymphopoietin, interleukin-33, and chemokines. He discusses diseases associated with eosinophilia and fungi, including allergic bronchopulmonary *Aspergillus*, severe asthma with fungal sensitization, allergic fungal sinusitis, and CRS. Treatment of certain eosinophil-associated diseases with anti-fungal agents has led to clinical improvement, whereas, in others, the response has been equivocal, especially in CRS, with the caveat that antifungal medications do not penetrate well into sinus cavities. Overall, this review is a summary of heretofore lacking information on immune responses to fungi and eosinophil participation and is a valuable summary of our current knowledge.

Interest in eosinophils and viruses stems from observations that respiratory syncytial virus (RSV) infection may be associated with eosinophilia. Further attention to a relationship emerged when a clinical trial of formalin-inactivated RSV vaccine resulted in strikingly more severe disease after subsequent RSV infection, with children developing *enhanced disease* and showing pronounced tissue eosinophilia. This raised the question whether eosinophils were responsible for the worsened outcome in the vaccinated children. The studies pertaining to these observations are reviewed in detail by Rosenberg and colleagues. Information supporting a protective role for eosinophils comes from studies of guinea pigs sensitized by allergen administration and then virus challenged. These animals showed a reduced parainfluenza/Sendai viral content, suggesting that the eosinophil, in an interleukin-5 (IL-5)-dependent manner, neutralized virus. The mechanisms by which this might occur still remain obscure. This chapter particularly is concerned with models of primary virus challenge in mice. Although the results from these models still leaves the role of the eosinophil in doubt, a caveat here is that the murine eosinophil seems to degranulate less readily than the human eosinophil and,

therefore, results in murine models may be misleading. In further exploring what is known about eosinophil-virus interactions, Rosenberg and colleagues allude to investigations with the pneumonia virus of mice (PVM), using human C-C motif chemokine 24 (CCL24/eotaxin-2)/mouse IL-5 double transgenic mice in which eosinophils in the respiratory tract demonstrate marked degranulation,[14] and state that they observed accelerated PVM clearance in this model. Hence, the human CCL24 (eotaxin-2)/mouse IL-5 double transgenic mice may provide a unique insight into the potential maximal effects of the activated eosinophil. Overall, this chapter summarizes our current understanding of the ability of the eosinophil to neutralize respiratory viruses and the mechanisms by which this might occur.

Taken together, these three chapters highlight important information of the eosinophil's role in both innate and adaptive immune responses.

REFERENCES

1. Bass DA. Behavior of eosinophil leukocytes in acute inflammation. I. Lack of dependence on adrenal function. *J Clin Invest* 1975;**55**(6):1229−36.
2. Bass DA. Behavior of eosinophil leukocytes in acute inflammation. II. Eosinophil dynamics during acute inflammation. *J Clin Invest* 1975;**56**(4):870−9.
3. Bass DA. Reproduction of the eosinopenia of acute infection by passive transfer of a material obtained from inflammatory exudate. *Infect Immun* 1977;**15**(2):410−6.
4. Bass DA, et al. Eosinopenia of acute infection: Production of eosinopenia by chemotactic factors of acute inflammation. *J Clin Invest* 1980;**65**(6):1265−71.
5. Baehner RL, Johnston Jr RB. Metabolic and bactericidal activities of human eosinophils. *Br J Haematol* 1971;**20**(3):277−85.
6. Mickenberg ID, Root RK, Wolff SM. Bactericidal and metabolic properties of human eosinophils. *Blood* 1972;**39**(1):67−80.
7. DeChatelet LR, et al. Comparison of intracellular bactericidal activities of human neutrophils and eosinophils. *Blood* 1978;**52**(3):609−17.
8. Migler R, DeChatelet LR, Bass DA. Human eosinophilic peroxidase: role in bactericidal activity. *Blood* 1978;**51**(3):445−56.
9. Yazdanbakhsh M, et al. Bactericidal action of eosinophils from normal human blood. *Infect Immun* 1986;**53**(1):192−8.
10. Bass DA, et al. Comparison of human eosinophils from normals and patients with eosinophilia. *J Clin Invest* 1980;**66**(6):1265−73.
11. Lehrer RI, et al. Antibacterial properties of eosinophil major basic protein and eosinophil cationic protein. *J Immunol* 1989;**142**(12):4428−34.
12. Gleich GJ, et al. Physiochemical and biological properties of the major basic protein from guinea pig eosinophil granules. *J Exp Med* 1974;**140**(2):313−32.
13. Linch SN, et al. Mouse eosinophils possess potent antibacterial properties in vivo. *Infect Immun* 2009;**77**(11):4976−82.
14. Ochkur SI, et al. Coexpression of IL-5 and eotaxin-2 in mice creates an eosinophil-dependent model of respiratory inflammation with characteristics of severe asthma. *J Immunol* 2007;**178**(12):7879−89.

Chapter 9.2

Eosinophil-Mediated Antibacterial Host Defense

Hans-Uwe Simon and Shida Yousefi

The primary function of eosinophils has previously been related to their interactions with helminthic parasites,[1] although this view has attracted some controversy. The cytoplasmic granules, which are believed to play an important role in host defense, consist of four distinct populations that can be identified by electron microscopy as primary granules, secondary granules, small granules, and lipid bodies.[2] The cytotoxic cationic proteins are stored in the secondary granules that consist of a core, which contains eosinophil granule major basic protein 1 (MBP-1), and a matrix composed of eosinophil cationic protein (ECP), eosinophil-derived neurotoxin, and eosinophil peroxidase (EPO).[3] MBP-1 is highly cytotoxic,[4] and because of its cationic nature, it affects the charge of surface membranes resulting in disturbed permeability, and disruption and injury of cell membranes.[5] Likewise, ECP can damage target cell membranes through the formation of pores or trans-membrane channels, but also has additional cytotoxic effects.[6] Eosinophils have also been implicated in antiviral defense mechanisms.[7–9]

Besides the antihelminthic and antiviral effects of at least some of the eosinophil granule proteins, antibacterial activities have also been demonstrated. By generating cytokines and chemokines, and by their ability to act as antigen-presenting cells, eosinophils may play different roles in antibacterial defense, although these topics are covered elsewhere in the book. In this chapter, we focus on our understanding of how eosinophils directly fight bacteria. For instance, ECP and MBP-1 can exhibit bactericidal activities by causing the permeabilization of the outer and inner membranes of *Escherichia coli*.[10] Moreover, eosinophil-derived reactive oxygen species, in combination with EPO, are efficient in destroying *E. coli*,[11] and eosinophil granules have also been implicated in the destruction of *Pseudomonas aeruginosa*.[12] The antibacterial properties of eosinophils have also been demonstrated in hypereosinophilic interleukin-5 (IL-5) transgenic mice or following the adoptive transfer of eosinophils in wild-type or eosinophil-deficient mice,[12,13] showing the importance of eosinophils in clearing bacteria *in vivo*. These data are supported by the observation that mice with congenital eosinophil deficiency (i.e., PHIL mice) show impaired bacterial clearance in an experimental model of *Pseudomonas* infection.[12]

The killing of bacteria might take place after phagocytosis (Fig. 9.2.1), which eosinophils are able to perform.[14] Subsequently, phagocytosis of gram-positive

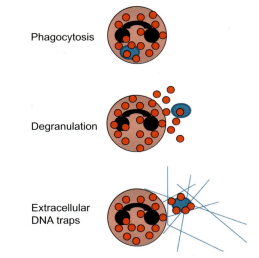

FIGURE 9.2.1 Antibacterial strategies used by eosinophils. *Phagocytosis*: Bacteria are ingested (blue). Granules (red) release cationic proteins into the phagosome. *Degranulation*: Granules and/or granule proteins are released in the extracellular space to kill the bacteria, but can also cause widespread tissue damage. *Extracellular DNA traps*: The incorporation of granule proteins into DNA traps, which also bind bacteria, likely increases the local concentration of antibacterial proteins and might limit tissue damage.

Staphylococcus aureus and gram-negative *E. coli* by eosinophils was demonstrated *in vitro*.[15] At least in equine eosinophils, eosinophil granules discharge their contents into the phagocytic vacuole,[14] providing one possible indication of how granule proteins could exhibit an antibacterial function. An alternative strategy to killing bacteria might be eosinophil degranulation[16] (Fig. 9.2.1). In such a scenario, eosinophil basic proteins could be released in the vicinity of infection. While the activation of eosinophils can occur via multiple different receptors, immunoglobulin A (IgA) receptors are particularly efficient in eliciting eosinophil degranulation on cross-linking.[17,18] IgA-mediated activation of eosinophils might be particularly important in the gastrointestinal mucosa, where secretory IgA is produced in high quantities.

Recently, a third strategy has suggested that extracellular DNA traps generated by eosinophils in the extracellular space are able to bind and kill bacteria was proposed (Fig. 9.2.1). Such mechanism might play a role in case of epithelial barrier defects of the gastrointestinal tract, avoiding the bacterial invasion of the body.[13] Epithelial barrier defects may occur due to inflammatory responses. But how are DNA traps formed? Under *in vitro* conditions, eosinophils need to be stimulated with IL-5 or interferon gamma (IFN-γ) for 20 min before being stimulated with lipopolysaccharide (LPS). However, nonbacterial triggers, e.g., complement component 5a or eotaxin, are also able to promote efficient DNA release from eosinophils.[13]

Although DNA seems to be required for efficient bacterial destruction,[13] it is unlikely that DNA carries out this function. Indeed, extracellular ECP and MBP-1 were detected as colocalizing with DNA, as assessed by double immunofluorescence and confocal microscopy. Thus, it is likely that bacterial death is actually mediated by granule proteins within extracellular DNA traps. Time-lapse confocal imaging allowed the analysis of the kinetics of DNA release in single cells. Strikingly, DNA release happened within 1 s. The mechanism(s) of DNA release appear(s) to differ from the secretion of granule proteins that occur either by classical exocytosis or piecemeal degranulation. Time-lapse confocal imaging revealed that DNA is released from perinuclear structures. Combined two-color DNA and mitochondrial staining suggested that IL-5-primed and LPS-stimulated eosinophils release mitochondrial DNA, which was subsequently confirmed by using molecular biological techniques.[13] Release of mitochondrial DNA was independent of cell death/apoptosis.

Extracellular DNA traps can also be generated by activated neutrophils.[19] The DNA here is associated with granule proteins, such as elastase or myeloperoxidase.[19] However, neutrophil DNA traps may additionally contain histones.[19] In contrast to eosinophils, much more information is available regarding the pathogens trapped and killed by neutrophil DNA traps; these include gram-positive and gram-negative bacteria, fungi, and parasites.[20] The release of DNA can occur within minutes[21,22] or hours;[23] in the latter case, cell death appears to be required.

Interestingly, eosinophil DNA traps were also seen in inflammatory skin diseases[24] and in bronchial asthma.[25] The primary function of these extracellular structures remains unclear under these conditions, although it is possible that eosinophils participate in anti-infection defense mechanisms in at least some of these subjects/diseases. On the other hand, the binding of released eosinophil cationic proteins to extracellular DNA may limit the collateral damage from granular contents in eosinophilic inflammatory diseases.

Taken together, there is accumulating evidence that eosinophils play a beneficial role in innate immune responses against bacteria. This suggests that therapies aiming to deplete eosinophils may cause increased susceptibility toward bacterial infections, although no such adverse effects were observed when treating patients with anti-IL-5 antibody.[26,27] Clearly, many additional questions remain. For instance:

- Do eosinophils play a role in the fight against pathogens in asthma and other allergic diseases?
- Under which conditions do eosinophil cationic proteins exhibit antimicrobial properties?
- What are the exact molecular mechanisms of extracellular DNA release and how long do DNA traps remain in tissues?
- How do DNA traps correlate with other markers of inflammation? Can they be used as biomarkers of

eosinophil activation? Do they prevent exaggerated eosinophil-mediated tissue pathology?

The mechanisms of indirect protection against bacteria (e.g., promotion of epithelial repair and bridging innate and adaptive immunity) also remain largely unexplored.

ACKNOWLEDGEMENTS

The Swiss National Foundation supports the laboratory research of the authors.

REFERENCES

1. Klion AD, Nutman TB. The role of eosinophils in host defense against helminth parasites. *J Allergy Clin Immunol* 2004;**113**:30—7.
2. Kariyawasam HH, Robinson DS. The eosinophil: the cell and its weapons, the cytokines, its locations. *Semin Respir Crit Care Med* 2006;**27**:117—27.
3. Peters MS, Rodriguez M, Gleich GJ. Localization of human eosinophil granule major basic protein, eosinophil cationic protein, and eosinophil-derived neurotoxin by immunoelectron microscopy. *Lab Invest* 1986;**54**:656—62.
4. Gleich GJ, Frigas E, Loegering DA, Wassom DL, Steinmuller D. Cytotoxic properties of the eosinophil major basic protein. *J Immunol* 1979;**123**:2925—7.
5. Kroegel C, Costabel U, Matthys H. Mechanism of membrane damage mediated by eosinophil major basic protein. *Lancet* 1987;**1**:1380—1.
6. Young JD, Peterson CG, Venge P, Cohn ZA. Mechanism of membrane damage mediated by human eosinophil cationic protein. *Nature* 1986;**321**:613—6.
7. Phipps S, Lam CE, Mahalingam S, Newhouse M, Ramirez R, Rosenberg HF, et al. Eosinophils contribute to innate antiviral immunity and promote clearance of respiratory syncytial virus. *Blood* 2007;**110**:1578—86.
8. Gleich GJ, Loegering DA, Bell MP, Checkel JL, Ackerman SJ, McKean DJ. Biochemical and functional similarities between human eosinophil-derived neurotoxin and eosinophil cationic protein: homology with ribonucleases. *Proc Natl Acad Sci USA* 1986;**83**:3146—50.
9. Rosenberg HF, Domachowske JB. Eosinophils, eosinophil ribonucleases, and their role in host defence against respiratory virus pathogens. *J Leukoc Biol* 2001;**70**:691—8.
10. Lehrer RI, Szklarek D, Barton A, Ganz T, Hammann KJ, Gleich GJ. Antibacterial properties of eosinophil major basic protein and eosinophil cationic protein. *J Immunol* 1989;**142**:4428—34.
11. Persson T, Andersson P, Bodelsson M, Laurell M, Malm J, Egesten A. Bactericidal activity of human eosinophilic granulocytes against *Escherichia coli*. *Infect Immun* 2001;**69**:3591—6.
12. Linch SN, Kelly AM, Danielson ET, Pero R, Lee JJ, Gold JA. Mouse eosinophils possess potent antibacterial properties in vivo. *Infect Immun* 2009;**77**:4976—82.
13. Yousefi S, Gold JA, Andina N, Lee JJ, Kelly AM, Kozlowski E, et al. Catapult-like release of mitochondrial DNA by eosinophils contributes to antibacterial defense. *Nat Med* 2008;**14**:949—53.
14. Archer GT, Hirsch JG. Motion picture studies on degranulation of horse eosinophils during phagocytosis. *J Exp Med* 1963;**118**:287—94.
15. Cline MJ, Hanifin J, Lehrer RI. Phagocytosis by human eosinophils. *Blood* 1968;**32**:922—34.

16. Melo RC, Spencer LA, Dvorak AM, Weller PF. Mechanisms of eosinophil secretion: large vesiculotubular carriers mediate transport and release of granule-derived cytokines and other proteins. *J Leukoc Biol* 2008;**83**:229–36.

17. Abu-Ghazaleh RI, Fujisawa T, Mestecky J, Kyle RA, Gleich GJ. IgA-induced eosinophil degranulation. *J Immunol* 1989;**142**: 2393–400.

18. Monteiro RC, Hostoffer RW, Cooper MD, Bonner JR, Gartland GL, Kubagawa H. Definition of immunoglobulin A receptors on eosinophils and their enhanced expression in allergic individuals. *J Clin Invest* 1993;**92**:1681–5.

19. Brinkmann V, Reichard U, Goosmann C, Fauler B, Uhlemann Y, Weiss DS, et al. Neutrophil extracellular traps kill bacteria. *Science* 2004;**303**:1532–5.

20. Papayannopoulos V, Zychlinsky A. NETs: a new strategy for using old weapons. *Trends Immunol* 2009;**30**:513–21.

21. Yousefi S, Mihalache C, Kozlowski E, Schmid I, Simon HU. Viable neutrophils release mitochondrial DNA to form neutrophil extracellular traps. *Cell Death Differ* 2009;**16**:1438–44.

22. Clark SR, Ma AC, Tavener SA, McDonald B, Goodarzi Z, Kelly MM, et al. Platelet TLR4 activates neutrophil extracellular traps to ensnare bacteria in septic blood. *Nat Med* 2007;**13**: 463–9.

23. Fuchs TA, Abed U, Goosmann C, Hurwitz R, Schulze I, Wahn V, et al. Novel cell death program leads to neutrophil extracellular traps. *J Cell Biol* 2007;**176**:231–41.

24. Simon D, Hoesli S, Roth N, Staedler S, Yousefi S, Simon HU. Eosinophil extracellular DNA traps in skin diseases. *J Allergy Clin Immunol* 2011;**127**:194–9.

25. Dworski R, Simon HU, Hoskins A, Yousefi S. Eosinophil and neutrophil extracellular DNA traps in human allergic asthmatic airways. *J Allergy Clin Immunol* 2011;**127**:1260–6.

26. Rothenberg ME, Klion AD, Roufosse FE, Kahn JE, Weller PF, Simon HU, et al. Treatment of patients with the hypereosinophilic syndrome with mepolizumab. *N Engl J Med* 2008;**358**:1215–28.

27. Straumann A, Conus S, Grzonka P, Kita H, Kephart G, Bussmann C, et al. Anti-interleukin-5 antibody treatment (mepolizumab) in active eosinophilic oesophagitis: a randomized, placebo-controlled, double-blind trial. *Gut* 2010;**59**:21–30.

Chapter 9.3

Interactions of Eosinophils with Respiratory Virus Pathogens

Helene F. Rosenberg, Kimberly D. Dyer and Joseph B. Domachowske

INTRODUCTION

Eosinophils, granule-bearing leukocytes found in peripheral blood and tissues, are best known for their roles in asthma, allergy, and other disorders in which they are recruited in response to cytokines released by T-helper type 2 (T_h2) lymphocytes. Eosinophils do not ordinarily come to mind when one thinks generally of a respiratory virus infection. However, at least for one important respiratory virus, the human respiratory syncytial virus (RSV), eosinophils and their unique secretory mediators have been detected in lung tissue in response to primary infection and as a feature of a characteristic hypersensitivity response to inactivated vaccines and vaccine components. Interestingly, eosinophil recruitment and accumulation are almost always perceived in a negative light, as it is assumed that these cells contribute to tissue damage, bronchoconstriction, and respiratory dysfunction via degranulation of their cationic secretory proteins, enzymes, and cytokines. However, recent descriptions of antiviral activity both *in vitro* and *in vivo* suggest that eosinophil function may encompass both of these functions, and present more of a double-edged sword. Clearly, we do not have a complete understanding of the role of eosinophils in disease caused by RSV; here we highlight many of the questions that remain to be explored.

HUMAN RESPIRATORY SYNCYTIAL VIRUS DISEASE

RSV infection is a near universal affliction of infancy and childhood, accounting for approximately 50% of all pneumonia and up to 90% of the reported cases of bronchiolitis in infancy. Of those infants infected during the first year of life, one-third develops lower respiratory tract disease and 2.5% are hospitalized, accounting for more than 90,000 children in the United States every year. In many previously healthy infants, RSV disease is a mild and self-limited infection involving the upper and lower respiratory tract, with varying degrees of peribronchiolar and interstitial inflammation. In others, disease progresses to severe bronchiolitis and pneumonia, including submucosal edema and bronchiolar obstruction requiring oxygen, and in the worst cases, mechanical ventilation. Infants at particularly high risk for severe disease include those born prematurely, infants and children with cardiac or pulmonary anomalies, and immunocompromised infants and children, although a recent study by Hall and colleagues[1] noted that a substantial proportion of children with serious RSV disease had no pre-existing predisposing condition. Prophylactic monoclonal antibody therapy is available for high-risk infants only, and no vaccine has been approved for use. RSV has also recently been recognized as an important pathogen in the institutionalized elderly. The clinical features and pathology of RSV disease have been reviewed extensively, and the reader is referred to these and other excellent sources of information.[2,3]

BASIC BIOLOGY OF HUMAN AND MOUSE EOSINOPHILS

Eosinophils are leukocytes of the granulocyte lineage, as are neutrophils and basophils. Eosinophils differentiate in the bone marrow from CD34 antigen-positive pluripotent progenitor cells and are released into the bloodstream in a more or less completely mature state. Under normal, homeostatic conditions, very few eosinophils can be detected in peripheral blood (only approximately 2–3% of total leukocytes), as the vast majority reside in the tissues, primarily in the gastrointestinal tract. In response to as yet incompletely characterized stimuli, typically observed in allergic states, during infection with helminthic parasites, and in some idiopathic hypereosinophilic states, T_h2 lymphocytes are activated, which results in the production of a specific subset of T_h2 cytokines, including interleukin-5 (IL-5). IL-5 has a unique impact on the eosinophil lineage, as it induces the expansion of eosinophil progenitors in the bone marrow, it primes eosinophils in the periphery, and it prolongs eosinophil survival in the tissues. Eosinophils are capable of responding to a wide variety of other stimuli and can undergo chemotaxis in response to eotaxin (CCL11), MIP1-alpha (CCL3) and RANTES (CCL5), which are chemoattractant cytokines that interact with eosinophils via specific cell surface receptors CCR3, CCR1 and CCR5, respectively. Interestingly, despite years of research, there is still no absolute consensus on eosinophil physiology and function, even in well-characterized disease states. For example, while eosinophils and eosinophil secretory mediators can promote destruction of helminthic eggs and larvae in experiments performed *in vitro*, experiments performed *in vivo* with cytokine-deficient and eosinophil-deficient mice have yielded complex and inconsistent results. Similarly, although the weight of evidence suggests that eosinophils contribute to the pathophysiology of allergy and asthma, a chronic respiratory disease in which broncho-constriction in response to environmental triggers is typically associated with production of T_h2 cytokines and recruitment of eosinophils to the airways, asthmatic responses are obviously negative sequelae of eosinophil function that alone cannot represent a direct evolutionary advantage to the host organism. Among the more recent hypotheses, several groups have focused on eosinophils as immunomodulatory mediators, as eosinophils can interact both directly and indirectly with T cells and mast cells, and can release a wide variety of preformed cytokines and other secretory mediators, primarily from cytoplasmic granules. Other chapters in this volume provide more extensive coverage of these subjects.

Mature eosinophils from all species are readily recognized by their eccentric bilobed nuclei and their characteristic red-staining cytoplasmic granules. As noted, human eosinophil granules are storage sites for cationic secretory mediators, including a unique eosinophil peroxidase, the eosinophil granule major basic protein 1 (MBP-1), eosinophil cationic protein (ECP/ribonuclease 3), eosinophil-derived neurotoxin (EDN/ribonuclease 2), and numerous enzymes and cytokines. Despite similar morphology, human and mouse eosinophils differ from one another and cannot be presumed to function identically in all circumstances. For instance, while there are several reports describing a high-affinity immunoglobulin E (IgE) receptor [high affinity immunoglobulin epsilon receptor subunit beta (FcεRI)] in human eosinophils from parasite-infected and asthmatic subjects,[4–6] this finding and its functional significance has been questioned,[7] and FcεRI has never been detected on eosinophils isolated from mice. Similarly, sialic acid-binding Ig-like lectin 8 (Siglec-8) can be detected on the surface of human eosinophils, while mouse eosinophils express the highly divergent functional orthologue, Siglec-F (Siglec-5). The mouse eosinophil ribonucleases are highly divergent orthologues of human ECP and EDN[8] and Charcot–Leyden crystal protein, a major human eosinophil component, cannot even be identified in the mouse genome. Mouse eosinophils also display a profoundly reduced propensity to degranulate and undergo differential chemotaxis to known exogenous stimuli (reviewed in[9]). As such, eosinophils from humans and mice may look similar to one another, but they may not be formally identical to one another in their actions and in their capacity to cause, to ameliorate, or even to serve as biomarkers for disease.

A number of recent studies have associated eosinophils and eosinophil degranulation products with various aspects of RSV infection in both human disease and in parallel mouse models, which need to be understood with the aforementioned caveats in mind. Here, we review some of these findings with an eye toward understanding what is and what is not known regarding the role of eosinophils, their role in vaccine-induced pathology, their interactions with respiratory virus pathogens, and the outcome of severe respiratory virus disease.

HYPERSENSITIVITY RESPONSES TO FORMALIN-INACTIVATED RESPIRATORY SYNCYTIAL VIRUS VACCINE—ARE THE EOSINOPHILS AT FAULT?

In the early 1960s, a number of children were enrolled in a clinical trial of a formalin-inactivated RSV vaccine. The negative outcomes of this trial, including records of the detailed responses of the vaccinated children who encountered natural RSV infection sometime thereafter, have been documented and reviewed.[10,11] Briefly, children immunized with the formalin-inactivated virus, upon encountering a natural RSV challenge, developed a hypersensitivity response, characterized by bronchoconstriction and severe pneumonia; this has been attributed in part to the development of non-neutralizing antibodies. Lung histology from two children that ultimately died as a result of this trial revealed the deposition of antibody–virus complexes and a pronounced tissue eosinophilia.[12] One or more of these features, which have collectively been termed *enhanced disease*, have been replicated and modeled with formalin-inactivated RSV in multiple species, including other primates, ferrets, cotton rats, and mice as well as with formalin-inactivated bovine RSV in cows.[13] Interestingly, hypersensitivity responses of this nature are not unique to formalin-inactivated RSV; there are a limited number of reports describing aberrant responses to formalin-inactivated measles vaccine.[14] This phenomenon has also been replicated experimentally to varying extents with formalin-inactivated versions of human metapneumovirus[15] and parainfluenza virus,[16] and in some reports even to carrier antigens.[17] There is also a recent report of a severe hypersensitivity reaction, including T$_h$2 cytokine-mediated eosinophil infiltration into the lung tissue, in BALB/c mice immunized with

a vaccinia virus construct expressing the severe acute respiratory syndrome (SARS) coronavirus (CoV) N nucleocapsid protein.[18] Our group has demonstrated that immunization of mice with formalin-inactivated pneumonia virus of mice (PVM) followed by intranasal virus challenge likewise results in pulmonary hypereosinophilia in the absence of a serum-neutralizing antibody response[19] (Fig. 9.3.1). PVM is a natural rodent pneumovirus pathogen that is related to RSV; PVM replicates extensively in mouse bronchiolar epithelial cells tissue, eliciting a profound and potentially lethal inflammatory response, similar to the more severe forms of RSV disease.[20]

Gene-deletion and cytokine depletion mouse model studies all point to T$_h$2 cytokines [interleukin-4 (IL-4), IL-5, and interleukin-13 (IL-13)] as crucial to eliciting pulmonary eosinophilia in response to formalin-inactivated RSV.[21,22] A recent study by Moghaddam and colleagues[23] suggested that the oxidation of RSV antigens resulting from formalin exposure elicits a T$_h$2 response *in vivo*. Delgado and colleagues[24] and Cyr and colleagues[25] have both reported that independent toll-like receptor (TLR) stimulation in conjunction with RSV antigens results in a rebalancing of the T$_h$1/T$_h$2 cytokine responses, thereby reducing pathology. In initial studies aimed at

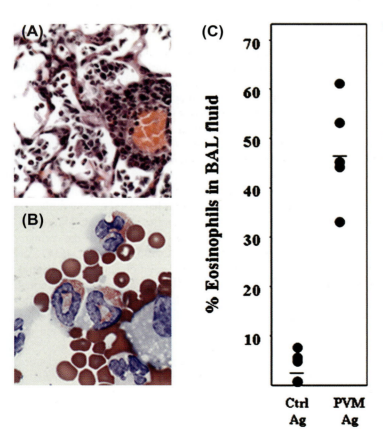

FIGURE 9.3.1 **Hypersensitivity responses in mice vaccinated with formalin-fixed pneumovirus antigens.** (A) Lung tissue from a mouse vaccinated with formalin-inactivated pneumonia virus of mice (PVM; a mouse pneumovirus related to human RSV) and then challenged intranasally with actively replicating virus. (B) Eosinophils detected in the bronchoalveolar lavage (BAL) fluid from the mouse described in (A). (C) Percentage eosinophils detected in the BAL fluid of mice vaccinated with formalin-inactivated PVM vs. control antigen. Ag, antigen; BAL, bronchoalveolar lavage; RSV, respiratory syncytial virus. *Reprinted from*[19].

exploring the molecular mechanism of T_h2-mediated immunopathology, pulmonary eosinophilia was observed in mice immunized with recombinant vaccinia virus expressing RSV-G protein followed by live RSV challenge in most, but not all published trials (reviewed in[11]), results which initially suggested that the pathology related to formalin-inactivation might be attributed mechanistically to aberrant reactivity to this one virus protein alone. Interestingly, although the end point—pulmonary eosinophilia—looks more or less the same, recent analysis indicates that the pulmonary eosinophilia that develops in response to RSV-G protein and the eosinophilia that develops in response to formalin-inactivated RSV proceed via different molecular mechanisms. Among other distinctions, eosinophilia in response to RSV-G protein is not dependent on IL-4 and requires the actions of RSV-G protein-specific Vbeta14$^+$ T cells; in contrast, eosinophilia in response to formalin-inactivated RSV antigens is IL-4 dependent, while not dependent on RSV-G protein-specific Vbeta14$^+$ T cells (reviewed in[11]).

Are Eosinophils Contributing to the Pathophysiology?

Much of the focus of the enhanced disease/hypersensitivity studies has been on the presence of eosinophils and the mechanisms of eosinophil recruitment to the lung tissue, yet it was never really clear whether or not eosinophils were directly responsible for the pathophysiological responses. In other words, it was unclear whether the vaccinated children became ill because of pulmonary eosinophilia, whether the eosinophils were engaged in altering future responses to virus infection, or whether eosinophilia was a neutral, secondary finding. These questions have been explored to some extent in mouse models using wild-type and eosinophil-deficient mice (including ΔdblGATA mice) immunized with vaccinia virus vectors expressing RSV-G and RSV-F proteins,[11,26] but, as noted above, these experimental systems are now recognized as mechanistically unrelated to the pathology induced by formalin-inactivated RSV antigens. As such, although findings address a role for eosinophils, and they likewise suggest that eosinophils actually may not be contributing to systemic disease—specifically, that clinical symptoms, weight loss, and respiratory dysfunction measurements may be unrelated to the presence of absence of pulmonary eosinophilia—these conclusions may not be directly relevant to the way in which eosinophils contribute to pathology in the setting of formalin-inactivated RSV antigens. The role of eosinophils in modulating the pathology induced by formalin-inactivated vaccine antigens has not been explored and might be addressed in mouse models of explicit eosinophil deficiency with formalin-inactivated RSV antigens.

Thus, although our long-standing prejudices might make it easy to conclude that eosinophils contributed directly to the lung and systemic pathology observed in the initial vaccine trials and in the subsequent mouse modeling experiments, it is important to recognize that the presence of eosinophils in the lung may or may not lead to these outcomes. The data from mouse models are inconclusive on this point. Furthermore, the presence of eosinophils alone, even in human conditions, does not necessarily imply severe respiratory pathology. For example, in eosinophilic bronchiolitis, patients complain of only minimal respiratory symptomatology despite pronounced pulmonary eosinophilia.[27] Thus, at current writing, while eosinophils may serve as important biomarkers for aberrant hypersensitivity reactions, we can reach no conclusions regarding their contributions to pathophysiology from the published experimental data.

EOSINOPHIL RECRUITMENT IN RESPONSE TO PRIMARY RESPIRATORY SYNCYTIAL VIRUS INFECTION—A CAUSE FOR ALARM?

Although respiratory virus infections are not among the diseases typically associated with T_h2 lymphocyte activation and profound pulmonary eosinophilia, eosinophils and/or eosinophil granule secretory proteins have been detected in lung washings or systemically in infants in need of supplemental oxygen secondary to severe RSV infection.[28-30] As mentioned earlier, it is not at all certain whether pulmonary eosinophilia is uniquely related to the RSV pathogen, or whether eosinophilia is observed in response to RSV because it is the predominant severe respiratory pathogen among very young infants. Although not reported as frequently, the eosinophil granule protein ECP has been detected in nasopharyngeal secretions in response to other respiratory virus infections, including influenza and parainfluenza.[31,32]

A number of recent studies have suggested that the age at which the individual experiences a first RSV infection has a profound impact on the nature of the primary response. In general, T_h2 cytokines (IL-4, IL-5) and evidence of eosinophilia (cells and/or degranulation products) are detected more readily in younger infants, although results are not completely consistent in all studies. For example, Kristjansson and colleagues[32] examined the responses of infants diagnosed with RSV and found that those who were less than 3 months of age at the time of first infection had higher levels of IL-4 in their nasopharyngeal secretions than children who were older, although no differences were observed in analogous levels of ECP. Likewise, Sung and colleagues[33] documented elevated levels of both IL-4 and IL-5 in serum samples of RSV-infected infants who were less than 18 months old at the time of primary infection than in older infants. Similarly, Kim and colleagues[34] examined eosinophils in the bronchoalveolar lavage fluid (BALF) from RSV-infected infants (ages 0.4–1.8 years) and found that the number of eosinophils detected in BALF correlated closely with IL-5 concentration, although interestingly, the age range of the group in which eosinophils were detected was not significantly different from the age range of the group in which eosinophils were absent.

Nasal eosinophilia has been detected in response to respiratory viruses other than RSV (including rhinoviruses and CoVs), although the circumstances tend to be limited and highly specific, such as in patients with pre-existing respiratory allergies.[35] Of particular interest, several groups have reported that influenza infection stimulates the production of the eosinophil chemoattractants eotaxin (CCL11) and CCL5 (RANTES) in normal nasal and airway epithelial cells in culture, which suggests the possibility of eosinophil recruitment.[36–39] The role of eosinophils in acute SARS-CoV remains completely unexplored, but the eosinophil secretory ribonuclease, eosinophil-derived neurotoxin (RNASE2/EDN), was among the 52 signature genes that discriminated between individuals recovering from severe SARS-CoV infection and healthy controls.[40]

Several large clinical studies have led to the consistent conclusion that infants who have recovered from severe RSV bronchiolitis are at significantly increased risk for both recurrent wheezing and childhood asthma.[41–43] Given the presumed role of eosinophils in the pathogenesis of acute allergic asthma, it seems reasonable to ask whether the eosinophils recruited to the lungs during severe primary RSV bronchiolitis might cause, or at least predict, the progression to wheezing. Causation is of course difficult to ascertain in human subjects; however, a prospective study by Pifferi and colleagues[44] demonstrated that infants who were less than 1 year old and who had elevated serum ECP levels during a primary RSV infection were nearly 10 times more likely to have developed symptoms of wheezing in later childhood than older children and children without elevated levels of serum ECP. However, Sigurs and colleagues,[45] following much the same methodology, found that serum ECP was not predictive of progression to wheezing. In a more recent prospective study, Castro and colleagues[46] examined the outcomes of RSV-infected, <1-year-old infants; in this study, 48% of those enrolled went on to develop allergic symptomatology by age 6, with a significantly higher prevalence of asthma developing among the children who were infected with RSV at a younger age (below 6 months). Among those developing asthma, there were no differences in peripheral blood eosinophil counts at the time of acute RSV infection, nor were the cytokine profiles (as determined by phorbol myristate acetate stimulation of isolated mononuclear cells) of children who developed allergic disease different from those who did not.

What Can We Learn from Mouse Models of Primary Virus Challenge?

Given the complexities of natural disease, and the fact that there is no human condition in which an individual is uniquely devoid of eosinophils, it is helpful (if not crucial) to have appropriate animal models to explore questions of association and causation. Inbred mice have been used extensively to study responses to RSV, although it is important to recognize that RSV inoculation of mice is formally a challenge-clearance model rather than an infection model, as RSV undergoes little if any replication in mouse lung tissue.

Schwarze and colleagues[47,48] have explored RSV challenge in mouse models; in these studies, the authors describe T_h2 cytokine-dependent recruitment of eosinophils and associated airways hyperreactivity, a finding that has implicated eosinophils in the pathophysiological mechanism. Eosinophilia was also noted in a mouse model of secondary RSV challenge; Culley and colleagues[49] detected eosinophil recruitment to the airways on secondary RSV challenge among mice undergoing primary challenge at 1 day of age; eosinophil recruitment declined dramatically if primary challenge was delayed until mice were 1 week old, although the authors found no statistically significant difference in systemic disease, measured as weight loss, between these two sets of challenged mice. Dakhama and colleagues[50] likewise found that the extent of airway hyperresponsiveness induced by a secondary challenge was directly dependent on the age of first virus challenge, similarly associated with augmented eosinophil recruitment when the primary challenge occurred in mice <1 week of age. Tasker and colleagues[51] performed a similar study, and identified T_h2 cytokine responses in neonatally primed mice that were associated with diminished virus replication in lung tissue. Finally, also noteworthy is the study by Harker and colleagues[52] in which mice were primed with recombinant RSV (rRSV) expressing T_h1 [interferon gamma (IFN-γ)] or T_h2 (IL-4) cytokines prior to RSV challenge. In contrast to mice primed with rRSV/IFN-γ, mice primed with rRSV/IL-4 sustained airway eosinophilia in response to subsequent RSV challenge. Although IL-4 clearly functions to suppress antiviral $CD8^+$ T cell function in other experimental settings,[53,54] challenge with RSV/IL-4 had no impact on the number of $CD8^+$ T cells recruited to the lung nor on the fraction producing IFN-γ when compared to mice challenged with wild-type RSV alone. The RSV/IL-4-primed mice had reduced lung virus titer and were protected against weight loss, a finding that correlated with the recruitment of eosinophils. As is clear from these findings, the precise role of eosinophils remains uncertain, but one thing that is clear is the fact that eosinophils do not universally provoke lung pathology and systemic disease.

The role of RSV and PVM in enhancing asthmatic-type responses via an interplay with known allergens has also been explored;[55–59] while the weight of evidence suggests that eosinophils play crucial roles in mouse models of asthma, what that precise role might be (promoting acute airways hyperreactivity vs. more chronic remodeling) is a complex and controversial issue that has been considered extensively by others and is beyond the scope of this chapter.

DO EOSINOPHILS PROMOTE ANTIVIRAL HOST DEFENSE?

One of the more curious aspects of eosinophil biology is, as discussed thus far in this review, that once they are detected, particularly in lung tissue, eosinophils are almost always considered as contributing in some negative way to the pathophysiology of disease. This is most intriguing, given our understanding of the role of their sister cell, the neutrophil, and the concept of the double-edged sword.[60] In other words, we know that neutrophils are recruited in response to bacterial and fungal infection, and serve to promote host defense against these invasive pathogens. However, if signals go awry, if neutrophil clearance does not proceed, and/or if neutrophil activation persists, pathology ensues. We were among the first groups to suggest that eosinophil function might encompass a positive, host-defense aspect as part of perhaps a more subtle double-edged sword,[61,62] and to consider the possibility that eosinophils may be recruited in part to promote primary antiviral host defense, perhaps in situations in which acquired immune responses are less than immediately effective.[63] Interestingly, eosinophilia has been reported in association with T cell dysfunction in human

immunodeficiency virus (HIV) infection;[64–66] this finding may in turn be related to the propensity for hypersensitivity reactions observed among HIV-infected patients.[67] However, correlations between eosinophilia and disease pathogenesis are often difficult to ascertain. In a primary study, Tietz and colleagues[64] found that elevated eosinophil counts among HIV-infected patients correlated with progression of disease and declining CD4[+] T cell counts, while Chorba and colleagues[68] found no correlation between HIV viral loads, CD4[+] T cell and eosinophil counts among more than 600 HIV-infected patients in sub-Saharan Africa, although concurrent helminthic parasite infection was clearly a confounding variable.

The first indication that eosinophils might have the means to function in promoting antiviral host defense came from a series of studies we performed in the late 1990s. In this work, we determined that eosinophils, acting at least in part via their secretory mediators, could reduce the infectivity of RSV for target epithelial cells in vitro;[69] Soukup and Becker[70] likewise demonstrated that eosinophils inhibit RSV infection in tissue culture. Shortly thereafter, Adamko and colleagues[71] demonstrated that eosinophils elicited by allergen sensitization served to limit virus replication and/or promote virus clearance in guinea pigs challenged with parainfluenza/Sendai virus. In a more recent study, Phipps and colleagues[72] demonstrated accelerated clearance of RSV from the lungs of the eosinophil-enriched IL-5 transgenic mice, and furthermore found that full antiviral activity was dependent on intact TLR signaling in eosinophils introduced exogenously (Fig. 9.3.2).

As discussed earlier, we have explored the responses of mice immunized with formalin-inactivated PVM followed by live virus challenge. In experiments using eosinophil-deficient ΔdblGATA mice, we found that eosinophils were not a crucial component of the (limited) protection resulting from this immunization strategy,[19] a finding perhaps related to the virulence of this pathogen in inbred strains of

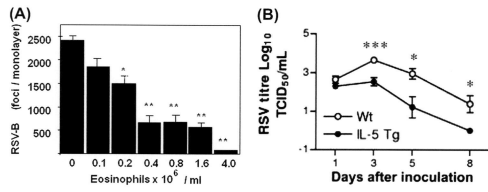

FIGURE 9.3.2 Eosinophils promote antiviral host defense. (A) Eosinophils reduce the infectivity of RSV for target epithelial cells in vitro. (B) Eosinophil-enriched interleukin-5 transgenic mice promote accelerated RSV clearance compared to wild-type mice, with reduced virus titers detected at all time points examined. IL-5, interleukin-5; RSV, respiratory syncytial virus; Wt, wild-type. Panels (A) and (B) reprinted from[69] and,[72] respectively.

mice, as well as its ability to infect eosinophils.[73] Nonetheless, we have recently observed accelerated clearance of PVM in eotaxin-2/IL-5 double-transgenic mice[74] in which eosinophils are undergoing profound and extensive degranulation (Percopo et al., unpublished data)

ARE EOSINOPHILS AMONG THE DIRECT TARGETS OF RESPIRATORY VIRUS INFECTION?

Our studies with PVM[75] and the examination of pathology specimens from RSV patients[76,77] demonstrate that pneumovirus replication *in vivo* takes place primarily in respiratory epithelial cells. However, it is clear that other cells, including human monocytes, support the replication of RSV and PVM in culture, and release proinflammatory cytokines in response to virus infection.[78–80] Given the questions regarding the role of eosinophils and their interactions with respiratory viruses, we set out to determine whether pneumoviruses could infect and replicate within eosinophils, and to determine what the outcome of this infection might be. Kimpen and colleagues[81,82] originally demonstrated that RSV could be taken up by purified human eosinophils, and virions were identified in phagolysosomal compartments, but virus replication was not examined. To explore PVM

replication in mouse eosinophils, we used our recently described method for generating sustained cultures of >95% pure mature eosinophils from unselected bone marrow progenitors.[83] With eosinophils generated by this culture method, we demonstrated a dramatic increase in virus titer within 7 days of inoculation (Fig. 9.3.3), associated with the replication-dependent release of infectious virions accompanied by the cytokine interleukin-6. Among others who have explored the direct interactions of pneumoviruses with eosinophils, Davoine and colleagues[84] determined that human eosinophils were unable to release granule proteins in response to RSV challenge without coincident exposure of virus to CD4+ T cells and antigen-presenting cells. Among the questions left to be explored are:

- How often are infected eosinophils detected *in vivo*?
- At what point during an acute infection are they detected and under what specific circumstances?
- Does virus infection and intracellular replication induce eosinophil apoptosis or disable eosinophils in some other, more subtle way, and thereby reduce their ability to promote virus clearance and antiviral host defense?

Answers to these questions may shed some light on the differential responses observed in the aforementioned experiments.

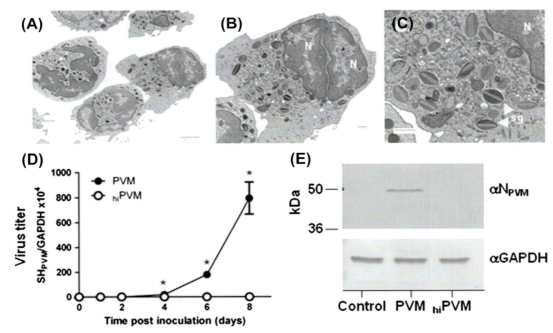

FIGURE 9.3.3 **Pneumovirus replication in eosinophils.** *(A−C)* Electron micrographs of cultured eosinophils derived from unselected mouse bone marrow; the symbol N denotes the characteristic bilobed nucleus, and sg (at arrows) denotes the cytoplasmic-specific granules. *(D)* Replication of PVM in cultured mouse eosinophils; filled symbols, replication-competent PVM; open symbols, heat-inactivated PVM. PVM is detected by quantitative RT-PCR targeting the virus *small hydrophobic (SH)* gene. *(E)* Detection of PVM replication in cultured mouse eosinophils with anti-PVM N protein-specific antibody. GAPDH, glyceraldehyde-3-phosphate dehydrogenase; PVM, pneumonia virus of mice; RSV, respiratory syncytial virus; RT-PCR, reverse transcription polymerase chain reaction. *Reprinted from*[73].

CONCLUSIONS AND FUTURE PERSPECTIVES

The manner in which eosinophils respond to and participate in respiratory virus infection is very far from clear. While pulmonary eosinophilia is a hallmark, or biomarker, of the aberrant hypersensitivity response to formalin-inactivated RSV, there is no clear indication that eosinophils actually contribute to the negative sequelae of disease. Likewise, while severe primary RSV is associated with pulmonary eosinophilia and progression to asthma, these two features have not been linked clearly to one another mechanistically or pathophysiologically. Finally, several groups have shown that eosinophils can promote virus clearance, but this interesting and positive feature of eosinophil function is not observed in all circumstances or in all situations. Among the possibilities that have yet to be explored, eosinophil function may be less dependent on numbers elicited, and may be more closely related to the quality and extent of activation, to the unique nature of the cytokines eliciting recruitment, and/or to the strength of the signals sustaining viability *in situ*. These are all issues that are worthy of consideration as we attempt to improve our understanding of the true nature of the eosinophilic leukocyte and strive to achieve some clarification and sense of balance between their perceived negative and their incompletely characterized positive contributions to homeostasis and host defense.

ACKNOWLEDGEMENTS

Research in Dr. Rosenberg's laboratory is supported by the Division of Intramural Research (DIR AI000941, AI000942, and AI000943) of the National Institute of Allergy and Infectious Diseases. Research in Dr. Domachowske's laboratory is supported by the Children's Miracle Network of Greater New York. This chapter is a revised version of a previously published review article.[85]

REFERENCES

1. Hall CB, Weinberg GA, Iwane MK, Blumkin AK, Edwards KM, Staat MA, et al. The burden of respiratory syncytial virus infection in young children. *N Engl J Med* 2009;**360**:588–98.
2. Stevens WW, Falsey AR, Braciale TJ. RSV 2007: recent advances in respiratory syncytial virus research. *Viral Immunol* 2008;**21**: 133–40.
3. Miyairi I, DeVincenzo JP. Human genetic factors and respiratory syncytial virus disease severity. *Clin Microbiol Rev* 2008;**21**:686–703.
4. Gounni AS, Lamkhioued B, Ochiai K, Tanaka Y, Delaporte E, Capron A, et al. High affinity IgE receptor on eosinophils is involved in defence against parasites. *Nature* 1994;**367**:183–6.
5. Barata L, Ying S, Humbert M, Barkans J, Meng Q, Durham SR, et al. Allergen-induced recruitment of FcεRI+ eosinophils in human atopic skin. *Eur J Immunol* 1997;**27**:1236–41.
6. Rajakulasingam K, Till S, Ying S, Humbert M, Barkans J, Sullivan M, et al. Increased expression of high affinity IgE (FcepsilonRI) receptor-alpha chain mRNA and protein-bearing eosinophils in human allergen-induced atopic asthma. *Am J Respir Crit Care Med* 1998;**158**:233–40.
7. Kita H, Kaneko M, Bartemes KR, Weiler DA, Schimming AW, Reed CE, et al. Does IgE bind to and activate eosinophils from patients with allergy? *J Immunol* 1999;**162**:6901–11.
8. Larson KA, Olson EV, Madden BJ, Gleich GJ, Lee NA, Lee JJ. Two highly homologous ribonuclease genes expressed in mouse eosinophils identify a larger subgroup of the mammalian ribonuclease superfamily. *Proc Natl Acad Sci USA* 1996;**93**:12370–5.
9. Lee JJ, Lee NA. Eosinophil degranulation: an evolutionary vestige or a universally destructive effector function? *Clin Exp Allergy* 2005;**35**:986–94.
10. Castilow EM, Varga SM. Overcoming T cell-mediated immunopathology to achieve safe RSV vaccination. *Future Virol* 2008;**2008**(3):445–54.
11. Castilow EM, Olson MR, Varga SM. Understanding respiratory syncytial virus (RSV) vaccine-enhanced disease. *Immunol Res* 2007;**39**:225–39.
12. Kim HW, Canchola JG, Brandt CD, Pyles G, Chanock RM, Jensen K, et al. Respiratory syncytial virus disease in infants despite prior administration of antigenic inactivated vaccine. *Am J Epidemiol* 1969;**89**:422–34.
13. Antonis AF, Schrijver RS, Daus F, Steverink PJ, Stockhofe N, Hensen EJ, et al. Vaccine-induced immunopathology during bovine respiratory syncytial virus infection: exploring the parameters of pathogenesis. *J Virol* 2003;**77**:12067–73.
14. Griffin DE, Pan CH, Moss WJ. Measles vaccines. *Front Biosci* 2008;**13**:1352–70.
15. de Swart RL, van den Hoogen BG, Kuiken T, Herfst S, van Amerongen G, Yüksel S, et al. Immunization of macaques with formalin-inactivated human metapneumovirus induces hypersensitivity to hMPV infection. *Vaccine* 2007;**25**:8518–28.
16. Ottolini MG, Porter DD, Hemming VG, Prince GA. Enhanced pulmonary pathology in cotton rats upon challenge after immunization with inactivated parainfluenza virus 3 vaccines. *Viral Immunol* 2000;**13**:231–6.
17. Piedra PA, Wyde PR, Castleman WL, Ambrose MW, Jewell AM, Speelman DJ, et al. Enhanced pulmonary pathology associated with the use of formalin-inactivated respiratory syncytial virus vaccine in cotton rats is not a unique viral phenomenon. *Vaccine* 1993;**11**: 1415–23.
18. Yasui F, Kai C, Kitabatake M, Inoue S, Yoneda M, Yokochi S, et al. Prior immunization with severe acute respiratory syndrome (SARS)-associated coronavirus (SARS-CoV) nucleocapsid protein causes severe pneumonia in mice infected with SARS-CoV. *J Immunol* 2008;**181**:6337–48.
19. Percopo CM, Phipps S, Foster PS, Domachowske JB, Rosenberg HF. *Pulmonary eosinophils and their role in immunopathology associated with formalin-inactivated pneumovirus vaccination, in review*; 2009.
20. Rosenberg HF, Domachowske JB. Pneumonia virus of mice: severe respiratory infection in a natural host. *Immunol Lett* 2008;**118**:6–12.
21. Connors M, Giese NA, Kulkarni AB, Firestone CY, Morse 3rd HC, Murphy BR. Enhanced pulmonary histopathology induced by respiratory syncytial virus (RSV) challenge of formalin-inactivated

RSV-immunized BALB/c mice is abrogated by depletion of inter-leukin-4 (IL-4) and IL-10. *J Virol* 1994;**68**:5321—5.

22. Castilow EM, Meyerholz DK, Varga SM. IL-13 is required for eosinophil entry into the lung during respiratory syncytial virus vaccine-enhanced disease. *J Immunol* 2008;**180**:2376—84.

23. Moghaddam A, Olszewska W, Wang B, Tregoning JS, Helson R, Sattentau QJ, et al. A potential molecular mechanism for hyper-sensitivity caused by formalin-inactivated vaccines. *Nat Med* 2006;**12**:905—7.

24. Delgado MF, Coviello S, Monsalvo AC, Melendi GA, Hernandez JZ, Batalle JP, et al. Lack of antibody affinity maturation due to poor Toll-like receptor stimulation leads to enhanced respiratory syncytial virus disease. *Nat Med* 2009;**15**:34—41.

25. Cyr SL, Angers I, Guillot L, Stoica-Popescu I, Lussier M, Qureshi S, et al. TLR4 and MyD88 control protection and pulmonary granulocytic recruitment in a murine intranasal RSV immunization and challenge model. *Vaccine* 2009;**27**:421—30.

26. Castilow EM, Legge KL, Varga SM. Cutting edge: Eosinophils do not contribute to respiratory syncytial virus vaccine-enhanced disease. *J Immunol* 2008;**181**:6692—6.

27. Scott KA, Wardlaw AJ. Eosinophilic airway disorders. *Semin Respir Crit Care Med* 2006;**27**:128—33.

28. Kristjánsson S, Wennergren D, Eriksson B, Thórarinsdóttir H, Wennergren G. U-EPX levels and wheezing in infants and young children with and without RSV bronchiolitis. *Respir Med* 2006;**100**:878—83.

29. Harrison AM, Bonville CA, Rosenberg HF, Domachowske JB. Respiratory syncytial virus-induced chemokine expression in the lower airways: eosinophil recruitment and degranulation. *Am J Respir Crit Care Med* 1999;**159**:1918—24.

30. Garofalo R, Kimpen JL, Welliver RC, Ogra PL. Eosinophil degranulation in the respiratory tract during naturally acquired respiratory syncytial virus infection. *J Pediatr* 1992;**120**:28—32.

31. Colocho Zelaya EA, Orvell C, Strannegård O. Eosinophil cationic protein in nasopharyngeal secretions and serum of infants infected with respiratory syncytial virus. *Pediatr Allergy Immunol* 1994;**5**:100—6.

32. Kristjansson S, Bjarnarson SP, Wennergren G, Palsdottir AH, Arnadottir T, Haraldsson A, et al. Respiratory syncytial virus and other respiratory viruses during the first 3 months of life promote a local TH2-like response. *J Allergy Clin Immunol* 2005;**116**:805—11.

33. Sung RY, Hui SH, Wong CK, Lam CW, Yin J. A comparison of cytokine responses in respiratory syncytial virus and influenza A infections in infants. *Eur J Pediatr* 2001;**160**:117—22.

34. Kim CK, Kim SW, Park CS, Kim BI, Kang H, Koh YY. Bronchoalveolar lavage cytokine profiles in acute asthma and acute bronchiolitis. *J Allergy Clin Immunol* 2003;**112**:64—71.

35. van Benten IJ, KleinJan A, Neijens HJ, Osterhaus AD, Fokkens WJ. Prolonged nasal eosinophilia in allergic patients after common cold. *Allergy* 2001;**56**:949—56.

36. Kawaguchi M, Kokubu F, Kuga H, Tomita T, Matsukura S, Kadokura M, et al. Expression of eotaxin by normal airway epithelial cells after influenza virus A infection. *Int Arch Allergy Immunol* 2000;**122**(Suppl. 1):44—9.

37. Kawaguchi M, Kokubu F, Kuga H, Tomita T, Matsukura S, Suzaki H, et al. Influenza virus A stimulates expression of eotaxin by nasal epithelial cells. *Clin Exp Allergy* 2001;**31**:873—80.

38. Matsukura S, Kokubu F, Noda H, Tokunaga H, Adachi M. Expression of IL-6, IL-8, and RANTES on human bronchial epithelial cells, NCI-H292, induced by influenza virus A. *J Allergy Clin Immunol* 1996;**98**(6 Pt 1):1080—7.

39. Matsukura S, Kokubu F, Kubo H, Tomita T, Tokunaga H, Kadokura M, et al. Expression of RANTES by normal airway epithelial cells after influenza virus A infection. *Am J Respir Cell Mol Biol* 1998;**18**:255—64.

40. Lee YS, Chen CH, Chao A, Chen ES, Wei ML, Chen LK, et al. Molecular signature of clinical severity in recovering patients with severe acute respiratory syndrome coronavirus (SARS-CoV). *BMC Genomics* 2005;**6**:132.

41. Mohapatra SS, Boyapalle S. Epidemiologic, experimental, and clinical links between respiratory syncytial virus infection and asthma. *Clin Microbiol Rev* 2008;**21**:495—504.

42. Pérez-Yarza EG, Moreno A, Lázaro P, Mejías A, Ramilo O. The association between respiratory syncytial virus infection and the development of childhood asthma: a systematic review of the literature. *Pediatr Infect Dis J* 2007;**26**:733—9.

43. Dakhama A, Lee YM, Gelfand EW. Virus-induced airway dysfunction: pathogenesis and biomechanisms. *Pediatr Infect Dis J* 2005;**24**(11 Suppl):S159—69.

44. Pifferi M, Ragazzo V, Caramella D, Baldini G. Eosinophil cationic protein in infants with respiratory syncytial virus bronchiolitis: predictive value for subsequent development of persistent wheezing. *Pediatr Pulmonol* 2001;**31**:419—24.

45. Sigurs N, Bjarnason R, Sigurbergsson F. Eosinophil cationic protein in nasal secretion and in serum and myeloperoxidase in serum in respiratory syncytial virus bronchiolitis: relation to asthma and atopy. *Acta Paediatr* 1994;**83**:1151—5.

46. Castro M, Schweiger T, Yin-Declue H, Ramkumar TP, Christie C, Zheng J, et al. Cytokine response after severe respiratory syncytial virus bronchiolitis in early life. *J Allergy Clin Immunol* 2008;**122**:726—33.

47. Schwarze J, Cieslewicz G, Hamelmann E, Joetham A, Shultz LD, Lamers MC, et al. IL-5 and eosinophils are essential for the development of airway hyperresponsiveness following acute respiratory syncytial virus infection. *J Immunol* 1999;**162**:2997—3004.

48. Schwarze J, Cieslewicz G, Joetham A, Ikemura T, Mäkelä MJ, Dakhama A, et al. Critical roles for interleukin-4 and interleukin-5 during respiratory syncytial virus infection in the development of airway hyperresponsiveness after airway sensitization. *Am J Respir Crit Care Med* 2000;**162**(2 Pt 1):380—6.

49. Culley FJ, Pollott J, Openshaw PJ. Age at first viral infection determines the pattern of T cell-mediated disease during reinfection in adulthood. *J Exp Med* 2002;**196**:1381—6.

50. Dakhama A, Park JW, Taube C, Joetham A, Balhorn A, Miyahara N, et al. The enhancement or prevention of airway hyperresponsiveness during reinfection with respiratory syncytial virus is critically dependent on the age at first infection and IL-13 production. *J Immunol* 2005;**175**:1876—83.

51. Tasker L, Lindsay RW, Clarke BT, Cochrane DW, Hou S. Infection of mice with respiratory syncytial virus during neonatal life primes for enhanced antibody and T cell responses on secondary challenge. *Clin Exp Immunol* 2008;**153**:277—88.

52. Harker J, Bukreyev, A, Collins PL, Wang B, Openshaw PJ, Tregoning JS. Virally delivered cytokines alter the immune response to future lung infections. *J Virol*; **81**:13105—11

53. Villacres MC, Bergmann CC. Enhanced cytotoxic T cell activity in IL-4-deficient mice. *J Immunol* 1999;**162**:2663−70.

54. Bot A, Holz A, Christen U, Wolfe T, Temann A, Flavell R, et al. Local IL-4 expression in the lung reduces pulmonary influenza-virus-specific secondary cytotoxic T cell responses. *Virology* 2000;**269**:66−77.

55. Schwarze J, Hamelmann E, Bradley KL, Takeda K, Gelfand EW. Respiratory syncytial virus infection results in airway hyper-responsiveness and enhanced airway sensitization to allergen. *J Clin Invest* 1997;**100**:226−33.

56. Barends M, de Rond LG, Dormans J, van Oosten M, Boelen A, Neijens HJ, et al. Respiratory syncytial virus, pneumonia virus of mice, and influenza A virus differently affect respiratory allergy in mice. *Clin Exp Allergy* 2004a;**34**:488−96.

57. Siegle JS, Hansbro N, Herbert C, Rosenberg HF, Asquith KL, Foster PS, et al. Early life viral infection and allergen exposure interact to induce an asthmatic phenotype in mice. *Respir Res* 2010;**11**:14.

58. Mäkelä MJ, Tripp R, Dakhama A, Park JW, Ikemura T, Joetham A, et al. Prior airway exposure to allergen increases virus-induced airway hyperresponsiveness. *J Allergy Clin Immunol* 2003;**112**:861−9.

59. Barends M, Van Oosten M, De Rond CG, Dormans JA, Osterhaus AD, Neijens HJ, et al. Timing of infection and prior immunization with respiratory syncytial virus (RSV) in RSV-enhanced allergic inflammation. *J Infect Dis* 2004b;**189**:1866−72.

60. Smith JA. Neutrophils, host defense, and inflammation: a double-edged sword. *J Leukoc Biol* 1994;**56**:672−86.

61. Rosenberg HF, Domachowske JB. Eosinophils, ribonucleases and host defense: solving the puzzle. *Immunol Res* 1999;**20**:261−74.

62. Rosenberg HF, Domachowske JB. Eosinophils, eosinophil ribonu-cleases, and their role in host defense against respiratory virus pathogens. *J Leukoc Biol* 2001;**70**:691−8.

63. Milner JD, Ward JM, Keane-Myers A, Paul WE. Lymphopenic mice reconstituted with limited repertoire T cells develop severe, multiorgan, Th2-associated inflammatory disease. *Proc Natl Acad Sci USA* 2007;**104**:576−81.

64. Tietz A, Sponagel L, Erb P, Bucher H, Battegay M, Zimmerli W. Eosinophilia in patients infected with the human immunodeficiency virus. *Eur J Clin Microbiol Infect Dis* 1997;**16**:675−7.

65. Cohen AJ, Steigbigel RT. Eosinophilia in patients infected with human immunodeficiency virus. *J Infect Dis* 1996;**174**:615−8.

66. Drabick JJ, Magill AJ, Smith KJ, Nutman TB, Benson PM. Hypereosinophilic syndrome associated with HIV infection. Mili-tary Medical Consortium for Applied Retroviral Research. *South Med J* 1994;**87**:525−9.

67. Phillips E, Mallal S. Drug hypersensitivity in HIV. *Curr Opin Allergy Clin Immunol* 2007;**7**:324−30.

68. Chorba TL, Nkengasong J, Roels TH, Monga B, Maurice C, Maran M, et al. Assessing eosinophil count as a marker of immune activation among human immunodeficiency virus-infected persons in sub-Saharan Africa. *Clin Infect Dis* 2002;**34**:1264−6.

69. Domachowske JB, Dyer KD, Bonville CA, Rosenberg HF. Recombinant human eosinophil-derived neurotoxin/RNase 2 func-tions as an effective antiviral agent against respiratory syncytial virus. *J Infect Dis* 1998;**177**:1458−64.

70. Soukup JM, Becker S. Role of monocytes and eosinophils in human respiratory syncytial virus infection in vitro. *Clin Immunol* 2003;**107**:178−85.

71. Adamko DJ, Yost BL, Gleich GJ, Fryer AD, Jacoby DB. Ovalbumin sensitization changes the inflammatory response to subsequent parainfluenza infection. Eosinophils mediate airway hyper-responsiveness, m(2) muscarinic receptor dysfunction, and antiviral effects. *J Exp Med* 1999;**190**:1465−78.

72. Phipps S, Lam CE, Mahalingam S, Newhouse M, Ramirez R, Rosenberg HF, et al. Eosinophils contribute to innate antiviral immunity and promote clearance of respiratory syncytial virus. *Blood* 2007;**110**:1578−86.

73. Dyer KD, Percopo CM, Fischer ER, Gabryszewski SJ, Rosenberg HF. Pneumoviruses infect eosinophils and elicit MyD88-dependent release of chemoattractant cytokines and interleukin-6. *Blood* 2009;**114**:2649−56.

74. Ochkur SI, Jacobsen EA, Protheroe CA, Biechele TL, Pero RS, McGarry MP, et al. Coexpression of IL-5 and eotaxin-2 in mice creates an eosinophil-dependent model of respiratory inflammation with characteristics of severe asthma. *J Immunol* 2007;**178**:7879−89.

75. Bonville CA, Bennett NJ, Koehnlein M, Haines DM, Ellis JA, DelVecchio AM, et al. Respiratory dysfunction and proin-flammatory chemokines in the pneumonia virus of mice (PVM) model of viral bronchiolitis. *Virology* 2006;**349**:87−95.

76. Welliver TP, Garofalo RP, Hosakote Y, Hintz KH, Avendano L, Sanchez K, et al. Severe human lower respiratory tract illness caused by respiratory syncytial virus and influenza virus is charac-terized by the absence of pulmonary cytotoxic lymphocyte responses. *J Infect Dis* 2007;**195**:1126−36.

77. Johnson JE, Gonzales RA, Olson SJ, Wright PF, Graham BS. The histopathology of fatal untreated human respiratory syncytial virus infection. *Mod Pathol* 2007;**20**:108−19.

78. Barr FE, Pedigo H, Johnson TR, Shepherd VL. Surfactant protein-A enhances uptake of respiratory syncytial virus by monocytes and U937 macrophages. *Am J Respir Cell Mol Biol* 2000;**23**:586−92.

79. Dyer KD, Schellens IM, Bonville CA, Martin BV, Domachowske JB, Rosenberg HF. Efficient replication of pneu-monia virus of mice (PVM) in a mouse macrophage cell line. *Virol J* 2007;**4**:48.

80. Krilov LR, McCloskey TW, Harkness SH, Pontrelli L, Pahwa S. Alterations in apoptosis of cord and adult peripheral blood mono-nuclear cells induced by in vitro infection with respiratory syncytial virus. *J Infect Dis* 2000;**181**:349−53.

81. Kimpen JL, Garofalo R, Welliver RC, Fujihara K, Ogra PL. An ultrastructural study of the interaction of human eosinophils with respiratory syncytial virus. *Pediatr Allergy Immunol* 1996;**7**:48−53.

82. Kimpen JL, Garofalo R, Welliver RC, Ogra PL. Activation of human eosinophils in vitro by respiratory syncytial virus. *Pediatr Res* 1992;**32**:160−4.

83. Dyer KD, Moser JM, Czapiga M, Siegel SJ, Percopo CM, Rosenberg HF. Functionally competent eosinophils differentiated ex vivo in high purity from normal mouse bone marrow. *J Immunol* 2008;**181**:4004−9.

84. Davoine F, Cao M, Wu Y, Ajamian F, Ilarraza R, Kokaji AI, et al. Virus-induced eosinophil mediator release requires antigen-presenting and CD4+ T cells. *J Allergy Clin Immunol* 2008;**122**:69−77.

85. Rosenberg HF, Dyer KD, Domachowske JB. Respiratory viruses and eosinophils: exploring the connections. *Antiviral Res* 2009;**83**:1−9.

Antifungal Immunity by Eosinophils: Mechanisms and Implications in Human Diseases

Hirohito Kita

INTRODUCTION

Eosinophils are implicated in the pathophysiology of allergic diseases, such as bronchial asthma and atopic dermatitis, and in host immunity to helminth infections.[1] During such inflammatory reactions, soluble mediators released by immune cells induce eosinophil recruitment from the bloodstream into the sites of inflammation, where as yet unknown stimuli trigger the release of proinflammatory mediators.[2] Marked extracellular deposition of released eosinophil granule proteins has been detected in specimens from patients with asthma, chronic rhinosinusitis (CRS), and atopic dermatitis.[3,4,5] However, the presence of eosinophils *per se*, as in normal intestinal mucosa,[6] does not lead to disease pathology. Thus, a fundamental and important question still remains: what *triggers* eosinophil activation and proinflammatory mediator release in human disease? The occurrence of eosinophilic inflammation in such disparate conditions as parasitic infections (presumably for the benefit of the human host) and hypersensitivity diseases (perhaps to the detriment of the patient) has become better understood as a consequence of recently reported information.

Fungi are ubiquitous in the environment, and as saprophytes or commensals, they may coexist without effect in the host with normal cellular immunity.[7] Nonetheless, these airborne fungi and their products may contribute to the development and exacerbation of human airway diseases. For example, fungal products induce immunological and inflammatory reactions, resulting in a T-helper type 2 (T$_h$2)-like cytokine response and the destruction of mucosal barrier functions.[8,9] Clinically, an association between fungal exposure and asthma has been widely recognized.[10] Therefore, antifungal immunity may provide a valuable clue toward improved understanding of eosinophil biology and the mechanisms of human diseases. This chapter describes the recent development of our understanding of the biological and immunological properties of eosinophils and the mechanisms of eosinophilic inflammation with specific focus on antifungal immune responses. The potential roles of eosinophils and antifungal immune responses in human diseases are discussed.

OVERVIEW OF ANTIFUNGAL IMMUNE RESPONSES

The host immune mechanisms against fungi range from protective mechanisms that were present early in the evolution of multicellular organisms (*innate immunity*) to adaptive mechanisms, which are specifically induced during infection and disease (*adaptive immunity*).[11] The life cycle of fungi starts with *conidia* (often called spores) that germinate and produce filaments (*hyphae*); these grow and branch in all directions to form a mass, collectively called the *mycelium* or colony. Colony formation only occurs if there are sufficient nutrients and moisture in the substrate. When the mycelium is mature, specialized hyphae called conidiophores are formed. On the conidiophores, thousands of conidia are produced, which can easily become airborne because of their small size and buoyancy.

Fungi are multicellular organisms and express various biological molecules during different stages of their lifecycle. The fungal cell wall is composed of several components, including a fibrillar layer, mannoprotein, chitin, and β-glucan.[12] Chitin is a (β1-4)-linked polymer of *N*-acetyl-D-glucosamine (NAG), and produced in the cytosol by chitin synthase. Glucan has a large group of D-glucose polymers having glycosidic bonds; of these, the most common glucans composing the fungal cell wall have the α-configuration, such as the (β1-3)- and (β1-6)-linked glucosyl units. Up to 60% of the dry weight of the cell wall of fungi consists of glucans.[13] Fungi also produce many proteases that are required for their germination, growth, and survival.[14]

Fungal cell wall components are recognized by cognate receptors, such as the soluble receptor, pentraxin-related protein PTX3 (PTX3),[15] and membrane-bound receptors, such as Toll-like receptors (TLRs). PTX3 binds to the yeast cells of *Paracoccidioides brasiliensis*, but not to *Candida albicans*, suggesting that PTX3 recognizes specific fungal species and morphological forms.[16] In *Aspergillus*-induced asthma, mannan-binding lectin A (MBL-A) appears to enhance the T$_h$2-type immune responses, suggesting that this opsonin enhances fungus-specific T-cell responses.[17] Fungal killing and host immunity often depend on multiple TLR- and non-TLR-mediated pathways.[18] Because the composition of the fungal cell wall is complex, the cell wall constituents can activate more than one TLR-dependent signaling pathway. For example, macrophage tumor necrosis factor (TNF-α) and interferon gamma (IFN-γ) secretion triggered by *C. albicans* yeast cells depend on

TLR2 and TLR4, whereas hyphal cells trigger only TLR2-dependent responses.[19]

The receptors for β-glucans are expressed by several immune cells including macrophages[20] and neutrophils.[21] The nonopsonic recognition of β-glucans by these cells has been ascribed to multiple receptors, including dectin-1,[22] lactosylceramide,[23] and CD11b/CD18 (integrin beta-2/CR3).[24] Dectin-1 possesses a single nonclassical C-type carbohydrate recognition domain (CRD) connected to the transmembrane region by a stalk, and a cytoplasmic tail, possessing an immunoreceptor tyrosine-based activation (ITAM) motif, which induces the production of reactive oxygen species and inflammatory cytokines.[25] Lactosylceramide (CDw17) is a glycosphingolipid found in the plasma membranes of many cells and was identified as a β-glucan receptor from the biochemical analyses of the interactions between poly-$^{1\text{-}6}$–D-glucopyranosyl-$^{1\text{-}3}$–D-glucopyranose glucan (PGG glucan) and isolated human leukocyte membrane components.[23] CD11b/CD18 (CR3) is a heterodimeric integrin, consisting of the α_M (CD11b) and β_2 (CD18) chains.[26] CD11b is an adhesion molecule but it also serves as a phagocytic receptor for iC3b-opsonized particles, including opsonized particulate glucans,[27] through its I-domain. Importantly, CD11b also possesses a lectin domain, which maps to a site C-terminus to the I-domain. The lectin domain recognizes selected monosaccharides and a variety of β-glucans, including zymosan, although the polymeric ligand with the highest affinity contained very little β-glucan and consisted mostly of mannose.[28]

EOSINOPHIL RESPONSES TO FUNGI AND FUNGAL PRODUCTS *IN VITRO*

Activated human eosinophils defend against large, non-phagocytosable organisms, most notably the multicellular helminthic parasites. Therefore, one could reasonably speculate that eosinophils may be involved in host defense and/or immune responses to other organisms with a large surface, for example filamentous fungi. We incubated human eosinophils with extracts from several environmental airborne fungi (*Alternaria alternata*, *Aspergillus versicolor*, *Bipolaris sorokiniana*, *C. albicans*, *Cladosporium herbarum*, *Curvularia spicifera*, and *Penicillium chrysogenum*).[29] *Alternaria* and *Penicillium* induced calcium-dependent exocytosis [e.g., eosinophil-derived neurotoxin (EDN) release] in eosinophils from normal individuals (Fig. 9.4.1). *Alternaria* also strongly induced other activation events in eosinophils, including increases in intracellular calcium concentration, cell surface expression of CD63 and CD11b, and production of interleukin-8 (IL-8). Interestingly, *Alternaria* did not induce neutrophil activation.

How do eosinophils recognize and respond to fungal products within the *Alternaria* extracts? An extensive analysis of eosinophil expression of TLRs and their responses to TLR ligands showed that eosinophils constitutively expressed TLR1, TLR4, TLR7, TLR9, and TLR10 mRNAs and that the TLR7 ligand, R848, can activate eosinophils.[2] However, eosinophil response to *Alternaria* is unlikely to be dependent on TLRs and their ligands.[30] Rather, the eosinophil-stimulating activity in *Alternaria* extract was highly heat-labile and had a molecular mass of about 60 kDa, suggesting the involvement of proteinaceous molecules and their receptors.

Proteases, especially serine proteases, activate hematological and interstitial cells through a family of G protein-coupled proteinase-activated receptors (PARs) and induce the production of several proinflammatory mediators.[31] Four members of this receptor family have been cloned and are designated PAR-1, PAR-2, PAR-3, and PAR-4. Protease cleavage of these receptors creates a neo-NH_2 terminus, which acts as a tethered ligand and activates the seven transmembrane segments of the PAR. Human eosinophils constitutively transcribe mRNA for PAR-2 and PAR-3, but not for PAR-1 and PAR-4.[32] Trypsin, an authentic agonist for PAR-2, induces superoxide anion production and degranulation of eosinophils; 5 nM trypsin induces responses that are 50−70% of those induced by 100 nM PAF, a positive control. Similarly, a cysteine protease, papain, induces isolated human eosinophils to degranulate and to produce superoxide anion.[33] *A. alternata* produces aspartate protease(s) and recognition of this protease(s) by PAR-2 is likely responsible for the activation of eosinophils in response to *Alternaria* extracts. Indeed, when stimulated with *Alternaria* extracts, eosinophils show an increased intracellular calcium concentration that is desensitized by peptide and protease ligands for PAR-2 and inhibited by PAR-2 antagonistic peptide(s). *Alternaria*-derived aspartate protease(s) cleaves PAR-2 to expose *neo-ligands*; these neo-ligands bind to the seven transmembrane segments of PAR-2 and activate eosinophil degranulation (Fig. 9.4.2). Treatment of *Alternaria* extract with aspartate protease inhibitors, which are conventionally used for human immunodeficiency virus 1 (HIV-1) and other microorganisms, attenuates the eosinophils' responses to *Alternaria*. Thus, fungal aspartate protease and eosinophil PAR-2 appear critical for the eosinophils' innate immune response to certain fungi, suggesting a novel mechanism for eosinophil activation, for pathological inflammation in asthma and for host−pathogen interaction.

Eosinophils respond not only to fungal extracts but also to living fungal organisms. Human eosinophils react vigorously to *A. alternata* grown in tissue culture dishes[34] (Fig. 9.4.3). Eosinophils release their cytotoxic granule proteins, such as EDN and eosinophil granule major basic protein 1 (MBP-1), into the extracellular milieu and onto

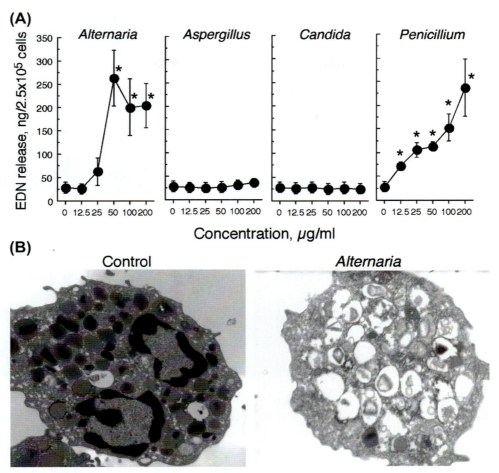

FIGURE 9.4.1   Eosinophil degranulation induced by fungal extracts. *(A)* Eosinophils were incubated with fungal extracts and the concentrations of eosinophil-derived neurotoxin (EDN) in the supernatants were measured by immunoassay. *(B)* Photoelectron micrographs of eosinophils stimulated with medium alone (control) or *Alternaria* extract. Note the granule fusions and electron lucent granules in eosinophils stimulated with *Alternaria*.

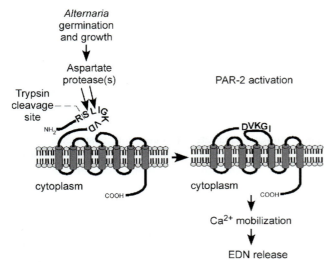

FIGURE 9.4.2   Mechanisms of PAR-2 activation and eosinophil EDN release in response to *Alternaria* extract. EDN, eosinophil-derived neurotoxin.

the surface of fungal organisms and kill the fungus. While human eosinophils do not express common fungal receptors such as dectin-1 and lactosylceramide as described earlier, eosinophils do express and use their integrin β2 molecule, CD11b, to adhere to a major fungus cell wall component, β-glucan. More specifically, the I-domain of CD11b is distinctively involved in how eosinophils interact with β-glucan. It has been known that eosinophils express CD11b constitutively and that the level of expression is increased by incubating eosinophils with various agonists, such as fMet-Leu-Phe (fMLP), granulocyte-macrophage colony-stimulating factor (GM-CSF), interleukin-5 (IL-5), or platelet-activating factor (PAF).[1] Furthermore, compared with neutrophils, eosinophils have a larger total pool of CD11b within the cells.[35] Historically, receptor ligands immobilized to relatively large surfaces [such as immunoglobulin G (IgG)-coated sepharose beads and parasites], though not particulate ligands (such as aggregated IgG and bacteria), are effective stimuli for eosinophil

FIGURE 9.4.3 **Interaction of eosinophils with living *Alternaria* organisms.** Isolated human eosinophils were incubated with germinating *Alternaria* for 24 h. The culture mix was analyzed by the following: *(A)* inverted light microscopy, *(B)* anti-MBP-1 staining and fluorescence microscopy, *(C)* scanning electron microscopy, and *(D)* transmission electron microscopy. Note that fungal hyphae are covered with eosinophils and their cellular components in *(C)*. In *(D)*, eosinophils adhere tightly to fungal hyphae and their granules are lost in >50% of the cells.

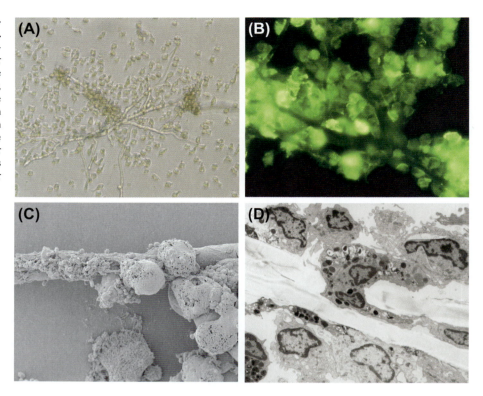

degranulation.[1] Eosinophil CD11b plays a pivotal role in eosinophil adhesion to these large surfaces and degranulation.[36] Thus, eosinophils likely use their versatile CD11b molecules to react to microorganisms (e.g., helminths, fungi) as well as other cellular targets with large surfaces.

Altogether, eosinophil response to filamentous fungi is mediated by at least two major pathways, recognition of fungal proteases by PARs, and cellular adherence to the cell wall β-glucan by integrin β2 integrin, CD11b/CD18. This two-pronged approach may have important implications in the mechanisms of antifungal immunity. *Drosophila* detects fungal infections by recognizing pathogen-associated molecular patterns (PAMPs) and by monitoring the effects of fungal virulence factors. Specifically, the receptor gram-negative bacteria-binding protein 3 on *Drosophila* recognizes β-glucan, and the secreted fungal virulence factor PR1 protease cleaves *Drosophila* serine protease Persephone, activating the downstream immune response to the fungi.[37] Similarly, eosinophil CD11b may recognize fungal structures by monitoring the presence of β-glucan. Eosinophil PARs may act like sensors to monitor fungal protease activities or putative virulence factors and provide a cue to immune and inflammatory systems once the activities reach a certain threshold. Therefore, recognition of fungi by both their structures and activities may be a conserved trait during the evolution of immunity. Such immune responses by eosinophils to fungi may benefit the host, but in turn, it may also play a role in the development

and/or exacerbation of eosinophil-related human airway diseases, such as asthma.

MECHANISMS INVOLVED IN FUNGUS-MEDIATED EOSINOPHILIC INFLAMMATION *IN VIVO*

Recently, major progress has been made regarding the mechanisms involved in eosinophilic inflammation in response to fungi *in vivo*. As described earlier, chitin is one of the major cell wall components and provides structural rigidity to fungi, as well as to crustaceans, helminths, and insects. Airway administration of chitin induces the accumulation of interleukin-4-expressing innate immune cells, including eosinophils and basophils.[38] Tissue infiltration of eosinophils is unaffected by the absence of TLRs but is reduced when the injected chitin is pre-treated with mammalian chitinase, acidic mammalian chitinase (AMCase), or when the chitin is injected into mice overexpressing AMCase. Chitin induces macrophage activation and the production of leukotriene B$_4$ (LTB$_4$), which is required for optimal recruitment of eosinophils. Similarly, cell wall preparations of *Aspergillus* isolated from house dust induce the recruitment of eosinophils into mouse lungs, and the effects are attenuated by enzymatic degradation of cell wall chitin and β-glucans.[39] Thus, chitin and likely β-glucan are recognized by innate immune cells in the airway and are implicated in eosinophil infiltration induced by fungal exposure.

Fungal proteases also likely play critical roles in eosinophilic airway inflammation induced by fungi. The protease activities in *Aspergillus* extracts correlate with their activities to induce immunoglobulin E (IgE) response and eosinophilic inflammation when they are administered to mouse lungs.[9] Furthermore, active proteases are present in almost all houses and many of these activities are derived from fungi, especially *Aspergillus niger*.[40] The conidia of *A. niger* readily establish airway mucosal infection, eosinophilic inflammation, and IgE antibody production, while protease-deficient *A. niger* has reduced ability to promote airway eosinophilia. Importantly, in mice infected with *A. niger*, IL-5 and interleukin-13 are required for optimal clearance of lung infection *in vivo*, suggesting T_h2-type airway inflammation and/or that eosinophils are protective against fungal infection *in vivo*.

The immunological pathways that link airway fungal exposure and development of T_h2-type eosinophilic inflammation have been an active area of research for many years. The airway epithelium is no longer just a structural barrier;[41] it can respond to environmental insults, such as allergens, microorganisms, cigarette smoke, and pollution, by secreting inflammatory mediators and antimicrobial peptides, and by recruiting immune cells.[42] Our current understanding emphasizes the importance of epithelial-derived cytokines, particularly the C-C chemokine receptor family, interleukin-25 and -33, and thymic stromal lymphopoietin (TSLP), in initiating T_h2-type airway inflammation.[43,44]

TSLP is an interleukin-7-like cytokine that is produced by epithelial cells in the lungs, gut, and skin.[45] Expression of TSLP in the airways of patients with asthma correlates with the severity of the disease, suggesting that it is involved in the development of allergic airway inflammation.[46] TSLP activates dendritic cells to polarize naive T cells toward the T_h2 cells that produce T_h2-type cytokines as well as TNF-α.[47] Mice expressing TSLP in the lungs develop a spontaneous airway inflammation with characteristics similar to human asthma.[48] Conversely, mice deficient in the TSLP receptor show decreased airway inflammation when they are challenged with allergens.[49] TSLP mRNA and protein are induced when a human airway epithelial cell line, BEAS-2B, is exposed to prototypic proteases, namely trypsin and papain. TSLP induction by trypsin requires intact protease activity and PAR-2. Importantly, when BEAS-2B cells or normal human bronchial epithelial cells are exposed to *Alternaria* extract, TSLP is also induced. The TSLP-inducing activity of *Alternaria* is partially blocked by treating the extract with a cysteine protease inhibitor, E-64, or by infecting BEAS-2B cells with small interfering RNA for PAR-2, suggesting roles for fungal protease(s) and their receptor, PAR-2.

Among the most potent molecules of the innate immune system are the interleukin-1 (IL-1) family members;[50] these cytokines, such as IL-1, interleukin-18, and interleukin-33 (IL-33), are evolutionarily ancient and are involved in regulating both innate and adaptive immune responses. IL-33 was first described as a nuclear factor abundantly expressed in the nucleus of endothelial and epithelial cells.[51] *In vivo* systemic administration of IL-33 to mice induces lung and gastrointestinal eosinophilia and increased levels of serum IgE and immunoglobulin A (IgA).[52] IL-33 drives the production of cytokines and chemokines by T_h2 cells, mast cells, basophils, eosinophils, natural killer T (NKT) cells, and natural killer (NK) cells.[52] Airway exposure of naive mice to *Alternaria* extracts induces the rapid release of IL-33 into the airway lumen, followed by innate T_h2-type responses.[53] Biologically active IL-33 is constitutively stored in the nuclei of human airway epithelial cells, and exposing these cells to *Alternaria* releases IL-33 extracellularly. Pharmacological inhibitors of purinergic receptors or deficiency in the *purinergic receptor P2Y, G-protein coupled, 2 (P2Y2)* gene abrogate IL-33 release and T_h2-type responses in the *Alternaria*-induced airway inflammation, suggesting essential roles for adenosine triphosphate and the P2Y purinoceptor 2 receptor in epithelial responses to airborne fungi. Altogether, these findings suggest that airway exposure to fungi or their products activate the innate immune cells (e.g., macrophages, epithelial cells) and facilitate their production of immunological molecules that are involved in eosinophilic inflammation, such as IL-33, LTB4, and TSLP. Fungal chitin and proteases likely are key components in triggering the immune response.

POTENTIAL ROLES FOR EOSINOPHIL ANTIFUNGAL IMMUNITY IN HUMAN DISEASES

Information regarding the roles of eosinophils in protecting against fungal infection in humans is scarce. Coccidioidomycosis, caused by the fungus *Coccidioides immitis*, may be accompanied by an increase in peripheral blood eosinophils.[54] Marked deposition of MBP-1 is observed in lesions of patients with paracoccidioidomycosis.[55] A number of studies suggest potential detrimental roles for eosinophils and fungi in T_h2-mediated airway diseases, such as allergic bronchopulmonary aspergillosis (ABPA), severe asthma associated with fungal sensitivity (SAFS), and CRS.

Allergic asthma is generally considered a chronic inflammatory disorder of the airways mediated by T_h2 cells, which drives the cardinal features of asthma, including airway eosinophilia, airway hyperreactivity, and excessive production of mucus. These pathological immune responses have long been recognized to mirror the beneficial immune responses to helminthic infection. Although the etiology of asthma could be complex, a number of epidemiological studies suggest that fungi are involved in asthma. For example, direct associations between increased fungal exposure and worsening of asthma have been demonstrated repeatedly.[56] Fungal sensitivity, particularly to *Aspergillus*, *Alternaria*, and *Cladosporium*, is associated with asthma.[57,58] In a large survey of US housing, exposure to *A. alternata* antigens was correlated with active asthma symptoms.[59] Sensitization to *Alternaria* at age 6 correlated with chronic asthma at age 22,[60] and sensitization to *Alternaria* and other species has been associated with severe and potentially fatal episodes of asthma.[56] In addition, epidemics of asthma caused by increased airborne fungal spores that occur during thunderstorms further illustrate the association between fungal exposure and worsening of asthma symptoms.[61]

ABPA is a T_h2 hypersensitivity lung disease caused by bronchial colonization of *A. fumigatus* and is characterized by exacerbations of asthma, recurrent transient chest radiographic infiltrates, and peripheral and pulmonary eosinophilia, especially during exacerbation.[62] The minimal criteria required for the diagnosis of ABPA are:

- asthma or cystic fibrosis with deterioration of lung function;
- immediate *Aspergillus* skin test reactivity;
- total serum IgE $\geq$ 1000 ng/ml (416 IU/ml);
- elevated *Aspergillus*-specific IgE and IgG antibodies; and
- chest radiographic infiltrates.

ABPA is likely the most common form of allergic bronchopulmonary mycosis (ABPM). Other fungi, including *Candida*, *Penicillium* and *Curvularia*, are occasionally responsible for a syndrome similar to ABPA.[63] The characteristics of ABPM include severe asthma, blood and pulmonary eosinophilia, markedly elevated IgE and specific IgE, bronchiectasis, and mold colonization of the airways.

The term SAFS has been coined to illustrate the high rate of fungal sensitivity in patients with severe asthma.[56] It is possible that ABPA represents one manifestation of a spectrum of fungus-associated airway diseases. From a fungal perspective, the human lung is not sterile. The conidia of *A. fumigatus*, *Penicillium* and *Cladosporium*, and presumably other fungi, are often present, and they induce an immune response when germination is initiated.[64] Excess mucus and airway architecture distortion may allow fungal germination and escape from host defense. In approximately 50% of patients with severe asthma, fungal cultures are positive without evidence of IgE fungal sensitization, suggesting that fungal colonization of the airways is common even in the absence of an allergic component.[65] There is a significant association between *A. fumigatus* IgE sensitization, colonization, and impaired postbronchodilator forced expiratory volume in one second (FEV_1).[66] Patients with SAFS respond to oral antifungal therapy, implicating fungi in the pathophysiology of asthma.[67]

CRS is chronic inflammation of the upper airway and is often associated with nasal polyps. CRS is defined as symptoms and signs of sinus inflammation persisting for more than 8–12 weeks.[68] In the past, a number of studies have attempted to elucidate the mechanisms of CRS. Several hypotheses concerning its pathogenesis have been proposed and tested, including chronic bacterial infection, inhalant or food allergies, and T-cell disturbance caused by aerodynamic factors.[69] Although its cause is still unclear, there are several distinctive features in CRS. For example, the histological hallmark of CRS is marked tissue eosinophilia, which is seen in almost all patients.[70,71] These eosinophils likely play a major role in the pathophysiology of CRS via the release of their cytotoxic granule proteins, such as MBP-1.[72] In fact, a highly significant correlation was noted between the extent of disease, as examined by computed tomography (CT) scan, and the peripheral blood eosinophil count.[73] Consequently, MBP-1 levels in mucus exceeding the concentrations that cause damage to the upper airway respiratory epithelium are detected in sinus mucus from patients with CRS.[4]

In 1981, Millar and colleagues first described sinus specimens from five patients with CRS that showed histological similarities to ABPA.[74] Since then the majority of reports have shown non-*Aspergillus* fungi, such as *Curvularia*, *Alternaria*, and *Bipolaris*, causing similar findings, promoting a change to the more general term, allergic fungal sinusitis (AFS). The diagnostic histological feature of AFS is the presence of *allergic mucin*, characterized by sloughed respiratory epithelial cells, accumulations of intact and degenerating eosinophils, Charcot–Leyden crystals, and cellular debris containing eosinophils and their products; noninvasive fungal hyphae are found within the *allergic mucin* and are best demonstrated with impregnated silver stain. AFS prevalence ranges from 4% to 7%[75] of all the CRS cases; however, it is also speculated that AFS is underdiagnosed because of confusion concerning the criteria for diagnosis[76] and less than adequate techniques to detect fungi. Indeed, fungi are ubiquitous and found in the mucus of a majority of patients with CRS as well as in healthy individuals.[77,78] Furthermore, in some patients with CRS, eosinophils sometimes form clusters around fungal organisms in the mucus (Fig. 9.4.4), reminiscent of the accumulation of eosinophils around helminth parasites. These findings suggest that the eosinophilic immune response to fungal organisms may be implicated in the pathophysiology of CRS.[77,78] Interestingly, systemic antifungal treatment in patients with SAFS not only

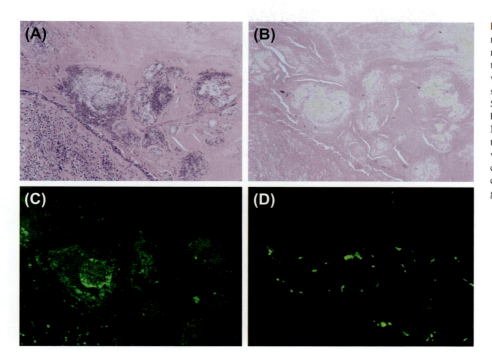

FIGURE 9.4.4 Eosinophilic inflammation and fungal localization in the mucus of a CRS patient. A surgical tissue specimen from a patient with CRS was stained by hematoxylin and eosin stain *(A)*, Gomori−Grocott Methenamine Silver GMS stain *(B)*, anti-MBP-1 antibody *(C)*, or anti-*Alternaria* antibody *(D)*. Note several clusters of eosinophils within the mucus *(A and C)*. Fungi were not visualized well by GMS stain *(B)* but were clearly shown by the antibody *(D)*. CRS, chronic rhinosinusitis; MBP-1, eosinophil granule major basic protein 1.

improved asthma symptoms but also rhinological symptoms.[67] Two other randomized controlled trials of CRS patients with antifungal agents have demonstrated a benefit while other trials have not.[79] The controversy over antifungal therapy for CRS and the involvement of fungi in CRS continues, requiring further investigations.

CONCLUSION AND FUTURE PERSPECTIVES

Evidence exists that eosinophils respond to fungal organisms and their products, resulting in the production of proinflammatory mediators and killing of fungi. Immune cells in the airways recognize carbohydrate molecules and proteases that are produced by fungal organisms and initiate T_h2 immune responses and eosinophilic inflammation. There are a number of epidemiological findings that link the exposure or sensitivity to fungi, eosinophilic airway inflammation, and human airway diseases. However, several major questions still remain. For example: are eosinophils protective against fungal colonization or infection in humans? How do human eosinophils become activated in diseased tissues or at the site of the immune response, resulting in tissue damage? Which molecules in fungi and host receptors play a pivotal role in antifungal immune responses? In SAFS or CRS, do antifungal immune responses and the aberrant activation of eosinophils play major roles in disease pathophysiology? Careful analysis of recent genetically engineered eosinophil-deficient mice *in vivo* and further characterization of immunobiology of eosinophils *in vitro* will be necessary to answer these critical

questions. Well-designed clinical studies and pharmacological trials in patients with SAFS and CRS will be necessary to dissect the roles for eosinophils and fungi in human diseases. Although these are challenging tasks, future studies have a promise to reveal the true importance of eosinophils in human health and disease.

REFERENCES

1. Gleich GJ, Adolphson CR. The eosinophilic leukocyte: structure and function. *Adv Immunol* 1986;**39**:177−253.
2. Kita H, Adolphson CR, Gleich GJ. Biology of eosinophils. In: Adkinson, Jr. NF, Bochner BS, Yunginger JW, Holgate ST, Busse WW, Simons FER, editors. *Middleton's Allergy: Principles and Practice*. 6th ed. Vol. 1. St. Louis: Mosby; 2003. p. 305−32.
3. Filley WV, Holley KE, Kephart GM, Gleich GJ. Identification by immunofluorescence of eosinophil granule major basic protein in lung tissues of patients with bronchial asthma. *Lancet* 1982;**2**:11−6.
4. Ponikau JU, Sherris DA, Kephart GM, Kern EB, Congdon DJ, Adolphson CR, et al. Striking deposition of toxic eosinophil major basic protein in mucus: implications for chronic rhinosinusitis. *J Allergy Clin Immunol* 2005;**116**:362−9.
5. Leiferman KM, Ackerman SJ, Sampson HA, Haugen HS, Venencie PY, Gleich GJ. Dermal deposition of eosinophil-granule major basic protein in atopic dermatitis. Comparison with onchocerciasis. *N Engl J Med* 1985;**313**:282−5.
6. Matthews AN, Friend DS, Zimmermann N, Sarafi MN, Luster AD, Pearlman E, et al. Eotaxin is required for the baseline level of tissue eosinophils. *Proc Natl Acad Sci USA* 1998;**95**:6273−8.
7. van Burik JA, Magee PT. Aspects of fungal pathogenesis in humans. *Annu Rev Microbiol* 2001;**55**:743−72.

8. Kauffman HF, Tomee JF, van der Werf TS, de Monchy JG, Koeter GK. Review of fungus-induced asthmatic reactions. *Am J Respir Crit Care Med* 1995;**151**:2109−15. discussion 2116.

9. Kheradmand F, Kiss A, Xu J, Lee SH, Kolattukudy PE, Corry DB. A protease-activated pathway underlying Th cell type 2 activation and allergic lung disease. *J Immunol* 2002;**169**:5904−11.

10. Bush RK, Prochnau JJ. Alternaria-induced asthma. *J Allergy Clin Immunol* 2004;**113**:227−34.

11. Romani L. Immunity to fungal infections. *Nat Rev Immunol* 2004;**4**:1−23.

12. Latge JP. The cell wall: a carbohydrate armour for the fungal cell. *Mol Microbiol* 2007;**66**:279−90.

13. Douwes J. (1–>3)-Beta-D-glucans and respiratory health: a review of the scientific evidence. *Indoor Air* 2005;**15**:160−9.

14. Reed CE. Inflammatory effect of environmental proteases on airway mucosa. *Curr Allergy Asthma Rep* 2007;**7**:368−74.

15. Diniz SN, Nomizo R, Cisalpino PS, Teixeira MM, Brown GD, Mantovani A, et al. PTX3 function as an opsonin for the dectin-1-dependent internalization of zymosan by macrophages. *J Leukoc Biol* 2004;**75**:649−56.

16. Garlanda C, Hirsch E, Bozza S, Salustri A, De Acetis M, Nota R, et al. Non-redundant role of the long pentraxin PTX3 in anti-fungal innate immune response. *Nature* 2002;**420**:182−6.

17. Kaur S, Thiel S, Sarma PU, Madan T. Mannan-binding lectin in asthma and allergy. *Curr Allergy Asthma Rep* 2006;**6**:377−83.

18. Roeder A, Kirschning CJ, Rupec RA, Schaller M, Korting HC. Toll-like receptors and innate antifungal responses. *Trends Microbiol* 2004;**12**:44−9.

19. van der Graaf CA, Netea MG, Verschueren I, van der Meer JW, Kullberg BJ. Differential cytokine production and Toll-like receptor signaling pathways by Candida albicans blastoconidia and hyphae. *Infect Immun* 2005;**73**:7458−64.

20. Yadav M, Schorey JS. The beta-glucan receptor dectin-1 functions together with TLR2 to mediate macrophage activation by mycobacteria. *Blood* 2006;**108**:3168−75.

21. Taylor PR, Brown GD, Reid DM, Willment JA, Martinez-Pomares L, Gordon S, et al. The beta-glucan receptor, dectin-1, is predominantly expressed on the surface of cells of the monocyte/macrophage and neutrophil lineages. *J Immunol* 2002;**169**:3876−82.

22. Taylor PR, Tsoni SV, Willment JA, Dennehy KM, Rosas M, Findon H, et al. Dectin-1 is required for beta-glucan recognition and control of fungal infection. *Nat Immunol* 2007;**8**:31−8.

23. Hahn PY, Evans SE, Kottom TJ, Standing JE, Pagano RE, Limper AH. Pneumocystis carinii cell wall beta-glucan induces release of macrophage inflammatory protein-2 from alveolar epithelial cells via a lactosylceramide-mediated mechanism. *J Biol Chem* 2003;**278**:2043−50.

24. Di Renzo L, Yefenof E, Klein E. The function of human NK cells is enhanced by beta-glucan, a ligand of CR3 (CD11b/CD18). *Eur J Immunol* 1991;**21**:1755−8.

25. Herre J, Marshall AS, Caron E, Edwards AD, Williams DL, Schweighoffer E, et al. Dectin-1 uses novel mechanisms for yeast phagocytosis in macrophages. *Blood* 2004;**104**:4038−45.

26. Vetvicka V, Thornton BP, Ross GD. Soluble beta-glucan polysaccharide binding to the lectin site of neutrophil or natural killer cell complement receptor type 3 (CD11b/CD18) generates a primed state of the receptor capable of mediating cytotoxicity of iC3b-opsonized target cells. *J Clin Invest* 1996;**98**:50−61.

27. Ross GD. Role of the lectin domain of Mac-1/CR3 (CD11b/CD18) in regulating intercellular adhesion. *Immunol Res* 2002;**25**:219−27.

28. Xia Y, Borland G, Huang J, Mizukami IF, Petty HR, Todd 3rd RF, et al. Function of the lectin domain of Mac-1/complement receptor type 3 (CD11b/CD18) in regulating neutrophil adhesion. *J Immunol* 2002;**169**:6417−26.

29. Inoue Y, Matsuwaki Y, Shin SH, Ponikau JU, Kita H. Nonpathogenic, environmental fungi induce activation and degranulation of human eosinophils. *J Immunol* 2005;**175**:5439−47.

30. Matsuwaki Y, Wada K, White TA, Benson LM, Charlesworth MC, Checkel JL, et al. Recognition of fungal protease activities induces cellular activation and eosinophil-derived neurotoxin release in human eosinophils. *J Immunol* 2009;**183**:6708−16.

31. Shpacovitch V, Feld M, Bunnett NW, Steinhoff M. Protease-activated receptors: novel PARtners in innate immunity. *Trends Immunol* 2007;**28**:541−50.

32. Miike S, McWilliam AS, Kita H. Trypsin induces activation and inflammatory mediator release from human eosinophils through protease-activated receptor-2. *J Immunol* 2001;**167**:6615−22.

33. Miike S, Kita H. Human eosinophils are activated by cysteine proteases and release inflammatory mediators. *J Allergy Clin Immunol* 2003;**111**:704−13.

34. Yoon J, Ponikau JU, Lawrence CB, Kita H. Innate antifungal immunity of human eosinophils mediated by a beta 2 integrin, CD11b. *J Immunol* 2008;**181**:2907−15.

35. Lundahl J, Hallden G, Hed J. Differences in intracellular pool and receptor-dependent mobilization of the adhesion-promoting glycoprotein Mac-1 between eosinophils and neutrophils. *J Leukoc Biol* 1993;**53**:336−41.

36. Kaneko M, Horie S, Kato M, Gleich GJ, Kita H. A crucial role for beta 2 integrin in the activation of eosinophils stimulated by IgG. *J Immunol* 1995;**155**:2631−41.

37. Gottar M, Gobert V, Matskevich AA, Reichhart JM, Wang C, Butt TM, et al. Dual detection of fungal infections in Drosophila via recognition of glucans and sensing of virulence factors. *Cell* 2006;**127**:1425−37.

38. Reese TA, Liang HE, Tager AM, Luster AD, Van Rooijen N, Voehringer D, et al. Chitin induces accumulation in tissue of innate immune cells associated with allergy. *Nature* 2007;**447**:92−6.

39. Van Dyken SJ, Garcia D, Porter P, Huang X, Quinlan PJ, Blanc PD, et al. Fungal chitin from asthma-associated home environments induces eosinophilic lung infiltration. *J Immunol* 2011;**187**:2261−7.

40. Porter P, Susarla SC, Polikepahad S, Qian Y, Hampton J, Kiss A, et al. Link between allergic asthma and airway mucosal infection suggested by proteinase-secreting household fungi. *Mucosal Immunol* 2009;**2**:504−17.

41. Tam A, Wadsworth S, Dorscheid D, Man SF, Sin DD. The airway epithelium: more than just a structural barrier. *Ther Adv Respir Dis* 2011;**5**:255−73.

42. Parker D, Prince A. Innate immunity in the respiratory epithelium. *Am J Respir Cell Mol Biol* 2011;**45**:189−201.

43. Saenz SA, Taylor BC, Artis D. Welcome to the neighborhood: epithelial cell-derived cytokines license innate and adaptive immune responses at mucosal sites. *Immunol Rev* 2008;**226**:172−90.

44. Locksley RM. Asthma and allergic inflammation. *Cell* 2010;**140**:777−83.

45. Hammad H, Lambrecht BN. Dendritic cells and epithelial cells: linking innate and adaptive immunity in asthma. *Nat Rev Immunol* 2008;**8**:193−204.

46. Ying S, O'Connor B, Ratoff J, Meng Q, Mallett K, Cousins D, et al. Thymic stromal lymphopoietin expression is increased in asthmatic airways and correlates with expression of Th2-attracting chemokines and disease severity. *J Immunol* 2005;**174**:8183—90.

47. Ito T, Wang YH, Duramad O, Hori T, Delespesse GJ, Watanabe N, et al. TSLP-activated dendritic cells induce an inflammatory T helper type 2 cell response through OX40 ligand. *J Exp Med* 2005;**202**:1213—23.

48. Zhou B, Comeau MR, De Smedt T, Liggitt HD, Dahl ME, Lewis DB, et al. Thymic stromal lymphopoietin as a key initiator of allergic airway inflammation in mice. *Nat Immunol* 2005;**6**:1047—53.

49. Al-Shami A, Spolski R, Kelly J, Keane-Myers A, Leonard WJ. A role for TSLP in the development of inflammation in an asthma model. *J Exp Med* 2005;**202**:829—39.

50. Sims JE, Smith DE. The IL-1 family: regulators of immunity. *Nat Rev Immunol* 2010;**10**:89—102.

51. Moussion C, Ortega N, Girard JP. The IL-1-like cytokine IL-33 is constitutively expressed in the nucleus of endothelial cells and epithelial cells in vivo: a novel 'alarmin'? *PLoS One* 2008;**3**. e3331.

52. Schmitz J, Owyang A, Oldham E, Song Y, Murphy E, McClanahan TK, et al. IL-33, an interleukin-1-like cytokine that signals via the IL-1 receptor-related protein ST2 and induces T helper type 2-associated cytokines. *Immunity* 2005;**23**:479—90.

53. Kouzaki H, Iijima K, Kobayashi T, O'Grady SM, Kita H. The danger signal, extracellular ATP, is a sensor for an airborne allergen and triggers IL-33 release and innate Th2-type responses. *J Immunol* 2011;**186**:4375—87.

54. Drutz DJ, Catanzaro A. Coccidioidomycosis. Part II. *Am Rev Respir Dis* 1978;**117**:727—71.

55. Wagner JM, Franco M, Kephart GM, Gleich GJ. Localization of eosinophil granule major basic protein in paracoccidioidomycosis lesions. *Am J Trop Med Hyg* 1998;**59**:66—72.

56. Denning DW, O'Driscoll BR, Hogaboam CM, Bowyer P, Niven RM. The link between fungi and severe asthma: a summary of the evidence. *Eur Respir J* 2006;**27**:615—26.

57. Arbes Jr SJ, Gergen PJ, Vaughn B, Zeldin DC. Asthma cases attributable to atopy: results from the Third National Health and Nutrition Examination Survey. *J Allergy Clin Immunol* 2007;**120**:1139—45.

58. Jaakkola MS, Ieromnimon A, Jaakkola JJ. Are atopy and specific IgE to mites and molds important for adult asthma? *J Allergy Clin Immunol* 2006;**117**:642—8.

59. Salo PM, Arbes Jr SJ, Sever M, Jaramillo R, Cohn RD, London SJ, et al. Exposure to Alternaria alternata in US homes is associated with asthma symptoms. *J Allergy Clin Immunol* 2006;**118**:892—8.

60. Stern DA, Morgan WJ, Halonen M, Wright AL, Martinez FD. Wheezing and bronchial hyper-responsiveness in early childhood as predictors of newly diagnosed asthma in early adulthood: a longitudinal birth-cohort study. *Lancet* 2008;**372**:1058—64.

61. Pulimood TB, Corden JM, Bryden C, Sharples L, Nasser SM. Epidemic asthma and the role of the fungal mold Alternaria alternata. *J Allergy Clin Immunol* 2007;**120**:610—7.

62. Greenberger PA. Allergic bronchopulmonary aspergillosis. *J Allergy Clin Immunol* 2002;**110**:685—92.

63. Lotvall J, Akdis CA, Bacharier LB, Bjermer L, Casale TB, Custovic A, et al. Asthma endotypes: a new approach to classification of disease entities within the asthma syndrome. *J Allergy Clin Immunol* 2011;**127**:355—60.

64. Aimanianda V, Bayry J, Bozza S, Kniemeyer O, Perruccio K, Elluru SR, et al. Surface hydrophobin prevents immune recognition of airborne fungal spores. *Nature* 2009;**460**:1117—21.

65. Denning DW, Park S, Lass-Florl C, Fraczek MG, Kirwan M, Gore R, et al. High-frequency triazole resistance found In non-culturable Aspergillus fumigatus from lungs of patients with chronic fungal disease. *Clin Infect Dis* 2011;**52**:1123—9.

66. Fairs A, Agbetile J, Hargadon B, Bourne M, Monteiro WR, Brightling CE, et al. IgE sensitization to Aspergillus fumigatus is associated with reduced lung function in asthma. *Am J Respir Crit Care Med* 2010;**182**:1362—8.

67. Denning DW, O'Driscoll BR, Powell G, Chew F, Atherton GT, Vyas A, et al. Randomized controlled trial of oral antifungal treatment for severe asthma with fungal sensitization: The Fungal Asthma Sensitization Trial (FAST) study. *Am J Respir Crit Care Med* 2009;**179**:11—8.

68. Spector SL, Bernstein IL, Li JT, Berger WE, Kaliner MA, Schuller DE, et al. Parameters for the diagnosis and management of sinusitis. *J Allergy Clin Immunol* 1998;**102**:S107—44.

69. Tomassen P, Van Zele T, Zhang N, Perez-Novo C, Van Bruaene N, Gevaert P, et al. Pathophysiology of chronic rhinosinusitis. *Proc Am Thorac Soc* 2011;**8**:115—20.

70. Harlin SL, Ansel DG, Lane SR, Myers J, Kephart GM, Gleich GJ. A clinical and pathologic study of chronic sinusitis: the role of the eosinophil. *J Allergy Clin Immunol* 1988;**81**:867—75.

71. Stoop AE, van der Heijden HA, Biewenga J, van der Baan S. Eosinophils in nasal polyps and nasal mucosa: an immunohistochemical study. *J Allergy Clin Immunol* 1993;**91**:616—22.

72. Motojima S, Frigas E, Loegering DA, Gleich GJ. Toxicity of eosinophil cationic proteins for guinea pig tracheal epithelium in vitro. *Am Rev Respir Dis* 1989;**139**:801—5.

73. Hoover GE, Newman LJ, Platts-Mills TA, Phillips CD, Gross CW, Wheatley LM. Chronic sinusitis: risk factors for extensive disease. *J Allergy Clin Immunol* 1997;**100**:185—91.

74. Millar JW, Johnston D., L. A, D., L. Allergic aspergillosis of the maxillary sinuses. *Thorax* 1981;**36**:710 (Abstract).

75. Corey JP, Delsupehe KG, Ferguson BJ. Allergic fungal sinusitis: allergic, infectious, or both? *Otolaryngol Head Neck Surg* 1995;**113**:110—9.

76. deShazo RD, Swain RE. Diagnostic criteria for allergic fungal sinusitis. *J Allergy Clin Immunol* 1995;**96**:24—35.

77. Ponikau JU, Sherris DA, Kern EB, Homburger HA, Frigas E, Gaffey TA, et al. The diagnosis and incidence of allergic fungal sinusitis. *Mayo Clin Proc* 1999;**74**:877—84.

78. Braun H, Buzina W, Freudenschuss K, Beham A, Stammberger H. 'Eosinophilic fungal rhinosinusitis': a common disorder in Europe? *Laryngoscope* 2003;**113**:264—9.

79. Rank MA, Adolphson CR, Kita H. Antifungal therapy for chronic rhinosinusitis: the controversy persists. *Curr Opin Allergy Clin Immunol* 2009;**9**:67—72.

Eosinophils: Mediators of Host-Parasite Interactions

Chapter 10.1

Introduction

David Abraham

HELMINTHS AND T-HELPER TYPE 2 RESPONSES

More than two billion people are infected with parasitic helminths and although the infections are generally not fatal, they are associated with high rates of morbidity. Helminths are a very diverse group of organisms with sizes ranging from microscopic to measurements in meters. The group is comprised of cestodes, trematodes, and nematodes with unique life cycles and habitats for each species within the phylum. Yet, it is remarkable how similar the immune responses are to these infections in human and animal hosts, especially considering the dissimilarities in life cycles and biology. With few exceptions, the immunological hallmark of the helminthic infection is the T-helper type 2 (T_h2) response, which is characterized in part by the presence of the cytokines interleukin-4 (IL-4), -5, -10 and -13 and of generalized eosinophilia. Each of the worms induces antigen-driven parasite-specific immune responses, yet the absolute common denominator that leads to a T_h2 response and eosinophilia has not been elucidated. It has been suggested that glycans, lipids and lipoproteins, proteases, or chitin may be the common type of molecule that induces the T_h2 response.[1] The shared characteristic of inducing eosinophilia between atopic diseases and helminth infections suggests that the two processes produce similar types of molecules that attract eosinophils. Alternatively, the two processes may produce similar types of tissue damage that attracts eosinophils for tissue remodeling and debris clearance.

HUMANS, HELMINTHS, AND EOSINOPHILIA

One of the diagnostic indicators of a helminth infection in humans is the presence of eosinophilia. Following helminth infection there is a dramatic increase in eosinophil numbers in the blood and these eosinophils then migrate to the site of infection where they degranulate, releasing eosinophil secondary granule products. The time of peak eosinophilia levels correlates with the specific developmental stages of the parasite. Interestingly, there is a spontaneous decrease in eosinophil numbers that occurs when migration of the larvae ceases and the parasites develop into sexually mature adults.[2] These correlations suggest that eosinophils are required for the control of the infections; however, these analyses are confounded in many cases by the potential for individuals to be infected acutely or chronically with multiple pathogens simultaneously, each containing various stages of parasite development. Histopathological evidence of eosinophils surrounding dying parasites in tissue biopsy specimens has suggested that eosinophils are active participants in killing the worms in the immune response. *In vitro* assays with human eosinophils have shown that eosinophils can kill helminths in collaboration with antibody and complement,[2] yet it remains impossible to ascribe a definitive positive or a negative role for eosinophils in the response of humans to helminth infections.

ANIMAL MODELS FOR THE STUDY OF EOSINOPHIL FUNCTION

Mouse models have been used to study the role of eosinophils in response to helminth infections. Several approaches have been used to deplete eosinophils from mice, each with their own strengths and weaknesses. IL-5 is the central regulator of eosinophilia and several approaches have been used to block this cytokine, including monoclonal antibody treatment and gene-deleted mice. These treatments have resulted in a significant decrease in eosinophil numbers and an increase in parasite survival for some infections and no effect in others. Alternatively, transgenic mice have been used to overexpress the *IL-5* gene and the increase in the number of eosinophils promoted the killing of some but not all of the tested species of worms. It must be emphasized that removal of IL-5 does not eliminate eosinophils, but rather limits their proliferation, survival, and recruitment to sites of infection.

Eosinophils in Health and Disease. http://dx.doi.org/10.1016/B978-0-12-394385-9.00010-9

Another approach taken to eliminate eosinophils was to alter the ability of mice to respond to the eosinophil chemokine eotaxin [C-C motif chemokine 11 (CCL11)], by eliminating the gene encoding the eotaxin receptor, C-C chemokine receptor type 3 (CCR3), or by treating mice with specific antireceptor antibody. Once again, mice treated in this fashion had decreased ability to control some helminth infections, yet were capable of controlling others. CCR3 is expressed on cell types other than eosinophils, so a direct eosinophil/helminth interaction cannot be concluded from these studies.

An important development in the tools available for the study of eosinophils in mice has been the development of mice in which the eosinophil lineage has been ablated. Studies of helminth infections in eosinophil-free transgenic (Tg)PHIL and ΔdblGATA mice have shown that eosinophils are required for the control of some worms and not others. Studies with the nematode *Strongyloides stercoralis* have demonstrated that PHIL mice develop both innate and adaptive immunity to the infection. However, if eosinophils isolated from normal mice are studied *in vitro* in culture or *in vivo* within diffusion chambers, it is clear that they are capable of killing the nematode larvae. TgPHIL mice controlled the infection through an eosinophil-independent mechanism that was neutrophil dependent.[3] Thus, in cases where there are redundant immune mechanisms to control an infection, eosinophil-free mice use collateral mechanisms to control the infection. The testing of isolated eosinophils *in vitro* and *in vivo* is required to answer the question of whether eosinophils have the potential to kill helminths.

The mechanisms through which eosinophils control helminth infections include both terminal effector functions and immunomodulatory capacities including the ability to produce cytokines, act as antigen-presenting cells, and modulate T-cell recruitment[4] (see Chapter 10.3 by Dr. Thomas Nutman). Mice deficient in eosinophil granule major basic protein 1 (MBP-1) and eosinophil peroxidase (EPO) have been generated and it is interesting to observe the range of susceptibilities to these molecules. Control of some nematodes required both MBP-1 and EPO, others only MBP-1, and in others neither molecule. Studies with *S. stercoralis*, infecting mice lacking either MBP-1 or EPO, demonstrated that neither molecule was required for resistance to the infection. However, if isolated eosinophils were studied *in vitro* in culture or *in vivo* within diffusion chambers it was shown that eosinophils required MBP-1 but not EPO to kill the parasites. It is assumed that killing of the parasites occurred in the MBP-1-deficient animals by neutrophils and therefore the requirement for MBP-1 from eosinophils was obscured.[3] It remains unclear why released eosinophil granule products do not universally kill all worms based on the nonspecific toxic nature of these molecules. Do parasites control the release of toxic molecules from the eosinophils or have the worms developed mechanisms to defuse them, as has been reported for some of the toxic pathways? Finally, it has also been proposed that there are eosinophil secondary granule product-independent mechanisms used by eosinophils to control the helminths.

One of the conundrums of the relationship between eosinophils and helminths is the observation that molecules from worms can directly recruit the cells. Eosinophils are directly recruited to the nematode *S. stercoralis* without the need for other host cell assistance. Chemoattractants derived from the worm and host species stimulate similar receptors and second messenger signals to induce eosinophil chemotaxis. The parasite extract stimulates multiple receptors on the eosinophil surface, which ensures a robust recruitment of eosinophils to the parasite. The redundancy of the chemotactic factors produced by the parasite and the multiple responding receptors on the eosinophils suggests chemotactic receptors on these pivotal cells may have evolved to ensure a robust response to this infection.[5] It is hard to understand why worms have evolved to produce chemoattractants for the cell that could kill them. One possible explanation is that worms recruit eosinophils to repair damage caused by the migrating parasites. It is to the advantage of the worm to keep its host alive and repairing tissues damaged by proteases released by the worm would have a selective advantage. An alternative explanation is that eosinophils have a beneficial role for helminths. Studies with the nematode *Trichinella spiralis* have shown that human eosinophils can kill the larvae *in vitro* and mice deficient in CCR3 or IL-5 had increased muscle larvae burdens in primary infections. However, eosinophil-free PHIL and ΔdblGATA mice infected with *T. spiralis* had significantly reduced numbers of parasites in the tissues. This was correlated with an increase in the T-helper type 1 (T_h1) response to the infection[6] (see Chapter 10.4 by Dr. Nebiat Gebreselassie and Dr. Judith Appleton). Therefore, eosinophils appear to have both a beneficial and a detrimental effect on the nematode *T. spiralis*. Furthermore, studying the filarial nematode *Litomosoides sigmodontis*, it was observed that eosinophils were required for immune clearance of the infection. However, eosinophils were also found to accelerate parasite development and reproduction[7] (see Chapter 10.2 by Dr. Katrin Gentil, Dr. Achim Hoerauf, and Dr. Laura Layland). It is therefore possible that worms recruit eosinophils to provide signals required to enhance parasite development and thereby evade potential killing mechanisms by the cells.

In conclusion, it is clear that the multifactorial nature of eosinophils and the inherent complexity of helminth infections will preclude a simple relationship between the cells and the pathogens. It has become clear that eosinophils can control some helminths through both terminal effector functions and immunomodulatory capacities.

Other helminths have developed mechanisms to evade eosinophil control processes and some helminths have even developed a need for eosinophils for their optimal development and survival.

REFERENCES

1. Anthony RM, Rutitzky LI, Urban Jr JF, Stadecker MJ, Gause WC. Protective immune mechanisms in helminth infection. *Nat Rev Immunol* 2007;**7**:975—87.

2. Klion AD, Nutman TB. The role of eosinophils in host defense against helminth parasites. *J Allergy Clin Immunol* 2004;**113**:30—7.

3. O'Connell AE, Hess JA, Santiago GA, Nolan TJ, Lok JB, Lee JJ, et al. Major basic protein from eosinophils and myeloperoxidase from neutrophils are required for protective immunity to *Strongyloides stercoralis* in mice. *Infect Immun* 2011;**79**:2770—8.

4. Shamri R, Xenakis JJ, Spencer LA. Eosinophils in innate immunity: an evolving story. *Cell Tissue Res* 2011;**343**:57—83.

5. Stein LH, Redding KM, Lee JJ, Nolan TJ, Schad GA, Lok JB, et al. Eosinophils utilize multiple chemokine receptors for chemotaxis to the parasitic nematode *Strongyloides stercoralis*. *J Innate Immun* 2009;**1**:618—30.

6. Fabre V, Beiting DP, Bliss SK, Gebreselassie NG, Gagliardo LF, Lee NA, et al. Eosinophil deficiency compromises parasite survival in chronic nematode infection. *J Immunol* 2009;**182**:1577—83.

7. Babayan SA, Read AF, Lawrence RA, Bain O, Allen JE. Filarial parasites develop faster and reproduce earlier in response to host immune effectors that determine filarial life expectancy. *PLoS Biol* 2010;**8**:e1000525.

Chapter 10.2

Eosinophil-Mediated Responses Toward Helminths

Katrin Gentil, Achim Hoerauf and Laura E. Layland

EXPOSURE AND PREVALENCE OF HELMINTHIC INFECTIONS

Helminths are extremely successful parasites that currently infect more than 2 billion people worldwide. Infections are usually chronic and often result in malnutrition, anemia, and increased susceptibility to other pathogens. They are also a major problem for domestic production both at home and abroad. Although most individuals are easily treated, they can be rapidly reinfected and there is growing concern that some parasites are becoming resistant to current chemotherapeutic methods, as has been observed for helminths parasitizing livestock.[1]

Helminths are divided into three families: cestodes such as *Taenia* and *Echinococcus*, which are more commonly known as tapeworms, trematodes (flukes), and nematodes (roundworms). Among the flukes, the schistosome family is endemic to Africa and Asia and morbidity arises from CD4$^+$ T cell-mediated granulomatous inflammation which are, at certain developmental stages, extremely rich in eosinophils.[2,3] Equally prevalent in the tropics are the filarial nematodes *Onchocerca volvulus*, *Loa loa*, *Wuchereria bancrofti*, and the *Brugia* species. These infections have a tremendous socio-economic impact, especially in women, since the disease can manifest into severe disfigurement.[4] In tropical countries, helminthic infections include the *Ascaris* species, and hookworms, pinworms, and whipworms. These generally occur in the gut and the prevalence of infection is high among the poor and immunocompromised individuals.[5]

A fascinating aspect of helminthology is deciphering how these infections are expelled from the body and the components of the immune system that are involved in effective immunity. Helminthic infections are renowned for modulating the host's response into producing a T-helper type 2 (T$_h$2) milieu that coincides with high levels of immunoglobulin E (IgE), T regulatory (T$_{reg}$) cells and other immune cells such as mast cells, basophils, and, of course, eosinophils.[6] Interleukin-4 (IL-4) and 13 (IL-13) appear to aid the host in antihelminthic immune responses and are produced not just by T$_h$2 and natural killer T (NKT) cells but also eosinophils. With a special focus on filarial and schistosomal infections, this chapter details the functions of eosinophils, including their role in pulmonary eosinophilia, their mediation of helminth destruction and, in contrast, their participation in parasite—host coevolution.

EOSINOPHILS IN FILARIAL INFECTIONS

The filarial nematodes *O. volvulus* and *W. bancrofti* are the causative agents of river blindness and lymphatic filariasis, respectively. Infective third-stage (L3) larvae are transmitted by vectors such as *Simulium* blackflies for *O. volvulus* and *Culex*, *Anopheles*, and *Aedes* mosquitos for *W. bancrofti*. After actively penetrating the wound site, the larvae undergo two molting steps over several weeks until they reach adulthood. Adult *O. volvulus* worms usually reside in the subcutaneous tissue, whereas *W. bancrofti* and *Brugia* species move throughout the lymphatic vessels. Released microfilariae [first-stage (L1) larvae] migrate through the skin (*O. volvulus*) or can be found in the bloodstream (*W. bancrofti*) and are taken up by the vector during a blood meal, completing the life cycle.[4,7] In the human host, adult worms can survive for more than a decade and the developing filarial-specific responses are characterized by strong T$_h$2 responses, especially increased levels of IL-5 and IgE. With the exception of *L. loa*, the previously mentioned filariae also harbor the endosymbiotic bacterium *Wolbachia*.[4,8] The dependency of filariae on this intracellular alphaproteobacterium was demonstrated through treatment with the antibiotic tetracycline, since depletion of *Wolbachia* leads to worm sterility and death.[7,9] These findings have provided a new avenue for treating individuals suffering from filarial disease including successful treatment regimens for *Mansonella perstans*, a filarial infection that elicits relatively mild symptoms but is impervious to treatment with conventional drugs such as ivermectin and diethylcarbamazine (DEC).[10] In addition, *Wolbachia* also contribute to innate immune responses due to the release of bacterial products from living or dead worms, and such responses are thought to drive beneficial adaptive scenarios for the helminth.[8] Although the basis of symbiosis is considered to be metabolic, other roles for the bacterium, such as immune evasion, have received little attention despite their propensity to attract neutrophils.[11] Interestingly, following antibiotic treatment, eosinophils are drawn to the worm and degranulate on the cuticle, providing support for the hypothesis that *Wolbachia* aid worm survival by diverting eosinophil responses.[12]

Eosinophils in Onchocerca Dermatitis and Keratitis

Infections with the filarial nematode *O. volvulus* affect approximately 37 million people, mainly in sub-Saharan Africa, but also in parts of Latin America and Yemen.[4,7] Onchocerciasis is manifested by dermatological symptoms and ocular disease. With regards to the latter, both the anterior (punctate keratitis) and posterior (chorioretinitis) segments of the eye are affected. Disease of the anterior segment can be studied using a mouse model in which chronic disease is mimicked by repeated sensitizations with parasite antigens followed by local injection of parasite-derived proteins into the corneal stroma. Studies with this model have shown that whereas neutrophil migration occurs early on after injection and does not require a filarial-specific adaptive immune response, eosinophil migration is delayed and is dependent on the presence of specific antibodies and functional T-cell responses.[13]

Eosinophil migration into the cornea depends on the expression of the chemokine eotaxin [C-C motif chemokine 11 (CCL11)], as well as on the adhesion molecules P-selectin and intercellular adhesion molecule 1 (ICAM-1).[14] Although the exact role of eosinophils during onchocercal keratitis remains unclear, they have been implicated in damaging the corneal structure, though they are also thought to be involved in clearing the injected proteins or the dying worms.[15] Active eosinophils in the cornea release eosinophil cationic protein (ECP) and eosinophil granule major basic protein 1 (MBP-1), which are thought to interfere with epithelial cell viability and wound healing.[15,16] These proteins are also increased in the serum and urine of *O. volvulus*-infected patients, confirming the findings in a mouse model in actual human patients.[17]

Several immunoregulatory mechanisms have been described in onchocerciasis including the induction of alternatively activated macrophages and T_{reg} cells, both of which downmodulate the host's immune system to ensure parasite survival.[6,18,19] Over a period of time, adult female worms become encased by host tissue forming a characteristic subcutaneous nodule (onchocercoma), which somewhat resembles a granuloma. Although the principal cells are macrophages (which fuse into giant cells), the surrounding infiltrate is rich in other cells including T_{reg} cells and eosinophils.[20] With regards to T_{reg} cells, several reports have described the ability of these cells to promote the induction of the noncomplement-fixing immunoglobulin G4 (IgG4) in an IL-10-dependent manner.[18,21] The benefits of such B-cell responses are apparent in infections where high IgE responses need to be counterregulated to avoid excessive immunopathology and is well illustrated in the two disease forms of *O. volvulus* infection. The *generalized* or hyporeactive condition is associated with high IgG4, elevated IL-10, positive T_{reg}, and a heavy worm burden, including elevated numbers of microfilariae. This contrasts with the high IgE-producing hyperreactive or *Sowda* patients, since in this situation the microfilariae are eliminated through strong, local, and systemic T_h2 immune responses.[18] However, these strong immune reactions are often associated with severe immunopathology and, interestingly, a strong eosinophil infiltration into the nodules.[22,23] IL-10-positive T_{reg} cells isolated and cloned from the nodules of generalized patients also show a greater propensity to induce IgG4 from B cells. When assessed by immunohistochemistry, abundant T_{reg} and IgG4$^+$ plasma cells were found in hyporeactive patients than in Sowda patients.[18,20,24] This correlates with a more recent study which showed that more transforming growth factor beta (TGF-β) and IgE$^+$ cells were found in the patients with the hyperreactive form.[25]

An infiltration of eosinophils into the nodules has been shown to be dependent on microfilarial release from adult worms[22] and these cells have been shown to systemically and actively attack microfilariae: a critical host defense step considering they represent the causative stage of cutaneous pathology.[23] Confirmation of their ability to destroy microfilariae was observed using a mouse model of onchocerciasis since this life stage was only cleared in an eosinophil-dependent manner.[26] Their ability to attack the parasite has also been shown to be required for the development of filarial immunity. *In vivo* immunization studies with irradiated L3 larvae lead to parasite killing during subsequent infection, but eosinophil depletion during challenge aborts this protective immunity.[27] However, if eosinophil migration is only partially abrogated, as observed when IL-12 is administered during priming or challenge infections, protective immunity remains intact.[28] As mentioned earlier, eosinophils migrate into the nodules on removal of *Wolbachia* by antibiotic treatment since there is a reduction in neutrophils.[11] If, however, the worms were killed directly by macro-filaricidal drugs, such as melarsomine, eosinophil infiltration into the nodules was not observed.[12] Eosinophil activity has also been associated with post-treatment reactions following treatment with a number of different anthelmintics including ivermectin, DEC, and amocarzine, all of which kill microfilariae and lead to the accumulation of eosinophils around the dying worms. For example, depending on the microfilarial load, acute inflammatory adverse reactions occur with varying intensity after the administration of DEC. This so-called Mazzotti reaction is characterized by acute papular onchodermatitis, lymph-adenitis, pruritus, rash and, in rare cases, hypotension. Initially, dermal eosinophils and mast cells degranulate with the simultaneous appearance of eosinophil micro-abscesses and microgranulomas, and microfilariae are mobilized from their usual resident sites, e.g., the skin. During the first 48 h, there is a transient eosinopenia and eosinophil degranulation products are deposited around the microfilariae. By the fourth day, there is an elevation in IL-5 which is followed by newly produced eosinophils and transient eosinophilia for 5−10 days.[29] Eosinophil migration to the skin is associated with increased expression of eotaxin and C-C motif chemokine 5 (CCL5/RANTES). CCL5 (RANTES) expression is first upregulated and mirrors eosinophil migration, whereas eotaxin is upregulated later and remains increased for an extended period of time.[30] In the tissues, eosinophils frequently degranulate and release granule proteins including ECP, eosinophil-derived neurotoxin (EDN), and eosinophil peroxidase (EPO).[31] Thus, during onchocercal infection, there are strong indications that eosinophils are involved in parasite killing due to their abundance in Sowda patients, who manage to clear parasites from their bloodstream; their abundance during post-treatment reactions and, in association, the abrogated clearance of microfilariae in the absence of eosinophils in a mouse model of filarial infection underlines the importance of eosinophils.

Eosinophils in Lymphatic Filariasis

Infections with *W. bancrofti* and *Brugia species* cause lymphatic filariasis, a disease characterized by swelling of the lymphatic vessels and lymph extravasation. Chronic vessel enlargement results in valve insufficiency and chronic lymphedema, primarily in the lower limbs and scrota of male patients.[4,7,18,32] A characteristic feature of *W. bancrofti* and *B. malayi* infection in endemic areas such as India, South East Asia, and Sri Lanka is the development of tropical pulmonary eosinophilia (formerly also known as tropical eosinophilia or pulmonary eosinophilia) which results from a hyperresponsiveness to the parasite. This syndrome is characterized by chronic cough and dyspnea, pulmonary eosinophilia, and fluffy infiltrates in the lungs, and encompasses type I, III and IV allergic reactions. In essence, microfilariae become *trapped* in the lung and during an acute attack, infiltrating eosinophils degranulate and parasite-specific IgE levels in the bronchoalveolar lavage fluid are significantly increased.[33] Thus, lung eosinophilia is actually a two-edged sword: on the one hand, eosinophils attack microfilaria in the lung, while on the other, eosinophils are involved in the destruction of lung tissue and asthmatic reactions.

This dual role of eosinophils during filarial infections was recently confirmed in a study by Babayan and colleagues,[34] where the authors could show that a lack of eosinophils results in delayed larval molting and worm growth following experimental infection with the rodent filaria *Litomosoides sigmodontis*. Infection with *L. sigmodontis* in a permissive mouse strain such as BALB/c mice allows the full development of the parasite's life cycle including patency, i.e., the release of microfilariae. Moreover, IL-5 has not only been shown to be essential for vaccine-induced protection but is the key cytokine for defending against invading L3 larvae, nodule formation, and microfilarial load.[35] Indeed, previous studies have shown that IL-5 deficiency during infection increases worm survival at an early stage of infection, and chronically infected IL-5-deficient mice also released larger numbers of microfilariae.[36] Increased eosinophilia following local subcutaneous injection of recombinant IL-5 at the site of larval inoculation also resulted in earlier fertility and increased microfilarial release.[34] During chronic infection, IL-5 leads to mobilization of eosinophils to the site of infection, but also affects neutrophil migration via the regulation of neutrophil-attracting chemokines. The source of IL-5-modulated chemokine release was found to be macrophages and due to decreased neutrophil infiltration, rather than eosinophil migration, granuloma formation was abrogated.[37] Similar to infections with *O. volvulus*, murine infections with *L. sigmodontis* also require eosinophil granule proteins (EPO, MBP-1) for parasite clearance since in their absence, worms develop faster and have an extended survival.[38] The effects of IL-5 itself and IL-5-induced eosinophilia have to be further identified using eosinophil-deficient mice.

While filarial infection may result in pulmonary eosinophilia and asthmatic symptoms, murine helminth infections also result in decreased airway inflammation and eosinophil migration when mice are challenged with the model allergen ovalbumin.[39] Reviewing the growing body of literature of eosinophils during filarial infection, it is clear that these cells drive diverse functions: the association of eosinophils with dying worms and post-treatment reactions; the involvement in the generation of *memory* to filarial infections; and the elicitation of pulmonary eosinophilia with asthmatic complications, all occurring in parallel to alleviate asthmatic responses against unrelated allergens.

EOSINOPHILS IN SCHISTOSOMA INFECTIONS

Although helminths can release endogenous substances that specifically attract eosinophils, the tissue accumulation of eosinophils is predominantly attributed to the direct chemotactic activities generated by the immediate and delayed immunological responses of the host such as CCL5 (RANTES), cytokines such as IL-5, eotaxin, and integrins. As mentioned earlier, migrating parasitic larvae often elicit a blood and tissue eosinophilia in the host. This peripherally induced eosinophilia has been linked with the overproduction of IL-5 and has been observed in both allergic and helminth-infected individuals.[33] Indeed, elevations in this T_h2 cytokine cause profound eosinophilia whereas deletion of the *interleukin 5 (colony-stimulating factor, eosinophil) (IL5)* gene has the opposite effect. Furthermore, the accumulation of eosinophils into the periphery is thought to be mediated by their attraction to IL-5 since this cytokine can induce their differentiation and release from the bone marrow.[40] In the end, this may have beneficial consequences for the host since eosinophils drawn into an inflammatory state produce the immunoregulatory cytokine IL-13. Such eosinophil activity is clearly seen in the immune responses that develop during schistosomiasis.[2] This chronic helminthic infection affects more than 200 million people worldwide and is endemic in 76 countries.[41] Three of these trematode blood flukes most commonly affect humans: *S. mansoni*, *S. japonicum*, and *S. haematobium* with the former pair eliciting liver and intestinal disease while *S. haematobium* affects the bladder. The mammalian host is initially infected by cercariae released from freshwater snails, which rapidly penetrate the skin and become schistosomula. Interestingly, in the presence of specific antibodies, eosinophils are able to kill schistosomula targets in a classical antibody-dependent cellular cytotoxicity (ADCC) mechanism.[42,43] This host mechanism has been described in several species and requires eosinophils to adhere to their targets. It is thought that the adhesion is mediated through the binding of specific antibodies to the parasite surface through their fragment antigen-binding (Fab) regions and to the eosinophil Fc receptor via their Fc fragments. However, it also appears that other eosinophil surface molecules could be involved, for example, in the case of IgE-dependent cytotoxicity both Fc-epsilon-RII and C-C chemokine receptor type 3 (CCR-3) are required. From epidemiological studies, such eosinophil-driven ADCC responses are thought to play a role in mediating immunity to reinfection.[44,45] However,

substantiating this hypothesis in the laboratory mouse remains difficult, as murine eosinophils express neither Fc-epsilon-RI nor II.[46]

On transformation, schistosomula migrate to the lung and may elicit pulmonary eosinophilia, which is similar to eosinophilic lung disease, induced by *Ascaris* and *Toxocara*, and tropical pulmonary eosinophilia, which was discussed earlier in the context of filarial infection. Pulmonary eosinophilia or Löffler syndrome was first described in 1932 and is a hypersensitivity response associated with elevated serum immunoglobulins (IgG and IgE) and mast cell degranulation.[33] Typical symptoms of pulmonary schistosomiasis include shortness of breath, wheezing, and nonproductive cough, and interestingly patients can be symptomatic with or without radiographic findings. Such techniques have also detected small ill-defined nodular lesions and these abnormalities sometimes appear after anthelmintic therapy indicating an immunological reaction to the dead parasite.[47] Interestingly, in a chemotherapeutic study of *S. mansoni*-infected fishermen, it was found that IL-5 was elevated 24 h after praziquantel treatment.[48,49] Gopinath and colleagues also reported a similar peak in lymphatic filariasis patients following DEC treatment.[50] Both treatment groups also reported an immediate dip in eosinophil numbers and it appears that the balance of IL-5 and eosinophils greatly depends on the intensity of infection prior to chemotherapy. Since praziquantel causes immediate worm destruction and thus the release of antigens, it is possible that the increase in IL-5 drives eosinophils to bind to destroyed worms and adhere to activated vascular endothelium: hence the drop in eosinophils.

During *S. mansoni* infection, the adaptive immune responses increases on the excretion of eggs from adult worm pairs and over a period of weeks these responses change from initial T_h1 reactions to a dominant T_h2 milieu.[2,51,52] As mentioned above, morbidity in schistosomiasis arises from T cell-mediated granulomatous inflammation that occurs due to eggs that become trapped in the liver and intestinal tissues.[2,3,41] Early histological observations illustrated the dual nature of eosinophils: immunohistochemistry showed MBP-1 around the eggs but also granule protein deposits in the tissue indicating these toxic molecules could be involved in egg destruction but also host tissue destruction.[53] Observations in *S. mansoni*-infected mice have shown three phases of granuloma development based on size, immunological content, and fibrotic appearance (Fig. 10.2.1). For example, *young* granulomas, those which are just beginning to form around the egg contain mainly T cells, macrophages, and some eosinophils. The latter population then proceeds to saturate the *middle-aged* granulomas and can represent up to 90% of the actual cellular content. Surprisingly, however, eosinophil ablation during

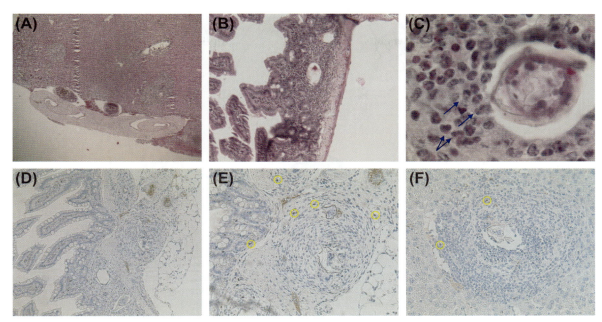

FIGURE 10.2.1   Eosinophil and T_{reg} activity in *Schistosoma mansoni*-induced granulomas. Nine weeks postinfection histological sections (4 μm) were prepared from the liver or intestines of *Schistosoma mansoni*-infected mice and stained with hematoxylin and eosin (upper panel) or immunohistochemically stained for forkhead box protein P3 (Foxp3) (lower panel). Images show *(A)* a schistosomal worm caught *in situ* in the liver; *(B)* a granuloma within the small intestine; *(C)* an enlarged image of a liver-bound granuloma (eosinophil infiltration is indicated by blue arrows). For double immunoenzymatic labeling of Foxp3/CD3, sections were incubated with a goat polyclonal antibody against the C-terminus of the Foxp3 protein and then a biotin-conjugated rabbit antigoat antibody and then processed using the EnVision peroxidase kit. Sections were then incubated with an antibody against CD3 and visualized by the alkaline phosphatase—anti-alkaline phosphatase method. Images show *(D)* an intestinal section of an infected mouse in which multiple granulomas can be seen; *(E)* enlarged image of *(D)* showing an influx of T_{reg} cells (in the yellow circles), within the granuloma. Finally, *(F)* shows that the absence of Foxp3⁺ T cells during infection (infected mice depleted of T_{reg} by α-CD25 blocking antibody) elicits uncontrolled immunopathology.[3,51] Microscopic analysis was performed using a Leitz microscope (Leica Microsystems, Wetzlar, Germany) and the histology preparations were obtained from the research group of Layland and Prazeres da Costa, MIH, München, Germany.

infection had no impact on granuloma size indicating that their primary function is perhaps along the lines of metabolic scavenging.[54] *Old* granulomas, those in which the eggs have been destroyed and display a strong fibrotic appearance, contain very few eosinophils and are instead dominated by an influx of forkhead box protein P3 (Foxp3)⁺ regulatory T cells.[3] These T_{reg} cells actually accumulated within the periphery of the granuloma indicating that their presence was one of preventing further granuloma expansion. Indeed, lack of antigen-specific T_{reg} cells during infection causes uncontrolled immunopathology and elevated T_h responses.[51] Nevertheless, there is a growing body of evidence which demonstrates that eosinophils can dampen T_{reg} activity or swing T_h responses at the site of inflammation. Such immunomodulatory capacity by eosinophils is thought to be modulated by the secretion of IL-10 and TGF-β.[55] Indeed, the production of these cytokines may even prevent naive T cell polarization, which would naturally have consequences on the developing schistosome-specific T_h1 or T_h2 responses or the shift from one phase to another. A summary of the relationship between eosinophils and schistosomal infection is depicted in Fig. 10.2.2.

EOSINOPHIL RESPONSES DURING OTHER HELMINTHI INFECTIONS

Besides tissue-dwelling nematodes, eosinophil activity is also prominent in gut-dwelling nematode infections, such as *Ancylostoma duodenale*, *Ascaris lumbricoides*, *Enterobius vermicularis*, *Strongyloides stercoralis*, *Trichinella spiralis*, and *Trichuris trichuria*. Although there are several overlapping features of eosinophil activity due to similar traits in helminthic infections (affected organs, larval migration patterns, etc.), defined functional effects of eosinophils remain unclear. As described earlier for schistosomes, much research has focused on the ability of eosinophils to mediate parasite elimination. As demonstrated with filarial models, evidence of effective killing by eosinophils has been observed in *S. stercoralis* infections,[56] intestinal parasites that can also survive in a free-living form in the environment. This is reflected in humans where patients suffering from severe strongyloidiasis show decreased eosinophil levels when compared with uninfected patients or individuals suffering from a mild infection. Such eosinophil-mediated parasite expulsion has also been observed in *T. spiralis*, *Nocardia brasiliensis*, and the

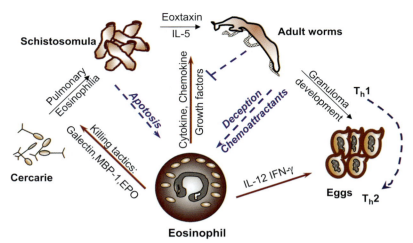

FIGURE 10.2.2 **Eosinophil activity during schistosomal infection.** *Schistosoma mansoni* infection begins with the penetration of the mammalian host by cercariae, which are released from the intermediate freshwater snail hosts. The cercariae transform into schistosomula and migrate to the lung where they may induce pulmonary eosinophilia. After a period of time, juvenile worms migrate to the liver where they mature into adult worms, pair up and produce offspring in the form of eggs. Throughout the infection there is a change from a prominent antigen-specific T_h1 phase to a dominant T_h2 response. To complete the life cycle, the eggs must penetrate the intestinal lumen so that they are excreted with the stool. Morbidity arises from CD4$^+$ T-cell mediated-granulomatous tissue. The image shows the activities and mechanisms employed by eosinophils during infection (dark brown lines). For example, eosinophils have been shown to kill schistosomula but their release of cytokines or chemokines to helminth-secreted products may regulate eosinophil-mediated tissue inflammation and attract other immune cells to the site of inflammation. On the other hand, helminths harbor specific molecules or methods to evade, trick, modulate, or paralyze eosinophil activity (blue lines). EPO, eosinophil peroxidase; IFN-γ, interferon gamma; IL-5, interleukin-5; IL-12, interleukin-12; MBP-1, eosinophil granule major basic protein 1; T_h1, T-helper type 1; T_h2, T-helper type 2.

murine model (*T. muris*) of the human whipworm infection *T. trichuris*,[57] which is currently estimated to infect 1 billion people worldwide through contaminated food and water. Worm expulsion often coincides with enhanced T_h2 responses such as IL-4, IL-5, and IL-13 and resolution of *T. muris* infection only occurs in a T_h2-dependent manner. In fact, it was reported that the release of helminth-induced IL-4 and IL-13 lead to goblet cell hyperplasia and increased turnover of the epithelial layer, which resulted in worm expulsion. Nevertheless, although eosinophils are present at the site of infection, no functional role for parasite expulsion has been described. For example, recent studies have shown that these cells infiltrate the local draining lymph nodes two weeks after infection and produce IL-4, but the value of this effect is required neither for worm expulsion nor T_h2 generation suggesting a distinct, yet undefined role of eosinophils in *T. muris* infection.[58]

In several of these helminthic infection models, eosinophil migration and recruitment has been found to require sufficient levels, or a combination of eotaxin (CCL11) and IL-5 but also by parasitic proteins and chitins. The necessity of IL-5 for eosinophil activity was demonstrated in *S. stercoralis*-infected IL-5$^{-/-}$ mice since these animals showed reduced levels of peripheral eosinophils and increased parasite survival, while peripheral eosinophilia in IL-5 transgenic mice resulted in increased parasite expulsion.[59] In addition, eosinophil recruitment during *T. muris* or *T. spiralis* infection depends on the expression of $\alpha_4\beta_7$/$\alpha_4\beta_1$ and β_7 integrin,[60] respectively. Of interest, lack of

eotaxin (CCL11) induces the upregulation of eotaxin-2 [C-C motif chemokine 24 (CCL24)] in the gut of nematode infected-mice, indicating a potential compensatory route for eosinophil migration. This is an interesting aspect since eotaxin (CCL11) is an agonist for CCR-3 and acts in synergy with IL-5 to recruit eosinophils to inflammatory sites.[57] During chronic infection with *Heligmosomoides polygyrus*, infiltrating eosinophils and mast cells are decreased as well as proinflammatory cytokine production. Nevertheless, infections with this helminth promote protective immune responses during experimental allergic airway inflammation since the helminth harbors specific components which actively decrease CCR-3 expression levels in lung eosinophils and downregulates eotaxin concentrations.[61] This protection, however, is limited to pulmonary allergy since the migration of eosinophils to allergic skin areas remains unaffected.[62] In contrast to *H. polygyrus*, *N. brasiliensis* migrates through various tissues and there are three distinct phases: an initial immune response in the skin, an intermediate response in the lung, and chronic responses in the gut. During the initial infection (first 30 min), eosinophils migrate to the site of infection in a complement-dependent manner that is no longer the case after 3 h. Interestingly, lack of eotaxin (CCL11) during the infection does not affect larval expulsion but instigates a decrease in eosinophil infiltration, which is not compensated by the administration of IL-5 in IL-5 transgenic mice.[63] However, complete eosinophil deficiency (ΔdblGATA mice) resulted in decreased secondary resistance to infection and increased worm burden during primary and secondary infection.[64] In contrast, a recent study of *T. spiralis* infection in eosinophil-deficient mice found decreased parasite burden and compromised survival.[65]

IL-5 transgenic mice are renowned for their numerous eosinophils, and in the absence of IL-5 or endothelial cells mature eosinophils only have a short half-life *in vitro*. Thus, although IL-5 is essential for their propensity, it remained unclear whether eosinophilia was maintained by the extended survival of eosinophils in the periphery or due to increased *de novo* generation in the bone marrow. Using the model of *N. brasiliensis* infection it was revealed that the influx of eosinophils into the lung during the intermediate phase was due to their increased survival rather than a constantly renewed production and was dependent on the ability of T cells to respond to IL-4.[66] This work confirms earlier studies with *T. spiralis*, which demonstrated that during acute infection, eosinophil output from the bone marrow was only increased three-fold, which is an insufficient amount to explain massive tissue eosinophilia.[67] Ohnmacht and colleagues also demonstrated that enhanced eosinophil survival during infection was due to preventing apoptosis, since the eosinophils had high expression of sialic acid-binding Ig-like lectin 5 (Siglec-5/Siglec-F): cross-linking of Siglec-5 (Siglec-F) can induce apoptosis in eosinophils.[66] In association, eosinophil survival is partially controlled by members of the apoptosis regulator Bcl-2 (Bcl-2) family and the antiapoptotic protein Bcl2 antagonist of cell death (Bad) is up-regulated by IL-5, again verifying the vital requirement of eosinophils for this cytokine. Remarkably, eosinophil survival was also enhanced in hypoxic conditions, which might explain why many eosinophils are found in the peritoneum since this is an area of low partial O_2 pressure. Therefore, one could hypothesize that from this reservoir eosinophils may recirculate to other tissues and act as antigen-presenting cells: this has been described in allergic airway inflammation where eosinophils mediated T-cell priming. In association, during *N. brasiliensis* infection, eosinophils were found within the subcapsular sinus and T-cell zone of mesenteric lymph nodes.[66] Furthermore, a series of studies using a mouse model of *S. stercoralis* infection confirmed the ability of eosinophils to present antigens even to naive T cells.[68,69] Moreover, antigen-specific immune responses to *S. stercoralis* were dependent on antigen presentation in major histocompatibility complex class II molecules on eosinophils and in these studies eosinophils were shown to be more effective at inducing T_h2 responses than dendritic cells.[60–62] In correlation, eosinophil-deficient mice have reduced T_h2 cytokine production and the increasing number of studies on these cells are clearly showing that eosinophils do not just possess a destructive dimension, but also have a broad regulatory capacity including the modulation of innate and adaptive immune responses.

CONCLUSION

Although still perceived as cells promoting tissue damage and pathology, it is becoming clear that eosinophils also possess immunoregulatory properties that may, in some cases, dictate the outcome of infection. With regard to helminthic infection, eosinophils seem to communicate with virtually all cells and their role in governing tissue immunomodulation describes yet another facet to their nature. Components that promote eosinophil function have also been isolated from other helminths, such as proteases from *Paragonimus westermani*, which attenuate the effector functions of eosinophils triggered by IgG and cathepsin L proteinase from *Fasciola hepatica*, which hinders antibody-mediated eosinophil attachment to newly released juveniles.

In addition, eosinophils are involved in direct parasite killing (*S. stercoralis*) and have an effect on worm expulsion (*N. brasiliensis*). Eosinophils are also involved in tissue damage during filarial infection (*O. volvulus*, *W. bancrofti*, *B. malayi*) and are active components in generating granulomas during schistosome infection and onchocercomas during onchocercal infection. In *T. spiralis*-infected mice, eosinophils seem to have a helminth-protective role rather than damaging the worms. Eosinophils are associated with faster worm development and increased microfilarial release during experimental filarial infection (*L. sigmodontis*), but also with microfilarial killing in a mouse model. Clearly, the plethora of data on eosinophil activity does not allow their activities to fall nicely under one umbrella, but it is clear that their activities either do not affect the parasite (*T. canis*), that they are protective (*N. brasiliensis*), or occasionally parasite-protective (*S. mansoni*). Lastly, one should not forget the manipulative ability of the helminths themselves, their trickery in attracting eosinophils through chemotactic molecules to aid their own survival, or the mechanisms used to evade eosinophil-mediated helminthotoxicity.

REFERENCES

1. Anziani OS, Suarez V, Guglielmone AA, Warnke O, Grande H, Coles GC. Resistance to benzimidazole and macrocyclic lactone anthelmintics in cattle nematodes in. *Argentina Vet Parasitol* 2004;**122**:303–6.
2. Pearce EJ, MacDonald AS. The immunobiology of schistosomiasis. *Nat Rev Immunol* 2002;**2**:499–511.
3. Layland LE, Mages J, Loddenkemper C, Hoerauf A, Wagner H, Lang R, et al. Pronounced phenotype in activated regulatory T cells during a chronic helminth infection. *J Immunol* 2010;**184**:713–24.

4. Taylor MJ, Hoerauf A, Bockarie M. Lymphatic filariasis and onchocerciasis. *Lancet* 2010;**376**:1175−85.

5. Hotez PJ, Gurwith M. Europe's neglected infections of poverty. *Int J Infect Dis* 2011;**16** [Epub ahead of print].

6. Allen JE, Maizels RM. Diversity and dialogue in immunity to helminths. *Nat Rev Immunol* 2011;**11**:375−88.

7. Hoerauf A, Pfarr K, Mand S, Debrah AY, Specht S. Filariasis in Africa-treatment challenges and prospects. *Clin Microbiol Infect* 2011;**17**:977−85.

8. Saint André A, Blackwell NM, Hall LR, Hoerauf A, Brattig NW, Volkmann L, et al. The role of endosymbiotic Wolbachia bacteria in the pathogenesis of river blindness. *Science* 2002;**295**:1892−5.

9. Debrah AY, Mand S, Marfo-Debrekyei Y, Batsa L, Albers A, Specht S, et al. Macrofilaricidal Activity in Wuchereria bancrofti after 2 Weeks Treatment with a Combination of Rifampicin plus Doxycycline. *J Parasitol Res* 2011;**2011**:201617.

10. Hoerauf A. Mansonella perstans—the importance of an endosymbiont. *N Engl J Med* 2009;**361**:1502−4.

11. Brattig NW, Büttner DW, Hoerauf A. Neutrophil accumulation around Onchocerca worms and chemotaxis of neutrophils are dependent on Wolbachia endobacteria. *Microbes Infect* 2001;**3**:439−46.

12. Hansen RD, Trees AJ, Bah GS, Hetzel U, Martin C, Bain O, et al. A worm's best friend: recruitment of neutrophils by Wolbachia confounds eosinophil degranulation against the filarial nematode Onchocerca ochengi. *Proc Biol Sci* 2011;**278**:2293−302.

13. Pearlman E, Hall LR, Higgins AW, Bardenstein DS, Diaconu E, Hazlett FE, et al. The role of eosinophils and neutrophils in helminth-induced keratitis. *Invest Ophthalmol Vis Sci* 1998;**39**:1176−82.

14. Kaifi JT, Diaconu E, Pearlman E. Distinct roles for PECAM-1, ICAM-1, and VCAM-1 in recruitment of neutrophils and eosinophils to the cornea in ocular onchocerciasis (river blindness). *J Immunol* 2001;**166**:6795−801.

15. Pearlman E, Hall LR. Immune mechanisms in Onchocerca volvulus-mediated corneal disease (river blindness). *Parasite Immunol* 2000;**22**:625−31.

16. Trocmé SD, Gleich GJ, Kephart GM, Zieske JD. Eosinophil granule major basic protein inhibition of corneal epithelial wound healing. *Invest Ophthalmol Vis Sci* 1994;**35**:3051−6.

17. Tischendorf FW, Brattig NW, Buttner DW, Pieper A, Lintzel M. Serum levels of eosinophil cationic protein, eosinophil-derived neurotoxin and myeloperoxidase in infections with filariae and schistosomes. *Acta Trop* 1996;**62**:171−82.

18. Adjobimey T, Hoerauf A. Induction of immunoglobulin G4 in human filariasis: an indicator of immunoregulation. *Ann Trop Med Parasitol* 2010;**104**:455−64.

19. Hoerauf A, Brattig N. Resistance and susceptibility in human onchocerciasis—beyond Th1 vs. Th2. *Trends Parasitol* 2002;**18**:25−31.

20. Korten S, Badusche M, Büttner DW, Hoerauf A, Brattig N, Fleischer B. Natural death of adult Onchocerca volvulus and filaricidal effects of doxycycline induce local FOXP3+/CD4+ regulatory T cells and granzyme expression. *Microbes Infect* 2008;**10**:313−24.

21. Satoguina JS, Adjobimey T, Arndts K, Hoch J, Oldenburg J, Layland LE, et al. Tr1 and naturally occurring regulatory T cells induce IgG4 in B cells through GITR/GITR-L interaction, IL-10 and TGF-beta. *Eur J Immunol* 2008;**38**:3101−13.

22. Wildenburg G, Kromer M, Buttner DW. Dependence of eosinophil granulocyte infiltration into nodules on the presence of microfilariae producing Onchocerca volvulus. *Parasitol Res* 1996;**82**:117−24.

23. Korten S, Wildenburg G, Darge K, Buttner DW. Mast cells in onchocercomas from patients with hyperreactive onchocerciasis (sowda). *Acta Trop* 1998;**70**:217−31.

24. Doetze A, Satoguina J, Burchard G, Rau T, Löliger C, Fleischer B, et al. Antigen-specific cellular hyporesponsiveness in a chronic human helminth infection is mediated by T(h)3/T(r)1-type cytokines IL-10 and transforming growth factor-beta but not by a T(h)1 to T(h)2 shift. *Int Immunol* 2000;**12**:623−30.

25. Korten S, Hoerauf A, Kaifi JT, Büttner DW. Low levels of transforming growth factor-beta (TGF-beta) and reduced suppression of Th2-mediated inflammation in hyperreactive human onchocerciasis. *Parasitology* 2011;**138**:35−45.

26. Folkard SG, Hogarth PJ, Taylor MJ, Bianco AE. Eosinophils are the major effector cells of immunity to microfilariae in a mouse model of onchocerciasis. *Parasitology* 1996;**112**(Pt 3):323−9.

27. Abraham D, Leon O, Schnyder-Candrian S, Wang CC, Galioto AM, Kerepesi LA, et al. Immunoglobulin E and eosinophil-dependent protective immunity to larval Onchocerca volvulus in mice immunized with irradiated larvae. *Infect Immun* 2004;**72**:810−7.

28. Hogarth PJ, Bianco AE. Interleukin-12 modulates T-cell responses to microfilariae but fails to abrogate interleukin-5-dependent immunity in a mouse model of onchocerciasis. *Immunology* 1999;**98**:406−12.

29. Ackerman SJ, Kephart GM, Francis H, Awadzi K, Gleich GJ, Ottesen EA. Eosinophil degranulation. An immunologic determinant in the pathogenesis of the Mazzotti reaction in human onchocerciasis. *J Immunol* 1990;**144**:3961−9.

30. Cooper PJ, Beck LA, Espinel I, Deyampert NM, Hartnell A, Jose PJ, et al. Eotaxin and RANTES expression by the dermal endothelium is associated with eosinophil infiltration after ivermectin treatment of onchocerciasis. *Clin Immunol* 2000;**95**:51−61.

31. Gutierrez-Pena EJ, Knab J, Buttner DW. Immunoelectron microscopic evidence for release of eosinophil granule matrix protein onto microfilariae of Onchocerca volvulus in the skin after exposure to amocarzine. *Parasitol Res* 1998;**84**:607−15.

32. Pfarr KM, Debrah AY, Specht S, Hoerauf A. Filariasis and lymphoedema. *Parasite Immunol* 2009;**31**:664−72.

33. Chitkara RK, Krishna G. Parasitic pulmonary eosinophilia. *Semin Respir Crit Care Med* 2006;**27**:171−84.

34. Babayan SA, Read AF, Lawrence RA, Bain O, Allen JE. Filarial parasites develop faster and reproduce earlier in response to host immune effectors that determine filarial life expectancy. *PLoS Biol* 2010;**8**:e1000525.

35. Martin C, Al-Qaoud KM, Ungeheuer MN, Paehle K, Vuong PN, Bain O, et al. IL-5 is essential for vaccine-induced protection and for resolution of primary infection in murine filariasis. *Med Microbiol Immunol* 2000;**189**:67−74.

36. Volkmann L, Bain O, Saeftel M, Specht S, Fischer K, Brombacher F, et al. Murine filariasis: interleukin 4 and interleukin 5 lead to containment of different worm developmental stages. *Med Microbiol Immunol* 2003;**192**:23−31.

37. Al-Qaoud KM, Pearlman E, Hartung T, Klukowski J, Fleischer B, Hoerauf A. A new mechanism for IL-5-dependent helminth

control: neutrophil accumulation and neutrophil-mediated worm encapsulation in murine filariasis are abolished in the absence of IL-5. *Int Immunol* 2000;**12**:899−908.

38. Specht S, Saeftel M, Arndt M, Endl E, Dubben B, Lee NA, et al. Lack of eosinophil peroxidase or major basic protein impairs defense against murine filarial infection. *Infect Immun* 2006;**74**: 5236−43.

39. Dittrich AM, Erbacher A, Specht S, Diesner F, Krokowski M, Avagyan A, et al. Helminth infection with Litomosoides sigmodontis induces regulatory T cells and inhibits allergic sensitization, airway inflammation, and hyperreactivity in a murine asthma model. *J Immunol* 2008;**180**:1792−9.

40. Sanderson CJ. Interleukin-5, eosinophils, and disease. *Blood* 1992;**79**:3101−9.

41. Wilson S, Vennervald BJ, Kadzo H, Ireri E, Amaganga C, Booth M, et al. Health implications of chronic hepatosplenomegaly in Kenyan school-aged children chronically exposed to malarial infections and Schistosoma mansoni. *Trans R Soc Trop Med Hyg* 2010;**104**: 110−6.

42. Capron M, Capron A. Effector functions of eosinophils in schistosomiasis. *Mem Inst Oswaldo Cruz* 1992;**87**(Suppl. 4):167−70.

43. Dombrowicz D, Quatannens B, Papin JP, Capron A, Capron M. Expression of a functional Fc epsilon RI on rat eosinophils and macrophages. *J Immunol* 2000;**165**:1266−71.

44. Davies SJ, Smith SJ, Lim KC, Zhang H, Purchio AF, McKerrow JH, et al. In vivo imaging of tissue eosinophilia and eosinopoietic responses to schistosome worms and eggs. *Int J Parasitol* 2005;**35**: 851−9.

45. MacDonald AS, Araujo MI, Pearce EJ. Immunology of parasitic helminth infections. *Infect Immun* 2002;**70**:427−33.

46. de Andres B, Rakasz E, Hagen M, McCormik ML, Mueller AL, Elliot D, et al. Lack of Fc-epsilon receptors on murine eosinophils: implications for the functional significance of elevated IgE and eosinophils in parasitic infections. *Blood* 1997;**89**:3826−36.

47. Kolosionek E, Crosby A, Harhay MO, Morrell N, Butrous G. Pulmonary vascular disease associated with schistosomiasis. *Expert Rev Anti Infect Ther* 2010;**8**:1467−73.

48. Fitzsimmons CM, Joseph S, Jones FM, Reimert CM, Hoffmann KF, Kazibwe F, et al. Chemotherapy for schistosomiasis in Ugandan fishermen: treatment can cause a rapid increase in interleukin-5 levels in plasma but decreased levels of eosinophilia and worm-specific immunoglobulin E. *Infect Immun* 2004;**72**:4023−30.

49. Reimert CM, Fitzsimmons CM, Joseph S, Mwatha JK, Jones FM, Kimani G, et al. Eosinophil activity in Schistosoma mansoni infections in vivo and in vitro in relation to plasma cytokine profile pre- and posttreatment with praziquantel. *Clin Vaccine Immunol* 2006;**13**:584−93.

50. Gopinath R, Hanna LE, Kumaraswami V, Perumal V, Kavitha V, Vijayasekaran V, et al. Perturbations in eosinophil homeostasis following treatment of lymphatic filariasis. *Infect Immun* 2000;**68**: 93−9.

51. Layland LE, Rad R, Wagner H, da Costa CU. Immunopathology in schistosomiasis is controlled by antigen-specific regulatory T cells primed in the presence of TLR2. *Eur J Immunol* 2007;**37**: 2174−84.

52. Taylor JJ, Krawczyk CM, Mohrs M, Pearce EJ. Th2 cell hyporesponsiveness during chronic murine schistosomiasis is cell intrinsic and linked to GRAIL expression. *J Clin Invest* 2009;**119**:1019−28.

53. Kephart GM, Andrade ZA, Gleich GJ. Localization of eosinophil major basic protein onto eggs of Schistosoma mansoni in human pathologic tissue. *Am J Pathol* 1988;**133**:389−96.

54. Swartz JM, Dyer KD, Cheever AW, Ramalingam T, Pesnicak L, Domachowske JB, et al. Schistosoma mansoni infection in eosinophil lineage-ablated mice. *Blood* 2006;**108**:2420−7.

55. Jacobsen EA, Taranova AG, Lee NA, Lee JJ. Eosinophils: singularly destructive effector cells or purveyors of immunoregulation? *J Allergy Clin Immunol* 2007;**119**:1313−20.

56. Rotman HL, Yutanawiboonchai W, Brigandi RA, Leon O, Gleich GJ, Nolan TJ, et al. Strongyloides stercoralis: eosinophil-dependent immune-mediated killing of third stage larvae in BALB/cByJ mice. *Exp Parasitol* 1996;**82**:267−78.

57. Dixon H, Blanchard C, Deschoolmeester ML, Yuill NC, Christie JW, Rothenberg ME, et al. The role of Th2 cytokines, chemokines and parasite products in eosinophil recruitment to the gastrointestinal mucosa during helminth infection. *Eur J Immunol* 2006;**36**:1753−63.

58. Svensson M, Bell L, Little MC, DeSchoolmeester M, Locksley RM, Else KJ. Accumulation of eosinophils in intestine-draining mesenteric lymph nodes occurs after Trichuris muris infection. *Parasite Immunol* 2011;**33**:1−11.

59. Herbert DR, Lee JJ, Lee NA, Nolan TJ, Schad GA, Abraham D. Role of IL-5 in innate and adaptive immunity to larval Strongyloides stercoralis in mice. *J Immunol* 2000;**165**:4544−51.

60. Bell LV, Else KJ. Mechanisms of leucocyte recruitment to the inflamed large intestine: redundancy in integrin and addressin usage. *Parasite Immunol* 2008;**30**:163−70.

61. Rzepecka J, Donskow-Schmelter K, Doligalska M. Heligmosomoides polygyrus infection down-regulates eotaxin concentration and CCR3 expression on lung eosinophils in murine allergic pulmonary inflammation. *Parasite Immunol* 2007;**29**:405−13.

62. Hartmann S, Schnoeller C, Dahten A, Avagyan A, Rausch S, Lendner M, et al. Gastrointestinal nematode infection interferes with experimental allergic airway inflammation but not atopic dermatitis. *Clin Exp Allergy* 2009;**39**:1585−96.

63. Knott ML, Matthaei KI, Foster PS, Dent LA. The roles of eotaxin and the STAT6 signalling pathway in eosinophil recruitment and host resistance to the nematodes Nippostrongylus brasiliensis and Heligmosomoides bakeri. *Mol Immunol* 2009;**46**:2714−22.

64. Fabre V, Beiting DP, Bliss SK, Gebreselassie NG, Gagliardo LF, Lee NA, et al. Eosinophil deficiency compromises parasite survival in chronic nematode infection. *J Immunol* 2009;**182**:1577−83.

65. Mearns H, Horsnell WG, Hoving JC, Dewals B, Cutler AJ, Kirstein F, et al. Interleukin-4-promoted T helper 2 responses enhance Nippostrongylus brasiliensis-induced pulmonary pathology. *Infect Immun* 2008;**76**:5535−42.

66. Ohnmacht C, Pullner A, van Rooijen N, Voehringer D. Analysis of eosinophil turnover in vivo reveals their active recruitment and prolonged survival in the peritoneal cavity. *J Immunol* 2007;**179**:4766−74.

67. Mori Y, Iwasaki H, Kohno K, Yoshimoto G, Kikushige Y, Okeda A, et al. Identification of the human eosinophil lineage-committed progenitor: revision of phenotypic definition of the human common myeloid progenitor. *J Exp Med* 2009;**206**:183−93.

68. Padigel UM, Lee JJ, Nolan TJ, Schad GA, Abraham D. Eosinophils can function as antigen-presenting cells to induce primary and secondary immune responses to Strongyloides stercoralis. *Infect Immun* 2006;**74**:3232−8.

69. Rodrigues RM, Silva NM, Goncalves AL, Cardoso CR, Alves R, Goncalves FA, et al. Major histocompatibility complex (MHC) class II but not MHC class I molecules are required for efficient control of Strongyloides venezuelensis infection in mice. *Immunology* 2009;**128**:e432—41.

Chapter 10.3

Immune Responses in Helminth Infections

Thomas B. Nutman

INTRODUCTION

Parasitic helminths are complex eukaryotic organisms, characterized by their ability to maintain long-standing, chronic infections in human hosts, sometimes lasting decades. Hence, parasitic helminths are a major health-care problem worldwide, infecting more than 2 billion people, mostly in resource-limited countries (Table 10.3.1). Common helminth infections, such as intestinal helminths, filarial, and schistosome infections are a major medical, social, and economic burden to the countries in which these infections are endemic. Chemotherapy, while highly successful in some areas, still suffers from the disadvantages of the length of treatment, the logistics involved in the distribution of drugs, and in some cases, the emergence of drug resistance. Vector control measures are at best an adjunct measure in the control of helminth infections but also suffer from the same social, logistic, and economic obstacles as mass chemotherapy. Therefore, the study of the immune responses to helminth infections attains great importance both in terms of understanding the parasite strategies involved in establishing chronic infection and in the delineation of a successful host immune response to develop protective vaccines against infection.

TABLE 10.3.1 Prevalence and Distribution of Common Helminth Infections

Types of Helminths	Numbers Infected (Millions)	Distribution
Intestinal helminths		
Ascaris lumbricoides	1221	Worldwide
Hookworm species[a]	740	Worldwide
Strongyloides stercoralis	75	Worldwide
Trichuris trichuria	795	Worldwide
Schistosomes		
Schistosoma mansoni	150	Africa, South America
Schistosoma japonicum	2	Eastern Asia
Schistosoma haematobium	50	Africa
Filarial parasites		
Onchocerca volvulus	39	Africa, Central and South America
Wuchereria bancrofti	120	Worldwide
Brugia malayi	10	Asia
Loa loa	13	Africa
Animal parasites[b]		
Taenia solium	1	Worldwide
Echinococcus granulosis	1.5	Worldwide
Toxocara canis	7	Worldwide

[a]*Both* Necator americanus *and* Ancylostoma duodenale.
[b]*These are characteristically parasites of animals that infect humans incidentally but cause serious diseases such as cysticercosis, hydatid disease, and visceral larva migrans.*

SPECTRUM OF HOST–PARASITE INTERACTIONS

While both protozoa and helminths can cause parasitic infections, the biology and the host response to each are extremely different. Protozoa are small, unicellular organisms that multiply intracellularly and pose an extreme immediate hazard to the host immune system. Helminths, in contrast, are large (often centimeters to meters in length), extracellular (the exception being *Trichinella spiralis*), and typically do not multiply in their vertebrate host and therefore do not present an immediate threat during the initial infection.

Helminths have characteristically complex life cycles with many developmental stages. Thus, the host is exposed during the course of a single infection to multiple life cycle stages of the parasites, each stage with both a shared and a unique antigenic repertoire. Thus, in *Schistosoma mansoni*, infection begins with penetration of the skin of humans exposed to infested waters by the free-swimming cercariae which then develop into tissue dwelling schistosomula. In the liver and mesenteric veins, schistosomula differentiate into sexually dimorphic adult worms, which mate and the resultant eggs produced migrate through tissues into the lumen of the intestine or bladder for environmental release. Similarly, in lymphatic filarial infection, the host is exposed to the infective stage larvae in the skin, lymph nodes, and lymphatics, to the adult worms in the lymph nodes and lymphatics, and finally to the microfilariae in the peripheral circulation. Hence, the host–helminth interaction is complex not only due to the multiple life cycle stages of the parasite, but also because of the tissue tropism of the different stages.

Antigenic differences among the life cycle stages can lead to distinct immune responses that evolve differentially over the course of a helminth infection. In addition, depending on the location of the parasite, the responses are compartmentalized (intestinal mucosa and draining lymph nodes in intestinal nematode infection or skin/subcutaneous tissue and draining lymph nodes in onchocerciasis) or systemic (lymphatic filariasis or schistosomiasis). Moreover, the migration patterns of the parasite might elicit varied cutaneous, pulmonary, and intestinal inflammatory pathologies as seen, for example, in *Ascaris* or *Strongyloides stercoralis* infection during their migratory phase. This is further complicated by the fact that human hosts are often exposed to multiple life cycle stages of the parasite at the same time. Thus, a chronically infected patient with lymphatic filariasis harboring adult worms and microfilariae might be exposed to insect bites transmitting the infective stage parasite. The immune response that ensues will not only be a reaction to the invading organism but will also bear an imprint of the previous exposures and the concurrent infection.

Helminth infections can elicit a spectrum of clinical manifestations mirroring diversity in host immune responses. For example, in lymphatic filariasis, most infected individuals remain clinically asymptomatic despite harboring significant numbers of worms, felt to reflect the induction of parasite-specific tolerance in the immune system. Others exhibit acute manifestations including fever and lymphadenopathy that is felt to reflect inflammatory processes induced by incoming larvae, dying worms, or superadded infections. Individuals who mount a strong but inappropriate immune response often have lymphatic damage and associated pathology—hydrocele and elephantiasis. Finally, a group of infected individuals mount exuberant immune responses often resulting in unusual pathology such as tropical pulmonary eosinophilia.[1] Thus, the clinical manifestations of lymphatic filariasis exemplify the spectrum of host–parasite interactions that occur during helminth infections.

Another hallmark of most helminth infections is their chronicity, with many helminths surviving in the host for decades. For example, adult schistosomes and filariae may survive in host tissues for as long as 30 years, producing eggs and larval stages throughout most of their lifespan. Similarly, *S. stercoralis* due to its ability to *autoinfect* can maintain its life cycle indefinitely. Chronic infections certainly reflect an adaptation that leads to *parasitism* in that causing mortality would prevent parasite transmission if the host were to die before larval release or egg production could occur. In addition to the long-lived nature of the infection, helminths frequently occur within a balanced host–parasite interface so that relatively asymptomatic carriers are available as reservoirs for ongoing transmission. When this balanced coexistence is interrupted, pathology—exemplified by cirrhosis and portal hypertension in schistosomiasis and elephantiasis associated with lymphatic filariasis—can ensue.

PROTOTYPICAL HOST RESPONSES TO HELMINTHS

The immunological hallmark of helminth infections is their ability to induce T-helper type 2 (T_h2) responses characterized by the presence of interleukin-4 (IL-4), -5 (IL-5), -9 (IL-9), -10 (IL-10), and -13 (IL-13), generalized eosinophilia, goblet and mucosal mast cell hyperplasia, and the production of immunoglobulin E (IgE) and G1 (IgG1) (in mice) and IgE and IgG4 (in humans). While the T_h2 responses induced by helminth parasites is a stereotypical response of the host, its initiation requires interaction with many different cell, most notably: stromal cells; dendritic cells and macrophages; eosinophils; mast cells; basophils; and innate helper cells (Fig. 10.3.1). These, in turn, can

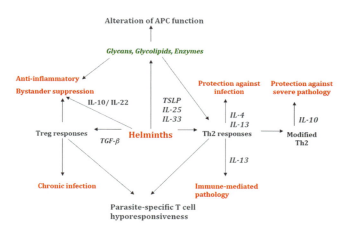

FIGURE 10.3.1 Outcomes and mediators in response to helminth infections. APC, antigen-presenting cell; IL-, interleukin; TGF-β, transforming growth factor-beta; T_h2, T-helper type 2 (cell); T_{reg}, T-regulatory (cell); TSLP, thymic stromal lymphopoietin.

induce and culminate in T_h2 responses. Over time, with chronic infection these prototypical T_h2 responses are modulated by both adaptive and natural regulatory T cells, alternatively activated macrophages, eosinophils, and likely other, heretofore, unidentified cell populations.

There are, however, exceptions to this canonical T_h2 response to helminth infections. Immune responses to *Schistosoma* cercariae are T-helper type 1 (T_h1) in nature and only on egg laying does a T_h2 response emerge.[2] Early immune responses of unexposed individuals to live filarial parasites are markedly T_h1-like and in murine infections, microfilariae of *Brugia malayi*, promote T_h1 responses.[3,4] Finally, immune responses to the gut nematode *Trichuris muris* can be T_h1 or T_h2 depending on the mouse strain used for infection.[5]

Helminth-Induced T-Helper Type 2 Polarization

It remains largely unknown why and how helminths induce T_h2 immune responses. Effector T cell responses against viruses, bacteria, and many protozoa are dependent on an initial signaling through pattern recognition systems that cause the activation of antigen-presenting cells (APCs) and release of interleukin-12 (IL-12), typically from macrophages and dendritic cells (DCs). To date, there has not been a system of pattern recognition defined for helminths or other T_h2-inducing molecules (e.g., allergens). Although it has been proposed that the T_h2 response induced by helminth parasites may be simply a default pathway in the absence of IL-12 production by APCs, there are multiple examples of helminth products[6] that actively promote T_h2 polarization through DCs. It had been shown that differentiation and maturation of DCs in the presence of helminth antigens can lead to robust T_h2 responses with inhibition of

IL-12 production.[7] This has been demonstrated using: (1) excretory–secretory (E–S) products from *Acanthocheilonema viteae*[8] and *Nippostrongylus brasiliensis*,[9] both rodent filarial parasites; (2) *S. mansoni* soluble egg antigen[10] or the schistosomal glycan lacto-N-fucopentaose III[11] or double stranded RNA[12]; and (3) live microfilariae from *B. malayi*.[13] More recently, some of these factors have been characterized at the molecular level. Omega-1, a secreted T2 ribonuclease of *S. mansoni* eggs has been shown to be an active component that drives T_h2 differentiation, in part by reducing the signaling strength between DCs and T cells.[14,15]

Basophils are also directly activated by such E–S molecules felt to be mediated by signaling through toll-like receptors (TLRs) and C-type lectin receptors.[16] *S. mansoni* eggs are known to contain a factor, termed IL-4-inducing principal of *S. mansoni* eggs, that has been shown to induce IL-4 release from human basophils in a non-IgE dependent manner.[17] Proteases from the hookworm *Necator americanus* or from *S. mansoni* cercariae[18] have been shown to activate human basophils to release histamine and IL-4. Alternatively, protease-mediated degradation of host tissue or other mechanisms of tissue injury could lead to the generation of factors that polarize a T_h2 immune response.[19]

Recently, much attention has been focused on innate cells that respond to interleukin-25 (IL-25), IL-33, and thymic stromal lymphopoietin (TSLP) to produce IL-4 and IL-13 that can in turn cause T_h2 polarization. IL-25, a broadly expressed cytokine of the IL-17 family, has been recognized lately as potent inducer of T_h2 immunity. IL-25-induced innate immune cells have been identified independently by a number of investigators and have been termed nuocytes,[20] innate type 2 helper cells (ih2),[21] or multipotent progenitor type 2 cells.[22] These appear very similar to *natural helper cells* that were described in adipose tissue and respond to a combination of IL-2 and IL-25.[23] The interrelationship among these cell types is complicated and not yet well elucidated.[24] IL-33, a member of the interleukin-1 (IL-1) family, binds to the IL-33 receptor (IL-33R/ST2) on the surface of many different cell types including T_h2 cells, mast cells, and basophils.[25,26] IL-33 [in concert with interleukin-18 (IL-18)] can activate murine basophils to release IL-4 and IL-13.[21] Exogenous IL-33, like IL-25, induces nuocytes and innate helper 2 (IH2) cells *in vivo*.[21,27]

TSLP, an interleukin-7 (IL-7)-like cytokine produced primarily by epithelial cells, induces the expression of tumor necrosis factor ligand superfamily member 4 (TNFL4) on DCs leading to their ability to promote T_h2 differentiation in naive T cells. TSLP also induces the production of chemokines (e.g., C-C motif chemokine 17) by DCs that recruit neutrophils, eosinophils, and T_h2 cells into the tissues.[28]

Nonprofessional Antigen-Presenting Cells and Sources of IL-4 for Helminth-Induced T-Helper Type 2 Polarization

Polarization of T_h2 cells *in vitro* generally requires the addition of exogenous IL-4. IL-4 binds to the type I IL-4 receptor on T cells consisting of the interleukin-4 receptor subunit alpha (IL-4RA) chain and the common gamma chain (γc) that is also used by the receptors for IL-2, IL-7, IL-9, interleukin-15 (IL-15), and interleukin-21 (IL-21). Binding of IL-4 to IL-4RA leads to the activation of signal transducer and activator of transcription 6 (STAT6) that in turn induces expression of GATA-binding factor 3 (GATA3), a master transcription factor for T_h2 cells. Based on the important role of IL-4 for *in vitro* polarization of T_h2 cells, it has been proposed that T_h2 polarization *in vivo* requires an innate cell type that rapidly releases IL-4. Basophils[29,30,31] and eosinophils[32] have been proposed to fulfill this function in the context of helminth infections, since they can rapidly produce IL-4 and have the capacity to present antigen to naive T cells in that both cell types can phagocytose and process antigen in the context of major histocompatibility complex class II (MHC-II) expressed on their surface.[29,30,31,32] Moreover, basophils have been shown to be recruited to lymph nodes after the infection of mice with *N. brasiliensis*[33,34] and injection of *S. mansoni* eggs,[30] as have eosinophils[35] (but not in the context of helminth infection).

REGULATION OF THE T CELL RESPONSE IN CHRONIC HELMINTH INFECTION

Generating a T_h2 response to parasite infections is essential, but controlling that (and non-T_h2 proinflammatory responses) is equally important. Some of this control is provided by regulatory T cells, alternatively activated macrophages, B cells, and eosinophils.

Regulatory T Cells

Regulatory T cells play an important role in the down-regulation of immune responses in infectious diseases, tumor immunology, and autoimmunity (reviewed in[36]). Natural regulatory T cells can turn off effector T cell responses mostly by a direct contact-mediated mechanism but can also act through suppressive cytokines. Adaptive regulatory T cells act through the induction of cytokines, particularly IL-10 and transforming growth factor-beta (TGF-β).[37] Evidence for the involvement of regulatory T cells in helminth-mediated downmodulation of the immune response has been accumulating in recent years. IL-10 and TGF-β, both associated with regulatory T cells, are elicited in response to helminth infections, and *in vitro* neutralization of IL-10 and TGF-β partially restores T-cell

proliferation and cytokine production in lymphatic filariasis.[38] Similar reversals of immunosuppression are observed in onchocerciasis and schistosomiasis with IL-10 producing natural regulatory T cells from egg-induced granulomas.[39,40] In addition, T-cell clones secreting IL-10 and TGF-β have been isolated in onchocerciasis.[40] Treatment of mice infected with *Litomosoides sigmodontis* with anti-CD25 and anti-GITR antibodies has been shown to render mice resistant to infection, implicating regulatory T cells in the susceptibility to this filarial infection.[41] Finally, in human lymphatic filariasis and in mouse models, increased expression of forkhead box protein P3 (FOXP3/Foxp3) has been shown to be associated with diminished responses in filarial infections.[42]

Hypo-Responsive T Cells

Effector T-cell responses can be turned off or modulated through a variety of mechanisms including through cytotoxic T-lymphocyte-associated antigen 4 (CTLA-4) and programmed cell death protein 1 (protein PD-1). Interestingly, increased expression of CTLA-4 and protein PD-1 has been demonstrated in filarial infections, while blocking CTLA-4 can partially restore a degree of immunological responsiveness in cells from infected individuals.[42,43] T cells from filaria-infected individuals exhibit classical signs of anergy, including diminished T-cell proliferation to parasite antigens, lack of IL-2 production, and increased expression of E3 ubiquitin ligases.[42]

In chronic helminth infection, T-regulatory (T_{reg}) cells also suppress host immune responses in humans, presumably allowing parasites to establish chronic infections.[44] Indeed, schistosome-infected individuals in Gabon had increased numbers of circulating T_{reg} cells compared with their uninfected neighbors.[45] T_{reg} cells from geohelminth-infected individuals in Indonesia, for example, were more effective at suppressing proliferation and interferon gamma (IFN-γ) production by effector T cells in response to malaria antigens and bacillus Calmette–Guérin (BCG) than T_{reg} cells from healthy individuals.[46] In studies from Mali, IL-10 mediated the diminished production of cytokines in response to malaria antigens has been demonstrated in helminth/malaria coinfected individuals as well.[47]

In human filariasis and in a mouse filarial infection model using *L. sigmodontis*, the chronic phase of infection is marked by T-cell anergy, loss of proliferative responses to parasite antigen challenge, reductions in effector cytokine levels, and elevated expression of inhibitory immune molecules, such as CTLA-4. In mouse models, parasite survival is linked to T_{reg} cell activity, and immunity to infection can be boosted only if T_{reg} cell depletion is accompanied by the delivery of CTLA-4-specific blocking antibodies or GITR-specific stimulatory antibodies to restimulate the anergized effector T-cell populations.[41] Similarly, anergic T cells are found in both humans and mice with schistosomiasis and, in the latter case, these T cells express the anergic protein gene related to E3 ubiquitin-protein ligase RNF128 (anergy) in lymphocytes.[48] Interestingly, anergy induction in murine schistosomiasis, as in filarial infections, is linked to a co-inhibitory signaling pathway, through PD-1/programmed cell death 1 ligand 1 (PD-L1) interaction.[49]

Antigen-Presenting Cells and Helminth Infections

DCs are the first APCs usually to encounter parasites, and helminth modulation of DC function has been characterized.[50] Filarial parasites induce downregulation of MHC-I and -II as well as cytokines and other genes involved in antigen presentation, thereby rendering DCs suboptimal in the activation of $CD4^+$ T cells.[51] Schistosomes induce similar effects on DCs with subsequent T_h2 polarization and inhibited responses to T_h1-inducing TLR ligands.[52] Experimental helminth infection with *B. malayi*, *N. brasiliensis* and *Toxocara canis*, respectively, has been shown to elicit macrophages with an alternative activation phenotype, leading to marked suppression of target cell proliferation and induction of T_h2 responses.[53]

Macrophages are the other important class of APCs that can serve as protective effector cells in bacterial and protozoan infections by their production of nitric oxide and other mediators. Helminth interaction with macrophages induces a population of macrophages preferentially expressing arginase instead of nitric oxide due to the increased activation of arginase-1 by IL-4 and IL-13. These macrophages, known as *alternatively activated macrophages*, are known to be important in wound healing and have been postulated to play a potential role in repairing wound damage that occurs during tissue migration of helminth parasites.[54] In addition, these macrophages mediate hyporesponsiveness in cognate T cells and play an additional role in the modulation of immune responses.[55] These macrophages, with downregulatory activity as evidenced by the production of IL-10 and TGF-β, are potent depressants of T cell activity. Alternatively activated macrophages are able to block the inflammatory proliferation of lymphocytes, while at the same time mediating immunity to tissue helminths and repairing tissue that has been damaged by the parasites.

Other Regulatory Cells

Other key immunoregulatory populations demonstrate that the immune system has also duplicated and diversified its regulatory mechanisms. For example, regulatory B cells are active in patients with multiple sclerosis whose remission is associated with helminth infection[56] and schistosome-infected mice are protected from anaphylactic shock[57] and airway allergy.[58] Moreover, *Heligmosomoides polygyrus*-infected mice generate regulatory B cells that can, on transfer to naive hosts, downmodulate both allergy and autoimmunity in a manner that is not IL-10 dependent.[59]

Innate effectors targeted by T_H2-type cytokines can also act as regulators. Eosinophils, the prototypical T_H2-type *effector cell*, produce TGF-β[60] and promote tissue remodeling.[61] Epithelial cells, particularly in the gastrointestinal tract and airways, are major producers of TGF-β and IL-10.

SPECIFIC MECHANISMS OF EVASION IN HELMINTH PARASITES

Helminths exert profound immunoregulatory effects on the host immune system with both parasite—antigen-specific and more generalized levels of immune suppression. It has been shown that patients with schistosomiasis or filarial infections have markedly diminished responses to parasite antigens and, in addition, some measurable attenuation in responses to bystander antigens and routine vaccinations. Thus, while host immunosuppression is usually antigen-specific, chronic infection can be associated with some spillover effects.[44] The long lifespan of helminth parasites is major evidence of successful immune evasion strategies of these parasites. Among the mechanisms used by parasites to avoid immune-mediated elimination are those of evasion and suppression, regulation, or blockade of immune effector pathways.

When the immune system fails to reject parasites and a chronic infection takes hold, the T cell compartment changes more with regard to its state of responsiveness than in its composition of $CD4^+$ cell subsets. Classic studies on schistosomiasis documented the attenuation of hepatic granulomas as the immune response subsides in chronically infected mice. A parallel is clearly seen in chronic human helminth infections, in which a T_h2 cell-dominated immune profile, with high levels of IL-4 and IL-10 production, is accompanied by a muted IL-5 and IL-13 response and an overall loss of T-cell proliferative responses toward parasite and bystander antigens.[44] This is suggested to represent a *modified T_H2-type response*, as discussed earlier. The host immune system can thus be in a state of effective tolerance even though many key markers of T_H2-type immunity are still evident.

Pathogens become sequestered when they enter into intra- or extracellular compartments, such as the central nervous system, that are not accessible to all components of the immune response. Encystation, as occurs in infections with the metacestodes *Echinoccous* spp. and *Taenia solium* or *Trichinella spiralis*, is considered a mechanism of protection against immune attack. It has been demonstrated, however, that parasite antigens and host antibodies can traverse the cyst membranes, indicating that the cyst environment is not completely immunologically inert. Similarly, in *Onchocerca volvulus* infection, the parasites are surrounded by a relatively avascular, acellular fibrotic capsule that serves as a barrier to the host response.[62] Tropism for immunological, relatively protected environments (e.g., the eye in *O. volvulus* infection and the intestinal lumen for intestinal nematodes) contributes to sequester the parasite from the host effector systems.

B. malayi secretes a cysteine protease inhibitor, cystatin, which interferes with both the cathepsins and asparigine endopeptidase involved in the processing of peptides for the MHC-II pathway. Similar inhibitors have been described for *O. volvulus* and *N. brasiliensis*. In addition, cystatins from helminth parasites can modulate T cell proliferation and elicit the upregulation of IL-10 expression.[63] Other modulators, such as prostaglandins and other arachidonic acid family members, such as prostaglandin-D2 (PGD2) and E-2 (PGE2), are known to inhibit IL-12 production by DCs. Aspartic proteases from *Ascaris lumbricoides* have been shown to block efficient antigen processing dependent on proteolytic lysosomal enzymes.[64]

Helminth parasites also use mechanisms involving cytokine mimicry and interference to establish chronic infection. Thus, parasites produce cytokine- and chemokine-like molecules to interfere with the function of the host innate immune products. The first helminth cytokines were found to be homologues of TGF-β expressed by *B. malayi* and both schistosomes and filarial parasites express members of the TGF-β receptor family. Similarly, *Echinococcus granulosus* expresses a TGF-β ligand, and thus all helminth groups might have the potential to exploit TGF-β-mediated immune suppression. Various helminths including *B. malayi* produce homologues of macrophage migration inhibitory factors (MIFs), which are known to activate an anti-inflammatory pathway through suppressor of cytokine signaling 1 (SOCS-1), a molecule involved in cytokine.[65] *T. muris* is known to express a homologue of IFN-γ, which can bind to the mammalian IFN-γ receptor *in vitro* and induce signaling.[64]

Similarly, helminth parasites use chemokine or chemokine receptor-like proteins to evade protective immunity. *Ascaris suum* is known to express a neutrophil chemoattractant with chemokine binding properties. *S. mansoni* eggs secrete a protein, *S. mansoni* chemokine binding protein (smCKBP), that binds the chemokines C-X-C motif chemokine 8 (CXCL8/IL-8) and C-C motif chemokine 3 (CCL3) and inhibits their interaction with host chemokine receptors and their biological activity, resulting in the suppression of inflammation.[66] Cells of the innate immune system use C-type lectin receptors such as

dendritic cell-specific ICAM-3-grabbing non-integrin 1 (DC-SIGN/CD209 antigen), liver/lymph node-specific ICAM-3-grabbing non-integrin (L-SIGN/CD299), the mannose receptor, the macrophage galactose-binding lectin, and other lectins to recognize particular glycan antigens from helminth parasites. Interestingly *S. mansoni* and *T. canis* express their own glycan and lectin antigens that act as molecular mimics of host antigens thereby interrupting or misdirecting the host inflammatory responses.[65]

Other parasite products mediate their effect by blocking effector functions, including the recruitment and activation of inflammatory cells, and limit the destructive potential of activated granulocytes or macrophages in the local extracellular milieu. For example, the host chemoattractant platelet-activating factor (PAF) is inactivated by a complementary enzyme PAF hydrolase secreted by *N. brasiliensis*.[64] Eotaxin, a potent eosinophil chemoattractant, is degraded by metalloproteases from hookworms. *Ancylostoma caninum* secretes a protein called neutrophil inhibitory factor that binds the integrins CD11b/CD18 and blocks the adhesion of activated neutrophils to vascular endothelial cells and also the release of H_2O_2 from activated neutrophils. *N. americanus* E−S products also bind to host natural killer (NK) cells and augment the secretion of IFN-γ, which might cross-regulate deleterious T_h2 responses.[67]

PROTOTYPIC EFFECTOR RESPONSES IN HELMINTH INFECTIONS

Eosinophilia

Blood and tissue eosinophilia is characteristic of helminth infection and is mediated by IL-5 [typically in concert with interleukin-3 (IL-3) and granulocyte-macrophage colony-stimulating factor (GM-CSF)]. Recruitment of eosinophils to the site of infection occurs very early in experimental helminth infection—as early as 24 h following exposure. Kinetics in humans has been harder to determine but it is postulated to occur as early as 2−3 weeks following infections as demonstrated in experimental infections of volunteers. Apart from the rapid kinetics of recruitment, eosinophils in the blood and tissue also exhibit morphological and functional changes attributable to eosinophil activation. These include decreased density, upregulation of surface activation molecules such as CD25, CD44, CD69, and human leukocyte antigen, DR subregion (HLA-DR), enhanced cellular cytotoxicity and the release of granular proteins, cytokines, leukotrienes, and other mediators of inflammation.[68] Activation of eosinophils requires T cells since, in the absence of T cells, eosinophils accumulate at the site of infection but do not degranulate or become cytotoxic.[68]

Basophil and Mast Cell Responses

Basophils are an important component of the immune response to helminth infections. Basophilia occurs in several animal models of helminth infections (though not in humans) and basophils can release IL-4 and histamine in response to parasite antigen stimulation. Moreover, basophils have the capacity to release large amounts of IL-4 and IL-13 rapidly, within minutes of surface IgE cross-linking. Basophils have been shown to be a major source of IL-4 and to accumulate in tissues in helminth infection, thereby contributing to both the initiation and maintenance of antigen-specific T_h2 responses.[69,70] Mast cells may also contribute to inflammatory reactions directed against invasive helminth parasites. These cells express the high affinity Fc-epsilon receptor-1 that is sensitized with parasite antigen-specific IgE and which can be triggered by parasite antigens. It has been postulated that cytokines and other mediators released by sensitized mast cells contribute to: (1) the recruitment and activation of effector eosinophils; (2) increased local concentrations of antibody and complement; and (3) mucus hypersecretion and increased peristalsis of the gastrointestinal tract that plays an important role in resistance to certain gastrointestinal nematode infections.[71]

Granulomatous Reactions

Although granulomatous reactions occur in many helminth infections (e.g., toxocariasis, *Angiostrongylus* infections, and lymphatic filariasis), parasitic granulomata have been best studied in *S. mansoni* infections, where granulomatous and fibrosing reactions against tissue-trapped eggs is orchestrated by $CD4^+$ T cells. Granulomas are composed primarily of eosinophils, lymphocytes, and macrophages and the fibrosis that results from the cellular response is the principal cause of morbidity in infected patients. The severity of the inflammatory process markedly varies both in humans and in experimental animal models, with severe pathology associated with T_h1 responses and milder pathology with T_h2-dominant responses.[72]

Studies in murine models of granuloma formation have demonstrated the important roles of IL-13 and tumor necrosis factor (TNF-α). IL-13 is a key mediator of chronic infection-induced liver pathology. IL-13 signaling is mediated by the type 2 IL-4 receptor, which consists of the IL-4RA and the interleukin-13 receptor subunit alpha-1 (IL-13RA1) chains. However, IL-13 can also bind to another receptor composed of the interleukin-13 receptor subunit alpha-2 (IL-13RA2), which acts to inhibit the actions of IL-13. IL-13RA2 appears to play an important role in preventing pathology since mice lacking IL-13RA2 fail to limit granuloma formation in the chronic phase of *S.*

mansoni infection. In addition, IL-13RA2-deficient mice develop severe IL-13-dependent fibrosis, portal hypertension, and succumb to infection.[73] Finally, TNF-α is necessary for egg laying and the excretion of eggs from the host. Additionally, TNF-α is necessary and sufficient to reconstitute granuloma formation in severe combined immunodeficient or immunodeficiency (SCID) mice.[74]

Lymphatic filariasis is associated with similar fibrotic reactions, wherein adult parasites residing in the afferent lymphatic channels and lymph nodes induce *scarring* felt to be partially responsible for the lymphedema and chyluria found in this condition.[75]

CONCLUSION

The complex interactions between the host and the various helminth parasites continue to be delineated. This expansion of knowledge has provided discoveries of novel cell types, new mediators of immune function and modulation, and the plasticity of those cells and processes already defined. Although helminths induce certain prototypic responses, they do not do so in isolation but through complicated, but potentially orchestrated, interactions. It will be the understanding of this complexity that will allow for appropriate approaches (be they vaccines, chemotherapy, immunomodulation, or prevention of pathology) toward these interesting fellow travelers.

REFERENCES

1. Ottesen EA, Nutman TB. Tropical pulmonary eosinophilia. *Annu Rev Med* 1992;**43**:417−24.
2. Pearce EJ, Caspar P, Grzych JM, Lewis FA, Sher A. Downregulation of Th1 cytokine production accompanies induction of Th2 responses by a parasitic helminth, Schistosoma mansoni. *J Exp Med* 1991;**173**:159−66.
3. Babu S, Nutman TB. Proinflammatory cytokines dominate the early immune response to filarial parasites. *J Immunol* 2003;**171**:6723−32.
4. Lawrence RA, Allen JE, Osborne J, Maizels RM. Adult and microfilarial stages of the filarial parasite Brugia malayi stimulate contrasting cytokine and Ig isotype responses in BALB/c mice. *J Immunol* 1994;**153**:1216−24.
5. Else KJ, Finkelman FD, Maliszewski CR, Grencis RK. Cytokine-mediated regulation of chronic intestinal helminth infection. *J Exp Med* 1994;**179**:347−51.
6. Robinson MW, Hutchinson AT, Donnelly S, Dalton JP. Worm secretory molecules are causing alarm. *Trends Parasitol* 2010;**26**:371−2.
7. Sher A, Pearce E, Kaye P. Shaping the immune response to parasites: role of dendritic cells. *Curr Opin Immunol* 2003;**15**:421−9.
8. Whelan M, Harnett MM, Houston KM, Patel V, Harnett W, Rigley KP. A filarial nematode-secreted product signals dendritic cells to acquire a phenotype that drives development of Th2 cells. *J Immunol* 2000;**164**:6453−60.

9. Holland MJ, Harcus YM, Riches PL, Maizels RM. Proteins secreted by the parasitic nematode Nippostrongylus brasiliensis act as adjuvants for Th2 responses. *Eur J Immunol* 2000;**30**:1977—87.

10. Cervi L, MacDonald AS, Kane C, Dzierszinski F, Pearce EJ. Cutting edge: dendritic cells copulsed with microbial and helminth antigens undergo modified maturation, segregate the antigens to distinct intracellular compartments, and concurrently induce microbe-specific Th1 and helminth-specific Th2 responses. *J Immunol* 2004;**172**:2016—20.

11. Okano M, Satoskar AR, Nishizaki K, Harn Jr DA. Lacto-N-fucopentaose III found on Schistosoma mansoni egg antigens functions as adjuvant for proteins by inducing Th2-type response. *J Immunol* 2001;**167**:442—50.

12. Aksoy E, Zouain CS, Vanhoutte F, Fontaine J, Pavelka N, Thieblemont N, et al. Double-stranded RNAs from the helminth parasite Schistosoma activate TLR3 in dendritic cells. *J Biol Chem* 2005;**280**:277—83.

13. Semnani RT, Liu AY, Sabzevari H, Kubofcik J, Zhou J, Gilden JK, et al. Brugia malayi microfilariae induce cell death in human dendritic cells, inhibit their ability to make IL-12 and IL-10, and reduce their capacity to activate CD4+ T cells. *J Immunol* 2003;**171**:1950—60.

14. Everts B, Perona-Wright G, Smits HH, Hokke CH, van der Ham AJ, Fitzsimmons CM, et al. Omega-1, a glycoprotein secreted by Schistosoma mansoni eggs, drives Th2 responses. *J Exp Med* 2009;**206**:1673—80.

15. Steinfelder S, Andersen JF, Cannons JL, Feng CG, Joshi M, Dwyer D, et al. The major component in schistosome eggs responsible for conditioning dendritic cells for Th2 polarization is a T2 ribonuclease (omega-1). *J Exp Med* 2009;**206**:1681—90.

16. Voehringer D. Basophils in immune responses against helminths. *Microbes Infect*; 2011.

17. Schramm G, Gronow A, Knobloch J, Wippersteg V, Grevelding CG, Galle J, et al. IPSE/alpha-1: a major immunogenic component secreted from Schistosoma mansoni eggs. *Mol Biochem Parasitol* 2006;**147**:9—19.

18. Machado DC, Horton D, Harrop R, Peachell PT, Helm BA. Potential allergens stimulate the release of mediators of the allergic response from cells of mast cell lineage in the absence of sensitization with antigen-specific IgE. *Eur J Immunol* 1996;**26**:2972—80.

19. Loke P, Gallagher I, Nair MG, Zang X, Brombacher F, Mohrs M, et al. Alternative activation is an innate response to injury that requires CD4+ T cells to be sustained during chronic infection. *J Immunol* 2007;**179**:3926—36.

20. Neill DR, Wong SH, Bellosi A, Flynn RJ, Daly M, Langford TK, et al. Nuocytes represent a new innate effector leukocyte that mediates type-2 immunity. *Nature* 2010;**464**:1367—70.

21. Price AE, Liang HE, Sullivan BM, Reinhardt RL, Eisley CJ, Erle DJ, et al. Systemically dispersed innate IL-13-expressing cells in type 2 immunity. *Proc Natl Acad Sci USA* 2010;**107**: 11489—94.

22. Saenz SA, Siracusa MC, Perrigoue JG, Spencer SP, Urban Jr JF, Tocker JE, et al. IL25 elicits a multipotent progenitor cell population that promotes T(H)2 cytokine responses. *Nature* 2010;**464**: 1362—6.

23. Moro K, Yamada T, Tanabe M, Takeuchi T, Ikawa T, Kawamoto H, et al. Innate production of T(H)2 cytokines by adipose tissue-associated c-Kit(+)Sca-1(+) lymphoid cells. *Nature* 2010;**463**:540—4.

24. Saenz SA, Noti M, Artis D. Innate immune cell populations function as initiators and effectors in Th2 cytokine responses. *Trends Immunol* 2010;**31**:407—13.

25. Oboki K, Ohno T, Kajiwara N, Arae K, Morita H, Ishii A, et al. IL-33 is a crucial amplifier of innate rather than acquired immunity. *Proc Natl Acad Sci USA* 2010;**107**:18581—6.

26. Oboki K, Ohno T, Kajiwara N, Saito H, Nakae S. IL-33 and IL-33 receptors in host defense and diseases. *Allergol Int* 2010;**59**: 143—60.

27. Neill DR, McKenzie AN. Nuocytes and beyond: new insights into helminth expulsion. *Trends Parasitol* 2011;**27**:214—21.

28. Soumelis V, Reche PA, Kanzler H, Yuan W, Edward G, Homey B, et al. Human epithelial cells trigger dendritic cell mediated allergic inflammation by producing TSLP. *Nat Immunol* 2002;**3**: 673—80.

29. Sokol CL, Chu NQ, Yu S, Nish SA, Laufer TM, Medzhitov R. Basophils function as antigen-presenting cells for an allergen-induced T helper type 2 response. *Nat Immunol* 2009;**10**:713—20.

30. Perrigoue JG, Saenz SA, Siracusa MC, Allenspach EJ, Taylor BC, Giacomin PR, et al. MHC class II-dependent basophil-CD4+ T cell interactions promote T(H)2 cytokine-dependent immunity. *Nat Immunol* 2009;**10**:697—705.

31. Tang H, Cao W, Kasturi SP, Ravindran R, Nakaya HI, Kundu K, et al. The T helper type 2 response to cysteine proteases requires dendritic cell-basophil cooperation via ROS-mediated signaling. *Nat Immunol* 2010;**11**:608—17.

32. Weller PF, Rand TH, Barrett T, Elovic A, Wong DT, Finberg RW. Accessory cell function of human eosinophils. HLA-DR-dependent, MHC-restricted antigen-presentation and IL-1 alpha expression. *J Immunol* 1993;**150**:2554—62.

33. Kim S, Prout M, Ramshaw H, Lopez AF, LeGros G, Min B. Cutting edge: basophils are transiently recruited into the draining lymph nodes during helminth infection via IL-3, but infection-induced Th2 immunity can develop without basophil lymph node recruitment or IL-3. *J Immunol* 2010;**184**:1143—7.

34. Ohnmacht C, Schwartz C, Panzer M, Schiedewitz I, Naumann R, Voehringer D. Basophils orchestrate chronic allergic dermatitis and protective immunity against helminths. *Immunity* 2010;**33**: 364—74.

35. Shi HZ, Humbles A, Gerard C, Jin Z, Weller PF. Lymph node trafficking and antigen presentation by endobronchial eosinophils. *J Clin Invest* 2000;**105**:945—53.

36. Rudensky AY. Regulatory T cells and Foxp3. *Immunol Rev* 2011;**241**:260—8.

37. von Boehmer H. Mechanisms of suppression by suppressor T cells. *Nat Immunol* 2005;**6**:338—44.

38. King CL, Mahanty S, Kumaraswami V, Abrams JS, Regunathan J, Jayaraman K, et al. Cytokine control of parasite-specific anergy in human lymphatic filariasis. Preferential induction of a regulatory T helper type 2 lymphocyte subset. *J Clin Invest* 1993;**92**:1667—73.

39. Baumgart M, Tompkins F, Leng J, Hesse M. Naturally occurring CD4+Foxp3+ regulatory T cells are an essential, IL-10-independent part of the immunoregulatory network in Schistosoma mansoni egg-induced inflammation. *J Immunol* 2006;**176**:5374—87.

40. Satoguina J, Mempel M, Larbi J, Badusche M, Loliger C, Adjei O, et al. Antigen-specific T regulatory-1 cells are associated with

immunosuppression in a chronic helminth infection (onchocerciasis). *Microbes Infect* 2002;**4**:1291—300.

41. Taylor MD, LeGoff L, Harris A, Malone E, Allen JE, Maizels RM. Removal of regulatory T cell activity reverses hyporesponsiveness and leads to filarial parasite clearance in vivo. *J Immunol* 2005;**174**: 4924—33.

42. Babu S, Blauvelt CP, Kumaraswami V, Nutman TB. Regulatory networks induced by live parasites impair both Th1 and Th2 pathways in patent lymphatic filariasis: implications for parasite persistence. *J Immunol* 2006;**176**:3248—56.

43. Steel C, Nutman TB. CTLA-4 in filarial infections: implications for a role in diminished T cell reactivity. *J Immunol* 2003;**170**: 1930—8.

44. Maizels RM, Yazdanbakhsh M. Immune regulation by helminth parasites: cellular and molecular mechanisms. *Nat Rev Immunol* 2003;**3**:733—44.

45. Maizels RM, Pearce EJ, Artis D, Yazdanbakhsh M, Wynn TA. Regulation of pathogenesis and immunity in helminth infections. *J Exp Med* 2009;**206**:2059—66.

46. Wammes LJ, Hamid F, Wiria AE, de Gier B, Sartono E, Maizels RM, et al. Regulatory T cells in human geohelminth infection suppress immune responses to BCG and Plasmodium falciparum. *Eur J Immunol* 2010;**40**:437—42.

47. Metenou S, Dembele B, Konate S, Dolo H, Coulibaly YI, Diallo AA, et al. Filarial infection suppresses malaria-specific multifunctional Th1 and Th17 responses in malaria and filarial coinfections. *J Immunol* 2011;**186**:4725—33.

48. Taylor JJ, Krawczyk CM, Mohrs M, Pearce EJ. Th2 cell hyporesponsiveness during chronic murine schistosomiasis is cell intrinsic and linked to GRAIL expression. *J Clin Invest* 2009;**119**: 1019—28.

49. Smith P, Walsh CM, Mangan NE, Fallon RE, Sayers JR, McKenzie AN, et al. Schistosoma mansoni worms induce anergy of T cells via selective up-regulation of programmed death ligand 1 on macrophages. *J Immunol* 2004;**173**:1240—8.

50. Semnani RT, Nutman TB. Toward an understanding of the interaction between filarial parasites and host antigen-presenting cells. *Immunol Rev* 2004;**201**:127—38.

51. Semnani RT, Law M, Kubofcik J, Nutman TB. Filaria-induced immune evasion: suppression by the infective stage of Brugia malayi at the earliest host-parasite interface. *J Immunol* 2004;**172**: 6229—38.

52. Pearce EJ, Kane CM, Sun J. Regulation of dendritic cell function by pathogen-derived molecules plays a key role in dictating the outcome of the adaptive immune response. *Chem Immunol Allergy* 2006;**90**:82—90.

53. Allen JE, Maizels RM. Diversity and dialogue in immunity to helminths. *Nat Rev Immunol* 2011;**11**:375—88.

54. Kreider T, Anthony RM, Urban Jr JF, Gause WC. Alternatively activated macrophages in helminth infections. *Curr Opin Immunol* 2007;**19**:448—53.

55. MacDonald AS, Maizels RM, Lawrence RA, Dransfield I, Allen JE. Requirement for in vivo production of IL-4, but not IL-10, in the induction of proliferative suppression by filarial parasites. *J Immunol* 1998;**160**:4124—32.

56. Correale J, Farez M, Razzitte G. Helminth infections associated with multiple sclerosis induce regulatory B cells. *Ann Neurol* 2008;**64**:187—99.

57. Mangan NE, Fallon RE, Smith P, van Rooijen N, McKenzie AN, Fallon PG. Helminth infection protects mice from anaphylaxis via IL-10-producing B cells. *J Immunol* 2004;**173**:6346—56.

58. Smits HH, Hammad H, van Nimwegen M, Soullie T, Willart MA, Lievers E, et al. Protective effect of Schistosoma mansoni infection on allergic airway inflammation depends on the intensity and chronicity of infection. *J Allergy Clin Immunol* 2007;**120**: 932—40.

59. Wilson MS, Taylor MD, O'Gorman MT, Balic A, Barr TA, Filbey K, et al. Helminth-induced CD19+CD23hi B cells modulate experimental allergic and autoimmune inflammation. *Eur J Immunol* 2010;**40**:1682—96.

60. Wong DT, Elovic A, Matossian K, Nagura N, McBride J, Chou MY, et al. Eosinophils from patients with blood eosinophilia express transforming growth factor beta 1. *Blood* 1991;**78**:2702—7.

61. Humbles AA, Lloyd CM, McMillan SJ, Friend DS, Xanthou G, McKenna EE, et al. A critical role for eosinophils in allergic airways remodeling. *Science* 2004;**305**:1776—9.

62. Mackenzie CD, Huntington MK, Wanji S, Lovato RV, Eversole RR, Geary TG. The association of adult Onchocerca volvulus with lymphatic vessels. *J Parasitol* 2010;**96**:219—21.

63. Hartmann S, Lucius R. Modulation of host immune responses by nematode cystatins. *Int J Parasitol* 2003;**33**:1291—302.

64. Else KJ. Have gastrointestinal nematodes outwitted the immune system? *Parasite Immunol* 2005;**27**:407—15.

65. Maizels RM, Balic A, Gomez-Escobar N, Nair M, Taylor MD, Allen JE. Helminth parasites—masters of regulation. *Immunol Rev* 2004;**201**:89—116.

66. Smith P, Fallon RE, Mangan NE, Walsh CM, Saraiva M, Sayers JR, et al. Schistosoma mansoni secretes a chemokine binding protein with antiinflammatory activity. *J Exp Med* 2005;**202**: 1319—25.

67. Loukas A, Constant SL, Bethony JM. Immunobiology of hookworm infection. *FEMS Immunol Med Microbiol* 2005;**43**:115—24.

68. Klion AD, Nutman TB. The role of eosinophils in host defense against helminth parasites. *J Allergy Clin Immunol* 2004;**113**:30—7.

69. Min B, Prout M, Hu-Li J, Zhu J, Jankovic D, Morgan ES, et al. Basophils produce IL-4 and accumulate in tissues after infection with a Th2-inducing parasite. *J Exp Med* 2004;**200**:507—17.

70. Mitre E, Taylor RT, Kubofcik J, Nutman TB. Parasite antigen-driven basophils are a major source of IL-4 in human filarial infections. *J Immunol* 2004;**172**:2439—45.

71. Pennock JL, Grencis RK. The mast cell and gut nematodes: damage and defence. *Chem Immunol Allergy* 2006;**90**:128—40.

72. Wynn TA, Thompson RW, Cheever AW, Mentink-Kane MM. Immunopathogenesis of schistosomiasis. *Immunol Rev* 2004;**201**: 156—67.

73. Mentink-Kane MM, Wynn TA. Opposing roles for IL-13 and IL-13 receptor alpha 2 in health and disease. *Immunol Rev* 2004;**202**: 191—202.

74. Amiri P, Locksley RM, Parslow TG, Sadick M, Rector E, Ritter D, et al. Tumour necrosis factor alpha restores granulomas and induces parasite egg-laying in schistosome-infected SCID mice. *Nature* 1992;**356**:604—7.

75. Figueredo-Silva J, Noroes J, Cedenho A, Dreyer G. The histopathology of bancroftian filariasis revisited: the role of the adult worm in the lymphatic-vessel disease. *Ann Trop Med Parasitol* 2002;**96**: 531—41.

Chapter 10.4

Eosinophils as Facilitators of Helminth Infection

Nebiat G. Gebreselassie and Judith A. Appleton

INTRODUCTION

Parasitic worms are complex organisms that are highly adapted to their hosts, commonly establishing chronic infections or demonstrating the capacity to infect the same individual repeatedly. Immune responses to parasitic helminths are characterized by the expansion of T-helper type 2 (T_h2) populations, the production of interleukin-4 (IL-4), -5 (IL-5), -9 (IL-9), -10 (IL-10), and -13 (IL-13), and are associated with basophilia, eosinophilia, and alternative activation of macrophages.[1] Investigation of the mouse model of *Schistosoma mansoni* has been central to the advancement of this field of study.[2] In addition, intestinal immune responses have been thoroughly studied in mice infected with nematodes that are natural parasites of rodents, including *Heligmosomoides polygyrus*, *Trichuris muris*, *Trichinella spiralis*, and *Nippostrongylus brasiliensis*. Parasite clearance from the intestine is abrogated in the absence of signal transducer and transcription activator 6 (Stat6), IL-4, and/or IL-13, confirming the importance of these mediators; however, due to differences in habitats and life cycles among these pathogens, the precise mechanisms of effective immunity vary. Immune responses and mechanisms of helminth clearance in extraintestinal sites have also been described, for example, in mouse models of infections caused by filarial nematodes,[3] such as *Strongyloides stercoralis*[4] and the muscle phase of *T. spiralis*.[5,6] Helminths employ a variety of evasive mechanisms that allow them to survive in their hosts for prolonged periods. In this chapter, we summarize the findings from studies of helminth infections in mice as we consider the role of the eosinophil in host defense and immune evasion, including recent evidence that eosinophils prevent nematode killing in extraintestinal sites.

IMMUNE EVASION AND REGULATION IN PARASITIC INFECTIONS

Evasion of immunity by parasites may result from simple avoidance of toxic host effectors or by more complex manipulations of the immune response. In the former category, some parasitic worms evade immune-mediated destruction by defending themselves against oxidants, e.g., *S. mansoni* sporocysts protect themselves against oxidative stress by producing antioxidant enzymes.[7] Extracellular pathogens may need to evade the effects of specific antibodies and often accomplish this by antigenic variation. While such variation is perhaps most elegantly executed by protozoan parasites like *Trypanosoma* and *Plasmodium*,[8] the different life stages of parasitic helminths display or secrete different immunogenic molecules, allowing the pathogen to stay ahead of the immune response. The surface glycoproteins of *T. spiralis* change with molting[9] and the highly immunogenic glycans synthesized in larvae are distinct from those in adults.[10,11] Other parasitic worms synthesize glycans that mimic those of their hosts, such as sialyl-Lewis[X12] or α-galactose,[13] a phenomenon that may enable the parasite to avoid detection by the host immune system.

An array of parasite products, including enzymes, enzyme inhibitors, and lectins have been described that have the potential to alter or regulate the immune response.[14] Downmodulation of T-cell activation or transition from T-helper type 1 (T_h1)-dominated responses to T-helper type 2 (T_h2) responses are characteristic of the transition from acute to chronic disease in several helminth infections.[1] In the example of murine schistosomiasis, a T_h1 response is mounted during the first 5 weeks, at which time egg laying commences, inducing a potent T_h2 response that dominates and sustains the chronic, patent phase of infection.[15] Investigations of this transition have revealed that schistosome eggs contain a glycan, lacto-N-fucopentaose III (LNFPIII) and a ribonuclease (omega-1), both of which enhance the activity of dendritic cells (DCs) in promoting T_h2 cell differentiation.[16–18]

While the influence of DCs is certainly a key factor in driving immunity against helminths, in recent years the potential for granulocytes, particularly basophils and eosinophils, to play regulatory roles in parasitic infection has come under scrutiny. Because the basophil produces large quantities of IL-4, it has been investigated in several rodent helminth infections, using newly available depletion techniques and gene knockout mice. Just as there are variations in T_h2-dependent effector mechanisms that clear the different parasitic worms, so there is also variation in the role of the basophil in generating protection. Depletion of basophils does not compromise T_h2 induction or parasite clearance in *S. mansoni*-infected mice[19] and appears to have a modest influence on T_h2 immune responses in *T. muris* infection.[20] In mice infected with *Litomosoides sigmodontis*, basophils have no apparent role in the protection against chronic infection but they are important for the generation of T_h2-dependent IL-4, immunoglobulin E (IgE), and eosinophilia.[21] In contrast, basophil deficiency does not prevent the development of T_h2 immunity in *N. brasiliensis*-infected mice, and IL-4 from both basophils and T cells is required for efficient worm expulsion.[22]

Available data are contradictory regarding a role for basophils in protective immunity against migrating larvae in a secondary infection.[22,23] Thus, in the context of helminth infection, basophils have a variable influence on T_h2 immunity and little impact on survival of the intestinal parasites studied thus far.

EOSINOPHILS IN IMMUNE REGULATION

Activated eosinophils are a heterogeneous population of cells, with variable granular, cytokine, chemokine, and surface phenotypes, compatible with a range of functional states. Eosinophils express major histocompatibility class II (MHC-II) and have been shown to function as antigen-presenting cells.[24,25] Specifically, intraperitoneal injection of eosinophils that were pulsed with *S. stercoralis* antigen stimulated antigen-specific primed T cells to increase T_h2 cytokine production when tested in restimulation assays.[25] Airway eosinophils have been shown to be capable of priming naive T cells and stimulating $CD4^+$ T cells to produce T_h2 cytokines.[24] Eosinophil-derived neurotoxin (EDN) in human eosinophils induces DC maturation via toll-like receptor 2 (TLR2) activation, resulting in an expansion of the T_h2 response.[26]

In addition, eosinophils influence the recruitment of lymphocytes to sites of inflammation. In murine models of asthma, T_h2 cytokine production is reduced in eosinophil-deficient mice, a result that derives from reduced T cell recruitment into the lung. Adoptive transfer of eosinophils or eosinophils and CD4 T cells reconstitutes disease in ΔdblGATA and PHIL mice, respectively.[27,28] Transfer of eosinophils deficient in IL-13 failed to restore disease, documenting IL-13 as a critical mediator of the regulatory effect of eosinophils in allergic airway disease.[27,29] Eosinophil-deficient mice also show reduced expression of chemokines such as C-C motif chemokine 7 (CCL7), 17 (CCL17), and 22 (CCL22), and the eotaxins CCL11 (eotaxin) and CCL24 (eotaxin-2) in response to allergy challenge.[27,28] Intranasal delivery of CCL11 restored T-cell infiltration and T_h2 cytokine responses[27] while blockade of CCL17 or CCL22 inhibited the recruitment of effector T cells,[27,28] implicating eosinophils in antigen-dependent T_h2 cell recruitment.

Eosinophils are present in certain healthy tissues, for example, the uterus, thymus, and adipose tissue. Recently, it was reported that IL-4 produced by eosinophils promotes alternative activation of macrophages in mouse adipose tissue, thereby promoting glucose tolerance and protecting against diet-induced obesity.[30] ΔdblGATA mice failed to support survival of plasma cells in the bone marrow, an effect that is attributed to a requirement for eosinophil-derived interleukin-6 and A proliferation-inducing ligand

(APRIL).[31] Thus, eosinophils are more than end-stage effector cells and can function in physiological and immunological regulation.

Eosinophilia is a prominent feature of helminth infection and the contributions of eosinophils to host defense have been investigated *in vitro* and *in vivo*. IL-5 is the central regulator of eosinophilia and several *in vivo* models have been developed to address the function of eosinophils by manipulating this cytokine at the gene and protein level. Evaluation of the influence of eosinophils in disease and parasite clearance has relied on knockout or overexpression of the gene encoding IL-5, *interleukin 5 (colony-stimulating factor, eosinophil) (IL5)*, or depletion of the cytokine with specific antibodies. In other models, the gene encoding the receptor for IL-5, *interleukin 5 receptor, alpha (IL5RA)*, was disrupted or the receptor was blocked by antibody. Results of these studies must be interpreted with caution, as murine IL-5 has effects on cells other than eosinophils, most notably B lymphocytes.[32] With rare exceptions,[4] the potential contribution of each of these IL-5 activities to the phenomena observed in IL-5 knockout or transgenic mice have not been elucidated. Furthermore, IL-5 deficiency does not eliminate eosinophils entirely, rather it limits their proliferation, survival, and recruitment to sites of infection. Other models employed to test the role of eosinophils are strains of mice engineered to be deficient in C-C chemokine receptor type 3 (CCR3) or eotaxin (CCL11).[33] Both have defective eotaxin-mediated recruitment of eosinophils, such that studies performed in these mice can address the requirement for eosinophils to enter sites of infection to affect parasite survival or immunity. Again, results must be interpreted with caution as CCR3 is expressed on cell types other than eosinophils, including mast cells, epithelial cells, and T_h2 cells.[34]

Investigations of the function of eosinophils have been aided significantly by development of two mouse strains in which the eosinophil lineage has been ablated. ΔdblGATA mice bear a deletion of the high-affinity double GATA site in the GATA-binding factor 1 (GATA-1) promoter.[35] In PHIL mice, a transgene encoding diphtheria toxin A was inserted downstream of the eosinophil peroxidase promoter.[36] Both constructs render mice free of eosinophils. These strains have been tested for susceptibility and responsiveness to several parasitic helminths, in some instances yielding results that contrast dramatically with findings from mice deficient in or overexpressing IL-5. Two other strains in which genes encoding the granular proteins eosinophil peroxidase (EPO) and eosinophil granule major basic protein-1 (MBP-1) have been knocked out are useful to test the significance of those eosinophil products.[37,38] In the next section of this chapter, we focus our attention on the results from experiments conducted in lineage-ablated and EPO or MBP-1 knockout mice. The findings provide evidence in support of eosinophil effector functions in parasite clearance in some instances, and evidence that the role of the eosinophil is negligible in others. In the case of *T. spiralis*, recent findings support a regulatory role for eosinophils that, rather than enhancing parasite clearance, serves to ensure parasite survival.

EOSINOPHILS IN PARASITIC INFECTION

Dramatic results obtained from studies in which eosinophils were shown to adhere to parasitic helminths, sometimes in the presence of antibody and/or complement, and to kill by discharging granules onto the parasite surface or releasing hydrogen peroxide *in vitro*, have informed the widely held belief that eosinophils function as cytotoxic effectors in host defense against helminths.[39] In contrast, experimental evidence obtained *in vivo* has not consistently supported a role for eosinophils in protection. The observations that are compatible with a regulatory role for eosinophils in helminth infections are discussed in the following sections.

Intestine-Dwelling Helminths

Data from experiments in which IL-5 was manipulated have yielded contradictory results for *H. polygyrus* and no phenotype for *Toxocara canis*.[40–42] Neither parasite has been evaluated in the context of eosinophil ablation. Intestinal infections with *N. brasiliensis*, *S. mansoni*, *T. muris*, and *T. spiralis* have been evaluated in PHIL and/or ΔdblGATA mice. Infection with *T. muris* progressed normally in ΔdblGATA mice,[43] with the only variation in the immune response being an increase in the production of interferon gamma (IFN-γ), suggesting that eosinophils are dispensable for parasite expulsion but contribute to the downmodulation of T_h1 responses. PHIL and ΔdblGATA mice infected with *S. mansoni* exhibited no obvious defects in immune response, worm burden, or egg deposition; there was also no eosinophil-dependent effect on granuloma size, fibrosis, and hepatic pathology.[44] Furthermore, eosinophil deficiency in PHIL and ΔdblGATA mice does not alter the T_h2 response induced by intestinal infection with *T. spiralis*.[5] Intestinal worm expulsion is normal in these strains. Liver injury caused by migrating *T. spiralis* newborn larvae is neither mediated nor prevented by eosinophils.[45] Eosinophil-ablated mice infected with *N. brasiliensis* show only a modest increase in fecal egg count and no alteration in intestinal worm burdens during a primary infection.[46] In aggregate, the studies described document that eosinophil deficiency does not markedly influence the outcome of intestinal helminth infection or the associated immune response. Worm infections have profound effects on tissue structures and functions, repair of intestinal tissues is crucial to recovery following infection. Eosinophilic mediators such as EPO, ribonucleases, and matrix metalloproteinases are known to contribute to debris clearance and lead the way to tissue remodeling in asthmatic patients.[47] Tissue remodeling following intestinal helminth infection is rarely investigated and the role of eosinophils in this process merits consideration.

Tissue-Dwelling Helminths

IL-5 is influential in the resistance to infection with *Angiostrongylus cantonensis*[48,49] and *Strongyloides ratti*, but is dispensable for immunity to *S. ratti* in secondary infection;[50] however, the relationship of IL-5 deficiency to eosinophil function has yet to be clarified in these infections. IL-5 and CCR3 are critical for the clearance of third-stage (L3) *Brugia pahangi* in primary infections in the mouse model, although EPO or MBP-1 do not contribute to this immunity.[51,52] Similarly, IL-5 and CCR3 are required for clearance of L3 *Onchocerca volvulus* from mouse skin, but EPO is dispensable.[53] The significance of these findings to natural infections with filarial worms is unclear, as the complete life cycles are not supported in the mouse.

In mice infected with *L. sigmodontis*, deficiency in either EPO or MBP-1 enhanced L3 establishment in association with reduced IL-4.[54] The findings are compatible with both effector and pro-T_h2 regulatory roles for eosinophil products in *L. sigmodontis* infection. More recently, it has been documented that *L. sigmodontis* development is transiently delayed in the absence of eosinophils or IL-5.[55] The mechanism underlying this effect is yet to be determined. A role for eosinophils in promoting protective T_h2 responses is evident in *N. brasiliensis*-infected mice in which eosinophil ablation in ΔdblGATA mice compromises the development of immunity that limits the early tissue migratory larval stage during secondary infections.[46] The function of eosinophils in the development of this immunity has not yet been elucidated.

More detailed understanding of regulatory and effector functions of eosinophils has been developed in a model of human *S. stercoralis* infection in which L3 are enclosed in a diffusion chamber prior to being implanted subcutaneously in mice. Experiments with mice deficient in C-X-C chemokine receptor type 2 (CXCR2) or depleted of CCR3 suggested that innate immunity to *S. stercoralis* in this model involves direct killing by eosinophils and neutrophils.[56] Furthermore, it has been shown that the parasite synthesizes proteins, as well as chitin, that are chemoattractants for eosinophils.[57] By producing these factors, *S. stercoralis* would recruit eosinophils and engage their activity early in infection. Adaptive, protective immunity in this model is dependent on a T_h2 response and on the production of immunoglobulin M (IgM). In secondary infection, the production of protective IgM is dependent on IL-5 and is restored when eosinophils are transferred to IL-5 knockout mice.[4] Independently, it has been reported that

IgM induced by alum-adjuvanted antigen is compromised in ΔdblGATA mice and can be restored by transfer of eosinophils.[24] The relationship between the enhancement of IgM production and the capacity of eosinophils to present antigens[58] is not obvious and the precise role of eosinophils in promoting IgM production remains to be elucidated.

The Muscle Phase of Trichinellosis

T. spiralis is a natural pathogen of rodents that is able to infect a broad spectrum of mammalian hosts including humans. Intestine-dwelling adults of *T. spiralis* produce newborn larvae that enter the bloodstream and colonize skeletal muscle. Establishment of intracellular muscle infection is coincident with an intestinal T_h2 immune response that causes the expulsion of adult worms. Blood and tissue eosinophilia are pronounced. Although there are multiple reports that human eosinophils kill *T. spiralis* larvae *in vitro*,[59-61] little or no alteration in parasite survival occurs in IL-5-deficient mice, IL-5 transgenic mice, or in mice depleted of eosinophils with anti-IL-5 antibodies.[62-64] In contrast, CCR3 knockout mice were reported to sustain enhanced muscle larvae burdens following primary infection with *T. spiralis*[65] and IL-5 knockout mice showed larger intestinal worm burdens and slower expulsion during secondary infection[62], two results that may implicate a host-protective role for eosinophils,

In stark contrast to these findings, when PHIL and ΔdblGATA mice are given primary infections with *T. spiralis*, newborn larvae colonize skeletal muscle normally but subsequently die in large numbers, reducing larvae burdens by 50–75%.[5] In the absence of eosinophils, infected muscle cells (called nurse cells) become infiltrated by leukocytes and larvae are destroyed (Fig. 10.4.1). Larval death correlates with enhanced IFN-γ and decreased IL-4

production by antigen-stimulated cells recovered from draining lymph nodes. In the absence of eosinophils, leukocytes at sites of infection produce inducible NO synthase (iNOS) and larval survival improves when mice are treated with specific iNOS inhibitors,[5] implicating NO in parasite clearance. Increasing NO production by introducing IL-10 deficiency into the PHIL background dramatically enhances NO production and increases parasite killing to over 90%.

The larvicidal effect of oxidative radicals on different stages of *T. spiralis* larvae has been evaluated *in vitro*. Comparison of the susceptibilities of different larval stages indicated that *T. spiralis* newborn larvae are vulnerable to oxidative damage,[66] while expressed sequence tag clusters encoding antioxidants (thioredoxin oxidase, peroxiredoxin, and glutathione peroxidase) were found in mature *T. spiralis* muscle larvae,[67,68] suggesting that this life stage actively protects itself from reactive oxygen (ROS) and nitrogen (RNS) species. Taken together, these findings suggest that there may be a window of susceptibility to nitrosative stress as larvae grow and mature in the muscle. NO-mediated larval clearing from muscles of eosinophil-ablated mice[5] indicates that eosinophils actively inhibit a T_h1 response in a way that promotes chronic *T. spiralis* infection and insures worm survival in the host population.

NO has been implicated in helminth killing in other murine models, including *B. malayi* and *S. mansoni*. Treatment of mice with an inhibitor of NO synthase abrogates resistance to *B. malayi*.[69,70] The parasites themselves are rich in cytoplasmic, cuticular, and secreted antioxidant enzymes, including superoxide dismutases (SODs) and glutathione peroxidase, which most likely help them survive in an unfavorable oxidative environment.[71] Studies conducted in mice infected with *S. mansoni* demonstrated that iNOS and NO production, induced by vaccination,

FIGURE 10.4.1 Photomicrographs of hematoxylin and eosin-stained sections of tongues from C57BL/6 and PHIL mice 22 days post-oral infection with *Trichinella spiralis*. *(A)* Tongue from infected C57BL/6 mouse. Leukocytes surround the capsule of nurse cells (arrow). Eosinophils are present in the infiltrate (arrowhead). Larvae are large and intact (small arrows). *(B)* Tongue from PHIL mouse. In the absence of eosinophils, the nurse cell (arrows) is infiltrated and normal architecture is destroyed. The larva is not visible in the infiltrated nurse cell, although a healthy larva and normal nurse cell are present in this field (small arrow). Scale bar = 100 μm.

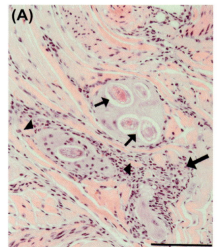
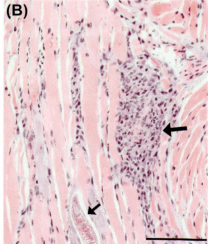

participate in the reduction of worm burdens.[72] Endothelial cells and macrophages, stimulated *in vitro* to produce NO, can kill *S. mansoni* larvae.[73] Susceptibility to the effects of NO has been shown to be larval age-dependent, with older larvae showing greater susceptibility.[74]

Taken together, these studies suggest that ROS and RNS play important host defense roles against tissue-dwelling parasitic worms, a striking contrast with the well-established role for T_h2 immune responses in the expulsion of intestinal worms. Although increased oxidative radical production in these systems may successfully limit parasite numbers, it will also likely cause immune-mediated tissue injury. In *T. spiralis* infection, the eosinophil does not appear to be critical for initiation of a T_h2 response in the intestinal phase of infection, but rather plays a critical role in promoting T_h2 responses during the transition to the chronic phase, when the infection can be transmitted to another host. The contradictory findings between models of IL-5 deficiency and eosinophil ablation indicate that enhanced eosinophilia is not required for regulatory impact to be manifest. Rather, it appears that steady-state numbers of eosinophils are critical. Parasite protection by eosinophils most likely benefits the host by maintaining a T_h2 response that prevents intestinal reinfection, which may overburden the host. In addition, preserving parasites will limit injury to skeletal muscle. Challenge experiments in eosinophil-ablated mice will determine if eosinophils have host protective roles in secondary infections or influence the development of immunological memory.

In an unexpected way, eosinophils may be influential in immunity to infections caused by protozoan parasites. It is well established that NO is an important mediator of immunity to *Toxoplasma gondii*.[75] Infected mice treated with the iNOS inhibitor aminoguanidine have increased numbers of tachyzoites, intracellular cysts, and an exacerbated brain inflammation.[76] Data on the role of eosinophils in *T. gondii* infection is contradictory, with enhanced survival of IL-5 knockout mice reported by one group[77] and increased brain cysts described by another.[78] Although the eosinophil dependence of the effect of IL-5 was not determined in either study, the results call for further investigation of the potential for eosinophils to influence infection by parasitic protozoa.

CONCLUSION AND FUTURE DIRECTIONS

Availability of eosinophil-ablated mice has advanced our ability to investigate the participation of these cells in parasitic infection and immunity. Just as there is variation among the mechanisms of immunity that clear worms, eosinophil function varies among parasitic worm infections. Although all intestinal helminths have not yet been

studied, it is generally true that primary intestinal immune responses do not appear to be influenced by eosinophils. In contrast, results obtained with tissue-dwelling nematodes more consistently show immune regulatory or effector influence for eosinophils. The properties and specific functions of eosinophils that enable them to execute these regulatory phenomena remain, in large measure, to be elucidated. Dissection of the functional attributes of eosinophils and identification of the cells with which they interact to exert their regulatory influence will be crucial next steps in determining how these findings can be applied in developing new tools to prevent and control the parasitic infections that continue to plague human and animal populations.

REFERENCES

1. Maizels RM, Pearce EJ, Artis D, Yazdanbakhsh M, Wynn TA. Regulation of pathogenesis and immunity in helminth infections. *J Exp Med* 2009;**206**:2059—66.
2. Pearce EJ, MacDonald AS. The immunobiology of schistosomiasis. *Nat Rev Immunol* 2002;**2**:499—511.
3. Fischer P, Supali T, Maizels RM. Lymphatic filariasis and Brugia timori: prospects for elimination. *Trends Parasitol* 2004;**20**:351—5.
4. Herbert DR, Lee JJ, Lee NA, Nolan TJ, Schad GA, Abraham D. Role of IL-5 in innate and adaptive immunity to larval Strongyloides stercoralis in mice. *J Immunol* 2000;**165**:4544—51.
5. Fabre V, Beiting DP, Bliss SK, Gebreselassie NG, Gagliardo LF, Lee NA, et al. Eosinophil deficiency compromises parasite survival in chronic nematode infection. *J Immunol* 2009;**182**:1577—83.
6. Beiting DP, Gagliardo LF, Hesse M, Bliss SK, Meskill D, Appleton JA. Coordinated control of immunity to muscle stage Trichinella spiralis by IL-10, regulatory T cells, and TGF-beta. *J Immunol* 2007;**178**:1039—47.
7. Mourao Mde M, Dinguirard N, Franco GR, Yoshino TP. Role of the endogenous antioxidant system in the protection of Schistosoma mansoni primary sporocysts against exogenous oxidative stress. *PLoS Negl Trop Dis* 2009;**3**:e550.
8. Stockdale C, Swiderski MR, Barry JD, McCulloch R. Antigenic variation in Trypanosoma brucei: joining the DOTs. *PLoS Biol* 2008;**6**:e185.
9. Jungery M, Clark NW, Parkhouse RM. A major change in surface antigens during the maturation of newborn larvae of Trichinella spiralis. *Mol Biochem Parasitol* 1983;**7**:101—9.
10. Morelle W, Haslam SM, Morris HR, Dell A. Characterization of the N-linked glycans of adult Trichinella spiralis. *Mol Biochem Parasitol* 2000;**109**:171—7.
11. Morelle W, Haslam SM, Olivier V, Appleton JA, Morris HR, Dell A. Phosphorylcholine-containing N-glycans of Trichinella spiralis: identification of multiantennary lacdiNAc structures. *Glycobiology* 2000;**10**:941—50.
12. Cummings RD, Nyame AK. Glycobiology of schistosomiasis. *FASEB J* 1996;**10**:838—48.
13. Duffy MS, Morris HR, Dell A, Appleton JA, Haslam SM. Protein glycosylation in Parelaphostrongylus tenuis—first description of the Galalpha1-3Gal sequence in a nematode. *Glycobiology* 2006;**16**:854—62.

14. Hewitson JP, Grainger JR, Maizels RM. Helminth immunoregulation: the role of parasite secreted proteins in modulating host immunity. *Mol Biochem Parasitol* 2009;**167**:1−11.

15. Pearce EJ, Caspar P, Grzych JM, Lewis FA, Sher A. Downregulation of Th1 cytokine production accompanies induction of Th2 responses by a parasitic helminth, Schistosoma mansoni. *J exp Med* 1991;**173**:159−66.

16. Okano M, Satoskar AR, Nishizaki K, Harn Jr DA. Lacto-N-fucopentaose III found on Schistosoma mansoni egg antigens functions as adjuvant for proteins by inducing Th2-type response. *J Immunol* 2001;**167**:442−50.

17. Steinfelder S, Andersen JF, Cannons JL, Feng CG, Joshi M, Dwyer D, et al. The major component in schistosome eggs responsible for conditioning dendritic cells for Th2 polarization is a T2 ribonuclease (omega-1). *J Exp Med* 2009;**206**:1681−90.

18. Faveeuw C, Mallevaey T, Paschinger K, Wilson IB, Fontaine J, Mollicone R, et al. Schistosome N-glycans containing core alpha 3-fucose and core beta 2-xylose epitopes are strong inducers of Th2 responses in mice. *Eur J Immunol* 2003;**33**:1271−81.

19. Phythian-Adams AT, Cook PC, Lundie RJ, Jones LH, Smith KA, Barr TA, et al. CD11c depletion severely disrupts Th2 induction and development in vivo. *J Exp Med* 2010;**207**:2089−96.

20. Perrigoue JG, Saenz SA, Siracusa MC, Allenspach EJ, Taylor BC, Giacomin PR, et al. MHC class II-dependent basophil-CD4+ T cell interactions promote T(H)2 cytokine-dependent immunity. *Nat Immunol* 2009;**10**:697−705.

21. Torrero MN, Hubner MP, Larson D, Karasuyama H, Mitre E. Basophils amplify type 2 immune responses, but do not serve a protective role, during chronic infection of mice with the filarial nematode Litomosoides sigmodontis. *J Immunol* 2010;**185**:7426−34.

22. Sullivan BM, Liang HE, Bando JK, Wu D, Cheng LE, McKerrow JK, et al. Genetic analysis of basophil function in vivo. *Nat Immunol* 2011;**12**:527−35.

23. Ohnmacht C, Schwartz C, Panzer M, Schiedewitz I, Naumann R, Voehringer D. Basophils orchestrate chronic allergic dermatitis and protective immunity against helminths. *Immunity* 2010;**33**:364−74.

24. Wang HB, Ghiran I, Matthaei K, Weller PF. Airway eosinophils: allergic inflammation recruited professional antigen-presenting cells. *J Immunol* 2007;**179**:7585−92.

25. Padigel UM, Hess JA, Lee JJ, Lok JB, Nolan TJ, Schad GA, et al. Eosinophils act as antigen-presenting cells to induce immunity to Strongyloides stercoralis in mice. *J Infect Dis* 2007;**196**:1844−51.

26. Yang D, Chen Q, Su SB, Zhang P, Kurosaka K, Caspi RR, et al. Eosinophil-derived neurotoxin acts as an alarmin to activate the TLR2-MyD88 signal pathway in dendritic cells and enhances Th2 immune responses. *J Exp Med* 2008;**205**:79−90.

27. Walsh ER, Sahu N, Kearley J, Benjamin E, Kang BH, Humbles A, et al. Strain-specific requirement for eosinophils in the recruitment of T cells to the lung during the development of allergic asthma. *J Exp Med* 2008;**205**:1285−92.

28. Jacobsen EA, Ochkur SI, Pero RS, Taranova AG, Protheroe CA, Colbert DC, et al. Allergic pulmonary inflammation in mice is dependent on eosinophil-induced recruitment of effector T cells. *J Exp Med* 2008;**205**:699−710.

29. Walsh ER, Thakar J, Stokes K, Huang F, Albert R, August A. Computational and experimental analysis reveals a requirement for eosinophil-derived IL-13 for the development of allergic airway responses in C57BL/6 mice. *J Immunol* 2011;**186**:2936−49.

30. Wu D, Molofsky AB, Liang HE, Ricardo-Gonzalez RR, Jouihan HA, Bando JK, et al. Eosinophils sustain adipose alternatively activated macrophages associated with glucose homeostasis. *Science* 2011;**332**:243−7.

31. Chu VT, Frohlich A, Steinhauser G, Scheel T, Roch T, Fillatreau S, et al. Eosinophils are required for the maintenance of plasma cells in the bone marrow. *Nat Immunol* 2011;**12**:151−9.

32. Erickson LD, Foy TM, Waldschmidt TJ. Murine B1 B cells require IL-5 for optimal T cell-dependent activation. *J Immunol* 2001;**166**:1531−9.

33. Pope SM, Zimmermann N, Stringer KF, Karow ML, Rothenberg ME. The eotaxin chemokines and CCR3 are fundamental regulators of allergen-induced pulmonary eosinophilia. *J Immunol* 2005;**175**:5341−50.

34. Ma W, Bryce PJ, Humbles AA, Laouini D, Yalcindag A, Alenius H, et al. CCR3 is essential for skin eosinophilia and airway hyper-responsiveness in a murine model of allergic skin inflammation. *J Clin Invest* 2002;**109**:621−8.

35. Yu C, Cantor AB, Yang H, Browne C, Wells RA, Fujiwara Y, et al. Targeted deletion of a high-affinity GATA-binding site in the GATA-1 promoter leads to selective loss of the eosinophil lineage in vivo. *J Exp Med* 2002;**195**:1387−95.

36. Lee JJ, Dimina D, Macias MP, Ochkur SI, McGarry MP, O'Neill KR, et al. Defining a link with asthma in mice congenitally deficient in eosinophils. *Science* 2004;**305**:1773−6.

37. Denzler KL, Borchers MT, Crosby JR, Cieslewicz G, Hines EM, Justice JP, et al. Extensive eosinophil degranulation and perox-idase-mediated oxidation of airway proteins do not occur in a mouse ovalbumin-challenge model of pulmonary inflammation. *J Immunol* 2001;**167**:1672−82.

38. Denzler KL, Farmer SC, Crosby JR, Borchers M, Cieslewicz G, Larson KA, et al. Eosinophil major basic protein-1 does not contribute to allergen-induced airway pathologies in mouse models of asthma. *J Immunol* 2000;**165**:5509−17.

39. Klion AD, Nutman TB. The role of eosinophils in host defense against helminth parasites. *J Allergy Clin Immunol* 2004;**113**:30−7.

40. Urban Jr JF, Katona IM, Paul WE, Finkelman FD. Interleukin 4 is important in protective immunity to a gastrointestinal nematode infection in mice. *Proc Natl Acad Sci USA* 1991;**88**:5513−7.

41. Takamoto M, Ovington KS, Behm CA, Sugane K, Young IG, Matthaei KI. Eosinophilia, parasite burden and lung damage in Toxocara canis infection in C57Bl/6 mice genetically deficient in IL-5. *Immunology* 1997;**90**:511−7.

42. Ovington KS, Behm CA. The enigmatic eosinophil: investigation of the biological role of eosinophils in parasitic helminth infection. *Mem Inst Oswaldo Cruz* 1997;**92**(Suppl. 2):93−104.

43. Svensson M, Bell L, Little MC, DeSchoolmeester M, Locksley RM, Else KJ. Accumulation of eosinophils in intestine-draining mesenteric lymph nodes occurs after Trichuris muris infection. *Parasite Immunol* 2011;**33**:1−11.

44. Swartz JM, Dyer KD, Cheever AW, Ramalingam T, Pesnicak L, Domachowske JB, et al. Schistosoma mansoni infection in eosinophil lineage-ablated mice. *Blood* 2006;**108**:2420−7.

45. Douglas DB, Beiting DP, Loftus JP, Appleton JA, Bliss SK. Combinatorial effects of interleukin 10 and interleukin 4 determine the progression of hepatic inflammation following murine enteric parasitic infection. *Hepatology* 2010;**51**:2162−71.

46. Knott ML, Matthaei KI, Giacomin PR, Wang H, Foster PS, Dent LA. Impaired resistance in early secondary Nippostrongylus brasiliensis infections in mice with defective eosinophilopoeisis. *Int J Parasitol* 2007;**37**:1367−78.

47 Jacobsen EA, Ochkur SI, Lee NA, Lee JJ. Eosinophils and asthma. *Curr Allergy Asthma Rep* 2007;**7**:18−26.

48. Yoshida T, Ikuta K, Sugaya H, Maki K, Takagi M, Kanazawa H, et al. Defective B-1 cell development and impaired immunity against Angiostrongylus cantonensis in IL-5R alpha-deficient mice. *Immunity* 1996;**4**:483−94.

49. Sasaki O, Sugaya H, Ishida K, Yoshimura K. Ablation of eosinophils with anti-IL-5 antibody enhances the survival of intracranial worms of Angiostrongylus cantonensis in the mouse. *Parasite Immunol* 1993;**15**:349−54.

50. Ovington KS, McKie K, Matthaei KI, Young IG, Behm CA. Regulation of primary Strongyloides ratti infections in mice: a role for interleukin-5. *Immunology* 1998;**95**:488−93.

51. Ramalingam T, Ganley-Leal L, Porte P, Rajan TV. Impaired clearance of primary but not secondary Brugia infections in IL-5 deficient mice. *Exp Parasitol* 2003;**105**:131−9.

52. Ramalingam T, Porte P, Lee J, Rajan TV. Eosinophils, but not eosinophil peroxidase or major basic protein, are important for host protection in experimental Brugia pahangi infection. *Infect Immun* 2005;**73**:8442−3.

53. Abraham D, Leon O, Schnyder-Candrian S, Wang CC, Galioto AM, Kerepesi LA, et al. Immunoglobulin E and eosinophil-dependent protective immunity to larval Onchocerca volvulus in mice immunized with irradiated larvae. *Infect Immun* 2004;**72**:810−7.

54. Specht S, Saeftel M, Arndt M, Endl E, Dubben B, Lee NA, et al. Lack of eosinophil peroxidase or major basic protein impairs defense against murine filarial infection. *Infect Immun* 2006;**74**:5236−43.

55. Babayan SA, Read AF, Lawrence RA, Bain O, Allen JE. Filarial parasites develop faster and reproduce earlier in response to host immune effectors that determine filarial life expectancy. *PLoS biology* 2010;**8**:e1000525.

56. Galioto AM, Hess JA, Nolan TJ, Schad GA, Lee JJ, Abraham D. Role of eosinophils and neutrophils in innate and adaptive protective immunity to larval strongyloides stercoralis in mice. *Infect Immun* 2006;**74**:5730−8.

57. Stein LH, Redding KM, Lee JJ, Nolan TJ, Schad GA, Lok JB, et al. Eosinophils utilize multiple chemokine receptors for chemotaxis to the parasitic nematode Strongyloides stercoralis. *J Innate Immun* 2009;**1**:618−30.

58. Padigel UM, Lee JJ, Nolan TJ, Schad GA, Abraham D. Eosinophils can function as antigen-presenting cells to induce primary and secondary immune responses to Strongyloides stercoralis. *Infect Immun* 2006;**74**:3232−8.

59 Hamann KJ, Barker RL, Loegering DA, Gleich GJ. Comparative toxicity of purified human eosinophil granule proteins for newborn larvae of Trichinella spiralis. *J Parasitol* 1987;**73**:523−9.

60. Bass DA, Szejda P. Mechanisms of killing of newborn larvae of Trichinella spiralis by neutrophils and eosinophils. Killing by generators of hydrogen peroxide in vitro. *J Clin Invest* 1979;**64**: 1558−64.

61. Buys J, Wever R, van Stigt R, Ruitenberg EJ. The killing of newborn larvae of Trichinella spiralis by eosinophil peroxidase in vitro. *Eur J Immunol* 1981;**11**:843−5.

62. Vallance BA, Matthaei KI, Sanovic S, Young IG, Collins SM. Interleukin-5 deficient mice exhibit impaired host defence against challenge Trichinella spiralis infections. *Parasite Immunol* 2000;**22**: 487−92.

63. Hokibara S, Takamoto M, Tominaga A, Takatsu K, Sugane K. Marked eosinophilia in interleukin-5 transgenic mice fails to prevent Trichinella spiralis infection. *J Parasitol* 1997;**83** 1186−9.

64. Herndon FJ, Kayes SG. Depletion of eosinophils by anti-IL-5 monoclonal antibody treatment of mice infected with Trichinella spiralis does not alter parasite burden or immunologic resistance to reinfection. *J Immunol* 1992;**149**:3642−7.

65. Gurish MF, Humbles A, Tao H, Finkelstein S, Boyce JA, Gerard C, et al. CCR3 is required for tissue eosinophilia and larval cytotoxicity after infection with Trichinella spiralis. *J Immunol* 2002;**168**:5730−6.

66. Kazura JW, Meshnick SR. Scavenger enzymes and resistance to oxygen mediated damage in Trichinella spiralis. *Mol Biochem Parasitol* 1984;**10**:1−10.

67. Mitreva M, Appleton J, McCarter JP, Jasmer DP. Expressed sequence tags from life cycle stages of Trichinella spiralis: application to biology and parasite control. *Vet Parasitol* 2005;**132**:13−7.

68. Mitreva M, Jasmer DP, Appleton J, Martin J, Dante M, Wylie T, et al. Gene discovery in the adenophorean nematode Trichinella spiralis: an analysis of transcription from three life-cycle stages. *Mol Biochem Parasitol* 2004;**137**:277−91.

69. Selkirk ME, Smith VP, Thomas GR, Gounaris K. Resistance of filarial nematode parasites to oxidative stress. *Int J Parasitol* 1998;**28**:1315−32.

70. Rajan TV, Porte P, Yates JA, Keefer L, Shultz LD. Role of nitric oxide in host defense against an extracellular, metazoan parasite, Brugia malayi. *Infect Immun* 1996;**64**:3351−3.

71. Dzik JM. Molecules released by helminth parasites involved in host colonization. *Acta Biochim Pol* 2006;**53**:33−64.

72. Wynn TA, Reynolds A, James S, Cheever AW, Caspar P, Hieny S, et al. IL-12 enhances vaccine-induced immunity to schistosomes by augmenting both humoral and cell-mediated immune responses against the parasite. *J Immunol* 1996;**157**:4068−78.

73. Oswald IP, Eltoum I, Wynn TA, Schwartz B, Caspar P, Paulin D, et al. Endothelial cells are activated by cytokine treatment to kill an intravascular parasite, Schistosoma mansoni, through the production of nitric oxide. *Proc Natl Acad Sci USA* 1994;**91**:999−1003.

74. Ahmed SF, Oswald IP, Caspar P, Hieny S, Keefer L, Sher A, et al. Developmental differences determine larval susceptibility to nitric oxide-mediated killing in a murine model of vaccination against Schistosoma mansoni. *Infect Immun* 1997;**65**:219−26.

75. Green SJ, Scheller LF, Marletta MA, Seguin MC, Klotz FW, Slayter M, et al. Nitric oxide: cytokine-regulation of nitric oxide in host resistance to intracellular pathogens. *Immunol Lett* 1994;**43**: 87−94.

76. Kang KM, Lee GS, Lee JH, Choi IW, Shin DW, Lee YH. Effects of iNOS inhibitor on IFN-gamma production and apoptosis of splenocytes in genetically different strains of mice infected with Toxoplasma gondii. *Korean J Parasitol* 2004;**42**:175−83.

77. Nickdel MB, Roberts F, Brombacher F, Alexander J, Roberts CW. Counter-protective role for interleukin-5 during acute Toxoplasma gondii infection. *Infect Immun* 2001;**69**:1044−52.

78. Zhang Y, Denkers EY. Protective role for interleukin-5 during chronic Toxoplasma gondii infection. *Infect Immun* 1999;**67**: 4383−92.

Eosinophil Cell–Cell Communication

Introduction

Calman Prussin

INTRODUCTION

Eosinophils, although best known as effector cells of T-helper type 2 (T_h2)-driven inflammation, play multiple roles in healthy cellular physiology, as well as in the setting of host defense and tissue inflammation. Under basal physiological conditions, eosinophils are a normal homeostatic constituent of the uterine endometrium and the gastrointestinal mucosa. In the settings of either helminth infection or allergen-driven T_h2 inflammation, eosinophils localize to the site of inflammation. Under both basal and pathological conditions, eosinophils are intimately associated with multiple other cellular constituents and are thus capable of cell–cell interactions.

A variety of heterotypic eosinophil cell–cell interactions are theoretically possible, including those occurring under basal physiological as well as pathological conditions. Theoretically, such cell–cell interactions can be bidirectional, with eosinophils as either the donor or the recipient of such cell–cell signaling. However, from an experimental point of view, a unidirectional hypothesis is the most easily addressed; for example, when investigating the neuropathic effects of eosinophil granule proteins. In this way, eosinophils may play a variety of roles as both initiators and effectors of T_h2-driven inflammation.[1]

Until recently, the study of heterotypic eosinophil–cell interactions was largely limited to *in vitro* experimental systems, in which well-defined cellular constituents could be cocultured to observe such interactions. The generation and availability of eosinophil-deleted mice is a major advance in characterizing such eosinophil–cell interactions *in vivo*. Eosinophil-deleted mice can be used to determine if eosinophils are required for specific pathological endpoints in an *in vivo* disease model. Once such changes have been shown, eosinophils can then be added back to eosinophil-deficient mice to reverse the effect, in a sense fulfilling Koch's postulates. The continued dissemination and sophistication of these models will further our understanding of eosinophil function *in vivo*. The generation of eosinophil-specific knockouts, either by Cre-Lox technology or by transferring eosinophils from specific gene-deleted mice into eosinophil-deficient mice, is an important next step in this regard.

EOSINOPHIL-DEPENDENT T CELL ACTIVATION AND RECRUITMENT

The best-characterized eosinophil cell–cell interaction is mediated by T cell expression of interleukin-5 (IL-5), which results in increased eosinophilopoiesis and eosinophil release from the bone marrow, and eosinophil activation and survival in tissues. Although other cell populations can express IL-5, including mast cells and eosinophils themselves, T_h2 cells remain the best-substantiated source of IL-5. Other T_h2 cytokines, such as IL-13 and to a lesser extent IL-4, indirectly promote eosinophil trafficking to tissues, largely through their induction of eotaxin [or chemokine (C-C motif) ligand] family chemokines (CCL11, CCL24, and CCL26) expression by endothelial and epithelial cells.

Recently, several groups have described innate non-T, non-B cells that express a T_h2 cytokine profile, variously named nuocytes, multipotent progenitor type-2, or natural helper cells.[2] Notably, upon activation with either IL-25 or IL-33, these cells produce large amounts of T_h2 cytokines and consequent eosinophilia. These findings suggest a possible innate cascade whereby epithelial cells secrete IL-25 and IL-33 upon activation by pattern recognition receptors, thus activating natural helper cell expression of T_h2 cytokines, and ultimately resulting in eosinophilic inflammation. How these various non-T, non-B cell populations are related and the mechanisms through which they promote eosinophilic inflammation is the subject of active investigation.

Through their direct and indirect responses to T_h2 cytokines, eosinophils have been characterized as *recipient* cells that respond to T_h2 cell signals. More recently, a number of reports have demonstrated that, conversely, eosinophils also act as *donors* and provide proinflammatory signals to T_h2 cells. As detailed by Jacobsen in Chapter 11.5 (T Cell Activation and Recruitment), eosinophils can

Eosinophils in Health and Disease. http://dx.doi.org/10.1016/B978-0-12-394385-9.00011-0

function as both initiators and potentiators of T_h2 dominant inflammation. Eosinophils are an important source of IL-25, which acts on T_h2 cells through the IL-17B receptor to potentiate T_h2 expression of IL-4, IL-5, and IL-13. Eosinophil-deleted mice have further elucidated the role of eosinophils in driving T_h2 cell differentiation and trafficking. In ovalbumin-induced models of asthma, efficient T_h2 cell trafficking to the inflamed lung requires eosinophil induction of C-C chemokine receptor type 3 (CCR3) ligands (CCL7, CCL11, and CCL24) and CCR4 ligands (CCL17 and CCL22). Thus, through multiple mediators and pathways, eosinophils and T_h2 cells may synergistically collaborate to induce inflammation.

EOSINOPHIL-MEDIATED ANTIGEN PRESENTATION

Although eosinophils are best known for their effector function, as presented by Akuthota and colleagues in Chapter 11.4 (Eosinophil-Mediated Antigen Presentation), a considerable literature has documented their capacity as antigen-presenting cells (APCs). Eosinophils are capable of all the requisite functions of APCs, being able take up, process and present antigens, express co-stimulatory molecules, and traffic to lymph nodes.

A fundamental, but unanswered question is whether eosinophils are specialized APCs for antigen presentation to T_h2 cells or more general APCs that can contribute to all immune responses. Obviously, from a teleological point of view, the former possibility is more attractive. Basophils, which have a lineage closely related to eosinophils, have recently gained notoriety for their newly established pro-T_h2 APC function.[3,4] Critics of eosinophils as APCs have suggested that since dendritic cells are so much more effective APCs for priming naive T cells, eosinophil APC activity is irrelevant. Even supposing that were true for naive cells, eosinophils could be APCs specialized for trafficking to inflammatory sites and activating memory T_h2 cells in the inflamed tissue. This is clearly an area of active investigation, in which eosinophil-deleted mice will contribute to our understanding of the antigen-presenting role for eosinophils *in vivo*.

EOSINOPHIL–NEURONAL INTERACTIONS

Eosinophils can be found in proximity to neurons in a variety of inflammatory diseases. Eosinophil–neuron interactions can be in either direction, with eosinophils acting as either the target or the source of the interaction. For example, neurons can directly or indirectly recruit eosinophils and, conversely, eosinophils can alternatively have trophic, stimulatory, or inhibitory effects on neurons.

How these eosinophil–neuronal interactions integrate into other pathological processes, their relative contribution to disease, and their physiological role are the subjects of active investigation.

EPITHELIAL CELL–EOSINOPHIL INTERACTIONS

One of the best-defined effector roles of eosinophils in allergic disease is through their degranulation and elaboration of highly cationic proteins, such as major basic protein, which is cytotoxic for epithelial cells. A functional consequence of this activity is indicated by the correlation between airway eosinophil number and bronchial epithelial cell desquamation. These and similar findings have been used to support the *eosinophil hypothesis in asthma*.

As detailed by Sexton and Walsh in Chapter 11.8 (Eosinophil–Airway Epithelial Cell Interactions), in contrast to having a role as simple targets for eosinophil granule proteins, epithelial cells have emerged as active collaborators in the induction of eosinophilic inflammation. In multiple allergic diseases, including asthma, allergic rhinitis, atopic dermatitis, and eosinophilic esophagitis, eosinophils are found in proximity to epithelial cells. This is not surprising, since epithelia are the initial site of aeroallergen and food allergen exposure. Epithelial cells are well placed to receive environmental signals from both innate and acquired immune sources and express a wide range of receptors, including toll-like, protease-activated, and cytokine receptors. Upon activation by multiple cytokines, particularly IL-13 and tumor necrosis factor (TN-α), epithelial cells express chemokines, notably CCR3 ligands, such as eotaxin family members, that attract eosinophils.

More recently, epithelial cells have emerged as the source of a variety of pro-T_h2 cytokines, including thymic stromal lymphopoietin, IL-25 and IL-33.[5,6] TSLP has been described as the *master T_h2 regulator* that induces dendritic cells to prime T cells to the T_h2 phenotype. IL-25 and IL-33 are expressed by epithelial cells and act to enhance T_h2 cytokine expression. Thus, epithelial cell-derived cytokines may act as a bridge between innate and acquired immunity.

EOSINOPHIL–MAST CELL INTERACTIONS

Mast cells and eosinophils are generally associated with early and late phase allergic responses, respectively. However, in chronic type 2 inflammation increased numbers of eosinophils and mast cells often cohabit in inflammatory tissue. A large body of evidence has shown that eosinophils and mast cells can each, through their elaboration of soluble mediators such as cytokines, eicosanoids, and granule proteins, modify the function and/or trafficking of the other cell type. Validating these findings *in vivo* has proven to be complicated and is the subject of active investigation. The complexity of such experiments is underscored by the fact that both cell types are immunologically active, and it is therefore often difficult to determine if their effects on each other are direct or are mediated through an intermediary cell type, such as a T cell.

PLASMA CELL/B CELL—EOSINOPHIL INTERACTIONS

Relatively little is known about B cell—eosinophil interactions. After antigen challenge, eosinophils can traffic to the T cell-rich areas of draining lymph nodes, but their ability to traffic to B cell-rich areas has not been demonstrated. Eosinophils have been identified as the alum-induced Gr-1$^+$ IL-4$^+$ cells found in the spleen that may mediate the alum adjuvant effect. Whether eosinophils play any role in later phases of the B cell response, such as modulating germinal center B cells or plasma cells, is the subject of active investigation. Interestingly, murine B cells express the IL-5 receptor and, furthermore, IL-5 is required for B-1 cell survival and can drive B-2 cells to undergo mu to gamma-1 class switch recombination. Notably, human B cells do not express the IL-5 receptor.

CONCLUSION

Under both homeostatic and inflammatory conditions, eosinophils interact with a large number and variety of cell populations. The wide range of heterotypic eosinophil cell—cell interactions underscores the diversity of potential eosinophil roles in health and disease.

REFERENCES

1. Barrett NA, Austen KF. Innate cells and T helper 2 cell immunity in airway inflammation. *Immunity* 2009;**31**:425—37.
2. Saenz SA, Noti M, Artis D. Innate immune cell populations function as initiators and effectors in Th2 cytokine responses. *Trends Immunol* 2010;**31**:407—13.
3. Nakanishi K. Basophils are potent antigen-presenting cells that selectively induce Th2 cells. *Eur J Immunol* 2010;**40**:1836—42.
4. Hammad H, Plantinga M, Deswarte K, Pouliot P, Willart MA, Kool M, et al. Inflammatory dendritic cells—not basophils—are necessary and sufficient for induction of Th2 immunity to inhaled house dust mite allergen. *J Exp Med* 2010;**207**:2097—111.
5. Saenz SA, Taylor BC, Artis D. Welcome to the neighborhood: epithelial cell-derived cytokines license innate and adaptive immune responses at mucosal sites. *Immunol Rev* 2008;**226**:172—90.
6. Bulek K, Swaidani S, Aronica M, Li X. Epithelium: the interplay between innate and Th2 immunity. *Immunol Cell Biol* 2010; **88**:257—68.

Chapter 11.2

Eosinophil and Nerve Interactions

Gregory D. Scott and Allison D. Fryer

INTRODUCTION

Over the past 10 years, it has become increasingly clear that physiological and pathological interactions between nerves and eosinophils exist. Eosinophils are typically not abundant in organs. However, in chronic inflammatory diseases associated with eosinophils, they are preferentially associated with nerves.[1,2] It has been established that nerves actively recruit and bind eosinophils and that this interaction produces a range of effects in both cell types. Tissues obtained from humans with chronic inflammatory diseases and from animal models contain eosinophils localized next to and inside nerves. Remarkably, the manifestations of some inflammatory diseases can be prevented by ablating eosinophils, by preventing eosinophil recruitment to diseased organs, or by separating eosinophils from nerves.

EOSINOPHIL RECRUITMENT TO NERVES

Studies that first highlighted neurons as important proinflammatory cells specifically in eosinophilic diseases were first performed in 1957. Asthma is a chronic inflammatory disease of the lungs that is characterized by infiltration of eosinophils and excessive bronchoconstriction. The vagus nerves contain the vast majority of sensory and parasympathetic neurons that cause bronchoconstriction. In an attempt to cure intractable asthma, surgeons severed the vagus nerves supplying the lungs. While asthma symptoms abated, an overlooked finding was that eosinophils in both the sputum and blood were reduced in these patients.[3] Thus, in addition to neurogenic inflammation, nerves may mediate eosinophil migration into the lungs. This may be part of a broader mechanism of eosinophil recruitment because in a study of leg nerve injury, transecting the sciatic nerve caused an influx of eosinophils within the nerve itself. Three to ten days after transection, the sciatic nerves were removed and disaggregated and all cells were counted. Eosinophils within the injured nerve comprised roughly half of the total inflammatory cells, far greater than the percentage of eosinophils in blood (<8%).

Eosinophils are also associated with the optic nerve in a mouse model of multiple sclerosis.[4] Similarly, in antigen-challenged guinea pigs and in humans with fatal asthma, eosinophils are located inside the nerve bundles, around the parasympathetic ganglia, and along nerve fibers (Fig. 11.2.1).[5] The density of eosinophils associated with nerves is greater than eosinophils associated with blood vessels or airway tissues. These four studies demonstrate that peripheral nerves can actively, and selectively, recruit eosinophils.[6]

Eosinophil Recruitment to Nerves: Role of Neuropeptides

Neuropeptides include the tachykinins, substance P, and neurokinin A, as well as calcitonin gene-related peptide (CGRP), vasoactive intestinal peptide (VIP), secretin, and secretoneurin. Neuropeptides are made by many cell types including neurons; predominantly, sensory neurons.[7,8,9] Neuropeptides regulate inflammatory responses and are associated with diseases involving eosinophil−nerve interactions, such as asthma, rhinitis, inflammatory bowel disease, atopic dermatitis, conjunctivitis, and myelitis.[9−12] The role of neuropeptides in eosinophil recruitment has been demonstrated *in vivo* and *in vitro*. *In vitro* studies have shown that isolated human eosinophils move up substance P, CGRP, secretoneurin, VIP, and secretin concentration gradients with half-maximal effective concentrations (EC_{50}) in the picomolar range.[13] *In vivo*, intradermal injections of substance P cause eosinophil recruitment to human skin.[14] Lesional pruritus nodularis, a skin disease characterized by itchy bumps on the arms and legs, is associated with both increased CGRP expression in neurons and increased nerve-associated eosinophils (Fig. 11.2.2).[15] Substance P, neurokinin A, and CGRP also prime human eosinophils isolated from allergic patients to migrate toward other chemotactic factors, such as platelet activating factor (PAF) and leukotriene B_4.[16] Thus, neuropeptides can prime and directly recruit eosinophils to nerves.

The pathology of other diseases also suggests that there may be a relationship between neuropeptides and eosinophil recruitment. In ulcerative colitis, both increased substance P and increased eosinophils are observed but have not been correlated.[17] Increased tear and plasma substance P production in vernal keratoconjunctivitis and seasonal allergic conjunctivitis also coincide with increased eosinophil accumulation in the finely innervated ocular surface mucosa.[18] Finally, cervical ripening in preparation for labor and delivery are characterized by increased neuronal expression of substance P, CGRP, and VIP, along with increased eosinophil numbers.[19,20] It is tempting to speculate that increased eosinophil presence in these inflammatory states is mediated by neuropeptide signaling. It will be important for future studies in each disease or biologic process to confirm whether neurons or another cell type are the source of neuropeptides.

Eosinophil Recruitment to Nerves: Role of Cytokines and Chemokines

Recruitment of eosinophils to nerves is an active process that requires chemotactic signals. These signaling proteins are cytokines capable of generating an immune response and also chemokines that cause leukocyte migration. In antigen-challenged guinea pigs and in humans with fatal asthma, there are more eosinophils associated with airway nerves than with anywhere else in the lungs.[5] Associated eosinophils are defined as those within 8 μm of a nerve, since 8 μm is the diameter of an eosinophil. Blocking the pleiotropic cytokine, tumor necrosis factor (TNF), in antigen-challenged guinea pigs prevents eosinophils localizing along airway parasympathetic nerves.[21] Eotaxin family members are potent chemokines responsible for eosinophil recruitment from blood vessels and bone marrow. Blocking the eotaxin receptor/C-C chemokine receptor type 3, CCR3, prevents eosinophil localization to airway nerves *in vitro* and *in vivo*.[22] CCR3 is a receptor for many cytokines in addition

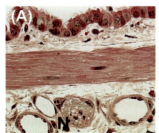 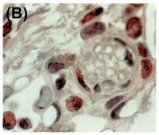 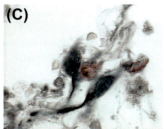 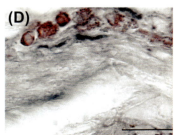

FIGURE 11.2.1 **Eosinophils are located inside and around airway parasympathetic nerves and ganglia in allergen-sensitized guinea pigs and humans who died of fatal asthma.** Photographs of glycol methacrylate-embedded sections of (*A*) healthy nonsensitized guinea pigs, (*B*) allergen-sensitized guinea pigs, and (*C, D*) human airways from patients who died of fatal asthma. Sections are stained with Luna's stain (*A*), hematoxylin and eosin (*B*), and major basic protein (*C, D*). Nerves are labeled in black and eosinophils are labeled red. Nerve bundle cross-section is labeled *N*. Scale bar represents 100 μm. (*Adapted from Costello et al.,[5] used with permission.*)

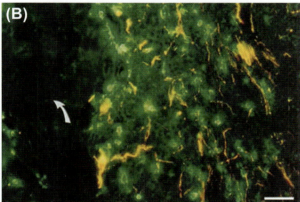

FIGURE 11.2.2 Dermis from a patient with prurigo nodularis showing increased density of calcitonin gene-related peptide (CGRP)⁺ nerves and increased eosinophil density next to CGRP⁺ nerves. Eosinophil (light green) density in the dermis of lesional skin from a patient with prurigo nodularis (B) is greater than the density of eosinophils (light green; straight arrow, A) in the dermis of healthy controls (A). Eosinophils in the dermis of patients with prurigo nodularis (B) are localized next to nerves and are less likely to be located in regions without nerves (curved arrow, B). Eosinophils in healthy skin (straight arrow, A) were not localized next to nerves (curved arrow, A). Skin from patients with prurigo nodularis (B) also contained increased nerve density. Scale bar represents 50 μm. The magnification of A is equivalent to that of B. *(Adapted from Liang et al.,[15] used with permission.)*

to eotaxin/C-C motif chemokine 11 (CCL11), including RANTES/CCL5 and MCP-3 (monocyte-specific chemokine 3).[23] Thus, while eotaxin has been identified in nerves, other chemokines, although they have not yet been discovered in nerves, may also be important for eosinophil recruitment to nerves. This is important because it appears that the presence of eotaxin alone is insufficient to recruit eosinophils to nerves. Although eotaxin is constitutively expressed in human and guinea pig airway parasympathetic nerves *in vitro*,[22] and eotaxin-3/CCL26 is expressed in airway nerves of allergen-sensitized rhesus monkeys, there is no association of eosinophils and nerves in the absence of antigen challenge.[24]

Eosinophil Recruitment to Nerves: Roles of Acetylcholine, Substance P, Leukotriene B₄, and Platelet Activating Factor

Nerves may also indirectly recruit eosinophils. The neurotransmitters acetylcholine and substance P stimulate macrophages and epithelial cells *in vitro* to secrete chemotactic factors for eosinophils.[25,26] In these studies, chromatography and antagonist experiments suggested that macrophage and epithelial cell-derived chemotactic factors were the lipid mediators leukotriene B₄ and PAF.[25,26] Another study demonstrated that the neuropeptides substance P, neurokinin A, and CGRP directly prime eosinophils to migrate toward PAF and leukotriene B₄.[16] This priming effect was blocked by inhibiting substance P signaling or CGRP receptors.[16] We speculate that neurotransmitters acetylcholine and/or substance P induce resident macrophages or epithelial cells to secrete chemotactic factors for eosinophils. A role for acetylcholine-mediated eosinophil recruitment is supported by the *in vivo* observation that, in a guinea pig model of allergic asthma, treating animals with tiotropium, an antagonist of acetylcholine M3 muscarinic receptors, prevented eosinophil influx into lungs.[27] Alternatively, like other inflammatory cells (neutrophils and macrophages), eosinophils express muscarinic receptors and it is possible that acetylcholine may directly recruit eosinophils to the nerves.[28]

Eosinophil Adhesion to Nerves: Role of Cell Adhesion Molecules

Following recruitment, eosinophils adhere to neurons using cell adhesion molecules (CAMs). Two CAM families participate in eosinophil–nerve binding: immunoglobulin superfamily CAMs (vascular cell adhesion molecule, VCAM; intercellular adhesion molecule, ICAM; and neural cell adhesion molecule, NCAM) and integrin CAMs (integrin α4β1/VLA-4 and leukocyte function-associated molecule 1; LFA-1). These two groups of adhesion molecules serve as ligands and counterligands for one another. In nerve cells, VCAM and ICAM expression augments leukocyte binding to nerves. Conversely, blocking VCAM, ICAM, LFA-1, and/or VLA-4 inhibits leukocyte binding to nerves,[29,30] showing that all of these CAMs are important for eosinophil adhesion to nerves.

CAM expression on nerves can be either constitutive or inducible. Although VCAM is constitutively expressed on parasympathetic nerves, the adhesion of eosinophils to nerves also requires ICAM expression.[30] TNF upregulates the VCAM and ICAM adhesion molecules on neuroblastoma and cortical neurons, and increases leukocyte adhesion to nerves.[29] In parasympathetic nerves isolated from trachea and maintained in cell culture, ICAM can be induced by TNF and interferon γ (IFN-γ) in a dose-related manner.[30] Dexamethasone, an anti-inflammatory corticosteroid, blocks TNF- and IFN-γ-induced ICAM expression in parasympathetic neurons from humans and guinea pigs, resulting in reduced eosinophil adhesion to parasympathetic nerves.[31] Interferon γ and TNF-induced ICAM expression was also blocked by an inhibitor of the proinflammatory transcription factor, nuclear factor κB (NFκB). Thus, not only IFN-γ and TNF but also NFκB can increase CAM expression in nerves that results in increased eosinophil adhesion to these nerves.

Chemokines and inflammatory cytokines increase eosinophil adhesion through CAMs and switch eosinophil CAM binding preference from VCAM to ICAM.[30] Eotaxin, though primarily considered to be key to eosinophil migration, also switches the predominant CAM binding in eosinophils from VCAM to ICAM.[32] Conversely, blocking the CCR3 eotaxin receptor prevented increased eosinophil binding to primary parasympathetic nerves after treatment with proinflammatory cytokines that increase ICAM expression.[22] However, not all nerves respond to inflammatory cytokines. For example, neuroblastoma cells do not recapitulate the reduction in eosinophil adhesion by dexamethasone as seen in primary parasympathetic neurons.[31]

Eosinophils are also abundant in the gut and localize to enteric nerves during inflammation. The role of ICAM and VCAM in enteric nerves is unknown, but expression of another cell adhesion molecule, NCAM, increased in the enteric nerves of rats infected by parasites.[33] Eosinophils localization also increased along the nerves (Fig. 11.2.3) in these infected animals.[33] Therefore, NCAM may also be important for eosinophil adhesion to nerves.[33]

There is some evidence that adhesion of eosinophils to nerves is also mediated by inflammation and CAMs *in vivo*. Antigen challenge of an allergen-sensitized guinea pig increases TNF in the airways[34] and increases ICAM and eotaxin expression by airway nerves.[31] Dexamethasone, etanercept, or a CCR3 eotaxin receptor antagonist blocked eosinophil association with nerves[21,22,31] by inhibiting ICAM, TNF, or eotaxin receptors, respectively, on airway nerves. These studies demonstrate eosinophil—nerve interactions are reversible at many different levels.

NERVE-MEDIATED ACTIVATION OF EOSINOPHILS

Eosinophil adhesion to ICAM and VCAM results in activation and eosinophil degranulation. This is true for eosinophil adhesion to nerves where mechanisms of eosinophil activation through nerve contact depend on CAMs. Eosinophil adhesion via ICAM and VCAM to cholinergic nerves activates eosinophils, as measured by release of eosinophil peroxidase and leukotrienes.[30,35] The requirement for physical contact between eosinophils and nerves is underscored by the observation that eosinophil activation was significantly decreased in coculture systems where direct eosinophil—nerve contact was prevented.[30] Eosinophils will degranulate in response to N-formyl-methionyl-leucyl-phenylalanine (fMLP), a secretagogue shown to cause release of eosinophil peroxidase and leukotrienes.[30] Eosinophil adhesion to nerves potentiates eosinophil degranulation in response to fMLP.[36] These studies show that neuronal adhesion, via ICAM and VCAM, directly induces eosinophil activation and also primes eosinophils for subsequent activation by other mediators (as with fMLP).

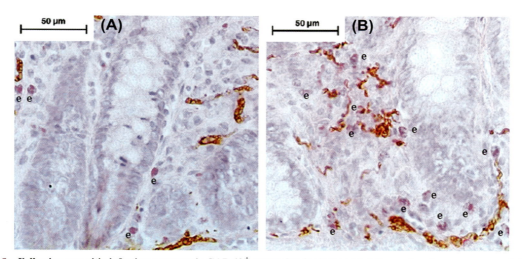

FIGURE 11.2.3 Following parasitic infection, rat enteric GAP-43⁺ nerve density is increased (brown) and nerves are associated with eosinophils (pink). *A,* Colonic tissue from healthy uninfected control rats. *B,* Colonic tissue from rats infected with the parasite, *Fasciola hepatica.* Nerves are stained for growth-associated protein-43 (GAP-43) using 3,3′-diaminobenzidine and eosinophils are labeled using chromotrope 2R. Eosinophils are labeled *e.* *(Adapted from O'Brien et al.,[33] used with permission.)*

In anesthetized guinea pigs, eosinophil activation is also inhibited by preventing CAM-mediated adhesion to nerves. In allergen-sensitized and challenged guinea pigs, administration of dexamethasone reduces ICAM expression specifically on airway nerves.[31] Dexamethasone treatment also reduces eosinophil association with airway nerves and prevents airway hyperreactivity in vivo.[37] Since airway hyperreactivity is mediated by eosinophil activation and degranulation, it is tempting to speculate that the protective effect of dexamethasone is due to decreased ICAM expression in nerves[31] and a consequent shift in eosinophils away from nerves[5] that results in reduced eosinophil activation and decreased airway hyperreactivity.[37]

Nerves can also activate eosinophils via release of neurotransmitters. Substance P releases eosinophil cationic protein and generates superoxide free radicals from isolated eosinophils in a dose-dependent manner, demonstrating that this neuropeptide induces eosinophil degranulation.[38] The C-terminal fragment of substance P mediates its stimulatory effect on eosinophils.[38] A similar role for substance P was shown in vivo, where administration of a stable substance P analog to guinea pigs led to acute airway hyperreactivity to electrical stimulation of the vagus nerves innervating the airway.[39] Substance P-mediated airway hyperreactivity was prevented by depleting inflammatory cells, thus demonstrating that hyperreactivity was not a direct effect of substance P on nerves but required inflammatory cells.[39] Furthermore, both an antibody to eosinophil major basic protein and heparin, which blocks major basic protein, prevented substance P-induced airway hyperreactivity, strongly supporting the hypothesis that substance P activates eosinophils in vivo.[39] The relevance of physiological levels of substance P to eosinophil activation and degranulation in vivo was demonstrated, since administration of a neurokinin-1 receptor antagonist prevented eosinophil-mediated airway hyperreactivity but did not change eosinophil influx into the lung or eosinophil accumulation around airway nerves.[40] Thus, neuronal substance P, in contrast to in vitro data[13] and in vivo data from skin eosinophils,[14] does not induce eosinophil recruitment to nerves, but does activate airway eosinophils in the lungs of antigen-challenged guinea pigs.

Eosinophils express a number of receptors for other neurotransmitters, including cholinoceptors, adrenoceptors, histamine receptors, and purinoceptors.[41,42] However, these receptors have not been specifically linked to nerve-induced eosinophil activation. Eosinophils from asthmatics contain increased purinoceptor mRNA and exhibit increased migration in response to adenosine 5'-triphosphate (ATP) than eosinophils from nonasthmatic patients.[41] Conversely, ATP-induced eosinophil migration is absent in eosinophils lacking purinoceptors.[41] Although nerves synthesize and release the nucleotide neurotransmitter, ATP, whether neuronal ATP is a physiological cause of eosinophil activation in vivo is unknown.[42]

Similarly, in vitro, histamine-induced eosinophil migration is mediated by histamine 4 (H4) receptors and is prevented by the H4 antagonist, thioperamide.[43] Although histamine is not a classical neurotransmitter, it is released by mast cells that are often found near to nerves.[10]

EOSINOPHIL-MEDIATED NEURAL PLASTICITY

Chronic inflammatory diseases with underlying eosinophil involvement are associated with symptoms such as excessive pain, cough, itch, enteric dysmotility, and bronchoconstriction, all of which could be caused by excessive neural activity.[10] Conversely, eosinophil degranulation is associated with ataxia and neuropathies that are associated with loss of neural activity. The impact of eosinophils on nerves has been studied by (1) measuring the physiological effects of eosinophil depletion or inhibition in vivo; (2) injecting eosinophil products into neuronal tissue compartments in vivo; (3) cataloging changes to nerves in eosinophilic diseases; and (4) coculturing eosinophil and nerves. The consequences of eosinophil—nerve interactions range from increased neuronal growth and increased activation to growth inhibition and nerve damage, all of which lead to aberrant neurotransmission.

Eosinophils Increase Neuronal Growth

In some inflammatory diseases the presence of eosinophils is associated with excessive neuronal growth. Increased eosinophil association with nerves following enteric parasitic infection coincides with an increased density of nerves expressing the growth and plasticity marker, growth-associated protein-43 (GAP-43)[33] (Fig. 11.2.3). Lesional skin from humans with prurigo nodularis contains both increased numbers of eosinophils and increased nerves compared to normal skin in the same patients[44] (Fig. 11.2.2). Histological studies of human acute appendicitis biopsies also found increased numbers of eosinophils along with a doubling of nerves and ganglion cells. In the appendicitis study, the increase in nerves was significant but may have been underestimated since nerve-specific immunolabeling was not performed and small unmyelinated and difficult-to-visualize fibers may have been missed.[45] Sensory nerves in airway tissues from patients with asthma were reported to be longer than nerves from nonasthmatic patients. This study, though intriguing, may be limited by the dependence on two-dimensional tissue sections to quantify the length of neurons, which are complex three-dimensional objects, branching and undulating through tissue. Thus, the distribution of potential neuron lengths that can be obtained in a single tissue section depends on the angle between the plane of the tissue section and plane of the nerve, as well as on the degree of nerve straightness or uniform directionality/anisotropy.[46] Since asthma is associated with eosinophil infiltration, and eosinophils can cause excessive nerve growth in cell culture,[47] eosinophil recruitment may be a mechanism for increasing nerve length. These three different diseases that affect three distinct organs (airway, gastrointestinal tract, and skin) demonstrate that there is an association between eosinophil recruitment and increased nerve density that remains to be confirmed. Whether increased nerve density is actually due to increased nerve cell numbers, branching, length, or undulation needs to be determined.

The physiological consequences of increased neural innervation have been studied in neuropathic pain. Hyperinnervation of peripheral tissues by sensory neurons increases their receptive field size and increases the likelihood of multiple receptive fields.[48] These changes result in sensitivity to otherwise subthreshold stimuli.[49] Similar changes occur in eosinophil-related diseases. This may explain how patients with eosinophilic disorders are sensitive to subthreshold mechanical and chemical stimuli that do not initiate a response in normal subjects. For example, patients with prurigo nodularis are hypersensitive to light touch,[44] and irritants that do not provoke an effect in normal lungs cause bronchoconstriction and coughing in patients with asthma.[10] Dexamethasone and steroids are used to treat all of these diseases. Although dexamethasone is an anti-inflammatory drug, it also decreases ICAM expression in nerves,[31] which would decrease eosinophil numbers around nerves. Inhibition of eosinophil and nerve interactions may be an additional mechanism for the effectiveness of dexamethasone and other steroids *in vivo*.

Eosinophils Regulate Neuropeptide Synthesis and Release

Eosinophils increase the synthesis and release of neuropeptides from nerves. Nasal biopsies from patients with seasonal allergic rhinitis, perennial allergic rhinitis, and aspirin-sensitive rhinitis contain more eosinophils and increased expression of VIP in mucosal nerves.[11,12,50] Nasal nerves from patients with seasonal allergic rhinitis also exhibit increased expression of substance P and neuropeptide Y.[12] Similarly, airway neurons in sensitized guinea pigs synthesize substance P *de novo* following antigen challenge.[10] *In vitro*, eosinophils isolated from patients with allergic rhinitis and/or asthma increase neuronal release of substance P when cocultured with sensory neurons from rats.[51] In addition, administration of eosinophil major basic protein increases substance P release from cultured sensory neurons.[51] Increased

substance P release is specific to eosinophil major basic protein, as the other eosinophil granule proteins, eosinophil cationic protein and eosinophil-derived neurotoxin, as well as the eosinophil products, leukotriene D_4, PAF, and hydrogen peroxide, do not cause substance P release.[51] Thus, eosinophils increase the synthesis and release of neuropeptides from nerves. This sets the stage for a positive feedback loop, with eosinophils increasing neuropeptide expression on nerves and neuropeptides subsequently recruiting additional eosinophils to the nerves.

Increased substance P is linked to an altered neuronal phenotype that may underlie the changed responses to stimuli characteristic of eosinophilic diseases, for example dermatitis-related itchiness in response to touch. Increased substance P expression by position sensing/proprioceptive neurons and mechanosensitive sensory neurons corresponds with a phenotypic change to a pain-sensing/nociceptive neuronal phenotype.[48] For example, adjuvant-induced inflammation causes large myelinated sensory neurons called A fibers to begin the *de novo* synthesis and release of substance P. The synthesis of substance P by A fibers coincides with a phenomenon called allodynia, in which A fibers produce increased sensations of pain due to otherwise innocuous mechanical stimuli.[48] Sensory A fibers responsible for perceiving mechanical stimuli in the lungs (e.g., stretch) are thought to undergo a similar phenotype switch in response to eosinophilic inflammation, and begin to synthesize substance P *de novo* following allergen sensitization.[10] This phenotypic change is believed to be one cause of the shortness of breath and inappropriate urge to cough seen in asthma patients.[9]

Eosinophil Adherence to Neurons Leads to Disinhibition or Activation of Neurons by Eosinophil Granule Proteins

Eosinophils are selectively recruited to airway nerves in asthma and in antigen-challenged guinea pigs.[5] These eosinophils are activated by major basic protein deposited along the nerves in the lungs of asthmatic humans and antigen-challenged guinea pigs.[5,52] Eosinophils are involved in airway hyperreactivity, since treatments that prevent eosinophil influx into lungs[53,54] or shift them away from nerves[21,22,37] prevent airway hyperreactivity in antigen-challenged guinea pigs. Similarly, a recent study of severe asthmatics shows that eosinophil depletion also prevents asthma exacerbations and reduces the need for steroids.[55,56]

Eosinophils cause airway hyperreactivity in antigen-sensitized and challenged guinea pigs by disrupting a cholinoceptor responsible for limiting acetylcholine release from parasympathetic neurons. Parasympathetic neurons induce airway smooth muscle contraction by releasing acetylcholine, which stimulates muscarinic receptors on airway smooth muscle. Acetylcholine also stimulates M2 muscarinic receptors located prejunctionally on airway parasympathetic neurons, which limits further release of acetylcholine and prevents excessive broncho-constriction.[57] Eosinophil major basic protein is an endogenous allosteric antagonist of M2 muscarinic receptors.[58] Release of major basic protein from eosinophils blocks M2 muscarinic receptors, resulting in disinhibition of acetylcholine release from parasympathetic neurons, thus leading to increased acetylcholine release and increased bronchoconstriction.

More recently, eosinophil granule proteins were shown to activate sensory neurons. Instillation of eosinophil cationic protein or eosinophil major basic protein into rat trachea *in vivo* increased baseline activity of unmyelinated airway sensory nerve fibers.[59] *In vitro*, administration of major basic protein to cultured sensory neurons also increased capsaicin-evoked, acid-evoked, and ATP-evoked action potentials.[60,61] In follow-up experiments, eosinophil major basic protein was shown to increase the decay and recovery times of action potentials via potassium channels.[61] This increased excitability of sensory neurons via eosinophil major basic protein or eosinophil cationic protein is a potential mechanism for sensory hypersensitivity associated with eosinophilic diseases.

Adhesion of eosinophils to nerves through ICAM and VCAM also activates nerves. Following whole eosinophil or eosinophil membrane adhesion to isolated nerves in culture, neurons activate the NFκB proinflammatory transcription factor.[62] This stimulatory effect is dependent on ICAM activation and involves intracellular kinase signaling and reactive oxygen species.[62] Eosinophil adhesion via the other major nerve CAM, VCAM, activates a different transcription factor, AP-1, which is also associated with neuroplasticity and inflammatory responses.[63] These data show that eosinophil adhesion to nerves via ICAM and VCAM is sufficient to change nerve function even in the absence of eosinophil degranulation.

Eosinophils Inhibit Nerve Growth and Cause Nerve Cell Damage

In some diseases and experimental systems, eosinophils inhibit neuronal growth and cause neuropathic damage. Case-control studies of demyelinating diseases (e.g., eosinophilic myelitis and neuromyelitis optica) demonstrate increased eosinophil numbers, as well as increased eosinophil cationic protein, eosinophil-derived neurotoxin, eotaxin, and IL-5 in cerebrospinal fluid. In eosinophilic myelitis, demyelination is associated with eosinophil cationic protein in the spinal cord, indicating the presence of activated eosinophils in the central nervous system.[64] Patients with another demyelinating disease,

neuromyelitis optica, also have increased eosinophil cationic protein as well as increased IL-5 and eotaxin in the cerebrospinal fluid and in active spinal cord lesions, which has been demonstrated histologically.[65,66] Eosinophil cationic protein levels in cerebrospinal fluid are also significantly elevated in cerebrovascular disease, acute central nervous system infections, and brain tumor malignancies.[67] The presence of eosinophil products in the cerebrospinal fluid is important because neurotoxic effects have been observed upon administration of eosinophils, or their granule proteins, to peripheral or to central neurons. Injection of eosinophils, eosinophil cationic protein, or eosinophil-derived neurotoxin into the cerebral cortex or cerebrospinal fluid of guinea pigs or rabbits has been shown to cause axonal damage, vacuolization, and destruction of central and peripheral neurons.[68] In a mouse model of multiple sclerosis, eosinophil influx into the optic nerve is one of the earliest events following induction of this auto-immune disease and preceded nerve demyelination and the development of motor paralysis.[4] Conversely, egress of eosinophils from the optic nerve precedes remyelination and disease remission. Eosinophils may also be more active in disease, since eosinophils from a patient with hyper-eosinophilia and peripheral neuropathy killed more peripheral nerves in coculture than did eosinophils from healthy subjects.[69] Thus, eosinophils have the capacity to

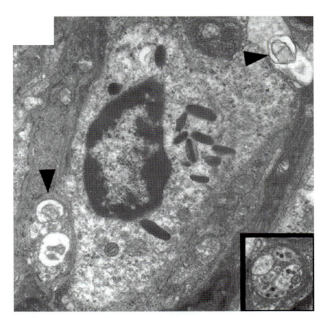

FIGURE 11.2.4 Eosinophils are adjacent to vacuolated and necrotic gut neurons in electron micrographs of the jejunum. The cell comprising almost the entire figure is an eosinophil, which contains characteristic electron dense granule proteins and is located in the reticular layer of the lamina propria (not shown). Arrowheads indicate enteric nerves, which are swollen and necrotic. Inset: a healthy enteric nerve bundle in a control mouse. Magnification ×16,900. *(Adapted from Hogan et al.,[71] used with permission.)*

damage and kill nerves in the central nervous system and are associated with demyelinating diseases.

Eosinophils may also mediate pathological and cytotoxic changes in the peripheral nervous system. Coculturing eosinophils with a cholinergic nerve cell line inhibited neurite outgrowth without increasing apoptosis.[70] *In vivo*, in a murine model of enteric eosinophilic inflammation, eosinophils were seen next to damaged axons, suggesting a neuropathic interaction with myenteric nerves[71] (Fig. 11.2.4). Eosinophils are found next to nerves in the gut; thus, if eosinophils inhibit peripheral neurons, the resulting disruption of myenteric control might cause gastric dysmotility characteristic of inflammatory bowel disease.

Eosinophils are therefore capable of causing trophic or pathological changes to nerves that increase nerve outgrowth, increase nerve activity, change neuropeptide content, and kill neurons (Fig. 11.2.5). Thus, it is odd that injured[6] and inflamed[4] nerves selectively recruit eosinophils, unless there are unrecognized beneficial interactions between eosinophils and nerves. The surrounding environment appears to influence the impact of eosinophils on nerves. For example, eosinophils mediate virus-induced airway hyperreactivity only in allergen-sensitized guinea pigs and not in nonsensitized guinea pigs.[72] Depletion of eosinophils is not protective in virus-infected guinea pigs unless the guinea pigs are also sensitized to a protein. Thus, the role of eosinophils is dependent upon atopic status.

Timing also determines the impact of eosinophils on nerves. In guinea pigs, airway hyperreactivity 1 day after ozone exposure is mediated by eosinophil recruitment to parasympathetic nerves (and the subsequent blockade of nerve M2 muscarinic receptors).[73] In contrast, 3 days after ozone exposure the role of eosinophils is changed and they become protective against ozone-induced airway hyperreactivity. The concentration and tissue distribution of eosinophil products also mediate their effect on nerves. For example, a low concentration of eosinophil major basic protein, as may be observed *in vivo*, does not kill isolated nerves in culture,[74] in contrast to the neurotoxicity caused by directly injecting major basic protein into cerebrospinal fluid.[74] Thus, the effect of eosinophils on nerves is dependent on atopic status, the time after a specific challenge, and the dosage and tissue distribution of eosinophils and their products.

EOSINOPHIL—NERVE INTERACTION AS A THERAPEUTIC TARGET

Recent data from animal models of allergic airway disease and human asthma show that inhibiting eosinophil—nerve interactions represents a potentially useful therapeutic target. Treatment of guinea pig asthma models with etanercept, a TNF blocker, or with dexamethasone, a corticosteroid, inhibited eosinophil localization next to nerves and prevented airway hyperreactivity without changing the

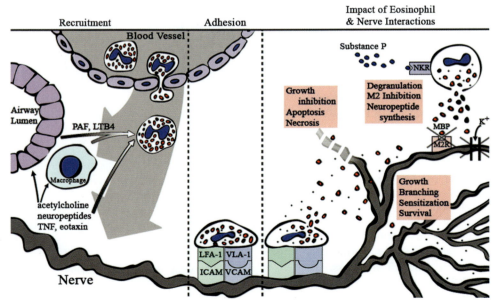

FIGURE 11.2.5   Summary schematic of eosinophil—nerve mechanisms. From left to right: eosinophil recruitment by nerves, adhesion of eosinophils to nerves, and the trophic and deleterious effects of eosinophils on nerves. Both direct and indirect mechanisms of eosinophil are shown. ICAM, intercellular adhesion molecule; LFA-1, leukocyte function-associated molecule 1; LTB4, leukotriene B4; M2R, M2 muscarinic receptor; MBP, major basic protein; NKR, neuromedin-K receptor; PAF, platelet activating factor; TNF, tumor necrosis factor; VLA-1, integrin α4β1; VCAM, vascular cell adhesion molecule.

overall quantity of eosinophils in the lungs.[21,37] Importantly, these medications are approved for human use and are efficacious in humans with asthma. In human asthma, dexamethasone reduces symptoms and reduces the relapse rate. Similarly, etanercept reduces hyperresponsiveness and exacerbations in severe steroid-dependent asthma.[21]

Newer muscarinic and neurokinin receptor antagonists show promise in recent animal and human trials.[7,75] Although older studies of nonselective anticholinergic drugs and selective neurokinin receptor antagonists concluded that these compounds were ineffective,[76] tiotropium, an anticholinergic agent selective for M3, or CS-003, a combined antagonist of NK-1, NK-2, and NK-3 receptors, were recently shown to reduce bronchoconstriction and airway hyperreactivity in humans with asthma.[7,75] It is tempting to speculate that these medications act in part by disrupting eosinophil–nerve interactions. A better understanding of eosinophil–nerve interactions within the gut, skin, eye, upper airway, lung, and other organs will not only extend our understanding of chronic inflammatory diseases but will also elucidate therapeutically useful signaling pathways.

CONCLUSION

In conclusion, nerves actively recruit eosinophils via the release of chemotactic factors and expression of adhesion molecules. Pathological nerve and eosinophil interactions are associated with deleterious effects that are dependent on atopic status, time, and dose/distribution of eosinophil products. An intriguing hypothesis is that there is also a physiological interaction between nerves and eosinophils that is as yet unappreciated.

REFERENCES

1. Rothenberg ME, Hogan SP. The eosinophil. *Annu Rev Immunol* 2006;**24**:147–74.
2. Leiferman KM, Gleich GJ, Peters MS. Dermatologic manifestations of the hypereosinophilic syndromes. *Immunol Allergy Clin North Am* 2007;**27**:415–41.
3. Balogh G, Dimitrov-Szokodi D, Husveti A. Lung denervation in the therapy of intractable bronchial asthma. *J Thorac Surg* 1957;**33**:166–84.
4. Milici AJ, Carroll LA, Stukenbrok HA, Shay AK, Gladue RP, Showell HJ. Early eosinophil infiltration into the optic nerve of mice with experimental allergic encephalomyelitis. *Lab Invest* 1998;**78**:1239–44.
5. Costello RW, Schofield BH, Kephart GM, Gleich GJ, Jacoby DB, Fryer AD. Localization of eosinophils to airway nerves and effect on neuronal M2 muscarinic receptor function. *Am J Physiol* 1997;**273**:L93–103.
6. Beuche W. Differential isolation of eosinophils and myelin phagocytes from mouse peripheral nerves during Wallerian degeneration

by uncoated and immunoglobulin-coated sheep red blood cells. *Brain Res* 1991;**558**:101–4.
7. Veres TZ, Rochlitzer S, Braun A. The role of neuro-immune crosstalk in the regulation of inflammation and remodelling in asthma. *Pharmacol Ther* 2009;**122**:203–14.
8. Undem BJ, Carr MJ. The role of nerves in asthma. *Curr Allergy Asthma Rep* 2002;**2**:159–65.
9. Barnes PJ. Neurogenic inflammation in the airways. *Respir Physiol* 2001;**125**:145–54.
10. Carr MJ, Undem BJ. Inflammation-induced plasticity of the afferent innervation of the airways. *Environ Health Perspect* 2001;**109**(Suppl. 4):567–71.
11. Fischer A, Wussow A, Cryer A, Schmeck B, Noga O, Zweng M, et al. Neuronal plasticity in persistent perennial allergic rhinitis. *J Occup Environ Med* 2005;**47**:20–5.
12. Heppt W, Dinh QT, Cryer A, Zweng M, Noga O, Peiser C, et al. Phenotypic alteration of neuropeptide-containing nerve fibres in seasonal intermittent allergic rhinitis. *Clin Exp Allergy* 2004;**34**:1105–10.
13. Dunzendorfer S, Meierhofer C, Wiedermann CJ. Signaling in neuropeptide-induced migration of human eosinophils. *J Leukoc Biol* 1998;**64**:828–34.
14. Smith CH, Barker JN, Morris RW, MacDonald DM, Lee TH. Neuropeptides induce rapid expression of endothelial cell adhesion molecules and elicit granulocytic infiltration in human skin. *J Immunol* 1993;**151**:3274–82.
15. Liang Y, Jacobi HH, Reimert CM, Haak-Frendscho M, Marcusson JA, Johansson O. CGRP-immunoreactive nerves in prurigo nodularis—an exploration of neurogenic inflammation. *J Cutan Pathol* 2000;**27**:359–66.
16. Numao T, Agrawal DK. Neuropeptides modulate human eosinophil chemotaxis. *J Immunol* 1992;**149**:3309–15.
17. Jonsson M, Norrgard O, Forsgren S. Substance P and the neurokinin-1 receptor in relation to eosinophilia in ulcerative colitis. *Peptides* 2005;**26**:799–814.
18. Mantelli F, Micera A, Sacchetti M, Bonini S. Neurogenic inflammation of the ocular surface. *Curr Opin Allergy Clin Immunol* 2010;**10**:498–504.
19. Collins JJ, Usip S, McCarson KE, Papka RE. Sensory nerves and neuropeptides in uterine cervical ripening. *Peptides* 2002;**23**:167–83.
20. Knudsen UB. Cervical ripening. A rat model for investigation of contractile and passive biomechanical properties, with focus on antigestagens, eosinophil granulocytes and mast cells. *Acta Obstet Gynecol Scand* 1996;**75**:88–9.
21. Nie Z, Jacoby DB, Fryer AD. Etanercept prevents airway hyperresponsiveness by protecting neuronal M2 muscarinic receptors in antigen-challenged guinea pigs. *Br J Pharmacol* 2009;**156**:201–10.
22. Fryer AD, Stein LH, Nie Z, Curtis DE, Evans CM, Hodgson ST, et al. Neuronal eotaxin and the effects of CCR3 antagonist on airway hyperreactivity and M2 receptor dysfunction. *J Clin Invest* 2006;**116**:228–36.
23. Erin EM, Williams TJ, Barnes PJ, Hansel TT. Eotaxin receptor (CCR3) antagonism in asthma and allergic disease. *Curr Drug Targets Inflamm Allergy* 2002;**1**:201–14.
24. Chou DL, Daugherty BL, McKenna EK, Hsu WM, Tyler NK, Plopper CG, et al. Chronic aeroallergen during infancy enhances

eotaxin-3 expression in airway epithelium and nerves. *Am J Respir Cell Mol Biol* 2005;**33**:1—8.

25. Sato E, Koyama S, Okubo Y, Kubo K, Sekiguchi M. Acetylcholine stimulates alveolar macrophages to release inflammatory cell chemotactic activity. *Am J Physiol* 1998;**274**:L970—9.

26. Koyama S, Sato E, Nomura H, Kubo K, Nagai S, Izumi T. Acetylcholine and substance P stimulate bronchial epithelial cells to release eosinophil chemotactic activity. *J Appl Physiol* 1998;**84**:1528—34.

27. Bos IS, Gosens R, Zuidhof AB, Schaafsma D, Halayko AJ, Meurs H, et al. Inhibition of allergen-induced airway remodelling by tiotropium and budesonide: a comparison. *Eur Respir J* 2007;**30**:653—61.

28. Gosens R, Zaagsma J, Meurs H, Halayko AJ. Muscarinic receptor signaling in the pathophysiology of asthma and COPD. *Respir Res* 2006;**7**:73.

29. Birdsall HH, Lane C, Ramser MN, Anderson DC. Induction of VCAM-1 and ICAM-1 on human neural cells and mechanisms of mononuclear leukocyte adherence. *J Immunol* 1992;**148**: 2717—23.

30. Sawatzky DA, Kingham PJ, Court E, Kumaravel B, Fryer AD, Jacoby DB, et al. Eosinophil adhesion to cholinergic nerves via ICAM-1 and VCAM-1 and associated eosinophil degranulation. *Am J Physiol Lung Cell Mol Physiol* 2002;**282**:L1279—88.

31. Nie Z, Nelson CS, Jacoby DB, Fryer AD. Expression and regulation of intercellular adhesion molecule-1 on airway parasympathetic nerves. *J Allergy Clin Immunol* 2007;**119**:1415—22.

32. Schleimer RP, Sterbinsky SA, Kaiser J, Bickel CA, Klunk DA, Tomioka K, et al. IL-4 induces adherence of human eosinophils and basophils but not neutrophils to endothelium. Association with expression of VCAM-1. *J Immunol* 1992;**148**:1086—92.

33. O'Brien LM, Fitzpatrick E, Baird AW, Campion DP. Eosinophil-nerve interactions and neuronal plasticity in rat gut associated lymphoid tissue (GALT) in response to enteric parasitism. *J Neuroimmunol* 2008;**197**:1—9.

34. Kelly DE, Denis M, Biggs DF. Release of tumour necrosis factor alpha into bronchial alveolar lavage fluid following antigen challenge in passively sensitized guinea-pigs. *Mediators Inflamm* 1992;**1**:425—8.

35. Kingham PJ, McLean WG, Sawatzky DA, Walsh MT, Costello RW. Adhesion-dependent interactions between eosinophils and cholinergic nerves. *Am J Physiol Lung Cell Mol Physiol* 2002;**282**:L1229—38.

36. Kingham PJ, Costello RW, McLean WG. Eosinophil and airway nerve interactions. *Pulm Pharmacol Ther* 2003;**16**:9—13.

37. Evans CM, Jacoby DB, Fryer AD. Effects of dexamethasone on antigen-induced airway eosinophilia and M^2 receptor dysfunction. *Am J Respir Crit Care Med* 2001;**163**:1484—92.

38. Iwamoto I, Nakagawa N, Yamazaki H, Kimura A, Tomioka H, Yoshida S. Mechanism for substance P-induced activation of human neutrophils and eosinophils. *Regul Pept* 1993;**46**:228—30.

39. Evans CM, Belmonte KE, Costello RW, Jacoby DB, Gleich GJ, Fryer AD. Substance P-induced airway hyperreactivity is mediated by neuronal M^2 receptor dysfunction. *Am J Physiol Lung Cell Mol Physiol* 2000;**279**:L477—86.

40. Costello RW, Fryer AD, Belmonte KE, Jacoby DB. Effects of tachykinin NK1 receptor antagonists on vagal hyperreactivity and neuronal M2 muscarinic receptor function in antigen challenged guinea-pigs. *Br J Pharmacol* 1998;**124**:267—76.

41. Muller T, Robaye B, Vieira RP, Ferrari D, Grimm M, Jakob T, et al. The purinergic receptor P2Y2 receptor mediates chemotaxis of dendritic cells and eosinophils in allergic lung inflammation. *Allergy* 2010;**65**:1545—53.

42. Burnstock G. Historical review: ATP as a neurotransmitter. *Trends Pharmacol Sci* 2006;**27**:166—76.

43. O'Reilly M, Alpert R, Jenkinson S, Gladue RP, Foo S, Trim S, et al. Identification of a histamine H4 receptor on human eosinophils—role in eosinophil chemotaxis. *J Recept Signal Transduct Res* 2002;**22**:431—48.

44. Johansson O, Liang Y, Marcusson JA, Reimert CM. Eosinophil cationic protein- and eosinophil-derived neurotoxin/eosinophil protein X-immunoreactive eosinophils in prurigo nodularis. *Arch Dermatol Res* 2000;**292**:371—8.

45. Singh US, Malhotra A, Bhatia A. Eosinophils, mast cells, nerves and ganglion cells in appendicitis. *Indian J Surg* 2008;**70**:231—4.

46. Ollerenshaw SL, Jarvis D, Sullivan CE, Woolcock AJ. Substance P immunoreactive nerves in airways from asthmatics and non-asthmatics. *Eur Respir J* 1991;**4**:673—82.

47. Kobayashi H, Gleich GJ, Butterfield JH, Kita H. Human eosinophils produce neurotrophins and secrete nerve growth factor on immunologic stimuli. *Blood* 2002;**99**:2214—20.

48. Sandkuhler J. Models and mechanisms of hyperalgesia and allodynia. *Physiol Rev* 2009;**89**:707—58.

49. Chu KL, Faltynek CR, Jarvis MF, McGaraughty S. Increased WDR spontaneous activity and receptive field size in rats following a neuropathic or inflammatory injury: implications for mechanical sensitivity. *Neurosci Lett* 2004;**372**:123—6.

50. Groneberg DA, Heppt W, Welker P, Peiser C, Dinh QT, Cryer A, et al. Aspirin-sensitive rhinitis-associated changes in upper airway innervation. *Eur Respir J* 2003;**22**:986—91.

51. Garland A, Necheles J, White SR, Neeley SP, Leff AR, Carson SS, et al. Activated eosinophils elicit substance P release from cultured dorsal root ganglion neurons. *Am J Physiol* 1997;**273**: L1096—102.

52. Evans CM, Fryer AD, Jacoby DB, Gleich GJ, Costello RW. Pretreatment with antibody to eosinophil major basic protein prevents hyperresponsiveness by protecting neuronal M2 muscarinic receptors in antigen-challenged guinea pigs. *J Clin Invest* 1997;**100**:2254—62.

53. Elbon CL, Jacoby DB, Fryer AD. Pretreatment with an antibody to interleukin-5 prevents loss of pulmonary M2 muscarinic receptor function in antigen-challenged guinea pigs. *Am J Respir Cell Mol Biol* 1995;**12**:320—8.

54. Fryer AD, Costello RW, Yost BL, Lobb RR, Tedder TF, Steeber DA, et al. Antibody to VLA-4, but not to L-selectin, protects neuronal M2 muscarinic receptors in antigen-challenged guinea pig airways. *J Clin Invest* 1997;**99**:2036—44.

55. Nair P, Pizzichini MM, Kjarsgaard M, Inman MD, Efthimiadis A, Pizzichini E, et al. Mepolizumab for prednisone-dependent asthma with sputum eosinophilia. *N Engl J Med* 2009;**360**:985—93.

56. Haldar P, Brightling CE, Hargadon B, Gupta S, Monteiro W, Sousa A, et al. Mepolizumab and exacerbations of refractory eosinophilic asthma. *N Engl J Med* 2009;**360**:973—84.

57. Fryer AD, Maclagan J. Muscarinic inhibitory receptors in pulmonary parasympathetic nerves in the guinea-pig. *Br J Pharmacol* 1984;**83**:973—8.

58. Jacoby DB, Gleich GJ, Fryer AD. Human eosinophil major basic protein is an endogenous allosteric antagonist at the inhibitory muscarinic M2 receptor. *J Clin Invest* 1993;**91**:1314—8.

59. Lee LY, Gu Q, Gleich GJ. Effects of human eosinophil granule-derived cationic proteins on C-fiber afferents in the rat lung. *J Appl Physiol* 2001;**91**:1318—26.

60. Gu Q, Wiggers ME, Gleich GJ, Lee LY. Sensitization of isolated rat vagal pulmonary sensory neurons by eosinophil-derived cationic proteins. *Am J Physiol Lung Cell Mol Physiol* 2008;**294**:L544—52.

61. Gu Q, Lim ME, Gleich GJ, Lee LY. Mechanisms of eosinophil major basic protein-induced hyperexcitability of vagal pulmonary chemosensitive neurons. *Am J Physiol Lung Cell Mol Physiol* 2009;**296**:L453—61.

62. Curran DR, Morgan RK, Kingham PJ, Durcan N, McLean WG, Walsh MT, et al. Mechanism of eosinophil induced signaling in cholinergic IMR-32 cells. *Am J Physiol Lung Cell Mol Physiol* 2005;**288**:L326—32.

63. Walsh MT, Curran DR, Kingham PJ, Morgan RK, Durcan N, Gleich GJ, et al. Effect of eosinophil adhesion on intracellular signaling in cholinergic nerve cells. *Am J Respir Cell Mol Biol* 2004;**30**:333—41.

64. Osoegawa M, Ochi H, Kikuchi H, Shirabe S, Nagashima T, Tsumoto T, et al. Eosinophilic myelitis associated with atopic diathesis: a combined neuroimaging and histopathological study. *Acta Neuropathol* 2003;**105**:289—95.

65. Lucchinetti CF, Mandler RN, McGavern D, Bruck W, Gleich G, Ransohoff RM, et al. A role for humoral mechanisms in the pathogenesis of Devic's neuromyelitis optica. *Brain* 2002;**125**:1450—61.

66. Correale J, Fiol M. Activation of humoral immunity and eosinophils in neuromyelitis optica. *Neurology* 2004;**63**:2363—70.

67. Hallgren R, Terent A, Venge P. Eosinophil cationic protein (ECP) in the cerebrospinal fluid. *J Neurol Sci* 1983;**58**:57—71.

68. Newton DL, Walbridge S, Mikulski SM, Ardelt W, Shogen K, Ackerman SJ, et al. Toxicity of an antitumor ribonuclease to Purkinje neurons. *J Neurosci* 1994;**14**:538—44.

69. Sunohara N, Furukawa S, Nishio T, Mukoyama M, Satoyoshi E. Neurotoxicity of human eosinophils towards peripheral nerves. *J Neurol Sci* 1989;**92**:1—7.

70. Kingham PJ, McLean WG, Walsh MT, Fryer AD, Gleich GJ, Costello RW. Effects of eosinophils on nerve cell morphology and development: the role of reactive oxygen species and p38 MAP kinase. *Am J Physiol Lung Cell Mol Physiol* 2003;**285**: L915—24.

71. Hogan SP, Mishra A, Brandt EB, Royalty MP, Pope SM, Zimmermann N, et al. A pathological function for eotaxin and eosinophils in eosinophilic gastrointestinal inflammation. *Nat Immunol* 2001;**2**:353—60.

72. Adamko DJ, Yost BL, Gleich GJ, Fryer AD, Jacoby DB. Ovalbumin sensitization changes the inflammatory response to subsequent parainfluenza infection. Eosinophils mediate airway hyperresponsiveness, m^2 muscarinic receptor dysfunction, and antiviral effects. *J Exp Med* 1999;**190**:1465—78.

73. Yost BL, Gleich GJ, Jacoby DB, Fryer AD. The changing role of eosinophils in long-term hyperreactivity following a single ozone exposure. *Am J Physiol Lung Cell Mol Physiol* 2005;**289**: L627—35.

74. Lee LY, Gu Q. Role of TRPV1 in inflammation-induced airway hypersensitivity. *Curr Opin Pharmacol* 2009;**9**:243—9.

75. Peters SP, Kunselman SJ, Icitovic N, Moore WC, Pascual R, Ameredes BT, et al. Tiotropium bromide step-up therapy for adults with uncontrolled asthma. *N Engl J Med* 2010;**363**:1715—26.

76. Westby M, Benson M, Gibson P. Anticholinergic agents for chronic asthma in adults. *Cochrane Database Syst Rev* 2004;**3**. CD003269.

Chapter 11.3

Eosinophils as Regulators of Gastrointestinal Physiological Homeostasis

Shauna Schroeder, Joanne C. Masterson, Sophie Fillon and Glenn T. Furuta

Time destroys the speculation of men, but it confirms nature.

Marcus Tullius Cicero

INTRODUCTION

The normal gastrointestinal (GI) mucosa contains varying numbers of eosinophils; this observation raises a number of interesting questions regarding their roles in intestinal health. Does the anatomical juxtaposition of eosinophils to the epithelia suggest a relationship that is protective to the mucosal barrier? Are eosinophils effector cells that are activated by other cells resident in the intestinal tract, such as fibroblasts, neurons, and others? Can eosinophils communicate with resident cells or recruit necessary immune cells to mucosal sites as in other organ systems? Since few studies have directly addressed these issues in the GI tract, extensions from other organ systems provide a rationale for considering these roles for eosinophils. In this chapter, we will begin to address these speculative roles of eosinophils in GI physiology.

Before going further, it is worth mentioning a few of the difficulties involved in studying eosinophils in the GI tract. Firstly, few animal models exist that replicate eosinophilic inflammation of the GI tract. These are often limited to sensitization and challenge models and to genetically engineered mice with poorly characterized GI eosinophilic phenotypes. Second, human studies are limited by difficulties in accessing tissues for analysis; archaic analytical techniques for assessing eosinophil inflammation; and a lack of standardized methodology to assess eosinophil activation *in vivo*. The GI tract is not readily accessible; for instance, to obtain samples for analysis, endoscopy and mucosal biopsy in humans and terminal examination in animal studies are required. Once samples are obtained, most studies manually count eosinophils following eosin or granule protein staining. The eosinophil activation state is difficult to determine and is usually based on measurements of eosinophil granule protein levels.[1] Novel insights into the role(s) of eosinophils in GI health and disease will require technical advancements in these areas.

To date, few studies have characterized numbers of eosinophils in the healthy GI tract. Studies to date have focused primarily on counting eosinophils in hematoxylin and eosin stained intestinal tissues. One study enumerated eosinophils in the intestinal tract mucosa from otherwise healthy infants and children who died suddenly without evidence of GI disease. Children ranged in age from 3 weeks to 17 years and demonstrated a gradient in eosinophil numbers ranging from none in the esophagus to a maximum of 50 per high power field (hpf) in the cecum. Another pediatric study showed virtually the same findings in mucosal biopsy samples.[2,3] One murine study identified a similar pattern of eosinophil density in the GI tract, with the stomach, cecum, and colon containing 36, 69, and 39 cells/mm,[2] respectively.[4] Thus, few studies have fully characterized normal GI eosinophil phenotypes.

ESOPHAGUS

Anatomy and Function

The esophagus is a muscular tube approximately 18−26 cm in length that serves to transfer food from the mouth to the stomach. Lining the mucosa of the esophagus is a non-keratinized stratified squamous epithelium that provides barrier function, antimicrobial protection, and structural integrity for the esophagus.[5] The submucosa is comprised of a dense network of connective tissue, blood vessels, nerves, and lymphatic channels, as well as glandular structures that secrete mucus, bicarbonate, and important factors for epithelial defense and repair, such as epidermal growth factor.[5] The proximal portion of the muscularis propria consists of striated muscle, whereas the distal two-thirds is composed of smooth muscle. The upper third of the esophagus is encased by a thick submucosal network of elastic and collagen called the serosa. Defining the proximal and distal ends of the esophagus are the upper and lower esophageal sphincters. These sphincters have high resting pressures that protect the lumen from oral and gastric contents and, once relaxed, allow the food bolus to pass from the upper pharynx to the stomach. Mast cells and lymphocytes are present in the normal esophageal mucosa.

Esophageal Eosinopenia

Two studies, and a wealth of clinical experience, have demonstrated that eosinophils are not present in healthy esophageal mucosa.[2,3] Despite the daily exposure of the epithelia to a number of stimuli often associated with accumulation of eosinophils, such as food and microbes, eosinophils are absent. Perhaps, as with the skin, the stratified squamous epithelium provides a sufficient barrier protection from the external environment. Other factors that may limit the exposure of the mucosa to these luminal products include antimicrobial peptides, bicarbonate, mucus, and peristalsis.

Therefore, the presence of eosinophils in the esophageal mucosa indicates a pathological state. To date, studies suggest that eosinophil numbers increase in the esophagus as a result of inflammatory stimuli. During disease states, including gastroesophageal reflux disease (acid) and eosinophilic esophagitis (allergy), eosinophil density increases and treatments that are not eosinophil-directed, such as proton pump inhibitors and antiallergic modalities, reduce eosinophilia, leading to symptom resolution.

STOMACH

Anatomy and Function

The stomach functions to initiate the digestive process and to deliver ingested nutrients via a rhythmic motion to the small intestine. Anatomically, it is divided into three regions, the cardia, corpus, and antrum, each with distinctive structures that promote specific functions. The cardia is located distal to the esophageal Z line, where the squamous epithelium of the esophagus gives way to the columnar epithelium of the stomach. The function of the cardia is to secrete bicarbonate and mucus, thus acting as an interface between the esophagus and stomach. The body or corpus of the stomach is grossly composed of rugae or folds and microscopically contains millions of glandular structures. The corpus glands include at least four cell types that secrete mucus (surface mucous cells), acid (parietal cells), serotonin (endocrine cells), and pepsin (chief cells). The antrum of the stomach ends in the pyloric os, which exits into the small intestine. This portion of the stomach contains neuroendocrine G cells that secrete the hormone gastrin, a protein that regulates acid production.[5]

A peak count of eight eosinophils per hpf in the antral gastric mucosa and 11 per hpf in fundic gastric mucosa was measured in children without GI disease.[2] A recent article by Genta counted gastric eosinophils and concluded that in the United States population normal gastric eosinophil counts are typically less than 38 eosinophils/mm.[2] The diagnostic histological pattern of eosinophilic gastritis showed an average density of >127 eosinophils/mm[2] or >30 eosinophils per hpf in at least five hpfs.[6]

In disease states, eosinophilic inflammation is characterized by increased numbers of eosinophils in the lamina propria; infiltrating eosinophils in the surface or glandular epithelium, submucosa, or muscularis; and other general features of inflammation, such as regenerative epithelia, and gland architectural changes, such as gland distortion and dropout.[7]

POTENTIAL ROLES OF EOSINOPHILS IN THE STOMACH

Host Microbial Defense

Infections of the stomach are rare. Due to the close anatomical juxtaposition of the eosinophil to the epithelial surface, and since eosinophils contain a number of antimicrobial proteins, eosinophils could potentially participate in an innate protective role against microbial infection. For instance, inflammatory patterns associated with *Helicobacter pylori* infections include the accumulation of neutrophils, macrophages, lymphocytes, and eosinophils, thus providing some circumstantial evidence of a role for eosinophils in this infection.[8,9]

Acid Production

A number of studies have shown that eosinophils can stimulate mast cell histamine secretion. Thus, eosinophil stimulation of mast cell-derived histamine could lead to histamine type 2 receptor activation on the basolateral surface of parietal cells, leading to proton pump secretion of acid.

Motility

Neurotrophins are a family of homologous growth factors that can mediate inflammatory signals among neurons, immune, and structural cells. They control the survival, differentiation, and maintenance of neurons in the peripheral and central nervous systems. Examples of these proteins include nerve growth factor (NGF), brain-derived neurotrophic factor (BDNF), and neutrophins 3 and 4.[10] Allergic inflammatory diseases, such as asthma and atopic dermatitis, demonstrate upregulated expression of these circulating neurotrophins on the mucosal surface.

Some of the first observations that neurotrophins may be involved in the pathogenesis of allergic diseases came from studies of vernal keratoconjuctivitis,[11] asthma, and atopic dermatitis, in which increased NGF levels were found in peripheral blood samples from affected patients.[12,13] Eosinophils can express cell surface receptors for all neurotrophins, as well as secreting NGF.[14,15] Increased levels of circulating NGF results in airway hyperresponsiveness in asthma by activating eosinophils and modifying cholinergic nerves.[12] Eosinophil-derived NGF may directly affect the nerve cholinergic phenotype, contributing to acetylcholine release, nerve hyperreactivity, and nerve remodeling.

Acetylcholine is a neurotransmitter that can exert its influence on the gastric mucosa, resulting in changes in muscle contraction, migrating motor complexes, and the secretion of other neurotransmitters and hormones that modify gastric motility. Other human and rodent studies have identified numerous other mediators derived from the eosinophil—mast cell axis with the capacity to regulate neuronal function, including vasoactive intestinal peptide, substance P, serotonin, histamine, and leukotrienes.[16—18] On the basis of these findings, one could speculate that the baseline presence of eosinophils in the gastric mucosa may play a regulatory physiological role in maintaining the neural tone required for gastric motility. These same paradigms could extend to functions in the small and large intestines.

SMALL AND LARGE INTESTINES

Anatomy and Function

The small intestine is composed of three organs: the duodenum, jejunum, and ileum. Each of these organs possesses different anatomical and functional features that facilitate nutrient digestion and absorption. The unique villus and microvillus surface of the small intestine provides a substantial surface area (approximately the size of a tennis court) to accomplish its tasks. In contrast, the large intestine (colon) serves to reabsorb water; it is presented with 10 liters of liquid and excretes approximately 200 mL of fecal matter on a daily basis. The colon is divided into several regions: cecum, ascending colon, transverse colon, descending colon, and sigmoid colon, each of which provides different functions. For example, the main roles of the ascending and transverse colons are reabsorption of digestive fluids and salvage of other dietary by-products. These products include short chain fatty acids produced by bacterial fermentation of carbohydrates, dietary fibers, and bile acids. While some absorption occurs in the descending and sigmoid colons, these portions act primarily as a storage reservoir for fecal waste.[19]

Epithelial Function

The epithelium is a cellular barrier between the external environment and the inside of the body. In the small and large intestines this single cellular layer (8—11 μm) serves not only as a physical barrier, but also possesses a number of innate mechanisms of defense including mucus, defensins, trefoil factor, and immunoglobulin A (IgA). In fact, the GI tract synthesizes and secretes the largest quantity of IgA in the body. The intestinal epithelium is designed to digest and absorb water and nutrients and to act as a barrier at the interface of the external environment and the rest of the body. Epithelial cells express several specialized properties that control absorption of luminal contents and secretion of water and electrolytes, including intercellular tight junctions, ion transport pumps (chloride channels), humoral influences (histamine and vasoactive intestinal peptide), and specific surface transport molecules (GLUT-2 and SGLT-1). Disturbances in the molecular regulation of small epithelial functions can lead to maldigestion or malabsorption, resulting in diarrhea and malnutrition. Colonic epithelial cells are actively involved in the absorption of fluid and other products but do not express absorptive transporters for conventional nutrients. Dysregulation of colonic epithelia, as in inflammatory bowel diseases, can lead to activation of proinflammatory pathways with the production of cytokines such as TNF-α.

POTENTIAL ROLES FOR EOSINOPHILS IN THE SMALL INTESTINE AND COLON

Interactions with Microbiota

Despite the daily exposure to a wide variety of microbes and microbial products, infections of the intestine and colon do not occur regularly. In fact, the small intestine and colonic epithelia interact with several kilograms of bacteria and over 500 different bacterial species. The exact quantity of microbes and their products varies throughout the length of the GI tract; 10^4/mL bacteria reside in the proximal small intestine and these numbers increase exponentially to reach a level of 10^{12}/mL in the distal small intestine and colon.[20,21] The reasons for the spatial organization of the intestinal microbiota may include differing oxygen concentrations, varying nutrients, intraluminal pH, mucus viscosity, and innate and acquired immune constituents (e.g., defensins, immunoglobulins, phosphatidylcholine, and trefoil peptides).[22] The concordant increase in mucosal eosinophils with increasing concentrations of luminal bacteria is important to this microbiological topography.

The microbiome is established in each host by 1 month of age and is influenced by the mode of delivery, mode of feeding (breast vs. formula) and use of antibiotics and probiotics. In health, these *biofilms*, or communities of bacteria, can associate with intestinal mucosa and may prevent pathogenic microorganism invasion, promote development and maturation of the mucosal immune system, and provide metabolic services (i.e., fermentation and deconjugation of bile acids). Homeostatic maintenance of intestinal barrier function against luminal bacteria is a key function of the GI innate immune system; in this light, some evidence allows for speculation on a role for eosinophils in this process. Eosinophils are anatomically juxtaposed to intestinal epithelia that line the luminal and crypt surfaces. The intestinal crypt is a sterile invagination composed of goblet, stem, and Paneth cells, which can secrete mucus and antimicrobial peptides such as cryptdins.

Could eosinophils signal to epithelial cells and help maintain the epithelial barrier? Eosinophils contain secretory granules, including major basic protein (MBP),

eosinophil peroxidase (EPO), eosinophil cationic protein (ECP), and eosinophil-derived neurotoxin (EDN), that could participate in developing a microbial shield.[23-27] Yousefi *et al.* demonstrated that eosinophils possess the ability to synthesize and release extracellular bactericidal traps, i.e., structures containing DNA in association with eosinophil granule proteins that are protective in murine models of bacterial infection.[28] Hu *et al.* described similar features in other inflammatory and infectious skin conditions.[29] Isolated mouse eosinophils possess properties that are directed against *Pseudomonas* species and that are effective *in vitro* and *in vivo*.[30] Bactericidal permeability-increasing protein (BPI) is an inducible antimicrobial polypeptide expressed in human neutrophils, eosinophils, and epithelial cells. BPI has a high affinity for lipopolysaccharides, direct antimicrobial activity, and the ability to neutralize the endotoxin activity of gram-negative bacteria, and is expressed on the basal surface of the esophageal epithelia.[31]

Role of Eosinophils in the Production of Mucus

Intestinal mucus barriers are the first line of defense against a wide range of potentially damaging agents of microbial, endogenous, and dietary origins that the intestinal epithelia may encounter. The secreted mucus barrier provides both a physical and chemical barrier to microbes, as well as keeping the mucosal surface well hydrated and providing lubrication that allows the continuous flow of luminal contents. Functional components of secreted mucus are polymeric glycoproteins termed mucins, for which there are approximately 20 different genes. The mucus barrier is over 800 μm thick and effectively serves as a barrier to separate the luminal bacterial content from the underlying epithelium.[32] A key function of mucus is to retain high concentrations of antimicrobial peptides, lectins, prokaryotic-associated molecular patterns (PAMP) and secretory IgA to prevent invasion by intestinal pathogens. Alternations in intestinal mucin composition can increase potential intestinal infections, potentially making tissue more sensitive to the development of inflammatory bowel disease.[33]

Translational and *in vitro* studies suggest that eosinophils may play a role in stimulating mucus production. For instance, mucosal eosinophils are juxtaposed to rhino-pulmonary epithelia that exhibit goblet cell hyperplasia and metaplasia, indicative of accelerated mucus production.[34] Exposure of human airway epithelial cells to supernatants derived from activated eosinophils leads to increased mucin production. Eosinophil-derived TGF-α and the activation of epidermal growth factor receptor appear to be crucial for this response.[35] In murine ovalbumin-induced airway eosinophilic inflammation, use of a C-C chemokine receptor type 3 (CCR3) monoclonal antibody reduces both eosinophil recruitment to the lung and mucus production.[36,37] In the absence of eosinophils, allergen challenge does not result in the accumulation of pulmonary mucus, airway hyperresponsiveness,[38] peribronchiolar collagen deposition, or increases in airway smooth muscle, suggesting that eosinophils contribute to airway dysfunction and remodeling.[39] Taken together, these studies provide strong support for a role for eosinophils in the regulation of mucus production and release. In the pulmonary system, where eosinophils do not normally reside, this may translate into a pathophysiological process. However, the intestine normally contains eosinophils and, under appropriate stimulation, eosinophils may activate mucus production and release, thus offering innate protection.

CONCLUSION

While the role of eosinophils in inflammatory diseases is being widely investigated, the fact that eosinophils are resident cells in the normal intestinal tract supports a role in maintaining health. Based on studies in other organ systems, such as the lung and skin, we speculate that eosinophils may play a role in physiological homeostasis by promoting intestinal mucosal health. Potential mechanisms include mucus stimulation, antimicrobial protein synthesis and release, and regulating intestinal motility. Future studies will determine the role of eosinophils in GI mucosal homeostasis.

REFERENCES

1. Protheroe C, Woodruff SA, de Petris G, Mukkada V, Ochkur SI, Janarthanan S, et al. A novel histologic scoring system to evaluate mucosal biopsies from patients with eosinophilic esophagitis. *Clin Gastroenterol Hepatol* 2009;**7**:749—55. e711.

2. DeBrosse CW, Case JW, Putnam PE, Collins MH, Rothenberg ME. Quantity and distribution of eosinophils in the gastrointestinal tract of children. *Pediatr Dev Pathol* 2006;**9**:210—8.

3. Lowichik A, Weinberg AG. A quantitative evaluation of mucosal eosinophils in the pediatric gastrointestinal tract. *Mod Pathol* 1996;**9**:110—4.

4. Mishra A, Hogan SP, Lee JJ, Foster PS, Rothenberg ME. Fundamental signals that regulate eosinophil homing to the gastrointestinal tract. *J Clin Invest* 1999;**103**:1719—27.

5. Barrett K. *Functional Anatomy of the GI Tract and Organs.* New York: McGraw-Hill; 2006.

6. Lwin T, Melton SD, Genta RM. Eosinophilic gastritis: histopathological characterization and quantification of the normal gastric eosinophil content. *Mod Pathol* 2010;**4**:556—63.

7. Collins MH. Histopathology associated with eosinophilic gastrointestinal diseases. *Immunol Allergy Clin North Am* 2009;**29**:109—17. x-xi.

8. Kim C, Kaufmann SH. Defensin: a multifunctional molecule lives up to its versatile name. *Trends Microbiol* 2006;**14**:428—31.

9. Kocsis AK, Ocsovszky I, Tiszlavicz L, Tiszlavicz Z, Mandi Y. Helicobacter pylori induces the release of alpha-defensin by human granulocytes. *Inflamm Res* 2009;**58**:241—7.

10. Nockher WA, Renz H. Neurotrophins in allergic diseases: from neuronal growth factors to intercellular signaling molecules. *J Allergy Clin Immunol* 2006;**117**:583—9.

11. Lambiase A, Bonini S, Micera A, Magrini L, Bracci-Laudiero L, Aloe L. Increased plasma levels of nerve growth factor in vernal keratoconjunctivitis and relationship to conjunctival mast cells. *Invest Ophthalmol Vis Sci* 1995;**36**:2127—32.

12. Bonini S, Lambiase A, Angelucci F, Magrini L, Manni L, Aloe L. Circulating nerve growth factor levels are increased in humans with allergic diseases and asthma. *Proc Natl Acad Sci USA* 1996;**93**: 10955—60.

13. Toyoda M, Nakamura M, Makino T, Hino T, Kagoura M, Morohashi M. Nerve growth factor and substance P are useful plasma markers of disease activity in atopic dermatitis. *Br J Dermatol* 2002;**147**:71—9.

14. Durcan N, Costello RW, McLean WG, Blusztajn J, Madziar B, Fenech AG, et al. Eosinophil-mediated cholinergic nerve remodeling. *Am J Respir Cell Mol Biol* 2006;**34**:775—86.

15. Labouyrie E, Dubus P, Groppi A, Mahon FX, Ferrer J, Parrens M, et al. Expression of neurotrophins and their receptors in human bone marrow. *Am J Pathol* 1999;**154**:405—15.

16. Curran DR, Morgan RK, Kingham PJ, Durcan N, McLean WG, Walsh MT, et al. Mechanism of eosinophil induced signaling in cholinergic IMR-32 cells. *Am J Physiol Lung Cell Mol Physiol* 2005;**288**:L326—32.

17. Jacoby DB, Gleich GJ, Fryer AD. Human eosinophil major basic protein is an endogenous allosteric antagonist at the inhibitory muscarinic M2 receptor. *J Clin Invest* 1993;**91**:1314—8.

18. Kingham PJ, McLean WG, Sawatzky DA, Walsh MT, Costello RW. Adhesion-dependent interactions between eosinophils and cholinergic nerves. *Am J Physiol Lung Cell Mol Physiol* 2002;**282**:L1229—38.

19. Madera JL, Anderson JM. *Epithelia biologic principles of organisation.* Philadelphia: Lippincott Williams & Wilkins; 2003.

20. Gibson GR, Roberfroid MB. Dietary modulation of the human colonic microbiota: introducing the concept of prebiotics. *J Nutr* 1995;**125**:1401—12.

21. Swidsinski A, Loening-Baucke V, Lochs H, Hale LP. Spatial organization of bacterial flora in normal and inflamed intestine: a fluorescence in situ hybridization study in mice. *World J Gastroenterol* 2005;**11**:1131—40.

22. Swidsinski A, Sydora BC, Doerffel Y, Loening-Baucke V, Vaneechoutte M, Lupicki M, et al. Viscosity gradient within the mucus layer determines the mucosal barrier function and the spatial organization of the intestinal microbiota. *Inflamm Bowel Dis* 2007;**13**:963—70.

23. Ackerman SJ, Gleich GJ, Loegering DA, Richardson BA, Butterworth AE. Comparative toxicity of purified human eosinophil granule cationic proteins for schistosomula of Schistosoma mansoni. *Am J Trop Med Hyg* 1985;**34**:735—45.

24. Lehrer RI, Szklarek D, Barton A, Ganz T, Hamann KJ, Gleich GJ. Antibacterial properties of eosinophil major basic protein and eosinophil cationic protein. *J Immunol* 1989;**142**:4428—34.

25. Pegorier S, Wagner LA, Gleich GJ, Pretolani M. Eosinophil-derived cationic proteins activate the synthesis of remodeling factors by airway epithelial cells. *J Immunol* 2006;**177**:4861—9.

26. Rosenberg HF, Domachowske JB. Eosinophils, eosinophil ribonucleases, and their role in host defense against respiratory virus pathogens. *J Leukoc Biol* 2001;**70**:691—8.

27. Wang J, Slungaard A. Role of eosinophil peroxidase in host defense and disease pathology. *Arch Biochem Biophys* 2006;**445**:256—60.

28. Yousefi S, Gold JA, Andina N, Lee JJ, Kelly AM, Kozlowski E, et al. Catapult-like release of mitochondrial DNA by eosinophils contributes to antibacterial defense. *Nat Med* 2008;**14**:949—53.

29. Simon D, Hoesli S, Roth N, Staedler S, Yousefi S, Simon HU. Eosinophil extracellular DNA traps in skin diseases. *J Allergy Clin Immunol* 2011;**127**:194—9.

30. Linch SN, Kelly AM, Danielson ET, Pero R, Lee JJ, Gold JA. Mouse eosinophils possess potent antibacterial properties *in vivo*. *Infect Immun* 2009;**77**:4976—82.

31. Canny G, Levy O. Bactericidal/permeability-increasing protein (BPI) and BPI homologs at mucosal sites. *Trends Immunol* 2008;**29**:541—7.

32. Pearson JP, Brownlee IA. The interaction of large bowel microflora with the colonic mucus barrier. *Int J Inflam* 2010;**2010**:321426.

33. McGuckin MA, Eri R, Simms LA, Florin TH, Radford-Smith G. Intestinal barrier dysfunction in inflammatory bowel diseases. *Inflamm Bowel Dis* 2009;**15**:100—13.

34. Ding GQ, Zheng CQ, Bagga SS. Upregulation of the mucosal epidermal growth factor receptor gene in chronic rhinosinusitis and nasal polyposis. *Arch Otolaryngol Head Neck Surg* 2007;**133**:1097—103.

35. Burgel PR, Lazarus SC, Tam DC, Ueki IF, Atabai K, Birch M, et al. Human eosinophils induce mucin production in airway epithelial cells via epidermal growth factor receptor activation. *J Immunol* 2001;**167**:5948—54.

36. Hur GY, Lee SY, Lee SH, Kim SJ, Lee KJ, Jung JY, et al. Potential use of an anticancer drug gefinitib, an EGFR inhibitor, on allergic airway inflammation. *Exp Mol Med* 2007;**39**:367—75.

37. Shen HH, Xu F, Zhang GS, Wang SB, Xu WH. CCR3 monoclonal antibody inhibits airway eosinophilic inflammation and mucus overproduction in a mouse model of asthma. *Acta Pharmacol Sin* 2006;**27**:1594—9.

38. Lee JJ, Dimina D, Macias MP, Ochkur SI, McGarry MP, O'Neill KR, et al. Defining a link with asthma in mice congenitally deficient in eosinophils. *Science* 2004;**305**:1773—6.

39. Humbles AA, Lloyd CM, McMillan SJ, Friend DS, Xanthou G, McKenna EE, et al. A critical role for eosinophils in allergic airways remodeling. *Science* 2004;**305**:1776—9.

Chapter 11.4

Eosinophil-Mediated Antigen Presentation

Praveen Akuthota, Jason J. Xenakis, Haibin Wang and Peter F. Weller

INTRODUCTION

Major histocompatibility complex class II (MHC-II)-restricted antigen presentation has traditionally been considered to be the province of dendritic cells, B cells, and macrophages.[1] These antigen-presenting cells (APCs) have been classified as *professional* APCs, with the ability to process and present extracellular antigens on MHC-II to T cells expressing antigen-specific T-cell receptors. As highlighted throughout this text, eosinophils have multi-faceted and complex immunoregulatory functions that may contradict established notions of immune function, and emerging evidence that eosinophils can function as professional APCs falls under this category.

This subchapter will discuss *in vitro* and *in vivo* evidence from human and murine studies supporting the notion that eosinophils have professional APC function. Firstly, MHC-II expression by eosinophils in various experimental and clinical settings will be detailed, followed by a discussion of key co-stimulatory molecules necessary for APC function that are also expressed by eosinophils. The subchapter will conclude with an account of the evidence that eosinophils can function as APCs *in vivo* and of their migration to regional lymph nodes for the purpose of antigen presentation in murine models of allergic inflammation.

EOSINOPHILS EXPRESS MAJOR HISTOCOMPATIBILITY COMPLEX CLASS II MOLECULES

Several studies of murine models of allergic airway inflammation or parasitic infection have subsequently described the expression of MHC-II by eosinophils recovered from the sites of induced inflammation. In ovalbumin (OVA) sensitization and challenge models of allergic inflammation, eosinophils from bronchoalveolar lavage fluid (BALF) express MHC-II proteins.[6–9] Moreover, eosinophils recovered from draining thoracic lymph nodes after OVA sensitization and challenge likewise express MHC-II proteins.[10] Compared to eosinophils from the lung itself, a higher proportion of eosinophils from draining lymph nodes display MHC-II expression.[10] Splenic eosinophils from interleukin-5 (IL-5) transgenic mice, which have high numbers of eosinophils due to IL-5 overexpression, demonstrate low-level expression of surface MHC-II molecules that is significantly upregulated following *in vitro* stimulation by granulocyte-macrophage colony-stimulating factor (GM-CSF).[11] Mice inoculated via the intraperitoneal route with the helminth *Brugia malayi* have increased MHC-II expression on eosinophils recovered from the peritoneum.[12] Similarly, murine eosinophils exposed *in vitro* to antigen from *Strongyloides stercoralis* also upregulate MHC-II.[13]

Human eosinophils express MHC-II molecules in a number of experimental and clinical settings in which they can be considered to be activated. Using an *in vitro* experimental migration model, human eosinophils that migrate through a monolayer of activated epithelial cells have increased HLA-DR expression as well as improved survival in culture.[14] This observed change in phenotype secondary to transepithelial migration is at least partially dependent on autocrine production of GM-CSF, as GM-CSF inhibition reverses the augmented survival phenotype. However, the impact of GM-CSF inhibition on MHC-II expression was not included in these studies. In patients with asthma, flow cytometry assays of sputum eosinophils show upregulation of HLA-DR and intercellular adhesion molecule-1 (ICAM-1).[15] Similarly, eosinophils from bronchoalveolar lavage fluid and tissue from nasal polyps in asthma have detectable HLA-DR expression.[16] HLA-DR expression has been reported on the surface of eosinophils recovered from the bronchoalveolar lavage fluid of patients with eosinophilic pneumonias.[17,18] Eosinophils from conjunctival samples in ocular allergic disease also have elevated HLA-DR expression.[19] Tissue eosinophils from esophageal biopsy specimens of children with eosinophilic esophagitis exhibit increased HLA-DR staining compared to control subjects.[20] Human eosinophils from bronchoalveolar lavage fluid of volunteers with allergic rhinitis who underwent pulmonary segmental allergen challenge have similarly been observed to have increased MHC-II expression.[21]

EXPRESSION OF CO-STIMULATORY MOLECULES

In addition to MHC-II molecules, the expression of co-stimulatory molecules is needed to provide second signals necessary for professional APC function. Co-stimulatory molecules expressed by eosinophils include tumor necrosis factor receptor superfamily member 5 (CD40) and T-lymphocyte activation antigens CD80 and CD86. CD40 provides a co-stimulatory signal to T cells through the CD40 ligand.[22] Both peripheral blood eosinophils from human donors with mild atopy and eosinophils isolated from nasal polyps express CD40.[23] Similarly, sputum eosinophils from asthmatic human subjects have detectable surface CD40.[24] Eosinophils from IL-5 transgenic mice also have CD40 surface expression that is upregulated after stimulation of eosinophils with GM-CSF.[11] In aeroallergen-challenged mice, CD40 has been detected on the surface of eosinophils isolated from harvested thoracic draining lymph nodes.[10] Of note, CD40 engagement by antibody cross-linking enhances eosinophil survival, with inhibition of apoptosis contributing to the improved survival.[23,24] Strong CD40 staining on eosinophils infiltrating biopsy specimens of cutaneous lesions from patients with hypereosinophilic syndrome has also been reported.[25]

Similar to MHC-II molecules, murine eosinophils recovered from BALF after aeroallergen challenge express the CD80 and CD86 co-stimulatory molecules.[6–9,11] CD80

and CD86 are members of the B7 family of ligands expressed on APCs that interact with CD28 on T cells, thus providing the second signal for antigen presentation.[26] Murine eosinophils harvested from the lung and spleen express CD80 and CD86, even without allergen stimulation.[9] Similar to the observations with CD40, CD80 and CD86 are upregulated on eosinophils isolated from thoracic lymph nodes of OVA-sensitized and challenged mice.[10] Eosinophils from the peritoneal cavities of IL-5 transgenic mice constitutively express CD80 and CD86, with increases surface expression in vitro following GM-CSF stimulation.[27] Murine eosinophils from the spleens of IL-5 transgenic mice demonstrate a fourfold increase in surface CD86 levels after in vitro exposure to antigens derived from S. stercoralis.[13] Peripheral blood eosinophils from hypereosinophilic human subjects express both CD86 and its ligand, CD28.[28] Eosinophils isolated from peripheral blood of normal subjects can also be stimulated to upregulate CD86 surface expression by incubation with interleukin-3 (IL-3).[29] Interestingly, apoptotic human eosinophils express CD28 and CD86 on their surface.[30]

EOSINOPHILS FUNCTION AS ANTIGEN-PRESENTING CELLS

As noted above, eosinophils express the requisite cell surface proteins for antigen presentation and co-stimulation. More importantly, they have been demonstrated in various experimental settings to function as APCs. In vitro coculture experiments featuring eosinophils and T cells have provided a substantial insight into the role of eosinophils as APCs. Murine eosinophils loaded with OVA or with antigens from the parasite Mesocestoides corti are able to induce the proliferation of T cell clones specific to OVA and M. corti antigens, respectively.[31] Furthermore, T cell proliferation is due to true MHC-II-restricted APC function, as antibody blocking of eosinophil MHC-II molecules specifically inhibits T cell proliferation.[31] Similar observations have been made regarding stimulation of antigen-specific T cells to produce IL-4 and IL-5 by S. stercoralis-pulsed murine eosinophils in an MHC-II-dependent fashion.[13] Importantly, S. stercoralis-loaded murine eosinophils can stimulate both primed and naive T cells, a hallmark of professional APCs.[13] In fact, the ability of eosinophils to act as professional APCs to naive T cells equivalent to that of dendritic cells in a single study using the S. stercoralis antigen.[13] Recent data have shown that rat eosinophils can phagocytose opsonized Cryptococcus neofomans, resulting in the upregulation of MHC-II and co-stimulatory proteins. These eosinophils induce MHC-II-dependent proliferation of splenic CD4$^+$ lymphocytes, consistent with an APC function.[32] Murine eosinophils isolated from the peritoneal exudates of IL-5 transgenic mice and subsequently stimulated with GM-CSF and loaded with OVA are able to induce the proliferation of lymph node T cells.[27] Importantly, antibody blocking of MHC-II or co-stimulatory CD80 and CD86 proteins inhibited the APC functionality of eosinophils in this experimental model.[27] In a similar experimental system using coculture of BALF eosinophils from OVA-challenged mice and sensitized CD4$^+$ T cells, eosinophils stimulated the production of the T-helper 2 (T$_h$2) cytokines IL-4, IL-5, and IL-13 by CD4$^+$ T cells, with antibody blocking of CD80 and CD86 inhibiting the cytokine production.[33] Another study demonstrated the ability of murine eosinophils loaded with OVA, bovine serum albumin, or human immunoglobulin to elicit *in vitro* proliferation of T cells in an antigen-specific fashion.[7] *In vitro* studies of human eosinophils have yielded concordant results. Human eosinophils have been found to process and present bee venom phospholipase A$_2$ antigen to specific T cells, likewise inducing their proliferation.[34] Peripheral blood human eosinophils exposed to tetanus toxoid similarly stimulate antigen-specific T cell proliferation via MHC-II-restricted antigen presentation.[4] Significantly, eosinophils in this experimental system are indeed processing antigen prior to presentation, as eosinophils fixed with paraformaldehyde after antigen loading have their APC function preserved, while those fixed prior to antigen exposure do not.[4] Rhinovirus antigens can also be presented by human eosinophils, resulting in the stimulation of rhinovirus-specific T cell clones.[35] Human eosinophils have additionally been observed to cause CD4$^+$ T cell proliferation *in vitro*, mediated by staphylococcal superantigens.[27,36]

In vivo and *ex vivo* experimental models have also contributed to our understanding of eosinophils as APCs. Responses to *S. stercoralis* antigen in mice have again proven to be a useful model. Murine eosinophils loaded with *S. stercoralis* antigen and subsequently administered to recipient mice induce proliferation of T cells harvested from the spleen and cultured *ex vivo*.[37] Additionally, both naive and memory T cells cultured *ex vivo* display increased production of the T$_h$2 cytokines, IL-4 and IL-5.[37] The production of T$_h$2 cytokines by splenocytes is dependent on the expression of MHC-II by the administered eosinophils. Furthermore, mice inoculated with antigen-loaded eosinophils produce both antigen-specific immunoglobulin M (IgM) and IgG.[37] Eosinophils have been demonstrated to act as professional APCs *in vivo* using the murine model of OVA-induced allergic airway inflammation. In one such experiment, GM-CSF-stimulated eosinophils loaded with OVA were endotracheally instilled into mice that had previously been inoculated with naive OVA-specific T cells.[11] After OVA challenge, OVA-specific T cells from harvested draining thoracic lymph nodes showed multiple effects resulting from MHC-II-restricted antigen presentation by eosinophils. T cells demonstrated increased activation associated with

upregulation of CD69, increased proliferation, and increased production of the T$_h$2 cytokine, IL-4.[11] In the same set of experiments, eosinophils and T cells were observed to interact physically using fluorescence microscopy techniques.[11] Similar to the *in vitro* data discussed previously, murine eosinophils have also been demonstrated to cause specific T cell proliferation *in vivo*, not only to OVA but to bovine serum albumin and human immunoglobulins as well.[7] Administration of antibodies to the co-stimulatory CD80 and CD86 proteins to mice sensitized to OVA and endotracheally instilled with OVA-elicited eosinophils has been demonstrated to reduce the production of IL-4, IL-5, and IL-13 from lymph node cells.[33]

Important to the APC function of eosinophils is their ability not just to present antigen but also to process it prior to presentation. While many of the studies of eosinophil APC function detailed above focused on the ability of eosinophils to process large proteins such as OVA for the purpose of presenting small peptides on MHC-II molecules, other studies directly address the requirement of peptide processing by eosinophils prior to antigen presentation. These reports depend on the inhibition of lysosomal function by chloroquine and ammonium chloride. Both agents are well-described inhibitors of APC function via preventing lysosomal acidification and, therefore inhibit peptide processing.[38,39] Chloroquine and ammonium chloride both inhibit presentation of the *M. corti* antigen by murine eosinophils *in vitro* when applied to eosinophils prior to antigen exposure.[31] In the murine OVA airway challenge model, in which OVA-loaded eosinophils and OVA-specific T cells were coadministered to mice, eosinophils exposed to ammonium chloride during their isolation [for the purposes of red blood cell (RBC) lysis] did not stimulate T cells.[11] However, when hypotonic saline was used for RBC lysis during eosinophil isolation, T cells displayed significant activation, proliferation, and T$_h$2 cytokine production.[11] Supporting data have also come from *in vitro* studies of human cells, in which bee venom-loaded eosinophils treated with chloroquine inhibited APC function toward specific T cells, while eosinophils not exposed to chloroquine caused robust and specific T cell proliferation.[34]

Also key to eosinophil APC function is their migration to regional lymph nodes after antigen exposure, where they may present antigen to T cells. In mice sensitized and challenged with OVA, eosinophils have been found to accumulate in the draining lymph nodes.[8] Furthermore, eosinophils that accumulate in the thoracic lymph nodes after aeroallergen administration have an APC phenotype, with upregulation of MHC-II, CD40, CD80, and CD86.[10] Eosinophils obtained from the airways of aeroallergen-challenged mice or from the peritoneum of IL-5 transgenic mice that are fluorescently labeled and reinstilled into the tracheas of normal mice have been observed to migrate to regional draining thoracic lymph nodes.[7] Eosinophil migration to regional lymph nodes appears to be independent of the effects of eotaxin, as there is no difference in the migration of eosinophils from C-C chemokine receptor 3 (CCR3) knockout mice in this experimental system.[7] Recent *in vitro* studies of human peripheral blood eosinophils have demonstrated that eosinophils pulsed with GM-CSF, IL-3, and IFN-γ show a chemotactic response toward C-C motif chemokine 21 (CCL21), one of the ligands of CCR7.[40] CCR7 is thought to have an important role in APC function by mediating the migration of APCs and T cells to the T-cell zones of lymph nodes, but has not been well described in eosinophils.[41] However, emerging data indicate that human eosinophils do indeed express CCR7 when stimulated with IL-5 and that they have chemotactic activity toward both CCR7 ligands (CCL19 and CCL21).[42]

Although data remain preliminary, recent reports indicate that localization of MHC-II to lipid rafts may have a role in eosinophil APC function.[43,44] Lipid rafts are cholesterol and sphingolipid enriched domains in cell membranes that facilitate the aggregation of protein complexes.[45] HLA-DR localizes to lipid rafts in GM-CSF-stimulated peripheral blood eosinophils, and disruption of lipid rafts inhibits superantigen-mediated T cell stimulation by eosinophils *in vitro*.[43] Localization of MHC-II proteins to lipid rafts appears to have an important, though not fully defined, role in antigen presentation by other APCs.[46,47] Another intriguing but ill-defined aspect of eosinophil APC function is the high level expression of the CD9 tetraspanin by human eosinophils.[48] CD9 associates with MHC-II in a variety of other APCs and is important in murine dendritic cells for the lateral aggregation of MHC-II molecules.[49]

CONCLUSION

As the evidence detailed above demonstrates, eosinophils express the requisite machinery for antigen presentation and can function as professional APCs (Fig. 11.4.1).[1] They increase MHC-II expression in response to a variety of *in vitro* and *in vivo* stimuli, with exposure to GM-CSF being a primary inducer of the APC phenotype. Eosinophils also express a variety of co-stimulatory molecules in response to activated conditions, including CD40, CD80, and CD86. The ability of eosinophils to provide the proper co-stimulatory *second* signal provides key support for the hypothesis that eosinophils are true professional APCs. Several *in vitro* experimental models using human eosinophils show that they can stimulate T cells in an antigen-specific manner. Murine models of allergic airway inflammation and parasitic infection provide additional demonstrations of eosinophil APC function. Data from murine models of asthma also suggest that eosinophils traffic to regional lymph nodes where they encounter

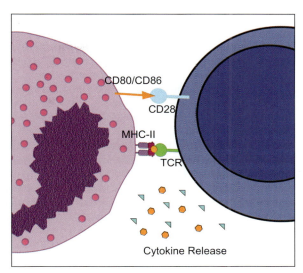

FIGURE 11.4.1 Eosinophil-mediated antigen presentation. Eosinophils express major histocompatibility complex MHC-II and co-stimulatory surface proteins, allowing them to interact with T cells in an antigen-specific manner. The antigen-presenting cell interaction of eosinophils with T cells results in cytokine release and T cell proliferation. MHC-II, major histocompatibility complex class II; TCR, T-cell receptor.

T cells for the purposes of antigen presentation. While the relative importance of eosinophil APC function with respect to that of dendritic cells and other APCs remains unknown, future studies may reveal that this function of eosinophils is important and nonredundant.

REFERENCES

1. Trombetta ES, Mellman I. Cell biology of antigen processing *in vitro* and *in vivo*. *Annu Rev Immunol* 2005;**23**:975—1028.
2. Koeffler HP, Billing R, Levine AM, Golde DW. Ia antigen is a differentiation marker on human eosinophils. *Blood* 1980;**56**:11—4.
3. Lucey DR, Nicholson-Weller A, Weller PF. Mature human eosinophils have the capacity to express HLA-DR. *Proc Natl Acad Sci USA* 1989;**86**:1348—51.
4. Weller PF, Rand TH, Barrett T, Elovic A, Wong DT, Finberg RW. Accessory cell function of human eosinophils. HLA-DR-dependent, MHC-restricted antigen-presentation and IL-1 alpha expression. *J Immunol* 1993;**150**:2554—62.
5. Luttmann W, Franz P, Schmidt S, Barth J, Matthys H, Virchow Jr JC. Inhibition of HLA-DR expression on activated human blood eosinophils by transforming growth factor-beta1. *Scand J Immunol* 1998;**48**:667—71.
6. MacKenzie JR, Mattes J, Dent LA, Foster PS. Eosinophils promote allergic disease of the lung by regulating CD4(+) Th2 lymphocyte function. *J Immunol* 2001;**167**:3146—55.
7. Shi HZ, Humbles A, Gerard C, Jin Z, Weller PF. Lymph node trafficking and antigen presentation by endobronchial eosinophils. *J Clin Invest* 2000;**105**:945—53.
8. van Rijt LS, Vos N, Hijdra D, de Vries VC, Hoogsteden HC, Lambrecht BN. Airway eosinophils accumulate in the mediastinal

lymph nodes but lack antigen-presenting potential for naive T cells. *J Immunol* 2003;**171**:3372—8.

9. Rose Jr CE, Lannigan JA, Kim P, Lee JJ, Fu SM, Sung SS. Murine lung eosinophil activation and chemokine production in allergic airway inflammation. *Cell Mol Immunol* 2010;**7**:361—74.

10. Duez C, Dakhama A, Tomkinson A, Marquillies P, Balhorn A, Tonnel AB, et al. Migration and accumulation of eosinophils toward regional lymph nodes after airway allergen challenge. *J Allergy Clin Immunol* 2004;**114**:820—5.

11. Wang HB, Ghiran I, Matthaei K, Weller PF. Airway eosinophils: allergic inflammation recruited professional antigen-presenting cells. *J Immunol* 2007;**179**:7585—92.

12. Mawhorter SD, Pearlman E, Kazura JW, Boom WH. Class II major histocompatibility complex molecule expression on murine eosinophils activated *in vivo* by Brugia malayi. *Infect Immun* 1993;**61**:5410—2.

13. Padigel UM, Lee JJ, Nolan TJ, Schad GA, Abraham D. Eosinophils can function as antigen-presenting cells to induce primary and secondary immune responses to Strongyloides stercoralis. *Infect Immun* 2006;**74**:3232—8.

14. Yamamoto H, Sedgwick JB, Vrtis RF, Busse WW. The effect of transendothelial migration on eosinophil function. *Am J Respir Cell Mol Biol* 2000;**23**:379—88.

15. Hansel TT, Braunstein JB, Walker C, Blaser K, Bruijnzeel PL, Virchow Jr JC, Virchow C, Sr. Sputum eosinophils from asthmatics express ICAM-1 and HLA-DR. *Clin Exp Immunol* 1991;**86**:271—7.

16. Walker C, Rihs S, Braun RK, Betz S, Bruijnzeel PL. Increased expression of CD11b and functional changes in eosinophils after migration across endothelial cell monolayers. *J Immunol* 1993;**150**:4061—71.

17. Beninati W, Derdak S, Dixon PF, Grider DJ, Strollo DC, Hensley RE, et al. Pulmonary eosinophils express HLA-DR in chronic eosinophilic pneumonia. *J Allergy Clin Immunol* 1993;**92**:442—9.

18. Okubo Y, Hossain M, Kai R, Sato E, Honda T, Sekiguchi M, et al. Adhesion molecules on eosinophils in acute eosinophilic pneumonia. *Am J Respir Crit Care Med* 1995;**151**:1259—62.

19. Hingorani M, Calder V, Jolly G, Buckley RJ, Lightman SL. Eosinophil surface antigen expression and cytokine production vary in different ocular allergic diseases. *J Allergy Clin Immunol* 1998;**102**:821—30.

20. Patel AJ, Fuentebella J, Gernez Y, Nguyen T, Bass D, Berquist W, et al. Increased HLA-DR expression on tissue eosinophils in eosinophilic esophagitis. *J Pediatr Gastroenterol Nutr* 2010;**51**:290—4.

21. Sedgwick JB, Calhoun WJ, Vrtis RF, Bates ME, McAllister PK, Busse WW. Comparison of airway and blood eosinophil function after *in vivo* antigen challenge. *J Immunol* 1992;**149**:3710—8.

22. Grewal IS, Flavell RA. The role of CD40 ligand in costimulation and T-cell activation. *Immunol Rev* 1996;**153**:85—106.

23. Ohkawara Y, Lim KG, Xing Z, Glibetic M, Nakano K, Dolovich J, et al. CD40 expression by human peripheral blood eosinophils. *J Clin Invest* 1996;**97**:1761—6.

24. Bureau F, Seumois G, Jaspar F, Vanderplasschen A, Detry B, Pastoret PP, et al. CD40 engagement enhances eosinophil survival through induction of cellular inhibitor of apoptosis protein 2 expression: Possible involvement in allergic inflammation. *J Allergy Clin Immunol* 2002;**110**:443—9.

25. Jang KA, Lim YS, Choi JH, Sung KJ, Moon KC, Koh JK. Hypereosinophilic syndrome presenting as cutaneous necrotizing eosinophilic vasculitis and Raynaud's phenomenon complicated by digital gangrene. *Br J Dermatol* 2000;**143**:641—4.

26. Sharpe AH, Freeman GJ. The B7-CD28 superfamily. *Nat Rev Immunol* 2002;**2**:116—26.

27. Tamura N, Ishii N, Nakazawa M, Nagoya M, Yoshinari M, Amano T, et al. Requirement of CD80 and CD86 molecules for antigen presentation by eosinophils. *Scand J Immunol* 1996;**44**:229—38.

28. Woerly G, Roger N, Loiseau S, Dombrowicz D, Capron A, Capron M. Expression of CD28 and CD86 by human eosinophils and role in the secretion of type 1 cytokines (interleukin 2 and interferon gamma): inhibition by immunoglobulin a complexes. *J Exp Med* 1999;**190**:487—95.

29. Celestin J, Rotschke O, Falk K, Ramesh N, Jabara H, Strominger J, et al. IL-3 induces B7.2 (CD86) expression and costimulatory activity in human eosinophils. *J Immunol* 2001;**167**:6097—104.

30. Seton K, Hakansson L, Carlson M, Stalenheim G, Venge P. Apoptotic eosinophils express IL-2R chains alpha and beta and co-stimulatory molecules CD28 and CD86. *Respir Med* 2003;**97**:893—902.

31. Del Pozo V, De Andres B, Martin E, Cardaba B, Fernandez JC, Gallardo S, et al. Eosinophil as antigen-presenting cell: activation of T cell clones and T cell hybridoma by eosinophils after antigen processing. *Eur J Immunol* 1992;**22**:1919—25.

32. Garro AP, Chiapello LS, Baronetti JL, Masih DT. Rat eosinophils stimulate the expansion of Cryptococcus neoformans-specific CD4(+) and CD8(+) T cells with a T-helper 1 profile. *Immunology* 2011;**132**:174—87.

33. Shi HZ, Xiao CQ, Li CQ, Mo XY, Yang QL, Leng J, et al. Endobronchial eosinophils preferentially stimulate T helper cell type 2 responses. *Allergy* 2004;**59**:428—35.

34. Hansel TT, De Vries IJ, Carballido JM, Braun RK, Carballido-Perrig N, Rihs S, et al. Induction and function of eosinophil intercellular adhesion molecule-1 and HLA-DR. *J Immunol* 1992;**149**:2130—6.

35. Handzel ZT, Busse WW, Sedgwick JB, Vrtis R, Lee WM, Kelly EA, et al. Eosinophils bind rhinovirus and activate virus-specific T cells. *J Immunol* 1998;**160**:1279—84.

36. Mawhorter SD, Kazura JW, Boom WH. Human eosinophils as antigen-presenting cells: relative efficiency for superantigen- and antigen-induced CD4+ T-cell proliferation. *Immunology* 1994;**81**:584—91.

37. Padigel UM, Hess JA, Lee JJ, Lok JB, Nolan TJ, Schad GA, et al. Eosinophils act as antigen-presenting cells to induce immunity to Strongyloides stercoralis in mice. *J Infect Dis* 2007;**196**:1844—51.

38. Loss Jr GE, Sant AJ. Invariant chain retains MHC class II molecules in the endocytic pathway. *J Immunol* 1993;**150**:3187—97.

39. Ziegler HK, Unanue ER. Decrease in macrophage antigen catabolism caused by ammonia and chloroquine is associated with inhibition of antigen presentation to T cells. *Proc Natl Acad Sci USA* 1982;**79**:175—8.

40. Jung YJ, Woo SY, Jang MH, Miyasaka M, Ryu KH, Park HK, et al. Human Eosinophils Show Chemotaxis to Lymphoid Chemokines and Exhibit Antigen-Presenting-Cell-Like Properties upon Stimulation with IFN-gamma, IL-3 and GM-CSF. *Int Arch Allergy Immunol* 2008;**146**:227—34.

41. Forster R, Davalos-Misslitz AC, Rot A. CCR7 and its ligands: balancing immunity and tolerance. *Nat Rev Immunol* 2008;**8**:362–71.
42. Akuthota P, Shamri R, Spencer LA, Weller PF. Human Eosinophils Express CCR7 and Exhibit Chemotaxis to CCR7 Ligands. *Am J Respir Crit Care Med* 2010;**181**:A2786.
43. Akuthota P, Spencer LA, Radke AL, Weller PF. MHC Class II and CD9 Localize to Lipid Rafts in Human Eosinophils. *Am J Respir Crit Care Med* 2009;**179**:A3697.
44. Akuthota P, Wang H, Weller PF. Eosinophils as antigen-presenting cells in allergic upper airway disease. *Curr Opin Allergy Clin Immunol* 2010;**10**:14–9.
45. Dykstra M, Cherukuri A, Sohn HW, Tzeng SJ, Pierce SK. Location is everything: lipid rafts and immune cell signaling. *Annu Rev Immunol* 2003;**21**:457–81.
46. Anderson HA, Hiltbold EM, Roche PA. Concentration of MHC class II molecules in lipid rafts facilitates antigen presentation. *Nat Immunol* 2000;**1**:156–62.
47. Eren E, Yates J, Cwynarski K, Preston S, Dong R, Germain C, et al. Location of major histocompatibility complex class II molecules in rafts on dendritic cells enhances the efficiency of T-cell activation and proliferation. *Scand J Immunol* 2006;**63**:7–16.
48. Matsumoto K, Bochner BS, Wakiguchi H, Kurashige T. Functional expression of transmembrane 4 superfamily molecules on human eosinophils. *Int Arch Allergy Immunol* 1999;**120**(Suppl. 1):38–44.
49. Unternaehrer JJ, Chow A, Pypaert M, Inaba K, Mellman I. The tetraspanin CD9 mediates lateral association of MHC class II molecules on the dendritic cell surface. *Proc Natl Acad Sci USA* 2007;**104**:234–9.

Chapter 11.5

Eosinophil Modulation of T Cell Activation and Recruitment

Elizabeth A. Jacobsen

INTRODUCTION

Eosinophils have immune-regulating activities capable of modulating both innate and adaptive immune cells. Studies in patients and animal models have demonstrated coordinated distributions and activities that link eosinophils and effector T cells in a variety of diseases, such as dermatitis, inflammatory bowel disease, and allergic asthma.[1] In the majority of these immunological diseases, effector CD4$^+$ T cell subtypes are the major source of cytokines characteristic of the T-helper 1 (T$_h$1)/T$_h$17/T$_h$2-polarized immune response. The importance of these T cells in propagating the inflammatory process has been demonstrated in T cell-deficient mice or mice impaired in

the production and recruitment of effector CD4$^+$ T cells. Failure to induce the activation and recruitment of effector T cells to the peripheral inflamed tissue results in attenuated pathophysiology. These polarized effector T cell responses are thought to occur via a succession of events:

1. Primary immune responses during a sensitization event(s) generating antigen-specific memory T cells that are imprinted for tissue specificity;
2. Secondary immune responses leading to the activation and proliferation of antigen-specific memory T cells upon exposure to antigen after the primary exposure;
3. Enforced or modified polarization of those T cells to induce a T$_h$1/T$_h$17/T$_h$2 response;
4. Recruitment of T cells to peripheral sites of inflammation.

This chapter will discuss the progression from naive CD4$^+$ T cell to recruited polarized effector CD4$^+$ T cell and the suggested roles of eosinophils in these steps.

PRIMARY IMMUNE RESPONSE

CD4$^+$ T Cell Trafficking from Thymus to Outlying Areas of the Body to Secondary Lymphoid Organs and Back to Outlying Areas

Movement of T cells throughout the body is a highly organized process and part of the tight regulation of the adaptive immune responses. Newly formed CD4$^+$ $\alpha\beta^+$ T cells exit the thymus, enter the bloodstream, and home to the secondary lymphoid organs (SLO), such as spleen, lymph nodes, and Peyer patches, as well as mucosal-associated lymphoid tissues (MALT). T cell migration out of the thymus and into and out of lymphatics and the periphery occurs through the alternating upregulation and downregulation of several cell surface molecules such as adhesion molecules [e.g., L-selectin (CD62L) and lymphocyte function-associated antigen (LFA-1, also known as αLβ2 and CD11a)], chemokine receptors [e.g., C-C chemokine receptor 7 (CCR7)], and sphingosine lipid Gαi-protein receptors (e.g., SIP$_1$).[2,3] T cells require the increased cell surface expression of SIP$_1$ for their release from the thymus (and from the SLO) to enter the bloodstream. Chemotactic migration and *docking* in the T cell zones of the SLO relies on the SLO-dependent expression of chemokines CCL19 and CCL21 that bind CCR7-positive T cells. Once the naive T cell enters the lymph venules, CD62L expression enables tethering and rolling of the lymphocyte, followed by binding of integrins [e.g., LFA-1 (αLβ2, CD11a)] to addressins [e.g., ICAM (CD54)] by high endothelial venules within the SLO.

Eosinophils express many of the same homing receptors as naive lymphocytes (e.g., CCR7),[4] likely enabling eosinophils to colocalize with T cells in T cell zones of the SLO.[5] Moreover, the expression of these cell surface molecules may in part explain eosinophil localization in the thymic corticomedullary junction, where naive T cells exit the thymus. Although eosinophils colocalize to similar regions as T cells, it is unknown whether eosinophils modulate naive T cell migration out of the thymus and into the SLO. Eosinophils possess a significant resource of effector functions that may activate postcapillary high endothelial venules, lymphatic vessels, other leukocytes, and stromal cells to signal for T cell recruitment to the SLO.

Activation of Naive T Cells: Primary Response

Exposure to antigen occurs primarily via antigen-presenting cells (APCs) in the periphery that migrate to the SLO to interact with the T cells. APCs process the exogenous antigen into a peptide fragment and present the peptide to the TCR via the major histocompatibility molecule II (MHC II), a required step for CD4$^+$ T cell activation. The three classical APCs are DCs, macrophages, and B-cells, although additional cells have been demonstrated to have antigen-presenting activities.[6] For example, intestinal epithelial M cells process antigen from the lumen, whereas DCs process antigen from the mucosal lining. In turn, these APCs present antigen to naive T cells in the Peyer patch or the mesenteric lymph node, respectively. In the skin, the Langerhans DCs migrate to local associated lymph nodes and lung APCs migrate to the caudal mediastinal lymph nodes (i.e., lung-draining lymph nodes or paratracheal lymph nodes). In addition to antigen presentation, full activation of naive CD4$^+$ T cells requires two additional signals, either from the APC, through multiple cell−cell contacts, or from a microenvironment that contributes to the response: stimulation through a co-stimulatory receptor (e.g., CD28)[3]; and polarization signals through cytokines [e.g., interleukin-4 (IL-4)],[7] coactivating molecules (e.g., OX-40),[3] or other mediators [e.g., toll-like receptors (TLRs)].[8] Together these signals determine the level of activation vs. tolerance and the type of cytokine production, i.e., T_h1 [e.g., IL-12 and interferon γ (IFN-γ)], T_h2 (e.g., IL-4, IL-13, and IL-25), and T_h17 (e.g., IL-17A and IL-17F).

Naive T Cell Activation by Eosinophils

Many elegant studies have demonstrated the ability of eosinophils to perform APC functions, leading to the activation of CD4$^+$ T cells [see[9] for review]. Mouse and human eosinophils increased MHC II expression when stimulated under various cell culture conditions, after allergen challenge in mouse models of asthma, and in various parasite infection models. In an attempt to clarify the role of eosinophils as APCs, investigators used several mouse models of allergen-induced asthma or parasite infection. In mouse models of allergen-induced asthma, four routes for naive CD4$^+$ T cell activation are often taken: (1) intraperitoneal injection of allergen with adjuvant (aluminum hydroxide, i.e., Imject Alum, Thermo Scientific); (2) intranasal exposure of allergen without adjuvant; (3) transfer of naive allergen-specific T cells; or (4) antigen-loaded APCs into naive recipient mice. This variety of methods has led to eosinophils being characterized as capable of migration to the draining lymph node with a kinetics similar to DC migration. Furthermore, these migrating eosinophils are capable of processing and presenting ovalbumin (OVA) to induce the activation of naive T cells.[5,10] Alternatively, variations in the ability[5,10] or lack of ability[11] of eosinophils to act as APCs to naive antigen-specific T cells have been ascribed to technical aspects of eosinophil isolation (e.g., the use of ammonium chloride for red blood cell lysis during eosinophil isolation). Alternatively, in a model of parasite infection, Padigel and colleagues demonstrated both in vitro[12] and in vivo[13] that eosinophils pulsed with nematode Strongyloides stercoralis antigen induced MHC II-dependent stimulation of IL-5 expression from naive T cells. Although these studies suggest a role for eosinophils in T cell priming, primary immune responses were unaffected in vivo when eosinophil-deficient PHIL mice were infected with S. stercoralis.[14] Thus, although eosinophils possess APC functions, their role as necessary APCs for naive T cell activation in vivo remains unclear. Nonetheless, the localization of eosinophils to T cell-rich regions in SLO suggests that eosinophils may have additional T cell immune modulating activities, such as polarization.

The abundance of cytokines and granule protein products, and the expression of co-stimulatory molecules suggest an unlimited potential for eosinophils to affect the activity and cytokine expression of CD4$^+$ T cells. Of particular focus is the role of eosinophils in polarizing T cells toward a T_h2 pathway during priming through expression of IL-4. For example, studies by Kool and colleagues showed that Gr1$^+$ IL-4 expressing cells infiltrate the peritoneal cavity 24 hours after injection of adjuvant (Imject Alum),[15] and Wang and colleagues verified that these Gr1$^+$ IL-4 expressing cells are eosinophils required for immunoglobulin M (IgM), but not IgG1 T_h2 production.[16] In a later study, McKee and colleagues confirmed that nearly all IL-4-expressing cells recruited to the peritoneal cavity are eosinophils; yet, upon using IL-5-deficient mice, which have reduced numbers of eosinophils, they found that eosinophils do not appear to be required for priming T cells.[17] They further confirmed that eosinophils are not necessary for activation of the humoral response by showing unaltered or modest changes in IgG1, IgG2, and IgE levels in both eosinophil-deficient PHIL and ΔdlbGATA mice. These results are similar to earlier studies of parasite infection. IL-4 expression and eosinophil infiltration were concurrent in *Schistosoma mansoni* infection[18]; yet, *Nippostrongylus brasiliensis* infection of ΔdblGATA mice demonstrated that eosinophils were not necessary for the primary immune response.[19] Collectively, these studies demonstrated that although eosinophils are recruited to sites with antigens and express IL-4, they appear to be unnecessary for priming the T_h2 immune responses.

SECONDARY IMMUNE RESPONSE: ACTIVATION AND PROLIFERATION OF T CELLS

Post Activation of Naive T Cells

Once naive T cells are activated, they undergo clonal proliferation and develop into effector or regulatory T cells that reduce SLO homing receptor expression and upregulate tissue-specific receptors; these events permit migration through the lymphatics, vasculature, and into the tissue.[3] Approximately 7—14 days after their first exposure to antigen, effector T cells undergo a contraction phase during which T cells die via apoptosis and activation-induced cell death. The termination of the effector T cell response is likely to occur through exhaustion (i.e., excessive activation) or Fas-mediated cell death. A few clonal cells survive this process by unknown mechanisms, although IL-7, IL-15, and TCR signaling likely play an important role in CD4+ memory T cell survival.[20] These central memory T cells increase their expression of CCR7, CD62L, and LFA-1 and reside in the SLO in a same manner to naive T cells, essentially *waiting* for a secondary exposure to antigen by APCs. A few effector memory cells have been shown to turn into central memory T cells that continue to reside in peripheral tissues (e.g., the lung and intestine) after inflammation. These cells gradually reduce their expression of IL-7 and IL-15 receptors, are capable of limited cytokine production, and do not proliferate extensively.[21] In addition to homing to the SLO or residing in tissues, they may home to the local MALT, which resembles a small local SLO in function.[22] The location and development of MALT varies depending on the tissue, developmental age, and chronic exposure to antigen, and varies between, for example, humans and mice. Thus, the majority of secondary immune responses that occur in eosinophil-associated diseases, such as asthma, require the expansion of SLO-associated central memory T cells.

Eosinophil Activation of Secondary Lymphoid Organ-Associated Central Memory T Cells

The role(s) of eosinophil in inducing proliferation of antigen-specific memory CD4+ T cells have been studied in two main ways: characterization of attenuated T cell responses in eosinophil-deficient/limited mice; and direct assessment of eosinophil APC activities leading to proliferation of memory T cells. Studies using double knockout mice deficient in eosinophil-specific cytokines and chemokines [e.g., IL-5 and eotaxin/C-C motif chemokine 11 (CCL11)/eotaxin-2 (CCL24), respectively], or using eosinophil-deficient mice [e.g., PHIL[23] and ΔdblGATA[24]] have demonstrated a critical role for eosinophils in memory T cell activation. Double knockout IL-5/eotaxin mice fail to induce T_h2 cytokine production (i.e., IL-13) from primed T cells, which is restored by transfer of eosinophils prior to allergen challenge,[25] thus indicating impaired T cell activation. In a spontaneous mouse model of inflammatory bowel disease (SAMP/Yit), where effector T cells are a key component of disease progression leading to significant pathology in the distal small intestine, eosinophil depletion results in a significant impairment of T cell activities.[26] Takedastu and colleagues isolated mesenteric lymph node CD4+ T cells from SAMP/Yit mice and transferred them into severe combined immunodeficient (SCID) mice that were treated with eosinophil-depleting antibody to IL-5. As a consequence of eosinophil depletion, the transferred disease-specific T cells failed to become activated and produce cytokines upon restimulation in culture.

Although cytokine and chemokine depletion provide insight into the roles of eosinophils in T cell activation, the genetic deletion of eosinophils in mice allows a direct cause—effect assessment of the role of eosinophils in T cell responses. In particular, the reduced tissue inflammation or parasite expulsion that was previously suggested to be a consequence of T cell-mediated effects on eosinophil destructive capabilities could be reevaluated through the use of eosinophil-deficient mice. For example, eosinophil-deficient mouse models fail to develop significant pathologies in a mouse model of acute asthma.[23,27] Previous studies had suggested that the reduced pathologies were a consequence of the reduced end-stage destructive effector cell activities of eosinophils that were regulated by T cells. Yet studies in PHIL[28] and ΔdblGATA[29] mice suggested that eosinophils are necessary for T cell activation in mouse models of acute asthma, although the literature overwhelmingly suggests that DCs are the cell type essential for this function (see for example[30]). In an acute OVA-induced asthma model, the role of eosinophils in activating T cells using ΔdblGATA mice was background strain-dependent, occurring only in C57BL/6 mice.[27,29] Conversely, challenging mice with a BALB/c background with *Aspergillus fumigatus* led to a reduction in effector T cell activities in both BALB/c ΔdblGATA and CCR3-deficient animals.[31] These results are similar to C57BL/6 PHIL mice that failed to accumulate wild-type levels of effector T cells (CD4$^+$ CD62L$^{-/lo}$ CD44^{hi}) in draining lymph nodes, as detected by flow cytometry and antigen-specific *in vitro* stimulation in an OVA-induced asthma model.[28] Data from eosinophil-dependent T cell activation in parasite infection models is less significant. Infection of ΔdblGATA or PHIL mice with *S. mansoni* did not alter the production of IL-5-expressing T cells in the liver (i.e., the site of infection) of these mice.[32]

The role of eosinophils as mediators of memory T cell activation depends in part on their migration into the SLO. Although eosinophils are capable of antigen presentation, recent studies in our laboratory using the acute OVA animal model of asthma demonstrate that MHC II expression on eosinophils is not necessary for T cell proliferation. Specifically, adoptive transfer of MHC II-deficient eosinophils into OVA-sensitized PHIL mice resulted in equivalent T cell proliferation in the lung draining lymph nodes during allergen challenge as occurs in wild-type mice.[33] Similarly, although MHC II-deficient eosinophils are incapable of *in vitro* stimulation of naive T cell cytokine production in a parasite model,[12] eosinophils are not essential for *in vivo* T cell activation in the same model.[14] It is our suggestion that eosinophils aid the induced proliferation of memory T cells through modulation of DC activities. Indeed, we have recently demonstrated that activated eosinophils are necessary for the accumulation of DCs within the lung-draining lymph node after allergen challenge in an acute model of asthma.[33] The mechanisms of such eosinophil-induced DC accumulation are unknown but several studies have suggested a role for eosinophils in DC recruitment and maturation. For example, Yang and colleagues demonstrated that eosinophil-derived neurotoxin (EDN) induced selective recruitment and release of inflammatory cytokines by dendritic cells.[34] *In vitro* eosinophils have been shown to enhance dendritic cell maturation,[35] viral antigen presentation,[36] and chemoattraction of CD11b^{hi} DCs.[37] Together these data suggest that rather than eosinophils being the APCs required to present antigen to memory T cells, it is likely that eosinophils have co-stimulatory activities leading to proliferation and polarization of memory T cells.

SECONDARY IMMUNE RESPONSE: POLARIZATION OF T CELLS

Memory T Cell Pool and Polarization

The paradigm suggesting that polarization of T cells during the primary immune response is retained following subsequent allergen exposures is undergoing reevaluation. Studies are now demonstrating that central memory T cells comprise a heterogeneous pool with the potential for a variety of responses that may be modified by several factors (described by Farber and colleagues[20]): (1) the initial polarization status of the T cell is maintained upon restimulation, possibly due to histone remodeling in the chromosomes; (2) the polarization status changes due to the cytokine/co-stimulatory signaling in the local microenvironment; and (3) a single pool of polarized T cells predominates due to enhanced production of that subclone from within the heterogeneous pool.

In addition, the ability of effector T cells to *switch* or transition to other phenotypes is well documented. In basic terms, CD4$^+$ T cells are defined by their cytokine expression: T_h17 (IL-17A and IL-17F), T_h1 (IFN-γ and IL-12), T_h2 cells (IL-4, IL-5, IL-13, and IL-25), and T regulatory cells [(IL-10 and transforming growth factor β (TGF-β)]. The regulation of T_h1 and T_h2 cells is balanced in part by expression of transcription factors T-bet and GATA-3, respectively. Furthermore, T regulatory and T_h17 cells are balanced by the interplay between transcription factors FOXP3 and RORγt, respectively. Depending on the presence of cytokines (e.g., IL-23, IL-6, and TGF-β) and their activation state, T regulatory cells may fail to become T_h17 cells.[38] Similarly, classical T_h2 or T_h1 cells may lack expression of phenotype-defining cytokines, such as IL-4 or IFN-γ, respectively.[39] Consequently, the ability to define diseases by changes to a particular polarization pathway is increasingly difficult, due to complex, delicately balanced interactions.

Eosinophil-Induced Polarization of T Cells During Secondary Immune Responses

Eosinophils have an almost unlimited potential to modulate polarization of memory T cells. For example, within their granules eosinophils contain numerous preformed and rapidly released cytokines and mediators, such as T_h2 cytokines (e.g., IL-4 and IL-25), T_h1 cytokines (e.g., IL-12), or suppressive mediators [e.g., TGF-β, IL-10, and indoleamine 2,3-dioxygenase (IDO); see Chapter 16.3]. In essence, eosinophils may suppress or enhance polarization of individual memory T cells within a mixed pool in a coordinated fashion. For example, release of IDO induces the death of T_h1 cells, allowing T_h2 responses to predominate.[40] Conversely, release of IDO in concert with release of TGF-β and weak co-stimulation of memory T cells could in essence result in expansion of induced T regulatory cells, rather than T_h1 or T_h2 cells. The activities of some of these cytokines may be modulated by DC activation of T cells. For example, release of IL-25 enhances the ability of DCs to polarize T cells toward a T_h2 phenotype.[41] Assessing the effect of these modifications on

DCs *in vivo* is critical, as DCs that have been characterized as T_h1-polarizing *in vitro* are rendered T_h2-polarizing once placed inside a mucosal environment that contains antigen-specific memory T cells.[7]

Other mechanisms for eosinophil-mediated activation of DCs include the release of endogenous TLR ligands that modulate activation and polarization of T cells by changing cytokine release from DCs.[8] Moreover, $CD4^+$ T cells in both humans and mice have been demonstrated to express all TLRs under various conditions,[42] and thus may be a direct target of eosinophils. Most TLRs bind foreign molecules, such as viral and bacterial particles, as a pathogen defense mechanism, but several TLRs, such as TLR2 and TLR4, recognize host molecules like hyaluronic acid and Hsp70, respectively. Endogenous release of eosinophil-derived TLR ligands may include eosinophil granule proteins and molecules released from dying cells, termed damage associated molecular patterns (DAMPs). In particular, Yang and colleagues have demonstrated that EDN activates TLR2 on DCs, resulting in an enhanced T_h2 polarized immune response.[34] Alternatively, high mobility group box 1 (HMGB1) released by apoptotic or necrotic eosinophils may create a T_h1 environment via binding TLR4 or TLR9.[8] Furthermore, several studies have demonstrated the release of nucleic acid material from eosinophils, such as the targeted release of mitochondrial DNA by eosinophils[43] and the release of DNA and RNA from eosinophil secretory granules,[44] which may have TLR ligand activities.

As mentioned previously, eosinophils constitutively express IL-4, but have not been proven to be the essential IL-4 producing cells that influence T_h2 polarization during immune reactions. Several studies have attempted to characterize the roles of MHC II presentation and T_h2 polarization (i.e., IL-4 release) by basophils, dendritic cells, or eosinophils. The problem with DCs being sufficient for both functions is that DCs do not express IL-4, which is necessary for most T_h2-polarizing conditions,[7] thus suggesting that eosinophils or basophils are also required.[6] Thus, delineating the role of these cells has required the use of various knockout mice, adoptive transfer techniques, and antibody or toxin-mediated depletion of cell subtypes. Together, several studies have shown that an alternative (i.e., non-DC) MCH II-expressing innate cell is required to regulate the T_h2/T_h1 polarized immune responses in models of pulmonary inflammation. For example, Niu and colleagues[45] demonstrated in a mouse model of OVA-induced asthma that T cell proliferation is induced by a DC-dependent mechanism, yet the cytokine profile is switched from that of T_h2 to T_h1 (T_h17 cells were not tested) when mice are deficient in an MHC II-expressing innate cell other than DCs. In a mouse model of helminth infection, T_h2 polarization, but not T cell proliferation, was shown to be dependent on basophil MHC II activities,[46] implicating

basophils as the candidate cell. Yet, the antibody used to deplete basophils [MAR-1 (FcRIα)] in this model was demonstrated by Hammad and colleagues to also deplete eosinophils.[47] Studies by our laboratory have demonstrated that DCs that are capable of inducing T_h2 responses in wild-type mice induce a markedly increased production of T_h1/T_h17 cytokines by T cells, in addition to a T_h2 response, in the absence of eosinophils (i.e., in PHIL mice).[33] These data suggest that eosinophils are necessary for suppressing T_h1/T_h17 responses in an acute mouse model of OVA-induced asthma, although their role in T_h2 polarization remains to be defined. Interestingly, in a skin sensitization model, IL-13/4 double knockout mice failed to recruit eosinophils and, instead, developed an elevated T_h17 response upon aerosol challenge, again suggesting a correlation between eosinophils and the suppression of T cell T_h1/T_h17 immune responses.[48] The mechanisms of eosinophil-mediated suppression remain to be clarified, but the release of IDO, TGF-β, or co-stimulatory activities may be relevant. These studies highlight unique inflammatory events that rely on these innate cells to influence the local cytokine milieu through both independent and redundant mechanisms.

Although eosinophils may regulate T cell polarization through release of molecular mediators, eosinophil–T cell physical interactions may amplify or attenuate these responses through the binding of co-stimulatory molecules or T cell-activating receptors. For example, eosinophils express CD40, CD80, and CD86 co-stimulatory molecules.[5,9,10] CD80 and CD86 (i.e., B7-1 and B7-2, respectively) bind to the CD28 receptor on T cells, which is the predominant T cell co-stimulatory receptor required for full activation of both naive and activated T cells. Conversely, two co-stimulatory receptors in the CD28 family that are found only on activated T cells include inducible co-stimulatory molecule (ICOS), which is upregulated on T cells in asthmatics, and cytotoxic T-lymphocyte antigen (CTLA)-4, an inhibitory coreceptor that is downregulated on asthmatic T cells.[49] As CTLA-4 is capable of binding CD80/86 on eosinophils, it is possible eosinophils downregulate the activity of a subset of memory T cells, such as T_h17/T_h1 cells, allowing the T_h2 immune response to predominate. ICOS generally binds B7RP1, rather than CD80/86, and has not been shown to be expressed on eosinophils to date. CD40L (or CD154)-deficient memory T cells proliferate in the absence of CD40 binding, although their cytokine profile is altered,[50] implying that eosinophils may suppress Th differentiation while allowing T cell proliferation in the SLO. In addition to co-stimulatory molecules, other cell–cell interactions, such as those that occur through the Jagged–Notch pathway, may mediate modulation of T cell activities. Human eosinophils constitutively express Jagged 1,[51] a Notch ligand that has been implicated in T cell polarization. Therefore, a variety of co-stimulatory and cell surface ligand/receptor molecules are potentially capable of modulating T cell activation and polarization, although eosinophil-specific activities need further elucidation.

SECONDARY IMMUNE RESPONSE: RECRUITMENT OF T CELLS

Effector T Cells to Sites of Peripheral Inflammation

Migration of T cells to peripheral sites of inflammation is an essential element of the tissue inflammatory process. Selective recruitment of T cells occurs through site-specific chemoattraction, cell adhesion molecule binding, and cellular activation, followed by extravasation into the tissue. Central memory T cells are imprinted during the primary immune response to express tissue-specific homing receptors.[3,52] That is, memory T cell activation results in changes in the expression of homing receptors, making them responsive to chemokines expressed by the tissue. In this process, CCR7 is downregulated and $S1P_1$ upregulated to release T cells from the SLO to enable contact with addressins, which are integrins expressed by site-specific cells. In general, T cells have been demonstrated to express various integrins, including VLA-4 (CD49d, $\alpha4\beta1$), LFA-1 ($\alpha L\beta2$ and CD11a), and integrin $\beta7$, that bind a variety of addressins. The regulation of these homing receptor—addressin interactions is highly complex and not completely understood. Some examples will be provided here to highlight the complexities of these interactions. LFA-1 expression is increased on effector $CD4^+$ T cells and its binding to addressin CD54 (ICAM) is necessary for effector T_h2 T cell recruitment to the lung after allergen challenge in a mouse model of asthma.[53] Conversely, $CD4^+$ T_h1 cells appear to rely on CD11a and CD49b interactions with CD54 or CD106 [(vascular endothelial protein-1 (VCAM-1)], respectively, to home to the lung.[53] The gastrointestinal tract, lungs, and skin all have T cell imprinting and express addressins. For example, unlike the lung, homing of T cells to the gastrointestinal tract depends on expression of integrins $\alpha4\beta7$ (LPAM-1) and $\alpha E\beta7$ (CD103) to bind mucosal addressin cell adhesion molecule-1 (MADCAM1) and E-cadherin, respectively.

Tightly coordinated T cell recruitment is also obtained through the expression of chemokines in peripheral tissues. Briefly, chemokines provide a molecular gradient that acts as a directional signal for the cells to move toward the tissue. An extensive array of chemokines has been demonstrated to coordinate the movement of $CD4^+$ T cells into tissues.[2,54] In general terms, the chemokine receptor expression on a T cell often corresponds to its polarization status. For example, T_h1 cells are recruited via CCR5 (CCL5) and CXCR3 (CXCL10), T_h2 cells via CCR8 (CCL17) and CCR4 (CCL22), and T_h17 cells via CCR6 (CCL20). CCR3 is a chemokine receptor found on human T_h2 T cells[55] that is not detected[31,56] or found to induce calcium influx in mouse T cells under a variety of conditions.[56] An additional T_h2-associated T cell receptor (TCR) is ST2, which binds the alarmin IL-33[57] and has been found to be expressed on murine T_h2 cells and human T_h2, T_h1, T_h17 cells. Although ST2 may activate polarized cells, its role in recruitment is unclear. T_h17 cells have been also shown to express both T_h1 and T_h2 classical chemokine receptors (i.e., CCR4 and CXCR3). Unfortunately, these T cell subtypes have been shown to express a variety of chemokine receptors concurrently, demonstrating that these classifications are imperfect guidelines. Moreover, other inflammatory models, such as CXCR4, CCR2, and CCR9, have all been shown to be essential for both T cell recruitment and T cell differentiation, although some have more tissue specificity than others.

Beyond chemokine signals, lipid mediators can also induce T cell movement into tissue.[58] For example, the cyclooxygenase pathway metabolite prostaglandin D2 binds the type-2 prostaglandin-2 receptor (CRTH2) on T_h2 cells and is considered to be a hallmark receptor for T_h2 cells in humans (it binds both T_h1 and T_h2 cells in mice).[59] CRTH2 acts as both a chemotactic receptor to induce recruitment of T cells and enhance T_h2 cell polarization. In parallel with the cyclooxygenase pathway is the 5-lipoxygenase pathway, in which arachidonic acid is metabolized into cysteinyl leukotrienes (LTC_4, LTD_4, and TLE_4) and leukotriene B_4 (LTB_4). LTB_4 is a classical activator and chemoattractant of innate cell inflammatory cells that express the LTB_4 receptors, BLT_1 and BLT_2.[60] However, several studies have demonstrated a role for LTB_4 in the recruitment of BLT_1-expressing T_h1 and T_h2 cells to the lung in mouse models of asthma.[61,62] Furthermore, LTB_4 specifically recruits effector T cells, as opposed to naive or memory T cells, and this response includes both $CD8^+$ and $CD4^+$ T cells depending on the model system assessed.

Eosinophil Recruitment of Effector T Cells

In essence, the role of eosinophils in effector T cell trafficking is a chicken-or-egg paradox. The original paradigm was that T cells regulated the recruitment of eosinophils and not vice versa, particularly as eosinophils were considered end-stage destructive cells recruited and regulated by T cells. First and foremost, T cell deficient animals fail to develop pulmonary inflammation in mouse models of asthma.[63] Thus, mouse models that resulted in the loss of eosinophil recruitment to sites of tissue inflammation (e.g., IL-5/eotaxin double knockout mice) also resulted in impaired T cell recruitment, thus indicating that the reduction in eosinophil numbers is T cell-dependent. Further supporting this concept were studies demonstrating that DCs, and not eosinophils, were required for T cell recruitment and T_h2 inflammation in mouse models of acute asthma.[30] As such, the possibility that eosinophils are modulating cells capable of regulating T cell recruitment was in doubt. However, recent studies have highlighted a significant role for eosinophils in T cell recruitment to the pulmonary compartment in mouse models of asthma.[28,29]

Eosinophil-mediated recruitment of T cells has been demonstrated both directly and indirectly through use of transgenic and knockout animals. Genetic manipulations in mouse models to overexpress cytokines necessary for eosinophilopoeisis (e.g., IL-5) and recruitment (e.g., eotaxin) have enabled an assessment of their roles in pathophysiology. In these models, the resulting recruitment and survival of eosinophils is a necessary component of T cell recruitment to the lung. In a mouse model that coexpresses IL-5 (from T cells) and eotaxin-2 (from lung Clara cells), eosinophils are recruited to the lung in naive animals and induce significant pulmonary histopathology, lung dysfunction [64] and T cell recruitment (our unpublished observations). Crossing these double transgenic mice to an eosinophil-deficient mouse (PHIL) demonstrated that both T_h2 pulmonary pathologies[64] and T cell recruitment (our unpublished observations) are dependent on the presence of eosinophils. Excessive eosinophil levels in circulation are insufficient for these responses (i.e., in transgenic IL-5 animals); rather, eosinophils must be recruited to the lung to induce tissue inflammation or T cell recruitment to that tissue. In a similar model, overexpression of IL-25 leads to hypereosinophilia, expression of eotaxin, and induced pulmonary inflammation dependent on IL-25-induced expression of IL-13.[65] Together, these data demonstrate that expression of cytokines and chemokines that activate eosinophils result in the recruitment of T_h2 effector cells. Conversely, mouse models that are deficient or devoid of T_h2 cytokines, such as IL-13/4, fail to recruit both eosinophils and T_h2 T cells, yet remain capable of recruiting T_h17 T cells.[48] These data suggest that eosinophils directly regulate T_h2 T cell recruitment and not T_h1/T_h17 T cell recruitment in models of lung and skin inflammation.

An alternative to inducing hypereosinophilia in transgenic mouse models is the use of gene knockout mice and eosinophil adoptive transfers. Transfer of eosinophils into mice has been used as a means of demonstrating eosinophil activation of T cells.[9,25] It is important to note that the origin of eosinophils for these assessments may lead to varying results in both T cell activation and inflammatory events. Nearly all models that rely on eosinophil adoptive transfer obtain eosinophils from two transgenic mice strains that overexpress IL-5 in T cells.[66,67] IL-5 overexpression results in tremendous splenomegaly in the mice, with

eosinophilopoeisis and possibly aberrant eosinophil activation occurring in the spleen.[68] Thus, eosinophils isolated from the spleen probably have a mixed phenotypic pool of progenitor eosinophils and activated eosinophils compared to blood-derived eosinophils. For example, transfer of spleen-derived eosinophils from naive IL-5 transgenic donor mice into the lungs of naive recipient wild-type mice induced pulmonary inflammation, even in the absence of lymphocytes.[69] Conversely, similar numbers of blood-derived eosinophils transferred into either naive IL-5-deficient[70] mice or sensitized eosinophil-deficient mice[28] fail to induce pulmonary inflammation or T cell recruitment.

In an effort to differentiate eosinophil-dependent and T cell-dependent recruitment to the lung, studies have used eosinophil-deficient mouse models as recipients in adoptive cell transfer experiments. As mentioned previously, ΔdblGATA1[29] and PHIL[28] mice that underwent OVA-induced asthma failed to induce effector T cell recruitment to the lung in a C57BL/6 background. In contrast, it was demonstrated in ΔdlbGATA mice in a BALB/c background that effector T cell responses were not disrupted, suggesting the possibility of strain differences. Yet similar studies with ΔdblGATA1 and CCR3-deficient BALB/c animals using *A. fumigatus* as an allergen showed that effector T cell activities were impaired in these mice.[31] The mechanism of failed effector T cell recruitment in both ΔdlbGATA and PHIL mice was suggested to be eosinophil dependent, although by different pathways. Jacobsen and colleagues[28] demonstrated that PHIL mice were impaired in their activation of memory T cells, and so bypassed the need for endogenous cells by transferring (intravenously) *in vitro* polarized OVA-specific T cells (i.e., OT-II T_h2-polarized T cells). When these cells were transferred into naive PHIL mice and then challenged with aerosolized OVA, the transferred T_h2 T cells did not accumulate in the lung and pulmonary inflammation was not induced. Conversely, intravenous transfer of both T_h2 T cells and blood-derived eosinophils into the lungs of PHIL mice resulted in T_h2 T cell accumulation in the lung upon allergen challenge. This recruitment of T_h2 cells was blocked by administration of antibodies that bind the chemokines known to bind the T_h2 chemokine receptors CCR4 and CCR8 (the CCL17 and CCL22 ligands). Walsh and colleagues[29] transferred eosinophils intravenously into sensitized ΔdblGATA1 mice and demonstrated that the presence of eosinophils is necessary and sufficient to rescue pulmonary inflammation in these mice upon allergen challenge. These studies showed that eotaxin levels are attenuated in these mice and that targeting cells that bind eotaxins (e.g., CCR3-expressing cells) restores T cell recruitment. Similarly, administration of eotaxin directly into the lungs of allergen-challenged ΔdblGATA1 mice restores the recruitment of T cells and pulmonary inflammation. The differences between eotaxin-dependent T cell recruitment and CCL17/CCL22-dependent recruitment are unclear. More importantly, as mentioned previously CCR3 has not been found to be expressed on mouse T cells *in vivo* or to induce activation of mouse T cells.[31,56] A subset of human T_h2 cells has been shown to express CCR3 and CCr3 is a marker for T_h2 polarization in those T cells; yet, their calcium mobilization to eotaxin is extremely low compared to CCL17 (sevenfold higher).[55] Thus, the report that 40% of all T cells in the pulmonary compartment express CCR3 (as determined by flow cytometry) in response to eotaxin administration[29] contradicts previous publications on CCR3 expression. Regardless, in humans CCR3 T cells may respond to eotaxins in conjunction with eosinophil eotaxin-dependent recruitment.

Although CCR3 may play a small role in T cell recruitment in humans, both CCR4 and CCR8 are well-characterized mediators of T_h2 T cell recruitment in both humans[71] and mice.[72] In particular, mice deficient in CCR4 or CCR8 (less so) are unable to induce wild-type levels of inflammatory events. The specificity of this process is demonstrated by the adoptive transfer of T_h2-polarized CCR4-deficient T cells into naive mice at the time of allergen challenge. These T cells fail to be recruited to the lung and as a consequence T_h2 inflammation is reduced, including a 50% reduction in BAL eosinophil levels.[72] Perros and colleagues[73] transferred peripheral blood monocytes from patients allergic to house dust mite into SCID mice treated with anti-human CCR4 antibody and challenged with house dust mites. The antibody blockade of CCR4 resulted in a complete inhibition of $CD4^+$ T cell recruitment to the lung and a lack of pulmonary inflammation. These studies demonstrate that T cell recruitment in allergic asthma models is dependent on CCR4 and its ligands, CCL17 and CCL22. Furthermore, eosinophil recruitment to the lung, although reduced, is independent of T cell recruitment, suggesting that eosinophils are not the end-stage mediators of T cell activities.

Although chemokine regulation of T cell recruitment has been most thoroughly studied in mouse models of asthma, eosinophils may mediate recruitment through additional activities. For example, eosinophils are potent producers of lipid mediators, such as prostaglandins and leukotrienes.[74] Thus, eosinophils may signal T cell recruitment via CRTH2 and BLT_1, although no direct role has been demonstrated to date. Additionally, eosinophil—T cell activation via TLR activation or co-stimulatory molecules may be a required component to activate cells sufficiently to modify expression of cell surface adhesion molecules and enable them to adhere, roll, and extravasate to the tissue. Finally, although eosinophil-deficient mice fail to recruit effector T_h2 cells to the lungs following allergen provocation, compensatory mechanisms are likely to replace eosinophils in certain inflammatory diseases (e.g., T_h1 or T_h17 inflammation) and parasite infections.

Similarly, it is possible that both T_h2 T cells and eosinophils are required for initial activation and that the immune responses following amplification cycles of repeated allergen provocation become eosinophil-dependent.[54]

CONCLUSION

Eosinophils are decidedly complex cells that have the capacity to alter the immune response in a variety of diseases, as well at homeostatic baseline. Over the last few decades a plethora of studies have demonstrated the ability of eosinophils to change the function of T cells, particular $CD4^+$ T_h2 polarized T cells. One may speculate that this pathway is a potentially important strategy to attenuate ongoing immune responses or switch the character of T cell responses (e.g., T_h2 to T_h1/T_h17) as a compensatory mechanism to fight the stressing agent (e.g., infection, cancer, or an exogenous allergen). The mechanisms of these eosinophil-mediated changes include T cell co-stimulation, polarization, and recruitment. In particular, eosinophils do not appear to be as important in primary immune responses, but rather directly or indirectly (through altering the function of DCs) modulate the polarization and recruitment of T cells to sites of T_h2 inflammation.

REFERENCES

1. Rothenberg ME, Hogan SP. The eosinophil. *Annu Rev Immunol* 2006;**24**:147—74.
2. Bromley SK, Mempel TR, Luster AD. Orchestrating the orchestrators: chemokines in control of T cell traffic. *Nat Immunol* 2008;**9**(9):970—80.
3. Marelli-Berg FM, Cannella L, Dazzi F, Mirenda V. The highway code of T cell trafficking. *J Pathol* 2008;**214**(2):179—89.
4. Jung YJ, Woo SY, Jang MH, Miyasaka M, Ryu KH, Park HK, et al. Human Eosinophils Show Chemotaxis to Lymphoid Chemokines and Exhibit Antigen-Presenting-Cell-Like Properties upon Stimulation with IFN-gamma, IL-3 and GM-CSF. *Int Arch Allergy Immunol* 2008;**146**(3):227—34.
5. Wang H-B, Ghiran I, Matthaei K, Weller PF. Airway eosinophils: allergic inflammation recruited professional antigen-presenting cells. *Journal of Immunology* 2007;**179**(11):7585—92.
6. Barrett NA, Austen KF. Innate cells and T helper 2 cell immunity in airway inflammation. *Immunity* 2009;**31**(3):425—37.
7. Pulendran B, Tang H, Manicassamy S. Programming dendritic cells to induce T(H)2 and tolerogenic responses. *Nat Immunol* 2010;**11**(8):647—55.
8. Piccinini AM, Midwood KS. DAMPening inflammation by modulating TLR signalling. *Mediators Inflamm* 2010;**2010**.
9. Akuthota P, Wang H, Weller PF. Eosinophils as antigen-presenting cells in allergic upper airway disease. *Curr Opin Allergy Clin Immunol* 2010;**10**(1):14—9.
10. MacKenzie JR, Mattes J, Dent LA, Foster PS. Eosinophils promote allergic disease of the lung by regulating $CD4^+$ Th2 lymphocyte function. *J Immunol* 2001;**167**(6):3146—55.

11. van Rijt LS, Vos N, Hijdra D, de Vries VC, Hoogsteden HC, Lambrecht BN. Airway eosinophils accumulate in the mediastinal lymph nodes but lack antigen-presenting potential for naive T cells. *J Immunol* 2003;**171**(7):3372—8.

12. Padigel UM, Lee JJ, Nolan TJ, Schad GA, Abraham D. Eosinophils can function as antigen-presenting cells to induce primary and secondary immune responses to Strongyloides stercoralis. *Infect Immun* 2006;**74**(6):3232—8.

13. Padigel UM, Hess JA, Lee JJ, Lok JB, Nolan TJ, Schad GA, et al. Eosinophils act as antigen presenting cells to induce immunity to Strongyloides stercoralis in mice. *The Journal of Infectious Diseases* 2007;**196**:1844—51.

14. O'Connell AE, Hess JA, Santiago GA, Nolan TJ, Lok JB, Lee JJ, et al. Major basic protein from eosinophils and myeloperoxidase from neutrophils are required for protective immunity to Strongyloides stercoralis in mice. *Infect Immun* 2011;**79**(7):2770—8.

15. Kool M, Soullie T, van Nimwegen M, Willart MA, Muskens F, Jung S, et al. Alum adjuvant boosts adaptive immunity by inducing uric acid and activating inflammatory dendritic cells. *J Exp Med* 2008;**205**(4):869—82.

16. Wang HB, Weller PF. Pivotal Advance: Eosinophils mediate early alum adjuvant-elicited B cell priming and IgM production. *J Leukoc Biol* 2008;**83**(4):817—21.

17. McKee AS, Munks MW, MacLeod MK, Fleenor CJ, Van Rooijen N, Kappler JW, et al. Alum induces innate immune responses through macrophage and mast cell sensors, but these sensors are not required for alum to act as an adjuvant for specific immunity. *J Immunol* 2009;**183**(7):4403—14.

18. Sabin EA, Kopf MA, Pearce EJ. Schistosoma mansoni egg-induced early IL-4 production is dependent upon IL-5 and eosinophils. *J Exp Med* 1996;**184**(5):1871—8.

19. Voehringer D, Reese TA, Huang X, Shinkai K, Locksley RM. Type 2 immunity is controlled by IL-4/IL-13 expression in hematopoietic non-eosinophil cells of the innate immune system. *J Exp Med* 2006;**203**(6):1435—46.

20. Lees JR, Farber DL. Generation, persistence and plasticity of CD4 T-cell memories. *Immunology* 2010;**130**(4):463—70.

21. Woodland DL, Kohlmeier JE. Migration, maintenance and recall of memory T cells in peripheral tissues. *Nat Rev Immunol* 2009;**9**(3):153—61.

22. Cesta MF. Normal structure, function, and histology of mucosa-associated lymphoid tissue. *Toxicol Pathol* 2006;**34**(5):599—608.

23. Lee JJ, Dimina D, Macias MP, Ochkur SI, McGarry MP, O'Neill KR, et al. Defining a link with asthma in mice congenitally deficient in eosinophils. *Science* 2004;**305**(5691):1773—6.

24. Pevny L, Simon MC, Robertson E, Klein WH, Tsai SF, D'Agati V, et al. Erythroid differentiation in chimaeric mice blocked by a targeted mutation in the gene for transcription factor GATA-1. *Nature* 1991;**349**(6306):257—60.

25. Mattes J, Yang M, Mahalingam S, Kuehr J, Webb DC, Simson L, et al. Intrinsic defect in T cell production of interleukin (IL)-13 in the absence of both IL-5 and eotaxin precludes the development of eosinophilia and airways hyperreactivity in experimental asthma. *J Exp Med* 2002;**195**(11):1433—44.

26. Takedatsu H, Mitsuyama K, Matsumoto S, Handa K, Suzuki A, Funabashi H, et al. Interleukin-5 participates in the pathogenesis of ileitis in SAMP1/Yit mice. *Eur J Immunol* 2004;**34**(6):1561—9.

27. Humbles AA, Lloyd CM, McMillan SJ, Friend DS, Xanthou G, McKenna EE, et al. A critical role for eosinophils in allergic airways remodeling. *Science* 2004;**305**(5691):1776—9.

28. Jacobsen EA, Ochkur SI, Pero RS, Taranova AG, Protheroe CA, Colbert DC, et al. Allergic Pulmonary Inflammation in Mice is Dependent on Eosinophil-induced Recruitment of Effector T Cells. *Journal of Experimental Medicine* 2008;**205**(3):699—710.

29. Walsh ER, Sahu N, Kearley J, Benjamin E, Kang BH, Humbles A, et al. Strain-specific requirement for eosinophils in the recruitment of T cells to the lung during the development of allergic asthma. *J Exp Med* 2008;**205**(6):1285—92.

30. Lambrecht BN, De Veerman M, Coyle AJ, Gutierrez-Ramos JC, Thielemans K, Pauwels RA. Myeloid dendritic cells induce Th2 responses to inhaled antigen, leading to eosinophilic airway inflammation. *J Clin Invest* 2000;**106**(4):551—9.

31. Fulkerson PC, Fischetti CA, McBride ML, Hassman LM, Hogan SP, Rothenberg ME. A central regulatory role for eosinophils and the eotaxin/CCR3 axis in chronic experimental allergic airway inflammation. *Proc Natl Acad Sci USA* 2006;**103**(44):16418—23.

32. Swartz JM, Dyer KD, Cheever AW, Ramalingam T, Pesnicak L, Domachowske JB, et al. Schistosoma mansoni infection in eosinophil lineage-ablated mice. *Blood* 2006;**108**(7):2420—7.

33. Jacobsen EA, Zellner KR, Colbert D, Lee NA, Lee JJ. Eosinophils regulate dendritic cells and Th2 pulmonary immune responses following allergen provocation. *Journal of Immunology*; 2011. In press.

34. Yang D, Chen Q, Su SB, Zhang P, Kurosaka K, Caspi RR, et al. Eosinophil-derived neurotoxin acts as an alarmin to activate the TLR2-MyD88 signal pathway in dendritic cells and enhances Th2 immune responses. *J Exp Med* 2008;**205**(1):79—90.

35. Lotfi R, Lotze MT. Eosinophils induce DC maturation, regulating immunity. *J Leukoc Biol* 2008;**83**(3):456—60.

36. Davoine F, Cao M, Wu Y, Ajamian F, Ilarraza R, Kokaji AI, et al. Virus-induced eosinophil mediator release requires antigen-presenting and CD4+ T cells. *J Allergy Clin Immunol* 2008;**122**(1):69—77. 77 e61—62.

37. Rose Jr CE, Lannigan JA, Kim P, Lee JJ, Fu SM, Sung SS. Murine lung eosinophil activation and chemokine production in allergic airway inflammation. *Cell Mol Immunol*(5):361—334, http://dx.doi.org/310.1038/cmi.2010.1031, 2010;**7**.

38. Damsker JM, Hansen AM, Caspi RR. Th1 and Th17 cells: adversaries and collaborators. *Ann N Y Acad Sci* 2010;**1183**:211—21.

39. Prussin C, Yin Y, Upadhyaya B. TH2 heterogeneity: Does function follow form? *J Allergy Clin Immunol*, http://dx.doi.org/10.1016/j.jaci.2010.1008.1031; 2010.

40. Odemuyiwa SO, Ghahary A, Li Y, Puttagunta L, Lee JE, Musat-Marcu S, et al. Cutting Edge: Human Eosinophils Regulate T Cell Subset Selection through Indoleamine 2,3-Dioxygenase. *J Immunol* 2004;**173**(10):5909—13.

41. Wang YH, Angkasekwinai P, Lu N, Voo KS, Arima K, Hanabuchi S, et al. IL-25 augments type 2 immune responses by enhancing the expansion and functions of TSLP-DC-activated Th2 memory cells. *J Exp Med* 2007;**204**(8):1837—47.

42. Kulkarni R, Behboudi S, Sharif S. Insights into the role of Toll-like receptors in modulation of T cell responses. *Cell Tissue Res* 2011;**343**(1):141—52.

43. Simon D, Hoesli S, Roth N, Staedler S, Yousefi S, Simon HU. Eosinophil extracellular DNA traps in skin diseases. *J Allergy Clin Immunol* 2010;**127**(1):194−9.

44. Behzad AR, Walker DC, Abraham T, McDonough J, Mahmudi-Azer S, Chu F, et al. Localization of DNA and RNA in eosinophil secretory granules. *Int Arch Allergy Immunol* 2010;**152**(1):12−27.

45. Niu N, Laufer T, Homer RJ, Cohn L. Cutting edge: Limiting MHC class II expression to dendritic cells alters the ability to develop Th2-dependent allergic airway inflammation. *J Immunol* 2009;**183**(3):1523−7.

46. Perrigoue JG, Saenz SA, Siracusa MC, Allenspach EJ, Taylor BC, Giacomin PR, et al. MHC class II-dependent basophil-CD4+ T cell interactions promote T(H)2 cytokine-dependent immunity. *Nat Immunol* 2009;**10**(7):697−705.

47. Hammad H, Plantinga M, Deswarte K, Pouliot P, Willart MA, Kool M, et al. Inflammatory dendritic cells—not basophils—are necessary and sufficient for induction of Th2 immunity to inhaled house dust mite allergen. *J Exp Med* 2010;**207**(10):2097−111.

48. He R, Oyoshi MK, Jin H, Geha RS. Epicutaneous antigen exposure induces a Th17 response that drives airway inflammation after inhalation challenge. *Proc Natl Acad Sci USA* 2007;**104**(40): 15817−22.

49. Botturi K, Lacoeuille Y, Cavailles A, Vervloet D, Magnan A. Differences in allergen-induced T cell activation between allergic asthma and rhinitis: Role of CD28, ICOS and CTLA-4. *Respir Res* 2010;**12**:25.

50. MacLeod M, Kwakkenbos MJ, Crawford A, Brown S, Stockinger B, Schepers K, et al. CD4 memory T cells survive and proliferate but fail to differentiate in the absence of CD40. *J Exp Med* 2006;**203**(4):897−906.

51. Radke AL, Reynolds LE, Melo RC, Dvorak AM, Weller PF, Spencer LA. Mature human eosinophils express functional Notch ligands mediating eosinophil autocrine regulation. *Blood* 2009;**113**(13):3092−101.

52. McKinstry KK, Strutt TM, Swain SL. The potential of CD4 T-cell memory. *Immunology* 2010;**130**(1):1−9.

53. Lee SH, Prince JE, Rais M, Kheradmand F, Ballantyne CM, Weitz-Schmidt G, et al. Developmental control of integrin expression regulates Th2 effector homing. *J Immunol* 2008; **180**(7):4656−67.

54. Medoff BD, Thomas SY, Luster AD. T cell trafficking in allergic asthma: the ins and outs. *Annu Rev Immunol* 2008;**26**:205−32.

55. D'Ambrosio D, Iellem A, Bonecchi R, Mazzeo D, Sozzani S, Mantovani A, et al. Selective upregulation of chemokine receptors CCR4 and CCR8 upon activation of polarized human type 2 Th cells. *J Immunol* 1998;**161**(10):5111−5.

56. Grimaldi JC, Yu NX, Grunig G, Seymour BW, Cottrez F, Robinson DS, et al. Depletion of eosinophils in mice through the use of antibodies specific for C-C chemokine receptor 3 (CCR3). *J Leukoc Biol* 1999;**65**(6):846−53.

57. Liew FY, Pitman NI, McInnes IB. Disease-associated functions of IL-33: the new kid in the IL-1 family. *Nat Rev Immunol* 2010;**10**(2):103−10.

58. Luster AD, Tager AM. T-cell trafficking in asthma: lipid mediators grease the way. *Nat Rev Immunol* 2004;**4**(9):711−24.

59. Schuligoi R, Sturm E, Luschnig P, Konya V, Philipose S, Sedej M, et al. CRTH2 and D-type prostanoid receptor antagonists as novel therapeutic agents for inflammatory diseases. *Pharmacology* 2010;**85**(6):372−82.

60. Ohnishi H, Miyahara N, Gelfand EW. The role of leukotriene B^4 in allergic diseases. *Allergol Int* 2008;**57**(4):291−8.

61. Miyahara N, Miyahara S, Takeda K, Gelfand EW. Role of the LTB4/BLT1 pathway in allergen-induced airway hyperresponsiveness and inflammation. *Allergol Int* 2006;**55**(2):91−7.

62. Tager AM, Bromley SK, Medoff BD, Islam SA, Bercury SD, Friedrich EB, et al. Leukotriene B4 receptor BLT1 mediates early effector T cell recruitment. *Nat Immunol* 2003;**4**(10): 982−90.

63. Gavett SH, Chen X, Finkelman F, Wills-Karp M. Depletion of murine CD4+ T lymphocytes prevents antigen-induced airway hyperreactivity and pulmonary eosinophilia. *Am J Respir Cell Mol Biol* 1994;**10**(6):587−93.

64. Ochkur SI, Jacobsen EA, Protheroe CA, Biechele TL, Pero RS, McGarry MP, et al. Co-Expression of IL-5 and Eotaxin-2 in Mice Creates an Eosinophil-Dependent Model of Respiratory Inflammation with Characteristics of Severe Asthma. *Journal of Immunology* 2007;**178**(12):7879−89.

65. Fort MM, Cheung J, Yen D, Li J, Zurawski SM, Lo S, et al. IL-25 induces IL-4, IL-5, and IL-13 and Th2-associated pathologies in vivo. *Immunity* 2001;**15**(6):985−95.

66. Dent LA, Strath M, Mellor AL, Sanderson CJ. Eosinophilia in transgenic mice expressing interleukin 5. *J Exp Med* 1990;**172**(5):1425−31.

67. Lee NA, McGarry MP, Larson KA, Horton MA, Kristensen AB, Lee JJ, et al. Expression of IL-5 in thymocytes/T cells leads to the development of a massive eosinophilia, extramedullary eosinophilopoiesis, and unique histopathologies. *J Immunol* 1997;**158**(3):1332−44.

68. Macias MP, Fitzpatrick LA, Brenneise I, McGarry MP, Lee JJ, Lee NA. Expression of IL-5 alters bone metabolism and induces ossification of the spleen in transgenic mice. *J Clin Invest* 2001;**107**(8):949−59.

69. Kanda A, Driss V, Hornez N, Abdallah M, Roumier T, Abboud G, et al. Eosinophil-derived IFN-gamma induces airway hyperresponsiveness and lung inflammation in the absence of lymphocytes. *J Allergy Clin Immunol* 2009;**124**(3):573−82. 582 e571−579.

70. Shen HH, Ochkur SI, McGarry MP, Crosby JR, Hines EM, Borchers MT, et al. A causative relationship exists between eosinophils and the development of allergic pulmonary pathologies in the mouse. *J Immunol* 2003;**170**:3296−305.

71. Hartl D, Griese M, Nicolai T, Zissel G, Prell C, Konstantopoulos N, et al. Pulmonary chemokines and their receptors differentiate children with asthma and chronic cough. *J Allergy Clin Immunol* 2005;**115**(4):728−36.

72. Mikhak Z, Fukui M, Farsidjani A, Medoff BD, Tager AM, Luster AD. Contribution of CCR4 and CCR8 to antigen-specific T(H)2 cell trafficking in allergic pulmonary inflammation. *J Allergy Clin Immunol* 2009;**123**(1):67−73 e63.

73. Perros F, Hoogsteden HC, Coyle AJ, Lambrecht BN, Hammad H. Blockade of CCR4 in a humanized model of asthma reveals a critical role for DC-derived CCL17 and CCL22 in attracting Th2 cells and inducing airway inflammation. *Allergy* 2009;**64**(7):995−1002.

74. Bandeira-Melo C, Bozza PT, Weller PF. The cellular biology of eosinophil eicosanoid formation and function. *J Allergy Clin Immunol* 2002;**109**(3):393−400.

Eosinophil–Mast Cell Interactions

Anastasya Teplinsky, Moran Elishmereni, Howard R. Katz and Francesca Levi-Schaffer

INTRODUCTION

Although mast cells and eosinophils have been classically associated with the early and late phases of the allergic process, respectively, both cells are present together in the inflamed tissue during the late stage, and when inflammation becomes chronic.[1] Their physical proximity within the inflamed tissue, together with their increased numbers and the persistence of these phenomena throughout the allergic inflammatory reaction, has led us to postulate that these cell types may communicate with each other and ultimately influence the outcome of allergic disease. Moreover, because eosinophils and mast cells are both involved in a number of nonallergic inflammatory diseases, and even in physiological settings at homeostasis, it is feasible to hypothesize that what has been found in allergy may also be valid in these other conditions. As for other cells of the immune system, eosinophils and mast cells can communicate via soluble mediators and through physical interactions.

In this subchapter, the interactions between eosinophils and mast cells are discussed. The most interesting feature is direct cell communication via unique mediators and/or ligands and receptors, based on specificity. Importantly, both eosinophils and mast cells have been shown to communicate with a number of other cells; for example, eosinophils interact with structural cells such as endothelial cells and fibroblasts[1] and can regulate T cell proliferation and activation.[2] Mast cells, too, are *promiscuous* cells, in that they interact with B cells, dendritic cells (DCs), fibroblasts, granulocytes, monocytes/macrophages, natural killer (NK) and NKT cells, and T cells.[3] Various aspects of these interactions are highlighted below.

EOSINOPHILS INFLUENCE MAST CELLS VIA SOLUBLE MEDIATORS

Human peripheral blood eosinophils produce stem cell factor (SCF), the major cytokine that regulates differentiation, maturation, and survival of mast cells; SCF can also prime and activate mast cells.[4] SCF is a hematopoietic cytokine that triggers its biologic effects by binding to its tyrosine kinase receptor, c-Kit, and is normally found in both soluble and transmembrane forms due to differential splicing and proteolytic cleavage.[5] SCF is detected in peripheral blood eosinophils from allergic patients, and colocalizes with major basic protein (MBP), which typically resides in the eosinophils' cytoplasmic secondary granules. This observation suggests that SCF can be released together with MBP and other granule-associated mediators upon eosinophils activation. In addition, SCF also colocalizes with c-Kit receptors on the eosinophils' plasma membrane. SCF membrane staining may be due to SCF produced by eosinophils being transported to the membrane from intracellular stores. Alternatively, SCF present in serum may have bound to c-Kit receptors prior to isolation of eosinophils.[4]

SCF can induce mediator release from human cutaneous mast cells[6] and enhance immunoglobulin E (IgE)-induced release of histamine and leukotriene C4 (LTC4), LTD4, and LTE4 from lung and intestinal mast cells.[7,8] Human lung mast cells are unresponsive to non-IgE-mediated activation when cocultured with fibroblasts *in vitro*, and release histamine and prostaglandin D2 (PGD2) in response to compound 48/30 and MBP through a mechanism involving the fibroblast-derived membrane form of SCF. Moreover, fibroblast-derived SCF also primes cord blood-derived mast cells (CBMC) so that they can be activated by MBP.[9] Because eosinophils produce SCF and release it upon incubation with the mast cell enzyme, chymase,[4] it is possible that they influence mast cells the same way, and promote their activation through nonimmunological routes. On the other hand, eosinophil cationic protein (ECP) and MBP cause direct activation of human heart mast cells *in vitro*, leading to the release of histamine, tryptase, and PGD2.[10] Recently, we found that MBP-induced CBMC activation is mediated at least in part by a physical interaction between MBP and CD29 on the mast cell surface (unpublished data).

Like mast cells, eosinophils also store and release nerve growth factor (NGF), which exerts its biological effects by interacting with two specific receptors: a tyrosine kinase high-affinity receptor (TrkA) and a low-affinity receptor (p75), which belongs to the tumor necrosis factor (TNF) receptor superfamily.[11] NGF receptors are expressed on both eosinophils and mast cells. In consequence, it is possible that activation and survival of these cells can occur via an autocrine/paracrine circuit mediated in part by NGF. Interestingly, NGF was shown to release eosinophil peroxidase (EPO) from peripheral blood eosinophils, which in turn activates rat peritoneal mast cells to release histamine and reduces apoptosis, thus suggesting a role of eosinophil-derived NGF in eosinophil–mast cell interactions.[12] On the other hand, another study reported that NGF in combination with SCF or with both SCF and interleukin-4 (IL-4) had no effect on the survival of mast cells isolated from normal intestinal tissue.[13]

In addition, human eosinophils and mast cells have the ability to convert arachidonic acid via the 5-lipoxygenase pathway to metabolites such as LTA4, LTC4, LTD4, and LTE4, which are chemoattractants not only for eosinophils but also for mast cells and neutrophils.[14,16,29] These mediators elicit a variety of responses in mast cells, including activation.[15] Since eosinophils express receptors for these metabolites, this implies that they are also involved in autocrine signaling functions.[14] Finally, another bidirectional interaction occurs between eosinophil MBP and heparanase, the latter an important enzyme involved in tissue remodeling and angiogenesis. MBP binds to and inhibits the activity of heparanase, which is produced by both eosinophils and mast cells.[16]

MAST CELLS INFLUENCE EOSINOPHILS VIA SOLUBLE MEDIATORS

Mast cells produce and secrete several cytokines that modulate eosinophil biology. Activated mast cells produce granulocyte-macrophage colony-stimulating factor (GM-CSF), IL-3 and IL-5, central cytokines involved in eosinophil growth, differentiation, chemotaxis, survival, activation and priming.[17] Mast cells synthesize and release TNF-α,[18] which promotes eosinophil survival, activation, and chemotaxis.[19,21,22] TNF-α secreted during the acute phase of allergy is responsible for eosinophil-mediated production of CXCL9 (a monokine induced by IFN-γ), CXCL10 (IFN-γ-inducible protein-10), CCL17 (a thymus and activation-regulated chemokine), and CCL22 (a macrophage-derived chemokine).[22] TNF-α activates the mitogen-activated protein kinase (MAPK) pathway in eosinophils.[20] Peripheral blood eosinophils cultured in the presence of mast cell sonicates showed enhanced viability, due in part to TNF-α binding to its receptors (TNF-RI and TNF-RII), leading to increased eosinophil production of GM-CSF. This effect was inhibited by blocking TNF-α and/or GM-CSF with neutralizing antibodies.[21]

Tryptase, a mast cell-specific neutral protease, can activate eosinophil by binding to the protease-activated receptor, PAR2.[23] Tryptase-induced release of EPO and β-hexosaminidase from peripheral blood eosinophils occurs in a time- and dose-dependent fashion.[24] In addition, tryptase promotes the production and release of cytokines, such as IL-6 and IL-8, in human peripheral blood eosinophils through the MAPK/AP-1 pathway.[23] Histamine, another prominent mast cell mediator, can elevate the expression of adhesion molecules and increase proinflammatory cytokine production by eosinophils.[25] Recent studies have also shown that histamine is an important chemoattractant for eosinophils that functions via the H4 receptor.[26]

Activated mast cells secrete PGD2, LTB4, LTC4, and LTE4,[27] which, in addition to being eosinophil chemoattractants, influence eosinophil adhesion and induce secretion of IL-8.[28] PGD2 acts through the prostaglandin D2 receptor 2 (the CRTH2 receptor, also known as the DP2 receptor) and the DP receptor (also referred to as DP1). Activation of both of these receptors releases mature eosinophils from the bone marrow and induces their migration. Through the CRTH2 receptor, PGD2 activates an eosinophil respiratory burst, causing the release of ECP and the synthesis of LTC4 during allergic inflammation. PGD2 also augments the responsiveness of eosinophils to other chemoattractants such as eotaxin, 5-oxo-ETE, and C5a. In contrast, PGD2 may inhibit eosinophil degranulation under C5a-activated conditions. In addition, PGD2 delays eosinophil apoptosis by activating the DP receptor.[29]

Another molecule proposed to be involved in eosinophil–mast cell interactions is the IL-7-like cytokine, thymic stromal lymphopoietin (TSLP). A recent report indicates a significant increase of TSLP mRNA expression in bronchial mast cells of asthmatic subjects.[30] Another study observed that human eosinophils constitutively express a functional heterodimeric TSLP receptor complex comprising the TSLP-binding chain, TSLPR, and the IL-7Ra chain. TSLP can significantly delay eosinophil apoptosis and upregulate cell surface expression of the CD18 adhesion molecule and the intercellular adhesion molecule (ICAM)-1, but downregulates L-selectin. TSLP also enhances eosinophil adhesion to fibronectin, and induces the release of CCL2, CXCL1, CXCL8, and IL-6.[31]

Another potential eosinophil–mast cell interaction is mediated by the major eosinophil chemokine eotaxin. Dramatic increases in the expression of *CCL24 [chemokine (C-C motif) ligand 24/eotaxin-2]* mRNA were detected in LPS-treated bone marrow mast cells after IgE stimulation.[32] Eotaxin also activates eosinophils *in vitro* and causes their accumulation in the lung *in vivo* in actively sensitized guinea pigs after aerosol allergen challenge.[33]

PHYSICAL INTERACTIONS BETWEEN EOSINOPHILS AND MAST CELLS

Physical contact is considered a more efficient, accurate, and reliable form of communication than paracrine interactions. In most cases, contact-dependent communication through receptor–ligand pairs constitutes a precise method of interaction for the efficient integration of multiple signals, whether activating, coactivating, or inhibiting.

Recently, and for the first time, we detected contact between eosinophils and mast cells in allergic inflammation-associated disorders. Histological and immunohistological staining of inflamed tissues from patients with nasal polyposis (a chronic inflammatory state often associated with asthma) and the bronchi of asthmatic patients revealed eosinophils and mast cells in close proximity. Some mast cells were partially degranulated, and released granules observed near to eosinophil. A murine chronic atopic dermatitis model, induced by epicutaneous sensitization with ovalbumin with or without staphylococcal enterotoxin B, showed similar associations: eosinophil—mast cell contacts were abundant, especially after the full 3-week sensitization period.[34] To examine the kinetics of this interaction, human peripheral blood eosinophils were cocultured with CBMC at a 1:1 ratio. The cells interacted to form a well-defined interface within minutes, and contacts lasted for several minutes, in contrast to the transient random homotypic binding observed in monocultures. Binding rates were highest in early cocultures. Complex multicellular eosinophil—mast cell aggregates were also found and consisted mostly of a single mast cell surrounded by several peripheral blood eosinophils.[34]

In electron microscopy (EM) imaging, we detected tight interactions between eosinophil—mast cell pairs along several membrane regions.[35] Cell—cell contact was also mediated by pseudopodia-like membrane structures extruded by both eosinophils and mast cells, linking the cells in a *synapse-like* manner (Fig. 11.6.1A). In some cases, mast cells appeared to deform and *embrace* the eosinophil, creating a wide contact region between the two cell types. Eosinophils, in turn, seemed more elongated, an established hallmark of activated eosinophils,[36] and exhibited larger peripheral vacuoles.

Further analysis revealed that the function and activation state of both cell types were altered as a result of their coculture. The impact of eosinophil—mast cell contact on cellular functions was first investigated by assessing their survival. In long-term cocultures (3 day), with or without SCF, CBMC significantly enhanced peripheral blood eosinophil viability. The mast cells' prosurvival signal for eosinophils was evident as soon as 48 h following the onset of coculture and lasted for at least 1 week. Intriguingly, this effect was proven to involve not only released mediators but also a contact-dependent mechanism. Mast cell-induced eosinophil survival was significantly higher in full-interaction cocultures, as opposed to transwell cocultures in which there was no cell—cell binding. Contact-mediated communication was able to override the soluble pathways, since the survival effect still occurred in full cocultures in which GM-CSF and TNF-α were blocked via neutralizing monoclonal antibodies. An important observation is that in cocultures treated with dexamethasone, which directly inhibits eosinophil survival, an effect that is partially blocked by GM-CSF and IL-3,[21] mast cells protected eosinophils from apoptosis.[34] In contrast, mast cells viability rates were slightly (but not significantly) higher following eosinophil coculture under all conditions. However, in cocultures with high eosinophil ratios (1:10), mast cells were slightly less viable, suggesting a possible toxic effect of eosinophils on mast cells in such settings.

Both cell types also showed altered activation morphology. EM revealed that cocultured eosinophils expressed significantly more vacuoles, which were at least twofold larger than those in eosinophil monocultures.[35]

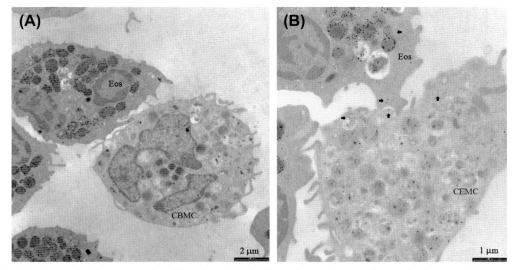

FIGURE 11.6.1 Physical contact between eosinophil and mast cells. Transmission electron micrograph of contacts between peripheral blood eosinophils (Eos) and human cord blood-derived mast cells (CBMC) in 2:1 ratio cocultures (60 min). Sections of cocultured cells were stained with specific gold-conjugated antibodies to eosinophil peroxidase (EPO). A, An eosinophil and a mast cell in a tight interaction characterized by extrusions formed by each cell. B, A few particles of EPO (arrows) are seen in the eosinophil and mast cell. *(Y. Minai-Fleminger, M. Eeshmereni, G. Zabucchi, and F. Levi-Schaffer; unpublished data.)*

In addition, altered lipid synthesis and accumulation in cocultured mast cells was revealed by the presence of significantly more lipid bodies, most of which were electron-dense, whereas those in monocultured mast cells were usually more translucent. Lipid bodies have been implicated in the biology and synthesis of eicosanoids, as they store the key substrate and enzyme for the generation of prostaglandins[37] and LTs, and contain arachidonic acid metabolites.[38] In addition, TNF-α and basic fibroblast growth factor are also known to be stored in mast cell lipid bodies.[39]

EM also revealed that specific mast cell mediators can be transferred into cocultured eosinophils, and vice versa. Tryptase was found to be directly translocated from mast cells to eosinophils only through tight contacts during cell—cell interactions, most likely via small vesicles that are probably generated by a piecemeal degranulation process that may also occur in mast cells. Tryptase was ultimately located in eosinophil multivesicular bodies associated with lysosomal compartments.[35] The pathophysiological significance of this exchange of specific mediators between eosinophils and mast cells is currently unknown.

In contrast, cocultured eosinophils release EPO into the extracellular medium through vesicles and following membrane fusion. Free EPO then binds to the mast cell surface and is engulfed in small vesicles (Fig. 11.6.1B). However, because mast cell monocultures treated with soluble EPO also become EPO^+, it is possible that physical communication is not necessary for EPO uptake. Indeed, larger amounts of EPO are released in cocultures vs. monocultures.[35]

Physical contact between eosinophils and mast cells can be mediated by the binding of receptors and ligands expressed on both cell types. Two potential candidates are the CD48 antigen CD2-family receptor (on mast cells) and its high-affinity ligand, natural killer cell receptor 2B4, expressed on eosinophils.[40] CD48 is a glycosylphosphatidylinositol-anchored cell-surface protein that induces numerous effects in eosinophils and mast cells as a co-stimulatory molecule.[40] 2B4 was initially thought to be an activating receptor in humans and mice, although inhibitory functions were detected in murine NK cells. The CD48—2B4 axis appears to play a role in physical interactions between eosinophils and mast cells. *In vitro* studies showed that the physical contact between eosinophils and mast cells is facilitated in part by CD48 binding to the 2B4 ligand, as antibody blockade of CD48 on mast cells and/or 2B4 on eosinophils in cocultures led to reduced abundant eosinophil—mast cell pairs.[34] The role of the CD48—2B4 axis was also demonstrated during the mast cell-sustained survival of eosinophils: antibody-mediated ligation of 2B4 on monocultured eosinophils induced their survival, and eosinophil—mast cell cocultures in which 2B4 was blocked by Fab fragments or recombinant human (rh) CD48 resulted in lower eosinophil survival rates.[34] The physiological relevance of CD48 in eosinophil—mast cell interactions is also supported *in vivo*: skin sections from mice with OVA/SEB-induced atopic dermatitis revealed CD48 to be located at the eosinophil—mast cell interface in several cases. These combined findings point to the CD48—2B4 axis as being one of the mechanisms underlying direct eosinophil—mast cell contact and the induced effects on eosinophil viability.[34]

Contact between these cells may also be facilitated by other surface molecules, for example CMRF35-like molecule 8 (CD300a), CD226 antigen (DNAM-1), intercellular adhesion molecule 1 (ICAM-1), integrin αvβ3, integrin alpha-L (LFA-1), and poliovirus receptor-related protein 2 (Nectin-2/CD112).[17] DNAM-1 is an adhesion molecule expressed on mast cells that regulates their activity state. One of its ligands is Nectin-2, which is expressed on eosinophils. A substantial increase in IgE-induced release of tryptase and IL-4 and eicosanoid synthesis was observed following DNAM-1 ligation on mast cells. Neutralizing this receptor in a peripheral blood eosinophil—CBMC coculture reduced eosinophil-increased IgE-mediated activation of mast cells.[41] A recent study suggests that another eosinophil—mast cell interaction is mediated by tumor necrosis factor ligand superfamily member 10 (TRAIL) and Trail receptor (TRAIL-R). Eosinophils express TRAIL-R[43] prosurvival molecules.[43] Further, human mast cell-surface expression of TRAIL-R transduces a proapoptotic signal[43] and mast cells have TRAIL in their cytoplasm.[44] Together, these findings suggest several routes mediate physical contact between eosinophils and mast cells, which could contribute to the modulation of allergic reactions.

EOSINOPHILS AND MAST CELLS ASSOCIATION IN HEALTH AND IN DISEASES OTHER THAN ALLERGY

Eosinophils and mast cells are constitutively located in both different and shared organs and tissues, and are found in the gastrointestinal (GI) tract, lymph nodes, mammary glands, spleen, thymus, and uterus.[45] Mast cells are predominantly found in the junctions between the body and the environment, namely in the skin and mucosal surfaces.[3] Most studies on the association between eosinophils and mast cells, other than in allergy, relate mainly to disease states rather than to normal physiology. It is well known that numerous eosinophils and mast cells are present in the gastrointestinal (GI) tract as normal constituents of the mucosal immune system.[46] Eosinophil and mast cell numbers are increased in several GI disorders, such as inflammatory bowel disease (Crohn's disease and ulcerative colitis).[47] Interestingly, while eosinophils and mast cells are present in high numbers in the submucosa of the colon of Crohn's disease patients, mast cells in the mucosa are degranulated and reduced in number, and increased numbers of eosinophils exhibit ultrastructural features of strong activation. These data indicate the possibility of cross-talk between eosinophils and mast cells in Crohn's disease.[46]

Increased numbers of eosinophils and mast cells are also found in gastric carcinomas and chronic gastritis. Both cell types release profibrogenic and proangiogenic factors, i.e., TGF-β and vascular endothelial growth factor, respectively, leading to fibrosis of the colon.[48,49] Thus, they may contribute to carcinogenesis.[49] Cell clusters containing a single mast cell and between one and three eosinophils can be found in gastric carcinomas. Moreover, eosinophils in contact with mast cells show signs of activation, such as alterations in the size and number of granules, cytoplasmic vacuoles, and scattered extracellular granules.[48]

Antral eosinophils and mast cells are significantly activated in pediatric functional dyspepsia.[50] In addition, in patients with food allergy, inflammatory mediators produced by eosinophils and mast cells such as ECP, eosinophil-derived neurotoxin (RNase2), histamine and its metabolite N-methylhistamine, tryptase, IL-5, and tumor necrosis factor are increased in gut lavage fluid, serum, stool, and urine.[51] These observations show that eosinophils and mast cells are in close proximity and are activated in this disease and suggest that they can influence each other.

The association of both eosinophils and mast cells with parasitic diseases is long established. For example, their numbers are significantly augmented in both the cecum and colon of mice infected with *Trichuris muris* helminthic parasites.[52] Another example is experimental ascariasis in Göttingen minipigs, characterized by a dramatic increase in eosinophils and mucosal-type mast cells in both the mucosa and the deeper layers of the intestinal wall. Interestingly, there is a simultaneous decrease of connective tissue-type mast cells.[46]

Eosinophilic esophagitis (EoE) is a clinical pathological syndrome characterized by the influx of numerous eosinophils into the esophageal epithelium. A recent study showed unequivocally that mast cells infiltrate esophageal smooth muscle in patients with EoE, where eosinophils also reside.[53] Other eosinophil-related diseases associated with the presence of mast cells are hypereosinophilic syndromes. In these, there is a common association with systemic mastocytosis (SM) and mast cell leukemia. Interestingly, eosinophilia in SM is of prognostic significance.[54] Eosinophils and mast cells are also present in

diseases associated with fibrosis such as asthma,[55] scleroderma, chronic graft-versus-host disease (cGVHD), and bleomycin-induced fibrosis,[56] as well as in physiological wound healing.[57,58] Both cells types, as mentioned above, have intrinsic profibrotic/proangiogenic properties and it is tempting to speculate that they can also act in concert to influence fibrosis.

CONCLUSION

Both eosinophils and mast cells can influence the properties of the other cell type both via released mediators and by physical contact mediated by receptor—ligand interactions. Eosinophil—mast cell pairs have been detected in allergic inflamed tissues, and *in vitro* they have been shown to have many avenues of communication. This communication appears important for the modulation of allergic inflammation, where eosinophils and mast cells are pivotal effector cells. In addition, evidence for their association in health and in other inflammatory diseases predicts that interactions between eosinophils and mast cells in nonallergic settings are worthy of future investigation.

ACKNOWLEDGEMENTS

F. Levi-Schaffer's research is supported by grants number 699/10 from the Israel Science Foundation and the Aimwell Charitable Trust, UK. F. Levi-Schaffer appreciates the kind partial financial support provided by Dr. Leone Nauri (Rome, Italy).

REFERENCES

1. Blanchard C, Rothenberg ME. Biology of the eosinophil. *Adv Immunol* 2009;**101**:81—121.
2. Shi HZ, Humbles A, Gerard C, Jin Z, Weller PF. Lymph node trafficking and antigen presentation by endobronchial eosinophils. *J Clin Invest* 2000;**105**:945—53.
3. Abraham SN, St John AL. Mast cell-orchestrated immunity to pathogens. *Nat Rev Immunol* 2010;**10**:440—52.
4. Hartman M, Piliponsky AM, Temkin V, Levi-Schaffer F. Human peripheral blood eosinophils express stem cell factor. *Blood* 2001;**97**:1086—91.
5. Ashman LK. The biology of stem cell factor and its receptor C-kit. *Int J Biochem Cell Biol* 1999;**31**:1037—51.
6. Columbo M, Horowitz EM, Botana LM, MacGlashan Jr DW, Bochner BS, Gillis S, et al. The human recombinant c-kit receptor ligand, rhSCF, induces mediator release from human cutaneous mast cells and enhances IgE-dependent mediator release from both skin mast cells and peripheral blood basophils. *J Immunol* 1992;**149**:599—608.
7. Bischoff SC, Schwengberg S, Raab R, Manns MP. Functional properties of human intestinal mast cells cultured in a new culture system: enhancement of IgE receptor-dependent mediator release and response to stem cell factor. *J Immunol* 1997;**159**:5560—7.

8. Bischoff SC, Dahinden CA. c-Kit ligand: a unique potentiator of mediator release by human lung mast cells. *J Exp Med* 1992;**175**:237–44.

9. Piliponsky AM, Gleich GJ, Nagler A, Bar I, Levi-Schaffer F. Non-IgE-dependent activation of human lung- and cord blood-derived mast cells is induced by eosinophil major basic protein and modulated by the membrane form of stem cell factor. *Blood* 2003;**101**:1898–904.

10. Patella V, de Crescenzo G, Marino I, Genovese A, Adt M, Gleich GJ, et al. Eosinophil granule proteins activate human heart mast cells. *J Immunol* 1996;**157**:1219–25.

11. Solomon A, Aloe L, Pe'er J, Frucht-Pery J, Bonini S, Bonini S, et al. Nerve growth factor is preformed in and activates human peripheral blood eosinophils. *J Allergy Clin Immunol* 1998;**102**:454–60.

12. Kawamoto K, Okada T, Kannan Y, Ushio H, Matsumoto M, Matsuda H. Nerve growth factor prevents apoptosis of rat peritoneal mast cells through the trk proto-oncogene receptor. *Blood* 1995;**86**:4638–44.

13. Lorentz A, Hoppe J, Worthmann H, Gebhardt T, Hesse U, Bienenstock J, et al. Neurotrophin-3, but not nerve growth factor, promotes survival of human intestinal mast cells. *Neurogastroenterol Motil* 2007;**19**:301–8.

14. Bandeira-Melo C, Weller PF. Eosinophils and cysteinyl leukotrienes. *Prostaglandins Leukot Essent Fatty Acids* 2003;**69**:135–43.

15. Jiang Y, Borrelli L, Bacskai BJ, Kanaoka Y, Boyce JA. P2Y6 receptors require an intact cysteinyl leukotriene synthetic and signaling system to induce survival and activation of mast cells. *J Immunol* 2009;**182**:1129–37.

16. Temkin V, Aingorn H, Puxeddu I, Goldshmidt O, Zcharia E, Gleich GJ, et al. Eosinophil major basic protein: first identified natural heparanase-inhibiting protein. *J Allergy Clin Immunol* 2004;**113**:703–9.

17. Minai-Fleminger Y, Levi-Schaffer F. Mast cells and eosinophils: the two key effector cells in allergic inflammation. *Inflamm Res* 2009;**58**:631–8.

18. Gordon JR, Galli SJ. Mast cells as a source of both preformed and immunologically inducible TNF-alpha/cachectin. *Nature* 1990;**346**:274–6.

19. Temkin V, Pickholtz D, Levi-Schaffer F. Tumor necrosis factors in a murine model of allergic peritonitis: effects on eosinophil accumulation and inflammatory mediators' release. *Cytokine* 2003;**24**:74–80.

20. Liu LY, Bates ME, Jarjour NN, Busse WW, Bertics PJ, Kelly EA. Generation of Th1 and Th2 chemokines by human eosinophils: evidence for a critical role of TNF-alpha. *J Immunol* 2007;**179**:4840–8.

21. Temkin V, Levi-Schaffer F. Mechanism of tumour necrosis factor alpha mediated eosinophil survival. *Cytokine* 2001;**15**:20–6.

22. Levi-Schaffer F, Temkin V, Malamud V, Feld S, Zilberman Y. Mast cells enhance eosinophil survival in vitro: role of TNF-alpha and granulocyte-macrophage colony-stimulating factor. *J Immunol* 1998;**160**:5554–62.

23. Temkin V, Kantor B, Weg V, Hartman ML, Levi-Schaffer F. Tryptase activates the mitogen-activated protein kinase/activator protein-1 pathway in human peripheral blood eosinophils, causing cytokine production and release. *J Immunol* 2002;**169**:2662–9.

24. Vliagoftis H, Lacy P, Luy B, Adamko D, Hollenberg M, Befus D, et al. Mast cell tryptase activates peripheral blood eosinophils to release granule-associated enzymes. *Int Arch Allergy Immunol* 2004;**135**:196–204.

25. Gelfand EW. Role of histamine in the pathophysiology of asthma: immunomodulatory and antiinflamatory activities of H1-receptor antagonists. *Am J Med* 2002;**113**(Suppl. 9A):2S–7S.

26. O'Reilly M, Alpert R, Jenkinson S, Gladue RP, Foo S, Trim S, et al. Identification of a histamine H4 receptor on human eosinophils—role in eosinophil chemotaxis. *J Recept Signal Transduct Res* 2002;**22**:431–48.

27. Stone KD, Prussin C, Metcalfe DD. IgE, mast cells, basophils, and eosinophils. *J Allergy Clin Immunol* 2010;**125**:S73–80.

28. Jacobsen EA, Ochkur SI, Lee NA, Lee JJ. Eosinophils and asthma. *Curr Allergy Asthma Rep* 2007;**7**:18–26.

29. Schuligoi R, Sturm E, Luschnig P, Konya V, Philipose S, Sedej M, et al. CRTH2 and D-type prostanoid receptor antagonists as novel therapeutic agents for inflammatory diseases. *Pharmacology* 2010;**85**:372–82.

30. Okayama Y, Okumura S, Sagara H, Yuki K, Sasaki T, Watanabe N, et al. FcepsilonRI-mediated thymic stromal lymphopoietin production by interleukin-4-primed human mast cells. *Eur Respir J* 2009;**34**:425–35.

31. Wong CK, Hu S, Cheung PF, Lam CW. Thymic stromal lymphopoietin induces chemotactic and prosurvival effects in eosinophils: implications in allergic inflammation. *Am J Respir Cell Mol Biol* 2009;**43**:305–15.

32. Nigo YI, Yamashita M, Hirahara K, Shinnakasu R, Inami M, Kimura M, et al. Regulation of allergic airway inflammation through Toll-like receptor 4-mediated modification of mast cell function. *Proc Natl Acad Sci USA* 2006;**103**:2286–91.

33. Griffiths-Johnson DA, Collins PD, Rossi AG, Jose PJ, Williams TJ. The chemokine, eotaxin, activates guinea-pig eosinophils in vitro and causes their accumulation into the lung in vivo. *Biochem Biophys Res Commun* 1993;**197**:1167–72.

34. Elishmereni M, Alenius HT, Bradding P, Mizrahi S, Shikotra A, Minai-Fleminger Y, et al. Physical interactions between mast cells and eosinophils: a novel mechanism enhancing eosinophil survival in vitro. *Allergy* 2011;**66**:376–85.

35. Minai-Fleminger Y, Elishmereni M, Vita F, Soranzo MR, Mankuta D, Zabucchi G, et al. Ultrastructural evidence for human mast cell-eosinophil interactions in vitro. *Cell Tissue Res* 2010;**341**:405–15.

36. Melo RC, Weller PF, Dvorak AM. Activated human eosinophils. *Int Arch Allergy Immunol* 2005;**138**:347–9.

37. Dvorak AM. Ultrastructure of human mast cells. *Int Arch Allergy Immunol* 2002;**127**:100–5.

38. Harris JR. Megakaryocytes, platelets, macrophages, and eosinophils. In: *Blood Cell Biochemistry*, **2**. New York: Plenum; 1991.

39. Dvorak AM. Ultrastructural studies of human basophils and mast cells. *J Histochem Cytochem* 2005;**53**:1043–70.

40. Elishmereni M, Levi-Schaffer F. CD48: A co-stimulatory receptor of immunity. *Int J Biochem Cell Biol* 2010;**43**:25–8.

41. Bachelet I, Munitz A, Mankutad D, Levi-Schaffer F. Mast cell costimulation by CD226/CD112 (DNAM-1/Nectin-2): a novel interface in the allergic process. *J Biol Chem* 2006;**281**:27190–6.

42. Daigle I, Simon HU. Alternative functions for TRAIL receptors in eosinophils and neutrophils. *Swiss Med Wkly* 2001;**131**:231—7.

43. Berent-Maoz B, Salemi S, Mankuta D, Simon HU, Levi-Schaffer F. Human mast cells express intracellular TRAIL. *Cell Immunol* 2010;**262**:80—3.

44. Berent-Maoz B, Piliponsky AM, Daigle I, Simon HU, Levi-Schaffer F. Human mast cells undergo TRAIL-induced apoptosis. *J Immunol* 2006;**176**:2272—8.

45. Hogan SP, Rosenberg HF, Moqbel R, Phipps S, Foster PS, Lacy P, et al. Eosinophils: biological properties and role in health and disease. *Clin Exp Allergy* 2008;**38**:709—50.

46. Beil WJ, McEuen AR, Schulz M, Wefelmeyer U, Kraml G, Walls AF, et al. Selective alterations in mast cell subsets and eosinophil infiltration in two complementary types of intestinal inflammation: ascariasis and Crohn's disease. *Pathobiology* 2002;**70**:303—13.

47. Jeziorska M, Haboubi N, Schofield P, Woolley DE. Distribution and activation of eosinophils in inflammatory bowel disease using an improved immunohistochemical technique. *J Pathol* 2001;**194**:484—92.

48. Caruso RA, Fedele F, Zuccala V, Fracassi MG, Venuti A. Mast cell and eosinophil interaction in gastric carcinomas: ultrastructural observations. *Anticancer Res* 2007;**27**:391—4.

49. Piazuelo MB, Camargo MC, Mera RM, Delgado AG, Peek Jr RM, Correa H, et al. Eosinophils and mast cells in chronic gastritis: possible implications in carcinogenesis. *Hum Pathol* 2008;**39**:1360—9.

50. Friesen CA, Lin Z, Singh M, Singh V, Schurman JV, Burchell N, et al. Antral inflammatory cells, gastric emptying, and electrogastrography in pediatric functional dyspepsia. *Dig Dis Sci* 2008;**53**:2634—40.

51. Bischoff S, Crowe SE. Gastrointestinal food allergy: new insights into pathophysiology and clinical perspectives. *Gastroenterology* 2005;**128**:1089—113.

52. Motomura Y, Khan WI, El-Sharkawy RT, Verma-Gandhu M, Grencis RK, Collins SM, et al. Mechanisms underlying gut dysfunction in a murine model of chronic parasitic infection. *Am J Physiol Gastrointest Liver Physiol*; 2010. In press.

53. Abonia JP, Blanchard C, Butz BB, Rainey HF, Collins MH, Stringer K, et al. Involvement of mast cells in eosinophilic esophagitis. *J Allergy Clin Immunol* 2010;**126**:140—9.

54. Bohm A, Fodinger M, Wimazal F, Haas OA, Mayerhofer M, Sperr WR, et al. Eosinophilia in systemic mastocytosis: clinical and molecular correlates and prognostic significance. *J Allergy Clin Immunol* 2007;**120**:192—9.

55. Puxeddu I, Levi-Schaffer F. Mast cells and eosinophils: the hallmark of asthma. *Paediatr Respir Rev* 2004;**5**(Suppl. A):S31—4.

56. Levi-Schaffer F, Weg VB. Mast cells, eosinophils and fibrosis. *Clin Exp Allergy* 1997;**27**(Suppl. 1):64—70.

57. Artuc M, Hermes B, Steckelings UM, Grutzkau A, Henz BM. Mast cells and their mediators in cutaneous wound healing—active participants or innocent bystanders? *Exp Dermatol* 1999;**8**:1—16.

58. Leitch VD, Strudwick XL, Matthaei KI, Dent LA, Cowin AJ. IL-5-overexpressing mice exhibit eosinophilia and altered wound healing through mechanisms involving prolonged inflammation. *Immunol Cell Biol* 2009;**87**:131—40.

Chapter 11.7

Plasma Cell/B Cell—Eosinophil Interactions

Van T. Chu and Claudia Berek

B CELL ACTIVATION AND THE DEVELOPMENT OF PLASMA CELLS

Immunization may induce the differentiation of B cells into antibody-secreting plasma cells by two alternative pathways.[1-3] Naive B cells (Fig. 11.7.1, B_{naive}) may be stimulated to differentiate directly into plasma cells so that immunoglobulin M (IgM)-secreting plasma cells are generated within the first days after immunization. These short-lived plasma cells are mainly found in the red pulp of the spleen. In contrast, long-lived *memory* plasma cells that protect the organism against future contact with the antigen are generated in the germinal center reaction and are induced by T cell-dependent antigens.

Three to four days after immunization with a T cell-dependent antigen, small groups of activated B cells accumulate at the border between the T cell and B cell zones of the follicles in splenic white pulp or within lymph nodes.[4-5] This marks the beginning of a germinal center reaction in which single germinal center B cells (Fig. 11.7.1, B_{GC}) give rise to large clones of antigen-specific cells. During this proliferative phase, a hypermutation mechanism is activated that diversifies the variable region genes of the heavy and the light chain of the B cell receptor (BCR) and generates B cell variants with different affinities for the antigen. Only those B cells with a relatively high affinity BCR are selected to differentiate into effector cells.

Germinal center B cells give rise to both, memory B cells (Fig. 11.7.1, Bmem) and plasmablasts. Some of the plasmablasts stay in the peripheral lymphoid organs, but the majority migrate to the bone marrow where they differentiate into mature plasma cells (Fig. 11.7.1).[6-7] During secondary immune responses, when memory B cells are reactivated, more plasma cells are generated and the influx of plasma cells to the bone marrow is even greater than in the primary responses (Fig. 11.7.1).[8-9] Within the bone marrow, plasma cells secreting antibodies of high affinity survive for months or even years.[10-11]

Eosinophils have long been regarded as cells of the innate immune system that play no significant role in adaptive immune responses. This subchapter will show that, at least in the murine immune system, eosinophils are essential in T cell-dependent immune responses, both supporting the immediate innate-type response of B cells as

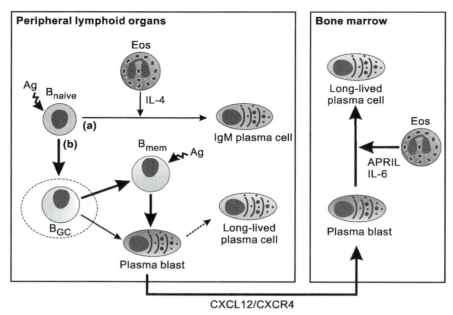

FIGURE 11.7.1   B cell activation and differentiation into plasma cells. A, Interleukin-4 (IL-4)-dependent direct differentiation of antigen-activated B cells into immunoglobulin M (IgM)-secreting plasma cells and B, germinal center formation and B cell differentiation. Antigen (Ag), eosinophils (Eos), germinal center B cells (B_{GC}), memory B cells (B_{mem}), naive B cells (B_{naive}). APRIL, tumor necrosis factor ligand superfamily member 13; CXCL12, stromal cell-derived factor 1/C-X-C motif chemokine 12; CXCR4, C-X-C chemokine receptor type 4; IL-6, interleukin-6.

well as being pivotal for the long-term maintenance of plasma cells in the bone marrow.

EOSINOPHILS SUPPORT B CELL ACTIVATION

Appropriate priming of naive B cells induces proliferation and differentiation into antibody-secreting plasma cells. As soluble antigens are rather ineffective in the activation of naive B cells, immune response are enhanced by the combination of antigen with adjuvants, such as complete Freund's adjuvants, or by absorption onto particulate aluminum hydroxide (alum).[12–13] Immunization with alum-precipitated antigens promotes interleukin-4 (IL-4) secretion, by Gr-1$^+$ granulocytes,[14] the majority of which are eosinophils.[15] During the early phase of the immune response, secretion of IL-4 by eosinophils promotes the differentiation of antigen-activated B cells into plasma cells. As a consequence, animals lacking eosinophils have a markedly decreased frequency of antigen-specific IgM plasma cells.[14–15] Thus the alum-induced activation of eosinophils is crucial for the initial rapid generation of IgM antibodies.

EOSINOPHILS ARE NOT REQUIRED FOR A NORMAL GERMINAL CENTER REACTION

A mutation in the promoter region of the *GATA1* gene (△dblGATA1) in the mouse results in the complete loss

of mature eosinophils.[16] As expected, immunization of △dblGATA1 mice with the T cell-dependent antigen results in a poor immediate B cell response.[14–15] Nevertheless, the subsequent development of germinal centers is normal and, just as in wild-type BALB/c mice, antigen-activated B cells differentiate into high-affinity effector cells.[17–18] Furthermore, in eosinophil-deficient △dblGATA1 mice and normal controls, comparable numbers of memory B and plasma cells are formed in spleen. Thus, a lack of eosinophils has no effect on germinal center formation and B cell differentiation.[18]

EOSINOPHILS ARE REQUIRED FOR THE RETENTION OF PLASMA CELLS IN THE BONE MARROW

An analysis of the secondary response to a T cell-dependent antigen showed that △dblGATA-1 mice respond to the challenge as well as wild-type BALB/c mice. In the spleens of BALB/c and △dblGATA1 mice similar numbers of plasmablasts are generated and they migrate to the bone marrow to the same extent. Nevertheless, significantly fewer antigen-specific plasma cells accumulate in the bone marrow of the mutant strain, suggesting that in the absence of eosinophils the retention of plasma cells in the bone marrow is impaired. This interpretation is supported by the finding that plasma cell accumulation in the bone marrow of △dblGATA1 mice can be enhanced by the adoptive transfer of mature wild-type eosinophils.[18]

EOSINOPHILS ARE IN CLOSE CONTACT WITH PLASMA CELLS IN THE BONE MARROW

Analysis of the bone marrow from BALB/c mice 6 days after secondary immunization shows that eosinophils and plasma cells are in close contact and that at any moment approximately 60% of plasma cells are interacting with eosinophils. A few days after a secondary challenge with antigen, the majority of plasma cells and eosinophils are seen in the vicinity of the venous sinusoids (Fig. 11.7.2). The underlying stromal reticular cells express the chemokine CXCL12 and, since both eosinophils and plasma cells express the CXCL12 receptor CXCR4, both cell types are attracted by stromal cells in the bone marrow.[19–20] As the retention of plasma cells in the bone marrow is dependent on eosinophils, this suggests that the interaction of newly generated plasmablasts with eosinophils is required for their development into mature long-lived plasma cells. Plasmablasts that do not receive appropriate signals from eosinophils may die by apoptosis or return to the circulation by leaving the bone marrow through the venous sinusoids.

By 60 days after secondary immunization, only the long-lived plasma cells are still alive. Staining of bone marrow tissue sections shows large clusters of these two cell types deep within the bone marrow parenchyma.[18] Expression of adhesion molecules, such as CD44, intercellular adhesion molecule-1 (ICAM-1), or vascular cell adhesion protein (VCAM-1), supports the tight contact between plasma cells and eosinophils and may promote interactions between different cellular components of the bone marrow plasma cell survival niche. The survival of eosinophils themselves may be prolonged by stromal cell secretion of IL-5, a cytokine required for their development and maintenance.[21–23] Furthermore, since antibody binding to the Fc receptor (FcR) has been shown to enhance the activation of eosinophils, IgG secretion by plasma cells may promote eosinophil cytokine production.[24]

EOSINOPHILS ARE ESSENTIAL FOR THE LONGEVITY OF PLASMA CELLS IN THE BONE MARROW

Bone marrow stromal reticular cells together with eosinophils provide a survival niche that supports the long-term maintenance of plasma cells.[11,18,25–26] The long-term survival of plasma cells is dependent on a number of different cytokines (Table 11.7.1) and in vitro experiments have shown that tumor necrosis factor ligand superfamily member 13 (APRIL; a proliferation-inducing ligand), IL-4, IL-5, IL-6, IL-10, and tumor necrosis factor α (TNF-α)[27–29] all contribute to the long-term maintenance of plasma cells in tissue culture. The crucial role of APRIL in the survival

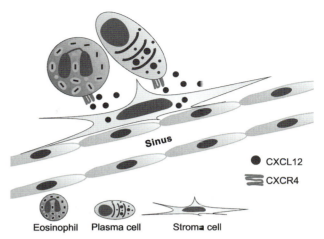

FIGURE 11.7.2 Stromal cell survival niche in the bone marrow. Eosinophils and plasma cells are attracted by stromal cell-secreted stromal cell-derived factor 1 (C-X-C motif chemokine 12; CXCL12). CXCR4, C-X-C chemokine receptor type 4.

of plasma cells in the bone marrow *in vivo* was shown by the finding that the frequency of bone marrow plasma cells is significantly reduced in APRIL-deficient mice.[29–30] Similarly, in mice deficient for the APRIL high-affinity receptor, BCMA (tumor necrosis factor receptor superfamily member 17), there is a significant reduction in the number of bone marrow plasma cells.[28,31] APRIL is continuously required for the survival of plasma cells, because blocking of APRIL by treatment with the APRIL decoy receptor TACI-Ig reduces the number of plasma cells in the bone marrow by more than 60%.[31] In vitro experiments have suggested that IL-6 may also be a crucial cytokine for long-term plasma cell survival. However, since the number of plasma cells are not reduced in the bone marrow of IL-6-deficient mice, other

TABLE 11.7.1 Plasma Cell Survival Factors

Ligand	Receptor
APRIL	TACI, BCMA, proteoglycan
IL-6	IL-6R
BAFF	BAFF-R, TACI, BCMA
IL-5	IL-5R
TNF-α	TNF-R
IL-10	IL-10R

APRIL, BCMA, BAFF, IL-5, interleukin-5; BCMA, tumor necrosis factor receptor superfamily member 17; IL-5R, IL-5 receptor; IL-6, interleukin-6; IL-6R, IL-6 receptor; IL-10, interleukin-10; TACI, tumor necrosis factor receptor superfamily, member 13B; TNFα, tumor necrosis factor α; TNF-R, TNF receptor.

cytokines, such as APRIL, may compensate for its loss *in vivo*.[25,27]

A number of cell types may provide the cytokines required for plasma cell survival in the bone marrow. Recently, the observation that macrophages and dendritic cells in the lymph node are the main sources of APRIL and IL-6 led to the suggestion that these cells provide an environment that supports plasma cell maintenance.[32] In the bone marrow, both macrophages and dendritic cells secrete APRIL and, to some extent, IL-6. Megakaryocytes have also been found in association with plasma cells in the bone marrow and they may well secrete survival factors, though it is questionable whether their level of IL-6 and APRIL expression is sufficient to support the long-term survival of plasma cells.[33] All of these different cell types may contribute to some extent to the microenvironment by which the long-term survival of plasma cells in the bone marrow is ensured (Table 11.7.1).

Recently, however, it was shown that eosinophils in the bone marrow are the key providers of plasma cell survival factors.[18] Mature eosinophils constitutively express APRIL and IL-6 and upon immunization the level of cytokine expression is significantly enhanced.[34] Once activated, eosinophils produce the additional plasma cell survival factors IL-5 and TNF-α.[21,35,36] The finding that in eosinophil-deficient ΔdblGATA-1 mice the level of APRIL expression in the bone marrow is significantly lower than in wild-type BALB/c mice underscores the central role of eosinophils as source of APRIL and IL-6.

In vitro coculturing of eosinophils and plasma cells showed that eosinophils prolong the survival of plasma cells by the secretion of APRIL and IL-6. To demonstrate whether eosinophils support plasma cell survival *in vivo*, BALB/c mice were immunized with a T cell-dependent antigen and boosted 6 weeks later. Three months later, when only long-lived plasma cells are present in the bone marrow, eosinophils were depleted by administration of Siglec-F-specific antibodies.[37,38] With the depletion of eosinophils, the bone marrow was no longer able to support plasma cell survival and there was a rapid decline in the number of plasma cells. Ten days after eosinophil depletion, the number of plasma cells in the bone marrow was reduced by 50% and nearly 20% of the remaining plasma cells were annexin V$^+$, suggesting that plasma cells undergo apoptosis in the absence of eosinophils.[39] These data demonstrate that eosinophils have a pivotal role in the long-term survival of plasma cells in the bone marrow.

Eosinophils have a half-life that is measured in days rather than years. Even if close contact with plasma cells were to prolong their survival, a constant new influx of eosinophils would be required to provide the continuous supply of survival factors necessary to support plasma cells. This implies that the bone marrow survival niche is a dynamic structure in which a constant turnover of eosinophils, and perhaps other components, maintains the environment necessary for the long-term survival of the plasma cell population.

CONCLUSION

Plasma cells and their secreted antibodies are of central importance in humoral immune protection but also play a critical role in diseases where antibodies have a pathological role. Since there is suggestive evidence to indicate that eosinophils may also play a critical role in the maintenance of long-lived plasma cells in humans,[40] anti-eosinophil strategies aimed at reducing plasma cells, and hence antibody levels, may offer a novel and effective therapeutic strategy for conditions ranging from asthmatic inflammation to autoimmune diseases or other conditions where plasma cells and their secreted antibodies are of relevance. We have only just started to unravel the outline of the bone marrow survival niche. A detailed understanding of its structure and function, and the therapeutic implications will be a fascinating and important story for the future.

ACKNOWLEDGEMENTS

We would like to thank R.S. Jack for helpful discussion. VTC is supported by the DFG grant Be-1171/2−1. The Deutsche Rheuma ForschungsZentrum is supported by the Berlin Senate of Research and Education.

REFERENCES

1. MacLennan IC, Toellner KM, Cunningham AF, Serre K, Sze DM, Zuniga E, et al. Extrafollicular antibody responses. *Immunol Rev* 2003;**194**:8−18.

2. Shapiro-Shelef M, Calame K. Regulation of plasma-cell development. *Nat Rev Immunol* 2005;**5**:230−42.

3. Smith KG, Light A, Nossal GJ, Tarlinton DM. The extent of affinity maturation differs between the memory and antibody-forming cell compartments in the primary immune response. *EMBO J* 1997;**16**:2996−3006.

4. Camacho SA, Kosco-Vilbois MH, Berek C. The dynamic structure of the germinal center. *Immunol Today* 1998;**19**:511−4.

5. Allen CD, Okada T, Cyster JG. Germinal-center organization and cellular dynamics. *Immunity* 2007;**27**:190−202.

6. Blink EJ, Light A, Kallies A, Nutt SL, Hodgkin PD, Tarlinton DM, et al. Early appearance of germinal center-derived memory B cells and plasma cells in blood after primary immunization. *J Exp Med* 2005;**201**:545−54.

7. Hargreaves DC, Hyman PL, Lu TT, Ngo VN, Bidgol A, Suzuki G, et al. A coordinated change in chemokine responsiveness guides plasma cell movements. *J Exp Med* 2001;**194**:45−56.

8. Benner R, Hijmans W, Haaijman JJ. The bone marrow: the major source of serum immunoglobulins, but still a neglected site of antibody formation. *Clin Exp Immunol* 1981;**46**:1−8.

9. Hauser AE, Debes GF, Arce S, Cassese G, Hamann A, Radbruch A, et al. Chemotactic responsiveness toward ligands for CXCR3 and CXCR4 is regulated on plasma blasts during the time course of a memory immune response. *J Immunol* 2002;**169**:1277−82.

10. Ahmed R, Gray D. Immunological memory and protective immunity: understanding their relation. *Science* 1996;**272**:54−60.

11. Radbruch A, Muehlinghaus G, Luger EO, Inamine A, Smith KG, Dorner T, et al. Competence and competition: the challenge of becoming a long-lived plasma cell. *Nat Rev Immunol* 2006;**6**:741−50.

12. Marrack P, McKee AS, Munks MW. Towards an understanding of the adjuvant action of aluminium. *Nat Rev Immunol* 2009;**9**:287−93.

13. Mannhalter JW, Neychev HO, Zlabinger GJ, Ahmad R, Eibl MM. Modulation of the human immune response by the non-toxic and non-pyrogenic adjuvant aluminium hydroxide: effect on antigen uptake and antigen presentation. *Clin Exp Immunol* 1985;**61**:143−51.

14. Jordan MB, Mills DM, Kappler J, Marrack P, Cambier JC. Promotion of B cell immune responses via an alum-induced myeloid cell population. *Science* 2004;**304**:1808−10.

15. Wang HB, Weller PF. Pivotal advance: eosinophils mediate early alum adjuvant-elicited B cell priming and IgM production. *J Leukoc Biol* 2008;**83**:817−21.

16. Yu C, Cantor AB, Yang H, Browne C, Wells RA, Fujiwara Y, et al. Targeted deletion of a high-affinity GATA-binding site in the GATA-1 promoter leads to selective loss of the eosinophil lineage in vivo. *J Exp Med* 2002;**195**:1387−95.

17. McKee AS, Munks MW, MacLeod MK, Fleenor CJ, Van Rooijen N, Kappler JW, et al. Alum induces innate immune responses through macrophage and mast cell sensors, but these sensors are not required for alum to act as an adjuvant for specific immunity. *J Immunol* 2009;**183**:4403−14.

18. Chu VT, Fröhlich A, Steinhauser G, Scheel T, Roch T, Fillatreau S, et al. Eosinophils are required for the maintenance of plasma cells in the bone marrow. *Nat Immunol* 2011;**12**:151−9.

19. Nagase H, Miyamasu M, Yamaguchi M, Fujisawa T, Ohta K, Yamamoto K, et al. Expression of CXCR4 in eosinophils: functional analyses and cytokine-mediated regulation. *J Immunol* 2000;**164**:5935−43.

20. Ma Q, Jones D, Springer TA. The chemokine receptor CXCR4 is required for the retention of B lineage and granulocytic precursors within the bone marrow microenvironment. *Immunity* 1999;**10**:463−71.

21. Dubucquoi S, Desreumaux P, Janin A, Klein O, Goldman M, Tavernier J, et al. Interleukin 5 synthesis by eosinophils: association with granules and immunoglobulin-dependent secretion. *J Exp Med* 1994;**179**:703−8.

22. Palframan RT, Collins PD, Severs NJ, Rothery S, Williams TJ, Rankin SM. Mechanisms of acute eosinophil mobilization from the bone marrow stimulated by interleukin 5: the role of specific adhesion molecules and phosphatidylinositol 3-kinase. *J Exp Med* 1998;**188**:1621−32.

23. Hogan MB, Piktel D, Landreth KS. IL-5 production by bone marrow stromal cells: implications for eosinophilia associated with asthma. *J Allergy Clin Immunol* 2000;**106**:329−36.

24. Bartemes KR, McKinney S, Gleich GJ, Kita H. Endogenous platelet-activating factor is critically involved in effector functions of eosinophils stimulated with IL-5 or IgG. *J Immunol* 1999;**162**: 2982−9.

25. Minges Wols HA, Underhill GH, Kansas GS, Witte PL. The role of bone marrow-derived stromal cells in the maintenance of plasma cell longevity. *J Immunol* 2002;**169**:4213−21.

26. Tokoyoda K, Egawa T, Sugiyama T, Choi BI, Nagasawa T. Cellular niches controlling B lymphocyte behavior within bone marrow during development. *Immunity* 2004;**20**:707−18.

27. Cassese G, Arce S, Hauser AE, Lehnert K, Moewes B, Mostarac M. et al. Plasma cell survival is mediated by synergistic effects of cytokines and adhesion-dependent signals. *J Immunol* 2003;**171**:1684−90.

28. Benson MJ, Dillon SR, Castigli E, Geha RS, Xu S, Lam KP, et al. Cutting edge: the dependence of plasma cells and independence of memory B cells on BAFF and APRIL. *J Immunol* 2008;**180**: 3655−9.

29. Belnoue E, Pihlgren M, McGaha TL, Tougne C, Rochat AF. Bossen C, et al. APRIL is critical for plasmablast survival in the bone marrow and poorly expressed by early-life bone marrow stromal cells. *Blood* 2008;**111**:2755−64.

30. Varfolomeev E, Kischkel F, Martin F, Seshasayee D, Wang H. Lawrence D, et al. APRIL-deficient mice have normal immune system development. *Mol Cell Biol* 2004;**24**:997−1006.

31. O'Connor BP, Raman VS, Erickson LD, Cook WJ, Weaver LK, Ahonen C, et al. BCMA is essential for the survival of long-lived bone marrow plasma cells. *J Exp Med* 2004;**199**:91−8.

32. Mohr E, Serre K, Manz RA, Cunningham AF, Khan M, Hardie DL, et al. Dendritic cells and monocyte/macrophages that create the IL-6/APRIL-rich lymph node microenvironments where plasma-blasts mature. *J Immunol* 2009;**182**:2113−23.

33. Winter O, Moser K, Mohr E, Zotos D, Kaminski H, Szyska M, et al. Megakaryocytes constitute a functional component of a plasma cell niche in the bone marrow. *Blood* 2010; **116**:1867−75.

34. Chu VT, Berek C. Immunization induces activation of bone marrow eosinophils required for plasma cell survival. *European Journal of Immunology* 2012;**42**:130−7.

35. Throsby M, Herbelin A, Pleau JM, Dardenne M. CD11c+ eosino-phils in the murine thymus: developmental regulation and recruit-ment upon MHC class I-restricted thymocyte deletion. *J Immunol* 2000;**165**:1965−75.

36. Rothenberg ME, Hogan SP. The eosinophil. *Annu Rev Immunol* 2006;**24**:147−74.

37. Song DJ, Cho JY, Lee SY, Miller M, Rosenthal P, Soroosh P, et al. Anti-Siglec-F antibody reduces allergen-induced eosinophilic inflammation and airway remodeling. *J Immunol* 2009;**183**: 5333−41.

38. Zimmermann N, McBride ML, Yamada Y, Hudson SA, Jones C, Cromie KD, et al. Siglec-F antibody administration to mice selec-tively reduces blood and tissue eosinophils. *Allergy* 2008;**63**: 1156−63.

39. Chu VT, Fröhlich A, Steinhauser G, Scheel T, Roch T, Filla-treau, S, Lee JJ, Löhning M, Berek, C. *Nature Immunology* (in press).

40. Juhlin L, Michaelsson G. A new syndrome characterised by absence of eosinophils and basophils. *Lancet* 1977;**1**:1233−5.

Chapter 11.8

Eosinophil–Airway Epithelial Cell Interactions

Darren W. Sexton and Garry M. Walsh

THE EPITHELIUM

The entire respiratory tract is lined with a continuous layer of epithelial cells, comprising eight different types of cells[1] that exhibit differing morphologies, functions, and responses to injury. Airway epithelial cells (AEC) comprise basal cells, brush cells, ciliated cells, Clara cells, mucous goblet cells, neuroendocrine cells, serous cells, and small mucous granule cells. There also exist a variety of intermediate or differentiating cells. The proximal airways are lined with pseudostratified columnar epithelium, consisting largely of basal and ciliated cells, and smaller numbers of goblet cells.[2,3] Glands are relatively common proximally, with a decrease in number toward the distal airway, where the epithelium also becomes much less columnar and there are fewer layers. Distal airways have reduced numbers of ciliated cells with a concomitant increase in basal and Clara cells as the epithelium starts to exhibit a more cuboidal nature, although basal cells are usually absent in the most distal airways. Eventually these cells merge with the type I and type II alveolar epithelial cells, which line the surface of respiratory bronchioles and alveoli.[4] For a single tissue type, epithelia boast a wide range of functions. Physically, the epithelium represents a barrier to nonself entities, whether allergens, inanimate, or living microorganisms. This physical defense is reinforced through the presence of epithelial cell-derived antimicrobial agents, such as defensins, ficolins, and surfactant proteins. Furthermore, it is now generally accepted that the epithelium plays a pivotal role in controlling many airway functions, including clearance of environmental agents, modulation of the inflammatory response to noxious stimuli, and regulation of the cellular activities necessary for responses to injury.[5]

Epithelial-Derived Cytokines

There is compelling evidence that AEC are integral components of both innate and adaptive immunity. AEC are capable of producing and responding to a variety of cytokines, eicosanoids, and growth factors that form a complex regulatory network of inflammatory responses. Production of cytokines by the epithelium is of particular interest with regard to allergic and asthmatic inflammation as these factors can influence inflammatory cells such as eosinophils, mast cells, and T-lymphocytes that are found infiltrating the airways in these disorders.[6] Cytokines derived from the airway epithelium can be functionally subdivided into chemotactic factors, colony stimulating factors, and proinflammatory pleiotropic cytokines and growth factors, each with the capacity to affect eosinophil function. Of equal importance are the chemokine receptors expressed on epithelial cells and eosinophils that facilitate cellular responses to cytokines and provide feedback mechanisms for cytokine production. Chemotactic factors govern the chemotaxis of various, and sometimes specific, inflammatory cells to the site of inflammation. The epithelium can produce members of both C-X-C (α) and C-C (β) chemokine subfamilies. These have come to prominence in recent years because of their potentially important role in allergic inflammation. Crucially, eosinophil populations in tissue can be small under homeostatic conditions and yet in disease states eosinophils numbers escalate, making recruitment, and the recruiting cells, central features of disease pathogenesis.

More recently, attention has focused on the eosinophil specific C-C chemokines, eotaxin/C-C motif chemokine 11 (CCL11), eotaxin-2/C-C motif chemokine 24 (CCL24), and eotaxin-3/C-C motif chemokine 26 (CCL26) that act predominantly via binding to the chemokine receptor (CCR3). The epithelium is a source of these chemokines and expression can be upregulated by proinflammatory cytokine exposure.[7] In particular, inducible eotaxin-3 expression has been closely linked to allergic inflammation, since it is induced (up to 100-fold) after allergen challenge in asthmatics.[8] siRNA knockdown of eotaxin-3 in the A549 type II alveolar cell line results in blockade of interleukin-4 (IL-4)-induced eotaxin-2 expression, as well as expression of C-C motif chemokines 5, 8, 13, and 15.[9] Supernatants from siRNA-treated cells also failed to activate or stimulate eosinophil chemotaxis.

IL-5 is well established as a crucial cytokine for eosinophil differentiation, survival, and activation. While AEC synthesis of IL-5 was reported by Salvi and colleagues in 1999,[10] in a recent study, Wu and coworkers[11] ovalbumin (OVA) challenged an IL-5 knockout mouse to highlight the significance of AEC-derived IL-5 to the pathogenesis of allergic airway inflammation. Bone marrow transfer between wild-type and IL-5 knockout mice distinguished the contribution made by hematopoietic cells and stromal cells and revealed that AEC are capable of IL-5 production at levels sufficient to induce eosinophil infiltration, mucous metaplasia, and OVA-specific immunoglobulin A (IgA) levels. Several epithelial-derived cytokines are crucial elements in the development of T_h2 inflammatory responses, including IL-33 and thymic stromal lymphopoietin (TSLP),[12,13] both of which are discussed below.

AIRWAY INFLAMMATION

Asthma is associated with structural and functional changes to the epithelial surface of the airways. Increasingly strong evidence indicates that the airway epithelium plays a critical role in the pathogenesis of asthma and may serve as an essential orchestrator of the events leading to persistent asthma. Histological alterations include goblet cell metaplasia, squamous metaplasia, thickening of the lamina reticularis, and the accumulation of both subepithelial and intraepithelial inflammatory cells.[14] Epithelial cell shedding is also an important feature, with extensive shedding being observed in the first pathological studies of fatal asthma. However, it is also commonly observed in mucosal biopsies obtained using flexible bronchoscopy in mild asthmatics, with ciliated columnar cells being detached from basal cells that remain firmly bound to the basement membrane. Asthmatic AEC also show evidence of fragility, with ciliated cells appearing swollen and vacuolated, often with loss of cilia.[15] Additional evidence of AEC damage in asthma comes from the study of bronchial AEC obtained by brushing. Clusters of sloughed AEC, known as Creola bodies, may be seen in bronchoalveolar lavage (BAL), histological specimens, and sputum samples from asthmatic subjects.[16]

AEC have considerable synthetic capability (Table 11.8.1) and their release of inflammatory mediators can be upregulated by stimuli associated with asthma (see Table 11.8.2 for epithelial stimuli that impact on eosinophil survival). Therefore, although it is difficult to prove in in vitro studies, it is highly likely that, in asthma, AEC are a source of mediators that direct the chemotactic recruitment of inflammatory cells, regulate the activity of these cells after their recruitment, and alter smooth muscle function. There are differences between normal and asthmatic AEC in both the type and quantity of inflammatory mediators released.[17] For example, in studies using monolayer cultures, AEC from asthmatic adults constitutively release greater amounts of IL-6, IL-8, granulocyte-macrophage colony-stimulating factor (GM-CSF), and soluble intercellular adhesion molecule-1 (sICAM-1) than AEC from control subjects, while RANTES (C-C motif chemokine 5) release is only detected in asthmatic AEC. Asthmatic AEC exhibit activation of the transcription factors activator protein-1 (AP-1), nuclear factor-κB (NF-kB), and signal transducer and activator of transcription-1 (STAT-1), which are involved in the synthesis of these inflammatory mediators.[18]

An area of speculative interest is calcium-activated chloride channel regulator 1 (hCLCA1). Originally classified as a chloride ion channel, this protein is now recognized to be an ion channel regulator. Although membrane bound, it also has a secreted component and thus a putative receptor has been suggested. hCLCA1 expression is known to be elevated in the epithelial cells of asthmatic patients. Indeed, when differential gene expression between epithelial cells from healthy controls and asthmatic patients was assessed, hCLCA1 was the most differentially expressed gene (6.2-fold).[19] Previously, Nakanishi and colleagues[20] had shown that overexpression or knockdown of the mouse homologue mCLCA3 (or gob-5) was sufficient to exacerbate or repress, respectively, mucous cell metaplasia and airway hyperresponsiveness in an OVA-challenge mouse model. While the function of the secreted fragment is unknown, it is interesting to speculate on a potential autocrine or paracrine effect of epithelial-derived hCLCA1. Intriguingly, Sexton and collaborators identified high hCLCA1 protein expression in human granulocytes (unpublished data). Thus, taken together, the interplay between epithelial and eosinophil-derived hCLCA1 may promote or exacerbate the asthma phenotype.

Repair of the epithelium is a normal process following inflammatory damage. Immediately postinjury, AEC at the edge of the damaged area dedifferentiate, flatten, and rapidly migrate to provide a covering layer over the denuded basement membrane.[21] This is followed by epithelial cell proliferation. AEC release of extracellular matrix (ECM) proteins, including fibronectin, is believed to be important in this process. Repairing AEC also directly influence adjacent subepithelial fibroblasts through the release of a variety of factors, including fibroblast growth factor (FGF), insulin-like growth factor (IGF), metalloproteases, and TGF-β that influence fibroblast migration, matrix production, and remodeling. Expression of epidermal growth factor receptor (EGFR) is markedly upregulated in the epithelium of adult asthmatics in proportion to disease severity and chronicity.[22] EGFR immunostaining in severe asthma can be observed throughout the repairing epithelium and also on the luminal surface, a feature not observed in normal epithelium. It should also be emphasized that the airway epithelium is a major target for inhaled anti-inflammatory drugs in asthma, including inhaled corticosteroids, the mainstay of modern asthma therapy. Although corticosteroids have effects on many inflammatory and structural cells involved in inflammation, AEC may be one of the most important targets for inhaled corticosteroids in asthma. Their most striking effect is to inhibit expression of multiple inflammatory genes, including adhesion molecules, cytokines, enzymes, and receptors, most likely via inhibition of the transcription factors NF-κB and AP-1, although they may also increase the transcription of genes encoding anti-inflammatory proteins.[23]

Brewster and coworkers[24] described a layer of subepithelial fibroblasts in the conducting airways that were

TABLE 11.8.1 Molecules Expressed/Released by AEC

Lipid Mediators		Cytokines		Chemokines	
PGE_2	LTB_4	IL-1	IL-10	IL-8	MCP-1/CCL2
PGI_2	LTC_4	TNFα	IL-11	GRO-α	MCP-4/CCL13
$PGF_{2\alpha}$	LTD_4	IL-5	IFNγ	GRO-γ	IL-16
PAF	LTE_4	IL-6	TSLP	RANTES/CCL5	Eotaxin
TXA_2	15-HETE	LIF	IL-33	MIP-1α/CCL3	Eotaxin-2/CCL24
TXB_2				MIP-2/ C-X-C motif chemokine 2	Eotaxin-3/CCL26

Growth Factors		ECM Proteins	Peptide Mediators
TGFα	G-CSF	Collagens 1 and 4	Endothelin family
TGFβ	M-CSF	Fibronectin	Vasopressin
EGF family	IGF	Laminin	Substance P
PDGF	FGF	Hyaluronan	Calcitonin gene-related peptide
GM-CSF	SCF	Tenascin	β-defensin-1
CSF-1		Nidogen-1	
		MMP 1,2,3,7,9	
		TIMP 1,2	

Adhesion Molecules	Integrins		Catabolic Enzymes
CD44	α2β1	α5β1	Neutral endopeptidase
ICAM-1	α3β1	αvβ1	Aminopeptidase N
Ep-CAM	α62β4	αvβ3	Acetylcholinesterase
Cadherin-1	α9β1	αvβ15	Histamine N-methyltransferase
Cadherin-2		αvβ6	
Cadherin-3		αvβ8	
HLA-DR			

Enzyme Inhibitors	Gases/ROS	Antibacterial Substances
Secretory leukoprotease inhibitor	NO	Lysozyme
Elafin	CO	Lactoferrin
Cystatin-C	H_2O_2	LC10
α1-antiprotease	Oxygen radicals	Defensins
α1-antichymotrypsin		Ficolin-2
		Ficolin-3
		Mucins

CCL, C-C motif chemokine; CO, carbon dioxide; CSF, colony stimulating factor; EGF, epidermal growth factor; FGF, fibroblast growth factor; GM-CSF, granulocyte-macrophage colony-stimulating factor; GRO, growth-regulated protein; 15-HETE, 15-hydroxyeicosatetraenoic acid; H_2O_2, hydrogen peroxide; IFN, interferon; IL, interleukin; IGR, insulin-like growth factor; LIF, leukemia inhibitory factor; LT, leukotriene; MMP, matrix metalloproteinase; NO, nitric oxide; PAF, platelet activating factor; PDGF, platelet-derived growth factor; PG, prostaglandin; SCF, stem cell factor; TGF, transforming growth factor; TNF, tumor necrosis factor; TSLP, thymic stromal lymphopoietin; TIMP, tissue inhibitors of MMPs; TX, thromboxane.

positioned in such a way as to allow close interactions among the epithelium, neural tissue, and ECM. This layer of thin mesenchymal cells was termed the attenuated fibroblast sheath. These fibroblasts maintain a constant spatial relationship with the epithelial basement membrane. The anatomical and functional interaction between the attenuated fibroblast sheath, epithelium, neural tissue, and ECM constitute the epithelial—mesenchymal trophic unit (EMTU). This complex unit was first identified as having an essential role in fetal lung development, during which interaction between these tissues allows the exchange of information and responses at different stages of airway growth and branching.[25] The pulmonary mesenchyme, via production of growth factors and other signaling molecules, such as FGF-7, FGF-10, NF-kB, and TGFβ2, is absolutely required for normal epithelial development.

Many of the transcription factors, growth factors and other signaling molecules involved in branching morphogenesis are also implicated in airway remodeling.[26] It has been proposed that the remodeling of the airways in asthma is a result of the EMTU remaining active after birth or becoming reactivated in susceptible asthmatics.[27] Increased susceptibility of AEC to injury and impaired epithelial

TABLE 11.8.2 Effect of Epithelial Cell Stimulation on Eosinophils

Stimulus	Epithelial Cell Response	Eosinophil Effect	Selected Refs
Fungus	TLR-mediated release of inflammatory mediators	Increased recruitment and eosinophil survival	107, 108
Allergen	Release of inflammatory mediators Protease activity of allergens can alter IL-13 receptors leading to increased bias toward inflammation	Increased recruitment and eosinophil survival	109, 110
Virus	TLR-mediated release of inflammatory mediators	Increased recruitment and eosinophil survival	111, 112
Steroids	Decreased inflammatory mediator release Increased phagocytosis	Apoptosis of eosinophils Phagocytosis by epithelial cells Luminal entry of tissue resident eosinophils	78, 79, 80, 83
Diesel exhaust particles	Decreased ciliary beat frequency Increased proinflammatory cytokines release	Decreased eosinophil clearance Increased recruitment and eosinophil survival	113, 114

IL-13, interleukin-13; TLR, Toll-like receptor.

proliferation lead to persistent mucosal injury and cause AEC to spend longer periods in a repair phenotype, resulting in increased production of proinflammatory mediators and increased secretion of profibrogenic growth factors that are capable of inducing proliferation of sub-epithelial fibroblasts and their differentiation into activated myofibroblasts. This theory is strongly supported by the myriad asthma susceptibility genes that have been identi-fied in epithelial cells and the fact that the strongest risk factors for asthma development, exacerbation or prolon-gation, such as air pollutants, biologically active allergens, environmental tobacco smoke (ETS), irritants, and respi-ratory viruses, primarily affect the EMTU and lead to disease in individuals in which the epithelial barrier func-tion is faulty.

The EMTU framework does not preclude the impact of other cell types on the epithelial and mesenchymal cell responses. Eosinophil-deficient mouse models have, thus, proven to be valuable models for investigating the effect of eosinophils on this system. The differing eosinophil-defi-cient mouse models that exist have been used in allergic asthma models and have yielded disparate results. Eosin-ophil involvement in asthma is, however, clearly estab-lished[28,29] and divergence between the models has been attributed to the differing mouse strain backgrounds, i.e., B6 mice vs. BALB/c. These models agree on a putative predominant role for eosinophils in airway remodeling and

disease severity.[30] This is reinforced by demonstrations that transforming growth factor-β2 (TGF-β2), which is princi-pally derived from tissue eosinophils, is the predominant isoform in severe asthma and is associated with augmented profibrotic responses.[31] The broader role of eosinophils in airway remodeling has been reviewed recently.[32] There is a crucial interaction between eosinophils and epithelial cells, with feedback between these cells purportedly being key to the chronicity and severity of a disease of global significance.

Inflammation Initiation

Innate Immune Response

Defective barrier function has been linked to diseases that involve eosinophilic inflammation in the airway and intestine.[32–36] Direct ingress of antigenic material into the body can alter the immune response to such antigens.[37] Epithelial cells are well equipped with pattern recognition receptors (PRRs). At different sites and in accordance with their differing cell types, these cells express mannose-binding lectins, NOD-like receptors (NLRs), pentraxins, retinoic acid-inducible gene 1 protein (RIG-I), and Toll-like receptors (TLRs), as well as dectin and HIN200 family members. These are capable of recognizing pathogen-associated molecular pattern proteins (PAMPs), alternatively called microbe-associated molecular patterns (MAMPS). The TLR family of PRRs has been extensively studied in recent years and are expressed on epithelial cells in the airways and gut. Polymorphisms in the genes that encode these proteins have been linked to eosinophil-associated diseases such as allergic airways disease,[38] Crohn disease, inflammatory bowel disease, and ulcerative colitis. Their significance to eosinophil interactions is their capacity to instigate proinflammatory cytokine responses in epithelial cells and thus initiate inflammation and consequent eosinophil recruitment, activation and survival. An interesting recent example of this process comes from research into the effects of chitin on nasal epithelial cell cultures. Results suggest that a novel innate immune pathway is capable of inducing acidic mammalian chitinase (chitinases and chitinase-like proteins are currently under investigation as antiallergy therapeutics[39]), as well as eotaxin-3 expression.[40] Other cytokines discussed in this subchapter are known to be inducible by PRR signaling.

Adaptive Immune Response

Adaptive immune responses in the airway are readily initiated by the resident antigen-presenting cells, which include dendritic cells, macrophages, and, indeed, eosinophils. However, epithelial cells express many of the cell-surface molecules associated with antigen presentation, such as MHC I and II and CD40.[41–43] B7 homologues are also expressed by AEC and together these data suggest a role for the AEC in maintaining and/or regulating the activation of antigen-specific lymphocytes that have migrated to the airways. With evidence now showing that eosinophils have the capacity to present antigen (reviewed in Walsh and August[44]) it is interesting to speculate on the influence that AEC may have on this activity. Evidence already exists to show that AEC can influence antigen-presenting cell (APC) function.[45] Epithelial cell-derived TSLP operates as a master switch for asthma or atopic dermatitis by inducing myeloid dendritic cells to initiate a T_h2 inflammatory response.[46] In addition, TSLP has also been shown to have chemotactic and prosurvival properties for eosinophils.[47] IL-33, which is also produced by mucosal epithelial cells, is similarly crucial to the T_h2 inflammatory response.[48] This IL-1 family member has been suggested to have a central role in inflammation, since it is among the primary signaling molecules released following epithelial cell damage.[49] IL-33 has an impact on eosinophil function through increased survival, degranulation, superoxide production, adhesion, and recruitment. More broadly, instillation of IL-33 in mice is sufficient to generate airway hyperresponsiveness and goblet cell hyperplasia and its overexpression results in spontaneous pulmonary eosinophilic inflammation.[50,51] Pecaric-

Petkovic and coworkers reported that mouse basophils and eosinophils are the only leukocytes directly targeted by IL-33; this places a great emphasis on epithelial and airway smooth muscle cells as sources of IL-33 in promoting the allergic airway response.[52]

Inflammation Resolution

Apoptosis Induction

Once generated in and released from the bone marrow, eosinophils may circulate for 8–12 h, but in tissues they are thought to survive for 8–12 days. Cytokines, growth factors, and interactions with the ECM play a role in this delay in eosinophil apoptosis. Apoptosis induction is a key facet of tissue homeostasis, and in resolving eosinophilic inflammation in asthmatic patients' apoptotic eosinophils, together with evidence of their engulfment by alveolar macrophages, are readily observed.[53] Indeed, in the human airways a decrease in the ratio of viable to apoptotic eosinophils in induced sputum samples correlates with reduced clinical symptoms in asthmatic subjects.[54] The processes that inhibit or initiate and control apoptosis in eosinophils have been comprehensively reviewed elsewhere.[55,56]

Phagocytosis of Apoptotic Eosinophils

Although much is said of the importance of apoptosis induction in homeostasis, its benefits arise only in conjunction with the phagocytosis of these apoptotic cells by neighboring cells. Failure to do so permits apoptotic cells to develop secondary necrosis, which results in injurious perpetuation of the inflammatory response. Our understanding of eosinophil accumulation and activation far outweighs that of the clearance mechanisms employed by the body for the safe removal of these cells (see reviews[57–60]).

Various studies in macrophages have elucidated the mechanisms for recognition and engulfment of apoptotic cells, including eosinophils. These include an uncharacterized lectin-dependent interaction[61]; a charge-sensitive process involving the CD36/vitronectin receptor complex interacting with unknown moieties on the surface of apoptotic neutrophils via a thrombospondin bridge[62,63]; stereo-specific recognition of phosphatidylserine (PS) exposed on the surface of apoptotic cells after loss of membrane asymmetry[64,65]; redistribution of PS on the phagocyte[66]; macrophage scavenger receptors,[67,68] CD14,[69,70] CD68,[71] the ABC1 transporter,[72] Dock180, β1 integrins,[73] and CD44[74]; homophilic CD31 ligation[75]; and optimization by MFG-E8, C1q, mannose-binding lectin (MBL), and E6 (reviewed by Elliott and Ravichandran[76]). It is now clear that a number of nonprofessional phagocytes, including dendritic cells, lung fibroblasts, and smooth muscle cells, also have the capacity to recognize

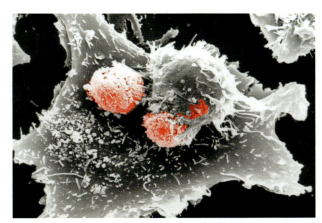

FIGURE 11.8.1 Phagocytosis of apoptotic human eosinophils (colored red) by human small airway epithelial cells.

and ingest apoptotic cells.[77] Although the bronchial epithelium is generally considered to be the target for cell damage and loss by eosinophil-derived mediators, we made the original observation that apoptotic eosinophils can be recognized and engulfed by both resting and cytokine-stimulated AEC.

In a series of papers, we characterized the capacity of normal human bronchial epithelial cells and small airway epithelial cells (SAEC) (Fig. 11.8.1) to recognize and to phagocytose apoptotic, but not freshly isolated, eosinophils.[78] Recognition and phagocytosis of apoptotic eosinophils is a specific event under the control of integrin, lectin, and phosphatidylserine membrane receptors. Importantly, we also demonstrated that the corticosteroid dexamethasone increases both the percentage of AEC that engulf apoptotic eosinophils and, in particular, in the number of apoptotic eosinophils ingested by each epithelial cell. These findings add a new dimension to the anti-inflammatory effects of corticosteroids. We also demonstrated that actin rearrangement is involved in the phagocytosis of apoptotic eosinophils by AEC and that the phagocytic capacity of cytokine-stimulated small and large AEC is approximately half that of human monocyte-derived macrophages. Intriguingly, AEC did not phagocytose apoptotic neutrophils, and the selective phagocytosis of apoptotic eosinophils was consistently observed on screening epithelial cell lines derived from alveolar (A549), colon (HT-29), and mammary (ZR-75—1) tissues.[79,80] Indeed, the preferential uptake of eosinophils was similarly observed in our monocyte-derived macrophages, which exhibited consistently higher phagocytic uptake of apoptotic eosinophils than apoptotic neutrophils. This latter observation suggests that clearance of apoptotic neutrophils by the bronchial epithelium does not represent a potential anti-inflammatory pathway in chronic obstructive pulmonary disease, in which neutrophil-induced inflammation

predominates.[81] What is clear is that given the extent of the lung epithelium and that LPS-dependent phagocytosis of apoptotic cells by alveolar macrophages is greatly impaired in patients with chronic asthma, AEC may prove to be vital to the clearance of apoptotic eosinophils.[82] While recognition and phagocytosis of apoptotic eosinophils may have important repercussions for the inflammatory status of airway epithelium, bulk clearance of eosinophils can be achieved by their passage into the airway lumen. AEC can orchestrate this process and, once in the lumen, the eosinophils are subject to the physical actions of epithelial mucociliary action (see review by Erjefalt[83]). Mucociliary action is, of course, perturbed by eosinophil-derived mediators.[84,85]

The effect of phagocytosis on the survival or cytokine profile of epithelial cells is as yet undetermined, but may provide interesting insights into the promotion of eosinophil survival, especially with regard to evidence highlighting the role of epithelial-derived cytokines in postponing eosinophil apoptosis.[36] Phagocytosis of apoptotic cells not only prohibits release of phlogistic agents, such as eicosanoids or proinflammatory cytokines,[87,88] but can actually suppress the inflammatory response. Engulfment of apoptotic eosinophils also induces an anti-inflammatory cytokine and mediator secretory profile in macrophages, i.e., IL-10, prostaglandin E2, and TGF-β, in direct contrast to ingestion of necrotic eosinophils, which results in a proinflammatory cytokine and mediator profile, i.e., release of thromboxane B2 and GM-CSF.[89] Whether an anti-inflammatory secretory profile can be established in AEC following ingestion of apoptotic eosinophils or as a result of ligation of their recognition receptors remains to be determined.

Eosinophil Apoptosis: Relationship with Prolonged Survival

It has been appreciated for many years that the culture of eosinophils *in vitro* with IL-3, IL-5, or GM-CSF enhances their survival for up to 2 weeks. Additionally, IL-13 has been shown to increase eosinophil survival,[89,90] which is further enhanced by coincubation with TNFα,[91] while IL-33 also prolongs eosinophil survival.[52] The significance of these findings is emphasized by ample evidence that these cytokines are present in asthmatic airway tissue[92] and that eosinophils isolated from patients with atopic dermatitis and, to a lesser extent, inhalant allergy, display enhanced survival compared with normal controls.[93] Vignola and coworkers[94] investigated the role of GM-CSF in prolonged survival of eosinophils from subjects with asthma and chronic bronchitis. They found that tissue production of this cytokine is greater in patients with asthma than in controls and those with chronic bronchitis. Significantly, GM-CSF production correlates with both the frequency of nonapoptotic eosinophils and macrophages and the severity of asthma. Although leukocyte infiltration is similar for both asthma and chronic bronchitis patients, the number of eosinophils is significantly lower in the latter and the percentage of apoptotic eosinophils in the tissues of patients with chronic bronchitis is greater than in asthma subjects. Moreover, Simon and colleagues[95] provided direct evidence, using human nasal polyps from nonatopic patients with bronchial asthma, that delayed eosinophil apoptosis can result in tissue eosinophilia and that nasal polyp tissue from allergic, but not nonallergic, subjects spontaneously release significant quantities of GM-CSF that enhances eosinophil survival *in vitro*.[96] Paracrine production of these cytokines by T_h2 cells, as part of an aberrant T_h2 response, is believed to be a significant part of the pathogenesis of asthma.[97] This hypothesis is given

credence in light of data revealing that mitogen-stimulated T cells from asthmatic subjects are resistant to Fas-mediated apoptosis, despite the presence of surface Fas expression[98] and the existence of clonal populations of abnormal T cells producing IL-5 in some patients with idiopathic eosinophilia.

Eosinophil interactions with the proteins of the ECM present in the asthmatic lung are also thought to make a significant contribution to their persistence by signaling through integrin receptors that mediate autocrine production of GM-CSF, IL-3 and IL-5. The finding that bronchial epithelial cells recognize and ingest apoptotic eosinophils is also interesting in the light of evidence demonstrating that under certain circumstances eosinophil persistence in the airways may also be enhanced by their interaction with epithelial cells[99] or epithelial cell-derived mediators.[100]

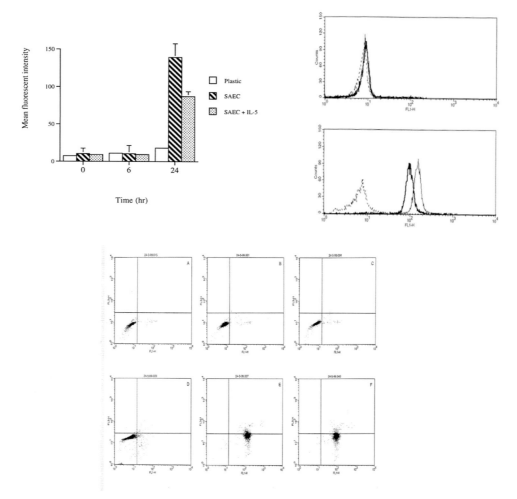

FIGURE 11.8.2   Fluorescence-activated cell sorting analysis. Three separate experiments revealing apoptosis induction in freshly isolated eosinophils after 1 h and 24 h coculture in the presence of small airway epithelial cell monolayers. Annexin V binding (FL-1) and propidium iodide (FL-3) exclusion were used to detect apoptosis and cell viability, respectively. A clear shift toward increased FL-1 without a similar rise in FL-3 reflects apoptosis induction without necrosis. Histograms more clearly indicate the degree of apoptosis induction and at 24 h one can see that interleukin-5 (10^{-10} M) treatment only partially inhibits the induced cell death.

Moreover, other resident lung cells, including mast cells[101] or IL-1β-stimulated airway smooth muscle cells,[102] elaborate GM-CSF, which in turn enhances eosinophil survival, while upper airway tissue eosinophils isolated from nasal polyps show enhanced survival when cultured *in vitro* compared with peripheral blood eosinophils.[103] Furthermore, eosinophils have been shown to express the CD40 membrane receptor, ligation of which results in enhanced eosinophil survival as a consequence of autocrine GM-CSF release. This study also demonstrated that tissue eosinophils resident in nasal polyp tissue have a high constitutive expression of CD40.[104] The ligand for CD40, CD40L, is expressed by CD4+ T cells that are also present in nasal polyp tissue, suggesting the intriguing potential for a further relationship between eosinophils and T cells.

AEC AND EOSINOPHIL INTERACTIONS

Elucidation of the effects of epithelial cells on eosinophils has relied largely on experiments using conditioned media from epithelial cell cultures. Much of the evidence suggests that factors in conditioned media can delay eosinophil apoptosis. These studies, however, failed to ascertain if direct cell—cell contact between the epithelial cells and eosinophils has any involvement in eosinophil survival. In addressing this issue, our data has revealed that cell—cell contact can have a dramatic effect on eosinophil survival.[105] Eosinophils cultured in fresh or conditioned media exhibit constitutive levels of apoptosis over a 24 h period, while those in equivalent media but in direct contact with the epithelial cell monolayer have significant apoptosis induction (Fig. 11.8.2). Instillation of IL-5 (10^{-10} M) at the beginning of coculture reduces annexin V positivity but does not restore constitutive levels of apoptosis to cultured eosinophils. Our findings are supported by a study that used bronchial biopsies from asthmatic patients to demonstrate that corticosteroids may permit the bronchial epithelium to exert a cytotoxic activity toward Fas-bearing cells, including eosinophils, by induction of Fas-L expression.[106]

CONCLUSION

Epithelial cells present at various sites of the body represent crucial sentinels of the body's defense mechanism, with a vast repertoire of sensory receptors and secretory molecules that act as effectors, activators, or recruiters of other cell types. Throughout the complex process of inflammation, epithelial cells can work with eosinophils to initiate and augment immune responses. In resolving inflammation, epithelial cells phagocytose apoptotic eosinophils with putative anti-inflammatory effects. Epithelial cells also interact indirectly with eosinophils through secreted mediators; each cell type is capable of effecting changes in the other. Our understanding of their cell—cell interactions is limited but eosinophils are no longer merely seen as the destructive partner in this relationship, although this capacity certainly exists. The epithelial cell cannot work in isolation but instead relies on specialist cell types such as the eosinophil. In normal host defense and in disease states, their interactions are multifaceted and much remains to be elucidated.

REFERENCES

1. Breeze RG, Wheldon EB. The cells of the pulmonary airways. *American Review of Respiratory Disease* 1977;**116**:705—77.
2. Plopper CG, Mariassy AT, Wilson DW, Alley JL, Nishio SJ, Nettesheim P. Comparison of nonciliated tracheal epithelial cells in six mammalian species: ultrastructure and population densities. *Exp Lung Res* 1983;**5**(4):281—94.
3. Mercer RR, Russell ML, Roggli VL, Crapo JD. Cell number and distribution in human and rat airways. *Am J Respir Cell Mol Biol* 1994;**10**(6):613—24.
4. Mason RJ, Apostolou S, Power J, Robinson P. Human alveolar type II cells: stimulation of DNA synthesis by insulin and endothelial cell growth supplement. *Am J Respir Cell Mol Biol* 1990;**3**(6):571—7.
5. Stick SM, Holt PG. The airway epithelium as immune modulator: the LARC ascending. *American Journal of Respiratory Cell and Molecular Biology* 2003;**28**(6):641—4
6. Saenz SA, Taylor BC, Artis D. Welcome to the neighborhood: epithelial cell-derived cytokines license innate and adaptive immune responses at mucosal sites. *Immunol Rev* 2008;**226**:172—90.
7. Garcia-Zepeda EA, Rothenberg ME, Ownbey RT, Celestin J, Leder P, Luster AD. Human eotaxin is a specific chemoattractant for eosinophil cells and provides a new mechanism to explain tissue eosinophilia. *Nat Med* 1996;**2**(4):449—56.
8. Berkman N, Ohnona S, Chung FK, Breuer R. Eotaxin-3 but not eotaxin gene expression is upregulated in asthmatics 24 hours after allergen challenge. *Am J Respir Cell Mol Biol* 2001;**24**:682—7.
9. Errahali YJ, Taka E, Abonyo BO, Heiman AS. CCL26-targeted siRNA treatment of alveolar type II cells decreases expression of CCR3-binding chemokines and reduces eosinophil migration: implications in asthma therapy. *J Interferon Cytokine Res* 2009;**29**(4):227—39.
10. Salvi S, Semper A, Blomberg A, Holloway J, Jaffar Z, Papi A, et al. Interleukin-5 production by human airway epithelial cells. *Am J Respir Cell Mol Biol* 1999;**20**:984—91.
11. Wu CA, Peluso JJ, Zhu L, Lingenheld EG, Walker ST, Puddington L. Bronchial epithelial cells produce IL-5: implications for local immune responses in the airways. *Cell Immunol* 2010;**264**(1):32—41.
12. Townsend MJ, Fallon PG, Matthews DJ, et al. T1/ST2-deficient mice demonstrate the importance of T1/ST2 in developing primary T helper cell type 2 responses. *J Exp Med* 2000;**191**:1069—75.
13. Coyle AJ, Lloyd CM, Tian J, Nguyen T, Erikkson C, Wang L, et al. Crucial role of the interleukin 1 receptor family member T1/ST2 in T helper cell type 2-mediated lung mucosal immune responses. *J Exp Med* 1999;**190**:895—902.

14. Jeffery PK. Comparative morphology of the airways in asthma and chronic obstructive pulmonary disease. *American Journal of Respiratory and Critical Care Medicine* 1994;**150**(5 Pt 2): S6−13.

15. Laitinen LA, Heino M, Laitinen A, Kava T, Haahtela T. Damage of the airway epithelium and bronchial reactivity in patients with asthma. *American Review of Respiratory Disease* 1985;**131**(4): 599−606.

16. Vignola AM, Chanez P, Siena L, Gagliardo R, Merendino AM, Bonsignore G, et al. Role of epithelial cells in asthma. *Pulmonary Pharmacology and Therapeutics* 1998;**11**(5−6):355−7.

17. Holgate ST. Has the time come to rethink the pathogenesis of asthma? *Curr Opin Allergy Clin Immunol* 2010 Feb;**10**(1):48−53.

18. Holgate ST, Lackie PM, Howarth PH, Roche WR, Puddicombe SM, Richter A, et al. Invited lecture: activation of the epithelial mesenchymal trophic unit in the pathogenesis of asthma. *International Archives of Allergy and Immunology* 2001;**124**(1−3):253−8.

19. Woodruff PG, Boushey HA, Dolganov GM, Barker CS, Yang YH, Donnelly S, et al. Genome-wide profiling identifies epithelial cell genes associated with asthma and with treatment response to corticosteroids. *Proc Natl Acad Sci USA* 2007;**104**(40): 15858−63.

20. Nakanishi A, Morita S, Iwashita H, Sagiya Y, Ashida Y, Shirafuji H, et al. Role of gob-5 in mucus overproduction and airway hyperresponsiveness in asthma. *Proc Natl Acad Sci USA* 2001;**98**(9):5175−80.

21. Knight DA, Holgate ST. The airway epithelium: structural and functional properties in health and disease. *Respirology* 2003;**8**(4):432−46.

22. Puddicombe SM, Polosa R, Richter A, Krishna MT, Howarth PH, Holgate ST, et al. Involvement of the epidermal growth factor receptor in epithelial repair in asthma. *FASEB Journal* 2000;**14**(10):1362−74.

23. Barnes PJ. Anti-inflammatory actions of glucocorticoids: molecular mechanisms. *Clinical Science* 1998;**94**(6):557−72.

24. Brewster CE, Howarth PH, Djukanovic R, Wilson J, Holgate ST, Roche WR. Myofibroblasts and subepithelial fibrosis in bronchial asthma. *American Journal of Respiratory Cell and Molecular Biology* 1990;**3**(5):507−11.

25. Warburton D, Schwarz M, Tefft D, Flores-Delgado G, Anderson KD, Cardoso WV. The molecular basis of lung morphogenesis. *Mechanisms of Development* 2000;**92**(1):55−81.

26. Demayo F, Minoo P, Plopper CG, Schuger L, Shannon J, Torday JS. Mesenchymal-epithelial interactions in lung development and repair: are modeling and remodeling the same process? *American Journal of Physiology—Lung Cellular and Molecular Physiology* 2002;**283**(3):L510−7.

27. Knight DA, Lane CL, Stick SM. Does aberrant activation of the epithelial-mesenchymal trophic unit play a key role in asthma or is it an unimportant sideshow? *Current Opinion in Pharmacology* 2004;**4**(3):251−6.

28. Humbles AA, Lloyd CM, McMillan SJ, Friend DS, Xanthou G, McKenna EE, et al. A critical role for eosinophils in allergic airways remodeling. *Science* 2004;**305**:1776−9.

29. Lee JJ, Dimina D, Macias MP, Ochkur SI, McGarry MP, O'Neill KR, et al. Defining a link with asthma in mice congenitally deficient in eosinophils. *Science* 2004;**305**:1773−6.

30. Wills-Karp M, Karp CL. Eosinophils in asthma: remodelling a tangled tale. *Science* 2004;**305**:1726−9.

31. Balzar S, Chu HW, Silkoff P, Cundall M, Trudeau JB, Strand M, et al. Increased TGF-beta2 in severe asthma with eosinophilia. *J Allergy Clin Immunol* 2005;**115**:110−7.

32. Venge P. The eosinophil and airway remodelling in asthma. *Clin Respir J* 2010;**4**(Suppl. 1):15−9.

33. Clayburgh DR, Shen L, Turner JR. A porous defense: the leaky epithelial barrier in intestinal disease. *Lab Invest* 2004;**84**(3): 282−91.

34. Su L, Shen L, Clayburgh DR, Nalle SC, Sullivan EA, Meddings JB, et al. Targeted epithelial tight junction dysfunction causes immune activation and contributes to development of experimental colitis. *Gastroenterology* 2009; **136**(2):551−63.

35. Swindle EJ, Collins JE, Davies DE. Breakdown in epithelial barrier function in patients with asthma: identification of novel therapeutic approaches. *J Allergy Clin Immunol* 2009;**124**(1): 23−34.

36. Bruewer M, Luegering A, Kucharzik T, Parkos CA, Madara JL, Hopkins AM, et al. Proinflammatory cytokines disrupt epithelial barrier function by apoptosis-independent mechanisms. *J Immunol* 2003;**171**(11):6164−72.

37. Clayburgh DR, Shen L, Turner JR. A porous defense: the leaky epithelial barrier in intestinal disease. *Lab Invest* 2004;**84**(3):282−91.

38. Iwamura C, Nakayama T. Toll-like receptors in the respiratory system: their roles in inflammation. *Curr Allergy Asthma Rep* 2008;**8**(1):7−13.

39. Sutherland TE, Maizels RM, Allen JE. Chitinases and chitinase-like proteins: potential therapeutic targets for the treatment of T-helper type 2 allergies. *Clin Exp Allergy* 2009; **39**(7):943−55.

40. Lalaker A, Nkrumah L, Lee WK, Ramanathan M, Lane AP. Chitin stimulates expression of acidic mammalian chitinase and eotaxin-3 by human sinonasal epithelial cells in vitro. *Am J Rhinol Allergy* 2009;**23**(1):8−14.

41. Jahnsen FL, Farstad IN, Aanesen JP, Brandtzaeg P. Phenotypic distribution of T cells in human nasal mucosa differs from that in the gut. *Am J Respir Cell Mol Biol* 1998;**18**:392−401.

42. Atsuta J, Sterbinsky SA, Schwiebert LM, Bochner BS, Schleimer RP. Phenotyping and cytokine regulation of the BEAS-2B human bronchial epithelial cell: demonstration of inducible expression of the adhesion molecule VCAM-1 and ICAM-1. *Am J Respir Cell Mol Biol* 1997;**17**:571−82.

43. Propst SM, Denson R, Rothstein E, Estell K, Schwiebert LM. Proinflammatory and Th2-derived cytokines modulate CD40-mediated expression of inflammatory mediators in airway epithelia: implications for the role of epithelial CD40 in airway inflammation. *J Immunol* 2000;**165**:2214−21.

44. Walsh ER, August A. Eosinophils and allergic airway disease: there is more to the story. *Trends Immunol* 2010;**31**(1):39−44.

45. Liu YJ. Thymic stromal lymphopoietin: Master switch for allergic inflammation. *J. Exp. Med* 2006;**203**:269−73.

46. He R, Geha RS. Thymic stromal lymphopoietin. *Ann N Y Acad Sci* 2010;**1183**:13−24.

47. Wong CK, Hu S, Cheung PF, Lam CW. Thymic stromal lymphopoietin induces chemotactic and prosurvival effects in eosinophils:

implications in allergic inflammation. *Am J Respir Cell Mol Biol* 2010;**43**(3):305–15.

48. Prefontaine D, Jessica N, Fazila C, Severine A, Abdelhabib S, Jamila C, et al. Increased IL-33 expression by epithelial cells in bronchial asthma. *J Allergy Clin Immunol* 2010;**125**:752–4.

49. Lloyd CM. IL-33 family members and asthma—bridging innate and adaptive immune responses. *Curr Opin Immunol* 2010;**22**:1–7.

50. Kondo Y, Yoshimoto T, Yasuda K, Futatsugi-Yumikura S, Morimoto M, Hayashi N, et al. Administration of IL-33 induces airway hyperresponsiveness and goblet cell hyperplasia in the lungs in the absence of adaptive immune system. *Int Immunol* 2008;**20**:791–800.

51. Zhiguang X, Wei C, Steven R, Wei D, Wei Z, Rong M, et al. Over-expression of IL-33 leads to spontaneous pulmonary inflammation in mIL-33 transgenic mice. *Immunol Lett* 2010;**131**:159–65.

52. Pecaric-Petkovic T, Didichenko SA, Kaempfer S, Spiegl N, Dahinden CA. Human basophils and eosinophils are the direct target leukocytes of the novel IL-1 family member IL-33. *Blood* 2009;**113**(7):1526–3.

53. Walsh GM. Defective apoptotic cell clearance in asthma and COPD—A new drug target for statins. *Trends in Pharmacological Sciences* 2008;**29**(1):6–11.

54. Duncan CJA, Lawrie A, Blaylock MG, Douglas JG, Walsh GM. Reduced eosinophil apoptosis in induced sputum correlates with asthma severity. *Eur Resp J* 2003;**22**:484–90.

55. Simon HU. Cell death in allergic diseases. *Apoptosis* 2009 Apr;**14**(4):439–46.

56. Walsh GM, Al-Rabia M, Blaylock MG, Sexton DW, Duncan CJA, Lawrie A. Control of eosinophil toxicity in the lung. *Current Drug Targets—Inflammation and Allergy* 2005;**4**:481–6.

57. Birge RB, Ucker DS. Innate apoptotic immunity: the calming touch of death. *Cell Death Differ* 2008;**15**(7):1096–102.

58. Fullard JF, Kale A, Baker NE. Clearance of apoptotic corpses. *Apoptosis* 2009;**14**(8):1029–37.

59. Fadeel B, Xue D, Kagan V. Programmed cell clearance: molecular regulation of the elimination of apoptotic cell corpses and its role in the resolution of inflammation. *Biochem Biophys Res Commun* 2010;**396**(1):7–10.

60. Elliott MR, Ravichandran KS. Clearance of apoptotic cells: implications in health and disease. *J Cell Biol* 2010;**189**(7):1059–70.

61. Duvall E, Wyllie AH, Morris RG. Macrophage recognition of cells undergoing programmed cell death (apoptosis). *Immunology* 1985;**56**:351–8.

62. Savill JS, Hogg Ren Y, Haslett C. Thrombospondin co-operates with CD36 and the vitronectin receptor in macrophage recognition of neutrophils undergoing apoptosis. *J Clin Invest* 1992;**90**:1513–22.

63. Savill JS, Dransfield I, Hogg N, Haslett C. Vitronectin receptor-mediated phagocytosis of cells undergoing apoptosis. *Nature* 1990;**343**:170–3.

64. Fadok VA, Savill JS, Haslett C, Braatton DL, Doherty DE, Campbell PA, et al. Differnet populations of macrophages use either the vitronectin receptor or the phosphatidylserine receptor to recognise and remove apoptotic cells. *J Immunol* 1992;**149**:4029–35.

65. Fadok VA, Voelker DR, Campbell PA, Cohen JJ, Bratton DL, Henson PM. Exposure of phosphatidyl-serine on the surface of apoptotic lymphocytes triggers specific recognition and removal by macrophages. *J Immunol* 1992;**148**:2207–16.

66. Marguet D, Luciani MF, Moynault A, Williamson P, Chimini G. Engulfment of apoptotic cells involves the redistribution of membrane phosphatidylserine on phagocyte and prey. *Nature Cell Biology* 1999;**1**:454–6.

67. Platt N, Suzuli H, Kurihara Y, Kodoma T, Gordon S. Role for the class A scavenger receptor in the phagocytosis of apoptotic thymocytes in vitro. *Proc Natl Acad Sci USA* 1996;**93**:12456–60.

68. Sambrano GR, Steinberg D. Recognition of oxidatively damaged and apoptotic cells by an oxidised low density lipoprotein receptor on mouse peritoneal macrophages: Role of membrane phosphatidylserine. *Proc Natl Acad Sci USA* 1995;**92**:1396–400.

69. Flora PK, Gregory CD. Recognition of apoptotic cells by human macrophages: Inhibition by a monocyte/macrophage-specific monoclonal antibody. *Eur J Immunol* 1994;**24**:2625–32.

70. Devitt A, Moffat OD, Raykundalia C, Capra JD, Simmons DL, Gregory CG. Human CD14 mediates recognition and phagocytosis of apoptotic cells. *Nature* 1998;**392**:505–9.

71. Ramprasad MP, Fischer W, Witzum JL, Sambrano GR, Quehenberger O, Steinberger D. The 94- to 97-kDa mouse macrophage membrane protein that recognises oxidised low density lipoprotein and phosphatidylserine rich liposomes is identical to macrosialin, the mouse homologue of human CD68. *Proc Natl Acad Sci USA* 1995;**92**:9580–4.

72. Luciani MF, Chimini G. The ATP binding cassette transporter ABC1 is required for engulfment of corpses generated by apoptotic cell death. *EMBO J* 1996;**15**:226–35.

73. Schwartz BR, Karsan A, Bombeli, Harlan JM. A novel β1 integrin mechanism of leukocyte adherence to apoptotic cells. *Journal of Immunology* 1999;**162**:4842–8.

74. Hart SP, Dougherty GJ, Haslett C, Dransfield I. CD44 regulates phagocytosis of apoptotic neutrophil granulocytes, but not apoptotic lymphocytes, by human macrophages. *J Immunol* 1997;**159**(2):919–25.

75. Brown S, Heinisch I, Ross E, Shaw K, Buckley CD, Savill J. Apoptosis disables CD31-mediated cell detachment from phagocytes promoting binding and engulfment. *Nature* 2002;**418**(6894):200–3.

76. Elliott MR. Ravichandran KS. Clearance of apoptotic cells: implications in health and disease. *J Cell Biol* 2010;**189**(7):1059–70.

77. Platt N, da Silva RP, Gordon S. Recognising death: the phagocytosis of apoptotic cells. *Trends Cell Biol* 1998;**8**:365–72.

78. Walsh GM, Sexton DW, Blaylock MG, Convery CM. Resting and cytokine-stimulated human small airway epithelial cells recognize and engulf apoptotic eosinophils. *Blood* 1999;**94**(8):2827–35.

79. Sexton DW, Al-Rabia M, Blaylock MG, Walsh GM. Phagocytosis of apoptotic eosinophils but not neutrophils by bronchial epithelial cells. *Clin Exp Allergy* 2004 Oct;**34**(10):1514–24.

80. Sexton DW, Blaylock MG, Walsh GM. Human alveolar epithelial cells engulf apoptotic eosinophils by means of integrin- and phosphatidylserine receptor-dependent mechanisms: a process upregulated by dexamethasone. *J Allergy Clin Immunol* 2001;**108**(6):962–9.

81. Walsh GM, McDougall CM. The resolution of airway inflammation in asthma and COPD. In: Rossi AG, Sawatzky D, editors. *Progress in Inflammation Research*. Birkhauser Verlag AG; 2008. p. 159–91.

82. Huynh ML, Malcolm KC, Kotaru C, Tilstra JA, Westcott JY, Fadok VA, et al. Defective apoptotic cell phagocytosis attenuates prostaglandin e2 and 15-hydroxyeicosatetraenoic acid in severe asthma alveolar macrophages. *Am J Respir Crit Care Med* 2005;**172**(8):972−9.

83. Erjefält JS. The airway epithelium as regulator of inflammation patterns in asthma. *Clin Respir J* 2010;**4**(Suppl. 1):9−14.

84. Devalia JL, Sapsford RJ, Rusznak C, Davies RJ. The effect of human eosinophils on cultured human nasal epithelial cell activity and the influence of nedocromil sodium in vitro. *Am J Respir Cell Mol Biol* 1992;**7**(3):270−7.

85. Hastie AT, Loegering DA, Gleich GJ, Kueppers F. The effect of purified human eosinophil major basic protein on mammalian ciliary activity. *Am Rev Respir Dis* 1987;**135**(4):848−53.

86. Daffern PJ, Jagels MA, Saad JJ, Fischer W, Hugli TE. Upper airway epithelial cells support eosinophil survival in vitro through the production of GM CSF and Prostaglandin E2: Regulation by glucocorticoids and TNFa. *Allergy and Asthma Proceedings* 1999;**20**(4):243−53.

87. Meagher LC, Savill JS, Baker A, Fuller RW, Haslett C. Phagocytosis of apoptotic neutrophils does not induce macrophage release of thromboxane B2. *J Leukoc Biol* 1992;**52**(3):269−73.

88. Stern M, Savill J, Haslett C. Human monocyte-derived macrophage phagocytosis of senescent eosinophils undergoing apoptosis: mediation by αvβ3/CD36/thrombospondin recognition mechanism and lack of phlogistic response. *Am J Path* 1996;**149**:911−21.

89. Luttmann W, Knoechel B, Matthys Foerster MH, Virchow Jr JC, Kroegel C. Activation of human eosinophils by IL-13. *J Immunol* 1996;**157**:1678−83.

90. Horie S, Okubo Y, Hossain M, Sato E, Nomura H, Koyama S, et al. Interleukin-13 but not interleukin-4 prolongs eosinophil survival and induces eosinophil chemotaxis. *Internal Med* 1997;**36**:179−85.

91. Luttmann W, Matthiesen T, Matthys H, Virchow Jr JC. Synergistic effects of interleukin-4 or interleukin-13 and tumor necrosis factor-α on eosinophil activation in vitro. *Am J Resp Cell Mol Biol* 1999;**20**(3):474−80.

92. Leung DYM. Molecular basis of allergic disease. *Mol Genetics Metab* 1998;**63**:177.

93. Wedi B, Raap U, Lewrick H, Kapp A. Delayed eosinophil programmed cell death in vitro: A common feature of inhalant allergy and extrinsic and intrinsic atopic dermatitis. *J Allergy Clin Immunol* 1997;**100**:536−43.

94. Vignola AM, Chanez P, Chiappara G, Siena L, Merendino A, Reina C, et al. Evaluation of apoptosis of eosinophils, macrophages and T lymphocytes in mucosal biopsy specimens of patients with asthma and chronic bronchitis. *J Allergy Clin Immunol* 1999;**103**:563−73.

95. Simon HU, Yousefi S, Shranz C, Schapowal A, Bachert C, Blaser K. Direct demonstration of delayed eosinophil apoptosis as a mechanism causing tissue eosinophilia. *J Immunol* 1997;**158**:3902−8.

96. Park H-S, Jung K-S, Shute J, Roberts K, Holgate ST, Djukanović R. Allergen-induced release of GM-CSF and IL-8 in vitro by nasal polyp tissue from atopic subjects prolongs eosinophil survival. *Eur Respir J* 1997;**10**:1476−82.

97. Umetsu DT, DeKruyff RH. Th1 and Th2 CD4+ cells in human allergic diseases. *J Allergy Clin Immunol* 1997;**100**:1−6.

98. Jayaraman S, Castro M, O'Sullivan M, Bragdon MJ, Holtzman MJ. Resistance to Fas-mediated T cell apoptosis in asthma. *J Immunol* 1999;**162**:1717−22.

99. Cox G, Ohtoshi T, Vancheri C, Denburg JA, Dolovich J, Gauldie J, et al. Promotion of eosinophil survival by human bronchial epithelial cells and its modulation by steroids. *Am J Resp Cell Mol Biol* 1991;**4**:525−31.

100. Peacock CD, Misso NL, Watkins DN, Thompson PJ. PGE2 and dibutyryl cyclic adenosine monophosphate prolong eosinophil survival in vitro. *J Allergy Clin Immunol* 1999;**104**:153.

101. Levi-Schaffer F, Temkin V, Malamud V, Feld S, Zilberman Y. Mast cells enhance eosinophil survival *in vitro*: role of TNF-α and granulocyte-macrophage colony-stimulating factor. *J Immunol* 1998;**160**:5554−62.

102. Hallsworth MP, Soh CPC, Twort CH, Lee TH, Hirst SJ. Cultured human airway smooth muscle cells stimulated by interleukin-1b enhance eosinophil survival. *Am J Resp Cell Mol Biol* 1998;**19**:910−9.

103. Ramis I, Finotto S, Dolovich J, Marshall J, Jordana M. Increased survival of nasal polyp eosinophils. *Immunol Letters* 1995;**45**:219−21.

104. Ohkawara Y, Lim KG, Xing Z, Glibetic M, Nakano K, Dolovich J, et al. CD40 expression by human peripheral blood eosinophils. *J Clin Invest* 1996;**97**:1761−6.

105. Sexton DW, Walsh GM. Induction of Apoptosis in Eosinophils via Co-culture with Airway Epithelium. In: *European Academy for Allergology and Clinical Immunology (EAACI)*. Berlin: Germany; 2001.

106. Druilhe A, Wallaert B, Tsicopoulos A, Lapa e Silva JR, Tillie-Leblond I, Tonnel AB, et al. Apoptosis, proliferation and expression of Bcl-2, Fas and Fas ligand in bronchial biopsies from asthmatics. *Am J Resp Cell Mol Biol* 1998;**19**:747−57.

107. Allard JB, Rinaldi L, Wargo MJ, Allen G, Akira S, Uematsu S, et al. Th2 allergic immune response to inhaled fungal antigens is modulated by TLR-4-independent bacterial products. *Eur J Immunol* 2009;**39**(3):776−88.

108. Iwamura C, Nakayama T. Toll-like receptors in the respiratory system: their roles in inflammation. *Curr Allergy Asthma Rep* 2008;**8**(1):7−13.

109. Daines MO, Chen W, Tabata Y, Walker BA, Gibson AM, Masino JA, et al. Allergen-dependent solubilization of IL-13 receptor alpha2 reveals a novel mechanism to regulate allergy. *J Allergy Clin Immunol* 2007;**119**:375−83.

110. Georas SN, Rezaee F, Lerner L, Beck L. Dangerous allergens: why some allergens are bad actors. *Curr Allergy Asthma Rep* 2010;**10**(2):92−8.

111. Busse WW, Lemanske Jr RF, Gern JE. Role of viral respiratory infections in asthma and asthma exacerbations. *Lancet* 2010;**376**(9743):826−34.

112. Torres D, Dieudonné A, Ryffel B, Vilain E, Si-Tahar M, Pichavant M, et al. Double-stranded RNA exacerbates pulmonary allergic reaction through TLR3: implication of airway epithelium and dendritic cells. *J Immunol* 2010;**185**(1):451−9.

113. Bayram H, Devalia JL, Sapsford RJ, Ohtoshi T, Miyabara Y, Sagai M, et al. The effect of diesel exhaust particles on cell function and release of inflammatory mediators from human

bronchial epithelial cells in vitro. *Am J Respir Cell Mol Biol* 1998;**18**(3):441—8.

114. Takizawa H, Ohtoshi T, Kawasaki S, Abe S, Sugawara I, Nakahara K, et al. Diesel exhaust particles activate human bronchial epithelial cells to express inflammatory mediators in the airways: a review. *Respirology* 2000;**5**(2):197—203.

Chapter 11.9

The Eosinophil and the Thymus

S.O. (Wole) Odemuyiwa, Meri K. Tulic, V. Olga Cravetchi, James J. Lee and Redwan Moqbel

INTRODUCTION

The thymus-derived lymphocyte (T cell) lineage is critical for defense against various pathogens and plays an important role in autoimmunity and allergic reactions. In general, T cells circulating in the peripheral blood are classified into CD4$^+$ or CD8$^+$ cells: CD8$^+$ cells specifically attack cells that express foreign antigens or host-derived neoantigens on their surfaces; CD4$^+$ cells secrete specific cytokines that orchestrate the adaptive immune response and regulate the activities of other immune cells. Over the last three decades, investigation of transcription factors, surface receptor expression patterns, and the diversity of cytokines secreted has revealed extensive heterogeneity among CD4$^+$ cells. The various subtypes of CD4$^+$ T cells include, among others, T-helper 1 (T$_h$1), T$_h$2, T$_h$17, and T-regulatory cells (T$_{reg}$) that are now well characterized. These different subsets of CD4$^+$ cells play diverse roles in both immune homeostasis and immunopathology. Thus, a good understanding of the mechanisms that control the generation and maintenance of these functional subtypes of T cells is critical for the understanding of allergic disease. The consistent association of eosinophils with allergic inflammation, the T$_h$2-promoting antigen-presenting function of eosinophils, and the tendency of eosinophils to home to the thymus suggest that this cell type may play a role during inflammation and immune ontogeny beyond what is currently known.

THE THYMUS

The thymus, a primary lymphoid tissue, is traditionally described as an organ chiefly concerned with the education of lymphocytes to tolerate *self* and prevent autoimmunity, while simultaneously recognizing and responding to *nonself* antigens to ensure protection from invading microorganisms. The educational activity in the thymus is an intricate process requiring the interaction of several infiltrating hematopoietic elements with one another, and with resident thymic stromal epithelial cells[3—4] (Fig. 11.9.1). However, the nature of the progenitors and other hematopoietic cells that infiltrate the thymus is currently unknown. Nonetheless, it is clear that while T cells ultimately complete their development in the thymus, they arrive at this organ as undifferentiated thymus-settling progenitors (TSP) arising from bone-marrow-derived hematopoietic stem cells.

When these cells first settle and populate the thymus, they express no known markers of lymphocyte differentiation. However, the mechanisms behind the sequential development and maturation of these cells into double-positive (DP; CD4$^+$ CD8$^+$) and single-positive (SP; CD4$^+$ or CD8$^+$ cells) thymocytes have been well characterized.[5—6] Nonetheless, the determinants of thymic settling, including the roles of other comigrating nonlymphocytic progenitors, are only recently becoming identified. We know that eotaxin is expressed by cells in the thymic medulla[7] and that eosinophils in the mouse thymus express co-stimulatory molecules CD80/CD86 and present antigens[8] that have been demonstrated to be essential for negative selection.[9] A major question is how infiltration of the thymus by mature granulocytes and *in situ* differentiation of granulocytes from stem cells in this organ affect the process and timing of lymphocyte development and maturation. It is also unknown whether infections that induce the upregulation of specific types of granulocytic responses during the window of thymocyte development could somehow regulate the type and functions of the eventual thymic immigrants. Indeed, it is well known that thymic settling is a highly selective process regulated by a panel of adhesion molecules, chemokines, and cytokines. This means that the temporal and spatial distribution of granulocytes that express this specific panel of proteins could be, at least, a major determinant of thymic settling.[1,10]

REGULATION OF SUBSET-SPECIFIC EMIGRATION FROM THE THYMUS

Interestingly, the need for different components of the thymic educational environment varies with the types of T cells that eventually emerge from this organ; for example, thymic selection of lymphocytes in the absence of thymic epithelial cells results in the emergence of fully functional CD8$^+$ but not CD4$^+$ lymphocytes.[11] The essential role of the thymic environment in the development of regulatory and helper T cells that constitute the bulk of the CD4$^+$ population was shown in an elegant study that demonstrated that extrathymic educated lymphocytes were poor at providing B cell help after infection with lymphocytic choriomeningitis virus or vesicular stomatitis virus.[12] The activities of cytotoxic CD8$^+$ lymphocytes were not affected in this model. These findings clearly demonstrate that T$_h$1

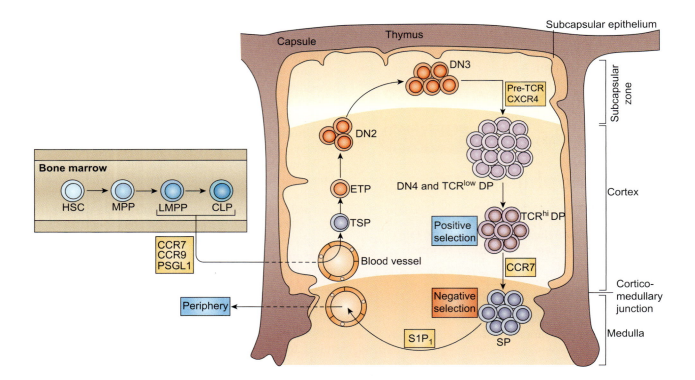

Nature Reviews| Immunology

FIGURE 11.9.1   **An overview of regulated migration events during T cell development.** Within the bone marrow, hematopoietic stem cells (HSCs) differentiate into multipotent progenitors (MPPs). A subset of MPPs, termed lymphoid-primed multipotent progenitors (LMPP), initiates transcription of recombination-activating genes. Subsequent lymphoid-primed bone marrow progenitors include common lymphoid progenitors (CLPs). All of these progenitors will make T cells if placed within the thymus. However, the ability of hematopoietic progenitors to migrate to the thymus is regulated by, and requires, C-C chemokine receptor 7 (CCR7; expressed by some LMPPs and highly expressed by CLPs) and (CCR9; expressed by subsets of LMPPs and CLPs), among other molecules. The identity of *in vivo* thymus-settling progenitors (TSPs) has not been precisely determined, but they probably include LMPPs and CLPs. TSPs enter the thymus near the corticomedullary junction to generate early T cell progenitors (ETPs; also known as mast/stem cell growth factor receptor kit[+] DN1 thymocytes). ETPs in turn generate DN2 and DN3 cells that migrate to the subcapsular zone. Expression of the pre-T cell receptor (pre-TCR) and CXC chemokine receptor 4 (CXCR4) on DN3 thymocytes induces cell proliferation, differentiation to the DN4, and subsequent differentiation to the DP stage. DP thymocytes that form appropriate interactions with self-peptide—MHC complexes on cortical thymic epithelial cells (a process termed positive selection) upregulate CCR7 expression and mature into SP mature T cells that migrate into the thymic medulla. Residence in the medulla (where negative selection occurs) is followed by emigration, which is regulated by the sphingosine-1-phosphate receptor 1 (S1P1). PSGL1, P-selectin glycoprotein ligand 1. *(Reproduced with permission from Love P.E. and Bhandoola A.,* Nature Immunology Reviews II *2011: 469—477.)*

and T$_h$2 responses, including diseases characterized by T$_h$1/T$_h$2 imbalance such as atopic asthma, may be especially sensitive to subtle changes to the thymic microenvironment during T cell education.

The need to expose lymphocytes to as many structural and nonstructural elements of self during their thymic educational program makes the thymic environment one of the most physiologically dynamic milieus in the body during this period. The major needs of the thymus during thymic education can be summarized as follows.

- A mechanism to remove the large number of cells undergoing apoptosis during negative selection, without inducing inflammation;
- A mechanism to ensure that all possible self-proteins are present and appropriately presented to lymphocytes in the thymus;

- The presence of an appropriate mix of chemokines, cytokines, and growth factors that ensure the proliferation of positively selected cells;
- A program that triggers thymic involution at the end of the period of active thymic selection.

Although the contributions of resident thymic cells and the dynamics of lymphocytic thymic infiltrates have been well characterized, the importance of nonlymphoid hematopoietic infiltrates into the thymus is often overlooked. Yet, a series of studies by Zinkernagel and Althage showed that the dogma of thymic epithelial cell preeminence in T cell selection only holds true under conditions of low T cell precursor frequency, as seen in the nude mouse,[13–15] where the thymus functions as a T cell receptor (TCR) rearrangement-promoting environment for T cells. Conversely, when there is a marked increase in precursor frequency, as

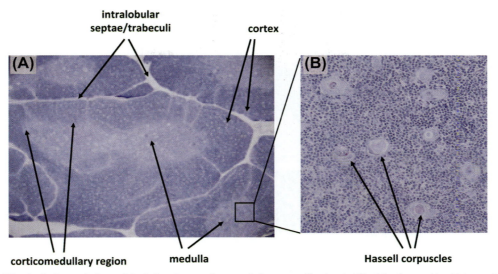

FIGURE 11.9.2 **Histological morphology of the infant human thymus.** *A*, Low magnification (×10) of the thymus identifying various regions of this primary lymphoid tissue. *B*, Higher magnification (×40) of the thymus showing Hassall corpuscles, a feature of the human thymus.

seen in mice carrying transgenic TCR, thymic epithelial cells become dispensable within the context of lymphocyte maturation. Crucially, however, nonthymic hematopoietic cells play a preeminent role in lymphocyte maturation under this type of condition.[16] Indeed, the maturation of gut intraepithelial lymphocytes, which normally receive their education in the gastrointestinal tract rather than the thymus, has confirmed these findings. It is of note that almost half of the total lymphocytes present in the body exist as gut intraepithelial lymphocytes.[17] Thus, under conditions of increased precursor frequency, the thymic microenvironment must necessarily include nonlymphocytic infiltrating cells of bone marrow origin to ensure the maturation of functional lymphocytes.

SPATIAL AND TEMPORAL DISTRIBUTION OF THYMIC-DWELLING EOSINOPHILS

Eosinophils naturally home to four organs following emigration from the bone marrow: the intestine, mammary glands, thymus, and uterus.[18] Although the seminal work of Lee and coworkers[19] showed that over 50% of eosinophils found in the thymus are products of *in situ* differentiation from thymus-dwelling hematopoietic stem cells, other studies suggest an element of chemotactic recruitment to this organ.[20–21] Regardless, immature cells are located in the intralobular septae and fibroreticular networks of the corticomedullary junction of preinvolutional thymi (Fig. 11.9.2). Eosinophils are frequently found at these locations in patients with active thymocyte development. In contrast, mature eosinophils show a less restricted distribution across the thymus. A similar, more

recent study from our laboratory focused not only on the spatial distribution of eosinophils but also on the time-dependent emigration of eosinophils from the peripheral blood into the thymus of children (Fig. 11.9.3).[22] We observed that the highest numbers of eosinophils are found in children less than 6 months old, while those older than 5 years have the least (Fig. 11.9.4). In children less than 6 months old, eosinophils, making up to 2% of all cells in the thymus, are found within intralobular septa and at corticomedullary regions. In addition, in children older than 5 years, who express significantly lower numbers of eosinophils, these cells are scattered throughout the thymus. Eosinophils in these older children are hypodense (Fig. 11.9.3B), suggesting activation and release of granule contents. The extrathymic origin of some of these thymus-dwelling eosinophils was confirmed in murine studies, in which eotaxin knockout mice were found to lack eosinophilic thymic infiltrates.[23–24] In the mouse thymus, the distribution of mature eosinophils is also compartment and age-dependent. Thymi consistently have the highest eosinophil density in corticomedullary junction from day 1 until day 21 postnatally. The maximal number of eosinophils is observed on days 11 and 21, which correlate with critical events in thymocyte development and differentiation.

The temporal and spatial distribution of the developing thymocytes in the thymus closely mirrors those of the eosinophils. The initial entry of double-negative progenitors into the thymus is via venules at the corticomedullary region. They then expand in number and migrate into the cortex, while moving between thymic epithelial cells; cells staining for both CD4 and CD8 first become apparent in the cortex at this time. These cells then begin to migrate back

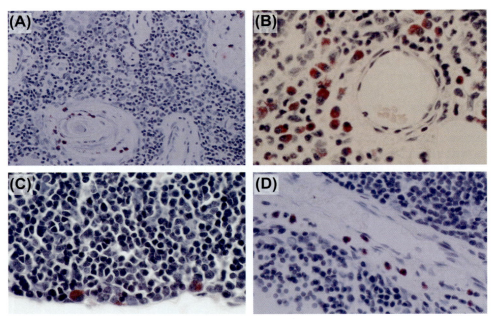

FIGURE 11.9.3   The presence of eosinophils in the infant human thymus. *A*, Eosinophils infiltrating the corticomedullary region of the thymus; *B*, eosinophils within the medulla of the thymus surrounding a blood vessel; *C*, eosinophils at the capsular edge of the thymic cortex; *D*, eosinophils in the interlobular septum (trabecula) connecting thymic lobes. Eosinophils are Luna stained (pink) against a hematoxylin background.

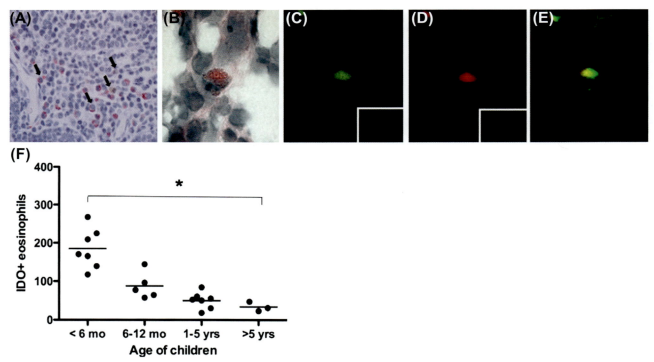

FIGURE 11.9.4   Thymic eosinophils contain express indoleamine 2,3-dioxygenase and their numbers decrease with age. *A*, Immunodetection of Luna[+] eosinophils (red) and express indoleamine 2,3-dioxygenase (IDO)[+] cells (brown) shows the majority of eosinophils containing IDO (colocalization shown by arrows). *B*, Confirmatory immunofluorescent staining using the BMK-13 monoclonal antibody (directed against human eosinophil major basic protein; MBP) conjugated to BODIPY FL (green, *C*) and IDO human antibody conjugated to tetramethylrhodamine B isothiocyanate (red, *D*), demonstrating colocalization of the two markers (*E*). Inserts illustrate the lack of staining in isotype controls for BMK-13 (IgG1 conjugated to BODIPY FL in *C*) or IDO (IgG1, IgG2a, IgG2b, and IgG3 conjugated to tetramethylrhodamine B isothiocyanate in *D*). The number of IDO[+] eosinophils in the thymus decreases with age (*F*). *$p = 0.01$ between 6-month (mo) and 5-year (yrs) age groups. *(Reprinted with permission from Tulic et al. Am J Pathol 2009: **175**: 2043−2052.)*

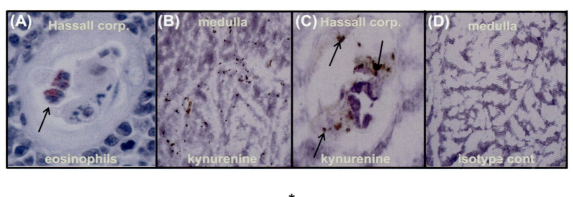

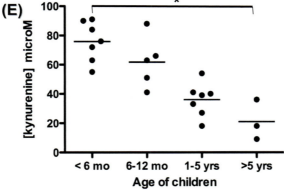

FIGURE 11.9.5 **Presence of eosinophils and L-kynurenine in infant thymus.** *A,* Eosinophils are detected within Hassall corpuscles in the early perinatal period, 4 weeks after birth. In the same 4-week-old neonate, kynurenine immunoreactivity is detected throughout the thymus (*B*), including Hassall corpuscles (*C*). Isotype control staining for kynurenine antibody is shown in *D.* The number of kynurenine+ cells in the thymus decreases with age (*E*). *(Reprinted with permission from Tulic et al.* Am J Pathol *2009: 175: 2043−2052.)*

toward the medulla, coinciding with the TCR becoming detectable on the cell surface.

The process of thymic selection that eventually results in the release of *educated* lymphocytes into the circulation begins with the interaction of the newly formed DP cells with epithelial and dendritic cells in the deeper cortex. Interestingly, however, the presence of thymocytes has been shown to be important in the maturational direction of the stromal thymic epithelial cells; this suggests a reciprocal codependence of thymic epithelial cells and thymocytes in terms of differentiation. Indeed, in the absence of thymocytes, thymic epithelial cells differentiate into well-polarized epithelial cells (like those lining the gut and respiratory tract) rather than nonpolarized epithelial cells characteristic of the thymic microenvironment where thymocyte selection usually takes place. The role of other infiltrates in this so-called *thymic cross-talk* is currently unknown.

Another unanswered question is whether eosinophils or other granulocytes have any role in thymic cross-talk. The spatial and temporal distribution of eosinophils in the thymus strongly suggests that this may indeed be the case. Thus, the educational program in the thymus may not be exclusively dependent on the teachers (i.e., thymic epithelial

cells and dendritic cells) and students (i.e., thymocytes) but also on other components (i.e., the school board) that may include eosinophils to ensure the success of this carefully orchestrated developmental process. It is conceivable that, while not necessarily preventing graduation from the thymus *school*, the absence of a balanced and functional *school board* will affect the quality of education and lymphocyte graduation rate from the thymus. An investigation of thymic epithelial cell polarization in the PHIL mouse should provide some answers to this important question.[25]

EOSINOPHILS AND LYMPHOCYTE SELECTION

Under homeostatic conditions, i.e., in the absence of inflammation, eosinophils are tissue-dwelling cells that are recruited to and maintained in the thymus and intestine under the influence of eotaxin.[26] The presence of eosinophils in these known sites of lymphocyte maturation under homeostatic conditions, along with macrophages, dendritic cells and lymphocytes, suggests a possible role for eosinophils in T cell maturation and selection. However, no active role in T cell selection was assigned to eosinophils until Throsby and colleagues[8] reported their work on a mouse model of acute negative selection. In this model of MHC-I-restricted negative selection of thymocytes, a dramatic elevation of eosinophil numbers was observed within 3 h of antigen exposure, suggesting active recruitment of eosinophils to the thymus during this process. Conversely, no such increase was observed in a model in which an MHC-II-restricted antigen was used.[8] This seminal study showed for the first time that eosinophils may play a role in T cell selection in the thymus. This work also demonstrated the complexity and heterogeneity of eosinophil function, i.e., that recruitment of eosinophils into tissues is not always an indication of ongoing inflammation. Thus, eosinophils are actively involved in the process that leads to the diversity of the TCR repertoire of lymphocytic thymic immigration. In addition, since these studies were conducted in models of high precursor frequency, it has become clear that the eosinophil is one cell type among many important hematopoietic cells that may be crucial for T cell development. Indeed, Zinkernagel and Althage[16] have previously shown that thymic epithelial cells are dispensable for T cell maturation under such conditions. Thus, if there are factors that increase precursor frequency during a crucial stage of immune development (for example, in children younger than 5 years of age), thymus-infiltrating eosinophils could significantly affect the TCR repertoire of circulating peripheral blood lymphocytes.

Following the study of Throsby et al.,[8] the next challenge was to delineate the mechanisms by which eosinophils may regulate thymocyte selection. A recent study from Kim and colleagues[10] showed that loss of eosinophils leads to a dramatic reduction in the efficiency of apoptotic cell clearance from the thymus during acute cell death characteristic of ongoing thymic selection. In our own study, we have shown that the neonatal thymus contains eosinophils (Fig. 11.9.3). These eosinophils were also shown to express indoleamine 2,3-dioxygenase (IDO) (Fig. 11.9.4) and a plethora of T_h2 markers.[22] Thymic IDO expression is predominantly associated with eosinophils and, importantly, a parallel age-related decline was seen between thymic IDO^+ eosinophils and T_h2 production in the thymus. In the same study, we also showed the active production of kynurenine (Fig. 11.9.5), a product of tryptophan catabolism catalyzed by IDO. In parallel mouse studies, eosinophils were also shown to be IDO^+; indeed, increases in IDO mRNA levels correlate with the influx of eosinophils into thymus. Notably, the recruitment of eosinophils into the thymus is regulated by eotaxin, which is constitutively expressed in the thymus.[26] These observations have led us to speculate that eosinophils, through the activity of T_h2-promoting enzymes and cytokines, may bias thymic immigration selected in their presence toward T_h2 rather than T_h1. We now know that the thymic microenvironment is skewed toward a T_h2-like pattern during the early neonatal

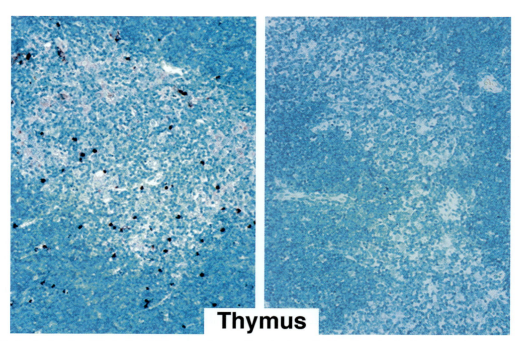

Thymus

FIGURE 11.9.6   Lack of eosinophilia in the PHIL mouse. Immunohistochemistry (dark purple-stained cells) with eosinophil-specific rabbit polyclonal antisera to MBP demonstrates that tissues or organs with prominent resident populations of eosinophils at baseline in wild-type mice are devoid of these granulocytes in PHIL mice. Scale bar, 100 μm. *(Reprinted with permission from: Lee et al., Science 2004; 305: 1773–1776.)*

period,[22] similar to previous findings in the periphery.[27] Furthermore, our most recent findings demonstrate the presence of functional T_{regs} in neonatal thymus, which are capable of downregulating thymic T cell and eosinophil production of the T_h2 signature cytokine, IL-13.[28] This inhibition is less effective in children who go on to develop allergic disease compared to those that remain healthy.

CONCLUSION

Surprisingly, the eosinophil has emerged as a crucial component of the thymic T cell educational and commitment program. The few studies examining the role of this cell type in the thymus have proposed that their presence in the thymus meets at least three of the four basic thymic needs enumerated above, i.e., removal of apoptotic cells; expression of appropriate cytokine mix; and presentation of appropriately expressed proteins to thymocytes (eosinophils are antigen-presenting cells, albeit weaker than professional APCs). Results of other *in vitro* and *in vivo* studies elucidating the immunological role of the eosinophil in peripheral tissues further strongly imply an expanded role for this cell type beyond our current understanding of its function in several tissues, including the thymus. The emergence of the PHIL mouse model (Fig. 11.9.6) and contemporary alternative eosinophil-deficient mouse strains have presented us with exciting tools to address these questions. However, we now know that the dogma that the eosinophil is a strict inflammation-related cell type will gradually disappear. The 21st century is a harbinger to new and exciting vistas about the expanding role of this enigmatic cell type.

REFERENCES

1. Zlotoff DA, Bhandoola A. Hematopoietic progenitor migration to the adult thymus. *Annals of the New York Academy of Sciences* 2011;**1217**:122—38.
2. Lee JJ, Dimina D, Macias MP, Ochkur SI, McGarry MP, O'Neill KR, et al. Defining a link with asthma in mice congenitally deficient in eosinophils. *Science* 2004;**305**:1773—6.
3. Haynes BF, Heinly CS. Early human T cell development: analysis of the human thymus at the time of initial entry of hematopoietic stem cells into the fetal thymic microenvironment. *J Exp Med* 1995;**181**:1445—58.
4. Haynes BF, Martin ME, Kay HH, Kurtzberg J. Early events in human T cell ontogeny. Phenotypic characterization and immunohistologic localization of T cell precursors in early human fetal tissues. *Journal of Experimental Medicine* 1988;**168**:1061.
5. Godfrey DI, Kennedy J, Mombaerts P, Tonegawa SAZ. Onset of TCR-β gene rearrangement and role of TCR-β expression during CD3−CD4−CD8−thymocyte differentiation. *J Immunol* 1994;**152**: 4783—92.
6. Gill J, Malin M, Sutherland J, Gray D, Hollander G, Boyd R. Thymic generation and regeneration. *Immunological Reviews* 2003;**195**:28—50.
7. Franz-Bacon K, Dairaghi DJ, Boehme SA, Sullivan SK, Schall TJ, Conlon PJ. Human thymocytes express CCR-3 and are activated by eotaxin. *Blood* 1999;**93**:3233—40.
8. Throsby M, Herbelin A, Pleau JM, Darcenne M. CD11c$^+$ eosinophils in the murine thymus: developmental regulation and recruitment upon MHC class I-restricted thymocyte deletion. *J Immunol* 2000;**165**:1965—75.
9. Nakamura T, Ohbayashi M, Toda M, Hall DA, Horgan CM, Ono SJ. A specific CCR3 chemokine receptor antagorist inhibits both early and late phase allergic inflammation in the conjunctiva. *Immunologic Research* 2005;**33**:213—21.
10. Kim HJ, Alonzo ES, Dorothee G, Pollard JW, Sant'Angelo DB. Selective depletion of eosinophils or neutrophils in mice impacts the efficiency of apoptotic cell clearance in the thymus. *PloS One* 2010;**5**. e11439.
11. Martinic MM, van den Broek MF, Rulicke T, Huber C, Odermatt B, Reith W, et al. Functional CD8$^+$ but not CD4$^+$ T cell responses develop independent of thymic epithelial MHC. *Proceedings of the National Academy of Sciences of the United States of America* 2006;**103**:14435—40.
12. Blais ME, Gerard G, Martinic MM, Roy-Proulx G, Zinkernagel RM, Perreault C. Do thymically and strictly extrathymically developing T cells generate similar immune responses? *Blood* 2004;**103**:3102—10.
13. Zinkernagel RM, Althage A, Waterfield E, Kindred B, Welsh RM, Callahan G, et al. Restriction specificities, alloreactivity, and allotolerance expressed by T cells from nude mice reconstituted with H-2-compatible or -incompatible thymus grafts. *The Journal of Experimental Medicine* 1980;**151**:376—99.
14. Zinkernagel RM, Althage A, Callahan G. Thymic reconstitution of nude F1 mice with one or both parental thymus grafts. *The Journal of Experimental Medicine* 1979;**150**:693—7.
15. Zinkernagel RM, Callahan GN, Althage A, Cooper S, Klein PA, Klein J. On the thymus in the differentiation of 'H-2 self-recognition' by T cells: evidence for dual recognition? *The Journal of Experimental Medicine* 1978;**147**:882—95.
16. Zinkernagel RM, Althage A. On the role of thymic epithelium vs. bone marrow-derived cells in repertoire selection of T cells. *Proceedings of the National Academy of Sciences of the United States of America* 1999;**96**:8092—7.
17. Lambolez F, Azogui O, Joret AM, Garcia C, von Boehmer H, Di Santo J, et al. Characterization of T cell differentiation in the murine gut. *The Journal of Experimental Medicine* 2002;**195**:437—49.
18. Rothenberg ME, Hogan SP. The eosinophil. *Annual Review of Immunology* 2006;**24**:147—74.
19. Lee I, Yu E, Good RA, Ikehara S. Presence of eosinophilic precursors in the human thymus: evidence for intra-thymic differentiation of cells in eosinophilic lineage. *Pathol Int* 1995;**45**:655—62.
20. Rothenberg ME, Luster AD, Lilly CM, Drazen JM, Leder P. Constitutive and allergen-induced expression of eotaxin mRNA in the guinea pig lung. *J Exp Med* 1995;**181**:1211—6.
21. Fulkerson WJ, Newman JH. Endogenous Cushing's syndrome complicated by Pneumocystis carinii pneumonia. *The American Review of Respiratory Disease* 1984;**129**:188—9.
22. Tulic MK, Sly PD, Andrews D, Crook M, Davoine F, Odemuyiwa SO, et al. Thymic indoleamine 2,3-dioxygenase-

positive eosinophils in young children: potential role in maturation of the naive immune system. *Am J Pathol* 2009;**175**:2043–52.

23. Fulkerson PC, Fischetti CA, McBride ML, Hassman LM, Hogan SP, Rothenberg ME. A central regulatory role for eosinophils and the eotaxin/CCR3 axis in chronic experimental allergic airway inflammation. *Proceedings of the National Academy of Sciences of the United States of America* 2006;**103**:16418–23.

24. Fulkerson PC, Zhu H, Williams DA, Zimmermann N, Rothenberg ME. CXCL9 inhibits eosinophil responses by a CCR3- and Rac2-dependent mechanism. *Blood* 2005;**106**:436–43.

25. van Ewijk W. The thymus: Interactive teaching during lymphopoiesis. *Immunology Letters* 2011;**138**:7–8.

26. Matthews AN, Friend DS, Zimmermann N, Sarafi MN, Luster AD, Pearlman E, et al. Eotaxin is required for the baseline level of tissue eosinophils. *Proc Natl Acad Sci* 1998;**95**:6273–8.

27. Prescott SL, Macaubas C, Holt BJ, Smallacombe TB, Loh R, Sly PD, et al. Transplacental priming of the human immune system to environmental allergens: universal skewing of initial T cell responses toward the Th2 cytokine profile. *Journal of Immunology* 1998;**160**:4730–7.

28. Tulic MK, Andrews D, Crook M, Charles A, Tourigny MR, Moqbel R, et al. Changes in thymic regulatory T cell maturation from birth to puberty: differences in atopic children. *J Allergy Clin Immunol* 2011;**129**:199–206.

Eosinophil-Mediated Tissue Remodeling and Fibrosis

Chapter 12.1

Introduction

Joanne C. Masterson and Glenn T. Furuta

EOSINOPHILS—HELPFUL OR HARMFUL?
GUILTY BY ASSOCIATION

On a minute-by-minute basis, mucosal tissues throughout the body reconstitute their barrier following injury. Whether in the rhinopulmonary, gastrointestinal, genitourinary tracts, eye, or skin, reconstitution of this barrier is characterized by a complex series of events that, in health, ultimately leads to restoration of function. In disease states, deregulation of these processes can lead to the deposition of excessive remodeling molecules, resulting in organ dysfunction.

The exact mechanisms that define wound healing and fibrosis are beyond the scope of this discussion and are presented in detail elsewhere.[1,2] Briefly, following the loss of the epithelial cell barrier the immediate response is to reconstitute the mucosal surface through processes that permit epithelial cells to migrate across the wound and thereby cover and protect the basement membrane. Secondary tissue responses then occur, including secretion of protective factors and plasma, recruitment of leukocytes, and activation of resident and recruited mesenchyme, resulting in, but not limited to, extracellular matrix deposition.[2,3] In addition, there is a potential role for epithelial—mesenchymal-like transitions in this process.[4,5] This new milieu allows for both protection and repair, leading to restoration of a normal functioning epithelium. However, these same processes are excessively activated during fibrotic diseases. Currently, the exact processes and signals that drive either healing or fibrosis in the mucosa are not certain.

Eosinophils have been associated with both processes. Eosinophils are topographically associated with remodeled tissues and are biological powerhouses that possess the ability to produce and release remodeling molecules within these locations. With regard to wound healing, the role of eosinophils is unknown. Nonetheless, a role in restitution may be speculated in the gastrointestinal tract, since a gradient of eosinophils is present within normal healthy mucosal surfaces. More is known about the potential role of eosinophils in fibrosis, where increased numbers of eosinophils occur in a number of different disease states, including asthma, eosinophilic esophagitis (EoE), and atopic dermatitis, and in some cases may reflect disease activity.[6] Within this chapter, the individual subchapters provide outstanding reviews and novel insights into mechanisms through which eosinophils may participate in fibrosis and tissue dysfunction.

EOSINOPHILS PRODUCE PROFIBROTIC MEDIATORS

Eosinophils are ripe with preformed mediators, such as eosinophil-derived granule proteins (EDGPs), that may participate in fibrosis. In Chapter 12.2, Marina Pretolani and colleagues identify a significant role for EDGPs in stimulating target cells that may contribute to a fibrogenic milieu. In Chapter 12.3, David Broide and colleagues provide both the rationale and data supporting the contribution of eosinophil-derived TFG-β1 to fibrosis in asthma and EoE. In addition, Elizabeth Kelly presents interesting data to support a potential role for tumor necrosis factor (TNF-α) in fibrosis that bears relevance to many diseases, in particular Crohn's disease.

These authors contribute cogent explanations determining how eosinophil-derived molecules may play a role in fibrosis and describe mouse studies that provide hope for therapeutic interventions. However, they also note that, to date, human studies in which eosinophils and their associated mediators are depleted have yielded conflicting results, and while anti-transforming growth factor β1 (anti-TGF-β1) studies may offer great therapeutic potential,[6] they have yet to be performed.

EOSINOPHILS CAN BE AFFECTED BY PROFIBROTIC MEDIATORS

Alternatively, eosinophils may be the cellular targets of profibrogenic molecules. For instance, eosinophils express

Eosinophils in Health and Disease. http://dx.doi.org/10.1016/B978-0-12-394385-9.00012-2

a number of receptors for cytokines associated with the remodeling process, including interleukin-13 (IL-13), TGF-β, and TNF-α. IL-13, which, while being primarily associated with mucus secretion and smooth muscle activation, can synergize with eotaxin to perpetuate eosinophil recruitment. In contrast to other target cells, TNF receptor activation in eosinophils leads to inhibition of apoptosis, thereby potentially elongating the time course of profibrotic outcomes resulting from the presence of eosinophils. Eosinophils may also respond to their cytokine microenvironment. As discussed in Chapter 12.4 by Elizabeth Kelly, TNF-α can synergize with either T-helper type 1 (T_h1; IFN-γ) or T_h2 (IL-4) cytokines to differentially induce the generation of T_h1 or T_h2 chemokines, thus potentially mediating the resolution or perpetuation of eosinophil-associated inflammation and fibrosis. As discussed by David Broide, TGF-β is commonly implicated in eosinophil-associated fibrotic diseases. However, less is known about the autocrine/paracrine effects of TGF-β on eosinophils themselves. Moreover, in studies by Seema Aceves, TGF-β/SMAD signaling in esophageal and pulmonary eosinophil-associated diseases report nuclear SMAD localization primarily in tissue resident cells, such as epithelial and fibroblast cells. Thus, while eosinophils have intact TGF-β signaling, the *in vivo* relevance is still uncertain.

MECHANISTIC INSIGHTS—IT ALL DEPENDS ON THE MICROENVIRONMENT

Considered in isolation, it would seem that eosinophils must participate in fibrosis and that therapeutic interventions could diminish this problematic response. However, neither *in vitro* nor *in vivo* studies can replicate the complexities of the local injury within the mucosal microenvironment. *Ex vivo* and clinical studies are chronically hampered by the inability to obtain samples most relevant to the disease process. For instance, obtaining samples of airway mucosa over the lifetime of a large population of asthma patients is impossible but would yield the most revealing information.

In this regard, Seema Aceves (Chapter 12.5) and Claudio Fiocchi (Chapter 12.6) detail their work and that of others that provide unique insights into the impact of eosinophils on intestinal fibrosis. Much of the work of these investigators focuses on the gastrointestinal tract and is based on mucosal samples obtained during endoscopy with biopsy. To date, best practice clinical care has recommended the use of endoscopy as a necessary diagnostic tool for patients with EoE, which has enabled research samples to be obtained as a part of a clinically indicated procedure. This opportunity has led to significant and rapid advances

into our understanding of how eosinophils impact fibrosis in the esophagus.

In the following subchapters, the authors are careful to point out that although the clinical endpoints of studies and the relevance of remodeling to these endpoints are often difficult to define, it remains of utmost importance. For instance, as discussed in Chapter 12.7 by Harsha Kariyawasam and colleagues, a defined clinical understanding of the precise role of tissue remodeling in asthma symptoms is currently lacking. In addition, Wynn suggests clarification of criteria for defining the characteristics of fibrosis and the clinical relevance of these criteria is of key importance for future research.[6] For instance, the morphological features of fibrosis in asthma and EoE include basement membrane and lamina propria thickening; however, many factors can contribute to these readouts, including sample size, tissue orientation, and the methodology used for obtaining measurements. In EoE, stricture and dysphagia are critical endpoints, but the best methods for measuring these endpoints are still to be defined. Furthermore, the identification of easily obtained and validated biomarkers that require limited invasive procedures will be invaluable.

THERAPEUTIC TARGETS

In the end, the main objective of research in this area is to identify novel therapeutic targets. To date, several potential candidates (EDGPs, IL-13, TGF-β, TGF-β receptors, and TNF-α) have been identified by this outstanding group of investigators. As research and clinical care advance, several important points are worth remembering. Firstly, different disease pathogeneses are likely to be associated with mucosal eosinophilia. For instance, in a significant number of patients with hypereosinophilic syndrome, have identified FiP1 mutations associated with eosinophilia, whereas mutations in eotaxin-3/C-C motif chemokine 26 (CCL26) are characteristic of EoE. This may be important when considering the selection of an eosinophil-specific target. Second, within the broad categorization of patients diagnosed with a specific disease, such as asthma and EoE, various phenotypes exist. Not all patients with either of these diseases go on to develop fibrosis and thus selection of the appropriate patient subtypes within studies and eventually for treatment with targeted novel agents will be critical. Third, practical aspects of novel therapeutics include mode of delivery, delivery vehicle, and methodologies/tools to monitor responses will need to be addressed. Finally, the safety of agents directed at eosinophils and relevant mediators will need to be carefully considered. To date, the overall well-being of eosinophil null mouse models is reassuring as to the safety of these types of interventions.

CONCLUSION

While the eosinophil is currently guilty by association, a growing body of evidence summarized within these chapters highlights a potential specific role for eosinophils in fibrotic disease processes, supported by the observation that eosinophils themselves are responsive to profibrotic mediators. Although knowledge is growing in this field, a better understanding of differences in tissue microenvironments, mechanisms of disease pathogenesis, and clinical/pathological phenotypes will lead to improved novel therapeutic targets. Continuing research into more efficient and tissue-specific modes of targeting delivery will improve treatment efficacy and advance our understanding of the role of eosinophils in fibrotic diseases.

REFERENCES

1. Erjefalt JS, Persson CG. Airway epithelial repair: breathtakingly quick and multipotentially pathogenic. *Thorax* 1997;**52**:1010–2.
2. Lee SB, Kalluri R. Mechanistic connection between inflammation and fibrosis. *Kidney Int* 2010;**78**(Suppl. 119):S22–6.
3. Strieter RM, Gomperts BN, Keane MP. The role of CXC chemokines in pulmonary fibrosis. *J Clin Invest* 2007;**117**:549–56.
4. Kalluri R. EMT: when epithelial cells decide to become mesenchymal-like cells. *J Clin Invest* 2009;**119**:1417–9.
5. Broekema M, Timens W, Vonk JM, Volbeda F, Lodewijk ME, Hylkema MN, et al. Persisting remodeling and less airway wall eosinophil activation in complete remission of asthma. *Am J Respir Crit Care Med* 2011;**183**:310–6.
6. Wynn TA. Common and unique mechanisms regulate fibrosis in various fibroproliferative diseases. *J Clin Invest* 2007;**117**:524–9.

Chapter 12.2

Potential Role of Eosinophil Granule Proteins in Tissue Remodeling and Fibrosis

Séverine Letuve and Marina Pretolani

INTRODUCTION

Regeneration of normal injured adult tissues involves the replacement of damaged cells and the production of new extracellular matrix (ECM). These events are accompanied by a robust inflammatory response surrounding the areas of damage. Chronic inflammatory diseases are frequently associated with an aberrant repair response, which results from the continuous exposure of wounded tissues to proinflammatory and fibrotic factors that are generated in excessive amounts by inflammatory and mesenchymal cells. This altered repair response favors the development of a fibrotic process, characterized by the accumulation of mesenchymal cells and by the abnormal deposition of ECM proteins, particularly of fibrillary collagen, in target tissues. Fibrosis leads to distortion of tissue architecture and is generally associated with organ failure and poor prognosis. Notably, therapy with corticosteroids and other immunomodulatory agents is largely ineffective in reversing these tissue alterations.

Fibroblasts are considered to be the main actors of the fibrotic process because of their ability to migrate, proliferate, produce ECM components, and differentiate into myofibroblasts that contract the wound.[1–3] Fibroblasts are activated and recruited to wound tissue areas by a variety of mediators, particularly transforming growth factor β1 (TGF-β1), which is produced by environmental cells and by fibroblasts themselves, resulting in an autocrine loop in which TGF-β1 perpetuates fibrosis by enhancing chemotaxis of mesenchymal cells, myofibroblast differentiation, and production of ECM proteins.[3]

Several reports have described the presence of eosinophilia and of degranulated eosinophils in areas of active fibrogenesis in a variety of inflammatory diseases, including asthma, allergic skin reactions, idiopathic pulmonary fibrosis, Crohn disease, retroperitoneal fibrosis, sclerosing cholangitis, eosinophilic esophagitis, and myocarditis.[4–10] The first report that eosinophils can directly participate in tissue fibrosis originated from a study demonstrating increased fibroblast proliferation in the presence of eosinophil extracts.[11] The eosinophil-derived fibrogenic factor(s) were not characterized at that time, but further studies demonstrated that eosinophils synthesize a wide array of cytokines, chemokines, and growth factors that are classically involved in the development and progression of fibrosis.[12] In particular, eosinophils are an important source of TGF-β in inflamed tissues,[13] and *in vitro* experiments demonstrated that this cytokine is predominantly responsible for the ability of activated eosinophils to stimulate proliferation, collagen synthesis, and lattice contraction in lung and skin fibroblasts.[14] This mechanism also operates *in vivo*, since eosinophil depletion from blood and tissues prevents antigen-induced hallmarks of fibrosis in human and mouse allergic skin and lung, particularly TGF-β1 production and ECM deposition.[15–18]

EFFECT OF EOSINOPHIL-DERIVED CATIONIC PROTEINS ON MESENCHYMAL CELLS

Compelling evidence has also demonstrated that eosinophils influence the function and activation state of different mesenchymal cell types involved in tissue repair and remodeling through cationic proteins (Table 12.2.1). In

TABLE 12.2.1 Effects of Eosinophil Granule Proteins on Cell Types Involved in Tissue Remodeling

Cell Type	Granule Protein	Effect	Reference
Human lung fibroblasts	MBP	Synthesis of cytokines of the IL-6 family (IL-6, IL-11, LIF)	19
	ECP	— Migration and collagen contraction — Synthesis of TGF-β1 — Proteoglycan accumulation	22, 23 24 27
Human intestinal myofibroblasts	MBP	Synthesis of IL-8	33
Human bronchial epithelial cells	MBP EPO	— Synthesis of MMP-9 — Inhibition of MMP-1 — Production of fibrogenic mediators — Synthesis of ECM proteins	41
Human umbilical vascular endothelial cells	MBP	— Cell proliferation — Capillary genesis and sprouting	42
Mouse cardiac fibroblasts	ECP	— Cell proliferation — Stress fiber formation	46
Rat natal cardiomyocytes	ECP	Cytoskeleton rearrangements associated with cell differentiation	46

Abbreviations: ECM, extracellular matrix; ECP, eosinophil cationic protein; EPO, eosinophil peroxidase; IL, interleukin; LIF, leukemia-inhibitory factor; MBP, major-basic protein; MMP, metalloproteinase; TGF-β1, transforming growth factor β1.

human lung fibroblasts, subcytotoxic concentrations of major basic protein (MBP) were initially shown to synergize with IL-1α and TGF-β1 to increase the synthesis of the inflammatory and fibrogenic cytokines interleukin (IL)-6, IL-11, and leukemia inhibitory factor.[19] The cationic charge and high arginine content of MBP largely account for the observed effects, since the anionic molecule heparin downregulates cytokine production, while the synthetic cationic polypeptide, poly-L-arginine, induces similar effects to MBP.[19] Notably, the stimulatory effect of MBP is not shared by eosinophil-derived neurotoxin (EDN) and is restricted to IL-6-related cytokines, since the synthesis of granulocyte-macrophage colony-stimulating factor (GM-CSF) is not similarly augmented.[19] In addition, MBP fails to upregulate type I collagen production and fibroblast proliferation.[19] In human myofibroblasts from intestinal origin, MBP, but not EDN, also initiates the synthesis of IL-8,[20] a proinflammatory cytokine that promotes airway smooth muscle cell migration and contraction.[21] These data thus identify another mechanism through which MBP may contribute to tissue remodeling and to respiratory dysfunction in asthma and in cystic fibrosis.

More recently, it was shown that the addition of conditioned medium from unstimulated eosinophils to lung fibroblasts enhances their migration and promotes collagen gel contraction.[22,23] These effects are dependent on eosinophil cationic protein (ECP) secretion, since they are almost completely suppressed by neutralizing anti-ECP antibodies and are reproduced by the addition of native, recombinant ECP.[22,23] Although ECP may promote fibroblast migration and collagen contraction directly, previous observations have demonstrated that this cationic protein induces TGF-β1 synthesis in lung fibroblasts, which may mediate the observed effects.[24]

ECP may also be involved in tissue remodeling through its effects on proteoglycan turnover. Proteoglycans are ECM components that maintain the architecture of connective tissues and ensure their optimal mechanical properties. Excessive proteoglycan deposition in proximal or distal airways is a hallmark of remodeling in asthma, emphysema, and pulmonary fibrosis, and correlates with the severity of respiratory dysfunction.[25,26] ECP has been shown to favor the intracellular accumulation of proteoglycan by preventing glycosaminoglycan degradation in human lung fibroblasts.[27] This property is unrelated to the cationic charge of ECP, as two other basic peptides, protamine and poly-L-lysine, fail to alter the intracellular levels of proteoglycan.[27] Moreover, ECP neither alters the expression of collagen and hyaluronan nor promotes fibroblast proliferation.[27]

Together, these findings suggest that eosinophil degranulation may participate in the onset of tissue fibrosis by stimulating mesenchymal cells to generate fibrogenic cytokines and growth factors; promoting fibroblast proliferation, migration and contraction; and altering the composition and turnover of the ECM. Such fibrogenic properties are mostly attributable to MBP and ECP, but not to EDN (Table 12.2.1). In addition, ECP-mediated effects seem to be unrelated to the cationic charge, while those attributable to MBP are usually inhibited by polyanions. It has been suggested that this disparity originates from differences in tertiary conformation of these proteins and in the relative proportions of arginine and other amino acids, which may influence their charge and interactions with the cellular environment.[28-30]

EFFECT OF EOSINOPHIL-DERIVED CATIONIC PROTEINS ON EPITHELIAL CELLS

Eosinophils preferentially accumulate in mucosal tissues, particularly in airways and the gut and, therefore, their primary targets are the epithelial barriers.[5,31] Indeed, several reports have shown that the function and integrity of bronchial and intestinal epithelial cells are markedly altered by cationic proteins (Table 12.2.1; 32-34). Until recently, most of the pathogenic effects of these proteins, at least on the airway epithelium, were thought to be restricted to their cytotoxic properties, probably involving their high arginine contents and cationic charge.[34-38] Accordingly, *in vitro* and *in vivo* exposure of airway epithelium to synthetic cationic polypeptides, such as poly-L-arginine, mimics many of the effects of eosinophil granule proteins, including bronchial hyperreactivity and increased epithelial permeability.[37-39] The respiratory epithelium is believed to be a key player in airway remodeling through the production of ECM components, fibrogenic cytokines and chemokines, growth factors, and matrix metalloproteinases (MMPs) that are responsible for the proliferation, migration, and activation of mesenchymal cells, for the differentiation of fibroblasts into myofibroblasts, and for ECM remodeling.[40] Eosinophils may activate the respiratory epithelium, particularly through the generation of TGF-β1, which influences epithelial integrity, adhesion, and migration, and the production of ECM components.[40] We recently demonstrated that eosinophils directly induce the production of remodeling factors by bronchial epithelial cells through their granule proteins.[41] Indeed, we found that incubation of primary cultured human bronchial epithelial cells with subcytotoxic concentrations of MBP or eosinophil peroxidase (EPO) augments the synthesis of the fibrogenic molecules endothelin-1, epidermal growth factor receptor, matrix metalloproteinase (MMP)-9, platelet-derived growth factor-3, TGF-α, and TGF-β1. We also observed that EPO and MBP promote the expression of the ECM proteins, fibronectin and tenascin, the accumulation of which has been documented in asthmatic airways.[26] In addition, we reported downregulation of MMP-1, which is primarily involved in the degradation of fibrillary collagen.[41] Some of these effects could be reproduced by poly-L-arginine, and coincubation with either poly-L-glutamic acid or heparin incompletely inhibited cellular effects induced by MBP and EPO, suggesting that their cationic charge is only partially responsible for the observed responses. Together, these findings identified eosinophil granule proteins as novel mediators that stimulate the bronchial epithelium to synthesize factors involved in both the fibrotic process and alterations of ECM homeostasis. To determine whether these mechanisms also operate *in vivo*, we examined the ability of EPO and MBP to promote hallmarks of fibrosis in the lungs of normal mice. We observed that endotracheal instillation of MBP, but not of EPO, to naive mice increased pulmonary expression of the transcripts encoding collagen type I α chain and IL-6 and reduced the levels of MMP-1 after 24 h (Pretolani, unpublished observations). These effects were followed, at 48 h, by the release of TGF-β1 in the bronchoalveolar lavage fluid, which was upregulated in the airway epithelium (Pretolani, unpublished observations). In addition, excessive collagen deposition was detected in the lung tissue and MMP-1 was downregulated in the airway epithelium (Fig. 12.2.1). In this setting, MBP-induced pulmonary fibrosis is unrelated to its cationic charge, since it was unaffected by poly-L-glutamic acid and its effects were not reproduced by poly-L-arginine.

Together, these observations suggest that TGF-β1 is involved in MBP-mediated pulmonary fibrosis and that this eosinophil granule protein facilitates collagen deposition by regulating collagen turnover.

OTHER CELL TARGETS OF EOSINOPHIL-DERIVED CATIONIC PROTEINS DURING TISSUE REMODELING

Eosinophil granule proteins may also induce tissue remodeling by favoring angiogenesis. Indeed, subcytotoxic concentrations of MBP induced cell proliferation, as well as capillary genesis and sprouting in rat aortic and human umbilical vascular endothelial cells.[42] This effect was again independent of the cationic charge of MBP.[42]

The fibrogenic properties of MBP have recently been demonstrated in Duchenne muscular dystrophy, a degenerative muscular disease characterized by invasion of injured muscle fibers by immune and inflammatory cells, including eosinophils.[43] This inflammatory process is associated with massive fibrosis of both skeletal muscles and the heart, which is the major cause of mortality in this disease.[44] *In vitro* experiments with isolated human myotubes demonstrated that MBP causes muscle cell lysis and *in vivo* experiments conducted in dystrophic mice showed that eosinophil

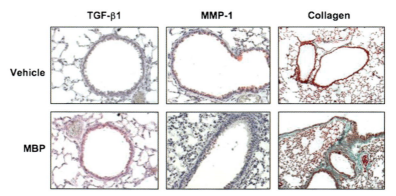

FIGURE 12.2.1 MBP promotes the hallmarks of lung fibrosis in mice: increased production of TGF-β1, MMP-1 and collagen. Naive mice received an endotracheal instillation of 50 μl of 0.025 μM sodium acetate (vehicle) or 15 μg MBP and were sacrificed 48 h later. Paraformaldehyde-fixed paraffin-embedded lung sections were incubated with polyclonal rabbit anti-TGF-β1 (transforming growth factor β1) or polyclonal chicken anti-MMP-1 (matrix metalloproteinase-1) antibody, followed by anti-rabbit, or anti-chicken secondary antibody conjugated to biotin, avidin-alkaline phosphatase complex, and Fast red substrate (red deposit), and counterstained using Mayer's hematoxylin. Collagen deposition was detected by staining tissue sections with Masson's trichrome stain (blue deposit). Images are representative of four distinct mice per group. Original magnification ×200.

depletion, or gene knockout of MBP, markedly reduced muscle cell lysis and collagen deposition in muscle and heart.[45] Thus, MBP may worsen the pathology of muscular dystrophy by promoting the lysis of dystrophic muscles and increasing skeletal and cardiac muscle fibrosis.[45]

ECP has been shown to stimulate both proliferation and stress fiber formation in mouse cardiac fibroblasts and to promote cytoskeleton rearrangements associated with differentiation in rat natal cardiomyocytes.[46] These findings suggest that ECP acts as a repair factor during heart diseases, such as myocardial infarction, and that its excessive production may contribute to cardiac hypertrophy and insufficiency. In this respect, a recent clinical study demonstrated that the assessment of serum levels of ECP provides valuable prognostic information in coronary atherosclerosis.[47]

INTERACTION BETWEEN EOSINOPHIL GRANULE PROTEINS AND EXTRACELLULAR MATRIX COMPONENTS

The capacity of ECP to selectively alter proteoglycan content, but not that of other ECM protein, suggests that this protein acts directly on mechanisms involved in proteoglycan catabolism. Proteoglycans consist of a core protein and one or more covalently attached glycosaminoglycan chains, such as chondroitin sulfate, dermatan sulfate, and heparan sulfate. Proteoglycans can thus be digested by heparanases, which are expressed by several types of immune and inflammatory cells and are known to influence a variety of normal and pathological processes, including tissue repair, inflammation, tumor growth and metastasis and angiogenesis.[48,49] MBP and, to a lesser extent, EDN are potent heparanase inhibitors, suggesting that these granule proteins may prevent proteoglycan

degradation and may in fact contribute to proteoglycan accumulation in fibrotic tissues.[28]

However, the net effect of eosinophil granule proteins on proteoglycan turnover is unclear, since eosinophils themselves express high levels of heparanase, in both active and latent forms.[28] In this study, however, no heparanase activity was detected in eosinophil extracts, an observation that was explained by the colocalization and binding of heparanase to intracellular MBP. In addition, through the presence of a heme group, EPO can catalyze the production of hypobromous acid that can react with glycosaminoglycans, leading to the formation of N-bromo derivatives and the subsequent breakdown of proteoglycans, as was reported for smooth muscle cell-derived ECM.[50]

CONCLUSION

Cationic protein-mediated effects may have important consequences on the severity of a variety of diseases characterized by eosinophil activation and tissue remodeling. Fibrosis of different organs, including gut, heart, lung, and skeletal muscle, significantly contributes to functional disabilities and to organ failure, and is potentially the proximal cause of death in several chronic diseases. Therefore, strategies specifically targeting eosinophil granule proteins may represent a novel therapeutic option for alleviating tissue fibrosis and its functional consequences in eosinophil-associated diseases.

REFERENCES

1. Araya J, Nishimura SL. Fibrogenic reactions in lung disease. *Annu Rev Pathol* 2010;**5**:77—98.
2. Brenner DA. Molecular pathogenesis of liver fibrosis. *Trans Am Clin Climatol Assoc* 2009;**120**:361—8.

3. Wynn TA. Cellular and molecular mechanisms of fibrosis. *J Pathol* 2008;**214**:199−210.

4. Noguchi H, Kephart GM, Colby TV, Gleich GJ. Tissue eosinophilia and eosinophil degranulation in syndromes associated with fibrosis. *Am J Pathol* 1992;**140**:521−8.

5. Bousquet J, Chanez P, Lacoste JY, Barnéon G, Ghavanian N, Enander I, et al. Eosinophilic inflammation in asthma. *N Engl J Med* 1990;**323**:1033−9.

6. Phipps S, Ying S, Wangoo A, Ong YE, Levi-Schaffer F, Kay AB. The relationship between allergen-induced tissue eosinophilia and markers of repair and remodeling in human atopic skin. *J Immunol* 2002;**1.169**:4604−12.

7. Birring SS, Parker D, McKenna S, Hargadon B, Brightling CE, Pavord ID, et al. Sputum eosinophilia in idiopathic pulmonary fibrosis. *Inflamm Res* 2005;**54**:51−6.

8. Xu X, Rivkind A, Pikarsky A, Pappo O, Bischoff SC, Levi-Schaffer F. Mast cells and eosinophils have a potential profibrogenic role in Crohn disease. *Scand J Gastroenterol* 2004;**39**:440−7.

9. Janin A. Eosinophilic myocarditis and fibrosis. *Hum Pathol* 2005;**36**:592−3.

10. Corradi D, Vaglio A, Maestri R, Legname V, Leonardi G, Bartoloni G, et al. Eosinophilic myocarditis in a patient with idiopathic hypereosinophilic syndrome: insights into mechanisms of myocardial cell death. *Hum Pathol* 2004;**35**:1160−3.

11. Pincus SH, Ramesh KS, Wyler DJ. Eosinophils stimulate fibroblast DNA synthesis. *Blood* 1987;**70**:572−4.

12. Kay AB, Phipps S, Robinson DS. A role for eosinophils in airway remodelling in asthma. *Trends Immunol* 2004;**25**:477−82.

13. Minshall EM, Leung DY, Martin RJ, Song YL, Cameron L, Ernst P, et al. Eosinophil-associated TGF-beta1 mRNA expression and airways fibrosis in bronchial asthma. *Am J Respir Cell Mol Biol* 1997;**17**:326−33.

14. Levi-Schaffer F, Garbuzenko E, Rubin A, Reich R, Pickholz D, Gillery P, et al. Human eosinophils regulate human lung- and skin-derived fibroblast properties in vitro: a role for transforming growth factor beta (TGF-beta). *Proc Natl Acad Sci USA* 1999;**96**:9660−5.

15. Cho JY, Miller M, Baek KJ, Han JW, Nayar J, Lee SY, et al. Inhibition of airway remodeling in IL-5-deficient mice. *J Clin Invest* 2004;**113**:551−60.

16. Humbles AA, Lloyd CM, McMillan SJ, Friend DS, Xanthou G, McKenna EE, et al. A critical role for eosinophils in allergic airway remodeling. *Science* 2004;**305**:1776−9.

17. Flood-Page P, Menzies-Gow A, Phipps S, Ying S, Wangoo A, Ludwig MS, et al. Anti-IL-5 treatment reduces deposition of ECM proteins in the bronchial basement membrane of mild atopic asthmatics. *J Clin Invest* 2003;**112**:1029−36.

18. Phipps S, Flood-Page P, Menzies-Gow A, Ong YE, Kay AB. Intravenous anti-IL-5 monoclonal antibody reduces eosinophils and tenascin deposition in allergen-challenged human atopic skin. *J Invest Dermatol* 2004;**122**:1406−12.

19. Rochester CL, Ackerman SJ, Zheng T, Elias JA. Eosinophil-fibroblast interactions. Granule major basic protein interacts with IL-1 and transforming growth factor-beta in the stimulation of lung fibroblast IL-6-type cytokine production. *J Immunol* 1996;**156**:4449−56.

20. Furuta GT, Ackerman SJ, Varga J, Spiess AM, Wang MY, Wershil BK. Eosinophil granule-derived major basic protein induces IL-8 expression in human intestinal myofibroblasts. *Clin Exp Immunol* 2000;**122**:35−40.

21. Govindaraju V, Michoud MC, Al-Chalabi M, Ferraro P, Powell WS, Martin JG. Interleukin-8: novel roles in human airway smooth muscle cell contraction and migration. *Am J Physiol Cell Physiol* 2006;**291**:C957−65.

22. Zagai U, Sköld CM, Trulson A, Venge P, Lundahl J. The effect of eosinophils on collagen gel contraction and implications for tissue remodelling. *Clin Exp Immunol* 2004;**135**:427−33.

23. Zagai U, Lundahl J, Klominek J, Venge P, Sköld CM. Eosinophil cationic protein stimulates migration of human lung fibroblasts in vitro. *Scand J Immunol* 2009;**69**:381−6.

24. Zagai U, Dadfar E, Lundahl J, Venge P, Sköld CM. Eosinophil cationic protein stimulates TGF-beta1 release by human lung fibroblasts in vitro. *Inflammation* 2007;**30**:153−60.

25. Faffe DS, Zin WA. Lung parenchymal mechanics in health and disease. *Physiol Rev* 2009;**89**:759−75.

26. Bergeron C, Boulet LP. Structural changes in airway diseases: characteristics, mechanisms, consequences, and pharmacologic modulation. *Chest* 2006;**129**:1068−87.

27. Hernnäs J, Sarnstrand B, Lindroth P, Peterson CG, Venge P, Malmstrom A. Eosinophil cationic protein alters proteoglycan metabolism in human lung fibroblast cultures. *Eur J Cell Biol* 1992;**59**:352−63.

28. Temkin V, Aingorn H, Puxeddu I, Goldshmidt O, Zcharia E, Gleich GJ, et al. Eosinophil major basic protein: first identified natural heparanase-inhibiting protein. *J Allergy Clin Immunol* 2004;**113**:703−9.

29. Fuchs SM, Raines RT. Pathway for polyarginine entry into mammalian cells. *Biochemistry* 2004;**43**:2438−44.

30. Hamann KJ, Barker RL, Ten RM, Gleich GJ. The molecular biology of eosinophil granule proteins. *Int Arch Allergy Appl Immunol* 1991;**94**:202−20.

31. Lamousé-Smith ES, Furuta GT. Eosinophils in the gastrointestinal tract. *Curr Gastroenterol Rep* 2006;**8**:390−5.

32. Barker RL, Gundel RH, Gleich GJ, Checkel JL, Loegering DA, Pease LR, et al. Acidic polyamino acids inhibit human eosinophil granule major basic protein toxicity. Evidence of a functional role for ProMBP. *J Clin Invest* 1991;**88**:798−805.

33. Furuta GT, Nieuwenhuis EE, Karhausen J, Gleich GJ, Blumberg RS, Lee JJ, et al. Eosinophils alter colonic epithelial barrier function: role for major basic protein. *Am J Physiol Gastrointest Liver Physiol* 2005;**289**:G890−7.

34. Frigas E, Loegering DA, Gleich GJ. Cytotoxic effects of guinea pig eosinophil major basic protein on tracheal epithelium. *Lab Invest* 1980;**42**:35−43.

35. Gleich GJ, Frigas E, Loegering DA, Wassom DL, Steinmuller D. Cytotoxic properties of the eosinophil major basic protein. *J Immunol* 1979;**123**:2925−7.

36. Gundel RH, Letts LG, Gleich GJ. Human eosinophil major basic protein induces airway constriction and airway hyperresponsiveness in primates. *J Clin Invest* 1991;**87**:1470−3.

37. Coyle AJ, Ackerman SJ, Irvin CG. Cationic proteins induce airway hyperresponsiveness dependent on charge interactions. *Am Rev Respir Dis* 1993;**147**:896−900.

38. Uchida DA, Ackerman SJ, Coyle AJ, Larsen GL, Weller PF, Freed J, et al. The effect of human eosinophil granule major basic protein on airway responsiveness in the rat in vivo. A comparison with polycations. *Am Rev Respir Dis* 1993;**147**:982−8.

39. Herbert CA, Edwards D, Boot JR, Robinson C. In vitro modulation of the eosinophil-dependent enhancement of the permeability of the bronchial mucosa. *Br J Pharmacol* 1991;**104**:391—8.

40. Davies DE. The role of the epithelium in airway remodeling in asthma. *Proc Am Thorac Soc* 2009;**6**:678—82.

41. Pégorier S, Wagner LA, Gleich GJ, Pretolani M. Eosinophil-derived cationic proteins activate the synthesis of remodeling factors by airway epithelial cells. *J Immunol* 2006;**177**:4861—9.

42. Puxeddu I, Berkman N, Nissim Ben Efraim AH, Davies DE, Ribatti D, Gleich GJ, et al. The role of eosinophil major protein in angiogenesis. *Allergy* 2008;**64**:368—74.

43. Cai B, Spencer MJ, Nakamura G, Tidball JG. Eosinophilia of dystrophin-deficient muscle is promoted by perforin-mediated cytotoxicity by T cell effectors. *Am J Pathol* 2000;**156**:1789—96.

44. Bernasconi P, Torchiana E, Confalonieri P, Brugnoni R, Barresi R, Mora M, et al. Expression of transforming growth factor-beta 1 in dystrophic patient muscles correlates with fibrosis. Pathogenetic role of a fibrogenic cytokine. *J Clin Invest* 1995;**96**:1137—44.

45. Wehling-Henricks M, Sokolow S, Lee JJ, Myung KH, Villalta SA, Tidball JG. Major basic protein-1 promotes fibrosis of dystrophic muscle and attenuates the cellular immune response in muscular dystrophy. *Hum Mol Genet* 2008;**17**:2280—92.

46. Fukuda T, Iwata M, Kitazoe M, Maeda T, Salomon D, Hirohata S, et al. Human eosinophil cationic protein enhances stress fiber formation in Balb/c 3T3 fibroblasts and differentiation of rat neonatal cardiomyocytes. *Growth Factors* 2009;**27**:228—36.

47. Niccoli G, Ferrante G, Cosentino N, Conte M, Belloni F, Marino M, et al. Eosinophil cationic protein: A new biomarker of coronary atherosclerosis. *Atherosclerosis* 2010;**211**:606—11.

48. Iozzo RV. Matrix proteoglycans: from molecular design to cellular function. *Annu Rev Biochem* 1998;**67**:609—52.

49. Vlodavski I, Friedman Y. Molecular properties and involvement of heparanase in cancer metastasis and angiogenesis. *J Clin Invest* 2001;**108**:341—7.

50. Rees MD, McNiven TN, Davies MJ. Degradation of extracellular matrix and its components by hypobromous acid. *Biochem J* 2007;**401**:587—96.

Chapter 12.3

Role of Eosinophils in Transforming Growth Factor β1-Mediated Remodeling

David H. Broide

INTRODUCTION

The presence of large numbers of eosinophils at sites of allergic inflammation associated with tissue remodeling (e.g., asthma, eosinophilic esophagitis, idiopathic hypereosinophilic syndrome, and nasal polyps) has led to investigations of whether eosinophils could be contributing to tissue remodeling.[1] These studies (detailed below) provide evidence

that eosinophils express significant levels of the pro-remodeling cytokine, transforming growth factor β1 (TGF-β1).[1–9] Moreover, eosinophil-deficient mice have reduced levels of TGF-β1 and allergen-induced remodeling,[10] thus providing a mechanistic link between eosinophil expression of TGF-β1 and remodeling in a mouse model. In addition, human studies in asthmatics provide evidence that eosinophils express TGF-β1 and that anti-interleukin-5 (anti-IL-5) therapy reduces numbers of eosinophils, as well as levels of TGF-β1 and extracellular matrix remodeling.[11] Thus, while TGF-β1 is well recognized as a profibrotic mediator,[12] its expression by eosinophils provides a mechanism for TGF-β1 to play a significant role in remodeling in diseases associated with significant eosinophilic tissue inflammation.

EOSINOPHILS AND TGF-β1

As the infiltration of eosinophils into tissues is often associated with extracellular matrix alterations, such as fibrosis, studies have examined whether eosinophils express TGF-β1. Initial studies by Wong and colleagues[2] demonstrated that eosinophils in the peripheral blood of patients with eosinophilia express TGF-β1, whereas eosinophils in the blood of normal donors contain little or no TGF-β1. Northern blot analysis detected the 2.5 kb TGF-β1 transcript in mRNA isolated from eosinophils purified from a patient with idiopathic hypereosinophilic syndrome. *In situ* hybridization and immunohistochemistry of leukocytes from patients with hypereosinophilic syndrome and patients with blood eosinophilia localized TGF-β1 mRNA and protein to eosinophils.[2] Subsequent studies examined whether tissue eosinophils in nasal polyps express TGF-β1,[3] and demonstrated that TGF-β1 mRNA is expressed in nasal polyp tissues and in sinus tissue from a patient with allergic rhinitis, but not in normal nasal mucosa. By combining chromotrope 2R staining to detect eosinophils with *in situ* hybridization to detect *TGFB1* mRNA, approximately 50% of eosinophils infiltrating the nasal polyp tissue were noted to express TGF-β1. Immunohistochemical localization of TGF-β1[3] in the sinus mucosa revealed an association with the extracellular matrix as well as localization to cells in the stroma.

Subsequent studies examined whether TGF-β1 was expressed by eosinophils in asthma using bronchial biopsy tissues from severe asthmatics, mild asthmatics, and normal subjects using *in situ* hybridization combined with histochemical staining.[4] The density of TGF-β1+ cells in severe asthmatic tissues (52 per mm²) was significantly greater than that in mild asthmatic tissues (1 per mm²) or normal tissues.[4] The vast majority of TGF-β1+ cells in bronchial tissues from severe asthma (99%) and mild asthma (100%) were identified as eosinophils.[4] In contrast, eosinophils constituted a small portion of TGF-β1+ cells (20%) in normal tissues. These results indicate that TGF-β1 mRNA is overexpressed in severe asthmatics and that the main source of TGF-β1 mRNA is eosinophils, suggesting that eosinophils through expression of TGF-β1 may play an important role in structural changes in the mucosa, such as subepithelial fibrosis in asthmatic airways.

The relationship between allergen-induced tissue eosinophilia and markers of tissue repair has also been investigated using the allergen-induced cutaneous late-phase reaction as a model of allergic inflammation.[13] These studies demonstrated that following the peak of the late-phase reaction (at 6 h), TGF-β1+ eosinophils, alpha smooth muscle actin-positive (ACTA+) myofibroblasts, tenascin immunoreactivity, and procollagen type 1+ cells 24–48 h persisted postchallenge. Direct evidence for repair markers induction by eosinophils was obtained by coculturing eosinophils and fibroblasts, which resulted in fibroblast alpha smooth muscle actin immunoreactivity that could be inhibited by neutralizing antibodies (Abs) to TGF-β1, as well as the production of tenascin transcripts and protein.[13] Thus, eosinophil-derived TGF-β1 contributes to repair and remodeling events in allergic inflammation in human atopic skin. Administration of anti-IL-5 reduces both levels of eosinophils and deposition of the extracellular matrix protein tenascin in allergen-challenged human atopic skin.[14]

TGF-β has three isoforms (TGF-β1, TGF-β2, and TGF-β3) and most studies of eosinophils and TGF-β have focused on eosinophil expression of TGF-β1. Fewer studies have examined eosinophil expression of TGF-β2 and TGF-β3. Studies on patients with severe asthma (with or without persistent eosinophilia), mild asthma, and normal subjects have demonstrated that TGF-β2 expression is increased in severe asthma.[15] Tissue eosinophils in patients with severe asthma, unlike eosinophils in other patient groups, expressed high amounts of TGF-β2.[15] Subjects with severe asthma also had the thickest subbasement membrane and highest tissue inhibitor of metalloproteinase-1 levels.

Eosinophils not only generate TGF-β but are also able to respond to TGF-β, as they express TGF-β type I and type II receptors, as well as intracellular TGF-β signaling components, including SMAD2, SMAD3 and SMAD4.[16] TGF-β induces phosphorylation of SMAD2 in eosinophils, which can be blocked using an inhibitor of TGF-β type I receptor kinase.[16,17] In addition, dominant-negative SMAD3 protein can suppress TGF-β-induced SMAD7 mRNA expression in eosinophils. Thus, eosinophils can generate TGF-β as well as being able to respond to TGF-β.

ACTIVATION OF TGF-β1 EXPRESSION BY EOSINOPHILS

Studies have examined which cytokines and mediators regulate the TGF-β1 expression by eosinophils.[7,8,9]

Experiments in which eosinophils isolated from the peripheral blood of healthy donors and cultured with IL-3, IL-4, or IL-5 have demonstrated that each of these cytokines upregulates TGF-β1 mRNA and protein expression by eosinophils in all donors.[7] Other studies have examined whether extracellular matrix glycosaminoglycans, such as hyaluronan, modulates eosinophil TGF-β production. Incubation of peripheral blood eosinophils with low molecular weight hyaluronan increases TGF-β mRNA expression and protein secretion by eosinophils.[8] These studies suggest that eosinophils exposed to cytokines or extracellular matrix proteins in tissues may also be activated by these signals to express TGF-β1 and thus contribute to tissue remodeling.[18] Studies have also demonstrated that TGF-β1 expression is induced in peripheral blood eosinophils cultured with leukotriene D4 (LTD4) in the presence of either IL-5 or granulocyte-macrophage colony-stimulating factor (GM-CSF), as assessed by real-time PCR.[19] LTD4 alone does not induce TGF-β1 expression. These results suggest that cysteinyl LT receptors may stimulate eosinophils to express TGF-β1 at sites of allergic inflammation if IL-5 or GM-CSF is present. In addition, LTs have been noted to play a role in airway remodeling in a mouse model.[20]

The peptidyl-prolyl isomerase Pin1 also influences levels of eosinophil TGF-β1 expression as it promotes the stability of TGF-β1 mRNA in human Eos.[9] Pin1 interacts with both protein kinase C (PKC)-α and protein phosphatase 2A, which together control Pin1 isomerase activity. Pharmacological inhibition of Pin1 in a rat asthma model selectively reduces eosinophilic pulmonary inflammation, TGF-β1 and collagen expression, and airway remodeling, thereby underscoring the importance of Pin1 to TGF-β1 expression *in vivo*. Furthermore, chronically challenged Pin1-deficient mice show reduced peribronchiolar collagen deposition compared with wild-type controls. These data suggest that pharmacological suppression of Pin1 may be a novel therapeutic option to prevent airway fibrosis in individuals with chronic asthma.

MOUSE MODELS OF ASTHMA AND TGF-β

IL-5-deficient mice have significantly reduced numbers of peribronchial eosinophils staining positive for major basic protein (MBP), which is paralleled by a similar reduction in the number of cells staining positive for TGF-β, suggesting that eosinophils are a significant source of TGF-β in the remodeled airway.[10] Eosinophil-deficient mice also have significantly reduced airway hyperreactivity.[21] As αvβ6 integrins are known to activate TGF-β, their levels of expression in have been examined in mouse models of allergen-induced airway remodeling.[10] Chronic ovalbumin

(OVA) challenge induces significantly higher levels of airway epithelial $\alpha v \beta 6$ integrin expression, as well as significantly higher levels of bioactive lung TGF-β in wild-type compared with IL-5-deficient mice.[10] Thus, increased airway epithelial expression of $\alpha v \beta 6$ integrin may contribute to the increased TGF-β activation. These results suggest an important role for eosinophils, $\alpha v \beta 6$ integrins, IL-5, and TGF-β in airway remodeling.

Intracellular signaling pathways that converge on Smad3 are used by both TGF-β and activin A, key cytokines implicated in the process of fibrogenesis.[12] Studies have demonstrated an important role for TGF-β and Smad signaling in mediating allergen-induced remodeling in mice.[22–24] Increased levels of activin A and increased numbers of peribronchial TGF-$\beta 1^+$ cells were detected in both wild-type and Smad3-deficient mice following repeated OVA challenge, as inactivation of Smad3 does not influence eosinophils trafficking to the lung.[22] In contrast, Smad3-deficient mice challenged with OVA had significantly less peribronchial fibrosis, reduced thickness of the peribronchial smooth muscle layer, and reduced epithelial mucus production compared with wild-type mice. Although the number of peribronchial myofibroblasts increased significantly in wild-type mice following OVA challenge, there was a significant reduction in the number of peribronchial myofibroblasts in OVA-challenged Smad3-deficient mice. This is likely due to the role of TGF-β in mediating myofibroblast differentiation in wild-type mice, which is inhibited in Smad3-deficient mice. There is no difference in levels of eosinophilic airway inflammation or airway responsiveness in Smad3-deficient compared with wild-type mice. These results suggest that Smad3 signaling is required for allergen-induced airway remodeling, as well as allergen-induced accumulation of myofibroblasts in the airway. However, Smad3 signaling does not significantly contribute to airway responsiveness.

COMBINED EFFECT OF ALLERGEN AND ENVIRONMENTAL TOBACCO SMOKE ON EOSINOPHIL TGF-$\beta 1$ EXPRESSION AND REMODELING

In addition to allergen challenge inducing expression of eosinophils and TGF-β, studies in mouse models have demonstrated that coexposure of mice to allergen and environmental tobacco smoke induces significantly enhanced airway remodeling compared to each stimulus alone.[25] Environmental tobacco smoke can increase asthma symptoms and the frequency of asthma attacks. However, mice exposed to chronic environmental tobacco smoke alone did not develop significant eosinophilic airway inflammation, airway remodeling, or increased airway hyperreactivity to methacholine.[25] In contrast, mice exposed

to chronic OVA allergen had significantly increased levels of peribronchial fibrosis and increased thickening of the smooth muscle layer, mucus, and airway hyperreactivity, which was significantly enhanced by coexposure to chronic environmental tobacco smoke.[25] Mice exposed to both chronic environmental tobacco smoke and chronic OVA allergen had significantly increased levels of eotaxin/C-C motif chemokine 11 (CCL11) expression in airway epithelium, which was associated with increased numbers of peribronchial eosinophils, as well as increased numbers of peribronchial cells expressing TGF-$\beta 1$. These studies suggest that chronic coexposure to allergen and environmental tobacco smoke significantly increases levels of allergen-induced airway remodeling (in particular, smooth muscle thickness) and airway responsiveness by upregulating expression of chemokines such as eotaxin in airway epithelium, with the resultant recruitment of cells expressing TGF-$\beta 1$ to the airway and enhanced airway remodeling. Administration of a TLR-9 agonist to mice simultaneously exposed to chronic environmental tobacco smoke and chronic OVA allergen significantly reduced levels of eosinophilic airway inflammation, mucus production, peribronchial fibrosis, peribronchial smooth muscle layer thickness, and airway responsiveness.[26,27] Reduced airway remodeling in mice treated with the TLR-9 agonist was associated with significantly reduced numbers of peribronchial MBP$^+$ and TGF-$\beta 1^+$ cells, and significantly reduced levels of lung T-helper type 2 (T$_h$2) cytokines (IL-5 and IL-13) and TGF-$\beta 1$.

CORTICOSTEROIDS, TGF-β, AND REMODELING

At present, there are conflicting results from investigations into the role of corticosteroids in inhibiting airway remodeling in humans with asthma. Mouse model studies of allergen-induced airway remodeling have demonstrated that corticosteroids significantly reduces allergen-induced peribronchial collagen deposition and total lung collagen levels, but does not reduce allergen-induced thickening of the peribronchial smooth muscle layer.[28] Levels of lung TGF-$\beta 1$ are significantly reduced in mice treated with systemic corticosteroids, and this is as associated with a significant reduction in peribronchial inflammatory cells expressing TGF-$\beta 1$, including eosinophils and mononuclear cells. Corticosteroids also significantly reduce the number of peribronchial myofibroblasts.[28] The reduction in peribronchial fibrosis mediated by corticosteroids is likely to be due to several mechanisms, including inhibition of TGF-$\beta 1$ expression, a reduction in the number of peribronchial inflammatory cells expressing TGF-$\beta 1$ (eosinophils and macrophages), and reduced accumulation of peribronchial myofibroblasts that contribute to collagen expression.[28]

Studies have also examined whether corticosteroids reduce levels of esophageal remodeling in human subjects with eosinophilic esophagitis, a disease associated with increased numbers of eosinophils expressing TGF-β1.[29,30] Eosinophilic esophagitis subjects were stratified based on the presence or absence of decreased epithelial eosinophilia following topical budesonide therapy.[31] Patients with residual eosinophil counts of less than seven eosinophils per high power field in the epithelial space (responders) demonstrated significantly reduced esophageal remodeling associated with decreased fibrosis, decreased TGFβ1 and SMAD2/3 positive cells, and decreased vascular activation in association with budesonide therapy.[31] Responders were more likely to have a TGF-β1 CC genotype at the -509 position in the TGF-β1 promoter. Thus, patients with eosinophilic esophagitis who demonstrate a reduction in eosinophils in response to topical corticosteroid therapy provide evidence of reduced TGF-β1 levels and reduced remodeling.

TARGETING EOSINOPHIL RECEPTORS AND SIGNALING PATHWAYS TO REDUCE BOTH LEVELS OF TGF-β AND REMODELING

In addition to using corticosteroids to reduce the numbers of eosinophils expressing TGF-β1, several alternative novel approaches to reduce levels of eosinophilic inflammation and remodeling have demonstrated efficacy in preclinical mouse models of allergen-induced remodeling. These approaches include targeting receptors that regulate eosinophil proliferation [IL-5 and sialic acid-binding Ig-like lectin 8 (Siglec-8)], intracellular signaling pathways important to eosinophil tissue recruitment [C-C chemokine receptor type 3 (CCR3) and phosphatidylinositol 3-kinase (PI3K)], and pathways that regulate T_h2 responses [nuclear factor κB (NF-κB) and Toll-like receptor 9 (TLR-9)]. IL-5 is an important eosinophil growth factor and studies in both mouse models of allergen-induced airway remodeling[10] and asthmatics have demonstrated an important role for IL-5 in regulating the number of eosinophils expressing TGF-β1.[11] An alternative approach to targeting IL-5 to regulate the number of eosinophils expressing TGF-β is to target Siglec-8 on eosinophils.[32] Siglec-8 is an innate immune receptor selectively expressed on eosinophils and mast cells,[32] and cross-linking Siglec-8 on human eosinophils induces apoptosis.[33] Siglec-F (a mouse paralogue of Siglec-8) is highly expressed on murine eosinophils,[34] and mouse models of asthma have demonstrated an important role for Siglec-F in mediating eosinophil survival *in vivo*.[34,35] Chronically allergen-challenged mice administered with an anti-Siglec-F Ab have significantly reduced levels of both peribronchial eosinophilic inflammation and subepithelial fibrosis, as assessed by trichrome staining and lung collagen levels.[34] Treatment with the anti-Siglec-F Ab reduced the number of blood, bone marrow, and tissue eosinophils, suggesting that the anti-Siglec-F Ab can reduce eosinophil production.[34] Administration of an anti-Siglec-F F(ab′)2 fragment also significantly reduced levels of eosinophilic inflammation in the blood and lungs. Fluorescence-activated cell sorting (FACS) analysis demonstrated increased numbers of apoptotic cells [annexin V+ CCR3+ bronchoalveolar lavage fluid (BALF) and bone marrow cells] in anti-Siglec-F Ab-treated mice challenged with OVA. Furthermore, treatment with the anti-Siglec-F Ab significantly reduces the number of peribronchial MBP+ TGF-β+ cells, suggesting that reduced levels of eosinophil-derived TGF-β in anti-Siglec-F Ab-treated mice contribute to a reduction in peribronchial fibrosis. Overall, these studies suggest that administration of an anti-Siglec-F Ab can significantly reduce levels of allergen-induced eosinophilic airway inflammation and features of airway remodeling, in particular subepithelial fibrosis, by reducing eosinophil production and increasing the number of apoptotic eosinophils in lung and bone marrow.

Studies using PI3Kγ-deficient mice exposed to chronic allergen challenge have examined whether targeting intracellular signaling molecules such as PI3kinase influences eosinophil inflammation and remodeling.[36] PI3Kγ-deficient mice challenged with OVA have significantly reduced numbers of BALF and peribronchial eosinophils compared with wild-type mice. However, there is no significant difference in the number of bone marrow or circulating peripheral blood eosinophils between wild-type and PI3Kγ-deficient mice, suggesting that eosinophil trafficking into the lung is reduced in PI3Kγ-deficient mice. In addition, PI3Kγ-deficient and wild-type mice have similar levels of IL-5 and eotaxin. The reduced eosinophil recruitment to the airway in PI3Kγ-deficient mice challenged with OVA is associated with significantly reduced numbers of TGF-β1+ peribronchial cells, Smad2/3+ airway epithelial cells, and Smad2/3+ peribronchial cells, as well as significantly reduced levels of peribronchial fibrosis.[36] In addition, the area of peribronchial alpha smooth muscle staining is significantly reduced in PI3Kγ-deficient compared with wild-type mice. Overall, this study demonstrated an important role for PI3Kγ in mediating allergen-induced eosinophilic airway inflammation and airway remodeling, suggesting that PI3Kγ may be a novel therapeutic target in asthma.

Other studies have demonstrated an important role for NF-kB-regulated genes in airway epithelium contributing to allergen-induced airway eosinophil recruitment, TGF-β expression, and remodeling.[37] Chronic allergen challenge of mice deficient in airway epithelial IKK-β specifically prevents the nuclear translocation of the Rel A NF-κB subunit in airway epithelial cells, resulting in significantly lower peribronchial fibrosis in CC10-Cre(tg)/ IKK-β(Δ/Δ) mice compared with littermate controls, as assessed by peribronchial trichrome staining and total lung collagen content. Levels of airway mucus, airway eosinophils, and peribronchial CD4$^+$ cells in chronic allergen-challenged mice are also significantly reduced upon airway epithelial IKK-β ablation.[37] The diminished inflammatory response is associated with reduced expression of NF-κB-regulated chemokines, including eotaxin and thymus- and activation-regulated chemokine, which attract eosinophils and T$_h$2 cells, respectively, into the airway. The number of peribronchial cells expressing TGF-β1, as well as TGF-β1 amounts in BALF, is also significantly reduced in mice deficient in airway epithelium IKK-β. Overall, these studies indicate an important role for NF-κB-regulated genes in the airway epithelium in allergen-induced airway remodeling, including peribronchial fibrosis and mucus production.

HUMAN STUDIES OF ASTHMA, EOSINOPHILIC ESOPHAGITIS, AND TGF-β

TGF-β expression has been examined in bronchial mucosal biopsies from asthmatics and normal human subjects and TGF-β immunoreactivity has been demonstrated to increase in the epithelium and submucosa of asthmatics.[6] A significant correlation was observed between the number of epithelial or submucosal cells expressing TGF-β in asthma and basement membrane thickness or fibroblast number. In contrast, no such correlation was found for epidermal growth factor (EGF) or GM-CSF. *In situ* hybridization of TGF-β1 mRNA confirmed results obtained by immunohistochemistry, while the combination of *in situ* hybridization and immunohistochemistry demonstrated that eosinophils and fibroblasts synthesize TGF-β in asthma.[6] These data suggest that TGF-β, but not EGF or GM-CSF, is involved in airway remodeling in asthma.

Minshall and coworkers[5] also demonstrated that asthmatic individuals exhibit higher expression of TGF-β1 mRNA and immunoreactivity in the airway submucosa than normal control subjects, and these increases are directly related to asthma severity. In asthmatic subjects, the presence of subepithelial fibrosis is associated with disease severity and correlates with reduced FEV$_1$ (forced expiratory volume in 1 s). Within asthmatic airways, EG2$^+$ eosinophils (expressing secreted eosinophil cationic protein) represent the major source of TGF-β1 mRNA and immunoreactivity. These results provide further evidence that TGF-β1 may play a role in the fibrotic changes occurring within asthmatic airways and that activated eosinophils are a major source of this cytokine.

Other studies have examined whether depletion of eosinophils with an anti-IL-5 antibody (Ab) reduces levels of eosinophil TGF-β1 expression and airway remodeling in mild asthmatics.[11] In these studies, bronchial biopsies were obtained before and after three infusions of a humanized anti-IL-5 monoclonal Ab in 24 atopic asthmatics in a randomized, double-blind, placebo-controlled study. Treating asthmatics with anti-IL-5 Ab specifically decreased airway eosinophil numbers and significantly reduced the expression of lumican, procollagen III, and tenascin in the bronchial mucosal reticular membrane. In addition, anti-IL-5 treatment was associated with a significant reduction in both the numbers of airway eosinophils and the percentage expressing TGF-β1 mRNA, and the concentration of TGF-β1 in BALF. Therefore eosinophils may contribute to tissue remodeling processes in asthma by regulating the deposition of extracellular matrix proteins.

In addition, the importance of eosinophils and eosinophil expressed TGF-β1 to remodeling in eosinophilic esophagitis has also been investigated. Eosinophilic esophagitis is associated with esophageal remodeling and results in esophageal stricture formation in a subset of patients. Esophageal biopsies in patients with eosinophilic esophagitis demonstrate both increased levels of subepithelial fibrosis and expression of TGF-β1 and its signaling molecule, phospho-SMAD2/3, compared with esophageal biopsies from patients with gastroesophageal reflux disease or normal controls.[29–31] In addition, esophageal biopsies from patients with eosinophilic esophagitis demonstrate increased vascular density and increased expression of the vascular endothelial adhesion molecule, vascular cell adhesion molecule-1. Thus, eosinophil expression of TGF-β1 may play an important role in esophageal remodeling and esophageal stricture formation in eosinophilic esophagitis.

TGF-β AND TOLERANCE

Although TGF-β1 is an important mediator of remodeling, T-regulatory (T$_{reg}$) cells also produce TGF-β and IL-10, which are important in inducing tolerance. The importance of eosinophils and TGF-β1 in airway remodeling is suggested from model studies in which airway remodeling is significantly reduced in mice treated with an anti-TGF-β1 Ab,[23] as well as in Smad3-deficient mice,[22] which have impaired TGF-β signaling. In addition, studies in IL-5-deficient mice,[10] as well as in human subjects with asthma treated with anti-IL-5,[11] demonstrate reduced numbers of eosinophils, reduced TGF-β1$^+$ eosinophils, and reduced airway remodeling. Although several studies show an important role for eosinophils, TGF-β1, and Smad signaling in airway remodeling,[22,23,24] other studies have demonstrated that administration of an anti-TGF-β1 Ab does not reduce remodeling[38] and increases airway hyperreactivity in mice.[39]

REMODELING PATHWAYS INDEPENDENT OF TGF-β AND/OR EOSINOPHILS

Conceptually, airway remodeling results from persistent inflammation and/or aberrant tissue repair mechanisms.[1] It is likely that several immune and inflammatory cell types and mediators are involved in mediating remodeling in allergic inflammation. In addition, different features of airway remodeling are likely to be mediated by different inflammatory pathways. Although this review has focused on the role of eosinophil-expressed TGF-β1 in remodeling, it is likely that both TGF-β-independent and eosinophil-independent pathways contribute to remodeling at sites of eosinophil inflammation. For example, eosinophil cationic protein is released by activated eosinophils and can stimulate TGF-β1 release in human lung fibroblasts.[40] Moreover, several important candidate mediators of remodeling have been identified in addition to TGF-β1, including a disintegrin, matrix metalloproteinase 9, metalloproteinase 33, T_h2 cytokines (IL-5 and IL-13), and vascular endothelial growth factor.[1]

CONCLUSION

There is significant evidence from both mouse and human studies to indicate that eosinophils express TGF-β1 and that TGF-β1 is expressed at increased levels at sites of remodeling associated with eosinophilic inflammation, and that depleting eosinophils reduces remodeling. Although several preclinical studies have demonstrated that inhibiting TGF-β reduces airway remodeling,[22–24] one preclinical study that failed to demonstrate that TGF-β inhibition reduces remodeling.[38] At present, there are no published data on the effects of anti-TGF-β treatment on eosinophilic inflammation and remodeling in human subjects. Therefore, further studies are needed to determine the safety and effectiveness of targeting eosinophils and/or TGF-β1 in diseases in which remodeling is associated with eosinophilic inflammation. The potential importance of the TGF-β pathway in asthma was suggested in a recent genome-wide association study that identified the TGF-β signaling molecule, *SMAD*, as one of five genes showing a high degree of linkage to asthma.[41] Identifying the subset of subjects with a genetic predisposition to remodeling may therefore lead to targeted therapy for a subset of subjects with eosinophilic inflammation who exhibit enhanced remodeling.

REFERENCES

1. Broide DH. Immunologic and inflammatory mechanisms that drive asthma progression to remodeling. *J Allergy Clin Immunol* 2008; **121**:560–70.

2. Wong DT, Elovic A, Matossian K, Nagura N, McBride J, Chou MY, et al. Eosinophils from patients with blood eosinophilia express transforming growth factor beta 1. *Blood* 1991;**78**:2702–7.

3. Ohno I, Lea RG, Flanders K, Clark DA, Banwatt D, Dolovich J, et al. Eosinophils in chronically inflamed human upper airway tissues express transforming growth factor beta 1 gene (TGF beta 1). *J Clin Invest* 1992;**89**:1662–8.

4. Ohno I, Nitta Y, Yamauchi K, Hoshi H, Honma M, Woolley K, et al. Transforming growth factor beta 1 (TGF beta 1), gene expression by eosinophils in asthmatic airway inflammation. *Am J Respir Cell Mol Biol* 1996;**15**:404–9.

5. Minshall EM, Leung DY, Martin RJ, Song YL, Cameron L, Ernst P, et al. Eosinophil-associated TGF-β1 mRNA expression and airways fibrosis in bronchial asthma. *Am J Respir Cell Mol Biol* 1997; **17**:326–33.

6. Vignola AM, Chanez P, Chiappara G, Merendino A, Pace E, Rizzo A, et al. Transforming growth factor-beta expression in mucosal biopsies in asthma and chronic bronchitis. *Am J Respir Crit Care Med* 1997;**156**:591–9.

7. Elovic AE, Ohyama H, Sauty A, McBride J, Tsuji T, Nagai M, et al. IL-4-dependent regulation of TGF-alpha and TGF-beta1 expression in human eosinophils. *J Immunol* 1998;**160**:6121–7.

8. Ohkawara Y, Tamura G, Iwasaki T, Tanaka A, Kikuchi T, Shirato K. Activation and transforming growth factor-beta production in eosinophils by hyaluronan. *Am J Respir Cell Mol Biol* 2000; **23**:444–51.

9. Shen ZJ, Esnault S, Rosenthal LA, Szakaly RJ, Sorkness RL, Westmark PR, et al. Pin1 regulates TGF-beta1 production by activated human and murine eosinophils and contributes to allergic lung fibrosis. *J Clin Invest* 2008;**118**:479–90.

10. Cho JY, Miller M, Baek KJ, Han JW, Nayar J, Lee SY, et al. Inhibition of airway remodeling in IL-5-deficient mice. *J Clin Invest* 2004;**113**:551–60.

11. Flood-Page P, Menzies-Gow A, Phipps S, Ying S, Wangoo A, Ludwig MS, et al. Anti-IL-5 treatment reduces deposition of ECM proteins in the bronchial subepithelial basement membrane of mild atopic asthmatics. *J Clin Invest* 2003;**112**:1029–36.

12. Massagué J. TGF beta in cancer. *Cell* 2008;**134**:215–30.

13. Phipps S, Ying S, Wangoo A, Ong YE, Levi-Schaffer F, Kay AB. The relationship between allergen-induced tissue eosinophilia and markers of repair and remodeling in human atopic skin. *J Immunol* 2002;**169**:4604–12.

14. Phipps S, Flood-Page P, Menzies-Gow A, Ong YE, Kay AB. Intravenous anti-IL-5 monoclonal antibody reduces eosinophils and tenascin deposition in allergen-challenged human atopic skin. *J Invest Dermatol* 2004;**122**:1406–12.

15. Balzar S, Chu HW, Silkoff P, Cundall M, Trudeau JB, Strand M, et al. Increased TGF-β2 in severe asthma with eosinophilia. *J Allergy Clin Immunol* 2005;**115**:110–7.

16. Kanzaki M, Shibagaki N, Hatsushika K, Mitsui H, Inozume T, Okamoto A, et al. Human eosinophils have an intact Smad signaling pathway leading to a major transforming growth factor-beta target gene expression. *Int Arch Allergy Immunol* 2007;**142**:309–17.

17. Liu T, Feng XH. Regulation of TGF-beta signaling by protein phosphatases. *Biochem J* 2010;**430**:191–8.

18. Burgess JK. The role of the extracellular matrix and specific growth factors in the regulation of inflammation and remodeling in asthma. *Pharmacol Ther* 2009;**122**:19–29.

19. Kato Y, Fujisawa T, Nishimori H, Katsumata H, Atsuta J, Iguchi K, et al. Leukotriene D4 induces production of transforming growth factor-β1 by eosinophils. *Int Arch Allergy Immunol* 2005;**137**: 17−20.

20. Henderson Jr WR, Tang LO, Chu SJ, Tsao SM, Chiang GK, Jones F, et al. A role for cysteinyl leukotrienes in airway remodeling in a mouse asthma model. *Am J Respir Crit Care Med* 2002;**165**: 108−16.

21. Lee JJ, Dimina D, Macias MP, Ochkur SI, McGarry MP, O'Neill KR, et al. Defining a link with asthma in mice congenitally deficient in eosinophils. *Science* 2004;**305**:1773−6.

22. Le AV, Cho JY, Miller M, McElwain S, Golgotiu K, Broide DH. Inhibition of allergen-induced airway remodeling in Smad 3-deficient mice. *J Immunol* 2007;**178**:7310−6.

23. McMillan SJ, Xanthou G, Lloyd CM. Manipulation of allergen-induced airway remodeling by treatment with anti-TGF-beta antibody: effect on the Smad signaling pathway. *J Immunol* 2005;**174**: 5774−80.

24. Gregory LG, Mathie SA, Walker SA, Pegorier S, Jones CP, Lloyd CM. Overexpression of smad2 drives house dust mite-mediated airway remodeling and airway hyperresponsiveness via activin and IL-25. *Am J Respir Crit Care Med* 2010;**182**:143−54.

25. Min MG, Song DJ, Miller M, Cho JY, McElwain S, Ferguson P, et al. Coexposure to environmental tobacco smoke increases levels of allergen-induced airway remodeling in mice. *J Immunol* 2007;**178**:5321−8.

26. Song DJ, Min MG, Miller M, Cho JY, Yum HY, Broide DH. Toll-like receptor-9 agonist inhibits airway inflammation, remodeling and hyperreactivity in mice exposed to chronic environmental tobacco smoke and allergen. *Int Arch Allergy Immunol* 2010;**151**:285−96.

28. Cho JY, Miller M, Baek KJ, Han JW, Nayar J, Rodriguez M, et al. Immunostimulatory DNA inhibits transforming growth factor-geta expression and airway remodeling. *Am J Respir Cell Mol Biol* 2004;**30**:651−61.

29. Miller M, Cho JY, McElwain K, McElwain S, Shim JY, Manni M, et al. Corticosteroids prevent myofibroblast accumulation and airway remodeling in mice. *Am J Physiol Lung Cell Mol Physiol* 2006;**290**:L162−9.

30. Aceves SS, Newbury RO, Dohil R, Bastian JF, Broide DH. Esophageal remodeling in pediatric eosinophilic esophagitis. *J Allergy Clin Immunol* 2007;**119**:206−12.

31. Aceves SS, Chen D, Newbury RO, Dohil R, Bastian JF, Broide DH. Mast cells infiltrate the esophageal smooth muscle in patients with eosinophilic esophagitis, express TGF-β1, and increase esophageal smooth muscle contraction. *J Allergy Clin Immunol* 2010;**126**: 1198−204.

32. Aceves SS, Newbury RO, Chen D, Mueller J, Dohil R, Hoffman H, et al. Resolution of remodeling in eosinophilic esophagitis correlates with epithelial response to topical corticosteroids. *Allergy* 2010;**65**:109−16.

33. Bochner BS. Siglec-8 on human eosinophils and mast cells, and Siglec-F on murine eosinophils, are functionally related inhibitory receptors. *Clin Exp Allergy* 2009;**39**:317−24.

34. Hudson SA, Bovin NV, Schnaar RL, Crocker PR, Bochner BS. Eosinophil-selective binding and proapoptotic effect in vitro of a synthetic siglec-8 ligand, polymeric 6'-sulfated sialyl lewis X. *J Pharmacol Exp Ther* 2009;**330**:608−12.

35. Song DJ, Cho JY, Lee SY, Miller M, Rosenthal P, Soroosh P, et al. Anti-Siglec-F antibody reduces allergen induced eosinophilic inflammation and airway remodeling. *J Immunol* 2009;**183**: 5333−41.

36. Zimmerman N, McBride ML, Yamada Y, Hudson SA, Jones C, Cromie KD, et al. Siglec-F antibody administration to mice selectively reduces blood and tissue eosinophils. *Allergy* 2008;**63**: 1156−63.

37. Lim DH, Cho JY, Song DJ, Lee SY, Miller M, Broide DH. PI3K gamma-deficient mice have reduced levels of allergen-induced eosinophilic inflammation and airway remodeling. *Am J Physiol Lung Cell Mol Physiol* 2009;**296**:L210−9.

38. Broide DH, Lawrence T, Doherty T, Cho JY, Miller M, McElwain K, et al. Allergen-induced peribronchial fibrosis and mucus production mediated by IkappaB kinase beta-dependent genes in airway epithelium. *Proc Natl Acad Sci USA* 2005;**102**: 17723−8.

39. Fattouh R, Midence G, Arias K, Johnson J, Walker T, Goncharova S, et al. Transforming growth factor-ß regulates house dust mite-induced allergic airway inflammation but not airway remodeling. *Am J Respir Crit Care Med* 2008;**177**:593−603.

40. Alcorn J, Rinaldi L, Jaffe E, van Loon M, Bates J, Janssen-Heininger Y, et al. Transforming growth factor-ß1 suppresses airway hyperresponsiveness in allergic airway disease. *Am J Respir Crit Care Med* 2007;**176**:974−82.

41. Zagai U, Dadfar E, Lundahl J, Venge P, Sköld CM. Eosinophil cationic protein stimulates TGF-β1 release by human lung fibroblasts in vitro. *Inflammation* 2007;**30**:153−60.

42. Moffatt MF, Gut IG, Demenais F, Strachan DP, Bouzigon E, Heath S, et al. A large-scale, consortium-based genomewide association study of asthma. *N Engl J Med* 2010;**363**:1211−21.

Chapter 12.4

Potential Role of Eosinophils and Tumor Necrosis Factor α in Tissue Remodeling

Elizabeth A.B. Kelly

INTRODUCTION

As outlined in previous chapters, eosinophils are thought to contribute to tissue remodeling in a variety of diseases through a complex array of mediators. The prototypic profibrotic factor, transforming growth factor β1 (TGF-β1), is expressed by tissue eosinophils in endobronchial and esophageal biopsies of patients with asthma and eosinophilic esophagitis, respectively (see Chapter 12.3). Furthermore, as described in Chapter 12.2, eosinophil granule proteins have profibrotic and proangiogenic activity. The present subchapter will examine the potential role of tumor necrosis factor α (TNF-α) in eosinophil-associated tissue remodeling. This area is not well established and will be explored from the two points of view (depicted in Fig. 12.4.1): TNF-α derived from eosinophils can contribute to aspects of remodeling, particularly profibrotic events; and eosinophils can be activated by TNF-α to release factors associated with remodeling. Eosinophils are considered to be involved in tissue remodeling in a variety of diseases, most prominently in eosinophilic esophagitis and airway diseases such as chronic obstructive pulmonary disease, idiopathic pulmonary fibrosis, and systemic sclerosis. This subchapter will focus primarily on airway remodeling in asthma.

CHARACTERISTIC FEATURES OF TISSUE REMODELING

The concept of tissue remodeling is described in depth elsewhere in this publication. Briefly, in asthma, the term *remodeling* generally refers to the following features.

- Changes in airway epithelium, including epithelial cell detachment from the basement membrane and increased numbers of goblet cells and their secretion of mucus.
- Thickening of the subbasement membrane area (lamina reticularis) due to deposition of extracellular matrix proteins such as collagen type I, III and V, fibronectin, proteoglycans, and tenascin by myofibroblasts.
- Expanded smooth muscle mass attributed to both cellular proliferation (hyperplasia) and enlargement (hypertrophy).

- Modification of the pulmonary vascular system including formation of new vessels (angiogenesis), increased vessel mass, and microvascular permeability leading to edema within the airway wall.

For an excellent review of these events and the underlying cellular and molecular mechanism, see Halwani.[1] Of the characteristics of tissue remodeling, eosinophils are most likely to contribute to subepithelial fibrosis and angiogenesis.

EOSINOPHILS ARE INVOLVED IN TISSUE REMODELING

The concept that eosinophils contribute to tissue remodeling evolved from observations of the presence of eosinophils in specific tissues associated with fibrotic diseases such as asthma,[2] eosinophilic esophagitis,[3] hepatic fibrosis,[4] systemic sclerosis,[5] and idiopathic pulmonary fibrosis.[6] A causative role for eosinophils was initially suggested by mouse studies in which eosinophils were depleted by elimination of interleukin-5 (IL-5), a cytokine required for eosinophil maturation and release from the bone marrow. For the most part, subsequent generation of IL-5 overexpressing transgenic mice and mice genetically deficient of eosinophils substantiated the association of eosinophils and fibrogenesis. For example, Humbles and coworkers generated an eosinophil lineage-deficient mouse strain, ΔdblGATA, which, when sensitized and chronically challenged with ovalbumin, did not develop antigen-induced eosinophilia.[7] Compared to

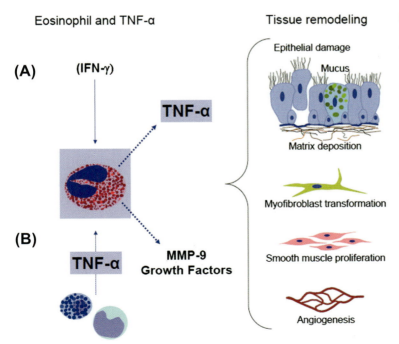

FIGURE 12.4.1 Tumor necrosis factor and eosinophils may function jointly to promote tissue remodeling. *A*, Eosinophil-derived tumor necrosis factor (TNF-α), released spontaneously or in response to factors such as interferon γ (IFN-γ), can directly influence tissue remodeling associated with epithelial cell damage and mucus production, myofibroblast differentiation and matrix generation, smooth muscle cell expansion, and vasculature changes such as angiogenesis. *B*, TNF-α derived from cells such as mast cells and monocytes can activate eosinophil generation of mediators, including matrix metalloproteinase 9 (MMP-9) and transforming growth factor β1 (TGF-β1) family growth factors that also contribute to changes in epithelial cells, matrix deposition, smooth muscle proliferation, and angiogenesis.

wild-type mice that do have an eosinophil response to antigen, ΔdblGATA mice have diminished subepithelial and submucosal fibrosis and a reduction in airway smooth muscle hyperplasia. Ochkur and colleagues developed a double transgenic mouse model in which IL-5 is expressed systemically from mature T cells and the eosinophil chemoattractant eotaxin-2/C-C motif chemokine 24 (CCL24) is localized to airway epithelial cells.[8] Interestingly, in the absence of antigen challenge, this mouse strain has airway eosinophilia and marked eosinophil degranulation with a corresponding increase in subepithelial collagen deposition, numbers of proliferating airway smooth cells, and mucus secretion. When these transgenic mice were crossed with the eosinophil-deficient PHIL mouse strain,[9] remodeling events were abolished. These studies demonstrate that, at least under certain conditions, eosinophils play a key role in tissue remodeling.

Utilization of anti-IL-5 antibody to reduce eosinophil numbers in humans has confirmed an association between eosinophils and tissue remodeling. In asthma, Flood-Page and colleagues[10] demonstrated that administration of a humanized anti-IL-5 monoclonal antibody ablated circulating eosinophils, but did not completely deplete eosinophils in bronchial tissues. The presence of eosinophils correlates with subepithelial deposition of lumican, procollagen type III, and tenascin in endobronchial biopsies.[10] The effect of anti-IL-5 treatment on remodeling in eosinophilic esophagitis has not been clarified. However, in a study of patients undergoing glucocorticoid treatment with budesonide, a subset of patients who showed decreased esophageal eosinophils had a concomitant reduction in fibrosis and vascular activation compared to patients who did not respond to therapy.[11]

Together, studies in humans and mice provide evidence that eosinophils are associated with airway remodeling, particularly the excessive deposition of extracellular matrix proteins. The mechanisms by which eosinophils contribute to fibrosis are not entirely clear, but they appear to involve eosinophil expression of TGF-β1 and, in some incidences, deposition of eosinophil granule proteins. Other mediators, such as TNF-α, have not been well characterized in these settings.

EOSINOPHIL-DERIVED MEDIATORS AND TISSUE REMODELING

Eosinophils have the potential to contribute to tissue remodeling through the release of mediators that activate stromal cells, such as airway smooth muscle, endothelial cells, fibroblasts, and goblet cells. For example, as discussed by Broide in Chapter 12.3, eosinophils are a major source of the prototypic profibrotic growth factor, TGF-β1.

The role of TGF-β1 in airway remodeling has been extensively discussed in a review by Makinde.[12] Briefly, TGF-β1 induces proliferation and synthesis of extracellular matrix synthesis by myofibroblasts and airway smooth muscle. As discussed by Bickham and Malter in Chapter 7.4, eosinophil TGF-β1 expression is post-transcriptionally regulated by a peptidyl-prolyl isomerase, Pin1. Pharmacological blockade of Pin1 in a rat model of antigen-induced airway inflammation reduced airway eosinophil TGF-β1 expression resulting in a concomitant reduction in peribronchial collagen deposition[13] and matrix metalloproteinase 9 (*Mmp9*) mRNA expression (S. Esnault, personal communication).

The role of eosinophils in other aspects of remodeling is less well defined. There is some evidence that eosinophils have the potential to stimulate mucus production.[9] Eosinophil-induced expression of mucin-5AC may occur through the release of TGF-α,[14,15] which is a ligand for epidermal growth factor receptors and is not a member of the TGF-β-superfamily.

As discussed in this publication, and as reviewed by Ribatti and coworkers,[15] eosinophils also release vascular endothelial growth factor (VEGF). In the context of tissue remodeling, VEGF is known principally for its effects on vascular endothelium and plays a key role in the induction of angiogenesis, vasodilation, and vascular permeability. VEGF also induces smooth muscle proliferation, stimulates mucus production, and amplifies T-helper type 2 (T$_h$2)-type responses through its effects on dendritic cells.

Finally, eosinophils synthesize and store TNF-α, which, as discussed below, can have direct effects on stromal cells and can activate eosinophils to produce factors that contribute to tissue remodeling.

EOSINOPHIL-DERIVED TNF-α AND TISSUE REMODELING

The proinflammatory cytokine TNF-α is produced as a transmembrane protein (mTNF-α). Soluble TNF-α is released following mTNF-α cleaved by the metalloproteinase ADAM17/TACE (TNF-α-converting enzyme). Both cell-associated mTNF-α and soluble (s) TNF-α are biologically active. Eosinophils express ADAM17 and release sTNF-α; however, the expression and function of mTNF-α on eosinophils is not well characterized. Using *in situ* hybridization, Costa and coworkers described the presence of *TNF* mRNA in unstimulated eosinophils from atopic and hypereosinophilic patients.[16] They also demonstrated the presence of TNF-α mRNA and protein in nasal polyps[16] and in colonic biopsies from patients with Crohn disease.[17] Extensive work by this group of investigators has clearly demonstrated that TNF-α is located in eosinophil-specific secondary granules[18] and can be spontaneously released upon *ex vivo* culture.[16] More recently, Spencer and colleagues[19] demonstrated that TNF-α protein is abundantly expressed in eosinophil lysates compared to cytokines such as IL-4 and IL-6. However, TNF-α is released in relatively small amounts, with a modest increase following eosinophil stimulation with interferon γ (IFN-γ).[19] Whether other factors, either alone or in combination, can trigger greater amounts of TNF-α has not been fully explored.

To date, there is no clear evidence that eosinophils directly contribute to tissue remodeling through the production of TNF-α. There are, however, reports that TNF-α is involved in remodeling, thus making it likely that eosinophil-derived TNF-α is a contributing factor. The role of TNF-α in lung fibrosis has been most extensively studied in mouse models of idiopathic pulmonary fibrosis (reviewed in 20). TNF-α expression in the lungs of these mice typically correlates with the degree of fibrosis, and elimination of TNF-α reduces fibrosis. *In vitro*, TNF-α can stimulate the proliferation of fibroblasts and airway smooth muscle and can stimulate differentiation of fibroblast into matrix-generating myofibroblasts by inducing TGF-β1 expression.[21,22] TNF-α can contribute to myofibroblast differentiation by promoting the transition of epithelial cells into myofibroblasts through a process known as epithelial—mesenchymal transition (EMT). EMT is delineated by the diminished expression of epithelial markers (cytokeratins and E-cadherin) and a corresponding acquisition of markers characteristic of myofibroblasts (alpha/aortic smooth muscle actin, collagen type I, fibronectin, and vimentin). TNF-α-induced EMT has been reported in the skin[23] and lung,[24] in which TNF-α can either induce EMT directly or augment TGF-β-induced EMT.[25]

EOSINOPHILS EXPRESS BIOLOGICALLY ACTIVE TNF RECEPTORS

Eosinophils express both types of TNF receptors:[26] TNF-RI (tumor necrosis factor receptor superfamily member 1A; CD120a/p55), which is the principal receptor for soluble TNF-α, and TNF-RII (tumor necrosis factor receptor superfamily member 1B; CD120b/p75a), which is the mTNF-α receptor. The most prominent signaling pathway induced by TNF-α is the nuclear factor κB (NF-κB) pathway. Intracellular NF-κB signaling involves the phosphorylation and subsequent degradation of the NF-κB-bound inhibitory molecule, IκBα, resulting in the release of the NF-κB heterodimer, allowing its translocation to the nucleus for induction of gene expression of a broad spectrum of proinflammatory cytokines/chemokines and adhesion molecules.[27]

TNF-α induces survival of eosinophils[28] and contributes to their migration by upregulating expression of adhesion molecules[29] and the formation of podosomes, which are dynamic adhesive structures that contain the metalloproteinase, ADAM8,[30] and are believed to be involved in eosinophil movement. TNF-α also induces eosinophil release of cytokines, such as IFN-γ, IL-4, and IL-6.[19] In addition to its singular effects on eosinophils, TNF-α synergizes with a variety of stimuli to enhance eosinophil function. For example, TNF-α interacts with fibronectin to promote eosinophil survival.[31] In addition, we have demonstrated that TNF-α is essential for IFN-γ-induced secretion of T_h1 chemokines (C-X-C motif chemokine 10 (CXCL-10)/IP-10 and CXCL-9/MIG) by highly purified human eosinophils and that it enhances IL-4-induced generation of T_h2 chemokines (CCL-17/TARC and CCL-22/MDC).[32]

TNF-α INDUCES EOSINOPHIL PRODUCTION OF REMODELING FACTORS

In addition to the direct effects of eosinophil-derived TNF-α on tissue remodeling, TNF-α can activate eosinophil functions that promote features of remodeling (Fig. 12.4.1B). Two principal factors will be discussed: MMP-9 and a member of the TGF-β superfamily, activin A. Both factors are present in the airway of asthmatics[33–35] and we have established that MMP-9 and activin are actively synthesized and secreted in abundant quantities by human eosinophils.

Ex vivo, eosinophils can release MMP-9 and migrate in response to 5-oxo-eicosatetraenoic acid,[36] complement factor C5a,[37] and platelet-activating factor.[38] TNF-α has been reported to induce the rapid release of MMP-9[39] through a p38 mitogen-activated protein kinase (MAPK)-dependent process and active synthesis of MMP-9[39,40] that requires NF-κB signaling. In light of the fact that TNF-α can provide potent synergistic signals through NF-κB-related mechanisms,[32] we evaluated the effect of T_h1, T_h2, and common β chain [granulocyte-macrophage colony-stimulating factor (GM-CSF), IL-3, and IL-5] cytokines on TNF-α-induced synthesis and release of MMP-9.[41] We found that neither IL-4 nor IFN-γ has an effect on TNF-α-induced eosinophil MMP-9. However, the βc-chain family family of cytokines interacted synergistically with TNF-α to promote eosinophil generation of profuse amounts of MMP-9.[2] Although all three β chain signaling cytokines have a striking synergistic effect with TNF-α, IL-3 is the most potent. The synergy between TNF-α and IL-3 is seen at the level of mRNA expression and is due, at least in part, to increased stabilization of MMP-9 mRNA.[42] Both NF-κB and the MAP kinase pathways involving extracellular signal-regulated kinases (ERKs) 1 and 2 and p38 but not c-Jun N-terminal kinase (JNK) are required for TNF-α/IL-3-induced MMP-9. Studies are ongoing to determine how eosinophil-derived MMP-9 is regulated post-transcriptionally and to determine whether IL-3 alters TNF-α-induced trafficking of proMMP-9 to the cell surface.

Similar to MMP-9, we have shown that activin A synthesis and secretion is induced by TNF-α in combination with IL-3.[43] However, in contrast to MMP-9, neither IL-5 nor GM-CSF significantly augments TNF-α-induced activin A (Fig. 12.4.2). This observation is somewhat surprising considering that signal transduction from the IL-3, IL-5, and GM-CSF receptors is commonly believed to be mediated only through βc (CD131) and the three cytokines tend to have similar effects on eosinophils.[44] IL-3 and TNF-α synergistically increase steady-state levels of activin A mRNA and prolonged mRNA stabilization. Studies are ongoing to determine the specific mechanisms contributing to TNF-α/IL-3-induced activin A mRNA stabilization in eosinophils.

POTENTIAL ROLES OF MMP-9 AND TGF-B-FAMILY GROWTH FACTORS IN AIRWAY REMODELING

By definition, remodeling involves *reorganization or renovation of existing tissues*. In order to be *remodeled*, tissue degradation or injury must occur first. MMPs are key players in this process and their release often precedes tissue remodeling, particularly in fibrotic diseases. In this setting, MMPs function to degrade the matrix and set the stage for deposition of extracellular matrix proteins, which often leads to disorganized tissue renovation. As previously discussed, eosinophils express MMP-9, which has specific proteolytic activity for denatured collagen and collagen type IV of the basement membrane.[45] However, in addition to its role in tissue injury, MMP-9 can contribute to tissue remodeling by releasing and activating growth factors that are sequestered in the matrix. MMP-9 can release TGF-β1 from the extracellular matrix by degrading matrix-bound latent TGF-β-binding

FIGURE 12.4.2 TNF-α synergizes with common β chain cytokines to induce secretion of pro-matrix metalloproteinase 9 and activin A from human blood eosinophils. Eosinophils from four atopic subjects were cultured for 72 h in medium alone or in the presence of 10 ng/mL of a common β chain cytokine (for GM-CSF, IL-3, or IL-5) ± TNF-α (10 ng/mL). Pro-matrix metalloproteinase 9 (pro-MMP-9) and activin A were measured by enzyme-linked immunosorbent assay. Data are expressed as means ± SE; $p < 0.05$, versus *medium, †TNF-α, or ‡respective single cytokine. GM-CSF, granulocyte-macrophage colony-stimulating factor; IL-3, interleukin-3; IL-5, interleukin-5; O.D., optical density; TNF-α, tumor necrosis factor.

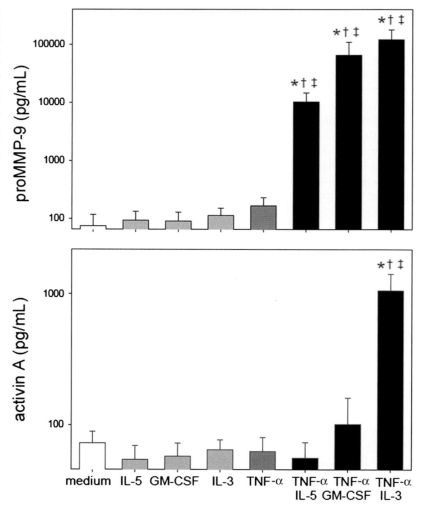

protein (LTBP-1).[46] Interestingly, MMP-9 can function in a noncatalytic manner by docking to cell surface molecules via its hemopexin domain. MMP-9 can bind to cell surface CD44 and then proteolytically activate TGF-β.[47] Additional *docking* molecules include α4β1 integrin; MMP-9-binding to these receptors can induce intracellular signaling.[48] These activities have been demonstrated primarily in cancer cells and require confirmation in other cell types. Studies in MMP-9 knockout mice have led to more questions than answers in terms of how MMP-9 affects antigen-induced airway fibrosis. Using repetitive antigen challenge to induce features of airway remodeling, Lim and coworkers[49] demonstrated modest but significant decreases in airway eosinophils and peribronchial fibrosis, but no changes in total TGF-β1 levels in MMP-9-knockout compared to wild-type mice. However, whether the decreased fibrosis was due to a reduction in MMP-9-mediated recruitment of airway inflammatory eosinophils and/or a decreased in MMP-9-mediated activation of TGF-β1 is not clear.

In contrast to MMP-9, which functions to degrade extracellular matrix and increase the bioavailability of profibrotic factors, activin A can directly activate stromal cells and regulate inflammation. Of note, activin A can activate fibroblast generation of TGF-β1 and, conversely, TGF-β1 can induce activin A.[50] Using primary bronchial fibroblasts from subjects with asthma, we have noted that activin A can induce fibronectin and VEGF generation in the absence of fibroblast proliferation (Fig. 12.4.3).

Activin A is a member of the TGF-β superfamily, meaning that it is structurally similar to TGF-β1 and signals through the common SMAD2/3 (mothers against decapentaplegic homologues 2 and 3) pathway. However, TGF-β1 and activin A bind to unique receptors comprised of different ligand-specific type I receptors and different type II serine/threonine kinase signaling receptors. Kariyawasam and colleagues demonstrated increased expression of SMAD2 and activin A type I and type II receptors in endobronchial biopsies from allergic asthma subjects following allergen inhalation.[51] In a mouse study of acute antigen-induced airway inflammation, bronchoalveolar lavage fluid levels of activin A increased in association with Th2-type responses.[50] Administration of follistatin, a natural inhibitor with specificity for activin A, reduced airway inflammation and mucus production.[50] Strikingly, follistatin diminished collagen deposition in a mouse model of bleomycin-induced airway fibrosis.[52] Taken together with our observations that TNF-α in combination with IL-3 is a potent inducer of eosinophil MMP-9 and activin A, these studies raise the possibility that eosinophils are involved in tissue remodeling through TNF-α-mediated activation of eosinophil-derived factors.

BLOCKING TNF-α *IN VIVO*

Several TNF-α-blocking agents have been developed for therapeutic use, including a decoy receptor, etanercept, which binds sTNF-α, but not mTNF-α, and specific monoclonal antibodies, infliximab, adalimumab, and certolizumab pegol, which inhibit the binding of both sTNF-α and mTNF-α to TNF receptors. These reagents have shown some benefit in treatment of diseases such as Crohn disease, juvenile arthritis, psoriasis, rheumatoid arthritis, and sarcoidosis; however, studies in asthma have not been encouraging (reviewed in[53]) and have recently

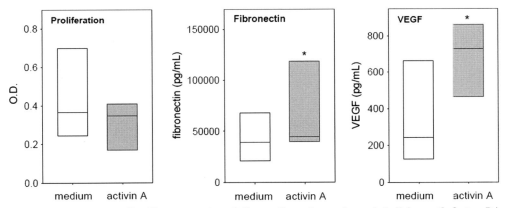

FIGURE 12.4.3 Activin A induces lung fibroblast generation of fibronectin and vascular endothelial growth factor. Primary human lung fibroblasts were obtained from endobronchial biopsies from four subjects with asthma. After serum starvation for 24 h to induce quiescence, cells were cultured in the presence of recombinant activin A (50 ng/mL) for an additional 5 days. Proliferation/survival was assessed by metabolism of 3-(4,5-dimethylthiazol-2-yl)-2,5-diphenyltetrazolium bromide (MTT); fibronectin and vascular endothelial growth factor (VEGF) were measured by enzyme-linked immunosorbent assay. Data are expressed as medians within 25% and 75% quartiles; *$p < 0.05$ compared to medium alone. O.D., optical density.

been halted due to increased incidence of respiratory severe adverse events in a large, multicenter trial. The primary outcome variables of these preliminary trials were treatment efficacy. Secondary outcome variables have not been widely published; in particular, the effect of anti-TNF-α on eosinophil function has not reported in human trials or murine studies.

To more clearly define the specific contribution of eosinophils to airway remodeling, particularly, the role of eosinophil-derived MMP-9 and activin A, we have initiated a study using anti-IL-5 (mepolizumab) as a tool to prevent allergen-induced airway eosinophilia in subjects with allergic asthma. This ongoing study should shed light on the effect of eosinophil depletion on allergen-induced initiation of factors such as MMP-9 and TGF-β family growth factors, which are associated with airway remodeling.

CONCLUSION

In summary, although eosinophil-derived TGF-β1 and specific granule proteins have received much attention as contributors to eosinophil-mediated tissue remodeling, the role of TNF-α should not be discounted. TNF-α and eosinophils may function jointly to promote tissue remodeling. As depicted in Fig. 12.4.1, eosinophil-derived TNF-α can directly affect stromal cells to influence remodeling events and TNF-α can activate eosinophil generation of factors such as MMP-9 and TGF-β1-family growth factors, which contribute to remodeling. This is an area of active ongoing investigation. It will be important to demonstrate the presence of TNF-α in association with eosinophil expression of MMP-9 and TGF-β-family growth factors in tissues from patients with diseases such as asthma, eosinophilic esophagitis, malignancies, and perhaps in healthy tissues, such as the endometrial lining of the uterus, which are associated with resident eosinophils.

ACKNOWLEDGEMENTS

The thoughtful comments and critical reading of this work by Mats Johansson, PhD, and Stephane Esnault, PhD, are greatly appreciated. Dr. Kelly is supported in part by NIH grants HL088594, HL080412, and HL069116.

REFERENCES

1. Halwani R, Al Muhsen S, Hamid Q. Airway remodeling in asthma. *Curr Opin Pharmacol* 2010;**10**:236—45.
2. Venge P. The eosinophil and airway remodelling in asthma. *Clin Respir J* 2010;**4**(Suppl. 1):15—9.
3. Aceves SS, Ackerman SJ. Relationships between eosinophilic inflammation, tissue remodeling, and fibrosis in eosinophilic esophagitis. *Immunol Allergy Clin North Am* 2009;**29**:197-xiv.
4. Bataller R, Brenner DA. Liver fibrosis. *J Clin Invest* 2005;**115**:209—18.
5. Atamas SP, White B. Cytokine regulation of pulmonary fibrosis in scleroderma. *Cytokine Growth Factor Rev* 2003;**14**:537—50.
6. Noble PW, Homer RJ. Idiopathic pulmonary fibrosis: new insights into pathogenesis. *Clin Chest Med* 2004;**25**:749—58. vii.
7. Humbles AA, Lloyd CM, McMillan SJ, Friend DS, Xanthou G, McKenna EE, et al. A critical role for eosinophils in allergic airways remodeling. *Science* 2004;**305**:1776—9.
8. Ochkur SI, Jacobsen EA, Protheroe CA, Biechele TL, Pero RS, McGarry MP, et al. Coexpression of IL-5 and eotaxin-2 in mice creates an eosinophil-dependent model of respiratory inflammation with characteristics of severe asthma. *J Immunol* 2007;**178**:7879—89.
9. Lee JJ, Dimina D, Macias MP, Ochkur SI, McGarry MP, O'Neill KR, et al. Defining a link with asthma in mice congenitally deficient in eosinophils. *Science* 2004;**305**:1773—6.
10. Flood-Page P, Menzies-Gow A, Phipps S, Ying S, Wangoo A, Ludwig MS, et al. Anti-IL-5 treatment reduces deposition of ECM proteins in the bronchial subepithelial basement membrane of mild atopic asthmatics. *J Clin Invest* 2003;**112**:1029—36.
11. Aceves SS, Newbury RO, Chen D, Mueller J, Dohil R, Hoffman H, et al. Resolution of remodeling in eosinophilic esophagitis correlates with epithelial response to topical corticosteroids. *Allergy* 2010;**65**:109—16.
12. Makinde T, Murphy RF, Agrawal DK. The regulatory role of TGF-beta in airway remodeling in asthma. *Immunol Cell Biol* 2007;**85**:348—56.
13. Shen ZJ, Esnault S, Rosenthal LA, Szakaly RJ, Sorkness RL, Westmark PR, et al. Pin1 regulates TGF-beta1 production by activated human and murine eosinophils and contributes to allergic lung fibrosis. *J Clin Invest* 2008;**118**:479—90.
14. Burgel PR, Lazarus SC, Tam DC, Ueki IF, Atabai K, Birch M, et al. Human eosinophils induce mucin production in airway epithelial cells via epidermal growth factor receptor activation. *J Immunol* 2001;**167**:5948—54.
15. Ribatti D, Puxeddu I, Crivellato E, Nico B, Vacca A, Levi-Schaffer F. Angiogenesis in asthma. *Clin Exp Allergy* 2009;**39**:1815—21.
16. Costa JJ, Matossian K, Resnick MB, Beil WJ, Wong DT, Gordon JR, et al. Human eosinophils can express the cytokines tumor necrosis factor-alpha and macrophage inflammatory protein-1 alpha. *J Clin Invest* 1993;**91**:2673—84.
17. Beil WJ, Weller PF, Peppercorn MA, Galli SJ, Dvorak AM. Ultrastructural immunogold localization of subcellular sites of TNF-alpha in colonic Crohn's disease. *J Leukoc Biol* 1995;**58**:284—98.
18. Beil WJ, Weller PF, Tzizik DM, Galli SJ, Dvorak AM. Ultrastructural immunogold localization of tumor necrosis factor-α to the matrix compartment of eosinophil secondary granules in patients with idiopathic hypereosinophilia syndrome. *J Histochem Cytochem* 1993;**41**:1611—5.
19. Spencer LA, Szela CT, Perez SA, Kirchhoffer CL, Neves JS, Radke AL, et al. Human eosinophils constitutively express multiple Th1, Th2, and immunoregulatory cytokines that are secreted rapidly and differentially. *J Leukoc Biol* 2009;**85**:117—23.

20. Moore BB, Hogaboam CM. Murine models of pulmonary fibrosis. *Am J Physiol Lung Cell Mol Physiol* 2008;**294**:L152–60.

21. Amrani Y, Panettieri Jr RA, Frossard N, Bronner C. Activation of the TNF alpha-p55 receptor induces myocyte proliferation and modulates agonist-evoked calcium transients in cultured human tracheal smooth muscle cells. *Am J Respir Cell Mol Biol* 1996;**15**:55–63.

22. Sullivan DE, Ferris M, Nguyen H, Abboud E, Brody AR. TNF-alpha induces TGF-beta1 expression in lung fibroblasts at the transcriptional level via AP-1 activation. *J Cell Mol Med* 2009;**13**:1866–76.

23. Yan C, Grimm WA, Garner WL, Qin L, Travis T, Tan N, et al. Epithelial to mesenchymal transition in human skin wound healing is induced by tumor necrosis factor-alpha through bone morphogenic protein-2. *Am J Pathol* 2010;**176**:2247–58.

24. Borthwick LA, Parker SM, Brougham KA, Johnson GE, Gorowiec MR, Ward C, et al. Epithelial to mesenchymal transition (EMT) and airway remodelling after human lung transplantation. *Thorax* 2009;**64**:770–7.

25. Camara J, Jarai G. Epithelial-mesenchymal transition in primary human bronchial epithelial cells is Smad-dependent and enhanced by fibronectin and TNF-alpha. Fibrogenesis. *Tissue Repair* 2010;**3**:2.

26. Matsuyama G, Ochiai K, Ishihara C, Kagami M, Tomioka H, Koya N. Heterogeneous expression of tumor necrosis factor-alpha receptors I and II on human peripheral eosinophils. *Int Arch Allergy Immunol* 1998;**117**(Suppl. 1):28–33.

27. Gupta S. A decision between life and death during TNF-alpha-induced signaling. *J Clin Immunol* 2002;**22**:185–94.

28. Temkin V, Levi-Schaffer F. Mechanism of tumour necrosis factor alpha mediated eosinophil survival. *Cytokine* 2001;**15**:20–6.

29. Hansel TT, DeVries JM, Carballido JM, Braun RK, Carballido-Perrig N, Rihs S, et al. Induction and function of eosinophil intercellular adhesion molecule-1 and HLA-DR. *J Immunol* 1992;**149**:2130–6.

30. Johansson MW, Lye MH, Barthel SR, Duffy AK, Annis DS, Mosher DF. Eosinophils adhere to vascular cell adhesion molecule-1 via podosomes. *Am J Respir Cell Mol Biol* 2004;**31**:413–22.

31. Esnault S, Malter JS. Hyaluronic acid or TNF-alpha plus fibronectin triggers granulocyte macrophage-colony-stimulating factor mRNA stabilization in eosinophils yet engages differential intracellular pathways and mRNA binding proteins. *J Immunol* 2003;**171**:6780–7.

32. Liu LY, Bates ME, Jarjour NN, Busse WW, Bertics PJ, Kelly EA. Generation of Th1 and Th2 chemokines by human eosinophils: evidence for a critical role of TNF-alpha. *J Immunol* 2007;**179**:4840–8.

33. Ohno I, Ohtani H, Nitta Y, Suzuki J, Hoshi H, Honma M, et al. Eosinophils as a source of matrix metalloproteinase-9 in asthmatic airway inflammation. *Am J Respir Cell Mol Biol* 1997;**16**:212–9.

34. Dahlen B, Shute J, Howarth P. Immunohistochemical localisation of the matrix metalloproteinases MMP-3 and MMP-9 within the airways in asthma. *Thorax* 1999;**54**:590–6.

35. Semitekolou M, Alissafi T, Aggelakopoulou M, Kourepini E, Kariyawasam HH, Kay AB, et al. Activin-A induces regulatory T cells that suppress T helper cell immune responses and protect from allergic airway disease. *J Exp Med* 2009;**206**:1769–85.

36. Langlois A, Chouinard F, Flamand N, Ferland C, Rola-Pleszczynski M, Laviolette M. Crucial implication of protein kinase C (PKC)-delta, PKC-zeta, ERK-1/2, and p38 MAPK in migration of human asthmatic eosinophils. *J Leukoc Biol* 2009;**85**:656–63.

37. DiScipio RG, Schraufstatter IU, Sikora L, Zuraw BL, Sriramarao P. C5a mediates secretion and activation of matrix metalloproteinase 9 from human eosinophils and neutrophils. *Int Immunopharmacol* 2006;**6**:1109–18.

38. Okada S, Kita H, George TJ, Gleich GJ, Leiferman KM. Migration of eosinophils through basement membrane components in vitro: role of matrix metalloproteinase-9. *Am J Respir Cell Mol Biol* 1997;**17**:519–28.

39. Wiehler S, Cuvelier SL, Chakrabarti S, Patel KD. p38 MAP kinase regulates rapid matrix metalloproteinase-9 release from eosinophils. *Biochem Biophys Res Commun* 2004;**315**:463–70.

40. Schwingshackl A, Duszyk M, Brown N, Moqbel R. Human eosinophils release matrix metalloproteinase-9 on stimulation with TNF-alpha. *J Allergy Clin Immunol* 1999;**104**:983–9.

41. Liu LY, Sedgwick JB, Jarjour NN, Kelly EA. Potential role of IL-3 and TNF-α in airway inflammation and remodeling: induction of eosinophil MMP-9 generation. *Am J Respir Crit Care Med* 2007;**177**:A30.

42. Kelly EA, Liu LY, Esnault S, Quinchia Johnson BH, Jarjour NN. Potent synergistic effect of IL-3 and TNF on matrix metalloproteinase 9 generation by human eosinophils. *Cytokine* 2012;**58**:199–206.

43. Liu LY, Kelly EA, Jarjour NN. Activin A is synthesized by human eosinophils and is selectively induced by IL-3 in combination with TNF-α. *Am J Respir Crit Care Med* 2010;**181**:A2799.

44. Martinez-Moczygemba M, Huston DP. Biology of common beta receptor-signaling cytokines: IL-3, IL-5, and GM-CSF. *J Allergy Clin Immunol* 2003;**112**:653–65.

45. Kelly EA, Jarjour NN. Role of MMPs in asthma. *Curr Opin Pulm Med* 2003;**9**:28–33.

46. Dallas SL, Rosser JL, Mundy GR, Bonewald LF. Proteolysis of latent transforming growth factor-beta (TGF-beta)-binding protein-1 by osteoclasts. A cellular mechanism for release of TGF-beta from bone matrix. *J Biol Chem* 2002;**277**:21352–60.

47. Yu Q, Stamenkovic I. Cell surface-localized matrix metalloproteinase-9 proteolytically activates TGF-beta and promotes tumor invasion and angiogenesis. *Genes Dev* 2000;**14**:163–76.

48. Redondo-Munoz J, Ugarte-Berzal E, Terol MJ, Van den Steen PE, Hernandez DC, Roderfeld M, et al. Matrix metalloproteinase-9 promotes chronic lymphocytic leukemia b cell survival through its hemopexin domain. *Cancer Cell* 2010;**17**:160–72.

49. Lim DH, Cho JY, Miller M, McElwain K, McElwain S, Broide DH. Reduced peribronchial fibrosis in allergen-challenged MMP-9-deficient mice. *Am J Physiol Lung Cell Mol Physiol* 2006;**291**:L265–71.

50. Hardy CL, O'Connor AE, Yao J, Sebire K, de Kretser DM, et al. Follistatin is a candidate endogenous negative regulator of activin A in experimental allergic asthma. *Clin Exp Allergy* 2006;**36**:941–50.

51. Kariyawasam HH, Pegorier S, Barkans J, Xanthou G, Aizen M, Ying S, et al. Activin and transforming growth factor-beta signaling pathways are activated after allergen challenge in mild asthma. *J Allergy Clin Immunol* 2009;**124**:454–62.

52. Aoki F, Kurabayashi M, Hasegawa Y, Kojima I. Attenuation of bleomycin-induced pulmonary fibrosis by follistatin. *Am J Respir Crit Care Med* 2005;**172**:713–20.

53. Dimov VV, Casale TB. Immunomodulators for Asthma. *Allergy Asthma Immunol Res* 2010;**2**:228–34.

Chapter 12.5

Esophageal Remodeling and Pediatric Eosinophilic Esophagitis

Seema S. Aceves

INTRODUCTION

The concept of tissue remodeling associated with allergic eosinophilic inflammation stems largely from studies in asthmatic patients.[1] Asthmatic airway remodeling consists of a number of histological features, including mucous gland metaplasia, subepithelial fibrosis, angiogenesis, and smooth muscle hypertrophy and hyperplasia. The clinical impact and natural history of these changes are difficult to assess in asthmatic children due to a paucity of airway biopsy specimens.[2] In contrast, eosinophilic esophagitis (EoE) is a new and increasingly well-described disease associated with food and aeroallergen triggers that requires repeated biopsy procurement for diagnosis and management.[3]

EoE is a clinicopathological entity defined as an acid independent pan-esophageal eosinophilic inflammation of >15 eosinophils/high power field (hpf) on hematoxylin and eosin stain at $400 \times$ light microscopy.[3] Typical accompanying histological features include eosinophil degranulation and clusters, as well as epithelial changes in basal zone hyperplasia and dilated intercellular spaces (Fig. 12.5.1). The pediatric symptom profile shifts with age from vomiting and failure to thrive to abdominal pain, dysphagia, food impactions, and stricture formation. Typical endoscopic EoE findings include concentric rings, esophageal pallor, linear furrows, white plaques, and at its most severe, strictures and narrowings. EoE remission and recrudescence following therapy necessitates repeated esophageal biopsy due to the current lack of surrogate disease markers. As such, EoE affords a unique opportunity to study the mechanisms, pathogenesis, and clinical impacts of eosinophil-associated esophageal remodeling in children.

MECHANISMS OF PEDIATRIC ESOPHAGEAL REMODELING

Epithelium

The normal esophageal epithelium contains a basal zone that comprises <20% of the total epithelial height and provides a constant renewal source for epithelial cells. In EoE, the epithelium becomes hyperplastic due to active proliferation of basal cells and >75% of the epithelial

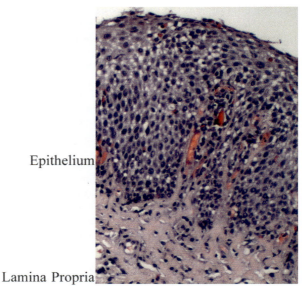

FIGURE 12.5.1 Representative image of an EoE biopsy showing basal zone hyperplasia, dilated intercellular spaces, and epithelial eosinophils in the epithelium. Eosinophils as well as thickened collagen bundles are present in the lamina propria.

height can consist of the basal zone[4,5] (Fig. 12.5.1). Eosinophil accumulation and activation in the esophagus can have direct effects on epithelial proliferation. Major basic protein (MBP) stimulates the production of fibroblast growth factor 9 (FGF-9), which in turn causes proliferation of the esophageal epithelial cell line, HET1A.[6] Consistent with a potential role in EoE pathogenesis, increased levels of FGF-9 are seen in the esophageal epithelium of pediatric EoE biopsies. Mice deficient in eosinophils due to mutations in the *Gata1* promoter or in interleukin-5 (IL-5) have significantly reduced basal zone thickness, further supporting the role of eosinophils in enhancing epithelial proliferation.[7] Although a pathogenic role for mucous production in EoE has not been described, similar to asthmatic epithelium, the esophageal epithelium has increased transcription of the *mucin 5AC* gene.

Lamina Propria

Fibrosis

A significant amount of tissue remodeling takes place in the subepithelial lamina propria (LP).[8,9] In non-diseased pediatric esophageal biopsy specimens, this area is relatively devoid of inflammatory cells and consists of a fine collagen network. Two patterns of collagen fibers have been described in nondiseased pediatric patients. The first is a lacy, reticular pattern and the second consists of coarser but still individually identifiable collagen fibers.[9] Importantly, there is no variation in collagen pattern by esophageal level, esophageal depth, or patient age. Therefore, LP fibrosis is neither a normal finding nor a usual consequence of the aging process.

In EoE, the LP often becomes fibrotic with increased collagen bundle width and tightly packed fibers (Fig. 12.5.1). This thickened collagen may serve as a structural trap for inflammation and both increased eosinophil density and free granules are found in the LP.[10] Among pediatric EoE patients, more than 50—76% have LP fibrosis, compared with children without EoE.[9,11] Adequate LP, defined in our laboratory as at least 3—5 hpf at 400× light microscopy, is present in approximately 40% of biopsies. As such, a thorough evaluation of esophageal LP findings requires a systemic analysis of a large number of biopsy specimens.

Although first described in adult patients with long-standing, stricture-associated EoE, much of our current understanding of EoE-induced tissue remodeling stems from pediatric studies.[7—9] IL-5 is increased in the esophagus of pediatric EoE subjects compared with non-diseased controls.[7] Animal model studies show that IL-5-deficient mice have diminished esophageal eosinophil accumulation and collagen deposition following allergen challenge, demonstrating that local IL-5 production is important for esophageal eosinophil trafficking and remodeling.[7] IL-13 is also increased in esophageal biopsy specimens from children with EoE.[12] Pulmonary IL-13 overexpression in mice causes significantly increased epithelial hyperplasia and LP collagen deposition.[13] This effect of IL-13 is eosinophil-independent, since eosinophil-deficient mice also have increased epithelial and subepithelial esophageal remodeling.

Degranulated eosinophils and mast cells are higher in EoE patients than in biopsies from either patients with reflux esophagitis or control subjects and the presence of eosinophil degranulation correlates with the severity of fibrosis.[4,5,14,15] IL-5 and eotaxin-3/C-C motif chemokine 26 (CCL26) are likely to be important and concerted stimulators of esophageal eosinophil activation and degranulation as both are elevated in pediatric EoE specimens.[7,16] Esophageal eosinophil and mast cell products include the profibrotic factor, transforming growth factor-β1 (TGF-β1). TGF-β1 transcript and protein is increased in pediatric EoE patients, suggesting that TGF-β1 provides an important pathway for fibrosis in EoE.[7,8] Consistent with this concept, pediatric EoE specimens have increased numbers of cells expressing the TGF-β1 heteromeric nuclear signaling protein, phosphorylated mothers against decapentaplegic homolog 2 and 3 (SMAD2/3), in the LP compared with healthy or reflux esophagitis subjects.[8]

TGF-β1 treatment of cultured esophageal fibroblasts causes increased release of the extracellular matrix protein periostin.[18] Periostin increases adhesion of eosinophils to the extracellular matrix loop. In this scenario, eosinophil- and mast cell-derived TGF-β1 released in the presence of T-helper type 2 (Th2) interleukins, such as IL-5 and IL-13, can promote extracellular matrix deposition and subsequent eosinophil accumulation, thus creating a positive feedback loop for perpetuating inflammation (Fig. 12.5.2).

As such, eosinophils and TGF-β1 function as both potential instigators and amplifiers of fibrosis in EoE.

Angiogenesis and Vascular Activation

Pediatric EoE patients have elevated numbers of LP blood vessels compared with subjects with reflux esophagitis or those with no esophageal pathology.[8] Few studies have documented the mechanisms by which this might occur but transgenic mice overexpressing respiratory Clara cell-derived IL-13 have increased numbers of vessels and increased esophageal circumference compared to wild-type mice.[13]

Th2-associated cytokines, such as IL-4 and tumor necrosis factor-α (TNF-α), induce vascular activation, defined as increased expression of adhesion molecules.[19] Children with EoE have increased expression of the endothelial activation marker, vascular cell adhesion molecule 1 (VCAM-1), compared with either nondiseased or reflux esophagitis subjects.[8] Increased expression of VCAM-1 provides a mechanism for eosinophil infiltration into the esophagus by allowing increased tethering of inflammatory cells to the endothelium of esophageal vessels (Fig. 12.5.2). The dilated intercellular spaces seen in EoE patient biopsies could reflect the presence of vascular leak with intercellular edema, a phenomenon that would be enhanced by adhesion molecules such as VCAM-1.[3,20]

Smooth Muscle

Due to the difficulty in obtaining esophageal biopsy samples that contain smooth muscle, there is a paucity of human data available on this important tissue compartment. However, endoscopic ultrasonographic studies in children have shown that the esophageal wall is thickened through the longitudinal and circular muscle layers of the esophagus in EoE subjects compared with nondiseased control subjects, suggesting that smooth muscle hypertrophy is prominent in EoE.[22,23] Diffuse esophageal thickening is found through the distal and proximal esophagus both in the mucosal layer and in the muscularis propria, consistent with the pan-esophageal distribution of EoE.[3,23] Significant smooth muscle dysmotility has been reported in EoE, with increased contractility, decreased relaxation, and loss of coordinated contraction of the concentric and longitudinal muscles.[25—28] This dysmotility may be a functional parallel to airway hyperresponsiveness in asthmatics.

Recent studies using pediatric biopsy specimens have demonstrated that the smooth muscle in EoE patients is infiltrated by tryptase+ mast cells that coexpress TGF-β1, in contrast to nondiseased control patients.[17] TGF-β1 stimulates esophageal smooth muscle cell contraction in vitro.[17] These data suggest a mechanism whereby inflammatory cell products influence smooth muscle function in EoE. Although eosinophils are also increased in the esophageal muscularis mucosa in EoE, their numbers are significantly lower than mast cells.[17]

CLINICAL IMPLICATIONS OF ESOPHAGEAL REMODELING IN EOE

Strictures

The most severe complication of EoE is esophageal stricture formation.[3] Once formed, strictures often require repeated dilation both in pediatric and adult EoE patients.[10,29–30] Pediatric EoE patients with long-standing or stricture-associated EoE often have the most severe features of remodeling.[8] Children with long-standing or stricture-associated EoE have elevated fibrosis scores, measured by either trichrome or hematoxylin/eosin staining, and the highest numbers of TGF-β1[+] cells, compared with normal control subjects, subjects with acid induced esophagitis, or pediatric EoE patients who do not have strictures.[8,21] The exact relationship between eosinophils, fibrosis, and progression to strictures remains to be clearly delineated. However, since the LP eosinophil appears to be an important source of TGF-β1, which has been linked to both fibrosis and strictures, therapies that reduce the numbers of LP eosinophils may be useful for controlling or reversing fibrosis.[21]

Dysphagia and Dysmotility

Dysphagia is a prominent complaint in school-aged and adolescent EoE patients.[3,31] Retrospective studies show that 42% of pediatric EoE patients with LP fibrosis complain of dysphagia and that dysphagia does not occur in the absence of fibrosis.[9] In addition, one study reported that 80% of fibrotic patients with dysphagia have food impactions.[9]

Although the clinical complaints in acid versus EoE are largely overlapping, clinical dysphagia and anorexia/early satiety can distinguish some children with EoE from those with reflux esophagitis.[32] An epithelial histology score that encompasses the remodeling features of dilated intercellular spaces and basal zone hyperplasia, as well as eosinophil features of clusters, degranulation state, and numbers correlates with endoscopic features of furrows/thickening, a potential macroscopic reflection of esophageal remodeling (Fig. 12.5.2). The only clinical features that correlate with the epithelial score are dysphagia and anorexia/early satiety. An LP score that accounts for both eosinophil numbers and fibrosis severity correlates specifically with furrow/thickening and dysphagia.[32] As such, both eosinophilic inflammation and remodeling can associate with clinical complaints and endoscopic findings in a subset of EoE children (Fig. 12.5.2).

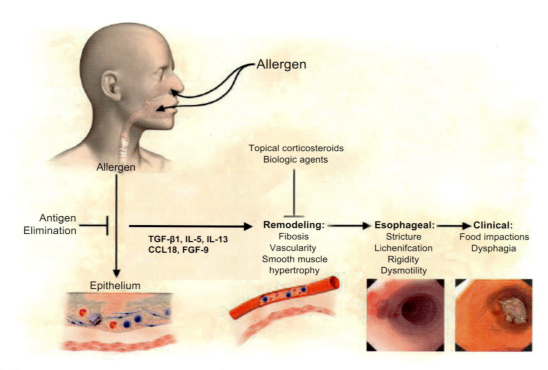

FIGURE 12.5.2 Inhaled or ingested allergen instigates an immune response involving pro-remodeling interleukins, chemokines, and growth factors. These factors promote subepithelial fibrosis, vascularity and smooth muscle changes that potentially lead to esophageal dysfunction and clinical outcomes of strictures and food impactions. CCL18, C-C motif chemokine 18; FGF-9, fibroblast growth factor receptor 9; IL-5, interleukin-5; IL-13, interleukin-13; TGF-β1, transforming growth factor β1. (*Reproduced with permission with modifications.[38]*)

Esophageal dysmotility has been noted in children with EoE and correlates with episodes of clinical dysphagia.[26] Since dysphagia correlates with LP and epithelial remodeling features, and since mast cell and eosinophil products affect smooth muscle cell contraction *in vitro*, it is possible to imagine a scenario whereby eosinophils, mast cells, and tissue remodeling can have significant clinical impacts in EoE (Fig. 12.5.2).

TREATMENT OF ESOPHAGEAL REMODELING IN EOE

EoE treatment modalities include elimination diets and swallowed topical corticosteroids.[3,33] Treatment with swallowed viscous budesonide or with swallowed fluticasone reduces the severity of fibrosis in children who have EoE resolution in the epithelial mucosa, as defined by decreased epithelial eosinophilia.[21,34] A combination of fluticasone and dietary elimination can reduce fibrosis scores in pediatric patients to the level of non-diseased control subjects.[11] However, the use of dietary elimination alone may not be as effective in reducing fibrosis, despite the resolution of epithelial eosinophilia.[34]

Topical viscous budesonide treatment in pediatric EoE reduces fibrosis as well as the numbers of TGF-β1 and pSMAD2/3$^+$ cells in the LP.[21,35] In addition, there is decreased expression of VCAM-1 and a reduction in intracellular space dilation.[21] However, decreased tissue remodeling occurs only in children with EoE who respond to therapy in the epithelial compartment, defined as a decrease of <7 eosinophils/hpf. In patients with continued EoE following therapy, there is continued, and sometimes progressive, esophageal remodeling.[21,35] The genetic and molecular differences that predict a clinical phenotype of corticosteroid responders versus nonresponders merits further study, but preliminary data suggest that a polymorphism in the *TGFB1* promoter aligns with therapeutic response.[21,35] Gene microarray data shows that fluticasone responsive EoE patients have increased expression of peptidyl-prolyl cis-trans isomerase FKBP5, a protein involved in steroid receptor function, as compared with control or active EoE subjects following therapy.[36] Whether FKBP5 reflects a steroid-inducible gene versus a predisposing factor for corticosteroid response remains to be understood.

Since IL-5 is pivotal for eosinophilopoesis and esophageal trafficking in animal models, it is reasonable to evaluate its utility in EoE patients. Adult trials have demonstrated improvements in remodeling-associated proteins, such as tenascin C and TGF-β1, following anti-IL-5 therapy.[37] However, the effects of IL-5 blockade on esophageal remodeling in pediatric EoE patients remains to be evaluated. Clinical trials using anti-IL-5 in pediatric EoE have been recently completed but remain unpublished to date. In addition, the use of other biologics, such as IL-13 blockade with agents such as pitrakinra, have not been evaluated to date but remain intriguing molecules that could affect esophageal remodeling in EoE patients.

CONCLUSIONS

Since EoE diagnosis and management requires repeated tissue procurement, this relatively new disease has already taught us important and novel lessons regarding eosinophil-associated tissue remodeling in children. Unlike other eosinophil-associated diseases, the clinical impact, natural history, and timing of remodeling in EoE can be assessed in young children. Tissue remodeling that includes angiogenesis, dilated intercellular spaces and subepithelial fibrosis, and epithelial basal zone hyperplasia are prominent features of pediatric EoE. These subepithelial changes can align with both endoscopic and clinical findings in children. Successful therapies for EoE, especially topical swallowed corticosteroids, can cause reversal and, at times, resolution of tissue remodeling. The effects of other potential therapies, including biologics, on esophageal remodeling remains to be assessed, as do many of the mechanisms that lead to fibrosis and angiogenesis in pediatric EoE.

REFERENCES

1. Broide DH. Immunologic and inflammatory mechanisms that drive asthma progression to remodeling. *J Allergy Clin Immunol* 2008; **121**:571–2.
2. Martinez FD. The origins of asthma and chronic obstructive pulmonary disease in early life. 2009;**1**:272–7.
3. Liacouras CA, Furuta GT, Hirano I, Atkins D, Attwood SE, Bonis PA, et al. Eosinophilic esophagitis: Updated consensus recommendations for children and adults. *J Allergy Clin Immunol* 2011;**128**(1):3–20.
4. Konikoff MR, Noel RJ, Blanchard C, Kirby C, Jameson SC, Buckmeier BK, et al. A randomized, double-blind, placebo-controlled trial of fluticasone propionate for pediatric eosinophilic esophagitis. *Gastroenterology* 2006;**131**:1381–91.
5. Noel RJ, Putnam PE, Collins MH, Assa'ad AH, Guajardo JR, Jameson SC, et al. Clinical and immunopathologic effects of swallowed fluticasone for eosinophilic esophagitis. *Clin Gastroenterol Hepatol* 2004;**2**:568–75.
6. Mulder DJ, Pacheco I, Hurlbut DJ, Mak N, Furuta GT, MacLeod RJ, et al. FGF9-induced proliferative response to eosinophilic inflammation in oesophagitis. *Gut* 2009;**58**:166–73.
7. Mishra A, Wang M, Pemmaraju VR, Collins MH, Fulkerson PC, Abonia JP, et al. Esophageal remodeling develops as a consequence of tissue specific IL-5-induced eosinophilia. *Gastroenterology* 2008; **134**:204–14.

8. Aceves SS, Newbury RO, Dohil R, Bastian JF, Broide DH. Esophageal remodeling in pediatric eosinophilic esophagitis. *J Allergy Clin Immunol* 2007;**119**:206−12.

9. Chehade M, Sampson HA, Morotti RA, Magid MS. Esophageal subepithelial fibrosis in children with eosinophilic esophagitis. *J Pediatr Gastroenterol Nutr* 2007;**45**:319−28.

10. Aceves SS, Newbury RO, Dohil R, Schwimmer J, Bastian JF. Distinguishing eosinophilic esophagitis in pediatric patients: clinical, endoscopic, and histologic features of an emerging disorder. *J Clin Gastroenterol* 2007;**41**:52−6.

11. Abu-Sultaneh SM, Durst P, Maynard V, Elitsur Y. Fluticasone and Food Allergen Elimination Reverse Sub-epithelial Fibrosis in Children with Eosinophilic Esophagitis. *Dig Dis Sci* 2010; **56**(1):97−102.

12. Blanchard C, Mingler MK, Vicario M, Abonia JP, Wu YY, Lu TX, et al. IL-13 involvement in eosinophilic esophagitis: transcriptome analysis and reversibility with glucocorticoids. *J Allergy Clin Immunol* 2007;**6**:1292−300.

13. Zuo L, Fulkerson PC, Finkelman FD, Mingler M, Fischetti CA, Blanchard C, et al. IL-13 induces esophageal remodeling and gene expression by an eosinophil-independent, IL-13R alpha 2-inhibited pathway. *J Immunol* 2010;**185**:660−9.

14. Mueller S, Aigner T, Neureiter D, Stolte M. Eosinophil infiltration and degranulation in oesophageal mucosa from adult patients with eosinophilic oesophagitis: a retrospective and comparative study on pathological biopsy. *J Clin Pathol* 2006;**59**:1175−80.

15. Kirsch R, Bokhary R, Marcon MA, Cutz E. Activated mucosal mast cells differentiate eosinophilic (allergic) esophagitis from gastro-esophageal reflux disease. *J Pediatr Gastroenterol Nutr* 2007;**44**: 20−6.

16. Blanchard C, Wang N, Stringer KF, Mishra A, Fulkerson PC, Abonia JP, et al. Eotaxin-3 and a uniquely conserved gene-expression profile in eosinophilic esophagitis. *J Clin Invest* 2006;**116**:536−47.

17. Aceves SS, Chen D, Newbury RO, Dohil R, Bastian J, Broide DH. Mast cells infiltrate the esophageal smooth muscle in eosinophilic esophagitis, express TGFβ1, and increase esophageal smooth muscle contraction. *J Allergy Clinical Immunol* 2010;**26**(6): 1198−204.

18. Blanchard C, Mingler MK, McBride M, Putnam PE, Collins MH, Chang G, et al. Periostin facilitates eosinophil tissue infiltration in allergic lung and esophageal responses. *Mucosal Immunol* 2008;**4**:289−96.

19. Hirata N, Kohrogi H, Iwagoe H, Goto E, Hamamoto J, Fujii K, et al. Allergen exposure induces the expression of endothelial adhesion molecules in passively sensitized human bronchus: time course and the role of cytokines. *Am J Respir Cell Mol Biol* 1998;**18**:12−20.

20. Sriramarao P, DiScipio RG, Cobb RR, Cybulsky M, Stachnick G, Castaneda D, et al. VCAM-1 is more effective than MAdCAM-1 in supporting eosinophil rolling under conditions of shear flow. *Blood* 2000;**95**:592−601.

21. Aceves SS, Newbury RO, Chen D, Mueller J, Dohil R, Hoffman H, et al. Resolution of remodeling in eosinophilic esophagitis correlates with epithelial response to topical corticosteroids. *Allergy* 2010;**65**:109−16.

22. Fox VL, Nurko S, Teitelbaum JE, Badizadegan K, Furuta GT. High-resolution EUS in children with eosinophilic 'allergic' esophagitis. *Gastrointest Endosc* 2003;**57**:30−6.

23. Dalby K, Nielsen RG, Kruse-Andersen S, Fenger C, Durup J, Husby S. Gastroesophageal reflux disease and eosinophilic esophagitis in infants and children. A study of esophageal pH, multiple intraluminal impedance and endoscopic ultrasound. *Scand J Gastroenterol* 2010;**45**:1029−35.

24. Binkovitz LA, Lorenz EA, DiLorenzo C, Kahwash S. Pediatric eosinophilic esophagitis: radiologic findings with pathologic correlation. *Pediatr Radiol* 2010;**40**:714−9.

25. Remedios M, Campbell C, Jones DM, Kerlin P. Eosinophilic esophagitis in adults: clinical, endoscopic, histologic findings, and response to treatment with fluticasone propionate. *Gastrointest Endosc* 2006;**63**:3−12.

26. Nurko S, Rosen R, Furuta GT. Esophageal dysmotility in children with eosinophilic esophagitis: a study using prolonged esophageal manometry. *Am J Gastroenterol* 2009;**104**:3050−7.

27. Lucendo AJ, Castillo P, Martín-Chávarri S, Carrión G, Pajares R, Pascual JM, et al. Manometric findings in adult eosinophilic oesophagitis: a study of 12 cases. *Eur J Gastroenterol Hepatol* 2007;**19**: 417−24.

28. Korsapati H, Babaei A, Bhargava V, Dohil R, Quin A, Mittal RK. Dysfunction of the longitudinal muscles of the oesophagus in eosinophilic oesophagitis. *Gut* 2009;**58**:1056−62.

29. Straumann A, Bussmann C, Zuber M, Vannini S, Simon HU, Schoepfer A. Eosinophilic esophagitis: analysis of food impaction and perforation in 251 adolescent and adult patients. *Clin Gastroenterol Hepatol* 2008;**6**:598−600.

30. Cohen MS, Kaufman AB, Palazzo JP, Nevin D, Dimarino Jr AJ, Cohen S. An audit of endoscopic complications in adult eosinophilic esophagitis. *Clin Gastroenterol Hepatol* Oct 2007;**5**(10): 1149−53.

31. Noel RJ, Putnam PE, Rothenberg ME. Eosinophilic esophagitis. *N Engl J Med* 2004;**26**:940−1.

32. Aceves SS, Newbury RO, Dohil MA, Bastian JF, Dohil R. A symptom scoring tool for identifying pediatric patients with eosinophilic esophagitis and correlating symptoms with inflammation. *Ann Allergy Asthma Immunol* 2009;**103**:401−6.

33. Aceves SS, Furuta GT, Spechler SJ. Integrated approach to treatment of children and adults with eosinophilic esophagitis. *Gastrointest Endosc Clin N Am* 2008;**18**:195−217.

34. Chehade M, Yershov O, Mayer L, Magid M, Morotti RA, Sampson H. Reversibility of esophageal fibrosis in response to various therapies in children with eosinophilic esophagitis. *Gastroenterology* 2009:A138.

35. Dohil R, Newbury R, Fox L, Bastian J, Aceves S. Oral viscous budesonide is effective in children with eosinophilic esophagitis in a randomized, placebo-controlled trial. *Gastroenterology* 2010;**139**: 418−29.

36. Caldwell JM, Blanchard C, Collins MH, Putnam PE, Kaul A, Aceves SS, et al. Glucocorticoid-regulated genes in eosinophilic esophagitis: a role for FKBP51. *J Allergy Clin Immunol* 2010;**125**: 879−88.

37. Straumann A, Conus S, Degen L, Felder S, Kummer M, Engel H, et al. Anti-interleukin-5 antibody treatment (mepolizumab) in active eosinophilic oesophagitis: a randomized, placebo-controlled, double-blind trial. *Gut* 2010;**59**:21−30.

38. Aceves Seema S. Tissue remodeling in patients with eosinophilic esophagitis: What lies beneath the surface? *Journal of Allergy and Clinical Immunology* 2011;**128**(5):1047−9.

Remodeling and Disease Pathology: Adult Eosinophilic Esophagitis

Florian Rieder, Claudio Fiocchi and Gary W. Falk

INTRODUCTION

Eosinophilic esophagitis (EoE) is a chronic inflammatory disorder of children and adults. The disease is defined by clinicopathological criteria that include symptoms of dysphagia and food impaction, esophageal biopsies with ≥ 15 eosinophils per high power field, and lack of response to antisecretory therapy.[1] In EoE, eosinophils are recruited into the esophagus, where they become activated and release a wide variety of inflammatory mediators.[2] One consequence of this inflammatory response is tissue remodeling, which is characterized histologically by sub-epithelial fibrosis; endoscopically by concentric rings, strictures and narrow caliber esophagus; and clinically by dysphagia and food impaction. This subchapter will discuss current knowledge about potential mechanisms of tissue remodeling, how to diagnose this complication, and how to manage it in adult patients.

KEY POINTS

- EoE is probably triggered by allergens that induce a T-helper (T_h2)-type immune response characterized by eosinophilic infiltration of the esophagus and organ remodeling.
- Tissue remodeling in EoE is a multifactorial process in which eosinophils, through the release of transforming growth factor β (TGF-β) and multiple degranulation products, interact with and induce the activation of local fibroblasts.
- Activated esophageal fibroblasts differentiate into myofibroblasts, which promote fibrosis by increasing production of collagen, fibronectin, and other extracellular matrix components.
- Esophageal fibrosis is a serious complication of EoE with distinct clinical, endoscopic, endosonographic, and histological features.
- Studies of adult patients with EoE are limited, particularly in regard to mechanistic investigations of EoE pathogenesis and fibrogenesis.
- Therapy with topical steroids, dietary manipulations, and anti-interleukin-5 (anti-IL-5) has the potential to improve symptoms and subepithelial fibrosis, but recurrence is common after discontinuation of therapy.
- Research in adult EoE should focus on the natural history of stricture formation, cellular and molecular mechanisms of fibrogenesis, and the development of novel antifibrotic therapies.

PATHOBIOLOGY

Inflammation-Associated Fibrosis

The concept that fibrosis is almost invariably the result of a chronic inflammatory process is firmly established, a paradigm that applies to both T_h1 and T_h2 immune responses.[3] While an inflammation-induced fibrotic response is part of the normal healing process, if excessive, deregulated, or unsuitably prolonged this response may result in pathological fibrosis with important structural and functional implications.[4] Many of the same cellular, molecular, and biological components can be found in a pathological fibrogenic process, regardless of the initiating factors or the type of underlying immune response. Injury, infection, or alterations of tissue homeostasis cause damage or activation of particular cells types, such as epithelial or endothelial cells, with release of a number of soluble mediators that initiate an inflammatory response. This involves the attraction, recruitment, activation and proliferation of different leukocytes at the affected site, accompanied by the local production of various bioactive molecules. This inflammatory and proliferative response initiates a process of regeneration, and the specific immune cells and mediators that participate in it depend on the type of T_h1, T_h2, T_h17, or mixed T_h response triggered by the initiating agent(s).[3] The regeneration phase is followed by actual remodeling and fibrogenesis, the major steps including activation of fibroblasts, their differentiation into myofibroblasts, and enhanced production of collagen, fibronectin and other extracellular matrix proteins, as well as angiogenesis.

Basic Components of Fibrogenesis

Traditionally, the key cellular mediator of fibrosis is the myofibroblast that, when activated, can differentiate and dedifferentiate into three interrelated cell types, i.e., the fibroblast, the myofibroblast, and the smooth muscle cell,[5] and acts as the primary collagen-producing cell.[6] TGF-β is still considered the prototypical and dominant profibrogenic molecule,[7] and of the three TGF-β isotypes (TGF-$\beta1$, 2, and 3), the $\beta1$ isoform is considered to be the main driver of fibrosis. Though still fundamentally correct, this restricted dual view is being progressively expanded to include several other components and pathways that range from innate to adaptive immune responses,[8] many of which are currently under investigation and are of potential relevance to the fibrogenic response of EoE. In addition to proliferation and migration of local fibroblasts, activated myofibroblasts may arise from bone marrow-derived cells, circulating fibrocytes, epithelial- and endothelial-to-mesenchymal transition, pericyte transdifferentiation, and local stellate cells; in addition to TGF-β, other cytokines and growth factors, such as connective tissue growth factor (CTGF), fibroblast growth factor (FGF), insulin-like growth factor 1 (IGF-1), interleukin-1β (IL-1β), IL-6, platelet-derived growth factor (PDGF), and reactive oxygen species (ROS) can activate fibroblasts[9,10] as well as microbial ligands that bind to Toll-like and NOD-line receptors.[11] Thus, although many of the same basic components are shared among different fibrogenic processes, it is becoming evident that fibrosis is a complex and diverse process, and that distinct events are implicated depending on the original trigger(s) and the cellular make-up of the organ affected. This is obviously important when considering the fibrogenic response of EoE, given its unique location, still unknown etiology, and incompletely defined Th2 immune pathogenesis.

T-Helper Type 2 Response and Fibrogenesis

Among Th2 immune responses, the best evidence for an association between allergic diseases and tissue remodeling can be found in human and animal studies of asthma. Eosinophils and Th2 cytokines play a central role in allergic airway remodeling.[12] This most likely results from the combined action of degranulation products and TGF-β secretion,[13,14] direct eosinophil–fibroblast interaction,[15] and secretion of IL-5, IL-13, and other mediators,[16,17] all of which leads to fibroblast activation, extracellular matrix deposition, and angiogenesis.[13,15,17,18] At the tissue level these events are translated into accumulation of fibroblasts and myofibroblasts beneath the subepithelial basement membrane, increased deposition of extracellular matrix proteins, formation of new blood vessels, and an increase in smooth muscle mass. The end result is thickening and stiffening of the bronchial wall.

Most of the cellular and molecular components and biological pathways activated in airways affected by a Th2 response are not unique to allergic asthma, and similar components and functional relationships among eosinophilic inflammation, tissue remodeling and fibrosis appear to be present in EoE.[19] Due to difficulties in accessing human esophageal tissue, and limitations imposed by the use of endoscopic biopsies that primarily consist of superficial mucosal layers, studies that have directly investigated mechanisms of fibrogenesis in humans are few in number and limited in reach. These deficiencies are partly compensated by studies carried out in animal

models of EoE, but these studies investigated more immune-related than remodeling-related pathogenic events.

Animal Studies

Animal studies clearly support the Th2 allergic nature of EoE,[20–22] and show that IL-5 is required for accumulation of eosinophils, as IL-5-deficient mice do not develop esophageal eosinophilic infiltration.[23] IL-5 also plays a key role in eosinophil-driven remodeling. Mishra and colleagues found an accumulation of collagen in the lamina propria of mice with experimental EoE compared to control mice.[20] This is associated with increased thickness of the basal layer, muscularis mucosa, and enhanced collagen deposition in elongated stromal papilla; moreover, they observed increased expression of *TGFB1* and *MUC5AC* mRNA. Furthermore, IL-5-deficient mice had reduced lamina propria collagen compared to wild-type mice. Esophageal remodeling can also occur through another Th2-dependent but eosinophil-independent pathway, as demonstrated in eosinophil-deficient, IL-13 transgenic mice.[24] In addition to eosinophils and Th2 cytokines, other types of immune cells and mediators are also involved in mouse models of EoE, including mast cells, T effector and T regulatory cells, and IL-15.[25–27] Whether these elements contribute to esophageal fibrosis remains to be investigated, but their presence highlights the complex nature of EoE pathogenesis and the associated fibrotic response.

Pediatric Studies

Studies of esophageal remodeling in EoE have been performed more frequently in children than in adults.[28–30] In a cohort of pediatric EoE patients, compared to gastroesophageal reflux disease (GERD) and control patients, Aceves and coworkers found that strictured areas displayed subepithelial fibrosis accompanied by an increased number of TGF-β1[+] cells (primarily eosinophils) in the lamina propria, increased expression of phosphorylated SMAD2/3 in the basal layer of the epithelium and lamina propria, and an increased number of activated blood vessels expressing vascular cell adhesion molecule 1 (VCAM-1) in the subepithelial space.[28] In a follow-up study, the same authors found that a beneficial response to topical steroid therapy is accompanied by a decrease in eosinophil numbers, edema, fibrosis scores, SMAD2/3[+] cells, TGF-β1[+] cells, and vascular cell adhesion molecule positive (VCAM[+]) blood vessels.[31] The strongest predictor of reduced remodeling factors is a drop in the number of eosinophils. In another pediatric EoE population, Chehade and coworkers also detected frequent esophageal subepithelial fibrosis, but this was not associated with the number of infiltrating eosinophils but rather with their degree of degranulation.[29] It is important to notice that these studies are all based on endoscopic mucosal biopsies, which are limited to sampling of the epithelial and subepithelial layers, so that information on deeper tissue layers, particularly the esophageal muscle, remains elusive.

Adult Studies

Investigation of esophageal remodeling in adult EoE patients has been quite limited, but it is reasonable to assume that the cellular and soluble mediators and related pathogenic mechanisms are similar to those of children with EoE.[32,33] While eosinophilic infiltration can be a feature of GERD and eosinophil numbers consistent with EoE can be found in this condition,[34,35] their effects may vary. In fact, at least one study claims that fibrosis of the esophageal lamina propria is selectively found in EoE but not in GERD.[36] Only preliminary reports of the potential mechanisms of fibrogenesis in human EoE are available. Using esophageal mucosal biopsies from adult EoE and control subjects, and primary human esophageal fibroblast and muscle cell lines, our laboratory found that active secretion of IL-5 and eotaxin, IL-6, and TGF-β1 occurs uniformly throughout the mucosa of EoE patients; secretion of collagen and fibronectin by esophageal mesenchymal cells is selectively promoted by eotaxin-3/C-C motif chemokine 26 (CCL26), IL-4, IL-6, IL-13, and TGF-β1 or by direct contact with activated eosinophils, whose binding to local mesenchymal cells is boosted by T$_h$2 cytokines or TGF-β1.[37] In addition, exposure to eosinophil-derived cytokines, degranulation products, or eosinophil sonicates increases collagen and fibronectin production by, and eosinophil binding to, esophageal mesenchymal cells through TGF-β1- and p38 mitogen-activated protein kinase (MAPK)-dependent pathways.[38] Thus, it appears that the T$_h$2 milieu of adult EoE drives a local fibrogenic response both directly through the action of cytokines and indirectly by promoting eosinophil binding and retention in the mucosa. Thus, based on the data described in this review a conceptual model of EoE-associated esophageal fibrosis can be proposed (outlined in Fig. 12.6.1).

CLINICAL PRESENTATION

EoE in adults is characterized by a variety of clinical symptoms, including chest pain, dysphagia, food impaction, and heartburn. Dysphagia and food impaction are the most common symptoms and a recent meta-analysis found dysphagia to be the presenting symptom in 93% and food impaction in 62% of adult patients.[39]

ENDOSCOPIC EVIDENCE OF TISSUE REMODELING

Patients with EoE have a wide range of endoscopic findings that may vary by age, and the disease does not have a uniform endoscopic appearance.[40] Such endoscopic findings may represent the macroscopic manifestation of tissue remodeling due to long-standing eosinophilic inflammation. They include concentric rings, crêpe-paper esophagus, focal strictures, linear furrows, narrow caliber esophagus, and diminished vascularity.

Linear furrows are some of the most characteristic findings of EoE. They appear as a linear pattern of grooves that typically involve the full length of the esophagus. They can be seen with conventional white light endoscopy and are further accentuated by narrow-band imaging. Observations with endoscopic ultrasonography suggest that furrows may be related to thickening of the mucosal and submucosal layers.[41] Furrows may be seen either alone or in conjunction with other endoscopic features of EoE. It has been estimated that linear furrows occur in 11—100% of EoE patients.[42—47]

Perhaps the most striking endoscopic finding of EoE is that of multiple concentric rings resembling the ringed, corrugated appearance of the *feline* esophagus. Rings may be the only finding or may be seen in conjunction with other endoscopic findings of EoE. Concentric rings may either be observed focally in esophageal segments or involve the entire length of the organ. The etiology of these esophageal rings is not completely understood and it has been suggested that they represent intermittent contraction of the deep muscle layer or possibly eosinophilic infiltration.[48] Interestingly, these rings may resolve with appropriate therapy.[47] Estimates of the prevalence of the ringed esophagus range from 19% to 88% of EoE patients.[42,45—51]

Focal strictures, in the absence of narrow caliber esophagus or concentric rings, may be encountered in 3—66% of EoE patients.[42,45—49,51] These strictures are typically encountered in the proximal esophagus, may be subtle, and are characterized by smooth circumferential narrowing of the esophagus. Strictures are more commonly seen in adults than in children.[42]

Small caliber esophagus is a unique finding in EoE and is defined as a narrow fixed internal diameter of the esophagus.[51] This feature may not be appreciated on endoscope insertion, but is characterized by extensive linear abrasions or mucosal rents best seen upon withdrawal of the endoscope or after dilation.[52] The prevalence of small caliber esophagus is estimated to be between 10% and 28%.[42,47,48,51]

The term *crêpe-paper* esophagus was first suggested by Straumann and coworkers in a 2003 case series of five men. In crêpe-paper esophagus, the mucosa of the esophagus is

SIDE UP

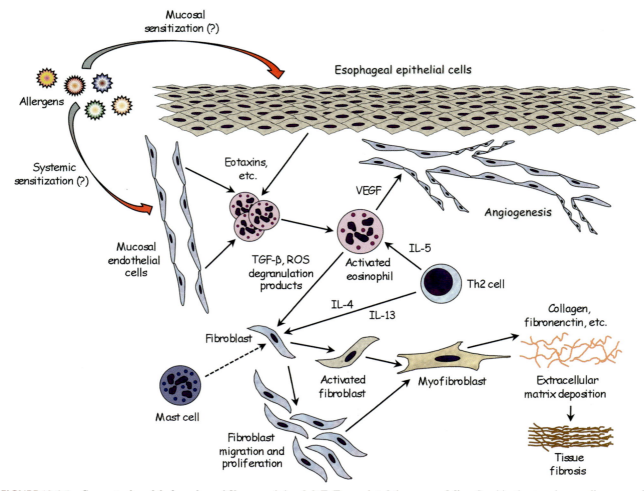

FIGURE 12.6.1   Conceptual model of esophageal fibrogenesis in adult EoE-associated tissue remodeling. Sensitization to unknown allergens could occur in EoE through mucosal or systemic routes, or both. This could set in motion an immune response leading to the generation of T-helper type 2 (T_h2) cells and the production of eosinophil chemoattractants, such as eotaxin, eotaxin-2/CCL24 (C-C motif chemokine 24), and eotaxin-3/CCL26, by esophageal epithelial cells and mucosal microvascular endothelial cells. The result is the recruitment of eosinophils into the epithelium and the mucosa and their subsequent activation by factors present in the local tissue microenvironment, such as T_h2-derived interleukin-5 (IL-5). Activated eosinophils release transforming growth factor β (TGF-β), reactive oxygen species (ROS), and a variety of granule-derived bioactive molecules, all of which act upon local fibroblasts that are also exposed to T_h2 cell-derived IL-4 and IL-13. Consequently, fibroblasts become activated, migrate, proliferate, and differentiate into myofibroblasts, which secrete increased amounts of extracellular matrix components, such as collagen and fibronectin, which initiate the fibrogenic process. These events are associated with an angiogenic response promoted by eosinophil-derived vascular endothelial growth factor (VEGF) and possibly other growth factors. A mast cell is shown to denote that these cells are also increased in number in EoE and likely contribute to the fibrogenic response and the associated motor abnormalities.

characterized as delicate, fragile, inelastic and prone to tear with little pressure.[53]

Finally, endosonography has been used in a limited number of EoE patients to date. Fox and colleagues found significant differences in total wall thickness (2.8 mm vs. 2.1 mm), combined mucosa and submucosa thickness (1.6 mm vs. 1.1 mm), and muscularis propria thickness (1.2 mm vs. 1.0 mm) between children with EoE and control children.[54] Wall thickening has also been described in an adult case report.[55] In summary, the diverse endoscopic

appearances of EoE clearly speak to an underlying change in the mucosal lining and probably in the deeper esophageal layers of these patients.

HISTOLOGY

EoE is typically identified by a rich mucosal infiltration by eosinophils, with fewer mast cells and T cells, accompanied by basal zone hyperplasia. Evidence for remodeling with subepithelial fibrosis starts in young children with EoE, extends to adults, and is clearly more common in biopsies obtained from patients with EoE than in biopsies from patients with GERD. Chehade and colleagues found subepithelial fibrosis in 12/21 children with EoE (57%) compared to 0/6 children with GERD.[29] In these pediatric patients, fibrosis was associated with dysphagia and eosinophil degranulation. These findings were confirmed recently in a study of 53 children with EoE, GERD, or normal histology;[56] 13/17 EoE patients (76.5%) had evidence of subepithelial fibrosis compared to 6.9% of GERD and 5.3% of control children.

Evidence for fibrosis in adults with EoE was first reported by Straumann and colleagues in a natural history study of 30 adults with persistent dysphagia.[57] In 6/7 patients with adequate biopsies that included the subepithelium, increased fibrous tissue was noted, accompanied by thickening and alteration of the architecture of the subepithelial stroma. Similar to the pediatric findings, Lee and coworkers found evidence of subepithelial sclerosis, defined as thickened or hyalinized collagen fibers within the stroma, immediately below the lamina propria in 17/23 (74%) adult patients with EoE, compared to only 1/20 (5%) patients with GERD.[58] In contrast, the normal subepithelial stroma is composed of loose fibroconnective tissue.

EFFECT OF THERAPY ON FIBROSIS

Information collected from clinical studies suggests that anti-IL-5, dietary manipulations, and topical steroids may reduce the extent of subepithelial fibrosis in EoE patients. Straumann and colleagues recently described the effect of oral budesonide given as 1 mg twice daily for 15 days in a randomized controlled trial of adults and adolescents.[59] Prior to therapy, fibrosis scores were assessed in EoE patients and compared to 20 healthy controls. It is important to note that only 253/565 biopsy specimens had adequate lamina propria tissue for analysis. Prior to therapy, fibrosis scores in the EoE patients were significantly increased compared to control subjects (2.43 vs. 0.33). Budesonide therapy reduced fibrosis scores by 0.42 points in the treatment group, with no change in the placebo group. Fibrosis scores in this short-term study did not, however, disappear. This study also was notable for finding that 70−80% of epithelial cells express TGF-β1 in active EoE, which decreases by 75% after budesonide treatment but remains unchanged after placebo treatment. Budesonide treatment is also associated with a decline in basal zone hyperplasia, esophageal eosinophil load, and fibrosis score (based on number of fibroblasts and density of collagen bundles). In addition, 8/9 individuals had resolution of linear furrows, although concentric rings persisted in 8/9 patients. All of these changes were accompanied by an improvement in symptoms of dysphagia.

Similar results were noted in children, for whom topical steroid therapy decreases fibrosis. This was first described in a case report of a 7-year-old boy treated with topical budesonide with a marked decrease in submucosal collagen deposition.[60] Subsequently, 11 pediatric EoE patients treated with fluticasone were found to have a reduction in eosinophil numbers accompanied by a decrease in fibrosis, from 71.5% of patients prior to therapy to 7.1% after therapy.[58] This decline was accompanied by an improvement in clinical symptoms in these patients. Fibrosis also decreased with dietary intervention.

A recent small trial with an anti-IL-5 monoclonal antibody resulted in decreased TGF-β1 and tenascin C expression in the esophageal epithelial layer of adult EoE patients after 13 weeks of therapy in conjunction with a decrease in eosinophil counts.[61] However, despite these findings, symptoms were unchanged due to short-term therapy, small sample size, or lack of efficacy of the biological agent.

CONCLUSIONS AND FUTURE DIRECTIONS

Since its emergence in the early 1990s, EoE has rapidly been recognized as a unique pathophysiological and clinical entity. The cellular and molecular mechanisms of this emerging disease are still being defined, but much remains to be done, particularly in regard to its fibrogenic response.[2] This is essential to learn how to manage this disease, given that the motility abnormalities associated with fibrosis are responsible for much of the patient complaints associated with EoE.[62] The key clinical issue in EoE is whether and how uncontrolled eosinophilic inflammation over the course of time leads to organ remodeling, fibrosis and luminal narrowing, with the functional correlates of dysphagia and food impaction.[63] However, it is still uncertain whether this outcome is inevitable or restricted to a subset of particularly susceptible individuals. Furthermore, it remains unclear whether treatment interferes with this pathway despite encouraging data to date. Finally, at what point tissue remodeling becomes irreversible and when treatment should start and end are critical but unanswered questions.

REFERENCES

1. Furuta GT, Liacouras CA, Collins MH, Gupta SK, Justinich C, Putnam PE, et al. Eosinophilic esophagitis in children and adults: a systematic review and consensus recommendations for diagnosis and treatment. *Gastroenterology* 2007;**133**:1342−53.
2. Mulder DJ, Justinich CJ. Understanding eosinophilic esophagitis: the cellular and molecular mechanisms of an emerging disease. *Mucosal Immunol* 2011;**4**:139−47.

3. Wynn TA. Fibrotic disease and the T(H)1/T(H)2 paradigm. *Nat Rev Immunol* 2004;**4**:583—94.

4. Fiocchi C, Lund PK. Themes in fibrosis and gastrointestinal inflammation. *Am J Physiol Gastrointest Liver Physiol* 2011;**300**(5):G677—83. May.

5. Powell DW, Mifflin RC, Valentich JD, Crowe SE, Saada JI, West AB. Myofibroblasts. I. Paracrine cells important in health and disease. *Am J Physiol* 1999;**277**:C1—19.

6. Wynn TA. Cellular and molecular mechanisms of fibrosis. *J Pathol* 2008;**214**:199—210.

7. Leask A, Abraham DJ. TGF-β signaling and the fibrotic response. *FASEB J* 2004;**18**:816—27.

8. Wick G, Backovic A, Rabensteiner E, Plank N, Schwentner C, Sgonc R. The immunology of fibrosis: innate and adaptive responses. *Trends Immunol* 2010;**31**:110—9.

9. Ghazizadeh M, Tosa M, Shimizu H, Hyakusoku H, Kawanami O. Functional implications of the IL-6 signaling pathway in keloid pathogenesis. *J Invest Dermatol* 2007;**127**:98—105.

10. Qi S, den Hartog GJ, Bast A. Superoxide radicals increase transforming growth factor-beta1 and collagen release from human lung fibroblasts via cellular influx through chloride channels. *Toxicol Appl Pharmacol* 2009;**237**:111—8.

11. Rieder F, Fiocchi C. Intestinal fibrosis in IBD—a dynamic, multifactorial process. *Nat Rev Gastroenterol Hepatol* 2009;**6**:228—35.

12. Humbles AA, Lloyd CM, McMillan SJ, Friend DS, Xanthou G, McKenna EE, et al. A critical role for eosinophils in allergic airways remodeling. *Science* 2004;**305**:1776—9.

13. Rochester CL, Ackerman SJ, Zheng T, Elias JA. Eosinophil-fibroblast interactions. *J Immunol* 1996;**156**:4449—56.

14. Pegorier S, Wagner LA, Gleich GJ, Pretolani M. Eosinophil-derived cationic proteins activate the synthesis of remodeling factors by airway epithelial cells. *J Immunol* 2006;**177**:4861—9.

15. Gomes I, Mathur SK, Espenshade BM, Mori Y, Varga J, Ackerman SJ. Eosinophil-fibroblast interactions induce fibroblast IL-6 secretion and extracellular matrix gene expression: implications for fibrogenesis. *J Allergy Clin Immunol* 2005;**116**:796—804.

16. Flood-Page P, Menzies-Gow A, Phipps S, Ying S, Wangoo A, Ludwig MS, et al. Anti-IL-5 treatment reduces deposition of ECM proteins in the bronchial subepithelial basement membrane of mild atopic asthmatics. *J Clin Invest* 2003;**112**:1029—36.

17. Borowski A, Kuepper M, Horn U, Knupfer U, Zissel G, Hohne K, et al. Interleukin-13 acts as an apoptotic effector on lung epithelial cells and induces pro-fibrotic gene expression in lung fibroblasts. *Clin Exp Allergy* 2008;**38**:619—28.

18. Aceves SS, Broide DH. Airway fibrosis and angiogenesis due to eosinophil trafficking in chronic asthma. *Curr Mol Med* 2008;**8**:350—8.

19. Aceves SS, Ackerman SJ. Relationships between eosinophilic inflammation, tissue remodeling, and fibrosis in eosinophilic esophagitis. *Immunol Allergy Clin North Am* 2009;**29**:197—211. xiii—xiv.

20. Mishra A, Hogan SP, Brandt EB, Rothenberg ME. An etiological role for aeroallergens and eosinophils in experimental esophagitis. *J Clin Invest* 2001;**107**:83—90.

21. Mishra A, Rothenberg ME. Intratracheal IL-13 induces eosinophilic esophagitis by an IL-5, eotaxin-1, and STAT6-dependent mechanism. *Gastroenterology* 2003;**125**:1419—27.

22. Blanchard C, Mingler MK, Vicario M, Abonia JP, Wu YY, Lu TX, et al. IL-13 involvement in eosinophilic esophagitis: transcriptome analysis and reversibility with glucocorticoids. *J Allergy Clin Immunol* 2007;**120**:1292—300.

23. Mishra A, Hogan SP, Brandt EB, Rothenberg ME. IL-5 promotes eosinophil trafficking to the esophagus. *J Immunol* 2002;**168**:2464—9.

24. Zuo L, Fulkerson PC, Finkelman FD, Mingler M, Fischetti CA, Blanchard C, et al. IL-13 induces esophageal remodeling and gene expression by an eosinophil-independent, IL-13R alpha 2-inhibited pathway. *J Immunol* 2010;**185**:660—9.

25. Levi-Schaffer F, Weg VB. Mast cells, eosinophils and fibrosis. *Clin Exp Allergy* 1997;**27**(Suppl. 1):64—70.

26. Mishra A, Schlotman J, Wang M, Rothenberg ME. Critical role for adaptive T cell immunity in experimental eosinophilic esophagitis in mice. *J Leukoc Biol* 2007;**81**:916—24.

27. Zhu X, Wang M, Mavi P, Rayapudi M, Pandey AK, Kaul A, et al. Interleukin-15 expression is increased in human eosinophilic esophagitis and mediates pathogenesis in mice. *Gastroenterology* 2010;**139**:182—93. e7.

28. Aceves SS, Newbury RO, Dohil R, Bastian JF, Broide DH. Esophageal remodeling in pediatric eosinophilic esophagitis. *J Allergy Clin Immunol* 2007;**119**:206—12.

29. Chehade M, Sampson HA, Morotti RA, Magid MS. Esophageal subepithelial fibrosis in children with eosinophilic esophagitis. *J Pediatr Gastroenterol Nutr* 2007;**45**:319—28.

30. Li-Kim-Moy JP, Tobias V, Day AS, Leach S, Lemberg DA. Esophageal subepithelial fibrosis and hyalinization are features of eosinophilic esophagitis. *J Pediatr Gastroenterol Nutr* 2011;**52**:147—53.

31. Aceves SS, Newbury RO, Chen D, Mueller J, Dohil R, Hoffman H, et al. Resolution of remodeling in eosinophilic esophagitis correlates with epithelial response to topical corticosteroids. *Allergy* 2010;**65**:109—16.

32. Furuta GT. Emerging questions regarding eosinophil's role in the esophago-gastrointestinal tract. *Curr Opin Gastroenterol* 2006;**22**:658—63.

33. Rothenberg ME. Biology and treatment of eosinophilic esophagitis. *Gastroenterology* 2009;**137**:1238—49.

34. DeBrosse CW, Collins MH, Buckmeier Butz BK, Allen CL, King EC, Assa'ad AH, et al. Identification, epidemiology, and chronicity of pediatric esophageal eosinophilia, 1982—1999. *J Allergy Clin Immunol* 2010;**126**:112—9.

35. Molina-Infante J, Ferrando-Lamana L, Ripoll C, Hernandez-Alonso M, Mateos JM, Fernandez-Bermejo M, et al. Esophageal eosinophilic infiltration responds to proton pump inhibition in most adults. *Clin Gastroenterol Hepatol* 2011;**9**:110—7.

36. Parfitt JR, Gregor JC, Suskin NG, Jawa HA, Driman DK. Eosinophilic esophagitis in adults: distinguishing features from gastroesophageal reflux disease: a study of 41 patients. *Mod Pathol* 2006;**19**:90—6.

37. Rieder F, Nonevski IT, Ouyang Z, West G, Scaldaferri F, Goldblum JR, et al. Integrated pathways of fibrogenesis in eosinophilic esophagitis: active secretion of T$_h$2 cytokines and TGF-β1, and binding of activated eosinophils promote collagen I and fibronectin production by human esophageal mesenchymal cells. *Digestive Disease Week Chicago* 2009;887 (abstract).

38. Rieder F, Nonevski IT, Ouyang Z, West G, Schirbel A, Goldblum JR, et al. Eosinophils and their products activate human esophageal mesenchymal cells—Implications for fibrogenesis in eosinophilic esophagitis. *Digestive Disease Week New Orleans, LA*; 2010. W1700 (abstract).

39. Sgouros SN, Bergele C, Mantides A. Eosinophilic esophagitis in adults: what is the clinical significance? *Endoscopy* 2006;**38**: 515—20.

40. Dellon ES, Gibbs WB, Fritchie KJ, Rubinas TC, Wilson LA, Woosley JT, et al. Clinical, endoscopic, and histologic findings distinguish eosinophilic esophagitis from gastroesophageal reflux disease. *Clin Gastroenterol Hepatol* 2009;**7**:1305—13. quiz 1261.

41. Fox VL, Nurko S, Furuta GT. Eosinophilic esophagitis: it's not just kid's stuff. *Gastrointest Endosc* 2002;**56**:260—70.

42. Gupta SK, Fitzgerald JF, Chong SK, Croffie JM, Collins MH. Vertical lines in distal esophageal mucosa (VLEM): a true endoscopic manifestation of esophagitis in children? *Gastrointest Endosc* 1997;**45**:485—9.

43. Muller S, Puhl S, Vieth M, Stolte M. Analysis of symptoms and endoscopic findings in 117 patients with histological diagnoses of eosinophilic esophagitis. *Endoscopy* 2007;**39**:339—44.

44. Cohen MS, Kaufman AB, Palazzo JP, Nevin D, Dimarino Jr AJ, Cohen S. An audit of endoscopic complications in adult eosinophilic esophagitis. *Clin Gastroenterol Hepatol* 2007;**5**:1149—53.

45. Remedios M, Campbell C, Jones DM, Kerlin P. Eosinophilic esophagitis in adults: clinical, endoscopic, histologic findings, and response to treatment with fluticasone propionate. *Gastrointest Endosc* 2006;**63**:3—12.

46. Gonsalves N, Policarpio-Nicolas M, Zhang Q, Rao MS, Hirano I. Histopathologic variability and endoscopic correlates in adults with eosinophilic esophagitis. *Gastrointest Endosc* 2006;**64**: 313—9.

47. Desai TK, Stecevic V, Chang CH, Goldstein NS, Badizadegan K, Furuta GT. Association of eosinophilic inflammation with esophageal food impaction in adults. *Gastrointest Endosc* 2005;**61**: 795—801.

48. Fox VL. Eosinophilic esophagitis: endoscopic findings. *Gastrointest Endosc Clin N Am* 2008;**18**:45—57. viii.

49. Potter JW, Saeian K, Staff D, Massey BT, Komorowski RA, Shaker R, et al. Eosinophilic esophagitis in adults: an emerging problem with unique esophageal features. *Gastrointest Endosc* 2004;**59**:355—61.

50. Kaplan M, Mutlu EA, Jakate S, Bruninga K, Losurdo J, Keshavarzian A. Endoscopy in eosinophilic esophagitis: 'feline' esophagus and perforation risk. *Clin Gastroenterol Hepatol* 2003;**1**: 433—7.

51. Straumann A, Spichtin HP, Bucher KA, Heer P, Simon HU. Eosinophilic esophagitis: red on microscopy, white on endoscopy. *Digestion* 2004;**70**:109—16.

52. Vasilopoulos S, Murphy P, Auerbach A, Massey BT, Shaker R, Stewart E, et al. The small-caliber esophagus: an unappreciated cause of dysphagia for solids in patients with eosinophilic esophagitis. *Gastrointest Endosc* 2002;**55**:99—106.

53. Straumann A, Rossi L, Simon HU, Heer P, Spichtin HP, Beglinger C. Fragility of the esophageal mucosa: a pathognomonic endoscopic sign of primary eosinophilic esophagitis? *Gastrointest Endosc* 2003;**57**:407—12.

54. Fox VL, Nurko S, Teitelbaum JE, Badizadegan K, Furuta GT. High-resolution EUS in children with eosinophilic 'allergic' esophagitis. *Gastrointest Endosc* 2003;**57**:30—6.

55. Bhutani MS, Moparty B, Chaya CT, Schnadig V, Logrono R. Endoscopic ultrasound-guided fine-needle aspiration of enlarged mediastinal lymph nodes in eosinophilic esophagitis. *Endoscopy* 2007;**39**(Suppl. 1):E82—3.

56. Abu-Sultaneh SM, Durst P, Maynard V, Elitsur Y. Fluticasone and food allergen elimination reverse sub-epithelial fibrosis in children with eosinophilic esophagitis. *Dig Dis Sci* 2011;**56**:97—102.

57. Straumann A, Spichtin HP, Grize L, Bucher KA, Beglinger C, Simon HU. Natural history of primary eosinophilic esophagitis: a follow-up of 30 adult patients for up to 11.5 years. *Gastroenterology* 2003;**125**:1660—9.

58. Lee S, de Boer WB, Naran A, Leslie C, Raftopoulos S, Ee H, et al. More than just counting eosinophils: proximal oesophageal involvement and subepithelial sclerosis are major diagnostic criteria for eosinophilic oesophagitis. *J Clin Pathol* 2010;**63**:644—7.

59. Straumann A, Conus S, Degen L, Felder S, Kummer M, Engel H, et al. Budesonide is effective in adolescent and adult patients with active eosinophilic esophagitis. *Gastroenterology* 2010;**139**: 1526—37. 1537 e1.

60. Maples KM, Henderson SC, Graham M, Irani AM. Treatment of eosinophilic esophagitis with inhaled budesonide in a 7-year-old boy with concomitant persistent asthma: resolution of esophageal submucosal fibrosis and eosinophilic infiltration. *Ann Allergy Asthma Immunol* 2007;**99**:572—4.

61. Straumann A, Conus S, Grzonka P, Kita H, Kephart G, Bussmann C, et al. Anti-interleukin-5 antibody treatment (mepolizumab) in active eosinophilic oesophagitis: a randomised, placebo-controlled, double-blind trial. *Gut* 2010;**59**:21—30.

62. Kwiatek MA, Hirano I, Kahrilas PJ, Rotle J, Luger D, Pandolfino JE. Mechanical properties of the esophagus in eosinophilic esophagitis. *Gastroenterology* 2011;**140**:82—90.

63. Liacouras CA, Furuta GT, Hirano I, Atkins D, Attwood SE, Bonis PA, et al. Eosinophilic esophagitis: Updated consensus recommendations for children and adults. *J Allergy Clin Immunol* Jul 2011;**128**(1):3—20.e6; quiz 21—22.

Chapter 12.7

Human Studies of Remodeling and Fibrosis in Asthma and Allergic Inflammation

Harsha H. Kariyawasam, Douglas S. Robinson and A. Barry Kay

INTRODUCTION

Eosinophils have traditionally been regarded as proinflammatory cells in the pathogenesis of allergic disease and asthma through the release of lipid mediators, cytokines, chemokines, and highly charged basic proteins.[1] However, in allergen challenge situations, such as the late-phase skin reaction, eosinophils persist long after the resolution of allergic inflammation.[2] This, and more recent observations in humans and experimental animals involving eosinophil depletion, has led to the view that eosinophils may also play a role in wound healing, remodeling, and the development of postinflammatory fibrosis, especially as tissue eosinophilia and eosinophil degranulation are associated with several fibrotic syndromes.[3–5] It is well documented that human eosinophils express the potent fibrogenic factor transforming growth factor-β1 (TGF-β1) and, in coculture systems, this cell type stimulates collagen synthesis, fibroblast proliferation, and lattice contraction.[6–8] In addition, the T-helper 2 (T$_h$2)-like cytokines, interleukin-4 (IL-4) and IL-13, also expressed by eosinophils, upregulate fibroblast chemokine and matrix protein expression[9,10] and induce a weak myofibroblastic phenotype.[10] Eosinophils also express other growth factors and cytokines that modulate mesenchymal cells, including fibroblast growth factor 2 (FGF-2),[11] nerve growth factor (NGF),[12] vascular endothelial growth factor (VEGF)[13] and IL-4.[14]

REPAIR AND REMODELING IN THE AIRWAYS

The response to airway tissue injury, whether from allergic inflammation or infection, involves the rapid activation of local host defense systems, along with recruitment and migration of activated inflammatory cells to the site of injury or antigen deposition. Subsequently, regulated induction of inflammatory resolution and healing of injured tissue occurs via controlled structural cell activation and regulated extracellular matrix (ECM) deposition. The goal of this repair response is the elimination of the inciting agent, resolution of inflammation, and the restoration of normal tissue architecture. As with other organs, such as the skin, airway eosinophils are implicated in several immunoregulatory and tissue repair processes. However, excessive eosinophil recruitment and deregulated function in allergic airway tissues may lead to an outcome of progressive and excessive tissue structural change, i.e., remodeling.

The precise clinical consequences of tissue remodeling in asthma symptomatology remains uncertain. Most studies have addressed this question using basement membrane thickening as a surrogate for remodeling in general. Whether this always reflects changes in airway smooth muscle mass, angiogenesis, and mucous gland hypertrophy is not known. It is generally assumed that remodeling events lead to a more ridged and thickened airway wall, with a consequent reduction in airway caliber and increased

or when ASM in increased, excessive BHR. Mathematical modeling predicts that these structural changes contribute to both obstructive and/or hyperresponsive phenotypes.[15] On the other hand, Broekema and colleagues[16] found that whereas airway eosinophils are elevated in current asthma compared to patients with either clinical or complete remission, basement membrane thickening is similar in all groups of patients.

Airway remodeling is seen in all layers of the asthmatic airway. Increased numbers of goblet cells and submucosal mucous gland hypertrophy contribute to excessive airway mucus. There are an excessive number of fibroblasts and myofibroblasts, both immediately below the thickened reticular basement membrane (RBM) and throughout the submucosa. Marked airway smooth muscle (ASM) hyperplasia and hypertrophy, along with excessive ECM deposition, is also present. TGF-β1 is the pivotal member of the TGF-β superfamily of ligands and is considered a potent inducer of tissue remodeling via both inhibitory and stimulatory effects on inflammatory and structural cells. A balance between production and degradation of ECM regulates turnover. In particular, metalloproteinase-9 (MMP-9) is important, as it can digest ECM collagen. An equal ratio of tissue inhibitors of MMPs (or TIMPs) serve to inhibit excessive MMP activity. Vascular and neuronal remodeling is also increasingly recognized as important in the asthmatic airway.

Eosinophilic bronchitis (EB) is a condition pathologically similar to asthma but clinically characterized by cough without reversible airway narrowing or airway hyperresponsiveness (AHR).[17] The most common outcome in EB is continuing disease and complete resolution is rare. Asthma and fixed airflow obstruction develop in relatively few patients.[18] Like asthma, RBM thickening is characteristic in EB.[19] Although vascular remodeling is similar to that seen in asthma,[20] a detailed study of other remodeling events in EB is incomplete. Thus, a more detailed study of this disease may reveal remodeling features that are eosinophil-dependent and identify those which are relevant to asthma.

EOSINOPHIL PRODUCTS AND REMODELING

Tissue eosinophilia with evidence of cell activation and degranulation is characteristic of several diseases related to organ fibrosis.[5] Our current understanding of the biology of eosinophils suggests multiple fibrogenic pathways by which these cells may potentially contribute to tissue remodeling in the allergic airway. In asthma, eosinophilia is associated with a more rapid decline in lung function[21] and exacerbation rate,[22] both determinants of asthma severity.

Several *in vitro* studies have highlighted potential mechanisms by which eosinophils may remodel tissue.

One possible mechanism is via direct eosinophil interaction with epithelium and submucosal structural cells. There is also extensive regulatory and modulatory *cross talk* between eosinophils and other inflammatory cells. Eosinophils may direct such cells to interact further with epithelium, fibroblasts, ASM and the ECM itself, leading to further induction of selected aspects of tissue remodeling. The airway epithelium is a significant source of immune factors and the resultant expression of adhesion and signaling receptors, along with the secretion of cytokines and growth factors, contributes to the maintenance of signaling and remodeling. Prestored eosinophil cytokines, chemokines, and growth factors release activate inflammatory and remodeling programs in epithelial cells that directly communicate with submucosal mesenchymal cells. The release of eosinophil cytotoxic mediators, free oxygen radicals, and collagenases from activated eosinophils impair epithelial integrity leading to a *chronic wound scenario* and prolonged epithelial activation, thus signaling to nearby mesenchymal cells. This structural relationship has been termed the epithelial—mesenchymal trophic unit (EMTU)[23]

EOSINOPHIL GRANULE PROTEINS

Eosinophil granule proteins are highly cationic compounds and can thus directly interact with cellular membranes, leading to disruption of the membrane integrity of bystander structural cells. Such direct injury activates inflammatory and remodeling programs in these cells. There is increasing recognition that eosinophils can target epithelium, leading to significant alterations in activation status and function but also compromising vital barrier functions.[24] Both major basic protein (MBP) and eosinophil peroxidase (EPO) lead to rapid but differential context-dependent expression and release of potent factors implicated in remodeling from airway epithelium,[25] including endothelin-1, MMP-9, and TGF-β1, along with the ECM proteins fibronectin and tenascin. Such epithelial modulation is dependent on the cationic charge of MBP and EPO, suggesting that epithelial injury activates epithelial remodeling programs. MBP binds the calcium-sensing receptor of esophageal epithelium leading to FGF-9 release and epithelial cell proliferation in eosinophilic esophagitis,[26] and it is likely that such mechanisms also apply to airway epithelium. Degranulated eosinophils are in close proximity to fibroblasts in the asthmatic airway and cellular cross talk effectively promotes fibroblast proliferation and activation to directly switch on specific cell functions related to remodeling. MBP can stimulate fibroblasts to produce the profibrogenic cytokines IL-6 and IL-11 both independently and synergistically with IL-1 and TGF-β1.[27] *In vitro* systems demonstrate that eosinophil cationic protein (ECP) promotes fibroblast migration.[28] ECP has

been shown to inhibit the degradation of proteoglycans,[29] and eosinophils store and release MMP-9 and TIMP-1/2. ECP can also release TGF-β1 from fibroblasts *in vitro*,[28] and there is extensive cross talk between these cells that modulates the functions of both cells.[30] ECM acts as an important reservoir for latent growth factors are released on demand and then activated. Eosinophils are an important source of heparanases,[31] which specifically cleave the heparan sulfate intrachain, leading to the release of ECM-resident heparin-binding growth and differentiation factors that modulate tissue remodeling events. MBP is a potent heparanase inhibitor and thus potentially modulates such remodeling events. In addition, MBP is cytostimulatory on basophils, mast cells, and neutrophils.[32,33] Given that these coinflammatory cells have an emerging important role in tissue remodeling, MBP activation of these cells probably also contributes to remodeling events.

The lipoxygenase pathway is particularly active in eosinophils. *In vitro* studies confirm increased structural cell activation, along with a marked proliferation of airway epithelium and fibroblasts in response to leukotriene (LT) C4,[34] while LTD4 participates with growth factors in the induction of ASM proliferation.[35]

Eosinophils are a rich source of an impressive array of cytokines and growth factors (Table 12.7.1), which all have the potential to directly modulate structural cells and inflammation. IL-4, IL-5, and IL-9 overexpression is associated with marked mucus metaplasia, while IL-5 and IL-9 overexpression leads to thickening of the basement membrane and AHR.[36,37] IL-11 overexpression is associated with smooth muscle hyperplasia.[38] IL-4 and IL-13 induce fibroblast differentiation into myofibroblasts and collagen induction, even in the absence of cellular inflammation. IL-13 overexpression leads to a dramatic fibrotic response in the lung, which may be related to the ability of IL-13 to induce the TGF-β1 production and to activate stored latent TGF-β1.[39] Eosinophils have been identified as an important source of IL-17 in the asthmatic airway,[40] and IL-17 can modulate remodeling at least via activation of structural cells.[41]

Eosinophils are an important airway source of TGF-β1.[42] They are also a significant source of antifibrotic growth factor bone morphogenetic protein 7 (BMP7) in asthma,[43] suggesting that eosinophils regulate fibrotic effects in an autocrine manner, although this mechanism may be defective in asthma. Other important growth factors produced by eosinophils include FGF-2 (a potent inducer of ASM proliferation),[11] along with the fibroblast and smooth muscle cell mitogen, heparin-binding epidermal growth factor-like growth factor (HB-EGF).[44] In a time course study, Kariyawasam and coworkers showed that cellular inflammation could be dissociated from remodeling.[45] Airway eosinophil numbers were significantly increased 24 h postallergen challenge, along with remodeling.

TABLE 12.7.1 Eosinophil-Derived Factors That Contribute to Tissue Remodeling

Eosinophil-Derived Mediator	Target Cell	Functional Effect
TGF-β, leukotrienes, IL-13, IL-1, IL-4, IL-5, IL-6, IL-9, IL-17, eotaxins/CCL11, CCL24, and CCL26, eosinophil granule proteins	Epithelium	Activation Hyperplasia Goblet cell hyperplasia Epithelial permeability
TGF-β, IL-13, IL-1, IL-4, IL-17, eotaxins/CCL11, CCL24, and CCL26, eosinophil granule proteins, FGF-2, HB-EGF, osteopontin	Fibroblast	Migration and proliferation Myofibroblast transformation ECM production
TGF-β, leukotrienes, IL-1, IL-13, IL-11, IL-17 eotaxins/CCL11, CCL24, and CCL26, eosinophil granule proteins, FGF-2, HB-EGF, osteopontin	Airway smooth muscle	Activation/dysfunction Hypertrophy Hyperplasia ECM production
VEGF, FGF-2, angiogenin, IL-1, IL-8, TGF-β, leukotrienes, eotaxins/CCL11, CCL24, and CCL26, eosinophil granule proteins	Endothelium	Activation Migration Proliferation
NGF	Neuron	Axonal growth

CCL, C-C motif chemokine; FGF-2, fibroblast growth factor 2; HB-EGF, heparin-binding EGF-like growth factor; NGF, nerve growth factor; TGF-β, transforming growth factor; VEGF, vascular endothelial growth factor.

However, the eosinophil number returned to baseline levels at 7 d, at which time remodeling was sustained and maximal. Importantly, increased numbers of inflammatory cells staining for TGF-β1 were still present after 7 d, with eosinophils accounting for 15.8% of submucosal cells expressing TGF-β1.[46] It should be emphasized that the effects of complex pleiotropic molecules such as TGF-β1 is context dependent and may be significantly modulated by concurrent tissue signals generated by the complex inflammatory milieu that characterizes asthma.

Postmortem histological specimens from asthma patients consistently demonstrate extensively degranulated eosinophils closely associated with neuronal fibers in ASM. Eosinophils produce NGF,[47] a neurotrophin that enhances neurogenic inflammation but also appears to have an important role in tissue repair and remodeling.[48]

In asthma, there is a strong association between airway vascularity and airway eosinophils.[49] Eosinophils produce a wide range of agents involved in angiogenesis (reviewed in 50). For example, VEGF is constitutively expressed in secondary eosinophil granules, with further upregulation and release in response to activation by granulocyte-macrophage colony-stimulating factor (GM-CSF) or IL-5.[13] Eosinophils also produce FGF-2.[11] FGF-2, like VEGF, is a powerful angiogenic factor that also activates endothelial cells to produce proteases that degrade blood vessel basement membranes.[51] Other eosinophil factors not generally considered to be angiogenic include eotaxin/C-C motif chemokine 11 (CCL11), GM-CSF, IL-8, NGF, and MBP.[50,52,53]

Osteopontin (OPN) is a novel cytokine implicated in allergen-induced airway remodeling,[54] with potent effects on fibroblast and ASM proliferation and migration but also proangiogenic activity.[54,55] OPN is constitutively expressed by eosinophils but is rapidly upregulated by GM-CSF and IL-5.[55]

The ability of eosinophils to *cross talk* with other inflammatory cells such as basophils, macrophages, mast cells, and neutrophils may contribute to tissue remodeling. For example, mast cells are potent sources of several cytokines, such as IL-13 and IL-4, and growth factors, such as TGF-β1 and activin-A, all implicated as potent inducers of airway remodeling (Fig. 12.7.1).

REPAIR AND REMODELING IN HUMAN SKIN

Using a human skin model of allergen provocation, Phipps and coworkers[56] showed that following the peak of the late-phase reaction (6 h), there were persistent TGF-β1+ eosinophils, alpha smooth muscle actin+ myofibroblasts, tenascin immunoreactivity, and procollagen-I+ cells 24–48 h postchallenge. Direct evidence of the generation of repair markers was obtained by coculture of eosinophils and fibroblasts. This resulted in aortic smooth muscle actin immunoreactivity that could be inhibited by neutralizing antibodies to TGF-β1, as well as production of tenascin transcript and protein products. IL-13 and TGF-β1 also induced tenascin expression. These studies suggest that

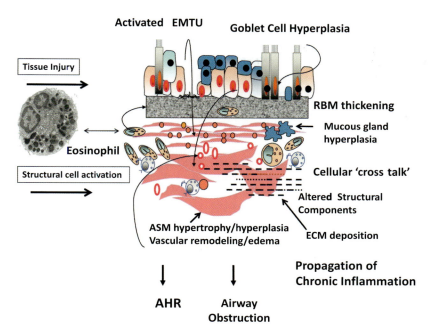

Activated EMTU **Goblet Cell Hyperplasia**

Tissue Injury

RBM thickening

Eosinophil

Structural cell activation

← **Mucous gland hyperplasia**

Cellular 'cross talk'

Altered Structural Components

ASM hypertrophy/hyperplasia Vascular remodeling/edema

ECM deposition

Propagation of Chronic Inflammation

↓ ↓

AHR **Airway Obstruction**

FIGURE 12.7.1 Diagrammatic representation of possible mechanisms underlying eosinophil-induced remodeling in the asthmatic airway. Epithelial injury and activation by eosinophil-derived mediators may promote further epithelial release of growth factors and cytokines. This, is turn, may lead to further inflammatory cell recruitment. In addition, epithelial signaling to the underlying fibroblast layer, collectively termed the epithelial–mesenchymal trophic unit (EMTU), promotes myofibroblast transformation and enhanced extracellular matrix (ECM) production. Eosinophils have the capacity to interact with other inflammatory cells (cross talk) and in this way may further regulate remodeling. This may lead to direct activation of other structural compartments, such as airway smooth muscle (ASM) and vascular tissue, by eosinophil products. Thus, the proinflammatory environment generated by chronic structural cell activation may sustain and propagate the inflammatory response triggered by allergens and other airway insults leading to sustained airway remodeling changes. AHR, alveolar hyperresponsiveness; RBM, reticular basement membrane.

IL-13 and TGF-β1, partly provided by eosinophils, contributed to repair and remodeling events in allergic inflammation in human atopic skin.

In order to test this hypothesis directly, skin eosinophil depletion was attempted using a humanized anti-IL-5 monoclonal antibody (mepolizumab).[57] Skin biopsies were performed in 24 atopic allergic subjects at allergen- and diluent-injected sites before and after (at 6 h and 48 h) three infusions of mepolizumab using a randomized double-blind, placebo-controlled design. Anti-IL-5 significantly inhibited eosinophil infiltration in 6-h and 48-h skin biopsies, as well as reducing the numbers of tenascin-immunoreactive cells at 48 h. In contrast, anti-IL-5 had no significant effect on the magnitude of the 6-h or 48-h late-phase cutaneous allergic reaction. We conclude that eosinophils are unlikely to account for the redness, swelling, and induration characteristic of the peak (6 h) late-phase cutaneous allergic reaction but probably contribute to the reduction in tenascin[+] cells at 48 h. This provides direct evidence that eosinophils may be involved in remodeling processes associated with allergic inflammation.

EOSINOPHIL DEPLETION STUDIES IN ASTHMA

Studies in human skin prompted us to determine the effect of eosinophil depletion on remodeling markers in mild asthma using mepolizumab. A previous study had shown that a single intravenous infusion to mild atopic asthmatics[58] had no appreciable effects on the late asthmatic

reaction or airway hyperresponsiveness. However, although anti-IL-5 almost totally ablated eosinophils in the blood and sputum,[59] we demonstrated that tissue eosinophil numbers were reduced rather than depleted.[59] Bronchial biopsies were obtained before and after three infusions of mepolizumab in 24 atopic asthmatics as described above. We observed only a 55% reduction in bronchial eosinophils, which is possibly a result of downregulated IL-5RA expression in airway eosinophils.[60]

Nevertheless, a significant reduction in markers of airway remodeling, defined by lumican, procollagen III, RBM, and tenascin, was observed, while type IV collagen levels remained unchanged.[61] There was also a significant reduction in the numbers and percentage of airway eosinophils expressing TGF-β1 mRNA and the concentration of TGF-β1 protein in bronchoalveolar fluid, thus confirming that eosinophils are a significant source of this profibrotic growth factor in asthma. In a cohort of patients with more severe asthma, the anti-IL-5 strategy was more effective in reducing disease exacerbation in asthmatics whose symptoms involved more airflow obstruction rather than AHR.[62] Biopsy studies are required to determine a role for eosinophils in this particular obstructive phenotype of remodeling. In fact, more detailed bronchoscopic longitudinal studies using such depletion strategies are required to define the role of eosinophils in activation and regulation of a range of remodeling pathways in asthma. A similar study in eosinophilic esophagitis supports these findings in the airways.[63] In eosinophilic esophagitis, mepolizumab also significantly reduces eosinophil numbers in esophageal tissues in adult patients with active disease and reverses changes in the expression of molecules associated with esophageal remodeling (i.e., tenascin and TGF-β in epithelial cells).

REFERENCES

1. Wardlaw AJ, Moqbel R, Kay AB. Eosinophils: biology and role in disease. *Adv Immunol* 1995;**60**:151–266.
2. Ying S, Robinson DS, Meng Q, Barata LT, McEuen AR, Buckley MG, et al. C-C chemokines in allergen-induced late-phase cutaneous responses in atopic subjects: association of eotaxin with early 6-hour eosinophils, and of eotaxin-2 and monocyte chemo-attractant protein-4 with the later 24-hour tissue eosinophilia, and relationship to basophils and other C-C chemokines (monocyte chemoattractant protein-3 and RANTES). *J Immunol* 1999;**163**:3976–84.
3. Todd R, Donoff BR, Chiang T, Chou MY, Elovic A, Gallagher GT, et al. The eosinophil as a cellular source of transforming growth factor alpha in healing cutaneous wounds. *Am J Pathol* 1991;**138**:1307–13.
4. Wong DT, Donoff RB, Yang J, Song BZ, Matossian K, Nagura N, et al. Sequential expression of transforming growth factors alpha and beta 1 by eosinophils during cutaneous wound healing in the hamster. *Am J Pathol* 1993;**143**:130–42.
5. Noguchi H, Kephart GM, Colby TV, Gleich GJ. Tissue eosinophilia and eosinophil degranulation in syndromes associated with fibrosis. *Am J Pathol* 1992;**140**:521–8.
6. Shock A, Rabe KF, Dent G, Chambers RC, Gray AJ, Chung KF, et al. Eosinophils adhere to and stimulate replication of lung fibroblasts in vitro. *Clin Exp Immunol* 1991;**86**:185–90.
7. Birkland TP, Cheavens MD, Pincus SH. Human eosinophils stimulate DNA synthesis and matrix production in dermal fibroblasts. *Arch Dermatol Res* 1994;**286**:312–8.
8. Levi-Schaffer F, Garbuzenko E, Rubin A, Reich R, Pickholz D, Gillery P, et al. Human eosinophils regulate human lung- and skin-derived fibroblast properties in vitro: a role for transforming growth factor beta (TGF-beta). *Proc Natl Acad Sci USA* 1999;**96**:9660–5.
9. Doucet C, Brouty-Boye D, Pottin-Clemenceau C, Canonica GW, Jasmin C, Azzarone B. Interleukin (IL) 4 and IL-13 act on human lung fibroblasts. Implication in asthma. *J Clin Invest* 1998;**101**:2129–39.
10. Richter A, Puddicombe SM, Lordan JL, Bucchieri F, Wilson SJ, Djukanovic R, et al. The contribution of interleukin (IL)-4 and IL-13 to the epithelial-mesenchymal trophic unit in asthma. *Am J Respir Cell Mol Biol* 2001;**25**:385–91.
11. Hoshino M, Takahashi M, Aoike N. Expression of vascular endothelial growth factor, basic fibroblast growth factor, and angiogenin immunoreactivity in asthmatic airways and its relationship to angiogenesis. *J Allergy Clin Immunol* 2001;**107**:295–301.
12. Solomon A, Aloe L, Pe'er J, Frucht-Pery J, Bonini S, Bonini S, et al. Nerve growth factor is preformed in and activates human peripheral blood eosinophils. *J Allergy Clin Immunol* 1998;**102**:454–60.
13. Horiuchi T, Weller PF. Expression of vascular endothelial growth factor by human eosinophils: upregulation by granulocyte macrophage colony-stimulating factor and interleukin-5. *Am J Respir Cell Mol Biol* 1997;**17**:70–7.
14. Moqbel R, Ying S, Barkans J, Newman TM, Kimmitt P, Wakelin M, et al. Identification of messenger RNA for IL-4 in human

15. eosinophils with granule localization and release of the translated product. *J Immunol* 1995;**155**:4939–47.
15. Moreno RH, Hogg JC, Pare PD. Mechanics of airway narrowing. *Am Rev Respir Dis* 1986;**133**:1171–80.
16. Broekema M, Timens W, Vonk JM, Volbeda F, Lodewijk ME, Hylkema MN, et al. Persisting Remodeling and Less Airway Wall Eosinophil Activation incomplete Remission of Asthma. *Am J Respir Crit Care Med*; 2010.
17. Brightling CE, Bradding P, Symon FA, Holgate ST, Wardlaw AJ, Pavord ID. Mast-cell infiltration of airway smooth muscle in asthma. *N Engl J Med* 2002;**346**:1699–705.
18. Berry MA, Hargadon B, McKenna S, Shaw D, Green RH, Brightling CE, et al. Observational study of the natural history of eosinophilic bronchitis. *Clin Exp Allergy* 2005;**35**:598–601.
19. Brightling CE, Symon FA, Birring SS, Bradding P, Wardlaw AJ, Pavord ID. Comparison of airway immunopathology of eosinophilic bronchitis and asthma. *Thorax* 2003;**58**:528–32.
20. Siddiqui S, Sutcliffe A, Shikotra A, Woodman L, Doe C, McKenna S, et al. Vascular remodeling is a feature of asthma and nonasthmatic eosinophilic bronchitis. *J Allergy Clin Immunol* 2007;**120**:813–9.
21. Broekema M, Volbeda F, Timens W, Dijkstra A, Lee NA, Lee JJ, et al. Airway eosinophilia in remission and progression of asthma: accumulation with a fast decline of FEV(1). *Respir Med* 2010;**104**:1254–62.
22. Haldar P, Brightling CE, Hargadon B, Gupta S, Monteiro W, Sousa A, et al. Mepolizumab and exacerbations of refractory eosinophilic asthma. *N Engl J Med* 2009;**360**:973–84.
23. Holgate ST, Lackie PM, Howarth PH, Roche WR, Puddicombe SM, Richter A, et al. Invited lecture: activation of the epithelial mesenchymal trophic unit in the pathogenesis of asthma. *Int Arch Allergy Immunol* 2001;**124**:253–8.
24. Furuta GT, Nieuwenhuis EE, Karhausen J, Gleich G, Blumberg RS, Lee JJ, et al. Eosinophils alter colonic epithelial barrier function: role for major basic protein. *Am J Physiol Gastrointest Liver Physiol* 2005;**289**:G890–7.
25. Pegorier S, Wagner LA, Gleich GJ, Pretolani M. Eosinophil-derived cationic proteins activate the synthesis of remodeling factors by airway epithelial cells. *J Immunol* 2006;**177**:4861–9.
26. Mulder DJ, Pacheco I, Hurlbut DJ, Mak N, Furuta GT, MacLeod RJ, et al. FGF9-induced proliferative response to eosinophilic inflammation in oesophagitis. *Gut* 2009;**58**:166–73.
27. Rochester CL, Ackerman SJ, Zheng T, Rankin JA, Elias JA. Major basic protein regulation of lung fibroblast cytokine production. Role of cytokine synergy and charge. *Chest* 1995;**107**:117S–8S.
28. Zagai U, Dadfar E, Lundahl J, Venge P, Skold CM. Eosinophil cationic protein stimulates TGF-beta1 release by human lung fibroblasts in vitro. *Inflammation* 2007;**30**:153–60.
29. Hernnas J, Sarnstrand B, Lindroth P, Peterson CG, Venge P, Malmstrom A. Eosinophil cationic protein alters proteoglycan metabolism in human lung fibroblast cultures. *Eur J Cell Biol* 1992;**59**:352–63.
30. Munitz A, Levi-Schaffer F. Eosinophils: 'new' roles for 'old' cells. *Allergy* 2004;**59**:268–75.
31. Temkin V, Aingorn H, Puxeddu I, Goldshmidt O, Zcharia E, Gleich GJ, et al. Eosinophil major basic protein: first identified natural heparanase-inhibiting protein. *J Allergy Clin Immunol* 2004;**113**:703–9.

32. O'Donnell MC, Ackerman SJ, Gleich GJ, Thomas LL. Activation of basophil and mast cell histamine release by eosinophil granule major basic protein. *J Exp Med* 1983;**157**:1981–91.

33. Moy JN, Gleich GJ, Thomas LL. Noncytotoxic activation of neutrophils by eosinophil granule major basic protein. Effect on superoxide anion generation and lysosomal enzyme release. *J Immunol* 1990;**145**:2626–32.

34. Leikauf GD, Claesson HE, Doupnik CA, Hybbinette S, Grafstrom RC. Cysteinyl leukotrienes enhance growth of human airway epithelial cells 1. *Am J Physiol* 1990;**259**:L255–61.

35. Panettieri RA, Tan EM, Ciocca V, Luttmann MA, Leonard TB, Hay DW. Effects of LTD4 on human airway smooth muscle cell proliferation, matrix expression, and contraction in vitro: differential sensitivity to cysteinyl leukotriene receptor antagonists 1. *Am J Respir Cell Mol Biol* 1998;**19**:453–61.

36. Temann UA, Geba GP, Rankin JA, Flavell RA. Expression of interleukin 9 in the lungs of transgenic mice causes airway inflammation, mast cell hyperplasia, and bronchial hyper-responsiveness. *J Exp Med* 1998;**188**:1307–20.

37. Lee JJ, McGarry MP, Farmer SC, Denzler KL, Larson KA, Carrigan PE, et al. Interleukin-5 expression in the lung epithelium of transgenic mice leads to pulmonary changes pathognomonic of asthma. *J Exp Med* 1997;**185**:2143–56.

38. Tang W, Geba GP, Zheng T, Ray P, Homer RJ, Kuhn III C, et al. Targeted expression of IL-11 in the murine airway causes lymphocytic inflammation, bronchial remodeling, and airways obstruction. *J Clin Invest* 1996;**98**:2845–53.

39. Lee CG, Homer RJ, Zhu Z, Lanone S, Wang X, Koteliansky V, et al. Interleukin-13 induces tissue fibrosis by selectively stimulating and activating transforming growth factor beta(1). *J Exp Med* 2001;**194**:809–21.

40. Molet S, Hamid Q, Davoine F, Nutku E, Taha R, Page N, et al. IL-17 is increased in asthmatic airways and induces human bronchial fibroblasts to produce cytokines. *J Allergy Clin Immunol* 2001;**108**:430–8.

41. Linden A. Interleukin-17 and airway remodelling. *Pulm Pharmacol Ther* 2006;**19**:47–50.

42. Kariyawasam HH, Robinson DS. The role of eosinophils in airway tissue remodelling in asthma. *Curr Opin Immunol* 2007;**19**:681–6.

43. Kariyawasam HH, Xanthou G, Barkans J, Aizen M, Kay AB, Robinson DS. Basal expression of bone morphogenetic protein receptor is reduced in mild asthma. *Am J Respir Crit Care Med* 2008;**177**:1074–81.

44. Powell PP, Klagsbrun M, Abraham JA, Jones RC. Eosinophils expressing heparin-binding EGF-like growth factor mRNA localize around lung microvessels in pulmonary hypertension. *Am J Pathol* 1993;**143**:784–93.

45. Kariyawasam HH, Aizen M, Barkans J, Robinson DS, Kay AB. Remodeling and airway hyperresponsiveness but not cellular inflammation persist after allergen challenge in asthma. *Am J Respir Crit Care Med* 2007;**175**:896–904.

46. Kariyawasam HJ, Barkans J, Aizen M, Kay AB, Robinson DS. Airway submucosal cells expressing TGF-beta1 are significantly increased at a time point when airway remodelling is maximal post-allergen. *Am J Respir Care Med* 2009;**179**:A1397.

47. Solomon A, Aloe L, Pe'er J, Frucht-Pery J, Bonini S, Bonini S, et al. Nerve growth factor is preformed in and activates human peripheral blood eosinophils. *J Allergy Clin Immunol* 1998;**102**:454–60.

48. Nassenstein C, Schulte-Herbruggen O, Renz H, Braun A. Nerve growth factor: the central hub in the development of allergic asthma? *Eur J Pharmacol* 2006;**533**:195–206.

49. Salvato G. Quantitative and morphological analysis of the vascular bed in bronchial biopsy specimens from asthmatic and non-asthmatic subjects. *Thorax* 2001;**56**:902–6.

50. Nissim Ben Efraim AH, Levi-Schaffer F. Tissue remodeling and angiogenesis in asthma: the role of the eosinophil. *Ther Adv Respir Dis* 2008;**2**:163–71.

51. Cross MJ, Claesson-Welsh L. FGF and VEGF function in angiogenesis: signalling pathways, biological responses and therapeutic inhibition. *Trends Pharmacol Sci* 2001;**22**:201–7.

52. Salcedo R, Young HA, Ponce ML, Ward JM, Kleinman HK, Murphy WJ, et al. Eotaxin (CCL11) induces in vivo angiogenic responses by human CCR3+ endothelial cells *J Immunol* 2001;**166**:7571–8.

53. Puxeddu I, Berkman N, Nissim Ben Efraim AH, Davies DE, Ribatti D, Gleich GJ, et al. The role of eosinophil major basic protein in angiogenesis. *Allergy* 2009;**64**:368–74.

54. Simoes DC, Xanthou G, Petrochilou K, Panoutsakopoulou V, Roussos C, Gratziou C. Osteopontin deficiency protects against airway remodeling and hyperresponsiveness in chronic asthma. *Am J Respir Crit Care Med* 2009;**179**:894–902.

55. Puxeddu I, Berkman N, Ribatti D, Bader R, Haitchi HM, Davies DE, et al. Osteopontin is expressed and functional in human eosinophils. *Allergy* 2010;**65**:168–74.

56. Phipps S, Ying S, Wangoo A, Ong YE, Levi-Schaffer F, Kay AB. The relationship between allergen-induced tissue eosinophilia and markers of repair and remodeling in human atopic skin. *J Immunol* 2002;**169**:4604–12.

57. Phipps S, Flood-Page P, Menzies-Gow A, Ong YE, Kay AB. Intravenous anti-IL-5 monoclonal antibody reduces eosinophils and tenascin deposition in allergen-challenged human atopic skin. *J Invest Dermatol* 2004;**122**:1406–12.

58. Leckie MJ, ten BA, Khan J, Diamant Z, O'Connor BJ, Walls CM, et al. Effects of an interleukin-5 blocking monoclonal antibody on eosinophils, airway hyper-responsiveness, and the late asthmatic response. *Lancet* 2000;**356**:2144–8.

59. Flood-Page PT, Menzies-Gow AN, Kay AB, Robinson DS. Eosinophil's role remains uncertain as anti-interleukin-5 only partially depletes numbers in asthmatic airway. *Am J Respir Crit Care Med* 2003;**167**:199–204.

60. Gregory B, Kirchem A, Phipps S, Gevaert P, Pridgeon C, Rankin SM, et al. Differential regulation of human eosinophil IL-3, IL-5, and GM-CSF receptor alpha-chain expression by cytokines: IL-3, IL-5, and GM-CSF down-regulate IL-5 receptor alpha expression with loss of IL-5 responsiveness, but up-regulate IL-3 receptor alpha expression. *J Immunol* 2003;**170**:5359–66.

61. Flood-Page P, Menzies-Gow A, Phipps S, Ying S, Wangoo A, Ludwig MS, et al. Anti-IL-5 treatment reduces deposition of ECM proteins in the bronchial subepithelial basement membrane of mild atopic asthmatics. *J Clin Invest* 2003;**112**:1029–36.

62. Pavord ID, Haldar P, Bradding P, Wardlaw AJ. Mepolizumab in refractory eosinophilic asthma. *Thorax* 2010;**65**:370.

63. Straumann A, Conus S, Grzonka P, Kita H, Kephart G, Bussmann C, et al. Anti-interleukin-5 antibody treatment (mepolizumab) in active eosinophilic oesophagitis: a randomised, placebo-controlled, double-blind trial. *Gut* 2010;**59**:21–30.

Eosinophils in Human Disease

Introduction

Erwin W. Gelfand

In healthy individuals, eosinophils represent a minor leukocyte subpopulation, accounting for less than 5% of total circulating white blood cells. Tissue compartments with abundant resident populations of eosinophils include bone marrow, primary, and secondary lymphoid tissues, the uterus, and most of the gastrointestinal tract (with the exception of the esophagus under homeostatic conditions). These tissues share features of substantial cellular turnover and regenerative capacity. To a large extent, eosinophils serve as effector cells, capable of inducing significant tissue damage as a result of their release of preformed cytotoxic mediators, including the granule proteins, major basic protein, and eosinophilic cationic protein. These mediators lead to the production of reactive oxygen species and generate an array of lipid mediators. The role of eosinophils was previously considered to be defensive in the setting of parasitic infections or offensive in the development of an allergic response to an environmental allergen. This binary expression of the role or function of eosinophils, especially in the context of human disease, has recently undergone considerable evolution. Eosinophilia and eosinophil products are now centrally positioned in ongoing immune responses through production of pivotal cytokines and chemokines, expression of features of antigen-presenting cells, ligation of Toll-like receptors, and the elicitation of T-helper (T_h2) immune responses. In these activities, eosinophils have been shown on the one hand to enhance local inflammatory responses, while on the other to dampen such responses. With this extensive array of activities, it is not surprising that a role for eosinophils has been demonstrated in normal tissue homeostasis and in many disease states. This chapter details the unique positions eosinophils play in a wide range of disease states, in both pathological and protective roles.

In Chapter 13.2, Per Venge begins by addressing the proteome of human eosinophils and the differences in molecular forms of many of the proteins in healthy and allergic subjects. Identification of the spectrum of proteins produced and the genetic polymorphisms of the major

secretory molecules is fundamental to our understanding of the role of eosinophils in health and disease and provides the potential for targeted regulation of specific functions.

Eosinophilia is a hallmark of allergic disorder characterized by the activation of selective hematopoietic processes during the onset and maintenance of allergic inflammation. The appearance of eosinophils in the circulation and in tissue involves processes in the bone marrow that lead to accumulation, differentiation, and proliferation of eosinophil lineage-committed hematopoietic progenitors at tissue sites. In Chapter 13.6, Gavreau and Denburg summarize mechanisms that lead to the accumulation of eosinophils and their progenitors in the airways of allergic asthmatics. The role of eosinophils in asthma is further developed by Thomas and Busse. Recognizing the controversial relationship between airway eosinophilia and asthma severity, they focus on the dynamic contribution of eosinophils to asthma. What emerges are two major roles, one in which the eosinophil serves as an effector cell in airway remodeling, the other as a biomarker for asthma exacerbations. The link between eosinophils and asthma is strengthened with the recognition that viruses are a primary cause of asthma exacerbations. In Chapter 13.7, Bivins-Smith and Jacoby explore the association of eosinophils with virus-induced asthma, especially eosinophil contact with airway nerves, which become activated and release mediators that cause dysfunction. In releasing excess acetylcholine, the altered airway nerves modulate airway smooth muscle responses and induce the development of bronchoconstrictive responses.

Moving from the airways to the skin, in Chapter 13.3 Simon and Simon discuss eosinophil infiltration of the skin in a wide variety of disorders, both allergic and nonallergic. However, it remains somewhat unclear what mechanisms are responsible for eosinophil recruitment and activation in the skin, especially as conditions and pathogenic roles vary from disorder to disorder. A similar scenario is observed in the various primary eosinophilic gastrointestinal disorders. In Chapter 13.8, Davis and Rothenberg identify common and uncommon features of these disorders and the surprising and somewhat frightening increases in prevalence of these conditions, especially eosinophilic esophagitis.

Beyond the association of eosinophilic infiltration of the airways, skin, or gastrointestinal tract, the entity *hyper-eosinophilic syndrome* has emerged, in which there is

Eosinophils in Health and Disease. http://dx.doi.org/10.1016/B978-0-12-394385-9.00013-4

significant peripheral blood eosinophilia in the absence of evidence for parasitic, allergic, infections, or other causes. In Chapter 13.9, Khoury and Klion define this rare group of disorders, taking advantage of advances in molecular diagnostics and the use of targeted therapies.

Importantly, eosinophils are now appreciated to be important players in newly recognized roles. In several cancers, eosinophilia is associated with the tumors. In Chapter, 13.10, Lofti and colleagues characterize tumor-associated eosinophils, with activities such as destructive effector functions potentially limiting tumor growth as well as immunomodulatory and remodeling activities, which may suppress immune responses. Eosinophils have been implicated in transplant rejection and have long been seen in graft-versus-host disease. In Chapter 13.13, Roufosse and colleagues review the *friend or foe* sides of eosinophil infiltration in solid organ and hematopoietic stem cell transplantation. In Chapter 13.14, Krahn and colleagues review the clinical and pathological features of the rare condition of eosinophil myositis. In the absence of known causes of eosinophilic myositis, such as parasite infections, systemic disorders or toxic causes, some cases of *idiopathic* eosinophilic myositis have been linked to calpain-3 mutations.

Over the last century, much has been learned about the functions of eosinophils, eosinophilia, and human disease. Indeed, from its initial appearance as a casual, innocent bystander in disease to its recognition as a major pathogenic effector cell causing the disease, the complexity of eosinophil activities at the molecular and cellular level is now being unraveled. An extension of this intense investigation is the recognition that targeting the eosinophil, its production and its accumulation in target tissues, is an important and necessary area of therapeutic exploration.

Chapter 13.2

Genomics and Proteomics of the Human Eosinophil

Per Venge

INTRODUCTION

Almost 40 years ago, the eosinophil granule major basic protein (MBP) was purified by Gleich and colleagues from guinea pig cells[1,2] and the eosinophil cationic protein (ECP/RNase3) purified from human leukemia cells by our group.[3,4] These achievements were the starting points for the study of the proteome of the human eosinophil, with the subsequent purification of MBP from human eosinophils[5] and the identification and purification of the two other major eosinophil granule proteins, eosinophil protein X/eosinophil-derived neurotoxin (EPX/EDN)[6,7] and eosinophil peroxidase (EPO).[8,9] Remarkably, these four highly basic proteins make up about 90% of the proteins contained in the secretory granules of the human eosinophil. The eosinophil is regarded to be a secretory cell and it was consequently assumed that the biological activities of the human eosinophil are governed to a large extent by the activities of these proteins. Thus, in attempts to understand the role of the human eosinophil in health and disease, a detailed study of the proteins and their genetics in relation to human disease therefore seemed logical. In this context, it should be emphasized that the granule content of human eosinophils is unique and shared only with other primates, since the duplicated gene products ECP and EPX/EDN are rapidly evolving and highly divergent orthologues are present in nonprimate mammalian species.[10] Such findings indicate that the interpretation of activities of eosinophils of other species should be extrapolated to humans with care. In this subchapter, I will describe some of the key activities of the four major granule proteins and the experience of assaying these proteins in various biological materials in human disease. I will also describe recent attempts to map the protein content further using modern proteomics techniques. At the end of the subchapter, I will summarize the genetic findings of the proteins and the associations of single nucleotide polymorphisms (SNPs) of the genes encoding these proteins with disease. More details of the biological activities of the four proteins are found throughout this volume.

THE FOUR MAJOR GRANULE PROTEINS OF HUMAN EOSINOPHILS

The four major granule proteins of human eosinophils will be considered in turn (Table 13.2.1).

Eosinophil Cationic Protein

ECP is a single chain, highly basic protein [isoelectric point (pI) ranging from 10.5 to 11] with apparent molecular masses ranging from 15.7 kDa to 22 kDa. The heterogeneity is largely due to glycosylation of the protein.[11,12] The gene encoding ECP comprises two exons and one intron and is located on the q arm of chromosome 14 (14q). Exon 2 is the coding DNA sequence for ECP. ECP is located in the secretory granules of human eosinophils and is unique to humans and primates. Minute amounts of ECP may be produced by monocytes and neutrophils under certain conditions.[13–15] However, most of the ECP located in neutrophils probably derives from the active uptake of ECP from the environment.[16] ECP belongs to the large family of RNases and is also named RNase3. In addition to being an RNase, ECP is a true multifunctional protein with both

TABLE 13.2.1 Some Characteristics of the Four Major Granule Proteins

Molecule	Chromosome	Protein Size (kDa)	pI	Major (Minor) Cellular Localization	Biological Activities
ECP (RNase3)	Chr 14q	15.5—22	10.5—11	Eosinophil (Neutrophil, Monocyte)	Cytotoxic, Noncytotoxic activities (RNase)
EPO	Chr 17q	~66	>11	Eosinophil	Peroxidase, Nonperoxidase activities
EPX/EDN (RNase 2)	Chr 14q	18.6	~9	Eosinophil (Neutrophil, Liver)	RNase, Alarmin
MBP	Chr 11q	13.8	10.8	Eosinophil, Placental cells (Basophil)	Cytotoxic, Noncytotoxic activities

ECP, eosinophil cationic protein; EDN, eosinophil-derived neurotoxin; EPO, eosinophil peroxidase; EPX, eosinophil protein X; MBP, major basic protein; pI, isoelectric point; RNase 2, non-secretory ribonuclease.

cytotoxic and noncytotoxic activities. The cytotoxic activities are determined by post-translational glycosylations and the majority, if not all, of ECP stored in the granules is richly glycosylated and noncytotoxic.[12,17,18] Upon release from the eosinophil, the molecule is deglycosylated and acquires cytotoxic capabilities.[19] Several SNPs have been found in the DNA sequence of ECP; however, only two are in the coding part of exon 2.[20,21] The most commonly found SNP is located at position 434, in which guanine (G) is replaced by cytosine (C). In Scandinavian populations 434G is most commonly found, whereas in African populations 434C is the most common.[22] Thus, in Scandinavia about 60% of the population carry the 434GG genotype and about 8% the 434CC genotype, whereas the reverse is the case in a Ugandan population. The 434G>C SNP results in an amino acid shift from arginine at position 96 to threonine and a fundamental change in biological activity, since the cytotoxic activity is lost.[18] Whether the loss in cytotoxic activity is due to the amino acid shift per se affecting the cytotoxic site of the molecule or due to the fact that the replacement of arginine with threonine potentially creates a new glycosylation site that might disguise another cytotoxic site is at present not entirely clear. Attempts to identify the bactericidal active sites were made by engineering recombinant protein and peptides.[23] These experiments suggest a location for the activity at the N-terminal portion of ECP. The presence in the ECP molecule of several active sites, possibly with different targets, seems likely. The SNP 277C>T in the coding region of the ECP gene is much less common and gives rise to a replacement of arginine at position 47 with cysteine. The possible functional consequences of this amino acid shift are unknown, but it is predicted to have a great impact on the molecular structure. Other biological activities of ECP, such as the RNase activity and the ability of ECP to activate fibroblasts, are not affected by the amino acid shift from

arginine to threonine, which shows that these capabilities of ECP are dissociated from each other.[17] Other SNPs in the ECP gene are associated with protein expression. Thus, the 562G>C SNP in the 3′UTR region was found to closely correlate with the cellular content of ECP.[24] The affected sequence is a binding site for the transcription factor, retinoic acid receptor (RXR), which in turn acts as a cofactor to the transcription factor, Sp1. Sp1 was shown to affect ECP synthesis by binding to the promoter region of the gene. Also, the intronic SNP c.-38A>C has been shown to relate to the ECP content of eosinophils. Thus, several parts of the ECP gene seem to affect ECP production. In a Japanese population, a promoter polymorphism -393C>T is common, but is not found at all in the Scandinavian population.[25] This mutation is closely related to serum levels of ECP, thus suggesting an impact on ECP production. The activities of ECP are counteracted by heparin, but also by a protease-modified α_2-macroglobulin.[11]

Eosinophil Protein X/Eosinophil-Derived Neurotoxin

EPX/EDN is a single chain protein of 18.6 kDa that shares 70% homology with ECP, since the formation of the two proteins is the result of a gene duplication 30—40 million years ago.[26,27] The ancestral gene was an RNase and this property has been conserved by the gene product EPX/EDN, but almost completely lost in the gene product ECP, which has instead acquired cytotoxic properties. The protein was independently described, purified, and named (EPX) by our group[7] and the group of Gleich (named EDN)[28] and both names are used in the literature. For the sake of clarity the name eosinophil derived neurotoxin (EDN) will be used throughout this volume, since this is the more commonly used name and also because the gene of eosinophil peroxidase (EPO) recently has been renamed EPX. The EDN gene is located close to the ECP gene at chromosome 14. Chromosome 14 is also the location of the genes of the RNase A superfamily to which ECP and EDN belong; hence, the alternative name of RNase2. EDN is also a highly basic protein (pI of about 9), although less so than ECP. EDN is produced in small amounts by macrophages and neutrophils, but also by liver cells. EDN is stored in the eosinophil within the secretory granules together with ECP, but is also stored in a separate compartment of easily mobilized secretory vesicles.[29] The biological activities of EDN are related to its RNase activity and involve antiviral properties. However, recent studies indicate several other activities of great interest. Thus, EDN has been added to the growing list of alarmins,[30] which are proteins that attract and enhance the activities of antigen-presenting cells, such as dendritic cells. Activation of cells through Toll-like receptor 2 (TLR2) further links the activity of EDN to components of innate immunity. The neurotoxic activity, which is the basis for the name EDN, suggests cytotoxic properties for EDN, although the cytotoxic activity of the molecule against any other cell is modest and mostly absent. In our previous studies, we could show some alterations in Purkinje cells in the cerebellum of rabbits following injection of EDN, thus resembling the Gordon phenomenon.[31] However, the injection of 100 times lower amounts of ECP had much more detrimental consequences, with the disappearance of Purkinje cells and the rapid development of ataxia and other neurological disturbances. The neurotoxic activities of the eosinophil proteins and the development of the Gordon phenomenon may therefore be the combined actions of the potent RNase EDN and the cytotoxic ECP. Four SNPs were identified in the EDN gene in a Scandinavian population, none of which gives rise to an amino acid shift. One SNP, 405G>C, is located in the intron and is closely related to the cellular content of EDN. This locus is the binding site for several different transcription factors that may be involved in the expression of EDN.

Eosinophil Peroxidase

EPO is a two chain heme-binding protein with one heavy chain of about 52 kDa and one light chain of about 14 kDa.[9] The gene is located on chromosome 17q31 and consists of 12 exons and 11 introns. The amino acid sequence shows an almost 70% homology with that of myeloperoxidase and also considerable homology with other members of the peroxidase family of proteins.[32] EPO is a highly basic protein with a pI of >11. It is located in the matrix of the secretory granules and is probably specific to eosinophil granulocytes, since no other locations have been identified in mature cells. EPO is difficult to extract from mixed blood leukocytes, since it has a high affinity for neutrophil membrane structures.[33] The biological activities of EPO are partly related to its peroxidase activity and partly to other properties of the

molecule. The peroxidase catalyzes halidation reactions leading to the formation of long-acting hypohalides, such as hypobromous acid, oxidation of thiocyanate, and nitration of tyrosine.[34,35] Such radicals may act on cellular membranes and take part in defense reactions against a variety of microbes. Numerous mutations and polymorphisms have been found in the EPO gene, five of which result in amino acid shifts. The possible consequence to functional activities of these amino acid shifts is unknown.

Major Basic Protein

Eosinophil MBP was named from findings in guinea pig eosinophils, since it appeared to make up the majority of the proteins contained in the secretory granules.[1,36–40] In human eosinophils the content of MBP is in the range of the other three major proteins, i.e., $5-10\ \mu g/10^6$ eosinophils. The mass of MBP is 13.8 kDa and its pI is 11.4. The MBP gene is located on chromosome 11q12 and consists of six exons and five introns. MBP is apparently produced as a much larger preproprotein, and an acidic portion of proMBP is cleaved off upon storage in the eosinophil granules. This acidic portion of proMBP may serve to protect cellular structures from its cytotoxic activities during synthesis and packaging. The larger proMBP, however, has been identified in immature bone marrow cells. proMBP has also been found in placental cells in complex with the metalloproteinase pappalysin-1, or pregnancy associated protein A (PAPP-A), and shown to inhibit the activities of PAPP-A. The MBP molecule makes up the typical crystals seen in the specific granules of human eosinophils. An MBP homologue was identified, characterized, and named MBP2.[40] This protein was purified from human eosinophils and has a molecular mass of about 13.5 kDa and a much lower isoelectric point of 8.7. The gene encoding hMBP2 is located in close proximity to the gene of MBP1 at chromosome 11q12 and has five exons. MBP1 is expressed in several cell types other than human eosinophils, such as basophils and placental cells, whereas hMBP2 seems to be located only in eosinophils. The biological activities of MBP are predominantly related to its cytotoxic capabilities, but numerous noncytotoxic activities have also been identified, many of which will be described throughout this volume. In the MBP genes, several mutations and polymorphisms have been identified, five of which may result in an amino acid shift. Consequences to the activities of MBP resulting from these amino acid shifts have not been described.

PROTEOMICS STUDIES OF HUMAN EOSINOPHILS

As discussed elsewhere in this volume, the human eosinophil is capable of producing and secreting a number of other proteins in addition to the major proteins described above. These include large numbers of adhesion molecules, chemokines, cytokines, and others. In an attempt to gain further insight into the biology of human eosinophils, modern proteomics techniques may be applied to map the major protein content of normal and diseased eosinophils. In this regard, several different approaches may be applied. One is the description of as many proteins as possible and another is the selected description of proteins based on criteria such as extraction procedures or detection methods, e.g., based on the identification of phosphorylated proteins only. One study incubated eosinophils with sonicates of mast cells and the cytokines granulocyte-macrophage colony-stimulating factor (GM-CSF) and tumor necrosis factor (TNF-α) and used [35S]methionine to monitor protein synthesis.[41] Extracts of eosinophils were run on two-dimensional (2-D) gels and the number of protein spots increased dramatically following these stimuli compared to control cells. In addition, the position of the spots differed depending on the stimuli used, which suggests that eosinophils respond differently to these stimuli. Unfortunately, no attempts were made to identify the proteins in these spots. Another study showed differences in 51 spots between healthy subjects and those affected by atopic dermatitis.[42] One such difference was downregulation of the Grb7 adaptor protein in cells from patients, which may relate to eosinophilia of the patients and antiapoptotic features of these cells. Overall, 1121 spots were identified in healthy subjects and 1310 spots in the eosinophils of atopic dermatitis patients, which emphasizes that circulating eosinophils of such patients are exposed to various stimuli that induce protein synthesis. One upregulated spot of particular interest in atopic dermatitis relates to increased expression of the low-affinity receptor for immunoglobulin E (IgE). A different approach involved

the study of phosphoproteins in an acute myelogenous leukemia (AML) eosinophil cell line after exposure to dexamethasone or IL-5.[43] Fourteen phosphoproteins showed significant changes, i.e., were either phosphorylated or dephosphorylated, after IL-5 and 12 after dexamethasone. Phosphorylation of the translation initiation factor elf-3 subunit was increased by IL-5 and was also found to be increased in patients with atopic dermatitis. Interestingly, phospho-apolipoprotein E (p-APOE) was induced in eosinophils by dexamethasone but was decreased by IL-5 treatment. p-APOE levels could therefore be used as an indicator of proliferation or apoptosis of eosinophils. A 2-D gel of a survey of proteins in whole eosinophil extracts and extracts of membrane fractions of eosinophils of healthy subjects and of eosinophils obtained from allergic subjects during a pollen season is shown (Fig. 13.2.1). Altogether more than 336 spots were identified by matrix-assisted laser desorption/ionization time-of-flight mass spectrometry (MALDI-TOF MS), representing 98 different proteins.[44] Among the proteins identified were the four major granule proteins described above and a number of other proteins hitherto not associated with human eosinophils. As expected, the proteins represent, to a large part, cytoskeleton-related proteins such as actin, but more than 11% of all proteins are of granular origin (Fig. 13.2.2). The study also showed large differences in the expected pIs of several proteins. Thus, the cytoskeleton-related proteins cofilin-1, profilin-1, adenylyl cyclase-associated protein 1 (CAP 1), and coronin-1A were all found to be significantly acidified, whereas EDN and MBP2 were much more basic than expected. The actual biological significance of alterations to cytoskeleton-related proteins is uncertain, but may relate to the well-documented migrating capacity of eosinophils. In eosinophils obtained from allergic subjects

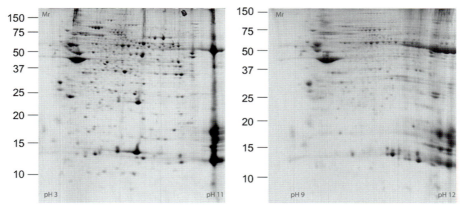

FIGURE 13.2.1   Two-dimensional separation of the proteins of eosinophils from healthy subjects. The molecular weight is given on the vertical axis. The left panel represents separation of proteins using a pH gradient 3–11 and the right panel separation from pH 9 to pH 12. A large number of proteins are gathered at the end, i.e., at about pH 11, of the left panel. These highly basic proteins were further separated and identified as shown in the right panel. Due to poor solubility, some proteins were not possible to separate using 2-D electrophoresis and their isoelectric points could therefore not be estimated. Among the insoluble proteins were eosinophil lysophospholipase (Charcot–Leyden protein), part of eosinophil peroxidase (EPO) and major basic protein (MBP). Mr, relative molecular mass.

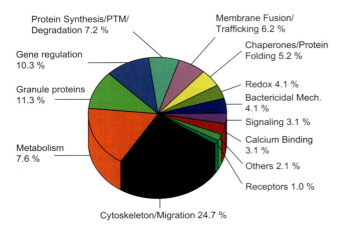

FIGURE 13.2.2 Distribution of the eosinophil proteins according to their biological functions. The largest proportion of proteins is related to the cytoskeleton and metabolism of the cell. However, as much as 11% of total proteins identified are granule secretory components. PTM, post-translational modification.

exposed to pollen, several intriguing changes were observed (Table 13.2.2). One such was a change of more than three units in the pI of two protein spots identified as the heavy chain of EPO due to heavy chain glycosylations. We speculate that such heavy glycosylation may interfere with the enzymatic activity of EPO. In support of such speculations are our previous findings that the peroxidase-dependent luminol-enhanced chemiluminescence reaction of blood eosinophils purified from allergic subjects during the pollen season is significantly reduced.[45] Altogether 12 spots were significantly changed in eosinophils from pollen-exposed allergic subjects, five of which were identified by MALDI-TOF MS. The other three identified spots were heat shock cognate protein 70 (hsc70) and the α and β subunits of CAP 1. As indicated above, CAP 1 subunits are involved in cell motility. The upregulation of these proteins, therefore, indicates that eosinophils from allergic subjects have an

increased potential to respond to chemoattractants, a capacity that is well documented, since eosinophils harvested from the blood of allergic subjects show increased migration toward several chemoattractants.[46] The upregulation of hsc70 has a number of biological implications of interest, such as changes in protein folding, intracellular protein transportation, and antigen presentation. The latter may lend further support to the eosinophil being actively involved in antigen presentation.[47]

ASSAYING EOSINOPHIL GRANULE PROTEINS IN DISEASE

The eosinophil marker that has become most widely used in the everyday clinical routine of the allergist is ECP, although several reports have shown that the measurement of EDN, EPO, or MBP may also be useful. The measurement of any of these eosinophil proteins may indicate the activity and turnover of the eosinophil granulocyte. Currently, ECP is measured in serum/plasma, but measurements in nasal lavage fluid, sputum, and possibly saliva are interesting alternatives, since ECP in these biological fluids more accurately reflects the local process. The advantages and disadvantages of measuring ECP in various biological fluids will be discussed below, and the current evidence that ECP may be a useful complement to the diagnostic armamentarium for monitoring and characterizing disease activity in the allergic patient. The emerging evidence of the clinical usefulness of urine measurement of EDN as alternative to serum ECP measurement to reflect eosinophil turnover and activity will also be considered.

Eosinophil Cationic Protein in Sputum and Other Secretions

Numerous reports show that assaying ECP in nasal lavage, saliva, and sputum has the potential to become a clinical

TABLE 13.2.2 Isoelectric Points and Molecular Weights (M_r) of the Four Major Granule Proteins of Healthy or Pollen-Allergic Subjects

Molecule	Theoretical pI/M_r	Experimental pI/M_r (Healthy Subjects)	Experimental pI/M_r (Allergic Subjects)
ECP	10.72/15.7	10.16–10.72/15.5–20.8	Similar to healthy
EPO, heavy chain	10.79/53.4	10.97/55.6	7.48/53.7
EPO, heavy chain	10.79/53.4	9.37/56.9	7.30/50.7
EPX/EDN (RNase 2)	9.2/15.5	9.67–10.65/17.2–25.3	Similar to healthy
MBP	10.8/13.8	12.6/ND	Similar to healthy

ECP, eosinophil cationic protein; EDN, eosinophil-derived neurotoxin; EPO, eosinophil peroxidase; EPX, eosinophil protein X; MBP, major basic protein; ND, not done; pI, isoelectric point; RNase 2, non-secretory ribonuclease.

instrument for characterizing and monitoring inflammatory processes in the airways.[11,48] This has been particularly shown in patients with asthma, chronic obstructive respiratory disease, and cystic fibrosis. In most cases, sputum has to be induced by hypertonic saline and cells in the sputum need to be separated from the supernatant in order to analyze mediators released from inflammatory cells. The relatively time consuming and complicated procedures required to achieve this are probably the main obstacles for a more widespread use of sputum measurement as a clinical tool. An alternative and much simpler procedure is the measurement of specific markers of various cells in whole sputum extracts. The numbers of eosinophil granulocytes in sputum have been estimated using ECP and several publications show that the numbers of eosinophils measured in this way correlate well with disease activity in asthma and are reduced as a consequence of corticosteroid treatment. An interesting alternative to sputum is saliva, since we showed recently that asthmatics have significantly raised levels of ECP in saliva that are reduced by corticosteroid treatment.[49] Still, however, we do not know what the ECP levels in saliva actually reflect, as they may be indicative of either systemic or local eosinophil activity. In addition, the measurements of specific cell markers in nasal lavage fluids or ear secretions have been widely used, and the usefulness of such measurements in the understanding of cellular involvement has been clearly indicated.[50] However, their clinical application is still not established.

Eosinophil Cationic Protein in Serum/Plasma

ECP may be measured in both serum and plasma.[11] If plasma is chosen, the blood should be anticoagulated with ethylenediaminetetraacetic acid (EDTA) or citrate in order to prevent spontaneous extracellular release of ECP and subsequent interaction with heparin. The levels of ECP in serum are consistently higher than in plasma, due to the fact that eosinophils in the test tube continue their extracellular release of ECP *ex vivo*. This is an active process that is both time and temperature dependent, which means that higher extracellular levels are achieved with increasing time and ambient temperature, and vice versa.[51] Thus, if ECP is measured in serum, strict standardization of the blood sampling procedure and handling of the blood sample are necessary in order to avoid unacceptable variations in ECP levels. Our recommendation is that blood should be taken in tubes with a gel separator and that coagulation is allowed for 1 h at room temperature (22°C) before centrifugation and separation of serum. Both plastic and glass tubes may used. However, differences in the material and the inclusion of coagulation activators in the tubes may affect measurements. This means that normal ranges of ECP have to be determined in each laboratory that does not follow the recommendations of the manufacturer. The levels of ECP in EDTA—plasma probably correctly reflect the circulating levels of ECP at the time of blood sampling. These levels are the consequences of production and elimination of ECP, i.e., local or systemic release of ECP to the circulation as well as variations in the turnover rate of ECP. Normally, turnover is quite rapid, with a half-life ($t_{1/2}$) of about 45 min. For several reasons, we can assume that the turnover is more rapid in subjects with ongoing inflammation. This means that an increased release of ECP to the circulation in certain diseases does not always lead to the anticipated increase in plasma levels, since the increased release may be partly or fully counteracted by an increased elimination rate. The dynamics of such counteracting principles may be the main explanation of the fact that in most cases clinical information obtained by EDTA—plasma measurements of ECP is less clear and less useful than the information obtained by serum measurements. In addition to the circulating levels, serum levels of ECP also reflect the secretory activity of the eosinophil population in the blood and, since the levels in serum are often 5—10 times those in plasma, we may draw the conclusion that it is above all the secretory activity of the eosinophils that determines ECP levels. The propensity of blood eosinophils to release ECP is increased in subjects exposed to allergen.[52] My own interpretation of serum ECP levels is that they reflect the propensity of the eosinophil population to release ECP in the local process, e.g., in the lung of asthmatics. The higher this propensity is, the more damage is inflicted on the patient. In order to eliminate the influence of eosinophil counts on the serum levels of ECP and thereby obtain a purer reflection of eosinophil activity, serum levels may be divided by the eosinophil count, thus forming an ECP/eosinophil ratio. In some studies, such a ratio was found to be more closely related to disease severity in asthma.

More than a thousand papers have been published dealing with the relation between ECP levels and allergic or other inflammatory diseases.[11,48,53,54] The majority of these publications indicate that ECP provides novel information about the process and that the information may be used in treatment stratification and monitoring of the disease, since ECP levels are closely related to exacerbation propensity and severity in diseases such as asthma and atopic dermatitis. Recent data also indicate that serum levels of ECP and EPO predict the further development of allergic disease.[55] Thus, serum levels of the two proteins were significantly elevated in a group of patients with allergic rhinitis who developed asthma-like symptoms 6 years later. The prediction was not seen for blood eosinophil counts or nasal lavage findings. Several publications, though, have questioned and sometimes rejected ECP as a clinically useful marker. One reason for this may be the simplified view that asthma is one disease and that the disease is caused by one cell, the eosinophil, and that one marker such as ECP will solve all clinical problems. This is obviously not true, since we know today that the involvement of eosinophils in the asthmatic process is very variable between individuals. Another cause of variations in the results is probably the lack of awareness of the importance of correct sample handling. Still another possibility may be related to the recent discovery of several genetic variants of ECP and the fact that these variants are related to the expression of allergic symptoms and serum levels of ECP. The levels of ECP in blood or any other biological fluid are not disease specific, but provide us with information about the extent to which eosinophils are involved in the particular disease process.

Eosinophil Protein X/Eosinophil-Derived Neurotoxin in Urine

In a search for noninvasive means to monitor eosinophil involvement in inflammatory diseases, urine measurement of EDN has emerged as a promising candidate, particularly in children.[56–58] In order to minimize the influence of differences in water dilution of the urine, EDN levels have to be adjusted to creatinine concentrations unless 24-hour samplings are carried out. It is also useful to bear in mind that all eosinophil markers, including the excretion of EDN, show circadian rhythms with the highest levels occurring at night.[59] Thus, for the sake of standardization of blood or urine measurements of eosinophil markers, sampling should always be carried out at the same time of the day. Several studies have shown that EDN in urine is elevated in asthma and atopic dermatitis, is related to disease activity, and reduced by anti-inflammatory treatment, i.e., corticosteroid treatment. Elevated levels of EDN in urine in wheezy children also seem to predict the development of asthma.

SINGLE NUCLEOTIDE POLYMORPHISMS IN THE MAJOR GRANULE PROTEINS AND THEIR ASSOCIATIONS WITH DISEASE

As mentioned above, two major granule proteins of human eosinophils are unique to humans and primates. Thus, knowledge of the role of these proteins in human disease cannot be extrapolated from mouse gene knockout experiments. The alternative is therefore to search for human counterparts to such knockouts, i.e., SNPs or mutations that have an impact on the biological activities of these proteins either with regard to their functional alterations or altered production. As shown above, only one SNP is known

to lead to a functional alteration in any of these proteins: the ECP434G>C gene polymorphism, in which the replacement of G with C in the DNA sequence results in the production of a noncytotoxic protein with the amino acid threonine at position 97 instead of arginine.[18]

ECP434G>C (rs2073342) Genotypes and Clinical Findings

In the first study examining the possible association of the ECP434G>C SNP with human disease, we found strong associations with the development of allergic symptoms, both in a group of 209 medical students and in a group of 79 subjects with allergic or nonallergic asthma.[21] In the group of medical students, the diagnosis of allergy was based on a self-assessment questionnaire and in the asthma group the distinction was based on a clinical diagnosis. We found that the genotype ECP GG, giving rise to the production of the cytotoxic species, was more common in those experiencing allergic symptoms than in nonallergic subjects or those with nonallergic asthma. However, most notable was the absence of any allergic manifestations in the two cohorts carrying the ECP CC genotype. The ECP CC genotype therefore seemed to exclude the development of allergic symptoms and provided strong support for a key role for eosinophils in allergy. In a larger community-based study on 574 randomly selected subjects in Estonia and Sweden [The European Community Respiratory Health Survey (ECRHS)], symptoms and signs of allergy were based on a structured interview.[60] The results of this study were much less clear, although the ECP434G>C genotypes (ECP434GG, ECP434GC, and ECP434CC) showed significant associations with various expressions of allergic symptoms. However, it also became clear that ethnicity, gender, and smoking habits are important confounders. One intriguing finding of the ECRHS study was the associations of the ECP434G>C genotypes with lung function with reduced lung functions found in both women and men carrying the ECP434G>C G-allele compared to those carrying the C-allele. If confirmed, these findings suggest a detrimental effect of the cytotoxic ECP on lung tissues. In a Norwegian–Dutch study, no associations between allergy and the ECP434G>C genotype were found.[61] In contrast, this study showed an association with nonallergic asthma, which was also the case in our ECRHS study. An association with allergic rhinitis of the ECP434G>C genotype was also negated in a Korean study.[62] The association between asthma/allergy and the ECP434G>C genotype is at present confusing and the seeming differences in findings not easily explained. Ethnicity and gender differences may have an impact, but the definition of the phenotypes studied is probably more important. It is important to clarify these relationships, since the cytotoxic activities of ECP could be targets for therapeutic interventions if such associations are confirmed and established.

ECP has the capacity to kill *Schistosoma mansoni* larvae. Knowing that the cytotoxic capacity is lost by an amino acid shift from arginine to threonine at position 97, we conducted a study on subjects living in Uganda in an endemic area of *S. mansoni* infections.[22] The ECP434G>C genotype distribution in this population was dominated by subjects carrying the ECP434 CC genotype, i.e., the opposite of the distribution found in non-African populations. Thus, the majority of people living in these endemic areas have a genotype that gives rise to the production of a noncytotoxic ECP and possibly a poorer defense against *S. mansoni* infection. We examined parasite egg excretion in feces, to reflect the level of defense against the infection, and indeed found higher numbers in subjects carrying the C-allele, thus suggesting the involvement of ECP in this defense reaction. We also found that subjects carrying the G-allele are much more prone to develop liver fibrosis, one of the serious consequences of the infection that affects about 10% of those infected by *S. mansoni*. Thus, it seems that the capacity to produce the cytotoxic ECP has some effect on defense against *S. mansoni*, but that the host's reaction to the infection is the major threat to the infected subject. In this regard, cytotoxic ECP may play a key role.

ECP 562G>C (rs2233860) Genotypes and Clinical Findings

As mentioned above, the ECP562G>C genotype is closely related to the cellular content of ECP, with the lowest levels found in those carrying the ECP CC genotype.[24] We found few, if any, associations between this genotype and the expression of allergic symptoms or asthma, but close relations to gender, with a higher prevalence of the G-allele in women, and relations to smoking habits.[60] Similar findings were seen for the 434G>C genotypes and may relate to the fact that these two genotypes are in strong disequilibrium. In the Korean study, the ECP562 G>C C-allele was found to be more prevalent in allergic rhinitis.[62] In the Norwegian–Dutch study, no apparent relations to allergy, asthma, and ECP levels were found, whether evaluated alone or as part of a haplotype.[61]

ECP c.-38A>C (rs2233859) Genotypes and Clinical Findings

The intronic SNP ECP c.-38A>C is located in a part of the gene that may be involved in regulating ECP production. The Norwegian–Dutch study showed a higher proportion of elevated serum levels of ECP and IgE, as well as higher proportions of subjects with asthma and bronchial hyperreactivity, in those carrying the adenine (A)-allele.[61] Most

of these associations, though, were only seen in the Dutch population. In a Japanese study, no association between the intronic SNP and serum ECP levels were found.[25] In our ECRHS study, we found an intriguing association between the ECP c.-38A>C genotypes and atopy.[60] Among males, but not among women, atopy was associated with the ECP c.-38(A>C) genotypes, with a significantly higher frequency of the CC genotype. In a logistic regression analysis, the ECP c.-38CC genotype was independently associated with an increased risk of atopy with an odds ratio of 1.9 and confidence interval (CI) of 1.2–3.1 when adjusted for gender, ethnicity, and smoking habits.

ECP -393C>T (rs11575981) Genotypes and Clinical Findings

The ECP -393C>T SNP is located in the promoter region of the ECP gene. This SNP has only been described in the Japanese population.[25] Interestingly, the -393C>T genotypes are related to serum levels of ECP, with undetectable levels in subjects carrying the TT genotype.

Eosinophil Protein X/Eosinophil-Derived Neurotoxin, Eosinophil Peroxidase, and Major Basic Protein Genotypes and Clinical Findings

One study examined a large number of EDN and ECP SNPs in a South Indian population with microfilaria infection and tropical pulmonary eosinophilia.[63] No associations between either of these conditions and the SNPs examined were seen. However, the South Indian population seemed to have unique SNPs and haplotypes of the EDN and ECP genotypes compared to Asian and Scandinavian populations. In two Japanese reports, SNPs in the coding parts of the EPO gene were found to be associated with cedar pollinosis.[64,65] In particular, an SNP in exon 7, resulting in the amino acid shift from proline to leucine and which might affect the activity of the protein, showed a strong association. In the second study, an association with a silent SNP in exon 6 was shown in addition to the exon 7 SNP. In a recent Czech study on allergic rhinitis, the exon 6 SNP was also found to be associated, whereas the exon 7 SNP was not present in this population.[66] These reports suggest the involvement of EPO in the allergic process although whether it involves the peroxidase activity of EPO or reflects other mechanisms is unknown. One report from Germany studied patients with atopic dermatitis and possible associations between nine different SNPs located in the four major eosinophil granule protein genes.[67] However, no associations with atopic dermatitis were found for any of the SNPs studied, despite the fact that eosinophils are regarded to be important effector cells in this disease. No other studies have investigated SNPs in the MBP gene in relation to human disease.

REFERENCES

1. Gleich GJ, Loegering DA, Maldonado JE. Identification of a major basic protein in guinea pig eosinophil granules. *J Exp Med* 1973;**137**:1459−71.
2. Gleich GJ, Loegering DA, Kueppers F, Bajaj SP, Mann KG. Physiochemical and biological properties of the major basic protein from guinea pig eosinophil granules. *J Exp Med* 1974;**140**:313−32.
3. Olsson I, Venge P. Cationic proteins of human granulocytes. I. Isolation of the cationic proteins from the granules of leukaemic myeloid cells. *Scand J Haematol* 1972;**9**:204−14.
4. Olsson I, Venge P. Cationic proteins of human granulocytes. II. Separation of the cationic proteins of the granules of leukemic myeloid cells. *Blood* 1974;**44 N0.2**:235−46.
5. Gleich GJ, Loegering DA, Mann KG, Maldonado JE. Comparative properties of the Charcot-Leyden crystal protein and the major basic protein from human eosinophils. *J Clin Invest* 1976;**57**:633−40.
6. Durack DT, Ackerman SJ, Loegering DA, Gleich GJ. Purification of human eosinophil-derived neurotoxin. *Proc Natl Acad Sci U S A* 1981;**78**:5165−9.
7. Peterson CGB, Venge P. Purification and characterization of a new cationic protein—eosinophil protein-x(EPX)—from granules of human eosinophils. *Immunology* 1983;**50**:19−26.
8. Olsen RL, Little C. Purification and some properties of myeloperoxidase and eosinophil peroxidase from human blood. *Biochem J* 1983;**209**:781−7.
9. Carlson MGCh, Peterson CGB, Venge P. Human eosinophil peroxidase: purification and characterization. *J Immunol* 1985;**134 N0.3**:1875−9.
10. Rosenberg HF, Dyer KD. Eosinophil cationic protein and eosinophil-derived neurotoxin. Evolution of novel function in a primate ribonuclease gene family. *J Biol Chem* 1995;**270**:21539−44.
11. Venge P, Byström J, Carlson M, Håkansson L, Karawacjzyk M, Peterson C, et al. Eosinophil cationic protein (ECP): molecular and biological properties and the use of ECP as a marker of eosinophil activation in disease. *Clin Exp Allergy* 1999;**29**:1172−86.
12. Eriksson J, Woschnagg C, Fernvik E, Venge P. A SELDI-TOF MS study of the genetic and post-translational molecular heterogeneity of eosinophil cationic protein. *J Leukoc Biol* 2007;**82**:1491−500.
13. Sur S, Glitz DG, Kita H, Kujawa SM, Peterson EA, Weiler DA, et al. Localization of eosinophil-derived neurotoxin and eosinophil cationic protein in neutrophilic leukocytes. *J Leukoc Biol* 1998;**63**:715−22.
14. Monteseirin J, Vega A, Chacon P, Camacho MJ, El BR, Asturias JA, et al. Neutrophils as a novel source of eosinophil cationic protein in IgE-mediated processes. *J Immunol* 2007;**179**:2634−41.
15. Byström J, Tenno T, Håkansson L, Amin K, Trulson A, Högbom E, et al. Monocytes, but not macrophages, produce the eosinophil cationic protein. *APMIS* 2001;**109**:507−16.
16. Byström J, Garcia RC, Håkansson L, Karawajczyk M, Moberg L, Soukka J, et al. Eosinophil cationic protein is stored in, but not produced by, peripheral blood neutrophils. *Clin Exp Allergy* 2002;**32**:1082−91.
17. Rubin J, Zagai U, Blom K, Trulson A, Engström A, Venge P. The coding ECP 434(G>C) gene polymorphism determines the cytotoxicity of ECP but has minor effects on fibroblast-mediated gel contraction and no effect on RNase activity. *J Immunol* 2009;**183**:445−51.
18. Trulson A, Byström J, Engström A, Larsson R, Venge P. The functional heterogeneity of eosinophil cationic protein is determined by a gene polymorphism and post-translational modifications. *Clin Exp Allergy* 2007;**37**:208−18.
19. Woschnagg C, Rubin J, Venge P. Eosinophil cationic protein (ECP) is processed during secretion. *J Immunol* 2009;**183**:3949−54.
20. Zhang J, Rosenberg HF. Sequence variation at two eosinophil-associated ribonuclease loci in humans. *Genetics* 2000;**156**:1949−58.
21. Jönsson UB, Byström J, Stålenheim G, Venge P. Polymorphism of the eosinophil cationic protein-gene is related to the expression of allergic symptoms. *Clin Exp Allergy* 2002;**32**:1092−5.
22. Eriksson J, Reimert CM, Kabatereine NB, Kazibwe F, Ireri E, Kadzo H, et al. The 434(G>C) polymorphism within the coding sequence of Eosinophil Cationic Protein (ECP) correlates with the natural course of Schistosoma mansoni infection. *Int J Parasitol*; 2007.
23. Torrent M, de la Torre BG, Nogues VM, Andreu D, Boix E. Bactericidal and membrane disruption activities of the eosinophil cationic protein are largely retained in an N-terminal fragment. *Biochem J* 2009;**421**:425−34.
24. Jönsson UB, Byström J, Stålenheim G, Venge P. A (G->C) transversion in the 3′ UTR of the human ECP (eosinophil cationic protein) gene correlates to the cellular content of ECP. *J Leukoc Biol* 2006;**79**:846−51.
25. Noguchi E, Iwama A, Takeda K, Takeda T, Kamioka M, Ichikawa K, et al. The promoter polymorphism in the eosinophil cationic protein gene and its influence on the serum eosinophil cationic protein level. *Am J Respir Crit Care Med* 2003;**167**:180−4.
26. Rosenberg HF, Domachowske JB. Eosinophil-derived neurotoxin. *Methods Enzymol* 2001;**341**:273−86.
27. Rosenberg HF. Eosinophil-derived neurotoxin/RNase 2: connecting the past, the present and the future. *Curr Pharm Biotechnol* 2008;**9**:135−40.
28. Durack DT, Ackerman SJ, Loegering DA, Gleich GJ. Purification of human eosinophil-derived neurotoxin. *Proc Natl Acad Sci USA* 1981;**78 N0.8**:5165−9.
29. Karawajczyk M, Sevéus L, Garcia R, Björnsson E, Peterson CG, Roomans GM, et al. Piecemeal degranulation of peripheral blood eosinophils: a study of allergic subjects during and out of the pollen season. *Am J Respir Cell Mol Biol* 2000;**23**:521−9.
30. Yang D, Chen Q, Su SB, Zhang P, Kurosaka K, Caspi RR, et al. Eosinophil-derived neurotoxin acts as an alarmin to activate the TLR2-MyD88 signal pathway in dendritic cells and enhances Th2 immune responses. *J Exp Med* 2008;**205**:79−90.
31. Fredens K, Dahl R, Venge P. The Gordon phenomenon induced by the eosinophil cationic protein and eosnophil protein-X. *J Allergy Clin Immunol* 1982;**70 N0.5**:361−6.
32. Davies MJ, Hawkins CL, Pattison DI, Rees MD. Mammalian heme peroxidases: from molecular mechanisms to health implications. *Antioxid Redox Signal* 2008;**10**:1199−234.

33. Zabucchi G, Menegazzi R, Soranzo MR, Patriarca P. Uptake of human eosinophil peroxidase by human neutrophils. *Am J Pathol* 1986;**124**:510–8.

34. Aldridge RE, Chan T, van Dalen CJ, Senthilmohan R, Winn M, Venge P, et al. Eosinophil peroxidase produces hypobromous acid in the airways of stable asthmatics. *Free Radic Biol Med* 2002;**33**:847–56.

35. Ulrich M, Petre A, Youhnovski N, Promm F, Schirle M, Schumm M, et al. Post-translational tyrosine nitration of eosinophil granule toxins mediated by eosinophil peroxidase. *J Biol Chem* 2008;**283**:28629–40.

36. Gleich GJ, Loegering DA, Frigas E, Filley WV. The eosinophil granule major basic protein: biological activities and relationship to bronchial asthma. *Monogr Allergy* 1983;**18**:277–83.

37. Hamann KJ, Barker RL, Ten RM, Gleich GJ. The molecular biology of eosinophil granule proteins. *Int Arch Allergy Appl Immunol* 1991;**94**:202–9.

38. Popken-Harris P, Thomas L, Oxvig C, Sottrup-Jensen L, Kubo H, Klein JS, et al. Biochemical properties, activities, and presence in biologic fluids of eosinophil granule major basic protein. *J Allergy Clin Immunol* 1994;**94**:1282–9.

39. Plager DA, Stuart S, Gleich GJ. Human eosinophil granule major basic protein and its novel homolog. *Allergy* 1998;**53**:33–40.

40. Plager DA, Adolphson CR, Gleich GJ. A novel human homolog of eosinophil major basic protein. *Immunol Rev* 2001;**179**:192–202.

41. Levi-Schaffer F, Temkin V, Simon HU, Kettman JR, Frey JR, Lefkovits I. Proteomic analysis of human eosinophil activation mediated by mast cells, granulocyte macrophage colony stimulating factor and tumor necrosis factor alpha. *Proteomics* 2002;**2**:1616–26.

42. Yoon SW, Kim TY, Sung MH, Kim CJ, Poo H. Comparative proteomic analysis of peripheral blood eosinophils from healthy donors and atopic dermatitis patients with eosinophilia. *Proteomics* 2005;**5**:1987–95.

43. Ryu SI, Kim WK, Cho HJ, Lee PY, Jung H, Yoon TS, et al. Phosphoproteomic analysis of AML14.3D10 cell line as a model system of eosinophilia. *J Biochem Mol Biol* 2007;**40**:765–72.

44. Woschnagg C, Forsberg J, Engström A, Odreman F, Venge P, Garcia RC. The human eosinophil proteome. Changes induced by birch pollen allergy. *J Proteome Res* 2009;**8**:2720–32.

45. Woschnagg C, Rak S, Venge P. Oxygen radical production by blood eosinophils is reduced during birch pollen season in allergic patients. *Clin Exp Allergy* 1996;**26**:1064–72.

46. Håkansson L, Carlson M, Stålenheim G, Venge P. Migratory responses of eosinophil and neutrophil granulocytes from patients with asthma. *J Allergy Clin Immunol* 1990;**85 N0.4**:743–50.

47. Wang HB, Ghiran I, Matthaei K, Weller PF. Airway eosinophils: allergic inflammation recruited professional antigen-presenting cells. *J Immunol* 2007;**179**:7585–92.

48. Venge P. Monitoring the allergic inflammation. *Allergy* 2004;**59**:26–32.

49. Schmekel B, Ahlner J, Malmström M, Venge P. Eosinophil cationic protein (ECP) in saliva: a new marker of disease activity in bronchial asthma. *Respir Med* 2001;**95**:670–5.

50. Hurst DS, Venge P. Evidence of eosinophil, neutrophil, and mast-cell mediators in the effusion of OME patients with and without atopy. *Allergy* 2000;**55**:435–41.

51. Björk A, Venge P, Peterson CG. Measurements of ECP in serum and the impact of plasma coagulation. *Allergy* 2000;**55**:442–8.

52. Carlson M, Håkansson L, Kämpe M, Stålenheim G, Peterson C, Venge P. Degranulation of eosinophils from pollen-atopic patients with asthma is increased during pollen season. *J Allergy Clin Immunol*, 89 N0.1. *Part* 1994;**1**:131–9.

53. Koh GC, Shek LP, Goh DY, Van BH, Koh DS. Eosinophil cationic protein: is it useful in asthma? A systematic review. *Respir Med* 2007;**101**:696–705.

54. Wolthers OD. Eosinophil granule proteins in the assessment of airway inflammation in pediatric bronchial asthma. *Pediatr Allergy Immunol* 2003;**14**:248–54.

55. Nielsen LP, Peterson CG, Dahl R. Serum eosinophil granule proteins predict asthma risk in allergic rhinitis. *Allergy* 2009;**64**:733–7.

56. Kristjánsson S, Strannegård IL, Strannegård Q, Peterson C, Enander I, Wennergren G. Urinary eosinophil protein X in children with atopic asthma: A useful marker of antiinflammatory treatment. *J Allergy Clin Immunol* 1996;**97**:1179–87.

57. Labbe A, Aublet-Cuvelier B, Jouaville L, Beaugeon G, Fiani L, Petit I, et al. Prospective longitudinal study of urinary eosinophil protein X in children with asthma and chronic cough. *Pediatr Pulmonol* 2001;**31**:354–62.

58. Gore C, Peterson CG, Kissen P, Simpson BM, Lowe LA, Woodcock A, et al. Urinary eosinophilic protein X, atopy, and symptoms suggestive of allergic disease at 3 years of age. *J Allergy Clin Immunol* 2003;**112**:702–8.

59. Wolthers OD, Heuck C. Circadian variations in serum eosinophil cationic protein, and serum and urine eosinophil protein X. *Pediatr Allergy Immunol* 2003;**14**:130–3.

60. Jönsson UB, Håkansson LD, Jogi R, Janson C, Venge P. Associations of ECP (eosinophil cationic protein)-gene polymorphisms to allergy, asthma, smoke habits and lung function in two Estonian and Swedish sub cohorts of the ECRHS II study. *BMC Pulm Med* 2010;**10**:36.

61. Munthe-Kaas MC, Gerritsen J, Carlsen KH, Undlien D, Egeland T, Skinningsrud B, et al. Eosinophil cationic protein (ECP) polymorphisms and association with asthma, s-ECP levels and related phenotypes. *Allergy* 2007;**62**:429–36.

62. Kang I, An XH, Oh YK, Lee SH, Jung HM, Chae SC, et al. Identification of polymorphisms in the RNase3 gene and the association with allergic rhinitis. *Eur Arch Otorhinolaryngol* 2010;**267**:391–5.

63. Kim YJ, Kumaraswami V, Choi E, Mu J, Follmann DA, Zimmerman P, et al. Genetic polymorphisms of eosinophil-derived neurotoxin and eosinophil cationic protein in tropical pulmonary eosinophilia. *Am J Trop Med Hyg* 2005;**73**:125–30.

64. Nakamura H, Miyagawa K, Ogino K, Endo T, Imai T, Ozasa K, et al. High contribution contrast between the genes of eosinophil peroxidase and IL-4 receptor alpha-chain in Japanese cedar pollinosis. *J Allergy Clin Immunol* 2003;**112**:1127–31.

65. Nakamura H, Higashikawa F, Miyagawa K, Nobukuni Y, Endo T, Imai T, et al. Association of single nucleotide polymorphisms in the eosinophil peroxidase gene with Japanese cedar pollinosis. *Int Arch Allergy Immunol* 2004;**135**:40–3.

66. Hrdlickova B, Izakovicova-Holla L. Association of single nucleotide polymorphisms in the eosinophil peroxidase gene with allergic

rhinitis in the Czech population. *Int Arch Allergy Immunol* 2009;**150**:184—91.

67. Parwez Q, Stemmler S, Epplen JT, Hoffjan S. Variation in genes encoding eosinophil granule proteins in atopic dermatitis patients from Germany. *J Negat Results Biomed* 2008;**7**:9.

Chapter 13.3

Eosinophils and Skin Diseases

Dagmar Simon and Hans-Uwe Simon

INTRODUCTION

Tissue eosinophilia with or without associated blood eosinophilia is observed in a number of skin disorders, including allergic diseases, autoimmune diseases, bacterial or viral infections, hematologic diseases, parasitic infestations, as well as in association with tumors.[1,2] The presence or absence of eosinophils in skin specimens is often used for differential diagnoses by dermatopathologists. For instance, eosinophils in the upper dermis might be a clue for the diagnosis of early lesions of bullous pemphigoid (BP), even in the absence of blisters. In addition, the detection of eosinophils might indicate the differential diagnosis of drug reactions, which often cannot be distinguished from other inflammatory skin diseases.

In hematoxylin and eosin (H&E) stained skin specimens, eosinophils are seen as round shaped cells stuffed with coarse eosinophil granules (Fig. 13.3.1A). Disrupted oval-shaped eosinophils may also be found, e.g., in subacute and chronic eczematous lesions (Fig. 13.3.1A, B).[3] Depending on the disease, eosinophils are located among other inflammatory cells in the perivascular areas (e.g., in eczema), between collagen bundles in the dermis (e.g., in eosinophilic cellulitis), or in the epidermis (e.g., in pemphigus foliaceus; Fig. 13.3.1B—D). Moreover, in eosinophilic pustular folliculitis, eosinophils enter the hair follicle.[4] Extracellular granular proteins can be detected in varying amounts either as separate deposits or as a thin coating on collagen bundles. The latter are called flame figures and are typically seen in eosinophilic cellulitis. Recently, deposition of eosinophil granule proteins in association with extracellular DNA traps was reported in several allergic, autoimmune, and infectious skin diseases.[5] Immunofluorescence staining using antibodies directed against eosinophil cationic protein (ECP) or major basic protein (MBP) allows a more sensitive detection of eosinophils and extracellular granular protein depositions than H&E staining.[6]

Under physiological conditions, eosinophils are usually not detectable in the skin. Therefore, primary causes (intrinsic) or stimuli (extrinsic) are required for initiating the increased production, recruitment, and/or survival of eosinophils under pathological conditions.[1] Myeloproliferative forms of hypereosinophilic syndromes (HES) represent intrinsic eosinophilic disorders,[7] in which mutations of multipotent or pluripotent hematopoietic stem cells occur, with subsequent increased eosinophil proliferation, often affect the skin (Table 13.3.1). Cutaneous manifestations vary from multiple erythematous papules, plaques, and nodules, to generalized erythematous maculopapular eruptions, often associated with pruritus.[8,9] Clonal eosinophilia can occur as a consequence of gene rearrangements that result in increased tyrosine kinase activity.[10] As a consequence, patients with hypereosinophilia due to fusion of the platelet-derived growth factor receptor α (*PDGFRA*) and FIP1 like 1 (*FIP1L1*) genes respond to imatinib therapy.[11]

Extrinsic eosinophilic disorders due to cytokine release by either T cells or tumor cells are more common than intrinsic HES forms due to genetic abnormalities within hematopoietic stem cells (Table 13.3.1). Cytokines involved in the development of skin eosinophilia are interleukin-3 (IL-3), IL-5, and granulocyte/macrophage colony-stimulating factor (GM-CSF). The expression of IL-5 in association with eosinophilic skin disorders has been reported for atopic dermatitis (AD), BP, cutaneous T cell lymphoma, episodic angioedema with eosinophilia, eosinophilic cellulitis, eosinophilic fasciitis, eosinophilic folliculitis, exanthematous drug reactions, hypereosinophilic syndrome with skin involvement, and urticaria.[2] IL-3 expression has been detected in blister fluids of BP and blood leukocytes of HES patients.[12—14] In Langerhans cell histiocytosis, as well as in AD, atopy patch test reactions, and cutaneous late phase reactions, the expression of both GM-CSF and IL-3 has been shown.[2] The chemokine eotaxin/C-C motif chemokine 11 (CCL11) is important for tissue recruitment and activation of eosinophils. Eotaxin expression has been observed in AD, autoimmune-blistering diseases like dermatitis herpetiformis and BP, drug reactions, eosinophilic folliculitis, and parasitic dermatoses, and also in lymphomas, e.g., cutaneous T-cell lymphoma and Hodgkin disease.[2]

The primary function of eosinophils was originally thought to be related to the protection against helminth parasites.[15] Recently, a novel mechanism of eosinophil function in innate immunity has been reported. By releasing mitochondrial DNA and granule proteins, eosinophils form extracellular structures that can bind and kill bacteria invading the gastrointestinal tract.[16] Such extracellular DNA structures generated by eosinophils have recently also been reported in inflammatory skin diseases.[5] Furthermore, eosinophils are presumed to cause tissue damage.[15,17] In addition, eosinophils play an important role in repair and remodeling processes, as well as in

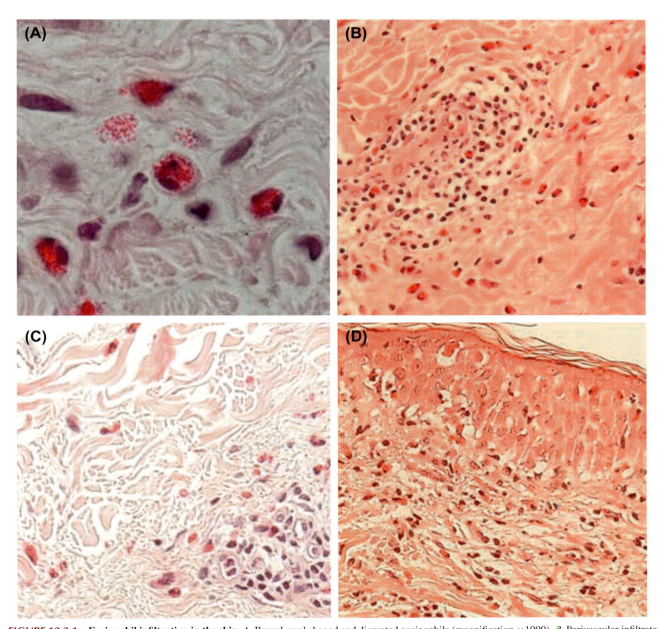

FIGURE 13.3.1   Eosinophil infiltration in the skin. *A*, Round, oval shaped and disrupted eosinophils (magnification ×1000). *B*, Perivascular infiltrate containing eosinophils in allergic contact dermatitis. *C*, Eosinophils between collagen bundles in eosinophilic cellulitis (Wells' syndrome). *D*, Eosinophilic spongiosis in pemphigus foliaceus.

immunomodulation.[18,19] The role of eosinophils under pathological conditions has mostly been studied in parasitic infections and bronchial asthma. In contrast, the role of eosinophils has not been explored in substantial depth in skin diseases. With regard to skin diseases, it can be assumed that eosinophils directly contribute to or amplify pruritus (itch) in the skin, by releasing neurotrophins (nerve growth factor and brain-derived neurotrophic factor) or indirectly by acting on mast cells.[20] Pruritus is associated with most eosinophilic skin diseases, in particular with

AD, BP, cutaneous T-cell lymphoma, and parasitic infections.

In the following sections, we summarize the current knowledge on the activation and function of eosinophils in selected eosinophilic skin disorders.

ATOPIC DERMATITIS

Tissue eosinophilia is a typical feature of eczema, in particular of AD. The numbers of eosinophils in the skin

TABLE 13.3.1 Selection of Eosinophilic Skin Disorders Associated with Eosinophilia[1,5]

Intrinsic Disorders	Extrinsic Disorders	
Mutations of Hematopoietic Stem Cells	Cytokines Released by	
	T cells	Tumor cells
Chronic eosinophilic leukemia	Allergic diseases: Atopic dermatitis, Urticaria, Drug reactions	Cutaneous T cell lymphoma
Acute myeloid leukemia		Langerhans cell histiocytosis
Chronic myeloid leukemia		B-cell lymphomas
Myelodysplastic syndromes		Hodgkin's lymphomas
Idiopathic hypereosinophilic syndromes	Autoimmune diseases: Bullous pemphigoid, Dermatitis herpetiformis, Pemphigus, Epidermolysis bullosa	Acute T-cell leukemia/lymphoma
	Infectious diseases: HIV, Ectoparasitosis, Insect bites, Erythema chronicum migrans, Erythema toxicum neonatorum	
	Hyper-IgE syndrome (Job syndrome)	
	Eosinophilic pustular folliculitis	
	Granuloma annulare	
	Angiolymphoid hyperplasia with eosinophilia	
	Localized scleroderma	
	Eosinophilic fasciitis	
	Eosinophilic cellulitis (Wells syndrome)	
	Hypereosinophilic syndromes	
	Inflammatory clonal T-cell disease	

are usually modest (2.8 cells/mm^2; range 0–90.3) and correlate with disease severity, as well as with the degree of spongiosis in acute exacerbations and marked epidermal hyperplasia in chronic stages.[21] Besides eosinophils, eosinophil-derived products, such as ECP, eosinophil-derived neurotoxin (EDN), and MBP, are present in increased amounts in the blood and the skin of AD patients.[22] Immunostaining with antibodies to ECP and MBP, as well as electron-microscopic evaluation revealed eosinophil granule proteins inside eosinophils, but also in the extracellular spaces, and the near absence of intact eosinophils, suggesting eosinophil degranulation and degeneration.[3,23] In AD, eosinophil production, differentiation, recruitment, survival, and activation are under tight control of cytokines, particularly GM-CSF, IL-3, and IL-5, and chemokines, including eotaxin and RANTES (C-C motif chemokine 5; CCL5), as well as adhesion molecules, complement factors and leukotrienes.[22] However, the pathogenic role(s) of eosinophils in AD have not yet been defined. The release of granule proteins suggests a role in host defense and/or tissue damage. Furthermore, eosinophils release a broad spectrum of mediators such as cytokines [GM-CSF, IL-1, IL-3, IL-4, IL-5, IL-8, IL-10, IL-13, and transforming growth factor (TNF)] and leukotrienes (in particular, the cysteinyl leukotrienes LTC$_4$, LTD$_4$, and LTE$_4$) and thus they can regulate immune responses or initiate tissue repair processes.[18] Improvement of AD upon both systemic and topical therapy is usually associated with a decrease in eosinophils and other inflammatory cells in the skin.[22] However, the administration of an anti-IL-5 antibody showed only moderate effects on clinical symptoms, although blood eosinophils were almost completely depleted.[24] Currently, it remains unclear whether anti-IL-5 antibody treatment reduces eosinophil tissue infiltration in lesional AD skin.

URTICARIA

Although the development of pruritic wheals in urticaria is attributed to the release of histamine by mast cells and basophils, other cell types, including eosinophils, neutrophils, and macrophages, and T cells, are also present in urticarial lesions. Eosinophils and extracellular deposits of eosinophil granule proteins have been described in chronic idiopathic urticaria, delayed pressure urticaria, and solar urticaria.[25] Extracellular deposits of MBP have been observed as granular deposits and coating tissue fibers in the dermis, as well as in small blood vessel walls.[26] ECP may stimulate histamine release by mast cells and basophils.[27] By generating eicosanoid mediators and secreting neuropeptides, such as substance P, eosinophils may contribute to vasodilation.[25] Vascular endothelial growth factor, which is elevated in the plasma of chronic urticaria patients and correlates with disease severity, has been reported to be predominantly expressed by eosinophils.[28] Recently, an involvement of the coagulation cascade in the pathogenesis of chronic urticaria has been suggested. Eosinophils that were shown to express tissue factor in urticarial lesions may activate the tissue factor pathway of coagulation, resulting in the generation of thrombin, which stimulates mast cells for histamine release.[29]

EOSINOPHILIC CELLULITIS (WELLS' SYNDROME)

Eosinophilic cellulitis is characterized by an intense infiltration of eosinophils, extracellular granule deposition, and flame figures in the dermis.[30] Recently, high numbers of eosinophils generating extracellular DNA traps in association with ECP have also been observed in this eosinophilic disorder.[5] Patients present with recurrent episodes of acute pruritic dermatitis and occasionally with blisters, painful edematous swellings, or persistent urticarial eruptions.[30] The cause is unknown, but some patients develop eosinophilic cellulitis in association with hematological disorders, infections, or anti-TNF-α therapy. Corticosteroids are usually helpful in this disease. In 37% of patients with HES, the skin is affected.[31] Skin manifestations of HES include blisters, eosinophilic cellulitis, erythematous macules, lichenoid eruptions or urticarial lesions, necrosis, papules or nodules, pruritus, purpura, and ulcerations. Cutaneous symptoms are usually present in a subgroup of patients with HES, in which IL-5-producing T cells exhibiting an abnormal immunophenotype have been identified.[13] Anti-IL-5 antibody therapy has been shown to improve skin symptoms in HES patients.[6]

EOSINOPHILIC PUSTULAR FOLLICULITIS

Eosinophilic Pustular Folliculitis (EPF) presents with recurring clusters of sterile follicular papules and pustules, predominantly on the face and trunk.[4] EPF may affect immunocompetent subjects (Ofuji disease), but is most commonly seen together with immunosuppression. EPF has been reported in association with infections, in particular acquired immunodeficiency syndrome (AIDS), autoimmune

diseases, and medications, as well as autologous peripheral blood stem cell and allogeneic bone marrow transplantation.[4] The histology shows a dense follicular and perifollicular infiltrate of eosinophils and scattered lymphocytes, and sometimes follicle destruction. A pathogenic role for eosinophils in response to fungi (Malassezia), Demodex mites, and bacteria has been suggested.[4]

BULLOUS PEMPHIGOID

The histopathology of BP reveals eosinophil infiltration in and below blisters and along the basement membrane,[32] as well as in nonblistering, urticarial, or eczematous lesions of BP.[33] Patients with active BP exhibit increased eotaxin and IL-5 levels, as well as eosinophil numbers in the blood compared with patients in clinical remission and healthy controls,[34–36] associated with significant eosinophil infiltration in the skin, as well as disease intensity.[37,38] Whether BP180 and/or BP230 autoantibodies of immunoglobulin G (IgG) or IgE types can activate eosinophils with or without preceding priming has not been investigated so far. Eosinophils exhibit CD16, the Fc-γ receptor, and degranulate upon stimulation with IgG immune complexes.[39] Furthermore, IL-5-primed eosinophils from these patients release granule proteins upon stimulation with complement C5a.[40] It has been hypothesized that in the presence of complement, eosinophils release enzymes and reactive oxygen onto the basement membrane, causing tissue destruction and blister formation in BP.[41] Eosinophil granule protein depositions have been observed in both blistering and evolving lesions.[42] Recently, extracellular DNA traps generated by eosinophils were also described in BP.[5] MMP9 has been reported to be expressed by eosinophils in lesional skin, as well as in blister fluids of BP.[43] Moreover, MMP9 cleaves the extracellular, collagenous domain of BP180 autoantigen *in vitro*.[43]

OTHER AUTOIMMUNE BULLOUS DISEASES

Eosinophilic spongiosis can be observed in early pemphigus including pemphigus foliaceus.[44] The presence of eosinophils may be due to IL-5 as part of the mixed T-helper 1/2 (T_h1)/T_h2 cytokine profile that has been found in pemphigus vulgaris.[45] Complement-fixing antibodies were shown to induce eosinophil infiltration in pemphigus.[46] Charcot–Leyden crystals have been observed in pemphigus vegetans.[47]

Eosinophils can be seen in the papillary dermis in dermatitis herpetiformis, although neutrophils and leukocytoclasis are more characteristic in this disorder.[52] Furthermore, both neutrophils and eosinophils are the predominant infiltrating cells in linear IgA bullous dermatosis.[48] Epidermolysis bullosa acquisita following

GM-CSF therapy has been related to eosinophil infiltration and deposition of eosinophil peroxidase (EPO) and MBP at the dermal–epidermal junction.[49]

CUTANEOUS DRUG ERUPTIONS

In drug reactions, the presence of eosinophils in the skin is quite a striking finding, despite various clinical and histopathological presentations [e.g., acute generalized exanthematous pustulosis (AGEP), erythema multiforme, maculopapular rashes, and pseudolymphomatous and granulomatous drug reactions].[50] Drug reaction with eosinophilia and systemic symptoms (DRESS) is a drug reaction with both blood and tissue eosinophilia and systemic symptoms. It presents with an acute, severe skin eruptions that may develop from a maculopapular rash into erythroderma, as well as with blood eosinophilia, fever, hepatitis, lymphadenopathy, and other organ involvement.[51] Eosinophils accompanied by other inflammatory cells are found in the skin and lymph nodes. Severe hepatitis, in which eosinophilic infiltration or granulomas as well as hepatocyte necrosis and cholestasis are striking features, may result in liver failure, accounting for the high mortality rate of 10%.[51] The treatment is based on high-dose corticosteroids. Drugs known to cause DRESS are anticonvulsants, anticancer drugs, antidiabetics, antimicrobial agents, nonsteroidal anti-inflammatory drugs, and sulfa drugs.[1]

LYMPHOMAS AND TUMORS

In Langerhans cell histiocytosis (LCH), among the infiltrate of Langerhans cells (LC), scattered or clusters of eosinophils can be found in the papillary and deeper dermis, respectively.[52] Since activated LC generate a broad spectrum of proinflammatory cytokines, e.g., chemokines and GM-CSF, they are able to recruit and activate eosinophils directly or via stimulation of other cell types. A predominant T_h2 type cytokine production by cutaneous T-cell lymphoma (CTCL) results in eosinophilia, extracellular granule protein depositions, as well as increased IL-5 levels in the skin and/or peripheral blood.[53] In angiolymphoid hyperplasia with eosinophilia, clonal T cell populations have been identified.[54] Currently, it is not known whether the eosinophils inhibit or modulate proliferation of the malignant cells.

SCLERODERMA-LIKE DISORDERS

Eosinophilia is a common histological finding in a number of diseases characterized by marked tissue fibrosis, such as cutaneous forms of systemic sclerosis, drug induced scleroderma-like illness (e.g., bleomycin), eosinophilia–myalgia syndrome, eosinophilic fasciitis, localized scleroderma, and toxic oil syndrome.[55] Elevated eosinophil granule protein levels have been detected in the serum of patients with diffuse cutaneous forms of systemic sclerosis, suggesting eosinophil activation and degranulation.[56] As they produce fibrogenic cytokines, e.g., transforming growth factor β (TGF-β), eosinophils are thought to contribute to skin fibrosis.[55]

CONCLUSION

Together, these examples show that most observations on eosinophils in skin diseases are rather descriptive. The exact mechanisms whereby eosinophils are recruited and activated, as well as their pathogenic role(s), are not yet fully understood. Depending on the skin disease, a role in host defense, immunoregulation and/or remodeling, and fibrosis can be assumed. Further research is therefore required in order to understand the function of eosinophils in skin diseases and to develop new therapeutic strategies.

ACKNOWLEDGEMENTS

Our research is supported by the Stanley Thomas Johnson Foundation and Swiss National Science Foundation, Bern, Switzerland.

REFERENCES

1. Simon D, Simon HU. Eosinophilic disorders. *J Allergy Clin Immunol* 2007;**119**:1291–300.
2. Simon D, Wardlaw A, Rothenberg ME. Organ-specific eosinophilic disorders of the skin, lung, and gastrointestinal tract. *J Allergy Clin Immunol* 2010;**126**:3–13.
3. Leiferman KM, Ackerman SJ, Sampson HA, Haugen HS, Venencie PY, Gleich GJ. Dermal deposition of eosinophil-granule major basic protein in atopic dermatitis: comparison with onchocerciasis. *N Engl J Med* 1985;**313**:282–5.
4. Nervi SJ, Schwartz RA, Dmochowski M. Eosinophilic pustular folliculitis: a 40 year retrospect. *J Am Acad Dermatol* 2006;**55**:285–9.
5. Simon D, Hoesli S, Roth N, Staedler S, Yousefi S, Simon HU. Eosinophil extracellular DNA traps in skin diseases. *J Allergy Clin Immunol* 2010;**127**(1):194–9.
6. Plötz SG, Simon HU, Darsow U, Simon D, Vassina E, Yousefi S, et al. Use of an anti-interleukin-5 antibody in the hypereosinophilic syndrome with eosinophilic dermatitis. *N Engl J Med* 2003;**349**:2334–9.
7. Simon HU, Rothenberg ME, Bochner BS, Weller PF, Wardlaw AJ, Wechsler ME, et al. Refining the definition of hypereosinophilic syndrome. *J Allergy Clin Immunol* 2010;**126**:45–9.
8. Simon HU, Plötz SG, Simon D, Dummer R, Blaser K. Clinical and immunological features of patients with interleukin-5-producing T cell clones and eosinophilia. *Int Arch Allergy Immunol* 2001;**124**:242–5.
9. Kaddu S, Zenahlik P, Beham-Schmid C, Kerl H, Cerroni L. Specific cutaneous infiltrates in patients with myelogenous leukemia: a clinicopathologic study of 26 patients with assessment of diagnostic criteria. *J Am Acad Dermatol* 1999;**40**:966–78.

10. Gotlib J, Cross NC, Gilliland DG. Eosinophilic disorders: molecular pathogenesis, new classification, and modern therapy. *Best Pract Res Clin Haematol* 2006;**19**:535–69.

11. Cools J, DeAngelo DJ, Gotlib J, Stover EH, Legare RD, Cortes J, et al. A tyrosine kinase created by fusion of the PDGFRA and FIP1L1 genes as a therapeutic target of imatinib in idiopathic hypereosinophilic syndrome. *N Engl J Med* 2003;**348**:1201–14.

12. Simon HU, Yousefi S, Dommann-Scherrer CC, Zimmermann DR, Bauer S, Barandun J, et al. Expansion of cytokine-producing CD4-CD8-T cells associated with abnormal Fas expression and hyper-eosinophilia. *J Exp Med* 1996;**183**:1071–82.

13. Simon HU, Plötz SG, Dummer R, Blaser K. Abnormal clones of T cells producing interleukin-5 in idiopathic eosinophilia. *N Engl J Med* 1999;**341**:1112–20.

14. Vassina EM, Yousefi S, Simon D, Zwicky C, Conus S, Simon HU. cIAP-2 and survivin contribute to cytokine-mediated delayed eosinophil apoptosis. *Eur J Immunol* 2006;**36**:1975–84.

15. Klion AD, Nutman TB. The role of eosinophils in host defense against helminth parasites. *J Allergy Clin Immunol* 2004;**113**:30–7.

16. Yousefi S, Gold JA, Andina N, Lee JJ, Kelly AM, Kozlowski E, et al. Catapult-like release of mitochondrial DNA by eosinophils contributes to antibacterial defense. *Nat Med* 2008;**14**:949–53.

17. Frigas E, Motojima S, Gleich GJ. The eosinophilic injury to the mucosa of the airways in the pathogenesis of bronchial asthma. *Eur Respir J* 1991;**13**:S123–35.

18. Jacobsen EA, Taranova AG, Lee NA, Lee JJ. Eosinophils: singularly destructive effector cells or purveyors of immunoregulation. *J Allergy Clin Immunol* 2007;**119**:1313–20.

19. Straumann A, Conus S, Gronzka P, Kita H, Kephart G, Bussmann C, et al. Anti-interleukin-5 antibody treatment (mepolizumab) in active eosinophilic esophagitis: A randomized, placebo-controlled, double-blind trial. *Gut* 2010;**59**:21–30.

20. Steinhoff M, Bienenstock J, Schmelz M, Maurer M, Wei E, Biro T. Neurophysiological, neuroimmunological, and neuroendocrine basis of pruritus. *J Invest Dermatol* 2006;**126**:1705–18.

21. Kiehl P, Falkenberg K, Vogelbruch M, Kapp A. Tissue eosinophilia in acute and chronic dermatitis: a morphometric approach using quantitative image analysis of immunostaining. *Br J Dermatol* 2001;**145**:720–9.

22. Simon D, Braathen LR, Simon HU. Eosinophils and atopic dermatitis. *Allergy* 2004;**59**:561–70.

23. Cheng JF, Ott NL, Peterson EA, George TJ, Hunkee MJ, Gleich GJ, et al. Dermal eosinophils in atopic dermatitis undergo cytolytic degeneration. *J Allergy Clin Immunol* 1997;**99**:683–92.

24. Oldhoff JM, Darsow U, Werfel T, Katzer K, Wulf A, Laifaoui J, et al. Anti-IL-5 recombinant humanized monoclonal antibody (mepolizumab) for the treatment of atopic dermatitis. *Allergy* 2005;**60**:693–6.

25. Staumont-Salle D, Dombrowicz D, Capron M, Delaporte E. Eosinophils and urticaria. *Clin Rev Allergy Immunol* 2006;**30**:13–8.

26. Peters MS, Schroeter AL, Kephart GM, Gleich GJ. Localization of eosinophil granule major basic protein in chronic urticaria. *J Invest Dermatol* 1983;**81**:39–43.

27. O'Donnell MC, Ackerman SJ, Gleich GJ, Thomas LL. Activation of basophil and mast cell histamine release by eosinophil granule major basic protein. *J Exp Med* 1993;**157**:1981–91.

28. Tedeschi A, Asero R, Marzano AV, Lorini M, Fanoni D, Berti E, et al. Plasma levels and skin-eosinophil-expression of vascular endothelial growth factor in patients with chronic urticaria. *Allergy* 2009;**64**:1616–22.

29. Cugno M, Marzano AV, Tedeschi A, Fanoni D, Venegoni L, Asero R. Expression of tissue factor by eosinophils in patients with chronic urticaria. *Int Arch Allergy Immunol* 2009;**148**:170–4.

30. Moossavi M, Mehregan DR. Wells' syndrome: a clinical and histopathologic review of seven cases. *Int J Dermatol* 2003;**42**:62–7.

31. Ogbogu PU, Bochner BS, Butterfield JH, Gleich GJ, Huss-Marp J, Kahn JE, et al. Hypereosinophilic syndrome: A multicenter, retrospective analysis of clinical characteristics and response to therapy. *J Allergy Clin Immunol* 2009;**124**:1319–25.

32. Blenkinsopp WK, Haffenden GP, Fry L, Leonard JN. Histology of linear IgA disease, dermatitis herpetiformis, and bullous pemphigoid. *Am J Dermatopathol* 1983;**5**:547–54.

33. Strohal R, Rappersberger K, Pehamberger H, Wolff K. Nonbullous pemphigoid: prodrome of bullous pemphigoid or a distinct pemphigoid variant? *J Am Acad Dermatol* 1993;**29**:2993–9.

34. Engineer L, Bhol K, Kumari S, Razzaque Ahmed A. Bullous pemphigoid: interaction of interleukin 5, anti-basement membrane zone antibodies and eosinophils. A preliminary observation. *Cytokine* 2001;**13**:32–8.

35. Frezzolini A, Teofoli P, Cianchini G, Barduagni S, Ruffelli M, Ferranti G, et al. Increased expression of eotaxin and its specific receptor CCR3 in bullous pemphigoid. *Eur J Dermatol* 2002;**12**:27–31.

36. Nakashima H, Fujimoto M, Asashima N, Watanabe R, Kuwano Y, Yazawa N, et al. Serum chemokine profile in patients with bullous pemphigoid. *Br J Dermatol* 2007;**156**:454–9.

37. Ameglio F, D'Auria L, Bonifati C, Ferrano C, Mastroianni A, Giacalone B. Cytokine pattern in blister fluid and serum of patients with bullous pemphigoid: relationships with disease intensity. *Br J Dermatol* 1998;**138**:611–4.

38. Wakugawa M, Nakamura K, Hino H, Toyama K, Hattori N, Okochi H, et al. Elevated levels of eotaxin and interleukin-5 in blister fluid of bullous pemphigoid: correlation with tissue eosinophilia. *Br J Dermatol* 2000;**143**:112–6.

39. Davoine F, Labonte I, Ferland C, Mazer B, Chakir J, Laviolette M. Role and modulation of CD16 expression on eosinophils by cytokines and immune complexes. *Int Arch Allergy Immunol* 2004;**134**:165–72.

40. Simon HU, Weber M, Becker E, Zilberman Y, Blaser K, Levi-Schaffer F. Eosinophils maintain their capacity to signal and release eosinophil cationic protein upon repetitive stimulation with the same agonist. *J Immunol* 2000;**165**:4069–75.

41. Sams Jr WM, Gammon WR. Mechanism of lesion production in pemphigus and pemphigoid. *J Am Acad Dermatol* 1982;**6**:431–52.

42. Borrego L, Maynard B, Peterson EA, George T, Iglesias L, Peters MS, et al. Deposition of eosinophil granule proteins precedes blister formation in bullous pemphigoid. Comparison with neutrophil and mast cell granule proteins. *Am J Pathol* 1996;**148**:897–909.

43. Ståhle-Bäckdahl M, Inoue M, Guidice GJ, Parks WC. 92-kD gelatinase is produced by eosinophils at the site of blister formation in bullous pemphigoid and cleaves the extracellular domain of recombinant 180-kD bullous pemphigoid autoantigen. *J Clin Invest* 1994;**93**:2022–30.

44. Brodersen I, Frentz G, Thomsen K. Eosinophilic spongiosis in early pemphigus foliaceus. *Acta Derm Venereol* 1978;**58**:368–9.

45. Rico MJ, Benning C, Weingart ES, Streilein RD, Hall RP. Characterization of skin cytokines in bullous pemphigoid and pemphigus vulgaris. *Br J Dermatol* 1999;**140**:1079–86.

46. Iwatsuki K, Tagami H, Yamada M. Pemphigus antibodies mediate the development of an inflammatory change in the epidermis. A possible mechanism underlying the feature of eosinophilic spongiosis. *Acta Derm Venereol* 1983;**63**:495–500.

47. Kanitakis J. Charcot-Leyden crystals in pemphigus vegetans. *J Cutan Pathol* 1987;**14**:127.

48. Caproni M, Rolfo S, Bernacchi E, Bianchi B, Brazzini B, Fabbri P. The role of lymphocytes, granulocytes, mast cells and their related cytokines in lesional skin of linear IgA bullous dermatosis. *Br J Dermatol* 1999;**140**:1072–8.

49. Ward JC, Gitlin JB, Garry DJ, Jatoi A, Luikart SD, Zelickson BD, et al. Epidermolysis bullosa acquisita induced by GM-CSF: a role for eosinophils in treatment-related toxicity. *Br J Haematol* 1992;**81**:27–32.

50. Ramdial PK, Naidoo DK. Drug-induced cutaneous pathology. *J Clin Pathol* 2009;**62**:493–504.

51. Bocquet H, Bagot M, Roujeau JC. Drug-induced pseudolymphoma and drug hypersensitivity syndrome (Drug Rash with Eosinophilia and Systemic Symptoms: DRESS). *Semin Cutan Med Surg* 1996;**15**:250–7.

52. Newman B, Hu W, Nigro K, Gilliam AC. Aggressive histiocytic disorders that can involve the skin. *J Am Acad Dermatol* 2007;**56**:302–16.

53. Ionescu MA, Rivet J, Daneshpouy M, Briere J, Morel P, Janin A. In situ eosinophil activation in 26 primary cutaneous T-cell lymphomas with blood eosinophilia. *J Am Acad Dermatol* 2005;**52**:32–9.

54. Kempf W, Haeffner AC, Zepter K, Sander CA, Flaig MJ, Mueller B, et al. Angiolymphoid hyperplasia with eosinophilia: evidence for a T-cell lymphoproliferative origin. *Hum Pathol* 2002;**33**:1023–9.

55. Foti R, Leonardi R, Rondinone R, Di Gangi M, Leonetti C, Canova M, et al. Scleroderma-like disorders. *Autoimmun Rev* 2008;**7**:331–9.

56. Cox D, Earle L, Jimenez SA, Leiferman KM, Gleich GJ, Varga J. Elevated levels of eosinophil major basic protein in the sera of patients with systemic sclerosis. *Arthritis Rheum* 1995;**38**:939–45.

Chapter 13.4

The Evolving Role of Eosinophils in Asthma

Alex Thomas and William W. Busse

INTRODUCTION: WHY HAS THE EOSINOPHIL HELD A POSITION OF PROMINENCE IN ASTHMA?

Asthma is characterized by variable airflow obstruction, bronchial hyperresponsiveness, and chronic airway inflammation, all of which are likely to be intertwined and interdependent. The immune processes involved in the development of these characteristics in asthma are complex, redundant, and interactive, making it difficult to specifically define which factor, or factors, are the principal contributors to these processes and the eventual pathophysiology of asthma. As has also become apparent, asthma is represented by multiple phenotypes in which the clinical profiles and patterns of inflammation have distinct, though overlapping, characteristics. To appreciate the mechanisms of disease in asthma and the role of eosinophils, it is helpful to explore the contribution of individual cells to the pathophysiology of asthma.

As early as the turn of the 20th century, the eosinophil was identified as a prominent cell associated with asthma.[1,2] For example, postmortem examinations of the lungs of patients who died from status asthmaticus showed, in many cases, sheets of eosinophils infiltrating the airways. These telling findings led to the long-held belief that eosinophils are an inherent characteristic and possibly an essential component of asthma, particularly in severe exacerbation of disease. As the role of airway inflammation became more fully defined, the assumption that eosinophils are a primary, if not *the* principal, contributor to asthma was integrated into concepts of asthma, at least until the last decade.[3]

This positioning into asthma was further supported by animal models, which strengthened the hypothesis that eosinophils are a principal contributor to inflammation, airflow obstruction, and airway hyperresponsiveness (AHR).[4] From animal models, it was possible to separate individual components that contribute to inflammation, and the cytokine, interleukin-5 (IL-5), was found to be responsible not only for terminal differentiation of eosinophils, but by ablating IL-5 with a specific monoclonal antibody, many of asthma-like airway responses to inhaled antigen were inhibited. Based upon these findings, eosinophils and IL-5 became a major pathway in allergic inflammation and a target to regulate and more mechanistically control asthma.

These theories would subsequently be tested with the administration of anti-IL-5 to patients with asthma. As expected, there was a significant reduction in circulating and sputum eosinophils but, surprisingly, little to no impact on features of asthma—airflow obstruction or AHR—and only a 50% reduction in bronchial mucosa eosinophils. Based upon these data, the role of the eosinophil in and contribution to asthma appeared less apparent. What has evolved since these dramatic shifts from an early appreciation of the eosinophil to asthma has been a more accurate and informed picture of this cell's involvement and contribution to asthma, which is the objective of our subchapter.

To most fully appreciate current views of eosinophils in asthma, we feel that it is helpful to trace the key observations that have attempted to define this cell's role in asthma. Retracing these discoveries has led to a more well-defined elucidation of the eosinophil's role in asthma.

WHAT HAS BEEN LEARNED FROM INVESTIGATIONS INTO THE EOSINOPHIL'S CONTRIBUTION TO ASTHMA?

Peripheral Blood Eosinophils

Clinical evidence to suggest that eosinophils are key players in asthma first emerged from examinations of peripheral blood samples, largely because of the ease and safety of such studies. Peripheral eosinophilia had been recognized as a feature of asthma for decades, with clinical correlations arising between airflow obstruction and the magnitude of peripheral blood eosinophilia.[5] In 1975, a pivotal study by Horn et al.[6] found that the total peripheral blood eosinophil count in a cohort of asthmatic patients was greater in those individuals with a more severe disease. This was an important early observation and provided a clue to what the eosinophil may contribute to asthma as an association arose between the degree of airflow obstruction and peripheral blood eosinophil counts. These and subsequent observations suggested that circulating eosinophils may be a key clinical feature of asthma and, because of these associations, contribute to airflow obstruction and hence disease severity.

In a subsequent study of 43 patients with chronic asthma, Bousquet et al.[7] assessed clinical symptoms of patients in relationship to peripheral blood eosinophil counts. These investigators found a positive correlation between the eosinophil count and disease severity. To extend the value of peripheral blood eosinophils to features of asthma, Taylor et al.[8] was also able to find correlations between peripheral blood eosinophilia and another key characteristic of asthma, bronchial hyperresponsiveness. Collectively, these studies were helpful in gaining an understanding of the role of eosinophils in asthma as they showed that the level of circulating eosinophils may serve as a biomarker not only for the presence of disease but also to indicate the degree of severity.

How Did the Eosinophil's Biology Further Link This Cell Type to Asthma?

The eosinophil was named for the ability of eosin to stain its basic granules and thus impart the cell's well-recognized red color.[5] These eosin-staining granules, however, were found to be more than a marker for this cell type. Eosinophil-derived granule products have been found to have multiple actions, some of which can produce pathological and physiological features associated with asthma, including injury to respiratory epithelium (desquamation) and AHR.

The eosinophil granules consist of four major cytotoxic cationic proteins: major basic protein (MBP), eosinophil peroxidase (EPO), eosinophil-derived neurotoxin (EDN) or nonsecretory ribonuclease, and eosinophil cationic protein (ECP).[9] MBP accounts for 55% of the eosinophil granule content, is toxic to parasites, and has similar cytotoxic effects on the respiratory epithelial cells of both animals and humans (Table 13.4.1).[10,11] In vitro studies have shown that the application of MBP to respiratory tissue, at concentrations consistent with values found in sputum and bronchoalveolar lavage (BAL) fluid of asthmatic patients (50–100 mg/mL), leads to almost complete erosion of tracheal epithelium and parallels the observed desquamation of respiratory epithelium in the bronchial mucosa of subjects with asthma. These effects of MBP suggest a potential mechanism by which eosinophils may contribute to airway injury, initiate repair (i.e., remodeling), and subsequently induce hyperresponsiveness.[12] To extend these observations, Flavahan et al. incubated guinea pig tracheal rings with MBP and found bronchial smooth muscle tension to acetylcholine was significantly enhanced compared to tissues not exposed to eosinophils.[13] In a dog model, the intraepithelial administration of MBP also increased the bronchoconstriction response.[14]

ECP also causes cytotoxic effects on respiratory tract cells. By adding small amounts of ECP to the respiratory tract, Dahl et al.[15] found damage and denudation of bronchial and tracheal epithelial cells, which paralleled observations in asthma. Elevated ECP values are found in BAL fluid of asthmatic subjects during the late-phase response (LPR) to an allergen challenge, as well as in lavage fluid of patients with chronic, persistent asthma.[16] In guinea pig models, the application of increasing concentrations of EPO to tracheal mucosa caused epithelial cell exfoliation, ciliostasis, and bleb formation.[17] Serum EDN levels are elevated in asthma and return to normal levels following treatment with prednisolone and the achievement of disease control.[18] Finally, when compared to normal subjects, EDN concentrations in urine are greater in asthmatic children.[19]

Eosinophils also produce other inflammatory products such as cysteinyl leukotrienes, platelet activating factor (PAF), reactive oxygen species (ROS), and substance P. In in vitro studies designed to assess eosinophil biology, PAF stimulated the release of granule products and led to superoxide anion generation.[20] PAF was also chemotactic for the eosinophil, as reflected by the induction of an eosinophilic infiltrate of the airway after local or systemic delivery of PAF in guinea pigs.[21,22] These early findings in animals suggested that the release of PAF by eosinophils could potentially create a self-perpetuating cycle to further the development of eosinophilic inflammation and hence airway injury. PAF also has direct proinflammatory properties. When administered to humans, PAF causes rapid bronchoconstriction, which can last up to 2 h.[23] Similar to changes that follow an airway allergen challenge, PAF causes a prolonged increase in bronchial responsiveness.[24]

TABLE 13.4.1 Eosinophil-Derived Inflammatory Mediators

		Cytotoxicity	Epithelial Damage	Airway Hyper-Responsiveness	Broncho-Constriction	Mucus Production, Vascular Leakage, Vasodilation	Eosinophil-Attraction Activation	Mast-Cell Proliferation
Granule proteins	MBP +	+	+	+				
	ECP +	+	+	+				
	EDN +	+	+	+				
	EPO +	+	+					
Lipid mediators					Leukotrienes C$_4$, D$_4$, E$_4$ +	+		
					Prostaglandin E2, I2 −	+		
					PAF +			
Cytokines						IL-3	+	+
						GM-CSF	+	
						IL-5	+	

GM-CSF, granulocyte-macrophage colony-stimulating factor; IL, interleukin; PAF, platelet-activating factor.

Although a role for PAF in human asthma has not been established, the biology of this eosinophil product is associated with airway changes reflective of asthma and also implies an important role for eosinophils in this process.

Eosinophils also generate the cysteinyl leukotrienes, LTC$_4$ and LTD$_4$, which can cause both an increase in vascular permeability and bronchoconstriction.[25] LTC$_4$ is secreted by eosinophils and is increased in asthmatic children.[26] Activated eosinophils can also release substance P, a neuropeptide, to increase vascular permeability and promote plasma extravasation.[27] Eosinophils, like other phagocytes, generate ROS, which are produced in greater concentrations in asthma.[28] In asthma, ROS can cause mucus secretion, which leads to increased vascular permeability and airway obstruction. When damaged by ROS, respiratory epithelium produces fewer bronchodilatory substances, such as prostaglandin E2 (PGE2) and nitric oxide, with the net result of increased airflow obstruction[29] (Fig. 13.4.1).

EOSINOPHILS GENERATE INFLAMMATORY CYTOKINES

Although T cell-derived cytokines may be the major source of factors that influence eosinophil development and function, eosinophils themselves can also produce several cytokines to perpetuate their own inflammatory biology and that associated with asthma. Transforming growth factor β (TGF-β) is a profibrotic cytokine produced by eosinophils and found in increased concentrations in asthma.[30] TGF-β stimulation of fibroblasts leads to a thickening of the reticular lamina of the airways, and thus provides a conceptual link between eosinophilic inflammation and structural remodeling of the airway in asthma.[30] Eosinophils also produce several other mediators/cytokines, such as IL-13, matrix metalloproteinase-9 (MMP-9), tissue inhibitor of metalloproteinases 1 (TIMP-1; or metalloproteinase inhibitor 1), and vascular endothelial growth factor (VEGF), which have also been implicated in matrix remodeling in asthma.[31]

IL-1, IL-2, IL-3, IL-4, IL-5, IL-6, IL-8, IL-10, and IL-16 are all produced by eosinophils in varying concentrations. IL-1, initially observed in mice and later found in hypereosinophilic patients, is associated with HLA-DR expression and is thought to contribute to the eosinophil's function as an antigen-presenting cell.[32] IL-3 and granulocyte macrophage colony stimulating factor (GM-CSF), are also produced by eosinophils, and can function in an autocrine fashion to prolong the cell's survival.[33] IL-4, which upregulates vascular cell adhesion molecule (VCAM) receptors on endothelial cells and is a cofactor in immunoglobulin E (IgE) isotype switching, is produced

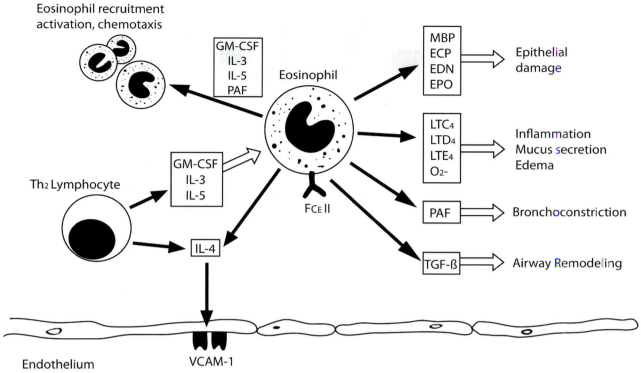

FIGURE 13.4.1 Eosinophil activities. Activated T-helper type 2 (T$_h$2) lymphocytes result in the generation of granulocyte macrophage colony stimulating factor (GM-CSF), interleukin-3 (IL-3), and IL-5. These cytokines have several proinflammatory properties, including activation, adhesion, chemotaxis, enhanced maturation, and prolonged survival. Antigen-induced stimulation of FCε receptor II also activates the eosinophil. IL-4, produced by T$_h$2 cells and the eosinophil, upregulates the vascular cell adhesion molecule 1 (VCAM-1) adhesion molecule and promotes adherence and migration through the vascular endothelium. The activated eosinophil is capable of mediator release. Granule constituents eosinophil cationic protein (ECP), eosinophil-derived neurotoxin (EDN), eosinophil peroxidase (EPO), and major basic protein (MBP) are capable of epithelial damage. Leukotrienes and reactive oxygen species (ROS) are capable of causing inflammation, edema and mucus secretion. Platelet activating factor (PAF) and transforming growth factor β (TGF-β) cause bronchoconstriction and airway remodeling, respectively. O$_2$−, peroxide; Fc$_E$II, Fc component of IgE.

by eosinophils from patients with atopic asthma.[34] IL-5, which regulates terminal differentiation and survival of eosinophils, is found in eosinophils obtained from BAL fluid of asthmatic patients.[35] IL-5 release also follows eosinophil stimulation by IgA, IgG, or IgE immune complexes.[36]

WHAT HAS BEEN LEARNED FROM THE STUDY OF SPUTUM AND BAL EOSINOPHILIA?

The presence of eosinophils in sputum has been a characteristic finding of asthma since the early 19th century.[2] Bousquet et al. increased sputum and peripheral blood eosinophils in asthma patients.[7] Sputum eosinophilia is also a feature of nonasthmatic eosinophilic bronchitis, a relatively nonspecific term. In contrast, sputum eosinophilia is not a feature of chronic obstructive pulmonary disease (COPD). In addition, previous studies have suggested a relationship between BAL fluid eosinophilia and

the level of bronchial responsiveness.[37] The concentrations of ECP and MBP in BAL fluid were also found to directly relate to the percentage of eosinophils, implying that eosinophils in asthmatic airways are activated and have undergone degranulation.[38] To extend these observations, Uchida et al.[4] evaluated the effect of MBP on airway responsiveness in an animal model. Within 1 h of direct instillation of MBP onto the trachea of rats, significant increases in airway responsiveness to methacholine occurred.

Finally, 22 subjects with asthma were identified, half of whom were treated with inhaled corticosteroids (ICS) and the other half given only bronchodilators, to be used as needed.[39] ICS use led to a fall in serum and BAL ECP levels but, interestingly, the eosinophil count was unchanged in both the corticosteroid or bronchodilator treated groups. Collectively, eosinophil-derived products can have a significant influence on the function and pathohistology of the airway and mirror many features of asthma.

WHAT HAS BEEN LEARNED FROM THE STUDY OF BRONCHIAL MUCOSAL EOSINOPHILS?

In the 1950s, a further relationship of asthma to eosinophils was supported by the finding of a marked infiltration of eosinophils in the lung tissue of patients who died suddenly of status asthmaticus.[40] In cases of death from acute, severe asthma, eosinophilic infiltrates were found in the lung parenchyma, bronchial lumen, and entire thickness of the bronchial wall.[41] Bronchial biopsies obtained by bronchoscopy from patients with mild disease also showed eosinophils as a prominent cellular infiltrate.[42] In addition to the presence of eosinophils and their granule products in airways of asthmatic patients, the airway histology showed epithelial desquamation, impaired ciliary function, basement membrane thickening and mucous plugs.[9,43] Expanding upon these findings, Ohashi et al.[44] found the eosinophilic infiltration of mucosal tissue was associated with opening of tight junctions of bronchial epithelial cells. These histological analyses showed eosinophil infiltration of the airways was associated with epithelial damage and possibly linked to subsequent airway hyper-responsiveness.[44]

WHAT HAS BEEN LEARNED ABOUT EOSINOPHIL INVOLVEMENT IN AIRWAY INFLAMMATION FROM INHALED ANTIGEN CHALLENGES?

When subjects with allergic asthma inhale allergens to which they are sensitized, there is an immediate, or early, response, characterized by an acute fall in forced expiratory volume in one second (FEV_1). Approximately 40% of subjects will go on to develop an LPR, which occurs 4–8 h after allergen exposure. In LPR, eosinophils are the prominent airway cellular infiltrate both in animal models and human subjects. To define the kinetics of eosinophil recruitment and eotaxin/C-C motif chemokine 11 (CCL11) generation, Humbles et al.[45] found the chemokine eotaxin was increased 2–3 h postallergen challenge in sensitized guinea pigs. By 12–24 h postallergen challenge, there was a significant increase in eosinophil numbers in the BAL fluid, but no further rise in eotaxin. These findings in the guinea pig suggest that eosinophil recruitment to the airway following an inhalation of allergen is, in part, regulated by eotaxins. Findings of an early recruitment of eosinophils and the later development of a late-phase airflow obstruction to allergen suggested that these events reflect processes in asthma that could provide insight into how eosinophils may contribute to asthma.

Histopathological findings of the airway in LPR are also similar to events found in patients with chronic asthma and persistent airway obstruction. In guinea pigs sensitized to ovalbumin, 17 h after an inhalation challenge with ovalbumin the cellular composition in BAL fluid is predominantly neutrophils.[46,47] By 72 h post antigen exposure, 50% of BAL cells are eosinophils. In addition, the subsequent eosinophilic infiltration of the peribronchial smooth muscle and epithelium persists for up to 7 days. Eosinophils in the BAL of immunized guinea pigs with an LPR were found to be activated, suggesting that these cells are primed during recruitment to the airway.[48]

Similarly in humans, de Monchy et al.[16] found significant eosinophilia only in patients who developed LPR to inhaled antigen. Moreover, the eosinophils in the BAL fluid had undergone degranulation, as evidenced by an elevated ECP:albumin ratio. There was also evidence that peripheral blood eosinophils and ECP concentrations are increased prior to antigen challenge in those subjects who eventually develop an LPR. Subsequently, Cockcroft et al.[49] evaluated the LPR in a population of asthma patients who had been given a single dose of inhaled beclomethasone dipropionate, inhaled salbutamol or inhaled cromoglycate in a randomized, double-blind, placebo-controlled crossover trial. While beclomethasone had no effect on the allergen-induced early pulmonary obstructive response, there was a significant inhibition of the LPR at 7 h and 30 h later. Collectively, these data suggest that allergen provocation of allergic inflammation is likely to be regulated by eosinophils and can be controlled by corticosteroids, which block both the late-phase rise in recruited eosinophils and the subsequent reduction in lung function. Thus, a logical conclusion from work at this time was that a direct association between eosinophil recruitment and altered lung function existed in asthma, a finding substantiated by Kidney et al.[50]

In 2000, Gauvreau et al.[51] extended these observations when she evaluated the development of the LPR and recruitment of eosinophils in asthma patients who were initially treated with inhaled budesonide for 1 week and then underwent an inhaled allergen challenge. Budesonide administration reduced the intensity of the LPR and also inhibited eosinophil recruitment to the airway, reflected by sputum eosinophils and the parallel increase in peripheral blood eosinophils 24 h after challenge. From these data, it was assumed that allergic activation of the airway and the subsequent development of the LPR were caused by eosinophils recruited to the lung.

WHAT HAS BEEN LEARNED ABOUT THE EOSINOPHIL AND ASTHMA FROM SEGMENTAL ANTIGEN CHALLENGES?

To extend observations from whole-lung allergen challenges, a number of groups developed the technique of using a bronchoscope to deliver antigen into single segments of the airway. To perform these studies, a bronchoscope is introduced into the lung and then wedged into an isolated airway segment, where a dose of allergen is introduced and a lavage performed immediately; the cells and lavage fluid obtained at this time represent events associated with the acute or early response. Bronchoscopy is repeated 24–48 h later, the challenged segment of the airway identified, and lavage performed: the airway events analyzed at this time model the LPR. This approach, while not allowing for measures of pulmonary function, provides a direct measure of cells, mediators, and retrieval of cells for *ex vivo* study to compare early- and late-phase allergic reactions.

When antigen is introduced to airway segments in allergic patients, there is a strong eosinophilic response 24–48 h postchallenge.[52] The analysis of BAL is characterized by large concentrations of granule proteins, ECP, EDN, EPO, and MBP, as well as LTC_4, suggesting that the recruited eosinophils are activated when they appear in the airway. Additionally, when eosinophils are retrieved from BAL fluid, they are phenotypically distinct from circulating cells and have greater superoxide anion release, collagen adherence and cell surface adherence receptors compared to peripheral blood eosinophils. These findings suggested that LPR-associated eosinophils, which are recruited to the lung during the late phase and are terminally differentiated, have a greater capacity to generate inflammation. BAL levels of IL-5 were also increased and correlated with the eosinophils, suggesting a key role for this cytokine in these processes. From these studies, a more expanded picture of the allergic inflammatory response was made, along with the identification of key mediators and evidence for an enhancement of the eosinophil's inflammatory potential.

Calhoun et al.[53] used segmental antigen challenges in subjects with allergic rhinitis who had been inoculated with rhinovirus to further evaluate the interactions between viral upper respiratory infections and allergic reactions, and gain insight into mechanisms of asthma exacerbations. During an acute rhinovirus infection, BAL fluid obtained 48 h after antigen challenge contained increased numbers of eosinophils. In some subjects, the increase in eosinophils persisted for up to 1 month following the acute viral infection and initial antigen challenge. This augmented antigen-induced eosinophil recruitment was thought to be a possible mechanism for an intensified inflammatory airway response during viral infections and to account for greater asthma symptoms at that time. Alternatively, virus-induced epithelial damage, with a subsequent increase in mucosal permeability, was hypothesized to increase allergen contact with immune cells, thus creating a greater inflammatory response.[54]

WHAT HAS BEEN LEARNED ABOUT EOSINOPHILS IN ASTHMA BY EVALUATING THE EFFECTS OF TREATMENT WITH CORTICOSTEROIDS?

The presence of eosinophils is regulated by apoptosis and modulated by cytokines, growth factors, and lipid mediators, which are released during allergic inflammation but suppressed by glucocorticoids.[55] Consequently, the persistence of airway eosinophilia in some patients with asthma was attributed to cells that had developed resistance to corticosteroids.[56] The reduction in airway eosinophils with corticosteroids and improved asthma control further supported a central role for eosinophils in asthma.[3] Under both circumstances, either a reduction or persistence of eosinophils, and their correlation with symptoms of asthma, supported a direct link of eosinophils to the pathobiology of asthma.

In 1991, Evans et al.[57] measured eosinophil numbers in 10 asthmatic subjects 14 days following the initiation of inhaled budesonide treatment. Airway responsiveness to methacholine was improved and was associated with a fall in peripheral eosinophils. These findings also suggested that eosinophil production, maturation, and differentiation may take place in the lung, as well as the bone marrow. As corticosteroid treatment reduced serum, sputum, and tissue eosinophils, and these reductions were associated with improved asthma symptoms, the eosinophil's place as a primary contributor to asthma appeared further substantiated. However, not all patients with asthma have eosinophilia, nor are all asthma patients responsive to corticosteroids. These observations, in the face of data already discussed, raised questions as to how essential the eosinophil is to all of the clinical features of asthma.

WHAT HAS BEEN LEARNED BY COMPARISONS OF EOSINOPHILIC AND NONEOSINOPHILIC ASTHMA?

While many studies of asthma confirm the presence of elevated eosinophils, circulating and BAL fluid eosinophilia are not always present in asthma.[40,58] Theories as to these differences included the possibility that eosinophils may be present only during an exacerbation and their absence may result from the actions of medication, particularly ICS.[59,60] These observations also led to an emerging theory that at least two asthma phenotypes exist, based on the presence or absence of tissue eosinophils.[48] Persistent eosinophilic inflammation, despite treatment, was found to be more common in adult onset asthma and less common in classic *allergic* asthma in patients receiving ICS.[61] While up to 40% of cases of severe asthma appear to start later in life, the presence of eosinophils in patients with late-onset asthma is also more variable. Interestingly and importantly, patients with asthma and existing eosinophilia had greater airway remodeling and more exacerbations, despite treatment.[62]

To begin to dissect and clarify these relationships, Woodruff et al.[63] used a number of innovative approaches for study. Firstly, to determine the molecular basis for asthma heterogeneity and the involvement of eosinophils, the effect of an underlying T-helper type 2 (T_h2)-mediated profile of inflammation was evaluated in a cohort of 42 patients with mild to moderate asthma and 28 healthy control subjects. The recruited subjects were stratified based on high or low expression of IL-13-inducible genes in samples of their airway epithelium. Using the response of their epithelium to stimulation with IL-13, investigators were able to classify the reaction as *Th-2-high* vs. *Th-2-low*. The T_h2-high asthma group was indistinguishable from the T_h2-low asthma group in relationship to demographics, lung function, and response to bronchodilators. However, the T_h2-high group had significantly greater AHR, total IgE levels, and BAL and peripheral blood eosinophils. The presence of airway remodeling was also evaluated in this study population, and both the reticular

basement membrane thickness and epithelial mucin stores were increased in subjects with a T_h2-high profile. Following this classification, subjects were randomized and treated with either inhaled fluticasone or placebo to determine if there was a difference in clinical response to ICS based upon their T_h2 profiles.

In the T_h2-high group, ICS improved the FEV_1, whereas no change occurred in the low T_h2 group. This study further supported the concept of heterogeneity in the pathogenesis and pathophysiology of asthma that can be characterized by the presence of T_h2-driven inflammation with eosinophilia and responsiveness to corticosteroids as markers of this profile. These findings also indicated that other phenotypes of asthma exist but their features were poorly understood.

WHAT HAS BEEN LEARNED ABOUT ASTHMA BY STUDYING INTERLEUKIN-5?

T_h2 cell-derived IL-5 has been identified as the major cytokine involved in terminal differentiation of eosinophils,

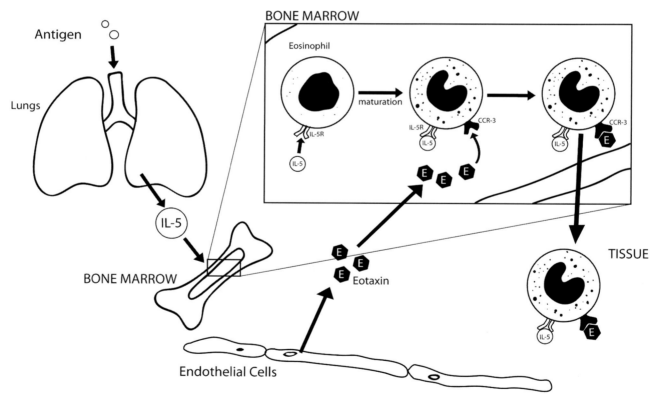

FIGURE 13.4.2   Interleukin-5. Interleukin-5 (IL-5) is a signaling molecule that stimulates eosinophil proliferation, maturation, and activation. An antigen-stimulated immune response in tissues leads to the secretion of IL-5 by cells such as eosinophils, mast cells, and T-helper type 2 (T_h2) cells. IL-5 then acts on the bone marrow to mobilize existing eosinophils and induce further eosinophil production. These maturing eosinophils become responsive to eotaxin/ C-C motif chemokine 11 (CCL11), produced by the endothelium, allowing for their exit from the bone marrow to tissue. CCR3, C-C chemokine receptor type 3; IL-5R, interleukin-5 receptor.

activation of mature eosinophils, and prolongation of eosinophil cell survival (Fig. 13.4.2).[64] IL-5 enhances eosinophil degranulation, chemotaxis, antibody-dependent cytotoxicity, and adhesion to endothelium.[65] Van Der Veen et al.[66] identified 22 patients with mild to moderate, dust mite-sensitive allergic asthma and found a significant correlation between the magnitude of the LPR, the allergen-specific proliferative response of peripheral T lymphocytes, and an increase in IL-5 *in vivo* following inhaled antigen. Extending beyond observational studies of an elevation of IL-5 in relationship to LPR, inhaled IL-5 was found to cause, or significantly contribute to, AHR.[67] Shi et al.[68] also demonstrated that inhaled IL-5 in asthma acted as an eosinophil chemoattractant and activator of the recruited eosinophils to the airway. In a subsequent, blinded, placebo-controlled crossover study of eight patients with allergic bronchial asthma, Shi et al.[68] found a significantly enhanced methacholine PC_{20} (the dose of the inhaled antagonist that provokes a 20% drop in FEV_1) responsiveness 24 h and 48 h after an inhalation of IL-5, as well as a significant increase in sputum eosinophils and ECP. Using a mouse model with IL-5 knocked out, inhaled allergen did not lead to eosinophilia or an increase in AHR.[69] IL-5 was also decreased during treatment with corticosteroids that improved asthma control.[70] From these and other data, IL-5 emerged as the dominant cytokine regulating the eosinophil's involvement in allergic airway disease.

WHAT HAS BEEN LEARNED ABOUT EOSINOPHILS IN ASTHMA FROM TREATMENT WITH ANTI-INTERLEUKIN-5?

The anti-IL-5 monoclonal antibody is an IgG antibody that binds with high affinity to free IL-5, thus preventing its binding to the IL-5 receptor on the surface of eosinophils and their progenitors. Van Oosterhouet et al.[71] used oval-bumin to challenge sensitized guinea pigs and induce airway eosinophilia, neutrophilia, and tracheal hyperreactivity. When the sensitized guinea pigs were treated with anti-IL-5 and then challenged with ovalbumin, airway eosinophilia, not neutrophil recruitment, was suppressed as well as the allergen challenge-induced increase in AHR. When sensitized guinea pigs were treated with the anti-IL-5 monoclonal antibody by Akutsu et al.,[72] there was a decrease in AHR, tracheal wall eosinophil infiltration, and the LPR. Animal studies in which anti-IL-5 decreased airway eosinophilia, AHR to allergen, and LPR provided further support for the hypothesis that eosinophils are central to many important aspects of the pathogenesis of asthma, and suggested that if eosinophil migration to the airway could be inhibited, signs and symptoms of asthma could be controlled or prevented.

WHAT HAS BEEN LEARNED ABOUT EOSINOPHIL INVOLVEMENT IN ASTHMA WITH ANTI-INTERLEUKIN-5 TREATMENT?

In 2000, Leckie et al.[73] conducted a randomized, double-blind, placebo-controlled trial in which a single infusion of either of two doses of anti-IL-5 humanized monoclonal antibody or placebo was administered to 24 men with mild allergic asthma. The study goal was to evaluate the effect of anti-IL-5 on the LPR to inhaled antigen, on the premise that this treatment would ablate eosinophil recruitment and hence the airway responses to allergen, including the development of the LPR and an associated increase in AHR. The investigators found anti-IL-5 to decrease blood and sputum eosinophils following inhaled allergen challenge. Surprisingly, anti-IL-5 treatment had no effect on either the LPR or postallergen increase in airway responsiveness to histamine. Thus, in striking contrast to animal studies, there was no significant effect of anti-IL-5 on the development of an LPR, despite the absence of eosinophils in the blood or the airways following antigen exposure. This study, though conducted in a small numbers of patients, prompted a total reevaluation of the eosinophil's role in asthma, including whether it had one at all.

Extending this study, Flood-Page et al.[74] evaluated anti-IL-5 monoclonal antibody treatment in asthma patients by examining its multidose effect on blood and sputum, as well as bone marrow and airway tissue eosinophils. In a randomized, double-blind, placebo-controlled study, 24 patients with mild asthma received three intravenous doses of mepolizumab (i.e., IL-5 monoclonal antibody) or placebo for 20 weeks. Within 4 weeks of the first anti-IL-5 dose, there was a significant decrease in peripheral blood eosinophils. At weeks 4 and 10 of the study, there was nearly a 100% reduction of eosinophils in blood and sputum samples following anti-IL-5 treatment. In contrast, eosinophils were reduced by only 52% in bone marrow aspirates. Similarly, bronchial mucosa eosinophils were reduced by 55% from baseline. Despite this reduction of intact eosinophils, staining of the bronchial biopsy for intracellular MBP was unchanged by mepolizumab. Despite these reductions in eosinophils, there was no change in AHR, exacerbations, FEV_1 values, peak flow measurements, or symptoms between the anti-IL-5 and placebo-treated groups.

While a reduction in blood and BAL fluid/sputum eosinophils occurred with anti-IL-5, there was only a 50% reduction in bone marrow and bronchial eosinophils.[74] This finding confirmed that, despite anti-IL-5 therapy, residual airway eosinophils persisted and the total amount of MBP present in airway tissues was unchanged. From these observations, it was also hypothesized that the residual eosinophil population in the airway continued to exist and release, or retain, granule proteins, despite anti-IL-5 treatment. Furthermore, it was proposed that these persistent effects may be responsible for the lack of improved asthma control, FEV$_1$, and AHR.

These findings with anti-IL-5 contrasted sharply with effects noted with oral corticosteroids, which caused an 80% decrease of bronchial mucosal eosinophilia and led to significant clinical improvements of asthma symptoms.[75] Given the possibility that, despite anti-IL-5 treatment, residual airway eosinophils may be sufficient to continue to exert their influence on clinical outcomes, including lung function and symptoms, the eosinophil should not have been excluded as a target for asthma therapy, although the outcome may not be symptoms or airflow obstruction.[74]

WHAT IS THE ROLE OF EOSINOPHILS IN EXACERBATIONS OF ASTHMA?

In 2008, Rothenberg et al.[76] reported the results of a randomized, double-blind, placebo-controlled trial of anti-IL-5 in 85 prednisone-dependent (20–60 mg/d) patients with hypereosinophilic syndrome. In addition to a significant lowering of peripheral blood eosinophilia, 84% of the mepolizumab-treated subjects were able to reduce their oral prednisone dose to $\leq$10 mg/d compared to only 43% in the placebo group. In addition, mepolizumab significantly reduced the likelihood of an exacerbation of their hypereosinophilic disease. While the effect of therapy on hypereosinophilic syndrome patients may not be extrapolated to asthma, this study did show that mepolizumab has the ability to decrease eosinophils, the prednisone requirement to maintain disease control, and exacerbations from the hypereosinophilic syndrome.

Expanding on this theme, Green et al.[3] examined a strategy designed to determine the effects of treatment directed toward reducing sputum eosinophil counts rather than administering a dose of ICS based on symptoms alone, i.e., a guidelines approach. Seventy-four patients with moderate to severe asthma were identified and randomly assigned to either a management strategy based upon British Thoracic Society guidelines or one using a dose of ICS that reduced sputum eosinophils to <3%. The sputum management strategy group had an average eosinophil count that was 63% lower than that of the guideline-managed group during the study. As an apparent consequence of this reduction in eosinophils, patients in the sputum management group had significantly fewer asthma exacerbations and were admitted less frequently to the hospital for asthma (Fig. 13.4.3). However, total asthma quality of life scores, mean peak flow measurements, postbronchodilator FEV$_1$, and the use of rescue

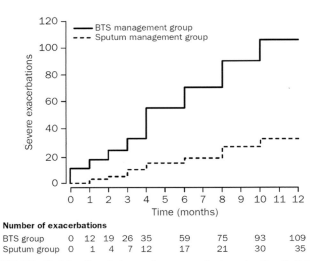

Number of exacerbations

BTS group	0	12	19	26	35	59	75	93	109
Sputum group	0	1	4	7	12	17	21	30	35

FIGURE 13.4.3 Cumulative asthma exacerbations in the British Thoracic Society (BTS) guideline management group and the sputum management group. *(Reproduced with permission from Green et al.[3])*

bronchodilators were not different in the two management groups.[3] While this study did not show a relationship between sputum eosinophils and variable airflow obstruction and other parameters, such as daily symptoms, it did support the notion that eosinophils play a central role in asthma by either serving as a marker for exacerbation risks or being possibly linked to an increased susceptibility for exacerbations.

It can be argued that previous studies of anti-IL-5 had examined patients with only mild asthma and with outcome measures that were not specifically related to ongoing eosinophilic inflammation. To extend upon this hypothesis, Haldar et al.[77] and Nair et al.[78] published results of studies designed to evaluate mepolizumab treatment in patients with severe asthma, persistent eosinophilia in sputum, and frequent exacerbations. In an approach similar to that of Green et al.,[3] Haldar et al.[77] evaluated the effect of a reduction in sputum eosinophils with anti-IL-5. In this randomized, double-blind, placebo-controlled trial, 61 patients were enrolled with asthma and sputum eosinophilia that was refractory to treatment and a history of recurrent exacerbations. Patients were given anti-IL-5 monoclonal antibody or placebo monthly for 1 year. The mepolizumab group had 57 exacerbations requiring prednisone (a mean of 2.0 exacerbations per year per subject) compared to 109 exacerbations (3.4 exacerbations per subject per year) in the placebo group (Fig. 13.4.4).

The mepolizumab treatment group also had a greater improvement in scores of the Asthma Quality of Life Questionnaire, with a mean improvement of 0.55 compared to 0.19 in the placebo group.[77] Similar to findings of Flood-Page et al.,[31,74] the anti-IL-5 treatment group had significantly lower eosinophil counts in BAL, blood, and bronchial wash, but less of a decrease in bronchial mucosa

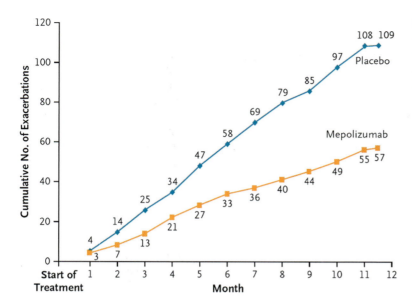

FIGURE 13.4.4 The cumulative number of severe exacerbations that occurred in each study group over the course of 50 weeks. The mean number of exacerbations per subject over the course of the 50-week treatment period was 2.0 in the mepolizumab group, compared with 3.4 in the placebo group (relative risk, 0.57; 95% confidence interval, 0.32 to 0.92; $p = 0.02$). *(Reproduced with permission from Haldar P, et al.[77])*

eosinophils. As noted in previous studies,[74] there were no significant changes from baseline in AHR, bronchodilator use, or FEV_1 in the mepolizumab group. Interestingly, when prednisolone was given after the mepolizumab treatment, there was an improvement in exhaled nitric oxide or lung function, suggesting these symptoms and pathways may be dissociated from eosinophilic inflammation, and are possibly mediated through other mechanisms.

WHAT EFFECTS DO EOSINOPHILS HAVE ON AIRWAY REMODELING?

Reticular basement membrane thickening in asthma is associated with the presence of bronchial mucosal eosinophils. Using allergen-sensitized mice, Humbles et al.[79] found eosinophil-deficient mice were protected from the development of peribronchiolar collagen deposition and increased airway smooth muscle mass following allergen challenge. However, increases in AHR and mucus secretion occurred in the eosinophilic deficient mice, just like wild-type mice. The authors concluded that eosinophils contribute substantially to airway remodeling, but are not obligatory for allergen-induced lung dysfunction.[79]

As noted earlier by Flood-Page et al.,[74] residual bronchial wall eosinophils persists despite treatment with anti-IL-5, even when blood and sputum eosinophils are nearly eliminated. In addition, anti-IL-5 has little effect on the MBP found in the airways, airflow obstruction, and AHR. These observations raise the possibility that bronchial mucosal eosinophils are a *privileged cell* or are in a *privileged location* when lodged in tissues, and their levels are not responsive to previously used methods of eosinophil

depletion. Furthermore, these findings suggest that another important contribution of eosinophils to the pathophysiology of asthma is airway remodeling, which is supported by other studies.

In a 2-year study, Sont et al.[80] evaluated an asthma treatment strategy aimed at reducing AHR compared to treatment that followed international guidelines and was based primarily on symptom and lung function assessments. In this randomized, prospective, parallel trial of 75 adults with mild to moderate asthma, 41 patients were placed in the reference strategy group, with a treatment based on current guidelines, while 34 patients were placed in the AHR group with a treatment strategy based on guidelines as well as a reduction in AHR. In addition to measuring bronchodilator use, FEV_1, peak expiratory flow (PEF) and symptoms, AHR was also quantified following methacholine challenges. Patients in the AHR strategy group received higher doses of ICS to reduce AHR. This use of higher doses of ICS and reduction in AHR was associated with a lower incidence of exacerbations and a greater improvement in lung function.

Of the 75 patients in the study,[80] 55 also underwent bronchial biopsies before and after their individual treatment approaches. In AHR strategy group subjects, who received an additional 400 µg/d of ICS, there was a significantly greater decrease in subepithelial reticular layer airway thickness and bronchial mucosa eosinophils compared to the reference group (Fig. 13.4.5). The study also showed that the decrease in mucosal eosinophils related to an accompanying improvement in AHR. When the investigators evaluated the relationship between improvements in AHR and changes in the histology of the airway biopsies, they found a correlation with the reduction in tissue eosinophils (Fig. 13.4.6). These findings

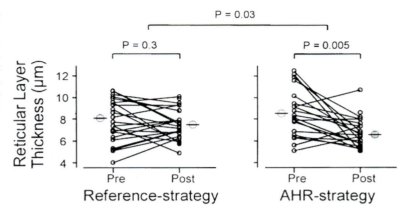

FIGURE 13.4.5 Individual changes in reticular layer thickness beneath the epithelium in bronchial biopsy specimens before and after treatment for 2 years according to the reference and AHR strategies. Bars indicate mean values at the visits for both strategies. There is a significant decrease in reticular layer thickness within the AHR strategy group, which is significantly greater than in the reference strategy group. (*Reproduced with permission from Sont JK, et al.[80]*)

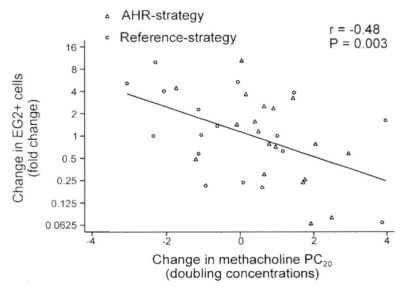

FIGURE 13.4.6 Relationship between changes in EG2[+] eosinophils and changes in methacholine PC_{20} (the dose of the inhaled antagonist that provokes a 20% drop in FEV_1, the forced expiratory volume in one second) during 2 years of treatment according to the reference and airway hyper-responsiveness (AHR) strategies. The greater the decrease in number of EG2[+] eosinophils, the greater the improvement in AHR to inhaled methacholine. EG2, antibody marker for activated eosinophils. (*Reproduced with permission from Sont JK, et al.[80]*)

further support for a role of eosinophils in airway remodeling.

The Haldar et al.[77] study also evaluated the effects of anti-IL-5 on features of airway remodeling. Airway wall thickness and airway wall area were evaluated by chest x-ray computed tomography (CT). Changes in these assessments of airway structure were made following a year of treatment in both the mepolizumab and placebo groups. The investigators found a significant reduction in airway wall thickness and total wall area (TA and WA) in the mepolizumab-treated group (Fig. 13.4.7).

Flood-Page et al.[31] also hypothesized that a reduction in bronchial wall eosinophils with anti-IL-5 would reduce markers of airway remodeling. Twenty-four mild atopic asthma patients received three monthly infusions of mepolizumab and had bronchial biopsies taken before and after each infusion. In an attempt to determine if bronchial mucosa eosinophilic inflammation was associated with an increased deposition of extracellular matrix (ECM) proteins, researchers found a positive correlation between the thickness, density and expression of the ECM protein tenascin, and bronchial mucosal eosinophil numbers. As noted previously, mepolizumab caused a significant, though incomplete elimination of bronchial mucosal eosinophils. Associated with these changes in eosinophils was a decrease in the thickness and density of tenascin, as well as density of the ECM proteins lumican and procollagen III, in the reticular basement membrane. In addition, there was a decrease in expression of both tenascin and lumican. BAL fluid in the mepolizumab-treated group also had a significant decrease in TGF-β. Not only did this study confirm that the expression of ECM proteins is greater in asthma, it also showed that a reduction in eosinophil numbers is associated with a decrease in ECM protein deposition in the airway.

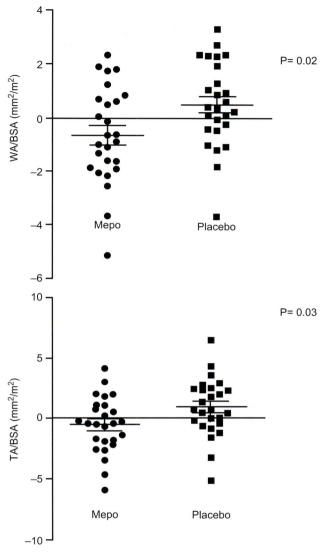

FIGURE 13.4.7 Mean change in CT measured wall area (WA) and total area (TA), corrected for body surface area (BSA) after 12 months of therapy with mepolizumab or placebo. Horizontal bars represent mean change from baseline and error bars +/−SEM. *(Reproduced with permission from Haldar P, et al.[77])*

Thus, the above studies provide convincing evidence for a relationship between the presence of eosinophils and features of airway remodeling. Firstly, an increased ICS dose leads to a decrease in bronchial mucosal eosinophils and airway thickness, as well as a reduction in bronchial hyperresponsiveness.[80] Anti-IL-5 studies built on such observations by showing that a decrease in bronchial mucosal eosinophils is associated with a reduction in airway thickness, as assessed by CT scans of the chest.[77] Finally, evidence for a decrease in TGF-β and ECM proteins following the use of anti-IL-5 provides further support of the link between bronchial mucosa eosinophilic inflammation and airway remodeling in asthma.[31]

CONCLUSION

Since the discovery of the eosinophil in the late 1800s, this cell has been considered to be a characteristic feature and possibly an essential and etiological component of the pathophysiology of asthma. These assumptions were supported by demonstrating tissue eosinophilia in postmortem analyses of the lungs of patients who died in status asthmaticus, as well as by finding parallel relationships between blood eosinophilia and asthma severity. The belief in the central role of eosinophil in asthma was further strengthened by animal models that convincingly found anti-IL-5 strategies to reduce AHR, blood and lung eosinophils, and the LPR to inhaled antigen. These studies identified eosinophils as a central contributor to asthma. Surprisingly, when anti-IL-5 was evaluated in patients with asthma, airway and peripheral eosinophils dramatically diminished, but there were no significant improvements in daily symptoms of asthma or airflow obstruction. What has emerged in the wake of anti-IL-5 studies is perhaps a more informed opinion on two major roles for eosinophils in asthma: as effector cells of airway remodeling and exacerbations.

REFERENCES

1. Wardlaw AJ, Brightling C, Green R, Woltmann G, Pavord I. Eosinophils in asthma and other allergic diseases. *Br Med Bull* 2000;**56**:985—1003.
2. Gibson PG, Girgis-Gabardo A, Morris MM, Mattoli S, Kay JM, Dolovich J, et al. Cellular characteristics of sputum from patients with asthma and chronic bronchitis. *Thorax* 1989;**44**:693—9.
3. Green RH, Brightling CE, McKenna S, Hargadon B, Parker D, Bradding P, et al. Asthma exacerbations and sputum eosinophil counts: a randomised controlled trial. *Lancet* 2002;**360**:1715—21.
4. Uchida DA, Ackerman SJ, Coyle AJ, Larsen GL, Weller PF, Freed J, et al. The effect of human eosinophil granule major basic protein on airway responsiveness in the rat in vivo. A comparison with polycations. *Am Rev Respir Dis* 1993;**147**:982—8.
5. Hirsch JG, Hirsch BI. Paul Ehrich and the discovery of the eosinphil. In: Mahmoud AAF, Austen KF, editors. *The Eosinophil in Health and Disease*. New York: Grune and Stratton; 1980. p. 2—23.
6. Horn BR, Robin ED, Theodore J, van Kessel A. Total eosinophil counts in the management of bronchial asthma. *N Engl J Med* 1975;**292**:1152—5.
7. Bousquet J, Chanez P, Lacoste JY, Barneon G, Ghavanian N, Enander I, et al. Eosinophilic inflammation in asthma. *N Engl J Med* 1990;**323**:1033—9.
8. Taylor KJ, Luksza AR. Peripheral blood eosinophil counts and bronchial responsiveness. *Thorax* 1987;**42**:452—6.
9. Gleich GJ, Adolphson CR. The eosinophilic leukocyte: structure and function. *Adv Immunol* 1986;**39**:177—253.
10. Frigas E, Loegering DA, Gleich GJ. Cytotoxic effects of the guinea pig eosinophil major basic protein on tracheal epithelium. *Lab Invest* 1980;**42**:35—43.
11. Frigas E, Loegering DA, Solley GO, Farrow GM, Gleich GJ. Elevated levels of the eosinophil granule major basic protein in the sputum of patients with bronchial asthma. *Mayo Clin Proc* 1981;**56**:345—53.
12. Gleich GJ, Frigas E, Loegering DA, Wassom DL, Steinmuller D. Cytotoxic properties of the eosinophil major basic protein. *J Immunol* 1979;**123**:2925—7.
13. Flavahan NA, Slifman NR, Gleich GJ, Vanhoutte PM. Human eosinophil major basic protein causes hyperreactivity of respiratory smooth muscle. Role of the epithelium. *Am Rev Respir Dis* 1988;**138**:685—8.
14. Brofman JD, White SR, Blake JS, Munoz NM, Gleich GJ, Leff AR. Epithelial augmentation of trachealis contraction caused by major basic protein of eosinophils. *J Appl Physiol* 1989;**66**:1867—73.
15. Dahl R, Venge P, Fredens K. The eosinophil. In: Barnes PJ, Rodger I, Thomson N, editors. *Asthma: Basic Mechanisms and Clinical Management*. London: Academic Press; 1988. p. 115—30.
16. De Monchy JG, Kauffman HF, Venge P, Koeter GH, Jansen HM, et al. Bronchoalveolar eosinophilia during allergen-induced late asthmatic reactions. *Am Rev Respir Dis* 1985;**131**:373—6.
17. Motojima S, Frigas E, Loegering DA, Gleich GJ. Toxicity of eosinophil cationic proteins for guinea pig tracheal epithelium in vitro. *Am Rev Respir Dis* 1989;**139**:801—5.
18. Robinson DS, Assoufi B, Durham SR, Kay AB. Eosinophil cationic protein (ECP) and eosinophil protein X (EPX) concentrations in serum and bronchial lavage fluid in asthma. Effect of prednisolone treatment. *Clin Exp Allergy* 1995;**25**:1118—27.
19. Lugosi E, Halmerbauer G, Frischer T, Koller DY. Urinary eosinophil protein X in relation to disease activity in childhood asthma. *Allergy* 1997;**52**:584—8.
20. Kroegel C, Yukawa T, Dent G, Venge P, Chung KF, Barnes PJ. Stimulation of degranulation from human eosinophils by platelet-activating factor. *J Immunol* 1989;**142**:3518—26.
21. Lellouchtubiana A, Lefort J, Simon MT, Pfister A, Vargaftig BB. Eosinophil Recruitment into Guinea-Pig Lungs after Paf-Acether and Allergen Administration—Modulation by Prostacyclin, Platelet Depletion, and Selective Antagonists. *American Review of Respiratory Disease* 1988;**137**:948—54.
22. Lellouchtubiana A, Lefort J, Pirotzky E, Vargaftig BB, Pfister A. Ultrastructural Evidence for Extravascular Platelet Recruitment in the Lung Upon Intravenous-Injection of Platelet-Activating Factor (Paf-Acether) to Guinea-Pigs. *Brit J Exp Pathol* 1985;**66**:345—55.
23. Rubin AHE, Smith LJ, Patterson R. The Bronchoconstrictor Properties of Platelet-Activating-Factor in Humans. *American Review of Respiratory Disease* 1987;**136**:1145—51.
24. Cuss FM, Dixon CM, Barnes PJ. Effects of inhaled platelet activating factor on pulmonary function and bronchial responsiveness in man. *Lancet* 1986;**2**:189—92.
25. Dahlen SE, Hedqvist P, Hammarstrom S, Samuelsson B. Leukotrienes are potent constrictors of human bronchi. *Nature* 1980;**288**:484—6.
26. Isono T, Koshihara Y, Murota S, Fukuda Y, Furukawa S. Measurement of immunoreactive leukotriene C4 in blood of asthmatic children. *Biochem Biophys Res Commun* 1985;**130**:486—92.
27. Garland A, Necheles J, White SR, Neeley SP, Leff AR, Carson SS, et al. Activated eosinophils elicit substance P release from cultured dorsal root ganglion neurons. *Am J Physiol* 1997;**273**:L1096—102.
28. Chanez P, Dent G, Yukawa T, Barnes PJ, Chung KF. Generation of oxygen free radicals from blood eosinophils from asthma patients after stimulation with PAF or phorbol ester. *Eur Respir J* 1990;**3**:1002—7.
29. Henricks PA, Nijkamp FP. Reactive oxygen species as mediators in asthma. *Pulm Pharmacol Ther* 2001;**14**:409—20.
30. Minshall EM, Leung DY, Martin RJ, Song YL, Cameron L, Ernst P, et al. Eosinophil-associated TGF-beta1 mRNA expression and airways fibrosis in bronchial asthma. *Am J Respir Cell Mol Biol* 1997;**17**:326—33.
31. Flood-Page P, Menzies-Gow A, Phipps S, Ying S, Wangoo A, Ludwig MS, et al. Anti-IL-5 treatment reduces deposition of ECM proteins in the bronchial subepithelial basement membrane of mild atopic asthmatics. *J Clin Invest* 2003;**112**:1029—36.
32. Weller PF, Rand TH, Barrett T, Elovic A, Wong DT, Finberg RW. Accessory cell function of human eosinophils. HLA-DR-dependent, MHC-restricted antigen-presentation and IL-1 alpha expression. *J Immunol* 1993;**150**:2554—62.

33. Kita H, Ohnishi T, Okubo Y, Weiler D, Abrams JS, Gleich GJ. Granulocyte/macrophage colony-stimulating factor and interleukin 3 release from human peripheral blood eosinophils and neutrophils. *J Exp Med* 1991;**174**:745−8.

34. Moqbel R, Ying S, Barkans J, Newman TM, Kimmitt P, Wakelin M, et al. Identification of messenger RNA for IL-4 in human eosinophils with granule localization and release of the translated product. *J Immunol* 1995;**155**:4939−47.

35. Broide DH, Paine MM, Firestein GS. Eosinophils express interleukin 5 and granulocyte macrophage-colony-stimulating factor mRNA at sites of allergic inflammation in asthmatics. *J Clin Invest* 1992;**90**:1414−24.

36. Kay AB, Barata L, Meng Q, Durham SR, Ying S. Eosinophils and eosinophil-associated cytokines in allergic inflammation. *Int Arch Allergy Immunol* 1997;**113**:196−9.

37. Kirby JG, Hargreave FE, Gleich GJ, O'Byrne PM. Bronchoalveolar cell profiles of asthmatic and nonasthmatic subjects. *Am Rev Respir Dis* 1987;**136**:379−83.

38. Wardlaw DF, Dunnett S, Gleich GJ. Eosinophils and mast cells in bronchoalveolar lavage and mild asthma: relationship to bronchial hyperreactivity. *Am Rev Respir Dis* 1988;**137**:62.

39. Adelroth E, Rosenhall L, Johansson SA, Linden M, Venge P. Inflammatory cells and eosinophilic activity in asthmatics investigated by bronchoalveolar lavage. The effects of antiasthmatic treatment with budesonide or terbutaline. *American Review of Respiratory Disease* 1990;**142**:91−9.

40. Earle BV. Fatal bronchial asthma; a series of fifteen cases with a review of the literature. *Thorax* 1953;**8**:195−206.

41. Kroegel C. The role of eosinophils in asthma. *Lung* 1990; **168**(Suppl):5−17.

42. Metzger WJ, Nugent K, Richerson HB, Moseley P, Lakin R, Zavala D, et al. Methods for Bronchoalveolar Lavage in Asthmatic-Patients Following Bronchoprovocation and Local Antigen Challenge. *Chest* 1985;**87**:S16−9.

43. Frigas E, Gleich GJ. The eosinophil and the pathophysiology of asthma. *J Allergy Clin Immunol* 1986;**77**:527−37.

44. Ohashi Y, Motojima S, Fukuda T, Makino S. Airway hyperresponsiveness, increased intracellular spaces of bronchial epithelium, and increased infiltration of eosinophils and lymphocytes in bronchial mucosa in asthma. *American Review of Respiratory Disease* 1992;**145**:1469−76.

45. Humbles AA, Conroy DM, Marleau S, Rankin SM, Palframan RT, Proudfoot AE, et al. Kinetics of eotaxin generation and its relationship to eosinophil accumulation in allergic airways disease: analysis in a guinea pig model in vivo. *J Exp Med* 1997;**186**: 601−12.

46. Dunn CJ, Elliott GA, Oostveen JA, Richards IM. Development of a prolonged eosinophil-rich inflammatory leukocyte infiltration in the guinea-pig asthmatic response to ovalbumin inhalation. *Am Rev Respir Dis* 1988;**137**:541−7.

47. Gulbenkian AR, Fernandez X, Kreutner W, Minnicozzi M, Watnick AS, Kung T, et al. Anaphylactic challenge causes eosinophil accumulation in bronchoalveolar lavage fluid of guinea pigs. Modulation by betamethasone, phenidone, indomethacin, WEB 2086, and a novel antiallergy agent, SCH 37224. *Am Rev Respir Dis* 1990;**142**:680−5.

48. Wenzel SE. Eosinophils in asthma—closing the loop or opening the door? *N Engl J Med* 2009;**360**:1026−8.

49. Cockcroft DW, Murdock KY. Comparative effects of inhaled salbutamol, sodium cromoglycate, and beclomethasone dipropionate on allergen-induced early asthmatic responses, late asthmatic responses, and increased bronchial responsiveness to histamine. *J Allergy Clin Immunol* 1987;**79**:734−40.

50. Kidney JC, Boulet LP, Hargreave FE, Deschesnes F, Swystun VA, O'Byrne PM, et al. Evaluation of single-dose inhaled corticosteroid activity with an allergen challenge model. *J Allergy Clin Immunol* 1997;**100**:65−70.

51. Gauvreau GM, Wood LJ, Sehmi R, Watson RM, Dorman SC, Schleimer RP, et al. The effects of inhaled budesonide on circulating eosinophil progenitors and their expression of cytokines after allergen challenge in subjects with atopic asthma. *Am J Respir Crit Care Med* 2000;**162**:2139−44.

52. Sedgwick JB, Calhoun WJ, Gleich GJ, Kita H, Abrams JS, Schwartz LB, et al. Immediate and late airway response of allergic rhinitis patients to segmental antigen challenge. *Am Rev Respir Dis* 1991;**144**:1274−81.

53. Calhoun WJ, Dick EC, Schwartz LB, Busse WW. A common cold virus, rhinovirus 16, potentiates airway inflammation after segmental antigen bronchoprovocation in allergic subjects. *J Clin Invest* 1994;**94**:2200−8.

54. Gern JE. Mechanisms of virus-induced asthma. *J Pediatr* 2003;**142**:S9−13. discussion S13−14.

55. Woolley KL, Gibson PG, Carty K, Wilson AJ, Twaddell SH, Woolley MJ. Eosinophil apoptosis and the resolution of airway inflammation in asthma. *Am J Respir Crit Care Med* 1996;**154**: 237−43.

56. Saeed W, Badar A, Hussain MM, Aslam M. Eosinophils and eosinophil products in asthma. *J Ayub Med Coll Abbottabad* 2002;**14**:49−55.

57. Evans PM, O'Connor BJ, Fuller RW, Barnes PJ, Chung KF. Effect of inhaled corticosteroids on peripheral blood eosinophil counts and density profiles in asthma. *J Allergy Clin Immunol* 1993;**91**:643−50.

58. Gleich GJ, Motojima S, Frigas E, Kephart GM, Fujisawa T, Kravis LP. The eosinophilic leukocyte and the pathology of fatal bronchial asthma: evidence for pathologic heterogeneity *J Allergy Clin Immunol* 1987;**80**:412−5.

59. Koch-Weser J. Beta adrenergic blockade and circulating eosinophils. *Arch Intern Med* 1968;**121**:255−8.

60. Ohman Jr JL, Lawrence M, Lowell FC. Effect of propranolol on the isoproterenol, responses of cortisol, isoproterenol, and aminophylline. *J Allergy Clin Immunol* 1972;**50**:151−6.

61. Miranda C, Busacker A, Balzar S, Trudeau J, Wenzel SE. Distinguishing severe asthma phenotypes: role of age at onset and eosinophilic inflammation. *J Allergy Clin Immunol* 2004;**113**:101−8.

62. Haldar P, Pavord ID, Shaw DE, Berry MA, Thomas M, Brightling CE, et al. Cluster analysis and clinical asthma phenotypes. *Am J Respir Crit Care Med* 2008;**178**:218−24.

63. Woodruff PG, Modrek B, Choy DF, Jia G, Abbas AR, Ellwanger A, et al. T-helper type 2-driven inflammation defines major subphenotypes of asthma. *Am J Respir Crit Care Med* 2009;**180**:388−95.

64. Sanderson CJ. Interleukin-5, eosinophils, and disease. *Blood* 1992;**79**:3101−9.

65. Walsh GM, Hartnell A, Wardlaw AJ, Kurihara K, Sanderson CJ, Kay AB. IL-5 enhances the in vitro adhesion of human eosinophils, but not neutrophils, in a leucocyte integrin (CD11/18) dependent manner. *Immunology* 1990;**71**:258−65.

66. van der Veen MJ, Van Neerven RJ, De Jong EC, Aalberse RC, Jansen HM, van der Zee JS. The late asthmatic response is associated with baseline allergen-specific proliferative responsiveness of peripheral T lymphocytes in vitro and serum interleukin-5. *Clin Exp Allergy* 1999;**29**:217−27.

67. Shi HZ, Xiao CQ, Zhong D, Quin SM, Liu Y, Liang GR, et al. Effect of inhaled interleukin-5 on airway hyperreactivity and eosinophilia in asthmatics. *Am J Respir Crit Care Med* 1998;**157**:204−9.

68. Shi H, Qin S, Huang G, Chen Y, Xiao C, Xu H, et al. Infiltration of eosinophils into the asthmatic airways caused by interleukin 5. *Am J Respir Cell Mol Biol* 1997;**16**:220−4.

69. Foster PS, Hogan SP, Ramsay AJ, Matthaei KI, Young IG. IL-5 deficiency abolishes eosinophilia, airways hyperreactivity, and lung damage in a mouse asthma model. *J Exp Med* 1996;**183**:195−201.

70. Robinson D, Hamid Q, Ying S, Bentley A, Assoufi B, Durham S, et al. Prednisolone treatment in asthma is associated with modulation of bronchoalveolar lavage cell interleukin-4, interleukin-5, and interferon-gamma cytokine gene expression. *Am Rev Respir Dis* 1993;**148**:401−6.

71. Van Oosterhout AJ, Ladenius AR, Savelkoul HF, Van Ark I, Delsman KC, Nijkamp FP. Effect of anti-IL-5 and IL-5 on airway hyperreactivity and eosinophils in guinea pigs. *Am Rev Respir Dis* 1993;**147**:548−52.

72. Akutsu I, Kojima T, Kariyone A, Fukuda T, Makino S, Takatsu K. Antibody against interleukin-5 prevents antigen-induced eosinophil infiltration and bronchial hyperreactivity in the guinea pig airways. *Immunol Lett* 1995;**45**:109−16.

73. Leckie MJ, ten Brinke A, Khan J, Diamant Z, O'Connor BJ, Walls CM, et al. Effects of an interleukin-5 blocking monoclonal antibody on eosinophils, airway hyper-responsiveness, and the late asthmatic response. *Lancet* 2000;**356**:2144−8.

74. Flood-Page PT, Menzies-Gow AN, Kay AB, Robinson DS. Eosinophil's role remains uncertain as anti-interleukin-5 only partially depletes numbers in asthmatic airway. *Am J Respir Crit Care Med* 2003;**167**:199−204.

75. Bentley AM, Hamid Q, Robinson DS, Schotman E, Meng Q, Assoufi B, et al. Prednisolone treatment in asthma. Reduction in the numbers of eosinophils, T cells, tryptase-only positive mast cells, and modulation of IL-4, IL-5, and interferon-gamma cytokine gene expression within the bronchial mucosa. *Am J Respir Crit Care Med* 1996;**153**:551−6.

76. Rothenberg ME, Klion AD, Roufosse FE, Kahn JE, Weller PF, Simon HU, et al. Treatment of patients with the hypereosinophilic syndrome with mepolizumab. *N Engl J Med* 2008;**358**:1215−28.

77. Haldar P, Brightling CE, Hargadon B, Gupta S, Monteiro W, Sousa A, et al. Mepolizumab and exacerbations of refractory eosinophilic asthma. *N Engl J Med* 2009;**360**:973−84.

78. Nair P, Pizzichini MM, Kjarsgaard M, Inman MD, Efthimiadis A, Pizzichini E, et al. Mepolizumab for prednisone-dependent asthma with sputum eosinophilia. *N Engl J Med* 2009;**360**:985−93.

79. Humbles AA, Lloyd CM, McMillan SJ, Friend DS, Xanthou G, McKenna EE, et al. A critical role for eosinophils in allergic airways remodeling. *Science* 2004;**305**:1776−9.

80. Sont JK, Willems LN, Bel EH, Van Krieken JH, Vandenbroucke JP, Sterk PJ. Clinical control and histopathologic outcome of asthma when using airway hyperresponsiveness as an additional guide to long-term treatment. The AMPUL Study Group. *Am J Respir Crit Care Med* 1999;**159**:1043−51.

Eosinophil-Targeted Treatment of Asthma

Parameswaran Nair

INTRODUCTION

The role of the eosinophils as key players in the pathophysiology of asthma has been debated, despite evidence that the cells are present and activated in the airway lumen and tissue[1] of patients with current asthma; are increased in number when asthma is uncontrolled[2] or severe[3] and decreased when asthma is controlled[4]; and treatment strategies that aim to control airway eosinophilia are significantly more effective and less expensive in improving asthma control[5,6] and decreasing asthma exacerbations compared to guideline-based clinical strategies.

Cynicism was fueled by observations that in murine models of allergic sensitization, airway hyperresponsiveness could be induced without eosinophils.[7] Skepticism grew stronger when therapy using monoclonal antibodies against interleukin-5 (IL-5), which has no known clinically relevant biologic activity other than targeting eosinophils, failed to demonstrate improvement in asthma outcomes despite decreasing airway and blood eosinophil numbers.[8] The molecule did not reduce allergen-induced airway constriction or hyperresponsiveness, airflow limitation, exacerbations, or symptoms. The likely explanations for this apparent paradox are inappropriate methodology, inadequate sample size,[9] or an inadequate reduction in bronchial mucosal eosinophil numbers.[10]

This subchapter will describe the clinical studies that demonstrated an improvement in asthma control using treatment strategies that aimed to normalize sputum eosinophil count using corticosteroids; to evaluate critically the clinical trials that failed to demonstrate an improvement in asthma using monoclonal antibodies directed against IL-5; and to present evidence from a prospective audit of clinical outcomes of patients managed by normalizing sputum cell counts.

EOSINOPHIL-BASED TREATMENT STRATEGIES USING CORTICOSTEROIDS

Airway eosinophilia can be reliably and relatively non-invasively assessed in sputum.[11] In clinical practice, approximately 30% of patients with asthma attending a tertiary clinic have eosinophilic bronchitis.[12] More severe asthma and more severe airflow limitation are associated with more intense sputum eosinophilia.[13] Two studies in adults and one study in children have evaluated the outcomes of titrating anti-inflammatory treatment with the intention of normalizing eosinophils in sputum. The first single center, 1-year trial that examined the effect of treating asthma to reduce eosinophils to 2% resulted in a significant reduction of severe exacerbations compared with a control group treated without sputum eosinophil counts.[5] The large number of exacerbations and their severity was probably a result of the policy at the time to reduce corticosteroid use further if control was maintained for 2 months. The second trial[6] was a multicenter trial conducted over 2 years, and differed in that the minimum dose of corticosteroid to maintain sputum eosinophils at 3% was determined first and then maintained for the duration of the study. Exacerbations were few and mild compared with the first study and were reduced by about 50% compared with the group treated with the same best-guideline approach to treatment without sputum cell counts. The active treatment reduced eosinophilic exacerbations but had no effect on neutrophilic exacerbations, which were regarded to be probably of viral cause. The benefits in both studies were achieved without any increase in corticosteroid dose over that required by the control group. In contrast, a similar study in children showed a nonstatistically significant effect on reducing exacerbations using a sputum strategy that aimed to keep eosinophil levels to below 2.5%.[14] The modest benefit was most likely due to the inadequate control of eosinophils in the treatment arm that was probably related to the inadequate dose

of inhaled corticosteroids allowed in the study. The effectiveness of using sputum eosinophils as a marker to decrease exacerbations in adults and children with moderate to severe asthma was recently confirmed in a systematic review and meta-analysis.[15] A critical review of the recently published clinical trial literature reveals that the reduction in exacerbation reported in all the most recent large clinical trials for either asthma or chronic obstructive pulmonary disease (COPD) for any new medication compared to placebo is significantly less than those reported for strategies employing the judicious use of currently available medications guided by cell counts in sputum (Table 13.5.1). In addition to a significant reduction of exacerbations at a reduced cost,[16] the adverse consequences of new therapies and suboptimal treatment may also be avoided.

ANTI-INTERLEUKIN-5 CLINICAL TRIALS

The beneficial effects of corticosteroids are not limited to decreasing eosinophil numbers in the airways.[17] They can also reduce the numbers of other cells, such as lymphocytes and mast cells, and decrease some markers of remodeling. Thus, it is not possible to conclude definitively the pathobiological role of eosinophils in asthma from those studies. Definitive proof would be obtained by reducing eosinophil numbers in the airway using treatments that directly target the eosinophils. Recently, the availability of monoclonal antibodies directed against IL-5 has provided us with the opportunity to examine this question. In two recently

TABLE 13.5.1 Relative Risk Reduction of Exacerbations for Various Interventions for Asthma and for Chronic Obstructive Pulmonary Disease

Asthma		COPD	
Intervention	RRR	Intervention	RRR
ICS + LABA (AJRCCM 2004;170:836—44)	10—15%	ICS + LABA (NEJM 2007;356:775—89)	25%
Tiotropium (NEJM 2010;363:1715—26)	~5%	Tiotropium (NEJM 2008;359:1543—54)	14%
Omalizumab (Ann Intern Med 2011;154:573—82)	25%	Tiotropium + ICS/LABA (NEJM 2011;364:1093—1103)	27%
ICS + Montelukast (BMJ 2003;327:891—7)	~1%	Roflumilast (Lancet 2009;374:695—703)	16—21%
Thermoplasty (AJRCCM 2010;181:116—24)	22%	Community-based multi-disciplinary programs (Thorax 2010;65:7—13)	~
Sputum strategy (ERJ 2006;27:483—94)	49%	Sputum strategy (ERJ 2007;29:906—13)	62%

COPD, chronic obstructive pulmonary disease; ICS, inhaled corticosteroid; LABA, long-acting β agonist; RRR, relative risk reduction.

TABLE 13.5.2 Response to Anti-Interleukin-5 and the Eosinophil Phenotype

Study	Intervention	Sputum Eosinophils at Study Entry	Success
Flood-Page et al. (AJRCCM 2007; 176: 1062–71)	Mepolizumab	5% of patients had >3% eosinophils	X
Kips et al. (AJRCCM 2003; 167: 1655–9)	Reslizumab	~30% of patients had >3% eosinophils	X
Haldar et al. (NEJM 2009; 360: 973–84)	Mepolizumab	all had >3% on one occasion in 2 years	√
Castro et al. (AJRCCM 2011; Aug 18 epub)	Reslizumab	all had >3% at randomization	√√
Nair et al. (NEJM 2009; 360: 985–93)	Mepolizumab	All had >3% on ≥3 occasions	√√√

published randomized controlled trials (RCTs) on the effect of mepolizumab[18,19] and a clinical trial on the effect of reslizumab,[20] these drugs reduced sputum eosinophils numbers to almost zero. This reduction was associated with a reduction of exacerbations compared with the placebo group in the first mepolizumab study[18] and a prednisone-sparing effect and improvement in clinical outcomes in a small sample number in the second.[19] In the larger reslizumab clinical trial,[20] the reduction in sputum eosinophils was associated with an improvement in the forced expiratory volume in one second (FEV_1) and in asthma controls over a 5-month period in patients with moderate to severe asthma. The results of these three studies contrasted with the negative results of five other trials, in which the effect of the antieosinophil drug was not examined in patients with asthma and current sputum eosinophilia. In two of the five studies that measured sputum eosinophils and in the three RCTs, the greater the certainty that an increase in eosinophils was persistent, the greater the success of the treatment (Table 13.5.2).

OBSERVATIONAL STUDY OF NORMALIZING SPUTUM EOSINOPHILS

Clinical outcomes are significantly improved when patients who require daily prednisone are monitored using sputum cell counts. Sixty-three patients with asthma (36 men; mean age, 52 years; mean BMI, 29.1) were followed for a median period of 7 years (range, 0.25–26 year).[21] Thirty-seven patients had associated chronic airflow limitation (post-bronchodilator FEV_1/vital capacity <70%). Twenty had never smoked. Forty-two percent were nonatopic. Significant comorbidities included gastroesophageal reflux disease (70%), nasal polyps and sinusitis (65%), and sensitivity to nonsteroidal anti-inflammatory drugs (28%).

Ethmoid and sphenoid sinusitis were the most important predictors of persistent airway eosinophilia.[22] At the time of their initial assessment, the majority of the patients were not on daily prednisone (median daily dose, 0 mg; sputum eosinophils: mean, 18.8%; median, 5.3%; minimum, 0%; maximum, 84%). Monitoring with the aim of keeping sputum eosinophils at <2% resulted in higher doses of corticosteroids (median daily dose of prednisone was 10 mg and for inhaled corticosteroids was 1000 µg of fluticasone equivalent), and this was associated with predictable significant adverse effects. Over the period of follow-up, despite decreasing the eosinophilic exacerbations to 0.2 years/patient, there were 22 noneosinophilic neutrophilic exacerbations. Overall, there was no significant loss of lung function over the period of follow-up (mean decrease in FEV_1, 35 mL/year).

OTHER ANTIEOSINOPHIL TREATMENT STRATEGIES

Corticosteroids are very effective in reducing eosinophil numbers and activation in the airways of most patients with asthma. Therefore novel treatment strategies such as anti-IL5 should probably be reserved for patients who require high doses of inhaled or regular ingested corticosteroids to control their airway eosinophilia and asthma. Although a larger number of antisense molecules, monoclonal antibodies, and small molecules are currently being evaluated to target a number of relevant cytokines or chemoattractants involved in eosinophil recruitment into the airway, such as eotaxin/ C-C motif chemokine 11 (CCL11), IL-4, and IL-13,[23] none of them have yet been demonstrated to be effective in suppressing an airway eosinophilia that persists despite being treated with prednisone. These are currently being investigated.

CONCLUSIONS

These studies illustrate a number of important principles. Firstly, they confirm that eosinophils are important in the pathophysiology of asthma. Second, since the nature of airway inflammation can change over time in the same patient,[24] antieosinophil treatment will be effective only when eosinophils are present in the airway. Third, luminal eosinophils represent biologically active cells. Fourth, clinical trials with small numbers of carefully characterized patients can demonstrate biologically relevant mechanisms, whereas large studies have not. Fifth, monitoring of exhaled nitric oxide is not effective in directing antieosinophil therapy in patients with severe asthma and airway eosinophilia.[25] Finally, since sputum cell count-based treatment strategies are more effective than any other currently available medications for the treatment of asthma or COPD, we are doing our patients with severe asthma and COPD a disservice by delaying the introduction of sputum cell counts in clinical practice.

ACKNOWLEDGEMENTS

Dr. Nair is supported by a Canada Research Chair in Airway Inflammometry. This chapter is dedicated to the fond memory of Professor Freddy Hargreave who tirelessly argued for the implementation of measurement of bronchitis by sputum examination in clinical practice.

REFERENCES

1. Bousquet J, Chanez P, Lacoste JY, Barnéon G, Ghavanian N, Enander I, et al. Eosinophilic inflammation in asthma. *N Engl J Med* 1990;**323**:1033–9.
2. Leuppi JD, Salome CM, Jenkins CR, Anderson SD, Xuan W, Marks GB, et al. Predictive markers of asthma exacerbation during stepwise dose reduction of inhaled corticosteroids. *Am J Respir Crit Care Med* 2001;**163**:406–12.
3. van Veen IH, ten Brinke A, Gauw SA, Sterk PJ, Rabe KF, Bel EH. Consistency of sputum eosinophilia in difficult-to-treat asthma: a 5-year follow-up study. *J Allergy Clin Immunol* 2009;**124**:615–7.
4. Lemière C, Ernst P, Olivenstein R, Yamauchi Y, Govindaraju K, Ludwig MS, et al. Airway inflammation assessed by invasive and noninvasive means in severe asthma: eosinophilic and non-eosinophilic phenotypes. *J Allergy Clin Immunol* 2006;**118**:1033–9.
5. Green RH, Brightling CE, McKenna S, Hargadon B, Parker D, Bradding P, et al. Asthma exacerbations and sputum eosinophil counts: a randomised controlled trial. *Lancet* 2002;**360**:1715–21.
6. Jayaram L, Pizzichini MM, Cook RJ, Boulet LP, Lemière C, Pizzichini E, et al. Determining asthma treatment by monitoring sputum cell counts: effect on exacerbations. *Eur Respir J* 2006;**27**:483–94.
7. Wills-Karp M, Karp CL. Eosinophils in asthma: remodeling a tangled tale. *Science* 2004;**305**:1726–9.
8. Leckie MJ, ten Brinke A, Khan J, Diamant Z, O'Connor BJ, Walls CM, et al. Effects of an interleukin-5 blocking monoclonal antibody on eosinophils, airway hyper-responsiveness, and the late asthmatic response. *Lancet* 2000;**356**:2144–8.
9. O'Byrne PM, Inman MD, Parameswaran K. The trials and tribulations of IL-5, eosinophils, and allergic asthma. *J Allergy Clin Immunol* 2001;**108**:503–8.
10. Flood-Page PT, Menzies-Gow AN, Kay AB, Robinson DS. Eosinophil's role remains uncertain as anti-interleukin-5 only partially depletes numbers in asthmatic airway. *Am J Respir Crit Care Med* 2003;**167**:199–204.
11. Nair P, Hargreave FE. Measuring bronchitis in airway diseases: clinical implementation and application. *Chest* 2010;**138** (2 Suppl):38S–43S.
12. D'silva L, Hassan N, Wang HY, Kjarsgaard M, Efthimiadis A, Hargreave FE, et al. Heterogeneity of bronchitis in airway diseases in tertiary care clinical practice. *Can Respir J* 2011;**18**:1–4–8.
13. ten Brinke A, Zwinderman AH, Sterk PJ, Rabe KF, Bel EH. Factors associated with persistent airflow limitation in severe asthma. *Am J Respir Crit Care Med* 2001;**164**:744–8.
14. Fleming L, Wilson N, Regamey N, Bush A. Use of sputum eosinophil counts to guide management in children with severe asthma. *Thorax* 2011;**67**(3):193–8.
15. Petsky HL, Cates CJ, Lasserson TJ, Li AM, Turner C, Kynaston JA, et al. A systematic review and meta-analysis: tailoring asthma treatment on eosinophilic markers (exhaled nitric oxide or sputum eosinophils). *Thorax* 2012 Mar;**67**(3):199–208.
16. D'silva L, Gafni A, Thabane L, Jayaram L, Hassack P, Hargreave FE, et al. Cost analysis of monitoring asthma treatment using sputum cell counts. *Can Respir J* 2008;**15**:370–4.
17. Chakir J, Loubaki L, Laviolette M, Milot J, Biardel S, Jayaram L, et al. Monitoring sputum eosinophils in mucosal inflammation and remodelling: a pilot study. *Eur Respir J* 2010;**35**:48–53.
18. Haldar P, Brightling CE, Hargadon B, Gupta S, Monteiro W, Sousa A, et al. Mepolizumab and exacerbations of refractory eosinophilic asthma. *N Engl J Med* 2009;**360**:973–84.
19. Nair P, Pizzichini MM, Kjarsgaard M, Inman MD, Efthimiadis A, Pizzichini E, et al. Mepolizumab for prednisone-dependent asthma with sputum eosinophilia. *N Engl J Med* 2009;**360**:985–93.
20. Castro M, Mathur S, Hargreave F, Boulet LP, Xie F, Young J, et al. for the Res-5-0010 Study Group. Reslizumab for Poorly Controlled, Eosinophilic Asthma: A Randomized, Placebo-Controlled Study. *Am J Respir Crit Care Med* 2011 Nov 15;**184**(10):1125–32.
21. Aziz-ur-Rehman A, Kjarsgaard M, Efthimiadis A, Hargreave FE, Nair P. An audit of prednisone-dependent asthma [abstract]. *Am J Respir Crit Care Med* 2008;**177**(suppl):A107.
22. Lui BT, Boylan C, Aziz-Ur-Rehman A, Sommer DD, Nair P. Paranasal sinus disease and sputum eosinophilia in prednisone dependent asthma. *J Otolaryngol Head Neck Surg* 2010;**39**:703–9.
23. Bochner BS, Gleich GJ. What targeting eosinophils has taught us about their role in diseases. *J Allergy Clin Immunol* 2010;**126**:16–25.
24. D'silva L, Cook RJ, Allen CJ, Hargreave FE, Nair P. Changing pattern of sputum cell counts during successive exacerbations of airway disease. *Respir Med* 2007;**101**:2217–20.
25. Nair P, Kjarsgaard M, Armstrong S, Efthimiadis A, O'Byrne PM, Hargreave FE. Nitric oxide in exhaled breath is poorly correlated to sputum eosinophils in patients with prednisone-dependent asthma. *J Allergy Clin Immunol* 2010;**126**:404–6.

Chapter 13.6

Eosinophils and Hemopoietic Processes in Allergic Asthma

Gail M. Gauvreau and Judah A. Denburg

EOSINOPHILS IN ASTHMA: AN OVERVIEW

Eosinophils, a prominent feature of asthma, are found in increased numbers in the circulation and airways in relation to the severity of asthma.[1,2] Inhaled allergen challenge in asthmatic subjects results in the appearance and accumulation of mature and immature eosinophils in the bone marrow, blood,[3] and airways.[4–6] The kinetics of eosinophilia are compartment-specific (Fig. 13.6.1),[4,7,8] and the number of eosinophils correlates with the severity of the late asthmatic reaction.[9] This will be covered in detail elsewhere in this book.

Eosinophils are terminally differentiated myeloid leukocytes that migrate to tissues as effector cells in a number of inflammatory processes, including allergic diseases and helminth infections.[10] The migration and accumulation of eosinophils is highly regulated via signaling of cytokines and chemokines through cell-surface

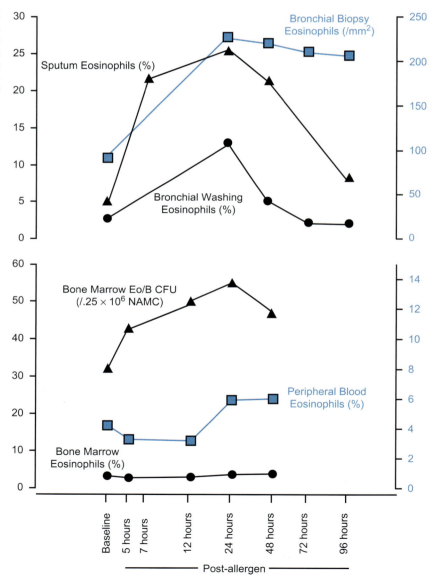

FIGURE 13.6.1   Kinetics of allergen-induced eosinophilia in airways, blood, and bone marrow. A rapid increase in airway eosinophils postallergen is mirrored by a decrease in mature eosinophils from the bone marrow and blood as these cells migrate to the airways, and an increase in eosinophil progenitors to replace the mature cells.[4,7,8] Eo/B CFU, eosinophil/basophil colony-forming units.

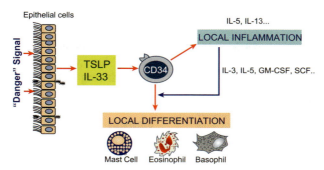

FIGURE 13.6.2 **TSLP effects on progenitors: a new vista on *in situ* hemopoiesis.** Thymic stromal lymphopoietin (TSLP) can induce T_h2 cytokines within cord blood CD34$^+$ progenitors, effectively enhancing inflammatory effector function of these progenitors, as well as providing lineage differentiating stimuli (via IL-5 acting on CD34$^+$ CD45$^+$ IL-5RA^{hi+} cells in an autocrine fashion) for tissue-resident (bronchial or nasal mucosal) CD34$^+$ cells. GM-CSF, granulocyte-macrophage colony-stimulating factor; IL-3, interleukin-3; IL-5, interleukin-5; IL-33, interleukin-33; SCF, stem cell factor. *(Delespesse and Allakhverdi, personal communication.)*

receptors, and by induction of adhesion molecule expression.[11] Since allergic asthma is primarily a T-helper type 2 (T_h2)-mediated disease, it is not surprising that cytokines driving eosinophilia are T_h2 cell products: specifically, granulocyte-macrophage colony-stimulating factor (GM-CSF), interleukin-3 (IL-3), and interleukin-5 (IL-5), which signal through specific high-affinity cell-surface receptors linked to a common β-chain—all of which can act as eosinophil growth factors that promote formation of eosinophil/basophil (Eo/B) colony-forming units (CFU) in functional assays.[12] The common β-chain is especially important for Eo/B proliferation, production of cytokines from eosinophils, and eosinophil migration to effector sites.[13] Of the three Eo/B differentiation-inducing cytokines, IL-5 is necessary for mobilization of eosinophil progenitors from the bone marrow and their terminal differentiation.[14] Other T_h2 cytokines, such as IL-4 and IL-13, are known to regulate transmigration of eosinophils from the vascular bed into the tissue compartments by augmenting expression of adhesion molecules on the endothelium[15] and inducing expression of potent eosinophil chemokines, such as eotaxin/C-C motif chemokine 11 (CCL11) and RANTES/CCL5 in the airways.[16,17] IL-9 has also been shown to play a supportive role in both mast cell and Eo/B differentiation.[18,19] Once in the inflamed tissue, eosinophils contribute to the manifestation of symptoms through release of granule proteins and proinflammatory mediators.[10]

HEMOPOIETIC ACTIVITY IN ALLERGIC ASTHMA

Hemopoietic progenitors, which are found in the circulation under steady-state and disease conditions, exist in the bone marrow at various stages of lineage commitment; in the latter compartment, these progenitors can be allowed or selected to differentiate in specific directions in response to various ambient stimuli, such as hemopoietic cytokines engaging specific cell-surface receptors. This hemopoietic activity occurs in the bone marrow proper under the influence of stromal cells and microvascular endothelial cells, and in tissues, also in response to epithelial cells. Each of these resident cell populations can respond to inflammatory stimuli by increasing transcription and translation of hemopoietic growth factors and/or cytokines. The bone marrow environment itself, as well as primed tissues, can each support and accelerate production of eosinophils for either release into the circulation or amplification of tissue eosinophilic inflammation, respectively.[20,21] The process of tissue amplification of eosinophilic inflammation through progenitor differentiation at the site has been termed *in situ hemopoiesis* (Figure 13.6.2).[22,23]

CD34-POSITIVE EOSINOPHIL/BASOPHIL PROGENITORS IN ALLERGIC ASTHMA

Hemopoietic progenitors express the cell stage-specific antigen, CD34, which is present at the highest levels on early hemopoietic myeloid progenitors and is progressively lost on terminally differentiating cells.[24] These CD34$^+$ hemopoietic progenitors contribute to the ongoing recruitment of eosinophils and basophils to sites of allergen challenge in allergic diseases, including asthma.[25] Indeed, the number of CD34$^+$ cells, as well as both mature and immature eosinophil cell numbers, are higher in blood and bone marrow of atopic subjects compared with nonatopic controls;[26,27] mild asthmatic subjects have fewer circulating CD34$^+$ cells than severe asthmatic subjects,[28] and CD34$^+$ cell numbers have been shown to correlate with the level of airflow obstruction.[28]

Interleukin-5 Receptor Subunit Alpha

After airway allergen inhalation challenge in atopic asthmatic subjects, CD34[+] cells from bone marrow synthesize mRNA and protein for membrane-bound IL-5 receptor subunit α (IL-5RA),[29] permitting them to further respond to the eosinophilopoietic cytokine, IL-5. Progenitor cells from the bone marrow of atopic subjects show increased responsiveness to IL-5,[26] probably due to increased levels of bone marrow CD34[+]IL-5RA[+] cells, a unique feature of atopic disease.[30] The phenotype of bone marrow CD34[+] cells from mild asthmatic subjects has been carefully examined by flow cytometry, demonstrating a higher proportion of CD34[+] IL-5RA[+] after allergen challenge in those subjects who developed airway eosinophilia and increased methacholine airway responsiveness.[29,31] Taken together with *in vitro* experiments demonstrating IL-5-induced expression of IL-5RA on human progenitor cells,[32,33] these data demonstrate eosinophil lineage skewing of CD34[+] cells in response to an allergic stimulus. Thus, increased production of Eo/B lineage-committed progenitors within the bone marrow, the subsequent development of blood and tissue eosinophilia,[29] and the maintenance of an allergic inflammatory response[26,34] have been linked. Reports of antigen-induced increases in CD34[+] IL-5RA[+] cell numbers in murine bone marrow coincident with enhancement of IL-5-dependent eosinophilopoiesis provide further evidence that eosinophil production occurs as a result of expansion of the relevant eosinophil progenitor population rather than exclusively from demargination or release of sequestered mature eosinophils into the circulation.[35–39] The rapid increase in expression of IL-5RA on CD34[+] cells from the bone marrow of atopic asthmatics[28,40] clearly demonstrates that progenitor cells in these subjects are *primed* to respond to IL-5; indeed, CD34[+] IL-5RA[+] cell numbers *circulate* at higher numbers in subjects with asthma compared to controls, and correlate with asthma severity.[28]

Although controversial, chemokines such as eotaxin may also play a role not only in the migration[41] but also in the differentiation of progenitor cells. Since CD34[+] progenitor cells are found to express C-C chemokine receptor 3 (CCR3),[42] which is upregulated in a Th2 environment, it is possible that signaling through this receptor also causes progenitor cells to differentiate into eosinophils independently of GM-CSF, IL-3, and IL-5.[43]

Colony-Forming Units

Colony assays are functional measures for the quantification of hemopoietic differentiation potential. Eo/B CFU are clusters of immature, nascent eosinophils and basophils derived from single progenitors[44–46] present in nonadherent mononuclear or purified CD34[+] cell populations seeded into a semisolid medium in the presence of hemopoietic growth factors. There are significantly greater numbers of Eo/B CFU in the peripheral blood of subjects with various allergic airway disorders, including asthma, nasal polyposis, and rhinitis.[47,48] Greater numbers of CD34[+] cells are detected in blood and bone marrow of atopic than nonatopic subjects, correlating positively with higher numbers of Eo/B CFU grown from the blood and marrow of these subjects.[26] The higher levels of CD34[+] cells and IL-5-responsive Eo/B CFU in atopic subjects indicate a role for Eo/B progenitors in allergic diseases, such as asthma.[26] The growth of Eo/B CFU is also influenced by exposure to specific antigen, which initiates a cascade of events with a stimulatory effect on the bone marrow to produce and release newly formed inflammatory cells. Importantly, allergic individuals can exhibit fluctuations in the numbers of circulating progenitors and tissue eosinophils during an allergen pollen season.[49,50] An initial increase in circulating Eo/B CFU at the beginning of seasonal allergen exposure is followed by a significant decline at the peak of the season, coincident with nasal symptoms and inflammation, probably reflecting the migration and differentiation *in situ* of progenitor cells from the blood to the inflamed tissue.[49–51] This increase in Eo/B CFU has also been demonstrated in allergic asthma exacerbation.[52,53] Supporting the latter *in vivo* observation, higher numbers of bone marrow[31] and circulating[54] Eo/B CFU can be measured 24 h after allergen inhalation. Studies conducted in animal models of allergic asthma that carefully investigate the trafficking of cells from bone marrow confirm that allergen-induced airway inflammation is associated with increased numbers of newly formed cells in the blood and airways.[34,55] In mouse models of allergic rhinitis and/or asthma, changes in bone marrow eosinophilopoiesis, accompanied by appropriate peaks of hemopoietic cytokines and chemokines, such as eotaxin, GM-CSF and IL-5, and changes in cell surface receptors for these factors, were shown to occur as early as 2 h after allergen challenge,[38,39,56–58] indicating that progenitor cell fluctuations and hemopoietic processes contributing to eosinophilic airways inflammation can occur quite rapidly in response to allergen challenge.

Studies to assess changes in cytokines levels within the bone marrow of sensitized mice or atopic asthmatics following allergen exposure have detected increases in levels of IL-5 consistent with the kinetics of eosinophil lineage commitment.[37–39,56,57,59] Using colony-forming assays, bone marrow progenitor cells from asthmatic subjects developing late-phase responses and airway eosinophilia postallergen challenge were shown to be more responsive to IL-5 than cells from subjects without these allergen-induced responses.[29] A subsequent investigation demonstrated that allergen inhalation by allergic asthmatic subjects induces a time-dependent change in the levels of growth factors and cytokines in the bone marrow: subjects with elevated levels of airway and circulating eosinophils postchallenge had increased numbers of IL-5-responsive progenitors at 12 h and 24 h postallergen (Fig. 13.6.1), coincident with increased IL-5 protein levels in the bone marrow.[4] As such, allergen-induced activation of an eosinophilopoietic process highlights the relationship between increased bone marrow IL-5 levels and the regulation of eosinophil production from bone marrow progenitor cells. Given the observed delayed interferon γ (INF-γ) increase in the marrow of these subjects,[4] it could be postulated that activated T cells migrate from the airways to the bone marrow and release cytokines such as IL-5 and IL-9 that may locally orchestrate activation of hemopoietic events during an allergic inflammatory response.[60–63] To this end, investigations of the cell-associated cytokine production within the bone marrow have shown that CD34+ cells[64,65] and T cells[61–63,65] produce IL-5 during the course of the allergic inflammatory response. Nascent eosinophils and basophils picked from Eo/B CFU have also been shown to constitute autocrine sources of GM-CSF and IL-5,[54,61–63,66,67] which could further amplify the process of differentiation and proliferation. More recently, it has been shown that CD34+ cells from blood and bone marrow release more IL-5 following stimulation with calcium ionophore and phorbol-12-myristate-13-acetate than do the equivalent number of CD3+ T-lymphocytes.[68]

IN SITU HEMOPOIESIS

Traditionally, the focal point of the differentiation and maturation process involving hemopoietic progenitors has been the bone marrow, under the regulation of its proper microenvironment. However, there is now an abundance of evidence demonstrating that eosinophil progenitors can traffic as fully or partly undifferentiated cells to allergic inflammatory tissues in the upper and lower airways, where they can and do differentiate into mature eosinophils under the control of local stimuli. Indeed, Eo/B CD34+ cells are found in tissues such as the bronchial mucosa,[69] nasal polyps,[70] and even in atopic dermatitis lesions.[71] Despite the importance of systemic and local IL-5 production for the differentiation of CD34+ cells in the bone marrow, it is apparent that Eo/B can differentiate similarly at sites of tissue inflammation. This has been demonstrated functionally in several studies. Firstly, inflamed tissue from subjects with allergic rhinitis and nasal polyposis has been shown to produce hemopoietic cytokines that promote the differentiation and maturation of Eo/B CFU.[23,72–78] Next, mononuclear cells extracted from nasal polyp tissue have the potential to produce Eo/B CFU in response to growth

factors such as IL-5.[23,70,72] This supports the concept of *in situ* hemopoiesis, in which locally elaborated growth factors can drive the differentiation and maturation of hemopoietic progenitors into mature eosinophils (Fig. 13.6.2). Finally, phenotypic evaluations have identified CD34+ cells in various airway compartments, including the mucosa of the upper and lower airways of subjects with allergic rhinitis and asthma, respectively.[60,69,70] The already elevated levels of CD34+ cells in induced sputum samples collected from mild asthmatic subjects are further increased after allergen inhalation challenges compared to healthy controls.[3] With respect specifically to *in situ* eosinophilopoiesis, increased numbers of cells double positive for CD34 and IL5RA mRNA are found in nasal biopsies[60] and lung biopsies of allergic asthmatic subjects compared to normal controls,[60,69] suggesting that IL-5-responsive CD34+ cells committed to the Eo/B lineage are present *within* the inflamed tissue. Furthermore, CD34+ cells appearing in sputum of allergic asthmatic subjects stain positive for IL-5 and IL-13 intracellularly, suggesting that they themselves may act as proinflammatory effector cells in the microenvironment of the inflamed tissue.[64] Autocrine production of growth factors in Eo/B CFU grown from the blood of allergic asthmatic subjects has likewise been documented, which suggests that cytokine expression by differentiating progenitors may provide an additional stimulus to enhance differentiation *in situ*.[54,66] A similar autocrine upregulation of both eosinophilopoietic factors and their receptors has recently been demonstrated for TLR-ligated CD34+ cells in cord blood.[79,80] *In situ* IL-5-dependent differentiation of eosinophil progenitors can be triggered in the nasal mucosa of allergic rhinitic donors: *ex vivo* nasal mucosa cultures stimulated with allergen were shown to contain reduced numbers of CD34+ IL5RA mRNA+ cells and increased numbers of major basic protein (MBP) immunoreactive cells, thus shifting cells locally from an immature to a mature phenotype.[60]

In a mouse model of allergen-induced airway inflammation, newly produced CD34+ CCR3+ cells, isolated from lung and stimulated with eotaxin-2/CCL24 or IL-5, differentiated into significant numbers of CFU; cell cycle analysis showed significant increases in the number of CD34+ CCR3+ proliferating cells in allergen-exposed animals.[81] These data support the notion of local Eo/B differentiation within tissues, suggesting that the CCR3/eotaxin pathway is also involved in the regulation of this allergen-driven *in situ* hemopoiesis, at least in mice. Recently, it has been shown that human airway smooth muscle of allergic asthmatics can also stimulate increased Eo/B differentiation *in vitro*,[82] adding to the complexity of systemic and local hemopoietic processes in the generation of eosinophilic tissue inflammation.

EOSINOPHIL PROGENITORS AND INNATE IMMUNITY

The evidence described above for *in situ* hemopoiesis[43] is also in keeping with a recent observation that IL-33/thymic stromal lymphopoietin (TSLP) is sufficient to induce the differentiation of CD34[+] progenitors into eosinophils, in addition to activating T_h2 and mast cells.[64] A major recent discovery is that TSLP promotes T_h2 pathways (via IL-25 and IL-33) that induce the development of a c-kit[int]Sca-1[+] multipotent progenitor (MMP; also known as 'nuocytes')[83] population in gut-associated lymphoid tissue;[84,85] this provides further support for *in situ* hemopoiesis, and promotes the idea that TSLP provides a critical link between adaptive and innate immunity in the process of hemopoietic CD34[+] cell differentiation within tissues (Fig. 13.6.2).

As mentioned above, recent data showing that CD34[+] IL-5[+] and CD34[+] IL-13[+] double-positive cells can be detected in sputum of asthmatic patients, with increases after allergen challenge,[64] are consistent with the concept of allergen activation, via TSLP, of multipotent as well as lineage-committed progenitors, some of which can thus potentially function as inflammatory effector cells without further differentiation. Preliminary data show that recombinant TSLP, with or without IL-33 as a coligand, can also upregulate IL-5RA on CD34[+] cells, rendering them more responsive to the effects of IL-5, with concomitant functional Eo/B CFU differentiation.[86] Moreover, stimulation of CD34-enriched human cord blood cells with the Toll-like receptor (TLR) agonists, lipopolysaccharide (LPS) or CpG-oligodeoxynucleotides (CpG ODN), induces increased expression of TLR-2, TLR-4, and TLR-9 on CD34[+] cells, as well as increases in GM-CSFRA, IL-3RA, and IL-5RA.[79,80] These observations demonstrate additional mechanisms through which innate immunity can regulate the responsiveness of Eo/B progenitors. CD34[+] cells stimulated with a combination of TLR agonists and hemopoietic cytokines have been shown to give rise to more Eo/B CFU responsive to IL-3 and IL-5 than through hemopoietic cytokine stimulation alone.[79,80] Collectively, these data point to TSLP—TLR—TLR ligand interactions that can result in autocrine *upregulation of T_h2 cytokines* (e.g., IL-5, IL-13) in CD34[+] cells,[61,64,65] and further support the idea that TSLP provides a critical link between adaptive and innate immunity in the process of hemopoietic CD34[+] cell differentiation within tissues (Fig. 13.6.2).

The Role of Interleukin-5 in Eosinophil Progenitor Recruitment

IL-5 is central for upregulation of myeloid progenitors in the bone marrow after airway allergen challenge,[38,87] and for trafficking from the marrow to the airways in several animal models of either upper or lower airways inflammation.[56,88,89] Airway, blood, and nasal eosinophilia are completely inhibited by either neutralizing the biologic effects of IL-5[90] or through deletion of the gene encoding IL-5.[62] Although the asthmatic lung can release abundant amounts of IL-5, studies in animal models have led to debate regarding the distribution and sources of IL-5 required to drive airway eosinophilia. Studies in mice have shown that circulating, rather than local, IL-5 in the lung is critical for the development of allergic airways eosinophilia.[91] This was investigated through systemic IL-5 gene transfer to IL-5-knockout mice, which effectively supported ovalbumin (OVA)-induced eosinophilia in the blood, bone marrow, and lung. This contrasts with IL-5 gene transfer into the airways, which did not induce eosinophilia following OVA challenge.[91] Thus, systemic IL-5 seems to be necessary for the development of eosinophilia in the mouse. However, this is more controversial in humans.

Further studies have demonstrated that inhalation of IL-5 by allergic asthmatic subjects leads to the development of peripheral blood and sputum eosinophilia,[92] and inhaled IL-5 induces airway eosinophilia accompanied by increased airway hyperresponsiveness (AHR).[93] In contrast, three subsequent studies reported no effect of IL-5 inhalation on eosinophil levels in blood, airways, or AHR in allergic asthmatics.[94–96] By inhalation, IL-5 was found to significantly *decrease* CD34[+] *IL5RA* mRNA[+] cells within the bronchial mucosa and *decrease* the percentage of CD34[+] CCR3[+] cells in the bone marrow of atopic asthmatics.[96] It has been hypothesized that lung-derived IL-5 provides a signal to a population of cells within the bronchial mucosa that traffic to the bone marrow, where they locally induce the efflux of CD34[+] CCR3[+] cells.[65,96] Elegant experiments involving perfusion of the femoral bone marrow of guinea pigs confirmed that whereas IL-5 induces chemokinesis of bone marrow eosinophils, mobilization of mature eosinophils by IL-5 occurs synergistically with, and requires, the presence of eotaxin.[41]

Roles of Eotaxins

Eotaxin is a potent and eosinophil-specific chemoattractant.[97–99] Eotaxin-1 is thought to be more important than eotaxin-2 for inducing mobilization of eosinophils and their progenitors from the bone marrow into the blood circulation.[41,100] A reduction of eosinophil numbers was observed in the lungs of mice treated with eotaxin-1 blocking antibodies[101,102] and in strain-specific eotaxin-1 knockout mice.[103,104] Eotaxin-1 is released from lung structural cells,[105] and is released at increased levels following allergen challenge.[106] Human endothelial progenitor cells also express eotaxin-1,[107] and have been shown to be rapidly mobilized to the lung after allergen challenge in sensitized mice[107] and atopic asthmatic subjects,[108] where they also can contribute to the development of lung eosinophilia through the expression and secretion of eotaxin-1.

More recently, both eotaxin-1 and eotaxin-2 have been shown to induce migration of murine bone marrow and blood CD34$^+$CCR3$^+$ cells using an *in vitro* transmigration assay.[81] These data suggest that the CCR3/eotaxin pathway is involved in the regulation of allergen-driven accumulation/mobilization of eosinophil lineage-committed progenitor cells in the lung. The receptor for eotaxin, CCR3, is upregulated on CD34$^+$ cells after allergen challenge, thereby facilitating eotaxin-mediated progenitor cell mobilization from the bone marrow to the peripheral circulation.[42,109]

Other Migration Signals

Studies in sensitized mice have indicated that T cells are the major gatekeepers regulating eotaxin and IL-5 levels, and thus eosinophilia. Reductions in airway, bone marrow, and peripheral blood eosinophil levels in CD4$^{-/-}$ and CD8$^{-/-}$ mice suggest that these T cell populations are critical regulators of allergen-induced eosinophilia. Furthermore, reduced serum IL-5 and bronchoalveolar lavage eotaxin-2 levels in CD4$^{-/-}$ mice suggest that CD4$^+$ T cells are obligatory for the development of allergen-induced airway eosinophilia.[110]

In addition to eotaxin and IL-5, many other mediators have been shown to induce responses in both mature eosinophils and Eo/B progenitors. Cysteinyl leukotrienes, which are released in the airways following perturbation by allergen, are chemoattractants for mature eosinophils,[111] but have also been shown to aid differentiation[112] and induce chemotaxis and *in vitro* transendothelial migration of Eo/B progenitors.[113] Preliminary work has shown that both IL-4 and IL-13 can prime hemopoietic progenitor cells in a transmigration assay,[114] and studies are under way to investigate whether progenitors are also responsive to the eosinophil chemoattractants prostaglandin D2 receptor 2 (CRTH2; reviewed in[115]) and peroxisome proliferator-activated receptor γ (PPARγ).[116,117] In the allergen challenge model, there is attenuation of CXCR4 (SDF-1α receptor) expression on bone marrow CD34$^+$ cells from mild asthmatic subjects, as well as a reduction in SDF-1α levels in the bone marrow. This demonstrates a mechanism whereby retention of progenitors in the bone marrow regulates their egress into the blood.[109]

EOSINOPHILS AND PROGENITORS AS TARGETS OF ASTHMA THERAPY

The discoveries that IL-5 is a specific eosinophil growth factor in humans and that eotaxins can selectively induce eosinophil recruitment, as described above, were instrumental for the development of drugs targeting the eosinophil. This is reviewed in detail elsewhere.[118]

Interleukin-5 Blockade

Anti-IL-5 treatment in mild atopic asthmatic subjects has been shown to induce a reduction of airway eosinophils, arrest bone marrow eosinophil maturation, and decrease eosinophil progenitors in the bronchial mucosa.[119,120] Though initial clinical trials of IL-5 blockade in patients with asthma were unsuccessful in demonstrating clinical efficacy,[119,121−124] a number of issues may have contributed to the failure of these studies, including lack of depletion of tissue eosinophils and their granule products; patient selection; and methodological problems.[125] Subsequent studies, conducted in a subgroup of asthmatic patients selected on the basis of having persistent eosinophilic asthma, have demonstrated that blocking IL-5 with the humanized IL-5 antibody, mepolizumab, has a steroid-sparing effect, reduces exacerbations, and improves quality of life for asthma patients.[126,127] The anti-IL-5 approach is troubled by the inability to completely abolish eosinophilia, consistent with murine observations (reviewed in Matthaei et al.[89]), since it is hypothesized that if eosinophilia were solely controlled by IL-5 then a more complete suppression of eosinophils would ensue from therapeutic anti-IL-5 interventions. *A propos* this possibility, however, MEDI-563 is a humanized anti-IL-5RA monoclonal antibody that binds IL-5RA with high affinity and mediates cell lysis via antibody-directed cell-mediated cytotoxicity. As such, MEDI-563 kills all cells bearing IL-5RA, including eosinophil progenitors, and has been shown to *eliminate* eosinophils from the circulation of subjects with mild asthma.[128,129] This antibody is currently under investigation in other clinical models of asthma.

That eosinophils can remain in the lung tissue of asthmatic subjects despite IL-5 blockade suggests that additional signals promote eosinophil survival. Indeed in IL-5-deficient mice, responses to GM-CSF and IL-3 are normal, despite absence of eosinophilia in the nasal mucosa and the bone marrow, significantly lower numbers of IL-5-responsive Eo/B CFU and maturing CFU eosinophils, and reduced expression of IL-5RA on bone marrow-derived CD34$^+$ CD45$^+$ progenitor cells.[56] These results indicate that *redundant* cytokine mechanisms can compensate for IL-5 deficiency and highlight the multifactorial nature of allergic inflammation, indicating that combined, as opposed to single-line, therapies may be more effective in the treatment of diseases such as asthma.

Antisense Therapy

One such therapy is currently being tested in clinical trials of allergic asthma. TPI-ASM8 is a combination of two antisense oligonucleotides, one blocking translation of the IL-3/IL-5/GM-CSF receptor common β chain, and the other blocking translation of CCR3. As such, TPI-ASM8 prevents expression of receptors for GM-CSF, IL-3, IL-5, eotaxin-1, and eotaxin-2, and will thus hypothetically inhibit many of the signals shown to be crucial for eosinophilopoiesis, as well as eosinophil migration, activation and survival. Following allergen challenge in mild atopic asthmatic subjects, inhaled TPI-ASM8 inhibited accumulation of mature eosinophils and CD34[+] IL-5RA[+] cells in the sputum, in addition to inhibiting the late asthmatic response.[130,131]

Steroid Therapy

Despite advances made in the development of eosinophil-specific therapies, corticosteroids remain the gold standard for the treatment of allergic inflammatory diseases like asthma.[132–134] Local delivery using inhalers is intended for topical treatment of the affected tissue; however, the anti-inflammatory actions of corticosteroids have been shown to extend beyond the environment of the airways, probably due to a small amount of systemic availability. These systemic effects are beneficial for regulating hemopoietic mechanisms that originate in the bone marrow. Stepwise withdrawal of inhaled corticosteroids results in a rapid and substantial increase in Eo/B progenitors assayed in peripheral blood, which returns to baseline when treatment is reinstated.[135] Only 1 week of treatment with inhaled corticosteroid is sufficient to significantly attenuate allergen-induced levels of circulating Eo/B CFU[66] and reduce baseline numbers of bone marrow Eo/B CFU[34] in mild allergic asthmatic subjects, further demonstrating the efficacy of corticosteroids on progenitors in peripheral blood. However, the inhaled steroid has no effect on the allergen-induced increase in the number of bone marrow CD34[+] cells, the increase in IL-5RA expression on these cells, or the number of Eo/B CFU. These findings suggest that topical corticosteroids may exert indirect suppressive effects on the differentiation of eosinophil progenitors.

CONCLUSION

Eosinophil progenitors are now emerging as effector cells that can migrate to inflamed tissue where they rapidly proliferate in response to allergic stimuli. Understanding communication between eosinophil progenitors and the innate immune system will require further exploration.

ACKNOWLEDGEMENTS

The authors would like to thank Drs Guy Delespesse and Zoulfia Allakhverdi, Laboratory on Allergy, CHUM Research Center, Notre Dame Hospital, Montreal, Canada, for providing input from their research into TSLP effects on progenitor cells (Fig. 13.6.2). Thanks also to Lynne Larocque for her expert assistance with the preparation of the manuscript. Funding for this research was provided by the Canadian Institutes of Health Research (CIHR), and the Allergy, Genes and Environment Network of Centres of Excellence (AllerGen NCE Inc.).

REFERENCES

1. Alfaro C, Sharma OP, Navarro L, Glovsky MM. Inverse correlation of expiratory lung flows and sputum eosinophils in status asthmaticus. *Ann Allergy* 1989;**63**:251–4.
2. Walker C, Kaegi MK, Braun P, Blaser K. Activated T cells and eosinophilia in bronchoalveolar lavages from subjects with asthma correlated with disease severity. *J Allergy Clin Immunol* 1991;**88**:935–42.
3. Dorman SC, Efthimiadis A, Babirad I, Watson RM, Denburg JA, Hargreave FE, et al. Sputum CD34+IL-5Rα+ cells increase after allergen: evidence for in situ eosinophilopoiesis. *Am J Respir Crit Care Med* 2004;**169**:573–7.
4. Dorman SC, Sehmi R, Gauvreau GM, Watson RM, Foley R, Jones GL, et al. Kinetics of bone marrow eosinophilopoiesis and associated cytokines after allergen inhalation. *Am J Respir Crit Care Med* 2004;**169**:565–72.
5. De Monchy JG, Kauffman HF, Venge P, Koeter GH, Jansen HM, Sluiter HJ, et al. Bronchoalveolar eosinophilia during allergen-induced late asthmatic reactions. *Am Rev Respir Dis* 1985;**131**:373–6.
6. Gauvreau GM, Evans MY. Allergen inhalation challenge: a human model of asthma exacerbation. *Contrib Microbiol* 2007;**14**:21–32.
7. Gauvreau GM, Newbold P, Perrett J, Craggs B, Foster M, Drong M, et al. Kinetics of cellular infiltration post allergen challenge in asthma subjects: differences between bronchial washings and lung tissue (abst). *Am J Respir Crit Care Med* 2007;**175**:A679.
8. Gauvreau GM, Watson RM, O'Byrne PM. Kinetics of allergen-induced airway eosinophilic cytokine production and airway inflammation. *Am J Respir Crit Care Med* 1999;**160**:640–7.
9. Bousquet J, Chanez P, Lacoste JY, Barneon G, Ghavanian N, Enander I, et al. Eosinophilic inflammation in asthma. *N Engl J Med* 1990;**323**:1033–9.
10. Weller PF. Eosinophils: structure and functions. *Curr Opin Immunol* 1994;**6**:85–90.
11. Rosenberg HF, Phipps S, Foster PS. Eosinophil trafficking in allergy and asthma. *J Allergy Clin Immunol* 2007;**119**:1303–10.
12. Clutterbuck EJ, Sanderson CJ. Regulation of human eosinophil precursor production by cytokines: a comparison of recombinant human interleukin-1 (rhIL-1), rhIL-3, rhIL-5, rhIL-6, and rh granulocyte-macrophage colony-stimulating factor. *Blood* 1990;**75**:1774–9.
13. Asquith KL, Ramshaw HS, Hansbro PM, Beagley KW, Lopez AF, Foster PS. The IL-3/IL-5/GM-CSF common receptor plays a pivotal role in the regulation of Th2 immunity and allergic airway inflammation. *J Immunol* 2008;**180**:1199–206.
14. Shalit M, Sekhsaria S, Malech HL. Modulation of growth and differentiation of eosinophils from human peripheral blood CD34+ cells by IL5 and other growth factors. *Cell Immunol* 1995;**160**:50–7.

15. Ying S, Meng Q, Barata LT, Robinson DS, Durham SR, Kay AB. Associations between IL-13 and IL-4 (mRNA and protein), vascular cell adhesion molecule-1 expression, and the infiltration of eosinophils, macrophages, and T cells in allergen-induced late-phase cutaneous reactions in atopic subjects. *J Immunol* 1997;**158**:5050–7.

16. Li L, Xia Y, Nguyen A, Lai YH, Feng L, Mosmann TR, et al. Effects of Th2 cytokines on chemokine expression in the lung: IL-13 potently induces eotaxin expression by airway epithelial cells. *J Immunol* 1999;**162**:2477–87.

17. Zimmermann N, Hogan SP, Mishra A, Brandt EB, Bodette TR, Pope SM, et al. Murine eotaxin-2: a constitutive eosinophil chemokine induced by allergen challenge and IL-4 overexpression. *J Immunol* 2000;**165**:5839–46.

18. Gounni AS, Gregory B, Nutku E, Aris F, Latifa K, Minshall E, et al. Interleukin-9 enhances interleukin-5 receptor expression, differentiation, and survival of human eosinophils. *Blood* 2000;**96**:2163–71.

19. Matsuzawa S, Sakashita K, Kinoshita T, Ito S, Yamashita T, Koike K. IL-9 enhances the growth of human mast cell progenitors under stimulation with stem cell factor. *J Immunol* 2003;**170**:3461–7.

20. Mohle R, Salemi P, Moore MA, Rafii S. Expression of interleukin-5 by human bone marrow microvascular endothelial cells: implications for the regulation of eosinophilopoiesis in vivo. *Br J Haematol* 1997;**99**:732–8.

21. Hogan MB, Piktel D, Landreth KS. IL-5 production by bone marrow stromal cells: implications for eosinophilia associated with asthma. *J Allergy Clin Immunol* 2000;**106**:329–36.

22. Denburg JA. Hemopoietic progenitors and cytokines in allergic inflammation. *Allergy* 1998;**53**(supp.145):22–6.

23. Denburg JA. Bone marrow in atopy and asthma: hematopoietic mechanisms in allergic inflammation. *Immunol Today* 1999;**20**:111–3.

24. Krause DS, Fackler MJ, Civin CI, May WS. CD34: structure, biology, and clinical utility. *Blood* 1996;**87**:1–13.

25. Sergejeva S, Johansson AK, Malmhall C, Lotvall J. Allergen exposure-induced differences in CD34+ cell phenotype: relationship to eosinophilopoietic responses in different compartments. *Blood* 2004;**103**:1270–7.

26. Sehmi R, Howie K, Sutherland DR, Schragge W, O'Byrne PM, Denburg JA. Increased levels of CD34+ hemopoietic progenitor cells in atopic subjects. *Am J Respir Cell Mol Biol* 1996;**15**:645–54.

27. Zeibecoglou K, Ying S, Yamada T, North J, Burman J, Bungre J, et al. Increased mature and immature CCR3 messenger RNA+ eosinophils in bone marrow from patients with atopic asthma compared with atopic and nonatopic control subjects. *J Allergy Clin Immunol* 1999;**103**:99–106.

28. Makowska JS, Grzegorczyk J, Cieslak M, Bienkiewicz B, Kowalski ML. Recruitment of CD34+ progenitor cells into peripheral blood and asthma severity. *Ann Allergy Asthma Immunol* 2008;**101**:402–6.

29. Sehmi R, Wood LJ, Watson R, Foley R, Hamid Q, O'Byrne PM, et al. Allergen-induced increases in IL-5 receptor α-subunit expression on bone marrow-derived CD34+ cells from asthmatic subjects. A novel marker of progenitor cell commitment toward eosinophilic differentiation. *J Clin Invest* 1997;**100**:2466–75.

30. Sehmi R, Howie K, Rerecich T, Watson RM, Foley R, O'Byrne PM, et al. Increased numbers of eosinophil progenitor cells (CD34+IL5Rα+) in the bone marrow of atopic asthmatic subjects (abst). *J Allergy Clin Immunol* 2000;**105**:S172.

31. Wood LJ, Inman MD, Watson RM, Foley R, Denburg JA, O'Byrne PM. Changes in bone marrow inflammatory cell progenitors after inhaled allergen in asthmatic subjects. *Am J Respir Crit Care Med* 1998;**157**:99–105.

32. Tavernier J, Van der Heyden J, Verhee A, Brusselle G, Van Ostade X, Vandekerckhove J, et al. Interleukin 5 regulates the isoform expression of its own receptor alpha-subunit. *Blood* 2000;**95**:1600–7.

33. Denburg JA, Sehmi R, Upham J. Regulation of IL-5R on eosinophil progenitors in allergic inflammation: role of retinoic acid. *Int Arch Allergy Immunol* 2001;**124**:246–8.

34. Wood LJ, Sehmi R, Gauvreau GM, Watson RM, Foley R, Denburg JA, et al. An inhaled corticosteroid, budesonide, reduces baseline but not allergen-induced increases in bone marrow inflammatory cell progenitors in asthmatic subjects. *Am J Respir Crit Care Med* 1999;**159**:1457–63.

35. Ohkawara Y, Lei XF, Stampfli MR, Marshall JS, Xing Z, Jordana M. Cytokine and eosinophil responses in the lung, peripheral blood, and bone marrow compartments in a murine model of allergen-induced airways inflammation. *Am J Respir Cell Mol Biol* 1997;**16**:510–20.

36. Gaspar Elsas MI, Joseph D, Elsas PX, Vargaftig BB. Rapid increase in bone-marrow eosinophil production and responses to eosinopoietic interleukins triggered by intranasal allergen challenge. *Am J Respir Cell Mol Biol* 1997;**17**:404–13.

37. Inman MD, Ellis R, Wattie J, Denburg JA, O'Byrne PM. Allergen-induced increase in airway responsiveness, airway eosinophilia and bone-marrow eosinophil progenitors in mice. *Am J Respir Cell Mol Biol* 1999;**21**:473–9.

38. Saito H, Howie K, Wattie J, Denburg A, Ellis R, Inman MD, et al. Allergen-induced murine upper airway inflammation: local and systemic changes in murine experimental allergic rhinitis. *Immunology* 2001;**104**:226–34.

39. Li J, Saito H, Crawford L, Inman MD, Cyr MM, Denburg JA. Hemopoietic mechanisms in murine allergic upper and lower airways inflammation. *Immunology* 2005;**114**:386–96.

40. Chou CL, Wang CH, Kuo HP. Upregulation of IL-5 receptor expression on bone marrow-derived CD34+ cells from patients with asthma. *Changgeng Yi Xue Za Zhi* 1999;**22**:416–22.

41. Palframan RT, Collins PD, Williams TJ, Rankin SM. Eotaxin induces a rapid release of eosinophils and their progenitors from the bone marrow. *Blood* 1998;**91**:2240–8.

42. Sehmi R, Dorman S, Baatjes A, Watson R, Foley R, Ying S, et al. Allergen-induced fluctuation in CC chemokine receptor 3 expression on bone marrow CD34+ cells from asthmatic subjects: significance for mobilization of haemopoietic progenitor cells in allergic inflammation. *Immunology* 2003;**109**:536–46.

43. Lamkhioued B, Abdelilah SG, Hamid Q, Mansour N, Delespesse G, Renzi PM. The CCR3 receptor is involved in eosinophil differentiation and is up-regulated by Th2 cytokines in CD34+ progenitor cells. *J Immunol* 2003;**170**:537–47.

44. Denburg JA, Richardson M, Telizyn S, Bienenstock J. Basophil/mast cell precursors in human peripheral blood. *Blood* 1983;**61**:775–80.

45. Denburg JA, Silver JE, Abrams JS. Interleukin-5 is a human basophilopoietin: Induction of histamine content and basophilic differentiation of HL-60 cells and of peripheral blood basophil-eosinophil progenitors. *Blood* 1991;**77**:1462—8.

46. Denburg JA, Telizyn S, Messner H, Lim B, Jamal N, Ackerman SJ, et al. Heterogeneity of human peripheral blood eosinophil-type colonies: Evidence for a common basophil-eosinophil progenitor. *Blood* 1985;**66**:312—8.

47. Denburg JA, Telizyn S, Belda A, Dolovich J, Bienenstock J. Increased numbers of circulating basophil progenitors in atopic patients. *J Allergy Clin Immunol* 1985;**76**:466—72.

48. Sehmi R, Wood LJ, Inman MD, Watson RM, O'Byrne PM, Lopez AF, et al. Increases in bone marrow derived CD34+ hemopoietic progenitor cells expressing the alpha-subunit of IL-3 receptors following allergen challenge in mild asthmatics (abst). *Am J Respir Crit Care Med* 1996;**153**:A880.

49. Otsuka H, Dolovich J, Befus AD, Telizyn S, Bienenstock J, Denburg JA. Basophilic cell progenitors, nasal metachromatic cells, and peripheral blood basophils in ragweed-allergic patients. *J Allergy Clin Immunol* 1986;**78**:365—71.

50. Linden M, Svensson C, Andersson M, Greiff L, Andersson E, Denburg JA, et al. Circulating eosinophil/basophil progenitors and nasal mucosal cytokines in seasonal allergic rhinitis. *Allergy* 1999;**54**:212—9.

51. Cyr MM, Baatjes AJ, Hayes LM, Crawford L, Denburg JA. The effect of desloratadine on eosinophil/basophil progenitors and other inflammatory markers in seasonal allergic rhinitis: a placebo-controlled randomized study (abst). *J Allergy Clin Immunol* 2002;**109**:S117.

52. Gibson PG, Dolovich J, Girgis-Gabardo A, Morris MM, Anderson M, Hargreave FE, et al. The inflammatory response in asthma exacerbation: changes in circulating eosinophils, basophils and their progenitors. *Clin Exp Allergy* 1990;**20**:661—8.

53. Gibson PG, Manning PJ, O'Byrne PM, Girgis-Gabardo A, Dolovich J, Denburg JA, et al. Allergen-induced asthmatic responses: Relationship between increases in airway responsiveness and increases in circulating eosinophils, basophils, and their progenitors. *Am Rev Respir Dis* 1991;**143**:331—5.

54. Gauvreau GM, O'Byrne PM, Moqbel R, Velazquez J, Watson RM, Howie KJ, et al. Enhanced expression of GM-CSF in differentiating eosinophils of atopic and atopic asthmatic subjects. *Am J Respir Cell Mol Biol* 1998;**19**:55—62.

55. Johansson AK, Sergejeva S, Sjostrand M, Lee JJ, Lotvall J. Allergen-induced traffic of bone marrow eosinophils, neutrophils and lymphocytes to airways. *Eur J Immunol* 2004;**34**:3135—45.

56. Saito H, Matsumoto K, Denburg AE, Crawford L, Ellis R, Inman MD, et al. Pathogenesis of murine experimental allergic rhinitis: a study of local and systemic consequences of IL-5 deficiency. *J Immunol* 2002;**168**:3017—23.

57. Saito H, Morikawa H, Howie K, Crawford L, Baatjes AJ, Denburg E, et al. Effects of a cysteinyl leukotriene receptor antagonist on eosinophil recruitment in experimental allergic rhinitis. *Immunology* 2004;**113**:246—52.

58. Southam DS, Widmer N, Ellis R, Hirota JA, Inman MD, Sehmi R. Increased eosinophil-lineage committed progenitors in the lung of allergen-challenged mice. *J Allergy Clin Immunol* 2005;**115**:95—102.

59. Hart TK, Cook RM, Zia-Amirhosseini P, Minthorn E, Sellers TS, Maleeff BE, et al. Preclinical efficacy and safety of mepolizumab (SB-240563), a humanized monoclonal antibody to IL-5, in cynomolgus monkeys. *J Allergy Clin Immunol* 2001;**108**:250—7.

60. Cameron L, Christodoulopoulos P, Lavigne F, Nakamura Y, Eidelman D, McEuen A, et al. Evidence for local eosinophil differentiation within allergic nasal mucosa: inhibition with soluble IL-5 receptor. *J Immunol* 2000;**164**:1538—45.

61. Minshall EM, Schleimer R, Cameron L, Minnicozzi M, Egan RW, Gutierrez-Ramos JC, et al. Interleukin-5 expression in the bone marrow of sensitized Balb/c mice after allergen challenge. *Am J Respir Crit Care Med* 1998;**158**:951—7.

62. Sitkauskiene B, Johansson AK, Sergejeva S, Lundin S, Sjostrand M, Lotvall J. Regulation of bone marrow and airway CD34+ eosinophils by interleukin-5. *Am J Respir Cell Mol Biol* 2004;**30**:367—78.

63. Sitkauskiene B, Radinger M, Bossios A, Johansson AK, Sakalauskas R, Lotvall J. Airway allergen exposure stimulates bone marrow eosinophilia partly via IL-9. *Respir Res* 2005;**6**:33.

64. Allakhverdi Z, Comeau MR, Smith DE, Toy D, Endam LM, Desrosier M, et al. CD34+ hemopoietic progenitor cells are potent effectors of allergic inflammation. *J Allergy Clin Immunol* 2009;**123**:472—8.

65. Wood LJ, Sehmi R, Dorman S, Hamid Q, Tulic MK, Watson RM, et al. Allergen-induced increases in bone marrow T lymphocytes and interleukin-5 expression in subjects with asthma. *Am J Respir Crit Care Med* 2002;**166**:883—9.

66. Gauvreau GM, Wood LJ, Sehmi R, Watson RM, Dorman SC, Schleimer RP, et al. The effects of inhaled budesonide on circulating eosinophil progenitors and their expression of cytokines after allergen challenge in subjects with atopic asthma. *Am J Respir Crit Care Med* 2000;**162**:2139—44.

67. Gauvreau GM, Lee JM, Watson RM, Irani AM, Schwartz LB, O'Byrne PM. Increased numbers of both airway basophils and mast cells in sputum after allergen inhalation challenge of atopic asthmatics. *Am J Respir Crit Care Med* 2000;**161**:1473—8.

68. Bossios A, Sjostrand M, Dahlborn AK, Samitas K, Malmhall C, Gaga M, et al. IL-5 expression and release from human CD34 cells in vitro; ex vivo evidence from cases of asthma and Churg-Strauss syndrome. *Allergy* 2010;**65**:831—9.

69. Robinson DS, Damia R, Zeibecoglou K, Molet S, North J, Yamada T, et al. CD34+/interleukin-5Rα messenger RNA+ cells in the bronchial mucosa in asthma: potential airway eosinophil progenitors. *Am J Respir Cell Mol Biol* 1999;**20**:9—13.

70. Kim YK, Uno M, Hamilos DL, Beck L, Bochner B, Schleimer R, et al. Immunolocalization of CD34 in nasal polyposis. Effect of topical corticosteroids. *Am J Respir Cell Mol Biol* 1999;**20**:388—97.

71. Mastrandrea F, Cadario G, Nicotra MR, Natali PG. Hemopoietic progenitor cells in atopic dermatitis skin lesions. *J Investig Allergol Clin Immunol* 1999;**9**:386—91.

72. Denburg JA, Dolovich J, Ohtoshi T, Cox G, Gauldie J, Jordana M. The microenvironmental differentiation hypothesis of airway inflammation. *Am J Rhinology* 1990;**4**:29—32.

73. Metcalf D. Clonal analysis of the response of HL60 human myeloid leukemia cells to biological regulators. *Leuk Res* 1983;**7**:117—32.

74. Ohnishi M, Ruhno J, Bienenstock J, Milner R, Dolovich J, Denburg JA. Human nasal polyp epithelial basophil/mast cell and

eosinophil colony-stimulating activity: the effect is T-cell-dependent. *Am Rev Respir Dis* 1988;**138**:560—4.

75. Ohnishi M, Ruhno J, Dolovich J, Denburg JA. Allergic rhinitis nasal mucosal conditioned medium stimulates growth and differentiation of basophil/mast cell and eosinophil progenitors from atopic blood. *J Allergy Clin Immunol* 1988;**81**:1149—54.

76. Cox G, Ohtoshi T, Vancheri C, Denburg JA, Dolovich J, Gauldie J, et al. Promotion of eosinophil survival by human bronchial epithelial cells and its modulation by steroids. *Am J Respir Cell Mol Biol* 1991;**4**:525—31.

77. Vancheri C, Gauldie J, Bienenstock J, Cox G, Scicchitano R, Stanisz A, et al. Human lung fibroblast-derived granulocyte-macrophage colony stimulating factor (GM-CSF) mediates eosinophil survival *in vitro*. *Am J Respir Cell Mol Biol* 1989;**1**:289—95.

78. Vancheri C, Ohtoshi T, Cox G, Xaubet A, Abrams JS, Gauldie J, et al. Neutrophilic differentiation induced by human upper airway fibroblast-derived granulocyte/macrophage colony-stimulating factor (GM-CSF). *Am J Respir Cell Mol Biol* 1991;**4**:11—7.

79. Reece P, Thanendran A, Crawford L, Tulic MK, Thabane L, Prescott SL, et al. Maternal allergy modulates cord blood hematopoietic progenitor Toll-like receptor expression and function. *J Allergy Clin Immunol* 2011;**127**:447—53.

80. Reece P, Crawford L, Baatjes A, Cyr M, Sehmi R, Denburg JA. Cord blood hemopoietic progenitor cell Toll-like receptor expression and function: a mechanism underlying allergic inflammation in early life (abst). *J Allergy Clin Immunol* 2010;**125**:AB105.

81. Radinger M, Bossios A, Sjostrand M, Lu Y, Malmhall C, Dahlborn AK, et al. Local proliferation and mobilization of CCR3(+) CD34(+) eosinophil-lineage-committed cells in the lung. *Immunology* 2011;**132**:144—54.

82. Fanat AI, Thomson JV, Radford K, Nair P, Sehmi R. Human airway smooth muscle promotes eosinophil differentiation. *Clin Exp Allergy* 2009;**39**:1009—17.

83. Neill DR, Wong SH, Bellosi A, Flynn RJ, Daly M, Langford TK, et al. Nuocytes represent a new innate effector leukocyte that mediates type-2 immunity. *Nature* 2010;**464**:1367—70.

84. Saenz SA, Siracusa MC, Perrigoue JG, Spencer SP, Urban Jr JF, Tocker JE, et al. IL25 elicits a multipotent progenitor cell population that promotes T(H)2 cytokine responses. *Nature* 2010;**464**:1362—6.

85. Ziegler SF, Artis D. Sensing the outside world: TSLP regulates barrier immunity. *Nat Immunol* 2010;**11**:289—93.

86. Hui CCK, Murphy DM, Thong B, Delespesse G, Denburg JA, Larché M. Induction of thymic stromal lymphopoietin (TSLP) in airway epithelium by recombinant allergens (abst). *J Allergy Clin Immunol* 2011;**127**:AB125.

87. Wood LJ, Inman MD, Denburg JA, O'Byrne PM. Allergen challenge increases cell traffic between bone marrow and lung. *Am J Respir Cell Mol Biol* 1998;**18**:759—67.

88. Egan RW, Athwahl D, Chou CC, Chapman RW, Emtage S, Jehn CH, et al. Pulmonary biology of anti-interleukin 5 antibodies. *Mem Inst Oswaldo Cruz* 1997;**92**(Suppl. 2):69—73.

89. Matthaei KI, Foster PS, Young IG. The role of interleukin-5 (IL-5) *in vivo*: studies with IL-5 deficient mice. *Mem Inst Oswaldo Cruz* 1997;**92**(Suppl. 2):63—8.

90. Kopf M, Brombacher F, Hodgkin PD, Ramsay AJ, Milbourne EA, Dai WJ, et al. IL-5-deficient mice have a developmental defect in CD5+ B-1 cells and lack eosinophilia but

have normal antibody and cytotoxic T cell responses. *Immunity* 1996;**4**:15—24.

91. Wang J, Palmer K, Lotvall J, Milan S, Lei XF, Matthaei KI, et al. Circulating, but not local lung, IL-5 is required for the development of antigen-induced airways eosinophilia. *J Clin Invest* 1998;**102**:1132—41.

92. Shi HZ, Li CQ, Qin SM, Xie ZF, Liu Y. Effect of inhaled interleukin-5 on number and activity of eosinophils in circulation from asthmatics. *Clin Immunol* 1999;**91**:163—9.

93. Shi HZ, Xiao CQ, Zhong D, Qin SM, Liu Y, Liang GR, et al. Effect of inhaled interleukin-5 on airway hyperreactivity and eosinophilia in asthmatics. *Am J Respir Crit Care Med* 1998;**157**:204—9.

94. Stirling RG, van Rensen EL, Barnes PJ, Chung KF. Interleukin-5 induces CD34(+) eosinophil progenitor mobilization and eosinophil CCR3 expression in asthma. *Am J Respir Crit Care Med* 2001;**164**:1403—9.

95. van Rensen EL, Stirling RG, Scheerens J, Staples K, Sterk PJ, Barnes PJ, et al. Evidence for systemic rather than pulmonary effects of interleukin-5 administration in asthma. *Thorax* 2001;**56**:935—40.

96. Menzies-Gow AN, Flood-Page PT, Robinson DS, Kay AB. Effect of inhaled interleukin-5 on eosinophil progenitors in the bronchi and bone marrow of asthmatic and non-asthmatic volunteers. *Clin Exp Allergy* 2007;**37**:1023—32.

97. Jose PJ, Griffiths-Johnson DA, Collins PD, Walsh DT, Moqbel R, Totty NF, et al. Eotaxin: a potent eosinophil chemoattractant cytokine detected in a guinea pig model of allergic airways inflammation. *J Exp Med* 1994;**179**:881—7.

98. Fulkerson PC, Fischetti CA, McBride ML, Hassman LM, Hogan SP, Rothenberg ME. A central regulatory role for eosinophils and the eotaxin/CCR3 axis in chronic experimental allergic airway inflammation. *Proc Natl Acad Sci U S A* 2006;**103**:16418—23.

99. Garcia-Zepeda EA, Rothenberg ME, Ownbey RT, Celestin J, Leder P, Luster AD. Human eotaxin is a specific chemoattractant for eosinophil cells and provides a new mechanism to explain tissue eosinophilia. *Nat Med* 1996;**2**:449—56.

100. Mould AW, Matthaei KI, Young IG, Foster PS. Relationship between interleukin-5 and eotaxin in regulating blood and tissue eosinophilia in mice. *J Clin Invest* 1997;**99**:1064—71.

101. Campbell EM, Kunkel SL, Strieter RM, Lukacs NW. Temporal role of chemokines in a murine model of cockroach allergen-induced airway hyperreactivity and eosinophilia. *J Immunol* 1998;**161**:7047—53.

102. Gonzalo JA, Lloyd CM, Wen D, Albar JP, Wells TN, Proudfoot A, et al. The coordinated action of CC chemokines in the lung orchestrates allergic inflammation and airway hyper-responsiveness. *J Exp Med* 1998;**188**:157—67.

103. Rothenberg ME, MacLean JA, Pearlman E, Luster AD, Leder P. Targeted disruption of the chemokine eotaxin partially reduces antigen-induced tissue eosinophilia. *J Exp Med* 1997;**185**:785—90.

104. Yang Y, Loy J, Ryseck RP, Carrasco D, Bravo R. Antigen-induced eosinophilic lung inflammation develops in mice deficient in chemokine eotaxin. *Blood* 1998;**92**:3912—23.

105. Smit JJ, Lukacs NW. A closer look at chemokines and their role in asthmatic responses. *Eur J Pharmacol* 2006;**533**:277—88.

106. Brown JR, Kleimberg J, Marini M, Sun G, Bellini A, Mattoli S. Kinetics of eotaxin expression and its relationship to eosinophil

accumulation and activation in bronchial biopsies and bron-choalveolar lavage (BAL) of asthmatic patients after allergen inhalation. *Clin Exp Immunol* 1998;**114**:137–46.

107. Asosingh K, Hanson JD, Cheng G, Aronica MA, Erzurum SC. Allergen-induced, eotaxin-rich, proangiogenic bone marrow progenitors: a blood-borne cellular envoy for lung eosinophilia. *J Allergy Clin Immunol* 2010;**125**:918–25.

108. Imaoka H, Babirad I, Watson RM, Obminski G, Strinich TX, Howie K, et al. Increased lung-homing of vascular endothelial progenitor cells in asthmatic subjects (abst). *Am J Respir Crit Care Med* 2010;**181**:A5624.

109. Dorman SC, Babirad I, Post J, Watson RM, Foley R, Jones GL, et al. Progenitor egress from the bone marrow after allergen challenge: role of stromal cell-derived factor 1alpha and eotaxin. *J Allergy Clin Immunol* 2005;**115**:501–7.

110. Radinger M, Bossios A, Alm AS, Jeurink P, Lu Y, Malmhall C, et al. Regulation of allergen-induced bone marrow eosinophilo-poiesis: role of CD4+ and CD8+ T cells. *Allergy* 2007;**62**: 1410–8.

111. Fregonese L, Silvestri M, Sabatini F, Rossi GA. Cysteinyl leuko-trienes induce human eosinophil locomotion and adhesion mole-cule expression via a CysLT1 receptor-mediated mechanism. *Clin Exp Allergy* 2002;**32**:745–50.

112. Braccioni F, Dorman SC, O'Byrne PM, Inman MD, Denburg JA, Parameswaran K, et al. The effect of cysteinyl leukotrienes on growth of eosinophil progenitors from peripheral blood and bone marrow of atopic subjects. *J Allergy Clin Immunol* 2002;**110**: 96–101.

113. Bautz F, Denzlinger C, Kanz L, Mohle R. Chemotaxis and trans-endothelial migration of CD34(+) hematopoietic progenitor cells induced by the inflammatory mediator leukotriene D4 are mediated by the 7-transmembrane receptor CysLT1. *Blood* 2001;**97**: 3433–40.

114. Punia N, Thomson J, Babirad I, Sehmi R. IL-4 and IL-13 prime the transmigrational responses of hemopoietic progenitor cells (abst). *Am J Respir Crit Care Med* 2010;**181**:A4034.

115. Pettipher R. The roles of the prostaglandin D(2) receptors DP(1) and CRTH2 in promoting allergic responses. *Br J Pharmacol* 2008;**153**(Suppl. 1):S191–9.

116. Ueki S, Kato H, Kobayashi Y, Ito W, Adachi T, Nagase H, et al. Anti- and proinflammatory effects of 15-deoxy-delta-prostaglandin J2(15d-PGJ2) on human eosinophil functions. *Int Arch Allergy Immunol* 2007;**143**(Suppl. 1):15–22.

117. Smith SG, Sehmi R, Howie K, Watson RM, Campbell H, Obminski G, et al. Effects of peroxisome proliferator-activated receptors (PPARs) on eosinophil migration (abst). *Am J Respir Crit Care Med* 2010;**181**:A2787.

118. Bochner BS, Gleich GJ. What targeting eosinophils has taught us about their role in diseases. *J Allergy Clin Immunol* 2010;**126**: 16–25.

119. Flood-Page P, Menzies-Gow A, Phipps S, Ying S, Wangoo A, Ludwig MS, et al. Anti-IL-5 treatment reduces deposition of ECM proteins in the bronchial subepithelial basement membrane of mild atopic asthmatics. *J Clin Invest* 2003;**112**:1029–36.

120. Menzies-Gow A, Flood-Page P, Sehmi R, Burman J, Hamid Q, Robinson DS, et al. Anti-IL-5 (mepolizumab) therapy induces bone marrow eosinophil maturational arrest and decreases eosinophil progenitors in the bronchial mucosa

of atopic asthmatics. *J Allergy Clin Immunol* 2003;**111**:714–9.

121. Leckie MJ, ten Brinke A, Khan J, Diamant Z, O'Connor BJ, Walls CM, et al. Effects of an interleukin-5 blocking monoclonal antibody on eosinophils, airway hyper-responsiveness, and the late asthmatic response. *Lancet* 2000;**356**:2144–8.

122. Kips JC, O'Connor BJ, Langley SJ, Woodcock A, Kerstjens HA, Postma DS, et al. Effect of SCH55700, a humanized anti-human interleukin-5 antibody, in severe persistent asthma: a pilot study. *Am J Respir Crit Care Med* 2003;**167**:1655–9.

123. Flood-Page PT, Menzies-Gow AN, Kay AB, Robinson DS. Eosinophil's role remains uncertain as anti-interleukin-5 only partially depletes numbers in asthmatic airway. *Am J Respir Crit Care Med* 2003;**167**:199–204.

124. Flood-Page P, Swenson C, Faiferman I, Matthews J, Williams M, Brannick L, et al. A study to evaluate safety and efficacy of mepolizumab in patients with moderate persistent asthma. *Am J Respir Crit Care Med* 2007;**176**:1062–71.

125. O'Byrne PM, Inman MD, Parameswaran K. The trials and tribu-lations of IL-5, eosinophils, and allergic asthma. *J Allergy Clin Immunol* 2001;**108**:503–8.

126. Nair P, Pizzichini MM, Kjarsgaard M, Inman MD, Efthimiadis A, Pizzichini E, et al. Mepolizumab for prednisone-dependent asthma with sputum eosinophilia. *N Engl J Med* 2009;**360**:985–93.

127. Haldar P, Brightling CE, Hargadon B, Gupta S, Monteiro W, Sousa A, et al. Mepolizumab and exacerbations of refractory eosinophilic asthma. *N Engl J Med* 2009;**360**:973–84.

128. Busse WW, Katial R, Gossage D, Sari S, Wang B, Kolbeck R, et al. Safety profile, pharmacokinetics, and biologic activity of MEDI-563, an anti-IL-5 receptor alpha antibody, in a phase I study of subjects with mild asthma. *J Allergy Clin Immunol* 2010;**125**:1237–44.

129. Kolbeck R, Kozhich A, Koike M, Peng L, Andersson CK, Damschroder MM, et al. MEDI-563, a humanized anti-IL-5 receptor alpha mAb with enhanced antibody-dependent cell-mediated cytotoxicity function. *J Allergy Clin Immunol* 2010;**125**:1344–53.

130. Gauvreau GM, Boulet LP, Cockcroft DW, Baatjes A, Cote J, Deschesnes F, et al. Antisense therapy against CCR3 and the common beta chain attenuates allergen-induced eosino-philic responses. *Am J Respir Crit Care Med* 2008;**177**:952–8.

131. Gauvreau GM, Pageau R, Seguin R, Carballo D, D'Anjou H, Campbell H, et al. Efficacy of increasing doses of TPI ASM8 on allergen inhalation challenges in asthmatics (abst). *Am J Respir Crit Care Med* 2010;**181**:A5669.

132. Bousquet J, Clark TJ, Hurd S, Khaltaev N, Lenfant C, O'Byrne P, et al. GINA guidelines on asthma and beyond. *Allergy* 2007;**62**:102–12.

133. Barnes PJ, Pedersen S, Busse WW. Efficacy and safety of inhaled corticosteroids. New developments. *Am J Respir Crit Care Med* 1998;**157**:S1–53.

134. Hargreave FE, Dolovich J, Newhouse MT. The assessment and treatment of asthma: a conference report. *J Allergy Clin Immunol* 1990;**85**:1098–111.

135. Gibson PG, Wong BJ, Hepperle MJ, Kline PA, Girgis-Gabardo A, Guyatt G, et al. A research method to induce and examine a mild exacerbation of asthma by withdrawal of inhaled corticosteroid. *Clin Exp Allergy* 1992;**22**:525–32.

Eosinophil Activities and Virus-Induced Asthma

Elizabeth R. Bivins-Smith and David B. Jacoby

VIRUS-INDUCED ASTHMA ATTACKS: AN OVERVIEW

Asthma is a chronic pulmonary disease characterized by airway remodeling, airway inflammation, and broncho-constriction. Patients with asthma cycle between periods of exacerbations, which result in significant morbidity, and recovery. Virus infections are the leading cause of asthma exacerbations in children and adults, yet specific treatment and prevention strategies for virus-induced asthma are limited. The tissue damage caused during virus-induced asthma is divided into two categories: damage caused by the virus itself and damage from the host immune response. Eosinophils play a significant role in both virus clearance and immunity-mediated tissue damage.

As many as 80% of childhood cases and 55% of adult cases of asthma attacks have an identifiable underlying virus infection.[1,2] Diagnosis of virus infection in acute asthma is often presumptive and based on patient history and physical examination, but other diagnostic laboratory techniques include serology, virus culture, and reverse-transcription polymerase chain reaction. Respiratory RNA viruses are the main types of virus that induce asthma attacks, with rhinovirus (the common cold virus) accounting for approximately two-thirds of viruses identified.[3] Coronavirus, influenza, parainfluenza, and respiratory syncytial virus (RSV) comprise the remainder of respiratory RNA viruses that induce asthma attacks.[1]

The immune response to viral infections is T-helper type 1 (T_h1)-driven and involves the production of proinflammatory cytokines such as interleukin-12 (IL-12) and interferon γ (IFN-γ). Studies show that T_h1 responses in patients with asthma are impaired, which results in decreased T_h1 cytokine production during viral infection.[4] Asthma also skews the immune system away from a T_h1 response toward a T_h2 immune environment through increased production of T_h2 cytokines, such as IL-4, IL-5, and IL-13. IL-5 is the central cytokine involved in eosinophil proliferation, maturation, and survival. When the immune environment of a virus-infected patient with asthma shifts from T_h1 to a T_h2 response, these patients experience more severe symptoms of infection and delayed pathogen clearance, which likely contribute to increased asthma exacerbations and hospital admissions.[4,5]

The association of eosinophils with virus-induced asthma is well documented. Eosinophil products are present in the sputum of patients with virus-induced asthma,[6] and histology studies show that in patients who have died from severe asthma, eosinophils are clustered around airway nerves (Fig. 13.7.1).[7] In the following sections, we will present the mechanisms of eosinophil pathophysiology in virus-induced asthma, focusing on virus-induced airway inflammation, viral activation of eosinophils, virus-induced eosinophil-mediated neural changes, and T cell–eosinophil interactions during virus infection. We then discuss the beneficial role of eosinophils in virus-induced asthma and potential targets for prevention and treatment of the disease.

EOSINOPHILS IN VIRUS-INDUCED AIRWAY INFLAMMATION

Respiratory virus infection induces airway inflammation in all individuals, but the inflammatory response is different in patients with asthma from in nonasthmatic individuals. Experimental rhinovirus infection of human subjects increases eosinophil numbers in the bronchial epithelium in both nonasthmatic and asthmatic volunteers. Eosinophils remained in airway tissues of virus-infected patients with asthma 6 weeks longer than in infected, nonasthmatic controls. These data suggest that virus-induced eosinophil influx into the airways occurs in individuals both with and without asthma and that the eosinophil response to viral infection is accentuated in patients with asthma.[8]

The airway epithelium is the primary site of respiratory virus infection. Virus-infected epithelial cells release a wide array of proinflammatory cytokines that recruit inflammatory cells into the airways. Among these proinflammatory cytokines are eotaxin/C-C motif chemokine 11 (CCL11), granulocyte-macrophage colony-stimulating factor (GM-CSF), MIP-1α/CCL3, and RANTES/CCL5, which are chemoattractants and activators of eosinophils.[9–16] Virus infection increased eosinophil inflammation in the airways of antigen-sensitized and antigen-challenged mice compared to sensitized-challenged, mock-infected animals. In addition, virus infection of epithelial macrophages induced the release of eotaxin, and the blockade of eotaxin bioactivity with a neutralizing antibody inhibited airway eosinophilia in sensitized-challenged, virus-infected but not sensitized-challenged, mock-infected mice. These data suggest that virus infection recruits eosinophils to the lungs of sensitized-challenged animals and that this is dependent on eotaxin.[17]

Virus infections of allergic individuals may lead to the development of eosinophilic inflammation as well as features of asthma. Calhoun et al. showed that patients with allergic rhinitis who were experimentally infected with rhinovirus had increased eosinophil influx into the airways following

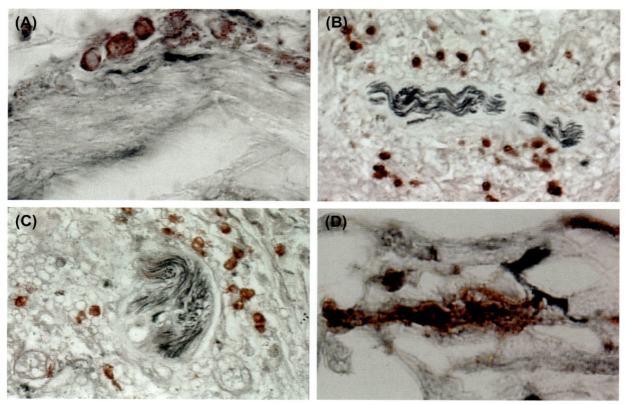

FIGURE 13.7.1 **Eosinophils are physically associated with airway nerves in asthma.** Postmortem airway tissues from patients with fatal asthma were stained with antibodies to the nerve-specific protein PGP9.5 (black) and to eosinophil MBP (red). *A*, Eosinophils line up along nerve fibers in airway smooth muscle layer. *B*, Eosinophils surround an airway nerve bundle, seen longitudinally. *C*, Eosinophils surround a nerve bundle, cut in cross section. Note extracellular MBP (arrow). *D*, A nerve fiber is encrusted with extracellular MBP.[7]

allergen challenge. These patients also had increased histamine release and edema from fluid leakage, which are consistent with asthma. These data suggest that virus infection may induce asthma in individuals with preexisting allergic respiratory diseases. The T_h2 immune environment that allergic rhinitis elicits closely resembles the cytokine profiles of asthma, so a subsequent viral infection may tip airway physiology toward an asthma phenotype.[18]

VIRUS-INDUCED EOSINOPHIL ACTIVATION

Eosinophils are activated during viral infection. Studies of patients with RSV bronchiolitis showed the presence of eosinophil cationic protein (ECP) and eosinophil-derived neurotoxin (EDN) in their lower airway secretions.[19,20] Other studies have shown that cell surface expression of CD11b, a marker of cellular activation, is upregulated in eosinophils of RSV-infected patients compared to uninfected volunteers.[21] These data suggest that eosinophils are activated and degranulate following viral infection.

Toll-like receptors (TLRs) are pattern recognition receptors that recognize pathogen-associated molecular patterns

(PAMPs). Many groups have shown that TLRs are present and functional in eosinophils. TLRs mediate both the innate and adaptive immune responses, and some believe they are involved in controlling sensitization. Studies suggest that polymorphisms in *TLR2*, *TLR4*, *TLR9*, and *TLR10* are associated with an increased risk of developing asthma.[22–25] Human eosinophils express *TLR1*, *TLR4*, *TLR7*, *TLR9*, and *TLR10* mRNA. TLR7 recognizes single-stranded RNA (ssRNA) found in respiratory RNA viruses. Nagase et al. showed that R848 (a synthetic ligand for TLR7 and TLR8) increases CD11b expression on human eosinophils, induces superoxide generation, and promotes cell survival.[26] Phipps et al. showed that mouse eosinophils express TLR7 on their cell surface and TLR3, TLR4, and TLR7 in endosomes. Furthermore, ssRNA treatment of mouse eosinophils increases eosinophil peroxidase (EPO) release, increases CD11b expression, and induces degranulation.[27]

Eosinophil TLR responses to stimulation may differ between atopic and nonatopic individuals. Mansson et al. showed that eosinophils treated with IL-5 and then stimulated with polyI:C (a TLR3 agonist) had increased IL-8 release compared to IL-5 treatment alone.[28] In addition, IL-5 pretreatment potentiates R837 (a TLR7-specific synthetic ligand)-induced IL-8 release. Moreover, eosinophils collected from atopic patients and stimulated with R837 release increased IL-8 compared to R837-stimulated eosinophils from nonallergic volunteers.[29] These data suggest that atopy affects eosinophil TLR responses and that aberrant virus-induced eosinophil activation in patients with asthma may exacerbate their airway disease.

Eosinophils can be infected and activated by respiratory viruses. Eosinophils can be activated directly by virus binding to receptors on either the cell surface or in endosomes, depending on the virus's mechanism of entry. RSV mediates entry into target cells by binding TLR4. Although it has not been demonstrated that RSV binds TLR4 on eosinophils, *in vitro* culture of RSV with human eosinophils prolongs cell survival.[30] Transmission electron microscopy studies showed that RSV is internalized by human eosinophils and identified virions in phagocytic vacuoles near the cell surface. In addition, infected eosinophils underwent piecemeal degranulation, suggesting that these cells were activated upon infection.[31]

In some cases, virus-induced eosinophil activation requires priming prior to infection and, in other cases, viruses prime eosinophils to respond to treatment with additional stimuli. Handzel et al. showed that rhinovirus binds intracellular adhesion molecule-1 (ICAM-1) on human eosinophils primed with GM-CSF and that GM-CSF treatment upregulates ICAM-1 on eosinophils.[32] Additional studies showed that RSV increased superoxide generation and CD11b in eosinophils that were first treated with platelet activating factor.[33] RSV infection of eosinophils potentiates phorbol-12-myristate-13-acetate (PMA)-induced superoxide generation and leukotriene C_4 release.[34] These data suggest that virus-induced eosinophil activation in asthma may lead to an exaggerated host immune response, resulting in airway inflammation and constriction.

VIRUS-INDUCED AIRWAY HYPERREACTIVITY: THE ROLE OF EOSINOPHILS

During virus-induced asthma, eosinophils cause dysfunction of parasympathetic nerve signaling. Airway parasympathetic nerves provide the dominant control over airway smooth muscle.[35] These neurons release acetylcholine (ACh), which binds M3 muscarinic receptors (M_3Rs) on smooth muscle to cause bronchoconstriction.[36] ACh release is normally limited via negative feedback inhibition of ACh on neuronal M2 muscarinic receptors (M_2Rs).[37] When neuronal M_2Rs are dysfunctional, ACh release by neurons is not inhibited, and increased levels of ACh binding to M_3Rs on airway smooth muscle cause airway hyperresponsiveness (AHR).

In antigen-sensitized, antigen-challenged guinea pigs, eosinophils mediate AHR.[38] Parasympathetic nerves recruit eosinophils to the nerves via eotaxin signaling.[39] Cholinergic nerves express ICAM-1 and vascular cell adhesion protein (VCAM), which bind to eosinophils.[40] Eosinophils bind to nerves and release major basic protein (MBP), which blocks M_2Rs on nerves. This causes loss of M_2R function, resulting in increased ACh release and bronchoconstriction.[41,42]

Viral infection of nonsensitized guinea pigs causes the loss of M_2R function and AHR, but this loss is not eosinophil-mediated (not prevented by antibody to IL5 or antibody to MBP).[43] In nonsensitized guinea pigs, the virus-induced loss of M2R function appears to be on the level of reduced M2R gene expression. This appears to be the result of production of TNF-α, which decreases M_2R mRNA stability.[44] In contrast, in sensitized guinea pigs, virus-induced loss of M_2R function and AHR is mediated by eosinophils. Adamko et al. showed that depletion of eosinophils with an anti-IL-5 antibody prevents virus-induced AHR in sensitized guinea pigs, but not in nonsensitized, virus-infected animals. Additional studies showed that blockade of MBP bioactivity inhibits virus-induced AHR and M_2R dysfunction in sensitized guinea pigs. Collectively, these data suggest that in the context of asthma, eosinophils are recruited to airway nerves, virus infection activates the eosinophils to release MBP, and MBP antagonizes M_2Rs, resulting in bronchoconstriction (Fig. 13.7.2).[45]

INTERACTIONS BETWEEN T LYMPHOCYTES AND EOSINOPHILS IN VIRUS-INDUCED ASTHMA

In addition to eosinophil—nerve interactions, eosinophils communicate with T cells in virus-induced asthma. As mentioned above, eosinophils mediate virus-induced AHR and M_2R dysfunction in sensitized guinea pigs but not in nonsensitized animals.[45] Adamko et al. demonstrated that depletion of $CD8^+$ T cells prevents virus-induced AHR and M_2R dysfunction in sensitized but not in nonsensitized guinea pigs. Sensitization increases the number of eosinophils in the airways, and virus infection of both nonsensitized and sensitized guinea pigs reduces the total number of eosinophils in the airways and eosinophils associated with nerves, suggesting that virus infection induces eosinophil degranulation. $CD8^+$ T cell depletion inhibits virus-induced eosinophil cytolysis, suggesting that $CD8^+$ T cells promote virus-induced eosinophil degranulation.[46] These data suggest that $CD8^+$ T cells interact with eosinophils in the airways of virus-infected, sensitized animals to promote AHR and M_2R dysfunction. Schwarze et al. showed that $CD8^+$ T cells are also necessary for AHR and lung eosinophilia during RSV infection of mice.[47]

FIGURE 13.7.2 **Viral infection activates airway eosinophils in a CD8$^+$ T cell-dependent process.** Activated eosinophils release major basic protein, which binds to inhibitory M2 muscarinic receptors on parasympathetic nerves. Blocking these receptors increases acetylcholine (Ach) release, causing bronchoconstriction. M3, M3 muscarinic receptor.

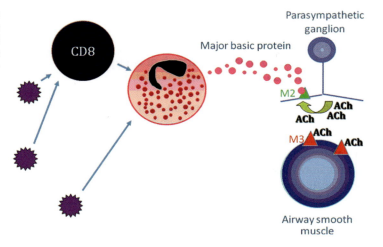

In vitro studies have shown that eosinophils directly interact with T cells to induce eosinophil degranulation and promote antiviral, T cell-mediated immunity. When human eosinophils are incubated with parainfluenza virus in the presence of T cells and antigen-presenting cells (APCs; macrophages or dendritic cells), they release EPO. UV-inactivated virus also induces EPO release from eosinophils, suggesting that this process occurs in the absence of viral replication. Additional studies showed that RSV is also able to induce EPO release from eosinophils in the presence of T cells and APCs.[48] Handzel et al. showed that eosinophils present rhinovirus antigens to CD4$^+$ T cells, inducing their clonal expansion and the release of IFN-γ.[32] These data suggest that eosinophils directly interact with CD4$^+$ and CD8$^+$ T cells, and that their interactions result in eosinophil degranulation, which may augment symptoms of asthma as well as initiate antiviral adaptive immunity. Despite the detrimental role eosinophils play in the pathophysiology of asthma, the antiviral effect of eosinophils may also be beneficial in the context of virus infection and asthma.

BENEFITS OF EOSINOPHILS IN VIRUS-INDUCED ASTHMA

The antibacterial and antiparasitic effects of eosinophils are well documented, yet their antiviral properties have been studied only recently. While eosinophils have historically been acknowledged for their pathophysiological role in asthma, recent studies suggest that these cells also play a beneficial antiviral role in virus-induced asthma. Adamko et al. showed that guinea pigs sensitized to ovalbumin had reduced parainfluenza titers in their lungs 4 days following infection compared to nonsensitized, virus-infected animals. Furthermore, depletion of eosinophils with an anti-IL-5 antibody increased viral levels in sensitized virus-infected animals to above those observed in nonsensitized infected

animals.[45] These data suggest that eosinophils serve an antiviral role in the context of virus infection and allergy.

Additional studies have investigated the mechanisms by which eosinophils promote virus clearance *in vivo*. Phipps et al. showed that IL-5 transgenic mice, which have increased levels of eosinophils, have improved RSV clearance compared to wild-type RSV-infected mice. MyD88 is the adaptor molecule for many TLR7 signaling pathways, and studies show that adoptive transfer of wild-type eosinophils, but not eosinophils deficient in MyD88, into the lungs of infected wild-type mice improves virus clearance and inhibits AHR. Blockade of nitric oxide synthase 2 (NOS-2; inducible nitric oxide synthase) activity inhibits RSV clearance *in vivo*. These data suggest that eosinophils contribute to RSV clearance in both a MyD88-dependent and NOS-dependent manner. Furthermore, these data suggest that eosinophils may aid the prevention of RSV-induced AHR.[27]

As mentioned above, eosinophils can be activated by viruses to degranulate. Upon degranulation, eosinophils release granule proteins and reactive oxygen species, which can cause damage to pathogen moieties. Several groups have shown that eosinophils increase superoxide generation in response to virus exposure.[33,34] Klebanoff et al. showed that human eosinophils stimulated with PMA are virucidal against human immunodeficiency virus (HIV) type 1 *in vitro* and that purified EPO, when incubated with hydrogen peroxide and a halide (its substrate), is also virucidal against HIV.[49] ECP and EDN contain intrinsic ribonuclease activity that can destroy the genomes of RNA viruses. Domachowske et al. showed that eosinophils and human recombinant EDN have antiviral activity toward RSV *in vitro*.[50] Others have demonstrated an antiviral effect of EDN against HIV *in vitro*.[51,52] Collectively, these data suggest that eosinophils inactivate RNA viruses by a number of different mechanisms involving granule proteins and reactive oxygen species.

While some studies have shown that eosinophils destroy virus, other groups have shown that eosinophils are productively infected by viruses. As mentioned previously, eosinophils can bind and internalize viruses. Kimpen et al. showed that RSV can bind to the eosinophil membrane and observed virus in the phagocytic vacuoles of eosinophils.[31,34] Dyer et al. showed that RSV and mouse pneumovirus productively infect human and mouse eosinophils, respectively, resulting in the release of infectious virions and proinflammatory cytokines *in vitro*.[53] In contrast, our laboratory has presented evidence suggesting that parainfluenza virus abortively infects human eosinophils. Parainfluenza virus can enter eosinophils and replicate the viral RNA genome, but infectious virus particles are not released.[54] A productive infection of eosinophils could potentially serve to induce cytokine release and attract other immune cells, while an abortive infection would remove extracellular virus from the local environment and/or promote the presentation of virus antigens to T cells. Collectively these data suggest that interactions between eosinophils and different types of viruses are varied and complex.

THERAPEUTIC TARGETS AND PREVENTION STRATEGIES IN VIRUS-INDUCED ASTHMA

Standard treatments for patients suffering from virus-induced asthma exacerbations consist of β-agonists in combination with anticholinergics and steroids. β-agonists relax smooth muscle by stimulating adenylate cyclase activity and closing calcium channels, while anticholinergics block the effects of the neurotransmitter ACh on airway smooth muscle constriction. Viruses and interferon downregulate inhibitory M_2Rs on parasympathetic nerves, thereby increasing ACh release, and steroids can reverse these effects.[55] Steroids can also reduce eosinophil influx into the lungs, ICAM-1 expression on nerves, and neuronal M_2R dysfunction.[56,57]

While eosinophil in sensitized animals can prevent virus-induced hyperreactivity,[45] the effects of eosinophil depletion in human virus-induced asthma attacks are less clear. While eosinophil depletion with anti-IL5 has limited effects on the response to antigen challenge in human subjects[58] and a clinical study of anti-IL5 had unimpressive effects in unselected asthmatics, when asthmatics with sputum eosinophilia were treated with anti-IL5, steroids were withdrawn without exacerbation in the majority of cases.[59] Whether eosinophil depletion can prevent virus-induced asthma exacerbation awaits further studies.

Another potential treatment for virus-induced asthma is the correction of deregulated cytokine production. As mentioned previously, the immune environment of patients with asthma is skewed toward a T_h2 phenotype, with decreased interferon production.[4] Treatments that promote a T_h1 phenotype, the immune response typically observed in viral infection, may prove clinically valuable.

The most direct method of treatment or prevention of virus-induced asthma is targeting the viral infection itself. However, such treatments are at present limited.

Rhinovirus

Treatment and prevention of rhinovirus, the cause of two-thirds of virus-induced asthma exacerbations, presents a significant challenge because over 100 serotypes of the virus exist. The varying antigenicity among serotypes further hinders vaccine design. As mentioned previously, rhinovirus uses ICAM-1 binding for entry into cells, thus the blockade of cellular infection by blocking virus binding to target cells presents one promising mechanism for drug design. While ribavirin, which is believed to interfere with viral RNA synthesis, has activity against a range of viruses, including rhinovirus and RSV, its clinical efficacy is limited and it is not widely used in the treatment of rhinovirus.

Respiratory Syncytial Virus

RSV vaccine development has been problematic, with paradoxical worsening of the clinical response being an issue. Palivizumab is a monoclonal antibody that prevents RSV infection by interfering with the RSV fusion protein and, thus, viral entry into cells. In addition, studies show that type IV phosphodiesterase inhibitors inhibit RSV-induced AHR and eosinophilia in the lungs.[60]

Influenza

Effective vaccines for influenza are recommended for patients with airway disease, including asthma. Although it is difficult to prove that influenza-induced asthma attacks are reduced by vaccination, this may be due to the relatively small number of asthma attacks caused by influenza. Treatment for influenza viral infections consists of neuraminidase inhibitors. Again, the role of neuraminidase inhibitors in preventing asthma attacks is difficult to demonstrate.

Coronavirus and Parainfluenza Virus

Currently no treatments or vaccines exist for parainfluenza virus and coronavirus. Improved immunological understanding of these viruses and of virus interactions with the host, and eosinophils specifically, will aid in the advancement of future vaccines and treatments.

CONCLUSION

Respiratory virus infections are the primary cause of asthma exacerbations in children and adults. In asthma, eosinophils are recruited to the airways as a part of the T_h2 immune response, where they contribute to asthma pathophysiology. However, eosinophils may play a dual role in virus-induced asthma; on the one hand recognizing viruses, releasing virucidal mediators, and presenting antigens, and on the other hand becoming activated and increasing airway reactivity. Current treatments for virus-induced asthma are not specific to virus infection and instead focus on smooth muscle relaxation and reduction of airway inflammation. Further investigation of the interactions between eosinophils and viruses is warranted and will likely lead to more targeted treatments and prevention.

REFERENCES

1. Johnston SL, Pattemore PK, Sanderson G, Smith S, Lampe F, Josephs L, et al. Community study of role of viral infections in exacerbations of asthma in 9–11 year old children. *BMJ* 1995;**310**:1225–9.

2. Atmar RL, Guy E, Guntupalli KK, Zimmerman JL, Bandi VD, Baxter BD, et al. Respiratory tract viral infections in inner-city asthmatic adults. *Arch Intern Med* 1998;**158**:2453–9.

3. Nicholson KG, Kent J, Ireland DC. Respiratory viruses and exacerbations of asthma in adults. *BMJ* 1993;**307**:982–6.

4. Papadopoulos NG, Stanciu LA, Papi A, Holgate ST, Johnston SL. A defective type 1 response to rhinovirus in atopic asthma. *Thorax* 2002;**57**:328–32.

5. Gern JE, Vrtis R, Grindle KA, Swenson C, Busse WW. Relationship of upper and lower airway cytokines to outcome of experimental rhinovirus infection. *Am J Respir Crit Care Med* 2000;**162**:2226–31.

6. Grunberg K, Smits HH, Timmers MC, de Klerk EP, Dolhain RJ, Dick EC, et al. Experimental rhinovirus 16 infection. Effects on cell differentials and soluble markers in sputum in asthmatic subjects. *Am J Respir Crit Care Med* 1997;**156**:609–16.

7. Costello RW, Schofield BH, Kephart GM, Gleich GJ, Jacoby DB, Fryer AD. Localization of eosinophils to airway nerves and effect on neuronal M2 muscarinic receptor function. *Am J Physiol* 1997;**273**:L93–103.

8. Fraenkel DJ, Bardin PG, Sanderson G, Lampe F, Johnston SL, Holgate ST. Lower airways inflammation during rhinovirus colds in normal and in asthmatic subjects. *Am J Respir Crit Care Med* 1995;**151**:879–886.

9. Griego SD, Weston CB, Adams JL, Tal-Singer R, Dillon SB. Role of p38 mitogen-activated protein kinase in rhinovirus-induced cytokine production by bronchial epithelial cells. *J Immunol* 2000;**165**:5211–5220.

10. Subauste MC, Jacoby DB, Richards SM, Proud D. Infection of a human respiratory epithelial cell line with rhinovirus. Induction of cytokine release and modulation of susceptibility to infection by cytokine exposure. *J Clin Invest* 1995;**96**:549–557.

11. Matsukura S, Kokubu F, Noda H, Tokunaga H, Adachi M. Expression of IL-6, IL-8, and RANTES on human bronchial epithelial cells, NCI-H292, induced by influenza virus A. *J Allergy Clin Immunol* 1996;**98**:1080–1087.

12. Noah TL, Becker S. Respiratory syncytial virus-induced cytokine production by a human bronchial epithelial cell line. *Am J Physiol* 1993;**265**:L472–8.

13. Becker S, Reed W, Henderson FW, Noah TL. RSV infection of human airway epithelial cells causes production of the beta-chemokine RANTES. *Am J Physiol* 1997;**272**:L512–20.

14. Olszewska-Pazdrak B, Casola A, Saito T, Alam R, Crowe SE, Mei F, et al. Cell-specific expression of RANTES, MCP-1, and MIP-1alpha by lower airway epithelial cells and eosinophils infected with respiratory syncytial virus. *J Virol* 1998;**72**:4756–64.

15. Kawaguchi M, Kokubu F, Kuga H, Tomita T, Matsukura S, Kadokura M, et al. Expression of eotaxin by normal airway epithelial cells after influenza virus A infection. *Int Arch Allergy Immunol* 2000;**122**(Suppl. 1):44–9.

16. Papadopoulos NG, Papi A, Meyer J, Stanciu LA, Salvi S, Holgate ST, et al. Rhinovirus infection up-regulates eotaxin and eotaxin-2 expression in bronchial epithelial cells. *Clin Exp Allergy* 2001;**31**:1060–6.

17. Nagarkar DR, Bowman ER, Schneider D, Wang Q, Shim J, Zhao Y, et al. Rhinovirus infection of allergen-sensitized and -challenged mice induces eotaxin release from functionally polarized macrophages. *J Immunol* 2010;**185**:2525–35.

18. Calhoun WJ, Dick EC, Schwartz LB, Busse WW. A common cold virus, rhinovirus 16, potentiates airway inflammation after segmental antigen bronchoprovocation in allergic subjects. *J Clin Invest* 1994;**94**:2200–8.

19. Harrison AM, Bonville CA, Rosenberg HF, Domachowske JB. Respiratory syncytical virus-induced chemokine expression in the lower airways: eosinophil recruitment and degranulation. *Am J Respir Crit Care Med* 1999;**159**:1918–24.

20. Garofalo R, Kimpen JL, Welliver RC, Ogra PL. Eosinophil degranulation in the respiratory tract during naturally acquired respiratory syncytial virus infection. *J Pediatr* 1992;**120**:28–32.

21. Lindemans CA, Kimpen JL, Luijk B, Heidema J, Kanters D, van der Ent CK, et al. Systemic eosinophil response induced by respiratory syncytial virus. *Clin Exp Immunol* 2006;**144**:409–17.

22. Eder W, Klimecki W, Yu L, von Mutius E, Riedler J, Braun-Fahrlander C, et al. Toll-like receptor 2 as a major gene for asthma in children of European farmers. *J Allergy Clin Immunol* 2004;**113**:482–8.

23. Raby BA, Klimecki WT, Laprise C, Renaud Y, Faith J, Lemire M, et al. Polymorphisms in toll-like receptor 4 are not associated with asthma or atopy-related phenotypes. *Am J Respir Crit Care Med* 2002;**166**:1449–56.

24. Lazarus R, Raby BA, Lange C, Silverman EK, Kwiatkowski DJ, Vercelli D, et al. TOLL-like receptor 10 genetic variation is associated with asthma in two independent samples. *Am J Respir Crit Care Med* 2004;**170**:594–600.

25. Lazarus R, Klimecki WT, Raby BA, Vercelli D, Palmer LJ, Kwiatkowski DJ, et al. Single-nucleotide polymorphisms in the Toll-like receptor 9 gene (TLR9): frequencies, pairwise linkage disequilibrium, and haplotypes in three U.S. ethnic groups and exploratory case-control disease association studies. *Genomics* 2003;**81**:85–91.

26. Nagase H, Okugawa S, Ota Y, Yamaguchi M, Tomizawa H, Matsushima K, et al. Expression and function of Toll-like receptors in eosinophils: activation by Toll-like receptor 7 ligand. *J Immunol* 2003;**171**:3977–82.

27. Phipps S, Lam CE, Mahalingam S, Newhouse M, Ramirez R, Rosenberg HF, et al. Eosinophils contribute to innate antiviral immunity and promote clearance of respiratory syncytial virus. *Blood* 2007;**110**:1578–86.

28. Mansson A, Fransson M, Adner M, Benson M, Uddman R, Bjornsson S, et al. TLR3 in human eosinophils: functional effects and decreased expression during allergic rhinitis. *Int Arch Allergy Immunol* 2010;**151**:118–28.

29. Mansson A, Cardell LO. Role of atopic status in Toll-like receptor (TLR)7- and TLR9-mediated activation of human eosinophils. *J Leukoc Biol* 2009;**85**:719–27.

30. Lindemans CA, Coffer PJ, Schellens IM, de Graaff PM, Kimpen JL, Koenderman L. Respiratory syncytial virus inhibits granulocyte apoptosis through a phosphatidylinositol 3-kinase and NF-kappaB-dependent mechanism. *J Immunol* 2006;**176**:5529–37.

31. Kimpen JL, Garofalo R, Welliver RC, Fujihara K, Ogra PL. An ultrastructural study of the interaction of human eosinophils with respiratory syncytial virus. *Pediatr Allergy Immunol* 1996;**7**:48–53.

32. Handzel ZT, Busse WW, Sedgwick JB, Vrtis R, Lee WM, Kelly EA, et al. Eosinophils bind rhinovirus and activate virus-specific T cells. *J Immunol* 1998;**160**:1279–84.

33. Tachibana A, Kimura H, Kato M, Nako Y, Kozawa K, Morikawa A. Respiratory syncytial virus enhances the expression of CD11b molecules and the generation of superoxide anion by human eosinophils primed with platelet-activating factor. *Intervirology* 2002;**45**:43–51.

34. Kimpen JL, Garofalo R, Welliver RC, Ogra PL. Activation of human eosinophils in vitro by respiratory syncytial virus. *Pediatr Res* 1992;**32**:160–4.

35. Nadel JA, Barnes PJ. Autonomic regulation of the airways. *Annu Rev Med* 1984;**35**:451–67.

36. Roffel AF, Elzinga CR, Zaagsma J. Muscarinic M3 receptors mediate contraction of human central and peripheral airway smooth muscle. *Pulm Pharmacol* 1990;**3**:47–51.

37. Fryer AD, Maclagan J. Muscarinic inhibitory receptors in pulmonary parasympathetic nerves in the guinea-pig. *Br J Pharmacol* 1984;**83**:973–8.

38. Fryer AD, Costello RW, Yost BL, Lobb RR, Tedder TF, Steeber DA, et al. Antibody to VLA-4, but not to L-selectin, protects neuronal M2 muscarinic receptors in antigen-challenged guinea pig airways. *J Clin Invest* 1997;**99**:2036–44.

39. Fryer AD, Stein LH, Nie Z, Curtis DE, Evans CM, Hodgson ST, et al. Neuronal eotaxin and the effects of CCR3 antagonist on airway hyperreactivity and M2 receptor dysfunction. *J Clin Invest* 2006;**116**:228–36.

40. Sawatzky DA, Kingham PJ, Court E, Kumaravel B, Fryer AD, Jacoby DB, et al. Eosinophil adhesion to cholinergic nerves via ICAM-1 and VCAM-1 and associated eosinophil degranulation. *Am J Physiol Lung Cell Mol Physiol* 2002;**282**:L1279–88.

41. Jacoby DB, Gleich GJ, Fryer AD. Human eosinophil major basic protein is an endogenous allosteric antagonist at the inhibitory muscarinic M2 receptor. *J Clin Invest* 1993;**91**:1314–8.

42. Evans CM, Fryer AD, Jacoby DB, Gleich GJ, Costello RW. Pretreatment with antibody to eosinophil major basic protein prevents hyperresponsiveness by protecting neuronal M2 muscarinic receptors in antigen-challenged guinea pigs. *J Clin Invest* 1997;**100**:2254–62.

43. Fryer AD, Yarkony KA, Jacoby DB. The effect of leukocyte depletion on pulmonary M2 muscarinic receptor function in parainfluenza virus-infected guinea-pigs. *Br J Pharmacol* 1994;**112**:588–94.

44. Nie Z, Jacoby DB, Fryer AD. Etanercept prevents airway hyperresponsiveness by protecting neuronal M2 muscarinic receptors in antigen-challenged guinea pigs. *Br J Pharmacol* 2009;**156**:201–10.

45. Adamko DJ, Yost BL, Gleich GJ, Fryer AD, Jacoby DB. Ovalbumin sensitization changes the inflammatory response to subsequent parainfluenza infection. Eosinophils mediate airway hyperresponsiveness, m(2) muscarinic receptor dysfunction, and antiviral effects. *J Exp Med* 1999;**190**:1465–78.

46. Adamko DJ, Fryer AD, Bochner BS, Jacoby DB. CD8+ T lymphocytes in viral hyperreactivity and M2 muscarinic receptor dysfunction. *Am J Respir Crit Care Med* 2003;**167**:550–6.

47. Schwarze J, Cieslewicz G, Joetham A, Ikemura T, Hamelmann E, Gelfand EW. CD8 T cells are essential in the development of respiratory syncytial virus-induced lung eosinophilia and airway hyperresponsiveness. *J Immunol* 1999;**162**:4207–11.

48. Davoine F, Cao M, Wu Y, Ajamian F, Ilarraza R, Kokaji AI, et al. Virus-induced eosinophil mediator release requires antigen-presenting and CD4+ T cells. *J Allergy Clin Immunol* 2008;**122**:69–77. 77 e1–2.

49. Klebanoff SJ, Coombs RW. Virucidal effect of stimulated eosinophils on human immunodeficiency virus type 1. *AIDS Res Hum Retroviruses* 1996;**12**:25–9.

50. Domachowske JB, Dyer KD, Bonville CA, Rosenberg HF. Recombinant human eosinophil-derived neurotoxin/RNase 2 functions as an effective antiviral agent against respiratory syncytial virus. *J Infect Dis* 1998;**177**:1458–64.

51. Lee-Huang S, Huang PL, Sun Y, Kung HF, Blithe DL, Chen HC. Lysozyme and RNases as anti-HIV components in beta-core preparations of human chorionic gonadotropin. *Proc Natl Acad Sci U S A* 1999;**96**:2678–81.

52. Rugeles MT, Trubey CM, Bedoya VI, Pinto LA, Oppenheim JJ, Rybak SM, et al. Ribonuclease is partly responsible for the HIV-1 inhibitory effect activated by HLA alloantigen recognition. *AIDS* 2003;**17**:481–6.

53. Dyer KD, Percopo CM, Fischer ER, Gabryszewski SJ, Rosenberg HF. Pneumoviruses infect eosinophils and elicit MyD88-dependent release of chemoattractant cytokines and interleukin-6. *Blood* 2009;**114**:2649–56.

54. Bivins-Smith E. Eosinophils mediate antiviral effects in vivo and directly kill parainfluenza virus in vitro. *6th Biennial Symposium International Eosinophil Society*; 2009.

55. Jacoby DB, Xiao HQ, Lee NH, Chan-Li Y, Fryer AD. Virus- and interferon-induced loss of inhibitory M2 muscarinic receptor function and gene expression in cultured airway parasympathetic neurons. *J Clin Invest* 1998;**102**:242–8.

56. Evans CM, Jacoby DB, Fryer AD. Effects of dexamethasone on antigen-induced airway eosinophilia and M(2) receptor dysfunction. *Am J Respir Crit Care Med* 2001;**163**:1484–92.

57. Nie Z, Nelson CS, Jacoby DB, Fryer AD. Expression and regulation of intercellular adhesion molecule-1 on airway parasympathetic nerves. *J Allergy Clin Immunol* 2007;**119**:1415–22.

58. Leckie MJ, ten Brinke A, Khan J, Diamant Z, O'Connor BJ, Walls CM, et al. Effects of an interleukin-5 blocking monoclonal antibody on eosinophils, airway hyper-responsiveness, and the late asthmatic response. *Lancet* 2000;**356**:2144–8.

59. Nair P, Pizzichini MM, Kjarsgaard M, Inman MD, Efthimiadis A, Pizzichini E, et al. Mepolizumab for prednisone-dependent asthma with sputum eosinophilia. *N Engl J Med* 2009;**360**:985–93.

60. Ikemura T, Schwarze J, Makela M, Kanehiro A, Joetham A, Ohmori K, Gelfand EW. Type 4 phosphodiesterase inhibitors attenuate respiratory syncytial virus-induced airway hyper-responsiveness and lung eosinophilia. *J Pharmacol Exp Ther* 2000;**294**:701–6.

Eosinophils and Gastrointestinal Disease

Benjamin P. Davis and Marc E. Rothenberg

INTRODUCTION

Primary eosinophilic gastrointestinal disorders (EGID) are diseases with eosinophilia in the absence of known causes (e.g., drug reactions, malignancy, and parasitic infections). These disorders include eosinophilic colitis, eosinophilic enteritis, eosinophilic esophagitis (EoE), eosinophilic gastritis, and eosinophilic gastroenteritis, a term used when more than one gastrointestinal (GI) segment is involved.[1,2] The symptoms of EGID include abdominal pain, diarrhea, dysphagia and food impaction, failure to thrive, gastric dysmotility, irritability, and vomiting. Both genetic and environmental factors have a role in EGID. Approximately 10% of patients with EGID have a first-degree relative that also has EGID. It appears that allergy may play a role. In fact, approximately 75% of EGID patients have atopy, allergen-free diets have been shown to be therapeutic, and tissue specimens show mast cell degranulation. Animal studies of EGID also provide evidence of allergic etiology.[2] Although food-specific immunoglobulin E (IgE) is common in EGID patients, food-induced anaphylactic responses occur in only a minority of patients. Thus, EGID appears to fall between pure IgE-mediated food allergy and cellular-mediated hypersensitivity disorders (Fig. 13.8.1). In fact, a recent study shows a higher incidence of EoE in patients with celiac disease, supporting a cell-mediated hypersensitivity or T-helper type 1 (T_h1)-mediated mechanism,[3] although another recent study demonstrates a lack of celiac disease-associated alleles in EoE.[4]

The chief differential diagnoses include gastroesophageal reflux disease (GERD), inflammatory bowel disease (IBD), and specific infections including parasites and *Helicobacter pylori*.[5,6] The prevalence of EGID has not been rigorously calculated, but they appear to be widespread and not uncommon.[1] For example, EoE has been reported in Australia,[7] Brazil,[8] England, Italy, Japan,[9] Spain, and Switzerland.[10] In one study approximately 10% of pediatric patients with GERD-like symptoms who were unresponsive to acid blockade had EoE.[11] Another study reported that 6% of children with chronic esophagitis have EoE.[1] Prevalence estimates vary from 1:70,000 adults in Australia to 1:2000 children in Cincinnati, USA. The increased prevalence of the disease is primarily due to increased recognition, as the disease accounted for approximately 30% of refractory chronic esophagitis in the 1980s and 1990s; however, a bona fide increase in disease incidence is also occurring.[12] It appears that EGID may be even more prevalent than pediatric IBD.

The blood eosinophil level in EoE patients is typically not dramatically elevated, although it averages twice the normal value.[13] The relative normal value of blood eosinophilia compared with esophageal eosinophilia highlights the tissue-specific pathogenesis of the disease. However, some EGID patients have peripheral eosinophilia high enough to meet criteria for hypereosinophilic syndrome (HES), defined by sustained peripheral blood eosinophilia (>1500 cells/mm^3) and end-organ involvement in the absence of known causes of eosinophilia.[14,15] While HES often involves the GI tract, the other organs typically affected in HES (heart and skin) are rarely involved in EGID.

EOSINOPHILS AND THE GASTROINTESTINAL TRACT

In most tissues, eosinophils are present in minute amounts. Organs with substantial eosinophils include GI tract, lymph nodes, spleen, and thymus. Interestingly, in a large study of healthy patients, on autopsy eosinophil degranulation was only seen in the GI tract. Eosinophils are normally present in the lamina propria of the colon, small intestine, and stomach, but are not normally present in the epithelium or Peyer patches, although they infiltrate these regions in EGID.[2] Eosinophil homing to the GI tract appears to occur independent of gut flora, as evidenced by prenatal, adult, and germ-free mice having eosinophils in similar locations and concentrations. Additionally, mice deficient in innate signaling responses (i.e., Myd88 deficient) have normal

FIGURE 13.8.1 The spectrum of eosinophilic gastrointestinal disorders. Eosinophilic gastrointestinal disorders (EGID) fall in the middle of the spectrum between immunoglobulin E (IgE)-mediated and non-IgE, cellular-mediated hypersensitivity disorders. EGID, eosinophilic gastrointestinal disorders; IBD, inflammatory bowel disease.

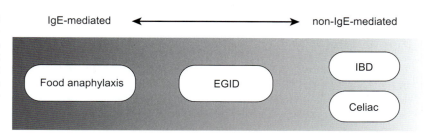

numbers of GI eosinophils. Together these data indicate that eosinophils respond to signals distinct from those of most other intestinal inflammatory cells that typically require gut flora-induced signaling. In fact, eotaxin/C-C motif chemokine 11 (CCL11) is the unique signal that induces localization of eosinophils to the GI tract.

In vitro, the granule components of eosinophils are toxic to many tissues, including intestinal epithelium. The eosinophil cationic proteins major basic protein (MBP), eosinophil peroxidase (EPO), and eosinophil cationic protein (ECP) are cytotoxic to epithelium at concentrations similar to those found in biological fluids from patients with eosinophilia. ECP can render cell membranes porous. MBP increases smooth muscle reactivity via vagal muscarinic M2 receptors and can trigger mast cell and basophil degranulation. In patients with eosinophilic gastroenteritis, MBP and ECP are deposited extracellularly in the small bowel, and ultrastructural changes in eosinophil secondary granules (indicating degranulation and mediator release) are found in duodenal biopsies. Furthermore, Charcot—Leyden crystals are commonly found in feces, and disease severity correlates with mucosal eosinophil numbers.

Eosinophils can secrete a number of different cytokines, suggesting that they may modulate multiple aspects of the immune response, as well as epithelial growth, fibrosis, and tissue remodeling. In addition, tissue eosinophils have distinct cytokine expression patterns under inflammatory versus noninflammatory conditions. For example, esophageal eosinophils from EoE patients express high levels of T_h2 cytokines and transforming growth factor β (TGF-β).[16] Other molecules secreted by eosinophils include halide acids, hydrogen peroxide, and leukotrienes, which increase vascular permeability and mucus secretion and stimulate smooth muscle contraction.

In IBD, eosinophils form only a small percentage of the infiltrating leukocytes, but their level has been proposed to be a negative prognostic indicator. Observations of eosinophilia in IBD suggest that eosinophils mediate axonal necrosis. Recently, it has been shown that eotaxin is upregulated in intestinal macrophages and epithelial cells in pediatric ulcerative colitis and thus may be a future target for therapy.[17]

EVALUATION OF EOSINOPHILIC GASTROINTESTINAL DISORDERS

Common symptoms of EGID patients include abdominal pain, diarrhea, dysphagia, failure to thrive, gastric dysmotility, hypoproteinemia, irritability, microcytic anemia, and vomiting. Patients with these refractory problems, especially individuals with a strong history of allergic diseases, peripheral blood eosinophilia, and/or a family history of EGID, should be evaluated for EGID. Evaluation for EGID starts with a comprehensive history and physical examination. Symptoms vary depending on the intestinal segment involved (e.g., abdominal pain and dysphagia are most common in eosinophilic gastroenteritis and EoE, respectively). In most EGID patients, peripheral eosinophilia is not present. Total IgE levels may help classify patients with atopy or those with occult parasite infection. In atopic EGID patients, food allergy is common and can be verified by skin prick and skin patch testing.[18] There are no pathognomonic symptoms or blood tests for diagnosing EGID. Therefore, the diagnosis of EGID is based on endoscopic biopsy procurement and appropriate clinical information.

The diagnosis of EGID is dependent on histological evaluation of biopsy samples, contingent upon the quantity, location, and characteristics of the eosinophilic inflammation. EGID patients often have a variety of endoscopic findings,[10] but it is not uncommon for the lumen to look normal endoscopically; thus, a microscopic evaluation of biopsy samples is required. EGID often has patchy involvement, requiring analysis of biopsies from multiple intestinal segments. While the normal esophagus is devoid of eosinophils, the rest of the GI tract contains readily detectable eosinophils. With the lack of diagnostic criteria, diagnosis of EGID depends on clinical and histopathological expertise. Diagnosis of EGID depends on several factors including the following: (1) eosinophil quantification, (2) the histological location of eosinophils, (3) associated histopathological abnormalities, and (4) the absence of pathological features suggestive of other primary disorders. Recently, gene expression profiles (transcriptomes) have been proposed to be helpful for the diagnosis of EoE; elevated levels of esophageal eotaxin-3/CCL26 in a single biopsy specimen has approximately 90% sensitivity.[19] Other disease processes, such as drug hypersensitivity, collagen vascular disease, malignancy, or infection, should be ruled out (Table 13.8.1). Evaluation for intestinal parasites via stool samples, colonoscopy aspirates, or antibody titers should be performed, especially if patients have high-risk exposure (e.g., drinking well water, living on a farm, or travel to endemic area). In one study of eosinophilic enteritis, the dog hookworm *Ancylostoma caninum* was identified as the cause in 15% of cases.[20] Infection with *Strongyloides stercoralis* should be excluded before treating for EGID, as immunosuppressants can be life threatening with this infection.[21]

EOSINOPHILIC ESOPHAGITIS

The esophagus is normally devoid of eosinophils.[1,2] Disorders associated with esophageal eosinophil infiltration include allergic vasculitis, bullous pemphigoid,

TABLE 13.8.1 Classification and Examples of Eosinophilic Gastrointestinal Disorders

Primary	Secondary
Eosinophilic esophagitis	Parasitic infection
Eosinophilic gastritis	Inflammatory bowel disease
Eosinophilic enteritis	Gastroesophageal reflux disease
Eosinophilic colitis	Churg-Strauss syndrome Scleroderma
Eosinophilic gastroenteritis	Drug induced Hypereosinophilic syndrome Celiac disease Leiomyomatosis Pemphigus vegetans

carcinomatosis, drug injury, EoE, eosinophilic gastroenteritis, esophageal leiomyomatosis, GERD, HES, IBD, myeloproliferative disorders, parasitic and fungal infections, periarteritis, pemphigus vegetans, recurrent vomiting, and scleroderma.[6,22] Eosinophil-associated esophageal disorders are classified as primary or secondary (i.e., due to another known disease process) (Table 13.8.1). The primary subtype includes the atopic, nonatopic, and familial variants, and the secondary subtype is subdivided into those with and without systemic eosinophilia. EoE is familial in about 10% of patients.

Etiology

The etiology of EoE is poorly understood, but food allergy has been implicated. Most EoE patients have specific IgE to foods and aeroallergens, but only a few have experienced anaphylaxis.[1] Esophageal eosinophilia may also be linked to allergic airway disease. Several independent studies have linked EoE to allergic etiology. For example, repeated delivery of allergens or interleukin-13 (IL-13) to murine lungs (via direct delivery or transgenic overexpression) induces experimental EoE; patients with allergic rhinitis have increased esophageal eosinophils;[23] intranasal delivery of indoor insect allergen induces EoE in mice;[24] patients with EoE commonly report seasonal variations in symptoms; and there is a seasonal variation in EoE diagnosis.[25,26]

In addition to eosinophils, T cell and mast cell numbers are increased in esophageal biopsies, suggesting a chronic T_h2-associated inflammation (Fig. 13.8.2).[27] Mast cells have been shown to be elevated in the esophagus and to degranulate in EoE, and they appear to be stimulated by the kit ligand. Carboxypeptidase A3 and tryptase appear to be good surrogate markers for mast cell involvement.[28] IL-13 is overproduced in the esophagus of EoE patients. In addition, in experimental systems, such as IL-13 lung transgenic systems, IL-13 induces eosinophilic esophagitis and tissue remodeling with features both dependent and independent of eosinophils.[29] Consistent with T_h2 activation, IL-5 overexpression induces EoE, and IL-5 neutralization completely blocks allergen- or IL-13-induced EoE in mice.[30] However, anti-IL-5 therapy in humans has not yet been shown to be effective at ameliorating clinical aspects of the disease, although esophageal eosinophilia improves.[31,32] Interestingly, a cytokine with more of a T_h1 skew, IL-15, appears to mediate $CD4^+$ T cells and to be involved in the pathogenesis of EoE.[33]

In a recent genome-wide microarray expression profile analysis of esophageal tissue[19] from patients with EoE and normal controls, an EoE transcriptome was found to contain changed expression of approximately 1% of the entire human genome. Interestingly, eotaxin-3 was the most prominently overexpressed gene in EoE patients, and levels correlated with disease severity. Furthermore, a single nucleotide polymorphism (SNP) in eotaxin-3 was overrepresented in EoE patients. Conversely, mice lacking the eotaxin receptor (C-C chemokine receptor type 3; CCR3) were protected from developing experimental EoE. Notably, eotaxin-3 is induced by IL-13. Two recent studies have implicated genetic susceptibility for EoE with genetic variants of the *TSLP* gene; *TSLP* encodes thymic stromal lymphopoietin, an epithelial gene product that targets dendritic cells and promotes their T_h2-polarizing ability. These studies included a genome-wide analysis study of common variants, as well as a large-scale candidate SNP approach. Notably, sex-associated TSLP receptor polymorphisms on Xp22.3/Yp11.3 may explain male predisposition to EoE.[34]

Clinical and Diagnostic Studies

Symptoms of EoE include dysphagia, epigastric or chest pain, GI obstructive problems, and vomiting.[35] Patients are predominantly male[35] and have robust esophageal eosinophilia,[2] extensive epithelial hyperplasia, and are commonly atopic compared to GERD patients. In adults with EoE, dysphagia and food impaction are common complaints.[36] One study presented a symptom scoring tool that can help identify patients with EoE and correlates with tissue inflammation.[37] However, another study demonstrated dissociation between symptom scores and histology in treated EoE patients.[38] At present, it is important to develop validated disease instruments so that these important parameters can be tracked, especially in clinical trials.

EoE is associated with esophageal dysmotility/dysphagia, which may be related to motor dysfunction of

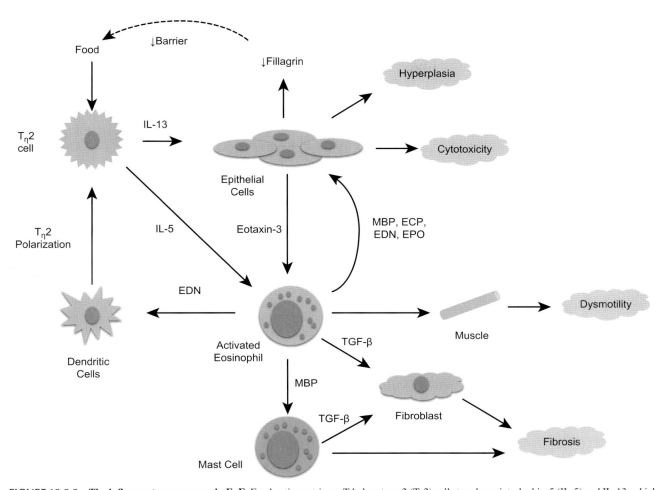

FIGURE 13.8.2   The inflammatory response in EoE. Food antigens trigger T-helper type 2 (T_h2) cells to release interleukin-5 (IL-5) and IL-13, which stimulate eosinophils and esophageal epithelial cells, respectively. IL-13 induces epithelial cells to produce eotaxin-3/C-C motif chemokine 26 (CCL26) and downregulate filaggrin. Reduced production of filaggrin might inhibit esophageal barrier function, which could perpetuate the inflammation by maintaining local food antigen uptake. IL-5 and eotaxin-3 activate eosinophils to release major basic protein (MBP) and eosinophil-derived neurotoxin (EDN), which activate mast cells and dendritic cells, respectively. Eosinophils and mast cells also produce TGF-β, which activates fibroblasts and muscle cells and contributes to dysmotility, fibrosis, and hyperplasia.[24,62] ECP, eosinophil cationic protein; EPO, eosinophil peroxidase; TGF-β, transforming growth factor β.

the esophagus rather than to physical narrowing.[39] Esophageal ultrasound shows dysfunctional muscularis mucosa in EoE.[40] Radiographic and endoscopic studies demonstrate furrowing, mucosal rings, polyps, strictures, ulcerations, and whitish papules.[41] It appears that a small diameter on barium esophagography can help diagnosis of EoE,[42] although many patients have normal barium assessments and need further evaluation by endoscopy.[43] Recently, it has been demonstrated that esophageal distensibility, defined as cross-sectional area following intraluminal pressure, is significantly reduced in EoE patients.[44]

The number and location of eosinophils helps distinguish EoE from GERD. Greater than 15 eosinophils per high-power field (hpf) and a lack of response to proton-pump inhibitors suggest EoE. Proximal and distal esophageal eosinophilia suggests EoE, whereas eosinophilia confined to the distal esophagus suggests GERD.[1] Histopathological changes in EoE include esophageal mucosa thickening with basal layer hyperplasia and papillary lengthening. Tissue eosinophil counts may underestimate eosinophil involvement, particularly with marked degranulation. Eosinophil-derived neurotoxin (EDN) staining of biopsy specimens may be useful for diagnosis and management.[45]

Clinical assessment of EoE includes analysis of food and aeroallergen sensitization and exclusion of GERD as well as other causes of eosinophils in the esophagus. However, EoE and GERD are not mutually exclusive and may coexist in the same patient (Table 13.8.2). Skin patch testing may facilitate identification of food allergy and lead to improved dietary therapy.[46]

TABLE 13.8.2 Comparison of EoE and GERD

Associated Features	EoE	GERD
Clinical		
Atopy	Yes	No
Food sensitivity	Yes	No
Gender preference	Male	No
Food impaction	Common	Uncommon
Procedural findings		
pH probe	Neutral	Acidic
Endoscopic furrowing	Yes	No
Endoscopic rings	Sometimes	No
Decreased luminal distention	Yes	Unknown
Radiographic small caliber	Sometimes	No
Histopathology		
Proximal disease	Yes	No
Distal disease	Yes	Yes
Epithelial hyperplasia	Severe	Moderate
Eosinophils/hpf	>15	<15 (typically)
Elevated eotaxin-3/ CCL26, carboxypeptidase A3	Yes	No
Treatment effectiveness		
H2-blockers	No	Yes
Proton-pump inhibitors	No	Yes
Glucocorticoids	Yes	No
Food elimination	Yes	No
Elemental diet	Yes	No

CCL26, C-C motif chemokine 26; hpf, high-power field.

Treatment

Specific food antigen and aeroallergen avoidance, identified by skin testing or history, is indicated for patients with atopic EoE. If feasible, an elemental diet is recommended, as it improves symptoms and reduces the number of eosinophils in the esophageal biopsies of patients with EoE (allergic or nonallergic subtypes). Elemental diet therapy frequently requires placement of a gastrostomy tube to achieve adequate caloric intake.

Glucocorticoids are also effective. Systemic steroids are used for acute exacerbations, and topical steroids provide day-to-day control.[47] For topical steroid delivery, the patient is instructed to swallow the dose to promote deposition on the esophageal mucosa. Topical fluticasone lowers the level of eosinophils, CD3[+] cells, and CD8[+] cells in the proximal and distal esophagus[48] and improves symptoms.[36] Side effects of inhaled glucocorticoids are less likely with swallowed

fluticasone, as this drug undergoes extensive first-pass metabolism in the liver. However, local esophageal candidiasis may occur.[48] In addition to inhaled steroids, an oral suspension of budesonide can be used.[49] Viscous budesonide improves symptoms and endoscopic/histological appearance and also appears to reach the distal esophagus.[50] A recent study shows FK506 binding protein 5 (FKBP5) transcript levels increase with glucocorticoid exposure, thus helping to distinguish responders from nonresponders and treated from untreated. This also provides the best evidence that swallowed steroids induce local effects in the esophagus.[51]

Although food antigen avoidance and glucocorticoids are the mainstay of treatment, additional therapies may be beneficial. In EoE patients for whom impaction has become an issue, esophageal dilation appears to be safe and effective.[52,53] Even if GERD is not present, neutralizing gastric acidity (with proton-pump inhibitors) may improve symptoms and esophageal pathology. On the horizon, an anti-IL-13 antibody is a promising future therapy. In an animal model of IL-13-induced esophageal eosinophilia, an anti-IL-13 antibody was effective, and clinical studies are now under way.[54] Additionally, anti-IL-5 antibodies prevent experimental EoE in mice[55,56] and appear to decrease eosinophilia of the human esophagus in early-stage clinical trials.[57] Human clinical trials are currently under way.

Prognosis

Although the natural history of EoE is not fully known, it is not uncommon for children with EoE to have a parent with a history of chronic esophageal strictures. Esophageal biopsies in some of these parents have revealed EoE.

The following symptoms occur in order of increasing age: feeding problems, vomiting, abdominal pain, dysphagia, and food impaction.[13] Thus, if left untreated, EoE is likely to progress to esophageal scarring and dysfunction. Recent data show that pediatric EoE patients, diagnosed by retrospective biopsy analysis, are at increased risk of developing persistent disease characterized by dysphagia, food impaction, a need for esophageal dilation, and food allergy.[12,58] Despite this, the development of Barrett esophagus in EoE has not been studied. However, it appears that EoE requires chronic treatment.

Having EoE increases the risk of developing other forms of EGID. Thus, routine surveillance guided by symptoms of the entire GI tract by endoscopy is recommended.

EOSINOPHILIC GASTRITIS AND GASTROENTERITIS

At baseline, the stomach and intestine have readily detectable eosinophils. Therefore, diagnosis of enteritis, eosinophilic gastritis, and gastroenteritis is more complex than diagnosis of EoE. These diseases are characterized by selective infiltration of eosinophils in the stomach and/or small intestine with variable involvement of the esophagus and/or large intestine. Eosinophilic enteritis, gastritis, and gastroenteritis can be divided into primary or secondary. The primary disorders have also been called idiopathic or allergic gastroenteropathy. Primary eosinophilic gastroenteritis is subdivided on the basis of the level of histological involvement into mucosal, muscularis, and serosal forms. Endoscopic biopsy can be normal in the latter two subtypes. Importantly, many other disorders feature eosinophil infiltration in the stomach, including allergic vasculitis, drug hypersensitivity, drug injury, HES, IBD, myeloproliferative disorders, parasitic and bacterial infections (e.g., *H. pylori*), periarteritis, and scleroderma.

Etiology

Because total IgE is elevated and food-specific IgEs are detectable in most EGID patients, an allergic mechanism is suspected. However, even though most patients have positive skin tests to a variety of food antigens, they do not have typical anaphylactic reactions, suggesting a delayed form of food hypersensitivity syndrome. In support of an allergic mechanism, mast cells are increased in EGID.[59] Also, eosinophilic gastroenteritis can be induced by feeding enteric-coated allergen beads to sensitized mice.[60] These mice develop eosinophil-associated GI dysfunction, including delayed food transit, gastromegaly, and weight loss.[61] In addition, duodenal lamina propria T cells in EGID preferentially secrete T_h2 cytokines (especially IL-13) when stimulated with milk proteins. Furthermore, elevated secretion of IL-4 and IL-5 by peripheral blood T cells has been observed in eosinophilic gastroenteritis. IgA deficiency can also be associated with eosinophilic gastroenteritis and may be related to the increased rate of atopy or to occult GI infection in these patients.

Clinical and Diagnostic Studies

In general, these disorders present with symptoms related to the degree and area of the GI tract affected. The mucosal form of eosinophilic gastroenteritis (the most common variant) is characterized by abdominal pain, bloody stools, diarrhea, failure to thrive, iron-deficiency anemia, malabsorption, protein-losing enteropathy, and vomiting. In the muscularis form, thickening of the bowel wall may result in GI obstructive symptoms. The serosal form is characterized by exudative ascites.

There are no standards for the diagnosis of eosinophilic gastritis or gastroenteritis, but the presence of increased eosinophils in biopsy specimens, infiltration of eosinophils within intestinal crypts and gastric glands, lack of involvement of other organs, and exclusion of secondary causes of eosinophilia support a diagnosis of eosinophilic gastroenteritis. Patients with eosinophilic gastritis may have micronodules (and/or polyposis), and these lesions often contain aggregates of lymphocytes and eosinophils. Food allergy and peripheral eosinophilia may be present but are not required for diagnosis.

Treatment

Eliminating foods implicated by skin prick or radioallergosorbent (RAST) testing has variable results, whereas complete resolution is generally achieved with amino acid-based elemental diets. Once remission has been achieved by dietary modification, specific food groups are reintroduced (at approximately 3-week intervals for each food group), and endoscopy is reperformed to identify sustained remission or disease flares.

Systemic or topical steroids are the main therapy when diet restriction has failed or is not feasible. For systemic steroid therapy, a course of 2–6 weeks of therapy with relatively low doses seems to work better than a 7-day course of glucocorticoid bursts. Various topical glucocorticoid preparations are designed to deliver drugs to specific segments of the GI tract [e.g., budesonide tablets (Entocort EC) targeted to the ileum and proximal colon]. As with asthma, topical steroids have a better risk–benefit ratio than systemic steroids. In cases refractory to or dependent on glucocorticoid therapy, parenteral nutrition or antimetabolite therapy (azathioprine or 6-mercaptopurine) may help.

Drugs such as cromoglycate, ketotifen, mycophenolate mofetil (an inosine monophosphate dehydrogenase inhibitor), suplatast tosilate, and various alternative therapies are not generally very useful, although successful long-term remission of eosinophilic gastroenteritis has been reported following montelukast treatment. Use of proton-pump inhibitors can improve symptoms and the degree of esophageal and gastric pathology, even if GERD is not present.

Prognosis

The natural history of eosinophilic gastritis, enteritis, and gastroenteritis is not well documented; however, they are often chronic, relapsing/remitting diseases. When the disease presents in infancy and specific food sensitization can be identified, remission is likely by late childhood. In food antigen-induced disease, abnormal levels of circulating IgE and eosinophils often serve as markers for tissue involvement/relapse. As noted earlier, due to concern for HES, routine surveillance of the cardiopulmonary system is recommended.

EOSINOPHILIC COLITIS

Colonic eosinophilia occurs in a variety of disorders, including allergic colitis of infancy, drug reactions, eosinophilic gastroenteritis, infections (e.g., helminths), IBD, and vasculitis (e.g., Churg–Strauss syndrome). Allergic colitis in infancy (or dietary protein-induced proctocolitis of infancy syndrome) is the most common cause of bloody stools in the first year of life. Similar to other EGID, these disorders are classified into primary and secondary.

Etiology

In contrast to other EGID, eosinophilic colitis is usually not IgE associated. Some studies point to a T cell-mediated process, though the exact mechanism is unclear. Allergic colitis of infancy may be an early expression of protein-induced enteropathy. Cow's milk and soy proteins are the foods most frequently associated, but other food proteins can also induce this disorder. Interestingly, this condition may more commonly occur in infants who are exclusively breastfed and can occur in infants fed with protein hydrolysate formulas. An association between allergic colitis and the later development of IBD has been reported but remains controversial.

Clinical and Diagnostic Studies

Similar to eosinophilic gastroenteritis, the symptoms of eosinophilic colitis vary depending on the degree and location of tissue involvement. Although diarrhea is a classic symptom, symptoms that can occur independent of diarrhea commonly include abdominal pain, anorexia, and weight loss. In infants, bloody diarrhea precedes diagnosis by several weeks, and anemia due to blood loss is not uncommon. Most infants affected do not have constitutional symptoms and are otherwise healthy. There is a bimodal age distribution, with the infantile form presenting with a mean age at diagnosis of approximately 60 days, and the second group presenting during adolescence and early adulthood.

There is no single gold standard diagnostic test, but peripheral blood eosinophilia or eosinophils in the stool are suggestive of eosinophilic colitis. On endoscopic examination, patchy erythema, loss of vascularity and lymphonodular hyperplasia are seen; findings are mostly localized to the rectum but can affect the entire colon. Histological examination often reveals preservation of mucosal architecture, with focal aggregates of eosinophils in the crypt epithelium, lamina propria, and muscularis mucosa and occasionally, multinucleated giant cells in the submucosa.

Treatment

Treatment of eosinophilic colitis varies according to the disease subtype. Clinical symptoms of eosinophilic colitis of infancy resolve within 72 hours of withdrawal of the offending protein. Treatment of eosinophilic colitis in older patients, for which IgE-associated triggers are rarely identified, usually requires medical management. Anti-inflammatory drugs, including aminosalicylates and systemic or topical steroids, are commonly used and appear to be efficacious, but clinical trials have not been conducted. Several forms of rectally delivered glucocorticoids treat the distal colon, though eosinophilic colitis typically also involves the proximal colon. In cases refractory or dependent on systemic glucocorticoid therapy, parenteral nutrition or antimetabolite therapy (azathioprine or 6-mercaptopurine) are alternatives.

Prognosis

Eosinophilic colitis presenting in the first year of life has a very good prognosis with the vast majority of patients able to tolerate the culprit food(s) by 1–3 years of age. The prognosis for eosinophilic colitis developing in adolescence or adulthood is less favorable. As with eosinophilic gastroenteritis, the natural history has not been studied, and this disease is considered to be a chronic relapsing/remitting disorder. Because eosinophilic colitis can often be a manifestation of other disease processes, ruling out autoimmune disease and other secondary causes of eosinophilia is recommended.

EVALUATION OF HYPEREOSINOPHILIC SYNDROME IN EOSINOPHILIC GASTROINTESTINAL DISORDERS PATIENTS

The term HES describes patients with systemic symptoms due to marked eosinophilia. Diagnostic criteria for HES include persistent eosinophilia of at least 1500 cells/mm^2 for a sustained duration; lack of known causes of eosinophilia; and symptoms and signs of organ system involvement. Patients with EGID and blood eosinophil counts greater than 1500/mm^2 meet these diagnostic criteria but generally do not have the high risk of life-threatening complications associated with classic HES (i.e., cardiomyopathy or central nervous system involvement).

As is true in any patient with prolonged and marked eosinophilia, HES patients are prone to develop eosinophilic endomyocardial disease with embolization. Thus, routine echocardiograms are warranted in patients with EGID and peripheral blood eosinophilia. Additionally, the diagnosis of HES should be considered in patients with

EGID who develop extra-GI manifestations. Additional testing may include bone marrow analysis (searching for myelodysplasia), serum tryptase and vitamin B12 levels (both moderately elevated in classic HES), and genetic analysis for the FIP1-like 1/platelet-derived growth factor receptor-α (*FIP1L1−PDGFRA*) fusion event.

A major advance in our understanding of HES has come about through treatment of HES with imatinib mesylate, a tyrosine kinase inhibitor. In many HES patients, treatment with imatinib mesylate dramatically reduces peripheral blood and bone marrow eosinophils, suggesting that certain HES patients express a kinase that is sensitive to imatinib mesylate. A fusion gene, the result of an 800-kb deletion in chromosome 4, yields a novel activated kinase (FIP1L1−PDGFR). FIP1L1−PDGFR is inhibited by imatinib *in vitro*, explaining the sensitivity of HES patients to imatinib. Those responsive to imatinib are typically 20−50-year-old males who present with marked peripheral eosinophilia (i.e., classic HES). These patients have been shown to meet minor criteria for systemic mastocytosis, having elevated levels of serum mast cell tryptase and high numbers of dysplastic mast cells in the bone marrow.

CONCLUSION

EGID are now being recognized more frequently. They have strong genetic and allergic components and share clinical and immunopathogenic features with asthma. EGID are associated with a variety of nonspecific common GI symptoms and laboratory findings, making their diagnosis dependent on microscopic examination biopsies and ruling out of other eosinophil-associated disease.

Eosinophils have potent proinflammatory effects mediated by their cytotoxic granule constituents and various lipid mediators and cytokines. In T_h2-associated GI inflammatory conditions, eosinophilia in the lamina propria is dependent on IL-5 and eotaxin. Esophageal eosinophilia can be induced experimentally by pulmonary deposition of aeroallergens or the T_h2 cytokine, IL-13.

Many new therapeutic approaches are now being developed for EGID, including eotaxin-3 receptor/CCR3 antagonists, eotaxin-3 blockers, humanized anti-IL-5, and IL-4/IL-13 inhibitors. However, although much progress has been made concerning GI eosinophils and EGID, there is still a paucity of knowledge compared with other cell types and GI diseases that may be even less common (e.g., IBD). A better understanding of the pathogenesis and treatment of EGID will emerge by combining comprehensive clinical and research approaches involving experts in the fields of allergy, gastroenterology, nutrition, and pathology.

REFERENCES

1. Fox VL, Nurko S, Furuta GT. Eosinophilic esophagitis: it's not just kid's stuff. *Gastrointest Endosc* 2002;**56**(2):260−70.
2. Rothenberg ME, et al. Pathogenesis and clinical features of eosinophilic esophagitis. *J Allergy Clin Immunol* 2001;**108**(6):891−4.
3. Leslie C, Mews C, Charles A, Ravikumara M. Celiac disease and eosinophilic esophagitis: a true association. *J Pediatr Gastroenterol Nutr* 2010;**50**(4):397−9.
4. Lucendo AJ, Arias Á, Pérez-Martínez I, López-Vázquez A, Ontañón-Rodríguez J, González-Castillo S, et al. Adult Patients with Eosinophilic Esophagitis Do Not Show an Increased Frequency of the HLA-DQ2/DQ8 Genotypes Predisposing to Celiac Disease. *Dig Dis Sci* 2011;**56**:1107−11.
5. Rothenberg ME, Mishra Anil, Brandt Eric B, Simon P. Hogan. Gastrointestinal eosinophils. *Immunol Rev* 2001;**179**:139−55.
6. Rothenberg ME, Mishra Anil, Brandt Eric B, Simon P. Hogan. Gastrointestinal eosinophils in health and disease. *Adv Immunol* 2001;**78**:291−328.
7. Croese J, Fairley SK, Masson JW, Chong AK, Whitaker DA, Kanowski PA, et al. Clinical and endoscopic features of eosinophilic esophagitis in adults. *Gastroint Endosc* 2003;**58**(4):516−22.
8. Cury EK, Schraibman V, Faintuch S. Eosinophilic infiltration of the esophagus: gastroesophageal reflux versus eosinophilic esophagitis in children—discussion on daily practice. *J Pediatr Surg* 2004;**39**(2):e4−7.
9. Fujiwara H, Morita A, Kobayashi H, Hamano K, Fujiwara Y, Hirai K, et al. Infiltrating eosinophils and eotaxin: their association with idiopathic eosinophilic esophagitis. *Ann Allergy Asthma Immunol* 2002;**89**(4):429−32.
10. Straumann A, Spichtin HP, Bucher KA, Heer P, Simon HU. Eosinophilic esophagitis: red on microscopy, white on endoscopy. *Digestion* 2004;**70**(2):109−16.
11. Markowitz JE, Liacouras CA. Eosinophilic esophagitis. *Gastroent Clin N Am* 2003;**32**(3):949−66.
12. DeBrosse CW, Collins MH, Buckmeier Butz BK, Allen CL, King EC, Assa'ad AH, et al. Identification, epidemiology, and chronicity of pediatric esophageal eosinophilia, 1982−1999. *J Allergy Clin Immunol* 2010;**126**(1):112−9.
13. Noel RJ, Putnam PE, Rothenberg ME. Eosinophilic esophagitis. *N Engl J Med* 2004;**351**(9):940−1.
14. Assa'ad AH, Spicer RL, Nelson DP, Zimmermann N, Rothenberg ME. Hypereosinophilic syndromes. *Chem Immunol* 2000;**76**:208−29.
15. Roufosse F, Cogan E, Goldman M. The hypereosinophilic syndrome revisited. *Annu Rev Med* 2003;**54**:169−84.
16. Straumann A, Kristl J, Conus S, Vassina E, Spichtin HP, Beglinger C, et al. Cytokine expression in healthy and inflamed mucosa: probing the role of eosinophils in the digestive tract. *Inflamm Bowel Dis* 2005;**11**(8):720−6.
17. Ahrens R, Waddell A, Seidu L, Blanchard C, Carey R, Forbes E, et al. Intestinal macrophage/epithelial cell-derived CCL11/eotaxin-1 mediates eosinophil recruitment and function in pediatric ulcerative colitis. *J Immunol* 2008;**181**(10):7390−9.
18. Spergel JM, Beausoleil JL, Mascarenhas M, Liacouras CA. The use of skin prick tests and patch tests to identify causative foods in eosinophilic esophagitis. *J Allergy Clin Immunol* 2002;**109**(2):363−8.

19. Blanchard C, Wang N, Stringer KF, Mishra A, Fulkerson PC, Abonia JP, et al. Eotaxin-3 and a uniquely conserved gene-expression profile in eosinophilic esophagitis. *J Clin Invest* 2006;**116**(2):536—47.

20. Walker NI, Croese J, Clouston AD, Parry M, Loukas A, Prociv P. Eosinophilic enteritis in northeastern Australia. Pathology, association with Ancylostoma caninum, and implications. *Am J Surg Pathol* 1995;**19**(3):328—37.

21. Liepman M. Disseminated Strongyloides stercoralis. A complication of immunosuppression. *JAMA* 1975;**231**(4):387—8.

22. Ahmad M, Soetikno RM, Ahmed A. The differential diagnosis of eosinophilic esophagitis. *J Clin Gastroenterol* 2000;**30**(3):242—4.

23. Onbasi K, Sin AZ, Doganavsargil B, Onder GF, Bor S, Sebik F. Eosinophil infiltration of the oesophageal mucosa in patients with pollen allergy during the season. *Clin Exp Allergy* 2005;**35**(11):1423—31.

24. Rayapudi M, Mavi P, Zhu X, Pandey AK, Abonia JP, Rothenberg ME, et al. Indoor insect allergens are potent inducers of experimental eosinophilic esophagitis in mice. *J Leukoc Biol* 2010;**88**(2):337—46.

25. Almansa C, Krishna M, Buchner AM, Ghabril MS, Talley N, DeVault KR, et al. Seasonal distribution in newly diagnosed cases of eosinophilic esophagitis in adults. *Am J Gastroenterol* 2009;**104**(4):828—33.

26. Prasad GA, Alexander JA, Schleck CD, Zinsmeister AR, Smyrk TC, Elias RM, et al. Epidemiology of eosinophilic esophagitis over three decades in Olmsted County, Minnesota. *Clin Gastroenterol Hepatol* 2009;**7**(10):1055—61.

27. Straumann A, Bauer M, Fischer B, Blaser K, Simon HU. Idiopathic eosinophilic esophagitis is associated with a TH2-type allergic inflammatory response. *J Allergy Clin Immunol* 2001;**108**(6):954—61.

28. Abonia JP, Blanchard C, Butz BB, Rainey HF, Collins MH, Stringer K, et al. Involvement of mast cells in eosinophilic esophagitis. *J Allergy Clin Immunol* 2010;**126**(1):140—9.

29. Zuo L, Fulkerson PC, Finkelman FD, Mingler M, Fischetti CA, Blanchard C, et al. IL-13 induces esophageal remodeling and gene expression by an eosinophil-independent, IL-13R alpha2-inhibited pathway. *J Immunol* 2010;**185**(1):660—9.

30. Mishra A, Rothenberg ME. Intratracheal IL-13 induces eosinophilic esophagitis by an IL-5, eotaxin-1, and STAT6-dependent mechanism. *Gastroenterology* 2003;**125**(5):1419—27.

31. Straumann A, Conus S, Grzonka P, Kita H, Kephart G, Bussmann C, et al. Anti-interleukin-5 antibody treatment (mepolizumab) in active eosinophilic oesophagitis: a randomised, placebo-controlled, double-blind trial. *Gut* 2010;**59**(1):21—30.

32. Conus S, Straumann A, Bettler E, Simon HU. Mepolizumab does not alter levels of eosinophils, T cells, and mast cells in the duodenal mucosa in eosinophilic esophagitis. *J Allergy Clin Immunol* 2010;**126**(1):175—7.

33. Zhu X, Wang M, Mavi P, Rayapudi M, Pandey AK, Kaul A, et al. Interleukin-15 expression is increased in human eosinophilic esophagitis and mediates pathogenesis in mice. *Gastroenterology* 2010;**139**(1):182—93. e7.

34. Sherrill JD, Gao PS, Stucke EM, Blanchard C, Collins MH, Putnam PE, et al. Variants of thymic stromal lymphopoietin and its receptor associate with eosinophilic esophagitis. *J Allergy Clin Immunol* 2010;**126**(1). p. 160—5.e3.

35. Orenstein SR, Shalaby TM, Di Lorenzo C, Putnam PE, Sigurdsson L, Mousa H, et al. The spectrum of pediatric eosinophilic esophagitis beyond infancy: a clinical series of 30 children. *Am J Gastroenterol* 2000;**95**:1422—30.

36. Konikoff MR, Noel RJ, Blanchard C, Kirby C, Jameson SC, Buckmeier BK, et al. A randomized double-blind-placebo controlled trial of fluticasone proprionate for pediatric eosinophilic esophagitis. *Gastroenterology* 2006;**131**:1381—91.

37. Aceves SS, Newbury RO, Dohil MA, Bastian JF, Dohil R. A symptom scoring tool for identifying pediatric patients with eosinophilic esophagitis and correlating symptoms with inflammation. *Ann Allergy Asthma Immunol* 2009;**103**(5):401—6.

38. Pentiuk S, Putnam PE, Collins MH, Rothenberg ME. Dissociation between symptoms and histological severity in pediatric eosinophilic esophagitis. *J Pediatr Gastroenterol Nutr* 2009;**48**(2):152—60.

39. Nurko S, Rosen R, Furuta GT. Esophageal dysmotility in children with eosinophilic esophagitis: a study using prolonged esophageal manometry. *Am J Gastroenterol* 2009;**104**(12):3050—7.

40. Fox VL, Nurko S, Teitelbaum JE, Badizadegan K, Furuta GT. High-resolution EUS in children with eosinophilic allergic esophagitis. *Gastrointest Endosc* 2003;**57**(1):30—6.

41. Fox VL. Pediatric endoscopy. *Gastrointest Endosc Clin N Am* 2000;**10**(1):175—94. viii.

42. White SB, Levine MS, Rubesin SE, Spencer GS, Katzka DA, Laufer I. The small-caliber esophagus: radiographic sign of idiopathic eosinophilic esophagitis. *Radiology* 2010;**256**(1):127—34.

43. Binkovitz LA, Lorenz EA, Di Lorenzo C, Kahwash S. Pediatric eosinophilic esophagitis: radiologic findings with pathologic correlation. *Pediatr Radiol* 2010;**40**(5):714—9.

44. Kwiatek MA, Hirano I, Kahrilas PJ, Rothe J, Luger D, Pandolfino JE. Mechanical properties of the esophagus in eosinophilic esophagitis. *Gastroenterology* 2011;**140**(1):82—90.

45. Kephart GM, Alexander JA, Arora AS, Romero Y, Smyrk TC, Talley NJ, et al. Marked deposition of eosinophil-derived neurotoxin in adult patients with eosinophilic esophagitis. *Am J Gastroenterol* 2010;**105**(2):298—307.

46. Spergel J, Rothenberg ME, Fogg M. Eliminating eosinophilic esophagitis. *Clin Immunol* 2005;**115**(2):131—2.

47. Liacouras CA, Spergel JM, Ruchelli E, Verma R, Mascarenhas M, Semeao E, et al. Eosinophilic esophagitis: A 10-year experience in 381 children. *Clin Gastroenterol Hepatol* 2005;**3**(12):1198—206.

48. Teitelbaum JE, Fox VL, Twarog FJ, Nurko S, Antonioli D, Gleich G, et al. Eosinophilic esophagitis in children: immunopathological analysis and response to fluticasone propionate. *Gastroenterology* 2002;**122**(5):1216—25.

49. Aceves SS, Dohil R, Newbury RO, Bastian JF. Topical viscous budesonide suspension for treatment of eosinophilic esophagitis. *J Allergy Clin Immunol* 2005;**116**(3):705—6.

50. Dohil R, Newbury R, Fox L, Bastian J, Aceves S. Oral viscous budesonide is effective in children with eosinophilic esophagitis in a randomized, placebo-controlled trial. *Gastroenterology* 2010;**139**(2):418—29.

51. Caldwell JM, Blanchard C, Collins MH, Putnam PE, Kaul A, Aceves SS, et al. Glucocorticoid-regulated genes in eosinophilic esophagitis: a role for FKBP51. *J Allergy Clin Immunol* 2010;**125**(4):879—88. e8.

52. Dellon ES, Gibbs WB, Rubinas TC, Fritchie KJ, Madanick RD, Woosley JT, et al. Esophageal dilation in eosinophilic

esophagitis: safety and predictors of clinical response and complications. *Gastrointest Endosc* 2010;71(4):706–12.

53. Bohm M, Richter JE, Kelsen S, Thomas R. Esophageal dilation: simple and effective treatment for adults with eosinophilic esophagitis and esophageal rings and narrowing. *Dis Esophagus* 2010;23(5):377–85.

54. Blanchard C, Mishra A, Saito-Akei H, Monk P, Anderson I, Rothenberg ME. Inhibition of human interleukin-13-induced respiratory and oesophageal inflammation by anti-human-interleukin-13 antibody (CAT-354). *Clin Exp Allergy* 2005;35(8):1096–103.

55. Mishra A, Hogan SP, Brandt EB, Rothenberg ME. An etiological role for aeroallergens and eosinophils in experimental esophagitis. *J Clin Invest* 2001;107(1):83–90.

56. Mishra A, Hogan SP, Brandt EB, Rothenberg ME. IL-5 promotes eosinophil trafficking to the esophagus. *J Immunol* 2002;168(5):2464–9.

57. Stein ML, Collins MH, Villanueva JM, Kushner JP, Putnam PE, Buckmeier BK, et al. Anti-IL-5 (mepolizumab) therapy for eosinophilic esophagitis. *J Allergy Clin Immunol* 2006;118(6):1312–9.

58. DeBrosse CW, Franciosi JP, King EC, Butz BK, Greenberg AB, Collins MH, et al. Long-term outcomes in pediatric-onset esophageal eosinophilia. *J Allergy Clin Immunol* 2011;128(1):132–8.

59. Brandt EB, Strait RT, Hershko D, Wang Q, Muntel EE, Scribner TA, et al. Mast cells are required for experimental oral allergen-induced diarrhea. *J Clin Invest* 2003;112(11):1666–77.

60. Hogan SP, Mishra A, Brandt EB, Foster PS, Rothenberg ME. A critical role for eotaxin in experimental oral antigen-induced eosinophilic gastrointestinal allergy. *Proc Nat Acad Sci USA* 2000;97:6681–6.

61. Aceves SS, Newbury RO, Dohil R, Bastian JF, Broide DH. A pathological function for eotaxin and eosinophils in eosinophilic gastrointestinal inflammation. *Nat Immunol* 2001;2(4):353–60.

62. Aceves SS, et al. Esophageal remodeling in pediatric eosinophilic esophagitis. *J Allergy Clin Immunol* 2007;119(1):206–12.

Chapter 13.9

Rare Hypereosinophilic Syndromes

Paneez Khoury and Amy D. Klion

INTRODUCTION

Although case reports describing eosinophilic infiltration of tissues accompanied by pronounced peripheral eosinophilia have existed since the 1950s, the concept of a hypereosinophilic syndrome (HES) was not introduced until 1975 in a landmark paper by Chusid and colleagues.[1] The original description of HES was based largely on clinical findings highlighting the heterogeneous presentations of HES and proposed the following diagnostic criteria:[1] blood eosinophilia >1500/mm³ for at least 6 months (or death before 6 months with signs and symptoms of eosinophilic

disease);[2] lack of evidence for allergic, infectious, parasitic, or other known causes of eosinophilia; and[3] presumptive signs of organ involvement. Since that time, the evolution of diagnostic testing has led to the identification of several HES variants with defined etiologies. Furthermore, the availability of novel targeted therapies precludes waiting 6 months for a diagnosis. As a result, several new conflicting classification schemes have been proposed.

The 2008 World Health Organization (WHO) classification specifically distinguishes between HES and eosinophilic disorders with proven clonal populations of myeloid or lymphoid origin.[2] Although correct from a diagnostic standpoint, there are many issues that are not addressed, namely the clinical overlap between patients with a detectable clonal population and those with similar disease in whom the mutation is not detectable with current techniques, and the heterogeneity of patients with non-clonal HES. In contrast, an HES classification scheme proposed at a 2005 NIH-sponsored workshop,[3] and subsequently refined by the same authors, groups clinically similar disorders with known and unknown etiologies into HES variants. Although useful from the standpoint of treatment approaches, this approach creates confusion with respect to diagnostic labels, since a patient with proven clonal eosinophilia is classified as having both HES and chronic eosinophilic leukemia (CEL).[3]

Despite these limitations, for the purposes of this subchapter, HES will be defined as (1) blood eosinophilia of >1500/mm³ on at least two separate occasions or evidence of prominent tissue eosinophilia associated with symptoms and marked tissue eosinophilia, and (2) exclusion of secondary causes of eosinophilia, such as drug hypersensitivity, hypoadrenalism, and parasitic infection.[4] The approach to classification, diagnosis, pathophysiology, and treatment of the various forms of HES and other rare manifestations of eosinophilic disease will be discussed.

EPIDEMIOLOGY

Hypereosinophilic syndromes are rare diseases, and correct estimates of incidence and prevalence are unavailable. By extrapolation from the Surveillance, Epidemiology and End Results (SEER) database of cancer statistics in the United States, the incident rate of HES has been estimated to be between 0.018 and 0.036 per 100,000 person-years in the period from 2001 to 2005. The calculated prevalence was estimated at 0.36–6.3 per 100,000.[5] These estimates are approximations and rely on coding of eosinophilia by individual physicians. In addition, the lack of specific codes for HES variants precludes a more specific characterization of incidence. Unified comprehensive patient databases would be useful in this regard. HES characteristically develops between the ages of 20 and 50 years; however, young children and the elderly can also be affected. A male predominance has been described, although this is due primarily to overrepresentation of males with the FIP1-like 1/platelet-derived growth factor receptor-α *(FIP1L1/PDGFRA)* mutation in most series.[6]

DIAGNOSING RARE HYPEREOSINOPHILIC SYNDROMES

Hypereosinophilia may be attributable to secondary causes requiring specific treatment. Therefore, a careful history and physical examination are of paramount importance for making the correct diagnosis (Fig. 13.9.1). The history should include the degree and duration of eosinophilia, with documentation if available. Specific symptoms related to individual organ systems should be elicited, as patients can present with a multitude of findings ranging from the most common complaints of skin involvement to more rare presentations of connective tissue or ocular involvement. Pertinent medication, occupational, and travel histories should also be obtained to exclude drug allergy, hypersensitivity reactions, and helminth infections, respectively. A family history of eosinophilia should also be investigated.

The initial physical exam should be complete, with a focus on examination for organ involvement commonly seen in eosinophilic syndromes. Notably, careful examination of the skin, abdomen (to assess organomegaly), cardiovascular system (to assess evidence of heart failure and valvular disease), lungs, neurological system (to exclude neuropathy) should be performed.

Finally, laboratory and diagnostic testing for eosinophilia should be directed by the history and prior physical findings, but should include basic testing to assess end-organ involvement (Fig. 13.9.1). Initial laboratory evaluation should include, at a minimum, a complete blood count (CBC) with a differential, chemistry panel to assess creatine kinase, creatinine, electrolytes, erythrocyte sedimentation rate (ESR), liver enzymes, quantitative immunoglobulin E (IgE), levels serum B12, and serum tryptase. Assessment of cardiac status should include echocardiogram, electrocardiogram (EKG), and measurement of serum troponin. Spirometry with assessment of lung volumes and diffusion capacity should be performed to assess occult pulmonary involvement. Often the initial evaluation will determine the need for further testing. For example, a patient with asthma and sinusitis should be screened for Churg-Strauss syndrome (CSS) with serum anti-neutrophil cytoplasmic antibody (ANCA). Electrolyte abnormalities with eosinophilia might necessitate a work-up for hypoadrenalism. Findings of anemia, neutropenia, or thrombocytopenia would initiate a search for hematological disorders, including CEL. A pertinent drug history

should prompt discontinuation of potentially offending agents to see if the eosinophilia resolves. Exceedingly uncommon in secondary eosinophilia, an elevated serum B12 or tryptase level should prompt evaluation for CEL and systemic mastocytosis (SM).

If no secondary or reactive cause is identified, one can reasonably proceed to further evaluation for HES and other rare eosinophilic disorders. Additional screening tests at this point should include computed tomography (CT) of the abdomen, chest, and pelvis, and bone marrow examination to assess cellularity, dysplastic changes of either eosinophils or mast cells, eosinophil precursors, mast cell numbers, and myelofibrosis. Cytogenetic analysis of dividing cells should be performed, as well as specific testing by real-time polymerase chain reaction (RT-PCR) or fluorescence *in situ* hybridization (FISH) for the most common genetic abnormalities associated with eosinophilia, including *PDGFR*-associated CELs. D816V *KIT* analysis should also be performed to exclude SM in patients with elevated serum tryptase and/or suggestive findings on bone marrow biopsy. T-cell clonality should be assessed by RT-PCR for T-cell receptor rearrangement and/or flow cytometry. Phenotypic assessment should also be performed to identify aberrant populations of T cells that may or may not be clonal and should include CD3, CD4, CD8, and if possible CD5 and CD7. Aberrant T-cell populations often exhibit elevated intracellular interleukin-4 (IL-4), IL-5, and IL-13 levels;[7] however, intracellular cytokine analysis is not routinely available except at specialized academic centers. Various biomarkers have been clinically correlated with particular variants of HES, including serum B12 and tryptase elevation in myeloproliferative forms of HES and elevated serum IgE and thymus and activation-regulated chemokine (TARC)/C-C motif chemokine 17 (CCL17) levels in lymphocytic variant HES and are discussed further in the next section. It is important to recognize that the diagnosis of HES is an iterative process and may be revisited if new clinical findings develop and as new biomarkers and diagnostic tests become available.

HYPEREOSINOPHILIC SYNDROME VARIANTS

As stated in the introduction, the current classification of HES variants is in evolution as specific etiologies become better defined. Despite the molecular uncertainties and current unknowns, it remains useful to subdivide patients into HES variants based on clinical characteristics and a combination of molecular and biologic data (Fig. 13.9.2), since this has important implications regarding prognosis, monitoring, and treatment. Unfortunately, the vast majority (60−70%) of patients with HES remain classified as idiopathic. In the following section, we discuss HES variants in

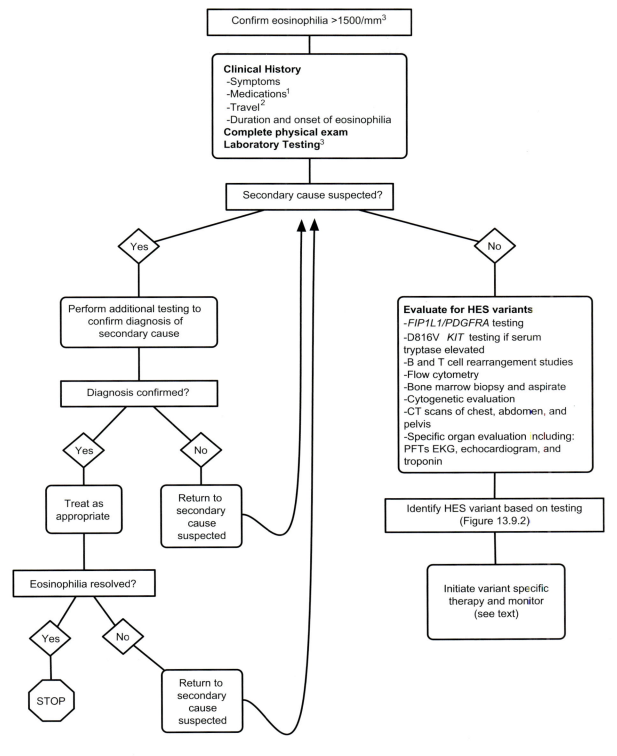

FIGURE 13.9.1 **Approach to diagnosis and treatment.** B12, vitamin B12; CK, creatine kinase; CT, computerized tomography; EKG, electrocardiogram; ESR, erythrocyte sedimentation rate; HES, hypereosinophilic syndrome; HIV, human immunodeficiency virus; IgE, immunoglobulin E; PFTs, pulmonary function tests; US, United States.

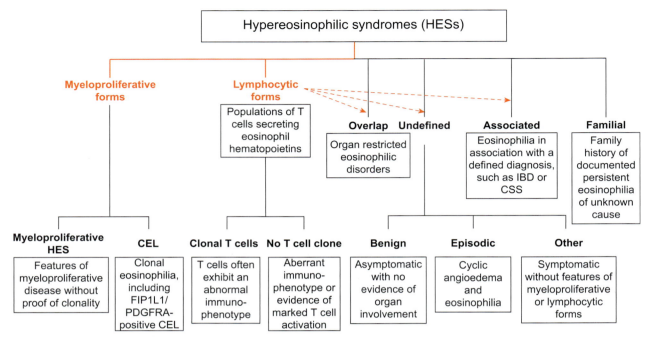

FIGURE 13.9.2 **Classification of hypereosinophilic syndrome.** Dashed arrows identify variants of hypereosinophilic syndrome (HES) that are thought to be T cell-mediated. *(Reproduced with permission from Elsevier[4]).*

further detail with a focus on clinical features, diagnosis, mechanisms, pathogenesis, and treatment. Areas with considerable controversy surrounding either diagnosis or treatment are also addressed.

Lymphocytic Variant Hypereosinophilic Syndrome

Mechanisms and Pathogenesis

Lymphocytic variant HES (L-HES) is defined by the expansion of an aberrant and/or clonal T-lymphocyte population with increased production of eosinophilopoietic cytokines in a patient who meets criteria for HES. Historically, the evidence for T-cell-mediated pathogenesis came to light when peripheral T-cell clones generated from a patient with HES were cultured with bone marrow progenitors, leading to outgrowth of eosinophilic colonies in culture.[8] Further studies demonstrated that IL-5 was overproduced; however, other cytokines such as IL-2, interferon γ (IFN-γ), and additional T-helper type 2 (T_h2) cytokines, IL-4 and IL-13, may also be increased in individual clones.[9]

Clinical Features and Diagnosis

L-HES affects males and females in equal proportions. Patients frequently have elevated serum IgE levels and skin involvement, ranging from urticaria and eczematous rashes to subcutaneous involvement with angioedema. Rare patients present with cyclic eosinophilia and angioedema in

the setting of an abnormal lymphocyte population (Gleich's syndrome). Lymphadenopathy and splenomegaly are uncommon except in the case of occult lymphoma, and bone marrow examination is generally normocellular with increased eosinophils. In contrast to the myeloproliferative variant, life-threatening end-organ involvement, including endomyocardial fibrosis, neurologic complications, and hypercoagulability are extremely rare in L-HES. Furthermore, L-HES is usually glucocorticoid-responsive, although moderate to high doses may be necessary to control symptoms.

Blood counts reflect elevated eosinophil levels; there may be lymphocytosis, but this is uncommon. There is often a polyclonal expansion of IgG or IgM, and markers of inflammation, such as ESR and C-reactive protein (CRP), and T-cell activation, such as CCL17, may be elevated. Thrombocytosis may also be present. The majority of aberrant T-cell populations in L-HES are detectable by flow cytometry using standard panels. Although $CD3^- CD4^+$ is the most common phenotype, $CD3^+ CD4^- CD8^-$ and $CD4^+ CD7^{dim/-}$ have also been described in some patients.[10] It is important to include appropriate markers to distinguish these cell populations from monocytes, especially if CD4 expression is decreased,[9] and to exclude T-cell lymphomas that may present with eosinophilia, including cutaneous T-cell lymphoma[11] and angioimmunoblastic T-cell lymphoma. Clonal populations of IL-5-producing cells with a normal surface phenotype have also been described in L-HES. Consequently, T-cell receptor

(TCR) rearrangement analysis and/or assessment of clonality by flow cytometry using TCR-Vβ staining should be performed. Demonstration of increased T$_h$2 cytokine production by the aberrant and/or clonal T-cell population by intracellular flow cytometry is also diagnostic, but is impractical at most centers.[12]

Therapy

The decision to treat L-HES patients depends on the nature and extent of disease. Patients with clinical manifestations attributable to the eosinophilia should be treated initially with corticosteroids (20—60 mg prednisone/d depending on the severity of the signs and symptoms), reducing slowly to the lowest dose that suppresses symptoms and eosinophilia. Steroid-sparing agents should be considered for patients with elevated eosinophil counts and symptoms requiring moderate to high dose (>10—15 mg prednisone equivalent daily) corticosteroids. Aims of therapy should be to reduce disease manifestations and prevent organ dysfunction. A number of steroid-sparing agents targeting T cells have been tried with variable success.[6] Among currently available agents, IFN-α (at a dose range of 1—3 mU daily) has shown the greatest efficacy. Due to in vitro data demonstrating decreased apoptosis of the clonal population in the presence of IFN-α, concomitant low dose corticosteroid therapy is recommended in patients with L-HES.[13] More recently, anti-IL-5 therapy with mepolizumab (available only for compassionate use through GlaxoSmithKline) has been shown to be safe and effective as a steroid-sparing agent in this subgroup of patients.[14]

It is important to monitor patients with a clonal T-cell population for progression to T-cell lymphoma throughout the course of their disease. This may be preceded by lymphadenopathy, expansion of the clone, and/or the development of cytogenetic abnormalities. Yearly bone marrow examination with karyotyping has been recommended,[9] although the impact of this approach on prognosis is unknown.

Myeloproliferative Variant Hypereosinophilic Syndrome

Mechanisms and Pathogenesis

The most common cause of marked eosinophilia with myeloproliferative features is an interstitial deletion in chromosome 4q12 between *FIP1L1* and the tyrosine kinase, *PDGFRA*, that results in the fusion gene product *FIP1L1/PDGFRA*.[15] Other *PDGFRA* fusion partners have been identified, but are less common. Colony forming assays using an *FIP1L1/PDGFRA* reporter showed that expression of the fusion gene in human hematopoietic progenitors induces differentiation of erythrocytes and neutrophils in addition to eosinophils,[16] and multiple

lineages, including B- and T-lymphocytes, monocytes, mast cells, and neutrophils, can express the fusion gene in affected patients.[17] Current data suggest, however, that the detrimental clinical effects are mediated primarily by eosinophils. In fact, the clinical presentation of *FIP1L/PDGFRA*-positive CEL is indistinguishable from that of a subset of idiopathic HES patients with myeloproliferative features.

A number of other myeloproliferative disorders (MPDs) can present with marked blood and bone marrow eosinophilia and have been defined at the molecular level. These include *PDGFRB*-associated chronic myelomonocytic leukemia (CMML), Janus kinase 2 (*JAK2*)-associated MPDs, 5q- syndrome and D816V *KIT*-associated SM. Although SM shares bone marrow features with *FIP1L1/PDGFRA*-positive CEL, the clinical presentation is often quite different from HES[18] and the eosinophils are not usually implicated in disease pathogenesis.

Clinical Features and Diagnosis

Myeloproliferative variant HES (MHES) is the most aggressive form of HES, with fatality rates of up to 50% at 3 years prior to the availability of imatinib therapy. Patients are predominantly male with a near 100% male predominance in *PDGFRA*-associated disease. MHES typically presents between the ages of 25 and 50 years, although children and the elderly can be affected. The clinical presentation ranges from fatigue and malaise to endomyocardial fibrosis and stroke. Splenomegaly is common and characteristic laboratory findings include anemia and thrombocytopenia, occasional neutrophilia, and elevation of serum B12 and tryptase levels. Bone marrow examination reveals a hypercellular marrow with increased and dysplastic eosinophils and eosinophil precursors, fibrosis, and, in most cases, a concomitant increase in atypical mast cells. Eosinophilia is unresponsive to corticosteroids in most patients with MHES.

Although the existence of a myeloproliferative (*leukemic*) variant of HES had long been recognized, the availability of molecular testing has revolutionized diagnosis of this syndrome. The *FIP1L1/PDGFRA* fusion gene can be detected either by RT-PCR or FISH in peripheral blood. Additional testing, including cytogenetics, FISH, and/or quantitative PCR, should be performed, as indicated, to identify other MPDs associated with peripheral eosinophilia.

Therapy

The discovery that the tyrosine kinase inhibitor, imatinib, can decrease eosinophilia and improve symptoms in patients with HES facilitated the discovery of the *FIP1L1/PDGFRA* fusion gene[15] and has dramatically improved prognosis in MHES. Some patients with *FIP1L1—PDG-FRA*-negative MHES also respond to imatinib. Therefore, a trial of imatinib in such patients is reasonable if they fail low dose corticosteroids, particularly in the setting of myeloproliferative features. Most *FIP1L1/PDGFRA*-positive patients achieve clinical and molecular remission within 1—2 weeks of beginning imatinib therapy and can be maintained on low-dose therapy (100 mg daily) for many years without disease progression. Although *FIP1L1/PDGFRA*-negative patients may require longer to respond, a trial of imatinib 400 mg daily for 4 weeks is sufficient to assess response. Side effects of therapy are comparable to that seen in the treatment of chronic myeloid leukemia (CML), but rarely lead to discontinuation of therapy, and the development of resistance appears to be extremely rare. Unfortunately, imatinib is not curative in MHES.[19]

Since the clinical symptoms may be rapidly progressive, any new diagnosis of HES with myeloproliferative features should be evaluated without delay so that treatment can be initiated rapidly. EKG, troponin, and echocardiogram should be performed prior to initiating therapy. If cardiac dysfunction is present or serum troponin is elevated, corticosteroids should be initiated concurrent with imatinib therapy, due to reports of necrotizing myocarditis after initiation of therapy in patients with preexisting cardiac involvement. Initial monitoring on imatinib therapy should include weekly complete blood counts, liver function tests, and serum troponin levels. Bone marrow examination should be repeated at 4—8 weeks, even in the setting of hematologic response, to exclude occult leukemia or lymphoma that may have been masked by the marked eosinophilia. Long-term monitoring should include echocardiograms every 3—6 months to assess progression of disease, as well as to monitor for possible treatment-related cardiomyopathy.

For patients with *FIP1L1/PDGFRA*-positive MHES who fail or are intolerant to imatinib therapy, a trial of one of the newer tyrosine kinase inhibitors with activity against the most common resistance mutations is indicated. For *FIP1L1/PDGFRA*-negative MHES patients who fail imatinib, a step-wise approach of commonly used medications for reduction of eosinophils should be employed, balancing toxicities of the drugs and the individual patient's comorbidities. Drugs that have been used successfully include hydroxyurea, IFN-α, anti-IL-5 therapy, and vincristine.[20] Patients with rapidly progressive or aggressive disease unresponsive to standard therapies should be considered for nonmyeloablative allogeneic bone marrow transplant, a strategy that has proven curative in a number of cases.[21]

Idiopathic Hypereosinophilic Syndrome

Mechanisms and Pathogenesis

By definition, the mechanism of pathogenesis in idiopathic HES is unknown. Similar to other forms of HES, pathogenesis is due to tissue infiltration by eosinophils, with deposition of granule proteins and release of inflammatory mediators.

Clinical Features and Diagnosis

Patients with idiopathic HES may have relatively few or no symptoms (benign eosinophilia) or a wide variety of manifestations attributable to eosinophilic infiltration of target organs. The most commonly affected organs are the gastrointestinal tract, lungs, and skin (Fig. 13.9.3).[6] Nonspecific symptoms, including arthralgias, fatigue,

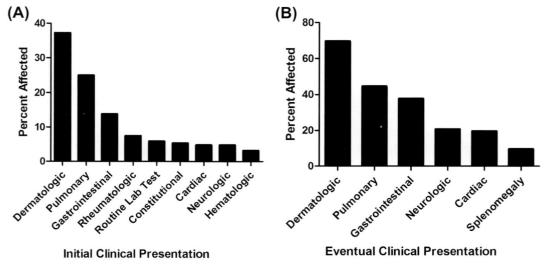

FIGURE 13.9.3 *A*, Presenting end-organ involvement. *B*, Eventual end-organ involvement.

malaise, and myalgias, are also common. Cardiac manifestations, including eosinophilic myocarditis and endomyocardial fibrosis, and neurological involvement each occur in 15–20% of patients with idiopathic HES and are major causes of morbidity and mortality in this patient group.

Two areas of confusion with regards to secondary forms of eosinophilia may delay treatment for true HES. The first involves appropriate testing to rule out parasitic disease. Prompt referral to an infectious disease specialist can target helminth infection testing based on the patient's travel history and likelihood of exposure. Whether to empirically treat for *Strongyloides* according to Centers for Disease Control and Prevention (CDC) guidelines is an area of controversy. Recommendations of the CDC are that all patients who are 'at risk of disseminated strongyloidiasis should be treated.' The drug of choice for treatment of uncomplicated strongyloidiasis is ivermectin. The second area of confusion is that of medications causing hypersensitivity syndromes such as drug reaction with eosinophilia and systemic symptoms (DRESS). Many drugs can cause eosinophilia starting from days to years after initiation. Simplification of drug regimens can vastly improve the ability to make the diagnosis of HES. Once the diagnosis of HES is made, the decision to treat is tailored to the likely etiology.

Therapy

Since patients with idiopathic HES are heterogeneous in their presentation, virtually all treatment must be individualized based on the presence of signs and symptoms and the likelihood of disease progression with end-organ involvement. Although the discovery of imatinib has dramatically altered prognosis in myeloproliferative HES, and in particular for patients with *FIP1L1/PDGFRA*-positive CEL, patients with idiopathic HES are much less likely to respond.

Corticosteroids remain first-line therapy for the majority of patients with HES, including L-HES, although many patients require moderate to high doses or develop significant steroid-related toxicity. Patients who fail to respond to corticosteroids or have significant side effects from prolonged high dose therapy should be considered for second-line therapy. A short (4–6 week) trial of imatinib, the only United States Food and Drug Administration (FDA)-approved drug for the treatment of HES, may be warranted in steroid-resistant patients.

Hydroxyurea is the most commonly used of the cytotoxic medications and has been used alone or in combination with IFN-α at doses ranging from 500 mg to 3 g daily. It can be associated with cytopenias or other adverse effects at high doses, thus limiting its use as a solitary agent. Furthermore, hydroxyurea may be associated with an increased risk of secondary malignancy with prolonged use. Of the immunomodulatory medications, IFN-α is most frequently used and at doses of 1–3 mU daily is effective in up to 30% of patients.[6] Prolonged remission of clinical symptoms and eosinophilia has rarely been reported after prolonged IFN-α use or when INF-α is used in combination with cytotoxic medications.[22] Other cytotoxic and immunomodulatory agents have been used with varied success and include cytosine arabinoside (ara-C),[23] vincristine,[20] alemtuzumab,[24,25] intravenous immunoglobulin,[26] cyclosporine,[27] cyclophosphamide,[28] azathioprine, and methotrexate. However, bone marrow transplantation remains the only curative therapy.[21]

Familial Eosinophilia

Clinical Features

Familial hypereosinophilia is an autosomal dominant disorder discovered in one kindred family. Eosinophilia is present at birth in affected individuals and remains remarkably stable over time. Although the index case and his sister presented with eosinophilic end-organ involvement (endomyocardial fibrosis and peripheral neuropathy) that progressed despite therapy, most affected family members have followed a benign course despite eosinophil counts ranging from 2000–5000/mm^3 over many years without treatment.[29] There have been isolated reports of additional families with eosinophilia, but no clear genetic inheritance pattern has been found. Common environmental exposures, including helminth infection, must be excluded prior to attributing the eosinophilia to a familial origin.

Mechanisms and Pathogenesis

The genetic defect in familial hypereosinophilic syndrome is not known. The gene has been mapped to an area on chromosome 5q harboring the cytokine gene cluster;[30] however, sequencing of a number of candidate genes in this region, including GM-CSF, IL-3, and IL-5, has failed to identify polymorphisms that could account for the affected phenotype. Additional sequencing is currently under way. Eosinophils from affected family members appear to be less activated than those from patients with HES, coincident with the general lack of eosinophil-mediated pathology.[29] No further family members have developed organ involvement, thus precluding further studies of disease pathogenesis in this disorder.

Organ-Restricted Eosinophilia

Single organ eosinophilic tissue infiltration has been described for nearly all organ systems, although isolated involvement of the skin, lung or gastrointestinal tract is most common. The mechanism whereby certain tissues are targeted preferentially over others is not well understood. Some organ-restricted disorders, such as eosinophilic esophagitis, have distinct clinical presentations and rarely progress to multiorgan involvement even in the presence of marked peripheral eosinophilia. Other disorders, such as chronic eosinophilic pneumonia (CEP), may overlap considerably in presentation with, or be the initial manifestation of, systemic HES. Treatment varies depending on the specific clinical manifestations, but local or systemic corticosteroids are often effective. It is beyond the scope of this subchapter to discuss all organ-restricted eosinophilic syndromes, several of which are covered in other subchapters. Consequently, only a few examples are provided to illustrate the diversity of syndromes seen.

Pulmonary Eosinophilia

Eosinophilia restricted to the lung, with or without peripheral eosinophilia, is associated with a wide variety of allergic, infectious, inflammatory, and neoplastic disorders, including asthma, allergic bronchopulmonary aspergillosis, drug hypersensitivity reactions, fungal pneumonia, helminth infection, and sarcoidosis. Pulmonary eosinophilia can also be the first indicator of systemic eosinophilic diseases, including CSS. When no trigger is identified and disease remains restricted to the lung, pulmonary eosinophilia can be classified into two distinct syndromes: acute eosinophilic pneumonia (AEP) and CEP.

AEP typically occurs in healthy men between the ages of 20 and 40 years without a history of asthma. Symptoms include the abrupt onset of dyspnea, fever, malaise, night sweats, nonproductive cough, and pleuritic chest pain, and respiratory failure requiring mechanical ventilation is common. Physical examination may reveal bibasilar rales or rhonchi. Radiological findings include diffuse alveolar infiltrates or reticular opacities in the early stages and bronchoscopy reveals increased absolute eosinophil counts as well as an increased percentage of eosinophils (>25%) relative to the total inflammatory cell content.[31] Whether an environmental exposure plays a role in the pathophysiology remains controversial. Patients typically respond well to high-dose corticosteroid therapy despite respiratory failure at presentation, and long-term sequelae are extremely uncommon.

CEP presents as a subacute illness, often mistaken for asthma in the early stages. It progresses to involve similar symptoms of cough and dyspnea, and is occasionally associated with constitutional symptoms of fever, sweats, and weight loss. Physical exam findings are similar to AEP, with wheeze and or crackles being present in most cases. Radiological findings typically show peripheral infiltrates. Bronchoalveolar lavage fluid (BALF) shows an eosinophil predominance.

While patients may respond initially to corticosteroids, many relapse and may require long-term corticosteroid treatment. Peripheral eosinophilia of $>1000/mm^3$ is common, but not universal.[32]

Eosinophilic Hepatitis

Eosinophilia of the liver may occur in primary liver diseases, such as primary biliary cirrhosis or autoimmune hepatitis, in the setting of hepatobiliary involvement by helminth infections, including clonorchiasis and schistosomiasis, or secondary to a wide variety of prescription and nonprescription drugs. Primary eosinophilic hepatitis with or without peripheral eosinophilia is, however, relatively rare. Early studies using immunohistochemistry implicated eosinophils in the pathogenesis of progression to chronic hepatitis by demonstrating major basic protein and activated, degranulated eosinophils in close proximity to affected hepatocytes.[33] Hepatic cholangiopathy, fibrosis and liver failure have been reported, although most patients in the literature have been responsive to corticosteroids or hydroxyurea.

Eosinophilic Cystitis

Eosinophilic cystitis is a rare disorder, primarily seen in children, that presents with dysuria, gross hematuria, or suprapubic pain. Urinalysis may reveal microscopic hematuria or pyuria. Cystoscopy often reveals bladder wall erythema, edema, or nodules, and the diagnosis is made histologically from biopsies of the bladder.[34] Infrequently, necrosis or fibrosis can be present with delayed diagnosis. It is important to exclude bladder cancer, since eosinophilic cystitis may be present in the setting of bladder cancer. Treatment with antihistamines, nonsteroidal anti-inflammatory medications, fulguration and steroids were reported to be successful in one series.[35]

Overlap Syndromes with Hypereosinophilic Syndrome

A number of systemic disorders have overlapping clinical presentations with HES. Of these, the two most common are SM and CSS. SM is a myeloproliferative disorder, most commonly associated with a D816V mutation in *KIT*. Patients typically present with symptoms related to mast cell tissue infiltration and mediator release, including anaphylaxis, diarrhea, flushing, and urticaria. Diagnosis is based on the presence of major and minor criteria according to the WHO classification.[2] D816V-positive SM with eosinophilia (SM-eo) is clinically distinct from *FIP1L1-PDGFRA*-positive CEL and can be distinguished using molecular and clinical findings.[18]

CSS is a distinct multisystem disorder that is characterized by the presence of eosinophilic vasculitis in the setting of asthma, pulmonary infiltrates, sinusitis with

TABLE 13.9.1 Selected Disorders Associated with Immune Deficiency or Dysregulation and Eosinophilia

Disorders Associated with Eosinophilia*
Autoimmune disorders
ALPS
Connective tissue disorders (e.g., dermatomyositis and rheumatoid arthritis)
Human immunodeficiency virus infection
Inflammatory bowel disease
Neoplasms (e.g., adenocarcinoma, lymphocytic leukemia, and lymphoma)
Primary immunodeficiencies
IPEX (immunodysregulation polyendocrinopathy enteropathy X-linked syndrome)
DIDS (Dock8 immunodeficiency syndrome)
Omenn syndrome
Hyper-IgE syndrome (Job syndrome)
Atypical DiGeorge syndrome
Sarcoidosis

This is only a partial list

polyps, and marked peripheral eosinophilia. ANCA may be present and appears to be associated with a more severe course, often with renal involvement. Definitive diagnosis can be made by tissue biopsy demonstrating granulomata and necrotizing eosinophilic vasculitis. Significant overlap in clinical presentation with idiopathic HES can cause diagnostic confusion, particularly since corticosteroids must often be initiated to prevent serious end-organ damage before appropriate biopsies can be obtained.[36]

Other Disorders Associated with Eosinophilia

Marked eosinophilia can be seen in association with immunodysregulation of varied etiologies, including primary immunodeficiencies, secondary immunodeficiencies [e.g., human immunodeficiency virus (HIV) infection], and autoimmune disorders (Table 13.9.1). In general, the peripheral eosinophilia seen in these disorders is not associated with characteristic end-organ manifestations. Exceptions include Omenn syndrome and Dock8 deficiency, which may present with eosinophilic tissue infiltration involving the skin or other organs.[37,38] In some disorders, including autoimmune lymphoproliferative syndrome (ALPS) and HIV, peripheral eosinophilia reflects more profound immunodysregulation and in some cases may portend a worse prognosis.[39,40]

NOVEL THERAPIES FOR HYPEREOSINOPHILIC SYNDROME

Improved understanding of the pathogenesis of eosinophilic disorders, including eosinophilic asthma, has led to the development of a number of novel agents, including agents targeting IL-5 and its receptor, although none are currently FDA approved. These are discussed in Chapter 15. Additional agents, including alemtuzumab, a monoclonal antibody that targets CD52 on aberrant T cells, and tyrosine kinase inhibitors with activity against imatinib-resistant *FIP1L1/PDGFRA*-positive mutations, have also been used with success in small numbers of patients with L-HES and CEL, respectively.[24]

CONCLUSION

The complexity of diagnosing and treating HES arises from the difficulty in excluding secondary forms of eosinophilia and knowing when to treat and with what medications. An understanding of underlying pathogenesis has improved genotype—phenotype classification of some eosinophilic disorders and reinforced the perception of HES as a heterogeneous group of rare disorders with differing pathogenesis, prognosis, and treatment options. Advancing knowledge of molecular pathogenesis and new genetic discoveries will provide the foundation for development of novel targeted therapies, as well as a promising outlook for the care of patients with HES.

REFERENCES

1. Chusid MJ, Dale DC, West BC, Wolff SM. The hypereosinophilic syndrome: analysis of fourteen cases with review of the literature. *Medicine (Baltimore)* 1975;**54**:1—27.
2. Swerdlow SH, International Agency for Research on Cancer, World Health Organization. *WHO classification of tumours of haematopoietic and lymphoid tissues.* Lyon, France: International Agency for Research on Cancer; 2008.
3. Klion AD, Bochner BS, Gleich GJ, Nutman TB, Rothenberg ME, Simon HU, et al. Approaches to the treatment of hypereosinophilic syndromes: a workshop summary report. *J Allergy Clin Immunol* 2006;**117**:1292—302.
4. Simon HU, Rothenberg ME, Bochner BS, Weller PF, Wardlaw AJ, Wechsler ME, et al. Refining the definition of hypereosinophilic syndrome. *J Allergy Clin Immunol* 2010;**126**:45—9.
5. Crane MM, Chang CM, Kobayashi MG, Weller PF. Incidence of myeloproliferative hypereosinophilic syndrome in the United States and an estimate of all hypereosinophilic syndrome incidence. *J Allergy Clin Immunol* 2010;**126**:179—81.
6. Ogbogu PU, Bochner BS, Butterfield JH, Gleich GJ, Huss-Marp J, Kahn JE, et al. Hypereosinophilic syndrome: a multicenter,

retrospective analysis of clinical characteristics and response to therapy. *J Allergy Clin Immunol* 2009;**124**:1319–1325, e1313.

7. Ravoet M, Sibille C, Gu C, Libin M, Haibe-Kains B, Sotiriou C, et al. Molecular profiling of CD3-CD4+ T cells from patients with the lymphocytic variant of hypereosinophilic syndrome reveals targeting of growth control pathways. *Blood* 2009;**114**: 2969–83.

8. Raghavachar A, Fleischer S, Frickhofen N, Heimpel H, Fleischer B. T lymphocyte control of human eosinophilic granulopoiesis. Clonal analysis in an idiopathic hypereosinophilic syndrome. *J Immunol* 1987;**139**:3753–8.

9. Roufosse F, Cogan E, Goldman M. Lymphocytic variant hypereosinophilic syndromes. *Immunol Allergy Clin North Am* 2007;**27**: 389–413.

10. Simon HU, Plotz SG, Dummer R, Blaser K. Abnormal clones of T cells producing interleukin-5 in idiopathic eosinophilia. *N Engl J Med* 1999;**341**:1112–20.

11. Edelman J, Meyerson HJ. Diminished CD3 expression is useful for detecting and enumerating Sezary cells. *Am J Clin Pathol* 2000;**114**:467–77.

12. Brugnoni D, Airo P, Rossi G, Bettinardi A, Simon HU, Garza L, et al. A case of hypereosinophilic syndrome is associated with the expansion of a CD3-CD4+ T-cell population able to secrete large amounts of interleukin-5. *Blood* 1996;**87**:1416–22.

13. Schandene L, Roufosse F, de Lavareille A, Stordeur P, Efira A, Kennes B, et al. Interferon alpha prevents spontaneous apoptosis of clonal Th2 cells associated with chronic hypereosinophilia. *Blood* 2000;**96**:4285–92.

14. Roufosse F, de Lavareille A, Schandene L, Cogan E, Georgelas A, Wagner L, et al. Mepolizumab as a corticosteroid-sparing agent in lymphocytic variant hypereosinophilic syndrome. *J Allergy Clin Immunol* 2010;**126**:828–835, e823.

15. Cools J, DeAngelo DJ, Gotlib J, Stover EH, Legare RD, Cortes J, et al. A tyrosine kinase created by fusion of the PDGFRA and FIP1L1 genes as a therapeutic target of imatinib in idiopathic hypereosinophilic syndrome. *N Engl J Med* 2003;**348**:1201–14.

16. Buitenhuis M, Verhagen LP, Cools J, Coffer PJ. Molecular mechanisms underlying FIP1L1-PDGFRA-mediated myeloproliferation. *Cancer Res* 2007;**67**:3759–66.

17. Robyn J, Lemery S, McCoy JP, Kubofcik J, Kim YJ, Pack S, et al. Multilineage involvement of the fusion gene in patients with FIP1L1/PDGFRA-positive hypereosinophilic syndrome. *Br J Haematol* 2006;**132**:286–92.

18. Maric I, Robyn J, Metcalfe DD, Fay MP, Carter M, Wilson T, et al. KIT D816V-associated systemic mastocytosis with eosinophilia and FIP1L1/PDGFRA-associated chronic eosinophilic leukemia are distinct entities. *J Allergy Clin Immunol* 2007;**120**: 680–7.

19. Klion AD, Robyn J, Akin C, Noel P, Brown M, Law M, et al. Molecular remission and reversal of myelofibrosis in response to imatinib mesylate treatment in patients with the myeloproliferative variant of hypereosinophilic syndrome. *Blood* 2004;**103**: 473–8.

20. Sakamoto K, Erdreich-Epstein A, deClerck Y, Coates T. Prolonged clinical response to vincristine treatment in two patients with idiopathic hypereosinophilic syndrome. *Am J Pediatr Hematol Oncol* 1992;**14**:348–51.

21. Ueno NT, Anagnostopoulos A, Rondon G, Champlin RE, Mikhailova N, Pankratova OS, et al. Successful non-myeloablative allogeneic transplantation for treatment of idiopathic hypereosinophilic syndrome. *Br J Haematol* 2002;**119**:131–4.

22. Butterfield JH. Treatment of hypereosinophilic syndromes with prednisone, hydroxyurea, and interferon. *Immunol Allergy Clin North Am* 2007;**27**:493–518.

23. Jabbour E, Verstovsek S, Giles F, Gandhi V, Cortes J, O'Brien S, et al. 2-Chlorodeoxyadenosine and cytarabine combination therapy for idiopathic hypereosinophilic syndrome. *Cancer* 2005;**104**:541–6.

24. Verstovsek S, Tefferi A, Kantarjian H, Manshouri T, Luthra R, Pardanani A, et al. Alemtuzumab therapy for hypereosinophilic syndrome and chronic eosinophilic leukemia. *Clin Cancer Res* 2009;**15**:368–73.

25. Pitini V, Teti D, Arrigo C, Righi M. Alemtuzumab therapy for refractory idiopathic hypereosinophilic syndrome with abnormal T cells: a case report. *Br J Haematol* 2004;**127**:477.

26. Orson FM. Intravenous immunoglobulin therapy suppresses manifestations of the angioedema with hypereosinophilia syndrome. *Am J Med Sci* 2003;**326**:94–7.

27. Donald CE, Kahn MJ. Successful treatment of hypereosinophilic syndrome with cyclosporine. *Am J Med Sci* 2009;**337**: 65–6.

28. Lee JH, Lee JW, Jang CS, Kwon ES, Min HY, Jeong S, et al. Successful cyclophosphamide therapy in recurrent eosinophilic colitis associated with hypereosinophilic syndrome. *Yonsei Med J* 2002;**43**:267–70.

29. Klion AD, Law MA, Riemenschneider W, McMaster ML, Brown MR, Horne M, et al. Familial eosinophilia: a benign disorder? *Blood* 2004;**103**:4050–5.

30. Rioux JD, Stone VA, Daly MJ, Cargill M, Green T, Nguyen H, et al. Familial eosinophilia maps to the cytokine gene cluster on human chromosomal region 5q31-q33. *Am J Hum Genet* 1998;**63**: 1086–94.

31. Philit F, Etienne-Mastroianni B, Parrot A, Guerin C, Robert D, Cordier JF. Idiopathic acute eosinophilic pneumonia: a study of 22 patients. *Am J Respir Crit Care Med* 2002;**166**:1235–9.

32. Marchand E, Reynaud-Gaubert M, Lauque D, Durieu J, Tonnel AB, Cordier JF. Idiopathic chronic eosinophilic pneumonia. A clinical and follow-up study of 62 cases. The Groupe d'Etudes et de Recherche sur les Maladies 'Orphelines' Pulmonaires (GER-M'O'P). *Medicine (Baltimore)* 1998;**77**:299–312.

33. Foong A, Scholes JV, Gleich GJ, Kephart GM, Holt PR. Eosinophil-induced chronic active hepatitis in the idiopathic hypereosinophilic syndrome. *Hepatology* 1991;**13**:1090–4.

34. Itano NM, Malek RS. Eosinophilic cystitis in adults. *J Urol* 2001;**165**:805–7.

35. Kilic S, Erguvan R, Ipek D, Gokce H, Gunes A, Aydin NE, et al. Eosinophilic cystitis. A rare inflammatory pathology mimicking bladder neoplasms. *Urol Int* 2003;**71**:285–9.

36. Wechsler ME. Pulmonary eosinophilic syndromes. *Immunol Allergy Clin North Am* 2007;**27**:477–92.

37. Zhang Q, Davis JC, Dove CG, Su HC. Genetic, clinical, and laboratory markers for DOCK8 immunodeficiency syndrome. *Dis Markers* 2010;**29**:131–9.

38. Villa A, Notarangelo LD, Roifman CM. Omenn syndrome: inflammation in leaky severe combined immunodeficiency. *J Allergy Clin Immunol* 2008;**122**:1082–6.

39. Kim YJ, Dale JK, Noel P, Brown MR, Nutman TB, Straus SE, et al. Eosinophilia is associated with a higher mortality rate among patients with autoimmune lymphoproliferative syndrome. *Am J Hematol* 2007;**82**:615—24.

40. Rosenthal D, LeBoit PE, Klumpp L, Berger TG. Human immunodeficiency virus-associated eosinophilic folliculitis. A unique dermatosis associated with advanced human immunodeficiency virus infection. *Arch Dermatol* 1991;**127**:206—9.

Chapter 13.10

Eosinophils and Cancer

Ramin Lotfi, Neal Spada and Michael Thomas Lotze

INFLAMMATION AND NECROSIS ARE MAJOR COMPONENTS OF THE EPITHELIAL TUMOR MICROENVIRONMENT

Advanced epithelial tumors typically undergo necrosis with subsequent release of damage-associated molecular pattern molecules (DAMPs).[1] Tumor cells are dependent on the host-created microenvironment, including endothelial cells, inflammatory cells, and stromal cells. This makes it difficult to cultivate more than a minority of tumor cells *in vitro* but, as is increasingly being understood, provides unique opportunities for cancer therapy. Thus, when evaluating a tumor, it is important to assess three elements within the microenvironment:

1. Factors released by tumor cells and their surrounding cells, consisting of epithelial and endothelial cells, fibroblasts, specialized local mesenchymal cells, and infiltrating leukocytes;
2. The quantity and quality of tumor-associated cells, specifically leukocytes;
3. Their state of activation.

The mammalian immune system reciprocally interacts within dynamic networks of tissue-associated nonimmune cells, enabling metabolic homeostasis, orderly scheduled cellular maturation and replacement, the timely eradication of effete cells, the repair of damage, and protection against pathogens. The simultaneous tolerance to self-antigens and reciprocal reactivity to new or occult antigens often occurs in settings of tissue damage and wound healing. When tissue homeostasis is perturbed, mast cells, granulocytes, and macrophages are recruited and rapidly release mediators such as cytokines, chemokines, matrix remodeling proteases and reactive oxygen species (ROS), and bioactive mediators such as histamine. These in turn induce mobilization and infiltration of additional leukocytes into damaged tissue (causing inflammation). Subsequently, the process of wound healing begins, characterized by phagocytosis of cellular debris and apoptotic cells, immune suppression, reepithelialization, and synthesis of extracellular matrix (ECM). Thus, inflammation is resolved, thereby restoring tissue homeostasis. Tumor cells, paradoxically growing in the setting of substantial necrosis and (chronic) inflammation, harness the collaborative capabilities (see below) of immune cells and local nonmutated but injured tissues to promote cell survival and proliferation, partly by releasing factors such as transforming growth factor β (TGF-β) and interleukin-10 (IL-10), leading to restoration of barrier function in epithelial tissues. The host therefore enables tumor cells to escape from eradication and to release tissue-healing factors, thereby providing neovascularization and subsequent nutritional supply to tumor cells. Wound healing and tumor stroma formation share many important properties. Nevertheless, wound healing is itself a self-limiting process, whereas tumors *addicted to death*[1] release DAMPs, thereby sustaining tissue proliferation, angiogenesis and continuous leukocyte recruitment.

Prolonged (chronic) inflammation is often associated with carcinogenesis. ROS generated largely intracellularly or at the cell membrane by NADPH oxidases can also promote mutagenic changes in cells when aerobic denaturation of extracellular DAMPs is ineffective.[2] Tumor necrosis factor α (TNF-α)[3] and matrix metalloproteinases promote recruitment of inflammatory cells and tissue remodeling, but also facilitate tumor metastasis. A state of metabolic symbiosis[4—8] between the tumor and the surrounding stroma, or within central hypoxic tumor cells and those at the growing rim of the tumor, allows regional variations in oxidative phosphorylation and autophagy that depend on nutrient supply and availability of molecular oxygen.

BIOLOGY OF EOSINOPHILS WITHIN NORMAL AND DAMAGED TISSUES

Compartments with abundant resident populations of eosinophils include tissues with substantial cellular turnover and regenerative capacity, such as the bone marrow, primary and secondary lymphoid tissues (e.g., lymph nodes, spleen, and thymus),[9] the uterus,[10] and almost the entire gastrointestinal tract (with the exception of the esophagus, except under abnormal states).[9,11] This link with cell turnover and tissue repair may also explain the presence of eosinophils at sites of wound repair[12] and the commonality of an eosinophil infiltrate among solid tumors.[13] Eosinophil localization to the lamina propria of the gastrointestinal tract is critically regulated by eotaxin/C-C motif chemokine 11 (CCL11), a chemokine

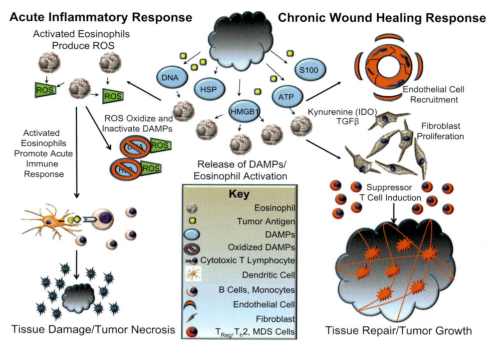

FIGURE 13.10.1 **Eosinophilic and neutrophilic granulocytes are recruited to neoplastic tissue by DAMPs.** DAMPs released by (necrotic) neoplastic cells recruit eosinophils and neutrophils. They in turn produce and release reactive oxygen species (ROS), inducing oxidation and degradation of tumor-related DAMPs. In addition, they initiate an acute inflammatory response to recruit, activate, and polarize B cells, cytotoxic lymphocytes, dendritic cells (DCs), and monocytes. Chronic exposure to active DAMPs promotes angiogenesis and immunosuppression by recruiting and activating endothelial cells, fibroblasts, and leukocytes such as T-regulatory (T_{reg}) cells, and myeloid-derived suppressor (MDS) cells, which are important for promoting wound healing. ATP, adenosine 5′-triphosphate; HMGB1, high mobility group box 1; HSP, heat shock protein; S100, protein S100 T_h2, T-helper type 2.

constitutively expressed throughout the gastrointestinal tract.[14] Nevertheless, eotaxin expression within the gastrointestinal tract (e.g., the esophagus) is by itself insufficient to induce eosinophil accumulation, because the esophagus is normally devoid of these granulocytes. This suggests the potential involvement of other eosinophil chemoattractants and activating factors that contribute to tissue-specific accumulation and degranulation. In particular, the correlation of eosinophil recruitment/activation with tissue damage and cell death associated with these inflammatory responses suggests that DAMPs may represent previously overlooked signaling molecules that elicit eosinophil agonist activities (Fig. 13.10.1). Consistently, high mobility group box 1 (HMGB1; a prototypic DAMP molecule) serves as a chemoattractant and survival factor for eosinophilic granulocytes.[15]

Eosinophilic granulocytes are found within necrotic tissues and the pseudocapsule surrounding tumors.[16] These immune cells contain, and can release, several cationic proteins that, in addition to their toxic tissue damaging character, are also potentially important for tissue remodeling and clearing cellular debris.[17] Eosinophils are thought to be, in part, responsible for the cell death and tissue damage commonly observed in disease states that are

associated with increased eosinophil numbers and tissue-specific eosinophil recruitment.[18]

Since the mid-1980s, eosinophils have been known to mediate their effects via at least three independent mechanisms in addition to the release of cytotoxic granule proteins, which enable them to modulate the intensity of inflammation, as well as to elicit cell death leading to the loss of tissue integrity:

1. Eosinophils may act as potent regulators of local inflammatory responses.[19]
2. Recruited eosinophils are a source of reactive oxygenated species[20] and established small molecule lipid mediators of inflammation. In particular, eosinophils generate cysteinyl leukotrienes (i.e., LTB_4, LTC_4, LTD_4, and LTE_4,[21,22] 5-HETE,[23] PGE_2,[24] and platelet-activating factor (PAF).[25] The capability of cysteinyl leukotrienes to mediate primary inflammatory responses such as edema,[26] the recruitment of other proinflammatory leukocytes,[27] and the induction of tissue histopathology[28] uniquely positions these molecules as mediators of inflammation.
3. Eosinophils are a prodigious source of cytokines associated with tissue repair and remodeling.

A growing body of literature suggests that both immuno-regulation and tissue repair/remodeling may represent important nonoverlapping eosinophil effector functions.[17] A quantitative assessment of eosinophil recruitment/accumulation in solid tumors showed that the tissue eosinophilia is apparently mediated by one or more factors released directly from necrotic tissues within the tumor. Studies linking eosinophil recruitment and activations with cell death and necrosis abound.[15,16,29] Thus, DAMPs released from damaged/dying epithelial cells may represent a previously underappreciated signaling event capable of mediating both eosinophil recruitment and the execution of effector functions leading to (and/or promoting) tissue repair and remodeling. In addition to their capacity to synthesize and release a variety of immunoregulatory molecules,[30] some studies have suggested that eosinophils may function as antigen presenting cells (APCs)[31] or enhance DC maturation.[19] Eosinophils may also affect local T cell responses by modulating the balance of T-helper type 1 (T_h1) and T_h2 immune responses [e.g., through eosinophil-derived indoleamine 2,3-dioxy-genase (IDO) production of kynurenine[32]]. Importantly, IDO appears to be essential for the induction of tolerance by tissue recruited T cells. Thus, similar to other eosinophil-mediated immunosupressive activities (e.g., the potential induction of T-regulatory cells through TGF-β production),[33] eosinophil-derived IDO may also play a crucial role in immunosuppression and potentially facilitate tissue repair and, as a by-product, tumor growth.

EOSINOPHILS AND THE IMMUNOTHERAPY OF PATIENTS WITH CANCER

The eosinophil plays a somewhat passive role in the tumor. Eosinophils are frequently observed following immunotherapy with IL-2,[34] IL-4,[35,36] GM-CSF,[37] repeated vaccination,[38] and antibodies to CTLA-4, but the significance of their appearance remains largely unknown. In particular, the antitumor effects of successful cytokine therapy of cancer with IL-2 has been associated with the identification of degranulating eosinophils within the tumor,[34] suggesting that eosinophil effector functions (e.g., direct or antibody-dependent tumor cell lysis or the immunoregulatory capacity of eosinophils modulating the local tumor microenvironment) may play a role in the anticancer activities mediated by systemic IL-2 administration.[34,39] However, despite the promise of these potential eosinophil-mediated antitumor activities, the presence of eosinophils is not an important prognostic indicator in high-dose IL-2-treated patients.

Mouse studies suggesting a link between eosinophils and the therapeutic value of antitumor responses associated with IL-4 administration[40] have led to clinical trials to evaluate these responses in cancer patients. In a phase I clinical trial of IL-4 administered to cancer patients, Sosman and colleagues[35] showed that IL-4 therapy induced systemic eosinophil degranulation associated with increased serum and urine major basic protein (MBP) levels. Moreover, the increase in serum MBP was IL-4 dose dependent. Unfortunately, the link of antitumor activities with eosinophil numbers in these patients is only correlative and similar to observations made in patients following IL-2 administration. No definitive conclusions can be made as to whether and how eosinophils modulate tumor growth.

Efforts to demonstrate experimentally a role for eosinophils in tumor immunity have also been fraught with complicating factors yielding qualified interpretations. Most notably, considerable excitement was generated by data from the elegant studies of Tepper and colleagues[36,40,41] that demonstrated in athymic nude mice that malignant cell lines transfected for constitutive expression of interleukin 4 (IL-4) elicited a tumor associated macrophage and eosinophil infiltrate that led to the attenuation of tumor growth. This provoked a series of tumor xenograft studies that attempted to define the cellular and molecular mechanisms of the apparent IL-4-mediated antitumor effect. Although these studies have shown that spontaneous tumors showed evidence of tumor regression, associated with tumor-infiltrating eosinophils, none of these studies has resolved the role(s) of eosinophils in tumor rejection reactions. Recently, colon tumor eradication by eosinophils in murine models suggested a more conventional cytolytic response.[42] In this study, coculture of eosinophils with colorectal tumor cells led to the release of eosinophil cationic protein and eosinophil-derived neurotoxin, as well as TNF-α secretion. Interestingly, this may be related to the ability of eosinophils to both lyse and promote clearance of *stressed* or damaged cells.[43] Eosinophils accumulate at unique sites in response to cell turnover, thus regulating tissue homeostasis, and are regulators of *Local Immunity and/or Remodeling/Repair* according to the LIAR hypothesis, suggesting a more nuanced and interesting role for these cells in damaged tissues and tumors.[44]

Within several tumor types, including gastrointestinal tumors, tumor-associated tissue eosinophilia (TATE) is associated with a significantly better prognosis.[30] The converse is true in other tumor types, such as differentiated oral squamous cell carcinoma[30] or Hodgkin lymphoma,[45,46] where eosinophils may be involved in promoting cancer cell growth. The mechanism by which eosinophils, in particular, are recruited into tumor tissues is largely unknown. We could characterize DAMPs, including the nuclear protein HMGB1, as candidate factors eliciting eosinophil chemotaxis into tumor tissue.[47] Thus, eosinophil activities are likely to have multiple roles, dictated by specific circumstances, which were originally adapted to maintain tissue homeostasis. Eosinophils are not only able to destroy tissue but are also attracted and activated by stressed and damaged cells, as we and others have demonstrated.[16,47] It is likely that stressed cells attract and activate eosinophils by expressing molecules such as NKG2D ligand 4 (LETAL), major histocompatibility complex class I chain related A (MICA), and MICB,[11] as well as other NKG2D ligands or UL16-binding proteins (ULBPs). These stress-associated molecules all serve as ligands for NKG2D, first described on natural killer (NK) cells[45] and subsequently on eotaxin-activated eosinophils[48] and T cells.[49] Thus, tumor-associated eosinophils appear to have at least two dominant nonoverlapping activities:

1. Destructive effector functions that may limit tumor growth and induce recruitment and activation of other leukocytes;
2. Immunoregulatory and remodeling activities that suppress immune response and release cytokines, thus promoting wound healing.

Consistent with the hypothesis that DAMPs initiate innate immune cell activation when encountering microbes or parasites,[50] eosinophils are often first responders to tissue damage and likely mediate some aspects of tissue remodeling and repair. The presence of DAMPs such as HMGB1 in the necrotic areas of tumors may, in part, elicit both eosinophil tissue recruitment and localized execution of effector functions such as degranulation. The available data, however, suggest that while all *threatened* epithelial cells, including cancer cells, release DAMPs, not all eosinophil tissue infiltration is associated with tumor eradication.[51,53] This conclusion again suggests that the relationship of eosinophils with the modulation of tumor onset/growth is complex and that expression of DAMPs, the emergent role of autophagy, and redox chemistry[54−68] is likely to be only one of several inflammatory mechanisms capable of eliciting eosinophil effector functions.

Interestingly, of all tested biologic activities, eosinophils respond most sensitively to the presence of DAMPs, including HMGB1, with generation of peroxide leading to oxidative degradation and thus inactivation of DAMPs.[15] These DAMPs with or without oxidation, play differential roles in the recruitment of mesenchymal stem cells as well as T regulatory cells which may confound and limit the antitumor response.[69−72]

CONCLUSION

In summary, while the role of eosinophils in tumor onset and growth is unresolved, recent studies suggest that eosinophils are a common and robust tumor infiltrate and that much interesting biology remains to be explored. Specifically, do eosinophil activities limit tumor growth through destructive effector functions or do eosinophil-derived immunoregulation and tissue repair/remodeling promote tumor growth and metastasis? How does the critical role of redox[49] inform eosinophil function in relation to DAMPs? The resolution of these questions could inform the initiation of eosinophil-based modalities and, in turn, novel therapeutic approaches to treat cancer patients.

REFERENCES

1. Zeh III HJ, Lotze MT. Addicted to death: invasive cancer and the immune response to unscheduled cell death. *J Immunother* 2005;**28**(1):1−9.
2. Cerutti PA. Prooxidant states and tumor promotion. *Science (New York, NY)* 1985;**227**(4685):375−81.
3. Moore RJ, Owens DM, Stamp G, Arnott C, Burke F, East N, et al. Mice deficient in tumor necrosis factor-alpha are resistant to skin carcinogenesis. *Nature medicine* 1999;**5**(7):828−31.
4. Sonveaux P, Vegran F, Schroeder T, Wergin MC, Verrax J, Rabbani ZN, et al. Targeting lactate-fueled respiration selectively kills hypoxic tumor cells in mice. *J Clin Invest* 2008 Dec;**118**(12):3930−42.
5. Semenza GL. Tumor metabolism: cancer cells give and take lactate. *J Clin Invest* 2008 Dec;**118**(12):3835−7.
6. Feron O. Pyruvate into lactate and back: from the Warburg effect to symbiotic energy fuel exchange in cancer cells. *Radiother Oncol* 2009 Sep;**92**(3):329−33.
7. Le A, Cooper CR, Gouw AM, Dinavahi R, Maitra A, Deck LM, et al. Inhibition of lactate dehydrogenase A induces oxidative stress and inhibits tumor progression. *Proc Natl Acad Sci U S A* 2010;**107**(5):2037−42. Feb 2.
8. Martinez-Outschoorn UE, Whitaker-Menezes D, Pavlides S, Chiavarina B, Bonuccelli G, Casey T, et al. The autophagic tumor stroma model of cancer or 'battery-operated tumor growth': A simple solution to the autophagy paradox. *Cell Cycle* 2010 Nov;**9**(21):4297−306.
9. Rothenberg ME, Mishra A, Brandt EB, Hogan SP. Gastrointestinal eosinophils. *Immunological Reviews* 2001;**179**:139−55.
10. Jones RL, Hannan NJ, Kaitu'u TJ, Zhang J, Salamonsen LA. Identification of chemokines important for leukocyte recruitment to the human endometrium at the times of embryo implantation and menstruation. *J Clin Endocrinol Metab* 2004 Dec;**89**(12):6155−67.
11. Matthews AN, Friend DS, Zimmermann N, Sarafi MN, Luster AD, Pearlman E, et al. Eotaxin is required for the baseline level of tissue eosinophils. *Proceedings of the National Academy of Sciences of the United States of America* 1998;**95**(11):6273−8.
12. Yang J, Torio A, Donoff RB, Gallagher GT, Egan R, Weller PF, et al. Depletion of eosinophil infiltration by anti-IL-5 monoclonal antibody (TRFK-5) accelerates open skin wound epithelial closure. *Am J Pathol* 1997 Sep;**151**(3):813−9.
13. Samoszuk M. Eosinophils and human cancer. *Histol Histopathol* 1997;**12**(3):807−12.

14. Rothenberg ME. Eotaxin. An essential mediator of eosinophil trafficking into mucosal tissues. *Am J Respir Cell Mol Biol* 1999;**21**(3):291–5.

15. Lotfi R, Herzog GI, DeMarco RA, Beer-Stolz D, Lee JJ, Rubartelli A, et al. Eosinophils oxidize damage-associated molecular pattern molecules derived from stressed cells. *J Immunol* 2009 Oct 15;**183**(8):5023–31.

16. Cormier SA, Taranova AG, Bedient C, Nguyen T, Protheroe C, Pero R, et al. Pivotal advance: eosinophil infiltration of solid tumors is an early and persistent inflammatory host response. *Journal of Leukocyte Biology* 2006;**79**(6):1131–9.

17. Lee JJ, Lee NA. Eosinophil degranulation: an evolutionary vestige or a universally destructive effector function? *Clin Exp Allergy* 2005;**35**(8):986–94.

18. Afshar K, Vucinic V, Sharma OP. Eosinophil cell: pray tell us what you do! *Current Opinion in Pulmonary Medicine* 2007;**13**(5): 414–21.

19. Lotfi R, Lotze MT. Eosinophils induce DC maturation, regulating immunity. *Journal of Leukocyte Biology* 2008;**83**(3):456–60.

20. Nagata M, Sedgwick JB, Bates ME, Kita H, Busse WW. Eosinophil adhesion to vascular cell adhesion molecule-1 activates superoxide anion generation. *J Immunol* 1995 Aug 15;**155**(4):2194–202.

21. Bandeira-Melo C, Woods LJ, Phoofolo M, Weller PF. Intracrine cysteinyl leukotriene receptor-mediated signaling of eosinophil vesicular transport-mediated interleukin-4 secretion. *J Exp Med* 2002 Sep 16;**196**(6):841–50.

22. Henderson WR, Harley JB, Fauci AS. Arachidonic acid metabolism in normal and hypereosinophilic syndrome human eosinophils: generation of leukotrienes B4, C4, D4 and 15-lipoxygenase products. *Immunology* 1984 Apr;**51**(4):679–86.

23. Turk J, Maas RL, Brash AR, Roberts LJ, Oates JA. Arachidonic acid 15-lipoxygenase products from human eosinophils. *J Biol Chem* 1982 Jun 25;**257**(12):7068–76.

24. Bruijnzeel P, Kok P, Kijne G, Verhagen J. Leukotriene synthesis by isolated granulocytes from intrinsic and extrinsic asthmatics and age-matched controls. *Agents Actions Suppl* 1989;**28**:191–4.

25. Lee T, Lenihan DJ, Malone B, Roddy LL, Wasserman SI. Increased biosynthesis of platelet-activating factor in activated human eosinophils. *J Biol Chem* 1984 May 10;**259**(9):5526–30.

26. Hay DW, Torphy TJ, Undem BJ. Cysteinyl leukotrienes in asthma: old mediators up to new tricks. *Trends Pharmacol Sci* 1995 Sep;**16**(9):304–9.

27. Ford-Hutchinson AW, Bray MA, Doig MV, Shipley ME, Smith MJ. Leukotriene B, a potent chemokinetic and aggregating substance released from polymorphonuclear leukocytes. *Nature* 1980 Jul 17;**286**(5770):264–5.

28. Busse WW. Leukotrienes and inflammation. *Am J Respir Crit Care Med* 1998 Jun;**157**(6 Pt 2):S210–3.

29. Stenfeldt AL, Wenneras C. Danger signals derived from stressed and necrotic epithelial cells activate human eosinophils. *Immunology* 2004;**112**(4):605–14.

30. Lotfi R, Lee JJ, Lotze MT. Eosinophilic granulocytes and damage-associated molecular pattern molecules (DAMPs): role in the inflammatory response within tumors. *J Immunother* 2007 Jan;**30**(1):16–28.

31. Akuthota P, Wang H, Weller PF. Eosinophils as antigen-presenting cells in allergic upper airway disease. *Curr Opin Allergy Clin Immunol* 2010 Feb;**10**(1):14–9.

32. Odemuyiwa SO, Ghahary A, Li Y, Puttagunta L, Lee JE, Musat-Marcu S, et al. Cutting edge: human eosinophils regulate T cell subset selection through indoleamine 2,3-dioxygenase. *J Immunol* 2004;**173**(10):5909–13.

33. Schramm C, Huber S, Protschka M, Czochra P, Burg J, Schmitt E, et al. TGFbeta regulates the CD4+CD25+ T-cell pool and the expression of Foxp3 in vivo. *Int Immunol* 2004 Sep;**16**(9):1241–9.

34. Huland E, Huland H. Tumor-associated eosinophilia in interleukin-2-treated patients: evidence of toxic eosinophil degranulation on bladder cancer cells. *J Cancer Res Clin Oncol* 1992;**118**(6):463–7.

35. Sosman JA, Bartemes K, Offord KP, Kita H, Fisher SG, Kefer C, et al. Evidence for eosinophil activation in cancer patients receiving recombinant interleukin-4: effects of interleukin-4 alone and following interleukin-2 administration. *Clin Cancer Res* 1995;**1**(8):805–12.

36. Tepper RI, Coffman RL, Leder P. An eosinophil-dependent mechanism for the antitumor effect of interleukin-4. *Science* 1992;**257**(5069):548–51.

37. Bristol JA, Zhu M, Ji H, Mina M, Xie Y, Clarke L, et al. In vitro and in vivo activities of an oncolytic adenoviral vector designed to express GM-CSF. *Mol Ther* 2003;**7**(6):755–64.

38. Schaefer JT, Patterson JW, Deacon DH, Smolkin ME, Petroni GR, Jackson EM, et al. Dynamic changes in cellular infiltrates with repeated cutaneous vaccination: a histologic and immunophenotypic analysis. *J Transl Med* 2010 Aug;**20**(8):79.

39. Rivoltini L, Viggiano V, Spinazze S, Santoro A, Colombo MP, Takatsu K, et al. In vitro anti-tumor activity of eosinophils from cancer patients treated with subcutaneous administration of interleukin 2. Role of interleukin 5. *International Journal of Cancer* 1993;**54**(1):8–15.

40. Tepper RI, Pattengale PK, Leder P. Murine interleukin-4 displays potent anti-tumor activity in vivo. *Cell* 1989 May 5;**57**(3):503–12.

41. Tepper RI. The eosinophil-mediated antitumor activity of interleukin-4. *The Journal of Allergy and Clinical Immunology* 1994; **94**(6 Pt 2):1225–31.

42. Legrand F, Driss V, Delbeke M, Loiseau S, Hermann E, Dombrowicz D, et al. Human Eosinophils Exert TNF-{alpha} and Granzyme A-Mediated Tumoricidal Activity toward Colon Carcinoma Cells. *J Immunol* 2010 Nov 10;**185**(12):7443–51.

43. Kim HJ, Alonzo ES, Dorothee G, Pollard JW, Sant'Angelo DB. Selective depletion of eosinophils or neutrophils in mice impacts the efficiency of apoptotic cell clearance in the thymus. *PLoS One* 2010 Jul 6;**5**(7):e11439.

44. Lee JJ, Jacobsen EA, McGarry MP, Schleimer RP, Lee NA. Eosinophils in health and disease: the LIAR hypothesis. *Clin Exp Allergy* 2010 Apr;**40**(4):563–75.

45. Costello RT, Fauriat C, Sivori S, Marcenaro E, Olive D. NK cells: innate immunity against hematological malignancies? *Trends in Immunology* 2004;**25**(6):328–33.

46. Enblad G, Molin D, Glimelius I, Fischer M, Nilsson G. The potential role of innate immunity in the pathogenesis of Hodgkin's lymphoma. *Hematology/oncology Clinics of North America* 2007;**21**(5):805–23.

47. Lotfi R, Herzog GI, DeMarco RA, Beer-Stolz D, Lee JJ, Rubartelli A, et al. Eosinophils oxidize damage-associated molecular pattern molecules derived from stressed cells. *J Immunol* 2009 Oct 15;**183**(8):5023–31.

48. Kataoka S, Konishi Y, Nishio Y, Fujikawa-Adachi K, Tominaga A. Antitumor activity of eosinophils activated by IL-5 and eotaxin

against hepatocellular carcinoma. *DNA Cell Biol* 2004;**23**(9):549—60.

49. Burgess SJ, Marusina AI, Pathmanathan I, Borrego F, Coligan JE. IL-21 Down-Regulates NKG2D/DAP10 Expression on Human NK and CD8+ T Cells. *J Immunol* 2006;**176**(3):1490—7.

50. Matzinger P. An innate sense of danger. *Ann N Y Acad Sci* 2002 Jun;**961**:341—2.

51. Coussens LM, Pollard JW. Leukocytes in mammary development and cancer. *Cold Spring Harb Perspect Biol* 2011 Mar 1;**3**(3). pii:a003285.

52. Curran CS, Evans MD, Bertics PJ. GM-CSF production by glioblastoma cells has a functional role in eosinophil survival, activation, and growth factor production for enhanced tumor cell proliferation. *J Immunol* 2011 Aug 1;**187**(3):1254—63.

53. Hudson SA, Herrmann H, Du J, Cox P, Haddad E, Butler B, et al. Developmental, malignancy-related, and cross-species analysis of eosinophil, mast cell, and basophil Siglec-8 expression. *J Clin Immunol* 2011 Dec;**31**(6):1045—53.

54. Kang R, Loux T, Tang D, Schapiro NE, Vernon P, Livesey KM, et al. The expression of the receptor for advanced glycation endproducts (RAGE) is permissive for early pancreatic neoplasia. *Proc Natl Acad Sci U S A* 2012 May 1;**109**(18):7031—6.

55. Buchser WJ, Laskow TC, Pavlik PJ, Lin HM, Lotze MT. Cell-mediated Autophagy Promotes Cancer Cell Survival. *Cancer Res*; 2012 Apr 17 [Epub ahead of print].

56. Weiner LM, Lotze MT. Tumor-cell death, autophagy, and immunity. *N Engl J Med* 2012 Mar 22;**366**(12):1156—8.

57. Livesey KM, Kang R, Vernon P, Buchser W, Loughran P, Watkins SC, et al. p53/HMGB1 Complexes Regulate Autophagy and Apoptosis. *Cancer Res* 2012 Apr 15;**72**(8):1996—2005.

58. Liang X, de Vera ME, Buchser WJ, Romo d, V, Loughran P, Beer-Stolz D, et al. Inhibiting Autophagy During Interleukin 2 Immunotherapy Promotes Long Term Tumor Regression. *Cancer Res*; 2012 Apr 3.

59. Kang R, Livesey KM, Zeh III HJ, Lotze MT, Tang D. Metabolic regulation by HMGB1-mediated autophagy and mitophagy. *Autophagy* 2011 Oct;**7**(10):1256—8.

60. Tang D, Kang R, Livesey KM, Kroemer G, Billiar TR, van HB, et al. High-mobility group box 1 is essential for mitochondrial quality control. *Cell Metab* 2011 Jun 8;**13**(6):701—11.

61. Kang R, Livesey KM, Zeh III HJ, Lotze MT, Tang D. HMGB1 as an autophagy sensor in oxidative stress. *Autophagy* 2011 Aug;**7**(8):904—6.

62. Tang D, Kang R, Livesey KM, Zeh III HJ, Lotze MT. High mobility group box 1 (HMGB1) activates an autophagic response to oxidative stress. *Antioxid Redox Signal* 2011 Oct 15;**15**(8):2185—95.

63. Amaravadi RK, Lippincott-Schwartz J, Yin XM, Weiss WA, Takebe N, Timmer W, et al. Principles and current strategies for targeting autophagy for cancer treatment. *Clin Cancer Res* 2011 Feb 15;**17**(4):654—66.

64. Kang R, Tang D, Lotze MT, Zeh III HJ. RAGE regulates autophagy and apoptosis following oxidative injury. *Autophagy* 2011 Apr;**7**(4):442—4.

65. Kang R, Zeh HJ, Lotze MT, Tang D. The Beclin 1 network regulates autophagy and apoptosis. *Cell Death Differ* 2011 Apr;**18**(4):571—80.

66. Tang D, Lotze MT, Kang R, Zeh HJ. Apoptosis promotes early tumorigenesis. *Oncogene* 2011 Apr 21;**30**(16):1851—4.

67. Kang R, Tang D, Livesey KM, Schapiro NE, Lotze MT, Zeh III HJ. The Receptor for Advanced Glycation End-products (RAGE)

protects pancreatic tumor cells against oxidative injury. *Antioxid Redox Signal* 2011 Oct 15;**15**(8):2175—84.

68. Tang D, Kang R, Zeh III HJ, Lotze MT. High-mobility group box 1, oxidative stress, and disease. *Antioxid Redox Signal* 2011 Apr 1;**14**(7):1315—35.

69. Bergmann C, Wild CA, Narwan M, Lotfi R, Lang S, Brandau S. Human tumor-induced and naturally occurring Treg cells differentially affect NK cells activated by either IL-2 or target cells. *Eur J Immunol* 2011 Dec;**41**(12):3564—73.

70. Lotfi R, Eisenbacher J, Solgi G, Fuchs K, Yildiz T, Nienhaus C, et al. Human mesenchymal stem cells respond to native but not oxidized damage associated molecular pattern molecules from necrotic (tumor) material. *Eur J Immunol* 2011 Jul;**41**(7):2021—8.

71. Wild CA, Brandau S, Lotfi R, Mattheis S, Gu X, Lang S, et al. HMGB1 is overexpressed in tumor cells and promotes activity of regulatory T cells in patients with head and neck cancer. *Oral Oncol* 2012 May;**48**(5):409—16.

72. Wild CA, Bergmann C, Fritz G, Schuler P, Hoffmann TK, Lotfi R, et al. HMGB1 conveys immunosuppressive characteristics on regulatory and conventional T cells. *Int Immunol*; 2012 Apr 3 [Epub ahead of print].

Chapter 13.11

Eosinophils and Chronic Rhinosinusitis

Robert P. Schleimer, Atsushi Kato and Robert Kern

INTRODUCTION

Eosinophils have been associated with numerous diseases affecting multiple organ systems and, as this volume attests, their activities have been implicated as a cause of tissue pathology in a wide range of these conditions. Eosinophils are strongly implicated in the pathogenesis of allergic diseases, including allergic rhinitis, asthma, atopic dermatitis, food allergy, and others. Although chronic rhinosinusitis (CRS) is not strictly an allergic disease, it is often associated with allergy and the appearance of intense eosinophilia in tissues from the upper airways and sinuses of patients suffering from this condition has been known for the better part of a century. The purpose of this subchapter is to review the evidence showing an association between the eosinophil and CRS; the mechanisms of recruitment of eosinophils into sinonasal tissue in CRS patients; the roles that eosinophils play in the pathogenesis of this disease; and the potential for treatment modalities based on targeting the eosinophil.

CRS is a disease of the upper airways and paranasal sinuses that affects 5—10% of the general population. It is typically characterized by one or more of the following

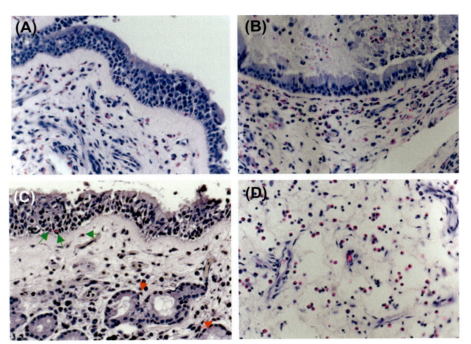

FIGURE 13.11.1 *A—D*, Hematoxylin and eosin stain of different views of typical nasal polyps from a patient with chronic rhinosinusitis. *C*, Green arrows denote clearly stained, *classic* eosinophils and red arrows refer to eosinophils that, while not as clearly visible as the eosinophils denoted by green arrows, nevertheless do have the pink granules surrounding their nuclei that are characteristic of eosinophilia. (magnification ×400) *(Courtesy of Roderick Carter.)*

symptoms: copious secretion of mucus, facial pressure and headache, fatigue, loss of smell, and nasal obstruction. The physical exam may be unremarkable, although in some cases nasal endoscopy may reveal the presence of purulent drainage and/or nasal polyps. Historically, CRS has been subdivided according to many of its pathophysiological features, including appearance of fungal mucins, comorbidity with asthma, formation of polyps, hyperplasia of connective tissue, and sensitivity to aspirin. More recently, partly for the sake of simplifying the study of this condition, it has generally been divided into CRS with nasal polyps (CRSwNP) and CRS without nasal polyps (CRSsNP), a terminology that we will use here. The vast majority of CRS cases are idiopathic with an unpredictable clinical course; however, a minority have a more characteristic clinical picture. Specifically, CRS with fungal mucins together with fungal atopy is often viewed as a distinct condition known as AFS (allergic fungal sinusitis). Patients with aspirin allergy, asthma, and nasal polyps have a syndrome known as Samter's triad, characterized by a severe form of recurrent polyposis. Lastly, CRS occurs in nearly all patients with cystic fibrosis, commonly with nasal polyposis, and this is frequently categorized as a discrete entity as well. In the United States, there are in excess of 300,000 surgeries performed annually to relieve the suffering of patients with CRS, and the discarded tissue from these procedures has provided access to tissue for those interested in studying the pathogenesis of the disease. The formation of nasal polyps occurs in roughly 15–20%

of patients with CRS and the inflammatory process within the polyp is particular intense and may differ from inflammation that occurs in CRSsNP. The level of eosinophilia is greater in CRSwNP and this form of the disease has been the most extensively studied. Consequently, we primarily focus the discussion in this review on the role of the eosinophil in CRS with nasal polyps.

ASSOCIATION OF EOSINOPHILS WITH SINONASAL INFLAMMATION IN CHRONIC RHINOSINUSITIS

The appearance of eosinophils in pathological conditions in the airways was noticed shortly after Paul Ehrlich developed the acidic dyes that continue to be used to detect these cells today.[1] CRS is one of the most intensely eosinophilic diseases, rivaling hypereosinophilic syndromes, eosinophilic esophagitis and eosinophilic gastroenteritis, in terms of tissue density of eosinophils (Fig. 13.11.1).[2,3] Bachert and his colleagues have shown that average eosinophil numbers are elevated in both CRSsNP and CRSwNP, but they are higher in CRSwNP and highest within nasal polyps themselves.[4] In our own studies, eosinophilia is the highest in patients with Samter's triad (see below—unpublished observations). Although some controversy exists as to whether there are relative differences in the eosinophilia of nasal polyps in populations of African, Caucasian, and Chinese descent, Bachert has reported low levels of eosinophilia in Chinese patients with CRS.[5–7] Such studies are often complicated by the fact that the degree of eosinophilia in sinonasal tissue of patients with CRS can be dramatically altered by therapies that patients receive prior to surgery, especially oral glucocorticoids, which can diminish the eosinophil number by an order of magnitude (see below). It has been suggested that some intranasal Chinese herbal remedies may also alter eosinophil numbers.[6] Along with elevated levels of eosinophils, nasal polyps also demonstrate elevated levels of B cells, dendritic cells, neutrophils, macrophages, mast cells, and T cells, and thus represent a tissue undergoing a robust immune and inflammatory response. Although mast cells are not the subject of this volume, it must be acknowledged that they are likely to play an important role in the pathogenesis of CRS, especially in the tissue swelling and formation of nasal polyps that occur.[8,9] Detection of eosinophils and eosinophilia in CRS has employed assays for eosinophil granule proteins, such as eosinophil cationic protein (ECP), as well as specific antibodies against eosinophil granule proteins, notably EG2, which is considered to specifically stain activated eosinophils.[10,11] Studies using these approaches have confirmed the presence of eosinophilia in CRS and provided evidence for activation of the tissue-resident eosinophils.[12–14] Persson and colleagues have emphasized the importance of cytolytic eosinophil degranulation and release of clusters of free granules. Evidence has been provided to demonstrate that eosinophils in sinonasal tissue of CRS patients undergo both piecemeal degranulation and cytolytic degranulation.[15,16] Recently, a method has been developed to purify eosinophils from human tissues, including nasal polyps, that should allow further studies into the activation state of eosinophils and gene expression in CRS.[17]

CRS is often comorbid with asthma, and one of the most severe forms of the disease, nasal polyps with asthma and aspirin sensitivity (referred to often as Samter's triad), manifests both severe polyposis and severe, often glucocorticoid-resistant, asthma.[18] CRS with nasal polyps is also comorbid with Churg-Strauss syndrome, a systemic vasculitic disorder in which eosinophils feature prominently.[19] Several groups have compared the level of eosinophils in sinonasal tissue with asthma severity, levels of eosinophils in bronchial lavage, or sputum samples from the lower airways. In one study, levels of eosinophils in lavage from the middle meatus correlated with FEV_1 (forced expiratory volume in one second) in asthmatics with CRS.[20] In another study, the presence of eosinophils in sinonasal tissue was higher in CRS patients undergoing sinus surgery who had comorbid asthma than in those who did not have asthma.[21] Mehta et al. found that the extent of CRS disease as measured by sinus computed tomography (CT) scores correlated with eosinophils in blood and sputum, suggesting that systemic elevation and activation of eosinophils may be a feature of CRS.[22] Patients with

CRSwNP were found to have increased asthma prevalence, as well as increased exhaled nitric oxide.[23]

The etiology of CRS has yet to be definitively ascribed to infection with any single pathogen or class of pathogens. It is clear that patients with CRS experience frequent acute, presumably infectious, exacerbations of disease, and CRS patients are generally treated frequently with antibiotics, suggesting that treating physicians suspect the presence of bacteria. Along these lines, it has been suggested that patients with CRS may have a defect in the innate immune barrier, making them more susceptible to infection or colonization with microorganisms in general.[24,25] Another line of investigation implicates fungi, or at least allergy to fungi, in CRS, based on the presence of activation of lymphocytes to express cytokines in response to fungal extracts.[26,27] While atopy to aeroallergens is frequently seen in patients with CRS, nearly half of CRS patients are nonatopic according to standard tests, and based on this present state of knowledge, CRS should not be viewed as a classical allergic disease.[28] Yet another line of investigation suggests that Staphylococcus aureus and the toxins it produces are important inciting factors in CRS.[29–32] According to the superantigen hypothesis of CRS, staphylococcus-derived superantigens drive a T-helper type 2 (T_h2)-skewed inflammatory response that is responsible for the eosinophilia observed in CRS (the mechanism of eosinophilia in CRS is discussed below). Superantigens activate large numbers of T cells by cross-linking class II MHC on antigen-presenting cells with specific Vβ regions on the T-cell receptor, leading to profound release of cytokines, in some cases skewed toward T_h2.[33] Reports on the extent of staphylococcal colonization in CRS have been variable but most have demonstrated normal or supranormal frequencies. Bachert et al. and Desrosiers et al. have reported that 50–90% of CRS patients with nasal polyps have colonization with staphylococcus.[34,35] Staphylococcal superantigens can also act as allergens, and several reports have detected staphylococcal superantigens in the airways of CRS patients and/or demonstrated the presence of functional immunoglobulin E (IgE) antibodies directed against the staphylococcal superantigens.[36–39] A recent study by Bachert et al. has implicated the presence of interleukin-5 (IL-5) and IgE directed to staphylococcal enterotoxins in CRS comorbid with asthma.[40] A longitudinal study demonstrated that high numbers of eosinophils in the nasal mucosa or in mucus collected from CRS patients are strong risk factors for recurrence of disease and the need for subsequent surgery in a 5-year follow-up study.[41] In summary, eosinophil numbers are increased in sinonasal tissue of patients with CRS and increased the most in those with nasal polyps, those with asthma, and especially those with all three conditions. The correlation of eosinophils with disease severity implicates them as either a biomarker of the pathogenic process or a mediating cell responsible for disease.

MECHANISMS FOR RECRUITMENT AND ACTIVATION OF EOSINOPHILS IN CHRONIC RHINOSINUSITIS

The molecular mechanisms by which eosinophils are recruited to tissues have been reviewed and are the topic of another subchapter in this volume. We restrict our comments therefore to the available evidence on recruitment of eosinophils in CRS. In general, several important processes occur. One is the priming and survival-promoting effects of cytokines on eosinophils, especially including granulocyte-macrophage colony-stimulating factor (GM-CSF) and IL-5, both of which also play a role in the generation of eosinophils. Another is the local expression of eosinophil-attracting chemokines by epithelium and other tissue cells. The expression of adhesion molecules by endothelium, especially vascular cell adhesion protein 1 (VCAM-1), is equally important in the rolling, arrest and transmigration of these cells into the affected tissue. An extensive literature that shows the elevation of these factors in sinonasal tissue of patients with CRS will be summarized next.

The primary receptor driving chemokine-mediated recruitment of eosinophils is C-C chemokine receptor 3 (CCR3), which has a number of ligands, notably RANTES (or C-C motif chemokine 5; CCL5), eotaxin (CCL11), eotaxin-2 (CCL24), eotaxin-3 (CCL26), and MCP-4 (CCL13), with less activity for MCP-1, MCP-2, and MCP-3.[42,43] Mice lacking CCR3 have severe, but not totally diminished, eosinophil infiltration in allergic inflammation, and CCR3 antagonists are under investigation in clinical trials. Beck et al. found elevated levels of RANTES mRNA and tissue staining in nasal polyp tissue but not in normal turbinate tissue.[44] RANTES expression is primarily found in epithelial cells, an observation leading to speculation that localization of eosinophils to the epithelium and lamina propria may be related to epithelial chemokine expression.[45–47] Similar findings have subsequently been made with other CCR3 ligands, including eotaxins 1–3 and MCP-4.[48–51] A careful study by Jahnsen et al. found that levels of mRNA for eotaxin-1, eotaxin-2, and MCP-4 are elevated in nasal polyp tissue and that levels of eotaxin-2 and MCP-4 in turbinates from CRSwNP patients are higher than in normal turbinates.[49] Taken together, it is clear that both surrounding tissue (i.e., turbinate) and affected tissue (nasal polyps, which commonly emerge from the ethmoid sinuses and other deeper tissues) in patients with CRS have elevated levels of CCR3 ligands that are likely to play a role in eosinophil recruitment to both of these regions in CRS. Regulation of chemokine expression by epithelium is complex, but both NF-κB and signal transducer and activator of transcription 6 (STAT6) play important roles in the response, and T$_h$2 cytokines including IL-4 and IL-13 are important inducers.[52,53] Other stimuli may also be important in activating CCR3 ligand expression in CRS, and chitin has recently been shown to induce eotaxin-3 in human sinonasal epithelial cells *in vitro*.[54]

Eosinophils not only respond to CCR3 agonists but also release eotaxin, eotaxin-2, eotaxin-3, and RANTES.[51,55] The release of CCR3 agonists from eosinophils may be involved in the local eosinophil accumulation in nasal polyps. Recently, our group has found that MPIF-1 (CCL23) is elevated in eosinophilic nasal polyps and that EG2$^+$ eosinophils are major MPIF-1-producing cells in nasal polyps.[55a] MPIF-1 is a ligand for CCR1 and is known to recruit monocytes, macrophages, and dendritic cells.[56,57] These findings indicate that activation of eosinophils may further enhance local inflammation via secondary recruitment of additional cells in nasal polyps.

Exposure of eosinophils to GM-CSF or IL-5 leads to several notable phenotypic changes that are likely to be relevant to their accumulation in CRS, including increased expression and function of adhesion molecules, increased transendothelial migration, and increased survival. The earliest studies on CRS were by Denburg and colleagues, and showed elevated levels of GM-CSF in eosinophils in nasal polyp tissue,[58,59] and that GM-CSF is prominently expressed by epithelial cells, in agreement with earlier *in vitro* studies showing that epithelial cells are a rich source of this cytokine.[60] Several studies have shown that conditioned medium from nasal polyps stimulated with antigen or from cultured nasal polyp epithelial cells can prolong eosinophil survival or activate eosinophils *in vitro* and that the activity can be significantly blocked by anti-GM-CSF antibodies.[61–64] Additional studies have shown that IL-5 is also an important eosinophil priming and survival factor in nasal polyp tissue.[13,65–67] This finding is supported by success in recent clinical trials by Bachert et al. using anti-IL-5 antibodies (see below). The relative importance of IL-5 and GM-CSF as priming cytokines in sinonasal tissue is unknown. Liu and Busse found that lung-migrating eosinophils have reduced expression of the IL-5 receptor in asthmatics and postulated that the main effects of IL-5 (and therefore probably anti-IL-5) may be in the bone marrow and circulation.[68] Whether the same is true in the upper airways and sinuses in patients with CRS requires further investigation.

In vitro, VCAM-1 is known to be an important and relatively selective endothelial adhesion molecule that mediates eosinophil rolling, firm adhesion, and transendothelial migration. Several groups have demonstrated increased endothelial expression of VCAM-1 in nasal polyps and shown that VCAM-1 levels correlate with the presence of eosinophils, leading to the hypothesis that recruitment of eosinophils in CRS is partially mediated by VCAM-1.[46,69–73] Using the Stamper–Woodruff assay, Symon, Wardlaw, and collaborators have demonstrated that eosinophils bind to nasal polyp tissue *in vitro* via interactions with P-selectin and on this basis suggested a role for P-selectin as well.[74] In general, the T cell cytokine milieu that most drives eosinophilic inflammation is a T_h2 pattern, including the VCAM-1 activators IL-4 and IL-13 as well as the eosinophil priming cytokine IL-5. Although most CRS is probably T_h2 driven, it has been suggested that in some cases of CRS, including nonallergic CRS, other T-cell cytokines may be involved.[75,76] At present, there is controversy about whether T_h17 are elevated in CRS. Saitoh et al. have reported increased IL-17 in CRS, while Peters et al. (our group) and Bachert and Geveart have not found evidence for a role of IL-17 in CRS.[77–79] As far as eosinophil recruitment and activation in CRS is concerned, there are reports of a possible role for complement activation, IL-33, and activation of protease-activated receptors (PAR), eicosanoids, and stem cell factor. In some cases, these pathways may activate local resident or infiltrating cells to express eosinophil priming and recruitment factors, including chemokines and cytokines such as GM-CSF.[80–85]

POTENTIAL ROLES OF EOSINOPHILS IN THE PATHOGENESIS OF CHRONIC RHINOSINUSITIS

The discussion above firmly establishes that eosinophils are present in high numbers in the sinonasal tissue of patients with CRS, especially the polypoid form, and that these tissue eosinophils are activated. Together, these two lines of investigation provide circumstantial evidence that eosinophils may play a role in alterations to nasal physiology, tissue structure, and clinical phenotype. This section discusses some features of CRS that may be mediated by activated eosinophils. In most cases, these discussions should be viewed as hypothetical or speculative, as there is no validated animal model of CRS and in most cases it is not possible to definitively assess mechanisms in human subjects (other than some recent studies with antibodies against IL-5 or IgE, discussed in the next section). In our view, it is clear that CRSsNP and CRSwNP are distinct disease entities that are primarily related by their frequent resistance to treatment, need for surgery (probably greater in CRSwNP), and eosinophilia (probably greater in CRSwNP). However, they are distinct in many of the clinical and inflammatory processes that we and others have been investigating at the molecular level.[86–89] As to the roles of eosinophils in pathogenesis, by far the most data are available for polypoid CRS, and we will restrict our comments here to CRSwNP, even though they may in many cases apply to CRSsNP. Although eosinophils are elevated in CRS and express the activation marker EG2 (see above), EG2 has not proven to be a reliable marker of degranulation; eosinophils in nasal polyps are undoubtedly activated,

as studies of eosinophil morphology by Erjefält and Persson have demonstrated extensive piecemeal degranulation and cytolytic degranulation in eosinophils in nasal polyp tissue.[16,90]

The most obvious and prominent feature of CRSwNP is the formation of the polyp itself, although details of the mechanism of this metamorphosis of sinonasal tissue are not well understood. Perhaps the best studies have been those by Bachert et al., who studied early-phase and established nasal polyps.[91] They found formation of a subepithelial eosinophilic cap at the upper surface of the tissue outgrowth and implicated fibronectin and deposition of albumin and extracellular matrix proteins in the early and late phase, respectively. The presence of IL-5 and eotaxin-2 correlated with the process, thus leading them to propose an important role for eosinophils.[91] Based on the deposition of albumin, we can presume that vascular leakage is occurring in CRS, although the stimuli for the leak are not well understood. Studies by Steinke, Borish, and others have shown that the 5-lipoxygenase (5-LO) pathway is activated in CRS, especially in aspirin-sensitive disease, and that eosinophils express 5-LO and LTC4 synthase within the polyp tissue.[92,93] Increased eicosanoid metabolism has been reported in CRS and correlates with ECP and IL-5.[82] Although eosinophil cyclooxygenase and lipoxygenase metabolites might be important in vascular leakage driving polyp formation, in general the clinical experience with inhibitors of both of these pathways has been disappointing and they are unlikely to be the primary mediators driving vascular leak in a forming polyp. The relative importance of mast cell and eosinophil mediators in polyp formation is unknown. Di Lorenzo et al. measured ECP, histamine, and tryptase in nasal lavage and found only tryptase and ECP to correlate with symptom scores.[94] Since mast cells are highly elevated in nasal polyps and are capable of releasing numerous mediators of vascular leakage, their potential role in polyp formation must be seriously considered.

Considerable evidence has accumulated to demonstrate elevations of both IgE and IgA in nasal polyp tissue, and it is not unreasonable to speculate that these antibodies play a role in activating mast cells and eosinophils, respectively, in CRSwNP. With respect to eosinophils, activation by IgE, if it occurs at all, most likely occurs indirectly as a result of the action of mediators released from basophils, mast cells, and other cells that clearly express a functional IgE receptor (e.g., inflammatory macrophage-like cells). Kita and colleagues have shown that IgG and IgA, but not IgE, can prolong eosinophil survival, and induce cytokine expression and effector function.[95,96] However, convincing data demonstrate that eosinophils are activated to degranulate by cross-linking receptors for IgA.[97,98] More uncertain is the nature of the antigen systems that might drive IgA-mediated degranulation in nasal polyp eosinophils. As discussed above, only half of patients with CRSwNP are atopic, and it is difficult to implicate typical aeroallergens as anything more than exacerbating factors. Recent studies by Tan and colleagues have demonstrated the presence of autoantibodies of both IgG and IgA isotypes in patients with recalcitrant CRS requiring revision surgery.[99] Suh et al. found correlations between both total IgG and total IgE with the number of EG2[+] cells, but not IgA or secretory IgA.[100] Clearly, further work is required to establish the importance of immunoglobulins in eosinophil activation in CRS.

In both asthma and CRS there is evidence for epithelial damage mediated by eosinophils. Gleich and colleagues demonstrated that the presence of eosinophils and extracellular levels of the toxic eosinophil granule protein major basic protein (MBP) correspond to regions of epithelial injury in patients with CRS.[101] This occurs in asthma as well, and there is ample evidence for a similar pathological process occurring in the two diseases. Chanez, Bousquet, and colleagues and Gaga et al. demonstrated a strong correlation between eosinophils in the nose and lung in asthmatics.[101–103] Bresciani et al. found that 70% of asthmatics have CRS, as determined using clinical and CT scores, and observed a correlation of clinical scores to blood eosinophils in those with mild and moderate asthma.[104] Eosinophil-derived granule mediators are well established to be toxic to epithelium, to activate mast cells and basophils, and to drive inflammation. An ultrastructural investigation of epithelial damage in asthmatic and nonasthmatic nasal polyps revealed reduced length of desmosomes in allergic CRSwNP and in asthmatics.[105] Several studies suggest that the epithelium in both asthma and CRS presents a poor barrier.[24,25,106] Whether damage from eosinophils occurs as a result of the cationic granular proteins, the respiratory burst and peroxidase activation, protein nitration, or another mechanism is worthy of further investigation.[107,108] Recent studies indicate reduced SPINK5 in epithelial tissue from CRS patients.[109] SPINK5 is a protease inhibitor that can regulate barrier function in the skin via preventing activation of PAR2 and subsequent induction of thymic stromal lymphopoietin (TSLP).[110] Briot et al. found that high levels of TSLP in the skin drive a highly eosinophil and mast cell rich inflammation even in the absence of T cells.[110] Kamekura et al. presented evidence that TSLP itself can induce claudins and occludins and enhance tight junction function in nasal epithelial cells *in vitro*, suggesting that its influences may be complex. TSLP is elevated in CRS, and whether TSLP regulates barrier function in CRS and/or contributes to eosinophilia needs further clarification.[111–113]

Studies in the lower airways indicate that eosinophilic inflammation is often associated with fibrotic changes, including the laying down of extracellular matrix proteins or repair proteins (e.g., lumican, procollagen, and tenascin—see Chapter 12).[114,115] Some nasal polyps are characterized by the deposition of collagen and other

matrix proteins, and it is possible that eosinophils may play a role in this process. Studies by Ohno et al. have demonstrated that transforming growth factors α and $\beta1$ (TGF-α and TGF-$\beta1$), as well as platelet-derived growth factor receptor (PDGF), are expressed by eosinophils in nasal polyps and suggest that eosinophil-derived growth factors may alter the structure of the affected nasal mucosa.[116,117] Eosinophil production of TGF in nasal polyps was confirmed by Elovic et al.[118] Ultrastructural studies in nasal polyposis using anti-IL-5 antibodies, such as the ones used by Flood-Page, Kay et al. in asthma, will be required to assess the role of eosinophils in remodeling of the sinonasal tissue in CRS.[114]

EOSINOPHILS AS A THERAPEUTIC TARGET FOR MANAGEMENT OF CHRONIC RHINOSINUSITIS

At present, there are no specific approved therapies for the treatment of CRS. In general, a significant number of patients that present with the diagnosis have failed treatment with antibiotics and/or intranasal glucocorticoids. Since nasal polyps commonly grow out of the sinuses, it is a challenge for intranasal glucocorticoid sprays to penetrate to the area of relevant inflammation. Treatment with oral steroids is often effective, especially in CRSwNP, and is frequently used to treat CRSwNP, although most patients and physicians prefer to avoid chronic treatment with oral glucocorticoids.

Glucocorticoids continue to be the most effective anti-inflammatory drugs available for a wide variety of autoimmune and allergic chronic inflammatory illnesses. Their mechanisms are pleiotropic and represent actions exerted upon numerous cell types known to participate in CRS, including epithelial, T_h1, and T_h2 cells. Other important cells, such as endothelial cells, mast cells, neutrophils, and T_h17 cells, are relatively unresponsive to glucocorticoid treatment and are unlikely to be important targets in CRS.[119] Among the steroid responsive cells are eosinophils, and the actions of glucocorticoids on eosinophils have been reviewed.[120,121] In 2007, Patiar and Reece performed a Cochrane review of the literature and identified only one randomized-controlled study comparing oral steroids with no intervention or placebo.[122,123] As found in earlier uncontrolled trials, oral treatment with prednisone or a similar systemic steroid shrank polyps, reduced symptoms, and improved olfaction in this study. Alobid et al. also found that continued treatment with intranasal steroid maintained the benefits of the 2-week treatment with oral steroids for nearly 1 year.[123] Subsequently, Hissaria et al. performed a controlled trial testing a short course of prednisolone in CRSwNP, and found improved symptom scores and a reduction in polyp size, as determined both subjectively and objectively by MRI.[124] Recently, Van Zele, Bachert, and collaborators performed a double-blind placebo-controlled multicenter trial of oral methylprednisolone in CRSwNP.[125] In addition to shrinking polyps, improving olfaction and reducing symptoms, the oral steroids improved congestion, nasal peak flow rates, postnasal drip, and rhinorrhea. Importantly, in this study methylprednisolone reduced ECP, IgE, and IL-5 in nasal secretions and decreased blood eosinophils, ECP, and soluble IL-5Ra in the serum.[125]

In vitro studies have identified numerous effects of glucocorticoids on eosinophils (for a summary, see[119]). Glucocorticoids diminish eosinophil survival *in vitro* by promoting apoptosis and this effect is blocked by survival-promoting cytokines such as GM-CSF or IL-5. Glucocorticoids also diminish production of these specific cytokines by many cell types, including epithelial cells and T lymphocytes. Steroids have a similar suppressive effect *in vitro* on the expression of cytokines that activate endothelial adhesion molecule expression [e.g., IL-1, IL-4, IL-13, and tumor necrosis factor (TNF)] and of chemokines known to cause eosinophil migration (e.g., CCR3 agonists). Thus, theoretically, glucocorticoids should diminish eosinophil recruitment to the sinuses and nasal cavity and hasten the disappearance of the eosinophils once they arrive in the sinonasal cavity. Numerous *in vivo* studies have addressed these theoretical effects of glucocorticoids by administering intranasal or oral glucocorticoids to patients with CRS prior to surgery or biopsy.

A general feature of these studies is that glucocorticoids uniformly reduce the number of eosinophils found in nasal polyps. Kanai et al. also found treatment with budesonide to reduce the *proportion* of activated (EG2[+]) eosinophils, along with reducing levels of total eosinophils, T cells, the ICAM-1 adhesion molecule, and HLA-DR, a marker of adaptive immune activation.[126] Similar findings were made by Hamilos and coworkers, who additionally found reduced P-selectin (but not IL-1β, TNF, or VCAM-1) in polyps from patients treated with intranasal fluticasone propionate.[127] They also found reduced numbers of cells expressing T_h2 cytokines (IL-4 and IL-13). Delbrouck et al. reported that budesonide decreases levels of both ICAM-1 and VCAM-1, and to a lesser extent P-selectin.[128] As mentioned above, Denburg, Dolovich, and collaborators identified GM-CSF as an important eosinophil-priming and survival-promoting cytokine in the nose of patients with allergic rhinitis and CRS. Nonaka et al. from this group found that budesonide reduces survival of peripheral blood eosinophils *in vitro* but not eosinophils extracted from nasal polyps, suggesting that the polyp eosinophils are primed and rendered glucocorticoid resistant as a result of the GM-CSF exposure within the polyp.[129] It is unclear whether GM-CSF or

IL-5 is the most important eosinophil survival cytokine in CRSwNP, and this is not a question that can be easily addressed, since both of these cytokines have effects on eosinophils outside of the sinonasal cavity. Nonetheless, Bolard et al. found a correlation between levels of IL-5 and eosinophils in the nasal secretions of patients with nasal polyps and that intranasal steroids reduce both of these parameters.[130] Whether glucocorticoid treatment directly induces apoptosis in tissue eosinophils or indirectly induces eosinophil death as a result of suppressing survival-promoting cytokines is still an open question. As to the expression of chemokines that attract eosinophils, Jahnsen et al. found that nasal polyp tissue expresses highly elevated levels of eotaxin, eotaxin-2, MCP-4 and RANTES, and that after treatment with oral glucocorticoid these factors are all profoundly reduced along with a reduction in eosinophil numbers in the tissue.[49] Importantly, in association with the reduction of eosinophils and of the cytokines that induce their accumulation, treatment of CRS with intranasal glucocorticoids also leads to a reduction in damage to the epithelium that is thought to be mediated by eosinophils and is accompanied by the restoration of epithelium integrity.[131]

Perhaps the most valuable data implicating eosinophils in the pathogenesis of CRSwNP is from the group of Gevaert, Bachert, and collaborators.[132–134] These researchers noted increased levels of IL-5 and soluble IL-5RA in patients with nasal polyposis and found that treatment of CRSwNP patients with an antibody against IL-5 led to a reduction in nasal polyp size, particularly in patients with elevated IL-5. The same patients had a corresponding reduction in the level of eosinophilia, as measured by total eosinophils in the blood and levels of ECP in both the serum and in nasal secretions.[132–134]

PROSPECTS FOR NEW THERAPIES TARGETING EOSINOPHILIC INFLAMMATION IN CHRONIC RHINOSINUSITIS

Studies using anti-IL-5 antibodies to reduce eosinophilic inflammation are still ongoing and this approach continues to show promise for the treatment of CRS. As kinase inhibitors that block the development of eosinophils are developed, it is possible that some will emerge that have some specificity for the eosinophilic lineage of cells and will thus have utility in a variety of diseases including CRS. Treatment of mice with antibodies against Fas caused widespread apoptosis of eosinophils but lacked specificity and thus killed many other important cell types.[135] Treatments based on Siglec-8 activation may induce apoptosis more selectively in eosinophils, and in the related allergic cells mast cells and basophils, and thereby provide some benefit without undue safety concerns[136] (see also Chapter 15.4). In addition, recent studies from our group demonstrate that B cells and immunoglobulins, especially IgA which is a potent eosinophil activator, are highly elevated in CRS.[137,138] Therapies targeting B cells, such as antibodies against B cell activating factor belonging to the TNF family (BAFF) or B cells (e.g., antiCD20), may prove to diminish the activation of eosinophils in CRS if IgA and secretory IgA are important in activating eosinophils in CRS.

REFERENCES

1. Mahmoud AAF, Austen KF. *The Eosinophil in Health and Disease*. New York: Grune and Stratton Inc. 1980:3—354.

2. Taillens JP. Nasal polyposis; value of blood eosinophils in surgical indications and in sinusal surgical therapy. *Rev Laryngol Otol Rhinol (Bord)* 1952;**73**:20—59.

3. Taillens JP. Polyposis of nasal sinuses; pathogenetic study with therapeutic conclusions. *Pract Otorhinolaryngol (Basel)* 1953;**15**:211—42.

4. Van Zele T, Claeys S, Gevaert P, Van Maele G, Holtappels G, Van Cauwenberge P, et al. Differentiation of chronic sinus diseases by measurement of inflammatory mediators. *Allergy* 2006;**61**:1280—9.

5. Zhang N, Holtappels G, Claeys C, Huang G, van Cauwenberge P, Bachert C. Pattern of inflammation and impact of Staphylococcus aureus enterotoxins in nasal polyps from southern China. *Am J Rhinol* 2006;**20**:445—50.

6. Lacroix JS, Zheng CG, Goytom SH, Landis B, Szalay-Quinodoz I, Malis DD. Histological comparison of nasal polyposis in black African, Chinese and Caucasian patients. *Rhinology* 2002;**40**:118—21.

7. Zhang N, Liu S, Lin P, Li X, van Bruaene N, Zhang J, et al. Remodeling and inflammation in Chinese versus white patients with chronic rhinosinusitis. *J Allergy Clin Immunol* 2010;**125**:507. author reply 507—508.

8. Drake-Lee AB, McLaughlan P, Baker TH. Histamine, asthma, and nasal polyps. *Lancet* 1982;**2**:213.

9. Pawliczak R, Kowalski ML, Danilewicz M, Wagrowska-Danilewicz M, Lewandowski A. Distribution of mast cells and eosinophils in nasal polyps from atopic and nonatopic subjects: a morphometric study. *Am J Rhinol* 1997;**11**:257—62.

10. Stoop AE, van der Heijden H, Biewenga J, van der Baan S. Eosinophils in nasal polyps and nasal mucosa: An immunohistochemical study. *J Allergy Clin Immunol* 1993;**91**:616—20.

11. Tai PC, Spry CJ, Peterson C, Venge P, Olsson I. Monoclonal antibodies distinguish between storage and secreted forms of eosinophil cationic protein. *Nature* 1984;**309**:182—4.

12. Appenroth E, Gunkel AR, Muller H, Volklein C, Schrott-Fischer A. Activated and non-activated eosinophils in patients with chronic rhinosinusitis. *Acta Otolaryngol* 1998;**118**:240—2.

13. Bachert C, Wagenmann M, Hauser U, Rudack C. IL-5 synthesis is upregulated in human nasal polyp tissue. *J Allergy Clin Immunol* 1997;**99**:837—42.

14. Sun DI, Joo YH, Auo HJ, Kang JM. Clinical significance of eosinophilic cationic protein levels in nasal secretions of patients with nasal polyposis. *Eur Arch Otorhinolaryngol* 2009;**266**:981—6.

15. Armengot M, Garin L, Carda C. Eosinophil degranulation patterns in nasal polyposis: an ultrastructural study. *Am J Rhinol Allergy* 2009;**23**:466—70.

16. Erjefält JS, Andersson M, Greiff L, Korsgren M, Gizycki M, Jeffery PK, et al. Cytolysis and piecemeal degranulation as distinct modes of activation of airway mucosal eosinophils. *J Allergy Clin Immunol* 1998;**102**:286—94.

17. Ben Efraim AH, Munitz A, Sherman Y, Mazer BD, Levi-Schaffer F, Eliashar R. Efficient purification of eosinophils from human tissues: a comparative study. *J Immunol Methods* 2009;**343**:91—6.

18. Samter M. Nasal polyps: their relationship to allergy, particularly to bronchial asthma. *Med Clin North Am* 1958;**42**:175—9.

19. Bacciu A, Buzio C, Giordano D, Pasanisi E, Vincenti V, Mercante G, et al. Nasal polyposis in Churg-Strauss syndrome. *Laryngoscope* 2008;**118**:325—9.

20. Ragab A, Clement P, Vincken W. Correlation between the cytology of the nasal middle meatus and BAL in chronic rhinosinusitis. *Rhinology* 2005;**43**:11—7.

21. Dhong HJ, Kim HY, Cho DY. Histopathologic characteristics of chronic sinusitis with bronchial asthma. *Acta Otolaryngol* 2005;**125**:169—76.

22. Mehta V, Campeau NG, Kita H, Hagan JB. Blood and sputum eosinophil levels in asthma and their relationship to sinus computed tomographic findings. *Mayo Clin Proc* 2008;**83**:671—8.

23. Guida G, Rolla G, Badiu I, Marsico P, Pizzimenti S, Bommarito L, et al. Determinants of exhaled nitric oxide in chronic rhinosinusitis. *Chest* 2010;**137**:658—64.

24. Kern R, Conley D, Walsh W, Chandra R, Kato A, Tripathi-Peters A, et al. Perspectives on the etiology of chronic rhinosinusitis: An immune barrier hypothesis. *Am J Rhinol* 2008;**22**(6):549—59.

25. Tieu DD, Kern RC, Schleimer RP. Alterations in epithelial barrier function and host defense responses in chronic rhinosinusitis. *J Allergy Clin Immunol* 2009;**124**:37—42.

26. Ponikau JU, Sherris DA, Kephart GM, Adolphson C, Kita H. The role of ubiquitous airborne fungi in chronic rhinosinusitis. *Clin Rev Allergy Immunol* 2006;**30**:187—94.

27. Shin SH, Ponikau JU, Sherris DA, Congdon D, Frigas E, Homburger HA, et al. Chronic rhinosinusitis: an enhanced immune response to ubiquitous airborne fungi. *J Allergy Clin Immunol* 2004;**114**:1369—75.

28. Pearlman A, Chandra R, Chang D, Conley D, Tripathi-Peters A, Grammer L, et al. Relationships between severity of chronic rhinosinusitis and nasal polyposis, asthma, and atopy. *American Journal of Rhinology & Allergy* 2009;**23**(2):145—8.

29. Bachert C, Gevaert P, van Cauwenberge P. Staphylococcus aureus superantigens and airway disease. *Curr Allergy Asthma Rep* 2002;**2**:252—8.

30. Bachert C, van Zele T, Gevaert P, De Schrijver L, Van Cauwenberge P. Superantigens and nasal polyps. *Curr Allergy Asthma Rep* 2003;**3**:523—31.

31. Bernstein JM, Ballow M, Schlievert PM, Rich G, Allen C, Dryja D. A superantigen hypothesis for the pathogenesis of chronic hyperplastic sinusitis with massive nasal polyposis. *Am J Rhinol* 2003;**17**:321—6.

32. Bernstein JM, Kansal R. Superantigen hypothesis for the early development of chronic hyperplastic sinusitis with massive nasal polyposis. *Curr Opin Otolaryngol Head Neck Surg* 2005;**13**:39–44.

33. Herman A, Kappler JW, Marrack P, Pullen AM. Superantigens: mechanism of T-cell stimulation and role in immune responses. *Annu Rev Immunol* 1991;**9**:745–72.

34. Van Zele T, Gevaert P, Watelet JB, Claeys G, Holtappels G, Claeys C, et al. Staphylococcus aureus colonization and IgE antibody formation to enterotoxins is increased in nasal polyposis. *J Allergy Clin Immunol* 2004;**114**:981–3.

35. Stephenson MF, Mfuna L, Dowd SE, Wolcott RD, Barbeau J, Poisson M, et al. Molecular characterization of the polymicrobial flora in chronic rhinosinusitis. *J Otolaryngol Head Neck Surg* 2010;**39**:182–7.

36. Perez-Novo CA, Kowalski ML, Kuna P, Ptasinska A, Holtappels G, van Cauwenberge P, et al. Aspirin sensitivity and IgE antibodies to Staphylococcus aureus enterotoxins in nasal polyposis: studies on the relationship. *Int Arch Allergy Immunol* 2004;**133**:255–60.

37. Rohde G, Gevaert P, Holtappels G, Borg I, Wiethege A, Arinir U, et al. Increased IgE-antibodies to Staphylococcus aureus enterotoxins in patients with COPD. *Respir Med* 2004;**98**:858–64.

38. Seiberling KA, Conley DB, Tripathi A, Grammer LC, Shuh L, Haines 3rd GK, et al. Superantigens and chronic rhinosinusitis: detection of staphylococcal exotoxins in nasal polyps. *Laryngoscope* 2005;**115**:1580–5.

39. Suh YJ, Yoon SH, Sampson AP, Kim HJ, Kim SH, Nahm DH, et al. Specific immunoglobulin E for staphylococcal enterotoxins in nasal polyps from patients with aspirin-intolerant asthma. *Clin Exp Allergy* 2004;**34**:1270–5.

40. Bachert C, Zhang N, Holtappels G, De Lobel L, van Cauwenberge P, Liu S, et al. Presence of IL-5 protein and IgE antibodies to staphylococcal enterotoxins in nasal polyps is associated with comorbid asthma. *J Allergy Clin Immunol* 2010;**126**:962–8. e966.

41. Matsuwaki Y, Ookushi T, Asaka D, Mori E, Nakajima T, Yoshida T, et al. Chronic rhinosinusitis: risk factors for the recurrence of chronic rhinosinusitis based on 5-year follow-up after endoscopic sinus surgery. *Int Arch Allergy Immunol* 2008;**146**(Suppl 1):77–81.

42. Combadiere C, Ahuja S, Tiffany HL, Murphy PM. Cloning and functional expression of CC CKR5, a human monocyte CC chemokine receptor selective for MIP-1α, MIP-1β, and RANTES. *J Leuk Biol* 1996;**60**:147–52.

43. Daugherty BL, Siciliano SJ, DeMartino JA, Malkowitz L, Sirotina A, Springer MS. Cloning, expression, and characterization of the human eosinophil eotaxin receptor. *J Exp Med* 1996;**183**:2349–54.

44. Beck LA, Stellato C, Beall LD, Schall TJ, Leopold D, Bickel CA, et al. Detection of the chemokine RANTES and endothelial adhesion molecules in nasal polyps. *J Allergy Clin Immunol* 1996;**98**:766–80.

45. Allen JS, Eisma R, LaFreniere D, Leonard G, Kreutzer D. Characterization of the eosinophil chemokine RANTES in nasal polyps. *Ann Otol Rhinol Laryngol* 1998;**107**:416–20.

46. Beck LA, Stellato C, Beall LD, Schall TJ, Leopold D, Bickel CA, et al. Detection of the chemokine RANTES and endothelial adhesion molecules in nasal polyps. *J Allergy Clin Immunol* 1996;**98**:766–80.

47. Meyer JE, Bartels J, Gorogh T, Sticherling M, Rudack C, Ross DA, et al. The role of RANTES in nasal polyposis. *Am J Rhinol* 2005;**19**:15–20.

48. Bartels J, Maune S, Meyer JE, Kulke R, Schluter C, Rowert J, et al. Increased eotaxin-mRNA expression in non-atopic and atopic nasal polyps: comparison to RANTES and MCP-3 expression. *Rhinology* 1997;**35**:171–4.

49. Jahnsen FL, Haye R, Gran E, Brandtzaeg P, Johansen F-E. Glucocorticosteroids inhibit mRNA expression for eotaxin, eotaxin-2, and MCP-4 in airway inflammation with eosinophilia. *J Immunol* 1999;**163**:1545–51.

50. Molinaro RJ, Bernstein JM, Koury ST. Localization and quantitation of eotaxin mRNA in human nasal polyps. *Immunol Invest* 2003;**32**:143–54.

51. Yao T, Kojima Y, Koyanagi A, Yokoi H, Saito T, Kawano K, et al. Eotaxin-1, -2, and -3 immunoreactivity and protein concentration in the nasal polyps of eosinophilic chronic rhinosinusitis patients. *Laryngoscope* 2009;**119**:1053–9.

52. Matsukura S, Stellato C, Plitt J, Bickel C, Miura K, Georas S, et al. Activation of eotaxin gene transcription by NF-kB and STAT6 in human airway epithelial cells. *J Immunol* 1999;**163**:6876–83.

53. Kuperman DA, Schleimer RP. Interleukin-4, interleukin-13, signal transducer and activator of transcription factor 6, and allergic asthma. *Curr Mol Med* 2008;**8**:384–92.

54. Lalaker A, Nkrumah L, Lee WK, Ramanathan M, Lane AP. Chitin stimulates expression of acidic mammalian chitinase and eotaxin-3 by human sinonasal epithelial cells in vitro. *Am J Rhinol Allergy* 2009;**23**:8–14.

55. Ying S, Meng Q, Taborda-Barata L, Corrigan CJ, Barkans J, Assoufi B, et al. Human eosinophils express messenger RNA encoding RANTES and store and release biologically active RANTES protein. *Eur J Immunol* 1966;**26**:70–6.

55a. Poposki JA, Uzzaman A, Nagarkar DA, Chustz RT, Pefers AT, Suh LA, et al. Increased expression of the chemokine CCk23 in eosinophilic chronic rhinosinusitiswith nasal polyps. *J Allergy Clin Immunol* 2011;**128**:73–81.

56. Nardelli B, Tiffany HL, Bong GW, Yourey PA, Morahan DK, Li Y, et al. Characterization of the signal transduction pathway activated in human monocytes and dendritic cells by MPIF-1, a specific ligand for CC chemokine receptor 1. *J Immunol* 1999;**162**:435–44.

57. Patel VP, Kreider BL, Li Y, Li H, Leung K, Salcedo T, et al. Molecular and functional characterization of two novel human C-C chemokines as inhibitors of two distinct classes of myeloid progenitors. *J Exp Med* 1997;**185**:1163–72.

58. Gauldie J, Cox G, Jordana M, Ohno I, Kirpalani H. Growth and colony-stimulating factors mediate eosinophil fibroblast interactions in chronic airway inflammation. *Ann N Y Acad Sci* 1994;**725**:83–90.

59. Ohnishi M, Ruhno J, Bienenstock J, Milner R, Dolovich J, Denburg JA. Human nasal polyp epithelial basophil/mast cell and eosinophil colony-stimulating activity. *Am Rev Respir Dis* 1988;**138**:560–4.

60. Churchill L, Friedman B, Schleimer RP, Proud D. Production of granulocyte-macrophage colony-stimulating factor by cultured human tracheal epithelial cells. *Immunology* 1992;**75**:189–95.

61. Hamilos DL, Leung DY, Wood R, Meyers A, Stephens JK, Barkans J, et al. Chronic hyperplastic sinusitis: association of tissue eosinophilia with mRNA expression of granulocyte-macrophage

colony-stimulating factor and interleukin-3. *J Allergy Clin Immunol* 1993;**92**:39–48.

62. Park HS, Jung KS, Shute J, Roberts K, Holgate ST, Djukanovic R. Allergen-induced release of GM-CSF and IL-8 in vitro by nasal polyp tissue from atopic subjects prolongs eosinophil survival. *Eur Respir J* 1997;**10**:1476–82.

63. Shin SH, Lee SH, Jeong HS, Kita H. The effect of nasal polyp epithelial cells on eosinophil activation. *Laryngoscope* 2003;**113**:1374–7.

64. Xaubet A, Mullol J, López E, Ferrer JR, Rozman M, Carrión T, et al. Comparison of the role of nasal polyp and normal nasal mucosal epithelial cells on in vitro eosinohpil survival. medication by GM-CSF and inhibition by dexamethasone. *Clin Exp Allergy* 1994;**24**:307–17.

65. Hamilos DL, Leung DY, Huston DP, Kamil A, Wood R, Hamid Q. GM-CSF, IL-5 and RANTES immunoreactivity and mRNA expression in chronic hyperplastic sinusitis with nasal polyposis (NP). *Clin Exp Allergy* 1998;**28**:1145–52.

66. Lamblin C, Bolard F, Gosset P, Tsicopoulos A, Perez T, Darras J, et al. Bronchial interleukin-5 and eotaxin expression in nasal polyposis. Relationship with (a)symptomatic bronchial hyperresponsiveness. *Am J Respir Crit Care Med* 2001; **163**:1226–32.

67. Simon H-U, Yousefi S, Schranz C, Schapowal A, Bachert C, Blaser K. Direct demonstration of delayed eosinophil apoptosis as a mechanism causing tissue eosinophilia. *J Immunol* 1997;**158**:3902–8.

68. Liu LY, Sedgwick JB, Bates ME, Vrtis RF, Gern JE, Kita H, et al. Decreased expression of membrane IL-5 receptor alpha on human eosinophils: II. IL-5 down-modulates its receptor via a proteinase-mediated process. *J Immunol* 2002;**169**:6459–66.

69. Beck LA, Schall TJ, Beall LD, Leopold D, Bickel C, Baroody F, et al. Detection of the chemokine RANTES and activation of vascular endothelium in nasal polyps. *J Allergy Clin Immunol* 1994;**93**:A234.

70. Corsi MM, Pagani D, Dogliotti G, Perona F, Sambataro G, Pignataro L. Protein biochip array of adhesion molecule expression in peripheral blood of patients with nasal polyposis. *Int J Biol Markers* 2008;**23**:115–20.

71. Eweiss A, Dogheim Y, Hassab M, Tayel H, Hammad Z. VCAM-1 and eosinophilia in diffuse sino-nasal polyps. *Eur Arch Otorhinolaryngol* 2009;**266**:377–83.

72. Hamilos DL, Leung DY, Wood R, Bean DK, Song YL, Schotman E, et al. Eosinophil infiltration in nonallergic chronic hyperplastic sinusitis with nasal polyposis (CHS/NP) is associated with endothelial VCAM-1 upregulation and expression of TNF-alpha. *Am J Respir Cell Mol Biol* 1996;**15**:443–50.

73. Jahnsen FL, Haraldsen G, Aanesen JP, Haye R, Brandtzaeg P. Eosinophil infiltration is related to increased expression of vascular cell adhesion molecule-1 in nasal polyps. *Am J Respir Cell Mol Biol* 1995;**12**:624–32.

74. Symon FA, Walsh GM, Watson SR, Wardlaw AJ. Eosinophil adhesion to nasal polyp endothelium is P-selectin-dependent. *J Exp Med* 1994;**180**:371–6.

75. Bachert C, Gevaert P, Holtappels G, van Cauwenberge P. Mediators in nasal polyposis. *Curr Allergy Asthma Rep* 2002;**2**:481–7.

76. Hamilos DL, Leung DY, Wood R, Cunningham L, Bean DK, Yasruel Z, et al. Evidence for distinct cytokine expression in allergic versus nonallergic chronic sinusitis. *J Allergy Clin Immunol* 1995;**96**:537–44.

77. Van Bruaene N, Perez-Novo CA, Basinski TM, Van Zele T, Holtappels G, De Ruyck N, et al. T-cell regulation in chronic paranasal sinus disease. *J Allergy Clin Immunol* 2008;**121**: 1435–41, 1441 e1431–1433.

78. Peters AT, Kato A, Zhang N, Conley DB, Suh L, Tancowny B, et al. Evidence for altered activity of the IL-6 pathway in chronic rhinosinusitis with nasal polyps. *J Allergy Clin Immunol* 2010;**125**:397–403. e310.

79. Saitoh T, Kusunoki T, Yao T, Kawano K, Kojima Y, Miyahara K, et al. Role of interleukin-17A in the eosinophil accumulation and mucosal remodeling in chronic rhinosinusitis with nasal polyps associated with asthma. *Int Arch Allergy Immunol* 2010;**151**:8–16.

80. Buysschaert ID, Grulois V, Eloy P, Jorissen M, Rombaux P, Bertrand B, et al. Genetic evidence for a role of IL33 in nasal polyposis. *Allergy* 2010;**65**:616–22.

81. Kowalski ML, Lewandowska-Polak A, Wozniak J, Ptasinska A, Jankowski A, Wagrowska-Danilewicz M, et al. Association of stem cell factor expression in nasal polyp epithelial cells with aspirin sensitivity and asthma. *Allergy* 2005;**60**:631–7.

82. Perez-Novo CA, Claeys C, Van Cauwenberge P, Bachert C. Expression of eicosanoid receptors subtypes and eosinophilic inflammation: implication on chronic rhinosinusitis. *Respir Res* 2006;**7**:75.

83. Shin SH, Lee YH, Jeon CH. Protease-dependent activation of nasal polyp epithelial cells by airborne fungi leads to migration of eosinophils and neutrophils. *Acta Otolaryngol* 2006;**126**:1286–94.

84. Van Zele T, Coppieters F, Gevaert P, Holtappels G, Van Cauwenberge P, Bachert C. Local complement activation in nasal polyposis. *Laryngoscope* 2009;**119**:1753–8.

85. Vandermeer J, Sha Q, Lane AP, Schleimer RP. Innate immunity of the sinonasal cavity: expression of messenger RNA for complement cascade components and toll-like receptors. *Arch Otolaryngol Head Neck Surg* 2004;**130**:1374–80.

86. Jankowski R, Bouchoua F, Coffinet L, Vignaud JM. Clinical factors influencing the eosinophil infiltration of nasal polyps. *Rhinology* 2002;**40**:173–8.

87. Polzehl D, Moeller P, Riechelmann H, Perner S. Distinct features of chronic rhinosinusitis with and without nasal polyps. *Allergy* 2006;**61**:1275–9.

88. Van Cauwenberge P, Van Hoecke H, Bachert C. Pathogenesis of chronic rhinosinusitis. *Curr Allergy Asthma Rep* 2006;**6**:487–94.

89. Schleimer RP, Kato A, Peters A, Conley D, Kim J, Liu MC, et al. Epithelium, inflammation, and immunity in the upper airways of humans: studies in chronic rhinosinusitis. *Proc Am Thorac Soc* 2009;**6**:288–94.

90. Erjefalt JS, Greiff L, Andersson M, Adelroth E, Jeffery PK, Persson CG. Degranulation patterns of eosinophil granulocytes as determinants of eosinophil driven disease. *Thorax* 2001;**56**:341–4.

91. Bachert C, Gevaert P, Holtappels G, Cuvelier C, van Cauwenberge P. Nasal polyposis: from cytokines to growth. *Am J Rhinol* 2000;**14**:279–90.

92. Adamjee J, Suh YJ, Park HS, Choi JH, Penrose JF, Lam BK, et al. Expression of 5-lipoxygenase and cyclooxygenase pathway enzymes in nasal polyps of patients with aspirin-intolerant asthma. *J Pathol* 2006;**209**:392–9.

93. Steinke JW, Bradley D, Arango P, Crouse CD, Frierson H, Kountakis SE, et al. Cysteinyl leukotriene expression in chronic hyperplastic sinusitis-nasal polyposis: importance to eosinophilia and asthma. *J Allergy Clin Immunol* 2003;**111**:342–9.

94. Di Lorenzo G, Drago A, Esposito Pellitteri M, Candore G, Colombo A, Gervasi F, et al. Measurement of inflammatory mediators of mast cells and eosinophils in native nasal lavage fluid in nasal polyposis. *Int Arch Allergy Immunol* 2001;**125**:164−75.

95. Bartemes KR, Cooper KM, Drain KL, Kita H. Secretory IgA induces antigen-independent eosinophil survival and cytokine production without inducing effector functions. *J Allergy Clin Immunol* 2005;**116**:827−35.

96. Muraki M, Gleich GJ, Kita H. Antigen-Specific IgG and IgA, but Not IgE, Activate the Effector Functions of Eosinophils in the Presence of Antigen. *Int Arch Allergy Immunol* 2011;**154**:119−27.

97. Abu-Ghazaleh RI, Fujisawa T, Mestecky J, Kyle RA, Gleich GJ. IgA-induced eosinophil degranulation. *J Immunol* 1989;**142**: 2393−400.

98. Pleass RJ, Lang ML, Kerr MA, Woof JM. IgA is a more potent inducer of NADPH oxidase activation and degranulation in blood eosinophils than IgE. *Mol Immunol* 2007;**44**:1401−8.

99. Tan BK, Li Q, Suh L, Kato A, Conley DB, Chandra R, et al. Evidence for intranasal antinuclear autoantibodies with patients with chronic rhinosinusitis with nasal polyps. *J Allergy Clin Immunol* 2011;**128**:1198−206.

100. Suh KS, Park HS, Nahm DH, Kim YK, Lee YM, Park K. Role of IgG, IgA, and IgE antibodies in nasal polyp tissue: their relationships with eosinophilic infiltration and degranulation. *J Korean Med Sci* 2002;**17**:375−80.

101. Harlin SL, Ansel DG, Lane SR, Myers J, Kephart GM, Gleich GJ. A clinical and pathologic study of chronic sinusitis: the role of the eosinophil. *J Allergy Clin Immunol* 1988;**81**:867−75.

102. Chanez P, Vignola AM, Vic P, Guddo F, Bonsignore G, Godard P, et al. Comparison between nasal and bronchial inflammation in asthmatic and control subjects. *Am J Respir Crit Care Med* 1999;**159**:588−95.

103. Gaga M, Lambrou P, Papageorgiou N, Koulouris NG, Kosmas E, Fragakis S, et al. Eosinophils are a feature of upper and lower airway pathology in non-atopic asthma, irrespective of the presence of rhinitis. *Clin Exp Allergy* 2000;**30**:663−9.

104. Bresciani M, Paradis L, Des Roches A, Vernhet H, Vachier I, Godard P, et al. Rhinosinusitis in severe asthma. *J Allergy Clin Immunol* 2001;**107**:73−80.

105. Shahana S, Jaunmuktane Z, Asplund MS, Roomans GM. Ultrastructural investigation of epithelial damage in asthmatic and non-asthmatic nasal polyps. *Respir Med* 2006;**100**:2018−28.

106. Holgate ST. Epithelium dysfunction in asthma. *J Allergy Clin Immunol* 2007;**120**:1233−44. quiz 1245−1236.

107. Gleich GJ. Mechanisms of eosinophil-associated inflammation. *J Allergy Clin Immunol* 2000;**105**:651−63.

108. Bernardes JF, Shan J, Tewfik M, Hamid Q, Frenkiel S, Eidelman DH. Protein nitration in chronic sinusitis and nasal polyposis: role of eosinophils. *Otolaryngol Head Neck Surg* 2004;**131**:696−703.

109. Richer SL, Truong-Tran AQ, Conley DB, Carter R, Vermylen D, Grammer LC, et al. Epithelial genes in chronic rhinosinusitis with and without nasal polyps. *Am J Rhinol* 2008;**22**:228−34.

110. Briot A, Deraison C, Lacroix M, Bonnart C, Robin A, Besson C, et al. Kallikrein 5 induces atopic dermatitis-like lesions through PAR2-mediated thymic stromal lymphopoietin expression in Netherton syndrome. *J Exp Med* 2009;**206**:1135−47.

111. Allakhverdi Z, Comeau MR, Jessup HK, Yoon BR, Brewer A, Chartier S, et al. Thymic stromal lymphopoietin is released by human epithelial cells in response to microbes, trauma, or inflammation and potently activates mast cells. *J Exp Med* 2007;**204**:253−8.

112. Kato A, Favoreto Jr S, Avila PC, Schleimer RP. TLR3- and Th2 cytokine-dependent production of thymic stromal lymphopoietin in human airway epithelial cells. *J Immunol* 2007;**179**:1080−7.

113. Liu T, Li TL, Zhao F, Xie C, Liu AM, Chen X, et al. Role of Thymic Stromal Lymphopoietin in the Pathogenesis of Nasal Polyposis. *Am J Med Sci* 2011;**341**(1):40−7.

114. Flood-Page P, Menzies-Gow A, Phipps S, Ying S, Wangoo A, Ludwig MS, et al. Anti-IL-5 treatment reduces deposition of ECM proteins in the bronchial subepithelial basement membrane of mild atopic asthmatics. *J Clin Invest* 2003;**112**:1029−36.

115. Laitinen LA, Laitinen A. Structural and cellular changes in asthma. *Eur Respir Rev* 1994;**4**:348−51.

116. Ohno I, Nitta Y, Yamauchi K, Hoshi H, Honma M, Woolley K, et al. Eosinophils as a potential source of platelet-derived growth factor B-chain (PDGF-B) in nasal polyposis and bronchial asthma. *Am J Respir Cell Mol Biol* 1995;**13**:639−47.

117. Ohno I, Nitta Y, Yamauchi K, Hoshi H, Honma M, Woolley K, et al. Transforming growth factor β1 (TGFβ1) gene expression by eosinophils in asthmatic airway inflammation. *Am J Respir Cell Mol Biol* 1996;**15**:404−9.

118. Elovic A, Wong DT, Weller PF, Matossian K, Galli SJ. Expression of transforming growth factors-alpha and beta 1 messenger RNA and product by eosinophils in nasal polyps. *J Allergy Clin Immunol* 1994;**93**:864−9.

119. Schleimer RP. Pharmacology of Glucocorticoids in Allergic Disease. In: *Middleton's Allergy*, 7th ed. New York: Mosby Elsevier; 2009.

120. Gleich GJ, Hunt LW, Bochner BSa, Schleimer RP. *Glucocorticoid effects on human eosinophils*. New York, NY: Marcel Dekker, Inc. 1997, pp.279−308.

121. Schleimer RP, Bochner BS. The effects of glucocorticoids on human eosinophils. *J Allergy Clin Immunol* 1994;**94**:1202−13.

122. Patiar S, Reece P. Oral steroids for nasal polyps. *Cochrane Database Syst Rev:CD005232*; 2007.

123. Alobid I, Benitez P, Pujols L, Maldonado M, Bernal-Sprekelsen M, Morello A, et al. Severe nasal polyposis and its impact on quality of life. The effect of a short course of oral steroids followed by long-term intranasal steroid treatment. *Rhinology* 2006;**44**: 8−13.

124. Hissaria P, Smith W, Wormald P, Taylor J. Short Course of Systemic Corticosteroids in Sinonasal Polyposis: A double-blind, randomized, placebo-controlled trial with evaluation of outcome measures. *J Allergy Clin Immunol* 2006;**118**:118−28.

125. Van Zele T, Gevaert P, Holtappels G, Beule A, Wormald PJ, Mayr S, et al. Oral steroids and doxycycline: two different approaches to treat nasal polyps. *J Allergy Clin Immunol* 2010;**125**: 1069−1076, e1064.

126. Kanai N, Denburg J, Jordana M, Dolovich J. Nasal polyp inflammation. Effect of topical nasal steroid. *Am J Respir Crit Care Med* 1994;**150**:1094−100.

127. Hamilos DL, Thawley SE, Kramper MA, Kamil A, Hamid QA. Effect of intranasal fluticasone on cellular infiltration, endothelial adhesion molecule expression, and proinflammatory cytokine mRNA in nasal polyp disease. *J Allergy Clin Immunol* 1999;**103**:79−87.

128. Delbrouck C, Kaltner H, Danguy A, Nifant'ev NE, Bovin NV, Vandenhoven G, et al. Glucocorticoid-induced differential expression of the sialylated and nonsialylated Lewis(a) epitopes and respective binding sites in human nasal polyps maintained under ex vivo tissue culture conditions. *Ann Otol Rhinol Laryngol* 2002;**111**:1097–107.

129. Nonaka R, Nonaka M, Takanashi S, Jordana M, Dolovich J. Eosinophil activation in the tissue: synthetic steroid, budesonide, effectively inhibits the survival of eosinophils isolated from peripheral blood but not nasal polyp tissues. *J Clin Lab Immunol* 1999;**51**:39–53.

130. Bolard F, Gosset P, Lamblin C, Bergoin C, Tonnel AB, Wallaert B. Cell and cytokine profiles in nasal secretions from patients with nasal polyposis: effects of topical steroids and surgical treatment. *Allergy* 2001;**56**:333–8.

131. Mastruzzo C, Greco LR, Nakano K, Nakano A, Palermo F, Pistorio MP, et al. Impact of intranasal budesonide on immune inflammatory responses and epithelial remodeling in chronic upper airway inflammation. *J Allergy Clin Immunol* 2003;**112**:37–44.

132. Gevaert P, Bachert C, Holtappels G, Novo CP, Van der Heyden J, Fransen L, et al. Enhanced soluble interleukin-5 receptor alpha expression in nasal polyposis. *Allergy* 2003;**58**:371–9.

133. Gevaert P, Hellman C, Lundblad L, Lundahl J, Holtappels G, van Cauwenberge P, et al. Differential expression of the interleukin 5 receptor alpha isoforms in blood and tissue eosinophils of nasal polyp patients. *Allergy* 2009;**64**:725–32.

134. Gevaert P, Lang-Loidolt D, Lackner A, Stammberger H, Staudinger H, Van Zele T, et al. Nasal IL-5 levels determine the response to anti-IL-5 treatment in patients with nasal polyps. *J Allergy Clin Immunol* 2006;**118**:1133–41.

135. Tsuyuki S, Bertrand C, Erard F, Trifilieff A, Tsuyuki J, Wesp M, et al. Activation of the fas receptor on lung eosinophils leads to apoptosis and the resolution of eosinophilic inflammation of the airways. *J Clin Invest* 1995;**96**:2924–31.

136. Kikly KK, Bochner BS, Freeman SD, Tan KB, Gallagher KT, D'Alessio K, et al. Identification of SAF-2, a novel Siglec expressed on eosinophils, mast cells, and basophils. *J Allergy Clin Immunol* 2000;**105**:1093–100.

137. Kato A, Peters A, Suh L, Carter R, Harris KE, Chandra R, et al. Evidence of a role for B cell-activating factor of the TNF family in the pathogenesis of chronic rhinosinusitis with nasal polyps. *J Allergy Clin Immunol* 2008;**121**:1385–92, 1392 e1381–1382.

138. Kato A, Xiao H, Chustz RT, Liu MC, Schleimer RP. Local release of B cell-activating factor of the TNF family after segmental allergen challenge of allergic subjects. *J Allergy Clin Immunol* 2009;**123**:369–75.

Chapter 13.12

Eosinophils and Vascular Healing

Mark C. Lavigne and Michael J. Eppihimer

The ability of organisms to repair themselves is an indispensable requirement for their survival. Most members of the animal kingdom inevitably encounter external or internal environmental pressures that can threaten their existence. Examples of these are, respectively, injury inflicted by a traumatic event, such as a stab wound, and parasites that inappropriately reside in various tissues. A complex series of reactions to these survival challenges has evolved among vertebrates to correct tissue damage traumatically imposed by a foreign object (i.e., a knife) or to completely remove an injurious agent (i.e., an infection). These shared mechanisms of self-preservation are collectively referred to as healing. In many individuals, healing events in the coronary vasculature occur and can be characterized by both types of reactions, including correction of vessel damage inflicted by circulating substances (i.e., coronary artery disease) and subsequent introduction of a foreign agent (i.e., a stent), as well as to mount an immune response (i.e., hypersensitivity) aimed at eliminating such a foreign body.

The most common type of healing occurrence in blood vessels is, paradoxically, also responsible for the most common type of coronary artery disease. Indeed, atherosclerosis is likely the result of a *response to injury* that occurs predominantly in medium-sized muscular vessels, including the left anterior descending, left circumflex, and right coronary arteries.[1] The injury is thought to be primarily inflicted on endothelial cells (ECs), which, from their position as a lining of the vessel lumen, normally provide a nonthrombogenic, nonadhesive surface despite their direct contact with flowing blood. Agents suspected of damaging ECs include elevated blood cholesterol levels and oxidized low-density lipoprotein. The outcome of vessel healing is often manifest as an eccentrically oriented lesion consisting of, in its most mature form, lipid-laden macrophages and smooth muscle cells (or so-called foam cells), and lymphocytes enclosed by a fibrous cap comprised of collagen, elastin, and proteoglycans.[1] Consequently, these lesions, or plaques, in their stable form can compromise critical oxygen supply to the energetic myocardium by significantly impeding blood flow and, in their unstable or vulnerable forms, can cause myocardial infarction and death by providing sites for platelet adherence and activation of clotting factors to promote formation of a fully occlusive thrombus. Although lifestyle changes and medications, including cholesterol synthesis inhibitors (or statins), effectively reduce the risk factors for developing atherosclerosis,[2] its place as a predominant cause of morbidity and mortality in industrialized nations has not wavered. Indeed, its recurrent detection is afforded by sophisticated angiographic evidence through catheterization of vessels suspected of harboring lesions. Once detected, a patient's physician has several choices for revascularizing occluded coronary vessels.

Since the late 1970s, a variety of minimally invasive, catheter-based procedures have evolved to displace occlusive atherosclerotic plaques in coronary vessels. Andreas Gruentzig and colleagues performed the first such procedure in 1977 by using a catheter to guide a balloon to the site of atherosclerosis.[3] Once positioned, the balloon was expanded to crush the plaque against the vessel wall. This procedure, referred to as percutaneous transluminal coronary angioplasty (PTCA), or simply angioplasty, was at least acutely successful for providing a less-invasive treatment, compared to traditionally used coronary artery bypass grafting (CABG), to restore adequate blood flow to the myocardium. Approximately 6 months following the procedure, however, approximately 35% of patients experienced a renarrowing (*restenosis*) of the blood vessel at the original atheromatous site.[4] Since then, two major modifications to PTCA have been made in an effort to reduce the incidence of restenosis, including balloon-mediated deployment of bare metal stents (BMS) and drug and polymer addition to such stents to form a device collectively referred to as a drug-eluting stent (DES). The former amendment to PTCA originated with the recognition that restenosis was at least partially attributable to elastic recoil of the vessel as an immediate reaction to balloon-induced vessel expansion, while the addition of drug came with an appreciation for the significant contributions of vascular smooth muscle cell (VSMC) proliferation[5] and migration[6] to restenosis following BMS insertion into coronary arteries. The cell proliferative component is similar to that which promotes benign tumor formation and, therefore, was as a good candidate for being disrupted by antiproliferative agents such as those on first-generation DES that emerged from Johnson and Johnson in 2003 (Cypher DES; Cordis, Miami Lakes, FL)[7] and Boston Scientific Corporation in 2004 (Taxus Express2, Maple Grove, MN),[8] and which included sirolimus and paclitaxel, respectively.

Vascular responses to stent implantation in coronary arteries occur in sequence to culminate in what is collectively referred to as *healing*. In the traditional sense, vascular wound healing is orchestrated by platelet- and clotting factor-mediated hemostasis, cytokine-/chemokine- and leukocyte-mediated inflammation, and VSMC- and fibroblast-mediated tissue remodeling.[9] Although *healing* naturally connotes a process that is favorable, it can be characterized by an exaggeration of the processes outlined above, leading to unfavorable consequences, such as reocclusion, or *restenosis* of the vessel and/or thrombosis. Restenosis can occur in conjunction with stent use.[10] Moreover, thrombosis rates associated with DES implantation are higher than those related to BMS deployment after 1 year,[11] possibly owing to insufficient EC coverage of the stent itself and of the local vascular area in which a DES resides.[12] The fate of the vessel wall localized to the area of stent implantation is determined, for the most part, by a balance between the interactions of cells and the soluble factors that they secrete. These events, in turn, may

be dictated by the circumstances of DES deployment itself, such as the extent of mechanical injury imposed by stent insertion.[13] Additionally, DES composition may play a significant role in determining the prognosis of DES use as a therapy to alleviate occlusive coronary vascular disease.

Along with their differential abilities to prevent restenosis and association with thrombosis, DES and BMS platforms may be distinguished by inflammatory cell infiltrates, based on the extent of eosinophil presence that follows deployment of each. Clinically used variations of DES currently feature controlled drug release from polymers that coat the metal stent skeleton. Twenty-eight days following their deployment, overlapping Taxus and Cypher DES each had more eosinophils associated with them compared to their respective overlapping BMS controls in rabbit iliac artery.[14] These DES are composed of different drugs and nonerodible polymers, but the same metals (316L stainless steel), suggesting that drugs and/or polymers may ultimately attract eosinophils to stent insertion sites. John and colleagues found that a critical amount of polymer may selectively incite eosinophil recruitment in rabbit iliac artery. Cypher was associated with significantly more luminal eosinophils than a polymer-free sirolimus-eluting stent, a polymer-free sirolimus–estradiol-eluting stent, or a BMS, but only when each stent platform was overlapped upon itself (i.e., one stent on top of another of the same kind) in the vessel.[15] Eosinophil recruitment following DES implantation in rabbits appears not to be specific to this species, since eosinophil accumulation was observed around Cypher stent struts, followed by less infiltration near both Taxus and BM stents, in porcine coronary arteries.[16] Table 13.12.1 provides a summary of studies citing the occurrence of eosinophil infiltration in response to a variety of catheter-based revascularization interventions, including purely balloon-mediated, or plain old balloon angioplasty (POBA), and BMS or DES implantation. Such reactions to foreign-body stent substances suggest an eosinophil-mediated hypersensitivity,[17] which has been implicated in thrombosis and restenosis following stent insertion into blood vessels, as described below.

Several instances of thrombosis occurring after BMS or DES implantation and their correlation with histological evidence of eosinophil infiltration have been recorded. Zavolloni and colleagues[18] reported that inflammatory infiltrate observed in thrombectomy material retrieved from right coronary artery previously implanted (10 years) with a BMS for myocardial infarction contained a prevalence of eosinophils. Data such as these contrast with previous reports that associated stent thrombosis and eosinophil recruitment particularly with DES rather than BMS. Virmani and colleagues provided histological evidence of thrombus formation in left circumflex coronary artery of

TABLE 13.12.1 Eosinophil Recruitment Following Catheter-Based Revascularization of Coronary Vessels

Species	Interventions	Extent of Eosinophil Inflammation	Reference Number
Rabbit	BMS and DES	Taxus > Cypher DES > BMS	14
Rabbit	BMS, DES (no polymer), DES (polymer)	DES (polymer) > DES (no polymer) and BMS	15
Pig	BMS and DES	Cypher > Taxus DES > BMS	16
Pig	POBA and BMS	BMS > POBA	22
Pig	POBA and BMS	BMS > POBA	23
Human	POBA and BMS	BMS > POBA	24

BMS, bare metal stent; DES, drug-eluting stent; POBA, plain old balloon angioplasty.

a 58-year-old male who had received overlapping Cypher stents in that vessel 18 months earlier.[19] Aneurysm formation and inflammatory prominence consisting of eosinophils, giant cells, lymphocytes, macrophages, and plasma cells within the vessel area localized to stent placements were evident. The authors referred to these phenomena as a 'hypersensitivity vasculitis,' which has since been described by other investigators following DES implantation in conjunction with repeated thromboses associated with implantation of Cypher stents into the left circumflex[20] and also in autopsy specimens.[21] The latter study is especially informative, given that it compared eosinophil presence among thrombotic events according to classification, and in the context of acute myocardial infarction or not, and time-frame, including early ($\leq$30 days) after BMS or DES [Cypher, Taxus, and Endeavor (Medtronic, Inc. Minneapolis, MN); drug is zotarolimus] deployment, late (31–365 days) after BMS deployment, and very late (>1 year) after DES deployment. Here, eosinophil accumulation and the fraction of leukocytes accounted for by eosinophils around Cypher stents was substantially greater than that measured in association with Taxus or Endeavor stents. Furthermore, eosinophils were most evident in thrombi that occurred very late following DES (predominantly of Cypher) deployment. Interestingly, vessel remodeling was exclusively related to very late stent thrombosis that occurred after DES, with its extent varying according to the number of eosinophils found in thrombi. The authors speculated that vessel remodeling likely caused stent malapposition and subsequent thrombosis. Taken together, these studies indicate eosinophilic reactions to BMS or DES deployment and suggest a role for

eosinophils in stent thrombosis. At least two questions may be drawn from these conclusions:

1. What factors related to coronary stents and/or their deployment attract eosinophils to stented vessel segments?
2. How might eosinophils contribute to stent-related thrombosis?

Eosinophilic inflammation occurs following both POBA and stent insertion. However, eosinophilic recruitment is more robust following stent placement,[22–24] suggesting that balloon-mediated vessel expansion, which is common to both POBA and stent placement, is not completely responsible for enhanced eosinophilic recruitment associated with stent placement. Indeed, stent components and prophylactic dual antiplatelet (medicinal) therapies (DAPT) that are prescribed for use after stent implantation, including the platelet adenosine diphosphate (ADP) receptor antagonist, clopidogrel (Plavix; Bristol-Myers Squibb, New York, NY, and Sanofi–Aventis, Paris, France) and acetylsalicylic acid (aspirin), an inhibitor of prostaglandin G/H synthase 1 (inhibits thromboxane production), can elicit hypersensitivity reactions.[25] The well-recognized role of eosinophils in allergic reactions[17] and their suggested role in a localized hypersensitivity reaction to stent placement[19] are in agreement with positive correlations between allergic sensitivity to metal stent components, including nickel[20,26] and molybdenum,[26] and eosinophil recruitment to stented vessel segments.[20,26] However, restenosis incidence may[27] or may not[28] positively correlate with allergic reactions to nickel and molybdenum. In addition, drug-eluting stent-related hypersensitivity reactions appear not to guarantee subsequent thrombosis, as revealed by a study[29] showing that a minority (approximately 1.5%) of total DES-specific hypersensitivity reactions (based on 262 hypersensitivity cases reported) were accompanied by thrombosis. These hypersensitive, thrombotic cases, however, were characterized by eosinophilic inflammation and incomplete stent coverage, or so-called delayed healing, 100% of the time. Overall, the majority of DES-specific hypersensitivity cases were associated with Cypher implantation (approximately 80%) and likely attributable to the poly n-butyl-methacrylate and polyethylene-vinyl acetate Cypher copolymer allergens,[25] but not to sirolimus, since the latter can reduce eosinophil infiltration.[30] Consistent with this, Finn and colleagues reported that 5/105 cases of late stent thrombosis were associated with hypersensitivity reactions, with four occurring after Cypher and one after Taxus implantation.[12] In these episodes, too, eosinophilic inflammation was always present. Taken together, these studies suggest that the risk of hypersensitivity to stent components is likely to be small among patients that have or will receive DES, with the onset of thrombosis occurring in such hypersensitivity cases also to be small. However, histological evidence shows that the incidence of such rare cases is consistently associated with eosinophil accumulation at stent sites and seems to be especially common with Cypher implantation.[12,29] At least one study found the combination of hypersensitivity and delayed healing (i.e., endothelialization) of stent struts to be a risk factor for developing late-stent thrombosis.[31] Identification of patients prone to each of these circumstances is not performed and may, in fact, be unpredictable considering that hypersensitivity reactions to stents appear not to be unequivocally associated with thrombosis[29] and, further, the incidence of eosinophil involvement in such reactions, which correlates with hypersensitivity-related thrombosis,[12,29] is also unclear. Thus, without objective, basic study- and clinical trial-based reasoning for differential postprocedural prescription to DES recipients, it is essential that all patients that receive DES comply with using DAPT medicines for their full recommended terms.

Currently, the exact duration of DAPT therapy following stent deployment is questionable, but would appear to be optimized by an awareness of when stent struts are completely *healed* (or endothelialized) such that the risk of platelet adhesion and activation is dramatically diminished. Unfortunately, the time required for adequate, antithrombotic endothelialization of stent struts and the local vascular wall likely varies among patients, because endothelialization rates among DES patients afflicted with complicating clinical backgrounds, such as diabetes,[32] may vary, especially compared to DES patients without complicating ancillary medical conditions. The concern, then, pertaining to eosinophil involvement in thrombus formation associated with stent deployment is that DAPT may be suspended before stent struts, whether covered by polymer or not, are masked enough by healing processes so as not to be a potentially chronic stimulus for eosinophil recruitment and activation. Indeed, the nonspecific nature of antiproliferative drugs currently included on clinically used DES inhibit not only VSMC proliferation and migration, two major contributors to restenosis,[5,6] but also attenuate the same properties of ECs.[33,34] Therefore, the risk of eosinophil-related thrombosis following DES deployment may be assessed by (1) whether the stent is BMS or DES and (2) evidence of allergies to stent components that may predispose the patient to a hypersensitivity reaction that involves eosinophils. Fortunately, research efforts have provided suggestive evidence concerning the mechanism of eosinophil recruitment to stented vessel segments and the mechanism through which eosinophils contribute to thrombus formation. These pieces of information may be used to direct future therapies intended to mitigate the purported role of eosinophils in promoting thrombosis following stent deployment.

Descriptions of two hypersensitivity reactions (I and IV) have been suggested to provide hypothetical explanations for eosinophilic involvement in inflammatory responses to stent insertion. As suggested by Virmani and coworkers, eosinophil recruitment to segments stented with Cypher and their association with thrombosis[19] in this context may be due to a type IV hypersensitivity reaction, in which T-helper lymphocyte liberation of T-helper type 2 (T_h2) cytokines and interleukins 4 and 13 attract eosinophils.[35] Another hypothesis[25] includes a two-phase cascade of cell- and soluble factor-mediated reactions being initially orchestrated by IgE-activated mast cells and the mediators that these cells elaborate (approximately 1—24 h post-stent deployment) to promote secondary infiltration of basophils, eosinophils, macrophages, neutrophils, and T lymphocytes (approximately 12—24 h post-stent deployment) that can remain chronically situated around stent struts. Thus, occurrence of the first phase of hypersensitivity would coincide with onset of acute stent thrombosis, while the latter time frame would fit the onset of both late and very late stent thromboses. Evidence of a role for eosinophils in thrombogenesis includes observations made in hypereosinophilic patients.[36] Furthermore, endothelial cells are likely targets for the highly basic-charged major basic and eosinophil cationic proteins (MBP and ECP) of eosinophils, considering that ECs express a negatively charged glycocalyx on their luminal surface. Once bound to ECs, these proteins may inflict damage or even kill these cells, as suggested by their cytotoxic capabilities.[17] Alternatively, direct activation of platelets by MBP and/or eosinophil peroxidase (EPO),[37] or disruption of thrombomodulin function by MBP,[38] may explain the contribution made by eosinophils to stent thrombosis. Eosinophil presence in the context of stent-related thrombosis cases that may be secondarily related to tissue remodeling and consequent stent malapposition[21] raises the possibility that eosinophils directly or indirectly possess tissue-remodeling properties. Both may be true, since eosinophils express matrix metalloproteinase-9 (MMP-9),[39] which is a collagenase capable of degrading type IV collagen, a major component of subendothelial layer basement membranes. Furthermore, eosinophils secrete interleukin-8,[17] a chemokine that has been shown to induce release of MMP-2 (a collagenase) and MMP-9 from ECs.[40] Eosinophils also express vascular endothelial growth factor (VEGF)[41] and heparanase.[42] These factors are likely to partially mediate the ability of eosinophils to promote angiogenesis,[43] by inducing EC growth (VEGF) and degradation of perlecan (heparanase), a heparan sulfate proteoglycan component of basement membranes. The relevance of this is that angiogenesis can occur in the context of the granulation stage of vascular healing following stent insertion. The cumulative occurrence of any thrombi that may form as a result of angiogenic events may culminate as a thrombus of significant size, with the ability to dramatically or completely block blood flow in the main stented vessel. Table 13.12.2 summarizes investigations that have linked stent-related thrombosis with eosinophilic inflammation. Kawano and coworkers reported that a patient who received a BMS to relieve total occlusion of the left coronary artery experienced repeated episodes of restenosis after stent implantation.[26] Examination of the restenotic lesion revealed granulation tissue with eosinophil infiltration. The patient

displayed positive reactions to allergic patch tests for nickel and molybdenum, both of which are components of the 316L stainless steel BMS that the patient received. Thus, through their direct association with granulation tissue formation, eosinophils may contribute to both thrombotic and restenotic mechanisms that pertain to stent deployment. Details concerning the former are outlined above, while the latter may be due to activation of platelets by eosinophilic proteins[37] and subsequent platelet degranulation to release promitogenic factors, such as platelet-derived growth factor and fibroblast growth factor, capable of stimulating VSMC migration and proliferation.

Restenosis of revascularization attempts by BMS or DES implantation are associated with eosinophil infiltration. This was documented nearly 20 years ago, when the transition from POBA to POBA with BMS deployment was being tested for clinical use. Karas and colleagues compared histological patterns of restenosis between POBA and insertion of a BM tantalum stent into swine coronary arteries.[22] Inflammation accompanied by greater VSMC proliferation was observed in association with in-stent restenosis. The inflammatory infiltrate consisted of eosinophils, macrophage-like histiocytes, and T lymphocytes surrounding stent struts. Consistent with this, a more recent report found VSMC proliferation to be primarily responsible for restenosis following stenting, but not following POBA, in swine.[23] Macrophages were selectively found in stented lesions and were accompanied by neutrophils and

TABLE 13.12.2 Association of Eosinophils with Stent Thrombosis

Species	Intervention(s)	Observations	Reference Number
Human	BMS	VLST, eosinophil infiltrate	18
Human	Overlapping Cypher	LST, eosinophils, T lymphocytes, aneurysm	19
Human	Overlapping Cypher	Repeated thrombosis, eosinophils	20
Human	Cypher and Taxus	VLST, eosinophils, vessel remodeling, stent malapposition Cypher > Taxus	21
Human	Cypher and Taxus	Occurrence of eosinophil inflammation in DES-specific, HS-associated thrombosis = 100%	29
Human	Cypher and Taxus	Occurrence of eosinophil inflammation in DES-related, HS-associated thrombosis = 100%	12

BMS, bare metal stent; DES, drug-eluting stent; HS, hypersensitivity; LST, late stent thrombosis (1 month–1 year postprocedure); VLST, very late stent thrombosis (>1 year postprocedure).

eosinophils. In another investigation, T lymphocytes were found in restenotic lesions of both POBA and BMS; however, significantly more VSMCs and eosinophils were associated with in-stent restenosis.[24] These studies suggest a positive correlation between inflammation characterized by eosinophil presence and VSMC proliferation and, similar to observations made relating eosinophils to stent thrombosis, a hypersensitivity reaction to a stent component that ultimately attracts eosinophils to the stented vessel segment. Such a relationship between the VSMC proliferative component of restenosis and eosinophil infiltration is significant, given the importance of VSMC growth in restenosis[5] and the fact that DES are loaded specifically with antiproliferative compounds primarily intended to block VSMC mitogenesis. Interestingly, other studies have shown the selective association of eosinophils, among other circulating cells including inflammatory and bone marrow-derived progenitor cells, with in-stent restenosis. Gabbasov and colleagues found that the number of osteonectin-positive progenitor cells, but not granulocytes, in blood were higher in patients afflicted with ischemic heart disease (IHD; $n = 38$) than in healthy individuals ($n = 17$).[44] In contrast, only elevations in eosinophils were detected in IHD patients that subsequently received Cypher DES and experienced restenosis ($n = 15$). Blood eosinophil levels were not increased in patients that did not undergo restenosis ($n = 23$). Together, histological and blood analyses have established a link between eosinophil presence and restenosis that is particularly associated with stent deployment, compared to POBA. These study observations suggest that eosinophils are equipped to contribute to the mechanism of restenosis. However, the question remains: are eosinophilic contents biomarkers of restenosis and thrombosis or do they actively contribute to these processes?

Niccoli and colleagues[45] showed that serum ECP levels prior to Cypher or Taxus implantation predicted whether patients would experience a major adverse cardiac event (MACE), including cardiac death, recurrent myocardial infarction, or target lesion revascularization (TLR), which was defined as being necessary if >50% stenosis occurred within 5 mm upstream or downstream of the stent. The majority (60%) of MACE onset occurred 180 days after stent insertion, while 27% of such cases happened more than 1 year following deployment. Clopidogrel was prescribed for 9 months and aspirin for a lifetime after stent insertion. Some patients had allergies, none of which were confirmed to stent components, such as metals or polymers. Furthermore, ECP levels were nearly equivalent between allergic and nonallergic patients that did not experience MACE. What factor(s) could preelevate ECP levels in individuals who had not yet been exposed to potential allergens contained in DES? Elevated ECP levels may be explained by prior observations indicating that eosinophil count positively correlated with IHD development[46] and that eotaxin/C-C motif chemokine 11 (CCL11) levels may play a role in atherosclerosis.[47] Taken together, these studies suggest that ECP may play a causative role in MACE, particularly TLR, which represented the majority of MACE cases in this study, since it was elevated before stent implantation. Related to this issue, experimental evidence implicates eosinophilic contents as having the potential to promote prorestenotic events during the healing process post-stent deployment. For example, as discussed earlier, activation of platelets by both MBP and EPO[37] may liberate growth factors from platelets that are capable of inducing migration and proliferation of VSMCs. By secreting eotaxin and transforming growth factor β (TBF- β),[17] eosinophils may directly stimulate VSMC

TABLE 13.12.3 Association of Eosinophils with Restenosis

Species	Intervention(s)	Observations	Reference Number
Pig	POBA and BMS	Eosinophils associated with BMS only; also histiocytes and macrophages present	22
Pig	POBA and BMS	Eosinophils associated with BMS only; neutrophils also present; VSMC hyperplasia	23
Human	POBA and BMS	Eosinophils, VSMCs, T lymphocytes: BMS > POBA	24
Human	BMS	Eosinophils, VSMC and matrix accumulation	26

BMS, bare metal stent; POBA, plain old balloon angioplasty; VSMC, vascular smooth muscle cell.

migration[48] and extracellular matrix production by fibroblasts and VSMCs,[49] respectively. Each of these manifestations of eosinophilic secretory products would contribute to restenotic lesion development. Masu and colleagues described that a heat-labile, unidentified constituent of <10 kDa in eosinophilic lysates can promote proliferation of airway smooth muscle cells (SMCs),[50] suggesting the ability of eosinophils to also stimulate vascular SMC growth. Table 13.12.3 provides a summary of studies that have reported an association between eosinophilic inflammation and restenosis. Further studies involving genetic deficiency or mRNA silencing may delineate which specific eosinophil component is responsible for this and other prorestenotic activities.

CONCLUSION

In summary, healing in coronary vessels can be a deleterious phenomenon in two instances, including forming atherosclerotic lesions following EC injury and forming thrombi and restenotic lesions following stent implantation. Ironically, treatments to alleviate atherosclerotic burden have evolved to include use of a minimally invasive therapeutic mode, namely deployment of BMS and DES, that potentially incites yet further, exaggerated, healing responses that can manifest as clinical concerns. This subchapter explains that healing events associated with stent deployment may be two-fold, including standard hemostatic, inflammatory, and tissue-remodeling phenomena that are likely common to all occurrences of stent implantation and, in a minority of individuals, superimposition of such standard healing responses by hypersensitivity reactions to the stent itself. To date, data suggest that both BM and polymer components of DES are candidates to stimulate involvement of factors and cells that mediate such hypersensitivity reactions, including eosinophils. Basic science studies have revealed the possibility that eosinophils contribute to thrombosis and restenosis associated with stent implantation, by virtue of the potentially prothrombotic effects of their granule proteins, such as ECP, EPO, and MBP, on platelets and thrombomodulin, and on SMC growth. As the incidence of hypersensitivity reactions as sequelae to stent implantation becomes more evident, routine prophylactic measures may be warranted, in addition to postprocedural DAPT prescription, in candidate stent recipients, to include prescreening individuals for allergies to stent components. Of course, the integrity of such tests is encumbered by the caveat that positive allergic patch tests may not predict the occurrence of stent-related thrombotic or restenotic events.[28] Alternatively, the use of next-generation bioabsorbable stents, which have relatively limited residency times in vessels compared to nonbioabsorbable stents such as Cypher and Taxus, would theoretically eliminate the stimulus for

eosinophilic responses relatively quickly and serve as a reasonable and useful way to reduce complications due to hypersensitivity in vulnerable patients. Clearly, observations made of eosinophils in the vicinity of stent struts in association with thrombosis and restenosis are highly suggestive of a role for these cells in such adverse events. Further work, perhaps involving eosinophil-deficient animals or cultured eosinophils deficient in granule proteins, is needed to more precisely define the role of eosinophils and the extent of medical attention deemed necessary to mitigate their presumed involvement in stent-associated thrombosis and in restenosis.

REFERENCES

1. Ross R. Rous-Whipple Award Lecture Atherosclerosis: A defense mechanism gone awry. *Am J Pathol* 1993;**143**:987–1002.
2. Mantell G. Lipid lowering drugs in atherosclerosis—the HMG-CoA reductase inhibitors. *Clin Exp Hypertens* 1989;**11**:927–41.
3. Gruentzig AR, Senning A, Siegenthaler WE. Nonoperative dilatation of coronary-artery stenosis: percutaneous transluminal coronary angioplasty. *N Engl J Med* 1979;**301**:61–8.
4. Holmes Jr DR, Vlietstra RE, Smith HC, Vetrovec GW, Kent KM, Cowley MJ, et al. Restenosis after percutaneous transluminal coronary angioplasty (PTCA): a report from the PTCA Registry of the National Heart, Lung, and Blood Institute. *Am J Cardiol* 1984;**53**:C77–81.
5. Komatsu R, Ueda M, Naruko T, Kojima A, Becker AE. Neointimal tissue response at sites of coronary stenting in humans. Macroscopic, histological and immunohistochemical analysis. *Circulation* 1998;**98**:224–33.
6. Casscells W. Migration of smooth muscle and endothelial cells: critical events in restenosis. *Circulation* 1992;**86**:723–9.
7. Regar E, Serruys PW, Bode C, Holubarsch C, Guermonprez JL, Wijns W, et al., RAVEL Study Group. Angiographic findings of the multicenter Randomized Study with the Sirolimus-Eluting Bx Velocity Balloon-Expandable Stent (RAVEL): sirolimus-eluting stents inhibit restenosis irrespective of the vessel size. *Circulation* 2002;**106**:1949–56.
8. Colombo A, Drzewiecki J, Banning A, Grube E, Hauptmann K, Silber S, et al., TAXUS II Study Group. Randomized study to assess the effectiveness of slow- and moderate release polymer-based paclitaxel-eluting stents for coronary artery lesions. *Circulation* 2003;**108**:788–94.
9. Scott NA. Restenosis following implantation of bare metal coronary stents: pathophysiology and pathways involved in the vascular response to injury. *Adv Drug Del Rev* 2006;**58**:358–76.
10. Auer J, Leitner A, Berent R, Lamm G, Lassnig E, Krennmair G. Long-term outcomes following coronary drug-eluting- and bare-metal-stent implantation. *Atherosclerosis* 2010;**210**:503–9.
11. Stone GW, Moses JW, Ellis SG, Schofer J, Dawkins KD, Morice MC, et al. Safety and efficacy of sirolimus- and paclitaxel-eluting coronary stents. *N Engl J Med* 2007;**356**:998–1008.
12. Finn AV, Nakazawa G, Joner M, Kolodgie FD, Mont EK, Gold HK, et al. Vascular responses to drug eluting stents. Importance of delayed healing. *Arterioscler Thromb Vasc Biol* 2007;**27**:1500–10.

13. Schwartz RS, Huber KC, Murphy JG, Edwards WD, Camrud AR, Vlietstra RE, et al. Restenosis and the proportional neointimal response to coronary artery injury: results in a porcine model. *J Am Coll Cardiol* 1992;**19**:267−74.

14. Finn AV, Kolodgie FD, Harnek J, Guerrero LJ, Acampado E, Tefera K, et al. Differential response of delayed healing and persistent inflammation at sites of overlapping sirolimus- or paclitaxel-eluting stents. *Circulation* 2005;**112**:270−8.

15. John MC, Wessely R, Kastrati A, Schomig A, Joner M, Uchihashi M, et al. Differential healing response in polymer- and nonpolymer-based sirolimus-eluting stents. *J Am Coll Cardiol Cardiovascular Interventions* 2008;**1**:535−44.

16. Wilson GJ, Nakazawa G, Schwartz RS, Huibregtse B, Poff B, Herbst TJ, et al. Comparison of inflammatory response after implantation of sirolimus- and paclitaxel-eluting stents in porcine coronary arteries. *Circulation* 2009;**120**:141−9.

17. Hogan SP, Rosenberg HF, Moqbel R, Phipps S, Foster PS, Lacy P, et al. Eosinophils: Biological properties and role in health and disease. *Clin Exp Allergy* 2008;**38**:709−50.

18. Zavolloni D, Bossi P, Rossi ML, Gasparini GL, Lisignoli V, Presbitero P. Inflammatory substrate with eosinophils may be present in bare-metal stent thrombosis. *J Cardiovasc Med* 2009;**10**:942−3.

19. Virmani R, Guagliumi G, Farb A, Musumeci G, Grieco N, Motta T, et al. Localized hypersensitivity and late coronary thrombosis secondary to a sirolimus-eluting stent. Should we be cautious? *Circulation* 2004;**109**:701−5.

20. Yokouchi Y, Oharaseki T, Ihara I, Naoe S, Sugawara S, Takahashi K. Repeated stent thrombosis after DES implantation and localized hypersensitivity to a stent implanted in the distal portion of a coronary aneurysm thought to be sequela of Kawasaki disease: autopsy report. *Pathol International* 2009;**60**:112−8.

21. Cook S, Ladich E, Nakazawa G, Eshtehardi P, Neidhart M, Vogel R, et al. Correlation of intravascular ultrasound findings with histopathological analysis of thrombus aspirates in patients with very late drug-eluting stent thrombosis. *Circulation* 2009;**120**:391−9.

22. Karas SP, Gravanis MB, Santoian EC, Robinson KA, Anderberg KA, King III SB. Coronary intimal proliferation after balloon injury and stenting in swine: and animal model of restenosis. *J Am Coll Cardiol* 1992;**20**:467−74.

23. Nakatani M, Takeyama Y, Shibata M, Yorozuya M, Suzuki H, Koba S, et al. Mechanisms of restenosis after coronary intervention. Difference between plain old balloon angioplasty and stenting. *Cardiovasc Pathol* 2003;**12**:40−8.

24. Rittersma SZH, Meuwissen M, van der Loos CM, Koch KT, de Winter RJ, Piek JJ, et al. Eosinophilic infiltration in restenotic tissue following coronary stent implantation. *Atherosclerosis* 2006;**184**:157−62.

25. Chen JP, Hou D, Pendyala L, Goudevenos JA, Kounis NG. Drug-eluting stent thrombosis. The Kounis hypersensitivity-associated acute coronary syndrome revisited. *J Am Coll Cardiol: Cardiovascular Interventions* 2009;**2**:583−93.

26. Kawano H, Koide Y, Baba T, Nakamizo R, Toda G, Takenaka M, et al. Granulation tissue with eosinophil infiltration in the restenotic lesion after coronary stent implantation—a case report. *Circ J* 2004;**68**:722−3.

27. Koster R, Vieluf D, Kiehn M, Sommerauer M, Kahler J, Baldus S, et al. Nickel and molybdenum contact allergies in patients with coronary in-stent restenosis. *Lancet* 2000;**356**:1895−7.

28. Hillen U, Haude M, Erbel R, Goos M. Evaluation of metal allergies in patients with coronary stents. *Contact Dermatitis* 2002;**47**:353−6.

29. Nebeker JR, Virmani R, Bennett CL, Hoffman JM, Samore MH, Alvarez J, et al. Hypersensitivity cases associated with drug-eluting coronary stents. A review of available cases from the Research on Adverse Drug Events and Reports (RADAR) project. *J Am Coll Cardiol* 2006;**47**:175−81.

30. Francischi JN, Conroy D, Maghni K, Sirois P. Rapamycin inhibits airway leukocyte infiltration and hyperreactivity in guinea pigs. *Agents Actions* 1993;**39**:C139−41.

31. Joner M, Finn AV, Farb A, Mont EK, Kolodgie FD, Ladich E, et al. Pathology of drug-eluting stents in humans. Delayed healing and late thrombotic risk. *J Am Coll Cardiol* 2006;**48**:193−202.

32. Cagliero E, Roth T, Taylor AW, Lorenzi M. The effects of high glucose on human endothelial cell growth and gene expression are not mediated by transforming growth factor-beta. *Lab Invest* 1995;**73**:667−73.

33. Parry TJ, Brosius R, Thyagarajan R, Carter D, Argentieri D, Falotico R, et al. Drug-eluting stents: sirolimus and paclitaxel differentially affect cultured cells and injured arteries. *Euro J Pharmacol* 2005;**524**:19−29.

34. Wessely R, Blaich B, Belaiba RS, Merl S, Gorlach A, Kastrati A, et al. Comparative characterization of cellular and molecular anti-restenotic profiles of paclitaxel and sirolimus. *Thromb Haemost* 2007;**97**:1003−12.

35. Grunig G, Warnock M, Wakil AE, Venkayya R, Brombacher F, Rennick DM, et al. Requirement for IL-13 independently of IL-4 in experimental asthma. *Science* 1998;**282**:2261−3.

36. Ogbogu P, Rosing DR, Horne III MK. Cardiovascular manifestations of hypereosinophilic syndromes. *Immunol Allergy Clin North Am* 2007;**27**:457−75.

37. Rohrbach MS, Wheatley CL, Slifman NR, Gleich GJ. Activation of platelets by eosinophil granule proteins. *J Exp Med* 1990;**172**:1271−4.

38. Mukai HY, Ninomiya H, Ohtani K, Nagasawa T, Abe T. Major basic protein binding to thrombomodulin potentially contributes to the thrombosis in patients with eosinophilia. *Br J Haematol* 1995;**90**:892−9.

39. Ohno I, Ohtani H, Nitta Y, Suzuki J, Hoshi H, Honma M, et al. Eosinophils as a source of matrix metalloproteinase-9 in asthmatic airway inflammation. *Am J Respir Cell Mol Biol* 1997;**16**:212−9.

40. Li A, Dubey S, Varney ML, Dave BJ, Singh RK. IL-8 directly enhanced endothelial cell survival, proliferation, and matrix metalloproteinases production and regulated angiogenesis. *J Immunol* 2003;**170**:3369−76.

41. Horiuchi T, Weller PF. Expression of vascular endothelial growth factor by human eosinophils: upregulation by granulocyte macrophage colony-stimulating factor and interleukin-5. *Am J Respir Cell Mol Biol* 1997;**17**:70−7.

42. Temkin V, Aingorn H, Puxeddu I, Goldshmidt O, Zcharia E, Gleich GJ, et al. Eosinophil major basic protein: first identified natural heparanase-inhibiting protein. *J Allergy Clin Immunol* 2004;**113**:703−9.

43. Puxeddu I, Alian A, Piliponsky AM, Ribatti D, Panet A, Levi-Schaffer F. Human peripheral blood eosinophils induce angiogenesis. *Int J Biochem Cell Biol* 2005;**37**:628−36.

44. Gabbasov ZA, Kozlov SG, Saburova OS, Titiov VN, Liakishev AA. Stromal progenitor cells and blood leukocytes after implantation of drug-eluting stents. *Kardiologiia* 2010;**50**:36–41.

45. Niccoli G, Schiavino D, Belloni F, Ferrante G, La Torre G, Conte M, et al. Pre-intervention eosinophil cationic protein serum levels predict clinical outcomes following implantation of drug-eluting stents. *Eur Heart J* 2009;**30**:1340–7.

46. Sweetnam PM, Thomas HF, Yarnell JW, Baker IA, Elwood PC. Total and differential leukocyte counts as predictors of ischemic heart disease: the Caerphilly and Speedwell studies. *Am J Epidemiol* 1997;**145**:416–21.

47. Emanuele E, Falcone C, D'Angelo A, Minoretti P, Buzzi MP, Bertona M, et al. Association of plasma eotaxin levels with the presence and extent of angiographic coronary artery disease. *Atherosclerosis* 2006;**186**:140–5.

48. Kodali RB, Kim WJ, Galaria II, Miller C, Schecter AD, Lira SA, et al. CCL11 (Eotaxin) induces CCR3-dependent smooth muscle cell migration. *Arterioscler Thromb Vasc Biol* 2004;**24**:1211–6.

49. Leask A, Abraham DJ. TGF-β signaling and the fibrotic response. *FASEB J* 2004;**18**:816–27.

50. Masu K, Ohno I, Suzuki K, Okada S, Hattori T, Shirato K. Proliferative effects of eosinophil lysates on cultured human airway smooth muscle cells. *Clin Exp Allergy* 2002;**32**:595–601.

Chapter 13.13

Eosinophils and Allograft Rejection

Florence Roufosse, Annick Massart, Fleur Samantha Benghiat, Philippe Lemaitre and Alain Le Moine

BRIEF OVERVIEW OF EOSINOPHIL BIOLOGY

Eosinophils are a minor leukocyte subset in healthy subjects, representing less than 5% of circulating white blood cells, and present in discrete locations, specifically the bone marrow, digestive tract, mammary glands, thymus, and uterus.[1,2] Eosinophils belong to the myeloid lineage, and their differentiation and proliferation in the marrow is controlled successively by specific transcription factors [including GATA binding protein 1 (GATA1) and transcription factor PU.1] and growth factors [granulocyte-macrophage colony-stimulating factor (GM-CSF), interleukin-3 (IL-3), and IL-5]. Among the latter, only IL-5 is specific for the eosinophil lineage in humans, as eosinophil precursors express the ligand-binding IL-5RA on their surface. Increased production of IL-5 *in vivo* by CD4+ T cells[3] or transformed cells (e.g., carcinomas[4] or Reed Sternberg cells[5]) results in increased eosinophilopoiesis and peripheral eosinophilia in blood and/or tissues. Various factors contribute to preferential eosinophil trans-endothelial migration and homing in tissues,[1] including adhesion molecules (vascular cell adhesion protein 1; VCAM-1), cytokines (IL-5), chemokines [specifically, eotaxins 1, 2 and 3, as well as RANTES (C-C motif chemokine 5) and monocyte chemoattractant protein (MCP)], and arachidonic acid metabolites [leukotriene B_4 (LTB_4), cysteinyl-leukotrienes, and prostaglandin D2 ($PGD2$)], in addition to the more general signals generated under conditions of cell stress and death.

Eosinophils were long considered as exclusively effector cells, able to induce significant tissue damage and dysfunction by releasing preformed, highly cytotoxic mediators, including the granule proteins, major basic protein and eosinophil cationic protein, producing reactive oxygen species, and generating arachidonic acid metabolites such as platelet activating factor (PAF) and LTC4 (reviewed in 1). These effector functions were considered potentially beneficial in the setting of parasitic infections, and harmful in the setting of allergic responses to environmental antigens. Recent studies have shattered this paradigm. It is now well established that eosinophils are active participants in ongoing immune responses through the production of cytokines and chemokines, and through previously unrecognized functions of their granule proteins [e.g., the ability of eosinophil-derived neurotoxin, a natural Toll-like receptor 2 (TLR2) ligand, to induce dendritic cell maturation and to promote antigen-specific T-helper type 2 (T_h2)-biased immune responses[6]], and that they possess many characteristics of antigen-presenting cells, enabling them to elicit antigen-specific CD4+ T cell responses. Indeed, eosinophils express major histocompatibility molecule (MHC) class II and co-stimulatory molecules (CD40, CD80, and CD86), and were shown to be able to process antigen (ovalbumin; OVA) and present it to naive CD4+ T cells in lymph nodes, in a murine model of allergic pulmonary inflammation.[7] Further upstream in allergic pulmonary inflammation, eosinophils have been shown to be required for the recruitment of antigen-specific effector CD4+ T cells to the lungs and the development of typical histopathological changes after allergen challenge in mice.[8] Their ability to induce production of the T_h2 chemokines thymus and activation-regulated chemokine (TARC) and macrophage-derived chemokine (MDC) in the lung is critical in this process.

In parallel with these proimmune functions, eosinophils also have the potential to modulate local inflammatory responses, for example dampening T_h1-dominated inflammation through release of IL-4 and IL-6.[9] Production of galectin-10 (also known as Charcot–Leyden crystal protein), IL-10, indoleamine 2,3-dioxygenase, and transforming growth factor β (TGF-β) may confer eosinophils with a regulatory role on effector T-cell responses.[10–12] Eosinophils also contribute to processes of remodeling and repair. Although eosinophils produce several potentially relevant factors, including fibroblast growth factor 2 (FGF-2), matrix metalloproteinase-9 (MMP-9), and vascular endothelial growth factor (VEGF), mechanistic studies establishing a causal role in remodeling are lacking; in contrast, several studies strongly suggest that eosinophil-derived TGF-β contributes to airway remodeling in allergic asthma.[13,14]

A series of observations on eosinophil behavior in health and disease unaccounted for by our current understanding of eosinophil biology, together with the paucity of experimental data supporting a causal relationship between eosinophil cytotoxicity and tissue damage and/or disease, have generated the *Local Immunity and/or Remodeling/Repair (LIAR) hypothesis*, which has recently been presented for scientific scrutiny.[15] The authors propose a central role for eosinophils in modulating LIAR, with recruitment of these cells to sites of cell death and turnover where stem cell activities are operative, in order to maintain tissue homeostasis. The accumulation and functions of eosinophils are dependent on various factors in the local microenvironment, including the presence of other specific immune effector cells, and of soluble growth factors liable to sustain eosinophil survival and activation. At physiological sites of high cell turnover, such as the endometrial lining and the gut, it is assumed that eosinophils dampen immune responses that could be triggered by such active metabolic activity. Similarly, eosinophils may inhibit the immune response elicited by tissue-infiltrating helminth parasites, favoring cohabitation between host and pathogen. In contrast, the local production of eosinophil growth-promoting and activating cytokines by other immune cells in allergic inflammation may favor positive feedback loops that sustain and amplify the immune response.[7,16,17]

Finally, in addition to the increasing complexity of eosinophil contributions to adaptive immunity, a role for eosinophils in innate immunity was recently suggested by a study showing that eosinophils express variable levels of CD3 and γ/δ T-cell receptors, which are involved in antimycobacterial and antitumor immune responses.[18] The inflammation that develops in solid organ transplants in the setting of an alloimmune response (i.e., transplant rejection) may contain, and in some instances be dominated by, eosinophilic infiltrates.[19] Whether eosinophils are directly involved in the damage to foreign tissue and thus actively contribute to rejection through release of their cytotoxic mediators, or are engaged in LIAR activities, is currently unknown. The local release of small molecule mediators, such as damage-associated molecular pattern molecules, by dying cells within the transplant may contribute to very early eosinophil recruitment.[15] Eosinophils could theoretically contribute to initiation of the allogeneic response by cross-presentation of foreign MHC antigens; they may also favor local recruitment of allospecific effector T cells, as is seen in allergic inflammation.

EOSINOPHILS IN EXPERIMENTAL ALLOGRAFT REJECTION

In mouse models of acute graft rejection in the setting of deficient CD8$^+$ T-cell effector functions, namely MHC class II-incompatible skin grafts[20] and fully histoincompatible cardiac transplants in CD8-deficient recipients,[21] marked eosinophilic infiltrates emerge that are dependent on IL-5 produced by antidonor CD4 T cells. However, IL-5 neutralization or silencing, and associated eosinophil depletion, fails to prevent rejection in these stringent alloreactive models, indicating at most a partial contribution of eosinophils to rejection.[19] This is not surprising, since allograft rejections are mediated by multiple redundant pathways.[22] Nevertheless, turning off T-cell cytotoxicity revealed a role for IL-5 and eosinophils in a model of acute and chronic rejection of MHC class II-incompatible skin grafts.[19,20,23] In the chronic rejection model, the T$_h$1 component of alloreactivity and T-cell cytotoxicity were dampened by repeated injections of anti-CD3 antibody. Similarly, treatment with anti-CD154 and a depleting CD8 monoclonal antibody Ab resulted in eosinophilic infiltration in a model of transplant arteriosclerosis.[24,25] In other experiments, the adoptive transfer of alloreactive noncytotoxic T$_h$2 clones into T cell-deficient mice induced the rejection of skin or cardiac allografts characterized by a dense eosinophil infiltrate.[26,27] Finally, IL-4 or IL-5 neutralization, as well as eosinophil depletion through repeated injections of anti-CCR3 (C-C chemokine receptor type 3) antibody, prevented the rejection of weakly immunogenic skin grafts bearing a single minor antigen disparity in MHC class I-deficient recipients.[28]

EOSINOPHILS IN SOLID ORGAN TRANSPLANTATION

Eosinophils and Kidney Transplantation

The role of eosinophils in renal transplantation has aroused little attention. Although publications described the presence of eosinophils during renal allograft rejection in the early 1980s,[29] the first systematic reviews on the subject only appeared in the mid-1980s.[30,31] This may be due to technical reasons. Indeed, conventional staining techniques, like hematoxylin—eosin (H&E), underestimate the presence of tissue eosinophils and do not usually detect their degranulation.[32] Motivated by a case report of marked hypereosinophilia and eosinophilic infiltration in a rejected renal allograft, in 1986 Weir and coworkers decided to investigate retrospectively a cohort of 132 renal transplant recipients (124 biopsies) with 187 episodes of acute

TABLE 13.13.1 Summary of Observations

Type of Transplantation	Supposed Role*	Marker of Severity	Predictive of Good or Bad Outcome	References
Kidney	Effector	Yes	Bad	30–36
Lung	Effector	Yes	Bad	37,40
Liver	Effector	Yes	Bad	41–50
HSC	?	Yes	Bad	55,56,59,72–74
		No	Good	57,58,61,63

*Based on each author's report, although the role of eosinophils might be more complex (see text).

rejection.[30] They concluded that increased eosinophils in the blood or renal biopsy represented an adverse prognostic factor for renal outcome. This was followed by a prospective study by Kormendi and coworkers analyzing cellular infiltrates using fine-needle aspiration in a cohort of 83 renal allograft recipients during the first month post-transplantation.[31] Tissue eosinophilia exceeding 4% was considered a useful cutoff with a predictive accuracy for serious or irreversible rejections of 71% (sensitivity: 78%; specificity: 91%, with a prevalence of acute rejections of 32.5%). In contrast, blood eosinophil counts were found to be less reliable. In another study, Ten and coworkers showed eosinophils in the kidney interstitium and in tubular casts.[33] Of note, eosinophil degranulation was evaluated by the extracellular localization of the eosinophil granule major basic protein (MBP) as revealed by immunofluorescence. In this small cohort of 16 patients, eosinophils and extracellular MBP were more frequently observed in acute rejection (94% and 87%, respectively) than in cyclosporine toxicity (6 patients; 17% and 17%), whereas both features were absent in controls (normal kidney donors). Similarly, urinary levels of MBP were also elevated in acute rejections and acute interstitial nephritis while they remained normal in cyclosporine nephrotoxicity. In another study, eosinophils and extracellular eosinophil cationic protein (ECP) were prominent features of acute vascular rejection rather than interstitial rejection, and eosinophil density increased in areas bordering necrotic tissue and in arteries with necrotic lesions.[34] A correlation between eosinophil infiltrates and rejection severity was also observed by Meleg and coworkers, who reviewed 29 allograft nephrectomies.[35] They concluded that a significant interstitial graft eosinophil infiltrate called SIGE was statistically associated with vascular rejections but not iatrogenic interstitial nephritis. This was already reported by Hongwei and coworkers, who also found a correlation between the density of the eosinophil infiltrate and the rate of graft loss by rejection.[36] All together, these observations reinforce the possibility of a nonincidental,

causative association between eosinophils and acute allograft rejection (Table 13.13.1).

Eosinophils are also linked to chronic allograft rejection characterized by interstitial fibrosis, obliterative arteriopathy, and tubular atrophy. Indeed, Nolan and coworkers reported the presence of eosinophils in 93% of renal allografts undergoing chronic rejection.[32] They were located in the intimal and adventitial space of the thickened arteries, as well as in the interstitium. Interestingly, in vitro experiments revealed that eosinophil by-products (see above) enhanced fibroblast and vascular smooth muscle cell proliferation in murine and human experiments, perhaps reflecting a pathogenic mechanism involved in obliterative arteriopathy. Although these data strongly suggest a pathogenic role for eosinophils and by-products in kidney allografts, their presence may be related to other functions, as evoked in the LIAR hypothesis.

Eosinophils and Lung Transplantation

Although eosinophils are described in chronic and acute lung allograft rejection, their role in these processes remains unclear.[37–39] Acute lung rejection classically occurs during the first year after transplantation and its diagnosis is essentially based on transbronchial biopsies (TBB). The current international guidelines, published in 2010 by the International Heart and Lung Transplantation Society (ISHLT),[40] establish acute lung rejection as the presence of, firstly, perivascular and interstitial mononuclear cell infiltrates (grade A) and, second, small-airway inflammation, namely lymphocytic bronchiolitis (grade B). Grading is scaled depending on the composition, extension, and intensity of the infiltrate. Eosinophils are not a feature of grade A1 (minimal rejection), but are found in grade A2 (mild vascular rejection) and, importantly, are considered to be a common finding in severe vascular rejection (grade A3). Regarding airway inflammation, eosinophils are considered to be occasional in low-grade (B1R) and common in high-grade (B2R) inflammation.

Chronic lung allograft rejection, synonymous with bronchiolitis obliterans syndrome (BOS), occurs in up to 50% of recipients. BOS is characterized by a persistent decrease in expiratory flow and is potentially life threatening, requiring retransplantation. Although eosinophils are not taken into consideration in the ISHLT working formulation for this condition, a recent prospective cohort study reported that recurrent tissue eosinophilia (with higher concentrations of IL-6 and IL-8) is significantly associated with an increased risk of developing BOS.[37]

Increased eosinophilia in lung transplant recipients should be interpreted with caution for two reasons.[39] Firstly, there are numerous nonrejection-related causes of graft eosinophilia. Among these, infectious diseases are dominant, including fungi (e.g., aspergillus and coccidioidomycosis), bacteria (e.g., tuberculosis), helminths (e.g., Toxocara canis and Ascaris lumbricoides), and even viruses (e.g., cocksackies). High-dose steroid treatment could be detrimental under these conditions. Drug reactions are also a common cause of pulmonary eosinophilia (e.g., antibiotics, diuretics, or methotrexate). The second reason is that blood eosinophilia does not always reflect bronchoalveolar lavage fluid or lung tissue eosinophilia. Lung biopsies are therefore crucial for assessing the role of eosinophils after lung transplantation.

Eosinophils and Liver Transplantation

There are many similarities regarding eosinophilia in liver transplantation compared with other transplanted organs already discussed. IL-5 and eosinophils were rapidly identified during liver allograft rejection.[41–45] There was a consensus for considering eosinophils and IL-5 as mediators of a nonclassical pathway of rejection (i.e., non-T_h1-mediated rejection).[46;47] Indeed, liver allografts with evidence of rejection showed concomitant intragraft IL-5 mRNA and activated eosinophils releasing MBP.[46] In pediatric recipients, elevated biliary and serum IL-5 correlate with rejection.[47] Along the same lines, another study reported that blood eosinophilia and serum ECP are early indicators of acute liver allograft rejection and precede alterations of conventional liver function tests by several days.[48] However, the use of increased serum ECP as a rejection marker is limited by its association with infections.[48] These pioneering findings were confirmed and refined by more recent studies that identified graft and blood eosinophilia as an independent highly specific marker of acute liver allograft rejection.[41,42,45,49,50] In addition, an elevated blood eosinophil count may predict the severity of rejection, just as in the case of lung and kidney transplantation.[49] However, the use of this potential marker of rejection is limited in patients with hepatitis C infection, and those treated with corticosteroids, as both these circumstances decrease eosinophil levels.[49]

EOSINOPHILS AND HEMATOPOIETIC STEM CELL TRANSPLANTATION

Inflammation plays a pivotal role in the complex pathogenesis of acute and chronic graft-versus-host disease (GVHD). Conventionally, acute GVHD (aGVHD) is described as a T_h1 disease associated with the release of interferon γ (IFN-γ), IL-2, IL-12, and TNF-α.[51–53] The gastrointestinal tract, liver, and skin are the most common targets of GVHD, and diarrhea, jaundice, and skin rash are its most common manifestations. Chronic GVHD has features resembling autoimmune and other immunological disorders, such as bronchiolitis obliterans (see above), chronic immunodeficiency, Sjögren syndrome, and systemic sclerosis. Although the pathophysiology has not been fully elucidated, chronic GVHD appears to be mediated by the overproduction of T_h2-type cytokines, namely IL-4 and IL-5.[53,54]

The role of eosinophils in acute and chronic GVHD is a matter for speculation. Thirty years ago, Shulman and coworkers showed for the first time that eosinophilia after allogeneic hematopoietic stem cell transplantation (HSCT) was often present at the time of diagnosis of chronic GVHD.[55] Afterwards, it was shown that eosinophilia could precede, sometimes by several months, the onset of chronic GVHD symptoms, and seemed to have a strong predictive value for the subsequent development of this condition.[56,57] Should this be confirmed by additional prospective trials, this observation may have significant clinical impact. More recently, eosinophilia was also observed among patients who developed aGVHD and, similar to chronic GVHD, was seen before the beginning of symptoms in some cases.[58,59] The pathophysiology behind eosinophilia in the setting of GVHD remains unclear. T_h2 cytokine production may be involved, as suggested by the finding that serum IL-5 concentrations are elevated in patients with symptoms of aGVHD.[60,61] However, no correlation with blood eosinophilia has been observed.[62]

Studies focusing on the prognostic importance of eosinophilia have produced conflicting data. In chronic GVHD, some retrospective data suggest a better outcome for HSCT recipients with eosinophilia, while others found no correlation, suggesting that eosinophilia may just be a bystander of cGVHD rather than a prognostic biomarker.[57,61,63] Among young patients with malignant diseases treated by HSCT, those with eosinophilia showed increased event-free survival and a lower relapse rate than those without eosinophilia, suggesting that eosinophils could be involved in the graft versus leukemia effect.[61] In aGVHD, observational studies showed that patients with eosinophilia after allogeneic HSCT have a milder disease than patients without eosinophilia.[58]

One could hypothesize that improved prognosis of aGVHD when eosinophils are present is related to the fact that this reflects T_h2, rather than T_h1 (classically considered as the effectors of aGVHD cell activation). In agreement, murine studies have shown that several cytokines that are known to favor T_h2 polarization of donor T cells, such as GM-CSF, IL-4, or rapamycin, can reduce aGVHD.[64-67]

In humans, it has been shown that an increased number of IL-4- and IL-10-producing cells are associated with reduced severity or absence of aGVHD,[68,69] also suggesting a possible protective role for T_h2 cytokines in aGVHD. Nevertheless, reports indicate that T_h2 subsets may actually cause aGVHD by targeting other organs than those targeted by T_h1 subsets.[70,71] This argues against a rigid paradigm according to which aGVHD is a T_h1 process and chronic GVHD a T_h2 process. In support of this hypothesis, prospective data have shown that bone marrow eosinophilia after HSCT may be a predictive marker of severe aGVHD.[72] Similarly, the presence of eosinophils in duodenal biopsy specimens taken during acute flares correlates with intestinal GVHD severity.[73] There is also evidence that eosinophils show signs of activation in both blood and target organs of patients during aGVHD flares.[74,75] The bad reputation of eosinophils in GVHD generated by these observations has led some investigators to test the efficacy of montelukast (an orally active leukotriene antagonist that inhibits eosinophils) as a supplement to standard therapy for patients with chronic GVHD, with promising preliminary results.[76] For the time being, the role of eosinophils in target organ damage still needs to be assessed in the setting of well-conducted experimental studies.[77]

CONCLUSION

In conclusion, compelling evidence links activated eosinophils with allograft rejection and graft-versus-host disease. In contrast to mast cells, they are not (yet) linked to transplantation tolerance.[78] Their presence seems to be correlated with the gravity of tissue damage, which may reflect effector functions contributing to rejection, or eosinophil accumulation in response to tissue damage, in agreement with the LIAR hypothesis. To date, the roles played by IL-5 and eosinophils in the redundant pathways of allograft rejection and the potential graft healing processes remain cryptic. The recent availability of genetically engineered mice lacking eosinophils (PHIL and ΔdblGATA) provides a unique opportunity to clarify their contribution to these processes.[79,80]

REFERENCES

1. Blanchard C, Rothenberg ME. Biology of the eosinophil. *Adv Immunol* 2009;**101**:81−121.
2. Rothenberg ME, Hogan SP. The eosinophil. *Annu Rev Immunol* 2006;**24**:147−74.
3. Romagnani S. Human TH1 and TH2 subsets: doubt no more. *Immunol Today* 1991;**12**:256−7.
4. Pandit R, Scholnik A, Wulfekuhler L, Dimitrov N. Non-small-cell lung cancer associated with excessive eosinophilia and secretion of interleukin-5 as a paraneoplastic syndrome. *Am J Hematol* 2007;**82**:234−7.
5. Samoszuk M, Nansen L. Detection of interleukin-5 messenger RNA in Reed-Sternberg cells of Hodgkin's disease with eosinophilia. *Blood* 1990;**75**:13−6.
6. Yang D, Chen Q, Su SB, Zhang P, Kurosaka K, Caspi RR, et al. Eosinophil-derived neurotoxin acts as an alarmin to activate the TLR2-MyD88 signal pathway in dendritic cells and enhances Th2 immune responses. *J Exp Med* 2008;**205**:79−90.
7. Wang HB, Ghiran I, Matthaei K, Weller PF. Airway eosinophils: allergic inflammation recruited professional antigen-presenting cells. *J Immunol* 2007;**179**:7585−92.
8. Jacobsen EA, Ochkur SI, Pero RS, Taranova AG, Protheroe CA, Colbert DC, et al. Allergic pulmonary inflammation in mice is dependent on eosinophil-induced recruitment of effector T cells. *J Exp Med* 2008;**205**:699−710.
9. Spencer LA, Szela CT, Perez SA, Kirchhoffer CL, Neves JS, Radke AL, et al. Human eosinophils constitutively express multiple Th1, Th2, and immunoregulatory cytokines that are secreted rapidly and differentially. *J Leukoc Biol* 2009;**85**:117−23.
10. Jacobsen EA, Taranova AG, Lee NA, Lee JJ. Eosinophils: singularly destructive effector cells or purveyors of immunoregulation? *J Allergy Clin Immunol* 2007;**119**:1313−20.
11. Odemuyiwa SO, Ghahary A, Li Y, Puttagunta L, Lee JE, Musat-Marcu S, et al. Cutting edge: human eosinophils regulate T cell subset selection through indoleamine 2,3-dioxygenase. *J Immunol* 2004;**173**:5909−13.
12. Kubach J, Lutter P, Bopp T, Stoll S, Becker C, Huter E, et al. Human CD4+CD25+ regulatory T cells: proteome analysis identifies galectin-10 as a novel marker essential for their anergy and suppressive function. *Blood* 2007;**110**:1550−8.
13. Kay AB, Phipps S, Robinson DS. A role for eosinophils in airway remodelling in asthma. *Trends Immunol* 2004;**25**:477−82.
14. Phipps S, Flood-Page P, Menzies-Gow A, Ong YE, Kay AB. Intravenous anti-IL-5 monoclonal antibody reduces eosinophils and tenascin deposition in allergen-challenged human atopic skin. *J Invest Dermatol* 2004;**122**:1406−12.
15. Lee JJ, Jacobsen EA, McGarry MP, Schleimer RP, Lee NA. Eosinophils in health and disease: the LIAR hypothesis. *Clin Exp Allergy* 2010;**40**:563−75.
16. MacKenzie JR, Mattes J, Dent LA, Foster PS. Eosinophils promote allergic disease of the lung by regulating CD4(+) Th2 lymphocyte function. *J Immunol* 2001;**167**:3146−55.
17. Sabin EA, Kopf MA, Pearce EJ. Schistosoma mansoni egg-induced early IL-4 production is dependent upon IL-5 and eosinophils. *J Exp Med* 1996;**184**:1871−8.
18. Legrand F, Driss V, Woerly G, Loiseau S, Hermann E, Fournie JJ, et al. A functional gammadeltaTCR/CD3 complex distinct from gammadeltaT cells is expressed by human eosinophils. *PLoS One* 2009;**4**, e5926.
19. Goldman M, Le Moine A, Braun M, Flamand V, Abramowicz D. A role for eosinophils in transplant rejection. *Trends Immunol* 2001;**22**:247−51.
20. Le Moine A, Surquin M, Demoor FX, Noel JC, Nahori MA, Pretolani M, et al. IL-5 mediates eosinophilic rejection of MHC

class II-disparate skin allografts in mice. *J Immunol* 1999;**163**:3778—84.

21. Braun MY, Desalle F, Le Moine A, Pretolani M, Matthys P, Kiss R, et al. IL-5 and eosinophils mediate the rejection of fully histo-incompatible vascularized cardiac allografts: regulatory role of alloreactive CD8(+) T lymphocytes and IFN-gamma. *Eur J Immunol* 2000;**30**:1290—6.

22. Le Moine A, Goldman M, Abramowicz D. Multiple pathways to allograft rejection. *Transplantation* 2002;**73**:1373—81.

23. Le Moine A, Flamand V, Demoor FX, Noel JC, Surquin M, Kiss R, et al. Critical roles for IL-4, IL-5, and eosinophils in chronic skin allograft rejection. *J Clin Invest* 1999;**103**:1659—67.

24. Ensminger SM, Spriewald BM, Witzke O, Morrison K, van Maurik A, Morris PJ, et al. Intragraft interleukin-4 mRNA expression after short-term CD154 blockade may trigger delayed development of transplant arteriosclerosis in the absence of CD8+ T cells. *Transplantation* 2000;**70**:955—63.

25. Ensminger SM, Spriewald BM, Sorensen HV, Witzke O, Flashman EG, Bushell A, et al. Critical role for IL-4 in the development of transplant arteriosclerosis in the absence of CD40-CD154 costimulation. *J Immunol* 2001;**167**:532—41.

26. Matesic D, Valujskikh A, Pearlman E, Higgins AW, Gilliam AC, Heeger PS. Type 2 immune deviation has differential effects on alloreactive CD4+ and CD8+ T cells. *J Immunol* 1998;**161**:5236—44.

27. Honjo K, Xu XY, Bucy RP. Heterogeneity of T cell clones specific for a single indirect alloantigenic epitope (I-Ab/H-2Kd54—68) that mediate transplant rejection. *Transplantation* 2000;**70**:1516—24.

28. Surquin M, Le Moine A, Flamand V, Rombaut K, Demoor FX, Salmon I, et al. IL-4 deficiency prevents eosinophilic rejection and uncovers a role for neutrophils in the rejection of MHC class II disparate skin grafts. *Transplantation* 2005;**80**:1485—92.

29. Shalev O, Rubinger D, Barlatzky Y, Kopolovic J, Drukker A. Eosinophilia associated with acute allograft kidney rejection. *Nephron* 1982;**31**:182—3.

30. Weir MR, Hall-Craggs M, Shen SY, Posner JN, Alongi SV, Dagher FJ, et al. The prognostic value of the eosinophil in acute renal allograft rejection. *Transplantation* 1986;**41**:709—12.

31. Kormendi F, Amend Jr WJ. The importance of eosinophil cells in kidney allograft rejection. *Transplantation* 1988;**45**:537—9.

32. Nolan CR, Saenz KP, Thomas III CA, Murphy KD. Role of the eosinophil in chronic vascular rejection of renal allografts. *Am J Kidney Dis* 1995;**26**:634—42.

33. Ten RM, Gleich GJ, Holley KE, Perkins JD, Torres VE. Eosinophil granule major basic protein in acute renal allograft rejection. *Transplantation* 1989;**47**:959—63.

34. Hallgren R, Bohman SO, Fredens K. Activated eosinophil infiltration and deposits of eosinophil cationic protein in renal allograft rejection. *Nephron* 1991;**59**:266—70.

35. Meleg-Smith S, Gauthier PM. Abundance of interstitial eosinophils in renal allografts is associated with vascular rejection. *Transplantation* 2005;**79**:444—50.

36. Hongwei W, Nanra RS, Stein A, Avis L, Price A, Hibberd AD. Eosinophils in acute renal allograft rejection. *Transpl Immunol* 1994;**2**:41—6.

37. Scholma J, Slebos DJ, Boezen HM, van den Berg JW, van der BW, de Boer WJ, et al. Eosinophilic granulocytes and interleukin-6 level in bronchoalveolar lavage fluid are associated with the development of obliterative bronchiolitis after lung transplantation. *Am J Respir Crit Care Med* 2000;**162**:2221—5.

38. Mogayzel Jr PJ, Yang SC, Wise BV, Colombani PM. Eosinophilic infiltrates in a pulmonary allograft: a case and review of the literature. *J Heart Lung Transplant* 2001;**20**:692—5.

39. Yousem SA. Graft eosinophilia in lung transplantation. *Hum Pathol* 1992;**23**:1172—7.

40. Stewart S, Fishbein MC, Snell GI, Berry GJ, Boehler A, Burke MM, et al. Revision of the 1996 working formulation for the standardization of nomenclature in the diagnosis of lung rejection. *J Heart Lung Transplant* 2007;**26**:1229—42.

41. Datta GS, Hudson M, Burroughs AK, Morris R, Rolles K, Amlot P, et al. Grading of cellular rejection after orthotopic liver transplantation. *Hepatology* 1995;**21**:46—57.

42. de Groen PC, Kephart GM, Gleich GJ, Ludwig J. The eosinophil as an effector cell of the immune response during hepatic allograft rejection. *Hepatology* 1994;**20**:654—62.

43. Foster PF, Sankary HN, Williams JW, Bhattacharyya A, Coleman J, Ashmann M. Morphometric inflammatory cell analysis of human liver allograft biopsies. *Transplantation* 1991;**51**:873—6.

44. Foster PF, Bhattacharyya A, Sankary HN, Coleman J, Ashmann M, Williams JW. Eosinophil cationic protein's role in human hepatic allograft rejection. *Hepatology* 1991;**13**:1117—25.

45. Foster PF, Sankary HN, Hart M, Ashmann M, Williams JW. Blood and graft eosinophilia as predictors of rejection in human liver transplantation. *Transplantation* 1989;**47**:72—4.

46. Martinez OM, Ascher NL, Ferrell L, Villanueva J, Lake J, Roberts JP, et al. Evidence for a nonclassical pathway of graft rejection involving interleukin 5 and eosinophils. *Transplantation* 1993;**55**:909—18.

47. Lang T, Krams SM, Berquist W, Cox KL, Esquivel CO, Martinez OM. Elevated biliary interleukin 5 as an indicator of liver allograft rejection. *Transpl Immunol* 1995;**3**:291—8.

48. Hughes VF, Trull AK, Joshi O, Alexander GJ. Monitoring eosinophil activation and liver function after liver transplantation. *Transplantation* 1998;**65**:1334—9.

49. Barnes EJ, Abdel-Rehim MM, Goulis Y, Abou RM, Davies S, Dhillon A, et al. Applications and limitations of blood eosinophilia for the diagnosis of acute cellular rejection in liver transplantation. *Am J Transplant* 2003;**3**:432—8.

50. Dollinger MM, Plevris JN, Bouchier IA, Harrison DJ, Hayes PC. Peripheral eosinophil count both before and after liver transplantation predicts acute cellular rejection. *Liver Transpl Surg* 1997;**3**:112—7.

51. Paczesny S, Hanauer D, Sun Y, Reddy P. New perspectives on the biology of acute GVHD. *Bone Marrow Transplant* 2010;**45**:1—11.

52. Ferrara JL, Levy R, Chao NJ. Pathophysiologic mechanisms of acute graft-vs.-host disease. *Biol Blood Marrow Transplant* 1999;**5**:347—56.

53. Tanaka J, Imamura M, Kasai M, Hashino S, Kobayashi S, Noto S, et al. Th2 cytokines (IL-4, IL-10 and IL-13) and IL-12 mRNA expression by concanavalin A-stimulated peripheral blood mononuclear cells during chronic graft-versus-host disease. *Eur J Haematol* 1996;**57**:111—3.

54. Martin PJ. Biology of chronic graft-versus-host disease: implications for a future therapeutic approach. *Keio J Med* 2008;**57**:177—83.

55. Shulman HM, Sullivan KM, Weiden PL, McDonald GB, Striker GE, Sale GE, et al. Chronic graft-versus-host syndrome

56. Kalaycioglu ME, Bolwell BJ. Eosinophilia after allogeneic bone marrow transplantation using the busulfan and cyclophosphamide preparative regimen. *Bone Marrow Transplant* 1994;**14**:113—5.

57. Kim DH, Popradi G, Xu W, Gupta V, Kuruvilla J, Wright J, et al. Peripheral blood eosinophilia has a favorable prognostic impact on transplant outcomes after allogeneic peripheral blood stem cell transplantation. *Biol Blood Marrow Transplant* 2009;**15**:471—82.

58. Imahashi N, Miyamura K, Seto A, Watanabe K, Yanagisawa M, Nishiwaki S, et al. Eosinophilia predicts better overall survival after acute graft-versus-host-disease. *Bone Marrow Transplant* 2010;**45**:371—7.

59. McNeel D, Rubio MT, Damaj G, Emile JF, Belanger C, Varet B, et al. Hypereosinophilia as a presenting sign of acute graft-versus-host disease after allogeneic bone marrow transplantation. *Transplantation* 2002;**74**:1797—800.

60. Fujii N, Hiraki A, Aoe K, Murakami T, Ikeda K, Masuda K, et al. Serum cytokine concentrations and acute graft-versus-host disease after allogeneic peripheral blood stem cell transplantation: concurrent measurement of ten cytokines and their respective ratios using cytometric bead array. *Int J Mol Med* 2006;**17**:881—5.

61. Sato T, Kobayashi R, Nakajima M, Iguchi A, Ariga T. Significance of eosinophilia after stem cell transplantation as a possible prognostic marker for favorable outcome. *Bone Marrow Transplant* 2005;**36**:985—91.

62. Imoto S, Oomoto Y, Murata K, Das H, Murayama T, Kajimoto K, et al. Kinetics of serum cytokines after allogeneic bone marrow transplantation: interleukin-5 as a potential marker of acute graft-versus-host disease. *Int J Hematol* 2000;**72**:92—7.

63. Aisa Y, Mori T, Nakazato T, Shimizu T, Yamazaki R, Ikeda Y, et al. Blood eosinophilia as a marker of favorable outcome after allogeneic stem cell transplantation. *Transpl Int* 2007;**20**:761—70.

64. Fowler DH, Kurasawa K, Smith R, Eckhaus MA, Gress RE. Donor CD4-enriched cells of Th2 cytokine phenotype regulate graft-versus-host disease without impairing allogeneic engraftment in sublethally irradiated mice. *Blood* 1994;**84**:3540—9.

65. Pan L, Delmonte Jr J, Jalonen CK, Ferrara JL. Pretreatment of donor mice with granulocyte colony-stimulating factor polarizes donor T lymphocytes toward type-2 cytokine production and reduces severity of experimental graft-versus-host disease. *Blood* 1995;**86**:4422—9.

66. Foley JE, Jung U, Miera A, Borenstein T, Mariotti J, Eckhaus M, et al. Ex vivo rapamycin generates donor Th2 cells that potently inhibit graft-versus-host disease and graft-versus-tumor effects via an IL-4-dependent mechanism. *J Immunol* 2005;**175**:5732—43.

67. Tawara I, Maeda Y, Sun Y, Lowler KP, Liu C, Toubai T, et al. Combined Th2 cytokine deficiency in donor T cells aggravates experimental acute graft-vs-host disease. *Exp Hematol* 2008;**36**:988—96.

68. Weston LE, Geczy AF, Briscoe H. Production of IL-10 by allo-reactive sibling donor cells and its influence on the development of acute GVHD. *Bone Marrow Transplant* 2006;**37**:207—12.

69. Takabayashi M, Kanamori H, Takasaki H, Yamaji S, Koharazawa H, Taguchi J, et al. A possible association between the presence of interleukin-4-secreting cells and a reduction in the risk of acute graft-versus-host disease. *Exp Hematol* 2005;**33**:251—7.

70. Nikolic B, Lee S, Bronson RT, Grusby MJ, Sykes M. Th1 and Th2 mediate acute graft-versus-host disease, each with distinct end-organ targets. *J Clin Invest* 2000;**105**:1289—98.

71. Yi T, Chen Y, Wang L, Du G, Huang D, Zhao D, et al. Reciprocal differentiation and tissue-specific pathogenesis of Th1, Th2, and Th17 cells in graft-versus-host disease. *Blood* 2009;**114**:3101—12.

72. Basara N, Kiehl MG, Fauser AA. Eosinophilia indicates the evolution to acute graft-versus-host disease. *Blood* 2002;**100**:3055.

73. Daneshpouy M, Socie G, Lemann M, Rivet J, Gluckman E, Janin A. Activated eosinophils in upper gastrointestinal tract of patients with graft-versus-host disease. *Blood* 2002;**99**:3033—40.

74. Rumi C, Rutella S, Bonini S, Bonini S, Lambiase A, Sica S, et al. Immunophenotypic profile of peripheral blood eosinophils in acute graft-vs.-host disease. *Exp Hematol* 1998;**26**:170—8.

75. Daneshpouy M, Facon T, Jouet JP, Janin A. Acute flare-up of conjunctival graft-versus-host disease with eosinophil infiltration in a patient with chronic graft-versus-host disease. *Leuk Lymphoma* 2002;**43**:445—6.

76. Or R, Gesundheit B, Resnick I, Bitan M, Avraham A, Avgil M, et al. Sparing effect by montelukast treatment for chronic graft versus host disease: a pilot study. *Transplantation* 2007;**83**:577—81.

77. Janin A, Ertault-Daneshpouy M, Socie G. Eosinophils and severe forms of graft-versus-host disease. *Blood* 2003;**101**:2073.

78. Lu LF, Lind EF, Gondek DC, Bennett KA, Gleeson MW, Pino-Lagos K, et al. Mast cells are essential intermediaries in regulatory T-cell tolerance. *Nature* 2006;**442**:997—1002.

79. Lee JJ, Dimina D, Macias MP, Ochkur SI, McGarry MP, O'Neill KR, et al. Defining a link with asthma in mice congenitally deficient in eosinophils. *Science* 2004;**17**:1773—6.

80. Yu C, Cantor AB, Yang H, Browne C, Wells RA, Fujiwara Y, et al. Deletion of a high-affinity GATA-binding site in the GATA-1 promoter leads to selective loss of the eosinophil lineage in vivo. *J Exp Med* 2002;**195**:1387—95.

Chapter 13.14

Eosinophils and Calpain-3 Mutation: A Genetic Cause Implicated in Idiopathic Eosinophilic Myositis

Martin Krahn, Marc Bartoli and Nicolas Levy

EOSINOPHILIC MYOSITIS

Eosinophilic myositis is a rare histopathological entity characterized by infiltration of skeletal muscle tissue by eosinophils, possibly in association with peripheral blood and/or bone marrow hypereosinophilia. These characteristics distinguish eosinophilic myositis from other idiopathic inflammatory myopathies, such as polymyositis and dermatomyositis. The diagnosis of eosinophilic myositis is

FIGURE 13.14.1 **Eosinophilic myositis caused by *CAPN3* mutations in a pediatric patient.** Depicted are histological findings (hematoxylin and eosin staining) on muscle biopsy samples evidencing mild myopathic changes with focal inflammatory lesions including abundant eosinophilic infiltration, involving necrotic fibers. *(Reproduced with permission from Krahn et al., 2006; Annals of Neurology; Wiley.)*

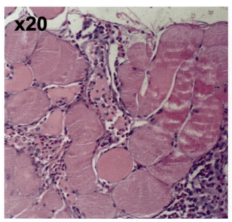
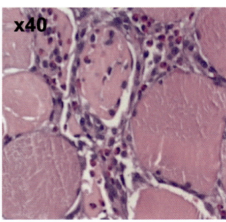

based on the histological examination of muscle biopsy sections. Infiltration of skeletal muscle tissue by eosinophils is an unusual event, observed especially during parasite infections (including *Taenia solium*, *Trichinella spiralis*, and sarcocystosis)[1] or more rarely bacterial infections (borreliosis). Several immune disorders, such as sarcoidosis or rheumatoid arthritis, may also be accompanied by eosinophilic infiltration of the skeletal muscle tissue. These forms must be identified as they may benefit from specific therapeutic management. Additionally, some toxic causes have been implicated in the formation of eosinophilic infiltrates in muscle tissue. Those include, in particular, the ingestion of certain plant oils (which caused Spanish toxic oil syndrome in 1981)[2] and the eosinophilia-myalgia syndrome, caused by the ingestion of L-tryptophan and presenting with a histological aspect resembling eosinophilic fasciitis, but associated with multisystemic manifestations.[3-4]

Idiopathic Eosinophilic Myositis

The exclusion of the aforementioned different etiologies determines the diagnosis of *idiopathic* eosinophilic myositis. Depending on the localization of the eosinophilic infiltrate, idiopathic eosinophilic myositis can be classified into three subgroups (reviewed in[5-6]): focal eosinophilic myositis, eosinophilic polymyositis, and eosinophilic perimyositis.

Focal Eosinophilic Myositis

Focal eosinophilic myositis includes eosinophilic infiltration of muscle tissue with invasion of muscle fibers, and is associated with necrotic fibers.[5] Clinically, myopathy preferentially affects the lower limbs, without involvement of skin or fascia. Eosinophilia is usually observed, and is associated with elevated serum creatine phosphokinase (CPK) levels. Spontaneous or corticosteroid-induced recovery may be observed, but with frequent relapses.

Eosinophilic Polymyositis

At the histological level, eosinophilic polymyositis combines diffuse eosinophilic infiltration of the muscle tissue and at perivenular locations, associated with necrotic fibers.[5,7] Unlike focal eosinophilic myositis, infiltration is instead located at the perimysium, without muscle fiber invasion. Clinical presentation associates myositis with severe systemic symptoms, including possible cardiac and skin involvement. Myopathy is preferentially proximal and high serum CPK levels are observed, reflecting extensive muscle damage. Corticosteroid treatment can allow recovery, but with possible relapses if not continued long term.

Eosinophilic Perimyositis

In eosinophilic perimyositis, infiltrates predominate at the superficial fascia and perimysium.[5,8] There is usually no damage to muscle fibers and, in particular, no necrosis. Clinically, a prodromal phase (abdominal pain, arthralgia, and fever) precedes muscle damage, which involves preferential impairment of lower limbs with localized induration. Eosinophilia is rare, and serum CPK levels are usually normal. The evolution can be spontaneously favorable.

Shulman's Syndrome

Shulman's syndrome,[9] or eosinophilic fasciitis, is a distinct entity characterized by eosinophilic infiltration of the deep fascia, without systemic manifestations, and about half the cases are responsive to corticosteroid treatment.[10]

A GENETIC CAUSE FOR A SUBSET OF IDIOPATHIC EOSINOPHILIC MYOSITIS

At the current state of knowledge, eosinophilic myopathies constitute a heterogeneous group of rare diseases without a causal factor being identified in most cases.

In 2006, our group reported an unexpected clinical observation: a 4-year-old boy was diagnosed with idiopathic eosinophilic myositis and muscle biopsy analysis revealed a calpain-3 protein defect.[11] Calpain-3 is a muscle-specific protein, belonging to the family of calpains, nonlysosomal calcium-dependent cysteine proteases.[12] The most well studied of the calpains are the ubiquitous heterodimeric calpains (μ-calpain and m-calpain). The human calpain-3 gene is located on chromosome 15q15.1—q21.1. The predominant product of this gene is encoded by 24 exons corresponding to a 3316 bp mRNA expressed mainly in adult skeletal muscles.[13]

The translation of the main calpain-3 gene (*CAPN3*) product leads to the formation of a 94 kDa protein comprising 821 amino acids and consisting of a short N-terminal region (domain I), a papain-type proteolytic domain (domains IIa and IIb), a C2-like domain (domain III), and a calcium-binding domain composed of five EF-hand motifs (domain IV). In addition, calpain-3 possesses three unique sequences not found in other calpains: the NS (N-terminal sequence), and the IS1 and IS2 (inserted sequences 1 and 2) sequences.[14] IS1 is a polypeptide of about 50 amino acids encoded by an alternative exon 6. It is composed of an α-helix flanked by loops that close the catalytic cleft, thus blocking access to substrates and inhibitors.[15] It also contains autolytic sites involved in the initiation of calpain-3 proteolysis by opening the catalytic cleft.[16] Immunolocalization studies carried out in human and mouse demonstrated that calpain-3 located in the N2A, M-line, and Z-line regions of the sarcomere. In addition to these localizations, calpain-3 has also been found at costameres and near the triad of the T-tubule.[17] Calpain-3 functions to promote proteolysis of several substrates located in the costameres, sarcolemma, and sarcomere and seems to be most important in fully differentiated fibers (for review see[14]). In adult fibers, calpain-3 participates in sarcomere adaptation by cleaving cytoskeletal proteins during muscular adjustment, in accordance with the distribution of known substrates.[17]

Mutations in the *CAPN3* gene cause the most prevalent form of autosomal recessive limb-girdle muscular dystrophy (LGMD), type 2A (LGMD2A),[18] also referred to as calpainopathy. Based on the calpain-3 protein defect identified in the boy with idiopathic eosinophilic myositis mentioned above, the *CAPN3* gene was analyzed, revealing mutations and thus a genetic cause associated with eosinophilic infiltration in this case. Nonspecific inflammatory features may be associated with a variety of muscle dystrophies, possibly leading to misdiagnosis of polymyositis in some cases (e.g., congenital muscle dystrophy, fascioscapulohumeral muscular dystrophy, and dysferlinopathies). However, eosinophilic infiltration had not been previously characterized as a component of inflammatory features in muscular dystrophies.

We subsequently identified five additional children diagnosed with idiopathic eosinophilic myositis (exemplified in one patient in Fig. 13.14.1) and identified *CAPN3* disease-causing mutations in all cases, either in a homozygous or compound heterozygous state. Following our publication of these pediatric cases, two adult cases of idiopathic eosinophilic myositis and *CAPN3* mutations were reported by Amato.[19] In 2009, Oflazer and colleagues reported another pediatric case of idiopathic eosinophilic myositis associated with *CAPN3* mutations,[20] in which a positive effect of immunosuppressive therapy was observed. Since our initial report, we have characterized five additional unrelated patients with idiopathic eosinophilic myositis and *CAPN3* mutations (one adult and four children, unpublished data). Importantly, except for the initially identified boy presenting with a calpain-3 defect, inclusion criteria for *CAPN3* mutation screening of the other cases were based only on the particular histopathological presentation, without any identified etiological factors.

Noteworthy, *CAPN3* mutations identified in patients with idiopathic eosinophilic myositis do not appear to constitute any particular mutational spectrum as compared to *typical* LGMD2A. Our findings demonstrate that at least a subset of idiopathic eosinophilic myositis has a genetic origin, caused by mutations in the *CAPN3* gene and with an autosomal recessive mode of inheritance. On the other hand, as eosinophilic infiltration is not known to be a typical feature of LGMD2A, it is possible that eosinophilia may be a transient feature in the natural course of this disease.

The explanation for eosinophilic infiltration correlating with defective calpain-3 needs to be further evaluated. T lymphocytes may be a key component in this process[21] as:

1. Together with macrophages, they are the main components of inflammatory lesions in the vicinity of damaged muscle fibers.
2. They play a central role in the chemoattraction of eosinophils by secreting interleukin-5, which induces local eosinophil accumulation[22] and they express calpain-3.[23]

Regarding the latter, no relevant function for calpain-3 in T lymphocytes has been identified to date. The presence of eosinophils has been previously reported to be a component of muscular dystrophy in mdx mice, promoted by perforin-dependent cytotoxicity of effector T cells.[24] In addition, eosinophils play a specific role in muscle fiber degradation, due to degradation of myofibrillar and membrane-associated proteins by eosinophil cationic protein.[25]

The identification of *CAPN3* mutations as the first genetic cause involved in eosinophilic myositis indicates that mutations in other genes may also be causal, or act as modifiers, of this pathophysiology. In this regard, a case of dystrophinopathy in which a muscle biopsy shows the appearance of eosinophilic myositis has been described,[26] and Baumeister and coworkers recently reported a case of idiopathic eosinophilic myositis caused by a homozygous mutation in the γ-sarcoglycan gene,[27] which is implicated in another form of LGMD. Interestingly, these reports suggest that early, but transient, eosinophilic infiltration may be a feature of a *wider* range of muscular dystrophies. The role of eosinophils in the natural history of diverse muscular dystrophies should therefore be further investigated, and it is possible that eosinophils may represent a novel therapeutic target.

REFERENCES

1. Bocanegra TS, Vasey FB. Musculoskeletal syndromes in parasitic diseases. *Rheum Dis Clin North Am* 1993 May;**19**(2):505—13.
2. Kilbourne EM, Rigau-Perez JG, Heath Jr CW, Zack MM, Falk H, Martin-Marcos M, et al. Clinical epidemiology of toxic-oil syndrome. Manifestations of a New Illness. *N Engl J Med* 1983 Dec 8;**309**(23):1408—14.
3. Hertzman PA, Blevins WL, Mayer J, Greenfield B, Ting M, Gleich GJ. Association of the eosinophilia-myalgia syndrome with the ingestion of tryptophan. *N Engl J Med* 1990 Mar 29;**322**(13):869—73.
4. Hollander D, Adelman LS. Eosinophilia-myalgia syndrome associated with ingestion of L-tryptophan: muscle biopsy findings in 4 patients. *Neurology* 1991 Feb;**41** (2(Pt 1)):319—21.
5. Hall FC, Krausz T, Walport MJ. Idiopathic eosinophilic myositis. *QJM* 1995 Aug;**88**(8):581—6.
6. Pickering MC, Walport MJ. Eosinophilic myopathic syndromes. *Curr Opin Rheumatol* 1998 Nov;**10**(6):504—10.
7. Layzer RB, Shearn MA, Satya-Murti S. Eosinophilic polymyositis. *Ann Neurol* 1977 Jan;**1**(1):65—71.
8. Serratrice G, Pellissier JF, Cros D, Gastaut JL, Brindisi G. Relapsing eosinophilic perimyositis. *J Rheumatol* 1980 Mar-Apr;**7**(2):199—205.
9. Shulman LE. Diffuse fasciitis with eosinophilia: a new syndrome? *Trans Assoc Am Physicians* 1975;**88**:70—86.
10. Barnes L, Rodnan GP, Medsger TA, Short D. Eosinophilic fasciitis. A pathologic study of twenty cases. *Am J Pathol* 1979 Aug;**96**(2):493—518.
11. Krahn M, Lopez de Munain A, Streichenberger N, Bernard R, Pécheux C, Testard H, et al. CAPN3 mutations in patients with idiopathic eosinophilic myositis. *Ann Neurol* 2006 Jun;**59**(6):905—11.

12. Goll DE, Thompson VF, Li H, Wei W, Cong J. The calpain system. *Physiol Rev* 2003 Jul;**83**(3):731—801.
13. Sorimachi H, Suzuki K. Sequence comparison among muscle-specific calpain, p94, and calpain subunits. *Biochim Biophys Acta* 1992 Nov 10;**1160**(1):55—62.
14. Duguez S, Bartoli M, Richard I. Calpain 3: a key regulator of the sarcomere? *FEBS J* 2006 Aug;**273**(15):3427—36.
15. Diaz BG, Moldoveanu T, Kuiper MJ, Campbell RL, Davies PL. Insertion sequence 1 of muscle-specific calpain, p94, acts as an internal propeptide. *J Biol Chem* 2004 Jun 25;**279**(26):27656—66.
16. Ono Y, Torii F, Ojima K, Doi N, Yoshioka K, Kawabata Y, et al. Suppressed disassembly of autolyzing p94/CAPN3 by N2A connectin/titin in a genetic reporter system. *J Biol Chem* 2006 Jul 7;**281**(27):18519—31.
17. Taveau M, Bourg N, Sillon G, Roudaut C, Bartoli M, Richard I. Calpain 3 is activated through autolysis within the active site and lyses sarcomeric and sarcolemmal components. *Mol Cell Biol* 2003 Dec;**23**(24):9127—35.
18. Richard I, Broux O, Allamand V, Fougerousse F, Chiannilkulchai N, Bourg N, et al. Mutations in the proteolytic enzyme calpain 3 cause limb-girdle muscular dystrophy type 2A. *Cell* 1995 Apr 7;**81**(1):27—40.
19. Amato AA. Adults with eosinophilic myositis and calpain-3 mutations. *Neurology* 2008 Feb 26;**70**(9):730—1.
20. Oflazer PS, Gundesli H, Zorludemir S, Sabuncu T, Dincer P. Eosinophilic myositis in calpainopathy: could immunosuppression of the eosinophilic myositis alter the early natural course of the dystrophic disease? *Neuromuscul Disord* 2009 Apr;**19**(4):261—3.
21. Brown Jr RH, Amato A. Calpainopathy and eosinophilic myositis. *Ann Neurol* 2006 Jun;**59**(6):875—7.
22. Murata K, Sugie K, Takamure M, Fujimoto T, Ueno S. Eosinophilic major basic protein and interleukin-5 in eosinophilic myositis. *Eur J Neurol* 2003 Jan;**10**(1):35—8.
23. De Tullio R, Stifanese R, Salamino F, Pontremoli S, Melloni E. Characterization of a new p94-like calpain form in human lymphocytes. *Biochem J* 2003 Nov 1;**375**(Pt 3):689—96.
24. Cai B, Spencer MJ, Nakamura G, Tseng-Ong L, Tidball JG. Eosinophilia of dystrophin-deficient muscle is promoted by perforin-mediated cytotoxicity by T cell effectors. *Am J Pathol* 2000 May;**156**(5):1789—96.
25. Sugihara R, Kumamoto T, Ito T, Ueyama H, Toyoshima I, Tsuda T. Human muscle protein degradation *in vitro* by eosinophil cationic protein (ECP). *Muscle Nerve* 2001 Dec;**24**(12):1627—34.
26. Weinstock A, Green C, Cohen BH, Prayson RA. Becker muscular dystrophy presenting as eosinophilic inflammatory myopathy in an infant. *J Child Neurol* 1997 Feb;**12**(2):146—7.
27. Baumeister SK, Todorovic S, Milic-Rasic V, Dekomien G, Lochmuller H, Walter MC. Eosinophilic myositis as presenting symptom in γ-sarcoglycanopathy. *Neuromuscul Disord* 2009 Feb;**19**(2):167—71.

Animal Models of Human Pathology

Introduction: The Trials and Tribulations of Linking Cellular and/or Molecular Pathways and the Pathologies Associated with Inflammation using Animal Models

James J. Lee

The utility of an animal model to understand unique cellular and molecular pathways, in fact the utility of any model system, lies in its ability to replicate (i.e., *to model*) the target(s) of interest and/or specific aspects of that target(s). Thus, the problem faced by research scientists working with animal models of human disease is getting it *right*, with *right* being defined as replicating human disease in an animal (e.g., the mouse) that often doesn't normally develop a given disease (e.g., asthma or atopic dermatitis). In addition, the underlying assumption in this pursuit is that what is being modeled is known in great detail. That is, that we understand and know all the things occurring in patients at a sufficient level to model these diseases in an animal. The unfortunate truth is that physician–scientists and basic researchers have yet to fully understand the details of the immune and inflammatory events associated with many diseases. For example, physiologists and clinicians have only a narrow understanding of how pulmonary pathologies associated with asthma relate to specific changes in lung function. This lack of clarity also extends to the potential roles (or lack thereof) of specific cell-mediated pathways, such as the pulmonary eosinophil infiltrate commonly linked with allergic respiratory inflammation. In this case, the countless correlative clinical studies linking airway eosinophilia with asthma, together with the accepted paradigm of eosinophils as destructive end-stage effector cells, led to the therapeutic option of treating asthma by targeting eosinophils and reducing their numbers in the lung. This hypothesis also found support in a variety of animal models, including studies in mice, that displayed improvement with targeting of the eosinophil infiltrate. However, despite these *positive* outcomes, a growing number of concurrent animal studies targeting eosinophils with potential agents that could be used in patients were suggesting a far less definitive association. This left investigators to quarrel over the specific utility of one model or another without even a clear definition of what eosinophils did (or did not do) in asthma patients. Clinical studies translating these potential therapeutic approaches to patients also provided only limited insights and served to complicate issues further. Specifically, early attempts targeting eosinophils in asthma patients using anti-interleukin-5 antibody approaches (e.g., mepolizumab[7]) were met with less than spectacular results. Thus, these early studies failed to provide much needed insights into the roles of eosinophils in asthma and additionally had the effect of crystallizing the view that animal models, particularly the mouse, were of limited value.

Our ongoing discussions regarding the utility of animal models to understand human disease have highlighted an underappreciated aspect of the problem faced by investigators: What is the true value of a mouse model and when is a given model useful? However, an answer to this question arose following many delightful and interesting conversations with a close friend and colleague (Dr. Mark Inman, McMaster University, Hamilton, Canada). We devised a simple analogy that originated with the desire of my young son Nicky (at the time 8 years old), who desperately wants to become a US fighter pilot and fly the F-35 Joint Strike Fighter to be stationed at Luke Air Force Base here in Phoenix, Arizona, USA. The first thing Nicky did upon hearing of the planes' new home in Arizona was to get me to buy one of those model airplane kits widely found in retail stores so he could build a replica of this plane, including painting it and adding those very sticky decals (i.e., the words feared by every father: 'some assembly required'). Nonetheless, Nicky did a great job with the project and when he showed me the finished F-35 model my reaction to him was to say that 'it was a perfect F-35 fighter.' The problem arose later in the day when Nicky came in from outside crying uncontrollably because his model plane was smashed beyond recognition. When asked what happened, I was both shocked at a child's naiveté yet exalted at the moment of clarity it provided. Nicky, upon hearing that his plane was a *perfect model*, had gone up to the roof of our house and thrown the plane skyward, fully expecting that such a perfect model would simply fly away like all of the other planes he sees at Luke Air Force Base. The disconnect for Nicky was that his plane only modeled the visual appearance (albeit to scale) of an F-35 fighter. That is, the limit of this model was that it only looked exactly like the plane and as such it could not defy gravity when thrown nor could it do anything else that the real F-35 can do. This was the moment of clarity and many subsequent discussions yielded a fundamental truth—a model's value is defined on the basis of what is intended to be replicated and how well you understand the character being modeled. The bottom line is that there are in fact many ways to model an F-35 that are all truly unique and accurate, yet each is perfectly flawed. For example, one can build a fighter out of Styrofoam for an exhibit in an air museum to model the plane's size and the details of its outward appearance. Alternatively, a large enclosed cockpit on hydraulic pistons controlled by a computer and sophisticated software (a flight simulator) could be used to train pilots because it provides an accurate model of how to fly the plane. A structural frame in a wind tunnel would provide an excellent model of the plane's aerodynamics and weapons systems at a proving ground afford accurate models of the plane's firepower. The point of these examples is that none of the *models* are complete and none of them is a good and/or realistic description of an F-35 Joint Strike Fighter. However, each model has immense value on its own and collectively they provide sufficient insights to accurately describe this aircraft. Mouse models of human diseases are no different! They are incredibly useful and provide valuable insights when the questions asked are narrow and specific. In addition, as with any model, the usefulness and/or value of a given mouse model is dependent on the background knowledge, descriptions, and details of the human disease being modeled. More importantly, while the value of a single mouse model is invariably limited, the larger collection of such models available to investigators may in that context provide a remarkably realistic view of human disease.

This chapter provides fascinating summations of unique mouse models of five human diseases, each of which is highlighted by the localized presence of eosinophils. Two fundamental assumptions were made by the contributing authors: (1) that the mouse models exploited are representative of human disease and (2) that eosinophils and their associated activities contribute to the pathologies occurring in patients and, in turn, the mouse models. The mouse models described by the contributing authors include both obvious and not so obvious candidate diseases that are accepted as being linked with eosinophils. For example, in Chapter 14.2 I describe the history and use of mouse models of asthma. As reviewed in this section, these respiratory models have both clarified and horrified

investigators. That is, these models have helped in many respects to move our understanding of eosinophil activities forward, while at the same time calling into question the role of eosinophils in this respiratory disease and the utility of the mouse as a model system of the lung. In Chapter 14.3, Dr. Carine Blanchard tackles mouse models of esophageal eosinophilia as representative of the human disease eosinophil esophagitis. As Dr. Blanchard notes, many models of eosinophilic esophagitis exist but herein lies the strength of this approach to understand the role(s) of eosinophils. Specifically, each model is representative of only a subset of the human conditions; however, together these models provide extraordinarily valuable insights from which investigators have developed (and continue to develop) a greater understanding of both this disease and the unique role(s) of eosinophil-mediated activities. Two more subchapters continue addressing the issues that surround eosinophils and their role(s) in the gastrointestinal tract. In Chapter 14.4, Furuta and colleagues note that even the colon of healthy subjects is home to a significant resident population of eosinophils. As such, they propose that eosinophils may have a unique role(s) here to maintain homeostasis. Moreover, they propose that these colonic eosinophils may also contribute to inflammatory bowel diseases linked with eosinophils. The issues surrounding food allergies and the potential role(s) of eosinophils in this disease are the topic of discussion in Chapter 14.5 by Waddel and Hogan. In this subchapter, the authors review the expanding number of mouse models of food allergy and review experimental evidence that suggests that eosinophils may have a causative role in disease pathogenesis extending from altered immune functions to intestinal pathophysiology. In Chapter 14.6, Rothenberg and colleagues provide an outstanding summation of the complexities associated with constructing and then characterizing mouse models of eosinophilic leukemia. Here, the goal isn't so much to convince the reader that eosinophils are a prominent disease component. Instead, this subchapter offers powerful insights regarding the utility of mouse models in the development of new therapeutic approaches. Finally, in Chapter 14.7 Odemuyiwa and Moqbel provide a comprehensive review of eosinophils from a comparative medicine perspective. In particular, these authors demonstrate that while a quick read of the literature could suggest that humans and mice are the only species with eosinophils, these granulocytes are widely distributed and are significant components of diseases occurring in a wide array of animals.

The breadth of knowledge and the quality of studies investigating the roles of eosinophils using animal models of human disease are hard to overestimate. It reminds me of an International Eosinophil Society meeting many years ago when, as a junior and somewhat young investigator, I was asked by a very prominent senior colleague whether animals even had eosinophils and as such were studies on them relevant to humans. I was taken aback by the question because clearly the answer was yes, but yet dumbfounded that this was even an issue. It appears that the intervening years have provided a far better response than anything I tried to say that day!

The Role of Eosinophils in Animal Models of Asthma: Causative Agents or Victims of Circumstance?

James J. Lee

The heterogeneous character of asthma symptoms and the complexity of both the innate and acquired inflammatory events occurring in the lung have complicated the definition of the underlying mechanisms leading to specific pathologies. In particular, although historically identified early as an accumulating leukocyte in the lungs of asthma patients, the role of eosinophils remains a hotly debated topic, the resolution of which does not appear imminent. This debate is ultimately fueled by the lack of consistency and efficacy between therapies targeting eosinophils in asthma patients and the resulting changes in lung function and related parameters. This uncertainty is highlighted in the current National Lung Heart and Blood Institute *Guidelines for the Diagnosis and Management of Asthma (EPR-3)* (http://www.nhlbi.nih.gov/guidelines/asthma/index.htm), which states that the role of eosinophils in asthma 'is undergoing a reevaluation based on studies with an anti-interleukin-5 (anti-IL-5) treatment that has significantly reduced eosinophils but did not affect asthma control.' This original study, and subsequent studies of higher power and more effective design that have followed, have elicited discussions in the intervening 5 years regarding appropriate study designs, parameters measured, the utility of animal models, and even the relevance of the eosinophil. Studies using eosinophil-dependent animal models of allergic respiratory inflammation have mirrored, and in some cases foretold, the ambiguous character of prospective patient studies. Specifically, subsets of animal model studies have supported a causative role for eosinophils, whereas other studies have suggested virtually no role in the pathologies linked with allergic respiratory inflammation. The case is made in this subchapter that, despite some understandable variability in results, studies of asthma using available eosinophil-dependent mouse models are convincingly consistent and demonstrate a direct link between the

presence of pulmonary eosinophils and several induced changes in the lung following allergen provocation. More importantly, these unique mouse model-based studies highlight significant roles for eosinophils in both the immune and remodeling/repair responses that are likely to underlie the pathologies occurring the lungs of asthma patients.

The accumulation of eosinophils in the lungs of asthma patients is an anchor phenomenon, with their differential recruitment to the airway mucosa and lumen occurring in >75% of reported cases,[1] even including patients with mild asthma.[2] Indeed, these granulocytes were identified as a significant inflammatory infiltrate in the earliest clinical studies of asthma.[3] Despite this established correlative history and accepted commonality of eosinophil accumulation in the lungs of both allergic asthmatics and animal models of asthma, the definition of a causative relationship between eosinophils and induced pulmonary pathologies has remained out of reach. That is, while many studies correlate asthma severity [e.g., pulmonary remodeling events and lung dysfunction, including airway hyper-responsiveness (AHR)] with either peripheral blood eosinophil numbers,[4] sputum and/or bronchoalveolar lavage (BAL) eosinophilia,[5] or airway/blood cationic eosinophil secondary granule protein levels,[6] these correlations are complex and hardly straightforward. The correlations have also tended not to be linear and published studies with conflicted results/conclusions are common even from the same investigators over time (e.g.,[7] vs.[8]). Moreover, exceptions to the rule that disassociate the presence of pulmonary eosinophils from pulmonary inflammation and induced lung dysfunction are well described.

WHY ARE EOSINOPHIL-MEDIATED MOUSE MODELS OF ASTHMA OF VALUE?

In vivo models of eosinophil effector functions in asthma have an inherit advantage over *ex vivo* experimentation using either isolated eosinophils or tissue/organ explants. They assess the integrated process vs. isolated components that on their own may, or may not, function in a similar fashion. Thus, while cellular and molecular studies of eosinophil activities have benefited tremendously from *ex vivo* cell culture experiments, this reductionist strategy has significant limitations that have prevented a detailed characterization of eosinophil activities in respiratory diseases.[9] To this end, animal models have been used to provide these much needed insights; the mouse, in particular, has risen to the forefront of these model systems. Although mouse models of asthma are limited because of their inability to model human lung structure and many aspects of function (see, for example,[10]), eosinophil-dependent mouse models have been sufficiently narrow and focused as to enable the underlying mechanisms of eosinophil effector functions in allergic respiratory disease to be defined. The utility and unique character of these eosinophil-dependent mouse models, including their relevance to clinical disease, also arise from two distinguishing features of the mouse relative to other animal models:

1. Genetic propinquity with humans and husbandry issues that allow for the maintenance of substantial cohorts/colonies of mice. That is, similar to comparisons between other mammalian species, the shared sequence complexity between mice and humans is startling—99% of all mouse genes have an identifiable human orthologue.[11] Moreover, maintaining colonies of mice is the most economical of the potential mammalian models and all mice in a given study may be of a single inbred strain. Thus, all subjects are genetically identical, eliminating subject-to-subject genetic variation.

2. The mouse offers a unique ability to manipulate its genome (i.e., alter gene expression) in a Mendelian-inheritable fashion through gene transfer technologies (i.e., transgenic and gene knockout/knock-in mouse strains).

LESSONS LEARNED FROM MOUSE MODELS DEFICIENT/DEVOID OF EOSINOPHILS

It continues to surprise me when I go to meetings or other venues to hear that eosinophils have been shown *not* to have a role in asthma. Invariably such discussions are also accompanied by a disparaging remark or two regarding the misguided use of mouse models to develop therapeutic strategies targeting eosinophils. The source of this negativity is ultimately traceable to the conclusion by many that while the blockade of eosinophils mediated by anti-IL-5 antibodies is widely effective in preventing the pulmonary pathologies in rodent models of allergic respiratory inflammation,[12–14] these strategies have no effect on allergic asthma symptoms in patients.[15,16] This perspective is pervasive and frustratingly annoying because the available clinical data, as well as results from experiments earlier using rodent/mouse models, are far from definitive and are not easily interpreted (as has been suggested otherwise).

The conclusion that eosinophils are not relevant in the development of asthma pathologies was fueled by the grandiose expectations that surrounded the initial studies targeting these cells in patients: for over 100 years, eosinophils have been uniquely linked with asthma[17] and therefore they must be important variable airflow components of the limitations associated with this lung disease. Thus, the lack of a clear and definitive link, no matter how poor the study,[18] was deemed as failing to support the

hypothesis of the importance of eosinophils and proved that they were not contributors to allergic pulmonary pathologies. The problem with such high expectations is that the only place to go is down. Specifically, these initial studies were based on the premise that eosinophils are the *central mediators* of asthma. As such, the lack of a clear and definitive answer with a cohort of asthma patients was surprising. However, given the complexity of this disease, the variability of patient populations, and the likely contributions of multiple immune/inflammatory events, a far more prudent premise would have been that eosinophils are simply one of many contributors to disease pathology. Moreover, in this specific paradigm eosinophils form a part of pathways that are both independent and, in some cases, overlapping with pathways mediated by many other immune/inflammatory cell types. The recent studies using the anti-IL-5 monoclonal antibody mepolizumab with specific subpopulations of asthma patients have supported this hypothesis. That is, stratifying asthma patients on the basis of defined clinical criteria indicating a role for eosinophils demonstrated an expected efficacy for this therapeutic approach.[5] More importantly, this efficacy was not all encompassing, but instead was limited to reducing exacerbation events leading to emergency room visits and steroid use.[5,19] Interestingly, a careful examination of the mouse model studies that were performed leading up to the development of drugs such as mepolizumab actually suggests that the results witnessed in the clinical trials with asthma patients should have been expected! In particular, these mouse model studies collectively should have suggested caution and lowered expectations for the development of this *wonder drug* to treat asthma:

1. Targeting IL-5 was successful in ameliorating lung dysfunction (i.e., AHR) only in some strains of mice and not in others (see, for example,[7] vs.[8]). In addition, a myriad of other studies using one mouse model or another also purportedly uncouple a *one-to-one* correlation of eosinophils with allergen-induced AHR (reviewed in[20,21]).

2. Allergen-sensitized/aerosol-challenged mice treated with anti-IL-5 antibodies consistently displayed decreases in goblet cell metaplasia/mucin accumulation via a mechanism(s) that was largely unexplained (see, for example,[12]). More importantly, other studies effecting eosinophil levels in the lung did not display such a correlative result (see, for example,[22]), suggesting a degree of complexity and, therefore, unpredictability of this antieosinophil approach.

3. Anti-IL-5 antibody treatment of mice effectively eliminated eosinophils from the blood of test animals but at most achieved only a 92% decrease of the induced pulmonary eosinophilia, regardless of the mouse strain used;[13,23] similar results were obtained with IL-5

knockout mice.[24] The point is that ablation of eosinophils was not complete in the mouse models and, depending on the mouse strain/model, this yielded variable levels of efficacy in the amelioration of symptoms.

In light of these data, it is also not surprising that an examination of asthma patients treated with mepolizumab showed that tissue eosinophils in the lung parenchyma of patients undergoing treatment are actually only reduced by approximately 50%.[25]

A more conservative conclusion from a summary of early studies with anti-IL-5 moieties in mouse models would be to suggest a more complex role for eosinophils in the immune/inflammatory events leading to allergic pulmonary inflammation. Fully understanding this complex role(s) was difficult, partly because existing strategies did *not* effectively eliminate eosinophils in the mouse models used. As noted earlier and summarized in Table 14.2.1, several independent studies in mice that

only partially ablated eosinophils resulted in variable effects on the pathologies following allergen provocation. Specifically, blockade of IL-5 (either with antibodies or through the use of IL-5 knockout mice) resulted in only a 92% decrease in eosinophils and, in turn, strain-dependent effects on allergen-induced AHR. Nonetheless, restoration of pulmonary eosinophilia in allergen-treated IL-5 knockout mice (background strain C57BL/6) by adoptive cell transfer recovered allergen-induced AHR in these mice.[24] The blockade of eosinophil trafficking to the lung using knockout mice deficient of the eosinophil agonist C-C chemokine receptor 3 (CCR3; background strain BALB/c) and two different allergen-challenge protocols (approximately 50%[26] or 10%,[27] relative to allergen-treated controls) provided mixed results. One study reported no effect on allergen-induced AHR[26] and the other reported its loss in allergen-challenged *Ccr3*[−/−] mice.[27] In contrast to the less than definitive results associated with the partial ablation of

TABLE 14.2.1 Allergen Challenge of Eosinophil-Deficient Mouse Models

Mouse Model	BAL Eosinophilia (% Relative to OVA-treated Wild-type Mice)	Background Strain	Goblet Cell Metaplasia (% Relative to OVA-treated Wild-type Mice)	AHR
Il5[−/−7,8]	~8	C57BL/6	~50	No
		BALB/c	~50	Yes
Il5[−/−] following restoration of pulmonary eosinophilia by endotracheal adoptive cell transfer[24]	~100	C57BL/6	~100	Yes
Ccr3[−/−29]	~50	BALB/c	?	Yes
Ccr3[−/−27]	~10	BALB/c	~70	No
Administration of an anti-CCR3 depleting monoclonal antibody[39]	<1	C57BL/6	~50	No
Ccl11[−/−]/*Ccl24*[−/−40,41]	<2	129	~50	?
Il5[−/−]/ *Ccl11*Eotaxin[−/−28]	<1	BALB/c	~50	No
PHIL[21]	~0	C57BL/6	~50	No
		BALB/c	~50	No
Δ*dblGATA1*[29]	~0	C57BL/6	~50	No
		BALB/c	~50	Yes
CD70-Tg[31]	<1	C57BL/6	~50	No
		BALB/c	~50	No
Mbp-1[−/−]/*Epo*[−/− ‡]	<1	C57BL/6	~50	No

? data not reported in the published manuscript.
‡ our unpublished observations; OVA, ovalbumin.

eosinophils, subsequent studies in which eosinophil ablation was complete (i.e., >98—99% relative to control mice) produced outcomes that were far more predictable. In these studies, a direct correlation emerged between the absence of eosinophils and the loss of allergen-induced pulmonary pathologies (Table 14.2.1). For example, concomitant systemic and local (i.e., lung) administration of a cell-depleting anti-CCR3 antibody to allergen-treated C57BL/6 mice resulted in both a >99% reduction of airway eosinophil numbers and the loss of allergen-induced AHR. A similar observation was observed in double-knockout BALB/c mice for cytokine IL-5 and the eosinophil agonist chemokine eotaxin (C-C motif chemokine 11; Ccl11) ($Il5^{-/-}Ccl11^{-/-}$).[28] In this allergen-challenge model, the loss of both IL-5 and eotaxin resulted in a >99% decrease in airway eosinophils and abrogation of allergen-induced AHR. Even the now *gold standard* mouse models devoid of eosinophils (i.e., PHIL[21] and ΔdblGATA[29]) display a surprising level of predictability correlating the loss of eosinophils with a reduction of lung pathologies. In PHIL mice, regardless of the background strain [i.e., C57BL/6[21] or BALB/c (our unpublished observations)] the complete loss of peripheral eosinophils (Fig. 14.2.1) resulted in the loss of allergen-induced AHR. For ΔdblGATA, a similar correlation was noted in mice with a C57BL/6 background,[30] although ΔdblGATA mice on BALB/c background still displayed allergen-induced AHR despite the loss of airway eosinophilia.[29] Recently, alternative models of absolute eosinophil

deficiency have also confirmed a link between eosinophils and allergen-induced lung dysfunction. Specifically, the elevated numbers of IFN-γ-producing T cells in CD70 overexpressing transgenic mice are accompanied by a collapse in eosinophilopoiesis and a nearly complete loss (>99% relative to control mice) of peripheral eosinophils.[31] In addition, our most recent studies with eosinophil granule protein knockout mice have shown that the presence of both major basic protein-1 (MBP-1) and eosinophil peroxidase was required for eosinophil development. That is, eosinophilopoiesis in the marrow of double knockout mice deficient in these granule proteins ($Mbp-1^{-/-}/Epo^{-/-}$) is abrogated (for reasons yet to be defined), resulting in a >99% loss of peripheral mature eosinophils even after allergen sensitization/aerosol challenge (Fig. 14.2.1). Significantly, allergen sensitization/aerosol challenge of either of these additional models (CD70 transgenic mice on both a C57BL/6 and BALB/c backgrounds and $Mbp-1^{-/-}/Epo^{-/-}$ mice on a C57BL/6 background) led again to a loss of AHR. Equally important are observations that, regardless of the method by which eosinophils are eliminated or even if ablation is only partial, every mouse model examined displays a 50% or more decrease in allergen-induced goblet cell metaplasia/airway epithelial cell mucin accumulation. Furthermore, data from several of these models directly link the presence of eosinophils to induced lung structural changes occurring in chronic allergen provocation.[29,32] Thus, while the mouse model data linking eosinophils and lung dysfunction (i.e., AHR) is

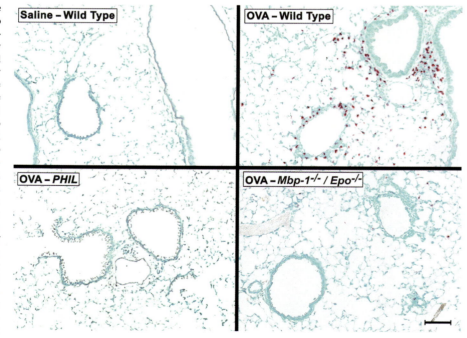

FIGURE 14.2.1 Transgenic and gene knockout mice have each been used to generate animal models devoid of circulating eosinophils. Immunohistochemistry using an anti-mouse MBP monoclonal antibody[35] was used to quantitatively demonstrate the level of eosinophil tissue infiltration occurring in the lungs of mice following allergen [ovalbumin (OVA)] sensitized and aerosol challenge. Relative to allergen naive saline-treated control animals (Saline—Wild Type) that fail to show evidence of eosinophil tissue infiltration, ovalbumin-treated mice (OVA—Wild Type) display a robust perivascular and peribronchial tissue eosinophilia. In contrast, this OVA-dependent pulmonary tissue eosinophil infiltrate is conspicuously absent in animals congenitally deficient of eosinophils either by transgenic expression of diphtheria toxin in marrow precursor cells (OVA-PHIL[21]) or double knockout mice deficient of the two abundant eosinophil secondary granule proteins, major basic protein-1 (MBP-1) and eosinophil peroxidase (EPO). OVA-$Mbp-1^{-/-}/Epo^{-/-}$ (our unpublished observations). Scale bar = 100 μm.

significant, the link between eosinophils and allergen-induced airway remodeling events is overwhelming.

LESSONS LEARNED FROM EOSINOPHIL-DEPENDENT TRANSGENIC MOUSE MODELS

A direct link between eosinophils and airway inflammatory events is also an extraordinarily consistent observation in the myriad of transgenic mouse models that our lab has created (Table 14.2.2). For example, IL-5-overexpressing transgenic mice using a T-cell promoter[33] elevates blood-circulating levels of eosinophils (approximately 1000-fold higher than control mice) but does not induce lung-associated eosinophilia and displays neither AHR nor goblet cell metaplasia.[34] In contrast, IL-5 overexpression directly in the lung results in both pulmonary eosinophilia and the advent of pulmonary

pathologies.[34] What was surprising about this result is that circulating IL-5 blood levels are equivalent in these transgenic strains of mice, suggesting a link with eosinophils and not simply IL-5 transgene expression. Since these early transgenic studies, we have created a more complex mouse model in an attempt to better replicate the pathologies observed in patients. This double transgenic model, systemically expressing IL-5 and locally expressing the eosinophil agonist chemokine eotaxin-2 (C-C motif chemokine 24; Ccl14) in the lung, induces a massive pulmonary eosinophilia that, as predicated, accompanies significant levels of lung dysfunction and airway histopathology.[35] Our subsequent studies combining these various models led to the generation of compound transgenic/gene knockout mice that modulate pulmonary eosinophil levels. Surprisingly, these data showed a tight correlation between airway eosinophilia and asthma-like pathologies. Specifically, triple transgenic mice (I5/hE2/PHIL), which maintain cytokine/chemokine

TABLE 14.2.2 Increased Airway Eosinophilia of Transgenic Mouse Models Correlates with Pulmonary Pathologies in Allergen-Naive Animals

Mouse Model	BAL Eosinophilia (% Relative to OVA-treated Wild-type Mice)	Background Strain	Goblet Cell Metaplasia (% Relative to OVA-treated Wild-type Mice)	AHR
T cell-derived, overexpression of IL-5 (Line: NJ.1638[33])	<1	C57BL/6	~0	No
		BALB/c	~0	No
Clara cell-derived, overexpression of IL-5 (Line NJ.1726[34])	~12	C57BL/6	~60	Yes
		BALB/c	~60	Yes
Double transgenic mice overexpressing IL-5 (T cell-derived) and eotaxin-2 (Clara cell-derived) (Line: I5/hE2[35])	>500	C57BL/6	~150	Yes
Eosinophil-less triple transgenic mice containing the PHIL transgene and overexpressing IL-5 (T cell-derived) and eotaxin-2 (Clara cell-derived) (Line: I5/hE2/PHIL[35])	~0	C57BL/6	~0	No
Compound I5/E2 double transgenic overexpressing IL-5 (T cell-derived) and eotaxin-2 (Clara cell derived)—double granule protein knockout mouse (Line: I5/hE2/Mbp-1$^{-/-}$/Epo$^{-/-\ddagger}$)	~0	C57BL/6	~0	No
Compound I5/E2 double transgenic overexpressing IL-5 (T cell-derived) and eotaxin-2 (Clara cell-derived)—CD4$^+$ T cell knockout mouse (Line: I5/hE2/CD4$^{-/-\ddagger}$)	~200	C57BL/6	50–100	Yes

$\ddagger$ our unpublished observations; OVA, ovalbumin.

transgene expression but are nonetheless eosinophil-less, showed that the lung dysfunction and histopathology in the parental *I5/hE2* double transgenic strain are dependent on the presence of eosinophils.[35] This was also true in the even more exotic eosinophil-less version of this double transgenic model that is also double knockout for the two abundant eosinophil granule proteins MBP1 and Epo (i.e., *I5/hE2/Mbp-1$^{-/-}$/Epo$^{-/-}$* mice)—no pulmonary eosinophils and no pathologies were observed, regardless of transgene expression. Equally interesting was the observation that while the induced lung pathologies require eosinophils, these pulmonary changes are only partially dependent on T-cell activities. Even in the absence of all CD4$^+$ T cells, the *I5/hE2/CD4$^{-/-}$* model maintains reduced yet significant airway eosinophilia that is similarly accompanied by lower yet significant changes in AHR and histopathology relative to the parental *I5/hE2* double transgenic model. Thus, consistent with established allergen provocation models of wild-type mice, eosinophils in these allergen naive transgenic models were shown to be direct contributors to the goblet cell metaplasia and AHR occurring in these models.

The degree of consistency linking eosinophils and allergic pulmonary inflammation displayed in mouse models (using both established allergen challenge models and transgenic models of cytokine/chemokine overexpression) is definitive. Yet, it is again surprising that these models have not changed the perception that eosinophils do not contribute to asthma pathologies, including commentaries suggesting that results simply reflect mouse model-to-model variations,[36] artifacts of the methodologies employed to generate the eosinophil-less condition,[37] or are simply a rodent-specific phenomenon.[38] I suggest that instead of creating complex excuses as to why mouse models are invalid, a far more parsimonious explanation is the obvious one—eosinophils are important contributors to a subset of symptoms arising from allergic respiratory inflammation. That is, the available mouse models modulating eosinophil numbers in ever increasingly reductionist approaches selectively targeting this cell demonstrate with few exceptions that eosinophils have a direct role in the development of specific allergic pulmonary pathologies. Moreover, these roles appear to be contributory and incremental, yet they are also significant in character, impinging on both lung function and the structural changes associated with chronic inflammation.

EPILOGUE

To say the least, it has been a tumultuous time for those of us studying eosinophils and their potential role in allergic asthma. The cycling pattern of *eosinophils are important* vs. *eosinophils are unimportant* has been unnerving, as we are all dependent on financial support from the National Institutes of Health (or similar governmental agencies across the globe), pharmaceutical interests, and/or both. Thus, when

eosinophils are considered important, times are good. However, when eosinophils are considered unimportant, times are bad … really bad!! Nonetheless, as one of the chief instigators of the entropy in this area, I can honestly say that the arguments and discussions debating the significance of eosinophils in asthma have been fruitful and informative. In addition, as the past President of the International Eosinophil Society and a life-long member of the *mouse models Я' us club*, I am pleased to thumb my nose at all detractors and say that the available data from mouse models collectively demonstrate (despite small variations and a few exceptions) that eosinophils are significant contributors to the pathologies resulting from allergic pulmonary inflammation and that this precept is gaining validity among clinical colleagues, which will only lead to improved care and quality of life for asthma patients.

ACKNOWLEDGEMENTS

I cannot say enough to thank my long-term collaborator, partner, and wife, Nancy Lee. My interests and desire to *conquer* the world of eosinophils have stemmed from this relationship and her belief in my abilities. I also wish to thank mentors and friends over the years who each helped me down the eosinophil-paved road that I have taken, including G. Gleich, M. McGarry, H. Rosenberg, R. Moqbel, M. Inman, E. Gelfand, and C. Irvin. Finally, the success that I have enjoyed in the lab, and in eosinophil research in general, is a direct consequence of the wonderfully dedicated folks in the integrated Lee Laboratories group. Their hard work and efforts do not go unnoticed and are the source of many of the studies noted in this subchapter.

REFERENCES

1. Tomassini M, Tsicopoulos A, Tai PC, Gruart V, Tonnel AB, Prin L, et al. Release of granule proteins by eosinophils from allergic and nonallergic patients with eosinophilia on immunoglobulin-dependent activation. *Journal of Allergy & Clinical Immunology* 1991;**88**(3 Pt 1):365–75.

2. Bousquet J, Chanez P, Lacoste JY, Barneon G, Ghavanian N, Enander I, et al. Eosinophilic inflammation in asthma [see comments]. *New England Journal of Medicine* 1990;**323**(15):1033–9.

3. Huber HL, Koessler KK. The Pathology of Bronchial Asthma. *Archives of Internal Medicine* 1922;**30**(6):689–760.

4. Stelmach I, Majak P, Grzelewski T, Jerzynska J, Juralowicz D, Stelmach W, et al. The ECP/Eo count ratio in children with asthma. *J Asthma* 2004;**41**(5):539–46.

5. Nair P, Pizzichini MM, Kjarsgaard M, Inman MD, Efthimiadis A, Pizzichini E, et al. Mepolizumab for prednisone-dependent asthma with sputum eosinophilia. *N Engl J Med* 2009;**360**(10):985–93.

6. Robinson DS, Assoufi B, Durham SR, Kay AB. Eosinophil cationic protein (ECP) and eosinophil protein X (EPX) concentrations in serum and bronchial lavage fluid in asthma. Effect of prednisolone treatment. *Clin Exp Allergy* 1995;**25**(11):1118–27.

7. Foster PS, Hogan SP, Ramsay AJ, Matthaei KI, Young IG. Interleukin 5 deficiency abolishes eosinophilia, airways hyperreactivity,

and lung damage in a mouse asthma model [see comments]. *J Exp Med* 1996;**183**(1):195—201.

8. Hogan SP, Matthaei KI, Young JM, Koskinen A, Young IG, Foster PS. A novel T cell-regulated mechanism modulating allergen-induced airways hyperreactivity in BALB/c mice independently of IL-4 and IL-5. *Journal of Immunology* 1998;**161**(3):1501—9.

9. Kuperman DA, Lewis CC, Woodruff PG, Rodriguez MW, Yang YH, Dolganov GM, et al. Dissecting asthma using focused transgenic modeling and functional genomics. *J Allergy Clin Immunol* 2005;**116**(2):305—11.

10. Epstein MM. Do mouse models of allergic asthma mimic clinical disease? *Int Arch Allergy Immunol* 2004;**133**(1):84—100.

11. Waterston RH, Lindblad-Toh K, Birney E, Rogers J, Abril JF, Agarwal P, et al. Initial sequencing and comparative analysis of the mouse genome. *Nature* 2002;**420**(6915):520—62.

12. Kung TT, Stelts DM, Zurcher JA, Adams GK, Egan RW, Kreutner W, et al. Involvement of IL-5 in a murine model of allergic pulmonary inflammation—prophylactic and therapeutic effect of an anti-IL-5 antibody. *Am J Respir Cell Mol Biol* 1995;**13**(3):360—5.

13. Egan RW, Athwahl D, Chou CC, Chapman RW, Emtage S, Jenh CH, et al. Pulmonary biology of anti-interleukin 5 antibodies. *Mem Inst Oswaldo Cruz* 1997;**92**(Suppl. 2):69—73.

14. Garlisi CG, Kung TT, Wang P, Minnicozzi M, Umland SP, Chapman RW, et al. Effects of chronic anti-interleukin-5 monoclonal antibody treatment in a murine model of pulmonary inflammation. *Am J Respir Cell Mol Biol* 1999;**20**(2):248—55.

15. Leckie MJ, ten Brinke A, Khan J, Diamant Z, O'Connor BJ, Walls CM, et al. Effects of an interleukin-5 blocking monoclonal antibody on eosinophils, airway hyper-responsiveness, and the late asthmatic response. *Lancet* 2000;**356**(9248):2144—8.

16. Kips JC, Anderson GP, Fredberg JJ, Herz U, Inman MD, Jordana M, et al. Murine models of asthma. *Eur Respir J* 2003;**22**(2):374—82.

17. Erlich P. Ueber die Specifischen granulationen des Blutes [in German]. *Arch Anant Physiol* 1879;**3**:571—9.

18. O'Byrne P,M, Inman MD, Parameswaran K. The trials and tribulations of IL-5, eosinophils, and allergic asthma. *J Allergy Clin Immunol* 2001;**108**(4):503—8.

19. Haldar P, Brightling CE, Hargadon B, Gupta S, Monteiro W, Sousa A, et al. Mepolizumab and exacerbations of refractory eosinophilic asthma. *N Engl J Med* 2009;**360**(10):973—84.

20. Lee NA, Gelfand EW, Lee JJ. Pulmonary T cells and eosinophils: coconspirators or independent triggers of allergic respiratory pathology? *Journal of Allergy and Clinical Immunology* 2001;**107**(6):945—57.

21. Lee JJ, Dimina D, Macias MP, Ochkur SI, McGarry MP, O'Neill KR, et al. Defining a link with asthma in mice congenitally deficient in eosinophils. *Science* 2004;**305**(5691):1773—6.

22. Henderson W, Chi EY, Albert RK, Chu SJ, Lamm WJE, Rochon Y, et al. Blockade of CD49D (alpha(4) integrin) on intrapulmonary but not circulating leukocytes inhibits airway inflammation and hyperresponsiveness in a mouse model of asthma. *Journal of Clinical Investigation* 1997;**100**(12):3083.

23. Hamelmann E, Cieslewicz G, Schwarze J, Ishizuka T, Joetham A, Heusser C, et al. Anti-interleukin 5 but not anti-IgE prevents airway inflammation and airway hyperresponsiveness. *Am J Respir Crit Care Med* 1999;**160**(3):934—41.

24. Shen HH, Ochkur SI, McGarry MP, Crosby JR, Hines EM, Borchers MT, et al. A causative relationship exists between

eosinophils and the development of allergic pulmonary pathologies in the mouse. *J Immunol* 2003;**170**:3296—305.

25. Flood-Page PT, Menzies-Gow AN, Kay AB, Robinson DS. Eosinophil's role remains uncertain as anti-interleukin-5 only partially depletes numbers in asthmatic airway. *Am J Respir Crit Care Med* 2003;**167**(2):199—204.

26. Humbles AA, Lu B, Friend DS, Okinaga S, Lora J, Al-Garawi A, et al. The murine CCR3 receptor regulates both the role of eosinophils and mast cells in allergen-induced airway inflammation and hyper-responsiveness. *Proc Natl Acad Sci USA* 2002;**99**(3):1479—84.

27. Ma W, Bryce PJ, Humbles AA, Laouini D, Yalcindag A, Alenius H, et al. CCR3 is essential for skin eosinophilia and airway hyper-responsiveness in a murine model of allergic skin inflammation. *J Clin Invest* 2002;**109**(5):621—8.

28. Mattes J, Yang M, Mahalingam S, Kuehr J, Webb DC, Simson L, et al. Intrinsic defect in T cell production of interleukin (IL)-13 in the absence of both IL-5 and eotaxin precludes the development of eosinophilia and airways hyperreactivity in experimental asthma. *J Exp Med* 2002;**195**(11):1433—44.

29. Humbles AA, Lloyd CM, McMillan SJ, Friend DS, Xanthou G, McKenna EE, et al. A critical role for eosinophils in allergic airways remodeling. *Science* 2004;**305**(5691):1776—9.

30. Walsh ER, Sahu N, Kearley J, Benjamin E, Kang BH, Humbles A, et al. Strain-specific requirement for eosinophils in the recruitment of T cells to the lung during the development of allergic asthma. *J Exp Med* 2008;**205**(6):1285—92.

31. de Bruin AM, Buitenhuis M, van der Sluijs KF, van Gisbergen KP, Boon L, Nolte MA. Eosinophil differentiation in the bone marrow is inhibited by T cell-derived IFN-gamma. *Blood* 2010;**116**(14):2559—69.

32. Cho JY, Miller M, Baek KJ, Han JW, Nayar J, Lee SY, et al. Inhibition of airway remodeling in IL-5-deficient mice. *J Clin Invest* 2004;**113**(4):551—60.

33. Lee NA, McGarry MP, Larson KA, Horton MA, Kristensen AB, Lee JJ. Expression of IL-5 in thymocytes/T cells leads to the development of a massive eosinophilia, extramedullary eosinophilopoiesis, and unique histopathologies. *J Immunol* 1997;**158**(3):1332—44.

34. Lee JJ, McGarry MP, Farmer SC, Denzler KL, Larson KA, Carrigan PE, et al. Interleukin-5 expression in the lung epithelium of transgenic mice leads to pulmonary changes pathognomonic of asthma. *Journal of Experimental Medicine* 1997;**185**(12):2143—56.

35. Ochkur SI, Jacobsen EA, Protheroe CA, Biechele TL, Pero RS, McGarry MP, et al. Co-Expression of IL-5 and Eotaxin-2 in Mice Creates an Eosinophil-Dependent Model of Respiratory Inflammation with Characteristics of Severe Asthma. *Journal of Immunology* 2007;**178**(12):7879—89.

36. Kips JC, O'Connor BJ, Langley SJ, Woodcock A, Kerstjens HA, Postma DS, et al. Effect of SCH55700, a humanized anti-human interleukin-5 antibody, in severe persistent asthma: a pilot study. *Am J Respir Crit Care Med* 2003;**167**(12):1655—9.

37. Wills-Karp M, Karp CL. Biomedicine. Eosinophils in asthma: remodeling a tangled tale. *Science* 2004;**305**(5691):1726—9.

38. Persson CG, Erjefalt JS, Korsgren M, Sundler F. The mouse trap. *Trends Pharmacol Sci* 1997;**18**(12):465—7.

39. Justice JP, Borchers MT, Crosby JR, Hines EM, Shen HH, Ochkur SI, et al. Ablation of eosinophils leads to a reduction of allergen-induced pulmonary pathology. *Am J Physiol Lung Cell Mol Physiol* 2003;**284**(1):L169—78.

40. Pope SM, Zimmermann N, Stringer KF, Karow ML, Rothenberg ME. The Eotaxin Chemokines and CCR3 are Fundamental Regulators of Allergen-Induced Pulmonary Eosinophilia. *J Immunol* 2005;**175**(8):5341–50.

41. Fulkerson PC, Fischetti CA, McBride ML, Hassman LM, Hogan SP, Rothenberg ME. A central regulatory role for eosinophils and the eotaxin/CCR3 axis in chronic experimental allergic airway inflammation. *Proc Natl Acad Sci USA* 2006;**103**(44):16418–23.

Chapter 14.3

Mouse Models of Esophageal Eosinophilia

Carine Blanchard

INTRODUCTION

Eosinophils are absent in normal human and mouse esophagi but are found at baseline levels in the other segments of the gastrointestinal tract. Eosinophilic esophagitis (EoE) is characterized by aberrant eosinophilic infiltrate in the esophageal epithelium [$\geq$15 eosinophils/high power field (hpf)], while the rest of the gastrointestinal tract is unchanged. At the beginning of the 21st century, the scientific literature started to recognize EoE as an allergic disease distinct from gastroesophageal reflux disease (GERD), as EoE patients are not responsive to antacid treatments. The development of EoE animal models has notably confirmed the allergic component of the disease and provided insight into the molecular mechanisms involved in esophageal eosinophilic infiltration.

FEATURES OF EOSINOPHILIC ESOPHAGITIS IN ANIMAL MODELS

Several models have been established to reproduce the human EoE and different strategies have been used. Exposure of rodents to allergens and overexpression of T-helper type 2 (T_h2) cytokines are the main methods used to induce esophageal eosinophilic infiltration.

Allergen-Induced Esophageal Eosinophilia Models

Aspergillus fumigatus *Model and Other Aeroallergens*

An extract of the fungus *A. fumigatus* is used to induce allergic airway inflammation in mice, which is characterized by elevated eosinophilic infiltration in the lung.[1] This model consists of nine intranasal instillations (three times a week for 3 weeks) with 50 µg *A. fumigatus* protein extract.[1] In 2001, Mishra and coworkers demonstrated that intranasal exposure to the aeroallergen *A. fumigatus* simultaneously induces eosinophilic airway and esophageal inflammation (without inducing lower gastrointestinal eosinophilia).[2] This increased eosinophilic infiltration—restricted to the esophagus (40 eosinophils/hpf)—represents an attractive model of EoE in mice. Eosinophils infiltrate mainly the lamina propria and only rare eosinophils are found in the epithelium of the esophagus. In these *A. fumigatus*-challenged mice, bromodeoxyuridine (BrdU)-positive cell numbers are increased in the basal epithelial cell layer, suggesting enhanced cell proliferation, while $CD4^+$ $CD45RB^{low}$ cell (T-regulatory) numbers are decreased, suggesting a role for these cells in maintaining esophageal homeostasis.[3] Due to its simplicity, reproducibility, and the fact that the induced eosinophilia is restricted to the esophagus without involving other parts of the gastrointestinal tract, the *A. fumigatus*-induced EoE model has then been widely used. Indeed, the quality of the allergen extract is a key factor in the induction and more *A. fumigatus* extract (up to 100 µg per challenge) has been used in recent studies.[4] Using arthropod allergen extracts (house dust mite and cockroaches) in the same model, Rayapudi and coworkers recently reported similar features to those observed in the *A. fumigatus*-induced EoE model, although with some variations in the levels of esophageal eosinophilic infiltration (from 18 to 33 eosinophils/mm^2), depending on the allergen used (cockroach and house dust mite allergens, respectively), thus demonstrating that other aeroallergens can induce EoE in mice.[5]

Ovalbumin-Induced Eosinophilic Gastroenteritis Including Esophageal Infiltration

In 2001, Hogan and coworkers described an interesting model of eosinophil-associated gastrointestinal disorder, involving the esophagus but also the stomach and the small intestine. This eosinophilic gastroenteritis (EGE) model is complex and requires oral administration of ovalbumin or placebo enteric-coated beads (20 mg) in ovalbumin/alum-sensitized mice.[6] In this experimental setting, mice developed eosinophilic infiltration in the esophagus (up to 26 eosinophils/mm^2), but other segments of the gastrointestinal tract, such as the stomach, are also involved. An interesting feature in this model is weight loss, a symptom associated with eosinophil-associated gastrointestinal disorder in human patients. Published in 2001, this model was innovative, complex, and interesting, but has been unused since then due to its technical complexity.

Peanut-Induced Esophageal Eosinophilia

Two recent animal studies have described the induction of esophageal eosinophilia with peanuts; one using an oral sensitization and the second using an intraperitoneal sensitization. Both models lead to high level of eosinophilia (over 100 eosinophils/mm^2). The work from Mondoulet et al., aimed at determining the efficacy of epicutaneous immunotherapy on several immunologic parameters including esophageal eosinophilia.[34] The authors showed a drastic decrease in esophageal eosinophilia and opened the possibility of a future therapeutic approach for EoE. The second work by Rajavelu et al. highlighted the earlier concept that a direct contact of the allergen with the esophageal mucosa was not driving the esophageal mucosal inflammation in this model.[35]

Epicutaneous Model

Atopic dermatitis and EoE share common features, such as eosinophil infiltration and squamous epithelial cell hyperplasia, suggesting a similar pathogenesis. Epicutaneous exposure to the allergens ovalbumin or *A. fumigatus* extract induces a strong systemic T$_h$2 response, accelerated bone marrow eosinophilopoiesis, atopic dermatitis-like skin inflammation, circulating eosinophilia, and eosinophilic infiltration in the gastrointestinal tract. In this model, eosinophils do not migrate into the esophagus unless sensitized mice are subsequently exposed to intranasal antigen.[7,8] Mice develop an eosinophilic infiltration in the esophagus comparable to that of the intranasal *A. fumigatus* model (30 or 35 eosinophils/mm^2 with *A. fumigatus* or ovalbumin, respectively) in addition to mild eosinophilia in the lungs. These findings could be particularly important for a better understanding of the pathogenesis of EoE, since a large fraction of patients with EoE have ongoing symptoms or a past history of atopic dermatitis.[9]

Collectively, these experimental systems have shown a connection between development of eosinophilic inflammation in the esophagus and in the respiratory tract, and highlight the potential for sensitization to occur via epicutaneous antigen exposure.[7,8] Additionally, some EoE patients may have seasonal variations in their symptoms,[10,11] providing clinical evidence to support a role for aeroallergen-driven eosinophil-associated responses in the esophagus.

Cytokine-Induced Esophageal Eosinophilia

Intratracheal Interleukin-13 (IL-13)

In several cell types or models, IL-13 is known to induce eosinophil chemoattractants, such as eotaxin (or C-C motif chemokine) genes. This experimental system consists of single or repeated intratracheal challenges of 10 μg IL-13 in anesthetized animals.[12] This EoE model is based on the high concordance of asthma and EoE in humans,[9] as well as the overlap between allergic airway inflammation and esophageal eosinophilia in the *A. fumigatus* model. In this

model, eosinophils infiltrate the esophagus, reaching up to 20 eosinophils/mm^2 after a single challenge and up to 50 eosinophils/mm^2 after five challenges,[12] levels comparable to those of previously described models. The advantage of this model is its prompt development of esophageal eosinophilic infiltration, but it recapitulates only a narrow part of the EoE disease, as it does not involve a sensitization step (see Table 14.3.1). In this model, human IL-13 can induce EoE and humanized anti-IL-13 antibody is able to decrease human IL-13-induced esophageal eosinophilia.[13]

Interleukin-5 Transgenic Mice

IL-5 is an important factor required for eosinophilic survival. In this mouse, IL-5 is overproduced under the control of the human T-cell surface antigen CD2 promoter.[4] This model is not a model of EoE per se, as eosinophil numbers are strongly increased in the esophagus (up to 120/mm^2) but are also increased systemically and in other parts of the gastrointestinal tract. This model, as well as other models involving systemic or local increased IL-5 levels, helped to delineate the role of IL-5 in esophageal eosinophilia in deficient animals.[14] Interestingly, a recent study has shown that these CD2-IL5 mice present with esophageal strictures. Using CD2-IL5 mice in a Δdbl-GATA1 background the authors have shown that these strictures depend on the presence of eosinophils.[36]

CC10/IL-13 Transgenic Mice

The allergic airway disease models continue to feed the EoE murine model field. Indeed, the CC10/IL-13 transgenic model is an additional example. In this model, the lung-specific CC10 (Clara Cell-Specific 10 kDa Protein) promoter drives the expression of reverse tetracycline transactivator (rtTA) and IL-13 is under the control of the Tet-on promoter, which itself is under the control of the rtTA/doxycycline complex. Double transgenic mice for CC10-rtTA and TET/on−IL-13 fed with doxycycline develop strong IL-13 expression in the lung, along with lung airway inflammation characterized by an eosinophilic infiltration. It is interesting to note that these mice develop enlargement of the esophagus, characterized by profound epithelial hyperplasia (see Fig. 14.3.1) and collagen deposition. In this EoE model, up to 400 eosinophils/mm^2 are seen and some are present in the epithelium of the esophagus.[15] Interestingly, in this model the remodeling is independent of the eosinophilia, as demonstrated in the eosinophil-deficient ΔdblGATA1 mice. Interestingly, CC10/IL-13 double transgenic mice experience weight loss in IL-13RA2-deficient background experience weight loss,[15] a symptom that can also be observed in the human disease.

Collectively, these experimental EoE systems have highlighted the key role of IL-13 and IL-5 in EoE pathogenesis and histological changes.

TABLE 14.3.1 Animal Models of Esophageal Eosinophilia

	Human EoE Disease	Allergen-induced Esophageal Eosinophilia					Cytokine-induced Esophageal Eosinophilia				
		Aspergillus	HDM/Cockroaches	Epicutaneous Sensitization	OVA-induced EGE	Peanut	IL-13	Intestinal IL5Tg	CD2-IL5tg	Systemic IL-5	CC10/IL-13Tg
Esophageal eosinophilia	+++	+++ 40/mm²	+++ 33 and 18/mm²	+++ 30/mm²	+++ 26/mm²	+++ >100/mm²	+++ 40/mm²	+++ 15/mm²	+++ 120/mm²	+++ 25/mm²	+++ 400/mm²
Epithelial eosinophilia	+++	Rare	Rare	Rare	Rare	Rare	Rare	Rare	Rare	Rare	++
Visible epithelial hyperplasia	+++	No	No	No	Unk	Unk	No	Unk	Unk	Unk	+++
Epithelial cell proliferation (molecular markers)	+++	+	Unk	Unk	Unk	Unk	+	Unk	Unk	Unk	+++
Collagen deposition	+++	+/−	Unk	Unk	Unk	Unk	Unknown	Unk	Unk	Unk	+++
Other eosinophilia		Lungs	Lungs	Blood, GI, lungs, skin	GI	Blood	Lungs	GI	GI	GI	Lungs
Sensitization route		Intranasal	Intranasal	Skin	Intraperitoneal	Intraperitoneal	None	None	None	None	None
Challenge/Inducer		Intranasal	Intranasal	Intranasal	Oral	Intragastric/intranasal	Intratracheal	Intestinal transgene	Systemic transgene	Miniosmotic pump	Lung transgene
Food or aeroallergens	Both	Aeroallergen	Aeroallergen	Both	Food	Food	None	None	None	None	None
Symptoms	++	−	−	−	Weight loss, and gastric dysmotility		−	−	Esophageal strictures and dysmotility	−	Esophageal strictures; weight loss and IL13RA2-deficient background

GI, gastrointestinal; Unk, unknown.

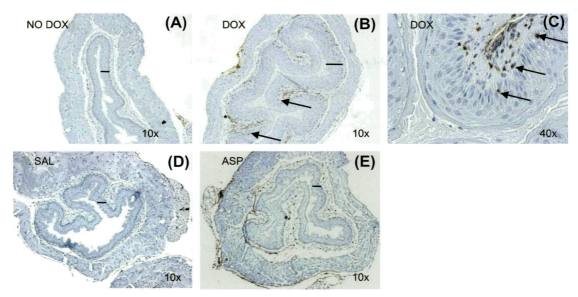

FIGURE 14.3.1 **Eosinophil staining in CC10/IL-13 transgenic mice and in *A. fumigatus* models.** Mouse esophageal cross sections were stained with anti-major basic protein (MBP) antibody and representative photomicrographs are shown. Eosinophil tissue accumulation in CC10/IL13 mice at baseline (*A*) was compared to the tissue eosinophilia occurring after IL-13 expression was induced in mice fed doxycycline (*B,C*). Wild-type mice were challenged with saline (*D*) or *A. fumigatus* extract (*E*). Eosinophils (*), epithelial eosinophils (arrows), and epithelial thickness (lines) are shown. Immunohistochemistry was performed in Cincinnati Children's Hospital and the rat anti-mouse MBP monoclonal antibody was provided by the laboratories of Drs Nancy and Jamie Lee.

Other Models

Other experimental systems—not directly linked to the EoE field—have described esophageal eosinophilic infiltration in animal models. Only some of these are described below. GERD is a complex disease that mimics EoE symptoms in humans and has histopathological findings that are sometimes very similar. In such cases, only a response to proton-pump inhibitors (PPI) can discriminate GERD from EoE. A recent murine model based on acid-induced esophageal pathological changes has shown interesting histological findings with eosinophilic infiltration in the epithelium of the esophagus and similar epithelial hyperplasia. It is expected that different etiology, with common histology, will provide a relatively similar molecular profile. However, the etiology is of primary importance since patients with high esophageal eosinophilic level may be GERD sufferers and respond to PPI. Similar results are observed between allergic and nonallergic EoE patients, whose transcriptome profiles are identical.[16] This indicates that the effector phase of the disease is the same between these phenotypes, regardless of which factors activated the inflammation; however, whether this holds in EoE vs. GERD with high eosinophilia is yet to be unraveled.

Other models have also described esophageal eosinophilia; for example, lymphopenic mice reconstituted with a limited repertoire of T cells develop severe, multiorgan, T_h2-associated inflammatory disease, including gastrointestinal eosinophilia (affecting the esophagus and stomach).[17]

An interesting publication recently reported that dogs can spontaneously develop EoE and potentially EGE with esophageal involvement.[18] Finally, C57BL/Ka mice with chronic proliferative dermatitis due to a mutation (cpdm/cpdm) develop esophageal and skin eosinophil infiltration,[19,20] suggesting a common etiology and/or pathogenesis for skin and esophageal eosinophilia.

It is interesting to note that all murine models of EoE have some inflammation in other organs. Collectively, these experimental systems have shown a strong connection between the development of eosinophilic inflammation in the esophagus, other gastrointestinal segments, the respiratory tract, and the skin, in response not only to external allergic triggers but also to intrinsic T_h2 cytokines, and highlight the potential for sensitization to occur via cutaneous antigen exposure.

ANIMAL MODELS AND MECHANISTIC INSIGHTS INTO EOE PATHOGENESIS

Eosinophil Trafficking to the Esophagus: A Role for Interleukin-5, Eotaxins, and C-C Chemokine Receptor 3

The study of eosinophilic infiltration in allergic airway diseases in the late 1990s preceded the findings in the esophagus of the next decade. Animal models and transgenic tools developed to study eosinophil trafficking have been quite helpful in uncovering the mechanisms involved in the esophagus. Several publications have reported the use of IL-5 or eotaxin-1 (or C-C motif chemokine 11; Ccl11)-deficient mice in the context of EoE.[2,8,12,14,21] In the *A. fumigatus* model, Mishra and coworkers demonstrated that eotaxin-1-deficient mice still show markedly elevated eosinophil levels in the esophagus following challenge compared with baseline wild-type controls or deficient mice treated with placebo.[2] This finding suggests that eotaxin-1 cooperates with other eosinophil chemo-attractants in the regulation of eosinophil trafficking to the esophagus. However, the esophagus of *A. fumigatus*-challenged IL-5-deficient animals is completely devoid of eosinophilic infiltration and epithelial cell hyperplasia. The same group showed similar results in the ovalbumin-coated bead EGE model. In this study, eotaxin-1 was shown to play a role in eosinophilic infiltration in the esophagus, but is not required since approximately 50% of wild-type levels of infiltration was still observed in deficient animals. Similarly, IL-5 is required to induce the esophageal eosinophilia and weight loss observed in this model. These two studies laid the foundations of the mechanism behind eosinophilic infiltration to the esophagus in murine models. Following studies revealed the major role of IL-5 using IL-5-deficient mice and/or the partial contribution of eotaxin-1 in eotaxin-deficient mice on esophageal eosinophilia in the CD2−IL5 mice, as well as in the IL-13 intratracheal and epicutaneous models.[8,22] Taken together, these studies highlight the dominant role of IL-5 in regulating eosinophil accumulation in the esophagus. Additionally, some (but not all) preliminary clinical studies have shown that administration of a humanized antibody against IL-5 reduces some symptoms/features of EoE.

In summary, these investigations have dissected the cellular and molecular mechanisms involved in eosinophil homing to the esophagus. These data demonstrate using independent approaches that IL-5 overexpression induces eosinophil trafficking to the esophagus. It is interesting to note that in these models eotaxin-1 seems to be involved but not required for the development of esophageal eosinophilia. However, C-C chemokine receptor 3 (CCR3), the eotaxin receptor, seems to be essential for eosinophilia in the esophagus, suggesting that CCR3 expression is needed for eosinophils to infiltrate the esophagus in the *A. fumigatus* model and/or that ligands other than eotaxin-1 are involved in eosinophil recruitment to the esophagus. In the cockroach and house dust mite models, Rayapudi and coworkers have recently shown that eotaxin-1 and eotoxin-2 (Ccl24) double-deficient mice do not develop esophageal eosinophilic infiltration, suggesting that both CCR3 ligands eotaxin-1 and eotoxin-2 are involved in EoE in mice. Interestingly, opposite results were obtained in CC10/IL-13 transgenic mice. In this study, Zuo and coworkers demonstrated that the eosinophilic infiltration in this model is primarily

eotaxin-1-dependent[15] suggesting that chemotactic signals may depend on the model used. Additionally, eosinophilic infiltration may be facilitated by other proteins, such as the IL-13-induced protein periostin, which increases eosinophilic adhesion to fibronectin and has been shown to cooperate in promoting esophageal eosinophilia in the *A. fumigatus* model.[23]

Immunologic Components

Studies in mouse models have shown that T_h2 signaling is required for induction of experimental EoE. In the epicutaneous sensitization model, mice genetically deficient in signal transducer and activator of transcription 6 (Stat6), which is involved in the IL4−IL13 pathway, or mice genetically deficient in IL-13, IL-4, and IL-5 have impaired induction of esophageal eosinophilia in response to allergen.[22,23]

The use of lymphocyte-deficient mice has defined the roles of T cells in the pathogenesis of EoE. T cell (forkhead box N1$^{-/-}$) and RAG1-deficient mice, but not B cell-deficient (IgH6) mice, fail to develop *A. fumigatus*-induced EoE, suggesting little or no role for B cells or antigen-specific antibodies in EoE induction.[24] Additionally, CD8$^+$ and CD4$^+$ T cells do not seem to be required for induction of *A. fumigatus*-induced experimental esophageal eosinophilia, indicating the involvement of a distinct cell type of the adaptive immune system.[24] While these results are certainly difficult to reconcile due differences in specific cell deficiency, they provide evidence that T_h2 cell-derived cytokines contribute to the pathogenesis of esophageal eosinophilia. Indeed, the involvement of invariant natural killer T has been recently suggested in CD1d-deficient mice.[35]

In the *A. fumigatus* model, Zhu and coworkers recently showed that IL-15 induces CD4$^+$ T-cell proliferation and T_h2 cytokine production, as well as eosinophil-specific chemokine mRNA expression by primary esophageal epithelial cells, thus indicating a functional IL-15 pathway in EoE.[4] In this study, mice deficient in the receptor chain of IL-15, the *Il15ra* gene, were protected from the development of experimental EoE.[4]

COMPARISON WITH HUMANS AND LIMITATIONS

These models have led to huge advances in the understanding of the etiology and pathogenic effectors involved in esophageal eosinophilia, but some may be disappointed by their divergences from human disease.

The increasing knowledge of human disease has emphasized the limitations of these different models. In humans, IL-13 is overexpressed in the esophagus of EoE patients and selectively induces the eosinophil chemo-attractant eotaxin-3 (CCL26) by a transcriptional mechanism in esophageal epithelial cells,[25] while eotaxin-1 and eotaxin-2 (CCL24) expression levels are not changed.[16] In mice, however, the eotaxin-3 gene is a pseudogene that does not encode a functional peptide;[26] depending on the model, eotaxin-1 alone or both eotaxin-1 and eotaxin-2 may be involved in esophageal eosinophilia. Additionally, eotaxin-1 and eotaxin-2 are increased in CC10/IL13 transgenic mice and IL-13 can directly increase eotaxin-1 and eotaxin-2 expression in murine esophageal rings.[27] In human EoE disease, eosinophils infiltrate the epithelium, and eosinophils are also seen in the lamina propria.[28] In contrast, most of these murine models induce only rare epithelial eosinophilia, mainly in the proliferative zone; and esophageal degranulation is globally less obvious than in the human disease. The processes that recruit eosinophils in the epithelium and the lamina propria may be different, involving, to some extent, the direct effect of IL-13 on eotaxin-1 and/or eotaxin-2 expression in the lamina propria. The absence of readout other than esophageal eosinophilic infiltration forces most of these studies to draw conclusions that are mainly based on eosinophil counts. A few other EoE features based on BrdU staining, collagen deposition, and T-cell subtypes and, recently, esophageal strictures have been described so far, but most have been only marginally used.

Of note, the esophageal epithelium of mice is cornified similar to the skin, while human esophageal epithelium is not, suggesting a difference in the expression of genes involved in protection of the epithelium at baseline. Notably, in contrast to atopic dermatitis that is associated with loss-of-function genetic variants of filaggrin, EoE is associated with impaired filaggrin mRNA expression.[29] IL-13 downregulates filaggrin expression in human esophageal epithelial cells,[29] providing a mechanism by which food antigen-elicited T_h2 cell adaptive immunity might impair the esophageal barrier function, and this may be involved in propagating local inflammatory processes (including sensitivity to acid) and in increasing antigen uptake by cells in the esophagus. These processes may be particularly important in humans. Transcriptome analysis has shown a nice overlap between the esophageal transcriptome in CC10/IL-13 transgenic mice and in the human disease.[15] In CC10/IL-13 transgenic mice, eosinophils migrate into the epithelium of the esophagus and visible hyperplasia is seen (Fig. 14.3.1), but no obvious reduction in filaggrin expression or in genes of the epidermal differentiation complex is observed according to microarray analysis. The *A. fumigatus*-induced EoE transcriptome includes only few epithelial genes compared to the CC10/IL-13 transgenic mice and the human transcriptome. A limitation to this comparison is the difference in tissue availability between mice and humans. In humans, only the epithelium and some of the lamina propria are obtained in biopsies, while in mice the whole esophagus, including the deeper muscle layers, can be used for analysis. The localization of inflammation to the lamina propria in *A. fumigatus*-challenged animals compared to the human disease, in which biopsies mainly sample the esophageal epithelium, may in part also explain these discrepancies.

Finally, the environmental and polygenic etiology of EoE is difficult to assess in humans or mice. The genetic components involved in EoE pathogenesis are being described: from the polymorphism in TGFβ1 promoter to eotaxin-3, filaggrin, or thymic stromal lymphopoietin (TSLP).[16,30–33] The animal models one the other hand, do not capture these subtle changes yet and are still just grossly representative of a deficient pathway or gene.

CONCLUSION

Models have been defined by George Edward Pelham Box as 'very useful but imperfect tools for science,' and it is agreed that the EoE models are useful and evidently not perfect. Only a few features of the human disease are conserved in some murine models, while others combine several conserved features. The combination of humanized transgenic mice expressing murine cytokines or epithelial cell components under the control of human promoters may be a step forward in the understanding of this disease. Knowledge gained using these animal models may not always be directly transferable to other T_h2 models or human diseases, such as asthma and atopic dermatitis, due to the tissue specificity, thus making the study of esophageal eosinophilia a unique opportunity to uncover distinct molecular and cellular pathogeneses. Esophageal immunologic response? Direct transport of inflammatory lung mediators through the trachea to the esophagus? Involvement of common draining lymph nodes? Nerve involvement? Reasons for tissue specificity of the esophageal eosinophilic infiltration? These are all questions that mouse models may help to answer.

REFERENCES

1. Mehlhop PD, van de Rijn M, Goldberg AB, Brewer JP, Kurup VP, Martin TR, et al. Allergen-induced bronchial hyperreactivity and eosinophilic inflammation occur in the absence of IgE in a mouse model of asthma. *Proc Natl Acad Sci USA* 1997;**94**:1344–9.

2. Mishra A, Hogan SP, Brandt EB, Rothenberg ME. An etiological role for aeroallergens and eosinophils in experimental esophagitis. *J Clin Invest* 2001;**107**:83–90.

3. Zhu X, Wang M, Crump CH, Mishra A. An imbalance of esophageal effector and regulatory T cell subsets in experimental eosinophilic esophagitis in mice. *Am J Physiol Gastrointest Liver Physiol* 2009;**297**:G550–8.

4. Zhu X, Wang M, Mavi P, Rayapudi M, Pandey AK, Kaul A, et al. Interleukin-15 expression is increased in human eosinophilic esophagitis and mediates pathogenesis in mice. *Gastroenterology* 2010;**139**:182–93.

5. Rayapudi M, Mavi P, Zhu X, Pandey AK, Abonia JP, Rothenberg ME, et al. Indoor insect allergens are potent inducers of experimental eosinophilic esophagitis in mice. *J Leukoc Biol* 2010;**88**:337–46.

6. Hogan SP, Mishra A, Brandt EB, Royalty MP, Pope SM, Zimmermann N, et al. A pathological function for eotaxin and eosinophils in eosinophilic gastrointestinal inflammation. *Nat Immunol* 2001;**2**:353–60.

7. Akei HS, Brandt EB, Mishra A, Strait RT, Finkelman FD, Warrier MR, et al. Epicutaneous aeroallergen exposure induces systemic TH2 immunity that predisposes to allergic nasal responses. *J Allergy Clin Immunol* 2006;**118**:62–9.

8. Akei HS, Mishra A, Blanchard C, Rothenberg ME. Epicutaneous antigen exposure primes for experimental eosinophilic esophagitis in mice. *Gastroenterology* 2005;**129**:985–94.

9. Noel RJ, Putnam PE, Rothenberg ME. Eosinophilic esophagitis. *N Engl J Med* 2004;**351**:940–1.

10. Wang FY, Gupta SK, Fitzgerald JF. Is there a seasonal variation in the incidence or intensity of allergic eosinophilic esophagitis in newly diagnosed children? *J Clin Gastroenterol* 2007;**41**:451–3.

11. Almansa C, Krishna M, Buchner AM, Ghabril MS, Talley N, DeVault KR, et al. Seasonal distribution in newly diagnosed cases of eosinophilic esophagitis in adults. *Am J Gastroenterol* 2009;**104**:828–33.

12. Mishra A, Rothenberg ME. Intratracheal IL-13 induces eosinophilic esophagitis by an IL-5, eotaxin-1, and STAT6-dependent mechanism. *Gastroenterology* 2003;**125**:1419–27.

13. Blanchard C, Mishra A, Saito-Akei H, Monk P, Anderson I, Rothenberg ME. Inhibition of human interleukin-13-induced respiratory and oesophageal inflammation by anti-human-interleukin-13 antibody (CAT-354). *Clin Exp Allergy* 2005;**35**:1096–103.

14. Mishra A, Hogan SP, Brandt EB, Rothenberg ME. IL-5 promotes eosinophil trafficking to the esophagus. *J Immunol* 2002;**168**:2464–9.

15. Zuo L, Fulkerson PC, Finkelman FD, Mingler M, Fischetti CA, Blanchard C, et al. IL-13 induces esophageal remodeling and gene expression by an eosinophil-independent, IL-13R alpha 2-inhibited pathway. *J Immunol* 2010;**185**:660–9.

16. Blanchard C, Wang N, Stringer KF, Mishra A, Fulkerson PC, Abonia JP, et al. Eotaxin-3 and a uniquely conserved gene-expression profile in eosinophilic esophagitis. *J Clin Invest* 2006;**116**:536–47.

17. Milner JD, Ward JM, Keane-Myers A, Paul WE. Lymphopenic mice reconstituted with limited repertoire T cells develop severe, multiorgan, Th2-associated inflammatory disease. *Proc Natl Acad Sci USA* 2007;**104**:576–81.

18. Mazzei MJ, Bissett SA, Murphy KM, Hunter S, Neel JA. Eosinophilic esophagitis in a dog. *J Am Vet Med Assoc* 2009;**235**:61–5.

19. Gijbels MJ, Elliott GR, HogenEsch H, Zurcher C, van den Hoven A, Bruijnzeel PL. Therapeutic interventions in mice with chronic proliferative dermatitis (cpdm/cpdm). *Exp Dermatol* 2000;**9**:351–8.

20. Gijbels MJ, HogenEsch H, Blauw B, Roholl P, Zurcher C. Ultrastructure of epidermis of mice with chronic proliferative dermatitis. *Ultrastruct Pathol* 1995;**19**:107–11.

21. Mishra A, Wang M, Pemmaraju VR, Collins MH, Fulkerson PC, Abonia JP, et al. Esophageal remodeling develops as a consequence of tissue specific IL-5-induced eosinophilia. *Gastroenterology* 2008;**134**:204–14.

22. Akei HS, Brandt EB, Mishra A, Strait RT, Finkelman FD, Warrier MR, et al. Epicutaneous aeroallergen exposure induces systemic TH2 immunity that predisposes to allergic nasal responses. *J Allergy Clin Immunol* 2006;**118**:62–9.

23. Blanchard C, Mingler MK, McBride M, Putnam PE, Collins MH, Chang G, et al. Periostin facilitates eosinophil tissue infiltration in allergic lung and esophageal responses. *Mucosal Immunol* 2008;**1**:289–96.

24. Mishra A, Schlotman J, Wang M, Rothenberg ME. Critical role for adaptive T cell immunity in experimental eosinophilic esophagitis in mice. *J Leukoc Biol* 2007;**81**:916–24.

25. Blanchard C, Mingler MK, Vicario M, Abonia JP, Wu YY, Lu TX, et al. IL-13 involvement in eosinophilic esophagitis: transcriptome analysis and reversibility with glucocorticoids. *J Allergy Clin Immunol* 2007;**120**:1292–300.

26. Pope SM, Fulkerson PC, Blanchard C, Akei HS, Nikolaidis NM, Zimmermann N, et al. Identification of a cooperative mechanism involving interleukin-13 and eotaxin-2 in experimental allergic lung inflammation. *J Biol Chem* 2005;**280**:13952–61.

27. Neilsen CV, Bryce PJ. Interleukin-13 directly promotes oesophagus production of CCL11 and CCL24 and the migration of eosinophils. *Clin Exp Allergy* 2010;**40**:427–34.

28. Aceves SS, Ackerman SJ. Relationships between eosinophilic inflammation, tissue remodeling, and fibrosis in eosinophilic esophagitis. *Immunol Allergy Clin North Am* 2009;**29**:197-xiv.

29. Blanchard C, Stucke EM, Burwinkel K, Caldwell JM, Collins MH, Ahrens A, et al. Coordinate interaction between IL-13 and epithelial differentiation cluster genes in eosinophilic esophagitis. *J Immunol* 2010;**184**:4033–41.

30. Blanchard C, Stucke EM, Burwinkel K, Caldwell JM, Collins MH, Ahrens A, et al. Coordinate interaction between IL-13 and epithelial differentiation cluster genes in eosinophilic esophagitis. *J Immunol* 2010;**184**:4033–41.

31. Aceves SS, Newbury RO, Chen D, Mueller J, Dohil R, Hoffman H, et al. Resolution of remodeling in eosinophilic esophagitis correlates with epithelial response to topical corticosteroids. *Allergy* 2010;**65**:109–16.

32. Sherrill JD, Gao PS, Stucke EM, Blanchard C, Collins MH, Putnam PE, et al. Variants of thymic stromal lymphopoietin and its receptor associate with eosinophilic esophagitis. *J Allergy Clin Immunol* 2010;**126**:160–5.

33. Rothenberg ME, Spergel JM, Sherrill JD, Annaiah K, Martin LJ, Cianferoni A, et al. Common variants at 5q22 associate with pediatric eosinophilic esophagitis. *Nat Genet* 2010;**42**:289–91.

34. Mondoulet L, Dioszeghy V, Larcher T, Ligouis M, Dhelft V, Puteaux E, et al. Epicutaneous immunotherapy (EPIT) blocks the allergic esophago-gastro-enteropathy induced by sustained oral exposure to peanuts in sensitized mice. *PLoS One* 2012;**7**(2). e31967. Epub 2012 Feb 21.

35. Rajavelu P, Rayapudi M, Moffitt M, Mishra A, Mishra A. Significance of para-esophageal lymph nodes in food or aeroallergen-induced iNKT cell-mediated experimental eosinophilic esophagitis. *Am J Physiol Gastrointest Liver Physiol* 2012;**302**(7):G645–54. 2012 Apr.

36. Mavi P, Rajavelu P, Rayapudi M, Paul RJ, Mishra A. Esophageal functional impairments in experimental eosinophilic esophagitis. *Am J Physiol Gastrointest Liver Physiol*; 2012 Feb 23 [Epub ahead of print].

37. Spergel JM, Mizoguchi E, Brewer JP, Martin TR, Bhan AK, Geha RS. Epicutaneous sensitization with protein antigen induces localized allergic dermatitis and hyperresponsiveness to methacholine after single exposure to aerosolized antigen in mice. *J Clin Invest* 1998 Apr 15;**101**(8):1614–22.

Chapter 14.4

Models of Inflammatory Bowel Diseases

Joanne C. Masterson and Glenn T. Furuta

INTRODUCTION

The term inflammatory bowel disease (IBD) characterizes a group of intestinal diseases that occur when a *genetically* susceptible individual encounters *environmental* cues thus resulting in a chronic deregulated *inflammatory* response in the colon. Two major forms of IBD, ulcerative colitis (UC) and Crohn disease (CD), manifest themselves by nonspecific symptoms, including abdominal pain and bloody diarrhea. While the presentations may be similar, a number of features are different between these two diseases (shown in Table 14.4.1). These are listed to emphasize the fact that no animal model of disease replicates the human condition, especially as it relates to the role of eosinophils in IBD. Thus, studies designed to understand IBD pathophysiology that use mouse models should always identify the limits and advantages of models as they relate to human disease. In addition, a number of lines of evidence support a potential role for eosinophils in the pathogenesis of IBD, although none has demonstrated that eosinophils are sufficient or necessary for the development or perpetuation of IBD.

PATHOPHYSIOLOGY OF INFLAMMATORY BOWEL DISEASE

Genes in Inflammatory Bowel Disease

A number of traditional genetic studies have highlighted particular genetic mutations associated with CD or UC, including those associated with APG16 autophagy 16-like (*ATG16L1*), interleukin-10 receptor (*IL-10R*), and nucleotidebinding oligomerization domain containing 2 (*NOD2*). More recent genome-wide association studies (GWAS) increased the number of gene targets associated with IBD to include *IL23R*, interleukin 12B (*IL12B*), macrophage stimulating 1 (*MST1*), NK2 homeobox 3 (*NKX2-3*), and signal transducer and activator of transcription 3 (*STAT3*).[1-3] In many circumstances, a clear link between the common genes identified and effective immune function has been elucidated. For instance, genes associated with barrier function have been identified in UC.[2] Animal models expressing these mutations have undergone pathophysiological analysis to determine the pathways affected and thus enable the generation of investigative and novel therapeutic approaches to the treatment of IBD.

Intestinal Microbiome and Inflammatory Bowel Disease

Simultaneous with the identification of IBD related genes, rapidly emerging studies cataloged the diverse microbiome associated with IBD and defined interactions between microbes and the underlying mucosa. The intestinal microflora consists of a large number of commensal bacteria. Because this commensal flora plays an important role in the metabolism of nutrients, as well as in the development of the mucosal immune system and immune tolerance, alterations in the microbiome may contribute to the development of IBD. Support for this hypothesis lies in finding that antibiotics may positively or negatively impact patients with IBD and can alter the course of mouse models of IBD. Very recent studies are identifying specific changes in microbiome diversity and abundance in IBD patients.[4] In particular, a common finding between each of these diseases is a reduced bacterial load and a skewed predominance of two particular phyla, *Bacteroidetes* and *Firmicutes*, in patients with CD and UC compared to normal individuals. However, whether these changes contribute to disease establishment or are a consequence of the heightened inflammatory environment has yet to be elucidated.

TABLE 14.4.1 Similarities and Differences in Two Major Forms of Inflammatory Bowel Disease

	Crohn Disease	Ulcerative Colitis
Symptoms	Abdominal pain, bloody diarrhea, growth failure	Abdominal pain, bloody diarrhea
Location	Any part of the GI tract—typically ileocolonic	Limited to colon
Pathology	Granuloma, transmural inflammation	Mucosal inflammation, no granuloma
Treatments	Aminosalicylates, immunosuppressives, targeted biologics	Aminosalicylates, immunosuppressives

GI, gastrointestinal.

Immune System and Inflammatory Bowel Disease

While the last few years have brought novel insights into the contribution of genetic and microbial factors to the pathogenesis of IBD, a large body of evidence has identified the role of deregulated immune responses in IBD. Despite daily encounters with food antigens, microbes, toxins, and a variety of other ingested products, the intestine is usually capable of maintaining a healthy state. The epithelial barrier, underlying mucosal immune system, and structural mesenchyme maintain the balance between responses to pathogenic and nonpathogenic commensal microbiome. Innate mechanisms contributing to this response include epithelial tight junctions, mucous, defensins, trefoil factors, and peristalsis. Acquired mucosal immune responses are orchestrated by a number of cellular and humoral factors including dendritic cells, lymphocytes, mast cells, and neutrophils. During disease states, imbalances in the cellular composition, as well as in the functionality of these cells, can lead to pathological injury and disease. In IBD a number of key cytokines including interferon γ (IFN-γ), interleukin-12 (IL-12), and tumor necrosis factor (TNF-α) as well as the regulatory cytokine IL-10 have been shown to contribute to this imbalance.

Roles for Eosinophils in Inflammatory Bowel Disease

While eosinophils are present in healthy intestinal mucosae, the precise features that define this normal pattern, including numbers, distribution, and state of activation, are uncertain. A wide variety of circumstantial evidence has determined that eosinophil numbers increase in tissue affected by IBD. For instance, initial studies identified increased numbers of ileal and colonic eosinophils in IBD patients. Local eosinophil activation is increased in IBD patients, as evidenced by increased eosinophil granule proteins within the mucosa, colonic lavage fluid, and fecal samples.[5−8] A straightforward interpretation of these studies would be that eosinophils and their inflammatory mediators are problematic; but one could speculate that eosinophils are protective and that the mediators released may have a beneficial impact. Since no eosinophil-targeted treatments have been used in IBD to date, this clinical question is difficult to answer, emphasizing the importance of appropriate mouse models. Extrapolating experiences from disease management and mouse models in which eosinophils can be the predominant inflammatory cell, such as asthma, suggests that eosinophils are pathological. If these findings can be replicated in mouse models, targeting of eosinophils in patients with IBD may provide a novel therapeutic benefit. Thus, studies in mouse models of IBD may offer an excellent venue to begin to determine the role of eosinophils in intestinal disease.

Mouse Models of Inflammatory Bowel Disease

Major advances in our understanding of the mechanisms of IBD have been gleaned from animal models that resemble this disease. These systems can be broadly subcategorized into spontaneous models, models induced by exogenous agents, genetically targeted models, disruption and cell transfer-induced models. In the context of each model and where relevant, we will indicate the presence of eosinophils and their potential role.

Spontaneous Disease Models

Spontaneous models include cotton-top tamarin monkeys,[9,10] and the SAMP1 lineage of mice.[11−14] C3H/HeJBir mice, a spontaneously derived substrain of C3H/HeJ mice, develop cecal and proximal colonic inflammation that is LPS-unresponsive as a result of Toll-like receptor 4 (TLR4) deficiency. Transmural inflammation is characterized by increased B- and T-cell responses, resulting in a T-helper type 1 (Th1)-dependent response to the commensal bacterial flagellin peptides, CBir1 and FlaX. Antibodies reactive to these commensal antigens are also elevated in patients with CD but not UC or normal individuals.[15] While acute and chronic inflammatory infiltrates are noted, there has been no study of eosinophils in this model.

One of the models that has led to significant insights into eosinophils in IBD is the SAMP1 mouse line. Inbreeding of the AKR mouse line for a senescence phenotype led to the first SAMP1 mouse strain, SAMP1/SkuSlc. Additional, subsequent inbreeding for particular phenotypes gave rise to the SAMP1/Yit and the SAMP1/YitFc mouse substrains. These mice develop a Crohn's-like transmural inflammation of the distal ileum. Inflammation in the SAMP1 lineages appears to be dependent on bacteria,[16] with identified roles for defects in epithelial barrier, expanded B cells, and increased T-cell activation.

Takedatsu et al. first described increased eosinophils in the ileal lamina propria of the SAMP1/YitFc substrain and the role of IL-5 in this response.[17] Recent work on the original strain of SAMP1 mice, the SAMP1/SkuSlc mouse, also noted ileal eosinophilia.[12] Studies directly targeting eosinophilia with anti-C-C chemokine receptor 3 (anti-CCR3) antibodies indicate a crucial role for eosinophils in the maintenance of inflammation and remodeling in SAMP1/SkuSlc mice.[18] Thus, eosinophils in the SAMP1 lineage contribute to spontaneous Crohn's-like ileal inflammation.

Chemically-Induced Models

Chemically-induced models of IBD constitute the other large group of IBD mouse models. In one model, mice are sensitized on the skin with agents associated with hypersensitivity responses such as 2,4,6-trinitrobenzenesulfonic acid (TNBS) solution[19] and oxazolone,[20] and then challenged rectally. This methodology induces a hypersensitivity response localized to the colon. In another model system, the intestinal mucosa is injured by an ingested agent, dextran sulfate sodium (DSS; a sulfated polysaccharide) or a rectally administered agent such as acetic acid.[21] The common mechanism in each of these models is disruption of the epithelial barrier with subsequent exposure of the intestinal immune system to luminal contents, including bacteria. Each model results in acute inflammation of the colon that begins within days of initiation. TNBS is characterized by a CD-like transmural immune infiltrate, consisting of lymphocytes, macrophages, and neutrophils, along the length of the colon. Colitis is defined by colon wall thickening, diarrhea, rectal prolapse, weight loss, and an IL-12-dependent inflammatory process that is mediated by a T_h1 cell response. Eosinophil involvement has not been reported in this model, possibly because inflammation results from a $CD4^+$ T_h1 cell-mediated injury.

In contrast to TNBS, oxazolone colitis is characterized by a distal colitis likened to human UC, due to its predominantly superficial inflammation and ulceration composed of a mixed granulocytic and lymphocytic infiltration. Similar to TNBS, oxazolone-mediated colitis is manifest by colon shortening, diarrhea, goblet cell depletion, and weight loss. While predominantly neutrophilic, the granulocyte infiltrate includes eosinophils. In contrast to TNBS colitis, oxazolone-induced inflammation is characterized by an IL-4- and IL-13-dependent T_h2 polarized immune response with increased levels of colonic IL-4, IL-5, IL-13, and transforming growth factor β (TGF-β). In one study, oxazolone induced intestinal permeability and weight loss was shown to be associated with eosinophilic inflammation. Major basic protein (MBP)-null mice exposed to oxazolone colitis are protected from weight loss and altered permeability, thus implicating MBP in a pathogenic role in this murine model of colitis.

As a consequence of its ease of use and reproducibility, DSS colitis is one of the most commonly used models to study colonic inflammation. This model of colitis is most severe in the distal colon and is characterized by crypt and epithelial cell erosion, goblet cell depletion, and granulocyte and mononuclear cell infiltration. DSS is administered in drinking water for 4−7 days, resulting in diarrhea, rectal bleeding, and weight loss. The development of DSS colitis is thought to be dependent on innate immune responses and it can be elicited in mice deficient in the humoral arm of the immune system. However, T cells can play a role in DSS colitis as both T_h1 and T_h2 cytokines are increased in affected mucosal tissue. Studies in mice deficient in IL-5 show no changes in eosinophil numbers recruited in DSS colitis.[22,23] However, DSS colitis studies of mice deficient in the chemoattractant eotaxin (C-C motif chemokine 11; Ccl11) identified a significant reduction in eosinophils, along with an improvement in disease activity indices. In the same study, a significant role for the eosinophil granule protein eosinophil peroxidase (EPO) was reported in DSS-induced colonic inflammation; mice deficient in EPO exhibit attenuated inflammation and disease activity.[22] In contrast to oxazolone colitis, MBP-null mice exposed to DSS colitis show greater weight loss and disease activity than control mice. Thus, these studies suggest alternative functions for eosinophil granule proteins, depending on the model of colitis chosen.

More recent studies of eosinophil-null mice (PHIL and ΔdblGATA mice) have reported significant protection against histological and disease activity in DSS-induced colitis.[24,25] Studies examining colonic chemotaxis of eosinophils used the DSS-induced model of colitis and pinpointed a critical role for intercellular adhesion molecule (ICAM) and $\beta2$ integrins in this process. However, authors found no role for $\beta7$ integrin in eosinophil recruitment toward DSS-induced large intestinal inflammation. Interestingly, an important role for $\beta7$ integrin had previously been described in small intestinal recruitment of eosinophils in numerous models of small intestinal inflammation, suggesting compartmentally specific eosinophil recruitment mechanisms in mice.[26−28]

Genetic Disruption

Targeting of IBD-related genes in mice, either with gene knockouts or transgenic approaches, offers an alternative strategy to understand the pathogenesis of intestinal inflammation. These include $G\alpha i2^{-/-}$, $IL-2^{-/-}$, $IL-10^{-/-}$, $TGF-\beta^{-/-}$, TNFΔARE, $mdr1a^{-/-}$ mice, and N-cadherin (cadherin-2) dominant-negative mutants (Table 14.4.2).

Clinical observations suggest that barrier dysfunction is present in patients with IBD, but whether this increased permeability is a result of ongoing inflammation or a predisposing factor to the development of IBD is not certain. In this regard, *in vitro* exposure of intestinal epithelial cells to the key IBD associated cytokine, TNF-α, results in increased epithelial permeability.[29]

With respect to mouse models, overexpression of TNF-α in TNFΔARE mice leads to chronic transmural ileal inflammation as well as barrier dysfunction. Alternative methods to determine the impact of barrier function in IBD may be found in genetic studies that suggest a role for polymorphisms in an adhesion molecule underlying UC.[2] Finally, barrier protective molecules may offer other insights into the pathogenesis of IBD. Intestinal trefoil factor (ITF) is an epithelial-derived protein that facilitates barrier regeneration. $ITF^{-/-}$ mice have increased baseline intestinal permeability that is amplified following hypoxic insult.[30] $ITF^{-/-}$ mice have significantly increased mortality from exacerbated DSS colitis as a result of delayed epithelial restitution.[31] Local delivery of ITF to mice

TABLE 14.4.2 Animal Models of Inflammatory Bowel Disease

Spontaneous	Chemical	Genetic	Transfer
Cotton-top tamarin	TNBS	IL-2$^{-/-}$	CD4$^+$ CD45RB^{high} into
C3H/HeJBir	Oxazolone	IL-2Rα$^{-/-}$	RAG$^{-/-}$ or SCID mice
SAMP1	DSS	IL-10$^{-/-}$	
	Acetic acid	TGF-β$^{-/-}$	
		TNFΔARE	
		Gαi2$^{-/-}$	
		N-cadherin dom. neg.	
		Mdr1a$^{-/-}$	
		ITF$^{-/-}$	
		TCRα$^{-/-}$	
		TCRβ$^{-/-}$	
		IL-7 transgenic	
		Stat4 transgenic	
		WASP-KO	
		MHCII$^{Δ/Δ}$	
		Tg$_e$26	

Models underlined denote those in which eosinophils have been identified. Was, Wiskott–Aldrich syndrome protein homologue. dom.neg, dominant negative; DSS, dextran sulfate sodium; GI, gastrointestinal; IL, interleukin; SCID, severe combined immunodeficient; TNBS, 2,4,6-trinitrobenzenesulfonic acid; TNF, tumor necrosis factor.

during DSS colitis and to mice deficient in IL-10 that develop chronic colitis results in attenuation of injury and inflammation.[32] While this is a model of epithelial injury and repair, no evaluation of eosinophils has been reported in these systems.

Mouse models exhibiting specific defects in epithelial barrier function as a result of genetic alteration include N-cadherin dominant-negative mice and mdr1a$^{-/-}$ mice.[33,34] N-cadherin acts to compete for E-cadherin (cadherin-1) during cell–cell adhesion in the epithelium. This competition results in areas in which low affinity cell adhesion occurs, resulting in a patchy intestinal inflammation, probably a response to exposure to luminal contents. Mdr1a$^{-/-}$ mice develop pancolitis, with epithelial disruption, lamina propria leukocyte infiltration, and mucosal thickening. Intestinal inflammation is primarily restricted to the large intestine and mice develop loose stools, anal mucus discharge, and colon shortening. This inflammation can be attenuated by treatment with antibiotics. In a detailed study of cellular infiltrates, chemokines, and cytokines, the presence of lymphocytes, macrophages, and neutrophils was observed in addition to a cytokine profile consistent with a T$_h$1 phenotype; however, no eosinophils were involved in this inflammation. Interestingly, the authors suggest that this may be the result of a significant reduction in eotaxin production in the mdr1a$^{-/-}$ mouse colons.[35]

Another theory into the mechanisms contributing to disease onset in IBD is that of a deregulated immune response, in particular deficiencies in the regulatory immune proteins TGF-β and IL-10. Mice deficient in TGF-β develop inflammation in multiple organs and succumb to this inflammation shortly after birth.[36] TGF-β is a ubiquitous growth factor and regulatory cytokine and acts on multiple cell types. Of interest in intestinal inflammation is the recent finding that mucosal macrophages are maintained in an anergic state despite the maintenance of their phagocytic and bactericidal function, a process dependent on TGF-β[37] Overexpression of a dominant-negative transforming growth factor-β type II (TGF-βRII) in T cells results in transmural inflammation of the colon involving lymphocytes, macrophages and plasma cells.[38] Epithelial overexpression of the same dominant-negative TGF-βRII receptor in the intestine results in the development of spontaneous colitis in mice maintained in conventional housing, as well as an increased susceptibility to DSS-induced colitis in these mice. Although eosinophils are potent producers of TGF-β no studies into the role of eosinophil-derived TGF-β in colitis have been performed.

IL-10$^{-/-}$ mice develop chronic transmural enterocolitis[39] with anemia, epithelial abnormalities, mucosal wall thickening, and weight loss, while eosinophils, lymphocytes, macrophages, neutrophils, occasional giant cells, and plasma cells are observed in the colon.[39] This colitis is described as a T$_h$1-type inflammation during early disease, with a switch to T$_h$2-type inflammation in late disease. Bacteria are important for the development of disease in these mice, with attenuation of inflammation found under specific pathogen-free conditions. Further studies identified an important role for the T$_h$2 cytokine IL-4 in the establishment of inflammation, as mice deficient in both IL-4 and IL-10 do not develop intestinal inflammation, including attenuation of colonic eosinophils. Eosinophils increase prior to the onset of colitis, suggesting that they may play a role in disease onset.[40] Despite this, direct targeting of eosinophils in IL-10$^{-/-}$ mice has not been examined, and may therefore shed light on the exact role of eosinophils in mouse colitis in the absence of regulatory IL-10 signaling.

TABLE 14.4.3 Comparison of Animal Models of Inflammatory Bowel Disease

Category	Model Name	GI Manifestations	Location	Pathology
Spontaneous	C3H/HeJ	Minimal	Cecal/proximal colon	Transmural
	SAMP1	Ileitis and ileal narrowing	Terminal ileum	Transmural
Chemical	TNBS	Diarrhea, rectal prolapse, weight loss	Pancolonic	Transmural
	Oxazolone	Diarrhea, weight loss,	Distal colon	Superficial
	DSS	Diarrhea, rectal prolapse, weight loss	Distal colon/pancolonic	Superficial
Genetic	TNFΔARE	None reported	Terminal ileum	Transmural
	Mdr1a$^{-/-}$	Anal mucus discharge, loose stool	Pancolonic	Superficial
	IL-10$^{-/-}$	Anemia, weight loss	Enterocolitis	Transmural
Transfer	CD45RB^{High}	Decreased activity, diarrhea and soft stool, piloerection, weight loss	Pan colonic/ileal	Transmural

DSS, dextran sulfate sodium; GI, gastrointestinal; IL-10, interleukin-10; TNBS, 2,4,6-trinitrobenzenesulfonic acid; TNF, tumor necrosis factor.

Cell Transfer Models

A model of increasing importance involves the transfer of CD45RB^{high} CD4$^+$ T cells to immune-deficient recombinase activating gene-deficient (RAG2$^{-/-}$) or severe combined immunodeficient (SCID) mice. Naive CD45RB^{high} CD4$^+$ T cells induce pancolitis and in some instances ileitis a number of weeks following transfer. Cotransfer of CD45RB^{low} CD4$^+$ T cells includes a population of regulatory cells and results in diminution of colitis. Colitis in this model results from an imbalance between effector immunity and regulatory immunity. CD45RB^{low} CD4$^+$ T-cell transfer models result in a slow-onset inflammation that may take up to 8 weeks to develop, resulting in a disease characterized by decreased activity, diarrhea and soft stool, piloerection, and weight loss. This transmural colitis involves a substantial mixed leukocyte infiltrate, including granulocytes, lymphocytes, macrophages, and monocytes. Disrupted epithelium, goblet cell depletion, and even crypt abscesses can sometimes be observed. Mice subjected to this model are described as having Crohn's colitis with a mixed T_h1/T_h17 cytokine profile.[41] Interestingly, IL-4 administration in this model led to accelerated development of the T_h1/T_h17 colitis, likely to be mediated through a combination of T cell and non-T cell responses.[42] Although granulocytes are noted in these colons, no direct examination of eosinophils has been performed in this model.

CONCLUSIONS

While a number of mouse models are available to study IBD, few of these have been used to determine the role of eosinophils in this process. Models that are primarily T_h2 skewed are most likely to have eosinophilia as histological and functional features, i.e., spontaneous SAMP1 ileitis and DSS- and oxazolone-induced colitis. In each of these models the reduction in eosinophil numbers has been associated with an improvement in histological and inflammatory measures of injury. In these mouse models of IBD, no direct or indirect correlation has been identified between the presence of eosinophils and any other factor, including recruited cell types or inflammatory profiles. Therefore, while current knowledge of the role of eosinophils in clinical IBD remains circumstantial, so too are the mechanisms and functions of eosinophils in murine models of IBD.

The models discussed here mimic the acute stage of colitis (Table 14.4.3). However, there remains a paucity in models that examine chronic and fibrosing forms of colitis and phenotypes of IBD. The SAMP1 mouse lines are chronic models of IBD that involve chronic transmural inflammation and ileal remodeling. An increasing number of studies are relying on repeated exposure to the chemical models (TNBS and DSS) to induce a remodeled colon and fibrosis. Together, these acute and chronic models may assist our understanding of the role of eosinophils in IBD, thus leading to novel therapeutic options.

REFERENCES

1. Barrett JC, Hansoul S, Nicolae DL, Cho JH, Duerr RH, Rioux JD, et al. Genome-wide association defines more than 30 distinct susceptibility loci for Crohn's disease. *Nat Genet* 2008;**40**:955–62.
2. Fisher SA, Tremelling M, Anderson CA, Gwilliam R, Bumpstead S, Prescott NJ, et al. Genetic determinants of ulcerative colitis include the ECM1 locus and five loci implicated in Crohn's disease. *Nat Genet* 2008;**40**:710–2.

3. Silverberg MS, Cho JH, Rioux JD, McGovern DP, Wu J, Annese V, et al. Ulcerative colitis-risk loci on chromosomes 1p36 and 12q15 found by genome-wide association study. *Nat Genet* 2009; **41**:216–20.

4. Frank DN, St Amand AL, Feldman RA, Boedeker EC, Harpaz N, Pace NR. Molecular-phylogenetic characterization of microbial community imbalances in human inflammatory bowel diseases. *Proc Natl Acad Sci USA* 2007;**104**:13780–5.

5. Berstad A, Borkje B, Riedel B, Elsayed S. Increased fecal eosinophil cationic protein in inflammatory bowel disease. *Hepatogastroenterology* 1993;**40**:276–8.

6. Carlson M, Raab Y, Peterson C, Hallgren R, Venge P. Increased intraluminal release of eosinophil granule proteins EPO, ECP, EPX, and cytokines in ulcerative colitis and proctitis in segmental perfusion. *Am J Gastroenterol* 1999;**94**:1876–83.

7. Luck W, Becker M, Niggemann B, Wahn U. In vitro release of eosinophil cationic protein from peripheral eosinophils reflects disease activity in childhood Crohn disease and ulcerative colitis. *Eur J Pediatr* 1997;**156**:921–4.

8. Wedemeyer J, Vosskuhl K. Role of gastrointestinal eosinophils in inflammatory bowel disease and intestinal tumours. *Best Pract Res Clin Gastroenterol* 2008;**22**:537–49.

9. Madara JL, Podolsky DK, King NW, Sehgal PK, Moore R, Winter HS. Characterization of spontaneous colitis in cotton-top tamarins (Saguinus oedipus) and its response to sulfasalazine. *Gastroenterology* 1985;**88**:13–9.

10. Sundberg JP, Elson CO, Bedigian H, Birkenmeier EH. Spontaneous, heritable colitis in a new substrain of C3H/HeJ mice. *Gastroenterology* 1994;**107**:1726–35.

11. Matsumoto S, Okabe Y, Setoyama H, Takayama K, Ohtsuka J, Funahashi H, et al. Inflammatory bowel disease-like enteritis and caecitis in a senescence accelerated mouse P1/Yit strain. *Gut* 1998;**43**:71–8.

12. McNamee EN, Wermers JD, Masterson JC, Collins CB, Lebsack MD, Fillon S, et al. Novel model of T(H)2-polarized chronic ileitis: The SAMP1 mouse. *Inflamm Bowel Dis* 2009.

13. Rivera-Nieves J, Bamias G, Vidrich A, Marini M, Pizarro TT, McDuffie MJ, et al. Emergence of perianal fistulizing disease in the SAMP1/YitFc mouse, a spontaneous model of chronic ileitis. *Gastroenterology* 2003;**124**:972–82.

14. Takeda T, Hosokawa M, Takeshita S, Irino M, Higuchi K, Matsushita T, et al. A new murine model of accelerated senescence. *Mech Ageing Dev* 1981;**17**:183–94.

15. Lodes MJ, Cong Y, Elson CO, Mohamath R, Landers CJ, Targan SR, et al. Bacterial flagellin is a dominant antigen in Crohn disease. *J Clin Invest* 2004;**113**:1296–306.

16. Olson TS, Reuter BK, Scott KG, Morris MA, Wang XM, Hancock LN, et al. The primary defect in experimental ileitis originates from a nonhematopoietic source. *J Exp Med* 2006;**203**:541–52.

17. Takedatsu H, Mitsuyama K, Matsumoto S, Handa K, Suzuki A, Funabashi H, et al. Interleukin-5 participates in the pathogenesis of ileitis in SAMP1/Yit mice. *Eur J Immunol* 2004;**34**:1561–9.

18. Masterson JC, McNamee EN, Jedlicka P, et al. CCR3 Blockade Attenuates Eosinophilic Ileitis and Associated Remodeling. *Am J Pathol* 2011;**179**:2302–14.

19. Morris GP, Beck PL, Herridge MS, Depew WT, Szewczuk MR, Wallace JL. Hapten-induced model of chronic inflammation and ulceration in the rat colon. *Gastroenterology* 1989;**96**:795–803.

20. Boirivant M, Fuss IJ, Chu A, Strober W. Oxazolone colitis: A murine model of T helper cell type 2 colitis treatable with antibodies to interleukin 4. *J Exp Med* 1998;**188**:1929–39.

21. Dieleman LA, Elson CO, Tennyson GS, Beagley KW. Kinetics of cytokine expression during healing of acute colitis in mice. *Am J Physiol* 1996;**271**:G130–6.

22. Forbes E, Murase T, Yang M, Matthaei KI, Lee JJ, Lee NA, et al. Immunopathogenesis of experimental ulcerative colitis is mediated by eosinophil peroxidase. *J Immunol* 2004;**172**:5664–75.

23. Stevceva L, Pavli P, Husband A, Matthaei KI, Young IG, Doe WF. Eosinophilia is attenuated in experimental colitis induced in IL-5 deficient mice. *Genes Immun* 2000;**1**:213–8.

24. Ahrens R, Waddell A, Seidu L, Blanchard C, Carey R, Forbes E, et al. Intestinal macrophage/epithelial cell-derived CCL11/eotaxin-1 mediates eosinophil recruitment and function in pediatric ulcerative colitis. *J Immunol* 2008;**181**:7390–9.

25. Vieira AT, Fagundes CT, Alessandri AL, Castor MG, Guabiraba R, Borges VO, et al. Treatment with a novel chemokine-binding protein or eosinophil lineage-ablation protects mice from experimental colitis. *Am J Pathol* 2009;**175**:2382–91.

26. Artis D, Humphreys NE, Potten CS, Wagner N, Muller W, McDermott JR, et al. Beta7 integrin-deficient mice: delayed leukocyte recruitment and attenuated protective immunity in the small intestine during enteric helminth infection. *Eur J Immunol* 2000;**30**:1656–64.

27. Brandt EB, Zimmermann N, Muntel EE, Yamada Y, Pope SM, Mishra A, et al. The alpha4bbeta7-integrin is dynamically expressed on murine eosinophils and involved in eosinophil trafficking to the intestine. *Clin Exp Allergy* 2006;**36**:543–53.

28. Mishra A, Hogan SP, Brandt EB, Wagner N, Crossman MW, Foster PS, et al. Enterocyte expression of the eotaxin and interleukin-5 transgenes induces compartmentalized dysregulation of eosinophil trafficking. *J Biol Chem* 2002;**277**:4406–12.

29. Bruewer M, Luegering A, Kucharzik T, Parkos CA, Madara JL, Hopkins AM, et al. Proinflammatory cytokines disrupt epithelial barrier function by apoptosis-independent mechanisms. *J Immunol* 2003;**171**:6164–72.

30. Furuta GT, Turner JR, Taylor CT, Hershberg RM, Comerford K, Narravula S, et al. Hypoxia-inducible factor 1-dependent induction of intestinal trefoil factor protects barrier function during hypoxia. *J Exp Med* 2001;**193**:1027–34.

31. Mashimo H, Wu DC, Podolsky DK, Fishman MC. Impaired defense of intestinal mucosa in mice lacking intestinal trefoil factor. *Science* 1996;**274**:262–5.

32. Vandenbroucke K, Hans W, Van Huysse J, Neirynck S, Demetter P, Remaut E, et al. Active delivery of trefoil factors by genetically modified Lactococcus lactis prevents and heals acute colitis in mice. *Gastroenterology* 2004;**127**:502–13.

33. Hermiston ML, Gordon JI. Inflammatory bowel disease and adenomas in mice expressing a dominant negative N-cadherin. *Science* 1995;**270**:1203–7.

34. Panwala CM, Jones JC, Viney JL. A novel model of inflammatory bowel disease: mice deficient for the multiple drug resistance gene, mdr1a, spontaneously develop colitis. *J Immunol* 1998;**161**:5733–44.

35. Masunaga Y, Noto T, Suzuki K, Takahashi K, Shimizu Y, Morokata T. Expression profiles of cytokines and chemokines in murine MDR1a$^{-/-}$ colitis. *Inflamm Res* 2007;**56**:439–46.

36. Shull MM, Ormsby I, Kier AB, Pawlowski S, Diebold RJ, Yin M, et al. Targeted disruption of the mouse transforming growth factor-beta 1 gene results in multifocal inflammatory disease. *Nature* 1992;**359**:693—9.

37. Smythies LE, Sellers M, Clements RH, Mosteller-Barnum M, Meng G, Benjamin WH, et al. Human intestinal macrophages display profound inflammatory anergy despite avid phagocytic and bactericidal activity. *J Clin Invest* 2005;**115**:66—75.

38. Gorelik L, Flavell RA. Abrogation of TGFbeta signaling in T cells leads to spontaneous T cell differentiation and autoimmune disease. *Immunity* 2000;**12**:171—81.

39. Kuhn R, Lohler J, Rennick D, Rajewsky K, Muller W. Interleukin-10-deficient mice develop chronic enterocolitis. *Cell* 1993;**75**:263—74.

40. Specht S, Arriens S, Hoerauf A. Induction of chronic colitis in IL-10 deficient mice requires IL-4. *Microbes Infect* 2006;**8**:694—703.

41. Ostanin DV, Bao J, Koboziev I, Gray L, Robinson-Jackson SA, Kosloski-Davidson M, et al. T cell transfer model of chronic colitis: concepts, considerations, and tricks of the trade. *Am J Physiol Gastrointest Liver Physiol* 2009;**296**:G135—46.

42. Fort M, Lesley R, Davidson N, Menon S, Brombacher F, Leach M, et al. IL-4 exacerbates disease in a Th1 cell transfer model of colitis. *J Immunol* 2001;**166**:2793—800.

Chapter 14.5

Eosinophil-mediated Events in Mouse Models of Food Allergy

Amanda Waddell and Simon P. Hogan

INTRODUCTION

The prevalence of food allergies is becoming significantly greater throughout the world. A European respiratory health survey in 15 countries reported food allergy or intolerance in 12% of their respondents.[1] It is estimated that food allergy occurs in approximately 4% of adults and about 6% of children > 3 years of age in the United States.[2–5]

Eosinophilic inflammation is a common feature of food allergies, including allergic colitis, allergic eosinophilic gastroenteritis, and early-phase immunoglobulin E (IgE)-mediated hypersensitivity.[6,7] The number of eosinophils and levels of free eosinophil secretory products [eosinophil cationic protein (ECP) and eosinophil peroxidase (EPO)] are increased in fecal and segmental perfusion fluid samples from patients with food allergy.[8–10] Furthermore, eosinophil levels are elevated in allergic eosinophilic gastroenteritis, and histological analyses reveal extracellular deposition of major basic protein (MBP) and ECP in the small bowel of patients with eosinophilic gastroenteritis.[11,12] Additionally, Charcot—Leyden crystals, remnants of eosinophil degranulation, are commonly found on microscopic examination of stool samples, and electron microscopy studies reveal ultrastructural changes in the secondary granules (indicative of eosinophil degranulation and mediator release) in duodenal samples from patients with eosinophilic gastroenteritis.[13,14]

While there is circumstantial evidence supporting a pathogenic role for eosinophils in food allergies, the contribution of eosinophils to disease pathogenesis is not yet fully understood. However, recent experimental analyses employing models of food allergies have provided some insight into the contribution of eosinophils and eosinophil-derived mediators to the clinical manifestations of disease.

MOUSE MODEL OF FOOD-INDUCED IGE-MEDIATED IMMEDIATE HYPERSENSITIVITY

Recently, investigators developed a mouse model of oral antigen-induced IgE-mediated immediate hypersensitivity to begin to decipher the cellular and molecular pathways involved in disease. In this model, mice are sensitized to ovalbumin (OVA) by intraperitoneal injection (i.p.) and 2 weeks later receive repeated intragastric delivery of relatively high amounts of soluble OVA (50 mg) Approximately 15—45 min following the fifth or sixth OVA challenge, mice develop secretory diarrhea and hypothermia. The development of anaphylactic symptoms is associated with increased levels of allergen-specific T-helper type 2 T_h2 cytokines, including interleukin-4 (IL-4), IL-9, and IL-13, the presence of OVA-specific IgE and IgG_1, increased Cl- secretory response, and induced eosinophilia and mastocytosis. Recent experimental studies in rodent models of anaphylaxis provide corroborative evidence demonstrating that IgE-mediated immediate hypersensitivity is critically dependent on FcεRI and mast cells.[15,16] Moreover, mice deficient in IL-4 or IL-4 receptor α ($IL\text{-}4RA^{-/-}$), mast cells, FcεRI, or IgE are protected against IgE-mediated anaphylaxis.[17]

Interestingly, there is a substantial body of emerging literature demonstrating bidirectional eosinophil and mast cell interactions. Human umbilical cord blood-derived mast cells can be activated by MBP to release granulocyte-macrophage colony stimulating factor (GM-CSF), histamine, IL-8, prostaglandin D2 (PGD2), and tumor necrosis factor (TNF-α).[18] Furthermore, eosinophil-derived MBP elicits mast cell exocytosis, cytokine production and eicosanoid generation, the latter two being prominent responses following FcεRI-dependent activation of mast cells.[18] Incubation of rat peritoneal mast cells with native ECP, EPO, and MBP (but not eosinophil-derived neurotoxin; EDN) results in concentration-dependent histamine release.[19] While experimental studies indicate a central role for mast cells in the effector phase of oral antigen-induced anaphylaxis, the contribution of eosinophils to the mast cell response is unclear.

Brandt and colleagues assessed oral antigen-induced anaphylaxis in IL-5 transgenic, wild-type (WT), and $Il5^{-/-}/Ccl11^{-/-}$ mice to define the role of eosinophils in oral antigen-induced IgE-mediated immediate hypersensitivity.[20] They demonstrated that repeated oral challenge of mice deficient in $Il5^{-/-}/Ccl11^{-/-}$ did not diminish oral antigen-induced IgE-mediated immediate hypersensitivity, suggesting that eosinophils do not contribute to the clinical symptoms of disease.[20] Furthermore, IL-5 transgenic mice, which have exaggerated levels of eosinophils, did not have any significant increase in the severity of oral antigen-induced IgE mediated anaphylactic reaction.[20] However, a recent study by Song and colleagues employing a similar model suggests a role for eosinophils in disease symptoms.[21] The investigators treated mice with a sialic acid-binding Ig-like lectin 5 (Siglec-F)-neutralizing antibody to deplete eosinophils and assessed the effect of eosinophil depletion on the development of the symptoms of oral antigen-induced anaphylaxis. Siglec-F is a sialic acid-binding immunoglobulin superfamily receptor that is highly expressed on eosinophils.[22] The authors demonstrated that administration of anti-Siglec-F neutralizing antibody significantly reduced levels of T_h2 cytokines, IgE, and intestinal eosinophils. Importantly, the reduction in eosinophil levels was associated with reduced intestinal permeability changes, normalization of intestinal villous crypt height, and restoration of weight gain. It is important to note that Siglec-F is also expressed on mast cells; while anti-Siglec F treatment did not inhibit intestinal mastocytosis, the authors could not exclude the possibility that anti-Siglec-F treatment was inhibiting mast cell activity.[21] Notably, *in vitro* studies with culture derived human mast cells demonstrate Siglec-8-mediated inhibition of FcεRI-dependent histamine release.[23,24] Collectively, these studies suggest that while eosinophils accumulate in compartments of the gastrointestinal tract in response to acute food-induced reactions, they are not required for the development of symptoms.

MOUSE MODEL OF ALLERGIC EOSINOPHILIC GASTROENTERITIS

We have previously developed an experimental model of food-induced chronic T_h2-inflammation that mimics certain pathophysiological features of allergic eosinophilic gastroenteritis.[25,26] In this model, mice are sensitized with an i.p. injection of the egg antigen OVA in the presence of adjuvant (alum) and then repeatedly challenged with oral OVA enteric-coated beads on days 12 and 15. The OVA antigen is encapsulated in a biodegradable particle that is resistant to degradation at gastric pH,

thereby protecting the antigen against gastric digestion. However, the particles are susceptible to degradation at pH 5.5, which facilitates the delivery and release of the allergen in a preserved native conformational state to the small intestine.[25] Administration of oral allergen to sensitized mice induces peripheral blood eosinophilia and expansion of a T_h2-biased immune response.[25,26] Histological examination of the jejunum reveals vascular congestion, edema, and a prominent cellular infiltrate comprised predominantly of eosinophils. Increased eosinophil numbers are also observed in various segments of the gastrointestinal (GI) tract, including the esophagus, stomach, small intestine, and Peyer patches, as has been similarly shown in patients with eosinophil-associated GI disorders.[27] The infiltrating eosinophils are interspersed throughout the reticular connective tissue of the lamina propria and mucosa and throughout the length of the lamina propria of the villi. In addition, eosinophils in this model are also in close proximity to damaged enteric nerves, which has been observed in eosinophil-associated GI diseases in humans.[28] Notably, oral antigen administration induced a number of pathological changes including cachexia, decreased villus:crypt ratio, gastric dysmotility, and vascular congestion and edema in the jejunum, all of which are similar to the pathogenic features observed in a variety of inflammatory GI disorders.[29]

The oral OVA-induced eosinophil infiltration into the GI tract is associated with elevated levels of chemokine (C-C motif) ligand 11 (Ccl11). Examination of $Il5^{-/-}$ and $Ccl11^{-/-}$ mice revealed that OVA-induced eosinophil infiltration into the GI tract is dependent on Ccl11, as the oral allergen-induced increase in eosinophil levels is ablated in $Ccl11^{-/-}$ mice. This conclusion is further supported by the demonstration that transgenic overexpression of CCL11 in $Ccl11^{-/-}$ mice significantly restores eosinophil levels in the small bowel.[30] The reduction in eosinophil recruitment to the GI tract in $Ccl11^{-/-}$ mice was associated with a marked reduction in the magnitude of antigen-induced pathology (failure to thrive and gastromegaly), supporting a critical role for eosinophils in pathogenesis of disease. Collectively, these results establish a crucial function for CCL11 in eosinophil recruitment into the small bowel and a role for eosinophils in allergic GI disease.

CONCLUSIONS

Clinical studies demonstrate increased eosinophil numbers and activation in the GI tract of patients with food allergies. Importantly, eosinophil level and activity are correlated with the clinical manifestations of disease. Experimental analyses have identified no major contribution of eosinophils in the IgE-mediated immediate hypersensitivity reaction. However, evidence suggests that eosinophils may have a pathogenic role in food-induced chronic T_h2 inflammatory responses, driving intestinal dysmotility and altered epithelial architecture. These changes may be brought about via eosinophil-mediated axonal necrosis and dysregulation of enteric nerve function; however, further experimental analyses are required to delineate the molecular basis of eosinophil-mediated dysfunction.

REFERENCES

1. Woods RK, Abramson MJ, Bailey H, Walter EH. International prevalences of reported food allergies and intolerances. Comparisons arising from the European Community Respiratory Health Survey (ECRHS). *Eur J Clin Nutr* 2001;**55**:298–304.

2. Sicherer SH, Munoz-Furlong A, Sampson HA. Prevalence of seafood allergy in the United States determined by a random telephone survey. *J Allergy Clin Immunol* 2004;**114**:159–65.

3. Sampson HA. Food Allergy. *J Allergy Clin Immunol* 2003;**111**: S540–7.

4. Sampson HA. Update on food allergy. *J Allergy Clin Immunol* 2004;**113**:805–19.

5. Munoz-Furlong A, Sampson HA, Sicherer SH. Prevalence of self-reported seafood allergy in the US. [abstract]. *J Allergy Clin Immunol* 2004;**113**(Suppl.):S100.

6. Torpier G, Colombel JF, Mathieu-Chandelier C, Capron M, Dessaint JP, Cortot A, et al. Eosinophilic gastroenteritis: ultrastructural evidence for a selective release of eosinophil major basic protein. *Clin Exp Immunol* 1988;**74**:404–8.

7. Machida HM, Catto SA, Gall DG, Trevenen C, Scott RB. Allergic colitis in infancy: clinical and pathologic aspects [see comments]. *J Pediatr Gastroenterol Nutr* 1994;**19**:22–6.

8. Bischoff SC, Grabowsky MS, Manns MP. Quantification of inflammatory mediators in stool samples of patients with inflammatory bowel disorders and controls. *Dig Dis Sci* 1997;**42**:394–403.

9. Bischoff SC, Mayer J, Nguyen QT, Stolte M, Manns MP. Immunohistological assessment of intestinal eosinophil activation in patients with eosinophilic gastroenteritis and inflammatory bowel disease. *Am J Gastroenterol* 1999;**94**:3521–9.

10. Schwab D, Raithel M, Klein P, Winterkamp S, Weidenhiller M, Radespiel-Troeger M, et al. Immunoglobulin E and eosinophilic cationic protein in segmental lavage fluid of the small and large bowel identify patients with food allergy. *Am J Gastroenterol* 2001;**96**:508–14.

11. Talley NJ, Kephart GM, McGovern TW, Carpenter HA, Gleich GJ. Deposition of eosinophil granule major basic protein in eosinophilic gastroenteritis and celiac disease. *Gastroenterology* 1992;**103**: 137–45.

12. Spry CJ, Tai PC, Barkans J. Tissue localization of human eosinophil cationic proteins in allergic diseases. *Int Arch Allergy Appl Immunol* 1985;**77**:252–4.

13. Klein NC, Hargrove RL, Sleisenger MH, Jeffries GH. Eosinophilic gastroenteritis. *Medicine* 1970;**2**:215–25.

14. Cello JP. Eosinophilic gastroenteritis: A complex disease entity. *Am J Med* 1979;**67**:1097–114.

15. Osterfeld H, Ahrens R, Strait R, Finkelman FD, Renauld JC, Hogan SP. Differential roles for the IL-9/IL-9 receptor alpha-chain pathway in systemic and oral antigen-induced anaphylaxis. *J Allergy Clin Immunol* 2010;**125**:469–76. e2.

16. Brandt EB, Strait RT, Wang Q, Hersko D, Muntel E, Finkelman FD, et al. Oral antigen-induced intestinal anaphylaxis requires IgE-dependent mast cell degranulation. *J Allergy Clin Immunol* 2003;**111**:S33919. Song, D. J., Cho, J. Y., Miller, M., Strangman, W., Zhang, M., Varki, A. and Broide, D. H. (2009) Anti-Siglec-F antibody inhibits oral egg allergen induced intestinal eosinophilic inflammation in a mouse model. Clin Immunol. 131, 157–169.

17. Finkelman FD. Anaphylaxis: lessons from mouse models. *J Allergy Clin Immunol* 2007;**120**:506–15.

18. Piliponsky A, Gleich G, Bar I, Levi-Schaffer F. Effects of eosinophils on mast cells: a new pathway for the perpetuation of allergic inflammation. *Mol Immunol* 2002;**38**:1369.

19. Zheutlin LM, Ackerman SJ, Gleich GJ, Thomas LL. Stimulation of basophil and rat mast cell histamine release by eosinophil granule-derived cationic proteins. *Source (Bibliographic Citation): J Immunol* 1984;**133**:2180–5.

20. Brandt EB, Strait RT, Hershko D, Wang Q, Muntel EE, Scribner TA, et al. Mast cells are required for experimental oral allergen-induced diarrhea. *J Clin Invest* 2003;**112**:1666–77.

21. Song DJ, Cho JY, Miller M, Strangman W, Zhang M, Varki A, et al. Anti-Siglec-F antibody inhibits oral egg allergen induced intestinal eosinophilic inflammation in a mouse model. *Clin Immunol* 2009;**131**:157–69.

22. Zhang JQ, Biedermann B, Nitschke L, Crocker PR. The murine inhibitory receptor mSiglec-E is expressed broadly on cells of the innate immune system whereas mSiglec-F is restricted to eosinophils. *Eur J Immunol* 2004;**34**:1175–84.

23. Yokoi H, Choi OH, Hubbard W, Lee HS, Canning BJ, Lee HH, et al. Inhibition of FcepsilonRI-dependent mediator release and calcium flux from human mast cells by sialic acid-binding immunoglobulin-like lectin 8 engagement. *J Allergy Clin Immunol* 2008;**121**:499–505 e1.

24. Yokoi H, Myers A, Matsumoto K, Crocker PR, Saito H, Bochner BS. Alteration and acquisition of Siglecs during in vitro maturation of CD34+ progenitors into human mast cells. *Allergy* 2006;**61**:769–76.

25. Hogan SP, Mishra A, Brandt EB, Foster PS, Rothenberg ME. A critical role for eotaxin in experimental oral antigen-induced eosinophilic gastrointestinal allergy. *Proc Natl Acad Sci USA* 2000;**97**:6681–6.

26. Hogan SP, Mishra A, Brandt EB, Royalty MP, Pope SM, Zimmermann N, et al. A pathological function for eotaxin and eosinophils in eosinophilic gastrointestinal inflammation. *Nat Immunol* 2001;**2**:353–60.

27. Kelly KJ. Eosinophilic gastroenteritis. *J Pediatr Gastroenterol Nutr* 2000;**30**(Suppl):S28–35.

28. Dvorak AM, Onderdonk AB, McLeod RS, Monahan-Earley RA, Antonioli DA, Cullen J, et al. Ultrastructural identification of exocytosis of granules from human gut eosinophils in vivo. *Int Arch Allergy Immunol* 1993;**102**:33–45.

29. Sampson HA. Food allergy. Part 1: immunopathogenesis and clinical disorders. *J Allergy Clin Immunol* 1999;**103**:717–28.

30. Mishra A, Hogan SP, Brandt EB, Wagner N, Crossman MW, Foster PS, et al. Enterocyte expression of the eotaxin and interleukin-5 transgenes induces compartmentalized dysregulation of eosinophil trafficking. *J Biol Chem* 2002;**277**:4406–12.

Models of Eosinophilic Leukemia

Yoshiyuki Yamada, Jose A. Cancelas and Marc E. Rothenberg

The World Health Organization (WHO) has published a set of criteria used to distinguish chronic eosinophilic leukemia (CEL) from classical hypereosinophilic syndromes (HES).[1] The diagnosis of CEL is met in patients who fulfill the criteria for HES if cytogenetic or molecular evidence of clonality of the myeloid lineage, especially involving eosinophils, is present, as well as an increase of blasts [>2% in peripheral blood or >5% but <19% in bone marrow (BM)]. The criteria for HES include persistent eosinophilia of 1500 eosinophils/mm^3 of blood for more than 6 months, lack of evidence of other known causes of secondary hypereosinophilia, and multiple organ involvement due to eosinophil infiltration.[2] In 2003, the FIP1 like 1 (S. cerevisiae)—platelet-derived growth factor receptor, α polypeptide (FIP1L1—PDGFRA) fusion gene was identified in patients initially diagnosed as having a myeloproliferative variant of HES who had a successful response to the administration of imatinib mesylate.[3] Since clonality of the disease in HES patients with the FIP1L1—PDGFRA fusion gene has been observed, the appropriate nomenclature for this disease is considered to be CEL. Since then, many other studies have reported the presence of the FIP1L1—PDGFRA fusion gene in a significant fraction of hypereosinophilia patients.[3—11] As a result, the FIP1L1—PDGFRA fusion gene has become the most

frequent clonal defect demonstrated in HES/CEL. Subsequently, other PDGFRA fusion gene variants have also been reported to be causes of CEL (Table 14.6.1).[10,12—15] In addition, PDGFRB or fibroblast growth factor receptor 1 (FGFR1) rearrangements have been described in myeloproliferative neoplasms (MPN) with eosinophilia.[16]

Based upon molecular information, a new category of myeloid neoplasms, 'myeloid and lymphoid neoplasms with eosinophilia and abnormalities of PDGFRA, PDGFRB, or FGFR1,' was added to the WHO criteria.[17] In addition, mouse models with these rearrangements have been reported.[18—21] Among them, a model of FIP1L1—PDGFRα-introduced mouse CEL is the only one that develops prominent eosinophilia.[19]

In this subchapter, we summarize the development and characteristics of animal models, focusing especially on those of FIP1L1—PDGFRα-associated diseases. In addition, we review the advantages of these models for dissecting the pathogenesis and exploring new therapeutic approaches.

MYELOPROLIFERATIVE NEOPLASM MODELS ASSOCIATED WITH FIP1L1—PDGFRα

Initially, Cools and coworkers used the classic BM transplantation method to simply introduce the FIP1L1—PDGFRA fusion gene into BM hematopoietic stem cells and progenitors (HSC/Ps) as a means of developing a mouse model.[22] The disease phenotype was a severe leukocytosis with mild eosinophilia in the peripheral blood and with a neutrophil-dominant myeloid tissue infiltration in multiple organs, similar to a p210-BCR—ABL-induced chronic myeloid leukemia

TABLE 14.6.1 Chronic Eosinophilic Leukemia with *PDGFRA* Rearrangement

Fusion Genes	Location	Break Points		Phenotype
		Partner Genes	PDGFRA	
FIPL1-PDGFRA	del(14q12)	exons 7—13, intron 16	exon 12	CEL, SM
BCR-PDGFRA	t(4;22)(q12;q11)	intron 7, exon 12, exon 1, exon 17	exon 12, exon 13	CEL, UMPD
KIF5B-PDGFRA	t(4;10)(q12;p11)	exon 3	exon 12	CEL
CDK5RAP2-PDGFRA	ins(9;4)(q33;q12q25)	exon 13	intron 9/exon 12	CEL
STRN-PDGFRA	t(2;4)(p24;q12)	exon 6	exon 12	CEL
ETVA6-PDGFRA	t(4;12)(q12;p13)	exon 6	exon 12	CEL

CEL, chronic eosinophilic leukemia; SM, systemic mastocytosis; UMPN, unclassified myeloproliferative neoplasms; PDGFRA, platelet-derived growth factor receptor.

(CML)-like disease. The disease was transplantable to secondary recipients. However, serial transplantation did not lead to a phenotype resembling HES/CEL.[19] In this mouse model, FIP1L1−PDGFRα expression promoted granulocytopoiesis that was not committed to eosinophil lineage cells, unlike patients with HES/CEL. This suggests that the induction of HES/CEL by FIP1L-1−PDGFRα may require a secondary event.

DEVELOPMENT OF FIP1L1−PDGFRα-INDUCED MOUSE CHRONIC EOSINOPHILIC LEUKEMIA

Attempts to seek the secondary events leading to induction of HES/CEL in the mouse eventually focused on inter-leukin-5 (IL-5), as this cytokine has been shown to be the most relevant cytokine in the pathogenesis of HES/CEL. Indeed, a subgroup of patients with HES displays aberrant, sometimes clonal, expansion of CD3$^-$ CD4$^+$ T-helper type 2 (T$_h$2) lymphocytes, which secrete large amounts of IL-5.[23] Notably, elevated levels of circulating IL-5 have also been shown in patients with HES/CEL, including FIP1L1−PDGFRα$^+$ patients,[24] and anti-IL-5 responder FIP1L1−PDGFRα$^+$ patients have been observed. Inter-estingly, it has been demonstrated that lymphoid lineage cells in patients with FIP1L1−PDGFRα$^+$ HES/CEL also express the *FIP1L1−PDGFRA* fusion gene,[25,26] implying that lymphocytes affect disease development. On the other hand, it has been shown that IL-5 transgenic mice display a blood eosinophilia, but not the tissue eosinophilia remi-niscent of HES/CEL.[27] In fact, neither IL-5 overexpression nor FIP1L1−PDGFRα alone induces substantial tissue eosinophilia. Finally, a humanized antibody against IL-5 lowers blood eosinophilia in all forms of HES, including patients with low serum levels of IL-5, thus indicating that even endogenous low levels of IL-5 regulate eosinophilia in HES.[28]

These collective findings led to the hypothesis that elevated levels of IL-5 or specific activation of the IL-5 pathway would induce the development of murine HES/CEL in conjunction with the expression of FIP1L1−PDGFRα in BM HSC/Ps in the model. To test this hypothesis, lethally irradiated wild-type mice were trans-planted with *FIP1L1−PDGFRA* fusion gene-transduced HSC/Ps derived from the BM of IL-5 transgenic mice that overexpress IL-5 in a T cell-dependent fashion.[19] The FIP1L1−PDGFRα-expressing cells engrafted and rapidly proliferated, resulting in the development of a myelopro-liferative neoplasm, characterized by eosinophilic infiltrate of multiple organs, leukocytosis with significant eosino-philia in the peripheral blood, and splenomegaly, thus resembling HES/CEL.[19]

BLOOD AND TISSUE EOSINOPHILIA IN THE FIP1L1−PDGFRα$^+$ HYPEREOSINOPHILIC SYNDROME/CHRONIC EOSINOPHILIC LEUKEMIA MOUSE MODEL

As mentioned above, the FIP1L1−PDGFRα$^+$ HES/CEL (F/P-CEL) mouse model demonstrates strikingly high eosinophilia in the blood, BM, spleen, and non-hematopoietic tissues. In addition to the large number of eosinophils, eosinophils in hematopoietic organs showed upregulation of α4-integrin and sialic acid-binding Ig-like lectin F (Siglec-F; or Siglec-5) in the F/P-CEL model mice, suggesting that they were activated.[19] Nonhematopoietic organs, as well as hematopoietic organs, also demonstrated eosinophil infiltrations. Eosinophils dominated the infil-tration in the perivascular regions of the liver, and in the peritubular regions of the kidney and lung. In addition, F/P-CEL mice showed a significantly higher eosinophil content in the myocardium and lamina propria of the small intestine.[9]

SPECIFICITY OF INTERLEUKIN-5 OVEREXPRESSION AND FIP1L1−PDGFRα IN PROMOTING HYPEREOSINOPHILIC SYNDROME/CHRONIC EOSINOPHILIC LEUKEMIA

Eosinophilia has been seen occasionally in MPN elicited by other oncogenes. For instance, there is an eosinophilic variant of BCR−ABL$^+$ CML.[29,30] In addition, CML patients with more than 5% of eosinophils are at risk of an eosinophilic blast crisis.[31] Thus, there is a possibility that other oncogenes, for example p210-BCR−ABL, may induce HES/CEL in the presence of IL-5 overexpression. To address this question, a combination of IL-5 over-expression and p210-BCR−ABL has been tested in mice. Interestingly, the eosinophilia that occurred when a mouse myeloproliferative disorder was induced by p210-BCR−ABL was not striking, even in the presence of transgenic IL-5 overexpression.[19] Moreover, FIP1L1−PDGFRA expression alone does not preferen-tially induce eosinophilia, when compared to the prolifer-ation of other myeloid lineages in this mouse model. Instead, an *ex vivo* analysis showed increased levels of expression and frequency of IL-5Rα in FIP1L1−PDGFRα$^+$ splenocytes compared to p210-BCR−ABL$^+$ and vector control splenocytes. Thus, a specific association of FIP1L1−PDGFRα expression and activation of the IL-5 in the development of mouse HES/CEL has been demon-strated (Figure 14.6.1).[19]

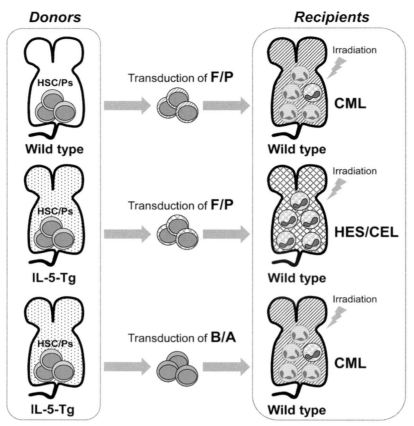

FIGURE 14.6.1 Disease phenotypes induced by the combination of donors, recipients, and fusion genes. Lethally irradiated wild-type mice transplanted with the *FIP1L1—PDGFRA* (F/P) fusion gene-transduced wild-type BM hematopoietic stem cells and progenitors (HSC/Ps) develop CML-like disease with mild eosinophilia. In contrast, the *FIP1L1—PDGFRA* fusion gene induces HES/CEL-like disease when used in BM HSC/Ps derived from the CD2-IL-5 transgenic (IL-5Tg) mouse. Unlike *FIP1L1—PDGFRA* fusion gene, *BCR—ABL* (B/A) cannot induce HES/CEL like disease even though BM HSC/Ps are harvested from the IL-5Tg mouse.

SHORT-TERM STEM CELLS OR EARLY MYELOID PROGENITORS MAY BE RESPONSIBLE FOR THE DEVELOPMENT OF FIP1L1—PDGFRα⁺ HYPEREOSINOPHILIC SYNDROME/ CHRONIC EOSINOPHILIC LEUKEMIA

In order to determine whether the mouse F/P-CEL model represents a stem cell or early progenitor-derived proliferative disease, serial transplantation from diseased mice was performed. A large number of splenocytes from primary CEL mice was required to develop a similar disease in secondary recipients and, at the same time, demonstrated FIP1L1—PDGFRα expression in circulating lymphocytes for up to several weeks after secondary transplantation. These data strongly suggest the involvement of a short-term repopulating stem cell or at least an early myeloid progenitor.[19]

SYSTEMIC MASTOCYTOSIS (SM) IN THE MOUSE FIP1L1—PDGFRα⁺ HYPEREOSINOPHILIC SYNDROME/ CHRONIC EOSINOPHILIC LEUKEMIA MODEL

FIP1L1—PDGFRα⁺ disorders are characterized by SM as well as clonal blood and tissue hypereosinophilia.[5] Although the most common SM, SM with D816V mutation,[32] occasionally shows eosinophilia,[33—35] it is clinically distinguishable from FIP1L1—PDGFRA-associated SM. FIP1L1—PDGFRA-associated SM shows lower tryptase levels in circulation, less aggregation of BM mast cells, more severe eosinophilia, higher serum vitamin B12 levels, and more frequent pulmonary and cardiac involvement, when compared with D816V SM with eosinophilia.[34] The mouse models induced by FIP1L1—PDGFRα were also shown to develop tissue mast cell infiltration and increased circulating mouse mast cell protease 1 (mMCP-1) levels, a systemic assay of mast cell content and degranulation in the mouse that resembles serum tryptase determination in SM patients.[36] Similar to patients with the *FIP1L1−PDGFRA* fusion gene, infiltration of morphologically aberrant tissue mast cells, characterized by abnormal shape with frequent cytoplasmic extensions (reminiscent of the *spindle shape* seen in patients), are seen in the hematopoietic organs, intestine, and skin of F/P-CEL mice. In addition, serum levels of mMCP-1 are extremely elevated in F/P-CEL mice. Moreover, *FIP1L1−PDGFRA* gene expression induced significantly higher mast cell infiltration in their skin and intestine, and even higher levels of mMCP-1 when IL-5 was overexpressed, suggesting that IL-5 may exacerbate mastocytosis in mouse F/P-CEL.[36] Mast cells expressing FIP1L1-PDGFRα are unique, as shown in experiments using blockade of v-kit Hardy-Zuckerman 4 feline sarcoma viral oncogene homologue (c-KIT) signaling in this model and *ex vivo* FIP1L1−PDGFRα[+] BM-derived mast cells. The c-KIT signaling pathway is pivotal for normal mast cell development and function. In normal mast cells, FIP1L1−PDGFRα-associated SM was still c-KIT dependent, as tissue mast cells and circulating levels of mMCP-1 were significantly decreased by the blocking the c-KIT signaling pathway using anti-c-KIT antibody. In addition, mast cell differentiation of FIP1L1−PDGFRα-expressing HSC/Ps is largely dependent on stem cell factor (SCF). Notably, prolonged survival and enhanced migration toward SCF is seen in FIP1L1−PDGFRα[+] mast cells. In addition, FIP1L1−PDGFRα and SCF signaling synergistically stimulate the Akt signaling pathway, indicating the collaboration of c-KIT and FIPL1L1/PDGFRα tyrosine kinase activities in activation of downstream signaling pathways.[36] FIP1L1−PDGFRα synergizes with the SCF/c-KIT pathway to promote mast cell development, activation, and survival, which results in the development of SM.

DEVELOPMENT OF NEW THERAPEUTIC APPROACHES FOR HUMAN HYPEREOSINOPHILIC SYNDROME/CHRONIC EOSINOPHILIC LEUKEMIA USING MOUSE MODELS INDUCED BY FIP1L1−PDGFRα

An FIP1L1−PDGFRα[+] MPN model (F/P-MPN), that is, a CML-like model without IL-5 overexpression, together with an F/P-CEL model would be very useful for exploring various therapeutic strategies. Indeed, the efficacy of the tyrosine kinase inhibitors, imatinib and nilotinib,[37] and the protein kinase C (PKC) inhibitor, PKC412, have been tested using the F/P-MPN model.[22] These models are promising tools for evaluating treatment regimens for patients expressing the imatinib-resistant FIP1L1−PDGFRA fusion

protein with a mutation at threonine 674 (analogous to the T315I mutation of BCR−ABL) located in the ATP-binding region of the PDGFRA protein.[3,38]

When focused on the impairment of eosinophilia as a therapeutic approach in HES/CEL, IL-5 is a potential therapeutic target for the treatment of HES/CEL. In fact, anti-IL-5 therapy induces a dramatic and sustained decrease in blood eosinophilia and decreased eosinophil activation in patients with a variety of eosinophilic disorders, including HES/CEL.[39] The efficacy of anti-IL-5 in HES/CEL is confirmed by evidence revealing that the development of mouse F/P-CEL is dependent on IL-5.[19]

In addition to anti-IL-5 therapy, an eosinophil-specific approach, anti-Siglec-8/Siglec-F therapy has been proposed.[40] Mouse Siglec-F on eosinophils is the functional paralogue of human Siglec-8.[41−43] On human eosinophils, Siglec-8 ligation specifically induces apoptosis.[44] The effect of mouse Siglec-F ligation was tested in F/P-CEL mice by the administration of anti-Siglec-F antibodies. The quantity of eosinophils in the peripheral blood of treated F/P-CEL mice was consistently and specifically decreased, while total leukocytes counts were unaffected by Siglec-F antibody administration.[40]

OTHER MODELS OF MYELOPROLIFERATIVE NEOPLASMS WITH EOSINOPHILIA IN A NEW WORLD HEALTH ORGANIZATION CLASSIFICATION

A new category in WHO criteria, 'myeloid and lymphoid neoplasm with eosinophilia and abnormalities in *PDGFRA, PDGFRB,* or *FGFR1*,' includes these three rearrangements.[16] Clinically, *PDGFRA* rearrangements affect the development of CEL as mentioned above, while *PDGFRB* rearrangements lead to chronic myelomonocytic leukemia (CMML) with eosinophilia, and *FGFR1* rearrangements cause of stem cell leukemia lymphoma syndrome.[45]

Mouse models of CEL induced by *PDGFRB* and *FGFR1* rearrangements have been reported, in addition to those for *PDGFRA*. *PDGFRB* rearrangement was initially reported to be a consequence of t(5;12)(q33;p13) translocation in a patient with CMML.[21] Subsequently, other PDGFRB fusion proteins have been reported, and the introduction of *TEL (ETV6)−PDGFRB*, which encodes a constitutively activated tyrosine kinase, induced MPN similar to p210-*BCR*−ABL in a mouse BM transplantation model.[46] Although human CMML occasionally has eosinophilia, mouse *PDGFRB* rearrangement models do not show apparent eosinophilia.

Stem cell leukemia/lymphoma caused by *FGFR1* rearrangement, also known as the 8p11 myeloproliferative syndrome (EMS), is clinically characterized by leukocytosis with eosinophilia, splenomegaly, and a short chronic phase followed by rapidly progressing acute leukemia.[47] In terms of mouse BM transplantation models, the clinical phenotypes are dependent on the fusion partner genes to the *FGFR1* gene. *BCR−FGFR1* induces CML-like MPN, whereas *ZMYM2 (ZNF198)−FGFR1* induces an EMS-like disease including eosinophilia.[20] More recently, humanized mouse BM transplantation models using immunodeficient animals have been established. Interestingly *BCR−FGFR1*-induced, as well as *ZMYM2 (ZNF198)−FGFR1*-induced, models demonstrate a marked increase in human granulocytopoietic cells with eosinophilia, indicating that the *FGFR1* rearrangement is primarily capable of inducing granulocytopoiesis with eosinophilia.[18]

A study in 2003 reported that *FIP1L1−PDGFRA* fusion gene expression is observed in approximately 60% of patients with a myeloproliferative variant of HES.[3] At that time, it was expected that the pathogenesis of the majority of CEL could be explained by this gene fusion. However, in a more recent report that included milder cases than earlier reports, the incidence of F/P-CEL occurred in 11% of the patients initially diagnosed as having HES.[7] This indicates that a large number of the HES/CELs have unknown molecular mechanisms. Progress in the molecular recognition of MPN may provide more information that would contribute to new therapeutic approaches. For this reason, it is important to facilitate the development of mouse models by introducing specific genetic rearrangements, mutations affecting protein tyrosine kinases, or related molecules leading to constitutively active signal transduction pathways.

REFERENCES

1. Bain B, Pierre R, Imbert M, Vardiman JW, Brunning RD, Flandrin G. Chronic eosinophilic leukaemia and the hypereosinophilic syndrome. World Health Organization of Tumours: Tumours of Haematopoietic and Lymphoid Tissues. In: Jaffe ES, Harris N, Stein H, Vardiman JW, editors. Lyon, France: IARC Press; 2001.

2. Chusid MJ, Dale DC, West BC, Wolff SM. The hypereosinophilic syndrome: analysis of fourteen cases with review of the literature. *Medicine (Baltimore)* 1975;**54**:1−27.

3. Cools J, DeAngelo DJ, Gotlib J, Stover EH, Legare RD, Cortes J, et al. A tyrosine kinase created by fusion of the PDGFRA and FIP1L1 genes as a therapeutic target of imatinib in idiopathic hypereosinophilic syndrome. *N Engl J Med* 2003;**348**:1201−14.

4. Bacher U, Reiter A, Haferlach T, Mueller L, Schnittger S, Kern W, et al. A combination of cytomorphology, cytogenetic analysis, fluorescence in situ hybridization and reverse transcriptase polymerase chain reaction for establishing clonality in cases of persisting hypereosinophilia. *Haematologica* 2006;**91**:817−20.

5. Klion AD, Noel P, Akin C, Law MA, Gilliland DG, Cools J, et al. Elevated serum tryptase levels identify a subset of patients with a myeloproliferative variant of idiopathic hypereosinophilic syndrome associated with tissue fibrosis, poor prognosis, and imatinib responsiveness. *Blood* 2003;**101**:4660−6.

6. La Starza R, Specchia G, Cuneo A, Beacci D, Nozzoli C, Luciano L, et al. The hypereosinophilic syndrome: fluorescence in situ hybridization detects the del(4)(q12)-FIP1L1/PDGFRA but not genomic rearrangements of other tyrosine kinases. *Haematologica* 2005;**90**:596−601.

7. Ogbogu PU, Bochner BS, Butterfield JH, Gleich GJ, Huss-Marp J, Kahn JE, et al. Hypereosinophilic syndrome: a multicenter, retrospective analysis of clinical characteristics and response to therapy. *J Allergy Clin Immunol* 2009;**124**:1319−25. e3.

8. Pardanani A, Ketterling RP, Li CY, Patnaik MM, Wolanskyj AP, Elliott MA, et al. FIP1L1-PDGFRA in eosinophilic disorders: prevalence in routine clinical practice, long-term experience with imatinib therapy, and a critical review of the literature. *Leuk Res* 2006;**30**:965−70.

9. Roche-Lestienne C, Lepers S, Soenen-Cornu V, Kahn JE, Lai JL, Hachulla E, et al. Molecular characterization of the idiopathic hypereosinophilic syndrome (HES) in 35 French patients with normal conventional cytogenetics. *Leukemia* 2005;**19**:792−8.

10. Score J, Curtis C, Waghorn K, Stalder M, Jotterand M, Grand FH, et al. Identification of a novel imatinib responsive KIF5B-PDGFRA fusion gene following screening for PDGFRA overexpression in patients with hypereosinophilia. *Leukemia* 2006;**20**:827−32.

11. Vandenberghe P, Wlodarska I, Michaux L, Zachee P, Boogaerts M, Vanstraelen D, et al. Clinical and molecular features of FIP1L1-PDFGRA (+) chronic eosinophilic leukemias. *Leukemia* 2004;**18**:734−42.

12. Baxter EJ, Hochhaus A, Bolufer P, Reiter A, Fernandez JM, Senent L, et al. The t(4;22)(q12;q11) in atypical chronic myeloid leukaemia fuses BCR to PDGFRA. *Hum Mol Genet* 2002;**11**:1391−7.

13. Curtis CE, Grand FH, Musto P, Clark A, Murphy J, Perla G, et al. Two novel imatinib-responsive PDGFRA fusion genes in chronic eosinophilic leukaemia. *Br J Haematol* 2007;**138**:77−81.

14. Trempat P, Villalva C, Laurent G, Armstrong F, Delsol G, Dastugue N, et al. Chronic myeloproliferative disorders with rearrangement of the platelet-derived growth factor alpha receptor: a new clinical target for STI571/Glivec. *Oncogene* 2003;**22**: 5702−6.

15. Walz C, Curtis C, Schnittger S, Schultheis B, Metzgeroth G, Schoch C, et al. Transient response to imatinib in a chronic eosinophilic leukaemia associated with ins(9;4)(q33;q12q25) and a CDK5RAP2-PDGFRA fusion gene. *Genes Chromosomes Cancer* 2006;**45**:950−6.

16. Tefferi A, Vardiman JW. Classification and diagnosis of myeloproliferative neoplasms: the 2008 World Health Organization criteria and point-of-care diagnostic algorithms. *Leukemia* 2008;**22**:14−22.

17. Bain BJ, Gilliland DG, Horny H-P, Vardiman JW, Swerdlow SH, Campo E, et al. *Myeloid and lymphoid neoplasms with eosinophilia and abnormalities of PDGFRA, PDGFRB and FGFR1. World Health Organization Classification of Tumours of Haematopoietic and Lymphoid Tissue.* Lyon, France: IARC Press; 2008.

18. Agerstam H, Jaras M, Andersson A, Johnels P, Hansen N, Lassen C, et al. Modeling the human 8p11-myeloproliferative syndrome in immunodeficient mice. *Blood* 2010;**116**:2103−11.

19. Yamada Y, Rothenberg ME, Lee AW, Akei HS, Brandt EB, Williams DA, et al. The FIP1L1-PDGFRA fusion gene cooperates with IL-5 to induce murine hypereosinophilic syndrome (HES)/ chronic eosinophilic leukemia (CEL)-like disease. *Blood* 2006;**107**:4071−9.

20. Roumiantsev S, Krause DS, Neumann CA, Dimitri CA, Asiedu F, Cross NC, et al. Distinct stem cell myeloproliferative/T lymphoma syndromes induced by ZNF198-FGFR1 and BCR-FGFR1 fusion genes from 8p11 translocations. *Cancer Cell* 2004;**5**:287−98.

21. Golub TR, Barker GF, Lovett M, Gilliland DG. Fusion of PDGF receptor beta to a novel ets-like gene, tel, in chronic myelomonocytic leukemia with t(5;12) chromosomal translocation. *Cell* 1994;**77**:307−16.

22. Cools J, Stover EH, Boulton CL, Gotlib J, Legare RD, Amaral SM, et al. PKC412 overcomes resistance to imatinib in a murine model of FIP1L1-PDGFRalpha-induced myeloproliferative disease. *Cancer Cell* 2003;**3**:459−69.

23. Bank I, Amariglio N, Reshef A, Hardan I, Confino Y, Trau H, et al. The hypereosinophilic syndrome associated with CD4+CD3− helper type 2 (Th2) lymphocytes. *Leuk Lymphoma* 2001;**42**:123−33.

24. Tefferi A, Pardanani A. Imatinib therapy in clonal eosinophilic disorders, including systemic mastocytosis. *Int J Hematol* 2004;**79**:441−7.

25. Tefferi A, Lasho TL, Brockman SR, Elliott MA, Dispenzieri A, Pardanani A. FIP1L1-PDGFRA and c-kit D816V mutation-based clonality studies in systemic mast cell disease associated with eosinophilia. *Haematologica* 2004;**89**:871−3.

26. Robyn J, Lemery S, McCoy JP, Kubofcik J, Kim YJ, Pack S, et al. Multilineage involvement of the fusion gene in patients with FIP1L1/PDGFRA-positive hypereosinophilic syndrome. *Br J Haematol* 2006;**132**:286−92.

27. Dent LA, Strath M, Mellor AL, Sanderson CJ. Eosinophilia in transgenic mice expressing interleukin 5. *J Exp Med* 1990;**172**:1425−31.

28. Rothenberg ME, Klion AD, Roufosse FE, Kahn JE, Weller PF, Simon HU, et al. Treatment of patients with the hypereosinophilic syndrome with mepolizumab. *N Engl J Med* 2008;**358**:1215−28.

29. Gotlib V, Darji J, Bloomfield K, Chadburn A, Patel A, Braunschweig I. Eosinophilic variant of chronic myeloid leukemia with vascular complications. *Leuk Lymphoma* 2003;**44**:1609−13.

30. Bennett JM, Catovsky D, Daniel MT, Flandrin G, Galton DA, Gralnick H, et al. The chronic myeloid leukaemias: guidelines for distinguishing chronic granulocytic, atypical chronic myeloid, and chronic myelomonocytic leukaemia. Proposals by the French-American-British Cooperative Leukaemia Group. *Br J Haematol* 1994;**87**:746−54.

31. Brito-Babapulle F. Clonal eosinophilic disorders and the hypereosinophilic syndrome. *Blood Rev* 1997;**11**:129−45.

32. Patnaik MM, Rindos M, Kouides PA, Tefferi A, Pardanani A. Systemic mastocytosis: a concise clinical and laboratory review. *Arch Pathol Lab Med* 2007;**131**:784−91.

33. Pardanani A, Reeder T, Li CY, Tefferi A. Eosinophils are derived from the neoplastic clone in patients with systemic mastocytosis and eosinophilia. *Leuk Res* 2003;**27**:883−5.

34. Maric I, Robyn J, Metcalfe DD, Fay MP, Carter M, Wilson T, et al. KIT D816V-associated systemic mastocytosis with eosinophilia and FIP1L1/PDGFRA-associated chronic eosinophilic leukemia are distinct entities. *J Allergy Clin Immunol* 2007;**120**:680−7.

35. Lawrence JB, Friedman BS, Travis WD, Chinchilli VM, Metcalfe DD, Gralnick HR. Hematologic manifestations of systemic mast cell disease: a prospective study of laboratory and morphologic features and their relation to prognosis. *Am J Med* 1991;**91**:612−24.

36. Yamada Y, Sanchez-Aguilera A, Brandt EB, McBride M, Al-Moamen NJ, Finkelman FD, et al. FIP1L1/PDGFRalpha synergizes with SCF to induce systemic mastocytosis in a murine model of chronic eosinophilic leukemia/hypereosinophilic syndrome. *Blood* 2008;**112**:2500−7.

37. von Bubnoff N, Gorantla SP, Thone S, Peschel C, Duyster J. The FIP1L1-PDGFRA T674I mutation can be inhibited by the tyrosine kinase inhibitor AMN107 (nilotinib). *Blood* 2006;**107**:4970−1. author reply 4972.

38. Ohnishi H, Kandabashi K, Maeda Y, Kawamura M, Watanabe T. Chronic eosinophilic leukaemia with FIP1L1-PDGFRA fusion and T674I mutation that evolved from Langerhans cell histiocytosis with eosinophilia after chemotherapy. *Br J Haematol* 2006;**134**(5):547−9.

39. Stein ML, Villanueva JM, Buckmeier BK, Yamada Y, Filipovich AH, Assa'ad AH, et al. Anti-IL-5 (mepolizumab) therapy reduces eosinophil activation ex vivo and increases IL-5 and IL-5 receptor levels. *J Allergy Clin Immunol* 2008;**121**:1473−83. 1483 e1−4.

40. Zimmermann N, McBride ML, Yamada Y, Hudson SA, Jones C, Cromie KD, et al. Siglec-F antibody administration to mice selectively reduces blood and tissue eosinophils. *Allergy* 2008.**63**:1156−63.

41. Aizawa H, Zimmermann N, Carrigan PE, Lee JJ, Rothenberg ME, Bochner BS. Molecular analysis of human Siglec-8 orthologs relevant to mouse eosinophils: identification of mouse orthologs of Siglec-5 (mSiglec-F) and Siglec-10 (mSiglec-G). *Genomics* 2003;**82**:521−30.

42. Zhang JQ, Biedermann B, Nitschke L, Crocker PR. The murine inhibitory receptor mSiglec-E is expressed broadly on cells of the innate immune system whereas mSiglec-F is restricted to eosinophils. *Eur J Immunol* 2004;**34**:1175−84.

43. Zhang M, Angata T, Cho JY, Miller M, Broide DH, Varki A. Defining the in vivo function of Siglec-F, a CD33-related Siglec expressed on mouse eosinophils. *Blood* 2007;**109**:4280−7.

44. Nutku E, Aizawa H, Hudson SA, Bochner BS. Ligation of Siglec-8: a selective mechanism for induction of human eosinophil apoptosis. *Blood* 2003;**101**:5014−20.

45. Pardanani A, Tefferi A. Primary eosinophilic disorders: a concise review. *Curr Hematol Malig Rep* 2008;**3**:37−43

46. Magnusson MK, Meade KE, Brown KE, Arthur DC, Krueger LA, Barrett AJ, et al. Rabaptin-5 is a novel fusion partner to platelet-derived growth factor beta receptor in chronic myelomonocytic leukemia. *Blood* 2001;**98**:2518−25.

47. Jackson CC, Medeiros LJ, Miranda RN. 8p11 myeloproliferative syndrome: a review. *Hum Pathol* 2010;**41**:461−76.

Chapter 14.7

Eosinophils in Veterinary Medicine

S.O. (Wole) Odemuyiwa

INTRODUCTION

The use of small animal models has significantly enhanced the understanding of the pathological mechanisms of several diseases. The bulk of such models is based on mice and, to a lesser extent, rats. These small rodents are popular because of the relatively low cost of housing, shorter gestation, and ease of genetic manipulations. However, these models may not fully mimic human diseases, especially when such conditions are subject to control by multiple genetic factors; this necessitates the induction of spontaneously arising diseases of humans in genetically modified animals, often providing a less-than-optimal model. The gene—gene and gene—environment interactions that form the basis of complex diseases are often poorly replicated in these models. This gap in translational medicine is significantly reduced when spontaneously occurring veterinary diseases that are similar to human diseases are studied in animals. For example, more than 40 naturally occurring diseases of dogs are due to mutations in genes that are homologous to those that cause similar diseases in human, making the dog an excellent natural disease model for such conditions.[1,2] In addition, the close contact of pets with humans suggests that shared environmental factors could trigger similar diseases in humans and their furry friends and thus provide unique epidemiological contexts for the study of both human and animal diseases.

Animal Models of Eosinophilic Diseases

As in many other diseases, the mouse and rat have been the main model animals used in the study of the role played by eosinophils in diseases of humans, especially eosinophilic airway inflammation and parasite elimination. These models have allowed a clear delineation of the environmental, genetic, and immunological bases of eosinophilic airway inflammation and have permitted an exquisite elucidation of the roles played by eosinophils. Today, the eosinophil is no longer seen as an evolutionary vestige with no known function. However, moving forward from here, a more complete understanding of the naturally occurring eosinophilic diseases of animals usually examined by veterinarians, including mice and rats, will allow maximal leveraging of the knowledge acquired from experimentally induced animal models toward clinical translation to the diagnosis and management of human and veterinary diseases.

Eosinophils in Animals: From Iguanas to Horses

In most animals, eosinophils comprise 1—10% of circulating leukocytes; however, over 20% of the blood leukocytes of cattle may be eosinophils. The proportion of leukocytes that are eosinophils could also vary according to the time of blood collection. For example, unlike humans, where the opposite is true, blood eosinophil counts in horses exhibit diurnal variation, with the highest counts in the morning and the lowest in the evening. The significance of these differences is unknown; however, in horses as in humans,

eosinophils are primarily tissue-dwelling cells with a rapid transit time between bone marrow, blood, and target tissues. Whether high circulating numbers in cattle and high early morning numbers in horses represent differences in tissue signals that recruit eosinophils to play a specific role in immune homeostasis in a species- and time-dependent manner is currently unknown.[3] Similarly, seasonal variations have been reported in the number of eosinophils circulating in reptilian blood: eosinophil numbers are highest in winter, when reptiles are usually in hibernation. Although it is conceivable that hibernation is associated with reduced stress, and thus a reduction in circulating endogenous steroids, it is unknown whether there is a specific eosinophil-dependent immunological state associated with hibernation in reptiles. A study of such a hypothetical state may provide important information on the immunological roles played by eosinophils in higher mammals.

The varied nature of eosinophils between species seen in veterinary practice is not limited to number but also pertains to morphology. The first confusion that confronts a veterinarian examining blood smears from several animals, after being trained in a human medical environment, is the marked variation in the colors and shapes of eosinophil granules. Using Romanowsky (Romanovsky)-type stains, eosinophils in animals can be differentiated by the colors of their granules—orange, red, or blue—suggesting that the accumulation of basic proteins that give the eosinophils their *eosinophilic* colors may be evolutionarily determined. The most striking color differences are seen in chelonians and crocodilians: for example, while eosinophil granules of most reptiles and birds stain red, those of the green iguana (*Iguana iguana*) stain blue and are difficult to differentiate from basophils, except for the presence of high nuclear-to-cytoplasmic ratio. Similarly, the presence of poorly stained vacuolated eosinophils has long been recognized in greyhounds. However, although recognized as a genetically determined trait, recent studies showed that these so-called gray eosinophils of greyhounds have the same cytochemical, functional, and ultrastructural characteristics as those of other breeds of dogs.[4]

Similar to their staining properties, the shape and size of eosinophil granules also vary across species. The ellipsoidal granules of horse eosinophils are often easily differentiated from those of other veterinary species, in which eosinophil granules may be either round or a mixture of round and ellipsoidal. The most striking difference is found in birds, where eosinophils have both large spherical primary granules and mature rod-shaped specific granules; however, the crystalline core characteristic of most mammals is absent in eosinophils from many avian species. Similarly, loggerhead sea turtles, but not green turtles, lack crystalline structures in their eosinophils.[5] An absence of Charcot—Leyden crystals has also been found in the eosinophils of some rodents. Whether these major variations in eosinophil ultrastructure translate to differences in eosinophil function in birds, turtles, and rodents versus other animals is also currently unknown.

Eosinophilic Diseases of Animals

Most veterinary clinicians become concerned about eosinophilia once the blood eosinophil count is raised above the upper reference value for the species in question and there is no historical basis to suspect helminth infection or allergy. Hypereosinophilia is often associated with a specific cause, usually an inflammatory reaction to helminth parasites or allergen. A response to parasites could lead to eosinophil numbers reaching *leukemoid reaction* levels, with *left shifts* toward the more immature metamyelocytes and myelocyte stages. Thus, terms like *eosinophil leukemoid reaction*, hypereosinophilia, and eosinophilic leukemia are often mentioned in the veterinary literature to describe varying degrees of eosinophilia in animals.[6] Generally speaking, eosinophilia in veterinary medicine is interpreted first on the basis of whether there is a known cause and if eosinophilia is diffuse or whether eosinophilia is localized to specific tissues without a readily identifiable cause. When peripheral eosinophilia is accompanied by diffuse infiltration of eosinophils into multiple tissues and organs, hypereosinophilic syndromes (HES) are described and these syndromes are quite similar to what is frequently reported in human patients.

In dogs and cats, mild to moderate eosinophilia is often associated with helminth infection and allergy, as is seen in humans. An interesting observation in animals, however, is the apparent predisposition of certain breeds to high blood eosinophil numbers and eosinophilic diseases relative to other breeds of the same species. For example, German shepherd and rottweiler dog breeds are well known for having high baseline eosinophil counts and a predisposition to eosinophilic diseases.[7] Whether this intraspecies variation is related to genetically determined atopy, defined as an intrinsic propensity for the elevated baseline production of IgE, as reported for atopic human beings, is currently unknown. In addition to genetic factors, however, there are a few specific eosinophilic diseases of domestic species ranging from laboratory to production and companion animals. These naturally occurring diseases are not only clinically important in animals but may also provide important directions in the study of the epidemiology, pathogenesis, and therapy of equivalent eosinophilic diseases in humans.

Spontaneously Occurring Eosinophilic Diseases of Laboratory Animals

The Matsumoto Eosinophilia Shinshu (MES) Rat

MES is a spontaneous, nonreactive disease characterized by the sudden appearance of blood eosinophilia at 9 weeks of age in rats with no history of allergy, hypersensitivity, or other eosinophil-associated conditions. The number of eosinophils continues to increase as the rat becomes older, and is accompanied by infiltration of eosinophils into several tissues and organs (especially the liver). This disease, first described by Matsumoto in the late 1990s, is found in a mutant Sprague-Dawley rat strain.[8] Histologically, it is characterized by a granulomatous reaction, with a center consisting of an intensely staining eosinophilic material that occurs in a radiate, star-like, or asteroid-shaped configuration. This central area is surrounded by histiocytes and a peripheral area of marked eosinophilic inflammation. It is noteworthy that such lesions, the so-called Splendore—Hoeppli phenomenon, have also been reported in HES in human patients.[9] In addition, eosinophils are found to infiltrate the aorta and to induce fibroid polyps in the stomach and interstitial eosinophilic pneumonia in the lungs. However, cardiac lesions commonly observed in human patients with HES are absent in MES rats. In addition, HES in MES rats is not associated with elevated immunoglobulin E (IgE) levels, as observed in human HES patients. Nonetheless, the efficacy of anti-interleukin-5 (anti-IL-5) treatment of these rats suggests an IL-5-dependent mechanism. It is now known that the MES phenotype in rats is due to a specific loss of function mutation in a gene located on rat chromosome 19. This mutation affects the gene encoding cytochrome b-245 light chain (*Cyba* or *p22-PHOX*), a protein component of the nicotinamide adenine dinucleotide phosphate (NADPH) oxidase system. An interesting finding in these rats is the increased phagocytic activity of peritoneal eosinophils during bacterial infection, which occurs as a consequence of nonfunctional NADPH oxidase system in neutrophils due to the inactivating mutation in the *Cyba* gene.[10] These findings suggest that eosinophils, in adequate numbers, possess enough residual phagocytic activity to replace neutrophils completely and to protect against infection with *Staphylococcus aureus*.

Eosinophilic Crystalline Pneumonia

Otherwise known as acidophilic macrophage pneumonia or crystalline pneumonitis, this is an idiopathic, sporadic pulmonary disease of inbred mice strains, such as C57BL/6 and 129Sv, and substrains derived from them. Subclinical to rapidly fatal disease of laboratory mice can occur alone or in combination with other pulmonary diseases, such as cryptococcosis, nematodiasis, pneumocystis pneumonia, and pulmonary adenomas. It has been described as a common background lesion in the C57BL/6 mouse and had a prevalence of 87% in 129S4/SvJae mice in one study. The importance of chronic antigen stimulation in the pathogenesis of this disease was demonstrated in some studies that reported its occurrence following a chronic albumin sensitization-challenge protocol used to induce eosinophilic airway inflammation in the C57BL/6 mouse. A defining histopathological picture of this disease is the occurrence of distinct intensely staining eosinophilic extracellular rectangular crystals that are also frequently found in the cytoplasm of macrophages and multinucleated giant cells in the airways of affected mice. Although these crystals have long been interpreted as being the Charcot— Leyden crystals of eosinophils, they were recently shown to primarily consist of chitinase-3-like protein 3, a nonenzymatic member of the chitinase family. This protein is usually found in neutrophils and macrophages; it is associated with the M2 (anti-inflammatory) phenotype linked to T-helper type 2 (T_h2) polarization when found in macrophages. Thus, eosinophilic crystalline pneumonia may represent a naturally occurring disease of C57BL/6 mouse characterized by T_h2 polarization.[11]

Eosinophilic Diseases of Domestic Animals

Like human patients, eosinophilia in domestic animals can be reactive, meaning that an increase in the number of circulating or tissue-infiltrating eosinophils may be a response to allergen challenge, insect bite, or infection, as is often reported for human patients. However, eosinophilia could also be a nonreactive process, as observed in malignant diseases and idiopathic HES. Although up to 10% of circulating leukocytes in some animals may be eosinophils, a proportion of leukocytes >5% in tissues or body fluid is often enough to suggest an eosinophilic inflammatory response. However, tissue eosinophilia is not usually diagnosed until up to 20% of infiltrating leukocytes consist of this cell type. Some eosinophilic diseases are well characterized in some veterinary species, while there are not enough case reports to describe the same diseases in other species. Thus, the description of eosinophilic diseases here will be limited to common diseases of domestic animals and those that may be similar to conditions encountered in humans. In addition, for ease of description, eosinophilic diseases will be considered according to whether eosinophils show localized infiltration of specific organs or whether eosinophilia is generalized. In most cases of tissue infiltration, the primary histopathological features of eosinophilic diseases are similar and often consist of a granulomatous reaction rich in eosinophils and usually with evidence of extensive collagenolysis or fibrosis.

Locally Infiltrative Eosinophilic Diseases of Domestic Animals

Eosinophilic Meningoencephalitis.
Specific infiltration of the nervous system is noted by the presence of eosinophils in cerebrospinal fluid (CSF), a location where they are rarely found in healthy animals. Eosinophilic pleocytosis is often not accompanied by blood eosinophilia at the time of CSF collection.

In general, mild CSF eosinophilia (<5% of nucleated cells) in dogs and cats is often associated with the placement of ventriculoperitoneal shunts following decompression surgery, as has been reported in pediatric patients undergoing similar surgical procedures.[12,13] Mild eosinophilia is also seen in infarction, neoplastic diseases affecting the central nervous system, and spinal cord compression. In addition, mild eosinophilia is occasionally a component of some infectious diseases, including bacterial encephalitis, canine distemper, cryptococcosis, rabies, and toxoplasmosis in dogs.

Eosinophilic meningitis (>10% eosinophils in CSF) is consistent with the presence of migrating helminth infection in different animal species: *Neospora* and *Prototheca* in dogs, *Dirofilaria immitis* and cuterebral myiasis in cats, and *Toxoplasma gondii* and *Hypoderma bovis* in horses. Infection of alpacas, llamas, moose, and sheep by *Parelaphostrongylus tenuis*, a strongyle that produces no disease in its natural deer host, is characterized by marked eosinophilic pleocytosis. The presentation of these parasitic diseases is often similar to that observed in parasite-induced eosinophilic meningoencephalitis of human patients with a history of travel to tropical destinations.

In addition, lead poisoning in cattle and salt poisoning in swine are all characterized by an elevated proportion of CSF eosinophils in affected animals.[14] Cases of idiopathic eosinophilic meningoencephalitis have also been reported in cats, cows, and dogs based on a failure to detect known infectious or toxic agents characteristically associated with eosinophilia.

Eosinophils in the Digestive System Feline Oral Eosinophilic Granuloma Complex.
This disease of domestic animals, mainly of cats, is an interesting epidemiological and clinical replica of eosinophilic ulcer of the oral mucosa described in human patients. Similar to the human disease, oral eosinophilic granuloma usually occurs once during the lifespan of the animal and the lesion is most often located on the tongue.

Eosinophilic Esophagitis in Dogs.
Eosinophilic esophagitis is not a common disease in animals. Esophagitis in dogs is most often due to reflux of gastric content into the esophagus following anesthesia, chronic vomiting, or hiatal hernia. As such, eosinophilic esophagitis is not usually considered a probable disease in dogs. In most such cases, a biopsy is not commonly performed to determine the cause of esophagitis, even when there is no clinical improvement following treatment with antireflux agents. However, a recent report described a case of eosinophilic

esophagitis following examination of biopsy samples in a dog that failed to respond to antireflux agents. Surprisingly, this dog demonstrated infiltration of eosinophils into esophageal tissue and showed significant clinical improvement following food allergen avoidance. This single case suggests possible similarities between eosinophilic esophagitis in dogs and humans. It will be interesting to see whether more reports are made in the future when an active search for such cases becomes a routine component of esophagitis in animals.[15]

Eosinophilic Diseases of the Gastrointestinal Tract (GIT). In discussing eosinophilic diseases of the GIT, attention should be drawn to the fact that the GIT forms the largest reservoir of immune and inflammatory cells in the body, a point often forgotten by most basic scientists and clinicians. The GIT is an immunologically active milieu of dendritic cells, eosinophils, lymphocytes (both B and T), macrophages, mast cells, and plasma cells, in addition to a wide variety of nonprofessional antigen-presenting cells. The most important role of this milieu is to act as an effective gatekeeper to tolerate nonpathogenic organisms and other nonself proteins, including food materials, while simultaneously ensuring an effective response to the presence and activities of pathogenic organisms in the microbe-rich lumen of the GIT.[16] Given this basic logic, and the fact that tolerance is also a form of immune response, the full complement of leukocytes important for the maintenance of immune tolerance should be present in the GIT of healthy animals under homeostatic conditions. In addition, following exposure to pathogenic organisms, the increased recruitment of specific cell types should be a plausible indication of their relative importance in the response of the GIT to specific pathogens and provide a better characterization of the pathomechanisms of gastrointestinal (GI) diseases. The caveat that comes with this line of reasoning, however, is quickly recognized by the pathologist examining histological sections of the GIT who quickly discovers that a *histologically normal* leukocyte composition of the GIT could be difficult to determine. Thankfully, a few studies in cats, dogs, humans, pigs, and rats have gone a long way toward elucidating the relative composition and distribution of leukocyte cell types in different segments of the GIT and providing important guidelines for the evaluation of GIT diseases.

For example, it is known that in the small intestine, $CD3^+$ (T cells) and $CD4^+$ (T_h cells) lymphocytes are predominantly located around the villi, with numbers decreasing toward the crypts. Conversely, the majority of B cells and plasma cells are present between the crypts, with fewer cells toward the villi. In addition, as one reaches the lamina propria, it becomes apparent that $CD4^+$ T cells outnumber $CD8^+$ T cells in this layer. This careful layering of lymphocytes is the first clue of the functional significance of leukocyte distribution in the GIT, including

a careful layering of the well-known proinflammatory and anti-inflammatory effectors of innate and adaptive immunity.[17] Interestingly, in studies examining eosinophils in the GIT of worm-free healthy cats, dogs, and horses, these cells were found to constitute over 1% of leukocytes found within the crypts and lamina propria, including intraepithelial locations, but not in the villi.[18] Thus, eosinophils, along with B and plasma cells, are found in a region of the intestine that may be most conducive for humoral, T_h2 or anti-inflammatory states under homeostatic conditions. It is instructive that mast cells tend to occupy a completely different histological niche from eosinophils.

The relatively high number of GI eosinophils in healthy animals makes an unequivocal characterization of eosinophilic GI diseases difficult in many species, and they are often conveniently lumped together with other conditions under the description of inflammatory bowel disease (IBD). However, the rapid and specific recruitment of eosinophils to the GIT following challenge with helminth antigens supports a reactive role for eosinophils in GI diseases. Nonetheless, it is possible to characterize distinct eosinophilic diseases of the GIT in some domestic animals. Future research endeavors that seek to determine the nature of eosinophil function under homeostatic conditions versus naturally occurring eosinophilic diseases of the GIT in domestic animals will go a long way toward further clarifying the functional diversity of eosinophils (homeostatic versus reactive).

Feline Intestinal Eosinophilic Sclerosing Fibroplasia. A disease of cats characterized by a trabeculated mass lesion made up of abundant eosinophilic infiltrates and large reactive fibroblasts with frequent intralesional bacteria was recently described. These masses were found along the length of the intestine, as well as in the area of the pyloric sphincter and the ileocecocolic junction. The disease is accompanied by peripheral eosinophilia and eosinophilic lymphadenitis in some cats. In spite of the presence of intralesional bacteria, the eosinophil-dependent nature of this disease is underscored by the dramatic response to corticosteroids and a comparably poor response to antibiotics.[19] The marked fibroplasia may be a result of excessive fibrogenic activity of eosinophils through the production of transforming growth factor-β (TGF-β).[20]

Canine Eosinophilic Enteritides. Diseases characterized by specific infiltration of the GIT by eosinophils in dogs include eosinophilic IBD, helminth infection (*Toxocara canis*, *Ancylostoma caninum*, and *Trichuris vulpis*), and neoplasia. Infiltration of GIT and other organs is seen in HES. Most of these diseases do not produce mass lesions but a marked thickening of the GIT is seen during exploratory laparotomy, imaging studies, or at necropsy. The advent of routine deworming has reduced the incidence of helminth-associated eosinophilic gastroenteritis in dogs.

Thus, eosinophilia in association with GI signs such as diarrhea or vomiting, coupled with evidence of thickened intestinal loops on imaging studies often requires further diagnostic investigations in dogs. Common findings in these cases include hypocholesterolemia, hypoproteinemia, mature neutrophilia, and peripheral eosinophilia, consistent with an inflammatory disease with protein-losing enteropathy. In these cases, the disease is described as diffuse eosinophilic gastroenterocolitis and is often caused by immune-mediated hypersensitivity reaction to food allergens.[21]

In some of these cases, however, abdominal palpation or rectal examination coupled with imaging studies reveal discrete intestinal masses in addition to diffuse thickening of intestinal loops. This condition has therefore been described as idiopathic eosinophilic gastrointestinal masses (IEGM). A distinctive feature of this condition in dogs is the preponderance of the disease in purebred animals. In contrast to IBD and diffuse eosinophilic gastroenteritis, where an increase in the number of eosinophils is usually confined to the mucosa and lamina propria, the lesions associated with eosinophilic masses in IEGM are most often transmural. Although it is quite reasonable to believe that IEGM may be a result of visceral larval migrans, the dramatic response to immunosuppressive doses of prednisone has been suggested as a possible indication of an immune-mediated disease.

Idiopathic Eosinophilic Enterocolitis (IEE) of Horses. Eosinophilic enteritis is an uncommon morphologic diagnosis that forms a component of the IBD complex with a predominance of eosinophils in the inflammatory infiltrate.[22,23] Eosinophilic enteritis is part of IBD in cats, dogs, cattle, horses, and humans.[24] In contrast to a diffuse eosinophilic infiltration of the GIT, however, a condition characterized by focal or multifocal eosinophilic infiltrates, sometimes accompanied by intestinal obstruction, is becoming increasingly important among horses.[25] This so-called focal eosinophilic enterocolitis of horses is similar both clinically and histologically to eosinophilic enteritis with intestinal obstruction in humans. In horses, chronic idiopathic IBD was recently identified as IEE, granulomatous enteritis, lymphocytic-plasmacytic enterocolitis, and multisystemic epitheliotropic eosinophilic disease (MEED).[22] The authors used IEE and MEED as morphological terms to describe diseases in which eosinophilic infiltration of intestinal and other tissues is prominent in horses.[26]

Eosinophils in the Respiratory Tract.
Diseases of the respiratory system are generally divided into airway or parenchymal diseases, depending on whether an increased number of eosinophils is found in the airways or within the lung parenchyma. Thus, airway disease is frequently categorized separately from interstitial lung disease. The most common eosinophilic airway disorder is asthma. As in human patients, asthma-like conditions in animals are characterized by bronchial hyperreactivity to cholinergic agents, chronic cough, eosinophilic infiltration of the airway wall, and reversible airway obstruction. Eosinophilic bronchopneumopathy is an example of interstitial lung disease in veterinary species.

Eosinophilic Airway Inflammation and Asthma in Animals. The characteristic airway inflammation of allergic asthma is frequently seen in animals. However, there is often a discrepancy between eosinophilic airway inflammation and airway hyperreactivity in the dog. *Asthma* in the dog is the best example of eosinophil infiltration into tissues that is not invariably associated with tissue injury. It is known that BAL fluid from dogs can contain up to 24% eosinophils without any evidence of lung injury or disease. In the dog, atopic dermatitis, often as a result of flea allergy, is usually associated with pulmonary infiltration by eosinophils and elevated eosinophils in BAL fluid. In these cases, however, the pathophysiological changes (e.g., reversible airway obstruction) associated with asthma are completely absent, thus making a clinical diagnosis of asthma difficult. The dog is, therefore, not a good animal model of human asthma. Indeed, asthma as a clinical disease is not widely recognized in the dog. In addition, unlike nonasthmatic eosinophilic inflammation in humans, eosinophilic bronchitis without involvement of the lungs is not common in dogs, although a few cases have been reported.

Similarly, the *asthmatic* condition of horses, recurrent airway obstruction (RAO), is a disease characterized by airway hyperreactivity, chronic cough, and neutrophilic, rather than eosinophilic, airway inflammation. Although some studies have suggested that RAO in horses is also a T_h2 driven phenotype, the absence of eosinophilia as a component of this disease is intriguing. Interestingly, while RAO affects older horses (>7 years), inflammatory airway disease (IAD) of younger horses is often characterized by a mixed inflammatory picture, with elevated proportions of mast cells and eosinophils coupled with reversible airway obstruction and hyperreactivity.[27,28] It is currently unknown whether young horses with IAD are more likely to develop RAO as they grow older. However, it is tempting to speculate that IAD may predispose to RAO later in life similar to the way that virus-induced airway hyperreactivity in children early in life predisposes to asthma in later life. RAO might be a more severe form of asthma, as seen in the neutrophil-rich phenotype of asthma in humans. Further studies of this phenomenon in horses could provide important data on the mechanisms behind the transition from *eosinophilic* to *neutrophilic* airway inflammation in some asthmatics.

Feline Asthma. Unlike the dog, and confusing observations in the horse, feline asthma is a distinct disease of the cat.[29] Any inflammatory disease distal to the bifurcation of the trachea is called feline bronchial disease. However, examination of the BAL fluid from cases of feline bronchial disease shows that there are at least two subsets of inflammatory diseases in the airways of cats. The preponderance of neutrophils is characteristic of chronic bronchitis, while eosinophilic inflammation is specifically observed in feline bronchial asthma. Although the two conditions are difficult to differentiate clinically, some authors have shown that the presence of airway hyperreactivity and reversible airway obstruction can differentiate feline bronchial asthma from chronic bronchitis. Feline bronchial asthma is one of the most common respiratory diseases of cats. It is characterized by clinical signs ranging from intermittent wheezing and coughing to episodes of severe dyspnea. Despite apparent similarities in the pathophysiology and clinical presentations of asthma in humans and cats, the immunological basis of feline asthma is still poorly understood. Up to 24% of BAL cells in some healthy cats can be eosinophils. In addition, bronchoscopy is often contraindicated and life threatening in cats with severe asthma, thus making it difficult to study the inflammatory component of the active disease in cats. However, the dependence on specific allergens and presence of allergen-specific IgE, along with the clinical usefulness of exhaled breath condensate for asthma diagnosis in cats as in humans, suggest that feline asthma may be a perfect natural model of human allergic asthma.

Eosinophilic Bronchopneumopathy. This disease has been called canine eosinophilic pneumonia, eosinophilic pneumonitis, HES, pulmonary hypersensitivity, pulmonary infiltrates with eosinophils, and pulmonary eosinophilic granulomatosis by different authors. In this interstitial lung disease, eosinophils are found to infiltrate the terminal bronchioles, the alveoli, and the blood vessels. The frequent involvement of bronchi justifies the term bronchopneumopathy in this disease. Like the human disease, the presence of bacteria, fungi, or parasites would make diagnosis easy. However, in most cases, the cause of eosinophilia is undetermined (i.e., idiopathic). If eosinophil infiltration of other organs is present, it is probably a component of an HES. Nonetheless, a diagnosis of eosinophilic bronchopneumopathy is often reserved for cases where the lung is specifically infiltrated. An attempt has been made by Peeters and Clercx to apply the human classification scheme for eosinophilic bronchopneumopathy to dogs.[30–32] Note that peripheral eosinophilia does not always accompany eosinophilic pneumonia. The release of toxic mediators from activated eosinophils results in extensive inflammatory lesions in the lung interstitium and the perivascular space during eosinophilic bronchopneumopathy of dogs.

Systemic Eosinophilic Diseases

In addition to the localized infiltration of specific organs by eosinophils, there are other conditions characterized by increased eosinophilia with diffuse infiltration of eosinophils into several organs. In these cases, generalized immune deregulation could lead to increased maturation and recruitment from the bone marrow into the circulation and thence into target tissues and organs. This is the case with HES and eosinophilic neoplasia/paraneoplastic syndrome.

Hypereosinophilic Syndromes. This is a rare syndrome characterized by marked eosinophilia with infiltration of multiple target organs leading to organ damage and dysfunction. The disease is often retrospectively diagnosed after prolonged eosinophilia, consistently negative fecal flotation tests, and the absence of any other evidence of helminth infection. This disease has been reported in dogs and cats; rottweilers and German shepherds show a marked predisposition to HES. However, specific genetic markers previously described in human patients have not been associated with the development of HES in dogs.

In dogs, the organs mainly affected are bone marrow, the GIT, and spleen. Unlike HES in humans the heart is not a target organ in the dog. In dogs and cats, HES is diagnosed based on evidence of peripheral eosinophilia and eosinophil infiltration into the bone marrow and organs in the absence of any other cause of eosinophilia. In the liver, several multifocal miliary pale lesions are found. In the bone marrow, there is an increase in the proportion of eosinophils; however, the proportion of myeloblasts is less than 2%.

In horses, a condition similar to HES is called MEED. This is the most diffuse eosinophilic disease of horses. It is characterized by the appearance of eosinophilic and lymphoplasmacytic granulomatous lesions in biliary and bronchial epithelium, the GIT, liver, lungs, pancreas, salivary glands, and skin.

Unlike locally infiltrative eosinophilic diseases, animals with HES are poorly responsive to corticosteroids. However, HES in dogs is relatively more benign than is observed in cats, with a survival time of 5 years being reported in a case treated with prednisolone. It is possible that this highly responsive form of HES in dog is related to what has been described as benign form of HES in human patients.

Eosinophilic Neoplasia and Paraneoplastic Syndrome. Eosinophils are often found in association with mast cell tumors as a result of abundant production of IL-3, IL-5, and GM-CSF in the tumor microenvironment. It has also been suggested that the production of IL-5 by neoplastic cells in intestinal lymphoma could lead to hypereosinophilia in cats, dogs, and horses.[33–35] In addition, eosinophilic infiltration is a prominent feature of certain carcinomas, probably due to the production of eosinophil chemoattractants by tumor cells. It is currently unknown whether the presence of eosinophils in any of these tumors is of prognostic importance. There is evidence to suggest that certain cases of HES in the cat may be secondary to systemic mast cell leukemia or lymphosarcoma.

CONCLUSION

Eosinophilia in animals is usually associated with helminth infection and hypersensitivity reactions. However, in cases where there is no evidence of helminths or allergen exposure, eosinophils can infiltrate tissues and lead to organ-specific dysfunction. In other cases, however, eosinophilia can be more systemic, with infiltration into multiple organs causing multisystem dysfunction (HES). Eosinophilia is also a frequent paraneoplastic phenomenon in mast cell tumors of animals, and occasionally with lymphoma, as a result of IL-5 production by neoplastic cells. The presence of clonality and markedly elevated bone marrow myeloblasts are used to distinguish HES from the rare cases of eosinophilic myeloid leukemia. As more cases are reported in the literature, eosinophilic diseases of animals could become important natural disease models for the study of these conditions in humans.

REFERENCES

1. Ostrander EA, Galibert F, Patterson DF. Canine genetics comes of age. *Trends Genet* 2000;**16**:117—24.
2. Rowell JL, McCarthy DO, Alvarez CE. Dog models of naturally occurring cancer. *Trends Mol Med* 2011;**17**:380—8.
3. McEwen BJ. Eosinophils: a review. *Vet Res Commun* 1992;**16**:11—44.
4. Iazbik MC, Couto CG. Morphologic characterization of specific granules in Greyhound eosinophils. *Vet Clin Pathol* 2005;**34**:140—3.
5. Casal AB, Freire F, Bautista-Harris G, Arencibia A, Oros J. Ultrastructural characteristics of blood cells of juvenile loggerhead sea turtles (Caretta caretta). *Anat Histol Embryol* 2007;**36**:332—5.
6. Lilliehook I, Tvedten H. Investigation of hypereosinophilia and potential treatments. *Vet Clin North Am Small Anim Pract* 2003;**33**:1359—78. viii.
7. Lilliehook I, Gunnarsson L, Zakrisson G, Tvedten H. Diseases associated with pronounced eosinophilia: a study of 105 dogs in Sweden. *J Small Anim Pract* 2000;**41**:248—53.
8. Matsumoto K, Matsushita N, Tomozawa H, Tagawa Y. Hematological characteristics of rats spontaneously developing eosinophilia. *Exp Anim* 2000;**49**:211—5.
9. Muto S, Hayashi M, Matsushita N, Momose Y, Shibata N, Umemura T, et al. Systemic and eosinophilic lesions in rats with spontaneous eosinophilia (mes rats). *Vet Pathol* 2001;**38**:346—50.
10. Mori M, Li G, Hashimoto M, Nishio A, Tomozawa H, Suzuki N, et al. Pivotal Advance: Eosinophilia in the MES rat strain is caused by a loss-of-function mutation in the gene for cytochrome b(-245), alpha polypeptide (Cyba). *J Leukoc Biol* 2009;**86**:473—8.
11. Hoenerhoff MJ, Starost MF, Ward JM. Eosinophilic crystalline pneumonia as a major cause of death in 129S4/SvJae mice. *Vet Pathol* 2006;**43**:682—8.
12. Windsor RC, Sturges BK, Vernau KM, Vernau W. Cerebrospinal fluid eosinophilia in dogs. *J Vet Intern Med* 2009;**23**:275—81.
13. Heidemann SM, Fiore M, Sood S, Ham S. Eosinophil activation in the cerebrospinal fluid of children with shunt obstruction. *Pediatr Neurosurg* 2010;**46**:255—8.
14. Lo Re 3rd V, Gluckman SJ. Eosinophilic meningitis. *Am J Med* 2003;**114**:217—23.
15. Mazzei MJ, Bissett SA, Murphy KM, Hunter S, Neel JA. Eosinophilic esophagitis in a dog. *J Am Vet Med Assoc* 2009;**235**:61—5.
16. German AJ, Hall EJ, Day MJ. Analysis of leucocyte subsets in the canine intestine. *J Comp Pathol* 1999;**120**:129—45.
17. Waly N, Gruffydd-Jones TJ, Stokes CR, Day MJ. The distribution of leucocyte subsets in the small intestine of healthy cats. *J Comp Pathol* 2001;**124**:172—82.
18. Packer M, Patterson-Kane JC, Smith KC, Durham AE. Quantification of immune cell populations in the lamina propria of equine jejunal biopsy specimens. *J Comp Pathol* 2005;**132**:90—5.
19. Sihvo HK, Simola OT, Vainionpaa MH, Syrja PE. Pathology in practice. Severe chronic multifocal intramural fibrosing and eosinophilic enteritis, with occasional intralesional bacteria, consistent with feline gastrointestinal eosinophilic sclerosing fibroplasia. (FIESF). *J Am Vet Med Assoc* 2011;**238**:585—7.
20. Craig LE, Hardam EE, Hertzke DM, Flatland B, Rohrbach BW, Moore RR. Feline gastrointestinal eosinophilic sclerosing fibroplasia. *Vet Pathol* 2009;**46**:63—70.
21. Lyles SE, Panciera DL, Saunders GK, Leib MS. Idiopathic eosinophilic masses of the gastrointestinal tract in dogs. *J Vet Intern Med* 2009;**23**:818—23.
22. Schumacher J, Edwards JF, Cohen ND. Chronic idiopathic inflammatory bowel diseases of the horse. *J Vet Intern Med* 2000;**14**:258—65.
23. Kleinschmidt S, Meneses F, Nolte I, Hewicker-Trautwein M. Characterization of mast cell numbers and subtypes in biopsies from the gastrointestinal tract of dogs with lymphocytic-plasmacytic or eosinophilic gastroenterocolitis. *Vet Immunol Immunopathol* 2007;**120**:80—92.
24. Craven M, Simpson JW, Ridyard AE, Chandler ML. Canine inflammatory bowel disease: retrospective analysis of diagnosis and outcome in 80 cases (1995—2002). *J Small Anim Pract* 2004;**45**:336—42.
25. Archer DC, Barrie Edwards G, Kelly DF, French NP, Proudman CJ. Obstruction of equine small intestine associated with focal idiopathic eosinophilic enteritis: an emerging disease? *Vet J* 2006;**171**:504—12.
26. Swain JM, Licka T, Rhind SM, Hudson NP. Multifocal eosinophilic enteritis associated with a small intestinal obstruction in a standardbred horse. *Vet Rec* 2003;**152**:648—51.
27. Bedenice D, Mazan MR, Hoffman AM. Association between cough and cytology of bronchoalveolar lavage fluid and pulmonary function in horses diagnosed with inflammatory airway disease. *J Vet Intern Med* 2008;**22**:1022—8.
28. Couetil LL, Rosenthal FS, DeNicola DB, Chilcoat CD. Clinical signs, evaluation of bronchoalveolar lavage fluid, and assessment of pulmonary function in horses with inflammatory respiratory disease. *Am J Vet Res* 2001;**62**:538—46.

29. Venema CM, Patterson CC. Feline asthma: what's new and where might clinical practice be heading? *J Feline Med Surg* 2010;**12**:681−92.

30. Clercx C, Peeters D. Canine eosinophilic bronchopneumopathy. *Vet Clin North Am Small Anim Pract* 2007;**37**:917−35. vi.

31. Clercx C, Peeters D, German AJ, Khelil Y, McEntee K, Vanderplasschen A, et al. An immunologic investigation of canine eosinophilic bronchopneumopathy. *J Vet Intern Med* 2002;**16**:229−37.

32. Peeters D, Day MJ, Clercx C. Distribution of leucocyte subsets in bronchial mucosa from dogs with eosinophilic bronchopneumopathy. *J Comp Pathol* 2005;**133**:128−35.

33. Barrs VR, Beatty JA, McCandlish IA, Kipar A. Hypereosinophilic paraneoplastic syndrome in a cat with intestinal T cell lymphosarcoma. *J Small Anim Pract* 2002;**43**:401−5.

34. Marchetti V, Benetti C, Citi S, Taccini V. Paraneoplastic hypereosinophilia in a dog with intestinal T-cell lymphoma. *Vet Clin Pathol* 2005;**34**:259−63.

35. La Perle KM, Piercy RJ, Long JF, Blomme EA. Multisystemic, eosinophilic, epitheliotropic disease with intestinal lymphosarcoma in a horse. *Vet Pathol* 1998;**35**:144−6.

Antieosinophil Therapeutics

Introduction

Alex Straumann

OVERVIEW

Eosinophil-targeted therapies exhibit particular features and difficulties, and before a proper treatment strategy can be established, several eosinophil- and disease-inherent characteristics must be taken into account.

Therapeutic Principles

At the outset, one must bear in mind that the pattern of eosinophilic inflammation involves, in addition to the eosinophils themselves, several other inflammatory cells, including basophils, mast cells, neutrophils, and various subsets of T cells.[1,2] Despite their prominent appearance, the precise contribution of eosinophils in generating symptoms and organ dysfunction has not yet been clearly and definitively elucidated. It is only in recent years that direct and selective clearance of eosinophils has become possible, and this has, in turn, led to an improved understanding of antieosinophil therapies and inspired new avenues for treatment. Upon this, a question arises: whether it is better to have a treatment with a high eosinophil specificity that leads to an isolated clearance of eosinophils, or a broader targeted therapy that results in the reduction of several involved inflammatory cells.

Though the question posed above cannot yet be answered for eosinophil-associated diseases in general or any one particular affliction, we present here three different antieosinophil approaches, classified according to their eosinophil specificity: some of the drugs discussed are well established (e.g., corticosteroids), but most are still in the experimental or early clinical use phase.

Firstly, a *nonspecific anti-inflammatory* approach is achieved with corticosteroids. This class of drugs interferes with multiple cellular and humoral sites in the inflammatory network. For decades, corticosteroids have been used as a first-line treatment for systemic eosinophilic diseases, such as hypereosinophilic syndromes,[3] as well as for localized maladies, including asthma,[4] atopic dermatitis,[5] or eosinophilic esophagitis.[6] Of note, almost all eosinophilic diseases are chronic inflammations and, as such, require long-term treatment. The main drawback of corticosteroids is that their prolonged use is limited by side effects. Two strategic approaches have been undertaken to manage this limitation: drugs evoking minimal systemic effects, e.g., budesonide, ciclesonide, and fluticasone, have been developed; and new formulations allowing a topical application, e.g., aerosols, enemas, and ointments, have been designed. These newer drugs and formulations are in some cases already well established for treating eosinophilic diseases (e.g., of the upper and lower airways and of the skin). In contrast, only initial results—albeit promising—are available for the treatment of eosinophilic inflammations of the digestive tract,[7,8] and further evaluation(s) are needed. In either case, options are suitable only for treating localized eosinophilic diseases, whereas the therapy of hypereosinophilic syndromes still requires corticosteroids with systemic efficacy.

A second approach addresses the fact that eosinophilic inflammations show mainly T-helper type 2 (T_h2)-type characteristics.[1,2] Focusing on the T_h2 inflammatory pathway is therefore a more selective therapeutic approach. Recently, a common receptor expressed on the surface of T_h2 cells such as basophils, eosinophils, and T_h2 lymphocytes, has been discovered.[9] The prostaglandin D2 receptor 2 (or so-called chemoattractant receptor-homologous model expressed on T_h2 cells; CRTH2) is a G protein-coupled receptor and mediates activation and recruitment of basophils, eosinophils, and T_h2 lymphocytes in response to prostaglandin D2.[9] It has been shown that a blockade of the CRTH2 receptor by an antagonist prevents the recruitment of several cellular and humoral key players in the T_h2 inflammatory network. This approach could therefore be considered as a *T_h2-specific therapeutic* principle. OC000459 is one example of a selective CRTH2 antagonist that blocks the ability of prostaglandin D2 to cause chemotaxis in addition to preventing activation of T_h2 lymphocytes and eosinophils. Several proof-of-concept trials using this compound to treat asthma, allergic rhinitis, and eosinophilic esophagitis are now under way. Initial results are expected soon, and should answer the question of whether or not this approach

Eosinophils in Health and Disease. http://dx.doi.org/10.1016/B978-0-12-394385-9.00015-8

is actually successful in the treatment of eosinophilic inflammations.

The third line of attack acknowledges that a highly *eosinophil-specific approach* would be desirable, but this strategy encompasses a further challenge: mature eosinophils originate from hematopoietic stem cells but do not display a cell-specific marker on their surface. As a consequence, no eosinophil-specific target exists and potential antieosinophil drugs must act indirectly, either by impairing their development or by accelerating the apoptosis of this myeloid cell. Interleukin-5 (IL-5) is a promising target for limiting eosinophil numbers, insofar as this cytokine interferes with several critical functions in eosinophil hematopoiesis, trafficking, and tissue accumulation.[1] Two humanized monoclonal antibodies—mepolizumab and reslizumab—both with the ability to specifically bind and inactivate human IL-5, have been developed. Another approach targeting IL-5 is the blockade of the corresponding receptor with an antagonist. In this approach, anti-IL-5 and anti-IL-5 receptor (IL-5R) drugs have been evaluated in several eosinophilic diseases, including asthma,[10] eosinophilic asthma,[11] eosinophilic dermatitis,[12] eosinophilic esophagitis,[13] and hypereosinophilic syndromes.[14] An alternative method suggested to achieve specialized cell depletion is to increase eosinophil apoptosis. Here, agonistic antibodies or artificial ligands that cross-link sialic acid-binding Ig-like lectin 8 (Siglec-8), which is expressed on the surface of eosinophils, lead to a selective induction of eosinophil death.[15] This recently recognized proapoptotic effect is mediated by caspase 3 activity (summarized in Table 15.1.1).[16]

Disease-Inherent Principles

Finally, one must keep in mind that several of the eosinophilic diseases are classified as *orphan* diseases. *Orphan* diseases are defined as disorders having a prevalence of less than 5 (in Europe) of less than 7.5 (in the USA) affected individuals per 10,000 inhabitants. Orphan diseases, e.g., hypereosinophilic syndromes, offer few financial incentives for pharmaceutical companies and fundraising for clinical trials is therefore difficult. The data on which therapeutic decisions must be made thus tend to be, in general, weak and limited to case reports and case series. In contrast, *nonorphan* diseases, e.g., asthma, have the potential for producing *blockbuster* drugs and pharmaceutical companies are eager to perform therapeutic studies on these. Higher standards are required to perform clinical trials and studies typically have a placebo-controlled, double-blind design. As a consequence, the resulting data are much more robust. In the following subchapters, when drugs and therapeutic options are discussed, one must keep in mind that the level of evidence is much lower for orphan than for nonorphan diseases.

TABLE 15.1.1 Treatment Approaches Based on Degree of Specificity

Approach/ Mechanism of Action	Advantage	Disadvantage
Nonspecific Approach/Interferes with Multiple Cellular and Humoral Sites		
Corticosteroids (e.g., prednisone, prednisolone, methylprednisolone)	Effective, well-known drug	Significant SE with long-term use
Topical corticosteroid formulations (e.g., budesonide, fluticasone, ciclesonide)	Little influence on the hypothalamic-pituitary-adrenal axis, no tapering before discontinuation	Not feasible for systemic disorders
T$_h$2-Specific Approach/Hinders Cellular Recruitment by Blocking CRTH2 Receptor		
Selective CRTH2 agents block chemotaxis ⇒ prevents eosinophil, basophil and T$_h$2 lymphocyte activation (e.g., OC000459)	So far minimal SE	Efficacy still open
Eosinophil-Specific Approach/Indirectly, Impairs Eosinophil Generation (a,b), or Accelerates Apoptosis (c)		
(a) Cytokine IL-5: disrupts hematopoiesis, eosinophil trafficking and tissue accumulation (e.g., mepolizumab)	Highly specific, minimal SE	Clinical efficacy questionable
(b) Anti-IL-5 and anti-IL-5R: blocks corresponding receptor with an antagonist	Minimal SE	Limited clinical experience
(c) Selective induction of apoptosis: e.g., cross-linking of Siglec-8 with agonistic antibodies or artificial ligands, mediated by caspase 3 activity	Promising approach	Still experimental

SE, side effects; T$_h$2, T-helper type 2.

In the following subchapters, several strategies for treating eosinophilic afflictions will be presented in detail. However, despite these varied therapeutic options and the enormous efforts already made, the question posed above remains relevant: is there a best approach to treating eosinophilic disorders, and if so, which one? Nonetheless, progress is being made and the pieces to the eosinophilic disease puzzle are now starting to shape a more complete picture. New treatment options discussed in this chapter will offer tangible hope for patients in the near future.

REFERENCES

1. Gleich GJ. Mechanisms of eosinophil-associated inflammation. *Journal of Allergy and Clinical Immunology* 2000;**105**:651—63.

2. Straumann A, Bauer M, Fischer B, Blaser K, Simon HU. Idiopathic eosinophilic esophagitis is associated with a T(H)2-type allergic inflammatory response. *Journal of Allergy and Clinical Immunology* 2001;**108**:954—61.

3. Klion AD, Bochner BS, Gleich GJ, Nutman TB, Rothenberg ME, Simon HU, et al, The Hypereosinophilic Syndromes Working Group. Approaches to the treatment of hypereosinophilic syndromes: a workshop summary report. *Journal of Allergy and Clinical Immunology* 2009;**117**:1292—302.

4. McFadden Jr ER, Gilbert IA. Asthma. *The New England Journal of Medicine* 1992;**327**:1928—37.

5. Bieber T. Atopic dermatitis. *The New England Journal of Medicine* 2008;**358**:1483—94.

6. Furuta GT, Liacouras CA, Collins MH, Gupta SK, Justinich C, Putnam PE, et al. Eosinophilic esophagitis in children and adults: a systematic review and consensus recommendations for diagnosis and treatment. *Gastroenterology* 2007;**133**:1342—63.

7. Aceves SS, Newbury RO, Chen D, Mueller J, Dohil R, Hoffmann H, et al. Resolution of remodeling in eosinophilic esophagitis correlates with epithelial response to topical corticosteroids. *Allergy* 2010;**65**:109—16.

8. Straumann A, Conus S, Degen L, Felder S, Kummer M, Engel H, et al. Budesonide is effective in adolescent and adult patients with active eosinophilic esophagitis. *Gastroenterology* 2010;**139**(5):1526—37.

9. Pettipher R, Hansel TT, Armer R. Antagonism of the prostaglandin D$_2$ receptors DP$_1$ and CRTH2 as an approach to treat allergic diseases. *Nature Reviews Drug Discovery* 2007;**6**:313—25.

10. Leckie MJ, ten Brinke A, Khan J, Diamant Z, O'Connor BJ, Walls CM, et al. Effects of an interleukin-5 blocking monoclonal antibody on eosinophils, airway hyper-responsiveness and the late asthmatic response. *Lancet* 2000;**356**:2144—8.

11. Haldar P, Brightling CE, Hargadon B, Gupta S, Monteiro W, Sousa A, et al. Mepolizumab and exacerbations of refractory eosinophilic asthma. *The New England Journal of Medicine* 2009;**360**:973—84.

12. Plötz SG, Simon HU, Darsow U, Simon D, Vassina E, Yousefi S, et al. Use of an anti-interleukin-5 antibody in the hypereosinophilic syndrome with eosinophilic dermatitis. *The New England Journal of Medicine* 2003;**349**:2334—9.

13. Straumann A, Conus S, Grzonka P, Kita H, Kephart G, Bussmann C, et al. Anti-interleukin-5 antibody treatment (mepolizumab) in active eosinophilic esophagitis: a randomised, placebo-controlled, double-blind trial. *GUT* 2010;**59**:21—30.

14. Rothenberg ME, Klion AD, Roufosse FE, Kahn JE, Weller PF, Simon HU, et al., and the Mepolizumab HES Study Group. Treatment of patients with the hypereosinophilic syndrome with mepolizumab. *The New England Journal of Medicine* 2008;**358**:1215—28.

15. Von Gunten S, Vogel M, Schaub A, Stadler BM, Miescher S, Crocker PR, et al. Intravenous immunoglobulin preparations contain anti-Siglec-8 autoantibodies. *Journal of Allergy and Clinical Immunology* 2007;**119**:1005—11.

16. Nutku E, Aizawa H, Hudson SA, Bochner BS. Ligation of Siglec-8: a selective mechanism for induction of human eosinophil apoptosis. *Blood* 2003;**101**:5014—20.

Chapter 15.2

Insights into the Pathogenesis of Asthma and Other Eosinophil-Mediated Diseases from Antagonists of Interleukin-5 and its Receptor

Pranab Haldar, Ian D. Pavord and Andrew J. Wardlaw

INTRODUCTION

It has long been known that peripheral blood and tissue eosinophilia can occur without any increase in the numbers of other leucocytes, for example in the context of infection with helminthic parasites. This suggested the existence of an eosinophil-specific growth factor in addition to nonspecific eosinophil growth factors such as interleukin-3 (IL-3) and granulocyte-macrophage colony-stimulating factor (GM-CSF). Two research groups identified this factor as IL-5, which in mice, but not humans, is also a B-cell growth factor.[1,2] Subsequent work has shown that although there is a low level of eosinophil production in the absence of IL-5, a nonleukemic peripheral blood eosinophilia is almost invariably the result of increased IL-5 production. As a result, IL-5 has been an obvious target for the treatment of diseases in which eosinophils are thought to play a pathogenic role, in particular hypereosinophilic diseases and asthma. IL-5 is secreted as a head-to-tail, intertwined, homodimeric protein that binds to a two-chain receptor consisting of a β chain, which is common to the GM-CSF and IL-3 receptor, and an α chain, which is specific to IL-5.[3] The receptor is expressed on eosinophils and to a lesser extent on basophils and a population of bone marrow precursors.[4] To date, two approaches taken to antagonize IL-5 have reached the clinic, both involving monoclonal antibodies: an antibody against IL-5 that prevents it binding to its receptor and an antibody to the α chain of the receptor, which prevents binding of IL-5 and also induces eosinophil cytotoxicity. Two humanized monoclonal antibodies against IL-5 have been developed for clinical trials: mepolizumab (GlaxoSmithKline), a fully humanized immunoglobulin G1 (IgG1) antibody, and SCH55700, which was originally developed by Schering-Plough, but is currently being taken forward by Cephalon as reslizumab. It is based on 39D10, a rat IgG2a antibody against human IL-5, and incorporates antigen recognition sites for human IL-5 in consensus human IgG4κ constant regions.[5,6] One monoclonal antibody against the receptor has been taken into patients, i.e., benralizumab (MEDI-563) from MedImmune.[7] The development of IL-5 antagonists has been important for two reasons: partly because of the possibility of new therapeutic agents to treat potentially life-threatening diseases, but also because of the insight they give into the role of eosinophils in disease. Their development is also an object lesson in the pitfalls that await companies trying to bring biological agents into the clinic. It is therefore important to briefly consider how eosinophils cause disease and the implications of this for the study design of clinical trials targeting eosinophils, in order to properly interpret the findings of these studies. This is particularly true in asthma, where much of the efforts of antieosinophil drugs have been focused and where the early studies of IL-5 antagonists did not meet expectations. Most of our knowledge about IL-5 antagonists has come from studies with mepolizumab and this subchapter will particularly focus on discussions of clinical trials involving this drug.

ROLE OF EOSINOPHILS IN ASTHMA AND OTHER EOSINOPHILIC DISEASES

Although there has been a longstanding debate about the extent to which eosinophils contribute to the pathogenesis of disease, evidence from hypereosinophilic syndrome (HES) strongly suggests that they can directly cause tissue damage in some circumstances. Hypereosinophilic diseases are heterogeneous with varying pathogenesis and severity, but there are common clinical features, such as endomyo-cardial (and other tissue) fibrosis, neurological and vascular damage, and the propensity to affect the lung and skin that together support the idea that eosinophils are directly responsible for disease rather than simply biomarkers of tissue damage.[8] This is not to say that the presence of eosinophils alone will necessarily lead to disease, as there are a number of conditions where there is marked peripheral blood eosinophilia but no evidence of significant tissue

damage, such as chronic infection with *Strongyloides* spp. and benign HES.[9] Clearly there are cofactors that come into play that determine disease severity. Most eosinophilic diseases generally follow a relapsing and remitting course characterized by exacerbations that can be life threatening. Good control of hypereosinophilic disease, including very eosinophilic forms of asthma, can usually be achieved with continuous oral steroids. Studies of patients with non-leukemic HES generally need to have a steroid-reducing design, but as exacerbations are unpredictable and often infrequent they also need to be long term, with reasonably large numbers of patients, so they are difficult to power and expensive. Clinicians and patients may also be unwilling to risk deterioration in control that can make recruitment difficult, especially in what are relatively rare conditions. It is also difficult to choose the correct outcome measure, as with HES each patient tends to have an individual pattern of organ damage. Using eosinophils as a biomarker of response is problematic where the primary effect of the drug (as is the case with all the antieosinophil therapies to date) is to reduce eosinophil numbers. Regulatory agencies want to see clear evidence of prevention of tissue damage for what are likely to be very expensive therapies. The challenges in designing studies to test the benefits in asthma are related more to the problems of developing drugs that block very specific pathways. A widely accepted, although oversimplistic, paradigm for the pathogenesis of asthma that dominated thinking in the 1990s was that allergen-stimulated T-helper type 2 (T_h2) cells recruited eosinophils into the lung and that these caused asthma.[10] This theory did not address the heterogeneity of asthma, particularly in its more severe manifestations, and the complex relationship between the inflammatory component of the disease and the airway dysfunction that defines it. It is increasingly clear that although inflammation and airway dysfunction are interlinked they can occur independently, particularly when considering eosinophilic inflammation.[11,12] Thus, an airway eosinophilia is neither necessary nor sufficient to cause variable airflow obstruction or increased airway hyperresponsiveness [AHR; caused by abnormal behavior of the airway smooth muscle (ASM)], which are the defining features of asthma and the pathophysiological abnormalities mainly responsible for day-to-day symptoms.[13-16] It is perhaps not surprising therefore that studies of IL-5 antagonists that used these parameters as primary outcomes, whether in allergen challenge studies or in clinical disease, did not show a benefit. In contrast, there is compelling evidence that eosinophils are closely linked to severe exacerbations of asthma that have a different pathophysiology to day-to-day asthma, being characterized by a progressive increase in airflow obstruction that is bronchodilator-resistant and probably caused in most cases by the airway lumen becoming blocked with inflammatory material and impacted mucous.[17,18] This is supported by the close link between asthma deaths that are the end result of severe exacerbations and florid eosinophilic inflammation.[19] It stands to reason, therefore, that if eosinophils are causative in asthma then severe exacerbations are the best place to look for a benefit of antieosinophil drugs. Associated with this is the importance of patient selection. Up to 50% of people with asthma do not have a significant airway eosinophilia at a given time point and are therefore unlikely to respond to an antieosinophil agent.[20,21,22] Although this figure is confounded by the effects of treatment (notably inhaled corticosteroids) and sputum eosinophil counts may underestimate the eosinophil load,[23] studies in steroid naive, noneosinophilic asthma have demonstrated this to be a stable phenotype that is clinically and physiologically indistinguishable from eosinophilic asthma.[24] In what is so far the largest clinical study of anti-IL-5, only 50% of participants had evidence of eosinophilic inflammation (with the caveat that only a minority had a sputum differential undertaken).[25] In addition, many patients with eosinophilic airway disease who have recurrent exacerbations as the major expression of their disease may not have marked features of variable airflow obstruction and are therefore excluded from clinical trials of asthma with a standard design. These problems are aggravated by the importance the regulatory agencies attach to forced expiratory volume 1 (FEV_1) as an outcome marker in studies of asthma, together with an apparent unwillingness to recognize exacerbations as a useful outcome measure. In addition, these agencies and to an extent the companies themselves are reluctant to restrict the use of a drug to a subgroup of patients, even when there is a clear rationale and biomarker for their identification. A combination of inappropriate study design and patient selection, together with an oversimplistic approach to the understanding of pathogenesis, has therefore greatly hampered the introduction of anti-IL-5 drugs and remains a significant barrier to their development.[26]

STUDIES OF MEPOLIZUMAB IN NONHUMAN PRIMATES

Experience with mepolizumab in animal models is limited by the species specificity of IL-5. In cynomolgus monkeys, the protein sequence of the cytokine differs from human IL-5 by two amino acids and this does not appear to affect mepolizumab efficacy. Reported outcomes of *in vivo* studies with this animal model closely resemble the experience in humans. Thus, there was profound depletion of peripheral blood and lavage eosinophils that remained suppressed after sequential antigen challenge and was dose dependent. Doses used ranged from 0.5 mg/kg to 50 mg/kg and none were associated with toxicity.[6] High-dose mepolizumab failed to abolish circulating eosinophils.

The drug had no significant effect on low basal counts of circulating eosinophils, possibly because of IL-3 or GM-CSF responsiveness in these cells. The blood eosinophil count remained significantly suppressed for 74 days after two-dose studies of mepolizumab. The biological half-life of the drug was therefore considerably longer than the pharmacological half-life (13 ± 2 days). No significant effect was seen in tissue eosinophil counts of either the lung or small intestine with high dose treatment. This suggested that peripheral blood eosinophil counts are a poor marker of the tissue response with mepolizumab. Mepolizumab concentrations were found to be decreased by 99.8% in lavage fluid relative to the bloodstream. This suggested poor tissue penetration, with the drug being retained primarily in circulation. Bioavailability and drug pharmacokinetics with subcutaneous delivery were comparable to intravenous administration. No anti-mepolizumab antibodies were detected after six doses at monthly intervals, suggesting that the drug lacks significant antigenicity and tachyphylaxis may not be a major problem with chronic therapy. Mepolizumab therapy did not alter AHR following allergen challenge in ascaris-sensitized monkeys. This observation was in keeping with the hypothesis that eosinophilic inflammation and airway dysfunction are independent processes that occur in parallel.

STUDIES OF MEPOLIZUMAB IN HUMANS

Immunobiological Effects of Mepolizumab in Human Asthma

Much of the information gathered about the biological effects of mepolizumab in asthma has been drawn from a series of publications that are based on a single, randomized, placebo-controlled trial of three doses of mepolizumab given at monthly intervals to subjects with mild, corticosteroid naive asthma.[27]

Mepolizumab and Eosinophil Counts in Different Tissue Compartments

The effect of mepolizumab on eosinophil numbers in asthma was characterized in a number of different tissue compartments, including the bone marrow, peripheral blood, proximal and distal airways, and within the bronchial submucosa. Within the bone marrow, three doses of mepolizumab achieved a 70% mean reduction in terminally differentiated bone marrow eosinophils, a 37% and 44% reduction in myelocytes and metamyelocytes, respectively, but had no effect on levels of early progenitors ($CD34^+$ $IL-5R^+$) or eosinophil/basophil colony forming units.[28] These results suggest that IL-5 is important in the later stages of eosinophil development in the bone marrow, inversely correlating with surface CD34 expression. Although IL-5R is expressed on early progenitors, the level of expression is low and accompanied by the expression of other hemopoietin receptors (IL-3 and GM-CSF) that are more important at this stage of development.

Studies indicated a gradient of antieosinophil efficacy with mepolizumab on mature eosinophils across different tissue compartments. Eosinophil suppression is most complete in the peripheral blood (>95% in all studies). This suppression is specific and not accompanied by a fall in the numbers of other leucocytes. Following three infusions of mepolizumab at monthly intervals, Flood-Page and coworkers reported progressively less efficacy in the airway (79% reduction of lavage eosinophil counts) and bronchial submucosa (55% reduction).[27] The relative resistance of tissue eosinophils to mepolizumab is likely to be multifactorial. Poor tissue penetration of the drug may be important, as the previously described studies in cynomolgus monkeys demonstrated a >99% lower mepolizumab concentration in lavage fluid than plasma. A second important reason is that migration of eosinophils into tissue is a multistep process that also involves chemoattractants and adhesion receptors. It is likely that IL-5 largely effects tissue localization indirectly by reducing the number of peripheral blood eosinophils available for recruitment.[29,30] What is striking is the marked inhibition by mepolizumab on migration into the airway lumen, despite a significant residual eosinophilia in the lamina propria. This remains unexplained.

Mepolizumab and Airway Structure

Reticular basement membrane (RBM) thickening is closely associated with eosinophilic inflammation and is ameliorated by regular inhaled corticosteroid use. Eosinophils are an important source of transforming growth factor β (TGF-β), a potent regulator of cell proliferation with profibrotic properties that are considered important in tissue repair.[31] It has therefore been hypothesized that persistent eosinophilic inflammation plays an important role in airway remodeling. Despite the modest effects on tissue eosinophilia, Flood-Page and colleagues showed that mepolizumab therapy is associated with a significant reduction in the extracellular matrix glycoproteins tenascin and lumican, together with a reduction in the thickness of the RBM.[32] The clinical significance of these structural changes is uncertain, as there was no associated improvement in either FEV_1 or AHR to histamine in study participants. A reduction in tenascin and TGF-β expression was also seen in patients given mepolizumab for eosinophilic esophagitis.[33] These results support a role for eosinophils in airway remodeling and indicate that the effect of mepolizumab on tissue eosinophils is sufficient to influence structural changes, either through quantitative suppression

of eosinophilic inflammation alone or possibly through additional effects on eosinophil activation.

Other Immunological Effects of Mepolizumab Therapy

Data on the effects of mepolizumab on other aspects of immune function is derived from studies in both asthma and other eosinophilic disorders. However, some discordance exists in the reported outcomes. In a study of patients with moderate asthma (receiving a daily dose of inhaled corticosteroid ≤1000 μg beclomethasone dipropionate equivalent), Buttner and colleagues reported no effect of three doses of mepolizumab on cytokine receptor expression, intracellular cytokine expression, markers of T-cell activation, or noneosinophil leucocyte numbers.[34] In another study of mepolizumab therapy administered to a heterogeneous population of patients with eosinophilic disease, Stein and colleagues reported an increase in the intracellular content of IL-5 in T cells after mepolizumab therapy.[35] A profound fall in peripheral blood eosinophil counts was observed in both studies; this was associated with a parallel fall in measured eosinophil cationic protein (ECP) levels by Buttner. However, no change was observed in the expression of other markers of eosinophil activation (CD11b and CD69), suggesting that the fall in ECP was due to reduced eosinophil numbers alone, with little additional effect of mepolizumab therapy on eosinophil activation. In contrast, Stein and colleagues reported a reduction in eosinophil shape change with eotaxin (C-C motif chemokine 11; CCL11) in vitro, suggesting impaired eosinophil activation after mepolizumab therapy. The authors also found an 18% increase in IL-5 R expression but no change in C-C chemokine receptor 3 (CCR3) expression with treatment. The increase in IL-5 R expression has not been corroborated in another study of mepolizumab in eosinophilic esophagitis. In this double-blind placebo-controlled study, two doses of mepolizumab (750 mg) were administered 7 days apart, followed by two further doses of 1500 mg at 4-weekly intervals if there was evidence of persistent tissue eosinophilia.[33] In addition to the expected fall in blood and tissue eosinophil counts, this study also reported no effect of mepolizumab therapy on the number of T cells and tryptase-positive mast cells in esophageal biopsies. A number of noteworthy points arise from these observations. The absence of an effect with mepolizumab on CCR3 expression is in keeping with the observations of Flood-Page and colleagues and supports the hypothesis that blockade of this receptor is also needed to effectively ameliorate tissue eosinophilia in disease. However, the relative efficacy of mepolizumab may vary with the severity and type of eosinophilic disease. In an open-label study of mepolizumab therapy for eosinophilic esophagitis, Stein and colleagues reported an impressive 89% reduction in tissue eosinophils after three doses.[36] The greater efficacy of mepolizumab observed in this disease may suggest differences in the relative importance of IL-5 and eotaxin for eosinophil trafficking to different organs. In keeping with this, Mishra and colleagues have shown IL-5 to be necessary and sufficient for the development of eosinophilic esophagitis in a mouse model.[37] A risk of rebound eosinophilic inflammation exists following cessation of therapy due to upregulated synthesis of IL-5 by T_h2 cells; upregulated expression of the IL-5R by eosinophils; elevated circulating levels of eotaxin and the theoretical availability of a circulating store of IL-5 in complex with a drug that may have impaired clearance kinetics. Although this has not been reported to date with mepolizumab, one study observed rebound eosinophilic inflammation to suprabasal levels following therapy with another anti-IL-5 agent, reslizumab (SCH55700), in patients with HES.[38] Although variability exists in the precise immunological outcomes after mepolizumab therapy, the evidence in these studies consistently describes immunological specificity. In particular, the absence of effect on T_h2 cells implies that mepolizumab therapy leads to uncoupling of eosinophil function from other processes of the T_h2 pathway. This is pertinent when considering the comparative clinical effects of corticosteroids and mepolizumab.

Studies of Clinical Effectiveness of Mepolizumab in Asthma

Initial Studies

Mepolizumab binds free IL-5 *in vivo* with high affinity, thus preventing interaction of the cytokine with its native cell surface receptor. The drug has a terminal half-life of 20 days following intravenous administration in humans and dosing is generally at 4-weekly intervals.[6] Early studies of mepolizumab therapy in asthma reported no effect of treatment on AHR, the late response to allergen challenge or lung function.[27,39] However, these studies were of short duration and performed in patients with very mild atopic asthma, a group with little scope or need for additional improvement. A pilot study of another anti-IL-5 compound (SCH55700) given to patients with severe asthma and evidence of lung function impairment also reported no significant improvement in FEV_1 with treatment.[5] In 2007, Flood-Page and colleagues performed the largest clinical study of mepolizumab to date in patients with moderate asthma, and with persistent symptoms despite regular inhaled corticosteroid therapy.[25] The study had a double-blind placebo-controlled parallel-group design, with treatment or placebo administered at monthly intervals for 3 months. The primary outcome of this study was a change in morning peak flow, and other outcomes measured included exacerbation events, lung function, quality of life,

and symptoms. The study identified no significant benefit with mepolizumab therapy in any of the measured outcomes. However, a 50% reduction in severe exacerbations was reported and although this did not reach statistical significance, the study was too short and inadequately powered for this endpoint. As noted above, the available evidence suggests that eosinophils do not have a direct causative role in the ASM dysfunction that causes variable airflow obstruction and AHR, whereas they are closely linked to exacerbations.[40] The general interpretation of the studies noted above is that eosinophils are not causing the tissue damage and clinical features of asthma. Indeed the small allergen challenge study by Leckie and coworkers[39] is one of the most highly cited papers in the asthma literature and is highly influential in the development of antieosinophil drugs, yet a proper interpretation of its message was very limited.[41] Two complementary studies have now examined the effect of mepolizumab in patients with refractory eosinophilic asthma using exacerbations as an outcome measure. Haldar and coworkers undertook a single-center clinical trial to evaluate the effect of mepolizumab on severe exacerbation frequency and Nair and colleagues examined the utility of mepolizumab as a steroid-sparing agent in patients with prednisolone-dependent eosinophilic asthma.[42,43] The two studies are discussed separately and their conclusions presented together.

Mepolizumab and Exacerbation Frequency in Refractory Eosinophilic Asthma

This was a single-center, investigator-led, randomized, double-blind, placebo-controlled, parallel-group study, undertaken using 61 patients with refractory eosinophilic asthma. The study hypotheses were:

1. Eosinophils are important effector cells in the pathogenesis of severe exacerbations in patients with refractory eosinophilic asthma and a history of recurrent severe exacerbations.
2. The inhibition of eosinophilic inflammation by mepolizumab would be associated with a reduction in severe exacerbation frequency.

Participants were well characterized and met American Thoracic Society criteria for refractory asthma; they had documented evidence of sputum eosinophilia >3% on at least one occasion in the previous 2 years and a history of at least two severe exacerbations requiring oral corticosteroid therapy in the previous 12 months. Mepolizumab or placebo was administered intravenously at 4-weekly intervals for 12 months. The primary outcome was severe exacerbation frequency; secondary clinical outcomes included change from baseline in asthma symptoms, asthma quality of life, AHR, and postbronchodilator FEV_1.

A gradient of antieosinophil efficacy with mepolizumab therapy was seen within different tissue compartments, consistent with earlier studies. The drug was most effective at suppressing eosinophil numbers in the blood (84% reduction) and airway lumen (74% reduction in sputum) and least effective in bronchial tissues (52% reduction). Of the clinical outcomes measured, mepolizumab therapy was associated with a 43% reduction in the rate of severe exacerbations ($p = 0.02$) and a small but clinically significant improvement in asthma quality of life (within group change of 0.55 points, $p = 0.02$). The reduction in exacerbations paralleled the reduction in sputum eosinophils, which was not seen in all subjects given mepolizumab. In keeping with earlier studies, no difference was observed between study groups in AHR, asthma symptoms, or lung function. Together, the results provide further evidence to support the view that eosinophilic inflammation is important in the pathogenesis of severe exacerbations, but not associated with clinical measures of symptoms or pulmonary physiology.

The clinical profile of severe exacerbations was similar in the study groups. Symptoms scores and measures of lung function, including postbronchodilator FEV_1 and peak flow recorded at the time of presentation, did not differ between the groups and there was no difference in the duration of prednisolone therapy needed for resolution. These data suggest that while effective in the prevention of exacerbations, mepolizumab therapy does not influence the clinical severity of events when they occur. In keeping with this, although sputum eosinophilia was significantly less prevalent at the time of exacerbation for participants receiving mepolizumab (59% vs. 35%, $p = 0.04$) the pattern of airway inflammation was not associated with any clinical measures of severity. From a health economic perspective, mepolizumab therapy was associated with a significant reduction in the cumulative dose of prednisolone and admission days to hospital over the period of the study.

A number of studies have demonstrated an association between eosinophilic airway inflammation and the clinical response to corticosteroid therapy in patients with airways diseases including asthma.[20,44,45] The generic nature of this association across disease boundaries suggests that the clinical response to corticosteroid therapy is not disease specific, but rather a function of antieosinophil activity, or at least closely related to this. In this study, all participants received a corticosteroid trial of prednisolone for 2 weeks at the beginning and end of the treatment phase. Interestingly, the authors found no difference in the clinical response to prednisolone, measured as change in lung function and symptom scores either between or within the study groups. Thus, the benefits of corticosteroid therapy are probably only partly due to a direct effect on eosinophils.

Mepolizumab in Prednisolone-Dependent Eosinophilic Asthma

The study by Nair and colleagues evaluated mepolizumab as a steroid-sparing therapy for eosinophilic patients requiring maintenance prednisolone to retain asthma control. As a group, these patients have the most severe asthma and suffer considerable additional long-term drug-induced iatrogenic morbidity. The study was a randomized double-blind placebo-controlled parallel-group design, enrolling 20 participants to receive five doses of intravenous mepolizumab or placebo at monthly intervals. Primary outcomes were the proportion of participants having exacerbations and the reduction in prednisolone dose achieved, expressed as a percentage of the protocol-defined projected target, in each group. Prednisolone tapering continued according to a prespecified protocol until the end of the study or until the participant had an eosinophilic exacerbation. Neutrophilic exacerbations were permitted and did not influence prednisolone withdrawal. The authors reported significantly fewer exacerbations in the group receiving mepolizumab (one event vs. 12 events in the placebo group; $p = 0.002$) and this was accompanied by a significantly greater reduction in prednisolone dose ($83.8 \pm 33.4\%$ vs. $47.7 \pm 40.5\%$ of maximum target-dose reduction). However, the final prednisolone dose did not differ significantly between the study groups.

Discussion

The two studies described above share similarities and some important differences. Both studies enrolled patients with eosinophilic asthma and poor control in the form of frequent exacerbations. Patients in the Nair study had disease of greater severity, as all required maintenance prednisolone, compared with 55% in the study population of Haldar and colleagues. Both studies reported a significant reduction in severe exacerbations with mepolizumab therapy. However, an important difference exists in the type of exacerbations described. In Haldar's study, no alteration of regular medication was permitted during the treatment phase. Exacerbations were therefore spontaneous events and were probably triggered by external factors. In contrast, exacerbations in the Nair study were primarily triggered by progressive withdrawal of glucocorticoid therapy, i.e., an induced loss of asthma control. Thus, while Haldar and coworkers have shown that add-on mepolizumab therapy effectively lowers the exacerbation risk for eosinophilic patients with a history of frequent severe exacerbations; Nair and colleagues have shown mepolizumab to be an effective steroid-sparing therapy for retaining asthma control and preventing exacerbations caused by inadequate maintenance treatment. This latter observation further suggests that, mechanistically, the antieosinophilic activity of corticosteroid therapy is important for preventing exacerbations.

A commentary accompanying these papers suggested that the phenotype of asthma responsive to mepolizumab therapy was rare.[46] This is not the case. While biological treatments are always going to be targeted at those people with disease uncontrolled by standard medication, this comprises up to 5% of the adult asthma population. In the study by Haldar and colleagues, about one-third of this group would potentially benefit from anti-IL-5 therapy. An important aspect of the studies using anti-IL-5 antibodies is that it appears to be very safe with relatively few adverse effects. A recent large multi-centre study in asthma utilising a similar study design to the work of Haldar and colleagues has also demonstrated a 50% reduction in severe exacerbations with mepolizumab (reference Mepolizumab for severe eosinophilic asthma (DREAM): a multicentre, double-blind, placebo-controlled trial 53.

Clinical Studies of Efficacy in Asthma with Other Anti-Iinterleukin-5 Agents

Interpretations of the role of eosinophils in asthma from studies of mepolizumab and reslizumab (SCH55700) are complicated by the relatively modest effect on tissue eosinophils. The lack of bronchodilator reversibility in eosinophilic exacerbations and the pathology of asthma deaths both strongly suggest that the primary abnormality leading to this expression of asthma is in the lumen of the airway, which becomes blocked with inspissated mucus and cell debris. Anti-IL-5 drugs are effective at blocking eosinophil migration into the lumen and may therefore be expected to prevent luminal inflammation. Variable airflow obstruction and AHR are due to abnormalities in the ASM and are therefore likely to be related to inflammatory events in the bronchial mucosa. It is therefore possible that treatments that ablate tissue eosinophils may have a greater effect on these more classical features of asthma. An argument against this is that glucocorticoids do reduce AHR and variable airflow obstruction in association with only a modest reduction in tissue eosinophils. However, these drugs may be having effects on eosinophil activation that are not seen with anti-IL-5. A compound that may answer the question of the importance of tissue eosinophils is benralizumab (MEDI-563), which has cytotoxic properties against eosinophils as well as blocking the binding of IL-5. Thus, it completely ablates blood eosinophilia, reduces numbers of eosinophil precursors, and is anticipated to reduce numbers in the bronchial mucosa. So far only a Phase I study has been reported, which was sufficiently encouraging to justify Phase II studies in asthma and chronic obstructive pulmonary disease, although some flu-like symptoms were reported.[7,47] Further support for the findings using mepolizumab comes from a study by Cephalon using the monoclonal antibody, reslizumab. In a Phase II study

reported in abstract form, they recruited 106 moderate eosinophilic asthmatics to a 4-month double-blind placebo-controlled study with improvement in asthma control score (ACQ) as the primary outcome. They confirmed the expected reduction in blood and sputum eosinophils. Improvements in ACQ just failed to reach significance, but there was a significant improvement in FEV_1 in the active group.[48] Improvements were more marked in patients with nasal polyposis. There were insufficient exacerbations to make any comment on this outcome, although a 50% reduction was seen in the active group. Patients with suboptimally controlled eosinophilic asthma do have impaired postbronchodilator FEV_1 and it is possible that the patients in this study at baseline were less well controlled than those in the study by Haldar and coworkers, who all had a course of prednisolone before being randomized.

Clinical studies of mepolizumab in other eosinophilic disorders

Mepolizumab and reslizumab have been used successfully as a therapeutic agent in FIP1L1—PDGFRα-negative idiopathic HES, a heterogeneous condition characterized by moderate to severe peripheral blood eosinophilia and end-organ damage associated with eosinophilic infiltration.[49,50] Both of these studies demonstrated a reduction in the requirement for oral steroids and the rate of exacerbations. However, regulatory approval for the use of mepolizumab in HES was not given, suggesting that the regulatory agencies at least were unconvinced that these studies had demonstrated prevention of tissue damage. This decision appears somewhat perverse to those clinicians looking after HES patients. In addition to HES, an open study on Churg-Strauss syndrome and a case report have shown the effectiveness of mepolizumab.[51,52] However evidence in eosinophil esophagitis is more mixed, perhaps because of the limited tissue response to the eosinophilia.[33,36]

CONCLUSIONS

IL-5 antagonists are a very promising new approach to the treatment of eosinophilic disease. IL-5 antagonists at least appear safe and well tolerated. Efficacy is likely to depend on the extent to which a condition is eosinophilic (i.e., the more eosinophilic, the more likely a successful outcome) and may treat only some aspects of the disease process. It is less likely to be effective where a marked tissue eosinophilia is the key hallmark of the disease, unless more effective agents for reducing the tissue eosinophilia become available. Most of the developments of anti-IL-5 pathway drugs have been in asthma. Asthma is a strikingly heterogeneous disorder and thoughtful consideration needs to be given to the pathological and clinical characteristics of disease most likely to respond to a given therapy. This should be based on the available scientific evidence and necessarily requires appropriate patient characterization or phenotyping, beyond currently accepted norms, prior to enrolment in clinical studies. Too often, pharmaceutical company sponsored clinical studies encourage a nonspecific approach to patient recruitment based on satisfactory fulfillment of accepted criteria for asthma. This practice is favored by governing bodies, partly for upholding the utilitarian philosophy of healthcare. However, as the clinical experience with mepolizumab illustrates, it is inappropriate for the study of highly specific therapies, and special consideration should be given to the target population for future Phase III studies. One consequence of this has been an increasing urgency to better identify and classify phenotypes of asthma that may predict response to therapy.[11] Clinical benefit with mepolizumab and probably other IL-5 antagonists is restricted to patients with eosinophilic asthma that exacerbate frequently and fail to achieve control with inhaled therapies. More generally, it is likely that the next generation of therapies for asthma will see a shift in the model of therapeutic management. A stepwise approach will be retained for mild to moderate disease. However, specific molecular therapies will play an important role in the management of severe asthma and the provision of such therapy will be phenotype specific.

REFERENCES

1. Campbell HD, Tucker WQ, Hort Y, Martinson ME, Mayo G, Clutterbuck EJ, et al. *Proc Natl Acad Sci USA* 1987;**84**:6629–33.
2. Kinashi T, Harada N, Severinson E, Tanabe T, Sideras P, Konishi M, et al. *Nature* 1986;**324**:70–3.
3. Miyajima A, Mui AL, Ogorochi T, Sakamaki K. *Blood* 1993;**82**:1960–74.
4. Yamada T, Sun Q, Zeibecoglou K, Bungre J, North J, Kay AB, et al. *J Allergy Clin Immunol* 1998;**101**:677–82.
5. Kips JC, O'Connor BJ, Langley SJ, Woodcock A, Kerstjens HA, Postma DS, et al. *Am J Respir Crit Care Med* 2003;**167**:1655–9.
6. Hart TK, Cook RM, Zia-Amirhosseini P, Minthorn E, Sellers TS, Maleeff BE, et al. *J Allergy Clin Immunol* 2001;**108**:250–7.
7. Kolbeck R, Kozhich A, Koike M, Peng L, Andersson CK, Damschroder MM, et al. *J Allergy Clin Immunol* 2010;**125**:1344–53. e2.
8. Ogbogu PU, Bochner BS, Butterfield JH, Gleich GJ, Huss-Marp J, Kahn JE, et al. *J Allergy Clin Immunol* 2009;**124**:1319–25. e3.
9. Klion A. *Annu Rev Med* 2009;**60**:293–306.
10. Kay AB. *Chem Immunol* 1998;**71**:178–91.
11. Haldar P, Pavord ID, Shaw DE, Berry MA, Thomas M, Brightling CE, et al. *Am J Respir Crit Care Med* 2008;**178**:218–24.
12. Pavord ID, Wardlaw AJ. *Clin Exp Allergy* 2010;**40**:62–7.
13. Brightling CE, Bradding P, Symon FA, Holgate ST, Wardlaw AJ, Pavord ID. *N Engl J Med* 2002;**346**:1699–705.
14. Crimi E, Spanevello A, Neri M, Ind PW, Rossi GA, Brusasco V. *Am J Respir Crit Care Med* 1998;**157**:4–9.
15. Rosi E, Ronchi MC, Grazzini M, Duranti R, Scano G. *J Allergy Clin Immunol* 1999;**103**:232–7.
16. Turner MO, Hussack P, Sears MR, Dolovich J, Hargreave FE. *Thorax* 1995;**50**:1057–61.
17. Green RH, Brightling CE, McKenna S, Hargadon B, Parker D, Bradding P, et al. *Lancet* 2002;**360**:1715–21.
18. Dougherty RH, Fahy JV. *Clin Exp Allergy* 2009;**39**:193–202.
19. Filley WV, Holley KE, Kephart GM, Gleich GJ. *Lancet* 1982;**2**:11–6.
20. Pavord ID, Brightling CE, Woltmann G, Wardlaw AJ. *Lancet* 1999;**353**:2213–4.
21. Green RH, Brightling CE, Woltmann G, Parker D, Wardlaw AJ, Pavord ID. *Thorax* 2002;**57**:875–9.
22. Gibson PG, Simpson JL, Saltos N. *Chest* 2001;**119**:1329–36.
23. Kulkarni NS, Hollins F, Sutcliffe A, Saunders R, Shah S, Siddiqui S, et al. *J Allergy Clin Immunol* 2010;**126**:61–9. e3.
24. Berry M, Morgan A, Shaw DE, Parker D, Green R, Brightling C, et al. *Thorax* 2007;**62**:1043–9.
25. Flood-Page P, Swenson C, Faiferman I, Matthews J, Williams M, Brannick L, et al. *Am J Respir Crit Care Med* 2007;**176**:1062–71.
26. O'Byrne PM. *Am J Respir Crit Care Med* 2007;**176**:1059–60.
27. Flood-Page PT, Menzies-Gow AN, Kay AB, Robinson DS. *Am J Respir Crit Care Med* 2003;**167**:199–204.
28. Menzies-Gow A, Flood-Page P, Sehmi R, Burman J, Hamid Q, Robinson DS, et al. *J Allergy Clin Immunol* 2003;**111**:714–9.
29. Wardlaw AJ. *J Allergy Clin Immunol* 1999;**104**:917–26.
30. Rosenberg HF, Phipps S, Foster PS. *J Allergy Clin Immunol* 2007;**119**. 1303-10; quiz 1311–2.
31. Wong DT, Elovic A, Matossian K, Nagura N, McBride J, Chou MY, et al. *Blood* 1991;**78**:2702–7.
32. Flood-Page P, Menzies-Gow A, Phipps S, Ying S, Wangoo A, Ludwig MS, et al. *J Clin Invest* 2003;**112**:1029–36.
33. Straumann A, Conus S, Grzonka P, Kita H, Kephart G, Bussmann C, et al. *Gut* 2010;**59**:21–30.
34. Buttner C, Lun A, Splettstoesser T, Kunkel G, Renz H. *Eur Respir J* 2003;**21**:799–803.
35. Stein ML, Villanueva JM, Buckmeier BK, Yamada Y, Filipovich AH, Assa'ad AH, et al. *J Allergy Clin Immunol* 2008;**121**:1473–83, 1483 e1–4.
36. Stein ML, Collins MH, Villanueva JM, Kushner JP, Putnam PE, Buckmeier BK, et al. *J Allergy Clin Immunol* 2006;**118**:1312–9.
37. Mishra A, Hogan SP, Brandt EB, Rothenberg ME. *J Immunol* 2002;**168**:2464–9.
38. Kim YJ, Prussin C, Martin B, Law MA, Haverty TP, Nutman TB, et al. *J Allergy Clin Immunol* 2004;**114**:1449–55.
39. Leckie MJ, ten Brinke A, Khan J, Diamant Z, O'Connor BJ, Walls CM, et al. *Lancet* 2000;**356**:2144–8.
40. Wardlaw AJ, Brightling CE, Green R, Woltmann G, Bradding P, Pavord ID. *Clin Sci (Lond)* 2002;**103**:201–11.
41. O'Byrne PM, Inman MD, Parameswaran K. *J Allergy Clin Immunol* 2001;**108**:503–8. t&artType=abs&id=a119149&target=.
42. Haldar P, Brightling CE, Hargadon B, Gupta S, Monteiro W, Sousa A, et al. *N Engl J Med* 2009;**360**:973–84.
43. Nair P, Pizzichini MM, Kjarsgaard M, Inman MD, Efthimiadis A, Pizzichini E, et al. *N Engl J Med* 2009;**360**:985–93.
44. Brightling CE, Monteiro W, Ward R, Parker D, Morgan MDL, Wardlaw AJ, et al. *Lancet* 2000;**356**:1480–5.
45. Brightling CE, Ward R, Wardlaw AJ, Pavord ID. *Eur Respir J* 2000;**15**:682–6.

46. Wenzel SE. *N Engl J Med* 2009;**360**:1026–8.

47. Busse WW, Katial R, Gossage D, Sari S, Wang B, Kolbeck R, et al. *J Allergy Clin Immunol* 2010;**125**:1237–44. e2.

48. Castro M, Mathur S, Hargreave FE, Xie F, Young J, Wilkins HJ, et al. Poster presented at the 2010 Annual Meeting of the American College of Allergy. *Asthma and Immunology* 2010.

49. Rothenberg ME, Klion AD, Roufosse FE, Kahn JE, Weller PF, Simon HU, et al. *N Engl J Med* 2008;**358**:1215–28.

50. Klion AD, Law MA, Noel P, Kim YJ, Haverty TP, Nutman TB. *Blood* 2004;**103**:2939–41.

51. Kim S, Marigowda G, Oren E, Israel E, Wechsler ME. *J Allergy Clin Immunol* 2010;**125**:1336–43.

52. Kahn JE, Grandpeix-Guyodo C, Marroun I, Catherinot E, Mellot F, Roufosse F, et al. *J Allergy Clin Immunol* 2010;**125**:267–70.

53. Pavord ID, Korn S, Howarth P, Bleecker ER, Buhl R, Keene ON, Ortega H. *Chanez PLancet* 2012 Aug 18;**380**(9842):651–9.

Chapter 15.3

Interleukin-5 Receptor-Directed Strategies

William W. Busse, Nestor A. Molfino and Roland Kolbeck

INTRODUCTION

Although interleukin-5 (IL-5) shares a number of functions with IL-3 and granulocyte-macrophage colony-stimulating factor (GM-CSF) on eosinophils, it acts primarily as an eosinophilopoietic cytokine. For example, IL-5 is the key cytokine in the terminal differentiation of eosinophils,[1] and its expression on progenitor cells is the first indication that bone marrow cells are committed toward the eosinophil lineage. Over the past 20 years, there has been considerable progress into understanding the regulation of eosinophil differentiation and survival as well as gaining insight into the expression of the IL-5 receptor (IL-5R), its function, and the associated signal transduction events that follow cell activation with IL-5. An appreciation of some basic features of the IL-5R helps to illustrate why it may be a valuable target to regulate eosinophils and how antibody treatment directed toward the IL-5R is distinct from efforts that focus on neutralization of IL-5 alone.

Using the promyelocytic leukemia HL-60 cell line, Plaetinck and coworkers[2] were able to identify high-affinity receptors on these cells that were specific for IL-5. Furthermore, by promoting the differentiation of HL-60 cells toward mature eosinophils, these investigators found increased IL-5 binding, thus suggesting that the eosinophil differentiation process, or cell maturation, was associated with enhanced expression of IL-5R. These initial observations were expanded by Chihara and coworkers[3] who used human eosinophils to evaluate whether IL-5Rs were found on primary human cells as well as on a cultured cell line. In their work, Chihara and coworkers[3] were able to show that IL-5 binds to human eosinophils, thus providing evidence for the presence of an IL-5R on these cells. Interestingly, phenotypically distinct eosinophils of the *hypodense* variety, presumably representing an upregulated subpopulation of cells, had increased binding to IL-5 and therefore more receptors. Finally, incubation of eosinophils with GM-CSF enhanced IL-5 binding, implying that activation of eosinophils by GM-CSF increased its surface expression of IL-5 receptors. Collectively, these early observations laid important groundwork in efforts to define the actions of IL-5 on eosinophils, the presence of its receptor, and evidence that the expression of IL-5R appeared to occur during differentiation and activation of eosinophils. As the IL-5R is central to IL-5 effects on eosinophils, this cell surface molecule was also identified as a potential therapeutic target.

Our understanding and appreciation of the IL-5R, and its complexity, has expanded considerably since these early studies. It is now known and appreciated that the biological effects of IL-5 occur through the recruitment of its cell surface receptor, which is composed of two subunits: an α chain and a βc (β common receptor) chain. The α subunit of this receptor is specific to IL-5, whereas βc is a common subunit that is shared with IL-3 and GM-CSF (Fig. 15.3.1).[4] Although binding of IL-5 is specific to the α subunit, this interaction is not sufficient to activate the cell and requires the recruitment of the βc chain to fully engage the cytokine and thereby initiate intracellular signaling and the generation of a physiological effect.[5] Once the α and βc chains of the cytokine receptor are engaged and form a heterodimer, there is cell signaling leading to eventual transcriptional activation of target genes and downstream biological effects (Figure 15.3.2).

The Janus kinase—signal transducer and activator of transcription (JAK—STAT) pathway is used by all cytokines, including IL-5, to transmit extracellular signals to promoters of target genes and is important for cytokine-induced proliferation and differentiation.[5] Tyrosine-protein kinases Lyn and SYK (Syk) are believed to be essential for activation of the antiapoptotic pathway signaled by βc activation.[6] The mitogen-activated protein kinase (MAPK) pathways, which include the extracellular signal-regulated kinase (ERK), regulate a number of key cellular functions, including cell survival, growth, and proliferation.[7]

These data pinpoint the importance of the IL-5R for directing the effect of IL-5 on eosinophils. Mapping studies have also found the IL-5R to be largely limited to eosinophils, indicating a specificity of this receptor to this cell type. Thus, to regulate many eosinophil actions, IL-5 is a—if not the—pivotal cytokine. Given the contributions of IL-5 to eosinophil biology, two approaches can be used to

FIGURE 15.3.1 Working model of receptor recruitment by interleukin-5 leading to receptor activation and signal triggering. Interleukin-5 (IL-5) initially binds to the IL-5 receptor α (IL-5Rα) to form a high-affinity 1:1 complex. Although IL-5 is a homodimeric protein, the stoichiometry of the IL-5—IL-5Rα interaction is 1:1. This IL-5—IL-5Rα complex is thought to bind to a preformed dimer of common receptor β (βc) to form a 1:1:2 intermediate complex. Subsequently, another IL-5—IL-5Rα complex binds to the initial 1:1:2 complex to form a 2:2:2 complex. It is proposed that this complex is further matured by disulfide bond formation between IL-5Rα and βc, and the final complex can induce cytoplasmic signaling through Janus kinase—signal transducer and activator of transcription (JAK—STAT) and mitogen-activated protein kinase (MAPK) pathways in eosinophils. *(Reproduced from[4] with permission.)*

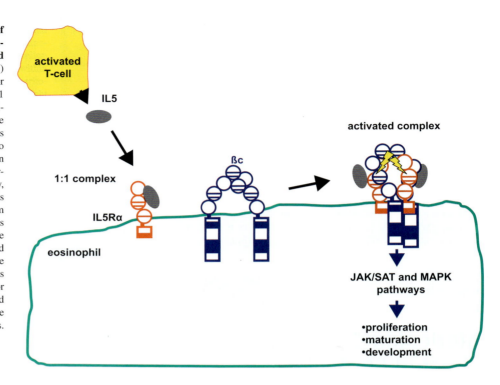

regulate IL-5 influence on eosinophils, with somewhat distinct outcomes and potential effects.

Antibodies to IL-5 have been developed and found to be effective in diminishing the presence of eosinophils in the circulation and airway fluids.[8] The effects of anti-IL-5 antibodies are less effective in diminishing eosinophil numbers when eosinophils are located in tissues, such as bone marrow or respiratory mucosa. Anti-IL-5 antibodies reduce the presence of eosinophils by preventing IL-5 from attaching to the cell, thereby diminishing the survival-enhancing activities of this cytokine. In a sense, anti-IL-5 antibodies diminish the presence of eosinophils by removing the cells' access to their cytokine *lifeline*.

Another approach to block the actions of IL-5 is via antibodies against IL-5R. With this approach, as is discussed below, the eosinophil itself becomes the target of the antibody and the resulting interaction between the IL-5R and its antibody can *kill* this cell (see below for further discussion). Cell death is mediated by two potential mechanisms:

1. The blockade is directed toward IL-5, with the reduction in eosinophils occurring as the consequence of IL-5 no longer being available, and the eosinophil undergoes death by cytokine starvation.
2. In a more direct mechanism, *antieosinophilic* IL-5R antibody mediates killing of the cellular target when the antibody binds to the receptor (Figure 15.3.3).

It is likely that each of these mechanisms will have a different outcome. Anti-IL-5 antibodies regulate the

IL-5 pathway and anti-IL-5R antibodies actively kill eosinophils. Therefore, the selection of approach to target IL-5-dependent processes will depend on which target is of greatest interest and importance: IL-5 or the eosinophil. In addition, anti-IL-5R-mediated cell killing also targets basophils, which are expected to be only marginally affected by targeting IL-5.

BENRALIZUMAB (ANTI-INTERLEUKIN-5 RECEPTOR α)

In Vitro Pharmacology

Benralizumab, formerly known as MEDI-563, is a humanized monoclonal antibody (mAb; IgG1κ) that was generated using conventional hybridoma technology.[9] Benralizumab binds to human IL-5Rα with high affinity and inhibits IL-5-dependent cell proliferation.[10] Benralizumab is produced in Chinese hamster ovary cells deficient in the enzyme α-1,6-fucosyltransferase (FUT8; Potelligent, BioWa, Inc., Princeton, NJ)[11] and as a result is not fucosylated on the biantennary oligosaccharide core. Fucose deficiency (A-Fuc) results in sixfold-enhanced binding affinity to low affinity immunoglobulin γ Fc region receptor III A (FcγRIIIa). Consequently, benralizumab induces killing of IL-5Rα-expressing eosinophils and basophils through enhanced antibody-dependent cell mediated cytotoxicity (ADCC; Figure 15.3.4) with picomolar potency, whereas the parent (fucosylated) anti-IL-5RA mAb at a 1000-fold

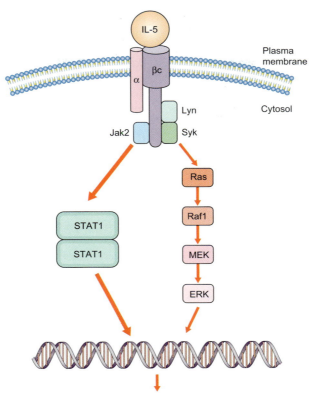

Transcriptional activation leading to antiapoptotic effects

FIGURE 15.3.2 **Signaling pathways leading from binding of interleukin-5 to its receptor in the membrane.** Following interleukin-5 (IL-5) binds to the IL-5 receptor (IL-5R), the β subunit of the receptor activates the Janus kinase/signal transducer and activator of transcription (Jak2— STAT1) pathway and the Ras—Raf1—MEK—ERK pathway, leading to transcriptional activation in the cell nucleus. Transcriptional activation is proposed to generate antiapoptotic effects in eosinophils. The murine pathways are shown: ERK, mitogen-activated protein kinase (extracellular signal-regulated kinase); Lyn, tyrosine-protein kinase Lyn; MEK, dual specificity mitogen-activated protein kinase kinase; Raf1, RAF proto-oncogene serine/threonine-protein kinase; Syk, tyrosine-protein kinase Syk. *[Reprinted from Moqbel et al. Biology of Eosinophils. In: Middleton's Allergy: Principles & Practice. 7th ed, 2009 (Vol 1, Ch 18), with permission from Elsevier; originally published in Lacy P, Becker AC, Moqbel R. The human eosinophil, in Wintrobe's Clinical Hematology, 11th edn, 2004 (Vol 1, Ch. 11) with permission. Copyright, Lippincott, Williams and Wilkins.]*

Regulation of IL-5 dependent pathways

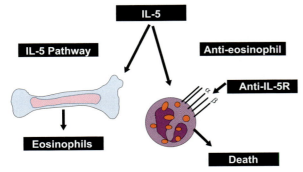

FIGURE 15.3.3 Differential pathways to regulate the eosinophil's presence: blockade of the interleukin-5 (IL-5) pathway or an *anti-eosinophilic* approach by targeting the cell via action on the IL-5 receptor (IL-5R).

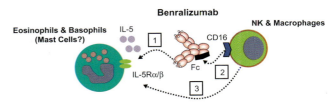

FIGURE 15.3.4 Benralizumab binds to interleukin-5 receptor α (IL-5Rα)-expressing target cells (basophils and eosinophils) in the circulation and tissues (1). Effector cells [macrophages and natural killer cells (NK)] express low affinity immunoglobulin γ Fc region receptor III A (FcγRIIIa/CD16), which binds to benralizumab Fc with high affinity (2). FcγRIIIa activation results in the release of cytotoxic mediators (granzyme, perforin, and proteases) that mediate target cell killing (3).

higher concentration is largely ineffective.[10] The dramatically enhanced ADCC potency of benralizumab is strictly dependent on the presence of effector cells, indicating a minor, if any, role for complement-mediated killing. Importantly, benralizumab induces eosinophil apoptosis without triggering the release of the eosinophilic granule proteins eosinophil-derived neurotoxin and eosinophil cationic protein (ECP) *in vitro*[10] and in patients with mild asthma.[12] The potent yet *silent* killing of eosinophils by benralizumab may be considered an important prerequisite for its safety profile, which, as assessed in nonhuman primates and clinical studies, continues to support ongoing development.

Pharmacology and Safety in Nonhuman Primates

Benralizumab binds to an epitope on domain 1 of IL-5Rα that is conserved in human and cynomolgus monkeys (Cynos) but not in rodents.[10] Therefore, lack of cross-reactivity of benralizumab has prohibited further investigations in murine models. However, cross-reactivity to Cynos eosinophils is fully maintained and has warranted pharmacological and safety studies in this species.

Benralizumab administration results in no sign of adverse events in Cynos that were injected four times intravenously with either vehicle or benralizumab (0.1, 1, 10, or 30 mg/kg) every 3 weeks, followed by an 18-day recovery period in the highest dose group. In all treatment groups, benralizumab depleted peripheral blood eosinophils to below the limit of detection throughout the recovery period. In addition, eosinophil precursors in the bone marrow of all animals receiving benralizumab were either markedly decreased (>90%) or absent in all groups at terminal necropsy (2–3 days after the last dose administration) and after the recovery period in the highest dose group.[10] The numbers of precursors of the megakaryocytic, myeloid (other than eosinophilic), and erythroid lineages remained within normal limits in animals receiving benralizumab. These data indicate potent and prolonged peripheral blood eosinophil and eosinophil precursor depletion in the bone marrow by benralizumab. The activity of benralizumab on bone marrow eosinophil precursors may partially explain its long-lasting effects on reducing peripheral eosinophil numbers in subjects with mild asthma.[12] Furthermore, benralizumab has been shown to reduce bronchoalveolar lavage (BAL) eosinophilia in Cynos allergic to *Ascaris suum* extract. Animals were injected with a single intravenous dose of benralizumab (1 mg/kg) on day 1 and received allergen challenges on days 2, 23, 60, 102 (for BAL), or 130 [for airway hyperresponsiveness (AHR)], and 161. Twenty-four hours after each allergen challenge, either BAL or AHR was measured.

All animals displayed substantial increases in BAL eosinophil numbers and enhanced AHR at baseline (24 hours before benralizumab administration). Benralizumab dramatically inhibited the increase in BAL eosinophilia on days 3, 24, 61, and 103 (>80%). There was also a trend toward decreased AHR in animals that received benralizumab that did not meet statistical significance.

Clinical Development

The pharmacodynamics, pharmacokinetics, and safety profile of single, escalating intravenous doses (0.0003–3 mg/kg) of benralizumab have been investigated in 44 subjects with mild atopic asthma.[12] Mean peripheral blood eosinophil levels decreased in a dose-dependent fashion. Ninety-four percent of subjects receiving ≥ 0.03 mg/kg exhibited peripheral blood eosinophil levels of $0.00-0.01 \times 10^3/\mu L$. Eosinopenia lasted at least 8 or 12 weeks with doses of 0.03–0.10 mg/kg or 0.3–3.0 mg/kg, respectively, with concomitant reductions in ECP levels from 21.4 ± 17.2 μg/L (baseline) to 10.3 ± 7.0 μg/L (24 hours postdose). The most frequently reported adverse events were reduced white blood cell counts (34.1%), nasopharyngitis (27.3%), and increased blood creatine phosphokinase (25.0%). Mean C-reactive protein levels increased approximately 5.5-fold at 24 hours postdose but returned to baseline by study end, and mean IL-6 levels increased approximately 3.9-fold to 4.7-fold at 6 hours and 12 hours postdose, respectively. In conclusion, single escalating doses of benralizumab had a safety profile that supports further development and resulted in marked reduction of peripheral blood eosinophil counts within 24 hours after dosing.

In a Phase IIa multiple ascending dose safety study, a subcutaneous formulation of benralizumab in adult asthmatic subjects resulted in pharmacokinetic/pharmacodynamic activities similar to those observed with intravenous dosing.[13,14] Benralizumab concentration–time profiles showed linear pharmacokinetics at the dose levels tested. The terminal half-life of benralizumab was approximately 18 days.

Another Phase Ib study collecting airway biopsy specimens in asthmatics and two additional Phase II studies of benralizumab in adult subjects with asthma are ongoing (ClinicalTrials.gov identifiers: NCT00659659, NCT00768079, and NCT01238861).

Other Products under Development Targeting Interleukin-5 Receptor α

The drug product TPI ASM8 contains two modified phosphorothioate antisense oligonucleotides designed to inhibit allergic inflammation by downregulating human C-C chemokine receptor 3 (CCR3) and the βc chain of IL-3, IL-5, and GM-CSF.[15]

A recent summary of presentations at the International Quality & Productivity Center[16] includes clinical results of TPI ASM8. At the highest dose level tested, this compound reduced early and late asthmatic responses by 32% and 49%, respectively.[16] A second study reported a significant reduction in an early-phase asthmatic response ($p = 0.04$) and a trend toward reduction in the late-phase asthmatic response ($p = 0.08$) among 17 subjects with mild atopic asthma who received 1500 μg of TPI ASM8 once daily for 4 days, compared with placebo.[15]

The relative contribution of IL-5R inhibition to the observed reduction of the asthmatic response by TPI ASM8 remains elusive.

CONCLUSIONS AND FUTURE DIRECTIONS

Treatment of asthma patients with anti-IL-5 mAbs has resulted in significant reductions in exacerbations.[17,18] One study has also shown improvements in lung function and quality of life.[19] Studies of differential effects of IL-5, IL-3, GM-CSF, and C-C motif chemokine 11 (CCL11/eotaxin) on the expression of these cytokines in human tissue eosinophils suggest that variable cross-regulation of the receptors by the cytokine ligands may cause eosinophils at sites of allergic inflammation to have reduced dependence on IL-5. These observations may explain the incomplete reduction in eosinophil numbers in lung tissue observed with anti-IL-5 therapies.

Monoclonal antibodies such as benralizumab that specifically target IL-5Rα rather than the cytokine itself and actively induce target cell killing through enhanced ADCC function may demonstrate that a more complete reduction of eosinophils results in a different clinical response, especially in the treatment of asthma and other diseases associated with elevated eosinophil numbers. In humans, in addition to eosinophils, basophils also express IL-5R[20,21] and in turn represent an additional target of benralizumab-based therapies. Specifically, evidence suggests that allergen provocation induces the accumulation of basophils in the airways of asthmatics[22] and that basophils are a powerful cell type capable of initiating IgE-mediated allergic inflammation in the skin.[23] Thus, the potential of IL-5R-targeted cell depletion strategies to reduce basophil numbers may represent an additional opportunity to ameliorate asthma pathogenesis beyond the eosinophil-targeting effects demonstrated to date.

REFERENCES

1. Clutterbuck EJ, Hirst EM, Sanderson CJ. Human interleukin-5 (IL-5) regulates the production of eosinophils in human bone marrow cultures: comparison and interaction with IL-1, IL-3, IL-6, and GMCSF. *Blood* 1989;**73**:1504–12.

2. Plaetinck G, Van der Heyden J, Tavernier J, Fache I, Tuypens T, Fischkoff S, et al. Characterization of interleukin 5 receptors on eosinophilic sublines from human promyelocytic leukemia (HL-60) cells. *J Exp Med* 1990;**172**:683–91.

3. Chihara J, Plumas J, Gruart V, Tavernier J, Prin L, Capron A, et al. Characterization of a receptor for interleukin 5 on human eosinophils: variable expression and induction by granulocyte/macrophage colony-stimulating factor. *J Exp Med* 1990;**172**:1347–51.

4. Ishino T, Harrington AE, Gopi H, Chaiken I. Structure-based rationale for interleukin 5 receptor antagonism. *Curr Pharm Des* 2008;**14**:1231–9.

5. Martinez-Moczygemba M, Huston DP. Biology of common beta receptor-signaling cytokines: IL-3, IL-5, and GM-CSF. *J Allergy Clin Immunol* 2003;**112**:653–65. quiz 666.

6. Yousefi S, Hoessli DC, Blaser K, Mills GB, Simon HU. Requirement of Lyn and Syk tyrosine kinases for the prevention of apoptosis by cytokines in human eosinophils. *J Exp Med* 1996;**183**:1407–14.

7. Watanabe S, Itoh T, Arai K. JAK2 is essential for activation of c-fos and c-myc promoters and cell proliferation through the human granulocyte-macrophage colony-stimulating factor receptor in BA/F3 cells. *J Biol Chem* 1996;**271**:12681–6.

8. Busse WW, Ring J, Huss-Marp J, Kahn JE. A review of treatment with mepolizumab, an anti-IL-5 mAb, in hypereosinophilic syndromes and asthma. *J Allergy Clin Immunol* 2010;**125**:803–13.

9. Koike M, Nakamura K, Furuya A, Iida A, Anazawa H, Takatsu K, et al. Establishment of humanized anti-interleukin-5 receptor alpha chain monoclonal antibodies having a potent neutralizing activity. *Hum Antibodies* 2009;**18**(1–2):17–27.

10. Kolbeck R, Kozhich A, Koike M, Peng L, Andersson CK, Damschroder MM, et al. MEDI-563, a humanized anti-IL-5 receptor alpha mAb with enhanced antibody-dependent cell-mediated cytotoxicity function. *J Allergy Clin Immunol* 2010 Jun;**125**(6):1344–53.

11. Yamane-Ohnuki N, Kinoshita S, Inoue-Urakubo M, Kusunoki M, Iida S, Nakano R, et al. Establishment of FUT8 knockout Chinese hamster ovary cells: an ideal host cell line for producing completely defucosylated antibodies with enhanced antibody-dependent cellular cytotoxicity. *Biotechnol Bioeng* 2004 Sep 5;**87**(5):614–22.

12. Busse WW, Katial R, Gossage D, Sari S, Wang B, Kolbeck R, et al. Safety profile, pharmacokinetics, and biologic activity of MEDI-563, an anti-IL-5 receptor alpha antibody, in a phase I study of subjects with mild asthma. *J Allergy Clin Immunol* 2010;**125**:1237–44.

13. Gossage D, Geba G, Gillen A, Le C, Molfino N. A multiple ascending subcutaneous (SC) dose study of MEDI-563, a humanized anti-IL5Ra monoclonal antibody, in adult asthmatics. *European Respiratory Society—20th Annual Congress* 2010. P1177.

14. Jin F, White W, Gossage D, Geba G, Molfino N. Multiple ascending subcutaneous (SC) dose study of MEDI-563: pharmacokinetics and immune response in adult asthmatics. *European Respiratory Society—20th Annual Congress* 2010. P4553.

15. Gauvreau GM, Boulet LP, Cockcroft DW, Baatjes A, Cote J, Deschesnes F, et al. Antisense therapy against CCR3 and the common beta chain attenuates allergen-induced eosinophilic responses. *American Journal of Respiratory and Critical Care Medicine* 2008;**177**:952–8.

16. Catley MC. Asthma & COPD—IQPC's Second Conference. *IDrugs* 2010;**13**:601–4.

17. Haldar P, Brightling CE, Hargadon B, Gupta S, Monteiro W, Sousa A, et al. Pavord. Mepolizumab and exacerbations of refractory eosinophilic asthma. *N Engl J Med* 2009;**360**:973–84.

18. Nair P, Pizzichini MM, Kjarsgaard M, Inman MD, Efthimiadis A, Pizzichini E, et al. Mepolizumab for prednisone-dependent asthma with sputum eosinophilia. *N Engl J Med* 2009;**360**:985–93.

19. Castro M, Mathur S, Hargreave F, Boulet LP, Xie F, Young J, et al. Reslizumab for poorly controlled, eosinophilic asthma: a randomized, placebo-controlled study. Res-5-0010 Study Group. *Am J Respir Crit Care Med* 2011 Nov 15;**184**(10):1125–32. Epub 2011 Aug 18.

20. Valent P. The phenotype of human eosinophils, basophils, and mast cells. *J Allergy Clin Immunol* 1994;**94**:1177–83.

21. Toba K, Koike T, Shibata A, Hashimoto S, Takahashi M, Masuko M, et al. Novel technique for the direct flow cytofluorometric analysis of human basophils in unseparated blood and bone marrow, and the characterization of phenotype and peroxidase of human basophils. *Cytometry* 1999;**35**:249–59.

22. Gauvreau GM, Lee JM, Watson RM, Irani AM, Schwartz LB, O'Byrne PM. Increased numbers of both airway basophils and mast cells in sputum after allergen inhalation challenge of atopic asthmatics. *Am J Respir Crit Care Med* 2000 May;**161**(5):1473–8.

23. Mukai K, Matsuoka K, Taya C, Suzuki H, Yokozeki H, Nishioka K, et al. Basophils play a critical role in the development of IgE-mediated chronic allergic inflammation independently of T cells and mast cells. *Immunity* 2005 Aug;**23**(2):191–202.

Chapter 15.4

Therapeutic Approaches Targeting Siglecs

Mi-Kyung Shin and Bruce S. Bochner

INTRODUCTION

In attempting to develop antieosinophil therapies to treat asthma and other disorders, the search for a truly unique, specific cell-surface phenotypic marker to function as a bull's-eye on eosinophils has been a long and elusive task.[1–3] Eosinophils express a variety of relatively selective markers including the C-C chemokine receptor 3 (CCR3), the interleukin-5 receptor, and most recently sialic acid-binding Ig-like lectin 8 (Siglec-8). All are useful markers of eosinophils but are not absolutely specific for these cells. Nevertheless, from the standpoint of developing future therapies that would be eosinophil specific, each has its own strengths, weaknesses, and noneosinophil expression patterns. This subchapter focuses on siglecs expressed by eosinophils, with the vast majority of the discussion concentrated on the one that is most eosinophil-specific, namely Siglec-8. The reader is referred to other papers on siglecs as therapeutic targets for a broader range of discussions.[4–7]

WHAT IS A SIGLEC?

Siglecs, a term coined in 1998,[8] are sialic acid-binding Ig-like lectins, all of which are I-type lectins belonging to the Ig superfamily.[9,10] All are single-pass type I transmembrane proteins with extracellular domains that share a high degree of similarity. Every siglec contains an N-terminal V-set Ig domain that binds various conformations of sialic acids, followed by variable numbers of C2-set Ig domains, a transmembrane region, and a cytosolic tail.[10] The V-domains typically have about 65–75 amino acid residues between the conserved disulfide bonds, and there are four β-strands in each β-sheet plus a short β-strand segment across the top of the domain. In C2-domains, the sequence between the disulfide bonds is shorter at 55–60 residues, yielding sheets with four or three β-strands.[11]

Most siglecs have immuno-receptor tyrosine-based motifs (ITIMs) in their intracellular domain, suggesting a role for siglecs in signaling events.[4,10,12–16] The best known is the membrane-proximal motif, with a typical 6-amino-acid sequence described as (I/L/V)xYxx(L/V), where I/L/V represents isoleucine, leucine, or valine, Y represents tyrosine and x represents any amino acid.[17] The tyrosine motif can be phosphorylated by Src-family tyrosine kinases that can interact with the tyrosine-protein phosphatase nonreceptor types 6 and 11 (Src homology 2 domain-containing phosphatase 1 and 2; SHP-1 and SHP-2), as well as with phosphatidylinositol 3,4,5-trisphosphate 5-phosphatase 1 (SH2-domain-containing inositol polyphosphate 5-phosphatase; SHIP-1). In general, transmembrane proteins that have this motif in their cytoplasmic domains are considered to have inhibitory functions. ITIMs in various siglecs can recruit SHP-1 when phosphorylated,[18] and modulate cellular activity in an inhibitory manner upon cross-linking with antibodies.[19] However, the exact pathways of signaling for siglecs, including Siglec-8 in eosinophils (see below), are still unclear and are part of ongoing investigations.

Siglecs can be divided into two primary subsets based on their sequence similarities and evolutionary conservation, namely the myeloid cell surface antigen CD33 (Siglec-3)-related siglecs and all the other siglecs.[15] Figure 15.4.1 shows schematic representations of human and mouse siglecs. In humans, CD33-related siglec proteins are highly structurally related and rapidly evolving; consequently, they differ significantly in composition between mammalian species.[20,21] In humans, this subset includes CD33 and Siglec-5, Siglec-6, Siglec-7, Siglec-8, Siglec-9, Siglec-10, Siglec-11, Siglec-14, and Siglec-16. In contrast, there are fewer siglec genes in the mouse, including *Cd33/Siglec3* (the only CD33-type siglec) and *Siglec12* (encoding Siglec-E), *Siglec5* (encoding Siglec-F), *Siglecg*, and *Siglech*, the latter representing paralogues rather than orthologues of various human

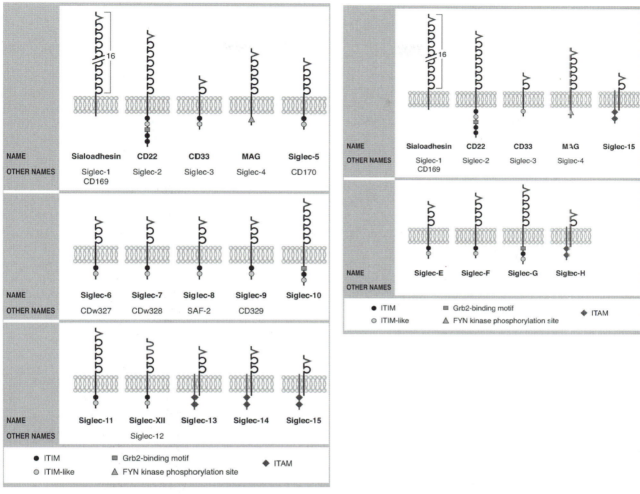

FIGURE 15.4.1 *A,* Nomenclature and key structural characteristics of human siglecs. Although 15 molecules are shown, Siglec-13 is present in nonhuman primates but not in humans. V structures indicate the arginine-containing V-set domains with lectin activity; these are followed by C2-type Ig repeat domains. U-shaped structures for Siglec-XII indicate the mutated V-set domains missing arginine that have lost their lectin activity. Also shown is DAP12 (TYRO protein tyrosine kinase-binding protein), illustrated as a shorter transmembrane structure coassociating with Siglec-13, Siglec-14, and Siglec-15. See key for symbols representing cytoplasmic signaling motifs. *B,* Nomenclature and key structural characteristics of mouse siglecs. V structures indicate the arginine-containing V-set domains with lectin activity; these are followed by C2-type Ig repeat domains. Also shown is DAP12, illustrated as a shorter transmembrane structure coassociating with Siglec-15 and Siglec-H. Whether Siglec-H, shown as having an arginine-containing V-set domains with lectin activity, can bind sialic acid ligands remains controversial (see text). Note that Siglecs-1—4 are conserved with humans, as is Siglec-15 (compare to *A*). While not orthologues, the closest functional paralogue of Siglec-E is Siglec-9, while Siglec-F resembles Siglec-8 and Siglec-G resembles Siglec-10. FYN, Grb2, growth factor receptor-bound protein 2; ITIM, immunoreceptor tyrosine-based inhibition motif; ITAM, immunoreceptor tyrosine-based activation motif; MAG, myelin-associated glycoprotein. *(Reproduced from[4] with permission.)*

siglecs. The remaining non-CD33-related human siglec proteins include sialoadhesin (Siglec-1), B-cell receptor CD22 (Siglec-2), myelin-associated glycoprotein (Siglec-4, the only nonleukocyte siglec), and Siglec-15 (the gene encoding the latter is uniquely located on chromosome 18q12.3), which are more distantly related, yet have well-conserved orthologues in all mammalian species examined so far.[10,14,22,23] Siglec-12 in humans no longer meets the definition of a siglec since it has lost the ability to bind sialic acid and is thus designated as Siglec-XII. Siglec-13 is present in nonhuman primates (baboons and chimpanzees) but not in humans.[20,22] Each siglec has a distinct expression pattern in different cell types, indicating that they perform highly specific functions on different cell types. Except for sialoadhesin and Siglec-15, all human siglec-encoding genes are found on human chromosome 19q13.3—q13.4, and of potential relevance to eosinophil siglecs is that genetic linkage studies of asthma and related phenotypes have previously implicated 19q13.33—q13.41 as disease loci.[24—26] Further encouragement in support of an important role of Siglec-8 in human eosinophil-associated diseases comes from a study of the genetic association between sequence variants in *SIGLEC8* and the diagnosis of asthma, where a significant association with asthma was observed for various single nucleotide polymorphisms among African American, Brazilian, and Japanese populations.[27]

SIALIC ACID-BINDING IG-LIKE LECTIN EXPRESSION ON EOSINOPHILS

Cellular Expression Patterns

Siglecs expressed by human eosinophils[28] are essentially limited to CD33 (weak), Siglec-7 (modest),[29] Siglec-8 (high)[30,31] and Siglec-10 (weak).[32] Expression of CD33 is higher on immature eosinophils and decreases, often to essentially undetectable levels, on mature circulating eosinophils.[33] Siglec-8 (originally called sialoadhesin factor-2 by one of the two groups who first discovered it from a cDNA library made from an unusual donor with extremely high eosinophil blood counts)[30,31] was initially reported to contain an unusually short cytoplasmic domain unlikely to be capable of signaling,[30,31] but subsequent studies identified a more typical form expressing two intracytoplasmic signaling motifs.[34–36] While highly and preferentially expressed on eosinophils,[37] it rapidly became clear that Siglec-8 was not specific to eosinophils since it was also expressed by human mast cells and, weakly but consistently, by human basophils.[30] Chimpanzees are predicted to express Siglec-8, but not any lower mammalians species.[10,20,22] All of the eosinophilic cell lines studied so far fail to express Siglec-8, suggesting that this is a late-differentiation antigen ([30] and unpublished observations). How Siglec-8 expression is regulated at the genetic level has not been studied.

With the discovery of Siglec-8, efforts were initiated to discover its counterpart in the mouse, where systems would be more amenable to *in vivo* manipulation for exploring its biology. It is now clear that Siglec-F, which only shares about 40% sequence identity with Siglec-8 and contains an extra Ig domain, is most prominently expressed by mouse eosinophils, although differences were subsequently found among patterns of cell-surface expression of Siglec-F and Siglec-8, with Siglec-F not being expressed on mouse mast cells and instead expressed on a wider range of cells, including alveolar macrophages and at very low levels on neutrophils and T cells, none of which express Siglec-8 in humans.[4,38–44] Eosinophils from interleukin-5 (IL-5) transgenic mice also possess mRNA for Siglec-E and Siglec-G,[44] but their surface protein expression of these molecules has not been well studied. Siglec-F is now routinely used as a reliable marker for detecting eosinophils in tissues[45–48] and during hematopoietic development as a late-stage marker of eosinophil maturation.[49]

Ligands

By definition, siglecs bind sialic acid, but the story is much more complicated than that, largely because the conformation of the sialic acid, along with other factors such as core protein expression, multivalency, and sulfation all contribute to binding specificity.[12,50,51] A major advance in our understanding of glycan-lectin binding occurred with glycan arrays developed by the National Institutes of Health-funded Consortium for Functional Glycomics (http://www.functionalglycomics.org).[52–55]

Using this approach, it was discovered that CD33 is highly *promiscuous* in that it recognizes α2,3-linked, α2,6-linked, and α2,8-linked sialic acids, while Siglec-7 prefers α2,8-linked sialic acids and Siglec-10 binds both α2,3-linked and α2,6-linked sialic acids.[10] Moreover, as astutely pointed out, the fact that Siglec-G (mouse Siglec-10) recognizes CD24 without defining the carbohydrate specificity of this interaction misses the probable enormous influence of the glycans on binding.[51] In contrast, Siglec-8 very specifically recognizes 6′ sulfated sialyl Lewis X (6′-su-sLe^x or NeuAcα2−3(6-O-SO$_3$)Galβ1−4(Fucα1−3) GlcNAc).[56] Neither sLe^x (a known ligand for E-selectin, P-selectin, and L-selectin[57]) nor 6su-sLe^x (a ligand for L-selectin[58]) bound Siglec-8, as glycan binding absolutely requires the presence of the sulfate on position 6 of the galactose residue. It has subsequently been shown that the fucose residue is not required for Siglec-8 binding ([53] and unpublished observations). Further evidence for the specificity of binding of this glycan comes from a report that incubation of whole human blood with a polyacrylamide polymer decorated with 6′-su-sLe^x resulted in binding to eosinophils that was Siglec-8-dependent. In contrast, no detectable binding to any other leukocyte subtype was detected (Figure 15.4.2), including lymphocytes and monocytes that express Siglec-7 and show some ability in other assays to bind 6′-su-sLe^x.[59] Finally, despite the structural and cellular differences mentioned above for Siglec-F, it is remarkable that Siglec-F, like Siglec-8, selectively recognizes 6′-su-sLe^x.[40] Therefore, because of its expression on eosinophils and preference for binding the same ligand, Siglec-F and Siglec-8 are best thought of as functionally convergent or *isofunctional* paralogues.[10]

The biochemical identity of natural tissue ligands for Siglec-8 and Siglec-F remains unknown. Recent strides have been made in characterizing glycan ligands in tissues for Siglec-F. Real-time polymerase chain reaction detected constitutive mRNA expression for the *Chst1* gene that encodes a putative essential sulfotransferase for ligand synthesis, carbohydrate sulfotransferase 1 (keratan sulfate Gal-6 sulfotransferase; KSGa16ST) in mouse lung.[60] Immunohistochemistry localized KSGa16ST expression in lung to airway epithelium. Siglec-F—Ig fusion protein selectively bound in a similar pattern, and binding was

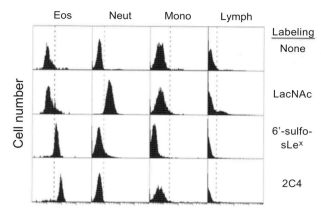

FIGURE 15.4.2 A 2000-kDa 6′-su-sLe^x polyacrylamide polymer selectively binds to eosinophils among leukocytes in whole blood. Attachment of the 6′-su-sLe^x polyacrylamide polymer and the 2C4 Siglec-8 monoclonal antibody to eosinophils (Eos) was readily apparent, whereas no labeling of monocytes (Mono), lymphocytes (Lymph), neutrophils (Neut), or basophils was observed. Results are representative of at least eight experiments with similar results. *(Reproduced from[59] with permission.)*

eliminated by sialidase, and was absent in tissues deficient in ST3 β-galactoside α-2,3-sialyltransferase 3 (*St3gal3*). In addition, binding of the Siglec-F—Ig fusion protein was completely inhibited if tissue sections were pretreated with a plant lectin that specifically recognizes α2,3-linked sialic acid. Thus, sialic acid-containing Siglec-F ligands, and enzymes required for their synthesis, are constitutively expressed in mouse lung, especially by airways epithelium, and *St3gal3* is required for constitutive Siglec-F ligand synthesis. These results set the stage for efforts to isolate and characterize this endogenous lung glycoprotein ligand, and to generate novel mouse models to explore the regulation of ligand expression, the biology of Siglec-8 and Siglec-F and its utility as a therapeutic target (also see end of next section).

Functions *in Vitro*

CD33 expression on eosinophils is low to absent, and its function on eosinophils has not been reported. Siglec-7 is expressed by eosinophils, but is most frequently thought of as an inhibitory receptor on natural killer cells involved in apoptosis.[61] Siglec-10 is expressed on monocytes, dendritic cells, particularly brightly on a CD16⁺ CD56[−] subpopulation of cells, and weakly on eosinophils and B cells.[32] Its mouse paralogue is Siglec-G, and it has been shown to play an important regulatory role in B cell biology.[62] Nothing has yet been reported regarding the function of either Siglec-7 or Siglec-10 on eosinophils.

In contrast, there is a great deal known about the function of Siglec-8 and Siglec-F on eosinophils.[63] Incubation of human eosinophils under Siglec-8 antibody cross-linking conditions resulted in pronounced cell death that was mediated through apoptosis.[36] Mechanistic studies implicated both caspases and reactive oxygen species (ROS) generation, resulting in mitochondrial injury in this type of cell death.[64,65] Indeed, recent studies have shown that Siglec-8 activation by specific antibodies potently induces ROS production mediated via phosphoinositide 3-kinases (Mi-Kyung Shin and colleagues, unpublished observations). It was subsequently discovered that intravenous immunoglobulin preparations that are used commercially contain autoantibodies to Siglec-8 at a sufficiently high titer so as to also induce eosinophil apoptosis and ROS production *in vitro*, especially in cytokine-primed cells.[66]

Initially confusing was the observation that Siglec-8-induced cell death could not be overridden by counterbalancing survival signals, such as those provided by the cytokines granulocyte-macrophage colony-stimulating factor (GM-CSF) and IL-5. Indeed, Siglec-8-induced death was enhanced by these cytokines, in that cells would die more readily even with less of a Siglec-8 engagement signal.[36] These results were subsequently confirmed in human eosinophils primed *in vivo* following allergen bronchoprovocation; primed cells no longer used caspases in the apoptosis process, instead relying exclusively on ROS generation and mitochondrial injury.[67] Overall, these data suggest that activated eosinophils might be particularly susceptible to approaches that engage Siglec-8.[16] Furthermore, polymers coated with 6′-su-sLe^x induced eosinophil apoptosis, albeit not as effectively as antibodies to Siglec-8.[59] Nevertheless, these studies provide proof of concept that Siglec-8 offers a means for selective targeting of eosinophils and activation of apoptosis in these cells.

Murine studies of Siglec-F have more or less confirmed those involving Siglec-8, in that antibody cross-linking causes apoptosis, albeit not as robustly as has been seen with human eosinophils and Siglec-8.[68–70] In addition to these effects on cell survival, many siglec proteins undergo endocytosis following engagement in an ITIM-dependent manner.[71] Siglec-F is reportedly internalized when engaged with antibody or glycan ligands, and this requires the presence of the cytoplasmic tyrosine motifs on Siglec-F.[72] In contrast to the way in which this occurs for another siglec, B-cell receptor CD22, it was reported that Siglec-F was endocytosed into lysosomal compartments independent of clathrin and dynamin. This phenomenon has not been studied in human eosinophils, but if present could suggest another way of delivering therapeutic agents into eosinophils.

Functions *in Vivo*

Zhang and coworkers successfully generated Siglec-F null mice.[41] While these mice were phenotypically normal, when put into allergen-sensitization and provocation models of asthma they display a more pronounced blood, bone marrow, and lung eosinophilia due to reduced rates of apoptosis.[41] Additional studies in normal and IL-5 transgenic mice, the latter displaying hypereosinophilia, showed that administration of a single dose of Siglec-F antibody results in selective reductions in eosinophil numbers in the blood and gastrointestinal tissues mediated via apoptosis. Eosinophil numbers in blood were decreased via apoptosis in multiple models: wild-type mice of two separate genetic backgrounds; a mouse model of hypereosinophilic syndrome/chronic eosinophilic leukemia (Figure 15.4.3); and a mouse model of gastrointestinal eosinophilia and diarrhea which, in the latter model normalized weight gain.[68,69] In a subsequent paper, Song and colleagues went on to show that mice sensitized and challenged repetitively for 1 month and then given anti-Siglec-F had decreased airways eosinophilia, and that the treatment almost completely normalized the amount of subepithelial fibrosis.[70] This latter finding is reminiscent of findings in asthma trials in humans with mepolizumab, an anti-IL-5 antibody, which nearly completely reduced to normal levels the deposition of several airway matrix proteins as assessed by bronchial biopsies.[73] Also associated with anti-Siglec-F administration was a reduction in mucin-containing cells and smooth muscle layer thickness, along with a significant reduction in the number of airway cells coexpressing major basic protein and transforming growth factor β (TGF-β). Since this latter finding was correlative, and since mouse alveolar macrophages and other cells can express Siglec-F,[42] these results strongly suggest, but do not prove, that

eosinophil-derived TGF-β is the cause of the observed peribronchial fibrosis. Finally, the anti-Siglec-F treatment had a modest, nonstatistically significant effect on airway hyperreactivity. Regardless, these studies demonstrate that targeting Siglec-F, and perhaps by analogy Siglec-8, can significantly reduce eosinophilic airway inflammation and airway remodeling, in particular subepithelial fibrosis. These conclusions are further supported by additional work from the Broide laboratory using Siglec-F-deficient mice in the same mouse chronic asthma model, where significantly increased amounts of airway mucin expression, fibronectin, lung eosinophils, peribronchial fibrosis, and TGF-β1+ cells were seen.[74] With chronic antigen challenge, they noted an increase in peribronchial smooth muscle thickness in these Siglec-F-deficient mice, but airway hyperreactivity was not increased to any great degree over the increase seen in wild-type mice. They also reported that mice given IL-4 or IL-13 intranasally, but not TNF-α, had increased expression of epithelial Siglec-F ligands. Thus, while not a perfect replica of the human Siglec-8 expression pattern, these data provide encouragement in support of the notion put forth by von Gunten and colleagues[16] that targeting an eosinophil-selective siglec could have profound effects on eosinophil numbers and airway disease activity.

CONCLUSION

While little is known about the function of CD33, Siglec-7, and Siglec-10 on eosinophils, there is now a significant body of literature surrounding Siglec-8 (human) and Siglec-F (mouse) expression and function on eosinophils. Siglec-8 ligation induces apoptosis in eosinophils via activation of caspases and/or ROS production leading to mitochondrial injury. The glycan 6'-su-sLe^x has been identified as

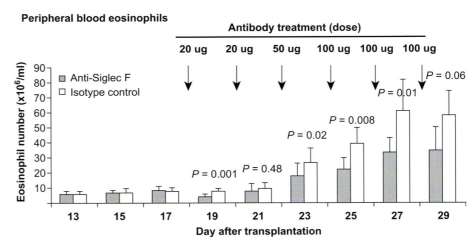

FIGURE 15.4.3 Selective decrease of circulating quantum of eosinophils by Siglec-F antibody administration in the hypereosinophilic syndrome/chronic eosinophilic leukemia mouse model. The hypereosinophilic syndrome/chronic eosinophilic leukemia was induced by transplanting lethally irradiated wild-type mice with FIP1-like 1–platelet-derived growth factor receptor-α-transduced bone marrow cells derived from interleukin-5 (IL-5) transgenic mice. Once the quantum of eosinophils started increasing above the IL-5 transgenic level, mice were treated with Siglec-F antibodies every 48 hours and the quantum of eosinophils in the peripheral blood was monitored. As the disease and number of eosinophils escalated, the dose of antibody was increased from 20 to 100 μg/mouse. *(Reproduced from[68] with permission.)*

a candidate ligand for both Siglec-8 and Siglec-F, and a 6′-su-sLe^x polymer selectively binds to eosinophils and induces their apoptosis. Siglec-F is the closest functional paralogue of Siglec-8 in the mouse. Antibody cross-linking of Siglec-F on mouse eosinophils *in vitro* induces apoptosis. Siglec-F-deficient mice have exaggerated eosinophilic responses during chronic allergic airways inflammatory reactions. Siglec-F antibody administered to mice reduces numbers of circulating and tissue eosinophils via apoptosis, as well as manifestations of tissue eosinophilia. Airway epithelial glycans containing α2,3-linked sialic acids serve as Siglec-F ligands, and enzymes required for the synthesis of 6′-su-sLe^x, are constitutively expressed in mouse lung. Binding of the Siglec-F—Ig fusion protein to mouse lung is eliminated by sialidase and is absent in tissues of *St3gal3*-deficient mice. Ongoing studies that will further our understanding of these molecules include the isolation and characterization of the natural Siglec-8/-F ligands; defining the role of ROS and SHP-1 in Siglec-8 versus Siglec-F signaling; and exploring the function of sialyl- and sulfotransferases in generating lung ligands *in vitro* and *in vivo*. The ultimate goal is that these studies will lead to the development of Siglec-8 targeting agents for imaging and treatment of eosinophil-associated diseases.

REFERENCES

1. Tachimoto H, Bochner BS. The surface phenotype of human eosinophils. *Chem Immunol* 2000;**76**:45−62.

2. Bochner BS. Verdict in the case of therapies versus eosinophils: the jury is still out. *J Allergy Clin Immunol* 2004;**113**:3−9.

3. Bochner BS, Gleich GJ. What targeting eosinophils has taught us about their role in diseases. *J Allergy Clin Immunol* 2010;**126**:16−25.

4. von Gunten S, Bochner BS. Basic and clinical immunology of Siglecs. *Ann NY Acad Sci* 2008;**1143**:61−82.

5. McMillan SJ, Crocker PR. CD33-related sialic-acid-binding immunoglobulin-like lectins in health and disease. *Carbohydr Res* 2008;**343**:2050−6.

6. O'Reilly MK, Paulson JC. Siglecs as targets for therapy in immune-cell-mediated disease. *Trends Pharmacol Sci* 2009;**30**:240−8.

7. Chen WC, Completo GC, Sigal DS, Crocker PR, Saven A, Paulson JC. *In vivo* targeting of B-cell lymphoma with glycan ligands of CD22. *Blood* 2010;**115**:4778−86.

8. Crocker PR, Clark EA, Filbin M, Gordon S, Jones Y, Kehrl JH, et al. Siglecs: a family of sialic-acid binding lectins. *Glycobiology* 1998;**8**:v−vi.

9. Powell LD, Varki A. I-type lectins. *J Biol Chem* 1995;**270**:14243−6.

10. Varki A, Angata T. Siglecs—the major sub-family of I-type lectins. *Glycobiology* 2006;**16**:1R−27R.

11. Williams AF, Barclay AN. The immunoglobulin super-family—domains for cell surface recognition. *Annu Rev Immunol* 1988;**6**:381−405.

12. Varki A. Glycan-based interactions involving vertebrate sialic-acid-recognizing proteins. *Nature* 2007;**446**:1023−9.

13. Crocker PR. Siglecs: sialic-acid-binding immunoglobulin-like lectins in cell-cell interactions and signalling. *Curr Opin Struct Biol* 2002;**12**:609−15.

14. Crocker PR. Siglecs in innate immunity. *Curr Opin Pharmacol* 2005;**5**:431−7.

15. Crocker PR, Paulson JC, Varki A. Siglecs and their roles in the immune system. *Nat Rev Immunol* 2007;**7**:255−66.

16. von Gunten S, Bochner BS. Expression and function of Siglec-8 in human eosinophils, basophils and mast cells. In: Pawankar R, Holgate S, Rosenwasser LJ, editors. *Allergy Frontiers: Classification and Pathomechanisms*. Tokyo: Springer; 2009. p. 297−313.

17. Vely F, Vivier E. Conservation of structural features reveals the existence of a large family of inhibitory cell surface receptors and non-inhibitory/activatory counterparts. *J Immunol* 1997;**159**:2075−7.

18. Ikehara Y, Ikehara SK, Paulson JC. Negative regulation of T cell receptor signaling by Siglec-7 (p70/AIRM) and Siglec-9. *J Biol Chem* 2004;**279**:43117−25.

19. Avril T, Attrill H, Zhang J, Raper A, Crocker PR. Negative regulation of leucocyte functions by CD33-related siglecs. *Biochem Soc Trans* 2006;**34**:1024−7.

20. Cao H, de Bono B, Belov K, Wong ES, Trowsdale J, Barrow AD. Comparative genomics indicates the mammalian CD33rSiglec locus evolved by an ancient large-scale inverse duplication and suggests all Siglecs share a common ancestral region *Immunogenetics* 2009;**61**:401−17.

21. Varki A. Colloquium paper: uniquely human evolution of sialic acid genetics and biology. *Proc Natl Acad Sci USA* 2010;**107** (Suppl. 2):8939−46.

22. Angata T, Margulies EH, Green ED, Varki A. Large-scale sequencing of the CD33-related Siglec gene cluster in five mammalian species reveals rapid evolution by multiple mechanisms. *Proc Natl Acad Sci USA* 2004;**101**:13251−6.

23. Crocker PR, Redelinghuys P. Siglecs as positive and negative regulators of the immune system. *Biochem Soc Trans* 2008;**36**:1467−71.

24. Venanzi S, Malerba G, Galavotti R, Lauciello MC, Trabetti E, Zanoni G, et al. Linkage to atopy on chromosome 19 in north-eastern Italian families with allergic asthma. *Clin Exp Allergy* 2001;**31**:1220−4.

25. Ober C, Cox NJ, Abney M, Di Rienzo A, Lander ES, Changyaleket B, et al. Genome-wide search for asthma susceptibility loci in a founder population. *Hum Mol Genet* 1998;**7**:1393−8.

26. The Collaborative Study on the Genetics of Asthma. A genome-wide search for asthma susceptibility loci in ethnically diverse populations. *Nat Genet* 1997;**15**:389−92.

27. Gao PS, Shimizu K, Grant AV, Rafaels N, Zhou LF, Hudson SA, et al. Polymorphisms in the sialic acid-binding immunoglobulin-like lectin-8 (Siglec-8) gene are associated with susceptibility to asthma. *Eur J Hum Genet* 2010;**18**:713−9.

28. Munitz A, Levi-Schaffer F. Inhibitory receptors on eosinophils: a direct hit to a possible Achilles heel? *J Allergy Clin Immunol* 2007;**119**:1382−7.

29. Munitz A, Bachelet I, Eliashar R, Moretta A, Moretta L, Levi-Schaffer F. The inhibitory receptor IRp60 (CD300a) suppresses the effects of IL-5, GM-CSF, and eotaxin on human peripheral blood eosinophils. *Blood* 2006;**107**:1996−2003.

30. Kikly KK, Bochner BS, Freeman S, Tan KB, Gallagher KT, D'Alessio K, et al. Identification of SAF-2, a novel siglec expressed

on eosinophils, mast cells and basophils. *J Allergy Clin Immunol* 2000;**105**:1093—100.

31. Floyd H, Ni J, Cornish AL, Zeng Z, Liu D, Carter KC, et al. Siglec-8: a novel eosinophil-specific member of the immunoglobulin superfamily. *J Biol Chem* 2000;**275**:861—6.

32. Munday J, Kerr S, Ni J, Cornish AL, Zhang JQ, Nicoll G, et al. Identification, characterization and leucocyte expression of Siglec-10, a novel human sialic acid-binding receptor. *Biochem J* 2001;**355**:489—97.

33. Wood B. Multicolor Immunophenotyping: Human Immune System Hematopoiesis. *Methods in Cell Biology* 2004;**75**:559—76.

34. Foussias G, Yousef GM, Diamandis EP. Molecular characterization of a Siglec-8 variant containing cytoplasmic tyrosine-based motifs, and mapping of the Siglec-8 gene. *Biochem Biophys Res Commun* 2000;**278**:775—81.

35. Aizawa H, Plitt J, Bochner BS. Human eosinophils express two Siglec-8 splice variants. *J Allergy Clin Immunol* 2002;**109**:176.

36. Nutku E, Aizawa H, Hudson SA, Bochner BS. Ligation of Siglec-8: a selective mechanism for induction of human eosinophil apoptosis. *Blood* 2003;**101**:5014—20.

37. Liu SM, Xavier R, Good KL, Chtanova T, Newton R, Sisavanh M, et al. Immune cell transcriptome datasets reveal novel leukocyte subset-specific genes and genes associated with allergic processes. *J Allergy Clin Immunol* 2006;**118**:496—503.

38. Angata T, Hingorani R, Varki NM, Varki A. Cloning and characterization of a novel mouse Siglec, mSiglec-F: differential evolution of the mouse and human (CD33) Siglec-3-related gene clusters. *J Biol Chem* 2001;**276**:45128—36.

39. Zhang JQ, Biedermann B, Nitschke L, Crocker PR. The murine inhibitory receptor mSiglec-E is expressed broadly on cells of the innate immune system whereas mSiglec-F is restricted to eosinophils. *Eur J Immunol* 2004;**34**:1175—84.

40. Tateno H, Crocker PR, Paulson JC. Mouse Siglec-F and human Siglec-8 are functionally convergent paralogs that are selectively expressed on eosinophils and recognize 6′-sulfo-sialyl Lewis X as a preferred glycan ligand. *Glycobiology* 2005;**15**:1125—35.

41. Zhang M, Angata T, Cho JY, Miller M, Broide DH, Varki A. Defining the *in vivo* function of Siglec-F, a CD33-related Siglec expressed on mouse eosinophils. *Blood* 2007;**109**:4280—7.

42. Stevens WW, Kim TS, Pujanauski LM, Hao X, Braciale TJ. Detection and quantitation of eosinophils in the murine respiratory tract by flow cytometry. *J Immunol Methods* 2007;**327**:63—74.

43. Guo JP, Nutku E, Yokoi H, Schnaar R, Zimmermann N, Bochner BS. Siglec-8 and Siglec-F: inhibitory receptors on eosinophils, basophils and mast cells. *Allergy Clin Immunol Inter—J World Allergy Org* 2007;**19**:54—9.

44. Aizawa H, Zimmermann N, Carrigan PE, Lee JJ, Rothenberg ME, Bochner BS. Molecular analysis of human Siglec-8 orthologs relevant to mouse eosinophils: identification of mouse orthologs of Siglec-5 (mSiglec-F) and Siglec-10 (mSiglec-G). *Genomics* 2003;**82**:521—30.

45. Voehringer D, van Rooijen N, Locksley RM. Eosinophils develop in distinct stages and are recruited to peripheral sites by alternatively activated macrophages. *J Leukoc Biol* 2007;**81**:1434—44.

46. Ohnmacht C, Pullner A, van Rooijen N, Voehringer D. Analysis of eosinophil turnover *in vivo* reveals their active recruitment to and prolonged survival in the peritoneal cavity. *J Immunol* 2007;**179**:4766—74.

47. Carlens J, Wahl B, Ballmaier M, Bulfone-Paus S, Forster R, Pabst O. Common gamma-chain-dependent signals confer selective survival of eosinophils in the murine small intestine. *J Immunol* 2009;**183**:5600—7.

48. Doherty TA, Soroosh P, Broide DH, Croft M. CD4+ cells are required for chronic eosinophilic lung inflammation but not airway remodeling. *Am J Physiol Lung Cell Mol Physiol* 2009;**296**:L229—35.

49. Dyer KD, Moser JM, Czapiga M, Siegel SJ, Percopo CM, Rosenberg HF. Functionally competent eosinophils differentiated ex vivo in high purity from normal mouse bone marrow. *J Immunol* 2008;**181**:4004—9.

50. Varki NM, Varki A. Diversity in cell surface sialic acid presentations: implications for biology and disease. *Lab Invest* 2007;**87**:851—7.

51. Varki A. Natural ligands for CD33-related Siglecs? *Glycobiology* 2009;**19**:810—2.

52. Blixt O, Head S, Mondala T, Scanlan C, Huflejt ME, Alvarez R, et al. Printed covalent glycan array for ligand profiling of diverse glycan binding proteins. *Proc Natl Acad Sci USA* 2004;**101**:17033—8.

53. Rapoport EM, Pazynina GV, Sablina MA, Crocker PR, Bovin NV. Probing sialic acid binding Ig-like lectins (siglecs) with sulfated oligosaccharides. *Biochemistry (Mosc)* 2006;**71**:496—504.

54. Campanero-Rhodes MA, Childs RA, Kiso M, Komba S, Le Narvor C, Warren J, et al. Carbohydrate microarrays reveal sulphation as a modulator of siglec binding. *Biochem Biophys Res Commun* 2006;**344**:1141—6.

55. Paulson JC, Blixt O, Collins BE. Sweet spots in functional glycomics. *Nat Chem Biol* 2006;**2**:238—48.

56. Bochner BS, Alvarez RA, Mehta P, Bovin NV, Blixt O, White JR, et al. Glycan array screening reveals a candidate ligand for Siglec-8. *J Biol Chem* 2005;**280**:4307—12.

57. Nimrichter L, Burdick MM, Aoki K, Laroy W, Fierro MA, Hudson SA, et al. E-selectin receptors on human leukocytes. *Blood* 2008;**112**:3744—52.

58. Rosen SD. Ligands for L-selectin: homing, inflammation, and beyond. *Annu Rev Immunol* 2004;**22**:129—56.

59. Hudson SA, Bovin N, Schnaar RL, Crocker PR, Bochner BS. Eosinophil-selective binding and pro-apoptotic effect in vitro of a synthetic Siglec-8 ligand, polymeric 6′-sulfated sialyl Lewis X. *J Pharmacol Exp Ther* 2009;**330**:608—12.

60. Guo JP, Brummet ME, Myers AC, Na HJ, Rowland E, Schnaar RL, et al. Characterization of expression of glycan ligands for Siglec-F in normal mouse lungs. *Am J Respir Cell and Molec Biol* 2011;**44**:238—46.

61. Falco M, Biassoni R, Bottino C, Vitale M, Sivori S, Augugliaro R, et al. Identification and molecular cloning of p75/AIRM1, a novel member of the sialoadhesin family that functions as an inhibitory receptor in human natural killer cells. *J Exp Med* 1999;**190**:793—802.

62. Hoffmann A, Kerr S, Jellusova J, Zhang J, Weisel F, Wellmann U, et al. Siglec-G is a B1 cell-inhibitory receptor that controls expansion and calcium signaling of the B1 cell population. *Nat Immunol* 2007;**8**:695—704.

63. Bochner BS. Siglec-8 on human eosinophils and mast cells, and Siglec-F on murine eosinophils, are functionally related inhibitory receptors. *Clin Exp Allergy* 2009;**39**:317—24.

64. Nutku E, Aizawa H, Hudson SA, Bochner BS. Function of Siglec-8 on human eosinophils. *Clin Exp Allergy Reviews* 2004;**4**(Suppl. 2):76—81.

65. Nutku E, Hudson SA, Bochner BS. Mechanism of Siglec-8-induced human eosinophil apoptosis: role of caspases and mitochondrial injury. *Biochem Biophys Res Commun* 2005;**336**:918–24.

66. von Gunten S, Vogel M, Schaub A, Stadler BM, Miescher S, Crocker PR, et al. Intravenous immunoglobulin preparations contain anti-Siglec-8 autoantibodies. *J Allergy Clin Immunol* 2007;**119**:1005–11.

67. Nutku-Bilir E, Hudson SA, Bochner BS. Interleukin-5 priming of human eosinophils alters Siglec-8 mediated apoptosis pathways. *Am J Respir Cell Mol Biol* 2008;**38**:121–4.

68. Zimmermann N, McBride ML, Yamada Y, Hudson SA, Jones C, Cromie KD, et al. Siglec-F antibody administration to mice selectively reduces blood and tissue eosinophils. *Allergy* 2008;**63**:1156–63.

69. Song DJ, Cho JY, Miller M, Strangman W, Zhang M, Varki A, et al. Anti-Siglec-F antibody inhibits oral egg allergen induced intestinal eosinophilic inflammation in a mouse model. *Clin Immunol* 2009;**131**:157–69.

70. Song DJ, Cho JY, Lee SY, Miller M, Rosenthal P, Soroosh P, et al. Anti-Siglec-F antibody reduces allergen-induced eosinophilic inflammation and airway remodeling. *J Immunol* 2009;**183**:5333–41.

71. Walter RB, Raden BW, Zeng R, Hausermann P, Bernstein ID, Cooper JA. ITIM-dependent endocytosis of CD33-related Siglecs: role of intracellular domain, tyrosine phosphorylation, and the tyrosine phosphatases, Shp1 and Shp2. *J Leukoc Biol* 2008;**83**:200–11.

72. Tateno H, Li H, Schur MJ, Bovin N, Crocker PR, Wakarchuk WW, et al. Distinct endocytic mechanisms of CD22 (Siglec-2) and Siglec-F reflect roles in cell signaling and innate immunity. *Mol Cell Biol* 2007;**27**:5699–710.

73. Flood-Page P, Menzies-Gow A, Phipps S, Ying S, Wangoo A, Ludwig MS, et al. Anti-IL-5 treatment reduces deposition of ECM proteins in the bronchial subepithelial basement membrane of mild atopic asthmatics. *J Clin Invest* 2003;**112**:1029–36.

74. Cho JY, Song DJ, Pham A, Rosenthal P, Miller M, Dayan S, et al. Chronic OVA allergen challenged Siglec-F deficient mice have increased mucus, remodeling, and epithelial Siglec-F ligands which are up-regulated by IL-4 and IL-13. *Respir Res* 2010;**11**:154.

Chapter 15.5

Targeting the FIP1L1–PDGFRα and Variant PDGFRα and PDGFRβ Fusion Kinases

Jan Cools

THE FIP1L1–PDGFRα FUSION PROTEIN

Until 2003, the hypereosinophilic syndromes remained a heterogeneous group of diseases characterized by the presence of unexplained persistent eosinophilia. Molecular work had clearly documented a clonal origin of the disease in specific cases, but it remained impossible to distinguish these clonal leukemias from secondary eosinophilia. These observations indicated that a malignant disease was present at least in a subset of the patients. The lack of molecular insight, however, blocked the development of specific diagnostic tests to identify these malignant cases. Similarly, due to the lack of understanding of the biology and heterogeneity of this disease, treatment for individuals with the hypereosinophilic syndrome did not make much progress and was mainly limited to the use of interferon and steroids.

It was the development of the kinase inhibitor imatinib for the treatment of breakpoint cluster region–c-abl oncogene 1, nonreceptor tyrosine kinase (BCR–ABL1)-positive chronic myeloid leukemia that eventually made it possible to uncover the presence of the oncogenic tyrosine kinase FIP1 like 1 (S. cerevisiae)–platelet-derived growth factor receptor,α peptide (FIP1L1–PDGFRα) in a subset of patients with persistent unexplained eosinophilia.[1] In 2003, work that I performed together with Gary Gilliland, Elizabeth Stover, Daniel De Angelo, and many collaborators, led to the discovery of a small deletion on the long arm of chromosome 4 in a subset of hypereosinophilia patients.[2] This deletion results in the generation and expression of a chimeric gene consisting of part of the *FIP1L1* gene and part of the *PDGFRA* gene (Figure 15.5.1). Subsequently, the *FIP1L1–PDGFRA* fusion gene was identified in a subset of hyper-eosinophilia patients in several studies, indicating that the deletion in chromosome 4q and the generation of the fusion gene is a recurrent, but still rare event, which has now been detected in more than a hundred patients worldwide.[3–6] In addition to patients with chronic eosinophilic leukemia, the fusion was also detected in some patients with mastocytosis and even in some T-cell lymphomas.[7–9] Functional studies confirmed that the FIP1L1–PDGFRα protein is an activated tyrosine kinase that activates proliferation and survival pathways and is sensitive to inhibition by imatinib, explaining the positive response to imatinib observed in a subset of patients with HES.[2,10] Most studies have shown a strong correlation between the presence of the *FIP1L1–PDGFRA* gene fusion and the response to imatinib therapy. All patients that express the *FIP1L1–PDGFRA* fusion gene respond to imatinib, but in some studies responses were also documented in some *FIP1L1–PDGFRA*-negative patients. Other studies have not confirmed this observation, but it remains possible that some of the *FIP1L1–PDGFRA*-negative patients may harbor *PDGFRA* mutations, or variant *PDGFRA* or *PDGFRB* fusions, that have remained undetected.[11–13]

FIGURE 15.5.1 The FIP1L1—PDGFRA fusion gene is generated by a chromosomal deletion. *Top,* Schematic overview of the deletion on chromosome 4 causing the fusion of part of the FIP1 like 1 (S. cerevisiae) (*FIP1L1*) gene to part of the platelet-derived growth factor receptor,α peptide (*PDGFRA*) gene. The fusion gene is created when the 800-kb region between *FIP1L1* and *PDGFRA* is deleted. *Bottom,* The deletion results in the expression of an *FIP1L1—PDGFRA* fusion transcript, which encodes an FIP1L1—PDGFRα fusion protein. CHIC2, cysteine-rich hydrophobic domain 2; FIP1L1, FIP1 like 1 (S. cerevisiae); GSH2, GS homeobox 2; LNX, ligand of numb-protein X; PDGFRα, platelet-derived growth factor receptor, α polypeptide gene; *PDGFRA*, platelet-derived growth factor receptor polypeptide gene

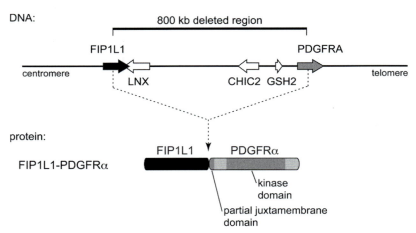

ADDITIONAL *PDGFRA* AND *PDGFRB* FUSIONS

In addition to the *FIP1L1—PDGFRA* fusion gene, which is the second most common fusion kinase in hematological malignancies (*BCR—ABL1* being the most common fusion gene), a few variant *PDGFRA* genes and numerous different *PDGFRB* fusion genes have been identified, of which only some are recurrent (Table 15.5.1). Of these, the transcription factor ETV6—platelet-derived growth factor

receptor, β peptide (*ETV6—PDGFRB/TEL—PDGFRB*) fusion gene was the first to be identified—in 1994 by Todd Golub and Gary Gilliland.[14] Since then, additional cases of the *ETV6—PDGFRB* fusion have been identified, as well as a large number of variant fusions involving *PDGFRB* (Table 15.5.2). Similarly, several variant *PDGFRA* fusion genes have been identified, including fusion partners such as *BCR* and *ETV6* (Table 15.5.1). Remarkably, while all these fusion genes have been associated with slightly variable myeloproliferative diseases, a unique feature of all *PDGFRA* and *PDGFRB* fusions is their association with eosinophilia. As a consequence, these malignancies are now renamed in the new World Health Organization (WHO) classification, being referred to as myeloid neoplasms associated with eosinophilia and abnormalities of *PDGFRA* or *PDGFRB* [the complete name of this entity is 'myeloid neoplasms associated with eosinophilia and abnormalities of PDGF receptor A and B (*PDGFRA* and *PDGFRB*), or FGF receptor 1 (*FGFR1*)[1]], since neoplasms with FGFR1 rearrangement are also associated with eosinophilia.

Importantly, all these PDGFRα and PDGFRβ fusion proteins act as oncogenic tyrosine kinases that can be effectively inhibited by the small molecule tyrosine kinase inhibitor imatinib. Thus, independent of the exact fusion that is expressed in the leukemia cells, the presence of a chimeric PDGFRα or PDGFRβ protein in patients with chronic eosinophilia predicts response to imatinib therapy.[2,12,15,16]

SIGNALING DOWNSTREAM OF FIP1L1—PDGFRα AND OTHER PDGFR FUSION PROTEINS

The FIP1L1—PDGFRα and all studied PDGFR fusion proteins are tyrosine kinases that are constitutively

TABLE 15.5.1 *PDGFRA* Fusion Genes Identified in Patients with Hypereosinophilia

Fusion Gene	Chromosomal Aberration	Number of Cases Described	Reference
FIP1L1 —PDGFRA	del[4](q12q12)	>100	2
BCR —PDGFRA	t[4;22](q12;q11)	5	42
KIF5B —PDGFRA	complex karyotype	1	43
CDK5RAP2 —PDGFRA	ins[9;4](q33;q12q25)	1	44
ETV6 —PDGFRA	t[4;12](q12;p13)	1	45
STRN —PDGFRA	t[2;4](p24;q12)	1	45

BCR, breakpoint cluster region; CDK5RAP2, CDK5 regulatory subunit associated protein 2; ETV6, transcription factor ETV6; FIP1L1, FIP1 like 1 (S. cerevisiae); KIF5B, kinesin family member 5B; PDGFRA, platelet-derived growth factor receptor,α peptide; STRN, striatin, calmodulin binding protein.

TABLE 15.5.2 Translocations Involving the *PDGFRB* Gene

Fusion Partners of *PDGFRB*	Chromosomal Location
TPM3	1q21
PDE4DIP	1q12
SPTBN1	2p21
WDR48	3p21
GOLGA4	3p21
PRKG2 [R]	4q21
HIP1	7q11
KANK1	9p24
CCDC6 (H4)	10q21
CAPRIN1 (GPIAP1)	11p13
ETV6 (TEL) [R]	12p13
BIN2	12q13
GIT2	12q24
SART3	12q24
NIN	14q24
TRIP11 (CEV14)	14q32
CCDC88C (KIAA1509) [R]	14q32
TP53BP1	15q15
NDE1	16p13
SPECC1 (HCMOGT-1)	17p11
RABEP1 (RABAPTIN-5)	17p13
MYO18A	17q11

[R]Indicates fusion partners that have been identified in more than one patient. BIN2, bridging integrator 2; CAPRIN1, cell cycle associated protein 1; CCDC6, coiled-coil domain containing 6; CCDC88C, coiled-coil domain containing 88C; ETV6, transcription factor ETV6; GIT2, protein-coupled receptor kinase interacting ArfGAP 2; GOLGA, golgin A; HIP1, huntingtin interacting protein 1; KANK1, KN motif and ankyrin repeat domains 1; MYO18A, myosin XVIIIA; NDE1, nudE nuclear distribution E homolog 1 (A. nidulans); NIN, ninein (GSK3B interacting protein); PDE4DIP, phosphodiesterase 4D interacting protein; PRKG2, protein kinase, cGMP-dependent, type II; RABEP1, rabaptin, RAB GTPase binding effector protein 1; SART3, squamous cell carcinoma antigen recognized by T cells 3; SPECC1, sperm antigen with calponin homology and coiled-coil domains 1; SPTBN1, spectrin, β, nonerythrocytic 1; TP53BP1, tumor protein p53 binding protein 1; TPM3, tropomyosin 3; TRIP11, thyroid hormone receptor interactor 11; WDR48, WD repeat domain 48.

active, without the need of ligand stimulation. In addition, recent work has illustrated that these proteins also escape normal degradation routes, another factor that assists in their oncogenic potential.[17] Similar to other activated tyrosine kinase fusions such as BCR−ABL1 and mutant fms-related tyrosine kinase 3 (FLT3), the FIP1L1−PDGFRα kinase is able to activate a diverse set of signaling proteins including signal transducer and activator of transcription 5A (STAT5A), phosphoinositide3-kinase (PI3K), dual specificity mitogen-activated protein kinase kinase (MEK/ERK), and mitogen-activated protein kinase (MAPK). This has been nicely demonstrated in human hematopoietic progenitor cell antigen CD34$^+$ cells, mouse bone marrow-derived pro-B-cell Ba/F3 cells stably expressing the *FIP1L1−PDGFRA* fusion, and human EOL-1 cells, a cell line derived from a patient with chronic eosinophilic leukemia.[2,10,18,19] Expression of FIP1L1−PDGFRα in the interleukin-3 (IL-3)-dependent Ba/F3 cells activates these signaling pathways and makes the cells independent of IL-3 for their proliferation and survival.[2] Similarly, inhibition of FIP1L1−PDGFRα kinase activity in EOL-1 cells leads to inhibition of proliferation and cell death.[10] Finally, expression of FIP1L1−PDGFRA in mouse bone marrow cells *in vivo* induces a myeloproliferative disorder.[20] Collectively, these data confirm that FIP1L1−PDGFRα kinase activity activates signaling pathways that drive the proliferation and survival of hematopoietic cells. The variant PDGFRα and PDGFRβ fusion proteins also share similar properties. Many of these variant fusions have also been studied in Ba/F3 cells or 32D cells (an IL-3 dependent myeloid cell line), and some also in the mouse bone marrow transplant model.

One important unanswered question is why the PDGFR fusion proteins cause eosinophilia. Indeed, PDGFR fusions can transform hematopoietic cells to cytokine-independent proliferation, but in simple mouse models these fusions cause a general myeloproliferative malignancy in which all myeloid cells are elevated and eosinophilia is not observed.[20] A possible explanation is that the fusion by itself is not sufficient to cause eosinophilia and that additional factors are required. A mouse model in which expression of the FIP1L1−PDGFRα fusion is combined with ectopic expression of IL-5 demonstrates a clear induction of eosinophilia, suggesting that IL-5 and/or other eosinophil growth-promoting factors or their receptors may cooperate with PDGFR fusion proteins in the human disease.[21] IL-5 may be produced by T cells, and T-cell abnormalities have indeed been observed in some *FIP1L1−PDGFRA* positive patients.[22] Alternatively, it was also shown that a variation in the IL-5 receptor (IL-5RA) that is present in the normal population is somehow associated with the severity of the disease in *FIP1L1−PDGFRA*-positive leukemias.[23] This variation leads to a higher expression of the IL-5R and may possibly sensitize the cells to IL-5 stimulation.

TARGETING FIP1L1—PDGFRα WITH KINASE INHIBITORS

As described above, the *FIP1L1—PDGFRA* fusion is sensitive to the kinase inhibitor imatinib, a characteristic that was central to the discovery of the *FIP1L1—PDGFRA* fusion gene.[1,2] In fact, the FIP1L1—PDGFRα kinase is around 200-fold more sensitive to imatinib than BCR—ABL, the target for which this inhibitor was designed.[2] The high sensitivity to imatinib brings a number of advantages: patients can be treated with low doses of imatinib (usually 100 mg/day), which comes with a lower risk of side effects, and a significantly lower cost of treatment.[1,15,24] In addition, reports have shown that during remission, patients can be treated with doses as low as 100 mg/week.[24] It is still unclear whether imatinib treatment can be stopped completely during remission, and especially at what time point this would be possible.[25] Despite the fact that most *FIP1L1—PDGFRA*-positive patients obtain a complete molecular remission after only few months of imatinib treatment,[26] trials have been undertaken to stop imatinib treatment. In most patients, however, reappearance of *FIP1L1—PDGFRA* transcripts has been observed at some point after termination of imatinib therapy.[25] These data seem to suggest that even though *FIP1L1—PDGFRA*-transformed cells are highly sensitive to imatinib, it remains difficult to completely eradicate the leukemia-initiating cells with imatinib treatment only. Further definitive studies are required, and there is hope that with close molecular monitoring, termination of imatinib therapy may be possible, at least for a subset of patients.

DEVELOPMENT OF RESISTANCE TO TYROSINE KINASE INHIBITOR THERAPY

Despite the high sensitivity of the FIP1L1—PDGFRα kinase to imatinib, the development of resistance to imatinib treatment has been observed in a few patients. We can conclude, however, from all available studies that resistance does not occur in patients with chronic disease. So far, all patients who developed resistance to imatinib had a more advanced disease and were those in whom treatment with imatinib was only started when the disease was at an advanced stage.[2,27−29] It is also not clear whether these patients had a clear phase of chronic disease, and it is most likely that the disease in these patients developed immediately as an acute leukemia.

In these patients, treatment with imatinib initially induced a complete hematological remission that, however, only lasted a few months. This is in sharp contrast with *FIP1L1—PDGFRA*-positive patients in the chronic phase of the disease, where remissions are stable and last for years, most likely for the rest of their lives.[2,15,26,30,31] Thus,

FIGURE 15.5.2 Mutations in FIP1L1—PDGFRα causing resistance to tyrosine kinase inhibitors. Mutations in the kinase domain of platelet-derived growth factor receptorα peptide (PDGFRα) can cause resistance to tyrosine kinase inhibitor therapy. The most frequent mutation observed to date is the threonine 674 to isoleucine (T674I) mutation in the adenosine triphosphate-binding pocket of the kinase domain. This mutation confers resistance to imatinib, but not to sorafenib. The aspartic acid 842 to valine (D842V) mutation close to the activation loop in the kinase domain causes resistance to both imatinib and sorafenib, but has only been identified in one patient after treatment with sorafenib.

to date there is no evidence that resistance development is a problem associated with life-long treatment of these patients, but is specifically associated with the advanced disease.

It is also remarkable that almost all patients who developed resistance to imatinib had acquired the same threonine 674 to isoleucine (T674I) mutation in the kinase domain of FIP1L1—PDGFRα[2,3] (Figure 15.5.2). In only one publication, resistance was reported to be caused by another mutation.[32] This preference for one particular resistance mutation was recently explained by a large-scale *in vitro* random mutagenesis profiling study. This study confirmed that upon treatment with imatinib, the T674I mutation was the only predominant resistance mutation occurring *in vitro*.[33] These data, together with the data from clinical studies with imatinib, suggest that only very few mutations are capable of providing resistance to imatinib in the FIP1L1—PDGFRα context, which may explain why this kinase is exquisitely sensitive to the drug. Thus, while many different mutations may slightly shift the sensitivity of FIP1L1—PDGFRA to imatinib, only the T674I and aspartic acid 842 to valine (D842V) mutations seem capable of providing clinical resistance (half-maximal inhibitory concentration, IC50 >1 μM)[27,33] (Figure 15.5.2).

In addition to imatinib, several other small molecule kinase inhibitors have been identified that have potent activity against PDGFRα and PDGFRβ. These include several drugs that are already approved by the United States Food and Drug Administration for the treatment of other leukemias or solid tumors, such as dasatinib, nilotinib, and sorafenib.[34−38] In addition, a large number of drugs that are still in preclinical development have been reported to be potent PDGFR inhibitors. This high number of alternative PDGFR inhibitors may be of benefit for the treatment of relapsed patients, although recent studies indicate the difficulty of treating relapsed patients. In the context of PDGFR fusions, relapses are rare and have only been observed in few patients with the *FIP1L1—PDGFRA* fusion. These leukemias were typically at a more advanced stage when imatinib treatment was initiated, and already had characteristics of acute leukemia. In one patient who developed imatinib resistance due to the T674I mutation in *FIP1L1—PDGFRA*, sorafenib treatment was shown to be effective *in vivo*, but only for a very short duration. A few weeks after initiation of sorafenib therapy, the leukemia cells had also acquired resistance to sorafenib, now due to a D842V mutation.[27] Remarkably, this was not caused by further mutation of T674I-mutant leukemia cells, but instead the D842V mutation was acquired in *FIP1L1—PDGFRA* cells lacking the T674I mutation. This example illustrates the high mutation rate of acute leukemias, and indicates that obtaining long-term responses may not be possible once resistance has started occurring.

EXPLORING OTHER WAYS TO TARGET FIP1L1—PDGFRα

When thinking about other ways to inhibit FIP1L1—PDGFRα kinase activity, an obvious strategy could be to target the FIP1L1 part of the fusion. It was previously shown that several fusion tyrosine kinases, such as BCR—ABL, ETV6—tyrosine-protein kinase JAK2 (JAK2), and zinc finger MYM-type protein 2 (ZNF198)—FGFR1, that these kinases were dependent on the presence of oligomerization domains within BCR, ETV6, or ZNF198 for the activation of kinase activity. Targeting the BCR coiled-coil domain has also been shown to be a valid strategy to decrease BCR—ABL1 kinase activity and transforming potential, and thus could also be a potential avenue for FIP1L1.

Unexpectedly, however, work with deletion variants of the *FIP1L1—PDGFRA* fusion gene revealed that the *FIP1L1* part is in fact completely dispensable for kinase activity, and for transformation of hematopoietic cells *in vitro* and *in vivo*: the FIP1L1 protein can be completely removed from the FIP1L1—PDGFRA fusion and the truncated protein retains all of its kinase and oncogenic activity.[39] The reason that the FIP1L1—PDGFRα protein is constitutively active lies not in any specific characteristic of the FIP1L1 protein, but rather in the interruption of the juxtamembrane domain of PDGFRα (Figure 15.5.3).[39] When the FIP1L1 gene becomes fused to the *PDGFRA* gene, part of exon 12 of *PDGFRA* is deleted. This results in truncation of the juxtamembrane domain, which is an inhibitory domain within PDGFRα that keeps the kinase inactive in the absence of ligand. In the absence of a functional juxtamembrane domain within the FIP1L1—PDGFRA fusion protein, the kinase is constitutively active and hence has transforming activity. The fusion with FIP1L1 is required to create this deletion and to provide a promoter for transcriptional

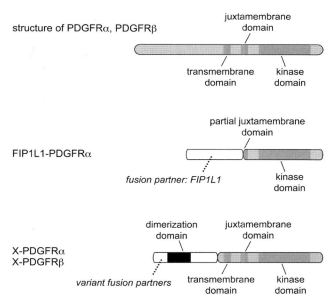

FIGURE 15.5.3 Comparison of the structure of PDGFR, FIP1L1—PDGFRα and variant PDGFR fusion proteins. The platelet-derived growth factor receptorα peptide (PDGFRα) and platelet-derived growth factor receptorβ peptide (PDGFRβ) proteins contain an extracellular ligand-binding domain, a transmembrane domain, a juxtamembrane domain, and a kinase domain. In the PDGFR fusion proteins, the ligand-binding domain is replaced by the fusion partner protein. In the case of the FIP1 like 1 (S. cerevisiae) (FIP1L1)—PDGFRα fusion, the transmembrane domain and at least part of the juxtamembrane domain are also removed. In case of other variant X—PDGFR fusions (where X represents the fusion partner), the transmembrane and juxtamembrane domains are usually present and the fusion partner contains a homodimerization domain. In the case of the FIP1L1—PDGFRα fusion, there is no known functional domain present in FIP1L1 contributing to kinase activation.

activity of the fusion gene and an ATG start codon for translation initiation. These findings are in line with genetic data obtained from the patients. Of all the different FIP1L1—PDGFRA fusion transcripts that have been identified in different patients, one constant finding is that the fusion results in the interruption or complete deletion of the juxtamembrane region of *PDGFRA*.[2,40]

In the context of the majority of variant PDGFRα and PDGFRβ fusions, the juxtamembrane region of PDGFR is usually present and intact. In these proteins, enforced homodimerization is required to activate the kinase domains and to overcome the inhibitory function of the juxtamembrane region (Figure 15.3.3). Thus, in these fusions, the partner protein is highly important and must be a protein with a homodimerization capacity, while the breakpoint in PDGFR can be more variable. Again, this is in line with biochemical and genetic data, which show that homodimerization is indeed observed in these fusion proteins and that the breakpoints in PDGFRA and PDGFRB are more variable than in the case of the *FIP1L1—PDGFRA* fusion.[40,41]

CONCLUSIONS

Independent of the mechanism of activation and the identity of the fusion partner, all PDGFRα and PDGFRβ gene fusions identified to date are highly sensitive to imatinib. These findings have opened a new approach for the treatment of eosinophilia caused by PDGFR oncogenes such as FIP1L1−PDGFRA, ETV6−PDGFRB, and the numerous variant PDGFRA and PDGFRB fusions. In addition to imatinib, other kinase inhibitors such as dasatinib, nilotinib, sorafenib—all of which have been developed for the treatment of other cancers—have all been shown to be active against PDGFR fusion proteins. This puts us in a remarkably advantageous position of having several targeted therapies available for a rare disease. Despite the fact that PDGFR fusions are rare in hematological malignancies, their detection is of the utmost importance, since their presence predicts the response to a nontoxic and highly effective therapy.

REFERENCES

1. Gleich GJ, Leiferman KM, Pardanani A, Tefferi A, Butterfield JH. Treatment of hypereosinophilic syndrome with imatinib mesilate. *Lancet* 2002;**359**:1577−8.

2. Cools J, DeAngelo DJ, Gotlib J, Stover EH, Legare RD, Cortes J, et al. A tyrosine kinase created by fusion of the PDGFRA and FIP1L1 genes as a therapeutic target of imatinib in idiopathic hypereosinophilic syndrome. *N Engl J Med* 2003;**348**:1201−14.

3. Gotlib J, Cools J. Five years since the discovery of FIP1L1-PDGFRA: what we have learned about the fusion and other molecularly defined eosinophilias. *Leukemia* 2008;**22**:1999−2010.

4. Pardanani A, Brockman SR, Paternoster SF, Flynn HC, Ketterling RP, Lasho TL, et al. FIP1L1-PDGFRA fusion: prevalence and clinicopathologic correlates in 89 consecutive patients with moderate to severe eosinophilia. *Blood* 2004;**104**:3038−45.

5. Vandenberghe P, Wlodarska I, Michaux L, Zachee P, Boogaerts M, Vanstraelen D, et al. Clinical and molecular features of FIP1L1-PDFGRA (+) chronic eosinophilic leukemias. *Leukemia* 2004;**18**:734−42.

6. Klion AD, Noel P, Akin C, Law MA, Gilliland DG, Cools J, et al. Elevated serum tryptase levels identify a subset of patients with a myeloproliferative variant of idiopathic hypereosinophilic syndrome associated with tissue fibrosis, poor prognosis, and imatinib responsiveness. *Blood* 2003;**101**:4660−6.

7. Pardanani A, Ketterling RP, Brockman SR, Flynn HC, Paternoster SF, Shearer BM, et al. CHIC2 deletion, a surrogate for FIP1L1-PDGFRA fusion, occurs in systemic mastocytosis associated with eosinophilia and predicts response to imatinib mesylate therapy. *Blood* 2003;**102**:3093−6.

8. Metzgeroth G, Walz C, Score J, Siebert R, Schnittger S, Haferlach C, et al. Recurrent finding of the FIP1L1-PDGFRA fusion gene in eosinophilia-associated acute myeloid leukemia and lymphoblastic T-cell lymphoma. *Leukemia* 2007;**21**:1183−8.

9. Maric I, Robyn J, Metcalfe DD, Fay MP, Carter M, Wilson T, et al. KIT D816V-associated systemic mastocytosis with eosinophilia and FIP1L1/PDGFRA-associated chronic eosinophilic leukemia are distinct entities. *J Allergy Clin Immunol* 2007;**120**:680−7.

10. Cools J, Quentmeier H, Huntly BJ, Marynen P, Griffin JD, Drexler HG, et al. The EOL-1 cell line as an in vitro model for the study of FIP1L1-PDGFRA-positive chronic eosinophilic leukemia. *Blood* 2004;**103**:2802−5.

11. Elling C, Erben P, Walz C, Frickenhaus M, Schemionek M, Stehling M, et al. Novel imatinib-sensitive PDGFRA activating point mutations in hypereosinophilic syndrome induce growth factor independence and leukemia-like disease. *Blood* 2011;**117**(10):2935−43.

12. Apperley JF, Gardembas M, Melo JV, Russell-Jones R, Bain BJ, Baxter EJ, et al. Response to imatinib mesylate in patients with chronic myeloproliferative diseases with rearrangements of the platelet-derived growth factor receptor beta. *N Engl J Med* 2002;**347**:481−7.

13. Erben P, Gosenca D, Muller MC, Reinhard J, Score J, Del Valle F, et al. Screening for diverse PDGFRA or PDGFRB fusion genes is facilitated by generic quantitative reverse transcriptase polymerase chain reaction analysis. *Haematologica* 2010;**95**:738−44.

14. Golub TR, Barker GF, Lovett M, Gilliland DG. Fusion of PDGF receptor beta to a novel ets-like gene, tel, in chronic myelomonocytic leukemia with t(5;12) chromosomal translocation. *Cell* 1994;**77**:307−16.

15. Baccarani M, Cilloni D, Rondoni M, Ottaviani E, Messa F, Merante S, et al. The efficacy of imatinib mesylate in patients with FIP1L1-PDGFRalpha-positive hypereosinophilic syndrome. Results of a multicenter prospective study. *Haematologica* 2007;**92**:1173−9.

16. David M, Cross NC, Burgstaller S, Chase A, Curtis C, Dang R, et al. Durable responses to imatinib in patients with PDGFRB fusion gene-positive and BCR-ABL-negative chronic myeloproliferative disorders. *Blood* 2007;**109**:61−4.

17. Toffalini F, Kallin A, Vandenberghe P, Pierre P, Michaux L, Cools J, et al. The fusion proteins TEL-PDGFRbeta and FIP1L1-PDGFRalpha escape ubiquitination and degradation. *Haematologica* 2009;**94**:1085−93.

18. Buitenhuis M, Verhagen LP, Cools J, Coffer PJ. Molecular mechanisms underlying FIP1L1-PDGFRA-mediated myeloproliferation. *Cancer Res* 2007;**67**:3759−66.

19. Goss VL, Lee KA, Moritz A, Nardone J, Spek EJ, MacNeill J, et al. A common phosphotyrosine signature for the Bcr-Abl kinase. *Blood* 2006;**107**:4888−97.

20. Cools J, Stover EH, Boulton CL, Gotlib J, Legare RD, Amaral SM, et al. PKC412 overcomes resistance to imatinib in a murine model of FIP1L1-PDGFRalpha-induced myeloproliferative disease. *Cancer Cell* 2003;**3**:459−69.

21. Yamada Y, Rothenberg ME, Lee AW, Akei HS, Brandt EB, Williams DA, et al. The FIP1L1-PDGFRA fusion gene cooperates with IL-5 to induce murine hypereosinophilic syndrome (HES)/chronic eosinophilic leukemia (CEL)-like disease. *Blood* 2006;**107**:4071−9.

22. Helbig G, Wieczorkiewicz A, Dziaczkowska-Suszek J, Majewski M, Kyrcz-Krzemien S. T-cell abnormalities are present at high frequencies in patients with hypereosinophilic syndrome. *Haematologica* 2009;**94**:1236−41.

23. Burgstaller S, Kreil S, Waghorn K, Metzgeroth G, Preudhomme C, Zoi K, et al. The severity of FIP1L1-PDGFRA-positive chronic eosinophilic leukaemia is associated with polymorphic variation at the IL5RA locus. *Leukemia* 2007;**21**:2428−32.

24. Helbig G, Stella-Holowiecka B, Majewski M, Calbecka M, Gajkowska J, Klimkiewicz R, et al. A single weekly dose of imatinib is sufficient to induce and maintain remission of chronic eosinophilic leukaemia in FIP1L1-PDGFRA-expressing patients. *Br J Haematol* 2008;**141**:200−4.

25. Klion AD, Robyn J, Maric I, Fu W, Schmid L, Lemery S, et al. Relapse following discontinuation of imatinib mesylate therapy for FIP1L1/PDGFRA-positive chronic eosinophilic leukemia: implications for optimal dosing. *Blood* 2007;**110**:3552−6.

26. Martinelli G, Malagola M, Ottaviani E, Rosti G, Trabacchi E, Baccarani M. Imatinib mesylate can induce complete molecular remission in FIP1L1-PDGFR-a positive idiopathic hypereosinophilic syndrome. *Haematologica* 2004;**89**:236−7.

27. Lierman E, Michaux L, Beullens E, Pierre P, Marynen P, Cools J, et al. FIP1L1-PDGFRalpha D842V, a novel panresistant mutant, emerging after treatment of FIP1L1-PDGFRalpha T674I eosinophilic leukemia with single agent sorafenib. *Leukemia* 2009;**23**:845−51.

28. Ohnishi H, Kandabashi K, Maeda Y, Kawamura M, Watanabe T. Chronic eosinophilic leukaemia with FIP1L1-PDGFRA fusion and T674I mutation that evolved from Langerhans cell histiocytosis with eosinophilia after chemotherapy. *Br J Haematol* 2006;**134**:547−9.

29. von Bubnoff N, Sandherr M, Schlimok G, Andreesen R, Peschel C, Duyster J. Myeloid blast crisis evolving during imatinib treatment of an FIP1L1-PDGFR alpha-positive chronic myeloproliferative disease with prominent eosinophilia. *Leukemia* 2005;**19**:286−7.

30. Klion AD, Robyn J, Akin C, Noel P, Brown M, Law M, et al. Molecular remission and reversal of myelofibrosis in response to imatinib mesylate treatment in patients with the myeloproliferative variant of hypereosinophilic syndrome. *Blood* 2004;**103**:473−8.

31. Jovanovic JV, Score J, Waghorn K, Cilloni D, Gottardi E, Metzgeroth G, et al. Low-dose imatinib mesylate leads to rapid induction of major molecular responses and achievement of complete molecular remission in FIP1L1-PDGFRA-positive chronic eosinophilic leukemia. *Blood* 2007;**109**:4635−40.

32. Salemi S, Yousefi S, Simon D, Schmid I, Moretti L, Scapozza L, et al. A novel FIP1L1-PDGFRA mutant destabilizing the inactive conformation of the kinase domain in chronic eosinophilic leukemia/hypereosinophilic syndrome. *Allergy* 2009;**64**:913−8.

33. von Bubnoff N, Gorantla SP, Engh RA, Oliveira TM, Thone S, Aberg E, et al. The low frequency of clinical resistance to PDGFR inhibitors in myeloid neoplasms with abnormalities of PDGFRA might be related to the limited repertoire of possible PDGFRA kinase domain mutations in vitro. *Oncogene* 2011;**30**(8):933−43.

34. Lierman E, Folens C, Stover EH, Mentens N, Van Miegroet H, Scheers W, et al. Sorafenib is a potent inhibitor of FIP1L1-PDGFRalpha and the imatinib-resistant FIP1L1-PDGFRalpha T674I mutant. *Blood* 2006;**108**:1374−6.

35. Lierman E, Lahortiga I, Van Miegroet H, Mentens N, Marynen P, Cools J. The ability of sorafenib to inhibit oncogenic PDGFRbeta and FLT3 mutants and overcome resistance to other small molecule inhibitors. *Haematologica* 2007;**92**:27−34.

36. von Bubnoff N, Gorantla SP, Thone S, Peschel C, Duyster J. The FIP1L1-PDGFRA T674I mutation can be inhibited by the tyrosine kinase inhibitor AMN107 (nilotinib). *Blood* 2006;**107**:4970−1. author reply 4972.

37. Stover EH, Chen J, Lee BH, Cools J, McDowell E, Adelsperger J, et al. The small molecule tyrosine kinase inhibitor AMN107 inhibits TEL-PDGFRbeta and FIP1L1-PDGFRalpha in vitro and in vivo. *Blood* 2005;**106**:3206−13.

38. Baumgartner C, Gleixner KV, Peter B, Ferenc V, Gruze A, Remsing Rix LL, et al. Dasatinib inhibits the growth and survival of neoplastic human eosinophils (EOL-1) through targeting of FIP1L1-PDGFRalpha. *Exp Hematol* 2008;**36**:1244−53.

39. Stover EH, Chen J, Folens C, Lee BH, Mentens N, Marynen P, et al. Activation of FIP1L1-PDGFRalpha requires disruption of the juxtamembrane domain of PDGFRalpha and is FIP1L1-independent. *Proc Natl Acad Sci USA* 2006;**103**:8078−83.

40. Walz C, Score J, Mix J, Cilloni D, Roche-Lestienne C, Yeh RF, et al. The molecular anatomy of the FIP1L1-PDGFRA fusion gene. *Leukemia* 2009;**23**:271−8.

41. Lierman E, Cools J. ETV6 and PDGFRB: a license to fuse. *Haematologica* 2007;**92**:145−7.

42. Baxter EJ, Hochhaus A, Bolufer P, Reiter A, Fernandez JM, Senent L, et al. The t(4;22)(q12;q11) in atypical chronic myeloid leukaemia fuses BCR to PDGFRA. *Hum Mol Genet* 2002;**11**:1391−7.

43. Score J, Curtis C, Waghorn K, Stalder M, Jotterand M, Grand FH, et al. Identification of a novel imatinib responsive KIF5B-PDGFRA fusion gene following screening for PDGFRA overexpression in patients with hypereosinophilia. *Leukemia* 2006;**20**:827−32.

44. Walz C, Curtis C, Schnittger S, Schultheis B, Metzgeroth G, Schoch C, et al. Transient response to imatinib in a chronic eosinophilic leukemia associated with ins(9;4)(q33;q12q25) and a CDK5RAP2-PDGFRA fusion gene. *Genes Chromosomes Cancer* 2006;**45**:950−6.

45. Curtis CE, Grand FH, Musto P, Clark A, Murphy J, Perla G, et al. Two novel imatinib-responsive PDGFRA fusion genes in chronic eosinophilic leukaemia. *Br J Haematol* 2007;**138**:77−81.

Emerging Concepts

Introduction

James J. Lee

EOSINOPHILS AND THE FUTURE: YOU NEVER KNOW WHERE THE NEXT IDEAS WILL COME FROM OR HOW FAR THEY WILL TAKE YOU!

My attraction to science and doing basic research comes from the misguided perspective, which usually develops early in life, that the world and events can be described from first principles. One simply had to devise ways to explain the series of absolute truths that surround us—the sky is blue, water is wet, and eosinophils are clearly the most important leukocytes in the blood (sorry … I simply couldn't help myself!). Unfortunately, too many of us never fully lose this simplistic early childhood view of the world and it interferes with our thinking as we try to develop hypotheses to account for experimental observations. That is, the eosinophil community at large (including me!) is often not immune to this bias. Many times the only eosinophil studies believed are the ones that reinforce previously held views and fundable research is usually only an incremental advance on previously established results. While this cynical view of the eosinophil research is clearly an exaggeration, sadly, there are more elements of truth here than most of us are willing to admit in public. On this point, the enduring and pervasive character of this *safe* and intellectually *soft* approach is surprisingly robust. This is especially true given the truly extraordinary scientific achievements in biomedical science that have occurred from thinking *outside the box*. One need look no further than the story surrounding the molecular tools that are now indispensable and commonplace elements of research activities in any medical center. Who would have thought that studies of bacterial mating types and host defense strategies against bacterial phages (see for example[1]) could possibly have led to the molecular biology revolution that is the driving force currently behind many basic as well as patient-based clinical studies. Even in those cases where the ideas and concepts were eventually refuted, the resulting discussions and debates substantially move areas of research forward. For example, in 1969 Britten and Davidson proposed a model for eukaryotic gene expression based on the presence of repetitive elements that could regulate *batteries* of genes in a coordinated fashion.[2] Clearly, subsequent studies showed the shortcomings of this hypothesis. Nonetheless, the *outside-the-box* character of this idea created the discourse and experimental strategies that have led to the currently accepted understanding of unique cis-acting regulatory sequence elements that drive the coordinated expression of multiple genes and/or gene families. I would suggest that this type of event (i.e., thinking outside the confines of established parameters) isn't an oddity at all and is instead all too often a moment of clarity. In this spirit, the coauthors of the subchapters within this 'Emerging Concepts' chapter describe research that neither reinforces current views of eosinophil effector functions nor are incremental advances of existing paradigms. As the heading suggests, *you never know where the next ideas will come from or how far they will take you!*

The subchapters focus on four novel areas of eosinophil research. In particular, these authors described new and unique areas of eosinophil research that may have significant implications on the emerging role(s) of eosinophils in health and disease. In Chapter 16.2, Foster and colleagues describe provocative studies that this group and others have performed in the area of gene regulation and allergic immune responses. Specifically, the subchapter focuses on the importance of the new area of microRNAs (miRNAs), which are small, highly conserved noncoding RNAs that were first shown in *Caenorhabditis elegans*,[3] as important regulators of nearly all cellular and molecular pathways associated with disease. The potential roles of miRNAs in eosinophil development are discussed, including their key position in the regulation of inflammatory events and the pathogenesis associated with eosinophilic diseases such as allergy, asthma, and cancer. In Chapter 16.3, David Traver and his colleague Keir Balla present an amazing and provocative story showing that eosinophils are not just mammalian blood cells found in human patients and mouse models. Instead, the studies presented in this subchapter suggest that eosinophils are evolutionarily conserved granulocytes that have likely been recognizable components in the blood of vertebrates for more than 400 million

Eosinophils in Health and Disease. http://dx.doi.org/10.1016/B978-0-12-394385-9.00016-X

years. Specifically, these authors describe their pioneering studies using zebrafish as a model system to demonstrate conserved elements of eosinophil differentiation, gene expression, and responses to pathological assaults such as helminth infection and antigen challenge.

Chapter 16.4 centers on the recent revelations regarding eosinophils as regulators of the tissue immune microenvironment, particularly as it relates to the establishment and propagation of T-helper type 2 (T_h2)-driven immunity. Moqbel and colleagues initially describe an emerging concept in which eosinophils help polarize the microenvironment toward T_h2 immune responses through the novel mechanism of tryptophan catabolism mediated by eosinophil-specific expression of indoleamine 2,3-dioxygenase. This subchapter outlines prospective mechanisms and proposes unique areas in which eosinophils may directly contribute to immune responses such as the selection of T cells and their maturation in the thymus. In Chapter 16.5, D. Smith and M. Comeau explore the emerging area of how local eosinophil effector functions are modulated by the group collectively known as the *epithelial-derived cytokines* (i.e., interleukin-25, interleukin-33, and thymic stromal lymphopoietin). This subchapter presents the available evidence suggesting that eosinophils are underappreciated targets of these cytokines, through modulating their production/maturation as well as their survival in tissues (i.e., their local accumulation). Moreover, these authors expand the novel concept of eosinophils as active participants establishing regulatory networks promoting T_h2 inflammation as opposed to passive participants simply mediating destructive end-stage effector activities.

Given the amazing lack of detailed knowledge that surrounds a leukocyte whose existence has been well documented for more than 130 years,[4] the potential importance and need to explore new areas of eosinophil biology cannot be overestimated. Thinking about old problems from new perspectives and risking failure form the core of any meaningful intellectual success, reminding me of the quote from Charles Darwin that I often use to encourage young trainees to think outside the box: *'False views, if supported by some evidence, do little harm, for everyone takes a salutary pleasure in proving their falseness.'*

REFERENCES

1. Morange M. *A History of Molecular Biology.* Cambridge, Mass: Harvard University Press; 1998.
2. Britten RJ, Davidson EH. Gene Regulation for Higher Cells: A Theory. *Science* 1969;**165**:349–57. 5789433.
3. Bartel DP. Micrornas: Genomics, Biogenesis, Mechanism, and Function. *Cell* 2004;**116**:281–97. 14744438.
4. Erlich P. Ueber Die Specifischen Granulationen Des Blutes [in German]. *Arch Anant Physiol* 1879;**3**:571–9.

Understanding the Role of MicroRNA in Regulating Immune Responses: A New Approach to Treating Eosinophilic Disorders and Allergic Inflammation?

Catherine Ptaschinski, Maximilian Plank, Joerg Mattes and Paul S. Foster

INTRODUCTION

MicroRNAs (miRNAs) are small, noncoding RNAs, approximately 22 nucleotides (nt) in length, and are increasingly recognized as important regulators of gene expression.[1] MiRNAs are involved in fine-tuning many biological processes, including apoptosis, development, cell identity, hematopoiesis, inflammation, and organogenesis.[2] Additionally, transcriptional profiles of miRNAs in cells and biological fluids are being used in the diagnosis and prognosis of disease, particularly in cancer.[3] Investigations into the role of miRNA in the regulation of cellular functions have also identified this class of molecule as a new target for potential therapeutics.[4] More than 1000 miRNAs have been discovered in humans, and it is hypothesized that miRNAs may regulate the expression of one-third of all genes.[5]

MICRORNA BIOGENESIS

MiRNAs are encoded either within the introns of protein-coding genes or as independent miRNA genes, and are usually transcribed by RNA polymerase II in addition to other transcription factors.[6] Alternatively, some miRNAs are grouped in clusters within the genome and are transcribed into a single transcript and then separately processed into functional miRNAs.[7] The biogenesis of miRNAs is detailed in Figure 16.2.1. Firstly, the primary transcript, known as the pri-miRNA, is processed by nuclear ribonuclease 3—microprocessor complex subunit DGCR8 (Drosha—DGCR8); these enzymes excise a stem-loop RNA structure known as the pre-miRNA from the primary transcript. The pre-miRNA is a double-stranded RNA hairpin approximately 70 nt long, which is exported from the nucleus by exportin-5 into the cytoplasm; this process in dependent on GTP-binding nuclear protein Ran—guanosine triphosphate. The pre-miRNA is spliced

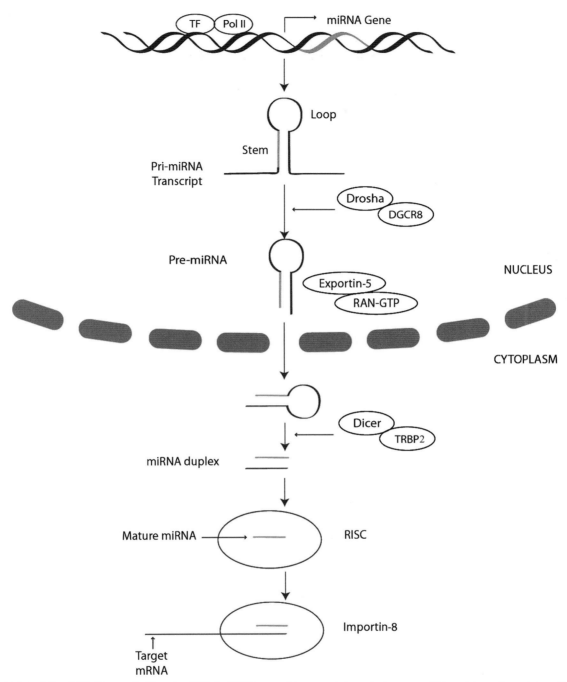

FIGURE 16.2.1 MicroRNA biogenesis. The microRNA (miRNA) gene, either as an independent gene or within an intron of a protein-coding gene, is transcribed by RNA polymerase II (Pol II) and other transcription factors (TF). The pri-miRNA is a processed by nuclear RNase III enzymes ribonuclease 3 (Drosha) and microprocessor complex subunit DGCR8 (DGCR8) to excise a pre-miRNA from the stem-loop structure. The pre-miRNA is a double-stranded hairpin that is exported from the nucleus by exportin-5, a process dependent on GTP-binding nuclear protein Ran—guanosine triphosphate (RAN-GTP). Once in the cytoplasm, the pre-miRNA is spliced by the RNase III enzyme endonuclease Dicer (Dicer), which works in concert with RISC-loading complex subunit TARBP2 (TAR RNA-binding protein 2;TRBP2). One strand of this miRNA duplex is incorporated into the RISC complex, and is targeting to mRNAs by importin-8.

by the cytoplasmic RNase III enzyme, endonuclease Dicer, into an miRNA duplex of approximately 22 nt, with help from RISC-loading complex subunit TARBP2 (TAR RNA-binding protein 2; TRBP2). This duplex is then unwound, and one strand is incorporated into the RNA-induced silencing complex (RISC), which is composed of the mature miRNA and several proteins, most importantly from the Argonaute family of RNA-binding RNA endonucleases, as well as other proteins like GW182 family members. This functional complex can then regulate mRNA levels, as the importin-8 protein can target Argonaute family proteins within the RISC complex to mRNA strands.[8]

MECHANISMS OF MICRORNA FUNCTION

MiRNAs primarily function by binding to the 3′ untranslated region (UTR) of a target mRNA.[9] A region of 2−8 nt at the 5′ end of the miRNA is known as the seed sequence, which is highly complementary to the target mRNA region, while the remaining nucleotide sequence is not fully complementary to the target.[10,11] Because miRNAs need only be partially complementary to their mRNA targets, one miRNA molecule can target many genes and thus a subset of miRNAs can modulate a range of specific biological processes.[12]

The manner by which miRNAs interfere with translation is not fully understood, although a number of mechanisms have been proposed. Initial studies indicated that the primary mechanism was suppression of translation, as the level of target mRNA-encoded proteins within cells was reduced but the stability of the mRNA transcript remained unchanged.[13] Recent data also suggest that miRNAs can reduce protein translation by blocking the polyribosome or by causing it to drop off during transcript elongation.[14,15] MiRNAs may also block the initiation of translation.[16]

In addition to blocking translation, miRNAs can also reduce gene expression by altering the stability of the mRNA target. The introduction of miRNAs into mammalian cell lines results in reduced levels of target transcripts, reflecting decreased stability of the mRNA.[17] It has also been demonstrated that miRNAs can directly mediate degradation of their target mRNAs.[18] The primary mechanism for this seems to be through deadenylation followed by removal of the mRNA 5′ cap.[19,20]

A third mechanism of miRNA-mediated gene regulation involves mRNA sequestration in processing (P) bodies. P bodies are structures involved in mRNA storage and degradation. The Argonaute proteins that bind to miRNAs in the RISC complex localize to P bodies, and miRNA-bound mRNA targets have been shown to accumulate in P bodies; this accumulation is miRNA dependent.[21] Deadenylation and decapping of mRNA can occur within P bodies, leading to degradation; alternatively, the inhibitory mechanism may simply involve mRNA sequestration from the translational machinery, as P bodies lack ribosomes. However, only a small number of Argonaute proteins localize to P bodies, and is it likely that multiple mechanisms are used by miRNAs to control gene expression.

Recent work has shown that miRNA stability is dependent on the abundance of target mRNAs and the extent of base pairing between the miRNA and its mRNA target.[22] Extensive complementarity of miRNA and target sites leads to 3′ end tailing and 3′ to 5′ trimming of the miRNA, which results in degradation of the miRNA. This may explain why miRNAs and their mRNA targets usually only display partial complementarity.

Multiple miRNA binding sites for the same or different miRNAs on target mRNAs are thought to mediate post-transcription regulation in a cooperative fashion.[23] The degree of post-transcriptional repression mediated by miRNAs is not absolute,[24,25] thus miRNAs are thought to function mainly by fine-tuning protein levels.

The bioinformatical identification of miRNA targets is rather difficult due to the partial complementarity of the miRNA−mRNA interactions. Target prediction software available today is reasonably accurate but a high false-positive rate remains (the three most commonly used programs are miRanda, PicTar, and TargetScan). The number of experimentally verified miRNA−mRNA target pairs is increasing but is still only a fraction of the predicted number.[26] Therefore it is important to verify the role of miRNAs experimentally, either by knocking down expression through the use of complementary antagomirs or by overexpressing miRNAs of interest.

MICRORNA REGULATION OF HEMATOPOIESIS

The role of miRNAs in hematopoiesis is currently an area of intense investigation. A subset of miRNAs has been identified that contribute to differentiation of $CD3^+$, $B220^+$, and $CD11b^+$ in long-term hematopoietic stem cells (HSCs).[27] Overexpression of one of these miRNAs, miR-125b, in bone marrow cells leads to a myeloproliferative disorder. Another study found that a cluster of miRNAs consisting of miR-99b, let-7e, and miR-125a is preferentially expressed in HSCs.[28] Specifically, miR-125a controls the size of the stem cell population by controlling HSC apoptosis. Furthermore, the miRNA processing enzyme Dicer is necessary for stem cell maintenance.

MiRNAs are thought to control differentiation of hematopoietic cells to all lineages, although the miRNAs involved in many of these pathways have yet to be identified. Nevertheless, several miRNAs have roles in lineage differentiation. Mast/stem cell growth factor receptor Kit (c-kit) is important in early progenitor proliferation and differentiation, and miR-221 and miR-222 promote decreased expression of the c-kit receptor, thus disrupting progenitor proliferation.[29] Other genes important in HSC homeostasis include the homeobox (HOX) genes, and miRNAs from the miR-196 and miR-10 families can directly repress HOX gene expression.[30]

Computational data also suggest that several miRNAs may be involved in the differentiation of multipotent progenitors into common lymphoid progenitors and myeloid progenitors.[31] For example, deletion of Argonaute 2 in the hematopoietic lineage leads to aberrant B cell development.[32] Additionally, deletion of Dicer in B cells leads to blocked transition of pro-B to pre-B lymphocytes and thereby impairs B cell survival and the production of a diverse antibody repertoire; this developmental defect is largely attributed to the loss of expression of the miR-17—92 cluster.[33] A conditional deletion of Dicer in T cells also affects lineage development of $\alpha\beta$ T cells.[34] MiRNAs can also have antagonizing effects during lymphocyte development: miR-150 blocks the progress of pro-B cells to pre-B cells,[35] while miR-34a promotes this transition.[36]

The granulocyte-macrophage progenitor differentiates into cells of the monocyte (macrophages and dendritic cells) and granulocyte lineages. The miR-17—92 cluster is implicated in the development of monocytes from this precursor; downregulation of miRNAs in this cluster leads to the expression of the macrophage colony stimulating factor receptor (M-CSFR) and thus monocyte lineage commitment.[37] MiR-223 is also expressed at low levels in this precursor population, but its expression increases as the cells differentiate into neutrophils and decreases as the cells develop into monocytes.[38] Furthermore, miR-223 is important not only in granulocyte development, but also in homeostasis; surprisingly, mice deficient in miR-223 have more neutrophils in their blood and bone marrow. However, these neutrophils have an aberrant morphology, indicating that miR-223, while dispensable for neutrophil differentiation, is necessary for normal maturation. The specific miRNAs involved in eosinophil differentiation have yet to be determined. However, recent investigations from our laboratory indicate that a select set of miRNAs is involved in the transition of eosinophil progenitors into mature cells. Importantly, these miRNAs are intimately regulated by signaling through the interleukin-5 (IL-5) receptor α chain as well as through the β common chain. Whether interleukin-5 signaling employs miRNAs for fine-tuning eosinophil development or whether they play an obligatory role remains undefined. Preliminary data indicate that targeting miRNA function with pharmacological antagonists can alter cell proliferation and survival.

MICRORNAS AND INNATE IMMUNITY

Eosinophils, neutrophils, and macrophages play critical roles in innate host defense. Toll-like receptors (TLRs) expressed on these cells are important components of the innate immune response and are activated by pathogen-associated molecular pattern molecules (PAMPs). Several miRNAs have been implicated in TLR signaling, with TLR4 being the best studied. TLR4 is expressed on eosinophils, macrophages, and neutrophils and has been shown to play crucial roles in activation of these cells in response to PAMPs. Notably, lipopolysaccharide (LPS) signaling through TLR4 is regulated by the let-7 family of miRNAs,[39] as well as by miR-9 in monocytes and neutrophils.[40] Regulation by miR-9 is dependent on myeloid differentiation primary response gene 88 (MyD88) and nuclear factor-κB (NF-κB); furthermore, *NFKB* has been identified as a target of miR-9. This suggests that induction of miR-9 following TLR4 signaling may serve as a negative-feedback mechanism to control NF-κB-mediated inflammation. Another potential negative-feedback loop following TLR4 stimulation involves miR-21.[41] MiR-21 targets programmed cell death 4 (neoplastic transformation inhibitor; *PDCD4*), which increases IL-10 expression and this in turn dampens NF-κB activation. A third example of an miRNA controlling innate immune responses is the increased expression of miR-146a following LPS activation of TLR4 on dendritic cells. This miRNA targets important components of the MyD88/NF-κB pathway such as the tumor necrosis factor (TNF) receptor-associated factor 6, E3 ubiquitin protein ligase (*TRAF6*) and interleukin-1 receptor-associated kinase 1 (*IRAK1*).[42] Finally, miR-147 is induced in macrophages following TLR4 stimulation, which then negatively regulates the production of inflammatory cytokines.[43] Because the inflammatory response initiated by LPS signaling through TLR4 can be lethal, it is critical that this response is tightly controlled, and miRNAs appear to play central roles in the control process.

The activation of macrophages through a variety of TLRs (TLR2, TLR3, TLR4, and TLR9) leads to the expression of miR-155.[44] These TLRs are the receptors for various viral and bacterial components, suggesting that miR-155 is important in controlling the host defense response of macrophages. This concept is also supported by data demonstrating that miR-155 is induced when macrophages are stimulated with the proinflammatory cytokines interferon β (IFN-β) and TNF-α. Thus, miR-155 is upregulated by multiple inflammatory signals, indicating its importance in innate immune regulation.

The contribution of miRNAs to host defense responses controlled by eosinophils is yet to be identified, but data from the aforementioned studies indicate that miRNAs are likely to play key roles in the activation of this cell type during infection (bacterial, parasitic, or viral) and allergic responses.

MICRORNAS AND ADAPTIVE IMMUNITY

The importance of miRNAs in shaping the adaptive immune response is also emerging. For example, miR-155 promotes the polarization of T-helper type 1 (T_h1)

cells,[45,46] and the T cells in mice deficient in miR-155 are skewed toward IL-4-producing T_h2 cells. Similarly, miR-326 is important in T_h17 differentiation:[43] silencing miR-326 leads to fewer T_h17 cells, while overexpression results in an increase in T_h17 cell numbers. The responsiveness of T cells to antigens is also regulated by miRNAs—miR-181a enhances T-cell receptor signaling by targeting phosphatases that would normally suppress this pathway.[47]

B cell development and function are controlled by miRNA expression. MiR-150 is expressed by mature B cells, but affects B-cell development when expressed earlier by blocking the differentiation of pro-B cells into pre-B cells.[35] Mice lacking miR-155 have impaired antibody responses, while activation of B cells through the B-cell receptor, tumor necrosis factor receptor superfamily member 5 (CD40) stimulation, or TLRs results in increased miR-155 expression.[45,46] Thus, it is becoming increasingly clear that miRNAs are critical regulators of the immune response.

MICRORNAS IN ALLERGIC DISEASES AND EOSINOPHILIA

The role of miRNAs in controlling the function of both innate and adaptive immune cells indicates that targeting these molecules may be effective in suppressing inflammation associated with chronic disease. Indeed, understanding the critical miRNA(s) involved in eosinophil, mast cell, and T_h2 cell development or activation in response to factors that trigger allergic inflammation may lead to a better understanding of how to control and treat allergic disorders.

MicroRNAs and Inflammatory Lung Diseases

Currently there are very limited data on the role of miRNA in the regulation of inflammatory diseases. Although data are emerging that show abnormal expression patterns in asthma and inflammatory diseases of the skin, most of the current knowledge on miRNA function has come from animal models. Two notable studies indicate that miR-126 and miR-21 may control T_h2 cell-induced pathology and eosinophil recruitment into inflamed tissue. In a mouse model of house dust mite-induced allergic airways disease and characterized by eosinophilia, a critical role for miR-126 in the regulation of allergic inflammation was identified.[48] Pharmacological antagonism of miR-126 function inhibited eosinophil recruitment into lung tissue, reduced mucus production, suppressed T_h2 cytokine production, and abolished airway hyperresponsiveness. Although the precise targets of miR-126 are yet to be elucidated,

miR-126 regulation is critically dependent on TLR4 and MyD88 signaling. Furthermore, miR-126 has also been shown to play an important a role in cystic fibrosis[49] by regulating epithelial responses. Collectively, the current evidence indicates that miR-126 plays an important role in the innate immune response in the lung.

In a second study, the expression of miR-21 was shown to be increased in both ovalbumin- and aspergillus-challenge mouse models of experimentally induced allergic asthma.[50] This increase was primarily attributed to high expression levels in macrophages and dendritic cells. IL-12A was found to be a primary target for miR-21 in these models, suggesting that increased expression of this miRNA could lead to a decrease in IL-12 production and promote the development of allergic inflammation of the airways, eosinophilia, and a T_h2 response. MiR-21 was also upregulated in the lungs of bleomycin-treated mice as well as in the lungs of patients with idiopathic pulmonary fibrosis.[31] In this case, the profibrogenic activity of TGF-β1 from primary pulmonary fibroblasts was shown to be dependent on miR-21 by potentially targeting *SMAD7*.

CONCLUSIONS

The role of miRNA in the regulation of the immune system is only beginning to emerge; however, current studies indicate that these molecules play important roles in regulating both the innate and adaptive arms of the host immune response. Evidence is also emerging that the expression of specific miRNA subsets are associated with the induction and progression of disease. Notably we have shown that miRNA regulate the severity of allergic inflammation and ongoing work demonstrates a role in eosinophil differentiation. Understanding the role of miRNA in eosinophilic disorders and other chronic inflammatory conditions may provide a new therapeutic approach to inhibiting inflammation and pathogenesis.

REFERENCES

1. Bartel DP. MicroRNAs: target recognition and regulatory functions. *Cell* 2009;**136**:215−33.
2. Schickel R, Boyerinas B, Park SM, Peter ME. MicroRNAs: key players in the immune system, differentiation, tumorigenesis and cell death. *Oncogene* 2008;**27**:5959−74.
3. Ferracin M, Veronese A, Negrini M. Micromarkers: miRNAs in cancer diagnosis and prognosis. *Expert Rev Mol Diagn* 2010;**10**:297−308.
4. Jackson A, Linsley PS. The therapeutic potential of microRNA modulation. *Discov Med* 2010;**9**:311−8.
5. Lewis BP, Burge CB, Bartel DP. Conserved seed pairing, often flanked by adenosines, indicates that thousands of human genes are microRNA targets. *Cell* 2005;**120**:15−20.

6. Lee Y, Kim M, Han J, Yeom KH, Lee S, Baek SH, et al. MicroRNA genes are transcribed by RNA polymerase II. *EMBO J* 2004; **23**:4051−60.

7. Lee Y, Jeon K, Lee JT, Kim S, Kim VN. MicroRNA maturation: stepwise processing and subcellular localization. *Embo J* 2002; **21**:4663−70.

8. Weinmann L, Hock J, Ivacevic T, Ohrt T, Mutze J, Schwille P, et al. Importin 8 is a gene silencing factor that targets argonaute proteins to distinct mRNAs. *Cell* 2009;**136**:496−507.

9. Hammond SM, Boettcher S, Caudy AA, Kobayashi R, Hannon GJ. Argonaute2, a link between genetic and biochemical analyses of RNAi. *Science* 2001;**293**:1146−50.

10. Mallory AC, Reinhart BJ, Jones-Rhoades MW, Tang G, Zamore PD, Barton MK, et al. MicroRNA control of PHABULOSA in leaf development: importance of pairing to the microRNA 5′ region. *EMBO J* 2004;**23**:3356−64.

11. Liu J. Control of protein synthesis and mRNA degradation by microRNAs. *Curr Opin Cell Biol* 2008;**20**:214−21.

12. Lewis BP, Shih IH, Jones-Rhoades MW, Bartel DP, Burge CB. Prediction of mammalian microRNA targets. *Cell* 2003;**115**: 787−98.

13. Wightman B, Ha I, Ruvkun G. Posttranscriptional regulation of the heterochronic gene lin-14 by lin-4 mediates temporal pattern formation in C. elegans. *Cell* 1993;**75**:855−62.

14. Nottrott S, Simard MJ, Richter JD. Human let-7a miRNA blocks protein production on actively translating polyribosomes. *Nat Struct Mol Biol* 2006;**13**:1108−14.

15. Petersen CP, Bordeleau ME, Pelletier J, Sharp PA. Short RNAs repress translation after initiation in mammalian cells. *Mol Cell* 2006;**21**:533−42.

16. Pillai RS, Bhattacharyya SN, Artus CG, Zoller T, Cougot N, Basyuk E, et al. Inhibition of translational initiation by Let-7 MicroRNA in human cells. *Science* 2005;**309**:1573−6.

17. Lim LP, Lau NC, Garrett-Engele P, Grimson A, Schelter JM, Castle J, et al. Microarray analysis shows that some microRNAs downregulate large numbers of target mRNAs. *Nature* 2005;**433**: 769−73.

18. Bagga S, Bracht J, Hunter S, Massirer K, Holtz J, Eachus R, et al. Regulation by let-7 and lin-4 miRNAs results in target mRNA degradation. *Cell* 2005;**122**:553−63.

19. Giraldez AJ, Mishima Y, Rihel J, Grocock RJ, Van Dongen S, Inoue K, et al. Zebrafish MiR-430 promotes deadenylation and clearance of maternal mRNAs. *Science* 2006;**312**:75−9.

20. Wu L, Fan J, Belasco JG. MicroRNAs direct rapid deadenylation of mRNA. *Proc Natl Acad Sci USA* 2006;**103**:4034−9.

21. Liu J, Valencia-Sanchez MA, Hannon GJ, Parker R. MicroRNA-dependent localization of targeted mRNAs to mammalian P-bodies. *Nat Cell Biol* 2005;**7**:719−23.

22. Ameres SL, Horwich MD, Hung JH, Xu J, Ghildiyal M, Weng Z, et al. Target RNA-directed trimming and tailing of small silencing RNAs. *Science* 2010;**328**:1534−9.

23. Doench JG, Petersen CP, Sharp PA. siRNAs can function as miRNAs. *Genes Dev* 2003;**17**:438−42.

24. Baek D, Villen J, Shin C, Camargo FD, Gygi SP, Bartel DP. The impact of microRNAs on protein output. *Nature* 2008;**455**:64−71.

25. Selbach M, Schwanhausser B, Thierfelder N, Fang Z, Khanin R, Rajewsky N. Widespread changes in protein synthesis induced by microRNAs. *Nature* 2008;**455**:58−63.

26. Rajewsky N. microRNA target predictions in animals. *Nature Genetics* 2006;**38**(Suppl):S8−13.

27. O'Connell RM, Chaudhuri AA, Rao DS, Gibson WS, Balazs AB, Baltimore D. MicroRNAs enriched in hematopoietic stem cells differentially regulate long-term hematopoietic output. *Proc Natl Acad Sci USA* 2010;**107**:14235−40.

28. Guo S, Lu J, Schlanger R, Zhang H, Wang JY, Fox MC, et al. MicroRNA miR-125a controls hematopoietic stem cell number. *Proc Natl Acad Sci USA* 2010;**107**:14229−34.

29. Felli N, Fontana L, Pelosi E, Botta R, Bonci D, Facchiano F, et al. MicroRNAs 221 and 222 inhibit normal erythropoiesis and erythroleukemic cell growth via kit receptor down-modulation. *Proc Natl Acad Sci USA* 2005;**102**:18081−6.

30. Yekta S, Shih IH, Bartel DP. MicroRNA-directed cleavage of HOXB8 mRNA. *Science* 2004;**304**:594−6.

31. Georgantas 3rd RW, Hildreth R, Morisot S, Alder J, Liu CG, Heimfeld S, et al. CD34+ hematopoietic stem-progenitor cell microRNA expression and function: a circuit diagram of differentiation control. *Proc Natl Acad Sci USA* 2007;**104**:2750−5.

32. O'Carroll D, Mecklenbrauker I, Das PP, Santana A, Koenig U, Enright AJ, et al. A Slicer-independent role for Argonaute 2 in hematopoiesis and the microRNA pathway. *Genes Dev* 2007;**21**: 1999−2004.

33. Koralov SB, Muljo SA, Galler GR, Krek A, Chakraborty T, Kanellopoulou C, et al. Dicer ablation affects antibody diversity and cell survival in the B lymphocyte lineage. *Cell* 2008;**132**:860−74.

34. Cobb BS, Nesterova TB, Thompson E, Hertweck A, O'Connor E, Godwin J, et al. T cell lineage choice and differentiation in the absence of the RNase III enzyme Dicer. *J Exp Med* 2005;**201**: 1367−73.

35. Zhou B, Wang S, Mayr C, Bartel DP, Lodish HF. miR-150, a microRNA expressed in mature B and T cells, blocks early B cell development when expressed prematurely. *Proc Natl Acad Sci USA* 2007;**104**:7080−5.

36. He L, He X, Lim LP, de Stanchina E, Xuan Z, Liang Y, et al. A microRNA component of the p53 tumour suppressor network. *Nature* 2007;**447**:1130−4.

37. Fontana L, Pelosi E, Greco P, Racanicchi S, Testa U, Liuzzi F, et al. MicroRNAs 17−5p-20a-106a control monocytopoiesis through AML1 targeting and M-CSF receptor upregulation. *Nat Cell Biol* 2007;**9**:775−87.

38. Johnnidis JB, Harris MH, Wheeler RT, Stehling-Sun S, Lam MH, Kirak O, et al. Regulation of progenitor cell proliferation and granulocyte function by microRNA-223. *Nature* 2008;**451**:1125−9.

39. Chen XM, Splinter PL, O'Hara SP, LaRusso NF. A cellular microRNA, let-7i, regulates Toll-like receptor 4 expression and contributes to cholangiocyte immune responses against Cryptosporidium parvum infection. *J Biol Chem* 2007;**282**:28929−38.

40. Bazzoni F, Rossato M, Fabbri M, Gaudiosi D, Mirolo M, Mori L, et al. Induction and regulatory function of miR-9 in human monocytes and neutrophils exposed to proinflammatory signals. *Proc Natl Acad Sci USA* 2009;**106**:5282−7.

41. Sheedy FJ, Palsson-McDermott E, Hennessy EJ, Martin C, O'Leary JJ, Ruan Q, et al. Negative regulation of TLR4 via targeting of the proinflammatory tumor suppressor PDCD4 by the microRNA miR-21. *Nat Immunol* 2010;**11**:141−7.

42. Taganov KD, Boldin MP, Chang KJ, Baltimore D. NF-kappaB-dependent induction of microRNA miR-146, an inhibitor targeted to

signaling proteins of innate immune responses. *Proc Natl Acad Sci USA* 2006;**103**:12481—6.

43. Du C, Liu C, Kang J, Zhao G, Ye Z, Huang S, et al. MicroRNA miR-326 regulates TH-17 differentiation and is associated with the pathogenesis of multiple sclerosis. *Nat Immunol* 2009;**10**:1252—9.

44. O'Connell RM, Taganov KD, Boldin MP, Cheng G, Baltimore D. MicroRNA-155 is induced during the macrophage inflammatory response. *Proc Natl Acad Sci USA* 2007;**104**:1604—9.

45. Thai TH, Calado DP, Casola S, Ansel KM, Xiao C, Xue Y, et al. Regulation of the germinal center response by microRNA-155. *Science* 2007;**316**:604—8.

46. Rodriguez A, Vigorito E, Clare S, Warren MV, Couttet P, Soond DR, et al. Requirement of bic/microRNA-155 for normal immune function. *Science* 2007;**316**:608—11.

47. Li QJ, Chau J, Ebert PJ, Sylvester G, Min H, Liu G, et al. miR-181a is an intrinsic modulator of T cell sensitivity and selection. *Cell* 2007;**129**:147—61.

48. Mattes J, Collison A, Plank M, Phipps S, Foster PS. Antagonism of microRNA-126 suppresses the effector function of TH2 cells and the development of allergic airways disease. *Proc Natl Acad Sci USA* 2009;**106**:18704—9.

49. Oglesby IK, Bray IM, Chotirmall SH, Stallings RL, O'Neill SJ, McElvaney NG, et al. miR-126 is downregulated in cystic fibrosis airway epithelial cells and regulates TOM1 expression. *J Immunol* 2010;**184**:1702—9.

50. Lu TX, Munitz A, Rothenberg ME. MicroRNA-21 is up-regulated in allergic airway inflammation and regulates IL-12p35 expression. *J Immunol* 2009;**182**:4994—5002.

Chapter 16.3

Eosinophils in the Zebrafish

Keir M. Balla and David Traver

EOSINOPHILS IN THE ZEBRAFISH

Eosinophils are ubiquitous among the five classes of vertebrates and at least three major phyla of invertebrates.[1] The remarkable conservation of this granulocyte subset suggests a selective importance for its roles in host defense and maintenance of immune homeostasis. Understanding these roles in terms of human health and disease has been a significant concern in research for more than 60 years,[2] and to this end animal models have been used nearly as long.[3] Together these efforts have delineated the diverse activities of eosinophils,[4] the bulk of which are implicated in host defense against parasitic helminth infection and exacerbation of allergic pathologies.[5,6] However, categorizing eosinophils primarily as effector cells during infection and allergic inflammation appears to be too narrow a definition, and in some aspects unfounded. In the attempt to establish definitively their role(s) in health and disease, eosinophils have been selectively ablated in different mouse models. Despite these efforts, the specific roles of eosinophils remain unclear, as eosinophil-less mice are not generally unhealthy, not obviously disadvantaged during helminth infection, and not less prone to allergic disease.[7–9] Therefore, a general reassessment of how and why eosinophils act may be necessary to understand their contributions to health. Eosinophils have been extensively studied in mammals, although a paucity of data exists on the biology of eosinophils in other vertebrates. New animal models for eosinophil biology could yield novel and relevant insights into their behavior. Recent work supports the zebrafish as a promising candidate for these studies.

Zebrafish are equipped with an immune system very much like that of mammals. Mammalian innate immune cells, such as eosinophils, macrophages, and neutrophils, as well as the B and T lymphocytes of adaptive immunity, share morphological and genetic features with their zebrafish counterparts.[10] Unlike in mammals, zebrafish develop externally and are transparent until beyond the establishment of adaptive immunity. In conjunction with their genetic amenability and high fecundity, zebrafish offer unique advantages for understanding the development and functions of the immune system during homeostasis and disease. For instance, specific leukocytes can be marked and distinguished from each other using multiple fluorophores, thus providing the ability to observe their activities in real time *in vivo*. This strategy has enabled progress in the understanding of macrophage and neutrophil behaviors macrophage during steady state and following pathogen challenge. Specifically, transgenic animals that have green fluorescent protein (GFP)-labeled neutrophils have provided insights into their migration patterns during inflammation.[11,12] Conversely, zebrafish infected with GFP-labeled mycobacteria have enabled the direct visualization of macrophage recruitment and aggregation during the pathogen-induced innate immune response.[13] Furthermore, mutants with subtle disruptions in genes important for leukocyte development or behavior can be identified with a high degree of precision *in vivo*.[14] Together, these advantages provide a strong prospect for the elucidation of eosinophil functions in the zebrafish.

While several advances in the realm of innate immunity have been made using the zebrafish, little attention has been made with regards to the presence or activities of eosinophils. The existence of granulocytes in fish equivalent to those found in mammals has long been a matter of controversy. This has been attributed to the lack of data establishing three basic types of criteria for comparison: functional, morphological, and ontogenetic.[15] The bulk of research investigating leukocytes in fish has defined eosinophils based on morphological criteria alone, although some evidence has also been provided correlating eosinophil presence with certain histopathological conditions, e.g., parasitic worm infections.[16] Underscoring the confusion involved in defining eosinophils based on morphology alone is the fact that there appear to be several species-dependent morphological traits, such as granule shape and appearance under electron microscopy (EM), peroxidase activity, and affinity for stains. At times, this has led to the designation of a combination mast cell/eosinophilic granule cell, or at other times a basophil/eosinophil, demonstrating the lack of defining features identified among those granulocytes in fish.[17,18] These interspecies differences do not reflect diversities inherent to fish alone, as histochemical reactions in mammalian granulocytes are also species-specific.[19] In order to make inferences about eosinophils across vertebrate species there must be a certain amount of consistency among their ontogenetic and functional features, in addition to the loose morphological similarities established thus far in teleosts. A comprehensive approach to this issue has recently been initiated in the zebrafish using the main advantages outlined above.

EOSINOPHIL MORPHOLOGY AND DISTRIBUTION

Eosinophils in the zebrafish, as in other teleosts, have been identified among other leukocytes by their affinity for histochemical stains and by EM (Figure 16.3.1). Specifically, zebrafish eosinophils stain positively with periodic acid-Schiff (PAS) but lack myeloperoxidase (MPO) activity and toluidine blue (TB) affinity, thus distinguishing them from neutrophils, which are negative for PAS but positive for MPO, and mast cells, which are positive for TB.[17,20–22] Spherical, electron-dense granules lacking paracrystalline inclusions comprise most of the cytoplasmic space of zebrafish eosinophils as detected in electron micrographs.[17,20,22] These basic properties have been used to identify their distribution in tissues during homeostasis: they comprise approximately 3% of the whole kidney marrow (WKM), which is the primary site of hematopoiesis in teleosts, are the most abundant leukocyte found in the intraperitoneal exudate (IPEX), and are found at low levels in the intestinal tract, peripheral blood, skin, spleen, and thymus.[17,20,23] Taken together, these observations establish some parallels between zebrafish and mammalian eosinophil morphology, ultrastructure, and a distribution that tentatively suggests an overlap in activities. However, that eosinophils from mammals and teleosts share histological features and tissue locality has been known for 100 years.[24] A definitive link in eosinophil function between species has been incomplete in the absence of more comprehensive comparisons.

THE EOSINOPHIL RESPONSE TO HELMINTH INFECTION

Hematologists have often looked to disease in order to understand the functions of leukocytes. Eosinophils were first observed at elevated levels in the circulation and peripheral tissues in humans infected with parasitic worms well over a century ago. Since then, it has largely been assumed that this association suggests a protective role for eosinophils against helminthic pathogens. Supporting evidence for this definition is the observation that circulating eosinophils increase upon infection by helminths and decrease as the infection is cleared, whereas other leukocytes of the innate and adaptive immune systems do not follow this pattern.[25] Whether or not teleosts exhibit such histopathologies has been of interest to eosinophil research since Metchnikoff proposed his theories of immunity,[24] and continues to be of interest now, with the aim of better understanding eosinophil-associated diseases using model organisms.

During a forward genetic screen at Oregon State University, the Spitsbergen laboratory discovered that part of their zebrafish colony was infected with helminth parasites.[26] Careful histological analyses of emaciated animals showed adult worms, larvae, and eggs within the intestines and peritoneal cavities. Ultrastructural analyses demonstrated the parasites to be *Pseudocapillaria tomentosa*, a capillarid nematode commonly found in a variety of fishes, and close relatives *of Capillaria philippinensis*, a parasite responsible for recent human epidemics in Asia.[27] Capillarid nematodes infect all classes of vertebrates, and often cause extensive tissue damage due to their invasive nature.[27] Whereas the life cycles of some capillarid nematodes such as oligochaete worms, require intermediate hosts, *P. tomentosa* is directly transmissible from infected to uninfected adult zebrafish.[26] This is likely due to the shedding of eggs and larvae through a fecal–oral transmission route.

Our laboratory was interested in determining the eosinophil response to helminthic *P. tomentosa* infection. Transverse sections were prepared from infected and uninfected adult zebrafish embedded in paraffin blocks. PAS-stained sections revealed low levels of eosinophils in the intestines of uninfected zebrafish, whereas infected intestines were robustly infiltrated by PAS⁺ eosinophils (Figure 16.3.2).[20] This change in the intestinal distribution of eosinophils mimics that observed during murine infection with intestinal parasites.[28] These observations therefore suggest that the eosinophil response to helminthic parasites is conserved from teleosts to mammals.

It was subsequently determined that infection could be transmitted experimentally by inoculating uninfected zebrafish in water containing infected fish separated by mesh netting, further demonstrating a direct mode of

FIGURE 16.3.1 The morphological and ultrastructural properties of zebrafish eosinophils are reminiscent of their mammalian counterparts. Eosinophils observed within whole kidney marrow (*A*, *C*) and the intraperitoneal exudate (*B*, *D*) stained with PAS under light microscopy (*A*, *B*) and with uranyl acetate and lead citrate under transmission electron microscopy (*C*, *D*).

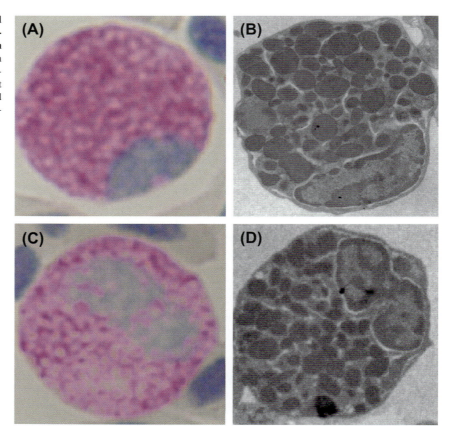

transmission that probably reflects its natural route. Further studies will use this tool to regulate infection and monitor eosinophil activities more thoroughly.

EOSINOPHIL ISOLATION AND *IN VIVO* VISUALIZATION

Leukocytes of the zebrafish hematopoietic system can be separated by their major lineages based upon light scatter characteristics measured by flow cytometry.[29] Furthermore, engineering transgenic animals that drive the expression of fluorescent proteins via lineage-affiliated gene regulatory elements permits the identification and isolation of distinct leukocyte subsets. For example, the *gata2:eGFP* (GATA-binding factor 2: enhanced green fluorescent protein) transgene is expressed highly by zebrafish eosinophils.[29] This reagent enables eosinophil isolation from various tissues, thus facilitating genetic and functional characterizations.[20] In addition to providing a tool for isolation, the *gata2:eGFP* transgene can be used to observe eosinophil activities *in vivo* in the transparent zebrafish embryo. Visualizing the caudal hematopoietic tissue of embryos 72 hours postfertilization by fluorescence microscopy

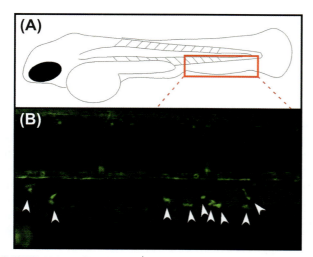

FIGURE 16.3.2 Gata2:eGFP⁺ eosinophils are visible in the 3-day-old zebrafish embryo. *A*, Cartoon representation of the 3-day-old embryo oriented for confocal imaging. Red box indicates the area depicted in B. *B*, Confocal image of the caudal hematopoietic tissue of a 3-day-old *gata2:eGFP* embryo. White arrowheads indicate mobile green fluorescent protein (GFP)⁺ eosinophils as determined by time-lapse microscopy.

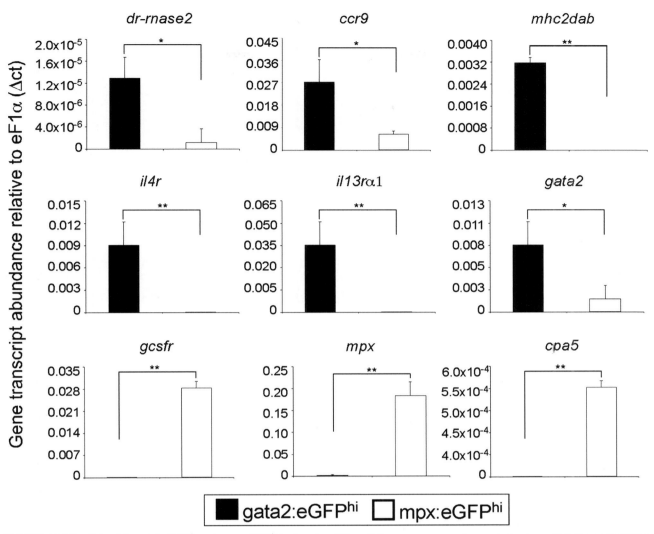

FIGURE 16.3.3 Zebrafish *gata2:eGFP*[+] and *mpx:eGFP*[+] leukocytes differentially express genes known to be important for eosinophil or neutrophil development, maintenance, and function. Eosinophils and neutrophils were isolated from whole kidney marrow by flow cytometry based on their expression of the *gata2:eGFP* (gata2:eGFP[hi]) or *mpx:eGFP* (mpx:eGFP[hi]) transgene, respectively. mRNA transcript abundance was measured by quantitative reverse-transcription polymerase chain reaction using the *eF1α* transcript as a reference. Black and white bars represent the level of gene expression measured in eosinophils and neutrophils, respectively. Data are shown as mean $\pm$ SD (n = 3–4). *$p < 0.05$, **$p < 0.001$. *ccr9*, chemokine (C-C motif) receptor 9; *cpa5*, carboxypeptidase A5; *dr-rnase 2*, ribonuclease 2 (*rnasel2*); *eF1α*, eukaryotic translation elongation factor 1α; *gata2*, GATA-binding factor 2; *gcsfr*, granulocyte stimulating factor receptor; *il4r*, interleukin-4 receptor; *il13rα1*, interleukin-13 receptor α1; *mhc2dab*, MHC class II DA-β; *mpx*, myeloperoxidase.

reveals the dynamic activities of GFP[+] leukocytes in the developing fish (Figure 16.3.3). These preliminary observations suggest that zebrafish could be used to discover roles of eosinophils during development and innate immunity that are not easily studied in mammals. Specifically, the eosinophil response to helminth pathogens can be visualized at all stages of infection. Additional transgenic tools could be engineered to modulate this response and to visualize the cooperative activities of different components of the immune system.

GENE EXPRESSION PROFILE OF EOSINOPHILS

The mammalian and zebrafish hematopoietic systems rely upon similar differentiation pathways to produce functional leukocytes. Analyzing the gene expression profiles of isolated immune cells can therefore be informative in terms of conservation. Eosinophils and neutrophils were separately isolated from kidney marrow, the main site of hematopoiesis in zebrafish, and mRNA transcript levels were

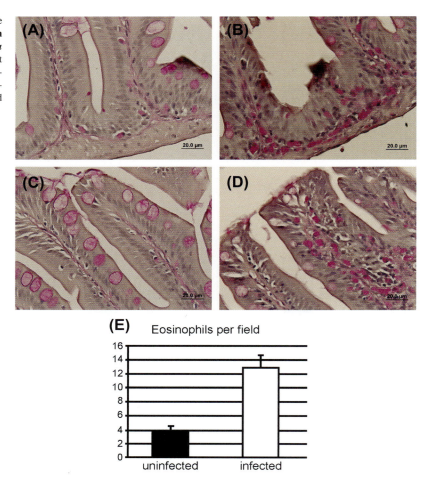

FIGURE 16.3.4 Periodic acid-Schiff-positive eosinophils infiltrate the intestine of zebrafish infected with intestinal helminth *Pseudocapillaria tomentosa.* Transverse sections of adult zebrafish gut tissue stained with periodic acid-Schiff from uninfected fish (*A, B*), and infected fish (*C, D*). E, Quantification of intestinal eosinophils in uninfected and infected zebrafish.

compared for genes involved in eosinophil or neutrophil development and function. Several genes thought to be important for eosinophil development, chemotaxis, activation, and function were expressed by *gata2:eGFP*[+] eosinophils, as compared to *mpx:eGFP*[+] neutrophils (Figure 16.3.4).

The endothelial transcription factor GATA-2 is essential for the differentiation and maintenance of murine eosinophils, in that it instructs progenitor cells to become eosinophils and regulates the expression of genes that are critical for eosinophil function.[30,31] Its importance in zebrafish eosinophils is demonstrated by its high steady-state expression and the specificity of the *gata2:eGFP* transgene.

Eosinophil ribonucleases are highly cationic proteins that are released during degranulation and are thought to mediate innate immune responses.[32] Three RNases have been identified in zebrafish. In conjunction with their bactericidal activity, their high rate of gene turnover during evolution implicates them in the role of host defense.[33] All of the zebrafish RNases are expressed by eosinophils, including one that is highly specific. It remains to be seen

whether any of these RNases might behave functionally like eosinophil-derived neurotoxin or eosinophil cationic protein.

C-C chemokine receptor type 3 (CCR3) is one of the most well-known chemokine receptors expressed by mammalian eosinophils, and is responsible for receiving the signal to migrate toward a gradient of eotaxin/chemokine (C-C motif) ligand 11 (CCL11). No clear orthologue of this receptor has been identified in zebrafish, although other chemokine receptors important for eosinophil activities are expressed by zebrafish eosinophils. C-C chemokine receptor type 9 (CCR9) is thought to mediate the trafficking of eosinophils to tissues, where they help regulate the adaptive immune response by acting as antigen-presenting cells.[34] Supporting this possibility is the observation that components of the major histocompatibility complex are highly expressed by zebrafish eosinophils. However, antigen-presenting cells were only recently defined in the zebrafish,[35] and it is not yet known whether eosinophils contribute to this process. The recently engineered transgenic zebrafish driving the expression of a fluorescent protein with the promoter for a major histocompatibility molecule class II component will serve as a useful tool both for visualization of eosinophils and evaluating their participation in antigen presentation (Wittamer V., Bertrand J., Traver D., submitted).

The interleukin-4 receptor (IL-4R) and IL-13R are related heterodimeric receptors that transduce a signal via signal transducer and activator of transcription 6 (STAT6) upon receiving their cognate interleukins.[36] The IL-4Rα dimerizes with the common γ chain (γc) in order to receive IL-4 signal. Alternatively, the IL-4Rα chain can dimerize with IL-13Rα1 to bind both IL-4 and IL-13.[36] IL-13 can also bind to IL-13Rα2 in the absence of a coreceptor, if not also interacting with IL-4α.[37] IL-4 and IL-13 are both released by T-helper type 2 (T$_h$2) cells in response to parasitic helminth infection and allergens, and lead to the expansion of eosinophils.[38] Furthermore, IL-13, IL-4α, and Stat6 are required for clearing infection by *Nippostrongylus brasiliensis*, a gastrointestinal parasite.[39] The polarization of different T-helper cell subsets has not been well characterized in the zebrafish, though orthologues to IL-4Rα, IL-13Rα1&2, and STAT6 have been identified in the genome. Interestingly, zebrafish eosinophils specifically express high levels of IL-4R and IL-13R. This could indicate that the eosinophil response to a T$_h$2 polarized environment is conserved from teleosts to mammals. Future tools will enable the dissection of this potential in more detail.

In agreement with cytochemical observations, zebrafish eosinophils do not express the myeloperoxidase gene. So far no clear orthologue to eosinophil peroxidase has been identified in the zebrafish genome. However, zebrafish eosinophils do exhibit peroxidase activity under EM.[22] This implies that a peroxidase gene other than myeloperoxidase must be responsible for the activity, and that an eosinophil-specific peroxidase has simply not yet been identified. Another gene expressed by neutrophils but not eosinophils is the receptor for granulocyte colony stimulating factor (*gcsfr*). Gcsf-r is essential for the differentiation of neutrophils from progenitor cells, as mice with nonfunctional receptors are neutropenic and highly susceptible to life-threatening infection, whereas eosinophil numbers are unaffected.[40] Additionally, work in our laboratory shows that recombinant zebrafish *gcsf* supports precursor cell differentiation to neutrophils but not eosinophils *in vitro* (D. Stachura, O. Svoboda, R.P. Lau, K.M. Balla, L.I. Zon, P. Bartunek, and D. Traver, submitted). The major growth factors for zebrafish eosinophils remain unknown, since orthologues of IL-5 or GM-CSF have not been identified in the genome. Previous studies have demonstrated that IL-5-deficient mice have normal baseline eosinophil levels and exhibit moderate eosinophilia in response to helminth infection, and it was hypothesized that IL-4 or IL-13 was compensating for the IL-5 deficiency in these mice.[41] Future studies with recombinant zebrafish IL-4 and IL-13

will determine whether or not they might serve as growth factors for zebrafish eosinophils.

Several studies have hypothesized that the eosinophil in teleosts represents a novel cell type that shares certain traits with mammalian eosinophils, basophils, and mast cells. Recent morphological and genetic evidence supports an alternative hypothesis that at least two different cell types exist. In addition to having distinct cytochemical properties, zebrafish mast cells have been defined genetically as expressing carboxypeptidase 5.[21] Zebrafish eosinophils do not express *cpa5*, and do not react with TB stain, thus distinguishing these two cell types on multiple levels. It is currently unknown whether or not basophils exist in the zebrafish. Further distinguishing eosinophils from mast cells or possibly basophils in the zebrafish will require a better understanding of their potential genetic and functional properties.

While the preliminary characterization of the *gata2:eGFP*$^+$ leukocyte gene expression profile supports the hypothesis that these leukocytes represent the equivalent of the mammalian eosinophil, it is limited in scope. To expand this classification, it will be useful to prepare eosinophils taken from different tissues before and after stimulation with allergens or helminth infection for microarray analysis in order to more comprehensively compare and contrast the biology of mammalian and teleostean eosinophils.

MODELING THE EOSINOPHIL RESPONSE TO ALLERGENS

Another range of maladies well known to involve eosinophil responses is found among allergic diseases. Papain is a cysteine protease commonly known to induce allergic responses involving eosinophils.[42] Eosinophils are also directly activated by papain *in vitro*.[43] It is thought that eosinophils might respond to papain as an allergen because it resembles cysteine protease factors released by parasitic helminths during infection.[42] Evidence that teleosts mount allergic responses involving eosinophils was recently reported. Systemic immunization with papain by intraperitoneal injection results in highly elevated eosinophil levels in the circulation, as determined by blood smears and *gata2:eGFP* leukocyte levels in circulating blood.[20] These observations further support the classification of zebrafish eosinophils as being conserved with their mammalian counterparts, and suggest the makings for an interesting novel model for studying the eosinophil response to allergens.

CONCLUSIONS

Mammalian models of eosinophil function have dominated research efforts seeking to improve our understanding of the perplexing nature and significance of their activities. Studies in these models have resulted in important advances materializing both as benefits to human health and more broadly in terms of understanding the various responses of the immune system to disease. However, certain essential characteristics of eosinophils remain ill defined. As eosinophils are found in most if not all vertebrates, nonmammalian vertebrate models of eosinophil biology may help resolve questions regarding their functions using their unique advantages. Teleosts have been candidate eosinophil models for decades, though until recently the inability to isolate them has resulted in a tenuous grasp of their resemblance to mammalian eosinophils. Recent studies show that eosinophils are conserved from teleosts to mammals in terms of their morphologies, genetic profiles, and responses to parasitic nematode pathogens and allergens. These characterizations comprise a platform from which further studies may elucidate questions regarding eosinophil activities *in vivo*. For example, the localization of labeled eosinophils could be followed through the entire animal during the immune response to helminth infection. Additionally, homeostatic activities of eosinophils might be better understood during development by observing the early colonization of the gut by gastrointestinal eosinophils. These studies will be enhanced by information gained from microarray analyses, in which gene functions crucial for eosinophil activities may be identified and targeted for manipulation to observe the biological outcomes of their deregulation. Finally, the precise roles of eosinophils in the response to helminth infestation may be illuminated through their conditional ablation using the recently developed nitroreductase transgenic system.[44,45] These and other advances now position the zebrafish as a new system to study the biology of eosinophils, in which future efforts will complement those in mammals.

REFERENCES

1. Lee JJ, Jacobsen EA, McGarry MP, Schleimer RP, Lee NA. Eosinophils in health and disease: the LIAR hypothesis. *Clin Exp Allergy* 2010;**40**:563—75.

2. Humphreys RJ, Raab W. Response of circulating eosinophils to norepinephrine, epinephrine and emotional stress in humans. *Proc Soc Exp Biol Med* 1950;**74**:302—3.

3. Bittner JJ, Halberg F, Vermund H. Daily eosinophil rhythm in mice bearing a transplanted mammary carcinoma. *J Natl Cancer Inst* 1956;**17**:139—44.

4. Hogan SP, Rosenberg HF, Moqbel R, Phipps S, Foster PS, Lacy P, et al. Eosinophils: biological properties and role in health and disease. *Clin Exp Allergy* 2008;**38**:709—50.

5. Cadman ET, Lawrence RA. Granulocytes: effector cells or immunomodulators in the immune response to helminth infection? *Parasite Immunol* 2010;**32**:1—19.

6. Walsh ER, Stokes K, August A. The role of eosinophils in allergic airway inflammation. *Discov Med* 2010;**9**:357—62.

7. Humbles AA, Lloyd CM, McMillan SJ, Friend DS, Xanthou G, McKenna EE, et al. A critical role for eosinophils in allergic airways remodeling. *Science* 2004;**305**:1776—9.

8. Lee JJ, Dimina D, Macias MP, Ochkur SI, McGarry MP, O'Neill KR, et al. Defining a link with asthma in mice congenitally deficient in eosinophils. *Science* 2004;**305**:1773—6.

9. Swartz JM, Dyer KD, Cheever AW, Ramalingam T, Pesnicak L, Domachowske JB, et al. Schistosoma mansoni infection in eosinophil lineage-ablated mice. *Blood* 2006;**108**:2420—7.

10. Stachura DL, Traver D. Cellular Dissection of Zebrafish Hematopoiesis. In: Westerfield M, Zon LI, Detrich HW, editors. *Essential Zebrafish Methods: Cell & Developmental Biology.* Oxford, UK: Academic Press; 2009.

11. Mathias JR, Perrin BJ, Liu TX, Kanki J, Look AT, Huttenlocher A. Resolution of inflammation by retrograde chemotaxis of neutrophils in transgenic zebrafish. *J Leukoc Biol* 2006;**80**:1281—8.

12. Renshaw SA, Loynes CA, Trushell DM, Elworthy S, Ingham PW, Whyte MK. A transgenic zebrafish model of neutrophilic inflammation. *Blood* 2006;**108**:3976—8.

13. Davis JM, Clay H, Lewis JL, Ghori N, Herbomel P, Ramakrishnan L. Real-time visualization of mycobacterium-macrophage interactions leading to initiation of granuloma formation in zebrafish embryos. *Immunity* 2002;**17**:693—702.

14. Walters KB, Green JM, Surfus JC, Yoo SK, Huttenlocher A. Live imaging of neutrophil motility in a zebrafish model of WHIM syndrome. *Blood* 2010;**116**:2803—11.

15. Ellis AE. The leucocytes of fish. *Journal of Fish Biology* 1977;**11**:453—91.

16. Chaicharn A, Bullock WL. The Histopathology of Acanthocephalan Infections in Suckers with Observations on the Intestinal Histology of two Species of Catostomid Fishes. *Acta Zoologica* 1967;**48**: 19—42.

17. Bennett CM, Kanki JP, Rhodes J, Liu TX, Paw BH, Kieran MW, et al. Myelopoiesis in the zebrafish, Danio rerio. *Blood* 2001; **98**:643—51.

18. Reite OB, Evensen O. Inflammatory cells of teleostean fish: a review focusing on mast cells/eosinophilic granule cells and rodlet cells. *Fish Shellfish Immunol* 2006;**20**:192—208.

19. Sieracki JC. The neutrophilic leukocyte. *Ann NY Acad Sci* 1955;**59**:690—705.

20. Balla KM, Lugo-Villarino G, Spitsbergen JM, Stachura DL, Hu Y, Banuelos K, et al. Eosinophils in the zebrafish: prospective isolation, characterization, and eosinophilia induction by helminth determinants. *Blood* 2010;**116**:3944—54.

21. Dobson JT, Seibert J, Teh EM, Da'as S, Fraser RB, Paw BH, et al. Carboxypeptidase A5 identifies a novel mast cell lineage in the zebrafish providing new insight into mast cell fate determination. *Blood* 2008;**112**:2969—72.

22. Lieschke GJ, Oates AC, Crowhurst MO, Ward AC, Layton JE. Morphologic and functional characterization of granulocytes and macrophages in embryonic and adult zebrafish. *Blood* 2001;**98**: 3087—96.

23. Siderits D, Bielek E. Rodlet cells in the thymus of the zebrafish Danio rerio (Hamilton, 1822). *Fish Shellfish Immunol* 2009;**27**: 539—48.

24. Drury AN. The eosinophil cell of teleostean fish. *J Physiol* 1915;**49**:349—440. 341.

25. Maxwell C, Hussain R, Nutman TB, Poindexter RW, Little MD, Schad GA, et al. The clinical and immunologic responses of normal human volunteers to low dose hookworm (Necator americanus) infection. *Am J Trop Med Hyg* 1987;**37**:126−34.

26. Kent ML, Bishop-Stewart JK, Matthews JL, Spitsbergen JM. Pseudocapillaria tomentosa, a nematode pathogen, and associated neoplasms of zebrafish (Danio rerio) kept in research colonies. *Comp Med* 2002;**52**:354−8.

27. Cross JH. Intestinal capillariasis. *Clin Microbiol Rev* 1992;**5**:120−9.

28. Rothenberg ME, Mishra A, Brandt EB, Hogan SP. Gastrointestinal eosinophils. *Immunol Rev* 2001;**179**:139−55.

29. Traver D, Paw BH, Poss KD, Penberthy WT, Lin S, Zon LI. Transplantation and in vivo imaging of multilineage engraftment in zebrafish bloodless mutants. *Nat Immunol* 2003;**4**:1238−46.

30. Iwasaki H, Mizuno S, Arinobu Y, Ozawa H, Mori Y, Shigematsu H, et al. The order of expression of transcription factors directs hierarchical specification of hematopoietic lineages. *Genes Dev* 2006;**20**:3010−21.

31. Qiu Z, Dyer KD, Xie Z, Radinger M, Rosenberg HF. GATA transcription factors regulate the expression of the human eosinophil-derived neurotoxin (RNase 2) gene. *J Biol Chem* 2009;**284**:13099−109.

32. Dyer KD, Rosenberg HF. The RNase a superfamily: generation of diversity and innate host defense. *Mol Divers* 2006;**10**:585−97.

33. Cho S, Zhang J. Zebrafish ribonucleases are bactericidal: implications for the origin of the vertebrate RNase A superfamily. *Mol Biol Evol* 2007;**24**:1259−68.

34. Jung YJ, Woo SY, Jang MH, Miyasaka M, Ryu KH, Park HK, et al. Human eosinophils show chemotaxis to lymphoid chemokines and exhibit antigen-presenting-cell-like properties upon stimulation with IFN-gamma, IL-3 and GM-CSF. *Int Arch Allergy Immunol* 2008;**146**:227−34.

35. Lugo-Villarino G, Balla KM, Stachura DL, Banuelos K, Werneck MB, Traver D. Identification of dendritic antigen-presenting cells in the zebrafish. *Proc Natl Acad Sci USA* 2010;**107**:15850−5.

36. Callard RE, Matthews DJ, Hibbert L. IL-4 and IL-13 receptors: are they one and the same? *Immunol Today* 1996;**17**:108−10.

37. Caput D, Laurent P, Kaghad M, Lelias JM, Lefort S, Vita N, et al. Cloning and characterization of a specific interleukin (IL)-13 binding protein structurally related to the IL-5 receptor alpha chain. *J Biol Chem* 1996;**271**:16921−6.

38. Coffman RL. Immunology. The origin of TH2 responses. *Science* 2010;**328**:1116−7.

39. Urban Jr JF, Noben-Trauth N, Donaldson DD, Madden KB, Morris SC, Collins M, et al. IL-13, IL-4Ralpha, and Stat6 are required for the expulsion of the gastrointestinal nematode parasite Nippostrongylus brasiliensis. *Immunity* 1998;**8**:255−64.

40. Lieschke GJ, Grail D, Hodgson G, Metcalf D, Stanley E, Cheers C, et al. Mice lacking granulocyte colony-stimulating factor have chronic neutropenia, granulocyte and macrophage progenitor cell deficiency, and impaired neutrophil mobilization. *Blood* 1994;**84**:1737−46.

41. Matthaei KI, Foster P, Young IG. The role of interleukin-5 (IL-5) in vivo: studies with IL-5 deficient mice. *Mem Inst Oswaldo Cruz* 1997;**92**(Suppl. 2):63−8.

42. Robinson MW, Dalton JP, Donnelly S. Helminth pathogen cathepsin proteases: it's a family affair. *Trends Biochem Sci* 2008;**33**:601−8.

43. Miike S, Kita H. Human eosinophils are activated by cysteine proteases and release inflammatory mediators. *J Allergy Clin Immunol* 2003;**111**:704−13.

44. Curado S, Stainier DY, Anderson RM. Nitroreductase-mediated cell/tissue ablation in zebrafish: a spatially and temporally controlled ablation method with applications in developmental and regeneration studies. *Nat Protoc* 2008;**3**:948−54.

45. Curado S, Anderson RM, Jungblut B, Mumm J, Schroeter E, Stainier DY. Conditional targeted cell ablation in zebrafish: a new tool for regeneration studies. *Dev Dyn* 2007;**236**:1025−35.

Chapter 16.4

Eosinophils as Immune Modulators: Roles of Indoleamine 2,3 Dioxygenase and Tryptophan Catabolites

Ramses Ilarraza, Solomon O. Odemuyiwa, Kanami Orihara, Narcy Arizmendi and Redwan Moqbel

CONVERGENCE OF METABOLIC PATHWAYS AND IMMUNE/INFLAMMATORY REGULATION: TRYPTOPHAN CATABOLISM BY INDOLEAMINE 2,3-DIOXYGENASE AND SUBSEQUENT KYNURENINE GENERATION

Tryptophan is an essential amino acid for *de novo* protein synthesis and other metabolic functions, and its catabolism yields the neuroactive derivatives, melatonin and serotonin,[1] as well as other bioactive by-products, such as niacin, which provide the backbone of the nicotinamide adenine dinucleotide electron transporters in the Krebs cycle. Humans have no metabolic pathway to synthesize tryptophan; therefore, it is obtained solely from the dietary intake. There are two known enzymes in charge of tryptophan degradation: hepatic tryptophan 2,3-dioxygenase and extrahepatically expressed and immunologically regulated indoleamine 2,3-dioxygenase (IDO). Tryptophan catabolism yields a series of biochemical intermediates collectively named kynurenines, which have various biological activities. In the reproductive system and central nervous system (CNS), tryptophan catabolites are employed for the consumption of superoxide ions, thereby playing a protective role.[2]

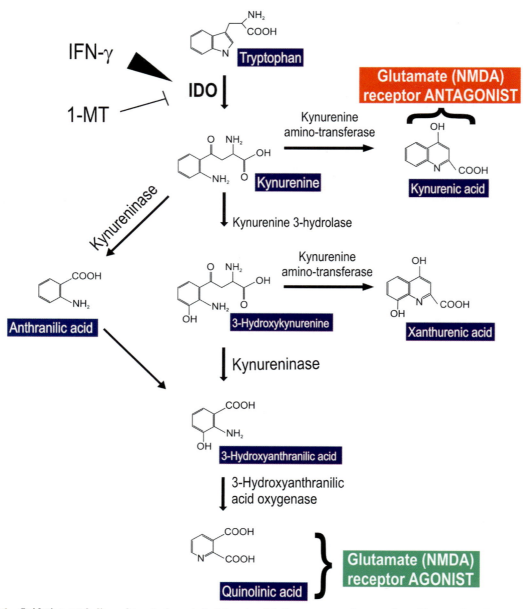

FIGURE 16.4.1 **Oxidative catabolism of tryptophan via indoleamine 2,3-dioxygenase and generation of kynurenines.** Indoleamine 2,3-dioxygenase (IDO) catalyzes the oxidation of the indole ring in tryptophan yielding unstable N-formyl-kynurenine, which is subsequently deformylated by formidases to yield kynurenines (not shown). IDO can be induced by interferon γ (IFN-γ), but its activity is blocked by 1-methyl-D-tryptophan (1-MT). Kynurenine is further degraded along one of two catabolic branches, one leading to the formation of kynurenic acid (KA), while the other generates 3-hydroxykynurenine and subsequently quinolinic acid (QA). QA and KA have opposing effects. QA is an agonist for glutamate (N-methyl-D-aspartate; NMDA) receptors, while KA is a glutamate/NMDA-R antagonist.

Besides its relevance in the CNS, in the past few decades, tryptophan metabolism has emerged as a relevant pathway also in the immune system (Figure 16.4.1). Within the confines of the immune system, several cell types have been shown to contribute to tryptophan degradation through the IDO pathway. We have provided evidence that human eosinophils constitutively express IDO.[3] Immature CCR-6$^+$ CD123$^+$ dendritic cells (DC) have also been shown to express IDO constitutively.[4,5] Other antigen-presenting cells (APC), including subsets of conventional DC bearing CD11c and CD8α markers, plasmacytoid DC[6] and macrophages, are also relevant sources of immunoactive IDO.[7] A role for IDO in the control of the immune regulation has been further confirmed with the use of IDO knockout mice[8] and models of allograft tolerance based on mice fetus immunization,[2] but the specific details of such roles are yet to be described.

IDO is expressed in many cell types, usually as an inducible enzyme with various immunoactive properties.[9] Biochemically, IDO catalyzes the oxidation of the indole ring in tryptophan down the kynurenine pathway,[10–12] thereby forming the relatively unstable N-formyl-kynurenine, which is subsequently deformylated by formidases to yield kynurenine (Figure 16.4.1). Kynurenine is further degraded along one of two catabolic branches, leading to the formation of 3-hydroxykynurenine (3-HK) and subsequently of quinolinic acid (QA) or kynurenic acid (KA). The latter two kynurenines (QA and KA) are a major focus of scientific interest, particularly because they have been shown to have opposite effects in some systems. Whereas QA is a known agonist for ionotropic glutamate receptors, specifically the N-methyl-D-aspartate (NMDA)-receptors (NMDA-R), KA is a glutamate/NMDA-R antagonist.[13–14] As a consequence of these interactions with NMDA-R, QA-induced continuous stimulation may lead to neurotoxicity, while KA-mediated NMDA-R antagonism is neuroprotective. The mechanisms controlling the KA/QA balance and its consequent effects are yet to be elucidated.

IDO is distributed ubiquitously in extrahepatic sites, with the lung and placenta having the highest activity.[15–17] IDO is an inducible enzyme, and its most potent known inducer is interferon γ (IFN-γ);[17] other triggers including Toll-like receptor agonists, lipopolysaccharide (LPS), tumor necrosis factor (TNF), and viral infections may also stimulate its expression.[18] During the onset of an infection or immune response (i.e., the presence of a *danger* signal) and the subsequent release of proinflammatory cytokines—including IFN-γ—many cell types react with robust IDO synthesis and tryptophan catabolism.[19] Studies using IFN-γ deficient mice revealed additional IFN-γ-independent IDO inducers, such as ligation of cytotoxic T-lymphocyte protein 4 (CTLA-4) and the co-stimulatory molecules, T-lymphocyte activation antigens CD80 and CD86[20–21] (B7 in mouse), both of which emphasize the relevance of IDO in the immune system.

NEURONAL AND IMMUNE REGULATORY ROLES OF INDOLEAMINE 2,3-DIOXYGENASE

While IDO has long been known to be an important contributor to tryptophan catabolism, mounting evidence suggests multiple roles for this enzyme in processes other than the merely amino acid degradation or the production of neurochemically-active mediators, such as serotonin (5-hydroxytryptamine; 5-HT) and melatonin (N-acetyl-5-methoxytryptamine; Figure 16.4.2). From the CNS standpoint, kynurenines have been of great interest, since they have been found to interact with a major type of neuronal receptor: the glutamate receptor, specifically, the ionotropic NMDA-R.[13] Besides their involvement in the neuronal system, one study ascribed an immune modulatory activity to the IDO pathway in trophoblasts during pregnancy and its protective role for the allogeneic fetus against maternal T cells,[2] involving a major role for IDO expression by DC,[22] which will be discussed later in this subchapter. Observations of this nature date from many years ago,[23] but their particular mechanisms of action have remained largely unknown until the past few years, when we started to elucidate the different factors involved in immune modulation exerted by IDO. Our own studies indicate that human eosinophils constitutively express bioactive IDO and that kynurenines derived from eosinophil-mediated tryptophan catabolism via IDO hampered the proliferation of T-helper type 1 (T_h1) but not T_h2 cells.[3]

Further observations on the effects of excessive tryptophan catabolism on live cells demonstrated a significant inhibition of T cell division and the induction of apoptosis.[24] However, the precise mechanisms underlying tryptophan catabolism-mediated cell changes and immune modulation are not well understood.

EOSINOPHILS EXPRESS INDOLEAMINE 2,3-DIOXYGENASE CONSTITUTIVELY

Eosinophils Regulate Immune Response via Inflammatory Mediators

Eosinophils are pleiotropic multifunctional leukocytes involved in the initiation and propagation of allergic inflammation,[25–27] with a major role in tissue remodeling in asthma and the modulation of innate and adaptive immunity.[28–31] A key feature that defines the role of eosinophils is that they synthesize, store and secrete multiple chemokines and cytokines,[32] which can be released in a controlled manner upon stimulation.[33–37] In addition to their individual roles as chemical mediator-producing immune cells, some studies have implicated the eosinophil in the immune regulation through their interaction with CD4$^+$ T cells,[38] as eosinophils express major histocompatibility complex class II (MHC-II) and the co-stimulatory molecules, CD28, CD40, CD80, and CD86.[39–40] Thus, eosinophils have the potential to exert immunoregulatory functions in the context of CD4$^+$ T cell responses,[41–44] through interactions between MHC-II molecules and T cell receptor, as well as the co-stimulatory effect of CD28, CD40, CD80, and CD86, in a manner that resembles that of APC. In addition to co-stimulation, the capacity of the eosinophil to release immune-modulating chemokines and cytokines confers

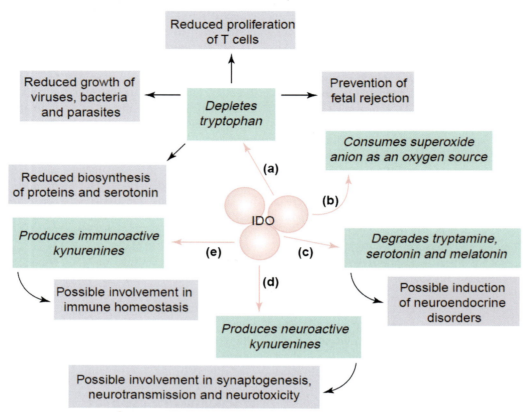

FIGURE 16.4.2   Multiplicity of effects of indoleamine 2,3-dioxygenase induction and possible role in physiopathological conditions. The scheme depicts the various physiological functions of the essential amino acid tryptophan, the possible consequences of its depletion, the effects of indoleamine 2,3-dioxygenase (IDO) activity on indole derivatives other than tryptophan, and the postulated biological functions of tryptophan breakdown products (kynurenines). By catalyzing the oxidative cleavage of the indole ring of tryptophan and thus depleting this essential amino acid, IDO impairs the biosynthesis of proteins and serotonin, limits microbial growth, and affects T-cell proliferation, with possible implications for cell-mediated immune responses, including fetal allograft rejection (a). As a consequence of inflammation (e.g., at sites of microbial infection), the superoxide anion is liberated and serves as a substrate for IDO. In the course of an inflammatory response, and possibly at the maternal−fetal interface, clearance of free radicals and inhibition of serotonin vasoactive properties could represent a common, protective mechanism (b). By acting on a variety of indole derivatives, IDO affects the biology of important regulatory molecules, including melatonin, serotonin, and tryptamine, with potential implications for neuroendocrine functions (c). Tryptophan degradation leads to the production of kynurenines, which display neuroactive and neurotoxic properties (d). Selected kynurenines also exhibit immunoactive properties and may contribute to the control of immune homeostasis (e). *(Reprinted from[89] with permission from Elsevier.)*

these eosinophil−T cell interactions an additional dimension. Considering that the outcome of T cell activation is, to a great extent, a result of the particular cytokine milieu,[45] eosinophils may hold powerful control mechanisms over the type of immune response generated.

Eosinophils produce and release a plethora of relevant immune modulators,[46] including:

- Chemokines: eotaxin (CCL11), IL-8 (CXCL8), C-C motif chemokine 3 (macrophage inflammatory protein MIP-1α; CCL3) and RANTES (C-C motif chemokine 5; CCL5);
- Cytokines: IL-1α, IL-2, IL-3, IL-4, IL-5, IL-6, IL-8, IL-9, IL-10, IL-12, IL-13, IL-16, IL-18, IFN-γ, granulocyte-macrophage colony-stimulating factor (GM-CSF), and TNF, among many others;

- Growth factors: transforming growth factors α (TGF-α) and TGF-β, heparin-binding epidermal growth factor (HB-EGF), nerve growth factor (NGF), platelet-derived growth factor β (PDGF-β), stem cell factor (SCF), and amphiregulin[47];
- Lipid mediators: platelet-activating factor (PAF), cysteinyl leukotrienes C_4 and D_4 (LTC_4, LTD_4), prostaglandins;
- Reactive oxygen species: superoxide anions, singlet oxygen;
- Cytotoxic granule proteins: major basic protein (MBP), eosinophil cationic protein (ECP), eosinophil-derived neurotoxin (EDN) and eosinophil peroxidase;
- Nitric oxide.

Eosinophils are Sources of Indoleamine 2,3-Dioxygenase in Different Immune Relevant Tissues

Besides their known immunomodulatory roles, human eosinophils constitute a relevant site of constitutive IDO bioactivity, shown to be present both at mRNA and protein levels[3] (Figure 16.4.3). We have identified eosinophils in

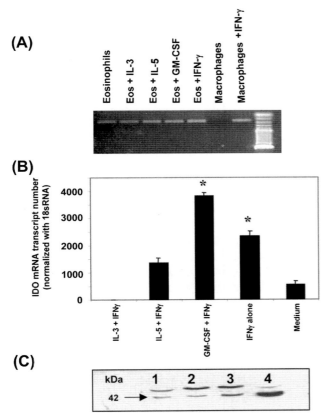

(A)

(B)

(C)

FIGURE 16.4.3 Eosinophils constitutively express indoleamine 2,3-dioxygenase. *A,* Indoleamine 2,3-dioxygenase (IDO) gene expression in human eosinophils (Eos) by reverse transcription polymerase chain reaction (RT-PCR) analysis of IDO mRNA expression in cytokine-treated Eos. IDO expression was detectable in nonstimulated and cytokine-treated human Eos and macrophages generated from the human THP-1 cell line by treatment with interferon γ (macrophages + IFN-γ); however, it was not detected in untreated THP-1 cells (macrophages). *B,* Real-time RT-PCR analysis of IDO mRNA expression. Following pretreatment with interleukin-3 (IL-3), IL-5, and granulocyte-macrophage colony-stimulating factor (GM-CSF; 10 ng/mL) or IFN-γ for 2 h, Eos were treated for an additional 2 h with IFN-γ (at 37°C). Following reverse transcription, PCR was performed with IDO-specific and 18S ribosomal RNA primers and quantified using the SYBR Green I dye. A standard curve was generated following the amplification of known starting copy numbers of IDO and 18S rRNA. Copy numbers of IDO mRNA were determined following normalization of the mRNA concentration with 18S rRNA ($n = 4$). *C,* Western blot analysis of Eos treated (48 h) with IL-3 (lane 1), IL-5 (lane 2), GM-CSF (lane 3), or IFN-γ (lane 4). Treatment with IFN-γ upregulated the 42-kDa form of IDO (lower band) and decreased the level of the upper band (approximately 45 kDa). *(Modified from Odemuyiwa et al.[3])*

the thymus as IDO-producing cells that may be relevant in T-cell clonal selection.[48] This will be discussed later in this subchapter. Interestingly, expression of IDO by eosinophils has also been found in non-small cell lung cancer,[49] where eosinophil IDO is thought to contribute to local immune suppression and consequent evasion of the immune response by tumor cells. Tryptophan catabolism via IDO, as a means to regulate T-cell responses, has grown in scientific interest during the past few years, and has yielded valuable insights into the role of amino acid metabolism and immune regulation. While mainly inducible, the finding of constitutive expression of IDO in eosinophils confers this cell type with added potential to be a relevant T-cell modulator.

INDOLEAMINE 2,3-DIOXYGENASE AS AN IMMUNE MODULATOR

How Indoleamine 2,3-Dioxygenase Came to be Recognized as an Important Component of Immunity (Role in Fetal Allograft Tolerance)

Munn and coworkers in 1998[2] analyzed the possibilities proposed by Peter Medawar in 1953[23] regarding the mechanisms that prevent fetal rejection. The elegant experiments of Billingham, Brent, and Medawar consisted of the transplantation of a mixture of adult allogeneic albino mice cells obtained from kidney, splenic, and testis tissue, into CBA mice fetuses *in utero*. Upon birth, the CBA offspring mice had the usual agouti color, but upon receiving skin grafts from the original donor mice (albino A-line), some were found to be tolerogenic, exhibiting characteristics of autografts, while retaining the albino color of the donor. Skin graft tolerance was found not to be hereditary, since the offspring of the tolerant mice were not tolerant themselves. Induction of tolerance was found to be fetal stage-related because when newborn mice, instead of fetuses *in utero*, were inoculated, most of the mice rejected allografts from CBA mice.

Medawar found that the tolerance generated during pregnancy was neither absolute nor irreversible. When adult tolerant mice received lymph node cells obtained from normal CBA mice that had been immunized with albino mice cells, all previously accepted allografts were rejected. These observations suggest that tolerance is insufficient to overcome the rejection induced by the transplanted immune cells, but is rather an *in utero*-induced phenomenon. The latter point gains relevance when taking into account that the fetus *per se* has the potential to induce maternal rejection due to the presence of paternal antigens, making it clear that a two-way tolerance mechanism has to exist in order to allow fetal survival. Therefore, the question of which mechanisms lead to fetal tolerance and survival was raised.

Three possibilities were discussed by Munn and colleagues to explain the outcome: anatomic separation between mother and fetus, antigenic immaturity of the fetus, and maternal immunologic tolerance. The first two possibilities were ruled out since there is evidence of active tolerance of paternal antigens (MHC class I) in the mother during pregnancy. Hence, an active mechanism for maternal tolerance was proposed by Munn and Mellor's group, involving IDO.[2] Munn and coworkers showed that IDO is induced in murine placenta at day 7.5 postcoitus, consistent with IDO expression in the syncytiotrophoblast in humans. Definitive evidence showing the involvement of IDO in maternal tolerance was obtained in experiments with the IDO inhibitor 1-methyl-D-tryptophan (1-MT), which induced allogeneic fetal rejection without impairing syngeneic fetal development. When Munn and coworkers used $Rag1^{-/-}$ mice, which are defective in lymphocyte development, it became evident that IDO-mediated tolerance involves such cells, since $Rag1^{-/-}$ mice did not exhibit 1-MT-induced rejection of allogeneic fetuses.

Originally, the effect of IDO in pregnancy-related immune tolerance was considered a consequence of a drop in tryptophan levels.[50] However, evidence exists for the existence of a more complex mechanism; this will be discussed in the next section.

Direct versus Indirect Effects of Indoleamine 2,3-Dioxygenase

Since its discovery, the role of IDO has been mostly attributed to the depletion of tryptophan.[51] However, in recent years it has become evident that IDO-mediated immune regulation extends beyond the depletion of this essential amino acid. Although tryptophan deficiency can indeed have an effect in cell metabolism, tryptophan catabolism by-products themselves also exert biological functions. Tryptophan catabolites have major relevance in other nonimmune systems, such as the CNS, where they have been widely shown to have both neuroprotective and neurotoxic roles, depending on the nature of the catabolite and the context of its action.[13,52]

To explain the effects of IDO in the immune system, two theories have been proposed.[53] As pointed out above, the first theory is based on the fact that humans have no pathway for tryptophan biosynthesis, so tryptophan can only be obtained from the diet. Hence, this theory is based on the concept that tryptophan depletion through its catabolism by IDO is the main contributor to suppression of T-cell proliferation via metabolic impairment (i.e., the inability to synthesize proteins with a high tryptophan content). While the evidence to support this theory is strong, it appears to fall short of explaining the observed phenomenon, since local tryptophan content is continuously replenished by blood flow. Therefore, the lowered tryptophan concentration hypothesis may not fully explain the role of this enzyme in immune modulation. The second theory postulates that tryptophan catabolites, mainly kynurenines, exhibit direct suppressive effects on certain immune cells via proapoptotic mechanisms. This theory will be discussed in a later section of this subchapter.

The Eosinophil is Involved in Regulating Clonal T Cell Depletion in the Thymus

Our discovery of IDO-expressing eosinophils in the human thymus is also quite intriguing.[48] The thymus is a key lymphoid organ involved in the maturation of T lymphocytes, positive selection of mature cells, and self-reactive clonal deletion.[54] It is in this organ that T-cell responses are largely determined since auto-reactive T-cell clones are deleted, while nonself reactive lymphocytes become mature as they continue their migration to the bloodstream. Therefore, it is probable that most of the features of adaptive immunity of an individual rely on the events that lead to T-cell selection or apoptosis, and maturation. Our group found that eosinophils are present in the thymus and that their number decrease with age, similar to thymus involution.[48] Supporting the role of the eosinophil in thymic selection, our evidence also showed that eosinophil IDO expression in the thymus contributes to local catabolism of tryptophan, and the by-products of such catabolism (kynurenines) decrease with age as well. Furthermore, thymic eosinophil IDO expression correlates positively with T_h2 markers [signal transducer and activator of transcription 6 (STAT6) and IL-4], and negatively with T_h1 cell marker T-box transcription factor TBX21 (T-bet).[48] Although these data still require expansion regarding the mechanistic aspects, both in human tissue and animal models, they strongly suggest an involvement of the eosinophil in both T lymphocyte clonal selection and subsequent phenotypic selection of $CD4^+$ T-cell subsets. These evidence, therefore, suggest that eosinophils may play an active role in negative selection of T_h1 cell clones in the thymus, via IDO and probably through kynurenines.

Eosinophil Indoleamine 2,3-Dioxygenase Suppresses T-Helper Type 1 but not T-Helper Type 2 Proliferation

Eosinophils are important sources of IDO in the immune system. Provided that tryptophan is available, they constitute a continuous, rather than inducible, source of bioactive kynurenines. In our studies, we found that eosinophil-derived, IDO-mediated tryptophan catabolism results in the

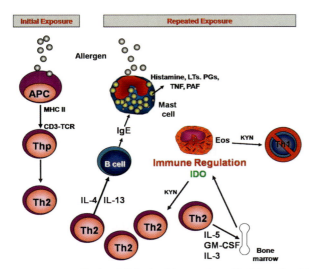

FIGURE 16.4.4 Eosinophil-derived interaction of indoleamine 2,3-dioxygenase mediates tryptophan catabolite interactions with T cells. While mast cells have been shown to be involved in allergic responses, via the release of a plethora of chemical mediators (growth factors and lipid mediators), it is now known that eosinophils have the capacity to exert powerful effector function on various tissues, which may contribute, for example, to airway hyperresponsiveness in asthma. Eosinophils constitutively express indoleamine 2,3-dioxygenase (IDO), a rate-limiting enzyme in the metabolism of the amino acid tryptophan. IDO, when stimulated intracellularly by interferon γ (IFN-γ), lowers the level of extracellular tryptophan and breaks it down into a series of by-products including kynurenines (KYNs), which have been shown to target T-helper type 1 (T_h1) cells for limited proliferation and apoptosis, while selectively allowing T_h2 proliferation. The precise mechanism and receptors involved in this distinct T_h1 versus T_h2 response remain unknown and under active study. APC, antigen-presenting cell; GM-CSF, granulocyte-macrophage colony-stimulating factor; IgE, immunoglobulin E; IL, interleukin; LT, leukotriene; PAF, platelet-activating factor; PG, prostaglandin; TCR, T-cell receptor; ThP, T helper progenitor cell; TNF, tumor necrosis factor.

generation of kynurenines, catabolites that exert differential apoptotic and antiproliferative effects over CD4+ T lymphocytes, i.e., affecting T_h1 but not T_h2 cells[3] (Figure 16.4.4). These data support the observation that T_h2 predominance, a major feature in allergic disease, may be more a consequence of T_h1 cell growth arrest rather than an increase in the proliferation or frequency of generation of T_h2 cells.[55-56] The basis for the preferential induction of apoptosis by kynurenines in T_h1 but not in T_h2 cells, and the exact mechanisms that lead to T_h2 predominance in the context of IDO, kynurenine activity, and tryptophan catabolism are currently unknown. However, these results identify the eosinophil as a contributor to the T_h2 phenotype associated with some allergic diseases, such as asthma and atopy. There is growing scientific interest in identifying such mechanisms for their understanding is likely to provide additional novel preventative and therapeutic targets in allergy and asthma.

KYNURENINES AS GLUTAMATE RECEPTOR AGONISTS AND ANTAGONISTS: THEIR INVOLVEMENT IN IMMUNE SYSTEM REGULATION

Mounting evidence supports the role of IDO in the polarization of immune responses.[53] However, information on the mechanisms leading to this outcome has never been fully convincing, clearly suggesting that more than one mechanism may be relevant to this phenomenon.[57] While evidence has been provided to support the hypothesis that IDO serves as a metabolic regulatory factor due to depletion of the essential amino acid tryptophan,[11] little information has emerged on the possible involvement of the by-products of tryptophan catabolism on this control mechanism. Although we acknowledge the importance of tryptophan availability for cellular viability, novel and exciting data have surfaced regarding kynurenine-mediated T-cell functional control.

Glutamate Receptors

It has long been known that kynurenines bind to and stimulate excitatory neuronal glutamate receptors, often resulting in a form of neuronal death termed excitotoxicity.[13,58] Glutamate is the major excitatory neurotransmitter in the brain,[59] activating cells through glutamate-gated ion channels (ionotropic glutamate receptors), as well as G-protein-coupled receptors (metabotropic receptors). The maintenance of homeostatic concentrations of glutamate is crucial for cognitive functions of neurons.[60] Conversely, high concentrations of glutamate act as a neurotoxin, resulting in neuronal death through excessive activation of glutamate receptors. The presence of glutamate in the synapse is tightly regulated under physiological conditions by active, ATP (adenosine-5'-triphosphat)-dependent transporters located in perisynaptic astroglial cells.[61] An intriguing unpublished observation from our group was the finding that eosinophils are capable of constitutively releasing glutamate,[62] which underscores the potential for interaction between eosinophils and other cell types via glutamate. In the airways, neurological studies from Haxhiu and coworkers[63] showed that glutamate, via glutamate receptor activation, leads to reflex responses and results in increased airway mucosal blood flow, submucosal gland secretion, and smooth muscle tone. Advances in the study of glutamate in extraneuronal systems have highlighted its potential relevance in the immune system. In this regard, one study[64] identified glutamate receptors, and in particular the activation of the NMDA-R, as major players in pathogenesis in lung injury and acute bronchial asthma. Mice injected intraperitoneally with glutamate exhibited elevated nitric oxide concentration, increased inducible nitric oxide synthase (iNOS) activity, and lung

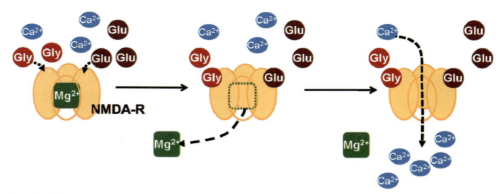

FIGURE 16.4.5 *N*-methyl-D-aspartate receptors are ionotropic cation channels. *N*-methyl-D-aspartate receptors (NMDA-Rs) are glutamate-gated Ca^{2+} channels that, under normal conditions, are blocked by Mg^{2+} ions. Removal of Mg^{2+} upon depolarization of the cell, following initial stimulation via AMPA-R, in the nervous system, allows the glutamate (Glu) to open the NMDA-R channel, causing an influx of Ca^{2+}. Gly, glycine.

injury.[65] All of these results indicate that glutamate may play a key role in hyperoxia-induced lung injury, possibly mediated by NO; however, the source of glutamate remains unidentified.

Ionotropic glutamate receptors are subdivided into three classes (based on their selective agonists): the α-amino-3-hydroxy-5-methyl-4-isoxazolepropionic acid receptor (AMPA-R), the kainate receptor, and the NMDA-R. While all three classes of ionotropic glutamate receptors are permeable to Na^+ and K^+, NMDA-R is the only ionotropic glutamate receptor that is also permeable to Ca^{2+}, via a high-conductance Ca^{2+} channel that is blocked by Mg^{2+} ions in a voltage-dependent manner (Figure 16.4.5). Due to the intrinsic implications of Ca^{2+} influx into the cell (i.e., second messenger role, T-cell proliferation induction, and toxicity), there is greater interest in this ion rather than in K^+ or Na^+.

Functional NMDA-R is a heterotetrameric structure comprised of individual subunits. Several NMDA-R configurations have been identified, which differ in their kinetic properties and sensitivity to various ligands. The functional properties of NMDA-R, including permeability to divalent ions and interactions with intracellular proteins,[66] are determined by the specific subunit assembly. In the immune system, human lymphoblasts and T cells have been shown to express NMDA-R, and stimulation of this receptor contributes to T cell activation and proliferation.[67-68]

In addition, NMDA-R is expressed by alveolar macrophages and other lung cells, with some evidence that different parts of the lung express distinct isoforms of one of the NMDA-R subunits, NR2, yielding a functionally distinct NMDA-R.[64] Administration of NMDA to rat lungs leads to nitric oxide-dependent pulmonary edema,[69] but may also increase NMDA-R expression.[64] NMDA-R activation also mediates capsaicin-induced injury, increased tracheal contractile responses to methacholine,[70] and sepsis-induced lung injury in guinea pigs.[71] The cells

activated by NMDA-R agonists that mediate these effects have not been extensively studied but are likely to be tissue-resident cells, including alveolar macrophages. Shang and coworkers[72] showed the expression of NMDA-R subtypes in alveolar macrophages from both newborn and adult rats; newborn rats exhibited differential expression of NMDA-R subunits compared to adults, raising the question of the presence of distinct NMDA-R arrays during developmental changes in the rat. NMDA-R was found also in the peripheral lung of rats as well as in the alveolar walls, bronchial smooth muscle, and bronchial epithelium.[73]

The NMDA Receptor and Excitotoxicity

Brain interstitial glutamate levels are tightly regulated under normal conditions, to maintain optimal synaptic function. During ischemic conditions, however, energy failure causes glutamate efflux sufficient to increase interstitial glutamate from less than $5\,\mu M$[74] to $100-200\,\mu M$.[75-76] This increase results in a phenomenon known as excitotoxicity (Figure 16.4.6), which results from excessive stimulation of NMDA-R and toxic intracellular Ca^{2+} accumulation. In the CNS, an initial depolarization event mediated by AMPA-R is required for NMDA-R-induced cytotoxicity; once the initial depolarization occurs, continuous and/or extensive NMDA-R activation leads to Ca^{2+} influx, elevating the concentration of this ion inside the cell. Elevated Ca^{2+} levels then trigger a cascade of toxic events that culminates in cell death. Overstimulation of the NMDA-R can lead to neuronal death under many acute and chronic conditions, and has been found responsible for neural loss in Alzheimer disease, epilepsy, Huntington chorea, ischemia, Parkinson disease, and AIDS encephalopathy, also known as HIV dementia.[77-78] Subsequent events lead to mitochondrial depolarization, permeability transition, and the activation of intrinsic apoptotic pathways characterized by mitochondrial release of the

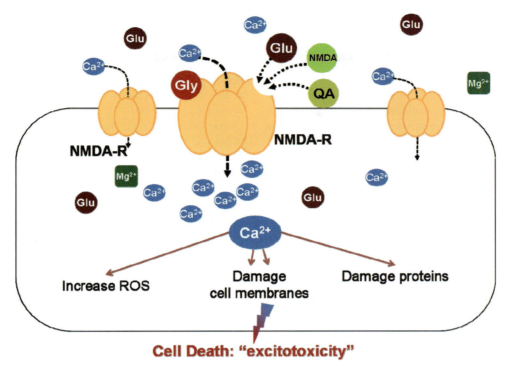

FIGURE 16.4.6 **Overstimulation of the *N*-methyl-D-aspartate receptor leads to excitotoxicity.** Binding of the major agonists, glutamate, *N*-methyl-D-aspartate (NMDA), and quinolinic acid, to NMDA receptors (NMDA-Rs), leads to Ca^{2+} influx. Under normal circumstances, the influx is controlled due to the intermittent and limited availability of glutamate (Glu). However, continuous and excessive stimulation of NMDA-R can induce neuronal cell death via excitotoxicity. Gly, glycine.

proapoptotic factor, cytochrome c, and downstream caspase-3 activation. Excitotoxic Ca^{2+} accumulation can also result in excessive lipase and protease activation, leading to cell membrane damage and necrotic-type neuron death.

Kynurenines as Surrogates for Glutamate: A Role of the Eosinophil and the Interaction of Kynurenines with T Cells

We have proposed that T-cell negative clonal selection is a process that may involve tryptophan catabolites produced by the initial activity of IDO on tryptophan, with eosinophils playing a key role both in peripheral and thymic T-cell selection.[48] In the thymus, the presence of IDO-expressing eosinophils, also capable of releasing glutamate,[62] would contribute to T-cell deletion via activation of NMDA-R by glutamate and QA. In the periphery, eosinophils, like APC, may be equally important sources of IDO and thus kynurenines, which in turn control the fate of T-cell polarization. It is likely that glutamate and/or QA bind to NMDA-R, thereby activating Ca^{2+} influx and Ca^{2+}-mediated signaling. Differential T-cell death would be a consequence of Ca^{2+}-mediated signaling in a process analogous to excitotoxicity, a mechanism not yet described in T lymphocytes.

Some of the downstream molecules of the kynurenine pathway, such as QA, 3-HK, and 3-hydroxyanthranilic acid (Figs 16.4.1 and 16.4.6), are known to be neurotoxic and are implicated in a range of neurodegenerative disorders.[79] While QA is an NMDA-R agonist, KA is known to be an NMDA-R antagonist.[13,58] Increased QA production occurs in the CNS and has been characterized in a number of neurological inflammatory disorders: beyond the CNS, substantial amounts of kynurenines and quinolinates have been detected in the lungs and lymphatic tissue, including our own observation in the human thymus, where kynurenines may influence the developing immune response.[48,80−81] Elevated levels of these by-products have also been reported in the serum of asthma patients as well as of those suffering from chronic bronchitis.[82]

Studies on excitotoxicity have been historically confined to neurons. It is a growing belief, however, that similar cytotoxic processes may occur in immune cells and result in apoptosis in a subset of T cells in a glutamate- and/or kynurenine-dependent fashion. However, data outside the field of the nervous system are scarce. Warraki and coworkers[82] first suggested a potential role for the kynurenine pathway in asthma and airway obstruction in humans. Furthermore, interesting evidence emerged from the work of Said and his lab at Stony Brook University Medical

Center, in New York. Said published elegant studies in which it was suggested that glutamate and NMDA may be key players in lung health.[69] His group also predicted the potential critical role of NMDA[64,69−70] in the airways and NMDA-R activation in pulmonary oxidative injury.[83] These researchers found that lung injury can arise from NMDA-R stimulation via excitotoxicity and subsequent generation of reactive oxygen species, in a nitric oxygen synthase-dependent manner.

INDOLEAMINE 2,3-DIOXYGENASE AND KYNURENINES IN ALLERGIC INFLAMMATION AND TOLERANCE: MODELS OF EOSINOPHILIC INFLAMMATION WITH A ROLE FOR INDOLEAMINE 2,3-DIOXYGENASE

IDO has been found to have T-cell immune-modulating properties in human subjects as well as in mouse models. Within this context, studies on allergic (eosinophilic) asthma patients have yielded some insight on such control. Maneechotesuwan and coworkers[84] showed that IDO activity in human sputum is decreased in house-dust mite (HDM)-sensitized patients with asthma, compared to non-asthmatic patients. The main HDM allergen (Der p 1) induced IDO activity, but the authors found no differences in kynurenine levels from HDM-sensitive or nonatopic subjects with asthma. This observation suggests that Der p 1 differentially modulates the expression and activity of IDO, depending on the sensitization status toward HDM. The existence of a T_h2-type response (IL-4-predominant) in HDM-sensitized individuals has been well documented, concurrently with suppression of IFN-γ secretion.[85] T cells from nonatopic individuals produced IL-10 and IFN-γ, but not T_h2-type cytokines,[86] while IFN-γ production was increased in response to HDM allergens.[85] This suggests the existence of a modulatory response mediated by both allergen exposure and the atopic or immunological status of the subjects.

Tryptophan catabolism by-products resulting from IDO activity, such as QA and KA, are of great relevance to the control of the immune response, especially in regards to T-cell viability and overall behavior. In particular, two studies, one from our lab[87] (Figure 16.4.7) and another by Taher and coworkers,[88] have provided strong evidence supporting the role of tryptophan catabolites in the development of tolerance using a mouse model. We showed that inhibition of IDO by 1-MT leads to the inhibition of tolerance (Figure 16.4.7A); accordingly, mice treated with 1-MT fail to develop tolerance in a protocol of desensitization with ovalbumin (OVA) that, under normal conditions, leads to tolerance induction (Figure 16.4.7B). Finally, addition of a mixture of kynurenines in the presence of 1-MT (i.e., bypassing the requirement for IDO) restored tolerance induction in this model (Figure 16.4.7C). In a similar manner, Taher and coworkers showed that the exogenous addition of kynurenine, 3-hydroxykynurenine, and xanthurenic acid, leads to the establishment of tolerance against OVA-induced hyperresponsiveness.

As discussed, we have proposed that kynurenines target CD4[+] T cells via NMDA-R and skew the balance toward T_h2 predominance. In T_h1 cells, glutamate may induce apoptosis (but definitely promotes growth arrest), in a pattern similar to neuronal excitotoxicity, while allowing the proliferation and expansion of T_h2 cells. These findings about the role of NMDA-R in IDO-mediated T-cell impairment introduce a new insight into the T_h cell bias in atopic asthma. Furthermore, the possibility of targeting the kynurenine pathway may provide opportunities for preventative and therapeutic measures within the context of immune regulation and inflammation in asthma and allergy, among other immune diseases.

CONCLUSIONS

The convergence of tryptophan oxidative catabolism via IDO, the generation of kynurenine by-products, and immune regulation is a relatively new but fascinating area of investigation particularly in neuroimmunology. This novel concept may have major and significant relevance to events leading to the development and regulation of T cell-dependent immunity and inflammation associated with atopy and asthma. The fact that eosinophils express IDO constitutively and have the ability to generate kynurenines in the presence of tryptophan is interesting and requires further study and probing. Beyond that, our observation that, similar to glutamate, eosinophil-derived kynurenines may have a direct influence on T_h subset cell survival (T_h2) provides evidence for a potentially critical role for the eosinophil in immune modulation. Together with other evidence accumulating from various studies, research efforts over the next few years will reveal the intersection of innate immunity and different cell types, including the eosinophil, with adaptive immunity. This is an exciting time to pursue these novel questions with a view to understanding the relationship between the mechanisms that regulate neuronal patterns of response and cell survival/function within the context of innate and acquired immune responses. From our perspective, the interaction between T cells with tryptophan catabolites via IDO (i.e., kynurenines), as well as glutamate and NMDA receptors, may provide the strongest expression of neuro-immunological processes in allergy and associated conditions, including asthma.

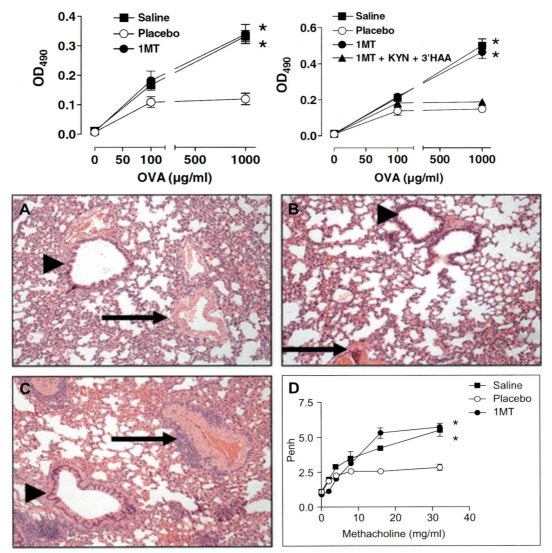

FIGURE 16.4.7 **Indoleamine 2,3-dioxygenase in tolerance to allergen sensitization.** *Top left,* inhibition of indoleamine 2,3-dioxygenase (IDO) activity prevents mucosal tolerance to ovalbumin. Following insertion of placebo control or 1-methyl-D-tryptophan (1-MT) pellet to inhibit systemic, mice were given 100 μg ovalbumin (OVA; 25 μl of a 100 μM solution) intranasally on three consecutive days. Control mice had no pellet inserted and were given saline intranasally. The development of tolerance was tested by intraperitoneal inoculation of OVA in Al(OH)$_3$. Five days after the immunization, splenic T cells were isolated and exposed to increasing concentrations of OVA and irradiated antigen-presenting cells *in vitro* for 4 day. T-cell proliferation was assessed using MTS [(3-(4,5-dimethylthiazol-2-yl)-5-(3-carboxymethoxyphenyl)-2-(4-sulfophenyl)-2H-tetrazolium] colorimetric cell proliferation read at 490 nm ($n = 3$). Successful tolerance was evidenced by failure of T-cell proliferation as seen with placebo pellets (open circles). Mice receiving repeated intranasal instillation of saline, rather than OVA, developed sensitization to OVA (closed squares). *$p < 0.05$, compared with mice given OVA with placebo pellet inserted. All results are shown as mean ± SEM. *Top right,* a mixture of kynurenine (KYN) and 3-hydroxyanthranilic acid (3'HAA) reconstituted tolerance to OVA in mice receiving 1-MT. Following implantation of a slow-release pellet of 1-MT, tolerance to OVA was induced in mice as described above, in the presence of a mixture of KYN and 3'HAA (15 mg/kg) administered intraperitoneally. Following intraperitoneal inoculation of OVA/Al(OH)$_3$, T-cell proliferation was used to evaluate sensitization to OVA. There was a significant increase ($n = 3$; $p < 0.05$) in T-cell proliferation in mice receiving 1-MT (closed circles) and saline (closed square). Conversely, mice receiving 1-MT/KYN/3-HAA, or placebo (open circles). *Bottom,* IDO regulates airway inflammation and hyperresponsiveness. Repeated intranasal instillation of OVA over a 3-day period led to the development of tolerance to OVA; this is evidenced by minimal perivascular (arrow) and peribronchial (arrowhead) inflammation in the airway of mice receiving placebo (*B*), similar to untreated mice (*A*). Conversely, mice receiving 1-MT pellets (*C*) developed extensive perivascular and peribronchial inflammatory response. These findings correlated with the results of whole body plethysmography showing a significant increase in airway responsiveness to methacholine ($n = 3$) in mice receiving 1-MT pellet or saline when compared with mice receiving placebo pellets (*D*), *$p < 0.05$. (*Modified from Odemuyiwa et al.*[87])

REFERENCES

1. Britan A, Maffre V, Tone S, Drevet JR. Quantitative and spatial differences in the expression of tryptophan-metabolizing enzymes in mouse epididymis. *Cell Tissue Res* 2006;**324**:301−10.

2. Munn DH, Zhou M, Attwood JT, Bondarev I, Conway SJ, Marshall B, et al. Prevention of allogeneic fetal rejection by tryptophan catabolism. *Science* 1998;**281**:1191−3.

3. Odemuyiwa SO, Ghahary A, Li Y, Puttagunta L, Lee JE, Musat-Marcu S, et al. Cutting edge: human eosinophils regulate T cell subset selection through indoleamine 2,3-dioxygenase. *J Immunol* 2004;**173**:5909−13.

4. Munn DH, Sharma MD, Lee JR, Jhaver KG, Johnson TS, Keskin DB, et al. Potential regulatory function of human dendritic cells expressing indoleamine 2,3-dioxygenase. *Science* 2002;**297**:1867−70.

5. Munn DH, Sharma MD, Hou D, Baban B, Lee JR, Antonia SJ, et al. Expression of indoleamine 2,3-dioxygenase by plasmacytoid dendritic cells in tumor-draining lymph nodes. *J Clin Invest* 2004;**114**:280−90.

6. Murray HW, Szuro-Sudol A, Wellner D, Oca MJ, Granger AM, Libby DM, et al. Role of tryptophan degradation in respiratory burst-independent antimicrobial activity of gamma interferon-stimulated human macrophages. *Infect Immun* 1989;**57**:845−9.

7. Pfefferkorn ER, Eckel M, Rebhun S. Interferon-gamma suppresses the growth of Toxoplasma gondii in human fibroblasts through starvation for tryptophan. *Mol Biochem Parasitol* 1986;**20**:215−24.

8. Xu H, Oriss TB, Fei M, Henry AC, Melgert BN, Chen L, et al. Indoleamine 2,3-dioxygenase in lung dendritic cells promotes Th2 responses and allergic inflammation. *Proc Natl Acad Sci USA* 2008;**105**:6690−5.

9. Mellor AL, Munn DH. Tryptophan catabolism and T-cell tolerance: immunosuppression by starvation? *Immunol Today* 1999;**20**:469−73.

10. Yoshida R, Imanishi J, Oku T, Kishida T, Hayaishi O. Induction of pulmonary indoleamine 2,3-dioxygenase by interferon. *Proc Natl Acad Sci USA* 1981;**78**:129−32.

11. Taylor MW, Feng GS. Relationship between interferon-gamma, indoleamine 2,3-dioxygenase, and tryptophan catabolism. *Faseb J* 1991;**5**:2516−22.

12. Takikawa O, Yoshida R, Kido R, Hayaishi O. Tryptophan degradation in mice initiated by indoleamine 2,3-dioxygenase. *J Biol Chem* 1986;**261**:3648−53.

13. Stone TW. Neuropharmacology of quinolinic and kynurenic acids. *Pharmacol Rev* 1993;**45**:309−79.

14. Stone TW. Subtypes of NMDA receptors. *Gen Pharmacol* 1993;**24**:825−32.

15. Shimizu T, Nomiyama S, Hirata F, Hayaishi O. Indoleamine 2,3-dioxygenase. Purification and some properties. *J Biol Chem* 1978;**253**:4700−6.

16. Yamazaki F, Kuroiwa T, Takikawa O, Kido R. Human indolylamine 2,3-dioxygenase. Its tissue distribution, and characterization of the placental enzyme. *Biochem J* 1985;**230**:635−8.

17. Carlin JM, Borden EC, Sondel PM, Byrne GI. Biologic-response-modifier-induced indoleamine 2,3-dioxygenase activity in human peripheral blood mononuclear cell cultures. *J Immunol* 1987;**139**:2414−8.

18. Munn DH, Sharma MD, Mellor AL. Ligation of B7−1/B7−2 by human CD4+ T cells triggers indoleamine 2,3-dioxygenase activity in dendritic cells. *J Immunol* 2004;**172**:4100−10.

19. Fujigaki S, Saito K, Sekikawa K, Tone S, Takikawa O, Fujii H, et al. Lipopolysaccharide induction of indoleamine 2,3-dioxygenase is mediated dominantly by an IFN-gamma-independent mechanism. *Eur J Immunol* 2001;**31**:2313−8.

20. Alberati-Giani D, Ricciardi-Castagnoli P, Kohler C, Cesura AM. Regulation of the kynurenine metabolic pathway by interferon-gamma in murine cloned macrophages and microglial cells. *J Neurochem* 1996;**66**:996−1004.

21. O'Connor JC, Andre C, Wang Y, Lawson MA, Szegedi SS, Lestage J, et al. Interferon-gamma and tumor necrosis factor-alpha mediate the upregulation of indoleamine 2,3-dioxygenase and the induction of depressive-like behavior in mice in response to bacillus Calmette-Guerin. *J Neurosci* 2009;**29**:4200−9.

22. Mahnke K, Schmitt E, Bonifaz L, Enk AH, Jonuleit H. Immature, but not inactive: the tolerogenic function of immature dendritic cells. *Immunol Cell Biol* 2002;**80**:477−83.

23. Billingham RE, Brent L, Medawar PB. Actively acquired tolerance of foreign cells. *Nature* 1953;**172**:603−6.

24. Fallarino F, Grohmann U, Vacca C, Bianchi R, Orabona C, Spreca A, et al. T cell apoptosis by tryptophan catabolism. *Cell Death Differ* 2002;**9**:1069−77.

25. Klion AD, Nutman TB. The role of eosinophils in host defense against helminth parasites. *J Allergy Clin Immunol* 2004;**113**:30−7.

26. Pritchard DI, Hewitt C, Moqbel R. The relationship between immunological responsiveness controlled by T-helper 2 lymphocytes and infections with parasitic helminths. *Parasitology* 1997;**115**(Suppl):S33−44.

27. Blanchard C, Rothenberg ME. Biology of the eosinophil. *Adv Immunol* 2009;**101**:81−121.

28. Kariyawasam HH, Robinson DS. The role of eosinophils in airway tissue remodelling in asthma. *Curr Opin Immunol* 2007;**19**:681−6.

29. Rothenberg ME, Hogan SP. The eosinophil. *Annu Rev Immunol* 2006;**24**:147−74.

30. Hogan SP, Rosenberg HF, Moqbel R, Phipps S, Foster PS, Lacy P, et al. Eosinophils: biological properties and role in health and disease. *Clin Exp Allergy* 2008;**38**:709−50.

31. Akuthota P, Wang HB, Spencer LA, Weller PF. Immunoregulatory roles of eosinophils: a new look at a familiar cell. *Clin Exp Allergy* 2008;**38**:1254−63.

32. Lacy P, Moqbel R. Eosinophil cytokines. *Chem Immunol* 2000;**76**:134−55.

33. Moqbel R, Lacy P. New concepts in effector functions of eosinophil cytokines. *Clin Exp Allergy* 2000;**30**:1667−71.

34. Lacy P, Mahmudi-Azer S, Bablitz B, Hagen SC, Velazquez JR, Man SF, et al. Rapid mobilization of intracellularly stored RANTES in response to interferon-gamma in human eosinophils. *Blood* 1999;**94**:23−32.

35. Melo RC, Spencer LA, Dvorak AM, Weller PF. Mechanisms of eosinophil secretion: large vesiculotubular carriers mediate transport and release of granule-derived cytokines and other proteins. *J Leukoc Biol* 2008;**83**:229−36.

36. Moqbel R, Coughlin JJ. Differential secretion of cytokines. *Sci STKE*; 2006:pe26.

37. Lacy P, Logan MR, Bablitz B, Moqbel R. Fusion protein vesicle-associated membrane protein 2 is implicated in IFN-gamma-induced

piecemeal degranulation in human eosinophils from atopic individuals. *J Allergy Clin Immunol* 2001;**107**:671–8.

38. Lucey DR, Nicholson-Weller A, Weller PF. Mature human eosinophils have the capacity to express HLA-DR. *Proc Natl Acad Sci USA* 1989;**86**:1348–51.

39. Ohkawara Y, Lim KG, Xing Z, Glibetic M, Nakano K, Dolovich J, et al. CD40 expression by human peripheral blood eosinophils. *J Clin Invest* 1996;**97**:1761–6.

40. Woerly G, Roger N, Loiseau S, Dombrowicz D, Capron A, Capron M. Expression of CD28 and CD86 by human eosinophils and role in the secretion of type 1 cytokines (interleukin 2 and interferon gamma): inhibition by immunoglobulin a complexes. *J Exp Med* 1999;**190**:487–95.

41. Harris N, Peach R, Naemura J, Linsley PS, Le Gros G, Ronchese F. CD80 costimulation is essential for the induction of airway eosinophilia. *J Exp Med* 1997;**185**:177–82.

42. Kuchroo VK, Das MP, Brown JA, Ranger AM, Zamvil SS, Sobel RA, et al. B7–1 and B7–2 costimulatory molecules activate differentially the Th1/Th2 developmental pathways: application to autoimmune disease therapy. *Cell* 1995;**80**:707–18.

43. Thompson CB. Distinct roles for the costimulatory ligands B7–1 and B7–2 in T helper cell differentiation? *Cell* 1995;**81**:979–82.

44. Spencer LA, Weller PF. Eosinophils and Th2 immunity: contemporary insights. *Immunol Cell Biol* 2010;**88**(3):250–6.

45. Curtsinger JM, Mescher MF. Inflammatory cytokines as a third signal for T cell activation. *Curr Opin Immunol* 2010;**22**:333–40.

46. Raap U, Wardlaw AJ. A new paradigm of eosinophil granulocytes: neuroimmune interactions. *Exp Dermatol* 2008;**17**:731–8.

47. Matsumoto K, Fukuda S, Nakamura Y, Saito H. Amphiregulin production by human eosinophils. *Int Arch Allergy Immunol* 2009;**149**(Suppl. 1):39–44.

48. Tulic MK, Sly PD, Andrews D, Crook M, Davoine F, Odemuyiwa SO, et al. Thymic indoleamine 2,3-dioxygenase-positive eosinophils in young children: potential role in maturation of the naive immune system. *Am J Pathol* 2009;**175**:2043–52.

49. Astigiano S, Morandi B, Costa R, Mastracci L, D'Agostino A, Ratto GB, et al. Eosinophil granulocytes account for indoleamine 2,3-dioxygenase-mediated immune escape in human non-small cell lung cancer. *Neoplasia* 2005;**7**:390–6.

50. Schrocksnadel H, Baier-Bitterlich G, Dapunt O, Wachter H, Fuchs D. Decreased plasma tryptophan in pregnancy. *Obstet Gynecol* 1996;**88**:47–50.

51. Belladonna ML, Puccetti P, Orabona C, Fallarino F, Vacca C, Volpi C, et al. Immunosuppression via tryptophan catabolism: the role of kynurenine pathway enzymes. *Transplantation* 2007;**84**: S17–20.

52. Moroni F, Russi P, Gallo-Mezo MA, Moneti G, Pellicciari R. Modulation of quinolinic and kynurenic acid content in the rat brain: effects of endotoxins and nicotinylalanine. *J Neurochem* 1991;**57**: 1630–5.

53. Moffett JR, Namboodiri MA. Tryptophan and the immune response. *Immunol Cell Biol* 2003;**81**:247–65.

54. Holt PG, Jones CA. The development of the immune system during pregnancy and early life. *Allergy* 2000;**55**:688–97.

55. Akdis M, Trautmann A, Klunker S, Daigle I, Kucuksezer UC, Deglmann W, et al. T helper (Th) 2 predominance in atopic diseases is due to preferential apoptosis of circulating memory/effector Th1 cells. *Faseb J* 2003;**17**:1026–35.

56. Akkoc T, de Koning PJ, Ruckert B, Barlan I, Akdis M, Akdis CA. Increased activation-induced cell death of high IFN-gamma-producing T(H)1 cells as a mechanism of T(H)2 predominance in atopic diseases. *J Allergy Clin Immunol* 2008;**121**:652–8. e1.

57. Mellor AL, Munn DH. IDO expression by dendritic cells: tolerance and tryptophan catabolism. *Nat Rev Immunol* 2004;**4**: 762–74.

58. Stone TW, Perkins MN. Quinolinic acid: a potent endogenous excitant at amino acid receptors in CNS. *Eur J Pharmacol* 1981;**72**:411–2.

59. Fonnum F. Glutamate: a neurotransmitter in mammalian brain. *J Neurochem* 1984;**42**:1–11.

60. Robbins TW, Murphy ER. Behavioural pharmacology: 40+ years of progress, with a focus on glutamate receptors and cognition. *Trends Pharmacol Sci* 2006;**27**:141–8.

61. Dirnagl U, Iadecola C, Moskowitz MA. Pathobiology of ischaemic stroke: an integrated view. *Trends Neurosci* 1999;**22**: 391–7.

62. Odemuyiwa SO, Ghaffari M, Lam V, Ilarraza R, Adamko D, Majaesic C, et al. Human Eosinophil-Derived Glutamate Modulates the Survival of Activated T Cells. *Journal of Allergy and Clinical Immunology* 2008;**121**:S17.

63. Haxhiu MA, Chavez JC, Pichiule P, Erokwu B, Dreshaj IA. The excitatory amino acid glutamate mediates reflexly increased tracheal blood flow and airway submucosal gland secretion. *Brain Res* 2000;**883**:77–86.

64. Dickman KG, Youssef JG, Mathew SM, Said SI. Ionotropic glutamate receptors in lungs and airways: molecular basis for glutamate toxicity. *Am J Respir Cell Mol Biol* 2004;**30**: 139–44.

65. Shen L, Han JZ, Li C, Yue SJ, Liu Y, Qin XQ, et al. Protective effect of ginsenoside Rg1 on glutamate-induced lung injury. *Acta Pharmacol Sin* 2007;**28**:392–7.

66. Cull-Candy S, Brickley S, Farrant M. NMDA receptor subunits: diversity, development and disease. *Curr Opin Neurobiol* 2001;**11**:327–35.

67. Miglio G, Varsaldi F, Lombardi G. Human T lymphocytes express N-methyl-D-aspartate receptors functionally active in controlling T cell activation. *Biochem Biophys Res Commun* 2005;**338**: 1875–83.

68. Lombardi G, Dianzani C, Miglio G, Canonico PL, Fantozzi R. Characterization of ionotropic glutamate receptors in human lymphocytes. *Br J Pharmacol* 2001;**133**:936–44.

69. Said SI, Berisha HI, Pakbaz H. Excitotoxicity in the lung: N-methyl-D-aspartate-induced, nitric oxide-dependent, pulmonary edema is attenuated by vasoactive intestinal peptide and by inhibitors of poly(ADP-ribose) polymerase. *Proc Natl Acad Sci USA* 1996;**93**:4688–92.

70. Said SI. Glutamate receptors and asthmatic airway disease. *Trends Pharmacol Sci* 1999;**20**:132–4.

71. da Cunha AA, Pauli V, Saciura VC, Pires MG, Constantino LC, de Souza B, et al. N-methyl-d-aspartate glutamate receptor blockade attenuates lung injury associated with experimental sepsis. *Chest* 2010;**137**:297–302.

72. Shang LH, Luo ZQ, Deng XD, Wang MJ, Huang FR, Feng DD, et al. Expression of N-methyl-D-aspartate receptor and its effect on nitric oxide production of rat alveolar macrophages. *Nitric Oxide* 2010;**23**:327–31.

73. Robertson BS, Satterfield BE, Said SI, Dey RD. N-methyl-D-aspartate receptors are expressed by intrinsic neurons of rat larynx and esophagus. *Neurosci Lett* 1998;**244**:77−80.

74. Benveniste H, Drejer J, Schousboe A, Diemer NH. Elevation of the extracellular concentrations of glutamate and aspartate in rat hippocampus during transient cerebral ischemia monitored by intracerebral microdialysis. *J Neurochem* 1984;**43**: 1369−74.

75. Sugahara M, Asai S, Zhao H, Nagata T, Kunimatsu T, Ishii Y, et al. Extracellular glutamate changes in rat striatum during ischemia determined by a novel dialysis electrode and conventional microdialysis. *Neurochem Int* 2001;**39**:65−73.

76. Yusa T. Effects of nitric oxide synthase inhibition on extracellular glutamate and cerebral blood flow during forebrain ischemia-reperfusion in rat in vivo. *J Anesth* 2000;**14**:24−9.

77. Zorumski CF, Olney JW. Excitotoxic neuronal damage and neuropsychiatric disorders. *Pharmacol Ther* 1993;**59**:145−62.

78. Choi DW. Excitotoxic cell death. *J Neurobiol* 1992;**23**:1261−76.

79. Heyliger SO, Mazzio EA, Soliman KF. The anti-inflammatory effects of quinolinic acid in the rat. *Life Sci* 1999;**64**: 1177−87.

80. Heyes MP, Chen CY, Major EO, Saito K. Different kynurenine pathway enzymes limit quinolinic acid formation by various human cell types. *Biochem J* 1997;**326**(Pt 2):351−6.

81. Moffett JR, Blinder KL, Venkateshan CN, Namboodiri MA. Differential effects of kynurenine and tryptophan treatment on quinolinate immunoreactivity in rat lymphoid and non-lymphoid organs. *Cell Tissue Res* 1998;**293**:525−34.

82. Warraki SE, el-Gammal MY, el-Asmar MF, Wahba N. Serum kynurenine in bronchial asthma and chronic bronchitis. *Chest* 1970;**57**:148−50.

83. Said SI, Pakbaz H, Berisha HI, Raza S. NMDA receptor activation: critical role in oxidant tissue injury. *Free Radic Biol Med* 2000;**28**:1300−2.

84. Maneechotesuwan K, Wamanuttajinda V, Kasetsinsombat K, Huabprasert S, Yaikwawong M, Barnes PJ, et al. Der p 1 suppresses indoleamine 2, 3-dioxygenase in dendritic cells from house dust mite-sensitive patients with asthma. *J Allergy Clin Immunol* 2009;**123**:239−48.

85. Hammad H, Charbonnier AS, Duez C, Jacquet A, Stewart GA, Tonnel AB, et al. Th2 polarization by Der p 1−pulsed monocyte-derived dendritic cells is due to the allergic status of the donors. *Blood* 2001;**98**:1135−41.

86. Bullens DM, De Swerdt A, Dilissen E, Kasran A, Kroczek RA, Cadot P, et al. House dust mite-specific T cells in healthy non-atopic children. *Clin Exp Allergy* 2005;**35**:1535−41.

87. Odemuyiwa SO, Ebeling C, Duta V, Abel M, Puttagunta L, Cravetchi O, et al. Tryptophan catabolites regulate mucosal sensitization to ovalbumin in respiratory airways. *Allergy* 2009;**64**: 488−92.

88. Taher YA, Piavaux BJ, Gras R, van Esch BC, Hofman GA, Bloksma N, et al. Indoleamine 2,3-dioxygenase-dependent tryptophan metabolites contribute to tolerance induction during allergen immunotherapy in a mouse model. *J Allergy Clin Immunol* 2008;**121**:983−91. e2.

89. Grohmann U, Fallarino F, Puccetti P. Tolerance, DCs and tryptophan: much ado about IDO. *Trends Immunol* 2003;**24**: 242−8.

Regulation of Eosinophil Responses by the Epithelial-Derived Cytokines TSLP, IL-25, and IL-33

Dirk E. Smith and Michael R. Comeau

THYMIC STROMAL LYMPHOPOIETIN AND EOSINOPHIL BIOLOGY

Thymic stromal lymphopoietin (TSLP) is a cytokine primarily expressed by epithelial and stromal cells in the periphery associated with a number of allergic inflammatory disorders. TSLP is a member of the interleukin 2 (IL-2) family of cytokines that also includes IL-4, IL-7, IL-9, IL-13, IL-15, and IL-21. TSLP is most closely related to IL-7 and mediates its biological activity through a heterodimeric receptor complex consisting of the shared IL-7 receptor subunit α (IL-7RA) and a TSLP-specific receptor chain (TSLPR).[1] TSLPR signaling is not well defined but is known to involve signal transducer and activator of transcription 5A (STAT5A) activation, which is reported to be dependent on the tyrosine-protein kinases JAK1 and JAK2[2] (Figure 16.5.1). TSLP was initially shown to play a role in the development of inflammatory responses through its activation of myeloid dendritic cells (DCs); it induces T-helper type 2 (T$_h$2) cell responses *in vitro* and is highly expressed in lesional skin samples from atopic dermatitis patients.[3] TSLP expression is also elevated in asthmatic lung biopsy samples and correlates with disease severity.[4] TSLP has also been found at increased levels in bronchoalveolar lavage (BAL) fluid from patients with asthma or chronic obstructive pulmonary disease, compared to normal controls.[5] Table 16.5.1 summarizes the cellular sources of TSLP commonly described. Transgenic overexpression of TSLP in the skin[6] or lungs[7] of mice leads to the induction of atopic dermatitis (AD) and asthma-like phenotypes, respectively. Taken together, these data link TSLP to human disease and suggest that deregulation of TSLP will promote pathological changes (summarized in Figure 16.5.2).

Several strains of transgenic mice have been developed that overexpress TSLP both locally and systemically. In these and other models, deregulated expression of TSLP invariably leads to profound disease. Interestingly, a common, prominent feature of these deregulated expression models is the development of eosinophilic inflammatory infiltrates, often associated with increases in

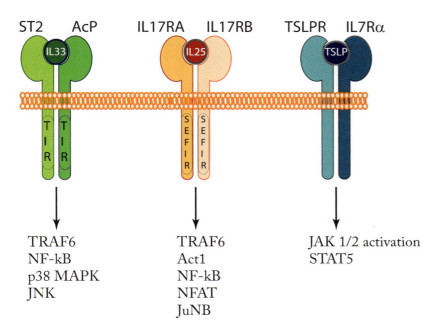

FIGURE 16.5.1 Representation of the unique heterodimeric receptors used by interleukin-25, interleukin-33, and thymic stromal lymphopoietin. Shown are the two receptor chains that comprise each receptor complex, the key intracellular domains recognized to be important for signaling, and the downstream factors that have been shown to be involved in mediating the intracellular response. AcP, interleukin-1 receptor accessory protein; Act1, adapter protein CIKS; IL7Rα, interleukin-7 receptor α; IL17RA/B, interleukin-17 receptor A,B; JAK, tyrosine-protein kinase JAK; JNK, C-Jun-amino-terminal kinase; JuNB, transcription factor jun-B; MAPK, mitogen-activated protein kinase; ST2, interleukin-33 receptor; NF-κB, nuclear factor κB; NFAT, nuclear factor of activated T-cells; STAT, signal transducer and activator of transcription; TRAF, TNF receptor-associated factor.

eosinophils-activating factors. Transgenic TSLP expression under the control of the lymphocyte-specific protein tyrosine kinase (*lck*) proximal promoter results in systemic inflammatory disease involving the kidney, liver, lungs, skin, and spleen. In these mice, eosinophils were a significant component of the mixed leukocyte infiltrate observed in the lungs that ultimately led to severe occlusion of the alveoli and death.[8] The differentiation of eosinophils from hematopoietic precursors and subsequent maturation, trafficking and activation is primarily regulated by interleukin-5 (IL-5),[9] and IL-5 is also a factor found at increased levels when TSLP is overexpressed and is induced both directly and indirectly by TSLP in responsive cell types.[10] Elevated levels of IL-5 in the serum, which the authors speculated were T_h2-cell derived, were reported in mice expressing TSLP under the control of the β-actin promoter. In this model, the TSLP-driven IL-5 levels possibly contribute to mobilization of myeloid progenitors, leading to extramedullary hematopoiesis and extensive accumulation of myeloperoxidase positive granulocytes in the spleens, similar to the phenotype of IL-5 transgenic mice.[11] Targeted expression of TSLP in the skin under the control of keratin, type II cytoskeletal 5 promoter in K5-TSLP transgenic mice leads to an AD-like phenotype that closely resembles the human disease. Abundant eosinophil accumulation coincident with increased IL-5 mRNA was observed within the inflammatory infiltrates of transgenic lesional skin.[6] Similar results were reported in keratin, type II cytoskeletal 14 (K14)-TSLP transgenic mice.[12] TSLP expression driven by the lung-specific surfactant protein C promoter (SPC-TSLP) results in an asthma-like phenotype, involving robust inflammatory cell infiltrates in the airways and associated airway remodeling.[7] Eosinophils were the predominant cell type among the cellular infiltrates in the lungs of SPC-TSLP mice and intracellular cytokine

TABLE 16.5.1 Cellular Sources and Inducing Stimuli of Interleukin-25, Interleukin-33, and Thymic Stromal Lymphopoietin

	Cellular Sources	Inducing Stimuli
TSLP	Basophils, epithelial cells, fibroblasts, keratinocytes, mast cells, smooth muscle cells	Allergens, IgE-receptor, cross-linking parasites?, pathogens, pollutants, proinflammatory cytokines, proteases, tissue damage, vitamin D3
IL-25	Basophils, epithelial, endothelial cells, eosinophils, mast cells, and T_h2 cells	Allergens, GM-CSF, IgE, IL-3, IL-5, parasites, proteases
IL-33	Adipocytes, cardiomyocytes, endothelial and epithelial cells, fibroblasts, keratinocytes, mast cells, and smooth muscle cells	IFN-γ, mechanical strain, necrosis, parasites, tissue damage, TNF (mechanisms of IL-33 release still poorly understood)

GM-CSF, granulocyte-macrophage colony-stimulating factor; IFN-γ, interferon γ; IgE, immunoglobulin E; IL-25, interleukin-25; IL-33, interleukin-33; T_h2, T-helper type 2; TNF, tumor necrosis factor α; TSLP, thymic stromal lymphopoietin.

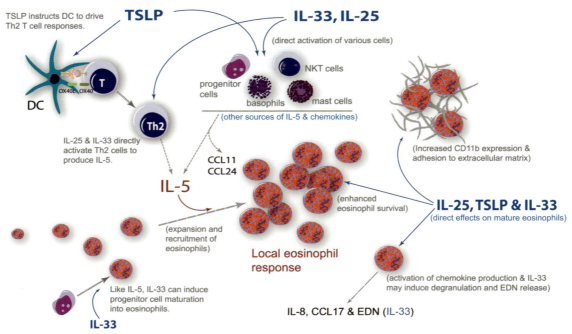

FIGURE 16.5.2 **Schematic of the mechanisms by which interleukin-25, interleukin-33, and thymic stromal lymphopoietin initiate, sustain and amplify local eosinophil responses.** Details of the various activities are described in the text. CCL, C-C motif chemokine; DC, dendritic cells; EDN, eosinophil-derived neurotoxin; IL, interleukin; NKT, natural killer T; TSLP, thymic stromal lymphopoietin.

staining revealed that IL-5-producing cells made up the highest percentage of T_h2 cytokine-producing T cells within the BAL fluid. Additionally, significant elevation of eotaxin/C-C motif chemokine 11 (CCL11), IL-5, and vascular endothelial protein 1 (VCAM-1) proteins were present in the BAL fluid, all factors known to facilitate the activation and recruitment of eosinophils to inflammatory sites.[13] Taken together, these independent models suggest that excess localized expression of TSLP promotes eosinophilic inflammation.

Similarly, significant eosinophilic inflammation is also observed in nontransgenic models associated with deregulated TSLP production. For example, nuclear receptor superfamily members are involved in the regulation of TSLP expression and keratinocyte-specific ablation of the retinoid X nuclear receptor-α (RXR-α), and RXR-β chains resulted in increased TSLP production in mice. This led to an AD-like phenotype, a response also seen in mice topically treated with RXR agonists, vitamin D3, or its low-calcemic analogues.[10,12,14] In all cases, and similar to TSLP transgenic mice, IL-5 transcripts in lesional skin and serum IL-5 levels were elevated. Reflective of endogenous TSLP expression models, exogenous TSLP administration to mice leads to allergic inflammatory responses involving eosinophils. Repeated intradermal administration of TSLP protein results in the development of a systemic T_h2 response characterized by increases in circulating immunoglobulin E, inflammatory cell infiltrates, and

increased T_h2 cytokines and chemokines in the skin. Interestingly, eosinophil-deficient △dblGATA mice failed to develop both local and systemic responses in this system, demonstrating the crucial contribution of eosinophils to TSLP-driven inflammatory responses.[15]

Beyond dendritic cells, expression of the functional TSLPR complex and TSLP activities have now been reported in a variety of innate and adaptive immune cell types, including mast cells, activated T cells, natural killer T (NKT) cells, CD34+ progenitor and, most recently, eosinophils.[10] Peripheral blood human eosinophils were found to express TSLPR and IL-7Rα transcripts and proteins and respond to TSLP in a dose-dependent and specific manner. TSLP-stimulated eosinophils demonstrated increased survival, surface expression of adhesion molecules, and adhesion to fibronectin. Interestingly, although TSLP induced the specific release of inflammatory cytokines and chemokines, it did not induce degranulation of eosinophils,[16] similar to the TSLP effects seen with mast cells.[17] The data obtained thus far with human cells suggest that TSLP may provide a signal to both increase the number of activated eosinophils that accumulate in tissues and extend the survival of these cells once there.

Differentiation and maturation of eosinophils from hematopoietic progenitor cell antigen CD34+ progenitor cells takes place primarily in the bone marrow. Progenitor cells are also found in circulation and, upon allergen exposure, they traffic to local tissues where they are capable of maturing into eosinophils, basophils, and mast cells, depending on their environment.[18,19] TSLPR is expressed on both mouse[20] and human[21] progenitor cells and stimulation of these cells by TSLP in combination with IL-1 or IL-33 (discussed below) and TNF induces the production of numerous cytokines and chemokines, including granulocyte-macrophage colony-stimulating factor (GM-CSF) and IL-5.[10,22] It is interesting to speculate that the induction of these cytokines may support the maturation of eosinophils from CD34$^+$ progenitor cells at sites of TSLP and IL-33 production in the periphery.

In addition to asthma and AD, TSLP is implicated in inducing eosinophilic esophagitis (EE). Human TSLP is located on chromosome 5q22.1, and a recent genome-wide association study identified 5q22 as a susceptibility locus for EE.[23] In a follow-up study by the same group, genetic variants within the TSLP gene were reported to be among the most dominant variants associated with EE, using a large panel of single nucleotide polymorphisms within allergy- and epithelial-associated genes. Additionally, in a sex-stratified analysis for TSLP and TSLPR polymorphisms performed to explore the mechanisms behind the higher prevalence of EE in males, a gender-specific association between single nucleotide polymorphisms (SNPs) in TSLP, as well as a nonsynonymous SNP in TSLPR, was identified. These data suggest that TSLP activities may contribute to EE and may partially explain the male bias seen in this disease.[24]

Multiple findings from several studies have now connected eosinophilic inflammatory responses with deregulated TSLP and have elucidated the mechanisms through which this cytokine promotes eosinophil function. Although TSLP is only one of many signals that individually or cooperatively are capable of promoting eosinophil mobilization and activation, in certain contexts it may be a pivotal factor. It remains to be determined whether intervention in the TSLP-mediated signaling pathway confers benefit in diseases where eosinophils are implicated.

INTERLEUKIN-25 AND EOSINOPHIL BIOLOGY

IL-25, also known as IL-17E, is a member of the structurally related IL-17 family of cytokines that also includes IL-17A, IL-17B, IL-17C, IL-17D, and IL-17F. Like many cytokines, IL-25 signals through a heterodimeric receptor complex consisting of IL-17RB and IL-17RA[25] (Figure 16.5.1). IL-25 activates nuclear factor of activated T-cells, cytoplasmic, calcineurin-dependent 1 (*NFATC1*) and jun B proto-oncogene (*JUNB*), and IL-17RB was recently shown to bind adapter protein CIKS through a structurally conserved

feature called the SEFIR (SEF/IL-17R) domain, present in the cytoplasmic tail.[26] Of the known IL-17 family members, IL-25 is the least similar at the level of conserved amino acids.[27] Likewise, the described biologic activities of IL-25 are divergent from other members of this family. IL-25 induces eosinophil accumulation in the lungs of mice,[28] unlike the neutrophil-rich inflammation seen in response to overexpression of IL-17A, IL-17C, or IL-17F. Daily infusion of IL-25 leads to robust systemic T_h2 responses in mice with eosinophilic infiltrates in various organs.[29] Splenomegaly occurs in IL-25 injected mice and is associated with increased numbers of splenic eosinophils. Additionally, dose-dependent increases in circulating eosinophils are observed, demonstrating the ability of IL-25 to mobilize eosinophils into the blood. Notably, inhibition of IL-5 completely prevents the eosinophilia, demonstrating that in this system the effects of IL-25 are indirectly mediated through the induction of T_h2-type cytokines. Additionally, the authors of this report noted a 20-fold higher number of myeloid colony-forming cells in the spleens of IL-25-treated mice, suggesting mobilization of hematopoietic progenitor cells,[29] similar to observations in mice overexpressing TSLP.[11] Consistent with studies in the mouse, IL-25 activates and enhances the viability of human eosinophils in a dose dependent manner,[30] suggesting that direct activities of IL-25 on eosinophils provide a further, conserved mechanism by which this cytokine promotes inflammatory responses. In addition to eosinophils, IL-25 has been reported to activate a variety of immune cell types including non-T non-B kit ligand (c-kit)$^+$ cells, T_h2 memory T cells, invariant natural killer T cells, monocytes, T cells and innate immune populations.[31] Although the cellular responses are varied, many involve the rapid production of IL-5 following stimulation, suggesting a common mechanism through which eosinophils may be mobilized to sites of IL-25-driven inflammation.

Numerous reports have demonstrated that IL-25 is a potent factor capable of promoting T_h2-cytokine-driven inflammatory responses in a variety of *in vivo* scenarios[31] and IL-25 is implicated in human diseases with allergic components, such as asthma and AD.[32] While primarily considered an epithelial-derived factor, the regulation of IL-25 protein production from these cells is poorly understood. Interestingly, a feature distinguishing IL-25 from other epithelial cell-derived cytokines including TSLP and IL-33 is that eosinophils are also a potential source of this cytokine. Peripheral blood eosinophils express IL-25 transcript and protein and activation of both normal and allergic donor cells with the eosinophilopoietic cytokines GM-CSF, IL-3, and IL-5 results in dramatic increases in IL-25 expression.[32] Recently, eosinophil-derived IL-25 was reported to play a role in Churg-Strauss syndrome (CSS), a rare disorder associated with hypereosinophilia in blood and tissues, severe asthma, and systemic vasculitis. Elevated serum IL-25 concentrations in CSS patients correlated with disease activity and eosinophil numbers, suggesting that IL-25 may play a critical role in promoting disease pathology.[33] As eosinophils have also been shown to respond to IL-25, the observation that they are capable of producing IL-25 suggests a potential feedback loop. Activated eosinophils may both produce and respond to IL-25, thereby self-sustaining functional responses and prolonged survival.

The aforementioned studies have revealed many similar and some unique eosinophil-focused attributes of IL-25, as compared with TSLP and IL-33 (discussed below). Future insights into eosinophil biology mediated by IL-25 are certain to be revealed by ongoing research on this epithelial- and eosinophil-derived cytokine.

INTERLEUKIN-33 AND EOSINOPHIL BIOLOGY

IL-33 is a member of the IL-1 family of cytokines[34] and is expressed broadly, especially within mucosal tissues in the gut and the lungs. Predominant sources include, but are not limited to, epithelial and endothelial cells, adipocytes and smooth muscle (as summarized in Table 16.5.1 and reviewed in[35]). IL-33 protein expression tends to be constitutive and localized within the nucleus, owing to nuclear localization and histone-binding motifs within the non-cytokine portion of the protein. This dual function nature of IL-33 suggests that it acts as a cytokine only when released from cells in response to tissue damage or distress, similar to other endogenous danger signals.[36] As such, soluble IL-33 acts directly on immune cells, including basophils, eosinophils, mast cells, NK, NKT cells, and T_h2 cells. It stimulates these cells by activating a receptor complex composed of the IL-33 receptor (known as ST2) and the IL-1 receptor accessory protein (IL-1RAcP), a second chain of the receptor that is required for signaling. ST2 activation leads to myeloid differentiation primary response protein MyD88- and interleukin-1 receptor-associated kinase 1 (IRAK-1)-dependent activation of nuclear factor (NF)-kB and mitogen-activated protein kinase (MAPK) responses (as shown in Figure 16.5.1). This mechanism of signal transduction distinguishes IL-33 from classical T_h2-associated cytokines. It also underlies the ability of IL-33 to both induce inflammatory response genes and amplify the effects of other eosinophil activation pathways.

Early observations hinted at a link between IL-33 and eosinophil biology. For example, ST2-deficient mice had reduced numbers of lung infiltrating eosinophils in response to parasite infection[37] and pulmonary challenge of naive animals with IL-33 led to infiltration of large numbers of eosinophils.[38] These findings suggested that eosinophils are targets of IL-33-mediated responses. One likely mechanism by which IL-33 enhances eosinophil accumulation is indirectly, through its ability to activate IL-5 production from a number of sources, including basophils, mast cells, NKT cells, and T_h2 cells. IL-33 also induces production of the eosinophil-recruiting C-C chemokine receptor type 3 (CCR3) chemokine CCL24 (C-C motif chemokine 24).[39]

Eosinophils are now recognized to express both chains of the IL-33 receptor and to be directly responsive to IL-33 stimulation *in vitro*. This may have escaped notice previously because ST2 surface expression on cultured eosinophils is only readily detectable after stimulation, such as by GM-CSF.[40] Eosinophils clearly express ST2 as confirmed by significant expression of ST2 mRNA, even in nonactivated cells.[40] IL-33 treatment leads to the rapid activation of extracellular signal regulated kinase (ERK), p38/MAPK, and NF-kB signaling molecules in human eosinophils.[41] As a result, IL-33 directly promotes several aspects of eosinophil-driven inflammation (summarized in Figure 16.5.2). For example, similar to IL-5, IL-33 enhances eosinophil survival[40,42] and also increases CD11b expression and adhesion to extracellular matrix proteins,[42] which could help retain eosinophils within local tissues. Our group has determined that IL-33 enhances eosinophil chemokinetic activity in a dose-dependent fashion (unpublished observations), again further enhancing local recruitment. IL-33 can also directly activate eosinophil effector functions. For example, IL-33 activates eosinophil superoxide production more quickly and more potently than IL-5, and in the same study, IL-33 induced release of the granule protein eosinophil-derived neurotoxin (EDN).[40] More study is needed in order to determine whether IL-33 is a robust inducer of degranulation; however, these observations suggest it could be an alternative trigger of acute activation. IL-33 also induces eosinophil production of chemokines, such as IL-8 and CCL17, thus enhancing their ability to orchestrate the recruitment of other inflammatory cell types. Interestingly, this activity is markedly enhanced when IL-33 is combined with any of the β-chain-utilizing cytokines, including GM-CSF, IL-3, and IL-5.[40,41] This means that IL-33 not only acts alone, but also provides an amplifying signal to other well-established forms of eosinophil activation.

The numerous direct and indirect effects of IL-33 on eosinophil recruitment and function suggest it could be a key factor required for optimal eosinophil responses in local tissues. This may be especially true in the asthmatic lung, where insults such as allergen exposure and ongoing inflammation may contribute to tissue damage and the release of IL-33. Furthermore, although IL-5 is a crucial eosinophil promoting cytokine, treatment of asthma patients with an anti-IL-5 antibody only reduces lung eosinophils by roughly 50%.[43] This may be related to the fact that eosinophils downregulate surface expression of the

IL-5 receptor when trafficking to the lung.[44] Also, animal studies have demonstrated that pulmonary eosinophilia can still occur in the complete absence of IL-5,[45] arguing that additional factors such as IL-33 may at times be driving eosinophil responses in the lung. Indeed, IL-33 acts on progenitor cells and is capable of inducing eosinophil differentiation from hematopoietic progenitor cells,[21,46] an observation that coincides with evidence for *in situ* eosinophilopoiesis in the asthmatic lung.[47] In addition, IL-33 activated eosinophils promoted a broad inflammatory response when transferred to the lungs of ST2-deficient mice, suggesting that IL-33 can drive pulmonary inflammation mediated solely by eosinophils.[46] A number of airway inflammation models have substantiated this role for IL-33 by demonstrating reductions in eosinophilic inflammation following ablation or disruption of the IL-33/ST2 pathway. Examples include pulmonary infection with respiratory syncytial virus,[48] a setting known to require a host eosinophil response,[49] as well as airway eosinophilia mediated by ovalbumin-specific T_h2 cells.[50]

Collectively, the emerging data suggest an evolutionarily conserved connection between eosinophil-mediated responses and the IL-33 cytokine axis. In particular, release of IL-33 from dead or damaged cells may provide an additional mechanism linking eosinophil accumulation at sites of cell turnover.[51] Just how large a role IL-33 plays relative to well-established modulators, such as GM-CSF and IL-5, remains to be determined, especially in humans. Thus far, however, genetic and disease sample analyses do point to a link. In particular, ST2 (*IL1RL1*) was one of only five gene polymorphisms reaching significance in a large-scale genome-wide association study designed to identify variants contributing to elevated eosinophil numbers across several human populations.[52] In addition, the concentration of circulating IL-33 is elevated in patients with hypereosinophilia and pulmonary eosinophilia.[53] Studies such as these will continue to strengthen our understanding of IL-33 and our appreciation for the role it plays in eosinophil responses.

CONCLUSION

Eosinophil differentiation, mobilization, and activation are all processes that play a critical role in protective immunity mediated by eosinophils. When these processes are deregulated, pathological inflammation and tissue destruction may result. Many factors are capable of influencing eosinophil responses both individually and in concert with each other. Besides the well-established eosinophilopoeitic cytokines GM-CSF, IL-3, and IL-5, novel cytokine pathways are increasingly recognized as important contributors. As summarized in this subchapter, the tissue-derived cytokines IL-25, IL-33, and TSLP all influence eosinophil biology both systemically and at the local level. Eosinophil function most likely evolved to maintain tissue homeostasis and assist with localized insults; thus, there is logic to the amplifying effect these tissue-derived cytokines have on eosinophil responses. While our understanding of the extent to which these factors may mediate eosinophil activities is in its infancy, clearly IL-25, IL-33, and TSLP are all capable of inducing or supporting established eosinophilic responses in certain contexts and continued study in this area is warranted.

REFERENCES

1. Park LS, Martin U, Garka K, Gliniak B, Di Santo JP, et al. Cloning of the murine thymic stromal lymphopoietin (TSLP) receptor: Formation of a functional heteromeric complex requires interleukin 7 receptor. *The Journal of Experimental Medicine* 2000;**192**:659−70.
2. Rochman Y, Kashyap M, Robinson GW, Sakamoto K, Gomez-Rodriguez J, Wagner KU, et al. Thymic stromal lymphopoietin-mediated STAT5 phosphorylation via kinases JAK1 and JAK2 reveals a key difference from IL-7-induced signaling. *Proceedings of the National Academy of Sciences of the United States of America* 2010;**107**(45):19455−60.
3. Soumelis V, Reche PA, Kanzler H, Yuan W, Edward G, Homey B, et al. Human epithelial cells trigger dendritic cell mediated allergic inflammation by producing TSLP. *Nature Immunology* 2002;**3**:673−80.
4. Ying S, O'Connor B, Ratoff J, Meng Q, Mallett K, Cousins D, et al. Thymic stromal lymphopoietin expression is increased in asthmatic airways and correlates with expression of Th2-attracting chemokines and disease severity. *J Immunol* 2005;**174**:8183−90.
5. Ying S, O'Connor B, Ratoff J, Meng Q, Fang C, Cousins D, et al. Expression and cellular provenance of thymic stromal lymphopoietin and chemokines in patients with severe asthma and chronic obstructive pulmonary disease. *J Immunol* 2008;**181**:2790−8.
6. Yoo J, Omori M, Gyarmati D, Zhou B, Aye T, Brewer A, et al. Spontaneous atopic dermatitis in mice expressing an inducible thymic stromal lymphopoietin transgene specifically in the skin. *The Journal of Experimental Medicine* 2005;**202**:541−9.
7. Zhou B, Comeau MR, De Smedt T, Liggitt HD, Dahl ME, Lewis DB, et al. Thymic stromal lymphopoietin as a key initiator of allergic airway inflammation in mice. *Nature Immunology* 2005;**6**:1047−53.
8. Taneda S, Segerer S, Hudkins KL, Cui Y, Wen M, Segerer M, et al. Cryoglobulinemic glomerulonephritis in thymic stromal lymphopoietin transgenic mice. *The American Journal of Pathology* 2001;**159**:2355−69.
9. Khaldoyanidi S, Sikora L, Broide DH, Rothenberg ME, Sriramarao P. Constitutive overexpression of IL-5 induces extramedullary hematopoiesis in the spleen. *Blood* 2003;**101**:863−8.
10. Comeau MR, Ziegler SF. The influence of TSLP on the allergic response. *Mucosal Immunology* 2010;**3**:138−47.
11. Osborn MJ, Ryan PL, Kirchhof N, Panoskaltsis-Mortari A, Mortari F, Tudor KS. Overexpression of murine TSLP impairs lymphopoiesis and myelopoiesis. *Blood* 2004;**103**:843−51.
12. Li M, Messaddeq N, Teletin M, Pasquali JL, Metzger D, Chambon P. Retinoid X receptor ablation in adult mouse keratinocytes generates an atopic dermatitis triggered by thymic stromal

lymphopoietin. *Proceedings of the National Academy of Sciences of the United States of America* 2005;**102**:14795−800.

13. Zhou B, Headley MB, Aye T, Tocker J, Comeau MR, Ziegler SF. Reversal of thymic stromal lymphopoietin-induced airway inflammation through inhibition of Th2 responses. *J Immunol* 2008;**181**:6557−62.

14. Li M, Hener P, Zhang Z, Kato S, Metzger D, Chambon P. Topical vitamin D3 and low-calcemic analogs induce thymic stromal lymphopoietin in mouse keratinocytes and trigger an atopic dermatitis. *Proceedings of the National Academy of Sciences of the United States of America* 2006;**103**:11736−41.

15. Jessup HK, Brewer AW, Omori M, Rickel EA, Budelsky AL, Yoon BR, et al. Intradermal administration of thymic stromal lymphopoietin induces a T cell- and eosinophil-dependent systemic Th2 inflammatory response. *J Immunol* 2008;**181**:4311−9.

16. Wong CK, Hu S, Cheung PF, Lam CW. Thymic stromal lymphopoietin induces chemotactic and prosurvival effects in eosinophils: implications in allergic inflammation. *American Journal of Respiratory Cell and Molecular Biology* 2009;**43**:305−15.

17. Allakhverdi Z, Comeau MR, Jessup HK, Yoon BR, Brewer A, Chartier S, et al. Thymic stromal lymphopoietin is released by human epithelial cells in response to microbes, trauma, or inflammation and potently activates mast cells. *The Journal of Experimental Medicine* 2007;**204**:253−8.

18. Cyr MM, Denburg JA. Systemic aspects of allergic disease: the role of the bone marrow. *Current Opinion in Immunology* 2001;**13**:727−32.

19. Fanat AI, Thomson JV, Radford K, Nair P, Sehmi R. Human airway smooth muscle promotes eosinophil differentiation. *Clin Exp Allergy* 2009;**39**:1009−17.

20. Hiroyama T, Iwama A, Morita Y, Nakamura Y, Shibuya A, Nakauchi H. Molecular cloning and characterization of CRLM-2, a novel type I cytokine receptor preferentially expressed in hematopoietic cells. *Biochemical and Biophysical Research Communications* 2000;**272**:224−9.

21. Allakhverdi Z, Comeau MR, Smith DE, Toy D, Endam LM, Desrosiers M, et al. CD34+ hemopoietic progenitor cells are potent effectors of allergic inflammation. *J Allergy Clin Immunol* 2009;**123**:472−8.

22. Allakhverdi Z, Smith DE, Comeau MR, Delespesse G. Cutting edge: The ST2 ligand IL-33 potently activates and drives maturation of human mast cells. *J Immunol* 2007;**179**:2051−4.

23. Rothenberg ME, Spergel JM, Sherrill JD, Annaiah K, Martin LJ, Cianferoni A, et al. Common variants at 5q22 associate with pediatric eosinophilic esophagitis. *Nature Genetics* 2010;**42**:289−91.

24. Sherrill JD, Gao PS, Stucke EM, Blanchard C, Collins MH, Putnam PE, et al. Variants of thymic stromal lymphopoietin and its receptor associate with eosinophilic esophagitis. *The Journal of Allergy and Clinical Immunology* 2010;**126**:160−5. e163.

25. Rickel EA, Siegel LA, Yoon BR, Rottman JB, Kugler DG, Swart DA, et al. Identification of functional roles for both IL-17RB and IL-17RA in mediating IL-25-induced activities. *J Immunol* 2008;**181**:4299−310.

26. Gaffen SL. Structure and signalling in the IL-17 receptor family. *Nature Reviews* 2009;**9**:556−67.

27. Moseley TA, Haudenschild DR, Rose L, Reddi AH. Interleukin-17 family and IL-17 receptors. *Cytokine Growth Factor Rev* 2003;**14**:155−74.

28. Hurst SD, Muchamuel T, Gorman DM, Gilbert JM, Clifford T, Kwan S, et al. New IL-17 family members promote Th1 or Th2 responses in the lung: in vivo function of the novel cytokine IL-25. *J Immunol* 2002;**169**:443−53.

29. Fort MM, Cheung J, Yen D, Li J, Zurawski SM, Lo S, et al. IL-25 induces IL-4, IL-5, and IL-13 and Th2-associated pathologies in vivo. *Immunity* 2001;**15**:985−95.

30. Cheung PF, Wong CK, Ip WK, Lam CW. IL-25 regulates the expression of adhesion molecules on eosinophils: mechanism of eosinophilia in allergic inflammation. *Allergy* 2006;**61**:878−85.

31. Monteleone G, Pallone F, Macdonald TT. Interleukin-25: A two-edged sword in the control of immune-inflammatory responses. *Cytokine Growth Factor Rev* 2010;**21**(6):471−5.

32. Wang YH, Angkasekwinai P, Lu N, Voo KS, Arima K, Hanabuchi S, et al. IL-25 augments type 2 immune responses by enhancing the expansion and functions of TSLP-DC-activated Th2 memory cells. *The Journal of Experimental Medicine* 2007;**204**:1837−47.

33. Terrier B, Bieche I, Maisonobe T, Laurendeau I, Rosenzwajg M, Kahn JE, et al. IL-25: a cytokine linking eosinophils and adaptive immunity in Churg-Strauss syndrome. *Blood* 2010;**116**(22): 4523−31.

34. Sims JE, Smith DE. The IL-1 family: regulators of immunity. *Nat Rev Immunol* 2010;**10**:89−102.

35. Oboki K, Ohno T, Kajiwara N, Saito H, Nakae S. IL-33 and IL-33 receptors in host defense and diseases. *Allergol Int* 2010; **59**:143−60.

36. Zhao W, Hu Z. The enigmatic processing and secretion of interleukin-33. *Cell Mol Immunol* 2010;**7**:260−2.

37. Townsend MJ, Fallon PG, Matthews DJ, Jolin HE, McKenzie AN. T1/ST2-deficient mice demonstrate the importance of T1/ST2 in developing primary T helper cell type 2 responses. *J Exp Med* 2000;**191**:1069−76.

38. Schmitz J, Owyang A, Oldham E, Song Y, Murphy E, McClanahan TK, et al. IL-33, an interleukin-1-like cytokine that signals via the IL-1 receptor-related protein ST2 and induces T helper type 2-associated cytokines. *Immunity* 2005;**23**:479−90.

39. Kurowska-Stolarska M, Stolarski B, Kewin P, Murphy G, Corrigan CJ, Ying S, et al. IL-33 Amplifies the Polarization of Alternatively Activated Macrophages That Contribute to Airway Inflammation. *J Immunol* 2009;**183**:6469−77.

40. Cherry WB, Yoon J, Bartemes KR, Iijima K, Kita H. A novel IL-1 family cytokine, IL-33, potently activates human eosinophils. *J Allergy Clin Immunol* 2008;**121**:1484−90.

41. Pecaric-Petkovic T, Didichenko SA, Kaempfer S, Spiegl N, Dahinden CA. Human basophils and eosinophils are the direct target leukocytes of the novel IL-1 family member IL-33. *Blood* 2009;**113**:1526−34.

42. Suzukawa M, Koketsu R, Iikura M, Nakae S, Matsumoto K, Nagase H, et al. Interleukin-33 enhances adhesion, CD11b expression and survival in human eosinophils. *Lab Invest* 2008; **88**:1245−53.

43. Flood-Page PT, Menzies-Gow AN, Kay AB, Robinson DS. Eosinophil's role remains uncertain as anti-interleukin-5 only partially depletes numbers in asthmatic airway. *Am J Respir Crit Care Med* 2003;**167**:199−204.

44. Liu LY, Sedgwick JB, Bates ME, Vrtis RF, Gern JE, Kita H, et al. Decreased expression of membrane IL-5 receptor alpha on human eosinophils: I. Loss of membrane IL-5 receptor alpha on airway eosinophils and increased soluble IL-5 receptor alpha in the airway after allergen challenge. *J Immunol* 2002;**169**:6452—8.

45. Domachowske JB, Bonville CA, Easton AJ, Rosenberg HF. Pulmonary eosinophilia in mice devoid of interleukin-5. *J Leukoc Biol* 2002;**71**:966—72.

46. Stolarski B, Kurowska-Stolarska M, Kewin P, Xu D, Liew FY. IL-33 Exacerbates Eosinophil-Mediated Airway Inflammation. *J Immunol* 2010;**185**:3472—80.

47. Dorman SC, Efthimiadis A, Babirad I, Watson RM, Denburg JA, Hargreave FE, et al. Sputum CD34+IL-5Ralpha+ cells increase after allergen: evidence for in situ eosinophilopoiesis. *Am J Respir Crit Care Med* 2004;**169**:573—7.

48. Walzl G, Matthews S, Kendall S, Gutierrez-Ramos JC, Coyle AJ, Openshaw PJ, et al. Inhibition of T1/ST2 during respiratory syncytial virus infection prevents T helper cell type 2 (Th2)- but not Th1-driven immunopathology. *J Exp Med* 2001;**193**:785—92.

49. Phipps S, Lam CE, Mahalingam S, Newhouse M, Ramirez R, Rosenberg HF, et al. Eosinophils contribute to innate antiviral immunity and promote clearance of respiratory syncytial virus. *Blood* 2007;**110**:1578—86.

50. Coyle AJ, Lloyd C, Tian J, Nguyen T, Erikkson C, Wang L, et al. Crucial role of the interleukin 1 receptor family member T1/ST2 in T helper cell type 2-mediated lung mucosal immune responses. *J Exp Med* 1999;**190**:895—902.

51. Lee JJ, Jacobsen EA, McGarry MP, Schleimer RP, Lee NA. Eosinophils in health and disease: the LIAR hypothesis. *Clin Exp Allergy* 2010;**40**:563—75.

52. Gudbjartsson DF, Bjornsdottir US, Halapi E, Helgadottir A, Sulem P, Jonsdottir GM, et al. Sequence variants affecting eosinophil numbers associate with asthma and myocardial infarction. *Nat Genet* 2009;**41**:342—7.

53. Kim HR, Jun CD, Lee YJ, Yang SH, Jeong ET, Park SD, et al. Levels of circulating IL-33 and eosinophil cationic protein in patients with hypereosinophilia or pulmonary eosinophilia. *J Allergy Clin Immunol* 2010;**126**:880—2. e886.

The findings presented in this book clearly support two significant conclusions. These conclusions challenge the previous status quo and require some significant further consideration as we proceed with eosinophil research, now and into the future.

EOSINOPHIL ACTIVITIES ARE COMPLEX AND DIVERSE

The change in the perceived role of eosinophil effector functions over the last two decades has been dramatic and worth noting again. In earlier times, eosinophils were considered to be end-stage cells that were recruited as cytotoxic effectors and thereby contributed to host defense and/or pathologies associated with disease. As detailed in the numerous contributions to this volume, we have a substantial understanding of the nuanced responses of eosinophils, including their modulation via cell-surface receptors, the variety of intracellular granule components, as well as the varied responses to extracellular ligands, all of which promote unique and situation-specific contributions to both homeostasis and disease. Adding to this complexity, we know from gene-deleted laboratory mice that the complete absence of eosinophils has not provided clear-cut evidence for an absolute, irrevocable, immutable homeostatic function for these leukocytes. Does this diminish the importance and significance of eosinophils? Absolutely not! A cell that is truly without beneficial pro-survival function for a given species would have been lost over the passage of evolutionary time. More appropriate,

the eosinophil is a specialized, multifunctional leukocyte whose true contributions to health and homeostasis are not yet fully understood.

ANTI-EOSINOPHIL STRATEGIES MAY NOT BE THE ONLY ANSWER

This point has its origins in the re-examination of the consensus view seen in the medical literature on how to treat patients with eosinophil-associated diseases—eliminate them and things will get better! We suggest that the collective stories outlined in this book provide a significant counterargument to this parochial point of view. Namely, eosinophils may be contributors to disease; certainly this is the case with respect to eosinophil leukemias and other primary eosinophil hematopoietic syndromes. However, it is important to recognize that eosinophil recruitment to tissue does not necessarily imply that eosinophils exclusively promote tissue damage, nor does it imply that eosinophils **alone** are promoting clinical symptoms. As noted above, eosinophil actions are complex. They can interact with other leukocytes and they have the immunomodulatory capacity and effector functions that can ameliorate and/or exacerbate inflammation. We suggest that the greater understanding of eosinophils and their associated activities, which will almost certainly occur in the near future, will continue to validate this new perspective. In turn, this greater understanding of both eosinophils and eosinophil-associated disorders will lead to an enlarged spectrum of therapeutic options.

Page references followed by "f" indicate figure, "t" indicate table, and by "b" indicate boxes.